Christiane Raab
Editor

Proceedings of the 9th International Conference on Maintenance and Rehabilitation of Pavements—Mairepav9

Springer

Editor
Christiane Raab
Swiss Federal Laboratories for Materials
Science and Technology
EMPA
Dübendorf, Switzerland

ISSN 2366-2557 ISSN 2366-2565 (electronic)
Lecture Notes in Civil Engineering
ISBN 978-3-030-48678-5 ISBN 978-3-030-48679-2 (eBook)
https://doi.org/10.1007/978-3-030-48679-2

This Springer imprint is published by the registered company Springer Nature Switzerland AG
The registered company address is: Gewerbestrasse 11, 6330 Cham, Switzerland

Preface

The Mairepav Conference series is the famous long-time flagship conference of the International Society of Maintenance and Rehabilitation of Transport Infrastructure (iSMARTi). The inaugural conference was held at Mackenzie Presbyterian University in Sao Paulo, Brazil, in 2000, and the series has steadily grown in number of participants and international visibility over the past 20 years, with installments hosted in various countries all over the world. The Mairepav Conference series is dedicated to the theme of maintenance and rehabilitation of infrastructure of the public and private transport, especially for roads and pavements.

This book gathers the proceedings of the Mairepav 9 Conference held at Empa (Swiss Federal Laboratories for Materials Science and Technology) in Dübendorf, Switzerland, in July 2020.

The contributions share the latest insights from research and practice in the maintenance and rehabilitation of pavements and discuss advanced materials, technologies and solutions for achieving an even more sustainable and environmentally friendly infrastructure. In this sense, it provides a state-of-the-art compendium and valuable source of knowledge for scientists, practitioners and students. The main topics are:

- Advanced Trends in Design, Rehabilitation and Preservation
- Management Systems and Life Cycle Analysis
- Sustainable Pavement Systems
- Recycling and By-products
- Advanced Pavement Materials and Technologies
- Evaluation of Pavement Performance
- Full Scale Studies Accelerated Pavement Testing
- Surface Characteristics and Road Safety

Infrastructure in general and pavements in particular provide a significant value worldwide. Since, the focus in most countries is not on construction of new infrastructure, maintenance and rehabilitation are receiving more and more importance and attention. Moreover, the development in maintenance and rehabilitation is driven by decreasing public budgets on the one hand and increasing traffic and user demand on the other hand. In addition, demands and requirements regarding digitalization, environmental issues and sustainable development have to be taken into consideration and have led to calls for new solutions. These solutions cover the whole range of sustainable materials, digital and big data technologies as well as life cycle considerations under an economic and environmental point of view. Therefore, it is high time that a new conference is held providing a platform for innovative developments and latest technologies for current and future applications for infrastructure and pavements.

In this sense, the 9th MAIREPAV Conference, July 1–3, 2020, in Switzerland, hosted for the first time by Empa Swiss Federal Laboratories for Materials Science and Technology, is of particular importance. All carefully peer-reviewed contributions present and discuss current knowledge and pioneering developments, clearly demonstrating that research and development in maintenance and rehabilitation of pavements is indispensable for providing durable and environmentally sustainable roads all over the world.

I would like to thank all the authors, peer reviewers, leading iSMARTi and scientific committee members as well as the Springer team who made this book possible and with this showed enormous effort and commitment in favor of the MAIREPAV9 Conference. This is particularly true in view of the extremely difficult situation caused by the COVID-19 pandemic. A situation, which clearly emphasizes that we are living in a global village, depending on each other in order to find effective problem solutions. I am particularly grateful to Manfred N. Partl, the former head of the Laboratory for Road Engineering/Sealing Components at Empa for his constant support. Moreover, I would like to acknowledge my assistant Michèle Köhli for her commitment in all MAIREPAV-related issues.

Dübendorf, Switzerland

Christiane Raab
Chair MAIREPAV9

Lecture Notes in Civil Engineering

Volume 76

Lecture Notes in Civil Engineering (LNCE) publishes the latest developments in Civil Engineering - quickly, informally and in top quality. Though original research reported in proceedings and post-proceedings represents the core of LNCE, edited volumes of exceptionally high quality and interest may also be considered for publication. Volumes published in LNCE embrace all aspects and subfields of, as well as new challenges in, Civil Engineering. Topics in the series include:

- Construction and Structural Mechanics
- Building Materials
- Concrete, Steel and Timber Structures
- Geotechnical Engineering
- Earthquake Engineering
- Coastal Engineering
- Ocean and Offshore Engineering; Ships and Floating Structures
- Hydraulics, Hydrology and Water Resources Engineering
- Environmental Engineering and Sustainability
- Structural Health and Monitoring
- Surveying and Geographical Information Systems
- Indoor Environments
- Transportation and Traffic
- Risk Analysis
- Safety and Security

To submit a proposal or request further information, please contact the appropriate Springer Editor:

- Mr. Pierpaolo Riva at pierpaolo.riva@springer.com (Europe and Americas);
- Ms. Swati Meherishi at swati.meherishi@springer.com (Asia - except China, and Australia, New Zealand);
- Dr. Mengchu Huang at mengchu.huang@springer.com (China).

All books in the series now indexed by Scopus and EI Compendex database!

More information about this series at http://www.springer.com/series/15087

Contents

Advanced Trends in Design, Rehabilitation and Preservation

Influence of the Aggregate Gradation on the Rutting Resistance of Bituminous Mixtures 3
B. S. Abhijith and J. Murali Krishnan

Experimental Investigation of Pothole Repair Materials 13
Debaroti Ghosh, Mugur Turos, and Mihai Marasteanu

A Durable Potholes Repair Method Using Polymer Modified Patching Material in Cold-Wet Weather 23
Sen Han, Jinping Xia, Hui Xu, and Hongwei Zhang

Maximising Stabilisation and Recycling Benefits for Sustainable Pavement Performance in New Zealand and Australia 35
Allen Browne

Evaluation of Warm Mix Asphalt Produced from Iraqi Materials 45
Noor J. Mahdi, Duraid M. Abd, and Taher M. Ahmed

Influence of Curing on the Mechanical Properties of Cement-Bitumen Treated Materials Using Foamed Bitumen: An Interlaboratory Test Program 55
Marco Pasetto, Emiliano Pasquini, Andrea Baliello, Simone Raschia, Amir Rahmanbeiki, Alan Carter, Daniel Perraton, Francesco Preti, Beatriz Chagas Silva Gouveia, Gabriele Tebaldi, Andrea Grilli, and Eshan V. Dave

Laboratory Tests for the Characterization of Cold Asphalt Patching Mixtures 67
Pier Paolo Riviera, Davide Dalmazzo, and Ezio Santagata

Performance Evaluation of Long-Life Pavements Using the Mechanistic-Empirical Asphalt Pavement Analysis (MEAPA) Web Application 79
M. Ghazavi, A. Seitllari, and M. E. Kutay

Management Systems and Life Cycle Analysis

The Challenges of Warm Mix Asphalt as a Mature Technology 93
Ali Jamshidi and Greg White

A Framework for Network Level Pavement Maintenance Planning for Low Volume Roads 103
H. R. Pasindu, R. M. K. Sandamal, and M. Y. I. Perera

P-F Curves in Modelling of Pavement Performance 115
Adam Zofka

Combined Life Cycle Cost Analysis and Life Cycle Assessment of Road Pavements 123
Egemen Okte and Imad L. Al-Qadi

Decision Support for New Holistic Uri Road Asset Management Process 133
Frank Schiffmann, Rade Hajdin, and Alfredo Serioli

Life Cycle CO_2 Analysis of Low Rolling Resistance Asphalt Pavements 143
A. Kawakami, M. Yabu, and H. Nitta

Accuracy Comparisons Between ASTM 1318-09 and COST-323 (European) WIM Standards Using LTPP WIM Data 155
Syed W. Haider and Muhammad Munum Masud

Detecting Significant Changes in Traffic Patterns for Pavement Design 167
Gopi K. Musunuru, Syed W. Haider, and Neeraj Buch

Development of Road Maintenance Management System for India's National Highway Network Using HDM-4 and Genetic Programming 177
Abhishek Sharma and Tanuj Chopra

Sustainable Pavement Systems

SENSO JOINT—An Innovative Sensor System for a Sustainable Joint Design of Concrete Pavements 191
Ch. Recknagel, S. Spitzer, J. Hoppe, N. Wenzel, and S. Pirskawetz

Numerical Evaluation of Crushing Resistance of Unbound Road Material 201
Erik Olsson, Denis Jelagin, and Manfred N. Partl

Developing and Modeling a Piezoelectric Energy Harvester (PEH) for Highway Pavements 211
Mohamadreza Khalili, Sara Ahmed, and A. T. Papagiannakis

Performance Optimization of Warm Recycled Mixtures 221
F. Cardone, F. Canestrari, X. Jiang, and G. Ferrotti

Development of Durable Pavement in Japan 231
Shigeki Takahashi, Shouichi Kanno, Yu Shirai, and Tamotsu Yoshinaka

Laboratory Evaluation of Recycled Asphalt Pavement Material in Warm-Mix Asphalt 243
Haritha Malladi, Abhilash Kusam, and Akhtarhusein A. Tayebali

Assessing Self-healing Asphalt by the Heating of Asphalt Mixtures 253
Caio Santos, Marina Cabette, Jorge Pais, Vitor Carvalho, and Paulo Pereira

Ultrasound Monitoring and Microwave Self-healing of Top-Down Cracks in Asphalt Pavements 263
Miguel A. Franesqui, Jorge Yepes, and Juan Gallego

Effects of Moisture and Aging on Asphalt Binder Adhesion Failure Using Pull-Off Tension Test 275
Muhammad Rafiq Kakar, Meor Othman Hamzah, and Christiane Raab

Effect of Addition of Plastic Fibres on Strength Characteristics of Subgrade Soil 285
Ashutosh Kaushal, Rajesh Pathak, and Tanuj Chopra

Modeling Rutting Behavior of Crumb Rubber Modified Binders Using Design of Experiments 295
Reza Azadedel, Nader Solatifar, and Maghsoud Rahbarnia

Self-healing Asphalt for Road Pavements 307
A. Tabaković and E. Schlangen

Recycling and By-products

Experimental Investigation on the Effect of Rejuvenator on the Use of a High Amount of Recycled Asphalt Binder 321
Di Wang, Maximilian Koziel, Augusto Cannone Falchetto, Chiara Riccardi, Martin Hugener, Laurent Porot, Yun Su Kim, Goshtasp Cheraghian, and Michael P. Wistuba

Rubber-Oil Distillation Bottoms Blends as a New Recycling Solution for Bitumen Extension 331
G. Tarsi, C. Sangiorgi, A. Varveri, and C. Oliviero Rossi

Improving the Sustainability of Semi-Dense Asphalt Pavements by Replacement of Recycled Concrete Aggregate Fractions 343
Peter Mikhailenko, Muhammad Rafiq Kakar, Zhengyin Piao, Moises Bueno, and Lily Poulikakos

Combined Effect of Warm Mix Processes and Multi-recycling on the Main Criteria of the French Asphalt Mix Design Method 353
P. Marsac, C. Petiteau, O. Burban, J. P. Terrier, G. Didelet, J. Demoncheaux, T. Lorino, and S. Pouget

Mechanical Behaviour of Cold Recycled Asphalt Mixtures for Binder Courses Produced with Bitumen Emulsion and High Strength Cement 365
Chiara Mignini, Fabrizio Cardone, and Andrea Graziani

Performance Evaluation of Hybrid EAF Slag and RAP in Pavement 375
Shih-Huang Chen, Hasnae Amal Smimine, Wei-Lun Tsai, Ching-Tsung Hung, Meng-Hsin Kuo, and Ching-Lien Zen

Effect of Waste Fillers on the Rutting and Fatigue Behavior of Asphalt Mastic and Mixes 385
Jayvant Choudhary, Brind Kumar, and Ankit Gupta

Evaluating the Properties of Bioasphalt Produced with Bio-oil Derived from Biodiesel Production 397
Caio Rubens Santos, Jorge C. Pais, Jorge Ribeiro, and Paulo Pereira

Evaluating the Characteristics of Crumb Rubber Modified Asphalt Binders Produced with Neat Bitumen—Case of Kuwait 409
Taha Ahmed, Dawoud Bahzad, Abdullah Al-Marshed, Zein-Eddine Merouani, and Mohamed Omar

Properties of Hot Mix Asphalt Containing Treated Recycled Concrete Aggregates Using SCB and ITS Tests 419
A. Kavussi, F. Kazemian, and M. Bayzidi

Bituminous Mixtures with High Environmental Compatibility: Laboratory Investigation on the Use of Reclaimed Asphalt and Steel Slag Aggregates 433
C. Nodari, M. Crispino, and E. Toraldo

Investigation of Selected Properties of Crumb Rubber Modified Bitumens with Different Rubber Contents 443
E. Manthos, J. Valentin, L. Benešová, D. Giannaka, P. Gravalas, and Ch. Tsakalidis

Performance Assessment of Rubberized Mixtures Containing Reclaimed Asphalt and a Viscosity Reduction Additive 457
Leonardo Urbano, Davide Dalmazzo, Pier Paolo Riviera, and Ezio Santagata

Evaluation of the Properties of Asphalt Concrete Modified with Crumb Rubber Using Marshall Test 469
Olumide M. Ogundipe, Omotola C. Aboloye, and Stephen O. Fatuase

Advanced Pavement Materials and Technologies

Influence of Source and Ageing on the Rheological Properties and Fatigue and Rutting Resistance of Bitumen Using a DSR 481
Mrinali Rochlani, Sabine Leischner, Gustavo Canon Falla, Puneet Goudar, and Frohmut Wellner

Permanent Deformation Characterisation of Gap-Graded and Continuous Graded Aggregate Blends for Bituminous Mixtures 493
V. T. Thushara and J. Murali Krishnan

Linear Viscoelastic Properties of a Half Warm Asphalt Mixture (HWMA) with Bitumen Emulsion 507
Silvia Angelone, Marina C. Casaux, Luis Zorzutti, and Fernando Martinez

Long Lasting Asphalt Materials with Highly Modified Asphaltic Binder 517
Laurent Porot, Erica Jellema, and David Bell

Mechanical Properties of Bio-Asphalt on Recycled Asphalt Pavement Binder 529
Atmy Verani Rouly Sihombing, Bambang Sugeng Subagio, Eri Susanto Hariyadi, and Anwar Yamin

The Effect of Fly Ash Based Geopolymer on the Strength of Problematic Subgrade Soil with High CaO Content 539
Nawfal Farooq Kwad, Ahmed H. Abdulkareem, and Taher M. Ahmed

Low Temperature Behavior of Asphaltite Modified Binders and Asphalt Concretes 553
Andrea Themeli, Emmanuel Chailleux, Cyrille Chazallon, and Nicolas Bueche

Repeatability Study on the Laboratory Production Process of Cement Bitumen Treated Materials with Foamed Bitumen 565
Simone Raschia, Amir Rahmanbeiki, Daniel Perraton, Alan Carter, Andrea Graziani, and Andrea Grilli

Effect of Global Thresholding Algorithms on Pervious Concrete Pore Network Properties Using XRCT-Based Digital Image Processing 575
Ajayshankar Jagadeesh, Ghim Ping Ong, and Yu-Min Su

Performance Characteristics of Nano-Modified Asphalt Mixtures 587
Lucia Tsantilis, Giuseppe Chiappinelli, Orazio Baglieri, Pier Paolo Riviera, Fabrizio Miglietta, and Ezio Santagata

Properties of Styrene-Isoprene-Styrene (SIS) Modified Asphalt Binder 597
Mithil Mazumder, Soon-Jae Lee, and Moon-Sup Lee

Effects of Binder Modification on Rutting Performance of Asphalt Binders 607
A. Seitllari, M. Ghazavi, and M. E. Kutay

Viscoelastic Response of Bitumen Emulsion Mastic with Various Active Fillers 617
Ahmed Al-Mohammedawi and Konrad Mollenhauer

Numerical Studies on Coir Geotextile Reinforced Flexible Pavement ... 627
V. Anusudha, V. Sunitha, Chithu Babu, Chetan R. Bhole, and Samson Mathew

Evaluation of Pavement Performance

Relating Asphalt Mixture Performance to Asphalt Mastic Rheology ... 639
Johannes Büchner and Michael P. Wistuba

Shear Bonding Performance of Reinforced Asphalt Pavements by Using Polyester Grids 651
Fabiana Leite-Gembus and Andreas Elsing

Discrete Element Simulations of 4-Point Bending Fatigue Tests of Asphalt Concrete Samples Reinforced by Fiberglass Grids 663
G. Liu, G. Koval, and C. Chazallon

Non-destructive Pavement Testing for Sustainable Road Management 675
G. Kneib

Precision Assessment of the Modified Wheel Tracking Device Based on Small-Scale Testing of New Zealand Hot Mix Asphalt 687
Abhirup Basu Roy-Chowdhury, Mofreh Saleh, and Miguel Moyers-Gonzalez

Simulation of Heavy Weight Deflectometer Test: Spectral Element Method vs Finite Element Method 699
Jean-Marie Roussel, Hervé Di Benedetto, Cédric Sauzéat, and Michaël Broutin

Three Dimensional Finite Element Model for Active Crack Control in Continuously Reinforced Concrete Pavement 709
Muhammad Kashif, Pieter De Winne, Ahsan Naseem, Nouman Iqbal, and Hans De Backer

Inference of Pavement Properties with Roadside Accelerometers 719
Julius Nielsen, Eyal Levenberg, and Asmus Skar

Behaviour of the Interface Bonding Between Asphalt Overlays and Rigid Pavements 729
K. Bayraktarova, M. Dimitrov, B. Hofko, and L. Eberhardsteiner

Estimation of Resilient Modulus for Fine-Grained Soils Using Ground Penetrating Radar 741
Logan Tihey and S. Sonny Kim

Impact of Construction Practices on Air Voids and Permeability of Asphalt Mixtures 751
Syed W. Haider, Michele Lanotte, Khurram Malik, and Aftab Quadri

Application of Dynamic Creep Testing to Investigate Permanent Deformation Characteristics of Asphalt Mixes 761
Amir Kavussi and Seyed Mohsen Motevalizadeh

A Performance Prediction Model for Continuously Reinforced Concrete Pavement Using Artificial Neural Network 771
Hakan Yasarer, Mohammad Najmush Sakib Oyan, and Yacoub Najjar

Full Scale Studies Accelerated Pavement Testing

Parameter Identification of Asphalt Pavements Subjected to Moving Loads 785
Zhaojie Sun, Cor Kasbergen, Karel N. van Dalen, Kumar Anupam, Athanasios Skarpas, and Sandra M. J. G. Erkens

Effects of Field Compaction Method on Water Permeability and Performance of Asphalt Concrete Pavements 795
Chinecherem Agbo Igboke, Eslam Elsayed, Yasser Hassan, and Abd El Halim Omar Abd El Halim

Cooling Time Requirements for Asphalt Pavement Repairs 805
L. Chu and T. F. Fwa

Cold Recycling in Germany—Current Experiences and Future Projects 813
B. Wacker, M. Kalantari, and M. Diekmann

Distributed Fiber Optic Strain Measurements in an Airfield Pavement 825
D. Hauswirth, F. Fischli, C. Rabaiotti, and A. M. Puzrin

Pavement Distress from Channelized and Lateral Wandering Loads Using Accelerated Pavement Tests 835
Martin Arraigada and Manfred N. Partl

Correlating Air Freezing Index and Frost Penetration Depth—A Case Study for Sweden 847
Sigurður Erlingsson and Denis Saliko

Simulating Deflection of a Jointed Rigid Pavement Under Rolling Wheel Deflectometer (RAPTOR) Loading 859
Pawan Deep, Mathias B. Andersen, Søren Rasmussen, Alessandro Marradi, Nick H. Thom, and Davide L. Presti

Study on the Asphalt Pavement Response in the Accelerated Pavement Testing Facility 871
Ruxin Jing, Aikaterini Varveri, Xueyan Liu, Athanasios Scarpas, and Sandra Erkens

In-Situ Measurement of Discontinuity Movements in Concrete Pavement Structures 881
Dongkyu Kim, Hyunsik Hwang, Christopher Jabonero, and Yoon-Ho Cho

Surface Characteristics and Road Safety

Pavement Surface Evaluation Interacting Vibration Characteristics of an Electric Mobility Scooter 893
Kazuya Tomiyama and Kazushi Moriishi

Acoustic Maintenance of Pavements by Large-Scale Grinding 901
Françoise Beltzung and Tobias Balmer

Framework for Pothole Detection, Quantification, and Maintenance System (PDQMS) for Smart Cities 913
Naga Siva Pavani Peraka, Krishna Prapoorna Biligiri, and Satyanarayana N. Kalidindi

Stochastic Prediction of Short-Term Friction Loss of Asphalt Pavements: A Traffic Dependent Approach 923
Christina Plati, Maria Pomoni, Andreas Loizos, and George Yannis

Tire Contact Stress Distribution Considering the Tire Inclination in Bend 933
Y. Oubahdou, E. Manyo, P. Reynaud, B. Picoux, J. Dopeux, and C. Petit

LCMS-2 Measurements of the Quality of Road Markings 943
Kars Drenth, Jun Yew Tan, Marc Drenth, and Ong Ju Kit

A Study on the Effect of Milling on Stress Distributions in Asphalt Pavements 953
Kaoutar Diouri, Rajae Bousselham, Anirban De, Adriana Hera, Tahar El-Korchi, and Rajib B. Mallick

Assessment of Preformed 3D-Thermoplastic Road Markings for Long-Term Durability, Skid Resistance and Texture Functionality 965
Kalpesh Purohit, Mujib Rahman, Andrew Price, and Alan Woodside

Multiple Linear Regression Models for Predicting Surface Damage Due to Repeated Dynamic Loading on Submerged Asphalt Pavement 975
Fauzia Saeed, Mujib Rahman, and Maher Mahmood

Advanced Trends in Design, Rehabilitation and Preservation

Influence of the Aggregate Gradation on the Rutting Resistance of Bituminous Mixtures

B. S. Abhijith and J. Murali Krishnan

Abstract Rutting in bituminous pavements predominantly depends on the aggregate gradation. In particular, for fine graded mixtures, the aggregate gradation below 2.36 mm play a critical role since more than 60% of aggregates will pass through this sieve. The present work is focused on quantifying the rut resistance of three different fine aggregate gradations. The control gradation is the well-graded gradation used in India, and the other two gradations were obtained following Bailey method. The Bailey gradations were designed in such a way that the gradation below 2.36 mm sieve size was varied while the gradation above this sieve size was kept the same as that of the control gradation. The mixtures were fabricated using an unmodified binder (binder content kept constant at 5.4%) at 6% air voids using a shear box compactor. The influence of aggregate gradation on the creep and recovery response of bituminous mixtures was investigated at two temperatures by applying 10,000 repetitions of a repeated trapezoidal loading and recovery cycles. From the resulting residual strain curve, the Bailey mixture designed towards the coarser side below 2.36 mm sieve was found to have a higher resistance to rutting when compared to the conventional bituminous concrete-middle gradation.

Keywords Aggregate gradation · Rutting · Bailey method · Creep and recovery test

1 Introduction

Premature failure by rutting is one of the primary distress observed in bituminous pavements in India. The major factors contributing to rutting is slow-moving traffic with overloaded axles, high pavement temperature, in addition to the bituminous mixture characteristics. Among the various components of bituminous mixtures, the rut resistance is found to depend on the aggregate gradation (Button et al. 1990).

B. S. Abhijith · J. Murali Krishnan (✉)
Department of Civil Engineering, Indian Institute of Technology Madras, Chennai 600036, India
e-mail: jmk@iitm.ac.in

B. S. Abhijith
e-mail: abi123bs@gmail.com

C. Raab (ed.), *Proceedings of the 9th International Conference on Maintenance and Rehabilitation of Pavements—Mairepav9*, Lecture Notes in Civil Engineering 76,
https://doi.org/10.1007/978-3-030-48679-2_1

In particular, much attention has been given for the aggregate gradation passing 4.75 or 2.36 mm sieve size (Kandhal and Cooley 2001). A set of control points were defined for aggregate gradations to be used in Superpave mixture design. The principle criterion introduced was the inclusion of a restricted zone in the aggregate gradation. This zone lies between the 4.75/2.36 and 0.3 mm sieve size. The allowable ratio between the percentage of the aggregate passing different sieve sizes within this zone was recommended to reduce the incidence of rut-prone mixtures.

In India, the wearing course has a nominal maximum aggregate size of 13.2 mm. A well-graded aggregate gradation from 13.2 to 0.075 mm is stipulated with a minimum binder content of 5.4% by weight of the mixture. In this mixture, the maximum percent passing 2.36 mm sieve size can be about 60% by weight of total aggregates, and possibly be the main reason for such premature rutting failure. Therefore, it is crucial to evaluate the influence of fine aggregate gradation on the rutting of bituminous mixtures. One way to achieve this objective is to design an optimal aggregate gradation for the finer portion of the blend using a rational approach towards blending aggregates of different sizes. The parameters associated with such approaches can then be used to relate to rutting in the bituminous mixture.

Among several approaches, Bailey method for aggregate gradation selection is a practical approach that has been successfully used for developing and analyzing bituminous mixtures (Daniel and Rivera 2009; Mohammad and Al-Shamsi 2007). The rutting resistance of the New Hampshire bituminous mixtures developed using the Bailey method was studied by Daniel and Rivera (2009). The original bituminous mixtures used in the field were re-designed using the principles of the Bailey method in the laboratory. Among the mixtures with same nominal maximum particle size, the Bailey mixtures were found to give the better rut resistant mixture. Mohammad and Al-Shamsi (2007) evaluated the rutting performance of three different bituminous mixtures designed using the Bailey method. Three mixtures had three different types of aggregates—limestone, sandstone, and granite. The rut depth of the bituminous mixtures was captured using a wheel tracking device, and it was found that all the mixtures had a higher rut-resistance.

In the current study, the Bailey method is used to design the finer portion of the aggregate gradation used for the wearing course in India. A detailed explanation of the Bailey method for gradation selection is given in Vavrik (2000). Bailey recommended ranges of aggregate ratios for fine graded mixtures were used to design the gradations. Three different bituminous mixtures were used to study the influence of fine aggregate gradation on rutting. Flow number test is used to evaluate the rut resistance of the bituminous mixtures. This test is recommended as a potential test to understand the rutting characteristics of the bituminous mixtures (Witczak et al. 2002).

2 Materials and Sample Preparation

The aggregates used were the crushed granite aggregates from Madurantakam located in Kanchipuram, India. The filler used was stone dust obtained from the same quarry.

Table 1 Properties of VG-30 binder as per IS:73 (2013)

Specification	Results	Limits
Penetration at 25 °C, 100 g, 5 s, 0.1 mm	47	45 min
Softening point (R & B), °C	50	47 min
Absolute viscosity at 60 °C, (Poise)	2735	2400–3600
Kinematic viscosity at 135 °C, (cSt)	553	350 min
The temperature at which $\lvert G^*\rvert/\sin\delta$ is min. 1 kPa (unaged)	64.9 °C	
The temperature at which $\lvert G^*\rvert/\sin\delta$ is min. 2.2 kPa (RTFO aged)	63.9 °C	
The temperature at which $\lvert G^*\rvert\sin\delta$ is max. 5000 kPa (PAV aged)	22 °C	
J_{nr} at 3.2 kPa (58 °C)	1.4249 1/kPa	

The specific gravity of the individual fractions was determined by ASTM C127, C128, and C188. The specific gravity for coarse aggregate, fine aggregates, and filler was found to be 2.717, 2.774, and 2.648, respectively. Viscosity graded VG-30 binder was used (see Table 1) for the fabrication of bituminous mixtures. A 5.4% binder content was used for the fabrication of all the bituminous mixtures following MoRTH (2013) guidelines.

The bituminous mixtures were compacted using a shear box compactor (ASTM D7981 2015). The mixtures were short term aged at a temperature of 135 °C for 4 h ± 5 min and 155 °C for 30 min before subjecting it to compaction (AASHTO R30 2015). All the mixtures were compacted to achieve a target air void of 6 ± 0.5%. The compacted beam specimen (450 × 150 × 170 mm) was then cored to get three cylindrical samples of 100 mm diameter and 150 mm height. The samples that met the air void criterion was used for further testing.

2.1 *Aggregate Gradation*

Three different aggregate gradations were used in the current study. All three gradations were selected in such a way that it meets the MoRTH (2013) requirements for bituminous concrete mixtures grade II. The control gradation, hereafter, referred to as BC-middle, is the mid-gradation used in India. The other two gradations were obtained using the principles of the Bailey method. The Bailey gradations were designed in such a way that the gradation below 2.36 mm sieve size was varied while the gradation above this sieve size was kept the same as that of the control gradation. The gradations Bailey-1 and Bailey-2 were designed using the values corresponding to the lower and upper side of the Bailey recommended range of aggregate ratio for fine graded mixtures. The recommended ranges for coarse aggregate ratio (CA) is 0.6 to 1.0, the fine aggregate coarser (FA_c) ratio and the fine aggregate finer (FA_f) ratio is 0.35 to 0.50.

3 Design of Bailey Gradation

According to the Bailey method, the type of mixtures is defined based on the unit weights of each individual stockpiles. For each stockpile, a loose unit weight (LUW) and rodded unit weight (RUW) is calculated (AASHTO T19 2004). The LUW is defined as the minimum density required for particle-to-particle contact and is considered as the dividing line between fine graded and coarse graded mixtures. The RUW is used to represent the amount of excess coarse aggregates in a unit volume after compaction.

Further, the LUW and RUW are used as a reference to select a chosen unit weight (CUW). The CUW is then used as a basis to select the desired type of mixtures. For a fine graded mixture, it is recommended to select a CUW as less than 90% of the LUW. In this study, as a first step, the BC-middle gradation was used as a reference to determine whether it is a fine graded or coarse graded mixture according to the Bailey method. The individual stockpiles were created in such a way that the targeted BC-middle gradation can be obtained. The specific gravities and unit weights of each stockpile were determined. By giving these as input to the Bailey method, the CUW was varied in such a way that the desired BC-middle gradation is obtained. The back-calculated CUW for BC-middle gradation was found to be 74.8% indicating that the gradation is fine graded.

After establishing that the BC-middle gradation is a fine gradation, the gradation for Bailey-1, and Bailey-2 were obtained by varying the percentage passing 2.36 mm in such a way that the lower limit and the upper limit of Bailey recommended aggregate ratios for fine graded mixes are met. A new sieve, 0.032 mm, was introduced in addition to the sieves recommended in MoRTH (2013) in order to control the Bailey aggregate ratios. It was also ensured that the gradations Bailey-1 and Bailey-2 fall within limits specified for bituminous mixtures grade II. The back-calculated chosen unit weight and the aggregate ratios are reported along with the individual stockpile details in Table 2. The aggregate gradation used along with the MoRTH (2013) gradation limits for bituminous mixtures is illustrated in Fig. 1.

4 Results and Discussion

4.1 Compaction Curve—Shear Compactor

4.1.1 Test Methodology

In the shear box compactor (ASTM D7981 2015), a constant compressive force and a cyclic shear force with a constant shear angle will be applied to the bituminous mixtures. Qiu et al. (2012) reported the compaction results obtained from the shear box compactor for bituminous mixtures with different gradations and binder types. The aggregate gradation was found to be more sensitive to the reduction in air voids

Table 2 Gradation of coarse, fine and filler aggregates

Sieve size (mm)	BC-middle			Bailey-1			Bailey-2		
	Coarse	Fine	Filler	Coarse	Fine	Filler	Coarse	Fine	Filler
	Cumulative percentage passing								
19.0	100	100	100	100	100	100	100	100	100
13.2/12.5	90	100	100	90	100	100	90	100	100
9.5	58	100	100	58	100	100	58	100	100
4.75	24	100	100	24	100	100	24	100	100
2.36	0	100	100	0	100	100	0	100	100
1.18	0	80.2	100	0	77.6	100	0	84.2	100
0.60	0	60.6	100	0	64.2	100	0	68.4	100
0.30	0	40.8	100	0	28.4	100	0	42.2	100
0.15	0	25.4	100	0	15	100	0	18.8	100
0.075	0	7.2	84	0	8	84	0	0	84
0.032	0	0	80	0	0	80	0	0	80
Final proportion (%)	50.02	45.61	4.38	49.96	44.79	5.25	50	38.13	11.88
	Unit weight (AASHTO T19 2004)								
LUW (kg/cm^3)	1523.63	–	–	1523.63	–	–	1523.63	–	–
RUW (kg/cm^3)	1653.98	1961.8	–	1653.98	1983.19	–	1653.98	1903.03	–
CUW (%)	74.80			75.17			73.45		
	Bailey aggregate ratios								
FG-CA	1			0.6			1.0		
FG-FA$_c$	0.5			0.35			0.50		
FG-FA$_f$	0.22			0.35			0.50		

Fig. 1 Aggregate gradation

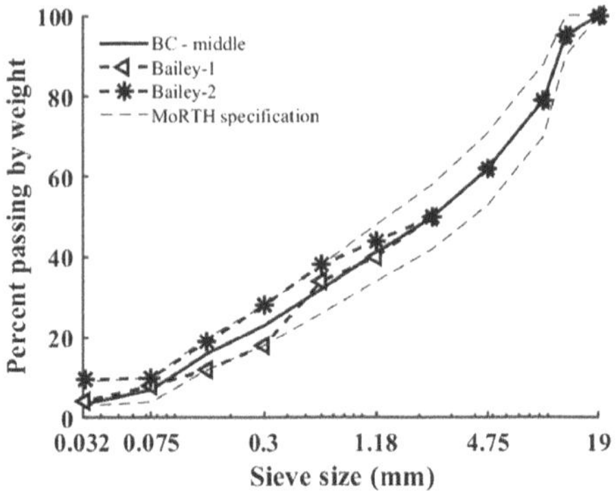

during compaction than the binder type. In this study, the influence of aggregate gradation on the compaction characteristics of bituminous mixtures was studied by applying vertical stress of 650 kPa and a shear angle of 4°.

4.1.2 Compactive Effort

The evolution of the height of the asphalt mixtures during compaction and the number of cycles of compaction the shear compactor takes to reach the required air voids for each type of gradation was studied. Figure 2 shows the evolution of height during compaction. The height is normalized with respect to the height corresponding to the first cycle. It can be observed from the figure that all the bituminous mixtures follow the same trend, and the aggregate gradations (BC-middle, Bailey-1, and Bailey-2) have negligible influence. Also, it was found that the number of cycles taken by all the mixtures to reach target air voids is in the range of 8–10, indicating that all the mixtures require nearly the same compaction effort.

Fig. 2 Compaction curve

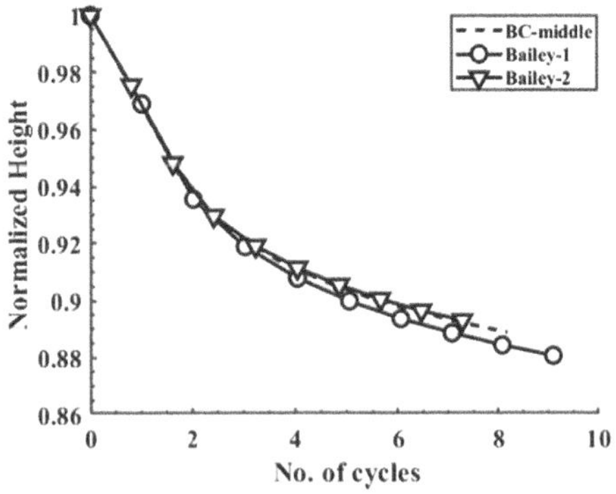

4.2 Flow Number Test

4.2.1 Test Methodology

Flow number test is a repeated creep and recovery test with a fixed rest period. In this study, the flow number test was used to characterize the influence of three different gradations on the rutting resistance of unmodified bituminous mixtures. The test was conducted using the asphalt mixture performance tester (AMPT). The test protocol consists of applying a repeated trapezoidal loading for 1 s, which is followed by 2 s rest period (Gayathri et al. 2016). In the trapezoidal loading, the deviatoric stress is linearly ramped in such a way that the desired stress is achieved in 0.1 s, and stress is held constant for 0.8 s followed by a ramp down in 0.1 s. The deviatoric stress (600 kPa) and confinement pressure (75 kPa) for the current investigation was chosen based on an earlier study by Gayathri et al. (2016) in which a three-stage creep curve was obtained. The test was run for 10,000 cycles or till actuator limit is reached. The applied load, confining pressure, temperature, and resulting axial deformation was recorded every 1/1000th s. The test was carried out at two different temperatures—45 and 55 °C.

4.2.2 Creep and Recovery Response

Figure 3 show the applied load and the resulting actuator deformation for one cycle. The response includes the residual deformation at the end of each cycle and total accumulated deformation at the end of the trapezoidal load. The evolution of total accumulated deformation with the number of cycles for three different bituminous mixtures at two different temperatures can be seen in Fig. 4. The repeatability of the test results was randomly checked by conducting experiments twice on BC-middle mixtures at 55 °C. The precision limits provided for the unconfined flow number test in AASHTO TP-79 (2011) was considered for checking repeatability. The deviation from the mean was found to be 4.35%, which is well within the

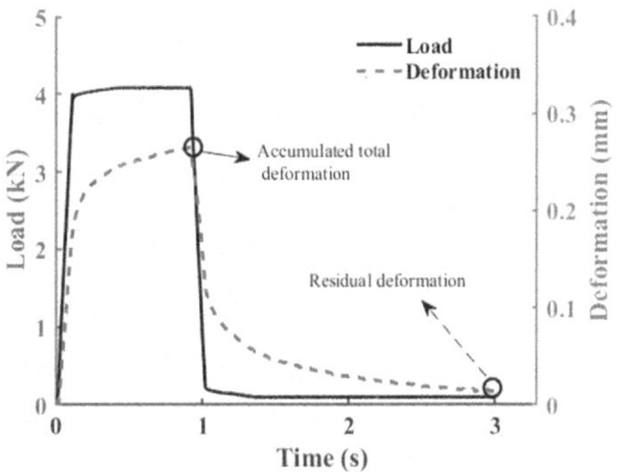

Fig. 3 Applied load and the resultant deformation for one cycle at 45 °C for Bailey-2 mixture (600 kPa deviatoric stress and 75 kPa confinement pressure)

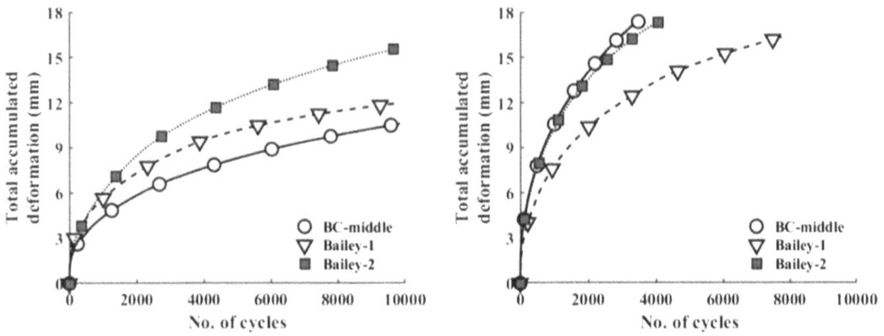

Fig. 4 Comparison of total accumulated deformation between different bituminous mixtures

acceptable range. The deformation curves at both 45 and 55 °C were found to have only primary stage until the end of the experiment. This behavior is one among four different types of deformation patterns observed for bituminous mixtures by Roy et al. (2016). Further, the influence of temperature on rutting characteristics was as expected. The bituminous mixtures tested at 55 °C were observed to reach the maximum deformation limit of the AMPT before 10,000 cycles without showing the tertiary stage.

Further, the influence of aggregate gradation on the deformation characteristics of bituminous mixtures were studied. At a test temperature of 45 °C, it was observed from the flow number test that the total accumulated deformation at the end of 10,000 cycles was more for Bailey-2 mixtures when compared to the other two mixtures. In comparison with BC-middle, the Bailey-2 mixtures exhibited nearly 1.5 times higher deformation. The ranking of bituminous mixtures at 45 °C is BC-middle > Bailey-1 > Bailey-2. It is worth noting here that the Bailey CA ratio and FAc ratio between the BC-middle and Bailey-2 mixtures are the same, whereas the FA_f ratio is different (0.22 for BC-middle and 0.50 for Bailey-2 mixtures). The bituminous mixture with lower FA_f ratio had a higher resistance to rutting.

Further, at a test temperature of 55 °C, all the mixtures reached the actuator deformation limit within 10,000 cycles of loading. Among these mixtures, Bailey-1 mixture was found to outperform when compared to the other two mixtures. The Bailey-1 mixture took nearly twice the number of cycles when compared to the other mixtures to reach the same total deformation. An interesting observation is that the Bailey-1 gradation falls below the BC-middle gradation towards the coarser side (see Fig. 1). This could possibly be the reason for higher resistance to rutting than the other two mixtures. The Bailey-1 gradation was designed by considering the lower side of the Bailey aggregate ratios for fine graded mixtures.

The influence of finer particles, particularly in the presence of confinement pressure, is found to play a critical role at 55 °C. For instance, the percentage passing 0.032 mm sieve size is not controlled in the BC-middle gradation and is performing better in terms of rutting when compared to the other mixtures at 45 °C. However, the same BC-middle mixture is found to have a low resistance to rutting among all the mixtures at 55 °C. Bailey-1 mixtures performed better than the other two mixtures in the presence of confinement pressure at 55 °C. The finer portion of the Bailey mixtures was effectively controlled by introducing a new sieve, 0.032 mm, in addition to the sieve sets recommended in the existing MoRTH (2013) specifications. In the case of Bailey-1 mixture, the percentage passing 0.032 mm sieve size was controlled in such a way that it gives an FA_f ratio of 0.35. Thus, from the current investigation, it can be concluded that the combination of percentage aggregate particles passing 0.032 mm sieve and the confinement pressure has minimal effect at 45 °C while the effect is significant at 55 °C

5 Conclusion

This study aimed to investigate the influence of fine aggregate gradation on the rutting potential of bituminous mixtures. Three different aggregate gradations were selected. The control gradation, BC-middle, is the mid-gradation (BC-grade II) used in India. The other two gradations, Bailey-1 and Bailey-2, were obtained using the principles of the Bailey method. The rutting performance at two different temperatures, 45 and 55 °C, was captured in terms of total accumulated deformation from the repeated trapezoidal creep and recovery test. The Bailey-1 mixtures performed satisfactorily than the other two gradations at both the temperatures.

The sensitivity of the amount and aggregates passing 0.032 mm sieve size on deformation characteristics was found to be significant at 55 °C, particularly in the presence of confinement pressure, as was seen for the Bailey-1 mixture. In most of the existing specifications for dense graded bituminous mixtures, the aggregate gradation is only specified till 0.075 mm. It is recommended to introduce 0.032 mm sieve since it helps in controlling the FA_f ratio for the fine graded mixtures effectively.

The bituminous mixtures in the current study were designed using the framework provided in MoRTH (2013) specifications for the dense graded bituminous mixtures. The binder content for all the mixtures was kept constant. In order to effectively quantify the effect of finer portion of the aggregate gradation on the deformation characteristics of the bituminous mixtures, an optimal binder content for each mixture has to be determined. The compactibility of such mixtures have to be studied before arriving at an optimal aggregate gradation. Further, this study was limited to one set of loading condition with the use of an unmodified binder. Currently, tests are being carried out at different levels of confinement pressure and deviatoric stress with different types of binders.

References

AASHTO T19 (2014) Standard method of test for bulk density ("unit weight") and voids in aggregate. American Association of State and Highway Transportation Officials, Washington, D.C.

AASHTO R30 (2002) Standard practice for mixture conditioning of hot mix asphalt. American Association of State Highway and Transportation Officials, Washington, D.C.

ASTM D7981 (2015) Standard practice for compaction of prismatic asphalt specimens by means of the shear box compactor. ASTM International, West Conshohocken

Button JW, Perdomo D, Lytton RL (1990) Influence of aggregate on rutting in asphalt concrete pavements. Transp Res Rec 1259:141–152

Daniel JS, Rivera F (2009) Application of the Bailey method to New Hampshire asphalt mixtures (No. FHWA-NH-RD-13733F)

Gayathri VG, Rajasekar YP, Roja KL, Krishnan JM (2016) Influence of confinement pressure on the development of three stage curve for bituminous mixtures. Transp Dev Econ 2(2):10

IS:73 (2013) Specifications for paving bitumen, Fourth Revision, Bureau of Indian Standards, New Delhi

Kandhal PS, Cooley LA (2001) The restricted zone in the superpave aggregate gradation specification. National Cooperative Highway Research Program (NCHRP) report no. 464, National Research Board & National Research Council, National Academy Press, Washington, D.C.

Mohammad LN, Al-Shamsi K (2007) A look at the Bailey method and locking point concept in superpave mixture design. In: Transportation research board circular E-C124: practical approaches to hot-mix asphalt mix design and production quality control testing, pp 12–32

MoRTH (2013) Specifications for road and bridge works. Section 500, Fifth Revision, Indian Roads Congress, New Delhi, India

Qiu J, Li N, Pramesti F, van de Ven M, Molenaar A (2012) Evaluating laboratory compaction of asphalt mixtures using the shear box compactor. J Test Eval 40(5):844–852

Roy N, Veeraragavan A, Krishnan JM (2016) Influence of confinement pressure and air voids on the repeated creep and recovery of asphalt concrete mixtures. Int J Pavement Eng 17(2):133–147

Vavrik WR (2000) Asphalt mixture design concepts to develop aggregate interlock. Doctoral dissertation, University of Illinois at Urbana-Champaign

Witczak MW, Kaloush K, Pellinen T, El-Basyouny M, VonQuintus H (2002) Simple performance test for superpave mix design. NCHRP report no. 465, Transportation Research Board, National Research Council, National Academy Press, Washington, D.C.

Experimental Investigation of Pothole Repair Materials

Debaroti Ghosh, Mugur Turos, and Mihai Marasteanu

Abstract Pothole repairs represent a major maintenance item in the budget of many highway agencies. Currently, there are no required specifications for patching materials. Although the appearance of potholes every spring is a major public relations concern, limited experimental work has been performed on pothole repair materials to evaluate their mechanical properties. The focus of this study is to investigate relevant mechanical properties of current pothole materials. A total of six materials consisting of both summer and winter mixtures were used in this study. Several issues were encountered during the sample preparation of cold mixtures, such as the need of significant curing to gain strength and stiffness at low temperatures. For the cold mixtures, only Indirect Tensile creep and strength testing were performed, while for the other mixtures fracture energy and toughness were also determined. Experiments were also performed to evaluate if the addition of graphene nano-platelets (GNP) to patching materials improve their properties. Based on the results, several recommendations were made to improve the durability of pothole repair materials.

Keywords Pothole repair · Experimental testing · Cold mixtures · Curing · Indirect tensile strength

1 Introduction

Pothole formation is mainly caused by the delayed response to fixing common pavement distresses in the initial phase of their development. The most common distress responsible for pothole formation is cracking that can be the result of different failure mechanisms. Despite considerable progress in the design and selection of pavement

D. Ghosh (✉)
Nichols Consulting Engineers, Sacramento, CA 95630, USA
e-mail: dghosh@ncenet.com

M. Turos · M. Marasteanu
University of Minnesota, Twin Cities, Minneapolis, MN 55455, USA
e-mail: turos001@umn.edu

M. Marasteanu
e-mail: maras002@umn.edu

C. Raab (ed.), *Proceedings of the 9th International Conference on Maintenance and Rehabilitation of Pavements—Mairepav9*, Lecture Notes in Civil Engineering 76,
https://doi.org/10.1007/978-3-030-48679-2_2

materials, pothole repair materials remain an area in which little progress was made. Most of the research reports published focus on the technological aspects of pothole repairs and less on mechanical properties (Biswas et al. 2016; Chatterjee et al. 2006; Chen et al. 2015; Dong et al. 2014; Evans et al. 1993; Maher et al. 2001; Liao et al. 2016; Rosales et al. 2007; Wilson and Romnie 1994).

A very comprehensive research effort on potholes was recently completed as part of a two-year European study called POTHOLE that involved seven countries (Nicholls et al. 2014). The main objective of the project was to address "the road agencies' need for durable construction and maintenance methods for the repair of damage occurring after hard winters due to repeated frost-thaw cycles." Currently, there are no required specifications for pothole repair materials in the US and Europe, making it very difficult for stakeholders to choose between the materials available on the market. In this investigation, experimental work is performed on several readily available materials to address this issue and a number of recommendations are made.

2 Experimental Investigation

The experimental investigation was conducted in two phases. In the 1st phase, cold mixtures used for pothole repairs were investigated, and several sample preparation methods were attempted to prepare testing specimens. The interface between the existing paving material and the pothole repair material was also investigated by preparing specimens with both hot-mix asphalt (representing existing pavement) and a winter mix (representing a pothole repair material). In the 2nd phase, a hot mix asphalt material used for summer repairs was tested, for comparison purposes. A new additive consisting of graphite nano-platelets (GNP) was added to the hot mix asphalt to determine if it improves mechanical properties.

2.1 Materials

A total of six pothole patching materials were tested in this investigation. A summary of the materials is presented in Table 1.

Figure 1 shows a few representative photos of the materials as well as of aggregate residue at the end of the ignition oven test.

Perma-Patch is a repair material with a proprietary formula composed of asphalt, a special aggregate, and pressure-sensitive plastics. This is the most common and frequently used pothole patching materials on the market.

UPM is a cold-mix repair material for asphalt and concrete pavements manufactured by Unique Paving Materials Corporation. Two types of UPM mixes were used in the experimental investigation: Winter Mix UPM, which is a winter grade

Table 1 List of materials tested

Patching materials	Acronym
DOT approved Perma-Patch from MnDOT	PP
Winter Mix from St. Paul Asphalt Plant	WM_SP
Winter Mix from Unique Paving Materials (UPM)	WM_UPM
Summer Mix from Unique Paving Materials (UPM)	SM_UPM
GAP Patching Mix	GAP
Summer Mix from St. Paul Asphalt Plant	SM_SP

Fig. 1 Pothole repair materials and aggregate residues

2 material specifically formulated to be used at temperatures below 40 °F; Summer Mix UPM, a summer grade 4 material, specifically formulated to be used between 60–80 °F.

The winter mix from St. Paul Asphalt Plant is designated as 6C and uses 12.7 mm minus aggregates. The oil content is 5.5% of SC-800 (slow cure). The asphalt binder is a PG 64-22.

GAP Mastic is a hot-applied, polymer modified asphalt mixed with engineered aggregates and modifiers designed to fill wide cracks and defects, to prevent water infiltration, and restore ride quality, according to the manufacturer. Unlike the other patching materials, GAP is a more expensive product that has specific preparation requirements.

The summer mix from St. Paul Asphalt Plant is a regular hot mix asphalt.

2.2 *Phase 1 Experimental Work*

One of the biggest challenges of testing the first five patching materials was the sample preparation. Different compaction techniques and conditioning methods were used to accomplish this goal. First, a gyratory compactor was used to prepare specimens for Indirect Tensile Strength (IDT) testing. However, the compaction of the repair materials at room temperature was not successful. Next, manual compaction was attempted by pouring a fixed amount of material into a mold and using a solid steel cylinder to compact the material at room temperature. Since this procedure is difficult and not repeatable, an MTS testing frame was used to simulate Marshall Compaction for repeatability. Several 100 compression cycles of a square wave between 0 and 20 kN at a frequency of 1 Hz were performed on each side of the mold. This compaction method was further used to obtain specimens for IDT testing.

2.2.1 IDT Testing

IDT testing was performed according to AASHTO T322-07. Unlike the preliminary attempts, in which the material was compacted at room temperature, cold conditioning of the materials was used before compaction to better simulate winter patching at low temperature. Test specimens were prepared as follows. First, a fixed amount of material was poured into steel molds. Molds were kept at 4 °C for 18 h in a refrigerator. After 18 h, molds were taken out of the refrigerator and compacted immediately. Specimens were extracted right away after compaction and conditioned for one hour at the test temperature (0 °C and −24 °C, respectively) in the MTS environmental chamber. Compaction was not performed on GAP samples, since the compaction of this material is not required in the field. Also, GAP needed some special sample preparation beforehand, since the binder and aggregate came separately for the materials. For GAP, the aggregates and the binder were kept in the bowl in the oven at 170 °C for 6 h and the heated aggregate and binder were mixed with the mixer for five minutes. The mixture was poured into the mold right away. Initially, the mixture was poured in hot molds and molds at room-temperature. With the hot molds it was observed that, due to a large amount of binder and lower aggregate cooling rate, the binder started to accumulate at the bottom of the specimen, which made extracting the specimens difficult. Consequently, cold or room-temperature molds were used.

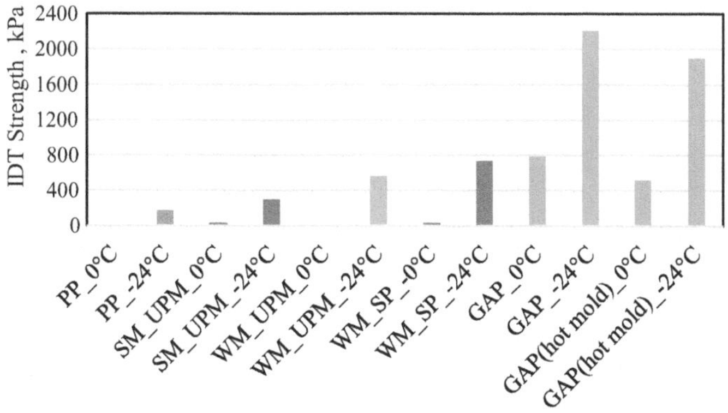

Fig. 2 IDT strength test for cold conditioning of patching materials compacted in steel molds at 0 and −24 °C

The strength results obtained following the procedure described above are shown in Fig. 2.

The GAP was found to have the highest strength followed by the winter mix from St Paul asphalt plant (WM_SP). As expected, the test temperature had a significant effect on the strength value. All patching materials had very low strength values at 0 °C. By comparison, a typical asphalt mixture has values in MPa rather than kPa. It was observed that for GAP, using a hot mold reduced the tensile strength at both testing temperatures. A much slower cooling rate of the material is the reason which allowed the binder to flow to the bottom of the testing specimen and making it weaker.

2.2.2 Bonding Experiment

Experiments were also performed to determine if qualitative information can be obtained regarding the bonding strength between the patching material and the existing asphalt mixture in the pavement. In this part, only two patching materials were used; Perma-Patch (PP) and Winter Mix_UPM (WM_UPM). Samples were prepared using half HMA and half patching material (PP and WM_UPM) to simulate the actual field condition and check the bonding at the interface of existing pavement and patching materials (Fig. 3). The diametric load was applied along the joint between the patching material and HMA and the crack propagation through the joint was observed. Patching materials were also tested individually to compare the test results with samples of half HMA and half patching materials. Specimen preparation and scheme photos of specimen preparation are shown in Fig. 3.

Since bonding between the cold patching mix and asphalt mix would not become effective until significant curing occurs, two accelerated methods of curing were used: in the first one the specimen was conditioned in the oven before compaction, and in the second method, the specimen was conditioned in the oven after compaction.

Fig. 3 Preparation of specimens with half HMA and half patching material bonding test

For conditioning specimens before compaction, materials were kept in the mold that was placed in an oven at 135 °C for 20 h, followed by compaction at room temperature. The compacted specimen was kept at room temperature for 5 days. After 5 days, the specimen was extracted and kept at test temperature (−24 °C) for 1 h before testing.

For conditioning specimens after compaction, the material was first compacted at room temperature and the compacted specimen was kept in the mold in an oven at 40 °C for 5 days. After 5 days, the specimen was extracted and kept at test temperature (−24 °C) for 1 h before testing.

Figure 4 shows the bond strength results. As expected, the HMA specimen used in the bonding experiments has the highest IDT strength. Conditioning samples in the oven before or after compaction did not make a significant difference in strength results. However, bonded samples consisting of half patching materials and half HMA were observed to have higher strength than the samples of the patching materials itself. Winter Mix from UPM was observed to have a strong interface bond with HMA (Fig. 4).

2.3 Phase 2 Experimental Work

In this phase of testing, graphite nano-platelets (GNP) were added to the Summer Mix from St. Paul. Both control mix (without GNP) and mix treated with GNP were

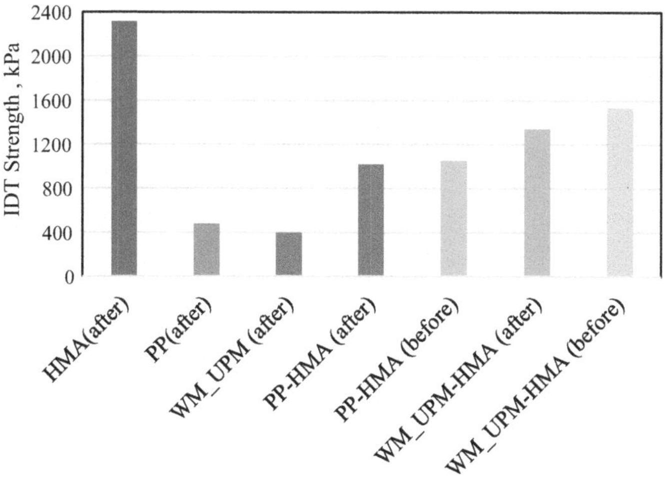

Fig. 4 IDT strength results for bonding specimens at −24 °C

tested using several testing methods. Three mixtures were created using the St Paul Summer Mix: a mixture containing 6% Micro 850 GNP, a mixture containing 6% 4827 GNP and a control mix, which contained no GNP additive.

The material was first heated in oven at 135 °C for 6 h. The mixer was then used for 1 min to mix the additive and heated asphalt material. The treated material was placed back into oven to keep warm until compaction. The material was compacted at 100 gyrations in Superpave gyratory compactor. The compacted mixture was then cured at room temperature for two days. Each cylinder was cut into two 38 mm thick samples for IDT testing and one 32 mm thick sample for SCB testing, that was further divided into two halves. Three tests were performed on the St Paul Summer Mix: IDT Creep Stiffness and IDT Strength according to AASHTO T322-07, and SCB fracture and toughness according to TP 105-13. The air void content was also investigated to check the compaction level.

Figure 5 shows air void content of the mixtures during compaction. The mixtures with GNP compact more quickly (with fewer gyrations), and the samples with 4827 GNP additive compact 1% more than the control mixture.

In the creep test, specimens were loaded diametrically using a vertical constant load of 0.8 kN/s. Horizontal and vertical deformation were measured using extensometers fixed near the center of the sample. Deformation measurements were then used to calculate creep stiffness. Three replicates were tested for each material at three temperatures, −24, −12, and 0 °C, for a total of 9 tests. Creep tests were followed by strength tests.

Creep stiffness curves were constructed for all three materials at all three testing temperatures. Average values from three replicates per material were used to construct curves . Figure 6 contains curves at −24, −12, and 0 °C.

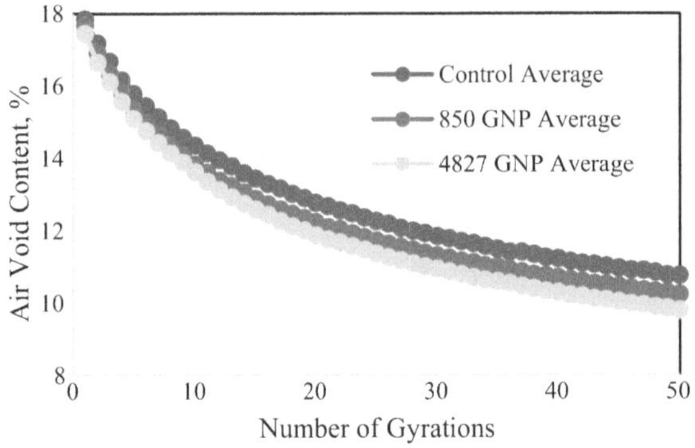

Fig. 5 Air void content of samples during compaction of Summer Mix

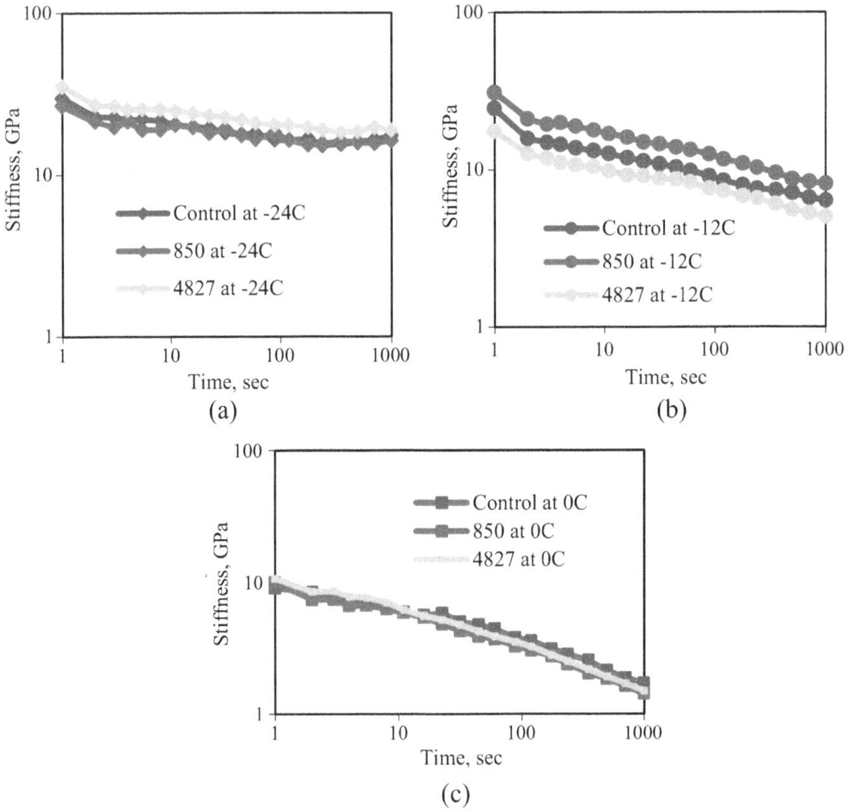

Fig. 6 Creep stiffness curves of Summer Mix at **a** −24 °C, **b** −12 °C and **c** 0 °C

Table 2 Tensile strength, SCB fracture energy and fracture toughness of St. Paul Summer Mix

Patching material	Tensile strength, kPa		Fracture energy, kJ/m^2		Fracture toughness, MPa * m$^{0.5}$	
	0 °C	−12 °C	−12 °C	−24 °C	−12 °C	−24 °C
Summer Mix without GNP	2601	2688	0.362	0.269	0.524	0.659
Summer Mix with 6% 850 GNP	2793	3263	0.377	0.270	0.572	0.673
Summer Mix with 6% 4827 GNP	2915	3418	0.371	0.341	0.633	0.716

Values for tensile strength are presented in Table 2. Each value is an average taken from three replicates. As shown, tensile strength increases in the samples containing GNP additive, and the 4827 GNP additive has a slightly higher increase in tensile strength than samples containing Micro 850 GNP additive.

SCB fracture testing was performed according to AASHTO TP 105-13. The results for fracture energy and fracture toughness are presented in Table 2. It can be observed that the fracture energy and fracture toughness increase in mixtures containing GNP additive, and the mixture containing the 4827 GNP additive has the highest increase in fracture toughness.

3 Conclusions and Recommendations

At the end of this study, several conclusions and recommendations were made. It was very difficult to prepare testing specimens since the patching materials are cold mixtures that do not gain strength and stiffness unless significant curing occurs. While cold mixtures can be poured and placed at typical winter temperatures, only minimal curing occurs at these temperatures, which makes these materials very weak, with characteristics more like a filling material.

GAP was the strongest materials tested in Phase 1 at both 0 and −24 °C. At −24 °C, GAP had strength values similar to regular hot mix asphalt.

The St Paul Summer Mix had properties similar to regular hot mix asphalt and addition of GNP was possible to this mix only. It was found that the addition of GNP increases compaction level. Tensile strength, fracture energy and fracture toughness all increased for the mixtures prepared with GNP additive. These preliminary results indicate that the addition of GNP can positively influence the performance and durability of summer patching materials.

The results obtained in this investigation confirm that, in general, cold mixtures used for winter patching have poor mechanical properties and are expected to require frequent re-repairs, especially in winters with many freeze-thaw cycles. Winter patching materials, such as GAP, that require an external source of heat have significantly better mechanical properties and are expected to last longer. However, they are more expensive and require specialized equipment and training.

A number of new materials and innovative technologies are available to provide more durable solutions for winter pothole repairs, which represent the most challenging situation: the repair materials need to be placed at cold and very cold temperatures at which curing and compaction is not possible unless additional heating is provided. Examples are the use of GAP, infrared heating, and use of taconite mixtures and microwave heating (Nazzal et al. 2014; Zanko 2015). The use of GNP additives shows promising benefits for summer patching since they increase the compaction level. The addition of GNP may provide benefits for winter pothole repair since they can be heated significantly by application of microwave energy, similar to the effect observed in taconite aggregates.

References

Biswas S, Hashemian L, Hasanuzzaman M, Bayat A (2016) A study on pothole repair in Canada through questionnaire survey and laboratory evaluation of patching materials. Can J Civ Eng 43(5):443–450

Chatterjee S, White RP, Smit A, Prozzi J, Prozzi JA (2006) Development of mix design and testing procedures for cold patching mixtures. Report no. FHWA/TX-05/0-4872-1, Texas Department of Transportation, Federal Highway Administration, Washington, D.C.

Chen JS, Ho HC, Liao MC, Wang TY (2015) Engineering properties of asphalt concrete patching mixtures. J Test Eval 44(1):534–542

Dong Q, Huang B, Zhao S (2014) Field and laboratory evaluation of winter season pavement pothole patching materials. Int J Pavement Eng 15(4):279–289

Evans LD, Mojab CG, Patel AJ, Romine AR, Smith KL, Wilson TP (1993) Innovative materials development and testing. Strategic Highway Research Program, SHRP-H-352, vol 1: Project Overview, National Research Council, Washington D.C.

Liao MC, Luo CC, Wang TY, Xie X (2016) Developing effective test methods for evaluating cold-mix asphalt patching materials. J Mater Civ Eng 28(10):04016108

Maher A, Gucunski N, Yanko W, Petsi F (2001) Evaluation of pothole patching materials. Report no. FHWA NJ 2001-02, Federal Highway Administration, U.S Department of Transportation, Washington D.C.

Nazzal M, Kim S, Abbas A (2014) Evaluation of winter pothole patching methods. Report no. FHWA/OH-2014/2, Ohio Department of Transportation, Office of Statewide Planning & Research, Cincinnati, OH

Nicholls C, Kubanek K, Karcher C, Hartmann A, Adesiyun A, Ipavec A, Komacka J, Nielsen E (2014) Durable pothole repairs. TRA-Transport Research Arena

Rosales VI, Prozzi J, Prozzi JA (2007) Mixture design and performance-based specification for cold patching mixtures. Report no. FHWA/TX-08/0-4872-2, Texas Department of Transportation, Federal Highway Administration, Washington, D.C.

Wilson TP, Romnie AR (1994) Pavement surface repair materials and procedures: training program-pothole repair. Report no. SS-20, Strategic Highway Research Program, Washington, D.C.

Zanko LM (2015) Evaluate and develop innovative pavement repair and patching: taconite-based repair options. Contract no. 99008, Work Order 51 Draft Final Report, Natural Resources Research Institute, University of Minnesota Duluth, MN

A Durable Potholes Repair Method Using Polymer Modified Patching Material in Cold-Wet Weather

Sen Han, Jinping Xia, Hui Xu, and Hongwei Zhang

Abstract There are two common deficiencies of the existing cold asphalt mixtures, one is the contradiction between workability or storage and strength, and the other is the problem of low initial strength after paving. In order to solve the problems, new polymer modified patching materials with dense gradation (PADG mixtures) are recommended for durable pothole repair in cold-wet weather, based on microcapsule technology and reinforcement action. In this paper, laboratory test and field investigation were performed to assess the performance of PADG mixtures, compared with dense graded cold mixtures (DG mixtures) and open graded cold mixtures (OG mixtures). It showed that the PADG mixtures solved the contradiction between workability, storability and strength of cold mixtures, and had better cohesion, stability, freeze-thaw resistance and durability in low temperature and wet conditions. Besides, the field survey over 13 months indicated that the pothole patches using PADG mixtures met the requirements of traffic safety and ride-ability. Therefore, combined with laboratory and field text results, it can be concluded that PADG mixtures are applicable to durable maintenance in adverse conditions of cold-wet weather.

Keywords Pothole repairs · Patching materials · Microcapsule · Polymer · Low temperature · Wet

S. Han (✉) · J. Xia
School of Highway, Chang'an University, Xi'an, Shannxi, China
e-mail: hyram_hs@chd.edu.cn

J. Xia
e-mail: 1297536011@qq.com

H. Xu
Gansu Province Highway Bureau, Lanzhou, Gansu, China
e-mail: 934381322@qq.com

H. Zhang
Inner Mongolia Transport Construction Engineering Quality Supervision Bureau, Huhehaote, China
e-mail: hhhtzhw@qq.com

C. Raab (ed.), *Proceedings of the 9th International Conference on Maintenance and Rehabilitation of Pavements—Mairepav9*, Lecture Notes in Civil Engineering 76,
https://doi.org/10.1007/978-3-030-48679-2_3

1 Introduction

Usually, the bowl-shaped holes on the surface of asphalt pavement called pothole problems, which adversely affect traffic safety and ride-ability (Dong et al. 2014; Liu et al. 2019a, b). Adverse weather and vehicular loads over asphalt pavements lead to the formation of bowl-shaped potholes, especially in winter or rainy seasons (Chatterjee et al. 2006; García 2017; Han et al. 2019). Water is the main factor for the formation of potholes, which penetrates into the internal pavement through cracking due to low-temperature deformation or fatigue (Wilson and Romine 1993). In the end, it would accelerate the failure of the pavement structure. Moreover, the construction and durability of pothole repairs are affected by low temperature and water (Prowell and Franklin 1996). In order to prevent further damage, we should take durable measures in time.

According to the construction technology, patching materials can be divided into three types: hot-mixed and hot-placed patching mixtures, hot-mixed and cold-placed patching mixtures, cold-mixed and cold-placed mixtures (Maher et al. 2001; María and Pimentel 2007). The cold-placed mixtures are more suitable in winter for the use of cutback asphalt and emulsified asphalt. However, construction temperature of these patching mixtures should be controlled rigidly (Maher et al. 2001). Generally, cold mixtures use continuous gradation and gap-graded gradation. The continuous gradation is good for pavement performance but has poor workability. The performance of gap-graded gradation mixtures is reversed (Chatterjee et al. 2006). Therefore, it is of vital importance to select the binder and gradation when designing the patching materials.

2 Objective and Scope of Study

The main objective of this paper is to propose a new polymer modified dense graded patching mixtures (PADG mixtures), and verify the practicability of PADG mixtures for timely and durable pothole repair in winter-rainy conditions. Past studies have summarized the main properties of cold mixtures, such as workability, storability, stability, anti-stripping, freeze-thaw resistance, durability, bonding performance and skid resistance (Dong et al. 2014; Prowell and Franklin 1996). The scope of this study consists of two parts, aiming to assess applicability of modified cold mixtures under adverse weather and heavy traffic load by laboratory tests and field surveys. The specifics of study scope are presented as follows:

Part A: Laboratory tests were conducted to evaluate the pavement performance of PADG mixtures, compared with the two typical patching materials of dense graded cold mixtures (DG mixtures) and open graded cold mixtures (OG mixtures), including workability, cohesion, storability, bonding performance in low temperature and wet conditions, and durability.

Part B: PADG mixtures were installed in the northeastern and southeastern part of China with complex weather. The field survey involved apparent damage, texture, skid resistance, seepage resistance and dishing depth.

3 Action Mechanism of PADG Mixtures

3.1 Film-Forming Capsule Mechanism

The capsule technology is realized by the film-forming component in additive of binder, in order to improve the workability and storability of PADG mixtures. Film-forming agent is a class of substances that contain conjugated olefinic bonds in the molecular structure. After exposure to air in a while, these substances undergo a series of oxidative cross-linking polymerization reactions due to the function of oxygen (Allen et al. 1991). Flexible solid film with spatial network structure was formed eventually, which has a certain compressibility and high density (Alanalp and Durmus 2018; Lv et al. 2016). If constant and sufficient pressure is loaded on the capsules, that would cause the film break down and the modifier and tackifier flow out of the capsules. As shown in Table 1, the microcapsules forming process is mainly described in the following four stages:

I. Introduction stage: An olefinic bond in the film-forming agent molecules breaks, resulting in a relatively stable new olefinic bonds and two free radicals. The free radicals would take away one electron from other substances to make themselves stable, since the electrons in the chemical bonds should appear in pairs.

Table 1 Principle of microcapsules forming procedure

Conjugated olefins:	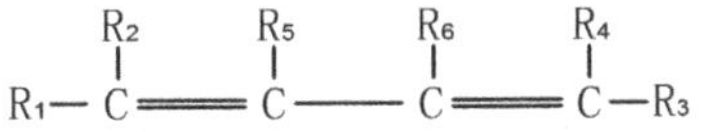
Introduction stage:	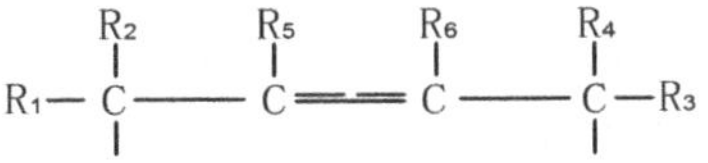
Oxidation stage:	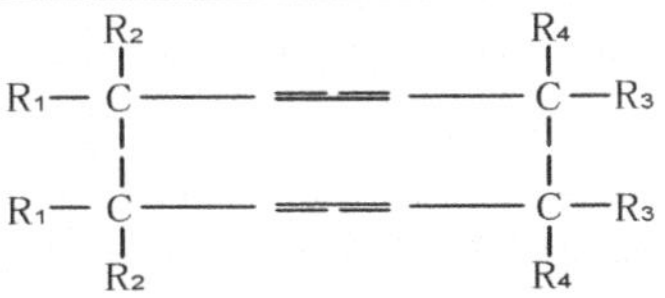
Film-forming stage	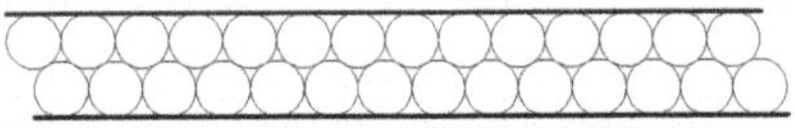
Microcapsules stage	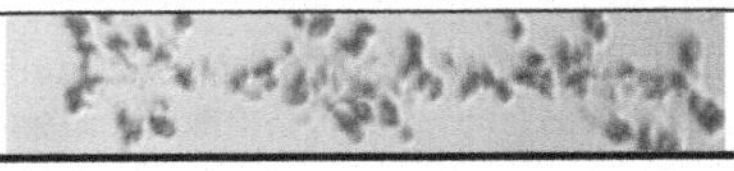

Note: R1, R2, R3, R4, R5 and R6 are functional groups in the film-forming agent molecules.

Ii. Oxidation stage: Reactivity of film-forming agent's molecular is enhanced with the increase of free radicals and the entire system is extremely unstable. Two major reactions occur and complete until the depletion of free radicals and the formation of a stable system. Namely, one is the combination of free radicals with oxygen, and the other is the inter-polymerization of free radicals.

Iii. Film-forming stage: Macromolecules from inter-polymerization of free radicals are further crosslinked into a spatial network structure to form film.

Iv. Microcapsules stage: the closed films are seen as microcapsules which can store modifier and tackifier, and spread in the additive evenly.

3.2 Reinforcement Mechanism

According to Mohr-Coulomb theory, the strength composition of cold mixtures is essentially identical to that of hot asphalt mixtures, which mainly depends on the cohesion from binder and the internal friction from the gradation and angularity of aggregates (Yu et al. 2004). However, the strength formation of the cold mixtures has delay and hysteresis characterises, compared with the traditional hot asphalt mixtures (Chatterjee et al. 2006; Estakhri et al. 1999). The strength formation of the cold mixtures can be divided into three stages:

i. Storage stage: Both of cohesion and internal friction of patching mixtures are smaller in this stage. Because some additive is isolated from air and asphalt by microcapsules, resulting in inhibition of binder curing. As for aggregates, the mixtures keep loose state without compaction.

ii. Construction stage: Cohesion and internal friction are enhanced significantly in this stage. PADG mixtures are compacted to form a denser structure compared with OG mixtures due to continuous gradation of aggregates, which benefits the increases of internal friction. At the same time, the microcapsules break and the binder begins to solidify due to sufficient external force, which can improve the cohesion of mixtures. PADG mixtures solve the disadvantages of DG mixtures based on microcapsule technology, i.e. poor workability and storability. It is conducive to initial internal friction and cohesion of patching materials.

iii. Road service stage: The strength continues to grow for denser structure and further curing under traffic loads in this stage. Both soft and hard segments from modifier are introduced into additive of PADG mixtures' binder, which is good for the cohesion of asphalt. And low-temperature deformation resistance and high temperature strength of mixtures can be effectively improved by controlling the ratio of two segments. In addition, the functional groups with low surface energy in tackifier can participate in the formation of the adhesive layer and reduce the surface energy, which improve the infiltration ability of binder. Thereby, the stripping resistance of binder from aggregate and adhesion between mixtures and pothole wall can be enhanced.

4 Test Program

4.1 Materials

As shown in the Fig. 1, PADG mixtures consist of 90# matrix asphalt, additive and aggregates and mineral fines. Among these, the functional components of additive involve modifier, tackifier, and film-forming agent, and solvent 1, solvent 2 and plasticizer are used as corresponding dissolvent.

Both aggregates and fines are limestone. OG mixtures and DG mixtures are produced in Canada and China, respectively, and are mature in engineering applications. The gradation of mixtures are plotted in Fig. 2.

4.2 Methodology

The test consists of two parts aiming to assess applicability of modified cold mixtures under adverse weather and heavy traffic load in the laboratory and field. The test process in laboratory and field were described in Table 2 and Table 3 respectively.

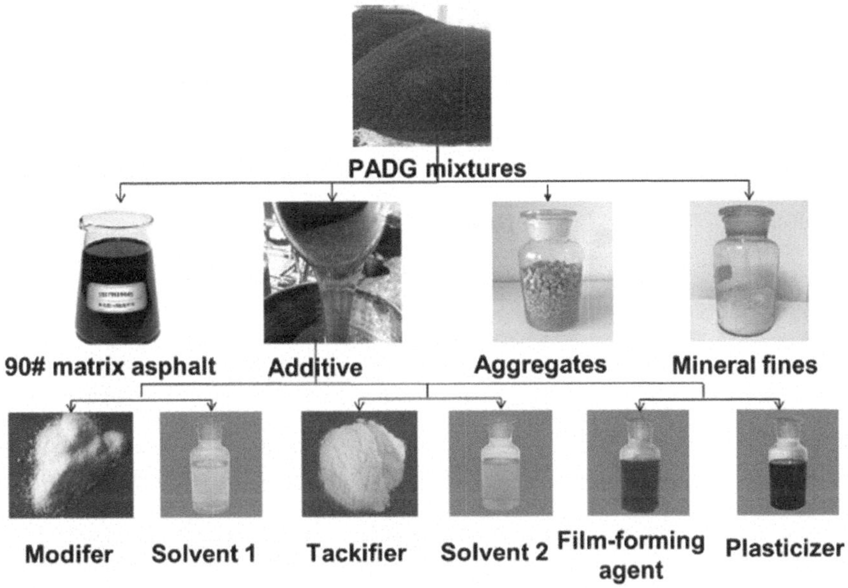

Fig. 1 Compositions of PADG mixtures

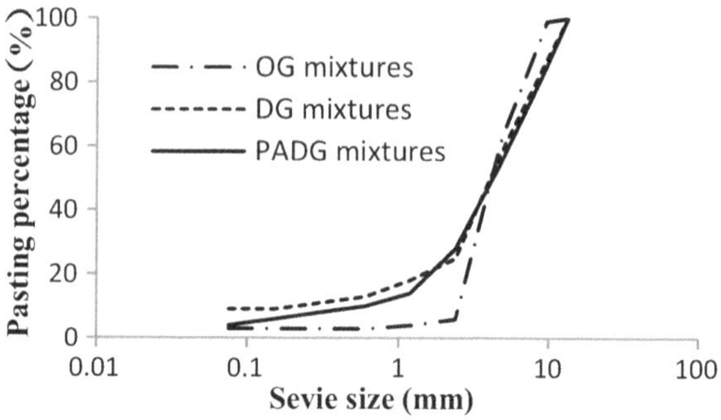

Fig. 2 Gradations of DG mixtures, OG mixtures and PADG mixtures

Table 2 Laboratory tests

Tests types	Properties	Testing process
Workability tests	Workability	3000 g of three cold mixtures were stored for 48 h under 100 kg loads at 5 and 20 °C. Torque when potholes were patched, were measured by the working stirring testers
Storability tests	Storability	3000 g of three cold mixtures were packed in sealed bags under 50 kg load at 5 °C for 3, 7, 30, 60, 120 and 240 days. Torque when potholes were patched, were measured by the working stirring testers
Cohesion tests	Cohesion	Following the testing procedure of the report of FHWA-RD-99-168(1)
Marshall tests	Initial stability	Following the testing procedure of ASTM D 1559 (9). Samples were placed in the air bath for 2 h at 25 °C
	Moulded stability	Following the testing procedure of ASTM D 1559 (9). Samples were placed in an oven at 100 °C for 48 h, then placed at 25 °C for 48 h, next removed to ventilated room for 2 h
	Stability forming speed	Following the testing procedure of ASTM D 1559 (9). Samples were placed in a ventilated room at 5 °C for 3, 7, 15, 30, 45 and 90 days

(continued)

Table 2 (continued)

Tests types	Properties	Testing process
	Low temperature stability	Following the testing procedure of ASTM D 1559 (9). Samples were placed in a refrigerator at −10 °C for 72 h
	Immersed stability	Following the testing procedure of ASTM D 1559 (9). One set of samples were immersed in a 10 °C water bath for 30 min and then placed in 25 °C ventilated room for 72 h. Another set of samples were laid in 25 °C ventilated room for 24 h after being soaked for 48 h
Freeze-thaw tensile strength tests	Durability (low temperature and wet conditions)	Following the testing procedure of Modified Lottman test AASHTO T-283 (10). One set of samples were stored at room temperature for 40 h. Another set of samples were placed in a refrigerator at a temperature of −18 °C for 16 h after saturation. Then samples were removed to a water bath at 10 °C for 24 h
Accelerated loaded tests	Durability (wet and accelerated loaded conditions)	To begin with, two 30 × 30 × 5 cm AC-13 slab specimens were stuck together and the middle section of the upper specimens was excised with the size of 30 × 13 × 5 cm. This simulated potholes were repaired using three patching materials and compacted 24 times with 9 kN. Finally, the composite plates were placed at 80 °C for 96 h and then stored at ambient temperature for 48 h. the wheel pressure of 0.7 MPa was loaded on the composite plates at 30 °C under immersed or dry conditions. The dishing depth of potholes patching in middle of composite plates was measured after 2400 times wheel loads

Table 3 Potholes patching field survey

Section	Weather[a]	Time[b] (months)	Traffic[c]	Test progress
BQ[d] 1	4 °C, cloudy	3	25600	Distress of pothole repair was investigated and field tests were also conducted as follows. DD[f] was obtained by measuring the maximum vertical gap between the surface of potholes patch and the bottom of 3 m straightedge. Besides, MTD[g], BPN[h] and SC[i] were conducted according to the standard sand patch method of ASTM E965-15 (11), British pendulum test procedure of ASTM E303-93(2013) (12) and standard water seepage testing method in China codes of JTG E20-2011 (13) respectively
BQ[d] 2	36 °C, sunny	5		
BQ[d] 3	28 °C, rainy	13		
HQ[e] 1	18 °C, rainy	3	22200	
HQ[e] 2	25 °C, rainy	5		
HQ[e] 3	23 °C, rainy	13		

Note [a]Installation weather; [b]Surveying Time after installation; [c]The Annual average daily traffic values; [d]BaQiao district ring expressway of Xi'an in China; [e]HangQian expressway of Hangzhou in China; [f]Dishing depth; [g]Mean texture depth; [h]British pendulum number; [i]Seepage coefficient

5 Test Results and Analysis

5.1 Laboratory Test Results Analysis

The following conclusions can be drawn from the laboratory test results:

(i) Workability: PADG mixtures and OG mixtures had better workability than DG mixtures. That's because the use of microcapsule technique in PADG mixtures and gap-grading in OG mixtures. In addition, the workability of patching materials deteriorated as the temperature decreases.

(ii) Storability: the workability of PADG mixtures and OG mixtures was good and kept relatively stable state. While the workability of DG mixtures varied over storage time.

(iii) Cohesion: retained percentage of all three cold mixtures is higher than 95%, which demonstrated that patching material could alleviate ravelling and stripping for enough cohesion. PADG mixtures with minimal mass loss had the best cohesion.

(iv) Stability: (a) The initial stability of PADG mixtures was better than that of the DG mixtures and OG mixtures because of dense gradation and additive. (b) PADG mixtures had the highest moulded stability among the three mixtures, followed by DG mixtures and OG mixtures. (c) The stability of PADG mixtures was always higher than DG mixtures and OG mixtures during the curing time of 90 days and reached a steady state earlier. (d) Even in low temperature environment, PADG mixtures had strong stability, indicating that it had sufficient strength to resist damage or deformation in winter. (e) The decrease of the stability of PADG mixtures was smaller than the other two mixtures after immersion in water. That's because water had negative effects on the bonding properties of binder in DG mixtures and OG mixtures.

(v) Durability: (a) The results of freeze-thaw tensile strength tests demonstrated that the patching materials were insensitive to water and low temperature. PADG mixtures had the highest indirect tensile strength, followed by DG mixtures and OG mixtures. (b) The rankings of dishing depth of patching materials was PADG mixtures < OG mixtures < DG mixtures in dry condition and PADG mixtures < DG mixtures < OG mixtures in immersion separately. It can illustrate that PADG mixtures can bear traffic loads even in humid environment. Meanwhile, it also can find that the dishing depth of OG mixtures is larger than that of DG mixtures after immersion. That's because massive pores in OG mixtures accelerated the seepage damage, which make the strength of OG mixture decline sharply.

5.2 *Field Test Results Analysis*

According to the field observation, patches retained satisfactory pavement performance without visible distress of bleeding, dishing, debonding, ravelling and pushing after 13 months, excluding HQ 3. As can be seen from Table 4, DD of HQ 3 was slightly larger than the smoothness requirement of 5 mm from FHWA, but pavement

Table 4 Test results in field

Section	Installation weather	L × W × D (cm × cm × cm)	Time after installation	DD[a] (mm)	SC[a] (ml/min)	MTD[a] (mm)	BPN[a]
BQ 1	4 °C, cloudy	47 × 73 × 7	3 months	2.23	141	0.65	66.0
BQ 2	36 °C, sunny	150 × 168 × 7	5 months	4.45	94	0.62	60.2
BQ 3	28 °C, rainy	102 × 130 × 7	13 months	4.91	68	0.54	53.1
HQ 1	18 °C, rainy	310 × 140 × 7	3 months	3.10	258	0.75	59.4
HQ 2	25 °C, rainy	50 × 210 × 7	5 months	4.51	137	0.68	61.0
HQ 3	23 °C, rainy	267 × 135 × 7	13 months	5.63	166	0.64	55.8

Note [a]L = Longitudinal length; W = Transverse width; D = Depth; DD = Dishing depth; SC = Seepage coefficient; MTD = Mean texture depth; BPN = British pendulum number evenness of the rest 5 installations was still appropriate for good ridestability. SC of all patches were smaller than the maximum requirement from China standard of 300 ml/min from JTG F40-2004 for dense internal structures. Potholes repair using new cold mixtures also can ensure traffic safety with adequate MTD and BPN. HQ 3 was installed near bridge expansion joint, where potholes patching withstood impact load and water easily seeped into the potholes through the interlayer between joint and pothole. Thus there were lightly ravelling and dishing in HQ3 section.

6 Conclusion

Laboratory tests and field surveys were performed to evaluate the performance of cold mixtures installed in low temperature and wet conditions. The main conclusion can be drawn as follows:

The workability, storability, cohesion, stability and durability of PADG mixtures were compared with commonly used DG mixtures and OG mixtures in the lab. As the microcapsules slowing down the cured reaction, the PADG mixtures had satisfactory workability and storability as the OG mixtures. All three mixtures had sufficient cohesion to resist raveling. PADG mixtures had strong stability in low temperature and immersion, which is of great significance for patches to resist the load in adverse weather. The test results of accelerated loaded tests demonstrated that PADG mixtures had good durability and was insusceptible to water.

Pothole repairs using PADG mixtures were patched in different climatic regions and surveyed over 13 months. Pothole patches had quality service capabilities without distress, except HQ 3 for the impact load and seepage near bridge expansion joint. The dishing depth, texture and skid resistance met the requirement for traffic safety and rideability. Besides, pothole patches can resist water damage effectively with low seepage coefficients. Based on the above conclusions, it is believed that PADG is practicable to apply for timely and durable potholes repair in winter-wet conditions.

Unfortunately, this paper only concentrated on the performance of cold mixtures in cold and wet conditions. Further studies involving properties of microcapsules in binder of PADG mixtures are expected to be conducted.

Acknowledgements This research was funded by National Natural Science Foundation of China (No. 51578076), the Fundamental Research Funds for the Central Universities, CHD (No. 300102219207) and Science and Technology Project of Shaanxi Provincial Communication Group (No. 15-04K).

References

Allen NS, Haque E, Howells EM, Kikkawa K, Minagawa M (1991) Synthesis and thermal and photo-oxidative behaviour of novel aminated 2-hydroxybenzophenone stabilizers in polyolefin films. Eur Polymer J 27(8):789–794

ASTM D (1989) Test method for resistance of plastic flow of bituminous mixtures using Marshall apparatus

AASHTO (2004) Standard method of test for resistance of compacted hot mix asphalt (HMA) to moisture-induced damage. T283, Washington, D.C.

ASTM E303-93(2013) (2013) Standard test method for measuring surface frictional properties using the British pendulum tester. West Conshohocken, PA

ASTM E965–15 (2015) Standard test method for measuring pavement macrotexture depth using a volumetric technique. ASTM International, West Conshohocken, PA

Alanalp MB, Durmus A (2018) Quantifying microstructural, thermal, mechanical and solid-state viscoelastic properties of polyolefin blend type thermoplastic elastomer compounds. Polymer 142:267–276

Chatterjee S, White RP, Smit A, Prozzi J, Prozzi JA (2006) Development of mix design and testing procedures for cold patching mixtures (No. FHWA/TX-05/0-4872-1)

Dong Q, Huang B, Zhao S (2014) Field and laboratory evaluation of winter season pavement pothole patching materials. Int J Pavement Eng 15(4):279–289

Dong Q, Yuan J, Chen X, Ma X (2018) Reduction of moisture susceptibility of cold asphalt mixture with Portland cement and bentonite nanoclay additives. J Clean Prod 176:320–328

Estakhri CK, Jimenez LM, Button JW (1999) Evaluation of Texas DOT item 334, hot-mix, cold-laid asphalt concrete paving mixtures. Texas Transportation Institute, Texas A & M University System

Han S, Liu M, Shang W, Qi X, Zhang Z, Dong S (2019) Timely and durable polymer modified patching materials for pothole repairs in low temperature and wet conditions. Appl Sci 9(9):1949

JTG E20-2011 (2011) Standard test methods of bitumen and bituminous mixtures for highway engineering. Ministry of Transport, Beijing

Kandhal PS, Mellott DB (1981) Rational approach to design of bituminous stockpile patching mixtures. Transportation Research Record (821)

Liu M, Han S, Shang W, Qi X, Dong S, Zhang Z (2019a) New polyurethane modified coating for maintenance of asphalt pavement potholes in winter-rainy condition. Prog Org Coat 133:368–375

Liu M, Han S, Han X, Qi X, Dong S (2019b) Microcapsule and polymer reinforcement techniques developed asphalt for use of pothole repairs in winter and rainy seasons. Cold Reg Sci Technol 167:102865

Lv L, Yang Z, Chen G, Zhu G, Han N, Schlangen E, Xing F (2016) Synthesis and characterization of a new polymeric microcapsule and feasibility investigation in self-healing cementitious materials. Constr Build Mater 105:487–495

Maher A, Gucunski N, Yanko W, Petsi F (2001) Evaluation of pothole patching materials (No. FHWA NJ 2001-02)

María J, Pimentel F (2007) Accelerated testing methodology for evaluating pavement patching materials. Presented at 86th annual meeting of the transportation research board, Washington, D.C.

Obaidi H, Gomez-Meijide B, Garcia A (2017) A fast pothole repair method using asphalt tiles and induction heating. Constr Build Mater 131:592–599

Prowell BD, Franklin AG (1996) Evaluation of cold mixes for winter pothole repair. Transp Res Rec 1529(1):76–85

Wilson TP, Romine AR (1993) Innovative materials development and testing. Volume 2: Pothole repair (No. SHRP-H-353)

Yu MH, Yu M, Yu MH (2004) Unified strength theory and its applications

Maximising Stabilisation and Recycling Benefits for Sustainable Pavement Performance in New Zealand and Australia

Allen Browne

Abstract The stabilisation of granular pavement materials and/or underlying soils is accepted practice in Australasia (i.e. Australia and New Zealand). Stabilisation in this context involves the mechanical introduction of reactive agents, typically lime, cement and foamed bitumen. The reuse/recycling of aged existing pavement materials is imperative for sustainable management of finite resources. Reduction in aggregate availability has hastened the need for development of insitu stabilisation to rehabilitate the pavement alongside other performance gains. Hiway Group have commissioned laboratory research and undertaken extensive field trials in partnership with industry and academic partners to lead industry adoption and confidence in sustainable recycling. This paper will outline a variety of proven approaches ranging from hot in-place asphalt recycling through aggregate stabilisation treatments employing waste materials such as ground steel slag to innovative processes to mitigate and control deleterious subgrade soils and low ground pressure fill drying methodologies. Case studies such as exhuming 30+ year-old pavements to evaluate durability of lime stabilised layers will be outlined through to recent research and field trials that successfully incorporate substantial proportions of waste plastic, glass, steel slag and concrete blended recycled aggregates. Examples of structural benefits will be detailed that have been monitored to substantiate performance and calibrate design parameters.

Keyword Sustainable pavement recycling stabilisation

1 Introduction

Hiway Stabilizers NZ (now Hiway Group) commenced in 1986 with a mission statement to integrate insitu stabilisation and recycling to civil works (especially roading) and since then has worked throughout the Pacific, established a strong presence in Australia and grown to be Australasia's largest stabilisation/insitu recycling contractor. Hiways' have maintained a close relationship with research and academic

A. Browne (✉)
Hiway Group, Silverdale 0944, New Zealand
e-mail: allen@hiways.co.nz

C. Raab (ed.), *Proceedings of the 9th International Conference on Maintenance and Rehabilitation of Pavements—Mairepav9*, Lecture Notes in Civil Engineering 76,
https://doi.org/10.1007/978-3-030-48679-2_4

entities through this time to provide robust and independent support for a myriad of innovative recycling initiatives over the last three decades.

Extraction of roading aggregate can only occur where the resource can be economically processed and transported to local markets. An additional 30 km travel distance typically doubles the cost of aggregate. Reprocessing recycled materials can consume substantially fewer resources than extracting and processing raw materials. The purpose of this paper is to outline commonly accepted best practice utilizing stabilisation in New Zealand/Australia, discuss some options to increase recycling and outline innovations that have been accepted and implemented by wider industry.

2 Industry Groups as a Conduit to Encourage Recycling

The NZ National Pavements Technical Group contains 14 pavement/materials specialists representing contracting, consulting, local authority, and national highway roading authorities. This group provides a formal value gateway role for the NZ Transport Agency by providing industry feedback on best practice, specification initiatives, innovation and research prioritization. Significant industry accomplishments include the development of stabilisation specifications, industry best practice notes, and most importantly a trusted platform for roading authorities to obtain industry feedback prior to a more general release.

Australia does not have a similar cross-industry focus group, and this is perhaps understandable with six states each having its own state road authority and different regional materials, design and construction protocols. An industry group AustStab (established in 1995), strive to raise the awareness of stabilisation in the industry, particularly recycling of pavements, formulation of specifications, technical notes and training courses covering all elements of stabilisation best practice.

3 Practices and Experience

3.1 Historic Subgrade Lime Stabilisation in New Zealand

Lime stabilisation was used sparingly in NZ before 1978, at which point a trial project was tendered for subgrade stabilisation on an Auckland Road upgrade. This project had the benefit of intensive post construction evaluation through 1979 to the late 1990's. The same section of road was re-tested in 2009 and a sequence of insitu and laboratory testing was undertaken to provide ongoing 'in-service' stabilised subgrade parameters. The original design life was 15-years, but 40 years later the stabilised subgrade is still performing exceptionally. The natural subgrade strength was soaked and remoulded subgrade CBR = 4, while the design strength of the subgrade stabilised with 4% lime (Calcium Hydroxide) was CBR = 25.

Annual performance evaluation through 1979 to 1983 revealed insitu CBR strength of CBR 100+. Further research testing (Transfund Research Report No. 127) in 1998 reported that "the strength and modulus properties of the stabilised subgrade are significantly superior to those of the original subgrade and this benefit has persisted for more than 20 years". Subsequent investigation in 2009 showed insitu CBR of 90+.

This 2009 research showed that despite the recognised process control limitations and stabilisation methodology of 1978 (i.e. variable layer thickness and intra-layer laminations), the layer properties and durability are excellent. The quality of spreading, stabilising and compaction plant has advanced tremendously over recent years and the level of assurance in obtaining dependable design parameters is very high relative to what was possible in the 1970's and 1980's.

A requirement raised by industry around soils stabilisation is expectation of permanence. Most importantly these improved material properties have been maintained for a sustained period of years with no relaxation thus confirming the durability of lime stabilised soils permanence of the pozzolanic reaction once lime demand and good mixing is ensured. One of the most significant opportunities for aggregate savings in pavements is the incorporation of a stabilised subgrade. NZ and Australia have seen industry advances such as a new Lime (hydrated or quicklime) subgrade stabilisation mix design and structural design specifications with significant improvement in construction protocols and quality assurance.

3.2 *Subgrade Lime Stabilisation in Australia*

Australian subgrade stabilisation employs several different design philosophies. The consistent initial methodology is to test for the lime 'demand' (i.e. quantity of lime required to attain the pH of 12.4) for ensuring a full and permanent pozzolanic reaction. This has led to a high level of success of the insitu lime stabilisation process but can require relatively large application rates. Some soils such as the highly expansive black cotton soils can require more than 6% application rate for lime demand.

In recent years two design protocols have been formalized in Austroads Part 4D (V2.1 April 2019). Method A requires lime application to achieve a laboratory 28-day Unconfined Compressive Strength of 1.0 to 2.0 MPa, while Method B requires the 7-day soaked CBR of the material to be tested and a design CBR derived of no more than half the laboratory mix design value. Both methods constrain the subgrade to a maximum strength of CBR = 15/Resilient Modulus = 150 MPa, and also limit the design top sublayer modulus to a Modular Ratio dependent on the support provided by the underlying 'unstabilised' material determined using AGPT Part 2 Eq. 39 as follows in Eq. (1) below:

$$E_{V\,\text{top sublayer}} = E_{V\,\text{underlying material}} \times 2^{(\text{thickness of each selected subgrade or stabilised layer}/150)} \tag{1}$$

Method A typically requires greater quantities of lime and is mainly used by Queensland DTMR. Method B is more commonly used by other local government or State Road Authorities. Some designers require a greater factor of safety, such as in Victoria where a reduction factor of 3 (i.e. Design CBR = 1/3 × Laboratory CBR) is typically required. Economic viability of subgrade stabilisation can thus be significantly influenced by geographic location and design protocol.

3.3 Employment of Subgrade Lime Stabilisation in New Zealand

Major projects incorporating lime stabilised subgrades have been an accepted part of New Zealand major project philosophy for new pavement construction for several decades in the North Island. The predominant soil types (Waitemata Group clayey silts and Northland Allochthon silty clays) have a naturally alkaline pH of 8.0 to 9.5 meaning only small quantities (1–2%) of lime are required to achieve pH of 12.4. While lime demand testing is recommended it is not regularly undertaken. New Zealand design protocols are to determine dependable soaked CBR strength with capacity reduction factor. This capacity reduction factor is typically 2 for laboratory to field (i.e. CBR of 20 in the laboratory allows a design subgrade CBR = 10) with a maximum permissible design subgrade of CBR = 15 as per the Austroads design guide.

A well-constructed stabilised subgrade will provide a strong and durable substrate that can replace a depth of aggregate virtually equivalent to the stabilised layer thickness. This also provides an improved construction anvil that is highly resistant to effects of moisture, preventing "aggregate punch" and upward migration of plastic fines into the overlying aggregate. A well designed stabilised subgrade can also deliver a 'perpetual' lower pavement system that can accommodate future rehabilitation at the end of the effective life of the upper aggregate courses.

3.4 Fill Drying

An extension of lime subgrade stabilisation is the use of low ground pressure (tracked) equipment to modify/strengthen weak soil layers to where they can be utilized rather than cut to waste. Industry has developed purpose-built equipment for spreading and stabilisation and this has seen the development of a variety of blends of lime and cement to optimize the strength and minimise moisture sensitivity.

Fill drying work, while traditionally used for roading infrastructure, has found high demand in the treatment of residential housing and commercial developments to heavy duty hardstands etc. Employing fill drying stabilisation provides many platform benefits beyond drying and strength gain with a stable platform reducing soil

Fig. 1 Example of low ground pressure spreader and stabilizer undertaking fill drying

particle migration into overlying aggregates, reduced plasticity/moisture sensitivity, and most importantly less demand on finite virgin aggregate resources (Fig. 1).

Fill drying can significantly decrease construction time where the conventional methods are cut-to-fill (or waste) or 'discing' to expose wet soils for air drying. Air drying is frequently compromised by rainfall. Lime fill drying will generate quicker drying from heat generated by hydration of quicklime (calcium oxide). A fill drying technique commonly used is to spread and mix unslaked quicklime (3 mm topsize) fines where moisture is consumed from hydration and heat. Refer to Eq. (2).

$$\text{Quicklime CaO} + \text{water H20} \Rightarrow \text{Calcium Hydroxide Ca(OH)2} + \text{HEAT} \tag{2}$$

The speed of curing and soil stiffness gain is enhanced by the heat of hydration with the added advantage of excess moisture consumption to dry the soils and aid in achieving optimum density. This process has been carried out for millions of cubic metres through the last two decades in New Zealand, and in recent years has been introduced to Victoria and NSW in Australia. The benefit to natural resources is the enormous reductions in virgin aggregate consumption, and the associated reduction in damage to the roading network otherwise caused by haulage of aggregates and building materials to the construction site.

4 Hot In-Place Asphalt Recycling (HIPAR)

HIPAR comprises reheating and gentle hot-milling of asphalt surfacing with the addition of a rejuvenating agent during inline pugmill remixing, then the recycled mix is paver laid in a single continuous process. Robust pavement structure is required where failure is within the aged oxidized, cracked or ravelled asphalt surfacing. The insitu remediation can be modified with a small quantity of 'make-up' asphalt to ensure the depth and geometric shape meet the needs of the traffic and asset lifecycle.

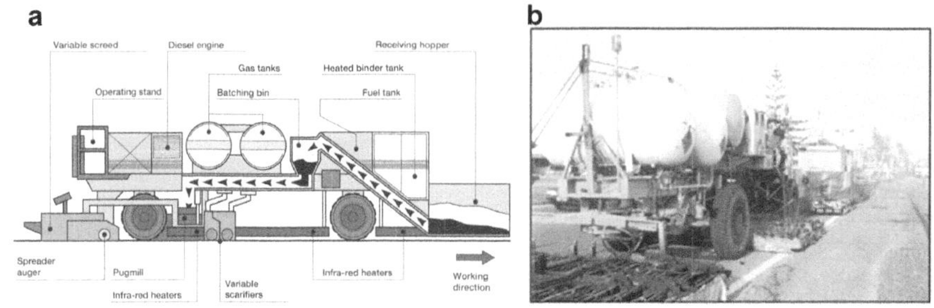

Fig. 2 **a** HIPAR schematic illustration. **b** Gas heaters operating in front of HIPAR

HIPAR was first utilized for the Queensland Department of Main Roads in 1990 and many of these sites are still monitored. The equipment is expensive and currently there is only one HIPAR resource in Australasia. Direct savings are typically 15–20% when compared to a conventional mill and resurface option and there are other environmental benefits such as zero waste leaving the site, reduced truck movements/damage to adjacent road networks, speed of operation reducing traffic disruption, strong bond with underlying asphalt due to heated contact. The stability of the 42-tonne remixer/paver provides for enhanced smoothness and ride properties.

The design process assesses the asphalt to be recycled and the suitability of the existing bitumen via Viscosity tests (Test Method QTMR Test Method No. Q337 previously, but now AS2341.5) which measure binder viscosity using slide plates. Australian practice has determined the onset of cracking due to bitumen hardening (ageing) is where bitumen achieves a critical viscosity of 7×10^6 Pas (6.86 log Pas). For fresh Class 320 bitumen the apparent viscosity is expected to be 4.3 log Pas, and any value exceeding this value will indicate the need for chemical analyses to establish what recycling agent is required to restore the properties of the aged binder (Fig. 2).

The bitumen rejuvenator is Rejuvenex 60E oil applied at ~0.7 L/m^2 for 50 mm treatment depth. Recent research simulates the HIPAR 125 °C heating, milling, 30-s pugmill mixing time then screw-fed paving. The Marshall specimens are prepared immediately (i.e. without any curing time). The specimens are tested to simulate early strength and are also placed in an 85 °C oven for accelerated curing of up to 40 days to simulate aging for testing of long-term fatigue properties.

Challenges for more widespread adoption include high cost of plant, specialized training for operators, selecting sites with failure mode comprising aged/oxidized wearing course (not structural). The recent prevalence of polymers is another consideration regarding management of fumes from heating modified binder asphalts.

Industry has been undertaking research using Epoxy in the place of rejuvenator to assess strength and fatigue capacity for HIPAR. This has been successful for open graded porous asphalts, but dense graded asphalts are more challenging with the quantity of epoxy required for full binder film coverage compromising volumetrics. Compounding this is the difficulty in achieving effective mixing of the epoxy with the hardened bitumen films in the recycled asphalt with the small amount of mixing time available. On this basis dense asphalt recycling with epoxy requires gradation adjustment and as such is better suited to exsitu/plant-based process where more mixing time and better feedstock control is possible.

5 Steel Slag Byproduct as Stabilizing Agent for Roading

During the 1990's NZ Industry pioneered the use of ground waste slag for use as a stabilisation binder. KOBM (Kontinuous Oxygen Blast Maxiite) is a by-product from the steel making process where Lime is added to the smelt to remove impurities from the steel products. The composition of ready-to-use KOBM binder (ground to a 3 mm topsize) is 45–50% Calcium Oxide, 13–15% Iron total, 8% Magnesium Oxide 6% Silicon Dioxide and then several other trace minerals.

The reaction occurs between the hydrated Calcium Oxide in the slag and the clay minerals in the roading aggregates. This plasticity reduction with only minor strength gain means that KOBM is typically used as a pretreatment binder for high plasticity aggregates prior to foamed bitumen stabilisation (employed since 2007) or used in a blended product with conventional binders such as cement (since 1997) (Fig. 3).

The recycling benefit of KOBM is in taking a processed by-product/waste material and utilizing it to modify poor quality aggregates to mitigate plasticity and improve shear strength/resilient modulus. Another advantage is that the KOBM is only a fraction of the cost of burnt lime and cement. KOBM is typically specified at 2–4% application rate (by mass) depending on the process and material properties.

NZ KOBM use follows extensive early independent research in the 1990's then ongoing field trials and monitored pavement rehabilitations. The high specific gravity of KOBM (relative to lime or cement) means increased cartage cost, but conversely provides a more stable product than lime as it is less susceptible to wind movement, so is the preferred pre-treatment product for urban or environmentally sensitive locations. Leachate testing suggests that constituents are not mobilized where KOBM is used with foamed bitumen or cement stabilisation. This is especially noted in foamed bitumen where permeability reduces by an order of magnitude and the fines are immobilized. Note, research also demonstrates that foamed bitumen is also employed for Coal Tar mitigation via immobilization ad reduced permeability.

KOBM is an alkaline material and therefore is handled with some caution. The high pH expedites the pozzolanic reaction for the liberated Calcium Oxide with the aluminates and silicates present in the clay fines. The KOBM is stored under cover as when exposed the Calcium Oxide carbonates to form Calcium Carbonate over

Client ID	Fe Total	CaO	SiO_2	TiO_2	Al_2O_3	MgO	P_2O_5	Na_2O	K_2O	S	V_2O_3	MnO
	%	%	%	%	%	%	%	%	%	%	%	%
KOBM 3mm 159/18	13.69	48.9	<6.83	2.76	1.74	8.03	1.4	<0.010	<0.010	0.13	3.26	2.36

Sieve Size (mm)	9.5	6.7	4.75	2.36	1.18	0.6	0.3	0.15	0.075
% Passing 159/18	100	100	100	87	67	57	49	37	21

Fig. 3 Steel Serve Glenbrook KOBM analytical report and particle size distribution summary

time—losing reactivity. The melter slag available from Australian steel mills is of a different composition to New Zealand, and currently high Calcium Oxide slag by-product is not available in commercial quantities.

6 Foamed Bitumen Incorporating Recycled Materials

Foamed bitumen ‘cold’ insitu recycling was undertaken in Australasia through the 80’s and 90’s but gained substantial traction with the evolution of purpose-built stabilisation equipment since the early 00’s. Foamed bitumen is seen as a cheaper alternative to structural asphalt or concrete for highly loaded expressways and arterials. Insitu treatment is also popular for level constrained urban rehabilitations.

Foamed bitumen in Australia gained additional support in recent years with the demonstration of resilience following the extensive Queensland flooding events of 2010/11 then 2013. Current research by Austroads and the Australian Road Research Board (ARRB) produced several encouraging performance metrics with test tracks constructed. A foamed bitumen stabilised materials test pavement consisting of three fatigue trial sections was constructed including a control section made with a class 3 20 mm crushed rock (granite), a section containing 80% previously cement treated material and section containing 50% RAP. All the test sections were stabilised with 3% foamed bitumen and 2% hydrated lime. These were designed to fail but performance was better than anticipated and the conclusion was that all foamed bitumen stabilised material iterations have performance better than expected. This initial evidence does suggest foamed bitumen stabilisation to be an excellent treatment for pavement rehabilitation, whether that be incorporating a pure host material, RAP or previously cement treated material. This is encouraging for confidence in utilizing more recycled material constituents for foamed bitumen treatment as opposed to cutting existing materials to waste and importing virgin aggregates.

Preference in some states for exsitu plant mixed foamed bitumen shows regional preference of associated active filler, with lime preferred in Queensland, NSW, Northern Territories and cement in Victoria, South Australia and Western Australia. The key is to ensure that active filler content is limited such that the bitumen contribution is dominant, and the desired ductile performance of the treated aggregate is maintained such that it does not become rigid/brittle and risk to crack failure.

Insitu foamed bitumen stabilisation permits full recycling of existing basecourse aggregates. In some materials plasticity mitigation is required before treatment, and in others mechanical stabilisation is required via supplementary fine or coarse aggregate fractions to optimize the particle size grading before foamed bitumen treatment.

Further to recycling existing aged aggregates, the preparation of high waste stream mix designs with aggregates comprising recycled crushed concrete and/or recycled asphalt has provided strong results and subsequent project utilisation. Furthermore, other laboratory and trial pavements have incorporated up to 35% recycled crushed glass where performance was exceptional, although the high proportion of crushed glass was prone to raveling when trafficked heavily before surfacing.

7 Dry Matting Recycling of Flushed Chip-Seal Surfacing Binder

HiTex dry matting is an alternative surface treatment for heavily flushed sprayed seal roads. Excess bitumen used to be burnt/set alight but this is no longer environmentally acceptable, so current approaches are to (1) 'watercut' (where there is a high risk of damaging seal coat with water-blaster overcutting), (2) profile surfacing to waste or (3) recycle surfacing into underlying basecourse via stabilisation.

The HiTex dry matting process utilizes the mobile HIPAR gas heater bank(s) in conjunction with sealing chip for volumetric improvement. The treatment methodology consists of spreading a light coating of 10 mm or (more typically) 14 mm sealing chip on the road to allow heater banks to pass over the heated flushed pavement. The ground speed is managed to optimize fluidity of the bitumen (Targeting 145–150 °C) to receive the additional sealing chip immediately behind the heater banks. The flushed portions of the pavement present liquefied bitumen while the non-flushed areas have lesser temperature due to the chip cover so the introduced chip bonds to the flushed portions and the chip placement/rolling "beds in" the introduced chip where required to correct wearing course volumetrics. Some additional chip may be hand spread for any localized more heavily flushed areas. Following compaction, the surplus chip is swept away, or where possible, it is reclaimed.

HiTex has been adapted for single lane treatment with smaller 0.6 to 1.0 m heater banks to permit focused treatment of pavements where wheeltrack flushing requires remediation. The large-scale heater banks can extend to 4.2 m width at full extension.

This treatment has been successfully undertaken for full pavement flushing through to more channelised wheeltrack flushing and can be employed as part of a poor texture mitigation process prior to full resurfacing, or as a holding maintenance treatment to extend timing until resurfacing is required. Some management of variable texture is required, and close attention to bitumen temperature is also required. The process is self-correcting with the introduced chip requiring liquefied bitumen (from the flushed portions of the pavement surface) to bond to otherwise it is removed.

8 Conclusions

There is widespread desire to incorporate recycling for sustainable pavements in Australasia for different applications as outlined in this paper. There are, however, many barriers to wider utilisation including lack of cross-industry consistency around recycling design protocols and specifications as well as limited desire of stakeholders to incorporate alternatives and/or manage perceived risk around recycling. Recycling, while usually providing cost savings, will inevitably involve different and typically higher risks and these need to be understood and categorized/managed rather than attempting to eliminate all risk.

Initiatives to encourage recycling practices include development of robust cross-industry recycling design guidelines and construction specifications, contract documents that incorporate minimum recycling project requirements or projects that contain incentives or attributes recognizing the use of recycling in tender evaluation.

Independent industry research and post construction evaluation is required to 'calibrate' the long-term performance of modified/recycled materials against laboratory mix design and construction quality assurance test data to provide confidence in dependable field performance. Validated performance criteria for innovative and non-conventional materials and processes substantially increase the desire for utilisation as an alternative, or ultimately being considered a part of conventional practice.

References

AustStab - Insitu stabilisation construction guide (and others). https://www.auststab.com.au/

Austroads guide to pavement technology part 4D stabilised materials: AGPT04D-19 Edition: 2.1, 12 April 2019. https://austroads.com.au/publications?facetScope=&f.Subject+Area%7CsubjectArea=Pavement&query=stabilised&sort=

Austroads 2011a: Review of structural design procedures for foamed bitumen pavements. AP-T188-11, Austroads, Sydney NSW, Australia

Bartley Consultants Limited (1998) Mechanistic design of pavements incorporating a stabilised subgrade. Transfund New Zealand research report no. 127

Depree C, Frobel T (2009) Reconstruction of coal tar contaminated roads by in-situ recycling using foamed bitumen stabilisation

Gray W, Frobel T, Browne A, Salt G (2011) Characterisation and use of stabilised basecourse materials in transportation projects in New Zealand. NZ Transport Agency research report 461

Grenfell J, Garton D, Yan Lu J, Jameson G, Bodin D (2019) Fatigue behaviour of foamed bitumen stabilised materials tested under the accelerated loading facility. In: 18th AAPA international flexible pavements conference 2019, Sydney, NSW, Australia

Hafeez I, Ozer H, Al-Qadi I (2014) Performance characterisation of hot in-place recycled asphalt mixtures. J Transp Eng 140(8):04014029

Ramanujam J (2004) Hot in place recycling of asphalt pavements. Department of Transport and Main Roads, Queensland

Thillairajah S (2006) Hot matting. RTA sprayed seal remedial treatment, bituminous sealing unit. Transport road sand maritime services, NSW, Sydney, Australia

Wilson P (1999) KOBM Binder and Durabind for successful aggregate modification. In: NZIHT Conference

Evaluation of Warm Mix Asphalt Produced from Iraqi Materials

Noor J. Mahdi, Duraid M. Abd, and Taher M. Ahmed

Abstract Warm mix asphalts (WMAs) have gained a widespread popularity as they have reduced emissions due to lower production temperature compared to that of traditional hot mix asphalt (HMA). In Iraq, such asphalt mixture is completely new while no project of WMA has constructed in the field so far although there is a huge demand to either re-construct or construct new highways. One of the considerations that should be taken into account in this regard is to get the advantages of installing WMA in Iraq. The aim of this study is to produce a warm mix asphalt using foaming additive, Zeolite, in both types natural and synthetic. The production temperatures of WMAs were at three levels 140, 130 and 120 °C while the traditional HMA was produced at 160 °C. This study includes producing HMA and WMAs from Iraqi materials, where one binder grade (40/50 pen) from one sources, it is Durah with one type of aggregate, limestone. Hydrated lime was also used as anti-stripping additive to study its effect on improving moisture damage of WMA produced at lower temperature. To study those properties, two techniques were used Marshall properties and Indirect Tensile test. The results showed there is no significant difference between the performances of WMAs compared to that of HMA taking in account level of reduction in the production temperature compared to that of traditional hot mix asphalt. Furthermore, the result of Indirect Tensile Strength ITS has shown the enhancement in moisture susceptibility of warm mix when hydrated lime was used.

Keywords WMAs · HMA · Zeolite · Hydrated lime

1 Introduction

Despite the production temperature of the hot asphalt mixture varies from 190 to 150 °C, is traditionally and commonly used in road's construction, however, it has become necessary to find a technique that reduces the production of asphalt mixture at high temperatures. A warm mix asphalt technique can be used to produce asphalt mixture lower than that required to produce traditional hot mix asphalt by 20–30 °C

N. J. Mahdi (✉) · D. M. Abd · T. M. Ahmed
Department of Civil Engineering, College of Engineering, University of Anbar, Ramadi, Iraq
e-mail: cedahi@uoanbar.edu.iq

C. Raab (ed.), *Proceedings of the 9th International Conference on Maintenance and Rehabilitation of Pavements—Mairepav9*, Lecture Notes in Civil Engineering 76,
https://doi.org/10.1007/978-3-030-48679-2_5

(Rubio et al. 2012). Great advantages in terms of environmental and economic benefits can be obtained using warm asphalt mixture as results of reducing gases and fuel consumption and improving workability of the asphalt mixture (Ghabchi et al. 2015). The mechanism of warm asphalt mixtures for reducing the production temperature depends on a group of additives that can be classified as follows: foaming additives, chemical additives and organic additives (Zaumanis 2010a). The foam technique can be classified into two categories: the first method is called a direct method which can be conducted by adding a quantity of water to the asphalt mixture or to the binder and, the letter is by adding additives containing water in its structure (Hurley and Prowell 2005). The zeolite is classified in two categories, natural and synthetic. The natural which is known as Clinoptilolite, is a fine powder and it is one of the most common natural zeolite containing a microstructure of tetrahedral of silica and alumina, and has high resistance to extreme temperatures. Furthermore it is chemically neutral basic structure (Clinoptilolite 1997). The synthetic zeolite is a soft white powder, composed of aluminum silicate that has a water content approximately of 18–22%. In general, Zeolite is capable of storing energy by 30% and decreases in production temperature by about 30 °C (Kristjánsdóttir et al. 2007; Smith 2009).

Many researchers studied properties and performance of asphalt mixture including foaming technology produced using zeolite, whether natural or synthetic. Hurley and Prowell (2005) reported that adding Aspha-Min (zeolite) to asphalt mixture improved temperature of mixing and compaction and resulted in decreasing the production temperature to 88 °C and reducing the percentage of air voids by 65%. They also reported that the addition of hydrated lime could maintain the issue of moisture damage, which may associate with addition of zeolite because of lower production temperature as result of incomplete aggregate drying. Gandhi (2008) highlighted that zeolite has a little effect on the viscosity of binder. The dosage of adding zeolite was also in question as reported also Gandhi (2008). He found that addition of zeolite whether natural or synthetic at different dosages 0.3–0.6% by total weight of asphalt mixture, reduced the production temperature by about 30 °C without any negative effect on the properties of asphalt mixture in terms of compatibility and water sensitivity. De Visscher et al. (2010) investigated the properties of warm asphalt mixture produced using for using Advera (Synthetic zeolite), Evotherm and Sasobit. Moisture susceptibility laboratory, dynamic modulus and rutting potential were in question. They concluded that the tensile strength ratio of warm asphalt mixture produced using Sasobit was higher than asphalt mixture containing Advera and Evotherm. Furthermore, the Sasobit had a superior performance on the rutting resistance of asphalt mixture while dynamic modulus of all mixtures were approximately same to that of control asphalt mixture. Zhang (2010) also investigated the performance of WMA produced using Aspha-min and Sasobit. Warm asphalt mixtures were manufactured at temperatures 125 and 135 °C. He conclude that the indirect tensile strength, resilient modulus and fatigue resistant of warm asphalt mixtures decreased while a positive increase in rutting resistance was noticed compared to that of control asphalt mixture. More importantly, Sasobit showed better than Aspha-min. Al-Jumaili et al. (2015) made a comparison this between the performance of natural zeolite and synthetic zeolite but not a clear conclusion was reported. Developed countries such as, USA,

United Kingdom, Germany, and Netherland have already used WMA instead of traditional hot mix asphalt to produce a more sustainable asphalt mixtures as warm additives which are used to produce such mixture, enhance the properties of asphalt mixtures. It is therefore that use of such mixture in Iraq will defiantly enhance the pavement structure by reducing deterioration and distresses in surface and wearing layers and the effect of transfer loads to subgrade layer. More importantly, In Iraq, the use of such kind of asphalt mixture is completely new therefore; there is a need to study the application of WMA in Iraq to produce a more sustainable asphalt mixture. Moreover, there is no demand to reconstruct due to some circumstances that happed in some parts of Iraq such as Anbar and Nineveh governorates. In addition, this research has considered another point which is the difference between natural and synthetic zeolite on the properties of asphalt mixture.

2 Experimental Work

2.1 Material and Mixes Design

The source of bitumen is the refinery of Durah in Baghdad. As mentioned previously, two types of Zeolite were used, synthetic and natural. Limestone was used in the asphalt mixture. Tables 1, 2 and 3 show the properties of bitumen, aggregate and Zeolite respectively.

Table 1 Properties of bitumen

Property	Value
Penetration at 25 °C (0.1 mm)	44
Ductility at 25 °C, 5 cm/min (cm)	>120
Flash point, °C	246
Specific gravity	1.047
Softening point	54
Thin-film oven test	
Retained penetration % of original	78
Ductility at 25 °C, 5 cm/min (cm)	88

Table 2 Properties of aggregate

Property	Coarse aggregate	Fine aggregate
Bulk specific gravity	2.607	2.620
Apparent specific gravity	2.665	2.664
Wear (Los Angeles) %	25	–
Chemical corrosion %	1.3	–

Table 3 Properties of zeolite

Content %						
SiO_2	Al_2O_3	Na_2O	CaO	MgO	Fe_2O_3	I.O.I
39.46	28.35	13.16	0.26	0.26	0.84	15.13

Physical properties			
Porosity size	Surface area	Absorption of water	Apparent density
0.32	7.7 m^2/gm	18.5	0.73 gm/cm^3

2.2 *Preparation of Asphalt Mixtures*

A control hot mix asphalt was manufactured at temperature of 160 °C after determining the optimal binder content based on Marshall Procedure. The binder content was 5%, which was considered for all asphalt mixtures under investigation in this research. Furthermore, four sets of warm asphalt mixtures were prepared at each production temperature of 140 °C, 130 °C and 120 °C respectively. The first group was the addition of synthetic zeolite and the second group was consisted the addition of a synthetic zeolite with the hydrated lime. The third group was the addition of natural zeolite while the last group was the addition of a natural zeolite with the addition of the hydrated lime at different production temperature in order study the effect of production on the performance of warm mix asphalt.

2.3 *Volumetric Characteristics*

Density and air voids of control hot asphalt mixture and warm asphalt mixture samples were measured in order to investigate the effect of zeolite type with and without hydrated lime and production temperature on the volumetric properties of those asphalt mixtures.

2.4 *Mechanical Characteristics*

The mechanical characteristics involved measuring the stability and flow for all asphalt mixtures and comparing the results with that of control hot mix asphalt.

In order to assess the performance of warm asphalt mixtures in terms of moisture damage, the indirect tensile test was performed. In this regards, a modification was made to Marshall Machine to use it as an indirect tensile test and has been adjusted and calibrated to read the amount of maximum load in at failure and the deformity resulting from the failure of the sample.

The condition and samples preparation were divided in two groups, the first groups was subjected to immersing in water for 24 h conditioned at 60 °C and then immersing the samples in water for two hours at a temperature of 25 °C. The letter group was an unconditional sample, which involved submerged the samples in water for 30 min at a temperature of 25 °C, then testing samples directly.

3 Results and Discussion

This study shows the comparison between the productions of warm asphalt mixture at different temperatures using a foaming additive zeolite, both types synthetic and natural with/without addition of hydrated lime to compare with that of hot mix asphalt. Although there was no significant enhancement in the volumetric properties in terms of density and flow for WMAs produced using synthetic and natural zeolite at different level of production temperature as seen in Fig. 1, the stability and flow of WMAs-Natural were same or better than that of control HMA as presented in Fig. 2. However, WMA-Natural produced at 120 °C was less than of control HMA. The reason behind that is at lower production temperature, 120 °C for example, the connection between particles, which comes from adhesion and cohesion forces, becomes week; therefore, the mixture is unstable to resist the deformation due to loading, there is a possibility to reduce the production temperature of asphalt mixture using synthetic-zeolite by 20–25 °C compared to that of control mix although the stability of WMA-Synthetic manufactured at 140 °C less than that of control HMA and that of WMA-synthetic manufactured at 130 °C as presented in Fig. 2. This issue is in question and currently under investigation.

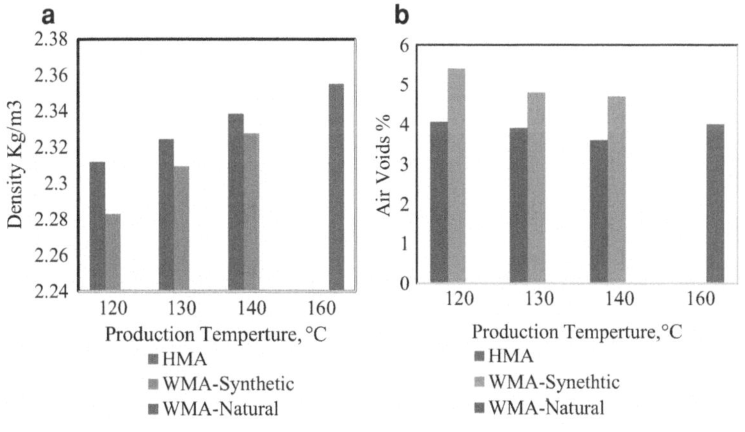

Fig. 1 Density (**a**) and air voids (**b**) of hot and warm asphalt mixtures without addition of hydrated lime

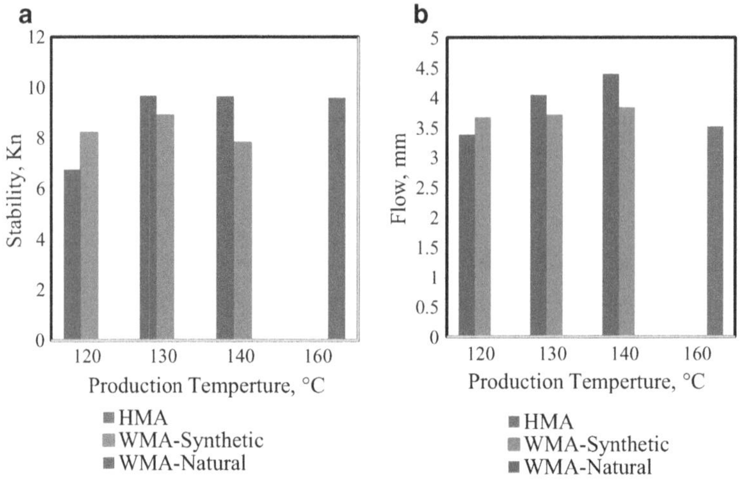

Fig. 2 Stability (**a**) and Flow (**b**) of hot and warm asphalt mixtures without addition of hydrated lime

In addition, it can be noticed that all WMAs manufactured using either synthetic or natural zeolite performed as same as that of control mix or even better that of control mix as presented in Fig. 3. Although there is a concern that the lower production temperature may make less adhesion between bitumen and aggregate as the letter may not be completely dried, the foaming process makes good connection between aggregate and bitumen. However, the lower production temperature make the mastic

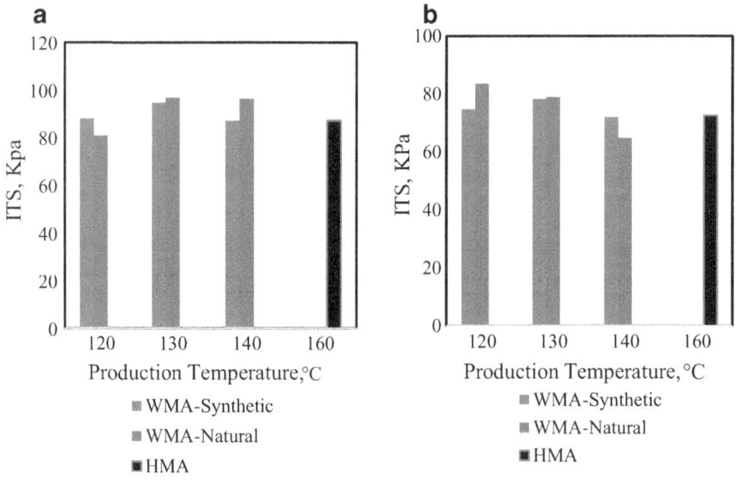

Fig. 3 Indirect tensile stress (ITS) (**a**) unconditioned samples (**b**) conditioned samples of hot and warm asphalt mixtures without addition of hydrated lime

phase in the asphalt mixture less stiff than required as highlighted by Abd et al. (2018). It therefore that it can be noticed, the production temperature has a significant of the stability of asphalt mixtures produced at temperature less than 140 °C while no moisture damage issue can observed in Fig. 4. Figure 3 shows the ITS results of unconditioned and conditioned samples.

In the scenario of addition-hydrated lime to WMAs to study its effect on improving the properties of WMAs manufactured at 130 and 120 °C, it can be noticed that there is a significant drop the volumetric and marshal properties of WMAs-Natural apart from WMA-Natural produced at 140 °C as illustrated in Figs. 5 and 6. This issue is because hydrated lime increases the viscosity of binder, which resulted a difficulty to mix the compounds adequately. It is however that in case of synthetic zeolite, WMA-synthetic produced at 130 °C performed as same as the control HMA. The

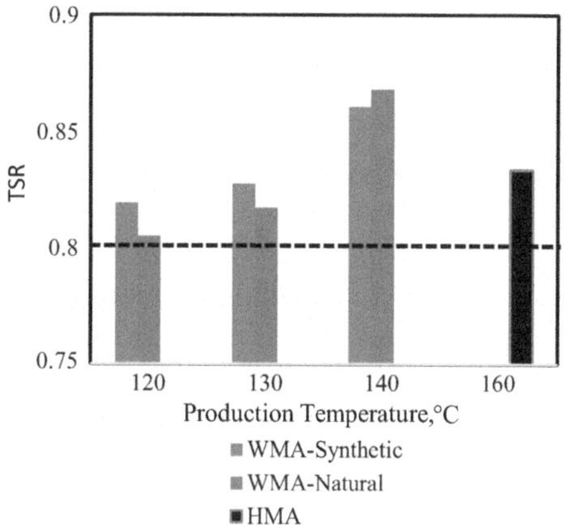

Fig. 4 Tensile strength ratio (TSR) of hot and warm asphalt mixtures without addition of hydrated lime

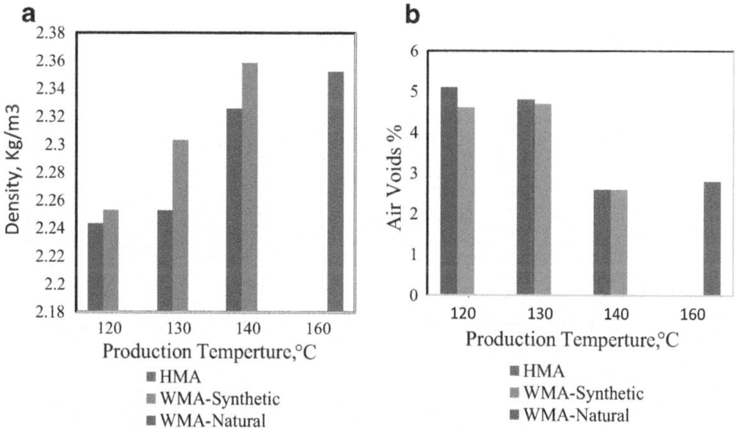

Fig. 5 Density (**a**) and air voids (**b**) of hot and warm asphalt mixtures with addition of hydrated lime

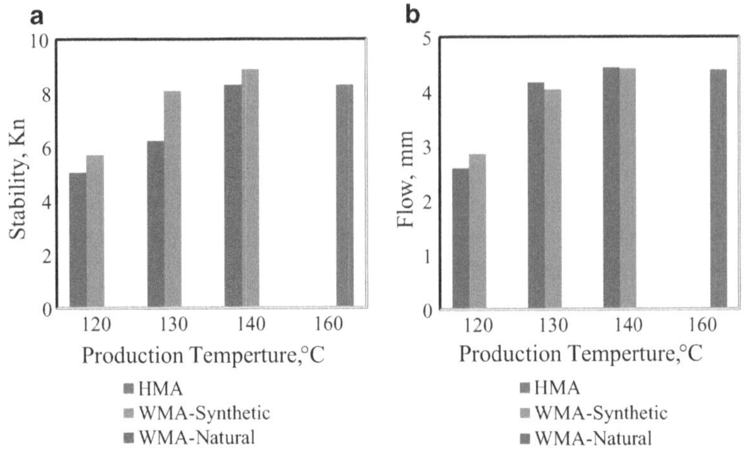

Fig. 6 Stability (**a**) and Flow (**b**) of hot and warm asphalt mixtures with addition of hydrated lime

reason, which needs further investigation, is the rate of smoothness of particles of synthetic and natural zeolite taking in account the availability of hydrated lime and their effect on the viscosity of binder.

In addition, although indirect tensile stress values of all WMAs excluding those manufactured at 140 °C, were less than that of control HMA as shown in Fig. 7, TRS values of all WMAs were higher than that of control mix and were higher than that of scenario without addition of hydrated lime as presented in Fig. 8. Therefore, in conclusion there is no need to add hydrated lime to improve moisture damage in case of using zeolite.

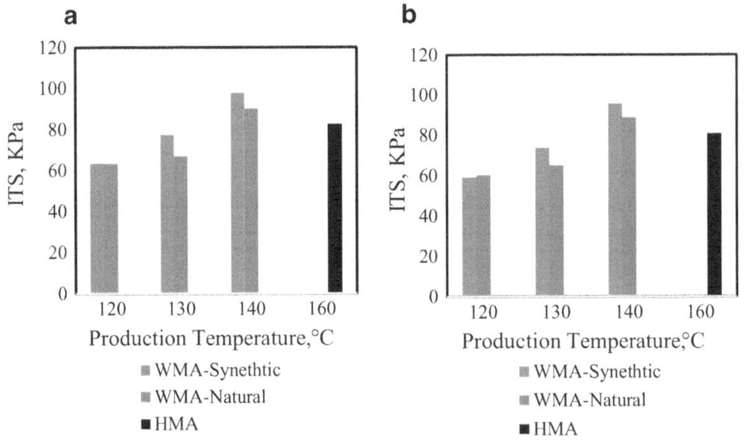

Fig. 7 Indirect tensile stress (ITS) (**a**) unconditioned samples (**b**) conditioned samples of hot and warm asphalt mixtures with addition of hydrated lime

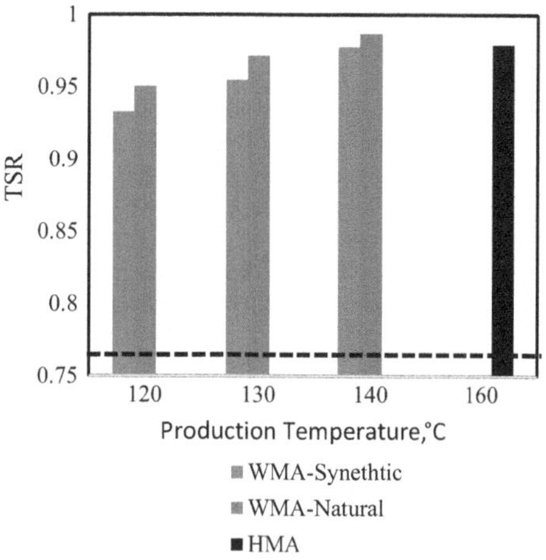

Fig. 8 Tensile strength ratio (TSR) of hot and warm asphalt mixtures with addition of hydrated lime

4 Conclusion

According to this research and the limitation of experimental work and the materials used, the main pinots can be drawn:

1. There is a possibility to produce WMA at production temperature lower than that of traditional hot mix asphalt by 40 °C using natural zeolite. It is however that further reduction in the production of WMA using such additive is not recommended.
2. The availability of zeolite with hydrated lime may increase the viscosity of binder and negatively affect foaming process, which in turn decreases the strength of asphalt mixture.
3. Susceptibility to moisture damage may not exist with using zeolite to produce asphalt mixture at lower temperature than that of control mix. However, production temperature plays a significant role in reflecting the overall performance of asphalt mixture, therefore there should be a balance in improving cohesive and adhesive force in the asphalt mixture.

References

Abd DM, Al-Khalid H, Akhtar R (2018) An investigation into the impact of warm mix asphalt additives on asphalt mixture phases through a nano-mechanical approach. Constr Build Mater 189:296–306

Al-Jumaili MAH et al (2015) Influence of selected additives on warm asphalt mixtures performance. Kufa J Eng 6(2):49–622

Clinoptilolite (1997) The mineral and locality database. http://www.mindat.org/min-1082.html

De Visscher J et al (2010) Asphalt production at reduced temperatures using zeolites and the impact on asphalt performance. Road Mater Pavement Des 11(1):65–81. https://doi.org/10.1080/14680629.2010.9690260
Gandhi T (2008) Effects of warm asphalt additives on asphalt binder and mixture properties
Ghabchi R, Singh D, Zaman M (2015) Laboratory evaluation of stiffness, low-temperature cracking, rutting, moisture damage, and fatigue performance of WMA mixes. Road Mater Pavement Des 16(2):334–357. https://doi.org/10.1080/14680629.2014.1000943
Hugener M, Emmenegger H (2008) Thin layer approximation and algebraic model for separated turbulent flows. In: 16th AIAA aerospace sciences meeting, Huntsville, Ala, 16–18 January
Hurley GC, Prowell BD (2005) Evaluation of Aspha-Min zeolite for use in warm mix asphalt. NCAT report (05–04)
Kristjánsdóttir Ó et al (2007) Assessing potential for warm-mix asphalt technology adoption. Transp Res Rec 2040(1):91–99
Raab C, Partl MN (2000) Random data: analysis and measurement procedures, 3rd edn. Wiley, Hoboken
Rubio MC et al (2012) Warm mix asphalt: an overview. J Clean Prod 24:76–84. https://doi.org/10.1016/j.jclepro.2011.11.053
Smith A (2009) Advera WMA zeolite. WMA Technical Working Group. Hunt Valley. PQ Corporation Presentation, Maryland. http://www.warmmixasphalt.com/submissions/82_20080629_Annette_Smith_Advera_Caltrans
Zaumanis M (2010a) W m a i
Zaumanis M (2010b) Warm mix asphalt investigation. Master of science thesis. Technical University of Denmark in cooperation with the Danish Road Institute, Department of Civil Engineering, Kgs. Lyngby
Zhang J (2010) Effects of warm-mix asphalt additives on asphalt mixture characteristics and pavement performance

Influence of Curing on the Mechanical Properties of Cement-Bitumen Treated Materials Using Foamed Bitumen: An Interlaboratory Test Program

Marco Pasetto, Emiliano Pasquini, Andrea Baliello, Simone Raschia, Amir Rahmanbeiki, Alan Carter, Daniel Perraton, Francesco Preti, Beatriz Chagas Silva Gouveia, Gabriele Tebaldi, Andrea Grilli, and Eshan V. Dave

Abstract The use of reclaimed asphalt (RA) in road pavements is continuously gaining interest thanks to the technical, economic and environmental advantages guaranteed by such sustainable practice. Cold recycling techniques compared to traditional asphalt mixes allow a significant reduction of energy, fume emissions, use of natural resources, etc. In this perspective, the Task Group 1 on "Cold Recycling" of the RILEM Technical Committee on "Asphalt Pavement Recycling" (TC 264-RAP) launched an interlaboratory test program (ITP) aimed at ensuring a better understanding for cold recycled mixtures. The paper presents the results collected by a restricted group of the participating laboratories testing cement-bitumen treated materials that included a single RA source and prepared with foamed bitumen. Gyratory compacted specimens were used to evaluate the influence of curing (free, partial or restricted-surface drying for 14 days at 40 °C at a relative humidity of 55 ± 5%). Stiffness was evaluated as a function of the curing stage and the corresponding water loss; strength was tested after 14 days of curing testing specimens in both dry and

M. Pasetto · E. Pasquini (✉) · A. Baliello
Department of Civil, Environmental and Architectural Engineering, University of Padova, Via Marzolo 9, 35131 Padua, Italy
e-mail: emiliano.pasquini@unipd.it

S. Raschia · A. Rahmanbeiki · A. Carter · D. Perraton
Construction Engineering Department, École de Technologie Supérieure (ÉTS), 1100 Notre-Dame, Ouest, Montreal, Canada

F. Preti · B. C. S. Gouveia · G. Tebaldi
Department of Engineering and Architecture, University of Parma, Parco Area delle Scienze 181/a, 43124 Parma, Italy

A. Grilli
Department of Economics, Science and Law, University of the Republic of San Marino, Via Consiglio dei Sessanta 99, 47891 San Marino, Republic of San Marino

E. V. Dave
Department of Civil and Environmental Engineering, University of New Hampshire, 33 Academic Way, Durham, NH 03824, USA

C. Raab (ed.), *Proceedings of the 9th International Conference on Maintenance and Rehabilitation of Pavements—Mairepav9*, Lecture Notes in Civil Engineering 76,
https://doi.org/10.1007/978-3-030-48679-2_6

wet conditions to also determine the water sensitivity. As expected, the different curing conditions clearly influenced the rate of water loss of tested samples with clear effects on mechanical properties and durability.

Keywords Cold recycling · Reclaimed asphalt · Strength · Stiffness · Water susceptibility

1 Introduction

The growing consciousness on the environmental sustainability is promoting the reuse of reclaimed asphalt (RA) in road pavements to avoid disposal and preserve natural raw materials. In this regard, cold recycling in road pavements is continuously gaining interest thanks to the technical, economic and environmental advantages that could be guaranteed (Grilli et al. 2018). Cold recycling of RA is usually performed in cement-bitumen treated materials (CBTMs) produced at ambient temperature; this construction technology allows a sensible reduction of energy consumptions and emissions with respect to traditional hot mix asphalt. Successful CBTMs applications have been worldwide demonstrated, particularly in the case of base and subbase layers (Cardone et al. 2015; Hugener et al. 2013). The presence of cement and water determines an evolutive behavior of CBTMs strictly related to the curing processes which in turn are clearly affected by the adopted construction procedures (Graziani et al. 2016). Significant efforts have also been made by researchers to properly characterize the mechanical properties of RA aggregates to be used in cold recycled mixes (Tebaldi et al. 2019). Given this background, Task Group 1 (TG1) on "Cold Recycling" of the RILEM Technical Committee TC 264-RAP on "Asphalt Pavement Recycling" launched an interlaboratory test program (ITP) investigating, among others, the effect of the curing conditions on the evolutive behavior as well as on the final properties of CBTMs prepared with different RAs and bituminous binders (foamed bitumen or bituminous emulsions). A total of 12 laboratories from 10 countries are actively involved in the TG1 activities.

2 Background and Research Objective

One of the crucial aspects related to the CBTM pavement layers is related to the construction procedures and, in particular, to the operations planned after the laying and compaction of such layers. The environmental site conditions, the possible different treatments of the upper surface as well as the scheduled time of the construction phases strongly affect the evolution of the moisture content within the mixtures with clear effects on the effective properties of CBTMs thus influencing the final behavior of the whole pavement. In this regard, the present paper illustrates a part of the above-mentioned ITP carried out in the framework of the activities of RILEM

Table 1 RA1 target gradation envelope

Size [mm]	20	16	8	4	2	1	0.5	0.25	0.125	0.063
Min. passing [%]	96	93	73	49	31	17	9.5	7.5	5	3
Max. passing [%]	100	100	79	53.5	37	25	17.5	10	5.5	4

TG1 TC 264-RAP and specifically aimed at investigating the influence of curing conditions on the performance of CBTMs. CBTMs were produced using the same physical raw materials: single-source of RA (RA1) and foamed bitumen (FB). The experimental results were collected by a restricted group of the TG1 participants, i.e. the University of Padova (Italy), the École de Technologie Supérieure (Canada) and the University of Parma (Italy), using a specific sample size compacted with the same technology: 150 mm cylinder from Shear Gyratory Compactor (SGC).

3 Experimental Approach

3.1 Materials and Mixtures

TG1 provided the same constituent raw materials (RA, filler, cement and bitumen for foaming) to all the involved laboratories along with a specific mix design to be followed. Based on the volumetric approach described in Grilli et al. 2012, the total solid part of the studied CBTMs was composed of RA, filler and cement. For this research, a RA1 with maximum aggregate size of 16 mm was used. The RA1 was characterized by a target gradation whose envelope is reported in Table 1. In the final aggregate composition, the RA1 was integrated with filler (particle size <0.063 mm) dosed at 5.5% of the total solid weight. A cement GU type (CSA A3000) with compressive strength at 28 days of 43.9 MPa dosed at 1.5% of the total solid weight was selected as co-binder. A bitumen for foaming was used at 3.0% by the total solid weight. Such a 2:1 bitumen to cement ratio avoided excessive brittleness of the mixture (Grilli et al. 2012). Figure 1 shows linear viscoelastic properties in terms of master curves of complex modulus G^* and phase angle δ of the base bitumen. The basic binder characteristics provided by the producer are shown in Table 2. Mix design also required 3.0% added water by the total solid weight. The overall mix design is summarized in Table 3.

3.2 Mixing, Compaction and Curing

After oven-drying at 40 °C, the selected RA1 was adequately mixed with filler and cement according to the mix design formulation achieving a homogeneous blend which was then accurately mixed with the added water. At the mixing time, the

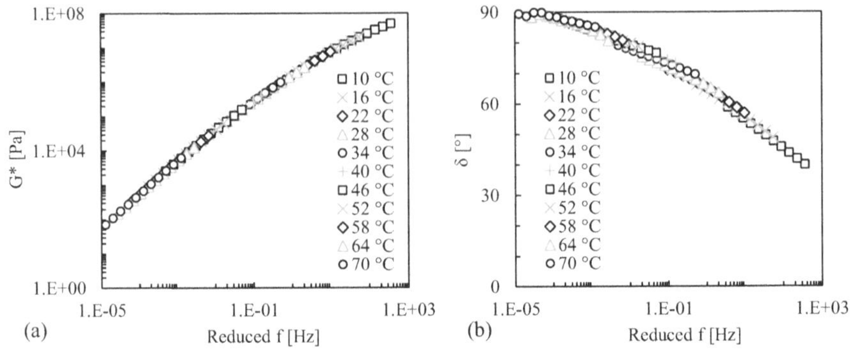

Fig. 1 Foamed bitumen master curves (34 °C): shear complex modulus G^* (**a**) and phase angle δ (**b**)

Table 2 Basic properties of bitumen for foaming

Property	Standard	Unit	Value
Penetration	EN 1426	0.1 mm	64
Softening point	EN 1427	°C	52

foamed bitumen was added to the wet aggregate blend immediately before the final mixing process carried out by an automatic equipment. In particular, the foamed bitumen was produced using a foaming unit adopting standardized temperature and water flow rate (170 °C and 1.9% of the bitumen content, respectively) based on a preliminary optimization study. In this regard, the spraying time was set to obtain an expansion ratio of 10 and a half-time of about 6 s. Lab_1 used a mortar mixer while pugmill mixing unit was used by Lab_2 and Lab_3.

The 150-mm diameter samples were compacted using an undrained SGC mold to target void content of 14.3% (pressure of 600 kPa; rotation speed of 30 rpm; angle of inclination of 1.25°). In this regard, the following Table 4 reports the actual average

Table 3 CBTM adopted mix design

Component	RA1	Filler	Cement	Bitumen	Water
Dosage [%]	93.0[a]	5.5[a]	1.5[a]	3.0[b]	3.0[b]

[a]of the total solid weight (RA1 + filler + cement)
[b]by the total solid weight (RA1 + filler + cement)

Table 4 Volumetric properties of CBTM samples

Property	Lab_1	Lab_2	Lab_3
Average bulk density (G_{mb}) [g/cm^3]	2.124	2.035	2.035
Maximum specific gravity (G_{mm}) [g/cm^3]	2.396	2.396	2.396

bulk densities (G_{mb}) measured by the different laboratories based on the geometric dimensions of specimens at the beginning of curing. The theoretical maximum specific gravity (G_{mm}) of the CBTM is also reported.

Samples were then subjected to a curing period of 14 days at 40 °C with a relative humidity of 55 ± 5% in three different conditions. In particular, the so-called Free-Surface Drying (FSD) allows free evaporation from the lateral and top specimens' surfaces whereas the Partial-Surface Drying (PSD) was achieved by sealing the lateral surface of cylindrical samples, thus allowing evaporation only from the top surfaces. Finally, Restricted-Surface Drying (RSD) was also evaluated by curing the specimens wrapped in sealed bags to avoid any free evaporation.

3.3 Experimental Plan and Testing Methods

Indirect Tensile Stiffness Modulus (ITSM) tests were carried out at 25 °C at different curing stages (from 1 to 14 days) in order to relate the evolution of the material properties with the corresponding Moisture Loss (ML). ML is the percentage of mass loss in accordance to the initial sample weight (fresh). At the end of the curing stage (14 days) ITSM was also measured at 2, 10 and 20 °C to evaluate the temperature sensitivity of the mixture. Tests were executed according to EN 12697-26/Annex C applying load pulses with 124 ms of rise-time at a target peak horizontal deformation of 7 μm (Poisson's ratio was fixed at 0.35). Stiffness properties at 25 °C at the different curing stages were also assessed using the Ultrasonic Pulse Velocity (UPV) non-destructive test according to ASTM C 597. The pulse waves propagation through the mix was used to estimate the dynamic modulus of elasticity at a resonant frequency of 54 kHz. Prior to the test, samples were conditioned 3 h at the testing temperature.

The cured specimens were also subjected to indirect tensile strength ITS tests (EN 12697-23). ITS was evaluated at 25 °C for all samples. Wet conditioning was

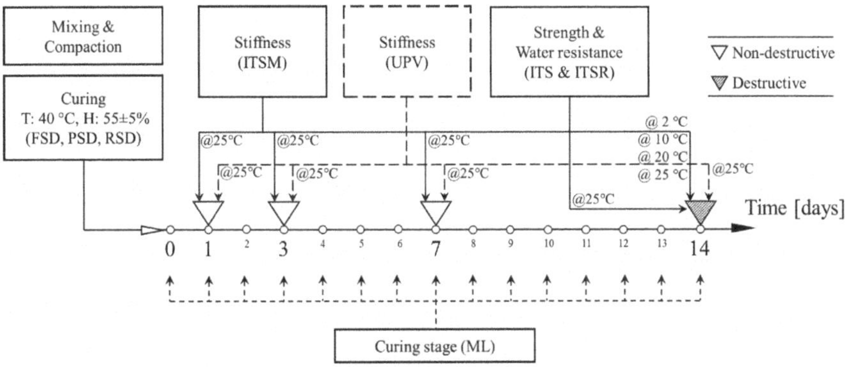

Fig. 2 Schematic summary of the experimental plan

3 days in water at 40 °C according to EN 12697-12. This allowed calculating the Indirect Tensile Strength Ratio (ITSR) as an indicator of the water resistance of the tested CBTMs. A summary of experimental plan is schematized in Fig. 2.

4 Results and Discussion

4.1 Moisture Loss

Figure 3 shows the evolution of the moisture loss ML in accordance to the curing time. Figure 3a depicts the results obtained by different laboratories in the case of FSD cured samples (6 specimens for Lab_1 and 10 specimens for Lab_2; Lab_3 did not assess ML) whereas Fig. 3b shows the comparison among the different curing procedures (FSD, PSD and RSD) by reporting the average data along with the corresponding error bars at 90% confidence level.

It can be clearly observed that the curing condition strongly influenced the ML evolution. Obviously, the higher the "restrictions" to the water evaporation is, the lower the moisture loss is during time. This happened till the end of the fixed curing phase. After 14 curing days at 40 °C, the residual water content by the mix weight was 0.60% for FSD, 1.25% for PSD and 1.93% for RSD samples bearing in mind that the total water content (w_{mix}) is 2.83% by mix weight. Indeed, specimens in a completely-cured state are expected to exhibit a ML equal to the w_{mix} minus the water of cement hydrated products, thus not countable for evaporation. Therefore, in the case that 0.53% of the total water would be part of the hydrate products (cement hydration degree of 90%), a maximum water loss could be estimated around 2.30% by mix weight (Cardone et al. 2015). In this respect, regardless the testing laboratories, FSD specimens (Fig. 3b) effectively approached such value also showing a quasi-constant behavior after 7 days curing at 40 °C (after one curing day the ML is already more than

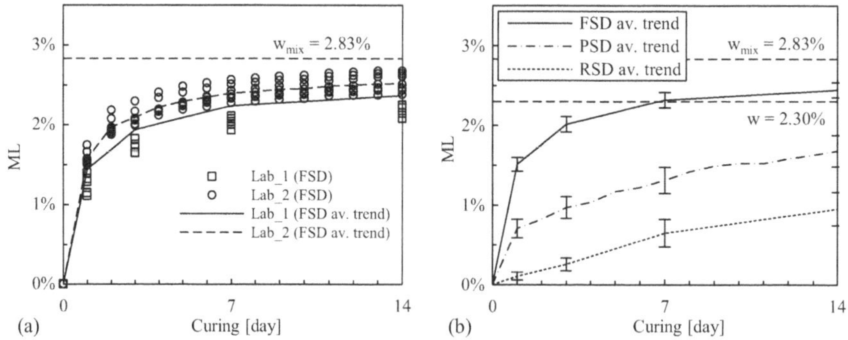

Fig. 3 Evolution of the ML in accordance to the curing time (40 °C): FSD samples (**a**); average trend of Lab_1 and Lab_2 for the three surface drying conditions tested (**b**)

50%). This finding suggests that the ML can be considered substantially concluded after 14 days in this curing condition. On the other hand, lower ML values and more progressive evolution of ML during the whole curing phase can be observed for PSD and RSD denoting that water is still remaining in samples even after 14 days curing in the established conditions. Comparable findings were obtained by other researchers testing similar materials under analogous curing conditions (Pasetto et al. 2019).

4.2 *Stiffness Properties*

The effect of curing on the CBTMs stiffness properties is depicted in Fig. 4 in terms of ITSM (Fig. 4a) and UPV (Fig. 4b) test results; for each data group the average ML is also reported. Results from both tests show similar trend and values (between 3000 and 7000 MPa) also consistent with field measurements from literature (Godenzoni et al. 2018; Graziani et al. 2017). As regards ITSM data, no difference was found between the two methods of drying. In contrast, UPV data seem to show some differences. Based on UPV data, RSD conditioning generally led to higher average stiffness values than the FSD during the whole curing process, even if the experimental points are scattered. In this regard, it is worth noting that the curing time affects the stiffness development of cold mixes due in particular to the cement hydration process. Higher water content into CBTMs cured in RSD conditions during the curing process could reasonably encourage cement hydration process that could be beneficial to stiffness increase.

At the same time, specimens cured in FSD condition could be characterized by an incomplete cement hydration and a slight oxidation process of the virgin binder. In this way, both FSD and RSD conditions were able to reach similar stiffness values, despite the very different residual water.

Figure 5 depicts the same stiffness data as a function of the corresponding ML. Results show that RSD samples were characterized by higher stiffness at a given

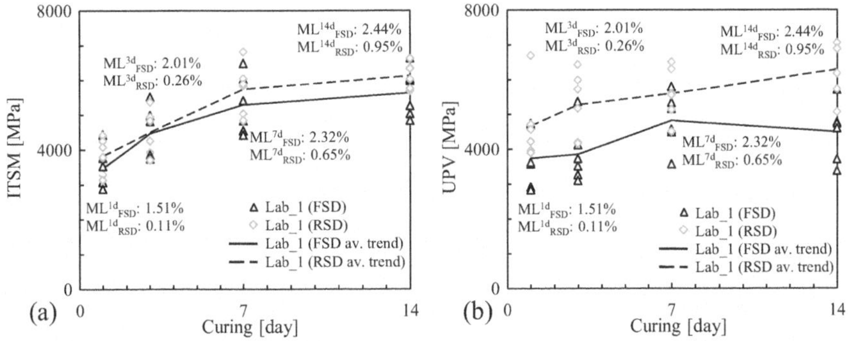

Fig. 4 ITSM (**a**) and UPV (**b**) stiffness test results vs. curing time (test temperature = 25 °C)

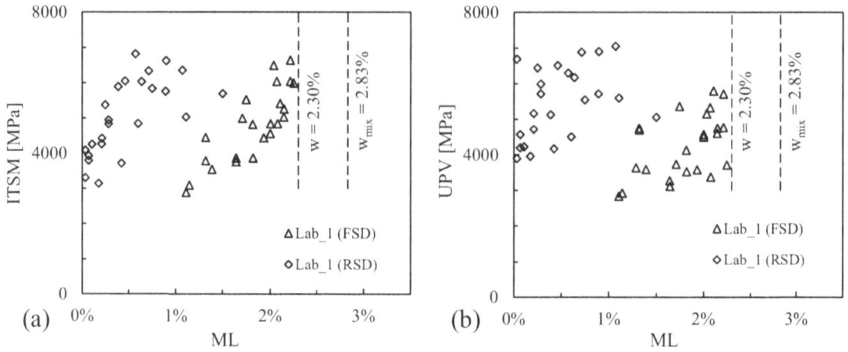

Fig. 5 ITSM (**a**) and UPV (**b**) test results vs. ML (tests temperature = 25 °C)

moisture loss level. Moreover, the rate of stiffness increase seems similar in both FSD and RSD curing conditions thus demonstrating the crucial role of the moisture loss. According to previous studies (Dulaimi et al. 2015), the experimental data also show that stiffness was still increasing at the end of the curing period (14 days) at the given curing conditions, suggesting a still incomplete development of the ultimate mix properties.

The influence of test temperature on the stiffness properties can be observed in Fig. 6. In particular, Fig. 6a plots ITSM at 2, 10 and 20 °C test temperature for 14 days cured FSD and PSD specimens. Again, PSD specimens were characterized by higher stiffness than FSD ones but the temperature sensitivity of the two samples appears almost equivalent. Moreover, Fig. 6b summarizes the stiffness properties of the selected CBTM tested at the end of curing by the different laboratories. Besides the already discussed aspects concerning the influence of test temperature, curing conditions and moisture loss, it is worth highlighting the higher stiffness measured

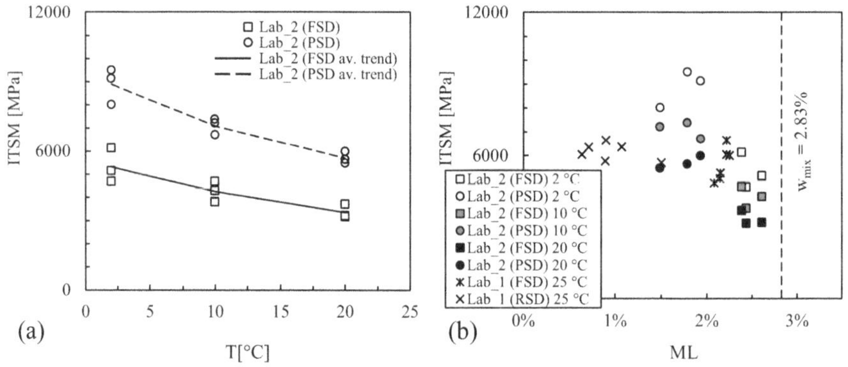

Fig. 6 ITSM test results after 14 days curing at 40 °C: influence of temperature (**a**) and ML (**b**)

by Lab_1 at 25 °C in FSD conditions with respect to that detected by Lab_2 at the same curing conditions but at lower temperature (20 °C). This fact can be likely explained by the higher bulk density of specimens tested by Lab_1 (Table 3).

4.3 Strength and Moisture Resistance

Strength and moisture resistance characteristics of 14 days cured samples are illustrated in Fig. 7. Firstly, it is worth noting that the measured ITS values were similar to those recently presented by other researchers investigating analogous materials in similar conditions (Gandi et al. 2019). Moreover, a good repeatability among the three involved laboratories can be detected by observing the average dry ITS of specimens cured in FSD conditions. In this sense, Table 5 reports the main outcomes of the ANalysis Of VAriance (ANOVA) carried out at 95% confidence level in order to assess the statistical significance of the different measured average values.

On the other hand, the different curing procedures did not seem to strongly influence the tensile strength of the CBTM tested in dry condition whereas a higher influence can be clearly observed for wet conditioned samples, thus deeply affecting the related moisture resistance in terms of ITSR. Reasonably, the greater wet ITS determined for PSD and RSD specimens could be ascribed to the higher residual moisture at the time of wet conditioning. In any case, it is worth specifying that the

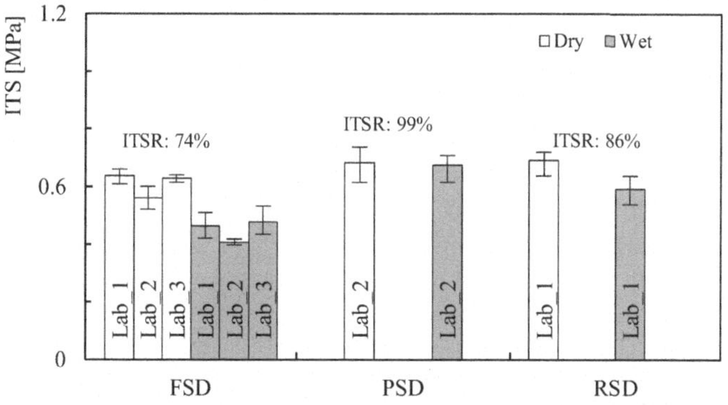

Fig. 7 ITS and ITSR results at 14 days

Table 5 ANOVA test: laboratory's reliability for dry ITS at 14 curing days (FSD samples)

Property	Lab_1 vs. Lab_2		Lab_1 vs. Lab_3		Lab_2 vs. Lab_3	
	Signif.?	*p*-value	Signif.?	*p*-value	Signif.?	*p*-value
ITS_{dry} (FSD)	Yes	0.046	No	0.552	Yes	0.049

water resistance of the tested material can be considered still acceptable in all the investigated conditions since an ITSR of 70% is commonly recognized as threshold value for material acceptance (Kennedy and Anangos 1984).

5 Summary and Conclusions

The present paper discusses a part of the results achieved during the first phase of the ITP carried out in the framework of the activities of TG1 on "Cold Recycling" of RILEM TC 264-RAP on "Asphalt Pavement Recycling". Specifically, the influence of different curing conditions (free, partial and restricted-surface drying) on the evolution of stiffness, strength and moisture resistance properties of CBTMs with foamed bitumen was analyzed. Results from three involved laboratories testing 150-mm diameter specimens prepared with SGC using common source of RA, cement and filler were used to accomplish this objective. Based on the experimental findings, the following main conclusions can be drawn:

- CBTMs mechanical properties strongly depend on curing and water evaporation conditions.
- Well-designed cured CBTMs are able to achieve noticeable stiffness and strength properties, particularly when the curing conditions allow for a "controlled" water evaporation with beneficial effects on cement hydration.
- The studied CBTMs in cured state demonstrate an acceptable water resistance, regardless the curing conditions.

Findings mainly highlighted that a "restrained" curing (e.g. early laying down of the upper layer during field construction) causes a higher amount of water in the mixture at the end of the process compared to the free-surface drying condition (i.e. extended field curing time without restrictions). Nevertheless, this is not affecting mechanical properties in terms of stiffness, strength and water sensitivity. More studies are encouraged to further investigate the role of water and cement in CBTM mixtures, especially when foamed bitumen is used.

Acknowledgements The authors wish to thank Nynas AB (Sweden) for providing the bitumen used for the foamed bitumen, and Cooperativa Braccianti Riminese (CBR—Italy) for providing the reclaimed asphalt used in this study.

References

Cardone F, Grilli A, Bocci M, Graziani A (2015) Curing and temperature sensitivity of cement-bitumen treated materials. Int J Pavement Eng 16(10):868–880

Dulaimi A, Al Nageim H, Ruddock F, Seton L (2015) A novel cold asphalt concrete mixture for heavily trafficked binder course. Int J Civ Environ Struct Constr Archit Eng 9(8):796–800

Gandi A, Cardenas A, Sow D, Carter A, Perraton D (2019) Study of the impact of the compaction and curing temperature on the behavior of cold bituminous recycled materials. J Traffic Transp Eng (English Edition) 6(4):349–358

Godenzoni C, Graziani A, Bocci E, Bocci M (2018) The evolution of the mechanical behaviour of cold recycled mixtures stabilised with cement and bitumen: field and laboratory study. Road Mater Pavement Des 19(4):856–877

Graziani A, Godenzoni C, Cardone F, Bocci M (2016) Effect of curing on the physical and mechanical properties of cold-recycled bituminous mixtures. Mater Des 95:358–369

Graziani A, Godenzoni C, Cardone F, Bocci E, Bocci M (2017) An application of the Michaelis-Menten model to analyze the curing process of cold recycled bituminous mixtures. Int J Pavement Res Technol 10:62–74

Grilli A, Graziani A, Bocci M (2012) Compactability and thermal sensitivity of cement–bitumen-treated materials. Road Mate Pavement Des 13(4):599–617

Grilli A, Cardone F, Bocci E (2018) Mechanical behaviour of cement-bitumen treated materials containing different amounts of reclaimed asphalt. Eur J Environ Civ Eng 22(7):836–851

Hugener M, Partl MN, Morant M (2013) Cold asphalt recycling with 100% reclaimed asphalt pavement and vegetable oil-based rejuvenators. Road Mater Pavement Des 15(2):239–258

Kennedy TW, Anangos JN (1984) Wet-dry indirect tensile test for evaluating moisture susceptibility of asphalt mixtures. Center for Transportation Research, Research Report 253-8, University of Texas, Texas, USA

Pasetto M, Baliello A, Giacomello G, Pasquini E (2019) Cold recycling pf reclaimed asphalt: analysis of alternative procedures. In: Bituminous mixtures and pavements VII. Taylor & Francis Group, London, pp 551–559

Tebaldi G, Dave E, Cannone Falchetto A, Hugener M, Perraton D, Grilli A, Lo Presti D, Pasetto M, Loizos A, Jenkins K, Apeagyei A (2019) Recommendation of RILEM TC237-SIB on fragmentation test for recycled asphalt. Mater Struct 52(4):52–82

Laboratory Tests for the Characterization of Cold Asphalt Patching Mixtures

Pier Paolo Riviera, Davide Dalmazzo, and Ezio Santagata

Abstract Cold asphalt patching mixtures are widely employed in pavement maintenance operations due to their technical and economical effectiveness. However, their selection is seldom based on the results of laboratory characterization tests, since there are no standard procedures which are recognized by the international community for such a purpose. The investigation described in this paper focused on the evaluation of the strength and stiffness properties of several cold asphalt patching products which were subjected to analysis in two characteristic compaction states: low compaction, as achieved after placement in the field, and high compaction, as reached under the action of traffic loads. Mechanical characteristics evaluated in the laboratory included indirect tensile strength, California Bearing Ratio, resilient modulus and quick shear strength. Analysis of the obtained results highlighted the existence of significant differences between the various products which were explained by referring to their composition and curing behavior.

Keywords Cold asphalt mixture · Patching · Indirect Tensile Strength · California Bearing Ratio · Resilient modulus · Quick shear strength

1 Introduction

One of the most common distresses occurring in asphalt pavements is represented by potholes. The presence of potholes greatly affects traffic safety, since they may act as a trigger for accidents. Furthermore, when potholes initially occur, they may have a significant impact on user costs as a result of the potential damage to vehicles. Finally, potholes and corresponding repair operations may be concurring factors for

P. P. Riviera (✉) · D. Dalmazzo · E. Santagata
Department of Environment, Land and Infrastructure Engineering, Politecnico di Torino, 24, Corso Duca degli Abruzzi, Turin 10129, Italy
e-mail: pierpaolo.riviera@polito.it

D. Dalmazzo
e-mail: davide.dalmazzo@polito.it

E. Santagata
e-mail: ezio.santagata@polito.it

C. Raab (ed.), *Proceedings of the 9th International Conference on Maintenance and Rehabilitation of Pavements—Mairepav9*, Lecture Notes in Civil Engineering 76,
https://doi.org/10.1007/978-3-030-48679-2_7

the creation of traffic jams, consequently increasing fuel consumption and release of air-borne pollutants (Dong et al. 2013; Naveen et al. 2018).

A pothole can be defined as a steep-sided bowl-shaped cavity of variable size affecting the pavement surface. It derives from the progressive damage of existing distresses and is mainly caused by the simultaneous action of several factors among which traffic and moisture are the most important (Miller and Bellinger 2003; Chen et al. 2016). Water seeping into the pavement through existing cracks tends to weaken asphalt layers which can therefore be easily deteriorated by heavy traffic and broken into pieces, thus leading to the creation of a pothole (Liao et al. 2016).

As a result of the frequent occurrence of potholes, one of the most typical pavement maintenance operations is represented by pothole repair, also known as "patching".

Patching activities, which are performed by employing different materials and techniques, are used for temporary and semi-permanent repair operations. In the first case, after being placed on the pothole, the patching material is compacted by means of shovels or by making use of truck tires (the so-called "throw-and-roll" technique). In the semi-permanent practice, the recommended procedure consists in a preliminary cleaning of the distressed area (by removing debris and dust) followed by the thorough compaction of the adopted patching material by means of appropriate equipment (Paige-Green et al. 2010; Biswas et al. 2018). In order to increase the durability of pavement areas subjected to rehabilitation, best practice entails the removal of the upper bituminous layers in a more extended surface (by saw cutting or milling), followed by application of a tack coat and by the laying of a traditional hot-mix asphalt (HMA) mixture compacted by means of steel drum rollers.

Different types of patching materials are available on the market and although it is common to consider HMA mixtures as the most durable, the environmental benefits and speed of use of cold asphalt patching mixes has promoted their widespread diffusion, especially for the repair of small potholes (McDaniel et al. 2014).

Cold patching mixtures are usually manufactured by making use of cutback asphalts or bituminous emulsions and can be easily stored for extended periods of time in appropriate bags or buckets. As reported in the literature, premature failure of cold patching repairs can be related to different factors that include poor short-term strength due to limited curing time, limited resistance to permanent deformation, and insufficient resistance to stripping and debonding in the presence of water (Prowell and Franklin 1996; Dong et al. 2014; Chen et al. 2016; Liao et al. 2016).

Several studies have been carried out in order to identify suitable laboratory test methods for the assessment of patching mixtures and for their relative ranking in the perspective of gathering information for their rational and cost-effective selection. So far, suggested laboratory tests have been developed by moving away from the classical procedures adopted for the characterization of HMA, and by consequently focusing on methods which can yield relevant information with respect to fundamental performance-related properties such as cohesion, workability and resistance to permanent deformation (Prowell and Franklin 1996; Estakhri and Button 1997; Chatterjee et al. 2006; Rosales-Herrera et al. 2007; Dong et al. 2014; Ferrotti et al. 2014; Gómez-Meijide and Pérez 2014; Biswas et al. 2016; Chen et al. 2016; Liao et al. 2016; Hasanuzzaman et al. 2017).

The goal of the experimental investigation described in this paper was to provide a contribution to the topic outlined above, with the consequent assessment of test procedures for the performance-related characterization of patching mixtures. Activities focused on the evaluation of the strength and stiffness properties of several cold asphalt patching products which were subjected to analysis in compaction states which are believed to be representative of the conditions encountered in service.

2 Experimental Investigation

2.1 *Materials*

Materials considered in the experimental investigation included four commercially available cold asphalt patching mixtures (CAPMs) deemed to be representative of the wide range of products normally employed for routine maintenance operations.

CAPM-1 is produced by using a polymer-modified bitumen with SBS and SBR derived from end-of-life tires, conveniently fluidized by means of selected vegetable oils, blended with high quality sand and gravel. CAPM-2 is composed of selected aggregates mixed with a patented binder that hardens in contact with water. CAPM-3 contains aggregates of unknown mineralogy and a fluxed bitumen. Finally, CAPM-4 is made by mixing basaltic aggregates with a patented bitumen emulsion. According to the information provided by suppliers and users, CAPM-1 and CAPM-3 have a positive performance record for temporary patching in urban areas, while CAPM-2 and CAPM-4 are promoted for use in heavy-duty semi-permanent applications.

In order to obtain accurate information on their composition, each CAPM was subjected to preliminary characterization for the evaluation of maximum density (ρ_{mv}, EN 12697-5, 2018), binder content (EN 12697-39, 2012), aggregate gradation (EN 12697-2, 2015) and particle density of extracted aggregates (ρ_p, EN 1097-6, 2013).

Results obtained in this phase of the investigation are provided in Table 1, which also shows the maximum aggregate size (D_{max}) and dust-to-binder ratio (DR, ratio between the content of particles finer than 63 μm and that of employed binder) of each CAPM. Aggregate gradations are displayed in Fig. 1, where they are also associated to their values of the uniformity coefficient C_u, defined as the ratio between

Table 1 Composition of investigated CAPMs

Mixture	ρ_{mv} (kg/m^3)	Binder content[a] (%)	ρ_p(−)	D_{max} (mm)	DR (−)
CAPM-1	2466	6.4	2.720	10	0.64
CAPM-2	2400	8.3	2.732	10	0.51
CAPM-3	2513	5.9	2.765	8	0.46
CAPM-4	2684	6.3	2.946	12.5	0.37

[a]By weight of dry aggregates

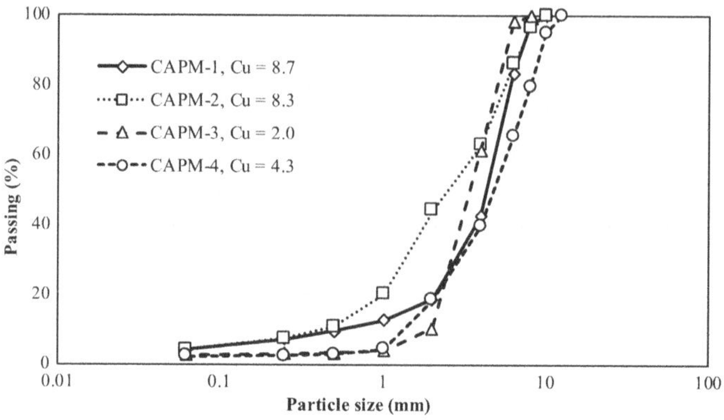

Fig. 1 Gradation curves of investigated CAPMs

the particle sizes corresponding, respectively, to a passing of 60 and 10% (EN ISO 14688-2, 2018).

It can be observed that the four mixtures exhibited a wide variation of ρ_{mv} as a result of different binder dosage and ρ_p of extracted aggregates. Measured binder content, which was comprised between 5.9 and 8.3% by weight of aggregates, was inclusive of the additives employed by producers in order to guarantee a satisfactory workability of the mixtures at low temperatures. Aggregates extracted from the mixtures were characterized by similar values of particle density, with the exception of those contained in CAPM-4, which showed the highest ρ_p as a result of their basaltic nature.

As clearly displayed in Fig. 1, significant differences were recorded among the various CAPMs in terms of the shape of aggregate gradation curves. In particular, CAPM-1 and CAPM-2 presented a medium-graded aggregate distribution (C_u comprised in the 6–15 range), while CAPM-3 and CAPM-4 were characterized by even-graded gradations (C_u lower than 6).

2.2 *Methods*

Specimens of each CAPM were prepared at room temperature by means of the gyratory shear compactor as per EN 12697-31 (2019). Although this method is normally used for the compaction of hot-mix asphalt, in the past it has been successfully adopted by the authors for cold asphalt mixtures (Santagata et al. 2007; Santagata et al. 2010).

Two compaction states of the CAPMs were considered in the investigation with the purpose of reproducing characteristic conditions which are encountered in service. Short-term compaction, achieved after placement and before opening to traffic, was mimicked by applying 50 gyrations. Long-term conditions, reached as a result of further densification due to traffic loads, were replicated by applying 180 gyrations.

In short-term low-compaction conditions, the function of a CAPM placed in a pothole is to prevent premature failure due to loss of material and excessive deformation. Thus, laboratory tests which were selected for the assessment of CAPMs in this state included Indirect Tensile Strength (ITS) and California Bearing Ratio (CBR) tests.

ITS tests were carried out at 20 °C as per EN 12697-23 (2017), with an imposed displacement rate of 50.8 mm/min until failure. In order to analyze the development of strength in time, specimens were cured in a climatic chamber set at 20 °C and were subjected to testing after curing periods of 1, 3, 7 and 28 days. For each curing condition, ITS tests were carried out on 3 specimens for each CAPM and average results were considered in the subsequent analyses.

Although CBR tests are normally used for subgrade soils, during the investigation they were employed for the assessment of the overall resistance to loading of CAPMs in low-compaction conditions, in which these mixtures may exhibit a very low cohesion, similar to that of granular materials. Specimens were prepared by means of the Proctor procedure (EN 13286-2, 2010) and were then subjected to loading as per EN 13286-47 (2012) with no surface surcharge and no preliminary soaking. Prior to testing, specimens were cured for 24 h at 20 °C in the abovementioned climatic chamber.

In long-term high-compaction conditions, a CAPM exhibits a response under loading in service which is similar to that of a standard asphalt mixture and depends upon both bulk structure and developed cohesion. Thus, laboratory tests which were considered appropriate to characterize CAPMs in these conditions included triaxial resilient modulus (M_r) and quick shear strength (QS) tests.

M_r tests were performed at 20 °C by following AASHTO T 307-99 (2017), which entails the application of various combinations of confining pressure (σ_c) and vertical deviatoric stress (σ_d) to slender cylindrical specimens (100 mm in diameter, 200 mm in height). Gyratory-compacted specimens were obtained by introducing material into the mold in three superposed layers, each of which was compacted to desired height and predefined target density (equal to that achieved with 180 gyrations in ITS specimens). Tests were carried out on 4 specimens for each CAPM and average results were considered in the subsequent analyses. In order to simulate long-term curing conditions, prior to testing specimens were cured for 28 days at 20 °C. For comparative purposes, tests were also carried out after a curing period of only 24 h.

After being tested for M_r evaluation, CAPM specimens were subjected to QS tests, in which a vertical strain rate of 0.01 min^{-1} was imposed in the absence of any confining pressure. Stress-strain were recorded during testing with the consequent identification, in peak load conditions, of quick shear strength (σ_{qss}) and strain at failure (ε_f).

3 Results

3.1 Volumetric Properties

Voids content of gyratory-compacted specimens prepared in the two characteristic compaction states and of Proctor-compacted specimens are listed in Table 2.

It can be noticed that significant differences were recorded when comparing the CAPMs with each other for any of the three selected compaction procedures. This is probably due to the combined effects of variations of the viscosity of employed binders, of the degree of uniformity of aggregate gradations, of the binder and filler dosages, and of the corresponding DR value. Mixtures characterized by medium-graded gradations (C_u values greater than 6) and higher DR values (CAPM-1 and CAPM-2) formed, as expected, structures which were denser than those exhibited by patching products with an even-graded gradation (C_u values lower than 6) and lower DR values (CAPM-3 and CAPM-4). Consistently with expectations, the lowest value of voids content was recorded for CAPM-2 which showed the highest binder dosage (8.3%) and filler content (4.2%).

When focusing on the results obtained on gyratory-compacted specimens, it can be observed that for most of the considered CAPMs (i.e. with the exception of CAPM-2) measured voids contents were significantly higher than those which are considered typical for standard hot asphalt mixes. However, values listed in Table 2 are consistent with those reported in literature for similar cold asphalt mixtures (Chatterjee et al. 2006; Gómez-Meijide and Pérez 2014; Hasanuzzaman et al. 2017).

An assessment of the tendency of the CAPMs to densify under the action of traffic loading in service can be made by comparing voids content values recorded at 50 and 180 gyrations. The absolute variation of voids content for all mixtures was comprised between 3.0 and 4.5%. However, it was found that the highest traffic densification potential was exhibited by the mixtures with medium-graded gradations, with a relative reduction in the voids content equal to 21.5 and 41.3% for CAPM-1 and CAPM-2, respectively. Smaller reductions were recorded in the case of the mixtures characterized by even-graded gradations, with values equal to 11.2 and 14.5% for CAPM-3 and CAPM-4, respectively.

Finally, it should be underlined that the voids content of specimens compacted by means of the Proctor procedure were in good agreement with those obtained with the gyratory compactor at 50 gyrations (i.e. in the so-called low-compaction state).

Table 2 Voids content of compacted CAPMs

Mixture	Voids content (%)		
	50 gyrations	180 gyrations	Proctor
CAPM-1	20.5	16.1	19.3
CAPM-2	7.5	4.4	8.8
CAPM-3	27.8	24.7	23.9
CAPM-4	22.8	19.5	28.8

Table 3 ITS and CBR of compacted CAPMs

Mixture	$ITS_{1\ day}$ (MPa)	$ITS_{3\ days}$ (MPa)	$ITS_{7\ days}$ (MPa)	$ITS_{28\ days}$ (MPa)	$CBR_{1\ day}$ (%)
CAPM-1	–	–	–	–	60
CAPM-2	0.37	0.42	0.42	0.45	62
CAPM-3	–	–	–	–	38
CAPM-4	0.02	0.03	0.07	0.26	42

3.2 *Low-Compaction Behavior*

Results obtained from ITS and CBR tests are presented in Table 3. No ITS data are available for CAPM-1 and CAPM-3 since the test specimens prepared with these mixtures, characterized by a very low cohesion, failed under the effect of their own weight as soon as they were positioned in the test jig. Such an outcome did not change even when considering specimens cured for a time period as long as 28 days.

The two CAPMs that could be tested in the indirect tensile configuration exhibited significantly different ITS values which were also characterized by a different sensitivity to curing time. Nevertheless, the range of recorded ITS values was consistent with that presented in literature for cold asphalt mixes having similar composition (Biswas et al. 2016; Rezaei et al. 2017).

As a result of its more closely packed aggregate skeleton and of its stiffer binder, CAPM-2 in the very early stages of curing (i.e. after only 1 day) achieved a strength which was one order of magnitude greater than that of CAPM-4 (0.37 MPa instead of 0.02 MPa). Over time the ITS of CAPM-2 increased only to a limited extent, with most of the changes of binder properties occurring in the first 3 days of curing. On the contrary, the time-dependent curing of the bituminous emulsion contained in CAPM-4 led to a build-up of strength which in post part took place after 7 days of curing, with an average ITS of 0.26 MPa reached after 28 days.

As indicated in Table 3, after only 24 h of curing the CBR index could be measured for all CAPMs due to the lateral restraint that is provided to test specimens by Proctor molds. Obtained results suggest that the CBR index is mainly influenced by the internal structure formed by mineral aggregates, since a clear distinction could be made between the denser mixtures with medium-graded gradations (CAPM-1 and CAPM-2, with CBR values greater than 60%) and those with even-graded gradations (CAPM-3 and CAPM-4, with CBR values of the order of 40%).

3.3 *High-Compaction Behavior*

Results obtained from M_r and QS tests performed at 20 °C on the considered CAPMs in their high-compaction state after 1 and 28 days of curing are presented in Figs. 2, 3 and Table 4. Data were recorded for all mixtures, thereby overcoming the problems which occurred for some of them during ITS low-compaction characterization.

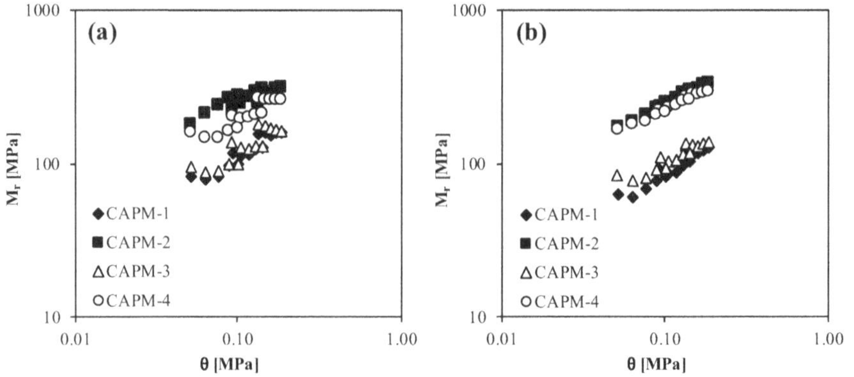

Fig. 2 Resilient modulus of CAPMs after 1 day **a** and 28 days **b** of curing

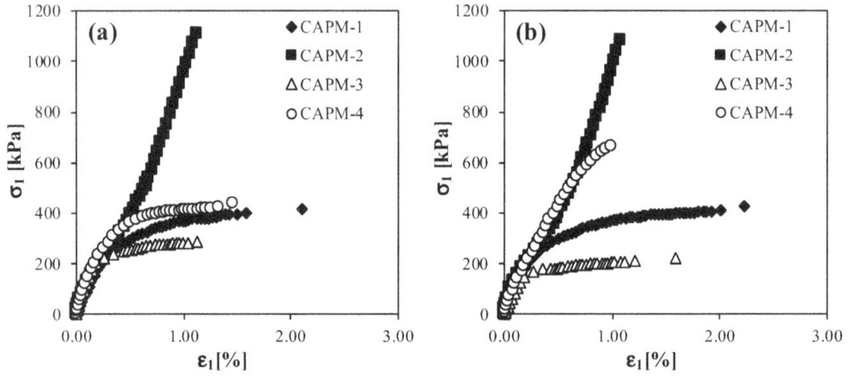

Fig. 3 Stress-strain curves obtained from quick shear tests after 1 day **a** and 28 days **b** of curing

Table 4 Average quick shear test parameters of compacted CAPMs

Mixture	Curing time (days)			
	1 day		28 days	
	σ_{qss} (MPa)	ε_f (%)	σ_{qss} (MPa)	ε_f (%)
CAPM-1	0.424	1.982	0.405	1.891
CAPM-2	1.092	1.146	1.082	1.027
CAPM-3	0.265	0.994	0.199	1.043
CAPM-4	0.446	1.516	0.682	1.075

As shown in Fig. 2, in which M_r is plotted as a function of the first stress invariant (θ), in both curing conditions the stiffest response was exhibited by the two CAPMs which were characterized, as previously shown, by a non-negligible ITS (CAPM-2 and CAPM-4, see Table 3). The other two mixtures (CAPM-1 and CAPM-3) showed significantly lower M_r values, comparable to those of compacted granular materials.

With respect to stress-dependency, Fig. 2 indicates that the stiffest mixture (CAPM-2) in both considered curing conditions displayed a true stress-stiffening behavior which is testified by the progressive M_r increase recorded for increasingly high θ values. On the contrary, as a result of their lower cohesion, at all levels of confining pressure (σ_c), the more deformable mixtures (CAPM-1 and CAPM-3) exhibited a slight reduction of M_r with increasing values of applied deviatoric stress (σ_d), thus combining stress-stiffening and stress-softening characteristics. Such an occurrence, which was more evident in short-term curing conditions, is proven by the existence of discontinuities in the plots provided in Fig. 2. CAPM-4, for which the build-up of ITS was found to occur after 7 days of curing (see Table 3), displayed an intermediate behavior, which was fully stress-stiffening after 28 days of curing but of the combined nature (stiffening and softening with respect to σ_c and σ_d, respectively) after only one day of curing.

Results obtained from QS tests (Fig. 3) were consistent with the outcomes of M_r tests and in line with those reported by Estakhri and Button (1997) for similar cold asphalt mixtures. As shown in Fig. 3, the behavior of CAPM-2 was different from that of all other mixtures, with a stiff quasi-linear response until brittle failure which can be explained by considering the dense aggregate structure and hardened bituminous binder. On the contrary, mixtures characterized by low cohesion (CAPM-1 and CAPM-3) exhibited a non-linear behavior, reaching failure at very high strains. In particular, CAPM-1 was found to be the most ductile, probably as a result of the presence of polymers in its binder. As a result of its cohesion build-up, CAPM-4 displayed an evolution in time, exhibiting a non-linear ductile response after 1 day of curing and a stiffer response, less ductile and closer to that of CAPM-2, after 28 days of curing.

Average QS parameters synthesized in Table 4 clearly indicate that most of the considered CAPMs did not exhibit any relevant changes in σ_{qss} and ε_f as a function of curing time. The only exception is constituted by CAPM-4, which displayed a 53% increase in strength and a 29% reduction in strain at failure when passing from 1 to 28 days of curing.

4 Conclusions

Based on the outcomes of the investigation discussed in this paper, it can be concluded that the experimental procedures selected for the characterization of cold asphalt patching mixtures provide meaningful results which highlight the differences between various products. Low-compaction properties, representative of those possessed by mixtures immediately after their placement in potholes, can be assessed for all products by means of CBR tests, while the applicability of indirect tensile strength tests is limited to mixtures in which the binding matrix has a sufficiently high cohesion. High-compaction properties, achieved in long-term conditions under traffic loading, can be described for all mixtures by means of resilient modulus and quick shear tests which provide information on stiffness and type of failure.

References

Biswas S, Hashemian L, Hasanuzzaman M, Bayat A (2016) A study on pothole repair in Canada through questionnaire survey and laboratory evaluation of patching materials. Can J Civ Eng 43(5):443–450

Biswas S, Hashemian L, Bayat A (2018) Investigation of pothole severity and maintenance methods in Canada through questionnaire survey. J Cold Reg Eng 32(2)

Chatterjee S, White RP, Smit A, Prozzi J, Prozzi JA (2006) Development of Mix Design and Testing Procedures for Cold Patching Mixtures. Federal Highway Administration, report n. FHWA/TX-05/0-4872-1

Chen J-S, Ho H-C, Liao M-C, Wang T-Y (2016) Engineering properties of asphalt concrete patching mixtures. J Test Eval 44(1):534–542

Dong Q, Onyango MA, Huang B (2013) Investigation on service time and effective cost of typical pothole patches in Tennessee. In: Proceedings of the international symposium of climatic effects on pavement and geotechnical infrastructure, Fairbanks, Alaska, USA, 4–7, August, pp 152–158

Dong Q, Huang B, Zhao S (2014) Field and laboratory evaluation of winter season pavement pothole patching materials. Int J Pavement Eng 15(4):279–289

Estakhri CK, Button JW (1997) Test methods for evaluation of cold-applied bituminous patching mixtures. Transp Res Rec 1590:10–16

Ferrotti G, Pasquini E, Canestrari F (2014) Experimental characterization of high-performance fiber-reinforced cold mix asphalt mixtures. Constr Build Mater 57:117–125

Gómez-Meijide B, Pérez I (2014) A proposed methodology for the global study of the mechanical properties of cold asphalt mixtures. Mater Des 57:520–527

Hasanuzzaman M, Hashemian L, Bayat A (2017) Laboratory investigations of cold mix asphalt for cold region applications. J Test Eval 46(1):20160263

Liao MC, Luo CC, Wang TY, Xie X (2016) Developing effective test methods for evaluating cold-mix asphalt patching materials. J Mater Civ Eng 28(10):04016108

McDaniel R, Olek J, Behnood A, Magee B, Pollock R (2014) Pavement Patching Practices, NCHRP Synthesis 463, Transportation Research Board

Miller JS, Bellinger WY (2003) Distress Identification Manual for the Long-Term Pavement Performance Program. Federal Highway Administration, report n. FHWA-RD-03-031

Naveen N, Mallesh Yadav S, Sontosh Kumar A (2018) A study on potholes and its effects on vehicular traffic. Int J Creat Research Thoughts 6(1):258–263

Paige-Green P, Maharaj A, Komba J (2010) Potholes: A Technical Guide to their Causes, Identification and Repair. CSIR Built Environment publication, ISBN 978-0-7988-5594-5

Prowell BD, Franklin AG (1996) Evaluation of cold mixes for winter pothole repair. Transp Res Rec 1529:76–85

Rezaei M, Hashemian L, Bayat A, Huculak B (2017) Investigation of rutting resistance and moisture damage of cold asphalt mixes. J Mater Civ Eng 29(10):04017193

Rosales-Herrera VI, Prozzi J, Prozzi JA (2007) Mixture Design and Performance-Based Specifications for Cold Patching Mixtures. Federal Highway Administration, report n. FHWA/TX-08/0-4872-2

Santagata E, Chiappinelli G, Riviera PP (2007) Experimental investigation for the analysis of cold-recycled bituminous mixtures. In: Proceedings of the 5th international conference on maintenance and rehabilitation of pavements and technological control (MAIREPAV5), Park City, Utah, USA, 8–10 August, pp 463–468

Santagata E, Chiappinelli G, Riviera PP, Baglieri O (2010) Triaxial testing for the short term evaluation of cold-recycled bituminous mixtures. Road Mater Pavement Des 11(1):123–147

Performance Evaluation of Long-Life Pavements Using the Mechanistic-Empirical Asphalt Pavement Analysis (MEAPA) Web Application

M. Ghazavi, A. Seitllari, and M. E. Kutay

Abstract The long-life pavements that last longer than 50 years without major structural rehabilitation and require only periodic surface renewal due to surficial distresses are known to have relatively low life cycle cost. The enhanced performances of such pavements (also known as perpetual pavements) are due to many factors such as improved structural design, better materials and construction practices. In Michigan, several long-life pavement sections were constructed in response to the Public Act 175 (2015) and the Roads Innovation Task Force (RITF) Report. The objective of this study was to evaluate the fatigue performance of one of the long-life pilot projects sections (US-131) and compare it with a standard design in Michigan. A new mechanistic-empirical analysis software (MEAPA) was used to predict the long-term performance of the pavements included in the study. The critical strains for bottom-up and top-down cracking, fatigue life (i.e., number of cycles to failure (Nf)) and predicted magnitudes of bottom-up and top-down cracking were compared. Even though it was observed that the long-life structure is expected to perform better than the standard section in terms of fatigue cracking, both sections are expected to perform well over 30-year design life.

Keywords Perpetual pavements · Fatigue performance · Mechanistic-empirical

1 Introduction

The terms of perpetual and long-life pavements have been introduced in the USA and Europe in the recent decades (Ferne 2006; Hall et al. 2007; Newcomb 2002; Pavements 2007). Such roadways are defined as a flexible or rigid pavement designed and built to last 30 to 50 years without any major structural rehabilitation or reconstruction and requiring only periodic surface renewal in response to distresses that are confined to the top surface. It is also recognized that many well-built, thick pavements that were categorized as either full-depth or deep-strength pavements had been in

M. Ghazavi · A. Seitllari · M. E. Kutay (✉)
Department of Civil and Environmental Engineering, Michigan State University, East Lansing, MI, USA
e-mail: kutay@egr.msu.edu

C. Raab (ed.), *Proceedings of the 9th International Conference on Maintenance and Rehabilitation of Pavements—Mairepav9*, Lecture Notes in Civil Engineering 76,
https://doi.org/10.1007/978-3-030-48679-2_9

service for decades with only minor periodic surface rehabilitation to remove defects and improve ride quality. Generally, surface rehabilitations are relatively easy and low-cost exercises. Therefore, it is a practical reality to consider long-life pavement concept.

Several states in the U.S. built long-life pavements by incorporating enhancements in structural design methods, materials evaluation and specification procedures, construction and maintenance practices (Harm 2001; Harvey 2015; Mahoney 2001; Newcomb et al. 2001; St Martin et al. 2001; Walubita et al. 2010). For instance, the California Department of Transportation (Caltrans) requires the use of 40-year service life for designing pavements along a highway corridor where the 20-year projected average annual daily traffic equals or exceeds 150,000 vehicles or average annual daily truck traffic equals or exceeds 15,000 trucks (Harvey 2015). Long-life concrete pavements have been quite attainable for a long time, as evidenced by the number of older pavements that have remained in service (Hall et al. 2007; Jackson et al. 2014).

In Michigan, four long-life pavement sections were constructed as a result of Public Act 175 (2015) and the Roads Innovation Task Force (RITF) Report (Bleech 2018): (i) 30-year HMA on US-131 in the Grand Region (constructed in 2017), (ii) 30-year concrete on I-69 in the Bay Region (constructed in 2018), (iii) 50-year concrete on US-131 in the Grand Region (constructed in 2018) and (iv) 50-year HMA on I-475 in the Bay Region (constructed in 2019). While designing these pilot projects, modifications were made to standard designs and materials to extend the design life via (i) increased layer thicknesses, (ii) improved materials, (iii) enhanced construction specifications and (iv) upgraded design aspects (e.g., increased drainage freeboard).

In this study, performance tests were conducted on material samples collected from 30-year HMA and the control section (standard design) at US-131 in the Grand Region. The testing matrix consisted of linear viscoelastic characterization of asphalt binders and mixtures and performing mechanistic-empirical analyses using the MEAPA web application, developed at Michigan State University for prediction of long-term performance of asphalt pavements.

The objective of this study was to evaluate the fatigue performance of the long-life pavement located on US-131 in the Grand Region (constructed in 2017) and the control section (standard design) via laboratory characterization and mechanistic-empirical analysis using MEAPA web application. Testing matrix included: (i) dynamic modulus (|E*|) and (ii) dynamic shear modulus (|G*|) for asphalt binder. The laboratory tests provided valuable Level 1 inputs for analysis using the MEAPA software.

Table 1 Asphalt mixture design characteristics

Test section	ID	PG	P_b (%)	NMAS (mm)	N_{des}	VMA (%)	VFA (%)
Long-life	GGSP (T)	70-28P	6.39	9.5	109	18.28	83.59
	4E30 (L)	70-28P	5.21	12.5	109	15.21	80.28
	3E30 (B)	64-28	4.90	19	109	14.29	79.01
Standard	5E10 (T)	64-28	5.90	9.5	96	16.14	81.41
	3E10 (L)	64-28	5.07	19	96	14.18	78.84
	2E10 (B)	58-28	4.48	25	96	13.08	77.06

Note: GGSP = Gap Graded Superpave, Ndes = number of design gyrations, P_b = binder content, NMAS = nominal maximum aggregate size, T = Top course, L = Leveling course, B = Base course

Table 2 Aggregate gradations of the selected asphalt mixtures

MIX/AGG GRADATION, %	2E10	3E10	3E30	4E30	5E10	GGSP
ASPHALT %	4.48	5.07	4.9	5.21	5.9	6.39
P 1″ (25.0 mm)	100	100	100	100	100	100
P 3/4″ (19.0 mm	89.2	100	100	100	100	100
P 1/2″ (12.5 mm)	76.5	87.6	89.8	93.2	100	100
P 3/8″ (9.5 mm)	70.6	79.2	80.4	87.1	96.7	93
P No. 4 (4.75 mm)	50.4	59.1	56.8	65.4	73.2	39.7
P No. 16 (1.18 mm)	27.6	27.2	27.7	32.4	38.1	18.2
P No. 200 (75 μm)	4.6	4.5	3.8	4.1	5.2	8.1

2 Materials and Methodology

2.1 Asphalt Mixture Volumetrics and Mix Design

Multiple loose mixture samples were collected from the test sections at US131. The Superpave mix design volumetric properties of the mixtures are presented in Table 1. The corresponding gradations of the selected mixtures are listed in Table 2. All mixtures were designed to an air void level of 3% according to the Superpave mix design procedure.

2.2 Asphalt Mixture and Binder Testing

Dynamic modulus (|E*|) and binder complex shear modulus (|G*|) data for each of the mixtures were measured in the laboratory and master curves were constructed in general accordance with AASHTO T 342-11 and AASHTO R 84-17. In this study, the original |E*| testing protocol (i.e., AASHTO T 342) was followed, primarily because MDOT's existing database is based on the AASHTO T 342 protocol and it includes measured |E*| data at −10 °C. The |E*| testing was conducted on triplicate samples. The target air void content of the samples was 7 ± 0.5%. The coefficient of variation of the test results was less than 15% for all test temperatures and frequencies.

The |E*| master curve were obtained at a single reference temperature (T_{ref}) by shifting horizontally the |E*| values collected during the test. The amount of shift is different at each temperature and defined by the so-called shift factor coefficients $a_T(T)$. The following second-order polynomial equation was used to develop the relationship between shift factors and the corresponding temperature:

$$log(a(T)) = a_1\left(T^2 - T_{ref}^2\right) + a_2\left(T - T_{ref}\right) \quad (1)$$

where T_{ref} is the reference temperature, T is the tested temperature, a_1 and a_2 are the polynomial fit coefficients for the temperature shift factor. During the shifting process, shift factors were varied until a good fit to the |E*| data at all the temperatures were obtained using the following sigmoidal function:

$$log\left(|E^*|\right) = b_1 + \frac{b_2}{1 + e^{-b_3 - b_4 log(f_R)}} \quad (2)$$

where b_1, b_2, b_3, b_4 are the sigmoid coefficients and f_R is the reduced frequency, which is the product of frequency (f) and the shift factor coefficient (i.e., $f_R = f * a(T)$).

The |G*| testing was conducted in general accordance with AASHTO T 315-12 for the binders of the asphalt mixtures, which included three Performance Grades (PGs): 70-28P (GGSP and 4E30 mixtures), 64-28 (3E30, 5E10 and 3E10 mixtures) and 58-28 (2E10 mixture). The results of the laboratory tests were input to the MEAPA software to describe the mixture properties as Level 1 input.

2.3 Structural Properties of the Pavement Sections

The structural properties of the pavement sections evaluated in this study are shown in Table 3. The MEAPA software was used to predict the long-term performance in terms of fatigue cracking of two pavement structures designed with six asphalt mixtures (see Table 1).

Level 1 inputs (dynamic modulus (|E*|), dynamic shear modulus (|G*|) and IDT strength) were measured and input as properties of the asphalt layers and binders, whereas Level 2 inputs were used for the base, subbase and subgrade layers (due to unavailability of measured resilient modulus data at time of preparation of this manuscript). Pavement structures were evaluated for a service life of 30 years as new flexible pavement. Same traffic level was used for both long-life pavement design and the standard design. Average annual daily truck traffic (AADTT), vehicle class

Table 3 Structural properties of the long-life and standard sections

Layer	Structure 1 - Long-life design		Structure 2 - Standard design	
HMA	GGSP	1.5 in (3.81 cm)	5E10	1.75 in (4.45 cm)
	4E30	2.5 in (6.35 cm)	3E10	3 in (7.62 cm)
	3E30	7.25 in (18.42 cm)	2E10	4.5 in (11.43 cm)
Base (Crushed gravel)	Thickness	12.0 in (30.48 cm)	Thickness	12.0 in (30.48 cm)
	Modulus	33,000 psi (227.5 MPa)	Modulus	33,000 psi (227.5 MPa)
Subbase (A-1-a)	Thickness	24.0 in (60.96 cm)	Thickness	24.0 in (60.96 cm)
	Modulus	20,000 psi (137.9 MPa)	Modulus	20,000 psi (137.9 MPa)
Subgrade (A-6)	Modulus	7000 psi (48.3 MPa)	Modulus	7000 psi (48.3 MPa)

distribution, growth factors, monthly adjustment factors were extracted from MDOT traffic data used for design of these test sections. Average annual daily truck traffic (AADTT) considered in this study was 2880, directional distribution was 51% and lane distribution was 96% in this study. Climatic data were obtained from the MEAPA climate database for the city of Grand Rapids, MI. The fatigue performance of the asphalt concrete layer including N_f (number of cycles to failure) and critical strains were the primary focus of this study.

2.4 Pavement Performance Prediction Using the Mechanistic-Empirical Asphalt Pavement Analysis (MEAPA) Web Application

The Mechanistic-Empirical Asphalt Pavement Analysis (MEAPA) is a web-based application to predict flexible pavement performance. Figure 1 illustrates the project detail page of the MEAPA web site. The MEAPA analysis engine includes implementation of the original formulations of the Mechanistic-Empirical Pavement Design Guide (MEPDG), with few improvements and simplifications on climatic model and improvements on the top-down cracking model. The rest of the formulations implemented in MEAPA are identical to those of MEPDG. The main analysis engines of MEAPA were coded in MATLAB, including the Layered Elastic Analysis algorithm, called MatLEA. The MatLEA is publicly available and formulations and computational steps are identical to those of the MnLayer software (Khazanovich and Wang 2007). The web-based application interface of MEAPA is composed of many sub algorithms written in several languages including JAVA, Python, JavaScript, HTML and CSS. In addition, the project data is saved in and retrieved from a MySQL database, which is integrated into the JAVA codes.

Fig. 1 Project detail page of MEAPA web application

Like in the Pavement ME software, prediction of pavement temperature with depth is based on the Climatic-Materials-Structural Model (CMS Model) originally developed at the University of Illinois (Dempsey 1969). Main improvement is how sunrise and sunset times are computed, which are important in calculation of net radiation flux at the pavement surface (Dempsey 1969).

The fatigue cracking model in MEAPA is based on the traditional fatigue life formulation implemented in the MEPDG (ARA Inc. ERES Consultants Division 2004; Seitllari and Kutay 2019; Seitllari et al. 2019):

$$N_f = C_H C \beta_{f1} k_{f1} \left(\frac{1}{\varepsilon_t}\right)^{\beta_{f2} k_{f2}} \left(\frac{1}{E}\right)^{\beta_{f3} k_{f3}} \tag{3}$$

$$C = 10^{4.84\left(\frac{V_{be}}{V_a + V_{be}} - 0.69\right)} \tag{4}$$

$$C_{H-bu} = \left(b_{bu1} + \frac{b_{bu2}}{1 + e^{(b_{bu3} - b_{bu} h_{ac})}}\right)^{-1} \tag{5}$$

where N_f = Number of cycles to failure, ε_t = critical tensile strain, E = equivalent modulus (at the given temperature/frequency), V_{be} = Effective asphalt content by volume, V_a = Percent air voids in the HMA mixture, k_{f1}, k_{f2}, k_{f3} = Global field calibration parameters (from the NCHRP 1-40D re-calibration $k_{f1} = 0.007566$, $k_{f2} = -3.9492$, and $k_{f3} = -1.281$), β_{f1}, β_{f2}, β_{f3} = Local or mixture specific field calibration constants; for the global calibration effort, these constants are initially set to 1.0, h_{ac} = height of the AC layer, $b_{bu1} = 0.000398$, $b_{bu2} = 0.003602$, $b_{bu3} = 11.02$, $b_{bu4} = 3.49$.

For top-down cracking, major tensile principal strain near the tire within top 0.5″ (1.27 cm) of the asphalt pavement layer was used in the fatigue life formulation (Eq. 3) instead of the horizontal tensile strain at the surface. This was because values of horizontal tensile strain at the surface were unrealistically low and it does not represent the true three-dimensional strain state near the surface. This is partially why an improved top-down cracking model was needed and this need was met with the NCHRP Project 01-52 (Lytton et al. 2018). However, NCHRP Project 01-52 project resulted in a procedure that uses Artificial Neural Networks (ANNs) and trained ANN is not publicly available (only available to AASHTO). This makes it impossible for the results of NCHRP Project 01-52 to be used by any institution other than AASHTO. The use of major principal strain near the pavement surface near the tire is the best intermediate solution until a more accurate and publicly available model is developed.

3 Results and Discussions

The dynamic modulus (|E*|) master curves of the mixtures and the dynamic shear modulus (|G*|) master curves of the binders are plotted in Fig. 2. In general, at lower temperatures and higher frequencies, 3E30, 2E10 and 3E10 have the highest modulus followed by 4E30, 5E10 and GGSP. On the other hand, at higher temperatures and lower frequencies, all mixtures had similar stiffness, except 5E10, which had relatively low stiffness. The |G*| master curves of different binders were similar at high frequency/low temperature combinations (i.e., at high reduced frequencies). This is expected because all binders have same PG. On the other hand, at high temperatures, as expected, the binders with high PGs have higher |G*|. MSCR (multiple

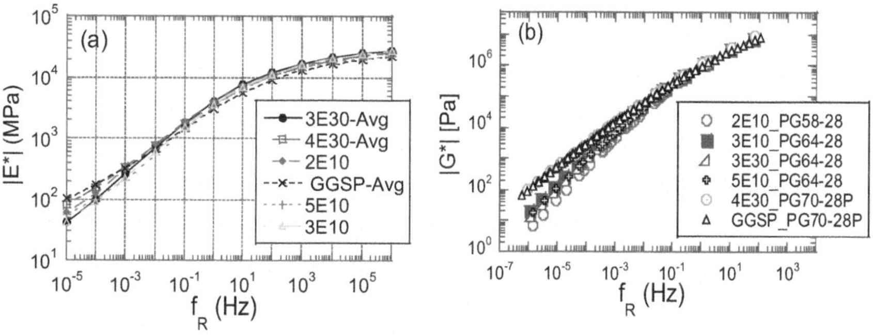

Fig. 2 **a** |E*| master curves and **b** |G*| master curves

stress creep recovery) grades for PG70-28P, PG64-28 and PG 58-28 are E-grade, H-grade and S-grade respectively and their percent recoveries are 78.8, 5.3 and 0.0%, respectively.

As part of the bottom-up cracking prediction, the MEAPA calculates the critical strains at the bottom of the AC layer for each month in each of the five temperature quintiles. Then these values are used in the fatigue life formulation shown in Eq. 3 to calculate Nf, which is subsequently used in Miner's law of damage growth using the traffic data. It is often of interest of the engineer to know the distribution of the strains throughout the analysis period for each month, and for each temperature quintile within a month. In this study, 30-year simulations of traffic were performed. As a result, for each of the single, tandem, tridem and quad axles, 30 years * 12 months * 5 quintiles = 1800 critical strains are computed by the layered elastic analysis program MatLEA. It is a bit hard to visualize this many data points in a single figure. Boxplots are quite useful for such data. The results of the critical strain boxplots for bottom-up fatigue cracking for long-life sections and standard sections for five temperature quantiles are presented in Fig. 3. On each box, the central mark is the median, the edges of the box are the 25th and 75th percentiles. The whiskers extend to the most extreme data points that the algorithm considers to be not outliers, and

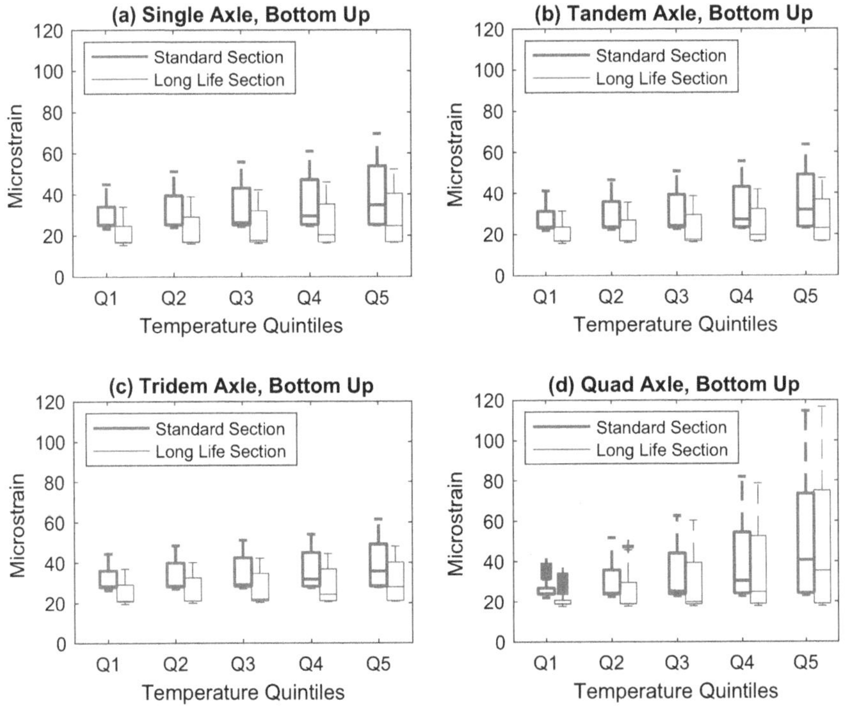

Fig. 3 Bottom-up cracking critical strains for long-life and standard sections for single, tandem, tridem and quad axles

the outliers are plotted individually. Each subfigure in Fig. 3 illustrates the data for different axles, and data is plotted for each temperature quintile within each subplot. As shown, for all axles and temperature quintiles, the critical strains for long-life section are lower than those of the standard section. This is due to structural and material differences between the standard and long-life sections. Comparing critical strains (median values for each combination) shows that strains for long-life section (design) are less than standard section (design) and this ratio is ranging from 0.67 to 0.71 for single axle, 0.72 to 0.73 for tandem axle, 0.75 to 0.78 for tridem axle and 0.78 times to 0.89 for quad axle in bottom-up cracking and 0.77 to 0.92 for single axle, 0.77 to 0.91 for tandem axle, 0.78 to 0.89 for tridem axle and 0.78 times to 0.87 for quad axle in bottom-up cracking.

Overall both standard and long-life sections produced quite low strain levels due to most axles and in most quintiles, except in quad axle, which is understandable. However, the axle load spectra did not include too many repetitions of quad axles in this road segment, therefore, overall accumulated bottom-up damage is quite low (see Fig. 5).

Top-down cracking critical strains for long-life and standard sections for single, tandem, tridem and quad axles are shown in Fig. 4. Similar to bottom-up cracking,

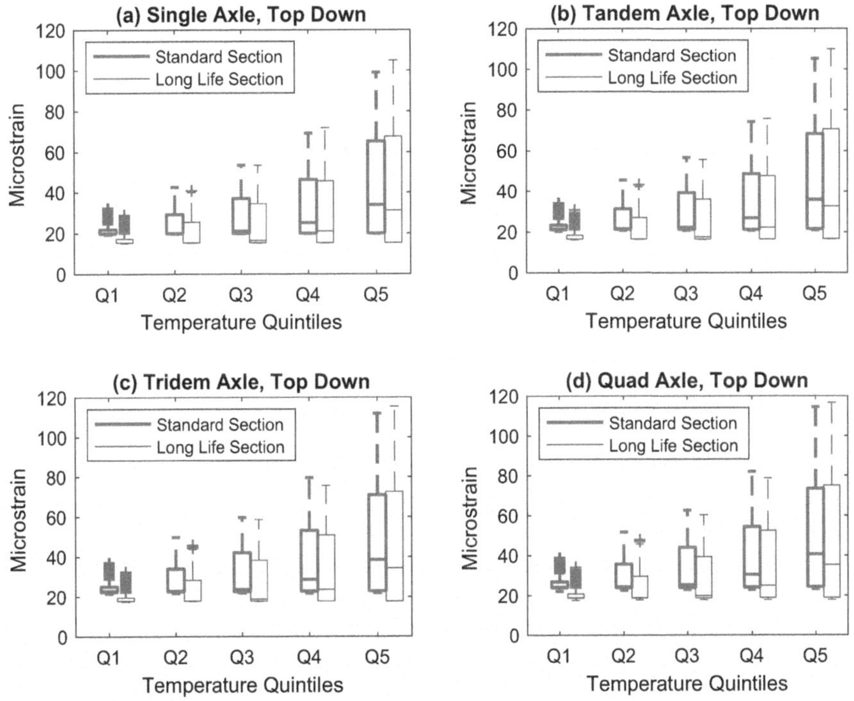

Fig. 4 Top-down cracking critical strains for long-life and standard sections for single, tandem, tridem and quad axles

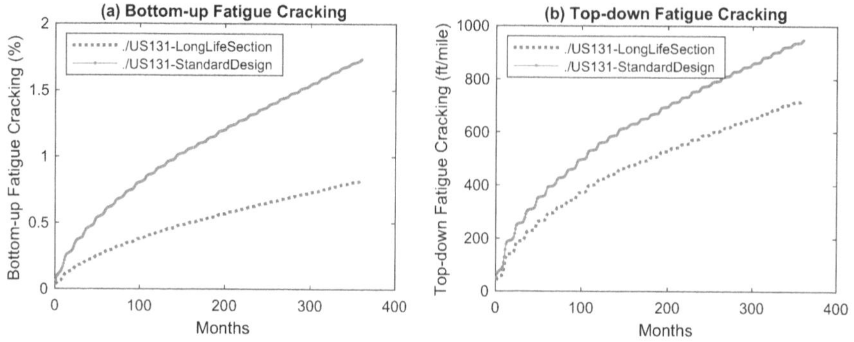

Fig. 5 Results of MEAPA predictions of **a** bottom-up and **b** top-down cracking for long-life and standard Sections (1 ft/mile = 0.19 m/km)

the long-life section resulted in lower critical strains near the surface. This is mostly due to the material-level differences rather than structural differences. On the other hand, the magnitudes of the top-down cracking strains were higher than those of the bottom-up cracking strains, which is expected for such thick structures. This explains why top-down cracking damage is accumulated at a point where actual cracking is predicted at the surface (see Fig. 5).

Results of MEAPA predictions of bottom-up and top-down cracking are shown in Fig. 5a and b, respectively. As shown, the long-life section is expected to perform better than the standard section. However, Fig. 5a shows that the total bottom-up cracking is expected to be below 2% in both sections at the end of the 30-year service life. This is significantly lower than the typical threshold of 25%. The top-down cracking magnitudes at the end of 30 years reached up to 1000 ft/mile(190 m/km), which is about half of the typical 2000 ft/mile (380 m/km) threshold. As such, both sections are expected to perform quite well in the field.

4 Conclusions

The objective of this study was to evaluate the fatigue performances of the long-life and standard (control) pavement sections constructed on route US-131 in the Grand Rapids, Michigan. A mechanistic-empirical pavement structural analysis methodology, called MEAPA, was used to conduct the performance analysis. The MEAPA, short for Mechanistic-Empirical Asphalt Pavement Analysis, is a web-based application to predict flexible pavement performance. The MEAPA uses the original formulations of the Mechanistic-Empirical Pavement Design Guide (MEPDG), with few improvements and simplifications on climatic model and improvements on the top-down cracking model. The rest of the formulations implemented in MEAPA are identical to those of MEPDG. The main analysis engines of MEAPA were coded in MATLAB, including the Layered Elastic Analysis algorithm, called MatLEA. Key inputs to MEAPA, i.e., mixture dynamic modulus (|E*|) and binder dynamic shear modulus (|G*|) were measured in the laboratory on the samples collected from the field. The analysis parameters included critical strains for bottom-up and top-down cracking, fatigue life (i.e., number of cycles to failure (Nf)) and predicted magnitudes of bottom-up and top-down cracking. It was observed that at all axles and temperature quintiles, lower critical strains and higher Nf values for long-life section were observed as compared to those of the standard section. Even though both top-down and bottom-up fatigue cracking values of the long-life section over 30-year analysis period was lower than those of the standard section, they were lower than the design thresholds. Therefore, both sections are expected to perform well in the field.

References

ARA Inc. ERES Consultants Division (2004) Guide for mechanistic–empirical design of new and rehabilitated pavement structures. Final report, NCHRP Project 1-37A - Appendix GG

Bleech C (2018) Roads innovation task force report. Michigan department of transportation

Dempsey BJ (1969) A heat-transfer model for evaluation frost action and temperature related effects in multilayered pavement systems. University of Illinois, Urbana-Champaign

Ferne B (2006) Long-life pavements—a European study by ELLPAG. Int J Pavement Eng 7(2):91–100

Hall K, Dawood D, Vanikar S, Tally Jr, R, Cackler T, Correa A et al (2007) Long-life concrete pavements in Europe and Canada

Harm E (2001) Illinois extended-life hot-mix asphalt pavements. Transp Res Circ 503:108–112

Harvey J (2015) New asphalt pavements research, and long-life (perpetual) asphalt. University of California Pavement Research Center, UC Davis

Jackson N, Puccinelli J, Mahoney J (2014). Using Existing Pavement in Place and Achieving Long Life. Transportation Research Board

Khazanovich L, Wang Q (Chuck) (2007) MnLayer: high-performance layered elastic analysis program. Transp Res Rec J Transp Res Board 2037(1):63–75. https://doi.org/10.3141/2037-06

Lytton RL, Luo X, Ling M, Chen Y, Hu S, Gu F (2018) A Mechanistic–Empirical Model for Top–Down Cracking of Asphalt Pavements Layers

Mahoney JP (2001) Study of long-lasting pavements in Washington State. Transp Res Circ 503:88–95

Newcomb D (2002) Perpetual pavements-a synthesis

Newcomb DE, Buncher M, Huddleston IJ (2001) Concepts of perpetual pavements. Transp Res Circ 503:4–11

Pavements LA (2007) Long-life asphalt pavements technical version, 1–24, June

Seitllari A, Kutay ME (2019) Development of 3-point bending beam fatigue test system and implementation of viscoelastic continuum damage (VECD) theory. J Assoc Asphalt Paving Technol

Seitllari A, Lanotte MA, Kutay ME (2019) Comparison of uniaxial tension-compression fatigue test results with SCB test performance indicators developed for performance-based mix design procedure. In: Nikolaides AF, Manthos E (eds.) Bituminous mixtures and pavements VII: Proceedings of the 7th international conference 'bituminous mixtures and pavements' (7ICONFBMP). CRC Press, Thessaloniki, Greece

St Martin J, Harvey JT, Long F, Lee EB, Monismith CL, Herritt K (2001) Long-life rehabilitation design and construction: I-710 Freeway, Long Beach, California. Transportation Research Circular (503)

Walubita LF, Scullion T, et al (2010). Texas perpetual pavements: new design guidelines

Management Systems and Life Cycle Analysis

The Challenges of Warm Mix Asphalt as a Mature Technology

Ali Jamshidi and Greg White

Abstract Laboratory tests and field investigations show promising results of structural performance of warm mix asphalt (WMA). Also, the lower environmental burdens and fuel requirement in WMA production increase sustainability in the asphalt industry. However, there are challenges that significantly affect WMA performance and marketing in the future. In this paper, all these challenges are discussed, and the trend of WMA technology is evaluated. The statistics indicate that the energy market plays a pivotal role in enthusiasm for WMA. Lastly, WMA requires further investigations to meet requirements of post-modern pavement, as a new concept of pavement design in the 21st century.

Keywords Energy crisis · Crude oil price · Greenhouse gas emission · Asphalt industry

1 Introduction

The WMA technology is a great step taken toward sustainable pavement construction. Lower environmental emissions and less energy requirement are main features of WMA. Therefore, WMA has the potential to build a consensus between environment and pavement engineers in eco-pavement design. This fragile consensus depends on WMA technologists paying attention to current and future technical challenges. It should be noted that the perception of WMA in the pavement and environment engineering is different. That is, the environmentalists recognize WMA as a sustainable technology due to reduced greenhouse gas emissions, while the asphalt manufacturers are very keen to produce cheaper asphalt via lower production temperatures. One of the current challenges in pavement technology is early failures due to global warming. For example, Mills et al. (2009) reported that the thermal cracking in the asphalt pavement decreases over the next 50 years in Canada, while rutting deteriorated earlier than expected time. The result is consistent with research carried out by Qiao et al. (2013) which showed that a 5% climate change significantly

A. Jamshidi (✉) · G. White
University of Sunshine Coast, Sippy Downs, QLD, Australia
e-mail: ajamshidi@usc.edu.au

C. Raab (ed.), *Proceedings of the 9th International Conference on Maintenance and Rehabilitation of Pavements—Mairepav9*, Lecture Notes in Civil Engineering 76,
https://doi.org/10.1007/978-3-030-48679-2_10

affects longitudinal cracking and rutting. In the United States, results of modelling integrated with into American Association of State Highway and Transportation Officials (AASHTO) and mechanistic empirical pavement design guide (MEPDG) software show that fatigue and rutting cracking increases by 2–9% and up to 40% depending on the climate zone Gudipudi et al. (2017).

This paper narrates how WMA has developed throughout the last 20 years. In addition, current and potential challenges for WMA in the pavement market are discussed. Finally, the potential of WMA for construction of multi-role infrastructure assets, based on post-modern pavement (PMP) concept, are evaluated.

2 History of WMA

Reduction of production and compaction temperatures for asphalt mixtures is not a new matter of research. For example, role of foamed bitumen in reduction of construction temperature was found in 1956 (Kristjansdottir 2006). Then in 1968, Mobil Oil Australia modified and patented the foam bitumen technology, which resulted in a practical procedure for marketing WMA technology (Muthen 1998). In the 1970s, the asphalt industry was challenged by the first oil shock that dramatically affected the cost of pavement construction. Consequently, alternative sources of pavement materials and low-energy asphalt mixes attracted significant attention. For example, the use of various waste materials, such as reclaimed asphalt pavement (RAP), reclaimed asphalt shingles (RAS) and cold asphalt mix technology were developed. In addition, environmental concerns were raised due to the anthropogenic greenhouse emissions associated with asphalt production. Consequently, the interest in sustainable pavement technology increased in order to meet stricter environmental regulations. To address this challenge, the asphalt industry developed new technologies and additives that reduced construction temperatures via various WMA technologies (Table 1).

Since 2000, the introduction and commercialization of WMA began across the world. The rapid rise in energy price associated with the second oil shortage from 2006 highlighted importance of WMA due to the reduced energy required for asphalt production. The second oil shock was significant because the peak point of global oil production was expected in the near future, or may have already occurred (Armstrong and Blundell 2007). The characteristics of the first oil shock in the 1970s was different to that in the 2000s. In the 1970s, the energy crisis was due to the Arab embargo on oil, while the more recent energy crisis is because of lack of oil supply. But both energy crises prompted the development of various WMA technologies to produce more energy-efficient asphalt mixes. Some companies invested enormously in the production of WMA additives. For example, Sasol Wax in the South Africa invested $360 million to pipe natural gas from Mozambique to Sasolburg for production of Sasobit® (Jamshidi et al. 2013). In other words, the increase energy price resulted in the rapid development of WMA solutions. For example, Fig. 1 shows that trend of research on WMA and crude oil price from 2002 to 2018. There is a direct correlation between number of papers published on WMA and crude oil price. For example, the

Table 1 List of the WMA technologies (Rubio et al. 2012)

WMA product	Mechanism	Dosage	Temperature reduction (°C)
Aspha-min	Water-containing technology using zeolites	0.3% by total weight of the mix	20–30
Asphaltan B	Refined montan wax with fatty acid amide for rolled asphalt	2%–4% by mass of asphalt binder	20–30
Advera	Water-containing technology using zeolites	0.25% by total weight of the mix	10–30
Aqua black	Water-based foaming process	2.5% water by weight of the asphalt binder	20–30
Double Barrel Green	Water-based foaming process	2% water by mass of asphalt binder	30–40
Hypertherm		0.2% by mass of asphalt binder	20–35
LT asphalt	Foam bitumen with hydrophilic additive	0.5% by mass of bitumen	40
WAM foam	Soft binder coating followed by foamed hard binder	2–5% water by mass	20–40
LEA	Hot course aggregate mixed with wet sand	3% water with fine sand; 0.4% bitumen weight	20–30
LEAB	Direct foam with binder additive Mixing of aggregates below water boiling point	0.1% of bitumen weight of coating and adhesion additive	20–40
Montan wax			
Sasobit	Fischer-Tropsch wax	Approx. 0.8–5% by weight of binder	20–30
Asphaltan B	Refined Montan wax with fatty acid amide for rolled asphalt	2.0–4.0% by mass of bitumen	20–30
Licomont BS	Fatty acid amide	3.0% by mass of bitumen	20–30
Evotherm	Water-containing technology	0.5% of mass of bitumen emulsion	20–40
Cecabase	Chemical-based	0.2–0.4% by mixture weight	20–30
Rediset	Cationic surfactants and organic additive	1.5–2% of bitumen weight	20–30

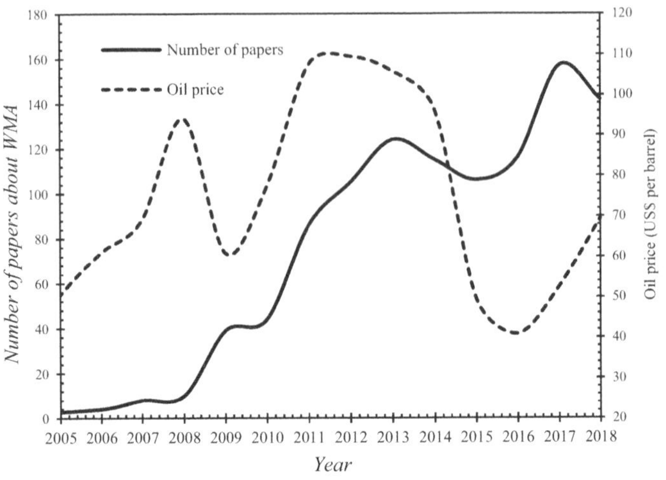

Fig. 1 The relationship between WMA papers and energy market. *Note* 1 the number of papers and technical documents (totally 1,086) were indexed by SCOPUS. *Note* 2 data of energy price is provided based on average annual OPEC statistics

oil price increased significantly from 2009 to 2012, and the number of papers also increased from 39 to 105, a 169% increase. After 2012, the oil price decreased gently, and the number of papers decreased over the same timeframe. In 2016, the oil price was at a low point and the number of papers decreased but increased again as the oil price increased in 2017 and 2018. As a result, the trend of research work depends on the energy market fluctuations.

Figure 2 shows a breakdown of major subjects of WMA in the documents indexed by the SCOPUS data base. Many papers touched on more than a single research topic. From Fig. 2, structural performance in terms of rutting and fatigue are the first and second most prevalent research subjects, respectively, followed by, mix durability

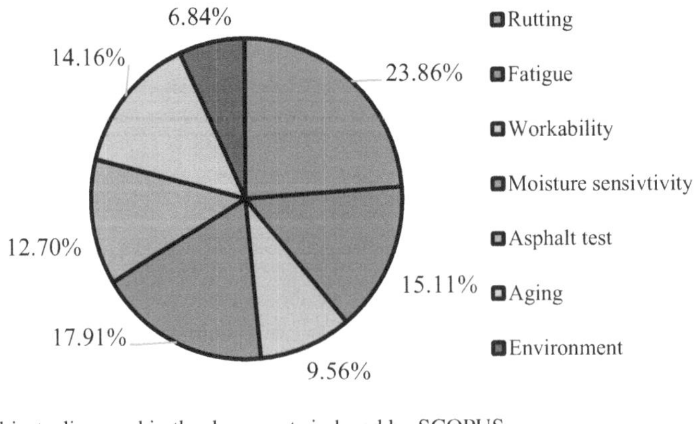

Fig. 2 Subjects discussed in the documents indexed by SCOPUS

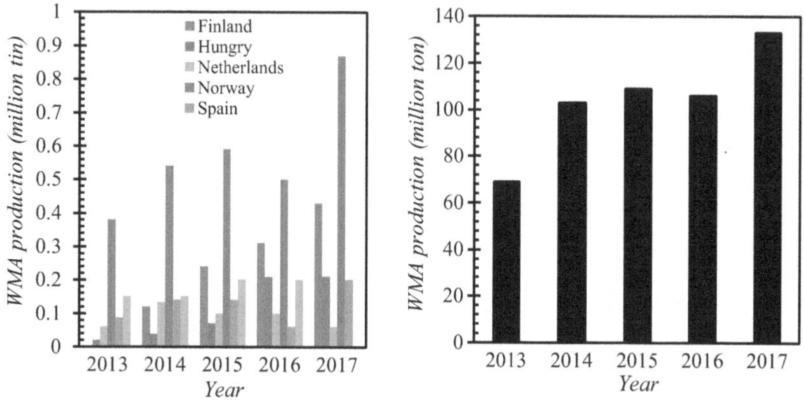

(a) WMA (b) WMA production in the United States

Fig. 3 WMA production in different countries, plotted based on data reported by EAPA (2018)

based on moisture sensitivity and aging. The figure also shows that only 6.84% of the research carried out related to environmental issues. As a result, structural stability and durability were main concerns of WMA researchers. Figure 3 shows that the WMA production in some European countries, South Africa, and the United States increased from 2012 to 2017, due to encouraging initial laboratory test results. In addition, promising results of field investigations in the Unites States and Europe spread the use of WMA into the Middle-East and Asian markets.

Obviously, the WMA performance should be equal or superior than the hot-mix asphalt (HMA). So, different manufacturers try to produce more efficient technologies and additives in terms of construction temperature reduction, structural strength, and durability. For example in France, every asphalt manufacturer and agency try to develop own WMA technology (D'Angelo et al. 2008).

3 Challenges and Opportunities of Asphalt Industry

3.1 Long-Term Performance

Cumulative structural and thermal strains/stresses are important factors for the long-term performance of pavements. Although the lower construction temperature of WMA results in less stiffness, the results of field investigation often showed satisfactory structural and functional performance. For example, Diefenderfer (2018) reported that the dynamic modulus, air void content and permeability of foamed WMA cored after 6 years were comparable with HMA samples. Also, the rut resistance of Sasobit® WMA mixtures, as indicated by dynamic modulus, was greater than or equal to that of HMA (Huerly et al. 2009). It seems that the structural performance of WMA is not a challenge. In addition, some additives, e.g., Sasobit®, can be used as compact aid which improves mix workability (Bonaquist 2011). However, the long-term performance of asphalt pavements depends on synergistic effects of structural loading and field ageing. Since the construction temperature of WMA is less than HMA, it is less prone to ageing. Furthermore, the ageing phenomenon in WMA can be different due to various additive types and production technology, binder type, ambient temperature, test temperature, and parameter chosen for analysis. Thus, it is necessary to develop prediction models for various WMA additives which can be used to better understand ageing process. In addition, the use of sustainable materials in the heavy-duty pavements, e.g., airports, is increasing (White et al. 2018). In this regard, shear stress resistance of pavement is important criterion (White 2017). Therefore, the less stiffness, due to the lowered construction temperature, may decrease shear stress resistance of the pavement. As a result, the long-term performance of WMA used in the heavy-duty pavements is matter of concern.

3.2 *Recyclability*

WMA usually contains some recycled or reused waste material, such as crumb rubber, RAPRAS, or waste glass. The WMA will be reclaimed as new source of RAP in future (RAP-WMA). However, the performance of new mixes containing RAP-WMA is less clear. It is expected that the engineering properties of such new mixes should be comparable with new WMA or HMA. However, further research on the recyclability of the WMA and HMA containing RAP-WMA is required. This is essentially to determining whether recycling RAP-WMA has economic justification or not.

3.3 *Selection of Appropriate WMA*

Since there are various types of WMA, selection of an appropriate WMA type is subjective, usually determined by local availability of equipment and/or materials. In this regard, it is necessary to compare the structural performance and durability of WMA samples produced using different WMA technologies. However, the comparison is not easy because the mechanism of each WMA technology is different (Table 1). To address this problem, Jamshidi et al. (2015) proposed criteria to compare various types of WMA technologies based on unaged and aged rheological characteristics. It should be noted that the selection of a particular WMA technology also depends on many other factors, including the initial cost, skill of paving crew, service condition, pavement application, availability of machinery and compatibility with the local materials.

3.4 Use of Local Materials, Practice Codes and Construction Technology

WMA technology should be compatible with local materials and current asphalt mixing plant facilities, which is an important factor for marketing WMA technology. Therefore, it is necessary to train the paving crews and engineers regarding its use. It is recommended that practice codes and guidelines for the use of various WMA technologies in local materials and construction practices be developed. It seems that the lack of courses in the field of WMA at universities and industry affects the basic knowledge of engineers to fully understand the complex problems of modelling and characterizing pavement materials and the principles associated with WMA technologies. Therefore, in parallel with the research on WMA, it is necessary to teach the principles of WMA technology. It is also recommended that the WMA manufactures provide periodical workshops to update the engineers dealing with pavement technology. As a result, a close collaboration between the WMA producers and universities can result in rapid development of WMA in future.

In addition, the use of current technology of the asphalt plant without any specific modification can be a motivation for WMA production. For example, Sasobit® can be blended with the hot asphalt binder at a terminal or in asphalt tank, introduced in a molten form, added with the aggregate materials, or pneumatically blown into a drum plant (Hurley et al. 2009). It requires simple modification that can be used in many asphalt mixing plants.

3.5 Costs of WMA

Although WMA technology reduces fuel requirement in the asphalt mixing plant, it may incur extra costs due to modification of the mixing plant and additive material supply. Therefore, WMA may only be an attractive option when the energy price is high. Table 2 shows the cost incurred using various WMA technology. It is clear that the cost can be different, depending on mix tonnage, shipping haul, WMA technology and industrial fuel type. From the table, the costs of HMA and WMA are comparable. However, more cost-effective WMA is more attractive for the asphalt manufacturers.

Table 2 A cost comparison of HMA and WMA production in Iceland (Diefenderfer 2018)

Cost ($US)	HMA	WAM foam	Aspha-Min	Sasobit
Additive cost	–	0.30	4	3.50
Energy	6.5	4.90	4.90	4.90
Reduction in energy cost compared HMA	–	1.60	1.60	1.60
Total	89	87.70	91.40	90.90

Since the WMA reduces the construction temperatures, more RAP or the other waste material can be used which results in more cost-effective mixes. As an example, the costs of WMA containing 30% RAP decreases by 32% (Onor and Sengoz 2015). Consequently, WMA may be better suited as an enabling technology for RAP usage, rather than a technology in its own right. In this regard, it is necessary to develop facilities for waste material management, such as RAP milling and tire shredding. However, development of such facilities may require massive investment, especially in the developing countries.

3.6 New Criteria for WMA Production

WMA technologists focus on the reduction of the construction temperatures, while the role of aggregate material in energy saving is neglected. Furthermore, many asphalt producers use WMA technologies as a compaction aid, producing asphalt at the same temperature as HMA. In both cases, the thermal characteristics of aggregate materials have a significant effect on the energy requirement and greenhouse gas (GHG) emissions. For example, the aggregate type, mechanical strength, gradation and specific gravity may be identical, but thermal properties, in terms of specific heat capacity, can be different. So, the selection of the appropriate aggregate source can result in huge energy saving. For example, Jamshidi et al. (2015) showed that energy requirement for HMA produced by the low-specific heat capacity can be lower than WMA using the high-specific aggregate. Therefore, the thermal properties of aggregate is proposed as new indicator for analysis environmental burdens in the Superpave mix design procedure (Jamshidi et al. 2013). It is therefore necessary for WMA technologists to pay more attention of the thermal properties of aggregate and filler. However, it must be noted that local availability and haulage costs often limit the choice of aggregate sources.

4 Future of WMA

The higher structural performance and lower environmental burdens are important factors for sustainable asphalt pavement design, but on their own, these factors are inadequate. Therefore, a new concept, called post-modern pavement (PMP) was developed to address the structural, sustainability, and socio-psychological requirements (Fig. 4). In other words, not only the high structural performance and sustainability are components, but also the social acceptability of WMA is important factor in future pavement technology. In other words, pavement based on PMP concept is a structural-environmental-social asset, under the policy of integration of the municipal infrastructure assets. However, the social benefits associated with reduced GHG emissions of WMA is not clear. To fill this gap, more research should be undertaken to highlight potential of WMA as PMP for the pavement end-users. For example,

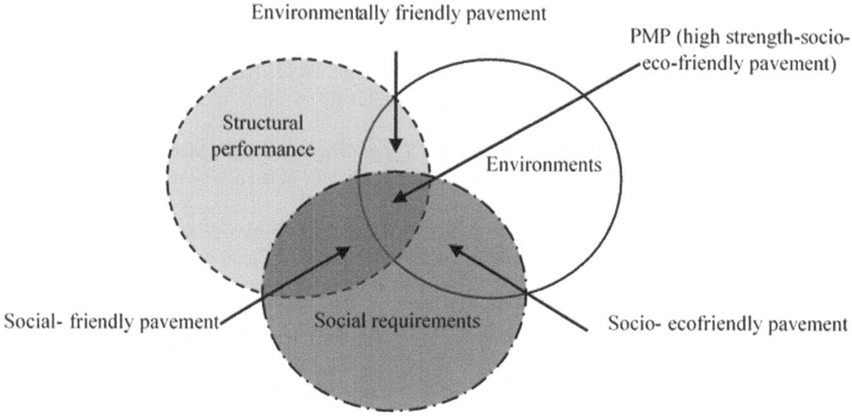

Fig. 4 PMP definition as an interface between structural, environmental, and social requirements (Jamshidi et al. 2019)

White et al. (2018) studied socio-environmental effects of airport pavement runway containing recycle materials and this approach could be extended to different, structurally equivalent, HMA and WMA mixtures.

5 Conclusion

The energy market oscillations play a more important role in WMA development than associated environmental benefits do. Also, the asphalt industry must understand and address the challenges and expectations of pavement and environment engineers to be successful in the future. This means that WMA technologists and suppliers will need to foster and promote the products meet future requirements. In conclusion, WMA is already sufficiently mature to support high strength-low energy asphalt pavement. However, more work is needed to disseminate WMA based on the PMP concept.

References

Armstrong FA, Blundell KM (2007) Energy… beyond oil. Oxford University Press, Oxford

D'Angelo J, Harm E, Bartoszek J, Baumgardner G, Corrigan M, Cowsert J, Harman T, Jamshidi M, Jones W, Newcomb D, Prowell B (2008) Warm-Mix Asphalt: European Practice (No. FHWA-PL-08-007). United States. Federal Highway Administration. Office of International Programs Washington, DC, USA

Diefenderfer SD (2018) Performance of virginia's early foamed warm mix asphalt mixtures. Transp Res Rec 2672(28):178–189

Gudipudi P, Underwood S, Zalghout A (2017) Impact of climate change on pavement structural performance in the United States. J Transp Res Part D 57:172–184
Hurley GC, Prowell BD, Kvasnak AN (2009) Michigan Field Trial of Warm Mix Asphalt Technologies: Construction Summary. National Center for Asphalt Technology, Auburn, AL, USA
Jamshidi A, Kurumisawa K, White G, Nishizawa T, Igarashi T, Nawa T, Mao J (2019) State-of-the-art of interlocking concrete block pavement technology in Japan as a post-modern pavement. Constr Build Mater 200:713–755
Jamshidi A, Golchin B, Hamzah MO, Turner P (2015) Selection of type of warm mix asphalt additive based on the rheological properties of asphalt binders. J Clean Prod 100:89–106
Jamshidi A, Hamzah MO, You Z (2013) Performance of warm mix asphalt containing sasobit: state-of-the-art. J Constr Build Mater. 38:530–553
Kristjansdottir O (2006) Warm mix asphalt for cold weather paving (No. WA-RD 650.1). MSC. thesis, University of Washington, Seattle, Washington, D.C, USA
Muthen KM (1998) Foamed Asphalt Mixes – Mix Design Procedure. CR-98/077. Sabita ltd and *CSIR* Transportek, South Africa
Mills BN, Tighe SL, Andrey J, Smith JT, Huen K (2009) Climate change implications for flexible pavement design and performance in Southern Canada. J Transp Eng 135(10):773–782
Bonaquist RF (2011). Mix design practices for warm mix asphalt, vol 691. Transportation Research Board. National Cooperative Highway Research Program (NCHRP), Report 691, Washington, USA
Oner J, Sengoz B (2015) Utilization of recycled asphalt concrete with warm mix asphalt and cost-benefit analysis. PLoS One 10(1):1–18
Qiao Y, Flintsch G, Dawson A, Parry T (2013) Examining effects of climatic factors on flexible pavement performance and service life. Transp Res Rec 2349:100–107
Rubio M, Carmen GM, Baena L, Moreno F (2012) Warm mix asphalt: an overview. J Clean Prod 24:76–84
White G, Fairweather H, Jamshidi A (2018) Sustainable runway pavement rehabilitation: a case study of an Australian airport. J Clean Prod 204:380–389
White G (2017) State of the art: interface shear resistance of asphalt surface layers. Int J Pavement Eng 18(10):887–901

A Framework for Network Level Pavement Maintenance Planning for Low Volume Roads

H. R. Pasindu, R. M. K. Sandamal, and M. Y. I. Perera

Abstract Low volume roads (LVRs) play a pivotal role in the economic development of rural areas especially by providing connectivity for the communities to access markets, education and social needs in an efficient manner. They serve as the link between the local road network to the arterial and collector road network designed at providing accessibility to residential, agricultural or industrial areas. Lack of funding, subjective and ad hoc decision making has resulted in an ineffcent utilization of resources in the local road agencies. Lack of a sound analytical process is a major impediment to maintain these roads in cost effective manner under the resource constraints prevalent. Existing pavement management systems (PMS) require extensive data collection and complex analysis processes, which makes them impractical to be deployed in local agencies. The core attributes of the proposed system are, reduced the data requirements, simplified the analytical tools and allowing users to customize considering the resource constraints. In this study, a relationship between International Roughness Index (IRI) and relevant distresses for LVR is established and based on that cost estimation model is developed for distress repair. Furthermore, the strategy which provide maximum condition for preventive maintenance is found by using decision tree approach in the network level optimization. A case study illustrated that the use of proposed PMS provides better overall network condition with compare to conventional decision making for same budget level.

Keywords Pavement management system · Low volume roads · Optimization · International Roughness Index · Cost estimation model

H. R. Pasindu (✉) · R. M. K. Sandamal · M. Y. I. Perera
Department of Civil Engineering, University of Moratuwa, Moratuwa 10300, Sri Lanka
e-mail: pasindu@uom.lk

R. M. K. Sandamal
e-mail: 198046r@uom.lk

M. Y. I. Perera
e-mail: 140460v@uom.lk

C. Raab (ed.), *Proceedings of the 9th International Conference on Maintenance and Rehabilitation of Pavements—Mairepav9*, Lecture Notes in Civil Engineering 76,
https://doi.org/10.1007/978-3-030-48679-2_11

1 Introduction

The classification of low volume road (LVR) is typically based on average daily traffic (ADT) and approximately less than 1000 vehicles/day, however the definition is differs from country to country significantly (Gamage et al. 2016). Low volume roads play an important role in providing accessibility to communities to fulfill their social and economic needs. Road pavement deteriorate with time, under traffic and environment effects and the condition of those pavements are generally in unsatisfactory condition, resulting in increase of vehicle operating cost, delay among several other issues to the road users. Therefore, it is imperative to maintain those roads at the satisfactory condition to meet the road user needs. In developing countries, the LVRs are maintained by local authorities (e.g. provincial councils, municipal councils) and typically, there is no comprehensive maintenance strategy in place mainly due to the lack of technical expertise and financial limitations. Conventional pavement management systems (PMS) require a certain level of technical knowledge to collect and analyze data, which may not always be available in local road agencies. Moreover, the data collection processes in PMS are time consuming and costly Therefore an alternative system with the capability to overcome those issues and be adoptive for the local authorities is required. In maintenance planning, several factors are to be considered, namely pavement structural and functional condition, social-economic importance, traffic volume etc. Moreover, in the decision making stage, selection of roads for different maintenance activities, strategies to be applied and post condition prediction must be accurately incorporated into the system to achieve optimum solution for the particular road network.

The main objective of this study is to develop a user-friendly and customizable system, with low data requirement with the capability of providing an optimal maitenance strategy within the budget and other constraints.

2 Pavement Management Systems for Low Volume Roads

Road asset management is a systematic process of maintaining, upgrading, and operating physical assets cost-effectively as defined by United State Department of Transportation (USDOT) in 1999 (Shah et al. 2017). It focuses on combining engineering principles with business practices and economic considerations, by providing a tool to facilitate a more organized, logical approach to decision-making. The existing PMS are consisted with different analytical tools, resource management methods, prioritization techniques. Ferreira et al. (2009) developed a PMS with road network database, quality evaluation tool, cost model, decision aid tool and a pavement performance model with incorporating Geographical Information System (GIS) for data collection.

Pavement condition evaluation is consisting of four aspects, i.e. distress condition evaluation, pavement roughness measurement, skid resistance and structural capacity evaluation (Hass et al. 1994). Among those aspects, pavement roughness is effective to use for network level evaluation due to its repeatability, productivity and ease of collection. Gamage et al. (2016), Islam et al. (2014) have developed models to evaluate road condition using roughness data based on a mobile phone application and a cost model is also developed to forecast maintenance cost with respect to International Roughness Index (IRI). Several researchers have shown that the applicability of roughness measurement is high in PMS for network level evaluation (Tai et al. 1998; Eriksson et al. 2008; Bisconsini et al. 2018; Buttlar and Islam 2014).

The maintenance and rehabilitation (M&R) strategies are incorporated in PMS based on the user defined threshold and trigger values. Prior to budget optimization process roads are categorized into routine maintenance, preventive maintenance, rehabilitation or reconstruction (Mane et al. 2016). In the budget optimization, performance jump for different operation types must be defined accurately. Islam et al. (2014) investigated pavement roughness improvement can be achieved by different maintenance treatment types such as slurry seal, chip seal, crack seal for flexible pavement. Moreover, Dwaikat and Haider (2012) estimated the pre-treatment pavement performance, jump and slope adjustment factors for various treatment types while showing that slurry and chip seal can be applied when rutting is minimal and thin overlay is cost-effective when the rutting is high.

Leanne et al. (2011) conducted a study on challenges and successes of implementing a PMS and identified that in the decision making, higher the number of decisions and higher the number of roads, make the output time longer. To reduce the output time and simplify the decision-making process, prioritization and optimization model can be implemented. Perera et al. (2019) shown that with the use of integer programming, average network condition can be reduced with the increasing of budget level and accuracy of optimization model. The challenges in optimizing large networks can be overcome with the using cost model, decision tree approach to solve the objective functions and prediction models with combining to a proper internal database (Mahoney et al. 1978; Swei et al. 2016). Scheinberg and Anastasopoulos (2010) shown that significant cost saving can be achieved with the use of year-by-year multi-constraint technique at network-level optimization using integer programming by a case study on the State of Virginia.

3 Proposed Pavement Management System

3.1 Condition Evaluation Method

In the system, pavement roughness is measured in terms of IRI and is used as the default road condition measurement parameter. Based on the results from the detail distress survey along with IRI measurement on selected low volume roads in Sri Lanka, a relationship is established between IRI and distresses as shown in Eq. (1). This suggests that IRI can accurately represent the relevant distress condition of rural

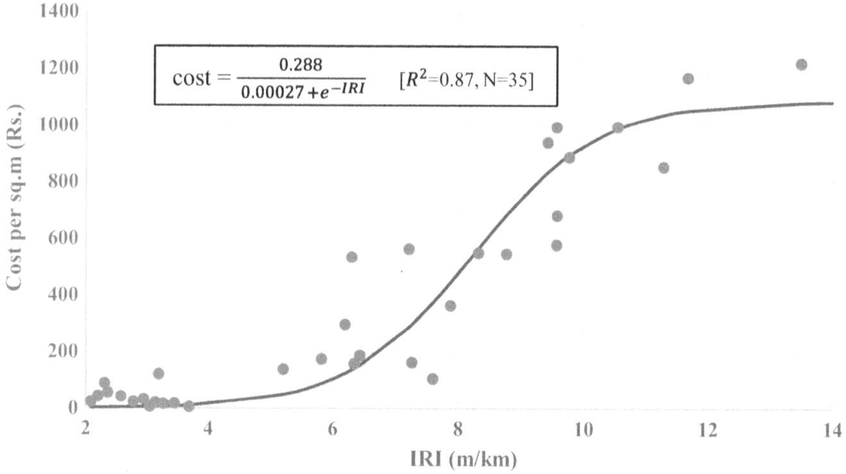

Fig. 1 IRI vs major distresses repair cost plot

roads. Moreover, by using these distress density a cost estimation model is developed for distress repair as shown in Fig. 1.

$$IRI = 2.90 + 0.16RAV\% + 0.29CRA\% + 0.40EDG\% \quad \left[R^2 = 0.75, N = 95\right] \tag{1}$$

where *RAV%* is the raveling area as a percentage of total pavement area, *CRA%* is the cracking area as a percentage of total pavement area, *EDG%* is the linear length of edge gap (more than 10 cm gap between carriageway and shoulder in both side of the pavement section) as a percentage of total length of road section.

Further, potholes are considered as a condition evaluation parameter due to its usage of identify priority roads to be repaired by local agencies. Pothole identification is relatively easy, and it can also be incorporated in the roughness measurement apps developed. Thus, it would not be a major issue for the local road agencies. In addition to roughness and number of potholes, basic road characteristics (e.g. pavement type, length of the section, average width) are the main component of the data required for the proposed system.

3.2 Screening and Selection for Corrective Maintenance

The software framework of the proposed PMS is developed by using Java programming language and MySQL software which provide the user interaction and data base respectively. In the system, following steps are performed in the screening and corrective maintenance prioritization process.

Step 1: Screening - If $IRI_i < IRI_{Prev}$ then road 'i' (i = 1, 2, 3…n) is filtered from the analysis step.
Step 2: Selection of roads for corrective maintenance - If $IRI_i > IRI_{Corr}$ then road 'i' (i = 1, 2, 3…n) is selected for the corrective maintenance/overlay in worst first ranking method until the allocated corrective maintenance budget is fully utilized.

Where IRI_i is the IRI of road 'i', IRI_{Prev} is the threshold IRI value used for screening, IRI_{Corr} is the threshold IRI value used for selecting roads for corrective maintenance, n is the number of roads in the network.

Based on the engineers' judgement and resource availability IRI_{Prev} and IRI_{Corr} can be defined in the selection criteria. The corrective maintenance cost is consisting of two components namely distress repair and overlay. The user must be input the unit value of overlay maintenance. Distress repair must be performed prior to the overlay and cost is based on IRI (see Fig. 1) and number of potholes per kilometer (NP). The most appropriate transfer function for cost estimation model is selected among the linear and non-linear regression analysis and among those sigmoid function is found to be the best representative of the data as shown in Fig. 1. Finally, by adding abovementioned two components, total cost of overlay is calculated.

3.3 *Preventive Maintenance Activities*

After selecting roads for corrective maintenance, the remaining road sections are performed under preventive maintenance for a given budget using the optimization approach. This optimization approach adopts a decision tree considering all possible combinations of operations. The operation type which are used for LVRs are identified by the opinion survey from engineers in local authorities. The representative cost values are shown in Table 1 using the Highway Schedule of Rates (HSR) (Provincial 2019).

Table 1 Selected preventive maintenance types used in decision tree approach with unit cost

Operation no. (r)	Preventive maintenance operation	HSR-2019 item no.	Cost (LKR.)
1	Minimum maintenance	DR-009 + DR-010 + MS1-021	33 per sq.m + 20 per L.m. + 40 per sq.m
2	Major distress repair + Pothole patching	Fig. 1 + MS1-007	Fig. 1 + 1404 per sq.m
3	Pothole patching + Double Bitumen Surface Treatment (DBST)	MS1-007 + S1-030B	1404 per sq.m + 477 per sq.m

Where DR is the drain work, MS is the miscellaneous, S is the surface treatment, sq.m is the square meter, L.m. is the linear meter, and LKR is the Sri Lankan rupee (1 US $ = 190 LKR)

In the decision tree approach, cost of each operation for respective section is calculate based on the pavement condition and road inventories as described below.

1. Minimum maintenance - Based on the length/area of the section and no distress repairing is considered in this operation.
2. Major distress repair - Based on the existing IRI value, cost is calculated for this operation. The unit value from Fig. 1 is multiply by the total area to calculate total cost for the entire section.
3. Pothole patching - NP is converted into the pothole repair cost by the system and a representative area of a pothole must be defined by the user.
4. DBST- Based on the HSR rates, using the unit values, total cost for DBST is calculated by multiplying by total area of the section.

IRI, NP, unit rates and road inventories (e.g. length, width etc.) used as the inputs and respective cost of operation for each section is calculated in the cost estimation model by following the described procedures. This would allow the cost estimation to be made for the maintenance activities based on IRI alone with NP, which would minimize the cost for extensive distress data collection.

3.4 Post Repair Condition Prediction

Performance jump (condition of each section after applying each repair) should be input by the user to the system prior to running optimization model. Minimum maintenance is not affected to change of IRI significantly while major distress repair is reduced roughness slightly. Applying DBST is restored IRI to the range of 2.5–3.4 m/km (Montenegro and Minc 1992).

3.5 Optimization Model

Step 3: Optimization Model for Preventive Maintenance Selection - Accepted preventive maintenance combination is selected by the combination which minimize overall network IRI under the budget constraint. The objective function and the constraint used in the system are represented in Eq. (2) to Eq. (4).

Objective function; $$Condition;\ Minimize\ Q = \frac{\sum_{s=1}^{n}\sum_{r=1}^{m} Qrs.Ls.Xrs}{L} \tag{2}$$

Constraints; $$Budget;\ B \geq \sum_{s=1}^{n}\sum_{r=1}^{m} Crs.Ls.Xrs \tag{3}$$

$$Annual\ operation; \sum_{r=1}^{m} Xrs = 1 \tag{4}$$

where Q is the average network IRI value, Qrs is the IRI of road section 's' after applying the operation 'r', Ls is the length of road section 's' in km, Xrs is the decision variable (if operation 'r' is applied to road section 's' then $Xrs = 1$ and operation 'r' is not applied to road section 's' then $Xrs = 0$, L is the total length of the road network, B is the total budget available for the financial year, Crs is the cost per km for applying operation 'r' to road section 's', n is the number of road sections in the network, and m is the number of operation used for road section 's'.

4 Illustrative Example

To illustrate the developed system, a sample of 27 road sections each having length of 1 km were selected. The budget for corrective maintenance works is Rs. 115 million and preventive maintenance budget is Rs. 15 million.

Step 1: Screening - IRI_{Prev} of 4 m/km were used in the case study and 9 sections were screened from the maintenance process based on the value of IRI_{Prev}.
Step 2: Selection of roads for corrective maintenance - 11 road sections were identified for the corrective maintenance under IRI_{Corr} value of 7 m/km and are prioritized based on IRI value. Total cost of corrective maintenance is Rs. 113.13Mn and out of 11 only 9 sections were selected due to budget constraint. The selected sections for corrective maintenance are shown in Table 2.
Step 3: Optimization Model for Preventive Maintenance Selection - The remaining 11 road sections are selected for the preventive maintenance process. Rs.15Mn budget allocation is used for the preventive maintenance and it was found that total cost of Rs. 14.62Mn and average network IRI of 3.98 m/km obtained from the optimization as shown in Table 3.

Table 2 Road sections prioritized under corrective maintenance

Rank	Road Id (s)	Existing IRI values (m/km)	Number of potholes per km (NP)	IRI after overlay (m/km)
1	C300/Sec004	10.3	41	2.5
2	C300/Sec003	9.1	37	2.4
3	C300/Sec002	9.0	21	2.4
4	C301/Sec004	8.8	32	2.4
5	C301/Sec003	8.6	17	2.3
6	C301/Sec002	8.5	07	2.3
7	D002/Sec001	8.1	32	2.2
8	D002/Sec002	7.5	21	2.1
9	C300/Sec001	7.4	15	2.1

Table 3 Summary of results in preventive maintenance process

Road Id (s)	Existing IRI values (m/km)	Number of potholes per km (NP)	Applied operation (r)	IRI after applying (m/km) (Qrs)	Cost (Crs) (Rs.)
C003/Sec001	5.7	10	3	3.5	1,697,600
C003/Sec002	5.4	10	3	3.5	1,697,600
C002/Sec002	5.3	7	3	3.4	1,689,170
C002/Sec001	5.2	7	1	5.2	311,000
C301/Sec001	7.1	14	3	3.8	1,708,840
C004/Sec002	4.4	3	1	4.4	311,000
D002/Sec004	7.3	22	3	3.9	1,731,320
D005/Sec001	6.6	17	3	3.8	1,717,270
D002/Sec003	6.1	21	3	3.7	1,728,510
D005/Sec002	5.0	4	1	2.7	311,000
D005/Sec003	5.9	17	3	3.6	1,717,270

Where r = 1 is the minimum maintenance, r = 2 is the major distress repair + pothole patching, r = 3 is the pothole patching + DBST

From the proposed PMS, overall network IRI of 3.16 m/km is obtained under total cost of Rs.127.75Mn (Budget constraint is Rs.130Mn).

The comparison of average network condition and total cost between different treatment strategies (minimum maintenance all, distress repair all, DBST all, overlay all) used for the selected road network and output from the developed PMS is shown in Fig. 2. In the PMS, 85% of total budget for corrective maintenance and other 15% for preventive maintenance is allocated for different budget limit.

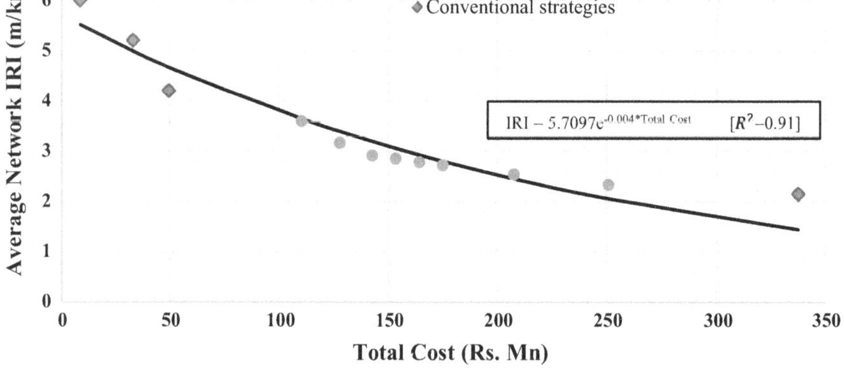

Fig. 2 Average network condition and total cost for different treatment strategies

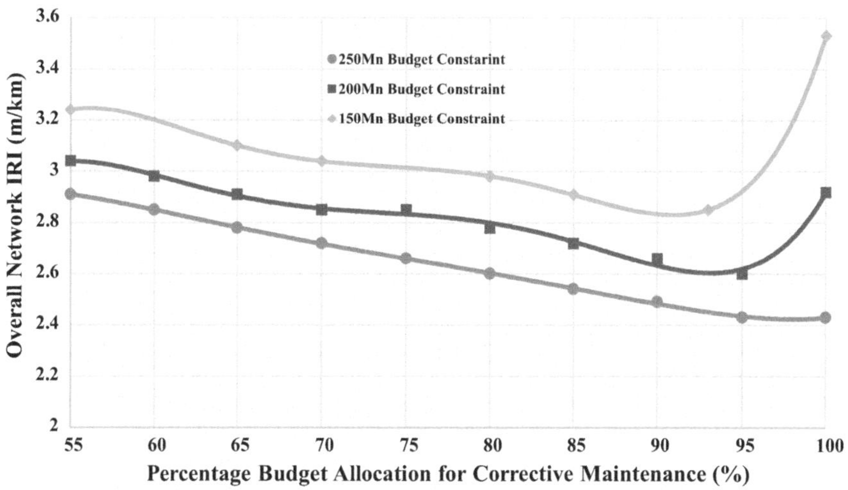

Fig. 3 Overall network condition under different budget allocation levels for corrective maintenance

Furthermore, for this network, total budget is fixed and the combination which gives the minimum overall IRI is found under that budget constraint is shown in Fig. 3.

From the comparison shown in Fig. 3, it can be concluded that when the budget limit is increasing, overall network IRI is reduced and the optimum combination is move towards the higher corrective maintenance percentage. Further, it refers that when the available budget is limited, by allocating relatively higher budget for the preventive maintenance will be provide a better overall network condition rather than allocating more to the corrective maintenance.

5 Conclusions

The research was focused into developing a system which requires minimum data and optimize the maintenance strategy within the available budget. This model can be customized based on road agency decision making strategy. By using the developed system's optimization model, an engineer could be able to decide the preliminary budget requirement for preventive and corrective maintenance. The research provides a decision support system which doesn't require intensive data collection and a systematic process which would yield better results than subjective decision making. Illustrated example shown that, the system has the capability to find the maximum overall network condition under the given budget. Moreover, for a given budget, the combination between preventive and corrective maintenance which gives the minimum network IRI can be found by running the analysis process for several iterations.

Acknowledgements The research was supported by SRC (Senate Research Committee) funding of grant no: SRC/LT 2018/19.

References

Bisconsini DR, Nunez JYM, Nicoletti R, Junior JLF (2018) Pavement roughness evaluation with smartphones. Int J Sci Eng Invest. 7:43–52

Buttlar W, Islam S (2014) Integration of smart-phone-based pavement roughness data collection tool with asset management system. McElhanney Consulting Service Ltd., Canada

Dwaikat MB, Haider SW (2012). Estimating optimum timings for maintenance treatments to alleviate pavement surface rutting. In: 91st Annual meeting of transport research board, Washington D.C.

Eriksson J, Girod L, Hull B, Newton R, Madeen S, Balakrishnan H (2008) The pothole patrol: using a mobile sensor network for road surface monitoring. In: 6th International conference on mobile systems, applications, and services, Breckenridge

Ferreira AJL, Meneses SCN, Vicente FAA (2009) Pavement Management System for Oliveira do Hospital, Portugal. Transportation 162, vol. 3, pp 157–169, June

Gamage D, Pasindu HR, Bandara SJ (2016) Pavement roughness evaluation method for low volume roads. In: 8th international conference on maintenance and rehabilitation of pavements, Singapore

Hass R, Hudson WR, Zaniewski JP (1994) Modern pavement management. Krieger Publishing Company, Florida

Islam S, Buttler WG, Aldunate RG, Vavrik WR (2014) Measurement of pavement roughness using android-based smartphone application. Transp Res Rec J Transp Res Board 2457:30–38

Leanne WL, Khaled H, Jason J (2011) Implementing a pavement management system for low volume roads: challenges and successes. In: Annual conference of transportation association in Canada, September, Alberta, Canada

Mahoney JP, Ahmed NU, Lytton RL (1978) Optimization of pavement rehabilitation and maintenance by use of integer programming. Transp Res Rec J Transp Res Board 674:15–22

Mane AS, Gujarathi SN, Arkatkar SS, Sarkar AK, Singh AP (2016). Methodology for pavement condition assessment and maintenance of rural road. In: National conference on fifteen years of PMGSY, Roorkee

Montenegro FM, Minc MJ (1992) Road condition and maintenance inputs for feasibility studies in developing countries. Transp Res Rec 1359:106–116

Perera MYI, Pasindu HR, Sandamal RMK (2019) Pavement maintenance management system for low volume roads in Sri Lanka. In: 5th international multidisciplinary engineering research conference, July, Moratuwa, Sri Lanka

Provincial Engineering Organization Western Province-Provincial Council (2019) Highway Schedule of Rates

Scheinberg T, Anastasopoulos PC (2010) Pavement preservation programming: a multi-year, multi-constraint optimization methodology. In: 89th Annual Meeting of Transport Research Board, Washington D.C.

Shah R, McMann O, Borthwick F (2017) Challenges and prospects of applying asset management principles to highway maintenance: a case study of the UK. Transp Res Part A Policy Pract 97:231–243

Swei O, Gregory J, Kirchain R (2016). Pavement management systems: opportunities to improve the current frameworks. In: 95th Annual meeting of transport research board, Washington D.C.
Tai Y, Chan C, Hsu JY (1998) Automatic road anomaly detection using smart mobile device. In: 15th Conference on artificial intelligence and applications, Wisconsin

P-F Curves in Modelling of Pavement Performance

Adam Zofka

Abstract Road asset management (RAM) is a systematic approach to maintain road network at sufficient functionality level in order to provide desirable level of services. Various strategies can be implemented within RAM but only proactive strategy ensures sustainable, optimal, and cost-effective decisions. One of the variations of the proactive strategy is reliability-centered maintenance (RCM). Integral part of the RCM framework are so called P-F curves that allow to knowingly consider the potential failure and loss of functionality of road asset. While this is not entirely new concept as comparing to models used for pavement performance, P-F curves emphasize critical stages of the development of pavement damage and concurrent continuous increase in probability of functional failure. This paper discusses several aspects of different strategies in maintenance management and associated deterioration models together with P-F curves highlighting the need for continuous improvement of processes involved in decision-making as well as verification of effectiveness and efficiency of actions applied onto road network.

Keywords Pavement performance · Deterioration models · Reliability-centered maintenance (RCM) · Proactive strategy

1 Maintenance Strategies and Models

Road asset management (RAM) allows for a systematic and comprehensive approach to implement the overall management strategy and to achieve strategic goals. Through a proper implementation of RAM for the road infrastructure, a reliable methodology is defined that will demonstrate the legitimacy of allocating funds for maintenance and will improve the decision-making process by introducing transparent rules and actions.

RAM is in fact a set of business processes for making decisions that encourage continuous improvement in asset management. Thanks to the implemented RAM,

A. Zofka (✉)
Road and Bridge Research Institute, IBDiM, ul. Instytutowa 1, 03-302 Warsaw, Poland
e-mail: azofka@ibdim.edu.pl

C. Raab (ed.), *Proceedings of the 9th International Conference on Maintenance and Rehabilitation of Pavements—Mairepav9*, Lecture Notes in Civil Engineering 76,
https://doi.org/10.1007/978-3-030-48679-2_12

road organization has efficient communication not only internally, but also with external stakeholders, e.g. politicians, media and users. RAM is a methodology that leads to more efficient and effective asset management based on good quality information (Taggart et al. 2017).

One of the fundamental elements of any RAM is management strategy. Generally, management strategies can be divided into two types, i.e. reactive and proactive. The latter can be further subdivided into:

- preventive,
- prognostic,
- fully proactive.

In fact, maintenance strategies can be even further categorized based on their scope, organizational effort, subject asset etc. One of the classifications is shown in Fig. 1 which is included in the EU standard (EN 13306: 2017 Maintenance—Maintenance terminology).

Historically management strategies have been progressing over last century starting with reactive approach and then moving to preventive, prognostic and finally a fully proactive approach. One of the variations of the proactive strategy is reliability-centered maintenance (RCM). In the RCM strategy, all failure modes are identified and effectiveness and efficiency of mitigations actions are constantly evaluated based on monitoring data. Further, RCM typically accommodates risk management processes to optimize maintenance actions and their timing for delivering appropriate level of service at acceptable reliability and costs (NASA 2008).

One of the most important aspects of a proactive strategy is the ability to forecast future demand and supply (in terms of pavement asset functionality and their service levels). Within RAM, such forecasts are used to prepare guidelines/specifications for design and maintenance, as well as for planning proactive activities on the road

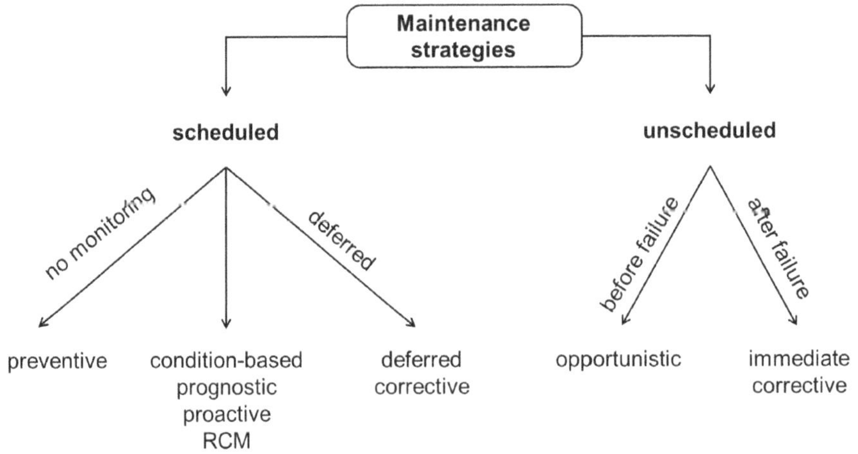

Fig. 1 Scheduled vs. unscheduled maintenance (EN 13306: 2017)

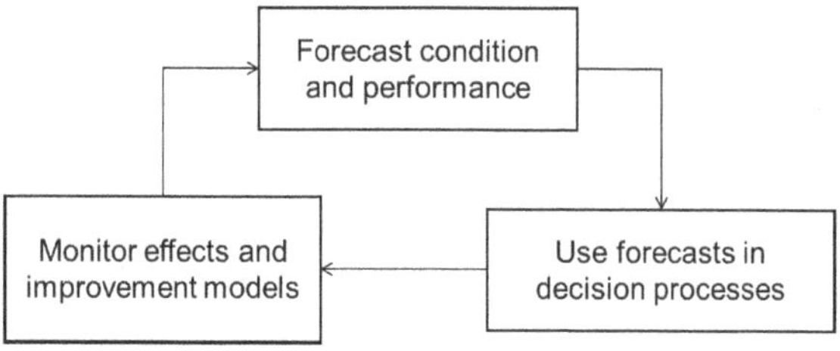

Fig. 2 Continuous improvement of forecasting models (AASHTO 2013)

network. In addition, the forecasted phenomena allow estimating the present value of costs related to alternative maintenance scenarios. After such cost estimation, it is possible to choose a scenario that is optimal for not only economic reasons, but also meets the strategic goals of the organization in terms of the goal level of services and asset functionality.

Depending on the type of asset, RAM advancement and needs in the decision-making process, prognostic models do not have to be complicated, especially if the modeled phenomenon (e.g. pavement damage) is simple, short-lived, of little importance or even unpredictable. Simple models or even lack of models is also advisable if the financial benefits of having information or knowledge do not justify investment in the development and maintenance of models. Such an assessment of the demand for forecasting models should be carried out during RAM gap analysis. All forecasting models should be periodically subjected to verification, calibration and validation in the process of continuous improvement. Such an approach is presented in Fig. 2.

The primary goal of forecasting models is to allow road administrator to value how its decisions affect the strategic goals of entire organization (ISO 55001 2014), (Baladi et al. 2017). In a proactive strategy, in order to control asset supply, the road administrator should forecast actions and processes that may conflict with its plans and fulfillment of strategic goals. The current state of practice among road administrators focuses on four key service attributes: network resilience, mobility, safety and other social aspects (e.g. environmental impact). Therefore, the goals and effects of forecasting models should first be determined based on RAM gap analysis, i.e., among other things, what value models bring to the decision-making process. At this stage, it is necessary to distinguish what the collected data on the asset performance bring and how the models are to use such data—that is, the interaction between the data and models is important.

It should be emphasized that strategic objectives of road administrator and key service attributes should be well communicated and aligned in order to deliver consistent decision process within organization and its stakeholder context. Such a cascade flow of decisions is presented in Fig. 3. It shows the connection between well-defined needs at the strategic level and low-level information on the asset performance at the project level (Gordon 2018).

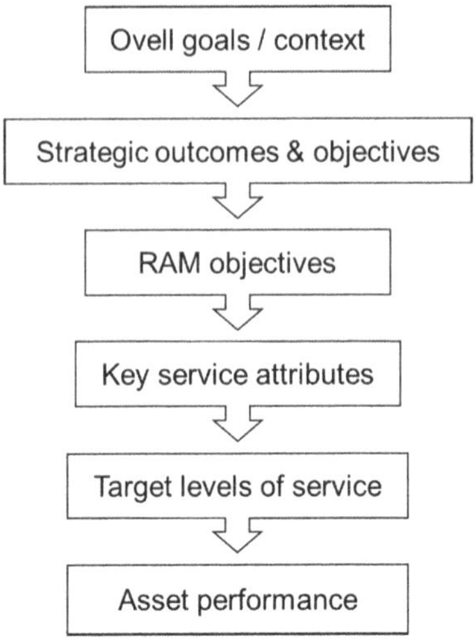

Fig. 3 Decision dependencies (Gordon 2018)

2 Deterioration Models in RAM

As discussed in previous section, modern RAM should comprise prognostic tools to support all key service attributes. One of the common aspects in prognostic tools is deterioration modelling of pavement condition which is a focus of this particular publication. However it should be stated that in advanced systems emphasis should be placed onto asset performance rather than solely on its condition.

Analytical business processes of the road administrator should use deterioration models to plan future activities on the network and to optimize their timing in order to deliver appropriate level of services and thus fulfill the overarching goals and objectives. But in order for the models to be effective element of the decision process, the models should accurately capture the deterioration and decline of asset condition. To improve the understanding of how and why asset condition worsens over time, special expert-based and data-based techniques can be used such as Fault Tree Analysis (FTA) supported by Analytical Hierarchy Process (AHP) (Schlotjes 2013). As shown in Fig. 2 it is also necessary to periodically monitor the actual deterioration of assets and compare this information with forecasted trends in order to enhance predictive capabilities of the models (AASHTO 2013).

There are various options of deterioration models in terms of their form, scope, type, dependent and independent variables. These options depend on many factor including among others model application, complexity of decision process, optimization engine, available data and information, as well as approach to risk management and reliability.

When choosing a pavement deterioration model, the following factors should be assessed (Chang 2017; Zofka 2019):

- the expected function and effect of the model in the decision-making process,
- availability and quality of current and historical data for dependent and independent variables,
- current and required data acquisition process (frequency and scope) to support decision-making process,
- form of mathematical notation and computational tools,
- capacity to meet the assumptions and boundary conditions,
- required precision and accuracy: it should be sufficient for the result to meet the expected effect of the model in the decision-making process,
- compatibility of selected models between assets and sharing of input data; different forecasting models for different assets due to, among others, different pace, scale and mechanism of degradation, other expected effects in the decision-making process.

While many forecasting models can be found in the literature related to road assets, they can be broadly grouped based on the following two criteria (Zofka 2018, 2019):

1) mechanism of damage development, i.e. mechanistic and empirical models,
2) type of dependent variable, i.e. deterministic (analytical) and stochastic (probabilistic).

All four model combinations, i.e. mechanistic and empirical combined with deterministic and stochastic, have been used to forecast the conditions of pavement or bridge assets. When choosing models, the aforementioned factors should be taken into account, as there is no universal model that would be suitable for situations. Proactive strategies such as mentioned earlier RCM use specific empirical models known as P-F that are further discussed in the following sections.

3 P-F Curves in RCM

Forecasting models according to the reliability centered maintenance (RCM) theory are called P-F curves (Moubray 1997). In the RAM literature it is not commonly emphasized that degradation and forecasting models have been introduced in other industries long before they were adopted to describe pavement condition. The classic P-F curve along with the basic concepts is shown in Fig. 4. P-F curve comprise two characteristics points. Point P (potential failure) indicates the point at which damage is detectable by the condition assessment technologies engaged by the road administrator. From this point, the probability of losing functionality increases significantly, until it reaches 100% at point F. So the point F (failure point) stands for complete loss of asset functionality, which does not necessary translate to a complete physical damage. Therefore location of point F strongly depends on definition of functionality and associated service attributes. Other interesting observation is that point P is directly related to the level of damage detection and their causes (LOD). Below

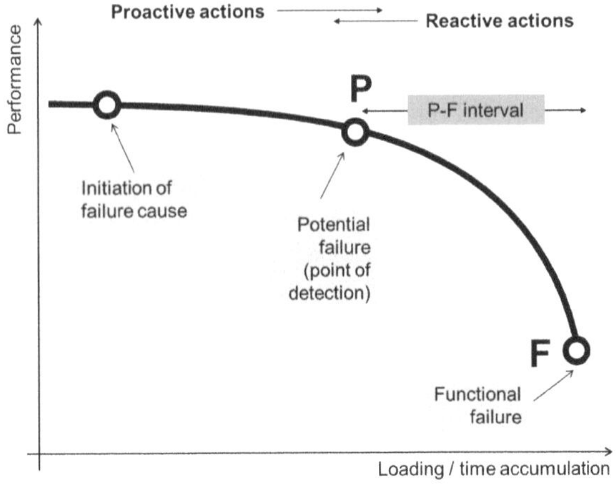

Fig. 4 Concept of P-F curve in RCM strategy (Moubray 1997)

point P, the LOD level is higher than the damage initiation threshold. Further, point P indicates a smooth transition between proactive and reactive actions as discussed earlier. So in fact it is desirable to pursue methods and techniques in order to move point P into the left making the P-F interval as large as possible. Such situation would lead to more reliable and predictable asset performance while leaving more options for proactive actions.

In a proactive strategy, asset condition assessment should be used to verify and validate assumptions, models and actions performed in the past. Current diagnostics technologies are not yet at the stage of proactive management of damage detection and their causes. For truly proactive use of diagnostics, there is a need for advanced methods of interpretation of relevant high-quality data and/or new diagnostic methods with low LOD, which will allow earlier detection of damage initiation. In other words, it is desirable to seek such condition assessment methods that feature LOD lower than the threshold of damage initiation or perhaps even lower than the threshold for the initiation of damage causes. Given the current progress of ICT technologies and machine learning algorithms, it seems that such condition assessment methods will be available in the near future.

4 P-F Curves in Pavement Maintenance Management

P-F curves does not change essentially the RAM processes however they allow to knowingly consider the potential failure and loss of functionality defined by P and F points, correspondingly. Furthermore, P-F concept allows simultaneously to model the performance and to assess the uncertainty of service delivery that is critical for the risk management.

P-F curves can be defined in time space or performance space i.e. independent variable which represents pavement performance (proxy of a certain level of service)

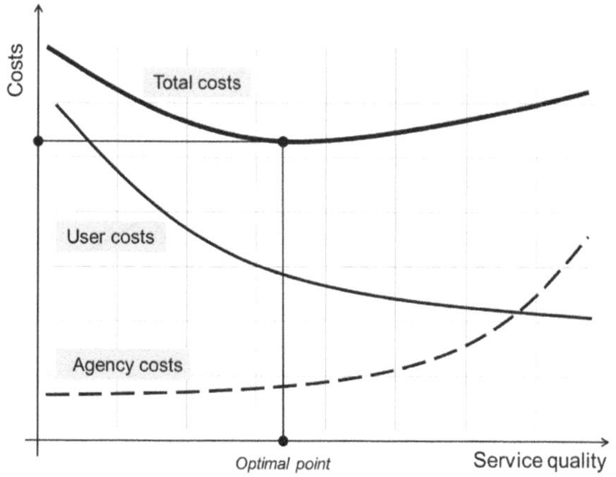

Fig. 5 Total costs as a function of service quality (AASHTO 2013)

versus time or accumulated loading (e.g. standard axles). In such a form, P-F curves (similarly to other forecasting models) are important part of optimized decision making (ODM) process. While they can be used to describe various independent variables they can be also utilized to determine derivative indicators such as benefit area of alternative maintenance strategies. Other example of application of P-F curve is demonstrated in Fig. 5. Here performance predictions are coupled with appropriate cost models to assess the effect of threshold quality level for the maintenance activities on the total cost that comprises user and agency contributions. Thus the optimal threshold for the service quality can be determined while minimizing the total costs.

5 Summary and Conclusions

P-F curves as well as other concepts from RCM strategy should be a part of modern RAM. RCM framework is already formulated and available in the relevant literature. These concepts are slowly adopted by few road administrators but the implementation rate should be accelerated especially on relatively new road networks. The sooner the RAM-RCM is implemented the greater the benefits are realized in the life-cycle horizon of road assets. This is shown in Fig. 6 that emphasizes the importance of timing for introducing systematic management of road network (especially concepts from the RCM strategy) and its relation to total agency costs.

P-F curves are advantageous tool in any proactive-type maintenance management. They allow to accurately consider life cycle options and asses the risk exposure. Therefore P-F curves should be used to schedule appropriate activities at the optimal time in order to provide desirable level of service through adequate performance and physical condition.

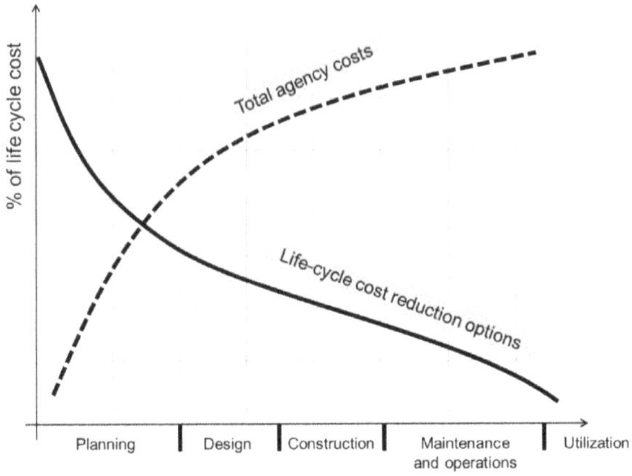

Fig. 6 Options for cost savings within asset life-cycle (AASHTO 2013)

Acknowledgements Study presented in this paper is a part of the research grant SIEC11.5t titled "Optimization of rehabilitation strategies for Polish national highway system to increase structural capacity to 11.5 tons". This grant is conducted within GOSPOSTRATEG program sponsored by the National Centre for Research and Development (NCBiR) in Poland.

References

AASHTO/FHWA (2013) AASHTO Transportation Asset Management Guide, A Focus on Implementation, Report Number FHWA-HIF-13-047, Washington, D.C.

Baladi GY, Dawson T, Musunuru G, Prohaska M, Thomas K (2017) Pavement Performance Measures and Forecasting and the Effects of Maintenance and Rehabilitation Strategy on Treatment Effectiveness (Revised): Report Number FHWA- HRT-17-095, McLean, VA

Chang CM, et al (2017) Consequences of delayed maintenance of highway assets; NCHRP Report 859. Transportation Research Board, Washington, D.C.

EN 13306: 2018-01 Maintenance – Maintenance terminology

Gordon M, Sharp K, Martin T (2018) Guide to Asset Management, Project Number AP1900, Sydney

International Organization for Standardization, Asset management systems – requirements (2014) ISO 55001:2014, Geneva

Moubray J (1997) Reliability-Centered Maintenance, Second Edition. ISBN-13 978-0831131463

National Aeronautics and Space Administration (NASA) (2008) RCM Guide: reliability–centered maintenance guide; for facilities and collateral equipment, NASA Report

Schlotjes MR (2013) The Development of a Diagnostic Approach to Predicting the Probability of Road Pavement Failure. ETheses Repository, 278

Taggart A, et al (2017) PIARC Asset Management Guide, Chapter 1, Paris

Zofka A (2018) Proactive pavement asset management with climate change aspects. IOP Conf Ser Mater Sci Eng 356. https://doi.org/10.1088/1757-899x/356/1/012005

Zofka A (2019) Proactive strategy for road asset management, Warsaw. ISBN 978-83-89252-29-6

Combined Life Cycle Cost Analysis and Life Cycle Assessment of Road Pavements

Egemen Okte and Imad L. Al-Qadi

Abstract Life cycle assessment (LCA) and life cycle cost analysis (LCCA) are the main pillars of pavement sustainability. LCA addresses environmental impacts of a pavement structure, and LCCA addresses life cycle costs. While both techniques go hand in hand with transportation agencies' decision making, they are usually used separately because there are no tools or methodologies that consider both under the same framework. This study introduces a LCCA add-on to a LCA tool developed by the Illinois Center for Transportation (ICT). The tool analyzes all pavement life stages, namely, construction and materials, maintenance, use and end of life. With the new add-on, the tool can calculate agency costs and user costs associated with both work zone and normal operating conditions as well as global warming potential and energy consumption. The developed tool makes the pay item framework easier to use for agencies and contractors. At the end, a case study is presented to illustrate the tool's capabilities.

Keywords LCA · LCCA · Tool · Pavement sustainability

1 Introduction

For any infrastructure that lasts for many decades, sustainability is a major concern both economically and environmentally. Pavements are no exception when it comes to long-lasting infrastructure. Within the scope of pavement engineering, sustainability implies the ability to achieve the engineering goals for which it was constructed, preserve and restore surrounding ecosystems, use resources economically and meet basic human needs such as health and safety (Van Dam et al. 2015). To achieve sustainability, life cycle impacts are a main concern for pavement engineers. Life cycle impacts are generally categorized as social, environmental and economic. Measuring the economic and environmental aspects of these impacts can be achieved by life

E. Okte (✉) · I. L. Al-Qadi
Illinois Center for Transportation, University of Illinois at Urbana–Champaign, 1611 Titan Drive, Rantoul, IL 61866, USA
e-mail: eokte2@illinois.edu

C. Raab (ed.), *Proceedings of the 9th International Conference on Maintenance and Rehabilitation of Pavements—Mairepav9*, Lecture Notes in Civil Engineering 76,
https://doi.org/10.1007/978-3-030-48679-2_13

cycle cost analysis (LCCA) and life cycle assessment (LCA), respectively. Materials production, pavement design and construction, pavement rehabilitation, preservation, maintenance, use, and end-of-life stages must be included for a complete analysis of sustainability (Harvey et al. 2010). Because these stages require separate analysis methods and strategies, tools were developed to help users assess the sustainability of their investment. Several LCA tools were developed for roadways (e.g., Al-Qadi et al. 2015; Huang et al. 2009; Athena Sustainable Materials Institute 2015), and several comprehensive LCCA tools were also developed (e.g., RealCost, 2004, Cal-B/C, 200 and APA LCCA, 2011). PTLaser and TCAce are examples of tools that conduct both LCA and LCCA outside of the pavement-engineering community (Norris 2001). However, to the best of the authors' knowledge, there are no tools that conduct both LCA and LCCA (agency and user costs) for pavement structures. This study introduces an LCCA add-on to a previously developed LCA tool by the Illinois Center for Transportation (ICT) (Al-Qadi et al. 2015). The developed LCA/LCCA toolkit uses a pay item framework to analyze all stages of material production and pavement life.

2 Pay Item Framework

Before the addition of the LCCA add-on, the tool was used as a stand-alone LCA (Al-Qadi et al. 2015). The tool takes advantage of the pay item framework to make it easier to use by agencies and contractors, because contracts are usually bid on a pay item basis. Figure 1 shows an example scheme of the pay item framework. Each pay item has general characteristics such as cost, unit and name. For each pay item, there can be associated materials, equipment and mix designs. Within those subsections, there are unit processes attached that capture the lifetime impacts of

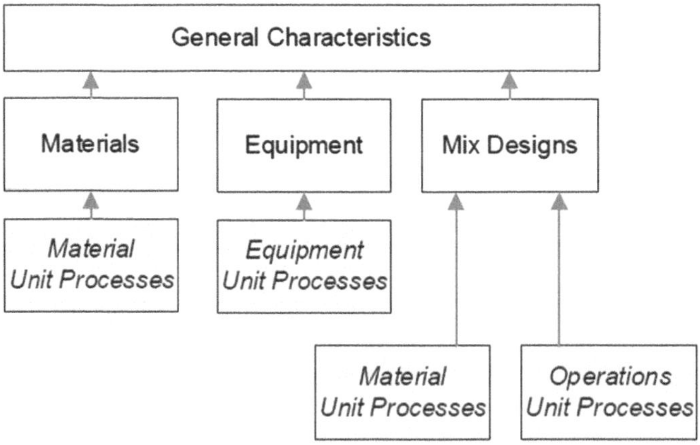

Fig. 1 Pay item framework example for an asphalt mix

that specific pay item from its production to pavement construction. Each pay item is broken down to smaller sections, and the analysis can be conducted by summing up all unit processes.

Each unit process within the pay item contributes to the overall cost and impact within the project life. An extensive database of pay items is built in the tool so that the users can select the pay items based on their project needs.

3 Tool Components

The tool is built to capture any pavement project from start to the end of analysis period. Therefore, the stages included in the tool are materials, construction, maintenance and rehabilitation, use, and end of life. Figure 2a provides an overview of the tool homepage. The main inputs include traffic information, number of lanes and year of construction. The materials and construction page lists all pay items used in the project as well as work-zone information. Figure 2b demonstrates a similar interface used for maintenance and rehabilitation. In the maintenance and rehabilitation interface, the user can see all activities related to the project from the start year to end year and all associated pay items. The user can also add work-zone activities to consider the impacts of work-zone closure and delays for both LCA and LCCA. In the interface for the use stage, the user can select and modify the International Roughness Index (IRI) and texture progression model parameters and see the plots of the progression as given in Fig. 2c. Finally, in the end-of-life stage, the user can allocate the landfill and recycling percentages of pay items or use presets.

4 LCCA Add-On

4.1 Net Present Value Approach

There are two common approaches in LCCA to compare money spent over time. One approach is equivalent annual cost (EAC), which converts cash flow to annual expenditures. The method used in this study is net present worth (NPV), which converts cash flow to a lump sum spent on the initial year of the project. Equation (1) gives the calculation for NPV.

$$NPV = \sum_{t=0}^{N} \frac{R_{t0}}{(1+i)^t} \quad (1)$$

where N is the analysis period, i is the real discount rate, and R_{t0} is the money spent on year t in year 0 dollars. Therefore, to compute NPV, the cost of the pay items used

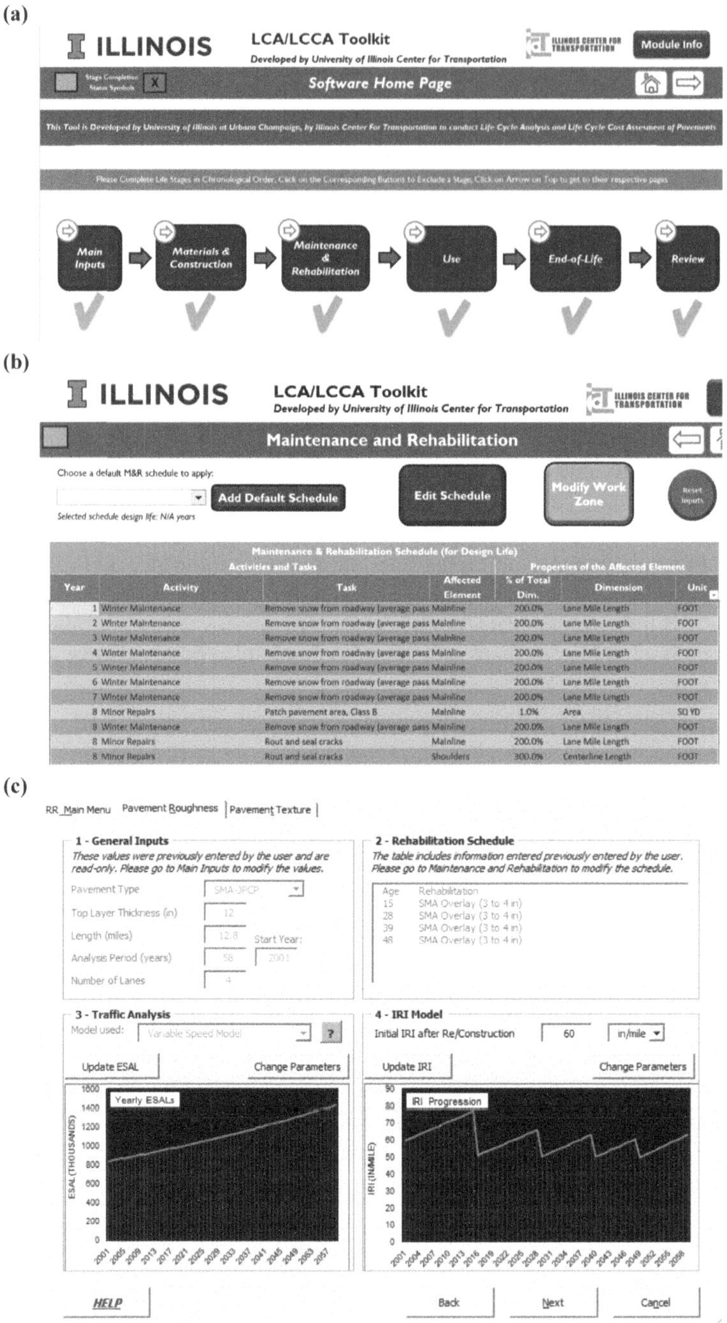

Fig. 2 The tool main page (**a**), maintenance and rehabilitation interface (**b**), and use stage interface (**c**)

in the project must be converted to the year 0. This can be done by discounting, as presented in Eq. 2.

$$R_0 = R_p \frac{CPI_p}{CPI_0} \tag{2}$$

where R_0 is the cost of the pay item at year 0 dollars (ex: 2015 dollars if construction year is 2015), R_p is the cost of the pay item reported at year p in year p dollars, CPI_p is the consumer price index at year p and CPI_0 is the consumer price index at construction year.

When comparing two or more projects, it is important to ensure that they have the same analysis period N. In real-world cases, it is possible for various projects to have different analysis periods. While this does not affect the calculation of user costs, agency costs must be reduced if a project's estimated lifetime exceeds the analysis period N. This can be done by the *serviceable life* approach and is discussed separately in agency- and user-cost computations.

4.2 Agency Costs

One of the advantages of working with pay items is that all pay items have associated costs and quantities. Therefore, if the reported year of the pay item is known, then the cost of a pay item at the construction year can be calculated using discounting and multiplying the unit cost by the quantity. Then, the agency cost at any year is simply the summation of all the unit costs of pay items used in that year multiplied by their respective quantities. If yearly routine maintenance is ignored, then agency cost is 0 or negligible compared to the years in which there is a scheduled maintenance activity such as mill and overlay. Finally, if the expected life of a maintenance activity exceeds the analysis period, then the cost of that maintenance activity can be reduced by the ratio of remaining analysis period to expected service life (Walls and Smith 1998).

4.3 User Costs

The user costs in LCCA can be divided into work-zone delay costs and vehicle operating costs. Work-zone delay costs are the costs arising from the expected delay of the users in work-zone conditions during maintenance or rehabilitation. Equation (3) gives the delay costs for a class A vehicle.

$$Delay\ Cost_A = Delay\ Time_A * Wage_A * \%_A Traffic * f_o * f_b \tag{3}$$

where $Wage_A$ is the hourly wage of a driver in a given vehicle type, $\%_A Traffic$ is the percentage of class A vehicles in traffic, f_o is the average number of individuals in class A vehicles and f_b is the business travel factor. If a vehicle's purpose of travel is personal and not business related, then their wages are reduced based on the type of road they travel on.

The *Delay Time* is computed by a two-phase traffic model reflecting normal and construction conditions. The details of the model are skipped for this study, but they can be found in Okte et al. (2019).

Operating condition costs are fuel consumption, tire wear and tear and repair-maintenance costs. These costs are all a function of pavement roughness and texture. However, pavement roughness contributes to more than 95%. Fuel consumption costs due to IRI are calculated using a Roughness - Speed impact model (RSI) developed by using EPA's motor vehicle emission simulator MOVES vehicle model emissions software and is given in Eq. 4 (Ziyadi et al. 2018).

$$\mathrm{RSI}_{t=0}^{\mathrm{Energy}} : E(v, \mathrm{IRI}) = \frac{p}{v} + (ka.\,\mathrm{IRI} + da) + b \times v + (kc.\,\mathrm{IRI} + dc) \times v^2 \quad (4)$$

where E is the energy consumed per vehicle miles travelled (VMT) in mega joules (MJ), v is the speed of the vehicle in mph and the rest are model coefficients depending on the vehicle class. Once the energy is computed, it is converted to gallons of either gasoline or diesel and multiplied by the unit price. The tire wear-tear and repair costs are adopted from NCHRP 720 for different vehicle classes (Chatti and Zaabar 2012). Overall, they contribute to less than 4% of the operating condition costs on highways, because they are less sensitive to pavement roughness, and the cost of fuel consumption per mile is higher than any other cost. The details of the calculation steps can be found in Okte et al. (2019).

Finally, it is unreasonable to assume that any agency is responsible for 100% of user costs. There is always a base cost of travel from point A to point B, even under perfect conditions. Therefore, user costs due to roughness are only considered if the IRI is above 40 in/mi. The cost at 40 in/mi is subtracted from the overall user cost because 40 in/mi is considered as a base level in this study.

5 Outputs of the Tool

Once the user enters all necessary information about the contract, traffic and maintenance of the road segment, the tool calculates the life cycle impacts and life cycle costs of the project. For illustration purposes, a 12.7-mi section of Illinois Tollway in Chicago is used with a 60-year analysis period. Average daily traffic is 70,000 vehicles with four lanes in each direction. A 3-in mill and overlay are assumed to be performed every 15 years. The discount rate is assumed to be 3%. Details of this section and its cost analysis can be found in Okte et al. 2019.

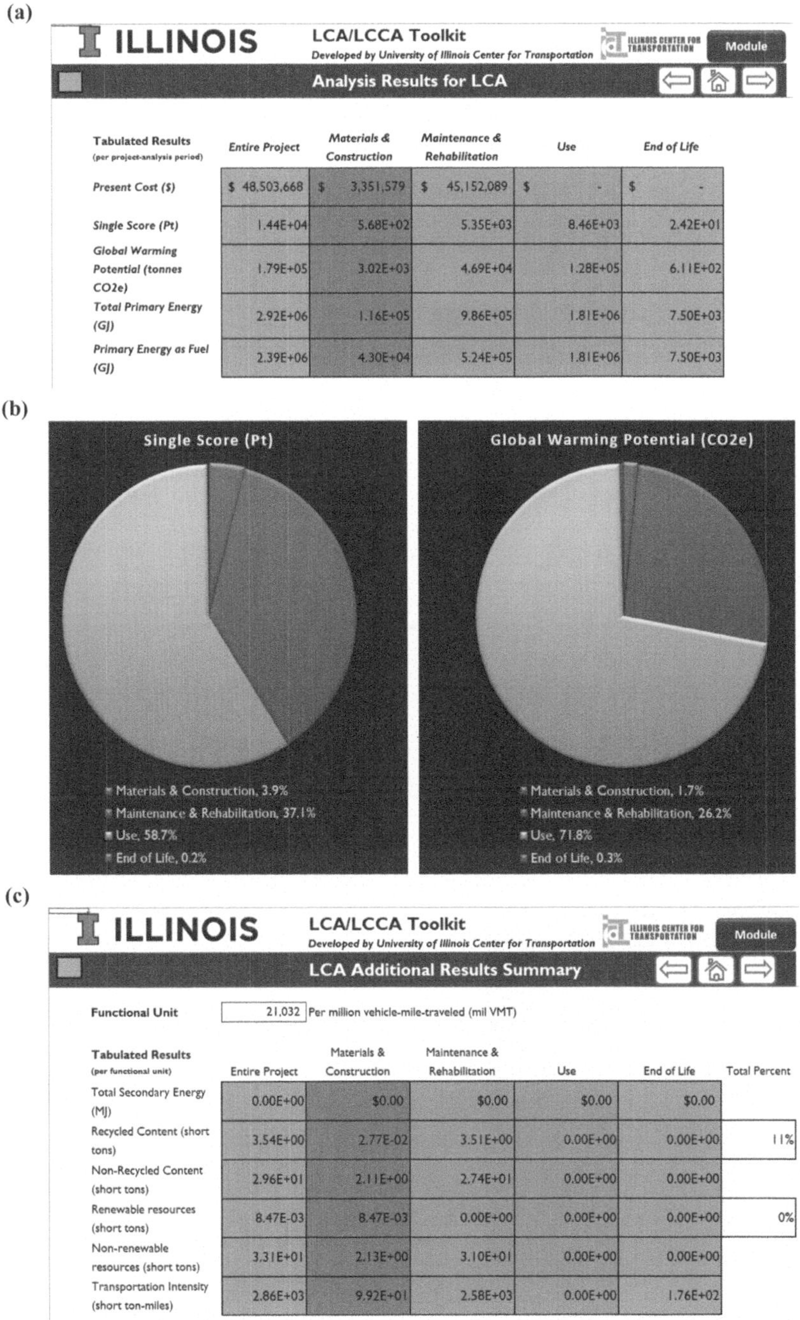

Tabulated Results (per project-analysis period)	Entire Project	Materials & Construction	Maintenance & Rehabilitation	Use	End of Life
Present Cost ($)	$ 48,503,668	$ 3,351,579	$ 45,152,089	$ -	$ -
Single Score (Pt)	1.44E+04	5.68E+02	5.35E+03	8.46E+03	2.42E+01
Global Warming Potential (tonnes CO2e)	1.79E+05	3.02E+03	4.69E+04	1.28E+05	6.11E+02
Total Primary Energy (GJ)	2.92E+06	1.16E+05	9.86E+05	1.81E+06	7.50E+03
Primary Energy as Fuel (GJ)	2.39E+06	4.30E+04	5.24E+05	1.81E+06	7.50E+03

Tabulated Results (per functional unit)	Entire Project	Materials & Construction	Maintenance & Rehabilitation	Use	End of Life	Total Percent
Total Secondary Energy (MJ)	0.00E+00	$0.00	$0.00	$0.00	$0.00	
Recycled Content (short tons)	3.54E+00	2.77E-02	3.51E+00	0.00E+00	0.00E+00	11%
Non-Recycled Content (short tons)	2.96E+01	2.11E+00	2.74E+01	0.00E+00	0.00E+00	
Renewable resources (short tons)	8.47E-03	8.47E-03	0.00E+00	0.00E+00	0.00E+00	0%
Non-renewable resources (short tons)	3.31E+01	2.13E+00	3.10E+01	0.00E+00	0.00E+00	
Transportation Intensity (short ton-miles)	2.86E+03	9.92E+01	2.58E+03	0.00E+00	1.76E+02	

Fig. 3 Summary of LCA results (**a**), graphical LCA results (**b**) additional LCA results (**c**)

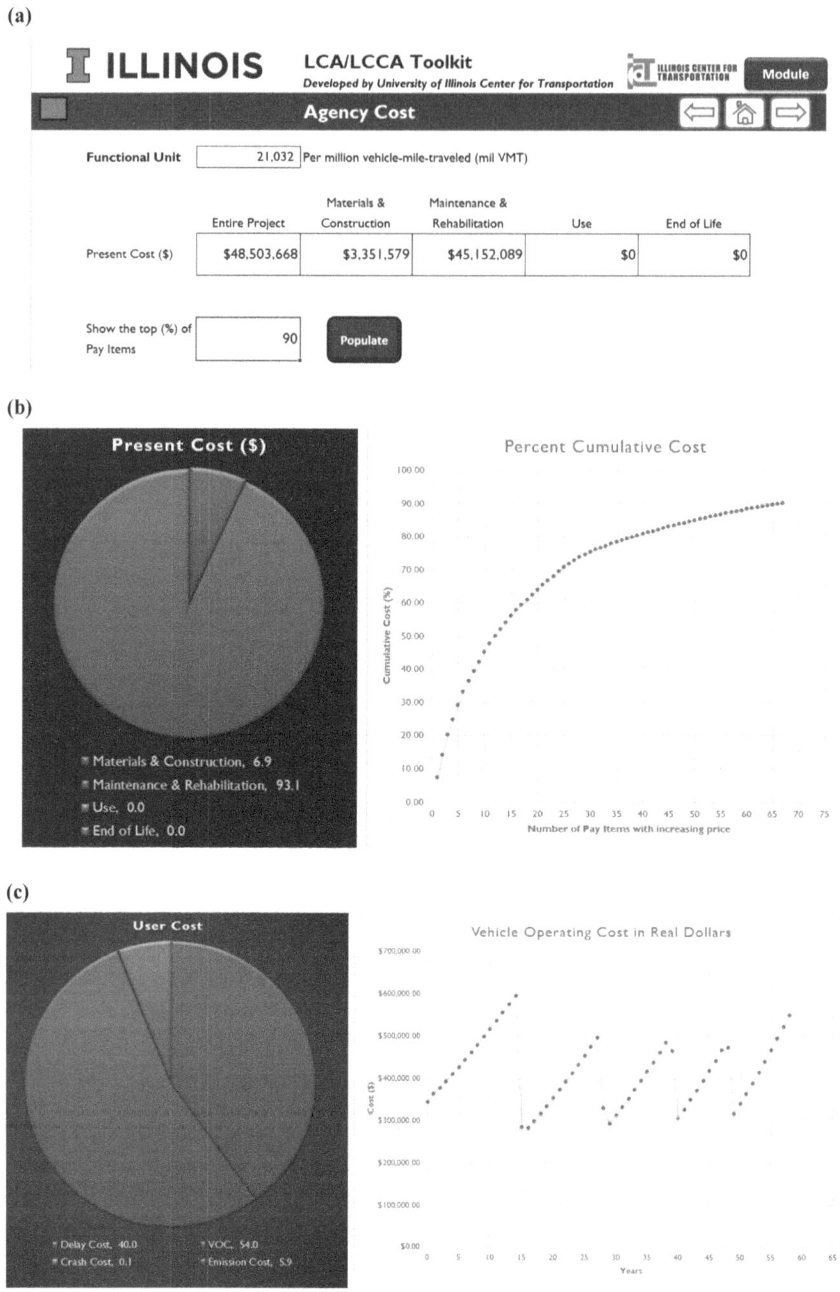

Fig. 4 Summary of LCCA results (**a**), graphical LCA results (**b**) additional LCA results (**c**)

5.1 LCA Outputs

The tool displays absolute and scaled results for LCA. The functional unit used for scaling is million vehicle miles travelled over the analysis period. Figure 3a presents the LCA analysis result page. All results are also displayed graphically in percentages (Fig. 3b). As expected with many LCA studies conducted on high-volume sections, the use stage is responsible for most of the global warming potential (GWP). Additionally, the tool is able to calculate information such as recycled and non-recycled content, renewable and non-renewable resources and transportation intensity (Fig. 3c).

5.2 LCCA Outputs

The results of LCCA are divided into agency costs and user costs. In the agency cost section, the user can see the cost of each stage (Fig. 4a). Additionally, percentages are displayed graphically. The user can also see the contribution of pay items to the total cost (Fig. 4b). In the user cost section, the user can see the difference between agency and user costs as well as the different types of user costs such as vehicle delay, operation, crash and emissions. Finally, Fig. 4c provides a view of vehicle-operating costs on a yearly basis. Vehicle-operating costs usually follow the roughness progression of the section because roughness is the main parameter that affects cost.

6 Summary and Conclusions

To perform pavement sustainability assessment, there is a need for tools that can conduct both LCA and LCCA. There are no tools in the literature that can both conduct LCA and LCCA with LCCA including both agency and user costs. This study presents a tool that can conduct both LCA and LCCA. With the use of this tool, users can generate any pavement section, decide on the maintenance schedule and compute the lifetime impacts and costs of their project. The tool covers all pavement stages. Material production and construction, maintenance and rehabilitation, use and end of life.

The developed tool also uses a pay item framework for easier use by contractors and agencies.

The tool has been used in multiple studies to assess the life cycle impacts and costs of roadway projects. Following its initial development, the tool was used to calculate the life cycle impacts of many scenarios for the Illinois Tollway (Al-Qadi et al. 2015). After the addition of the LCCA add-on, it was also used to calculate the life cycle costs of additional Illinois Tollway sections (Okte et al. 2019). Reader is advised to further read those studies regarding the performance of the developed tool.

Acknowledgements The support of Illinois Tollway for supporting work on LCA is acknowledged. The valuable assistance of Hasan Ozer, Rebekah Yang, and Qingwen Zhou is appreciated.

References

Al-Qadi IL, Yang R, Kang S, Ozer H, Ferrebee E, Roesler JR, Salinas A, Meijer J, Vavrik WR, Gillen SL (2015) Scenarios developed for improved sustainability of Illinois tollway: life-cycle assessment approach. Transp Res Rec 2523(1):11–18

Athena Sustainable Materials Institute (2015) Athena Impact Estimator for Highways, Ottawa, Ontario, Canada

Chatti K, Zaabar I (2012) Estimating the effects of pavement condition on vehicle operating costs. NCHRP Report 720. Transportation Research Board of the National Academies

Harvey J, Kendall A, Lee IS, Santero N, Dam TV, Wang T (2010) Pavement Life Cycle Assessment Workshop: Discussion Summary and Guidelines. University of California Pavement Research Center

Huang Y, Bird R, Heidrich O (2009) Development of a life cycle assessment tool for construction and maintenance of asphalt pavements. J Clean Prod 17(2):283–296

Norris GA (2001) Integrating life cycle cost analysis and LCA. Int J Life Cycle Assess 6(2):118–120

Okte E, Al-Qadi IL, Ozer H (2019) Effects of pavement condition on LCCA user costs. Transp Res Rec 2673(5):339–350

Van Dam TJ, Harvey J, Muench ST, Smith KD, Snyder MB, Al-Qadi IL, Ozer H, Meijer J, Ram PV, Roesler JR, Kendall A (2015) Towards Sustainable Pavement Systems: A Reference Document (No. FHWA-HIF-15-002). Federal Highway Administration

Walls III J, Smith MR (1998) Life-Cycle Cost Analysis in Pavement Design-Interim Technical Bulletin (No. FHWA-SA-98-079). Federal Highway Administration

Ziyadi M, Ozer H, Kang S, Al-Qadi IL (2018) Vehicle energy consumption and an environmental impact calculation model for the transportation infrastructure systems. J Clean Prod 174:424–436

Decision Support for New Holistic Uri Road Asset Management Process

Frank Schiffmann, Rade Hajdin, and Alfredo Serioli

Abstract By maintaining Uri road infrastructure, the cantonal road agency is ensuring safety, reliability and driving comfort on the roads for passengers and freight transport and thus, contributes to economic and social added value. The canton Uri has one of the first road agencies with an infrastructure division responsible for all road infrastructure objects. This includes inventory, condition inspections, maintenance planning and necessary data management. Uri maintenance and rehabilitation processes for pavements are embedded in a holistic road asset management process supported by an essential IT infrastructure.

One of the key applications is a flexible decision support tool "infFaros Uri" which was customized and developed together with the infrastructure division of canton Uri to meet their needs. infFaros can handle road sections and bridge together allowing synergy effects and corridor planning. It supports various methodologies including probabilistic and deterministic deterioration, cost/benefit and cost/effectiveness decision-making as well as consideration of four-year maintenance and rehabilitation plan and evaluation of its impact. By linking short- and long-term maintenance and rehabilitation planning using existing cantonal transportation data complex questions can be answered and decision making can be efficiently supported by scenario comparisons. infFaros is a modern web-based application. The paper presents both the technical background of the software and use of it to support four-year maintenance and rehabilitation planning within new Uri road asset management process.

Keywords Road infrastructure · Asset management · Asset management system · Decision process · Decision making

F. Schiffmann (✉) · R. Hajdin
Infrastructure Management Consultants LLC, Bellerivestrasse 209, 8008 Zurich, Switzerland
e-mail: frank.schiffmann@imc-ch.com

A. Serioli
Canton Uri, Road Infrastructure Division, Klausenstrasse 2, 6460 Altdorf, Switzerland

C. Raab (ed.), *Proceedings of the 9th International Conference on Maintenance and Rehabilitation of Pavements—Mairepav9*, Lecture Notes in Civil Engineering 76,
https://doi.org/10.1007/978-3-030-48679-2_14

1 Introduction

The Swiss road network connects regions, cities and agglomerations and raise the growth potential of Swiss economy as well as private, public and social benefit. In addition to national roads, the cantonal roads in particular, are of great importance for the regional economy and cultural development. Swiss road authorities by maintaining and preserving their road infrastructure are expected to provide a safe, efficient and reliable road network for passengers and freight transport. This is a challenging task in time of increasing mobility demand, already existing maintenance backlog and decreasing financial resources. The classical silo-based approaches fall short on this complex duty. Road infrastructure asset management as a holistic framework is helping to establish a management process for road authorities to manage their road infrastructure (PIARC 2012). Within these framework processes should be adopted that enable agency to answer the following five key question (EPA/FHWA 2009):

- What is the current state of our assets?
- What is the required level of service?
- Which assets are critical to sustained performance?
- What are the best investments options available?
- What is the best funding strategy from a long-term point of view?

The infrastructure division of the canton of Uri is about to implement road infrastructure asset management process in their decision making. This process is supported by a modular designed road infrastructure asset management system which helps to collect, store, manage and analyze all kinds of necessary road infrastructure data (see Fig. 1).

One of the key module applications within Uri road infrastructure asset management system is infFaros Uri. The name infFaros—meaning "inf" for infrastructure and "faros" is Greek for lighthouse—stands for an integral asset management analysis tool. It is a modern web-based application with an integrated GIS browser. User can analyze and present any available road infrastructure data from other sources, simulate various decision scenarios and evaluate on screen or use different reports to support decision-making process within Uri road authority.

infFaros makes use of recent research in the field of road information systems to handle data flow from available databases (Bernard et al. 2015) and data historization (Rosenthaler et al. 2015). Since the input data is fundamental for realistic simulation, systematic data management of source data is one of the important tasks for Uri infrastructure division. infFaros gets data by the use of automated interfaces from available data source for analysis and presents these together with simulation results within the browser. To avoid redundancy and inconsistency infFaros will not alter source data.

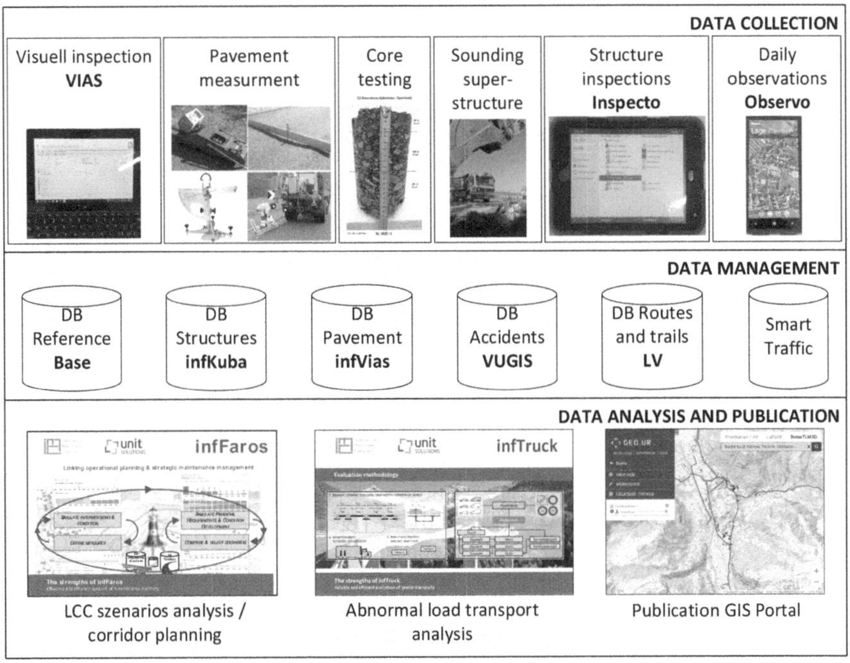

Fig. 1 Modular Uri road infrastructure asset management system

2 Uri Road Infrastructure Input Data

2.1 *Inventory and Condition*

Uri's cantonal road network includes approximately 2865 engineering structures and 154 roadway kilometers of cantonal roads (see Fig. 2). This network connects all municipalities of that mountainous canton with an area of 1077 km^2 and 36000 inhabitants. Different standards for traffic and maintenance needs lead to 4 road classes with specified level of service. The big challenge in managing this small road network is the topographical diversity with an altitude range from 435 m up to 2436 m above sea level. Seasonal road closures and consideration of natural hazards and their risk assessment is part of the management process but is not described within that paper.

Survey and inspection procedures for road pavement sections and bridges of Uri infrastructure division strongly comply with relevant Swiss standard regulation. Inspections are performed on basis of a yearly defined inspection program for about 20% of infrastructure objects. This definition leads to a five-year frequency for roads pavement sections and structures.

While engineering structures on that routine monitor basis are visually inspected and rated in condition states within CS1 (best) and CS5 (worst) using mobile devices

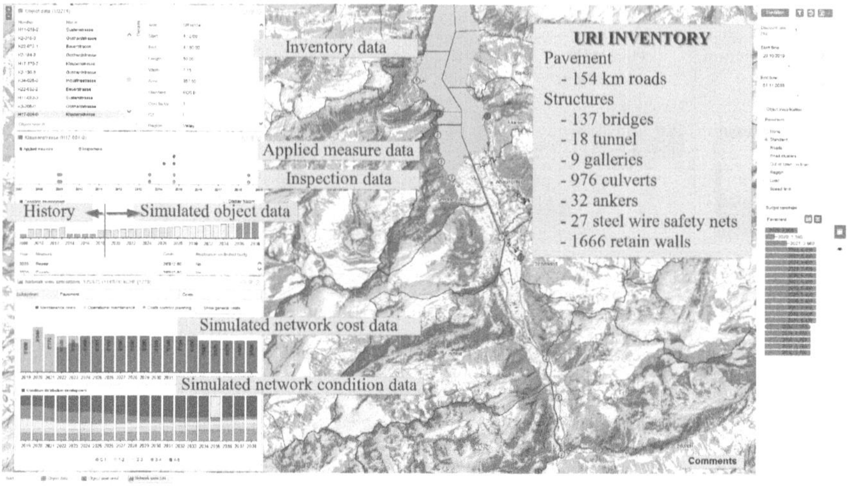

Fig. 2 Presenting data within infFaros Uri

and application Inspecto according to (SIA 469 1997). Road pavement sections are standardized assessed by surface condition (VSS 40 925 2019). The recorded data is normalized according to COST 354 (2008) to a scale from 0 (good) to 5 (poor). This includes the following road surface characteristics (see Fig. 1):

- Pavement distresses via visual inspection (VIAS conform to VSS 40 925)
- Longitudinal evenness by measurement (Goniograph conform to VSS 40 517)
- Transverse evenness by measurement (4m-board conform to VSS 40 518)
- Skid resistance by measurement (SRT-Pendulum conform to VSS 40 512)
- Bearing capacity by measurement (Benkelman-beam conform to VSS 640 330)

Additionally, data of core testing, sounding shafts of road superstructure and plate load test of unbounded layers is collected networkwide where it appears to be necessary. This inspection procedure is successfully and fully applied since 2008 and data is available for planning and analysis. By collecting condition data of road infrastructure objects, the decision makers are getting a recent snapshot of object characteristics (see Fig. 3). Generally, road infrastructure objects deteriorate over time, which means especially condition data refer to a single time instance. So temporal aspects need to be considered in managing all road infrastructure data.

The current focus in Uri for specific deterioration analysis is on pavement distresses and their development over time because of decision process to localize relevant objects for interventions. The deterioration analyses result in a model to report on present road condition data. This is necessary because of the yearly inspection procedure for only one part of the network. Data of objects which are not inspected in the current year refer to past years and need to be updated using a deterioration model. This calibration is done for all network-wide condition data, to identify network hot spots for maintenance needs.

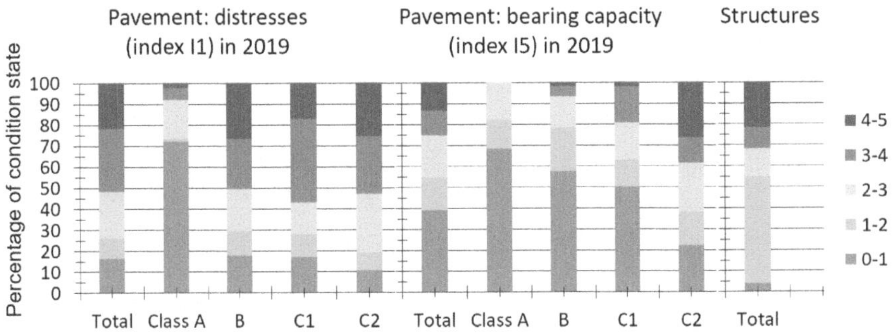

Fig. 3 Road network condition of canton of Uri (I1 and I5 for pavement/structures)

2.2 *Applied Interventions*

Besides condition calibration from last known inspection recently applied maintenance interventions are considered. After end of project, approval date, intervention type and scope as well as costs and duration must be recorded as part of the data collection process. On one hand, this data allows for a more precise determination of current network condition state. On the other hand, an analysis of this data collection enables the calibration of the cost model, relevant trigger values for maintenance interventions and effects of certain intervention types on deterioration.

3 Uri Decision Model

3.1 *Deterioration Model*

Infrastructure is deteriorating over time due to physical or chemical processes as a result of different impacts from e.g. heavy traffic load, winter maintenance or climatic conditions. This deterioration can be simulated by deterministic or probabilistic models. The deterioration of pavement is mainly modeled deterministically (e.g. Maerschalk et al. 2013) whereas for bridges the Markov Chain probabilistic model is common (e.g. Roelfstra et al. 2004) (see Fig. 4).

Currently in Uri, the dimensionless I1 data serve as parameters for the pavement surface condition. To consider not only surface but also condition of pavement structure, the state of the bearing capacity is currently used in decision making on the basis of the index I5. Further existing condition data in combination with further technical parameter, e.g. freight traffic volume and material specifications, are still being analyzed for ongoing deterioration model refinement.

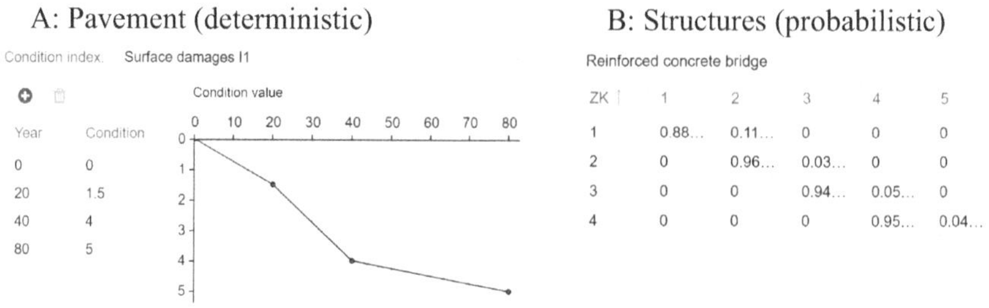

Fig. 4 Uri deterioration models: a piecewise linear function/B Markov chain

3.2 Intervention Types and Maintenance and Rehabilitation Strategy

In order to fulfill maintenance needs, interventions need to be applied on assets or infrastructure objects. The questions are: Where is a need for maintenance intervention? When should this be applied? What type of intervention would be the optimal one? For that reason, (1) the impact an intervention on condition must be specified and (2) a maintenance strategy needs to be determined. A maintenance strategy defines in which condition state a specific intervention type will be triggered.

For assets with deterministic modeled deterioration the following parameters specify the intervention type:

- a reset value representing short-term impact for each different deteriorating condition parameter,
- an application range for that intervention type.

Impact of intervention types and strategy for probabilistic modelling are specified by efficiency matrices for a specific object type. This matrix contains:

- a condition distribution after applying an intervention type and
- applicable intervention types for each condition state

3.3 Decision Making

With the definition of these fundamentals—such as:

- knowledge of asset inventory and actual condition state,
- modelling the deterioration of assets,
- setting of intervention types as well as
- the cost modelling –

It is now possible to schedule interventions on assets over a given time frame. Different decision options in terms of intervention plans (e.g. 20 years) for each asset in the road network will be evaluated under defined yearly budget restrictions. The decision modell within infFaros Uri is based on cost-benefit-analysis (e.g. Adey et al. 2010). If wanted a cost-efficiency-analysis would also be possible which is quite commonly used for pavement (e.g. Maerschalk et al. 2013). In both cases a reference case needs to be defined in order to calculate benefit of an intervention plans. Benefit of an intervention plan is always a relative benefit to a defined reference plan. Whereas the cost benefit analysis postulates a monetized benefit function, cost efficiency analysis is based on a non-monetized benefit definition, e.g. area under curve. A detailed explanation on that provides (Adey et al. 2010). The decision model is prioritizing with incremental benefit cost ratio IBCR algorithm. With this widely used optimization algorithm infFaros finds the intervention plan by maximizing net benefit for each asset under budget restriction (e.g. Rafi et al. 2005).

In addition to that, depending on agency data set, their requirements and needs, the decision model can use manual defined prioritization. Currently besides IBCR, the single asset condition state is part of the prioritization rules. All pavement assets with condition I1 ≥ 4.5 are prioritized for maintenance and rehabilitation.

4 Uri Strategy Definition

One of the main first tasks to develop the holistic Uri road asset management process was to define a new strategy "Strategie Strasse 2019" for the road agency. The strategy is mainly based on the following core elements:

- Road classification
- Budget definition and related road condition
- Four-year maintenance and rehabilitation plan

The strategy aims to spend the scarce financial resources for road infrastructure efficiently to gain a maximum of benefit for Uri population and economy.

4.1 Road Classification

The cantonal roads of Uri are divided into four different road classes (EQS) A to C2 according to the development needs of the municipalities and their population and economy. The road class is linked to specifications on how a road section is designed, maintained and operated. This includes specifications for the needed standard for road profile and structural design, for the existing traffic and for maintenance and rehabilitation as well as operational needs.

4.2 Budget Definition and Related Road Condition

In order to maintain the existing cantonal road network efficiently, long-term maintenance planning is necessary. Otherwise, cantonal roads will deteriorate rapidly and the need for rehabilitation will exceed the canton's financial resources. The possible annual intervention in Uri road network are also limited to avoid obstruction of traffic. The financial needs within a defined period depend on road condition at the beginning, the deterioration and tolerated condition at the end of a defined period. The question rises: what condition state is acceptable and affordable under given lack of financial resources with regard to a 20 year time period. For pavement the focus should be on the percentage of road sections in poor condition state as a critical risk parameter regarding traffic safety and a representation of maintenance backlog. The decision process of Uri road authority leads to a budget decision the 20 year time period. infFaros Uri did support the decision makers with consequence simulation for different scenarios (see Fig. 5).

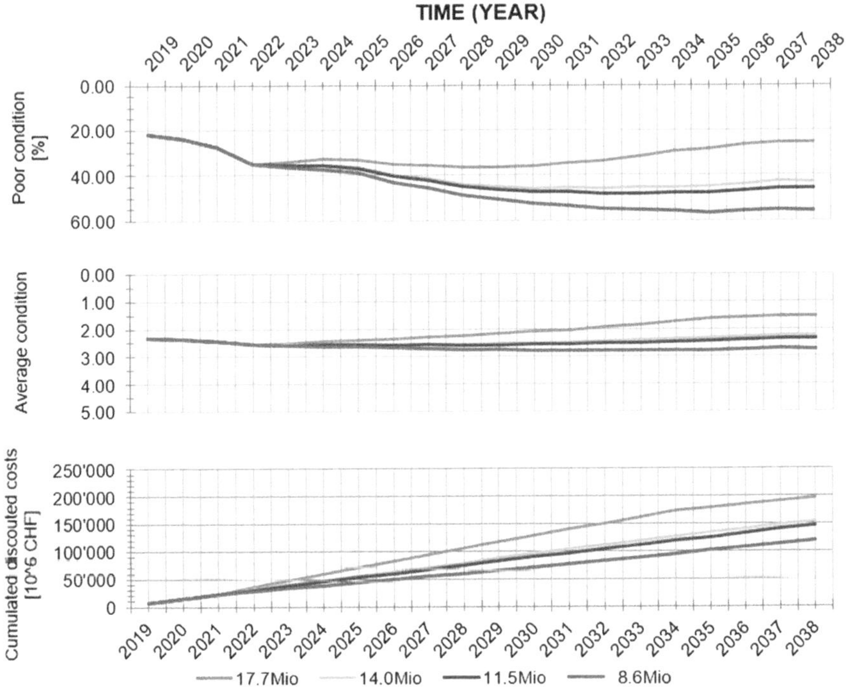

Fig. 5 Comparison of different Uri budget scenarios for a 20 year time period

Table 1 Additional financial expenditure due to four-year maintenance and rehabilitation plan (11.5 mio CHF scenario) compared to defined rehabilitation strategy

Total additional costs	Early intervention	Late intervention
17 Mio	13 years (average)	−2 years (average)

4.3 Four-Year Maintenance and Rehabilitation Plan

Four-year maintenance and rehabilitation plan includes new road construction, intervention for road extension and rehabilitation. The financial needs for these interventions are mainly beyond rehabilitation. In some cases, intervention on road sections are not yet necessary in terms of pure rehabilitation, but for reasons to keep traffic flowing or increase of safety level it is of high priority. The already existing lack of financial resources will increase for the necessary rehabilitation. This must be considered when defining the annual overall budget. Table 1 shows the consequences by comparing the 11.5 mio CHF scenario with the defined rehabilitation strategy and resulting rehabilitation costs.

Early interventions are applied in average 13 years before triggered by rehabilitation strategy (this counts for about 10% of the interventions). Late interventions are in average late by 2 years compared to rehabilitation strategy (this counts for about 70% of the interventions). Due to the four-year maintenance and rehabilitation plan in total 17 mio CHF will be spent more compared to normal rehabilitation interventions.

5 Conclusion and Outlook

During the strategy development infFaros Uri already could create benefit by numerous scenario simulations. The results did show consequences in terms of condition, intervention plan and resulting costs. These were the basis for the communication and decision process within the cantonal authority of Uri. It could be stated what are the financial needs under different defined LOS. Compared to real decided budget in the past with 8.6 mio CHF, new budget will be 11.5 mio CHF to avoid the worst deterioration. Still, the budget is not enough to hold current condition state and discussion is open if predicted LOS is preferred by the public in future.

infFaros Uri will help to keep track on that issue by different reporting functions which will be customized next. This includes a yearly network status report followed by proof of strategy report.

Further data analysis of available sources of road infrastructure should lead to an enhancement of existing decision model by precising the deterioration model for certain road classes. The next steps for infFaros development will be the integration of work zone optimization and risk assessment to be able to consider full risk allocation from “inside” (e.g. road agencies, user, society) and “outside” (e.g. natural hazards).

References

Adey BT, Herrmann T, Tsafatinos K, Lüking J, Schindele J, Hajdin R (2010) Methodology and base cost models to determine the total benefits of preservation interventions on road sections in Switzerland, Structure and Infrastructure Engineering, 23 June 2010

Bernard E, Marschal C, Hajdin R (2015) Interfaces between RIS and Business Itelligence Systems (Schnittstellen aus den Auswertesystemen SIS-DHW). Report 1508, VSS, UVEK, FEDRO, Bern

COST 354 (2008) COST action 354 performance indicators for road pavements. Final report

Environmental Protection Agency and Federal Highway Administration EPA/FHWA (2009) The Fundamentals of Asset Management. Training Course Materials. Environmental Protection Agency, Washington, DC

Maerschalk G, Krause G, Socina M, Köhler M, Stöckner M (2013) Data and methods for a systematic maintenance management of urban roads (Daten und Methoden für ein systematisches Erhaltungsmanagement innerörtlicher Strassen). Wirtschaftsverlag NW, Verlag für neue Wissenschaft, Bremerhaven, Germany

Rafi A, Hajdin R, Welte U (2005) Optimization processes in road infrastructure asset management (Optimierungsprozesse im Management der Strassenerhaltung MSE). Report 1109, VSS, UVEK, FEDRO, Bern

Roelfstra G, Hajdin R, Adey BT, Brühwiler E (2004) Condition evolution in bridge management systems and corrosion induced deterioration. J Bridge Eng 2(9):200–208 ASCE

Rosenthaler C, Koch R, Hajdin R, Botzen M (2015) Aspects of time and historization (Zeitaspekte und Historisierung). Report 1516, VSS, UVEK, FEDRO, Bern

SIA 469 (1997) Conservation of structures; communication, conservation objectives, conservation measures and activities, construction files (Erhaltung von Bauwerken; Verständigung, Erhaltungsziele, Erhaltungsmassnahmen und -tätigkeiten, Bauwerksakten). Swiss Standard SN 588 469, SIA, Zurich, Switzerland

VSS 40 925 (2003) Pavement management: condition survey and index evaluation (Erhaltungsmanagement der Fahrbahnen (EMF); Zustandserhebung und Indexbewertung). Swiss Standard VSS 40 925, VSS, Zurich, Switzerland

VSS 40 517 (2019) Characteristics of road surfaces: Longitudinal evenness (Eigenschaften der Fahrbahnoberfläche: Längsebenheit). Swiss Standard VSS 40 517, VSS, Zurich, Switzerland

VSS 40 518 (2019). Characteristics of road surfaces: Transverse evenness (Eigenschaften der Fahrbahnoberfläche: Querebenheit). Swiss Standard VSS 40 518, VSS, Zurich, Switzerland

World Road Association PIARC (2012) Asset Management Manual; A Guide for Practioners, Paris, France

VSS 40 330 (2019) Characteristics of road surfaces: measurement of load-bearing capacity - deflection measurements (Eigenschaften der Fahrbahnoberflächen; Messverfahren der Tragfähigkeit – Deflexionsmessungen). Swiss Standard VSS 40 330, VSS, Zurich, Switzerland

VSS 40 512 (2019) Characteristics of road surfaces: Skid resistance measurements (Eigenschaften der Fahrbahnoberflächen; Griffigkeitsmessungen). Swiss Standard VSS 40 512, VSS, Zurich, Switzerland

Life Cycle CO_2 Analysis of Low Rolling Resistance Asphalt Pavements

A. Kawakami, M. Yabu, and H. Nitta

Abstract Reducing rolling resistance of pavement surfaces contributes to improved fuel efficiency for automobiles and reduced CO_2 emissions in exhaust gases. The authors have developed actual low-rolling-resistance asphalt pavements and shown that such pavements can reduce CO_2 emissions from automobiles by 1.3–6.6% through a driving test. In contrast, in order to evaluate the CO_2 emissions of pavement technologies, which will contribute to the realizing of a sustainable society in the future, it is important to conduct the life cycle assessment of these pavement technologies, including CO_2 emissions, due to the production and transportation of pavement materials, pavement construction work, and the recycling as well as the reconstruction of pavements. Thus, for the purpose of clarifying the environmental improvement effect of low-rolling-resistance asphalt pavements, this research paper reports on the life cycle CO_2 emissions assessment conducted for studying not only CO_2 emissions due to the production and transportation of pavement materials, pavement construction work, reconstruction but also reductions in CO_2 emissions through the improvement in fuel efficiency of automobiles by the low-rolling-resistance asphalt pavements. As a result of the life cycle CO_2 emissions assessment, it is clarified that the low-rolling-resistance asphalt pavements can reduce CO_2 emissions by 70% compared to conventional pavement (porous asphalt pavement).

Keywords Life cycle assessment · Carbon dioxide · Rolling resistance · Asphalt pavement

A. Kawakami (✉) · M. Yabu
Pavement Research Team, Road Technology Research Group, Public Works Research Institute, 1-6 Minamihara, Tsukuba-shi, Ibaraki 305-8516, Japan
e-mail: kawakami@pwri.go.jp

H. Nitta
Innovative Materials and Resources Research Center, Public Works Research Institute, 1-6 Minamihara, Tsukuba-shi, Ibaraki 305-8516, Japan

C. Raab (ed.), *Proceedings of the 9th International Conference on Maintenance and Rehabilitation of Pavements—Mairepav9*, Lecture Notes in Civil Engineering 76,
https://doi.org/10.1007/978-3-030-48679-2_15

1 Introduction

In Japan, the gross emissions of greenhouse gases was 1,292 million tons (CO_2 equivalent) in 2017. Although the gross emissions have been on a decreasing trend since 2013, the value in 2017 was 1.3% larger than that in 1990. In the gross emissions in 2017, the gross CO_2 emissions was 1,190 million tons, in which the industrial and transportation sectors accounted for 34.7 and 17.9% respectively. Thus, the pavement sector has also been requested to contribute to reducing CO_2 emissions when pavements are constructed, maintained and put in service. In Japan, the types of pavement technologies expected to contribute to reducing CO_2 emissions include: those for lowering the temperature to produce hot asphalt mixtures, producing asphalt at cold temperature, recycling pavement materials and extending lives of pavements (Japan Road Association 2009). Thus, the authors have conducted life cycle assessments focusing on the technologies for recycling pavement materials in plants and in situ (Kawakami et al. 2010), concrete pavements with extended lives (Kawakami et al. 2011a), and producing asphalt at warm temperatures (Kawakami et al. 2011b).

Recently, reducing the rolling resistance of pavements has attracted attention as a new CO_2 reduction measures in the pavement sector. The rolling resistance of pavements can be reduced by improving the texture and roughness of pavement surfaces. Reducing rolling resistance also contributes to improving the fuel efficiency of automobiles and thereby reducing CO_2 emissions in exhaust gases. Actually, the authors have developed low-rolling-resistance asphalt pavements (hereinafter referred to as the "LRRAPs") and clarified that the LRRAPs can reduce CO_2 emissions of automobiles by 1.3–6.6% through the driving test (Kawakami et al. 2016). The pavement technology for reducing rolling resistance is expected to be used as measures to save energy and reduce greenhouse effect gas emissions and thereby contributing to achieving the sustainable development goals (SDGS) promoted by the United Nation.

For evaluating the CO_2 emissions of pavement technologies that will contribute to realizing a sustainable society in the future, it is important to conduct a life cycle assessment of these pavement technologies, including CO_2 emissions due to the production and transportation of pavement materials, pavement construction work, and the recycling as well as reconstruction of pavements. Thus, for the purpose of clarifying the environmental improvement effect of the LRRAPs, this research paper reports on the life cycle CO_2 emissions assessment conducted for studying not only CO_2 emissions due to the production and transportation of pavement materials, pavement construction work, reconstruction but also the reductions in CO_2 emissions through the improvement in fuel efficiency of automobiles with the LRRAPs.

2 Assessment Method

2.1 Life Cycle CO_2 Emissions Calculation Method

Research into the life cycle assessment (LCA) of asphalt pavements has been conducted worldwide by, for example, Santero et al. (2010), Ram et al. (2011) and Harvey et al. (2020). Also, many LCA tools have been developed for modeling LCA such as PaLATE, Athena Pavement LCA and VTTI/UC. However, the LCA requires calculations of the amounts of resources and energy to be consumed and environmental loads to be discharged. Because these amounts vary depending on each countries or regions economic conditions and level of environmental efforts, it is important to use environmental load intensities suitable for the respective countries.

Regarding research into the LCA of pavements in Japan, the authors have conducted the LCA of pavements using the intensities for energy consumption and discharge amounts of environmental loads (Kawakami et al. 2010). In this research, the LCA of pavements are conducted in a manner that focuses on consumption energy, CO_2, SOX, NOX, and SPM as environmental load items and uses the LIME (life-cycle impact assessment method based on endpoint modeling) coefficients which integrate these environment load items in terms of costs (Itsubo and Inaba 2005). These LIME coefficients have also been set taking into consideration the situations in Japan. Recently, along with the enhancement of global warming countermeasures, the LCA of pavements has been conducted for pavements with extended working lives and the technologies for producing asphalt at warm temperatures with environment load items limited to those related to CO_2 emissions. Also, for the purpose of standardizing the calculation methods for evaluating the life cycle CO_2 emissions in pavements, the Japan Road Association has published the "Calculation Guidebook for Reducing the Environmental Loads of Pavements" (2015). The Calculation Guidebook adopts a hybrid method which uses both input-output data and bottom-up data as the method for calculating life cycle CO_2 emissions. This research also used the hybrid method for evaluating the life cycle CO_2 emissions of asphalt pavements in a manner that quantitatively estimates CO_2 emissions in each stage of the life cycle of asphalt pavements such as the production and transportation of pavement materials, pavement construction work, recycling (reconstruction).

This research also incorporated the CO_2 emissions reduction effect of the LRRAPs from automobiles in the evaluation of the life cycle CO_2 emissions of asphalt pavements. The LCA of roads in consideration of the CO_2 emissions from automobiles was proposed as the expanded life cycle CO_2 evaluation (ELC-CO_2) (Nakamura et al. 1998). Thus, this research quantitatively estimated the reduction effect of CO_2 emitted from automobiles with the LRRAPs applied to general highways by: calculating the life cycle CO_2 emissions due to the production and transportation of materials, construction, and reconstruction of the LRRAPs; and using actual measurements of the reduction effect of CO_2 emissions from automobiles through the LRRAPs.

2.2 Estimations of the Life Cycle CO_2 Emissions of Asphalt Pavements

The life cycle CO_2 emissions of asphalt pavements were estimated by the calculation method following the guidebook (Japan Road Association 2015). Also, CO_2 emissions were estimated for each life stage of asphalt pavements: the production and transportation of pavement materials, pavement construction work, recycling (reconstruction of base course).

The structure of asphalt pavements used for the estimations in this research was based on N6 in the Japanese pavement design traffic volume classification. N6 corresponds to a typical traffic volume on national roads in urban districts in Japan; i.e., large daily automobile traffic of not less than 1,000 and less than 3,000 in one lane. The cross-sectional view of the asphalt pavement to estimate CO_2 emission in this study is shown in Fig. 1. The cross-sectional structure of asphalt pavement comprises: 10 cm of asphalt mixture as surface and binder layers; 8 cm of bituminous stabilized materials and 20 cm of crushed stone for mechanical stabilization as an upper base course layer; and 35 cm of crusher-run as a lower base course layer. The estimations were based on the assumption that asphalt pavements were subjected to reconstruction of their upper base course, surface and binder layers in the 10th year after new construction in their life-spans.

Table 1 shows the list of unit CO_2 emissions for asphalt pavement materials. For the LRRAPs, the unit of Type H polymer-modified bitumen is used because the LRRAPs use it to prevent the deformation of pavement surfaces. The life cycle CO_2 emissions of porous asphalt pavement (PA) using the same Type H polymer-modified bitumen as the LRRAPs was also estimated for comparison.

In addition, the estimations of the life cycle CO_2 emissions were based on the common distance of 20 km required for transporting general materials such as aggregate and 10-ton dump trucks as the common mode of transportation of the materials in all cases. The distance for transporting construction machines was also commonly

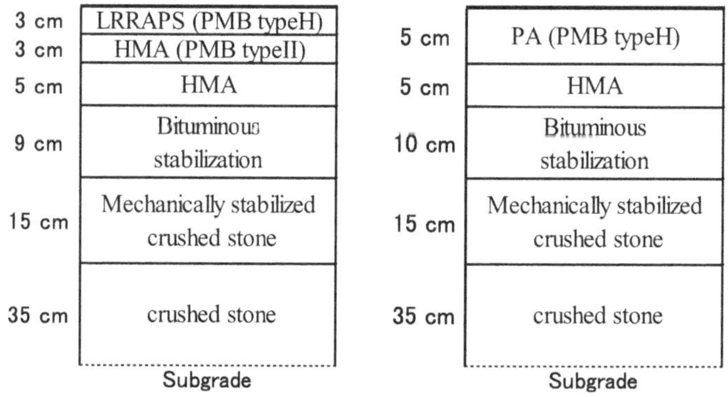

Fig. 1 Cross-sectional view of asphalt pavement used for the estimations

Table 1 Unit CO_2 emissions

Material	Unit	CO_2 emissions ($kg\text{-}CO_2$/unit)	Material	Unit	CO_2 emissions ($kg\text{-}CO_2$/unit)
Aggregate	t	7.98	PMB type II	t	488.82
Sand	t	11.54	PMB type H	t	670.40
Screenings	t	7.98	Electric power	kWh	0.46
Filler (Lime)	t	5.41	A-type heave oil	L	2.91
Bitumen (60/80)	t	107.56	Light oil	L	4.19
Asphalt emulsion	L	5.44			

set at 20 km. However, a common distance of 240 km was set as an exception for bitumen transportation by lorries taking into consideration the decrease in the number of bitumen refineries in Japan.

2.3 Estimations of CO_2 Emissions from Automobiles

The CO_2 emissions vary depending on the types of automobiles and their traffic. This research applied the CO_2 emissions factor equations, as shown in Eqs. (1) and (2) by two types of automobiles used in the road environmental assessment for the calculation of the CO_2 emissions intensities of automobiles for the estimations (Dohi et al. 2012). The estimations assumed that object roads were located in Ibaraki Prefecture in the Tokyo metropolitan area with the average travel speed and traffic of automobiles as well as the average number of lanes on these roads set with reference to the road traffic census. The types of roads used in the estimations were: an arterial high-standard highway; a general highway; a principal prefectural road; and a general prefectural road (as shown in Table 2).

The CO_2 emissions from automobiles can be calculated by multiplying the emissions factors by traffic and distances. Then, on the basis of the average traffic lanes, daily CO_2 emissions from automobiles per lane were calculated for these types of roads.

Table 2 Daily traffic per lane (Ibaraki Prefecture)

Road classification	Daytime average speed per 12 h [km/h]	Average traffic volume per 24 h (Vehicles/day)		Average lanes (weighted average)
		Light vehicles	Heavy vehicles	
High-standard highway	81.3	21,985	8,359	4.63
General highway	37.2	13,797	3,651	2.85
Principal prefectural road	37.5	7,128	1,124	2.34
Prefectural road	36.8	4,686	636	2.14

$$EF_L = 1501.20185/V - 2.40935 \times V + 0.02115 \times V^2 + 174.47635. \quad (1)$$

$$EF_H = 908.52069/V - 23.49899 \times V + 0.18396 \times V^2 + 1364.81344 \quad (2)$$

where EF_L is the CO_2 emissions factor [g-CO_2/km/vehicle] for light vehicle and EF_H is for the heavy vehicle. V is the running velocity.

2.4 Actual Measurement of CO_2 Emissions Reduction Effect of LRRAPs

In the estimations of the CO_2 emissions reduction effect of the LRRAPs in the life cycle CO_2 emissions analysis, the actual measured data of CO_2 emissions from automobiles obtained through the driving test previously conducted by the authors (Kawakami et al. 2016) were used.

The actual measurements of CO_2 emissions from automobiles were obtained through the driving test in which four types of pavements having different surface properties as shown in Table 3 and a travel distance of 300 m each were tested. The driving test was conducted with the test vehicle driven at constant speeds on test pavements. In the test, exhaust gas concentrations were measured with the exhaust gas measurement system as shown in Table 4. The test vehicle runs on gasoline and has 3,500 cc displacement and a fuel consumption rate of 8.20 km/L in the ten-fifteen mode. The exhaust gas concentrations were measured when the vehicle was driven at the speeds of 20, 40, 60 and 80 km/h for respective compositions of CO, CO_2, THC, and NOX with a recording cycle of 10 Hz. In the results of this measurement,

Table 3 Pavement surface properties of test section

Pavement surface	PA (13)	DG (13)	LRRAPs type 1	LRRAPs type 2
Length (m)	300	300	300	300
Max. aggregate size (mm)	13	13	5	5
Evenness σ (mm)	0.87	0.75	0.58	0.55
Mean profile depth (mm)	1.88	0.49	0.74	0.66

Table 4 Devices of measuring emissions and fuel consumptions

Measurement item	Measuring device	Model
Emissions	Vehicle-mounted emission measuring system	OBS-2200
Fuel consumption rate	Fuel meter	FC-9521
Tire temperature	Stationary non-contact thermometer	CS-30TAC
Test speed	GPS speedometer	LC-8100
Wind direction and velocity	Vane anemometer	WMT52

it was revealed that Type 1 and Type 2 LRRAPs had the CO_2 emissions reduction effects of 1.3–5.3% and 1.8–6.6% with respect to the PA respectively.

Here, the fuel consumption of an automobile changes depending on the gradient (vertical alignment) and degree of curvature (horizontal alignment) of the roads, however it was difficult to reflect these factors in predicting the CO_2 emission. Moreover, considering that the LRRAPs could not be ruled out of the possibility for the aging degradation of CO_2 emissions reduction effects even though they use polymer modified bitumen to prevent the deformation of pavement surfaces, the LRRAPs' effects on the reduction of the CO_2 emissions from automobiles were regarded to be 1% and 2% in this life cycle CO_2 analysis compared to a maximum of 6% in our experiments.

3 Research Results

3.1 Estimations of the Life Cycle CO_2 Emissions of Asphalt Pavements

Figure 2 shows the estimated CO_2 emissions when the LRRAPs and the PA are newly constructed. For both types of pavements, it can be said that the majority of the CO_2 emissions are generated through the production of materials and the CO_2 emissions through the construction and machine transportation are small. This is because a large amount of energy needs to be consumed when producing bitumen and heating asphalt mixtures as calculated in the previous research (Kawakami et al. 2016). In the comparison of the LRRAPs with the PA, the LRRAPs generate 2.5% larger CO_2 emissions than the PA. This is because the LRRAPs require larger amounts of bitumen and a longer paving time (double layered surface) than the PA.

Figure 3 shows the estimated life cycle CO_2 emissions of the LRRAPs and the PA for 40 years since their new construction with the reconstruction of the layers above

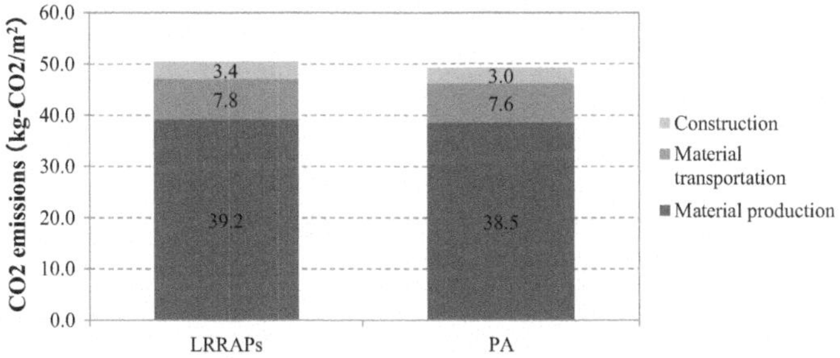

Fig. 2 CO_2 emissions of LRRAPs and PA in new construction

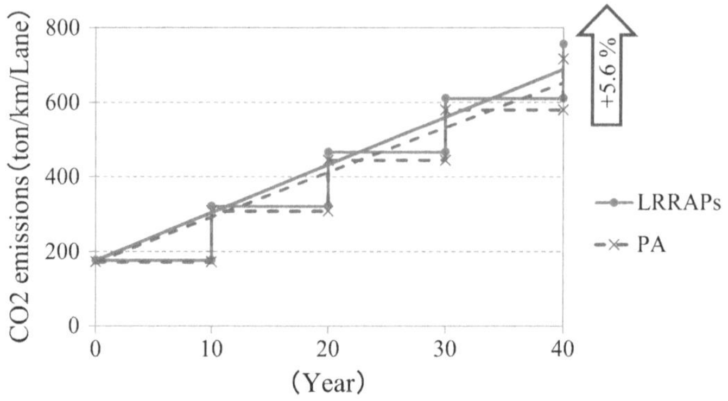

Fig. 3 Estimated life cycle CO_2 emissions of the LRRAPs and PA

base course in the 10th years. These estimations are based on the cross-sectional structure as shown in Fig. 1, a road width of 3.5 m and a distance of 1 km.

As a result, the life cycle CO_2 emissions of the LRRAPs is 5.6% larger than that of the PA. This is because the LRRAPs generate slightly larger CO_2 emissions than the PA in the new construction and reconstruction of the layers above base course.

3.2 Estimations of CO_2 Emissions from Automobiles

The estimations of the CO_2 emissions from automobiles are based on actual traffic on an high-standard highway; a general highway; a principal prefectural road; and a general prefectural road in Ibaraki Prefecture. Figure 4 shows the estimation results together with the life cycle CO_2 emissions of the LRRAPs. As shown in Fig. 3, since the CO_2 emissions of LRRAP are almost equivalent results for the PA, the value of LRRAP was used here.

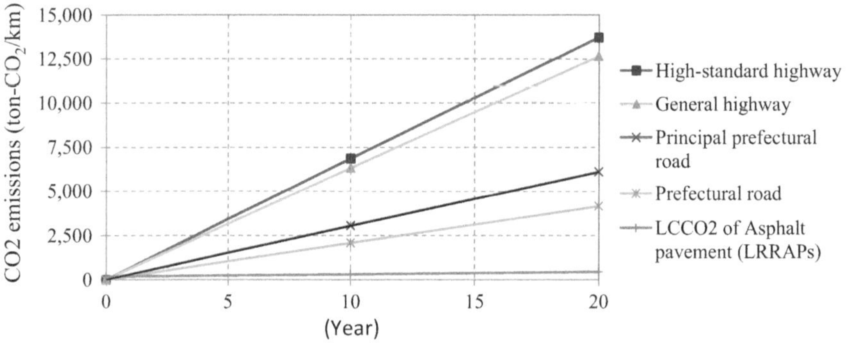

Fig. 4 CO_2 emissions from automobiles and life cycle CO_2 emissions of asphalt pavements

As can be seen in the Fig. 4, the CO_2 emissions from automobiles are significantly larger than the life cycle CO_2 emissions of asphalt pavements. This is because the CO_2 emissions from automobiles per unit road length are increased with increasing traffic. Thus, the CO_2 emissions reduction through cutting the CO_2 emissions from automobiles even by 1 or 2% considerably exceeds the increase in the CO_2 emissions generated through the new construction of an entire road with the LRRAPs.

3.3 *Life Cycle CO_2 Emissions of the LRRAPs*

The effect of the LRRAPs on the reduction in the CO_2 emissions from automobiles was compared with the life cycle CO_2 emissions of the LRRAPs. The comparison was based on the estimations of the CO_2 emissions on a 1-km section of a general highway with the LRRAP in Ibaraki Prefecture as mentioned above. Figure 5 shows the comparison result. In the figure, the legend "present condition" means the life cycle CO_2 emissions on a road with PA without the effect of the reduction in the CO_2 emissions from automobiles. Also, "LRRAPs (1.0%)" and "LRRAPs (2.0%)" mean the life cycle CO_2 emissions of the LRRAPs minus the reduction amount of the CO_2 emissions from automobiles by LRRAPs.

With the comparison result in Fig. 5, this research is able to quantitatively show that, in the case where the LRRAPs' effect on the reduction of the CO_2 emissions from automobiles is 2%, the LRRAPs have the CO_2 emissions reduction effect of about 70% with respect to the life cycle CO_2 emissions of PA. Also, it is estimated that the LRRAPs can still maintain the CO_2 emissions reduction effect of about 30% with respect to the life cycle CO_2 emissions of PA even in the case where the aging

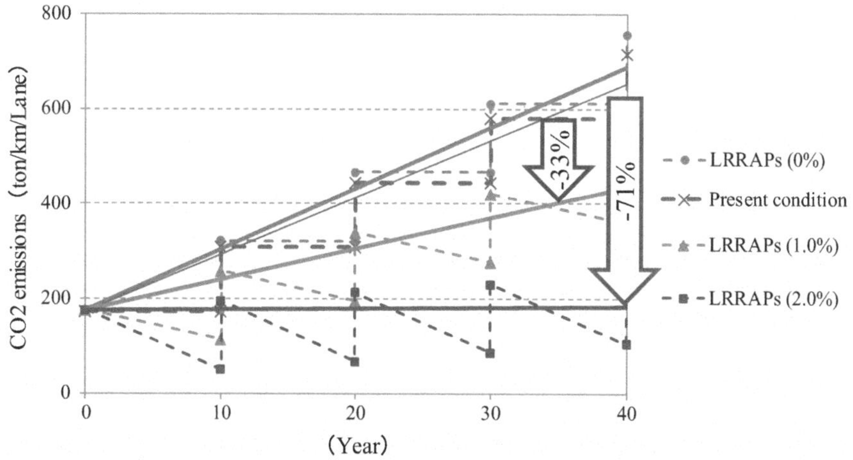

Fig. 5 Life cycle CO_2 emissions in consideration of the LRRAPs' effects on the reduction of CO_2 emissions from automobiles

degradation of LRRAPs causes their effect on the reduction of the CO_2 emissions from automobiles to be reduced to 1%. Because the LRRAPs showed the maximum fuel efficiency improvement ratio of 6.6% in the verification test, it is considered that the LRRAPs have large fuel efficiency improvement effects. Thus, this research quantitatively proves that the LRRAPs when applied to general highways can produce CO_2 emissions reduction effects.

4 Conclusion

(1) Because the CO_2 emissions from automobiles are significantly larger than the life cycle CO_2 emissions of asphalt pavements, the CO_2 emissions from automobiles per 1 km increase along with the increases in traffic on arterial high-standard highways and general highways. Thus, applying the LRRAPs to the actual roads is considered to produce large CO_2 emissions reduction effects.
(2) In the case where the LRRAPs' effect on reducing CO_2 emissions from automobiles is 2%, the LRRAPs can contribute to about 70% reduction with respect to the life cycle CO2 emissions of the PA (present condition). Also, even in the case where the LRRAPs' effect on the reduction of the CO_2 emissions from automobiles is 1%, the LRRAPs can still contribute to about 30% reduction with respect to the life cycle CO_2 emissions of the PA. Thus, this research quantitatively proves that the LRRAPs when applied to general highways can produce CO_2 emissions reduction effects.

As above, the LRRAPs' life cycle CO_2 emissions reduction effects can be quantitatively proven through the expanded life cycle CO_2 evaluation (ELC-CO_2) with the CO_2 emissions from automobiles incorporated into the life cycle CO_2 emissions of the LRRAPs.

References

Dohi M, Sone S, Takimoto M, Ogawa T, Namikawa Y (2012) Grounds for the Calculation of Motor Vehicle Emission Factors using Environment Impact Assessment of Road Project (Revision of FY 2010), Technical note of National Institute for Land and Infrastructure Management, No. 671
Harvey J, Kendall A, Butt A, Saboori A, Lozano M, Ostovar M (2020) Supply curves using LCA and LCCA for conceptual evaluation of proposed policies to improve the environment. In: ISAP APE 2019. LNCE, vol 48, pp 144–152
Itsubo N, Inaba A (2005) LIME, a Life Cycle Environmental Impact Assessment Method - Assessment Methods and Databases for LCA. Environmental Accounting and Environmental Efficiency. Japan Environmental Management Association for Industry
Japan Road Association (2009) Guideline on Environment-friendly Pavement Technologies. Maruzen Publishing Co., Ltd.
Japan Road Association (2015) Calculation Guidebook for Reducing the Environmental Loads of Pavements. Maruzen Publishing Co., Ltd.

Kawakami A, Nitta H, Kanou T, Kubo K (2010) Environmental loads of pavement recycling methods in Japan. In: TRB 89th Annual meeting, compendium of papers, No. 10-1487

Kawakami A, Nitta H, Kubo K (2011a) Life-cycle CO_2 of long-life pavement and technologies for extending the life of pavement. In: TRB 90th Annual meeting, compendium of papers, No. 11-1864

Kawakami A, Nitta H, Kubo K (2011b) Estimation of the CO_2 emissions in producing HMA and WMA in Japan. In: 24th World road congress

Kawakami A, Ishigaki T, Shirai Y, Terada M, Kubo K (2016) Demonstration on fuel consumption reduction performance of low rolling resistance asphalt pavement. In: 2016 ISAP symposium

Nakamura H, Kato H, Maruta H, Futamura T (1998) Life Cyde Assessment of CO_2 Emission from Inter-city Motorways by Road Cross-sectional Structure, Environmental systems research. Japan Society of Civil Engineers, pp 261–270

Ram P, Van Dam T, Meijer J, Smith K (2011) Sustainable Recycled Materials for Concrete Pavements. Michigan Department of Transportation, Lansing, MI

Santero N, Harvey J (2010) Consideration of time-dependent factors in the environmental assessment of longer life pavements. In: Proceedings, International conference on sustainable concrete pavements, Sacramento, CA

Accuracy Comparisons Between ASTM 1318-09 and COST-323 (European) WIM Standards Using LTPP WIM Data

Syed W. Haider and Muhammad Munum Masud

Abstract Weigh-in-Motion (WIM) is a primary technology used for monitoring and collecting vehicle weights and axle loads on roadways. Highway agencies collect WIM data for many reasons, including highway planning, pavement and bridge design, freight movement studies, motor vehicle enforcement, and regulatory studies. The process of weighing vehicles in motion estimates static truck weight by the wheel [single or tandem axles] or gross vehicle weight (GVW) as vehicles drive over sensors installed in a roadway or under a bridge. Two primary protocols are currently used across the globe to assess the accuracy of a WIM system, (a) ASTM E1318-09 and (b) European WIM accuracy protocols. The quality and accuracy of the data largely depend on the characteristics of the WIM equipment, calibration/validation, site characteristics, and data reporting. This paper compares the WIM sensor accuracies in the LTPP data for both protocols. The results show that there are minor differences in calculated accuracies.

Keywords Weigh-in-Motion (WIM) · Sensors · ASTM type I and COST323

1 Introduction

Weigh-in-Motion (WIM) is a primary technology used for monitoring and collecting vehicle weights and axle loads on roadways (Selezneva and Mcdonnell 2017). Highway agencies collect WIM data for many reasons that include highway planning, pavement and bridge design, freight movement studies, motor vehicle enforcement, and regulatory studies. The new mechanistic-empirical pavement design guide

S. W. Haider (✉)
Civil and Environmental Engineering Department, Michigan State University, 428 S. Shaw Lane, Room 3546, East Lansing, MI 48824-1226, USA
e-mail: syedwaqa@egr.msu.edu

M. M. Masud
Civil and Environmental Engineering Department, Michigan State University, 428 S. Shaw Lane, Room 1206, East Lansing, MI 48824-1226, USA
e-mail: masudmuh@msu.edu

C. Raab (ed.), *Proceedings of the 9th International Conference on Maintenance and Rehabilitation of Pavements—Mairepav9*, Lecture Notes in Civil Engineering 76,
https://doi.org/10.1007/978-3-030-48679-2_16

(Pavement-ME) also requires WIM data for predicting pavement distresses. Inaccurate WIM data may result in significant over- or underestimation of the pavement performance period and hence, lead to higher costs and premature failure, respectively. Therefore, the data collected at WIM systems must be accurate and consistent (Papagiannakis and Quinley 2008).

The process of weighing vehicles in motion estimates the static truck weight (SA or TA and GVW) as vehicles drive over sensors installed in a roadway or under a bridge. It measures the transient dynamic tire forces applied by the tires as a vehicle passes over the sensors installed in-road or on structural bridge members and estimates the corresponding static loads of the vehicle when it is at rest (FHWA 2009). Several WIM technologies exist to measure the applied forces and predict static weight. Additionally, two protocols are currently used across the globe to assess the accuracy of a WIM system, (a) ASTM E1318-09 and (b) European WIM accuracy protocols. The quality and accuracy of the data largely depend on the characteristics of the WIM equipment, calibration/validation, site characteristics, and data reporting.

Therefore, the primary objectives of the paper are to (a) describe WIM system accuracy and consistency, (b) review of ASTM and European WIM accuracy protocols, and (c) evaluate the LTPP WIM data accuracy by using these protocols. The objectives were accomplished by synthesizing and analyzing the WIM data available in the LTPP database.

2 Background

The accuracy of the WIM systems is a primary concern for its manufacturers and users. The users desire different levels of accuracy according to the application of data usage (Jacob 2000). The primary goal of a WIM station is to accurately measure the dynamic loads exerted by moving vehicle wheels on the road pavement and estimate static axles and GVW. The accuracy of weighing results obtained from WIM systems broadly impacts the legal load enforcement, i.e., the elimination of overloaded vehicles from the highways. It also facilitates the pavement repair planning and development of efficient traffic models (Gajda et al. 2013). WIM systems also provide statistical knowledge of the traffic loads that are used for traffic flow optimization and the design of road infrastructure (Burnos and Rys 2017).

Establishing a baseline for assessing the impact of multiple factors on WIM data accuracy would require an understanding of measurement accuracy and consistency concepts. Figure 1 shows the target analogy to visualize the differences between accuracy and consistency. Accuracy is the conformity of results to the true value, i.e., the absence of bias. Bias is a tendency of an estimate to deviate in one direction from the actual value. The target analogy in Fig. 1 is a practical way of understanding how accuracy can be quantified by measurement bias. Consistency or precision is related to the repeatability of a process. The variability of repeat measurements can characterize precision under carefully controlled conditions. Figure 1 also illustrates that it is possible to be consistent (or precise, as applied to target shooting) without

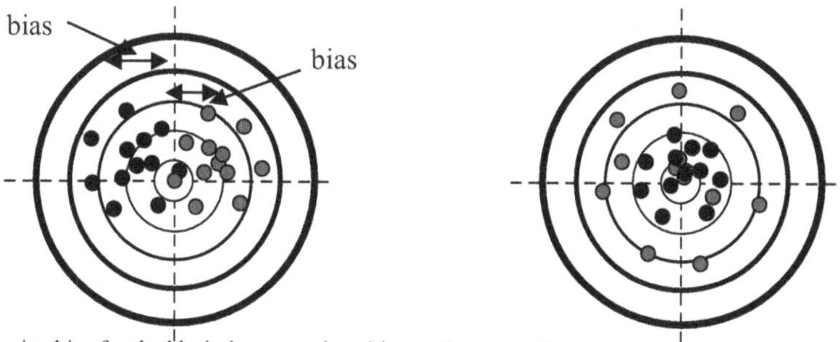

(a) Negative bias for the black dataset and positive bias for the red data set, similar precision for both data sets

(b) Good (black) and poor (red) precision with no bias

Fig. 1 Target analogy for precision and bias

being accurate or accurate without being consistent (low precision). Ideally, we would like a measurement process to be both accurate and consistent.

The WIM system accuracy is measured in terms of the relative difference between WIM and static weights. The following equation can express the relative WIM error:

$$\varepsilon = \frac{\text{WIM weight} - \text{Static weight}}{\text{Static weight}} \times 100 \tag{1}$$

This relative error is commonly referred to as measurement error for a WIM scale. Further, this accuracy will vary for different types of WIM sensor technologies. For a well-calibrated WIM system, typical WIM measurement error follows a normal distribution with a zero mean (no bias) and a standard deviation (Haider et al. 2011) as shown in Eq. (2):

$$\varepsilon = \frac{X' - X}{X} \sim N\left(0, \sigma_{\varepsilon}^{2}\right) \tag{2}$$

where

$X' =$ load measured on a WIM scale for an axle configuration
$X =$ load measured on a static scale for the same axle configuration
$\sigma_{\varepsilon} =$ standard deviation (SD) characterizing the accuracy of the WIM scale

3 WIM Performance Requirements

Two specifications are currently being followed for WIM systems calibration and accuracy. American Society for Testing and Materials International Standard, ASTM E1318-09 (ASTM 2009) is mainly adopted in the US and the European Road Specification COST-323 (Jacob 2000; Jacob and O'Brien 1998) is used in the European countries. Accuracy criteria for both the protocols are briefly presented in this section.

3.1 ASTM E1318-09 (2017)

This specification describes the WIM system as "*the process of measuring the dynamic tire forces of a moving vehicle and estimating the corresponding tire loads of the static vehicle. WIM systems installed on the highways can estimate the gross weight of moving vehicle as well as the portion of this weight (called load in this standard), that is carried by the tires of each wheel assembly, axle, and axle group on the vehicle.*" Table 1 provides a summary of performance specifications for different WIM types as per ASTM E1318-09.

3.1.1 Procedure for Static Weight (Reference Values) Measurements

Two test vehicles are used to get the reference values, required to perform calibration, type-approval test, and for on-site acceptance/verification test. Each vehicle is weighed for a minimum of three times with brakes released to measure tire loads for the wheel(s) on each end of every axle on the static vehicle. The arithmetic mean rounded to the nearest 100 lb for all-wheel load, axle load, tandem-axle load, and

Table 1 Functional performance requirements for WIM systems

Function	Tolerance for 95% compliance[a]				
	Type I	Type II	Type III	Type IV	
				Value $\geq$ lb[b]	±lb
Wheel load	±25%	–	±20%	5000	300
Axle load	±20%	±30%	±15%	12000	500
Axle-group load	±15%	±20%	±10%	25000	1200
Gross-vehicle weight	±10%	±15%	±6%	60000	2500
Speed				±1 mph	
Axle-spacing and wheelbase				±0.5 ft	

[a]95% of the respective data items produced by the WIM system must fall within tolerance. [b]Lower values are not usually a concern for enforcement

GVW values are calculated from the individual observations on test vehicles. Difference in percent (truncated to an integer) from each value is also calculated and compared to the mean with following specified limits for each applicable load and test vehicle: gross vehicle weight = ±2%, tandem axle load = ±3%, axle load = ±4%, and wheel load = ±5%. If the reference values exceed then the prescribed limits, corrections are made by re-weighing the test vehicles at least three more times. Further discussion on reference values is available elsewhere (ASTM 2009).

3.1.2 Procedure for Calculating the Percent of Non-compliance

WIM system performance is ascertained by comparing the reference values with the WIM values for all the data items listed in Table 1. The following relationship is used in the specification to calculate the difference or percent difference in the WIM system value and the corresponding reference value:

$$D = (C - R) \tag{3}$$

The difference, d, in loads and weights (%):

$$d = 100[(C - R)/R] \tag{4}$$

where,

D = The difference in speed (mph), axle-spacing (ft), and wheelbase (ft)
d = The difference in the value of the data item produced by the WIM system and the corresponding reference value expressed as a percent of the reference value.
C = Value of the data item produced by the WIM system.
R = The corresponding reference value for the data item

3.2 *European Road Specification (COST 323)*

This specification mainly addresses the issues associated with high-speed WIM systems, i.e., the WIM systems installed on one or more traffic lanes and operated under normal traffic conditions. However, this specification may also be adapted for low-speed WIM systems, i.e., the WIM system installed in a specifying weighing area outside the traffic lanes under controlled traveling conditions.

According to this specification, under defined operating conditions (moving traffic, tire loads, etc.) the accuracy of a WIM system may only be defined statistically by a confidence interval of the relative error of a unit (an axle, an axle group or a gross weight) defined as by Eq. (1). Such a confidence interval centered on the static load/weight is $[-\delta; +\delta]$, where δ is the tolerance for a confidence level π (for example, 90 or 95%). According to this specification, the main requirements and

Table 2 Accuracy classes definition, [value of δ i.e., confidence interval width (%)]

Function	Accuracy classes						
	A(5)	B + (7)	B (10)	C (15)	D + (20)	D (25)	E[a]
Gross weight (GW)	5	7	10	15	20	25	>25
Group of axle (AoG)	7	10	13	18	23	28	>28
Single axle (SA)	8	11	15	20	25	30	>30
Axle of group (GA)	10	15	20	25	30	35	>35

[a]Class E is defined for the WIM systems which do not meet the class D (25) requirements

applications of the WIM system can be classified into three accuracy classes with an increasing level of statistical accuracy:

- **Statistics** (δ up to 20 to 30%): This criterion *[for class D + (20) or D (25)]* is described for economic and technical studies of freight transport, general traffic evaluation on roads and bridges, and for collecting statistical data.
- **Infrastructure and preselection** (*δ* up to 10 to 15–20%): This *criterion [for class B + (10) or C (15)]* is described for detailed analysis of traffic and vehicle classification, design, and maintenance of roads and bridges, and preselection for enforcement.
- **Legal purposes** (*δ* up to 5 to 10%): This *criterion [for class A (5) or B + (7)]* is described for enforcement and industrial applications, but only if the legislation allows the use of WIM for this purpose.

A typical standardized table of δ values taken from European specification is shown in Table 2 (Jacob and O'Brien 1998).

3.2.1 Test Conditions and Confidence Levels (π_o)

This specification allows the user to set a test plan by selecting an appropriate combination of repeatability/reproducibility and environmental conditions.

As per the specification, Table 3 provides the minimum levels of confidence (π_o) for different tests and environmental conditions. As compared to environmental repeatability (**I**), smaller π_o values are required for limited (**II**) and full (**III**) environmental reproducibility conditions.

3.2.2 Accuracy Assessment of the WIM System

The accuracy test on a WIM system is carried out using a pre-weighed or post-weighed vehicle as per European standard. The sample statistics, including mean (bias) $\boldsymbol{m}$, standard deviation $\boldsymbol{s}$, and the number values $\boldsymbol{n}$ are calculated and used as per the specification. "A lower bound π of the probability that an individual error falls within the specified interval [−δ; +δ] is calculated and compared to the specified π_o."

Table 3 Minimum % levels of confidence π_o of the centered confidence intervals case

Test conditions		Sample size (n)					
		10	20	30	60	120[a]	∞
Full repeatability	I	95.0	97.2	97.9	98.4	98.7	99.2
	II	93.3	96.2	97.0	97.8	98.2	98.9
	III	91.4	95.0	96.0	97.0	97.6	98.5
Extended repeatability	I	90.0	94.1	95.3	96.4	97.1	98.2
	II	87.5	92.5	93.9	95.3	96.1	97.5
	III	84.7	90.7	92.4	94.1	95.1	96.8
Limited reproducibility	I	85.0	90.8	92.5	94.2	95.2	97.0
	II	81.9	88.7	90.7	92.7	93.9	96.0
	III	78.6	86.4	88.7	91.1	92.5	95.0
Full reproducibility	I	80	87.4	89.6	91.8	93.1	95.4
	II	76.6	84.9	87.4	90.0	91.5	94.3
	III	73.0	82.3	85.1	88.1	89.9	93.1

[a]Sample sizes (n) not mentioned in the table may be interpolated

According to the statistics provided in the standard, an upper bound on the customer risk, π, for an $\alpha = 0.05$, is given by:

$$\pi = \Phi(u_1) - \Phi(u_2) \tag{5}$$

$$u_1 = \frac{\delta - m}{s} - \frac{t_{n-1,0.975}}{\sqrt{n}} \tag{6}$$

$$u_2 = \frac{-\delta - m}{s} + \frac{t_{n-1,0.975}}{\sqrt{n}} \tag{7}$$

The *function* Φ is the cumulative distribution function of a student variable and $t_{n-1,0.975}$ is a student variable with $(n - 1)$ degrees of freedom. For a sample size greater than 60, the cumulative function Φ in the above equation can be approximated by the cumulative distribution of a standard normal variable. Following criteria are used for the acceptance of WIM systems:

- If $\pi \geq \pi_o$, the system is accepted in the accuracy class of tolerance for the criterion considered.
- If $\pi < \pi_o$, the system cannot be accepted in the proposed accuracy class, and the acceptance test is repeated with a lower accuracy class, i.e., a larger value of δ.

A concept of reduction factor k was also introduced in the specification to cater for different testing conditions, i.e., initial verification after calibration, an in-service check of the WIM system (Jacob and O'Brien 1998).

4 LTPP WIM Data Extents

The WIM data were obtained from the LTPP database standard release 33.0 (July 2019). The data elements were organized in various data tables to create a relational database. The following items were identified and reviewed for the data analyses:

- Section inventory (Sate code, WIM ID, and site location)
- WIM calibration data (SA, TA, and GVW bias and SD, truck type and passes, LTPP data quality, calibration technique)
- WIM equipment data (sensor type [bending plate (BP), load cell (LC), quartz piezo (QP), piezo cable (PC)], calibration technique and reason)
- LTPP climatic regions [dry freeze (DF), dry-no freeze (DNF), wet freeze (WF), wet-no freeze (WNF)]

Table 4 presents the climate and sensor type distribution of WIM sites and associated records available in the LTPP database. It can be noted that the majority of the WIM sites have PC sensors followed by BP and QP in the LTPP WIM database. Also, the majority of the WIM sites are in a wet climate. Only a few sites (9) with LC sensors are located in the WF climate.

Figure 2 presents the distribution of sites with the typical WIM sensor type for each State in the LTPP database. A majority of WIM sites have PC followed by BP and QP sensors in North America. The dataset was further filtered to get more accurate calibration data. The primary purpose of the WIM calibration is to remove any systematic bias of WIM equipment. The calibration data considered for the comparisons were obtained by applying a filter on GVW bias of $\leq \pm 5$ collected by at least 10 or more runs of a Class-9 truck.

Furthermore, WIM data should be available for at least 1 year after a successful calibration or validation. It is necessary to prove that data quality stays consistent overtime during a year after calibration. Table 5 presents the summary of LTPP WIM data complying with the above criteria.

Table 4 Distribution of WIM sites and records by sensor type and climate

Climatic region	Sensor type				Total
	BP	LC	PC	QP	
DF	3[a] (16[b])	–	44 (249)	3 (15)	50 (280)
DNF	14 (46)	–	24 (81)	5 (18)	43 (145)
WF	14 (69)	9 (118)	58 (349)	26 (146)	107 (682)
WNF	17 (66)	–	137 (396)	15 (56)	169 (518)
Total	48 (197)	9 (118)	263 (1075)	49 (235)	369 (1625)

[a]No of WIM sites, [b]No of WIM records

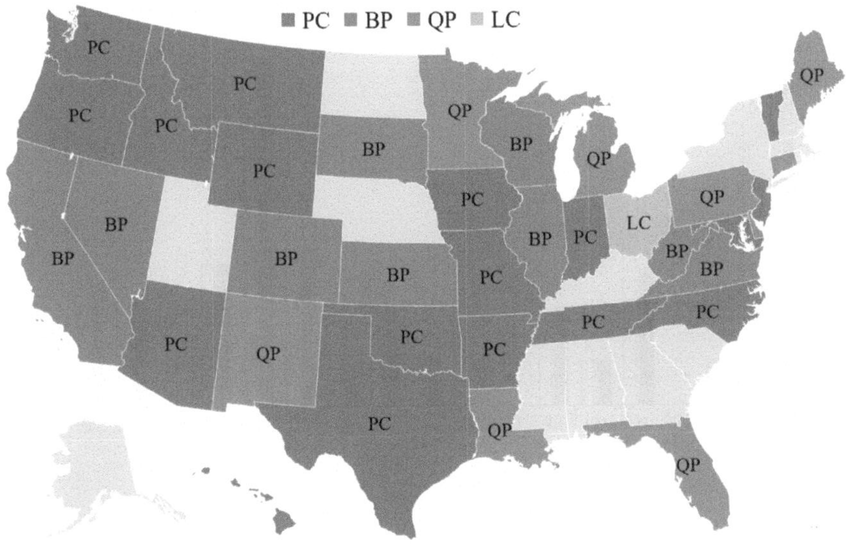

Fig. 2 Distribution of LTPP WIM site location with different sensors in the US

Table 5 WIM data accuracy criteria and minimum calibration requirements

Climatic region	Sensor type				Total
	BP	LC	PC	QP	
DF	1[a] (9[b])	–	2 (113)	3 (9)	6 (131)
DNF	4 (21)	–	9 (9)	5 (12)	18 (42)
WF	5 (22)	4 (8)	26 (65)	19 (77)	54 (172)
WNF	5 (29)	–	21 (40)	8 (28)	34 (97)
Total	15 (81)	4 (8)	58 (227)	35 (126)	112 (442)

[a]No of WIM sites, [b]No of WIM records

5 Accuracy Comparisons ASTM E1318-09 and COST-323

The WIM accuracy was calculated based on ASTM E1318-09 and COST-323 WIM protocols. The limited reproducibility (R1), environmental repeatability (I), and in-service test conditions were selected to calculate WIM system accuracy using COST-323. Additionally, the ASTM Type I and B accuracy classes were compared for GVW. These test conditions are the closest to WIM calibration guidelines provided in the ASTM E1318-09 protocol. Figure 3 presents the comparison of WIM accuracies based on both protocols. The results show that out of available 442 LTPP WIM records used in this analysis, 282, and 297 calibration records met the ASTM Type-I, and B (10) accuracy class for ASTM E1318-09 and COST-323, respectively. There were 15 records for PC sensors which met the COST-323 B accuracy class but did

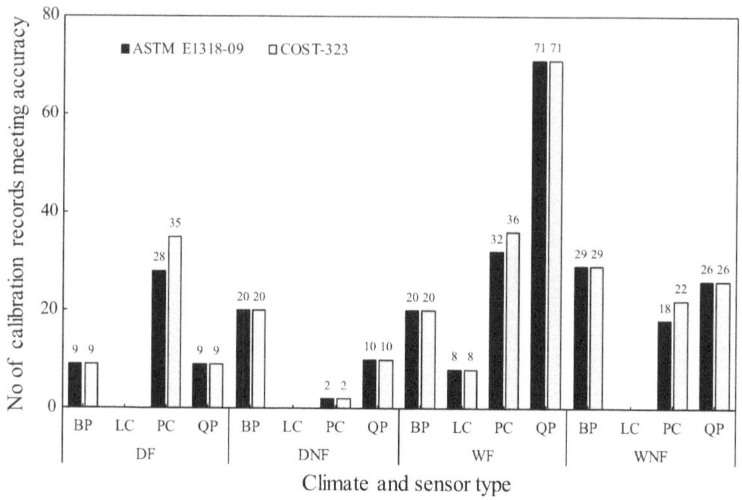

Fig. 3 Comparisons between ASTM E1318-09 and COST-323 WIM accuracy protocols

not meet ASTM E1318-09 Type I accuracy. This difference is caused mainly because of the flexibility in the confidence levels based on the sample size (i.e., number of runs). These records did not meet the ASTM Type I accuracy requirement of less than 10% with a small margin. Also, the number of LTPP WIM records meeting either of the two accuracy protocols was the highest for BP sensors located in different climates, followed by QP. While PC sensors showed the lowest number of WIM records meeting the accuracy.

6 Conclusions

- In the LTPP database, the majority of WIM measurement accuracy information is available for the sites with PC sensors (263 sites) followed by QP (49 sites) and BP sensors (48 sites). Only 9 sites with LC sensors had WIM available measurement accuracy information.
- Out of available 442 LTPP WIM records used in the analysis, 282, and 297 calibration records met the ASTM Type-I, and B (10) accuracy class for ASTM E1318-09 and COST-323, respectively. There is good agreement between ASTM E1318-09 and COST-323 calculated accuracy requirements.
- The number of LTPP WIM records meeting either of the two accuracy protocols were the highest for BP sensors located in different climates followed by QP. The PC sensors showed the lowest number of WIM records meeting the accuracy.
- The number of runs and selection of test conditions may cause a minor difference in the sensor accuracies between the two protocols.

References

ASTM (2009) Standard specification for highway weigh-in-motion (WIM) systems with user requirements and test methods E 1318-09. In: 2007 Annual book of ASTM standards. Edited by ASTM Committee E17-52 on Traffic Monitoring. ASTM International, USA

Burnos P, Rys D (2017) The effect of flexible pavement mechanics on the accuracy of axle load sensors in vehicle weigh-in-motion systems. Sensors 17(9):2053

FHWA (2009) LTPP Field Operations Guide for SPS WIM Sites Version 1.0 Draft Office of Infrastructure Research. Development, and Technology, Federal Highway Administration, McLean, Virginia, 173

Gajda J, Sroka R, Zeglen T, Burnos P (2013) The influence of temperature on errors of wim systems employing piezoelectric sensors keywords: piezoelectric sensors, temperature influence, temperature error of wim systems, error correction. Metrol Measur Syst 20(2):171–182

Haider SW, Harichandran RS, Dwaikat MB (2011) Impact of systematic axle load measurement error on pavement design using mechanistic-empirical pavement design guide. J Transp Eng 138(3):381–386

Jacob B (2000) Assessment of the accuracy and classification of weigh-in-motion systems part 1: statistical background. Int J Heavy Veh Syst 7(2–3):136–152

Jacob B, O'Brien EJ (1998) European specification on weigh-in-motion of road vehicles (COST323). In: Proceedings of second European conference on weigh-in-motion of road vehicles, Held Lisbon, Portugal, 14–16 September 1998

Papagiannakis AT, Quinley R (2008) High speed weigh-in-motion system calibration practices. Transportation Research Board

Selezneva O, Mcdonnell AM (2017) Weigh-in-Motion. Advancing highway traffic monitoring through strategic research

Detecting Significant Changes in Traffic Patterns for Pavement Design

Gopi K. Musunuru, Syed W. Haider, and Neeraj Buch

Abstract The mechanistic-empirical pavement design guide (AASHTOWARE Pavement-ME) incorporates mechanistic models to estimate stresses, strains, and deformations in pavement layers using site-specific climatic, material, and traffic characteristics. These traffic characteristics include monthly adjustment factors (MAF), hourly distribution factors (HDF), vehicle class distributions (VCD), axle groups per vehicle (AGPV), and axle load distributions for different axle configurations. Site-specific traffic inputs (Level 1) were generated for each of the 41 WIM sites after extensive QC checks. The averages from nearby sites (regional) with similar traffic characteristics (groups or clusters) can be used as Level 2 data or Level 3 data when Level 1 data are unavailable. Multiple approaches were used to develop Level 2 and Level 3 traffic input levels. These developed traffic inputs at different levels need to be updated every few years due to several reasons, including the change in land use nearby the WIM locations, economic conditions resulting in the change in traffic patterns. Equations were developed to identify these changes in traffic patterns that would cause significant changes in design lives. Once these patterns are identified, the traffic inputs can be updated so that the pavement sections would not be over-designed or under-designed.

Keywords Pavement-ME · Traffic inputs · Pavement design

1 Introduction

In the AASHTO 93 pavement design procedure, the truck traffic is converted to an equivalent number of 18-kip single-axle loads (ESALs) using the load equivalency factors (LEFs) developed based on Present Serviceability Index (PSI) concept. Several studies have found that the multiple failure modes of pavement structures cannot be explained by this single value (Zhang et al. 2000; Carvalho et al. 2006).

G. K. Musunuru (✉)
Fugro USA Land Inc., Gainesville, FL, USA
e-mail: g.musunuru@fugro.com

S. W. Haider · N. Buch
Department of Civil Engineering, Michigan State University, East Lansing, MI, USA

C. Raab (ed.), *Proceedings of the 9th International Conference on Maintenance and Rehabilitation of Pavements—Mairepav9*, Lecture Notes in Civil Engineering 76,
https://doi.org/10.1007/978-3-030-48679-2_17

The mechanistic-empirical pavement design guide (AASHTOWARE Pavement-ME) addresses these limitations by incorporating mechanistic models to estimate stresses, strains, and deformations in pavement layers using site-specific climatic, material, and traffic characteristics (NCHRP Project 1-37A 2004). These structural responses are used to predict different performance measures for each pavement type using empirical models (i.e., transfer functions). Therefore, the use of ESALs to characterize traffic loadings is not compatible with the Pavement-ME. This new analysis and design approach require several types of traffic data. These traffic inputs include:

- Annual average daily truck traffic (AADTT),
- Vehicle class distribution (VCD),
- Monthly adjustment factors by vehicle class (MAF),
- Hourly truck volume distribution factors (HDF),
- Number of axle groups per vehicle (AGPV), and
- Axle load distributions by vehicle class and axle group.

The Pavement-ME addresses the unavailability of detailed traffic data over the years; the hierarchical input levels are used depending on the level of detail of the available traffic data (NCHRP Project 1-37A 2004). These input levels range from site-specific input values to "best-estimate" or default values and are classified as follows:

- Level 1 – There is a very good knowledge of past and future traffic characteristics. At this level, it is assumed that the past traffic volume and weight data have been collected along or near the roadway segment to be designed.
- Level 2 – There is a modest knowledge of past and future traffic characteristics. At this level, only regional truck volume and weight data may be available for the roadway in question.
- Level 3 – There is inadequate knowledge of past and future traffic characteristics. At this level, the designer will have little truck volume information. In this case, a statewide or some other default value must be used.

In the State of Michigan, the traffic inputs listed above were developed for use in the Pavement-ME by analyzing the permanent traffic recorder (PTR) traffic volumes and WIM axle load data in the year 2009. Axle weight and vehicle classification data were obtained from 44 WIM and classification stations located throughout the State of Michigan to develop Level 1 (site-specific) traffic inputs. Cluster analyses were conducted to group sites with similar characteristics to develop Level 2 (regional) inputs. Finally, data from all sites were averaged to establish the statewide Level 3 inputs. The traffic characterization was based on data collected from 2005 to 2007. However, the developed traffic inputs at all levels need to be updated every few years due to the following reasons (Buch et al. 2009; Haider et al. 2011):

a. Addition of new classification and WIM sites at different geographical locations or changes in the status of the existing site (e.g., down- or up-grading from WIM to classification or vice versa).
b. Significant changes in land use near the existing WIM locations.

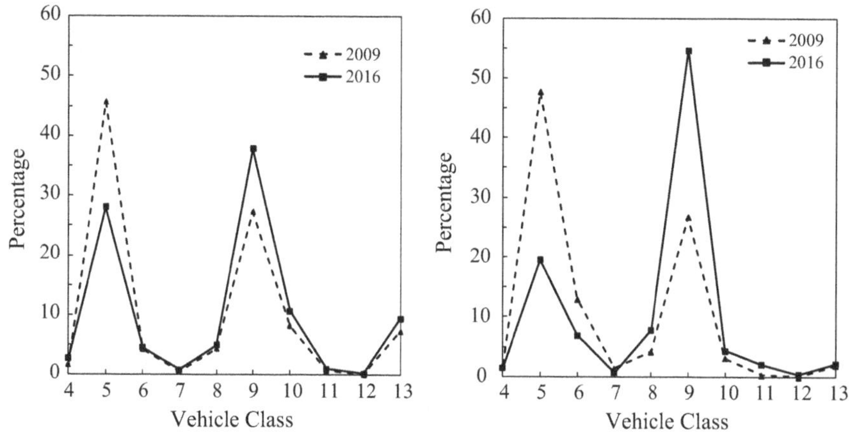

(a) VCD at site 4129 (b) VCD at site 9759

Fig. 1 Temporal changes in vehicle class distribution at the selected sites

c. WIM technology improvements (e.g., less accurate piezo-polymer sensors are replaced with more accurate piezo-quartz sensors).

During the last seven (7) years, new traffic data were collected, reflecting the recent economic growth, additional, and downgraded WIM sites. Figure 1 shows the changes in VCD traffic data between the years 2009 and 2016 for selected sites in Michigan. Consequently, the current traffic inputs needed to be re-evaluated and developed with the latest traffic data collected at all the PTR locations.

The current Level 2 methodology used in Michigan has some practical limitations (freight data availability issues for cluster assignments). Hence, one of the goals of the study was to develop an alternative, simplified methodology for the generation of Level 2 inputs. One such method is to use available MDOT traffic inventory data (AADTT, VCD, and road information) to group the PTR sites. Once the groups are identified, the traffic inputs based on the averages of sites in each group are established. The inputs developed using clustering methodology and the alternative simplified methodology are termed as "Level 2A" and "Level 2B", respectively. Both these input levels were tested for pavement design accuracy. The methodology that balances accuracy and practicality was recommended for adoption and future updates.

2 Generation of Multiple Traffic Input Levels

Site-specific traffic inputs (Level 1) were generated for each of the 41 WIM sites (a few WIM sites were eliminated from 2009 list) after extensive QC checks. The development of regional inputs (Level 2) is crucial when site-specific data are not available. The averages from nearby sites (regional) with similar traffic characteristics (groups or clusters) can be used as Level 2 data. Two approaches were adopted for developing Level 2 inputs (a) cluster analyses (Level 2A), and (b) grouping roads with similar attributes (Level 2B). Level 3 data were further divided into Levels 3A and 3B—3A represents an average of freeways and non-freeways, and 3B represents the overall statewide average for traffic inputs.

2.1 Level 2A Traffic Inputs

Level 2A inputs were generated using cluster analysis, a data mining technique that identifies homogeneous subsets of data (also known as clusters) within a dataset using only the information found in the data. It uses a mathematical similarity of two data objects to group them. The Euclidean distance and Ward's method were used as the similarity measure and the linkage method, respectively. After evaluating several criteria to find the optimal number of clusters in the dataset, the Calinski-Harbasz criterion (Calinski and Harabasz 1974) and gap criterion methods (Tibshirani et al. 2001), and engineering judgment were used to determine the optimal number of clusters for each traffic input. The detailed procedure used for obtaining the optimum number of clusters for each traffic input in this paper can be found elsewhere (Haider et al. 2018).

2.2 Level 2B Traffic Inputs

Grouping roads by attributes are more subjective and involve identifying roads that are expected to behave similarly (i.e., similar traffic patterns). Attributes of the roadways (e.g., road class—freeway vs. non-freeway) are used to identify groups that have similar traffic patterns. Such groups, based on these attributes, are easy to interpret by the users. A minimum of three to six groups are required, but more groups may be appropriate if significant regional differences exist (TMG 2016). The advantages of this methodology are that the creation of groups is intuitive, and a new site can be assigned to an existing group easily based on its attributes. The drawback of this process is that it is not entirely objective (involves subjective decisions that may not explain the variability of traffic patterns within a group). Also, the challenge lies in identifying a combination of attributes that can be used to group the PTR locations. The attributes used to classify groups need not necessarily be the same for all the traffic inputs. The traffic patterns at the PTR locations should be similar within a group and should be different between the groups. An algorithm was developed to help identify an optimal attribute combination. Some of the attributes contained in the MDOT's sufficiency database for different PTR locations are as follows:

- Functional classification (freeway vs. nonfreeway)
- Development type (urban vs. rural)

Table 1 Attributes used for grouping PTR sites for Level 2B inputs

Input	Attribute 1	Attribute 2
VCD	VC9 levels	Development type
HDF	AADTT level	Development type
MAF	VC9 levels	Development type
SALS	COHS	Development type
TALS	Number of lanes	Development type
TRALS	COHS	Development type
QALS	COHS	Development type

- AADTT level (<1,000, 1,000 to 3,000, >3,000)
- Corridors of highest significance (COHS) (national, regional, and statewide)
- Number of lanes (2, 3, and 4)
- Road type (non-freeway divided, non-freeway undivided, and freeway)
- Vehicle Class 9 (VC9) distribution levels (<45, 45 to 70, >70)

Several attributes can be chosen at a time to divide the PTR locations into groups. Each attribute has sublevels (e.g., functional classification has two sublevels: freeway and non-freeway), and hence a combination of attributes has different sublevel combinations. The Level 1 traffic inputs of all PTR sites belonging to a combination of sublevels were averaged to develop Level 2B inputs. Pairwise Euclidean distances between each sublevel combinations were calculated to identify the combination of attributes that show different traffic patterns. The higher the number of attributes used for grouping, the lower the number of PTR locations in each sublevel combination. Hence it was more appropriate to use only two attribute combinations to form road groups for developing Level 2B inputs. After careful evaluation of the results, the following attribute combinations (Table 1) were chosen based on the availability of the sublevel combinations and the distances between them. The Level 1 traffic data of all PTR sites in each of the sublevel combinations for the attributes chosen were averaged to obtain the Level 2B inputs. The detailed procedure used to develop the Level 2B traffic inputs can be found elsewhere (Haider et al. 2018).

3 Significant Traffic Inputs

The Pavement-ME traffic inputs were generated for Levels 1, 2A, 2B, 3A, and 3B. Level 1 inputs should always be used for design purposes wherever possible as it is the actual traffic data specific to the site. When Level 1 inputs are unavailable, either Level 2 or Level 3 inputs should be used. The results of the sensitivity analyses based on statistical and practical significance were used to decide on the appropriate traffic input level. Subsequently, the appropriate input levels can be selected for a site for which Level 1 traffic data are not available. Baseline flexible and rigid pavement designs used for the sensitivity analyses were established as per MDOT guidance. Pavement-ME Version 2.3 was used for the sensitivity analysis. For both the flexible and rigid pavement, locally calibrated performance models were used (Haider et al. 2014, 2017). The pavement design life was assumed to be 20 years with 95% design reliability. For each of the 41 WIM locations, the hot mix asphalt (HMA) surface layer thickness was designed to achieve a 20-year design life for bottom-up fatigue cracking threshold of 20% for flexible pavements for Level 1 inputs. For these designs, the rut depth values at the end of 20 years were also recorded.

Similarly, the slab thickness was designed to achieve a 20-year design life for the international roughness index (IRI) threshold of 172 in./mile (2.73 m/km) for rigid pavements. Faulting and transverse cracking values were also recorded at the end of 20 years. For both the flexible and rigid pavement designs, one traffic input was changed at a time to appropriate cluster or road groups for the site of the PTR (levels 2A and 2B) to determine the effect of that input on the design life. Levels 3A and 3B inputs for each design (one input at a time) were also evaluated in the Pavement-ME to determine their impact on the design life. The time for the distress values (for Levels 2 and 3) to reach the threshold values in the Level 1 designs were documented. The differences in design lives between different input levels were quantified for further analyses. Statistical analyses could detect differences in the datasets, but the differences might not have much practical significance or vice versa. Hence in addition to finding the likelihood of a value (significance value a) outside the 95% confidence interval (CI), a life difference value of 2 years between Level 1 and the level being evaluated was also adopted as an indicator of significant difference. The life differences in pavements between Level 1, and Levels 2A and 2B for each WIM location were examined. If there was at least one WIM location in any cluster or groups with a life difference of at least 2 years, then that cluster/group was considered sensitive. Table 2 shows input levels recommended for each traffic input for different pavement types.

Table 2 Recommended traffic input levels

Input	Flexible pavements	Rigid pavements
VCD	2B	2B
HDF	n/a	2B
MAF	3A	3A
SALS	3A	3A
TALS	2B	2B
TRALS	3A	3A
QALS	3A	3A

4 Identifying the Changes in Traffic Patterns

In this study, 5 years of data (2011–2015) were averaged to obtain Level 1 inputs. However, there will be inherent variability in traffic data from year to year due to the change in economic growth, additional, and downgraded WIM sites. Therefore, there is a need to identify the change in traffic patterns to update the traffic inputs so that the pavement sections would not be over-designed or under-designed. To this effect, the changes in design live data when the traffic inputs were changed from Level 1 to Level 2A or Level 2B or Level 3A tabulated as part of the sensitivity analyses were used. The data were used to model the effect of change in VCD values on the predicted design lives by the Pavement-ME. Figures 2 and 3 show the predicted life differences using these models compared to the estimated life differences using the Pavement-ME for flexible and rigid pavements. For example, for rutting in flexible pavements, the relationship between change in VCD values and the estimated life differences are as follows:

$$\begin{aligned} \Delta LifeRut = & -0.210 - 0.087 \times VC8 - 0.139 \times VC9 - 0.204 \times VC10 \\ & - 0.141 \times VC11 - 0.446 \times VC137 \end{aligned} \quad (1)$$

Bottom-up fatigue cracking in the flexible pavement:

$$\Delta LifeCrack = -0.273 - 0.070 \times VC9 - 0.084 \times VC10 - 0.370 \times VC13 \quad (2)$$

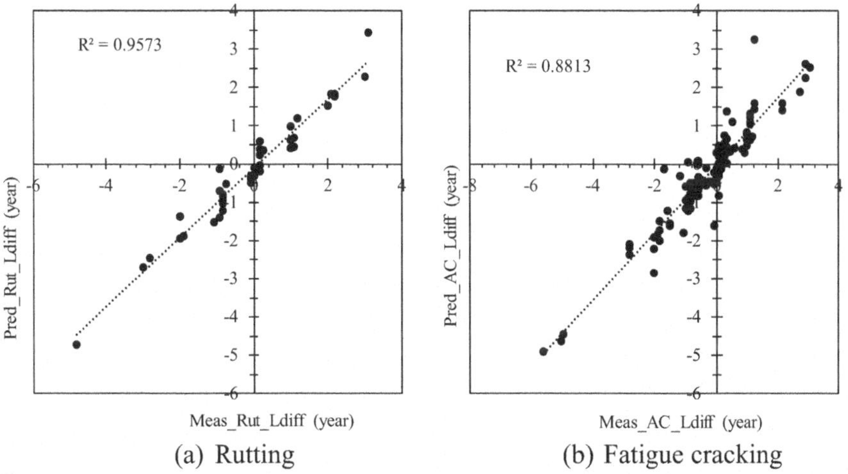

Fig. 2 Predicted vs. measured life differences for flexible pavements

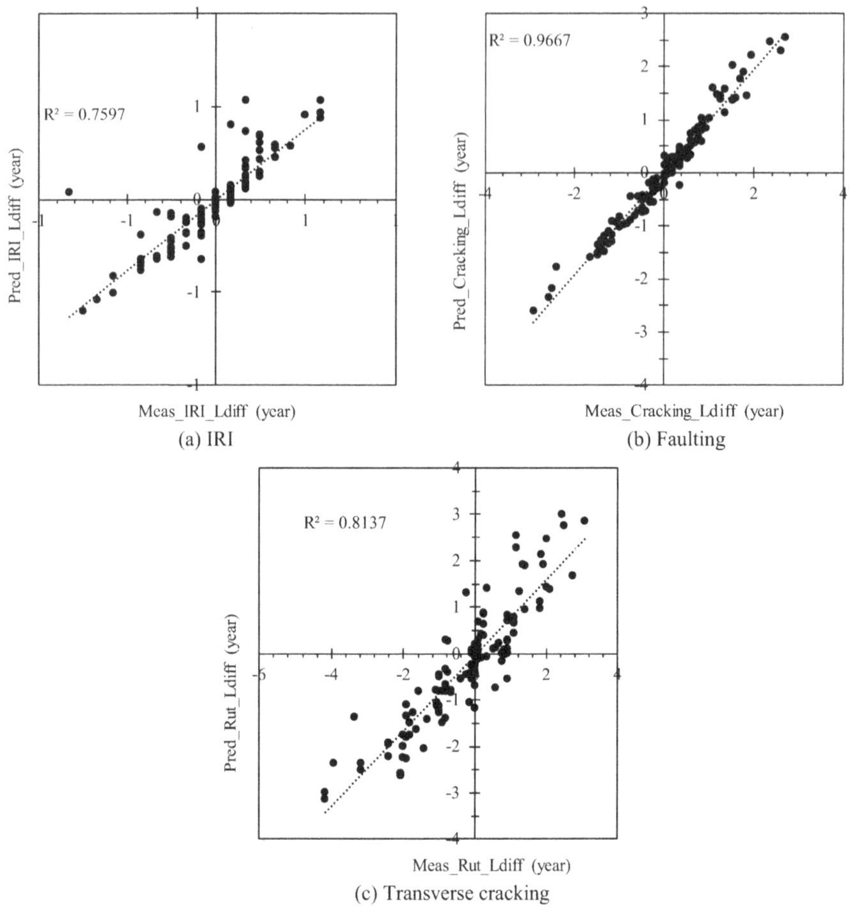

Fig. 3 Predicted vs. measured life differences for rigid pavements

IRI in the rigid pavements:

$$\Delta LifeIRI = -0.022 - 0.025 \times VC9 - 0.016 \times VC10 - 0.045 \times VC11 - 0.050 \times VC13 \quad (3)$$

Faulting in the rigid pavements:

$$\Delta LifeFault = 0.0220 - 0.112 \times VC9 - 0.071 \times VC10 - 0.088 \times VC11 - 0.219 \times VC13 \quad (4)$$

Table 3 Estimated life differences due to changes in traffic patterns

Site	Flexible		Rigid		
	Rutting	Fatigue cracking	IRI	Faulting	Transverse cracking
4129	4.64	1.90	0.78	3.47	4.54
9579	2.78	1.45	0.40	1.86	2.50

Note: Design lives in years

Transverse cracking in the rigid pavements:

$$\Delta LifeCrack = -0.224 - 0.158 \times VC9 - 0.224 \times VC10 - 0.245 \times VC13 \quad (5)$$

where, VC5, VC8, VC9, VC10, VC11, and VC13 are the change in their VCD between the years.

The short-term counts from the PTR sites can be used as inputs into these equations to check if there are any substantial differences in design life predictions. If the life differences are considered significant enough by the highway agencies at a PTR location for at least 3 years, then the new 3 years traffic data should be used to update the traffic inputs. Otherwise, the new data should be combined with the available traffic database. The change in axle load spectra can also be included in the above equations by converting the entire axle load spectra into a single factor called the load spectra factor. Additional variables may be incorporated into the equations based on the needs of the traffic agencies.

For illustration purposes, the changes in traffic patterns at the PTR sites shown in Fig. 1 are used in the equations developed to estimate the design live changes. The estimated change in design lives are listed in Table 3 for different distresses for both flexible and rigid pavements. Since life differences are significant enough, and if it continues for at least three years, then updating the traffic inputs is warranted.

5 Conclusions

The AASHTOWARE Pavement-ME uses traffic inputs including monthly adjustment factors (MAF), hourly distribution factors (HDF), vehicle class distributions (VCD), axle groups per vehicle (AGPV), and axle load distributions for different axle configurations to estimate stresses, strains, and deformations in pavement layers. Site-specific traffic inputs (Level 1) were generated for each of the 41 WIM sites after extensive QC checks. Multiple approaches were used to develop Level 2 and Level 3 traffic input levels which can be used when Level 1 inputs are unavailable. The developed traffic inputs need to be updated frequently due to the inherent variability in traffic data from year to year due to reflecting the change in economic growth, additional and downgraded WIM sites. To this effect, the changes in design live data when the traffic inputs were changed from Level 1 to Level 2A or Level 2B or Level 3A were modeled to see the effect of change in vehicle class distribution values on the predicted design lives by Pavement-ME. The equations developed can be used to identify these changes in traffic patterns that would cause significant changes in design lives. Once these patterns are identified, the traffic inputs can be updated so that the pavement sections would not be over-designed or under-designed.

Acknowledgements The authors acknowledge the Michigan Department of Transportation for funding the study.

References

Buch N, Haider SW, Brown J, Chatti K (2009) Characterization of truck traffic in Michigan for the new mechanistic-empirical pavement design guide. MI Report # RC-1537. Michigan Department of Transportation, Lansing

Calinski T, Harabasz J (1974) A dendrite method for cluster analysis. Commun Stat Theory Methods 3:1–27

Carvalho R, Schwartz C (2006) Comparisons of flexible pavement designs: AASHTO empirical versus NCHRP project 1-37A mechanistic-empirical. Transp Res Rec J Transp Res Board 1947:167–174

Haider SW, Buch N, Chatti K, Brown J (2011) Development of traffic inputs for the mechanistic-empirical pavement design guide in Michigan. Transp Res Rec J Transp Res Board 2256:179–190

Haider SW, Buch N, Brink W, Chatti K, Baladi G (2014) Preparation for implementation of the mechanistic-empirical pavement design guide in Michigan part 3: local calibration and validation of the pavement-ME performance models. Report No. RC-1595. Michigan State University, East Lansing, MI

Haider SW, Musunuru G, Kutay ME, Lanotte MA, Buch N (2017) Recalibration of mechanistic-empirical rigid pavement performance models and evaluation of flexible pavement thermal cracking model. Report Number SPR-1668. Michigan State University, Department of Civil and Environmental Engineering, East Lansing, MI

Haider SW, Musunuru G, Buch N, Selezneva O, Desaraju P, Li J (2018) Updated analysis of Michigan traffic inputs for pavement-ME design. Final Report No. SPR-1678. Michigan Department of Transportation

NCHRP Project 1-37A (2004) Guide for Mechanistic-Empirical Design of New and Rehabilitated Pavement Structures. Transportation Research Board of the National Academies, Washington, D.C.

Tibshirani R, Walther G, Hastie T (2001) Estimating the number of clusters in a data set via the gap statistic. J R Stat Soc Ser B (Stat Methodol) 63:411–423

Traffic Monitoring Guide (2016) FHWA, Washington, D.C.

Zhang Z, Leidy J, Kawa I, Hudson W (2000) Impact of Changing traffic characteristics and environmental conditions on flexible pavements. Transp Res Rec J Transp Res Board 1730:125–131

Development of Road Maintenance Management System for India's National Highway Network Using HDM-4 and Genetic Programming

Abhishek Sharma and Tanuj Chopra

Abstract With the increasing traffic loads on National highways of India, pavements are deteriorating at a faster rate leading to premature failure. In addition to this lack of scientific road management system leads to lower levels of serviceability and unreliable road network in the long term. Therefore, in order to maintain the highway network in good condition the road administration should focus on long lasting and economical road maintenance solutions. The research study focuses on the development of road management system for high volume roads using calibrated Highway Development & Management (HDM-4) model. Long-term pavement performance under various maintenance strategies has been measured in terms of roughness progression using HDM model. Genetic Programming (GP) system has been configured to develop four distress prediction models i.e., roughness, ravelling, cracking and rutting. Adequacy of GP models has been measured using simple linear regression analysis. Statistical significance of roughness model has been evaluated using student's t-test. Variability in the output results of the two deterministic models i.e., HDM and GP has been computed by comparing the difference between predicted and observed roughness behaviour. Prediction models play a crucial role in development of Road Maintenance and Management System (RMMS) for systematic technical as well as economic appraisal of road projects. Future modelling of pavement behaviour related to various maintenance activities will assist the highway planners and road agencies in timely monitoring and conditioning of roads by adopting suitable management framework.

Keywords HDM-4 model · GP system · RMMS · Prediction models · Roughness progression

---

A. Sharma (✉) · T. Chopra
Thapar Institute of Engineering and Technology, P.O Box 32, Bhadson Road, Patiala, Punjab 147004, India
e-mail: abhishek.sharma44267@gmail.com

C. Raab (ed.), *Proceedings of the 9th International Conference on Maintenance and Rehabilitation of Pavements—Mairepav9*, Lecture Notes in Civil Engineering 76,
https://doi.org/10.1007/978-3-030-48679-2_18

1 Introduction

India is a diverse country with second largest road network out of which national highways holds a share of 1.80% (MORT&H). Road projects require large capital investment and therefore timely allocation and utilization of funds is necessary for conditioning of roads and providing safe passage at appropriate speed and optimized road user cost. Due to lack of decentralisation of funds in India, the funds allocated for maintenance does not exceed 60% of the normal requirements of roads (MORT&H 2001b). This gap between the requirements and allocation has been accumulating over the years and is a concern for the long-term pavement performance.

To enhance the pavement performance, conditioning of road pavements needs to be done at regular intervals. It has been observed that delay in maintenance and rehabilitation of roads leads to faster pavement deterioration, increase in vehicle operating costs and accident costs thereby reducing the reliability of road network. Further if the pavement maintenance work is delayed even after noticeable deterioration, there comes a break point after which only reconstruction or extensive rehabilitation can be done costing many times more than routine maintenance works. Timely implemented maintenance and rehabilitation strategies bags a remarkable serviceability level for the road agencies. The main goal of highway engineers should be to accomplish a healthy pavement condition index in terms of both functional as well as structural condition keeping in mind other aspects such as economic, safety and environment. Road maintenance and investment projects both require strategic planning and enormous funding. Therefore, a decision support system needs to be adopted by transportation agencies for selection of optimum maintenance works within the budget constraint (Jorge 2012).

2 Development of RMMS Using HDM-4 Model

In this study RMMS has been developed using HDM-4 model for National Highways road network across various geographical locations in India to analyze and prepare road management scheme for analysis period of 20 years. Road condition data pertaining to 29 road sections has been acquired from literature survey and government publications (Deori 2016). Development of RMMS is a hierarchical process which requires a systematic methodology and the modelling has been performed using HDM-4 tool. The main concept behind implementing HDM-4 model in this study is because it assists the road agencies and other concerned authorities involved in road management and investment alternatives with smart effective pavement conditioning strategies along with optimized economic solutions for the entire pavement life cycle. In the early sixties of 19th century, World Bank developed and tailored it as a strategic pavement management analysis tool dedicated to road projects. Due to its wide range of applicability on various geographical as well as climatic conditions this model has been trending in many countries whenever there is a need of project appraisal system.

The fundamental concept behind an adequate and reliable RMMS is to calibrate accurately the in-built distress models in HMD-4 to the existing local conditions. If, the calibration factors related with distress models are not configured precisely then due to large variation in predicted and observed distress behavior, deterioration model obtained will be unable to simulate more realistic local conditions. The default values of calibration factors of in-built distress models in HDM equals to unity. In this study, calibration factors have been carefully scrutinized from past research works before development of RMMS for the selected National Highway (NH) network. Chopra et al. (2018) determined calibration coefficients for NH road networks identical to physical attributes and environmental conditions of selected road sections and therefore have been considered in the present study. Calibration coefficients adopted in the research study are 2.4, 0.6, 0.4, 0.4 for cracking distress, ravelling distress, rutting distress and roughness distress respectively. Validation of distress models developed using HMD-4 has been performed by simply comparing the similarity between the predicted values (given by HDM-4 model) and observed values (field survey data) to check the applicability and adequacy of the model for the specified conditions. Table 1 depicts field data pertaining to identified NH road sections showing various road physical characteristics such as pavement age, AADT (Annual average daily traffic), in-service layers thickness, distresses and BBD (Benkelman Beam Deflection).

2.1 *Long Term Pavement Performance Analysis Using HDM-4 Model*

Roughness is a reliable and powerful index to assess average pavement condition of road section at any instant (Chandra 2012). After assigning the maintenance strategies for the candidate road sections, project analysis application has been run to analyse the effect of these maintenance works on long-term pavement performance in terms of variation in average roughness with time. Roughness progression graphs for each section has been obtained and the work effects have been analysed. These graphs serve as a tracker of work items and verify the intervention criteria at trigger points.

It can be observed from the Fig. 1 and Fig. 2 that maintenance alternatives (Alt 1, Alt 2, Alt 3 and Base alt corrospond to Resealing, Resealing and Thin Overlay, Major Overlay and Routine Maintenance, respectively) have been correctly triggered corresponding to the intervention limits established for the entire NH road network for the analysis period of 20 years (Jain 2013). In some of the road sections Alternative 3 i.e., Major Overlay intervenes at roughness value much higher than the specified critical limit (IRI > 6 m/km) due to the 'AND' condition imposed as the total damaged area limit of 20% has not been exceeded. Therefore, the maintenance work operation will come into effect only when both the intervention criteria have been surpassed at the same instance.

Table 1 Road section attributes

Section name	Description	Pavement age (years)	AADT	Surface type	Surface Thickness (mm)	Roughness (m/km) 2011	Cracking (%)	Ravelling (%)	Rutting (mm)	BBD (mm)
NE-1GJ01	Ahmadabad-Vadodhara	8	8387	BC	50	2.3	2	2.3	2.2	0.49
NE-1GJ02	Vadodhara-Ahmadabad	8	8304	BC	50	2.5	7	5	3.5	0.37
NE-1GJ03	Ahmadabad-Vadodhara	8	8387	BC	50	2.5	1	1	2.7	0.24
NE-1GJ04	Vadodhara-Ahmadabad	8	8304	BC	50	2.5	1.5	1	2.2	0.35
NH-14GJ05	Radhanpur-Deesa	3	3715	BC	50	2.7	27.07	23	2.4	0.56
NH-2UP01	Allahabad-Khaga	6	4789	BC	50	2.5	7	4	3	0.55
NH-2UP02	Allahabad-Varanasi	6	4906	BC	50	2.8	1	2	2.7	0.67
NH-37AS01	Nagaon-Guwahati	1	4316	BC	40	2.5	0	0	2	0.64
NH-4KA01	tumkur-banglore	8	16789	BC	40	3.2	4.57	2	2.6	0.75
NH-4KA02	dharwad-belgaon	3	5088	BC	40	3.7	5	2	2.1	0.7
NH-4KA03	tumkur-sira	7	11456	BC	40	3.9	19.86	15	3.6	0.75
NH-4MH01	mahrastra.border-belgaon	7	5029	BC	40	3	0	0	2.1	0.8
NH-5AP01	Vizag–Srikakuram	7	6771	BC	50	2.5	1.16	0	1.6	0.68
NH-5AP02	Vizag–Srikakuram	7	6771	BC	50	2.8	0.66	0	2.7	0.75
NH-5AP03	NH-5-vizag port	5	5682	BC	50	2.9	10.87	8	1.9	0.72
NH-5AP04	NH-5-vizag port	5	5682	BC	50	2.9	11	9	1.7	0.88
NH-7MH02	hyderabad-nagpur	3	4288	BC	40	2.5	0	0	2	0.75
NH-7MH03	hyderabad-nagpur	3	4288	BC	50	2.4	0	1	2	0.65
NH-73 01	Roorkee-Saharanpur	9	7840	AC	25	3.6	4	5	7	0.65
NH-73 02	Roorkee-Saharanpur	7	7750	AC	25	2.4	1	1	2	0.8

(continued)

Table 1 (continued)

Section name	Description	Pavement age (years)	AADT	Surface type	Surface Thickness (mm)	Roughness (m/km) 2011	Cracking (%)	Ravelling (%)	Rutting (mm)	BBD (mm)
NH-73 03	Roorkee-Saharanpur	9	7670	DBSD	25	6.9	15	10	8	1.02
NH-73 04	Roorkee-Saharanpur	9	7650	DBSD	25	6.3	10	7	7	0.95
NH-74 01	Haridwar-Najibabad	9	5480	DBSD	25	4.5	6	5	7	0.95
NH-74 02	Haridwar-Najibabad	9	5520	DBSD	25	4.7	5	4	6	1.1
NH-72 01	Poanta Sahib-Dehradun	11	3650	SBSD	15	5.9	7	10	5	0.85
NH-72 02	Poanta Sahib-Dehradun	9	5720	AC	25	5.1	10	5	8	0.7
NH-72 03	Poanta Sahib-Dehradun	9	5870	AC	25	5.7	12	5	6	0.75
NH-72 04	Poanta Sahib-Dehradun	11	7280	DBSD	25	5.2	10	8	10	0.9
NH-72 05	Poanta Sahib-Dehradun	11	6840	DBSD	25	6.2	12	9	9	0.68

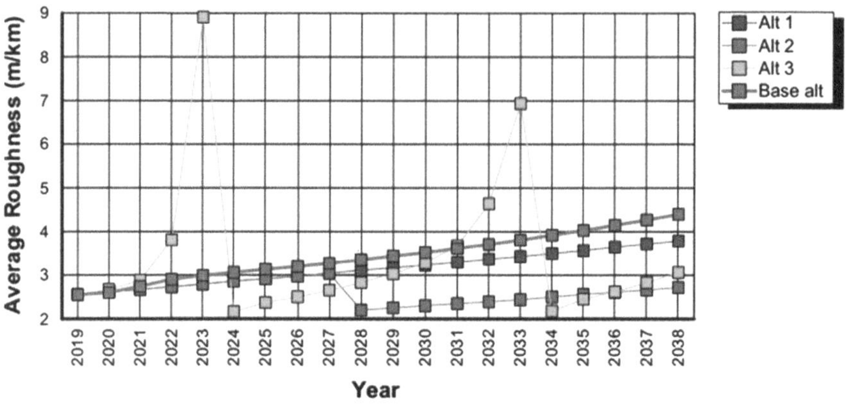

Fig. 1 Average roughness trend of NE-1 GJ01 for maintenance standards

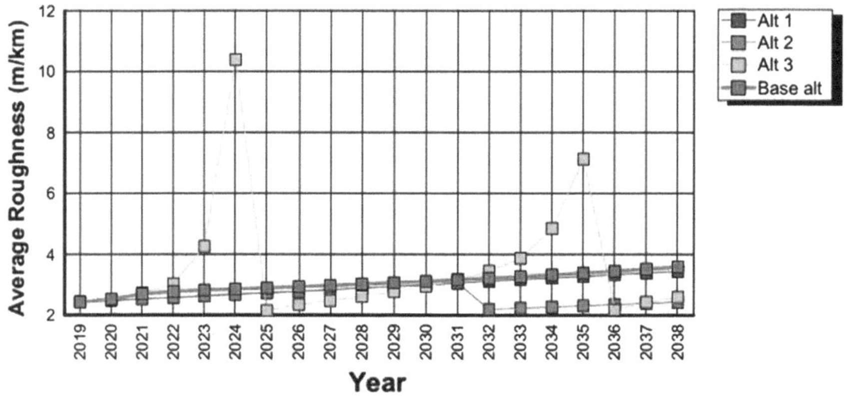

Fig. 2 Average roughness trend of NH-2UP 01 for maintenance standards

3 Development of Prediction Models Using Genetic Programming System

Pavement deterioration prediction models have been developed in this study using Genetic Programming (GP). Four distress models have been generated using GP to predict pavement distresses viz. cracking, ravelling, rutting and roughness for the National Highway road network. Development of this model will assist the road agencies in planning a structured road management program with cost effective solutions for preservation of pavement service life and providing a reliable road network to the users. Genetic programming has been performed using GP KERNEL software program to develop prediction models for the specified NH network. Statistical evaluation of GP model has been done by regression analysis technique by checking the goodness of fit between the predicted and observed roughness values.

GP is a domain-independent method which on the Darwinian principles of randomly occurring operations creates an automatic working computer program of new generation to solve problems by genetic mutation of population of computer programs (Koza 1994). The whole process involves numerous iterations of input population using the analogy of genetic operations such as mutation, crossover, reproduction, gene deletion and duplication. The five paramount steps required to generate GP model are to (a) specify parameters of the various terminals of stem linked with the program to be developed; (b) specify basic functions for stem of the program to be developed; (c) specify the accuracy measure; (d) set the conditions required prior to running the program; and (e) fix criteria for termination and classification of results obtained from the run (Babovic 2000).

In the present study, four distress models linked to National Highway network have been generated using genetic programming system. The input population to the computer program consists of dataset pertaining to physical characteristics of candidate road sections such as pavement condition, pavement age, traffic (AADT). The dataset congregated is of two consecutive years i.e. 2011 and 2012 and modelling of this training dataset has been done using GP to predict pavement distresses for the year 2012. GP gives numerous prediction equations depending upon the attributes of framework in which it has run and based upon fitness measure (such as R^2 and RMSE) the best possible prediction model has been selected corresponding to various distresses as shown in Table 2.

3.1 Validation of GP Model

Aggarwal et al. (2004) gathered pavement condition data for the identified National Highway network in the states of Uttaranchal and Uttar Pradesh to develop pavement management system. To check the adequacy and reliability of these distress models, validation dataset of year 2012 has been used and goodness of fit of these models measured simply by plotting scatter graphs between observed distress values and predicted distress values. Validation is essential as it examines the applicability of model for the particular conditions. It can be observed from post-regression results that there is slight difference between the observed values and predicted values. This difference is quiet rational due to the variation in physical characteristics such as pavement history, traffic conditions, geographical locations of selected NH road sections. Also due to the modelling technique adopted and certain assumptions like normal construction quality of bituminous surfacing at optimum conditions has led to noticeable deviation in this quantitative assessment of distress models. Table 3 depicts the post-regression results in terms of coefficient of determination (R^2) and root mean square error (RMSE) related to various distress models which indicate the adequacy of GP models.

Table 2 Pavement deterioration prediction models for national highway roads

Distress Models	GP Equations
Cracking ($R^2 = 0.99$)	$CA_j = \left(\left(\left(\sqrt{\left(\left(\sqrt{e^{AGE}} \times BBD\right) \times BBD \times \sqrt{\sqrt{CA_i} \times BBD}\right) + \sqrt{BBD} \times \log\left(TRAFFIC \div AGE + CA_i^2\right) + \sqrt{CA_i}\right) \times \sqrt{e^{BBD}}}\right)\right) + CA_i\right)$ where CA_j corresponds to cracking predicted for next year, CA_i is cracking of previous year, AGE is pavement age (years) from last reconstruction/resurfacing work done till analysis period, BBD is Benkelman Beam Deflections and TRAFFIC corresponds to annual average daily traffic (AADT) in analysis year
Ravelling ($R^2 = 0.97$)	RAV_j = (((log(age) * tanh((tanh((RAV_i * (tanh(log(tanh(tanh(tanh(tanh(tanh((exp(tanh(RAV_i)) - tanh(pow(traffic, 2)))))))))) + RAV_i))) * RAV_i))) + RAV_i) + (log(age) * tanh((tanh((tanh(tanh((RAV_i * RAV_i))) * age)) * RAV_i)))) where RAV_j corresponds to ravelling predicted for next year, RAV_i is ravelling of previous year, AGE is pavement age (years) from last reconstruction/resurfacing work done till analysis period and TRAFFIC corresponds to annual average daily traffic (AADT) in analysis year
Rutting ($R^2 = 0.79$)	RUT_j = (RUT_i + tanh(tanh(tanh(tanh(pow((tanh(0.37790966) + exp(((tanh((exp(sqrt(sqrt(BBD))) * exp((BBD * age)))) – tanh((tanh(tanh(RUT_i))/exp(BBD)))) * (BBD - exp(sqrt(pow((BBD/traffic), 2))))))), 2)))))) where RUT_j corresponds to rutting predicted for next year, RUT_i is rutting of previous year, AGE is pavement age (years) from last reconstruction/resurfacing work done till analysis period, BBD is Benkelman Beam Deflections and TRAFFIC corresponds to annual average daily traffic (AADT) in analysis year
Roughness ($R^2 = 0.95$)	IRI_j = (IRI_i – (IRI_i + (IRI_i – (−3.174337e – 002 * ((1.41491628 – (IRI_i + (((IRI_i + IRI_i) - IRI_i) + sqrt(sqrt(IRI_i))))) + (IRI_i - (pow(−0.3, 2) * ((sqrt(traffic) + (age + pot hole)) * −3.31783e – 002))))))) * −3.273277e – 002)) where IRI_j corresponds to roughness predicted for next year, IRI_i is roughness of previous year, AGE is pavement age (years) from last reconstruction/resurfacing work done till analysis period and TRAFFIC corresponds to annual average daily traffic (AADT) in analysis year

Table 3 Post-regression results

S. No.	Distresses models	Linear regression model	R^2	RMSE
1	Cracking	$y = 0.7672x + 3.5187$	0.8582	0.347
2	Ravelling	$y = 0.8133x + 1.2516$	0.7558	0.148
3	Rutting	$y = 1.1442x - 0.5736$	0.9386	0.054
4	Roughness	$y = 0.9704x - 0.0091$	0.9722	0.024

x: represents observed distress values of given year
y: represents predicted distress values of same year

3.2 *Comparative Analysis*

A comparison has been done to analyse the capabilities of the two prediction models i.e., HDM-4 and GP in terms of the roughness parameter for the candidate road sections. Table 4 represents average predicted roughness values using both HDM-4 and GP models which have been compared with the observed field values for 18 identified road sections. It can be concluded that there is marginal variation in predicted

Table 4 Variability of HDM & GP models in terms of average roughness (IRI) prediction

Sections	HDM model (mm)	GP model (mm)	Observed IRI (mm)	HDM variability (%)	GP variability (%)
NE-1GJ01	2.47	2.42	2.40	1.47	1.42
NE-1GJ02	2.54	2.63	2.60	1.54	1.63
NE-1GJ03	2.55	2.63	2.60	1.55	1.63
NE-1GJ04	2.55	2.63	2.71	1.55	1.63
NH-14GJ05	2.78	2.85	2.80	1.78	1.85
NH-2UP01	2.59	2.64	2.60	1.59	1.64
NH-2UP02	2.87	2.96	3.10	1.87	1.96
NH-37AS01	2.55	2.64	2.90	1.55	1.64
NH-4KA01	3.38	3.36	3.40	2.38	2.36
NH-4KA02	3.79	3.91	3.80	2.79	2.91
NH-4KA03	3.81	4.11	4.10	2.81	3.11
NH-4MH01	3.08	3.17	3.10	2.08	2.17
NH-5AP01	2.56	2.63	2.70	1.56	1.63
NH-5AP02	2.87	2.95	2.90	1.87	1.95
NH-5AP03	3.03	3.06	3.00	2.03	2.06
NH-5AP04	3.03	3.06	3.00	2.03	2.06
NH-7MH02	2.56	2.64	2.80	1.56	1.64
NH-7MH03	2.47	2.53	2.60	1.47	1.53

roughness values mainly due to the difference in the architecture and working principles of these two software programs. Also it has been observed that the fitness of GP model increases with incorporation of more control parameters affecting pavement performance. Therefore, it shows that there is no such restriction on consideration of significant input parameters for analysis in GP.

Comparison of predicted roughness values obtained from HDM-4 and GP models with the observed field values for all road sections has been done to check the functional capabilities and adequacy of the models. Variations in roughness values predicted by HDM model for all road sections vary in the range of 1.47% to 2.8% whereas for GP model the variability range lies from 1.41% to 3.1%, which is very much acceptable.

4 Conclusions

It has been concluded that a scientific approach system should be adopted before planning road maintenance strategy for providing long lasting management and investment solutions to concerned road agencies. HDM tool used in the study has developed a distress prediction model for all the identified flexible pavement sections of NH network for an analysis period of 20 years. The maintenance strategies assigned to the road network has been traced out from graphical representations and results show that the alternatives intervene precisely as per the defined critical limits of control parameters i.e., roughness and total damaged area. Pavement performance in terms of distress patterns and rate of deterioration is a function of the physical characteristics of road such as structural condition, traffic loading, pavement age, climatic conditions etc. and previous maintenance work records. Genetic programming system has been tailored to generate distress prediction models which defines the quantitative assessment of future pavement performance. Statistical evaluation results prove that GP prediction models are highly accurate and applicable for National highway roads having flexible pavements and high traffic loading conditions. Validation of GP model has been done in terms of fitness measures such as R^2 and RMSE. R^2 obtained from graphs are of the values 0.85, 0.75, 0.93 and 0.97 for cracking, ravelling, rutting, Roughness models respectively which indicate high adequacy of these models.

References

Chandra S, Sekhar CR, Bharti AK, Kangadurai B (2012) Relationship between pavement roughness and distress parameters for Indian highways. J Transp Eng 139(5):467–475

Chopra T, Parida M, Kwatra N, Chopra P (2018) Development of pavement distress deterioration prediction models for urban road network using genetic programming. Advances in Civil Engineering

Deori S, Choudhary R, Tiwari D, Gangopadhyay S (2016) Calibration of HDM-4 models for Indian conditions of flexible pavement having modified bitumen in wearing course. Int J Pavement Eng 19(9):772–785

Jain K, Jain SS, Chauhan MS (2013) Selection of optimum maintenance and rehabilitation strategy for multilane highways. Int J Traffic Transp Eng 3(3):269–278

Jorge D, Ferreira A (2012) Road network pavement maintenance optimisation using the HDM-4 pavement performance prediction models. Int J Pavement Eng 13(1):39–51

Babovic V, Keijzer M (2000) Evolutionary algorithms approach to induction of differential equations. In: Proc. of Genetic and Evolutionary Computation Conference (GECCO2000), pp 251–258

Aggarwal S, Jain SS, Parida M (2004) Development of pavement management system for Indian national highway network. J Indian Road Cong 65(2):271–326

Koza JR (1994) Genetic programming as a means for programming computers by natural selection. Stat Comput 4:(2)

Sustainable Pavement Systems

SENSO JOINT—An Innovative Sensor System for a Sustainable Joint Design of Concrete Pavements

Ch. Recknagel, S. Spitzer, J. Hoppe, N. Wenzel, and S. Pirskawetz

Abstract Inacceptable capability and durability of joint sealing systems but also inadequate traffic performance (noise emission; overrolling comfort) up to traffic safety aspects reflect the still enormous demand for data-based description of concrete pavements performance under heavy loading conditions. Especially the deformation behavior of concrete pavement slabs in the joint region in consideration of new pavement construction types and improved concrete mixtures meanwhile established but also under the steeply rising traffic loads is not sufficiently explored. To create a data basis for advanced design rules, evaluation methods and product standards - and with it to improve quality, durability and finally sustainability of pavements - an innovative 3-D sensor system SENSO JOINT adapted to german roadworking requirements and suitable for heavy-duty operating conditions was developed. The contribution introduced describes the development of an extensive technical solution based on the analysis of decisive loads, interactions and boundary conditions. Based on calibration data, results of laboratory testing and finally field-testing on different concrete pavement construction types the outcome of a multi-level evaluation process shall introduce the potential of the new sensor system.

Keywords Concrete pavements · Sustainable joint sealing · Load quantification · Innovative measurement technology · Field validation · Performance design for joint systems

1 Introduction

Current studies predict a further increase of heavy-goods vehicles (HGV) of up to +39% over the next decade (BVWP 2030 2016) at simultaneously more difficult climatic conditions (Wellner et al. 2017). Highly stressable concrete pavements may be structural and technological answers to withstand higher axle loads, increased amounts of overrunning and higher climatic loads. Therefore, improvements in concrete technology and new design concepts were already transferred in innovative

Ch. Recknagel (✉) · S. Spitzer · J. Hoppe · N. Wenzel · S. Pirskawetz
Federal Institute for Materials Research and Testing, Unter den Eichen 87, 12205 Berlin, Germany
e-mail: Christoph.Recknagel@bam.de

C. Raab (ed.), *Proceedings of the 9th International Conference on Maintenance and Rehabilitation of Pavements—Mairepav9*, Lecture Notes in Civil Engineering 76,
https://doi.org/10.1007/978-3-030-48679-2_19

concrete pavements with optimized static and surface performance. Unfortunately, both the joint design as well as the materials improvement for sustainable joint sealing systems doesn't keep pace with these developments. Joint design rules as well as materials and joint system validation criteria (EN 14188 2004) are still determined by an empiric methodology reflecting the knowledge of the early 1970's. Increasing malfunction response regards to inadequate functionality and insufficient durability are the practical expression of the outdated joint sealing technology. The essential performance requirements on joint sealing systems under the changing use conditions (pavement construction; traffic increase; climate change) cannot ensured by the actual regulated methodology. Studies (Vater et al. 1997–2013) give indication for a necessarily method change towards a performance-related investigation and approval process to overcome insufficient performance. The identification of decisive climatic and mechanical loading parameters and their real impact on joints in heavy-loaded concrete pavements are of key importance to realize this approach (Fig. 1).

There is especially a gap of knowledge with regards to nature, size and amount of relevant deformations in the joint region of carriageway slabs induced by external impacts like climatic and/or traffic loads (Spitzer et al. 2019). The knowledge so far is mainly determined by sporadic data of climatic-induced deformations in longitudinal roll axis (x-axis) of elder concrete constructions. Measurement tools used were often ordinary folding rules or steel scales with very limited technical resolution and unsatisfying accuracy. After supersession by first digital extensometers the data basis remains still restricted in their technical message by discontinuity of the measurements. Correlation to the ambient climatic conditions and the temperature conditions inside the concrete pavement are only inadequately considered. First assumptions

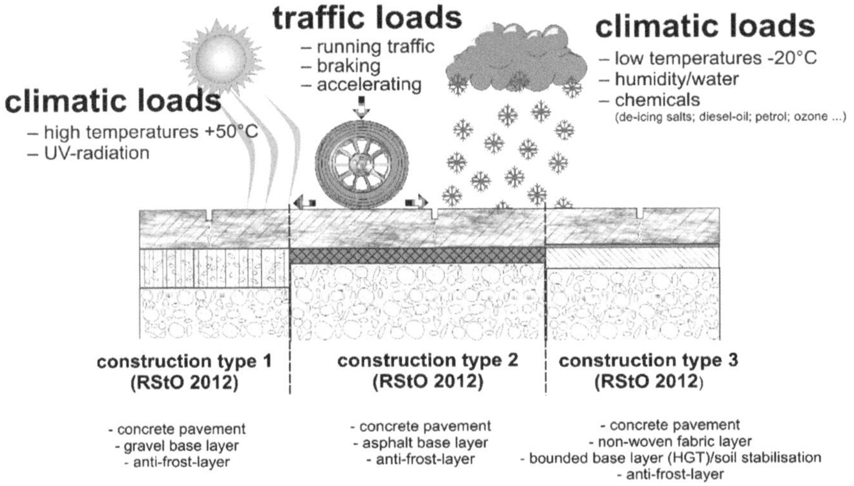

Fig. 1 Modern construction types for actual concrete pavements (e.g. exposed aggregate concrete) in Germany and decisive external impacts on pavement joints

Fig. 2 Application concept for an innovative sensor system

for an improved data acquisition especially with regards to traffic induced deformations were presented in (Hean and Partl 2004) but still restricted by reference points outside the wheel tracks. That's why there is still no significant knowledge about traffic induced deformations in the joint region available. Indications for deformations in other main spatial axes (vertical and/or orthogonal to traffic direction) are not investigated. To master these challenges within the SENSO JOINT project a new application concept for the innovative sensors system was developed (see Fig. 2).

The significance of defined decisive load parameters to realize a performance related investigation and evaluation methodology for joint sealing systems is motivation for our innovative SENSO JOINT sensor system. To explore deformation nature and size between adjacent slabs of modern concrete pavements a suitable sensor system should realize these main service features (Spitzer et al. 2019) for an improved approach try to reality:

- three-axial sensor system with axis-adapted measurement ranges and accurate resolution of
    - x-axis: ±5 mm with 0.01 mm
    - y-axis: ±2 mm with 0.01 mm
    - z-axis: ±2 mm with 0.01 mm
- continuous operation at adaptable measurement frequencies up to 1000 Hz and automated data management
- direct link to ambient and pavement temperature distribution
- safe operating conditions:
    - at temperatures: −20 °C ≤ T ≤ +50 °C
    - at humidity: 20% r.h. ≤ H ≤ 95% r.h.
    - water-resistant

  - pressure resistant up to compression stresses of 1.5 N/mm^2 (overrunning wheel loads)
  - serviceable under basically as well as acidic environment

- reusable miniaturized construction
- simple, robust and fast installation technology.

2 Design of the Sensor System

The three-dimensional relative movement of two carriageway slabs and therefore the deformation of the joint between these slabs is measured by a sensor system which consists of three single sensors. These sensors are protected by a housing that is embedded in one of the slabs near the joint under monitoring.

The measurement principle is shown in Fig. 3a. One sensor, a LVDT, measures changes in distance of the slabs in x-axis. A spherical joint connects the core of the LVDT to a stable rod that is fixed in the opposite concrete slab. The LVDT is gimbal-mounted, allowing zero-backlash rotations of the LVDT around the y- and z-axis. Two rotary encoders measure the inclination of the LVDT. The result is a representation of the relative movement of the slabs in spherical coordinates that is transferred in cartesian coordinates of x-axis, y-axis and z-axis acc. to Fig. 2. All sensors measure absolute positions and therefore the relative position of the adjacent slabs can be calculated anytime if the initial position in known, even if the measurement was interrupted meanwhile.

The sealed, cylindrical housing of the sensor system has an outer diameter of 100 mm and a height of 100 mm. It is made of stainless steel and constructed to bear the loads of directly overrunning wheels of heavy trucks. The system is designed to be mounted in boreholes near the joint. After sensor installation the borehole is grouted by a fast concrete repair system. Figure 3b–d shows the installation of a 3D-sensor system.

A monitoring station includes the 3D-Sensors, sensors for temperature distribution inside the concrete slabs, climate sensors, a data acquisition system and other supporting technical equipment. All data are recorded with a sampling frequency of 2 kHz and sent to the operator via mobile communications network.

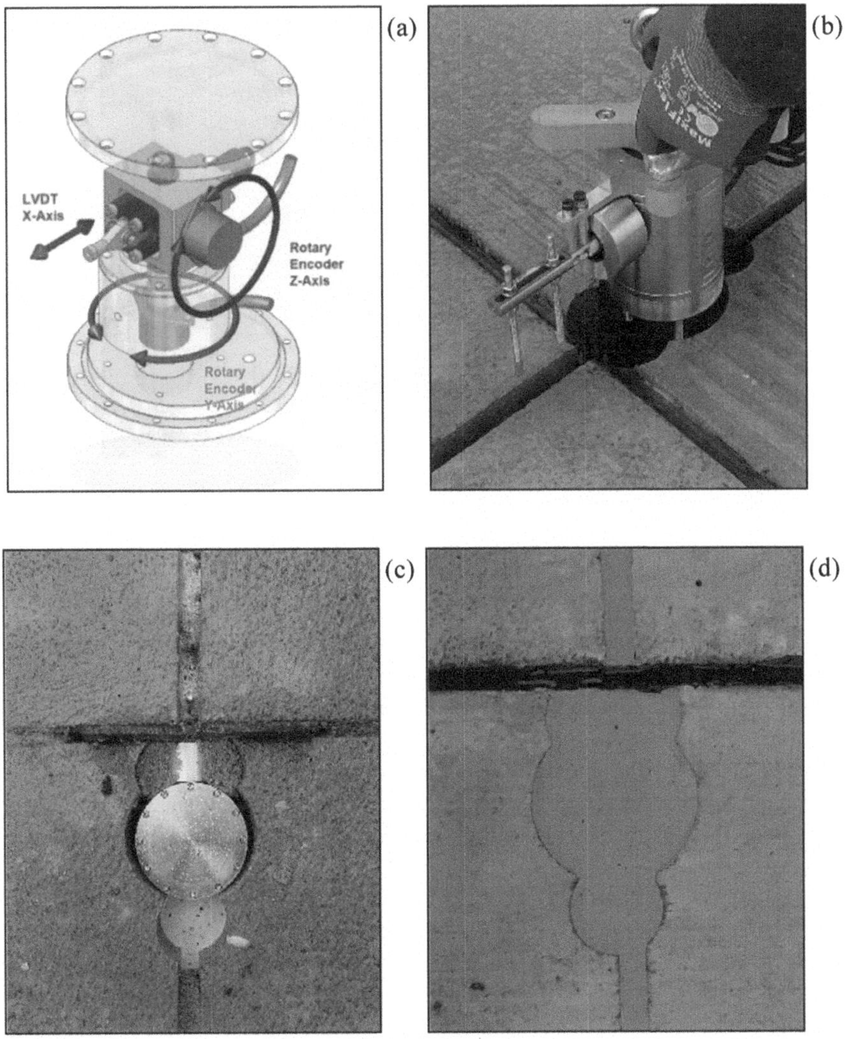

Fig. 3 **a** Measurement principle, **b** Sensor system installation, **c** embedded Sensor System before **d** after grouting

3 Validation of the Sensor System

3.1 Laboratory Validation

Prior to complex field tests the sensitivity and accuracy of the sensor system was tested under laboratory conditions. After an individual calibration at room temperature (validation step 1), all LVDT's and rotary sensors were tested in a second

Fig. 4 Setup for validation step 4

validation step separately in a temperature range of −20 °C to 50 °C to specify the temperature dependencies. In a third step, every single sensor was calibrated in the assembled sensor system. During this calibration it was ensured that only the sensor under test was moved whereas the other two sensors were on their zero positions. The sensor system was mounted on a high precision 3D calibration device and moved to predefined positions in the 3D measurement range. The positions in spherical coordinates measured by the sensor system were converted to cartesian coordinates and compared to the real position indicated by the calibration device. This test confirmed the required three-dimensional accuracy and sensitivity under static conditions in a temperature range of −20 °C to 50 °C.

For a dynamic validation six 3D-sensor systems were finally installed on a two-axial servo hydraulic testing system (validation step 4—see Fig. 4). At temperatures ranging from −20 °C to 50 °C the sensors were slowly displaced to predefined positions. At these positions the sensors were additionally deflected dynamically with a sine function of 5 Hz and an amplitude of ±0.1 mm. The test confirmed that all sensor systems fulfill the requirements on functionality, accuracy and sensitivity under realistic conditions as far as possible to simulate in laboratory scale.

3.2 Validation in Situ

In 2018 the sensor systems were tested on the "Demonstration, Investigation and Reference Area" (duraBASt) of the German Federal Highway Research Institute (BASt) near Cologne. The test included installation of the sensors and of the data acquisition system under real conditions as well as the response to climatic and traffic load. Six sensor systems were installed on joints between concrete slabs of a

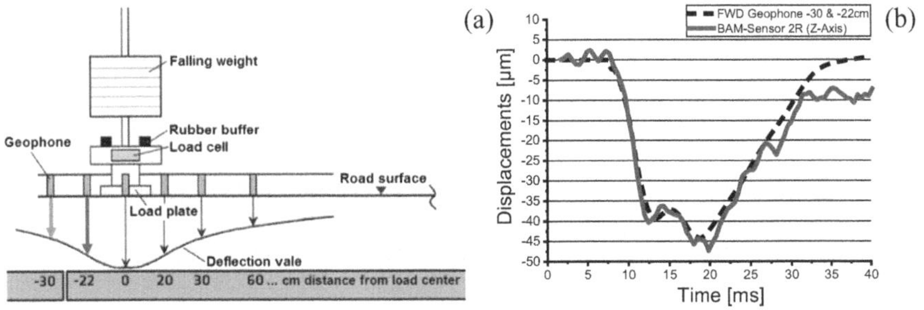

Fig. 5 Setup of the FWD-test in the joint region between two slabs **a** and the response of the BAM sensor system in comparison to the deflection measured by the FWD system **b**

test track, directly in the wheel track as well as in the middle between the tracks as schematically described in Fig. 2. The slabs have a dimension of 5 m × 4.25 m × 0.23 m and are bedded on an asphalt base layer acc. to construction type 2 (Fig. 1).

The dynamic response of the Sensor systems was first practical validated by the results of Falling Weight Deflectometer (FWD) test near the joint region (Fig. 5a). The FWD-test is a standardized method for determining soil stiffness based on dynamic deflection measurements (Březina et al. 2017). Figure 5b shows the relative movement in vertical direction (Z–Axis) of two adjoining slabs during the test. The blue curve, measured by the new BAM-Sensor system, is in good correlation with the black curve which is the output of the FWD test (difference measurement signal geophone at position −30 resp. −22 cm from load impact point FWD).

Secondly the sensor response to traffic load was tested with a 2 axis truck with a total weight of around 20 t. The truck over-rolled the embedded sensors several times under controlled uniform conditions with speeds of 2, 30 and 70 km/h. As an example, Fig. 6 shows the deformation of a joint measured by BAM sensor "2 R" in the wheel track. The truck rolled over the joint with a speed of 2 km/h. The crossing of the front axle with a load of approximately 7.3 t causes an opening of the joint (X–Axis) of about 17 µm. In vertical direction (Z–Axis) the sensor first detects a deformation of about 17 µm, indicating a downward movement of the slab in front of the joint. It is followed by a sudden deformation of −25 µm when the axle has crossed the joint, indication a downward movement of the slab behind the joint. The crossing of the rear axle with a load of 12.1 t caused similar but higher deformations. The ratio of the deformation amplitudes caused by front and rear axles is about 1.7 and corresponds to the load ratio of the axles of 7.3:12.1 $\approx$ 1.7. No horizontal deformation was detected orthogonal to the driving direction (Y–Axis). Similar joint deformations were measured at higher speed. Peak amplitudes and time between the peaks decrease with increasing speed of the truck. Before and after direct truck over-rolling (see Fig. 6 course of BAM-Sensor 2R (X-Axis)), the sensor system also detects relative slab movements indicating the interaction of the adjacent slabs. With it the mechanical functionality of the pavement construction can be characterized. For this pavement at duraBASt (not under regular traffic) no horizontal deformation

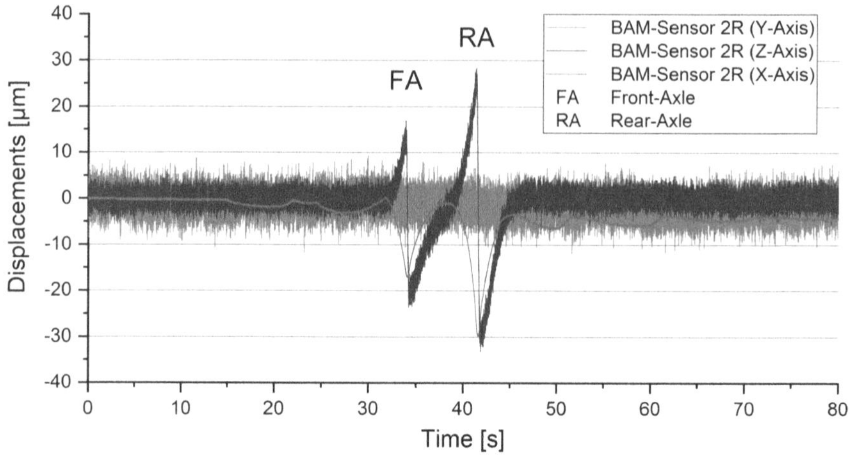

Fig. 6 Deformation of adjacent concrete slabs while being over-rolled

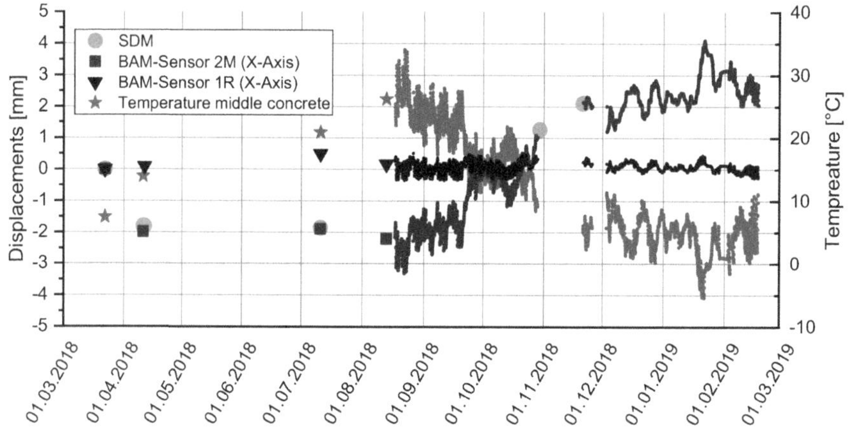

Fig. 7 Joint movements over a one year period with validating external measuring system (SDM)

in joint longitudinal axis (Y-axis) orthogonal to the driving direction was detected at uniform and straight over-rolling conditions (see Fig. 6).

Thirdly the long-term field performance of the sensors was tested over a period of 12 months. During the first six month the local technical infrastructure was available only temporary. The absolute opening of the joint (X–Axis) was measured additionally five times in situ with a digital extensometer (SDM). The comparison of these manual measurements with the results of the sensor system shows a good correlation. Moreover, it was validated that shut down and restart of the system resulting from interruption of technical logistics doesn't influence the measurement results.

The joint deformations measured on joint 2 (BAM sensor 2 M) show a strong correlation to temperature. Starting from an initial joint width, within the monitoring period March 2018 the joint was compressed by −3.3 mm in late summer in 2018 at a temperature of 34 °C inside the concrete pavement and expanded by +4.2 mm at a temperature of −6 °C in the following winter. In contrast, nearly no deformation was detected on joint 1 (BAM sensor 1 R). FWD-results and visual observation shows that this joint is not working correctly, and these two slabs are still connected. Therefore, the pavement structure does not work as designed (Fig. 7).

4 Conclusions

An innovative sensor system to measure seasonal as well as traffic-induced multiaxial deformations also directly in heavy-loaded pavements was developed and approved in various validation steps. The laboratory validation shows high accuracy at all use scenario between −20 and +50 °C. The in situ validation approves the practicability of the sensor system design, the developed field installation technology and the sensor system performance under practical conditions. After one field trial the sensor systems are easily removable and reusable for other field trials.

With the innovative sensor system, it is for the first time possible to measure three-dimensional movements of adjacent slabs in real-time. First measurements encourage the need for an improved exploration of joint as well as the pavement performance in the joint region by innovative sensor solution. The sensor system opens new opportunities for a scientifically exploration of the real performance of pavement construction types and will support further improvements in evaluation methodology as well as design of joint sealing systems, improved joint performance and finally for more sustainability in concrete pavement design.

Acknowledgements This presentation is based on the research project carried out at the request of the Federal Ministry of Transport and Digital Infrastructure, represented by the Federal Highway Research Institute (BASt), under research project No. FE 08.0244/2015/FRB "Improvement of joint design—Determination of real loads of joint sealing systems in heavy-loaded concrete pavements". The authors wish to thank the BASt for the interesting task and financial support.

References

BVWP 2030 (2016). Bundesverkehrswegeplan. Federal Ministry of Transport and Digital Infrastructure. Referat G 12; 10115 Berlin. www.BVWP2030.de

Wellner F, Kayser S, Clauß M (2017) Projizierter Klimawandel und Dimensionierung von Straßenbefestigungen. Forschung Straßenbau und Straßenverkehrstechnik, Heft 1128; 120 p. Fachverlag NW im Carl Schünemann Verlag GmbH. ISBN 978-3-95606-322-0

EN 14188 2004. part 1 to 3. Joint fillers and sealants. Part 1: Specifications for hot applied sealants. Part 2: Specifications for cold applied sealants. Part 3: Specifications for preformed joint seals. www.beuth.de

Vater E-J, Recknagel Ch, Pirskawetz St, Huth Ch (1997–2013) Alterungsverhalten von Fugenfüllsystemen in Verkehrsflächen, Bearbeitungsteile 1 bis 5. Bundesanstalt für Straßenbau (BASt); Forschungsvorhaben FE 08.146 G95G. Bundesanstalt für Straßenbau (BASt); Forschungsvorhaben FE 08.169/2001

Spitzer St, Wenzel N, Pirskawetz St, Hoppe J, Recknagel Ch (2019) Optimierung von Fugenkonstruktionen—Ermittlung realer Beanspruchungen des Fugensystems. Bundesanstalt für Straßenbau (BASt). Forschungsvorhaben FE 08.0244/20015/FRB

Hean S, Partl MN (2004) Long-term behavior of polymer bitumen joint sealants on a trial road section. In: Proceedings of the 3rd eurasphalt and eurobitume congress, May 2004 in Vienna (Austria), pp 546–559

Březina I, Grošek J, Janků M (2017) Measurement of deflections and determination of jointed plain concrete pavements stiffness by falling weight Deflectometer. Procedia Eng 190:162–169 Elesevier—Structural and Physical Aspects of Construction Engineering at www.Sciencedirect.com, 2017 Elsevier Ltd. https://doi.org/10.1016/j.proeng.2017.05.322

Numerical Evaluation of Crushing Resistance of Unbound Road Material

Erik Olsson, Denis Jelagin, and Manfred N. Partl

Abstract The service life of a road is to great extent controlled by the performance of the unbound layers. Assessing the constitutive behavior of these layers is thus imperative for a sustainable pavement design. Adequate description and measurement of unbound materials resistance to aggregate crushing is an issue, in terms of the measured response coupled to intrinsic properties of the aggregates and unbound materials gradation. In this study, a new Discrete Element Method (DEM)—based model is developed to investigate aggregate damage in unbound road materials. In order to get better insight into micro-mechanics of aggregate crushing, the developed model incorporates granular mechanics-based particle contact and damage laws. By numerical analysis with the DEM, several unbound granular materials have been examined investigating the effect of the materials gradation and aggregates toughness properties on their performance in crushing tests. The capability of the model to capture the effect of unbound material properties on its crushing performance, is evaluated based on comparison with experimental findings.

Keywords Unbound road material · Particle behavior · Crushing resistance · Discrete element modelling

1 Introduction

Aggregate degradation in unbound road layers is known to affect significantly functional performance and service life of roads. In particular, as discussed in e.g. (Saeed et al. 2001) aggregate degradation in unbound layers may contribute to a higher rate

E. Olsson (✉)
Department of Solid Mechanics, KTH Royal Institute of Technology, Teknikringen 8D, 100 44 Stockholm, Sweden
e-mail: erolsson@kth.se

D. Jelagin (✉) · M. N. Partl
Department of Civil and Architectural Engineering, KTH Royal Institute of Technology, Brinellvägen 23, 100 44 Stockholm, Sweden
e-mail: jelagin@kth.se

C. Raab (ed.), *Proceedings of the 9th International Conference on Maintenance and Rehabilitation of Pavements—Mairepav9*, Lecture Notes in Civil Engineering 76,
https://doi.org/10.1007/978-3-030-48679-2_20

of rutting, fatigue cracking and surface corrugation in flexible pavements. Accordingly, ensuring adequate aggregate resistance to degradation through crushing and abrasion is crucial for good pavement performance.

In order to address this issue, a number of empirical tests to characterize aggregates toughness have been proposed, such as the Los Angeles abrasion test. Summaries of the existing test methods for aggregate toughness/abrasion characterization may be found in the literature (Bjarnason et al. 2000; Saeed et al. 2001). While the existing test methods provide a relatively simple way for determining empirical indices capturing aggregates crushing and/or abrasion resistance, their relationship with the aggregates performance in road structures is not necessarily strong, cf. e.g. (Wu et al. 1998). This study aims to contribute to establishing a stronger link between fundamental mechanical properties of aggregates and their crushing performance in unbound road layers.

In this study, a new Discrete Element Method (DEM) based model is developed to investigate aggregate damage in aggregate crushing tests. The developed model builds up on a DEM modelling approach developed by the authors (Celma Cervera et al. 2017; Olsson et al. 2019a, b) and incorporating particle contact and damage laws based on granular mechanics. It allows thus to capture unbound materials crushing performance at arbitrary loading and geometries. Accordingly, the proposed DEM model may provide a better quantitative insight into aggregate breakage in unbound layers in roads and railways. Presently, models capability to quantitatively capture the effects of aggregates gradation and fracture toughness on unbound materials performance in aggregate crushing tests is evaluated, based on the experimental findings reported in Bjarnason et al. (2000). Feasibility of using DEM modelling for establishing a quantitative link between aggregate performance in crushing tests and aggregate degradation in unbound layers is illustrated and discussed.

2 Problem Formulation and Methodology

In unbound road layers, external forces applied to the layer are distributed in aggregate skeleton through the stone-to-stone contact points. Aggregates in pavement structures have to withstand a large number of load cycles during construction as well as under repeated traffic loads. The magnitude of pressure applied to unbound layers during their service life depends on the pavement structure and materials used as well as traffic type and speed and may reach approximately 1.6 MPa for the cases of heavy construction traffic travelling directly on unbound layer, cf. (Bjarnason et al. 2000). Furthermore, even higher pressure pulses may be applied to unbound material during compaction (Holtz and Kovacs 1981). Aggregate propensity to breakage in unbound layers will, in addition to aggregate properties, be controlled by the magnitude and distribution of contact forces at stone-to-stone contact points. Accordingly, in order to obtain the best computational insight into the aggregate crushing process, models should explicitly account for stone-to-stone interactions, particle rearrangements and discuss stone breakage as a function of fundamental materials properties. The

Discrete Element Method (DEM), developed by Cundall and Strack (1979) is the natural tool for investigating such conditions and hence the main goal of the present paper is to develop such a modelling framework based on DEM.

The DEM model used in this work is an extension of the granular mechanics based DEM framework developed for studying deformation and damage of both unbound and asphalt materials. The framework is implemented in an in-house C++ code and uses a standard DEM formulation with explicit time integration. More details of the framework and implementation can be found in Celma Cervera et al. (2017); Olsson et al. (2019a, b); Olsson and Jelagin (2018).

In DEM, each physical particle (stone) is represented by one or more DEM particles and the movements of the particles are determined by explicit integration of Newton's second law given forces acting on the particles. In the present case, with unbound stone materials, the contributing forces are gravity and contact forces between the stones. Hence, in order to get reliable predictions from DEM models the contact forces need to be accurately calculated.

The contact behavior of a particular material used in road construction, Arlanda granite, was investigated experimentally using instrumented indentation testing in Celma Cervera et al. (2017). A spherical indenter was used having a radius of 6.25 mm. In that work, two damage mechanisms were identified, cyclic damage leading to energy dissipation and macroscopic cracking leading to a possible fragmentation of the stone. The latter damage mechanism will be used as a damage criterion and will be discussed later in this section. To be able to transfer the results from the indentation testing to contact between two spherical stone particles, the effective contact radius of the contact pair, shown in Fig. 1 is defined according to Eq. (1) as

$$\frac{1}{R_0} = \frac{1}{R_1} + \frac{1}{R_2} \tag{1}$$

Based on the results in (Celma Cervera et al. 2017), a model for the normal force, F_N, as function of the indentation depth, h, was derived by Olsson et al. (2019a) and the resulting relationship is sketched in Fig. 2(a). It consists of three parts. Ini-

Fig. 1 Sketch of the two particle contact problem with quantities needed to define the force-displacement relationship

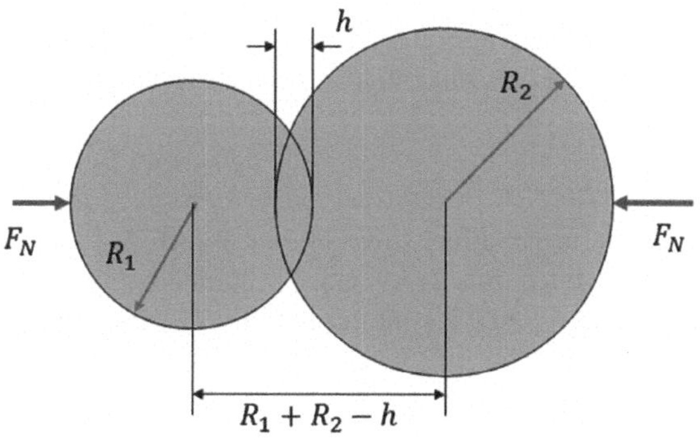

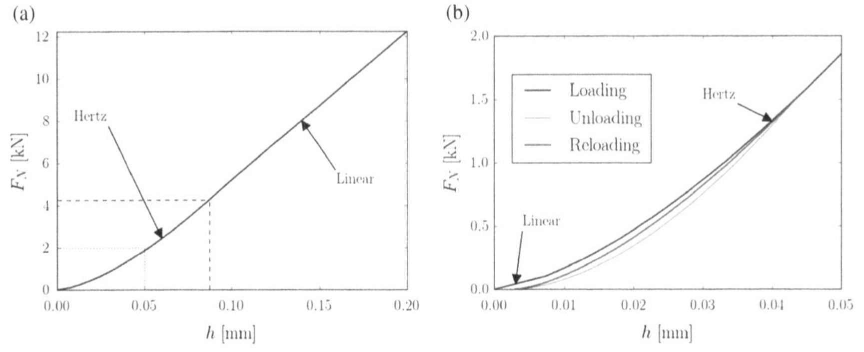

Fig. 2 **a** The force displacement relationship at loading. **b** A magnification of the force-displacement relation in **b** showing the cyclic behavior

tially, the force—indentation depth relation is linear accounting for crushing of small asperities. In the indentation experiments where $R_0 = 6.25$ mm, this region occurs for forces below 100 N and is scaled in the simulations with R_0^2. This region is visible in the magnification in Fig. 2(a). At higher forces, the Hertz (1882) contact theory is used assuming that the stones behave elastically. Due to shear driven damage below the contact, deviation from the elastic Hertzian behavior is seen at large indentation depths. This is modelled by an additional linear region that in the indentation experiments starting at $h = 0.087$ mm which is scaled with R_0 in the simulations. Unloading of the contact is modelled to give a hysteresis leading to cyclic energy dissipation by having different stiffnesses at loading and unloading. This is seen in Fig. 2(b). The benefit with this rather complex force-displacement relationship is that it results in energy dissipation without introducing any unphysical damping. This model was further confirmed using FE simulations in Olsson et al. (2019a).

Another important mechanism for energy dissipation and for restricting the movements of the stones is friction. Presently, the cyclic stick-slip model in Thornton et al. (2011) is used. At full slip, a coefficient of friction of 0.7 is used both for stone-to-stone contacts and for stone-surface contacts. This rather high value accounts for surface irregularities.

The main focus of this paper is damage of the aggregates in the road material and thus a suitable criterion for damage has to be used given the contact force between the aggregates. For each contact a fracture force can be computed according to Eq. (2)

$$F_C = \sigma_F R_0^2 \tag{2}$$

where σ_F is the failure stress of the contact. As there is a larger probability of finding a critical defect in a large loaded volume than in a small volume, a Weibull weakest-link model is used for the probability distribution of the failure stress according to Eq. (3).

$$F = 1 - \exp\left[-\left(\frac{\sigma_F}{\sigma_W}\right)^m \frac{R_0^3}{V_{Ref}}\right] \tag{3}$$

where σ_W is the Weibull stress controlling the strength of the material and m is a parameter controlling the scatter. For dimensional consistency, a reference volume V_{ref} is introduced, here defined being the radius of the indenter used in Celma Cervera et al. (2017) cubed, $V_{ref} = (6.25\,\text{mm})^3$. Using this model, a contact with a large effective contact radius R_0 will have a lower fracture stress than a contact with small effective contact radius. If the contact force exceeds the fracture force F_c, it is registered that the aggregate has fractured at the position of the contact and used for post-processing of the number of cracks.

3 Computational Study

The focus of this paper is a numerical investigation of a modified version of the proctor test for studying aggregate damage. The experimental basis for the investigation is found in Bjarnason et al. (2000). In their experiments, a cylinder having a radius of 2 inches (50.8 mm) and height of 116.43 mm is used. The cylinder is filled with five layers of stone material and a cylindrical hammer with radius 50 mm is released 25 times on each layer. The hammer has a weight of 4.54 kg and is released from a height 457 mm above the stone layer. During testing, the cylinder containing the stones is rotated so that the hammer impacts the stone surface evenly.

The gradations before and after the tests are determined and an index, the modified Bg index, is calculated by using Eq. (4)

$$\text{Bg} = \sum_{i=1}^{sieves-1} \langle \Delta S_a^i - \Delta S_b^i \rangle \tag{4}$$

where ΔS_a^i is the difference in volume of material passing sieve size i and sieve size $i + 1$ after the test and ΔS_b^i is the corresponding quantity after the test. These quantities are visualized in Fig. 3(a) and <> is the Macaulay brackets resulting in that only positive differences are included in the summation. The larger the Bg value, the more fragmentation of the stones during testing. This method was used to investigate 4 materials, prepared with 4 different aggregate types from Icelandic quarries, in Bjarnason et al. (2000) with results presented in Fig. 3(b).

Two different gradations have been the focus presently, presented in Fig. 4(a). One contains only stones having sizes between 8 to 16 mm and one follows the theoretical Fuller curve given by Eq. (4)

$$F = \left(\frac{d}{d_{max}}\right)^{0.5} \tag{5}$$

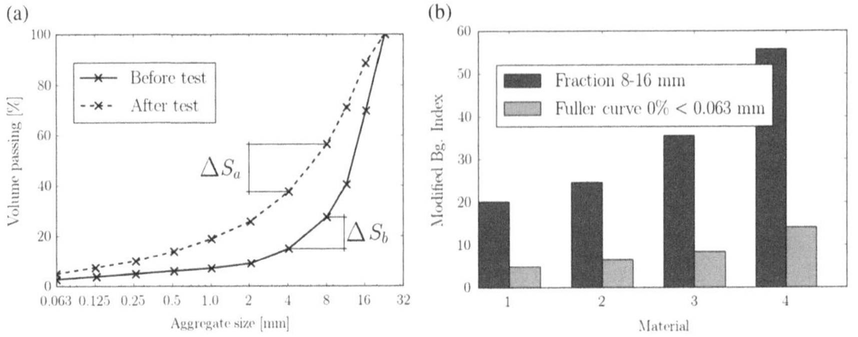

Fig. 3 **a** Definition of quantities used for the Bg. index evaluation. **b** Bg. index values obtained experimentally for 4 different materials

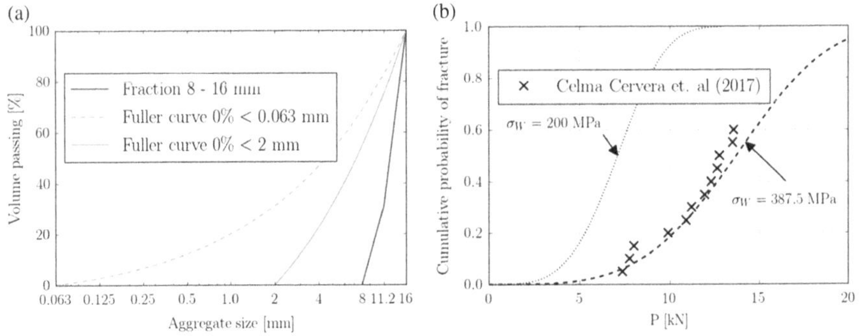

Fig. 4 **a** Gradations used in the experiments and in the simulations. **b** A Weibull model fitted to the experimental results in Celma Cervera et al. (2017) together with predictions of a weaker material

where F is the cumulative mass distribution and d_{max} the maximum aggregate size. A cut-off was used in experiments excluding fine material smaller than 0.063 mm. To have a feasible number of particles in the DEM simulations, the cut-off is there set to 2 mm, also included in Fig. 4(a). The number of particles in the simulations for all five layers is 583 particles for the 8–16 mm gradation and 18,304 particles for the Fuller curve.

The experimental results in Celma Cervera et al. (2017), presented in Fig. 4(b) are used for determining parameters for the weakest-link fracture model in Eq. (2). A Weibull distribution has been fitted to the experimental outcome giving the parameters $\sigma_W = 386.5$ MPa and $k = 3.87$ which matches the experiments well. To investigate the effect of material properties, a weaker material is defined having the same exponent k but with a Weibull stress of $\sigma_W = 200$ MPa. Its predicted behavior at the fracture tests in Celma Cervera et al. (2017) is shown as a dotted line in Fig. 4(b).

4 Results

In Fig. 5(a) the computationally predicted number of fractured particles is shown as a function of number punch impacts; vertical dashed lines in the Fig. 5(a) mark moments when the next layer of particles is added to the specimen. Results in Fig. 5(a) are presented for the two gradations and two aggregate qualities, and, as may be seen, in all 4 cases examined, particle crushing curves have approximately the same form, with number of fractured particles increasing more or less linearly with blows. As may also be noticed, tests performed on 8–16 mm particles result in approximately 3 times higher number of fractured particles as compared to the corresponding measurements for the gradation following the Fuller curve. Furthermore, reducing Weibull stress 386.5 MPa to 200 MPa increases particle crushing by approximately a factor of 2. Comparing the results with the ones reported in Fig. 3(b), it may be concluded that the developed DEM model, captures at least qualitatively, the effect of unbound material gradation on aggregate damage. Effect of aggregate toughness on the material performance in crushing test may further be effectively captured by adjusting material's Weibull stress. Distribution of fractured particles at the end of the test, i.e. after 125 blows, is shown in Fig. 5(b) for the case of 8–16 mm gradation.

In order to examine the effect of test parameters on materials performance further, two additional simulations of crushing tests for 8–16 mm gradation are performed as follows: (i) after the specimen is loaded according to the test procedure described above, additional 75 blows are applied to the specimen surface; (ii) all the unbound material is filled into the container as a single layer and complete specimen is loaded with 125 blows. Number of fractured particles as function of blows is shown in Fig. 6(a,b) for the simulation cases (i) and (ii) respectively. As may be seen in Fig. 6(a), particle fracture rate decreases somewhat at the end of the simulations, in particular for the case of weak aggregates ($\sigma_W = 200$ MPa). Single layer specimen case reported in Fig. 6(b) results in somewhat smaller number of particle fractures as compared to the results presented in Fig. 5(a). It is worth noting, however, that for

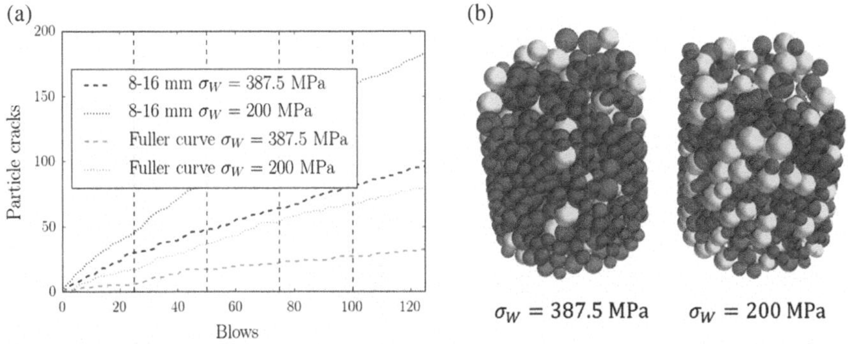

Fig. 5 **a** Number of fractured particles as a function of number of blows; **b** distribution of fractured particles (red) at the end of the test, 8–16 mm gradation

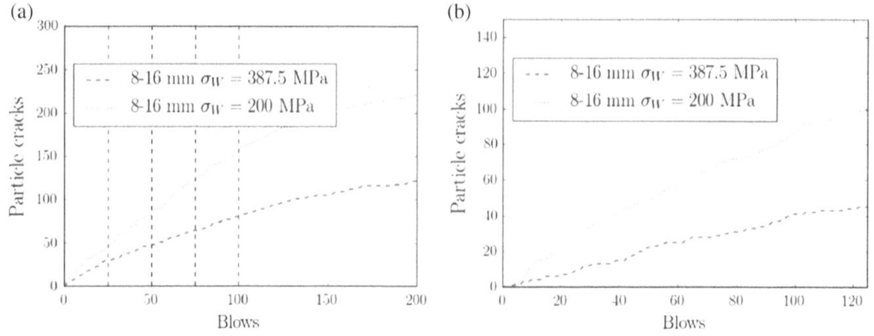

Fig. 6 Number of fractured particles as a function of number of blows, 8–16 mm gradation, **a** prolonged test; **b** single layer specimen

both cases in Fig. 6 weaker aggregates accumulate approximately twice the number of fractured particles as compared to the strong ones. The effect of aggregate properties on material simulated crushing performance is thus more or less identical for the 3 loading scenarios, examined in Figs. 5(a) and 6(a,b).

To compare the Bg index value from the experimental data with the simulations it is assumed that each crack in a stone divides it into two pieces of equal volume. A modified gradation is then calculated by assuming that these pieces are themselves spherical with a diameter used as size characteristics in the gradation curves. The results of these calculations are shown in Fig. 7(a) where it is clearly seen that the 8-16 mm fraction leads to larger change of gradation after the testing than for the Fuller curve gradation. For the case of Fuller curve gradation, it may also be noted in Fig. 7(a), that, large aggregates of 8–16 mm in size are primarily damaged in the crushing test.

From the calculated gradations it is possible to calculate the modified Bg index in the same way as in the experimental results presented in Fig. 3(b); the result of this calculation is shown in Fig. 7(b). As may be observed, the calculated values

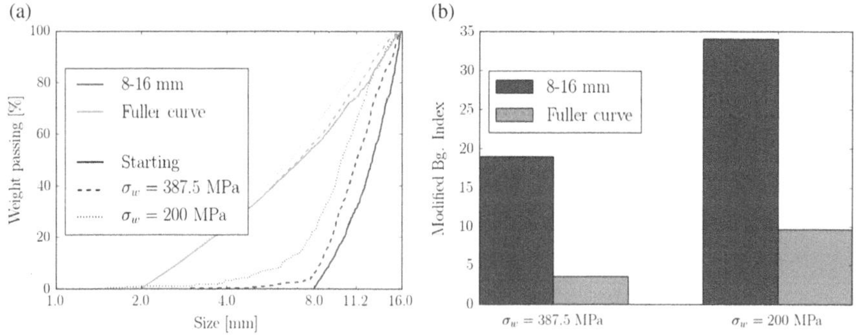

Fig. 7 **a** Simulated gradations before and after the testing. **b** Calculated Bg. index values from the simulations

are in the same range as the experimental ones. Furthermore, computational results obtained for $\sigma_W = 386.5$ MPa and $\sigma_W = 200$ MPa agree well quantitatively with the experimental ones, in Figs. 3(b) for the materials 1 and 4 correspondingly.

Full quantitative agreement is not expected however, as the failure law parameters used in DEM simulations are obtained from experiments performed by Celma Cervera et al. (2017), performed on a different aggregate type. Presently, emphasis lies on the evaluation of the ability of the model to capture relative effects of gradation and aggregate properties on the unbound material performance in crushing test. In this context, it seems to be instructive to compare the ratios between the Bg values reported in Figs. 3(b) and 7(b), for the different gradations and material types. Based on this comparison it may be concluded that both the effect of gradation and of aggregate properties are captured well.

5 Conclusions

The new DEM-based model is proposed to investigate unbound material performance with respect to aggregate crushing. The model is used to evaluate performance of unbound materials in aggregate crushing test at two different gradations and two different aggregate qualities. Based on comparison with the experimental findings, it is shown that the developed model captures successfully the effect of unbound materials gradation and of aggregate fracture toughness on materials crushing performance. It is also found numerically that, for both gradation examined, primarily the largest aggregates get fractured during the test. Based on a brief parametric study, the models capability for examining the influence of testing parameters on the specimens crushing is illustrated.

The model, developed in this study, relies on granular mechanics-based contact and damage laws for aggregates and allows thus to quantify the effect of fundamental properties of unbound materials on their performance with respect to aggregate crushing. Accordingly, the model can be used to evaluate unbound materials crushing performance at arbitrary geometries and loading scenarios, such as e.g. the ones arising in unbound layers in road and railway structures.

Acknowledgements The authors are indebted to Prof. Sigurdur Erlingsson for providing data and details from the crushing experiments.

References

Bjarnason G, Petursson P, Erlingsson S (2000). Aggregates resistance to fragmentation, weathering and abrasion, In: Unbound Aggregates in Road Construction (UNBAR)

Celma Cervera C, Jelagin D, Partl MN, Larsson P-L (2017) Contact-induced deformation and damage of rocks used in pavement materials. Mater Des 133:255–265

Cundall PA, Strack ODL (1979) A discrete numerical model for granular assemblies. Géotechnique 29:47–65
Hertz H (1882) Über die Berührung fester elastischer Körper. J. für die reine und Angew. Math 92:156–171
Holtz RD, Kovacs WD (1981) An introduction to geotechnical engineering. Prentice-Hall, Upper Saddle River
Olsson E, Jelagin D (2018) A contact model for the normal force between viscoelastic particles in discrete element simulations. Powder Technol 342:985–991
Olsson E, Jelagin D, Forquin PA (2019a) Computational framework for analysis of contact-induced damage in brittle rocks. Int J Solids Struct
Olsson E, Jelagin D, Partl MN (2019b) New discrete element framework for modelling asphalt compaction. Road Mater Pavement Des 1–13
Saeed A, Hall JW, Barke, WR (2001) Performance-related tests of aggregates for use in unbound pavement layers, NCHRP Report. National Academy Press, Washington, D.C
Thornton C, Cummins SJ, Cleary PW (2011) An investigation of the comparative behaviour of alternative contact force models during elastic collisions. Powder Technol 210:189–197
Wu Y, Parker Jr, F, Kandhal K (1998) Aggregate Toughness/Abrasion Resistance and Durability/Soundness Tests Related to Asphalt Concrete Performance in Pavements

Developing and Modeling a Piezoelectric Energy Harvester (PEH) for Highway Pavements

Mohamadreza Khalili, Sara Ahmed, and A. T. Papagiannakis

Abstract This paper describes a piezoelectric energy harvester (PEH) that uses PZT or lead zirconate titanate (Pb[Zr(x)Ti(1 − x)]O3) elements to convert mechanical energy from moving vehicles to electricity. The PEH consists of a rectangular metal housing supported by four PZT stacks located at its corners. It is sized to capture the right wheel path of the weaving traffic. Each PZT stack consists of six PZT cylindrical elements connected in parallel. Extensive laboratory tests were performed to properly characterize and model the stacks. An electromechanical model of the stacks was developed to translate its mechanical properties to electrical properties. This model was implemented into Matlab/Simulink in order to optimize the power harvesting circuitry. The output power depends on the applied stress and the loading frequency. The power output of one of the PZT stacks is in the order of 10 mW and 1.4 W for a car axle and a truck axle, respectively. In addition to generating power, the PZT stacks can be used as axle load sensors by utilizing the output voltage to back calculate the stress using the developed model.

Keywords Piezoelectric · Energy harvesting · Pavement · Electromechanical

1 Introduction

Energy harvesting consists of tapping into ambient energy and convert it into electrical energy suitable to power electrical devices. Ambient energy can be in various forms including mechanical, thermal and solar. Depending on the energy source, harvester designs involve piezoelectric (Erturk and Inman 2011), thermoelectric (Tahami et al. 2019), electromagnetic (Gholikhani et al. 2019) or photoelectric (Guilar et al. 2009) technologies. Each technology has its pros and cons. Solar cells scavenge solar

M. Khalili (✉) · A. T. Papagiannakis
Civil and Environmental Engineering Department, University of Texas at San Antonio, 1 UTSA Circle, San Antonio, TX 78249, USA
e-mail: reza.khalili@my.utsa.edu

S. Ahmed
Electrical and Computer Engineering, University of Texas at San Antonio, 1 UTSA Circle, San Antonio, TX 78249, USA

C. Raab (ed.), *Proceedings of the 9th International Conference on Maintenance and Rehabilitation of Pavements—Mairepav9*, Lecture Notes in Civil Engineering 76,
https://doi.org/10.1007/978-3-030-48679-2_21

energy and have high efficiency. They are easy to use and modular with no emissions, noise or moving parts. Solar cells are ideal for certain applications, however their cost and dependence on light and available space are their limitations. Thermoelectric devices that harvest thermal energy need high temperature gradients (>10 °C) and have efficiency limitations (Cook-Chennault et al. 2008). Piezoelectric harvesters are currently the simplest method to harvest power from structural vibrations.

Piezoelectric ceramics are easy to use and modular with no emissions, noise and moving parts. They are widely used as stress sensors as well. These materials are produced synthetically for engineering applications. The most popular piezoelectric ceramic is lead zirconate titanate (Pb[Zr(x)Ti(1 − x)]O3) which is commonly known as PZT. It was first developed at the Tokyo Institute of Technology in 1950s (Erturk and Inman 2011). PZT crystals have high material strength and long service life. They are also resistant to humidity and temperatures well over 100 °C. Moreover, they can be fabricated easily into various shapes (Kour and Charif 2016).

Piezoelectric materials exhibit electric polarization when they are mechanically strained. Their electric polarization is proportional to the applied strain so they can be also used as sensors when properly characterized.

Piezoelectric materials has been used for power harvesting from human motions in common wearables, like shoes and backpacks (Feenstra et al. 2008). There have been studies that could harvest 1.3 mW at 3 V from embedded PZT wafers in shoes when walking at a rate of 0.8 Hz (Platt et al. 2005a, b). Delivering sustainable electric power to micro-electro-mechanical systems (MEMS) or a wireless network system by harvesting ambient energy is valuable since it can remove batteries and thus eliminating the need for replacing and maintaining the batteries especially in remote and hard to reach sensor nodes, safety-monitoring devices and embedded sensors in structures (Batra and Alomari 2017). These battery-less solutions will limit or eliminate the battery disposal which is environmentally beneficial.

There are approximately four million lane-miles in the USA, experiencing over 3.2 trillion vehicle-miles of travel per year according to the Federal Highway Administration ("Press Release: 3.2 Trillion Miles Driven On U.S. Roads In 2016, 2/21/2017| Federal Highway Administration"). This is a significant untapped source of energy. There have been many recent efforts to harvest the mechanical energy generated from vehicles passing over road pavements. Jung et al. developed a PEH module measuring 15 cm by 15 cm that produced a power density of 8 W/m^2 (Jung et al. 2017). A study at Virginia Tech evaluated nine different PEH designs and reported an average output of 3.1 mW per vehicle pass. It was concluded that the parallel electric connection between the piezoelectric disks is more efficient compared to the series connection. This configuration allows for lower output voltage peaks and more importantly, lower matching impedance in order to maximize power (Platt et al. 2005a, b); (Xiong 2014). Xiong et al. (Xiong and Wang 2016) developed and field tested a PEH, which produced an average power of 3.106 mW. It was concluded that this power was sufficient to operate roadside microelectronics. Sun et al. (2013) modeled and compared PZT stacks in series and in parallel and suggested that parallel configuration produces higher electric current and power. Yang et al. (2017) developed a PEH prototype casing made of MC Nylon® containing 9 PZT stacks which produced an open-circuit voltage of 280 V. Jiang et al. (2014) also used PZT stacks made of 36 layers of piezoelectric disks connected in parallel. They generated 85 mW under an excitation force amplitude of 1,360 N at 6 Hz. Roshani et al. (2016), tested piezoelectric material in the laboratory and concluded that the effect of ambient temperature has no significant effect on the power output of a PEH prototype.

2 Objectives

The particular objectives of this paper are:

1. Design and fabricate a PEH
2. Characterize its electromechanical properties through laboratory testing
3. Quantify the amount of electrical power that can be generated under traffic.

3 Harvester Design and Fabrication

The PEH housing was conceived as a steel box (AISI 1018 steel having yield stress of 53.7 ksi) with a top plate supported in each corner by a PZT stack (Fig. 1). Each PZT stack was built from six individual piezoelectric disks with diameters of 25 mm and a thickness of 6.8 mm which are connected electrically in parallel together. This housing is to be embedded into the roadway pavement layers in a way that its top plate is flush with the pavement surface (Fig. 2). Its dimensions and design were dictated by the need to safely carry the entire load of half a truck axle (i.e., 1 wheel path) and transmit it to the PZT stacks. Figure 1b shows the fabricated box, housing the four PZT stacks. For each post special supports were designed to hold the PZT stacks in place.

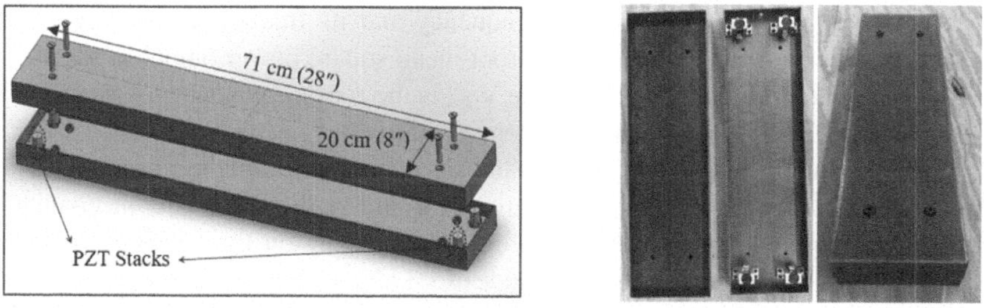

(a) (b)

Fig. 1 **a** PEH Housing schematic stacks **b** Fabricated PEH box

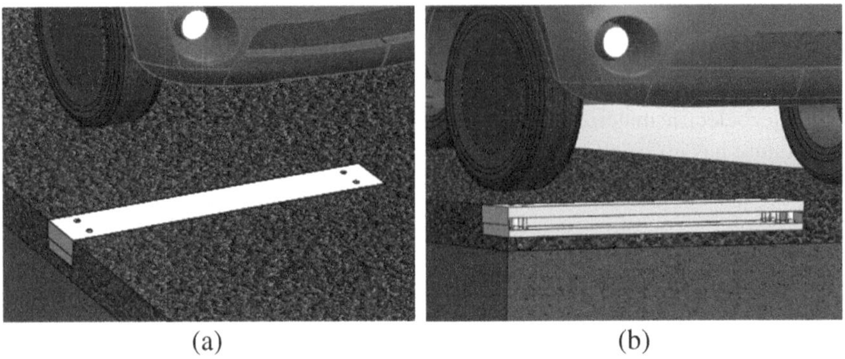

Fig. 2 PEH embedded in the pavement **a** top view **b** section view

4 Laboratory Testing

The mechanical loads were applied on the PZT stacks with a servo-hydraulic Universal Testing Machine (UTM) which is capable of applying dynamic loads with a wide range of amplitudes and frequencies. The loading frequencies applied ranged from 2.5 to 62 Hz. The higher test frequency was close to the loading frequency expected from vehicles moving at highway speeds, (e.g. at 96 km/h, a 0.2 m diameter circular tire contact takes 0.015 s to traverse the 0.20 m width of the PEH housing, which translates to a frequency of 66 Hz). Load magnitudes ranged from 4.45 kN to 44.5 kN, representing a typical half-car axle and legally-loaded half-truck axle, respectively. The corresponding stresses on each of the four PZT posts were 2.245 kPa (1.1 kN) and 22.453 kPa (11 kN), respectively.

The voltage output measurements across a variable resistor connected to PZT stack were recorded using an oscilloscope along with a voltage probe with a 500X attenuation rate.

The output voltage of a PZT stack under a sinusoidal mechanical load is a sinusoidal signal recorded by the oscilloscope which its amplitude is a function of both mechanical load amplitude and frequency and its frequency is the same as the mechanical loading frequency. The root mean square power output of the PZT stack is computed as $P = V_{rms}^2/R_L$ where V_{rms} is the root mean square of the output voltage signal and R_L is the resistance connected to the PZT stack.

Figure 3a illustrates the effect of mechanical load amplitude on the power output of the PZT stacks. The harvested power increases significantly with higher mechanical load levels. Figure 3b evaluated the effect of frequency on the harvested power. Having the same mechanical load amplitude of 1.1 kN, the harvested power increases drastically for higher frequencies. This shows more energy can be harvested from vehicles with higher speeds.

According to the power-transfer theorem also known as "Jacobi's law", the maximum amount of power output from a source occurs when the external electrical impedance matches the internal impedance of the source. Figure 3a shows that this

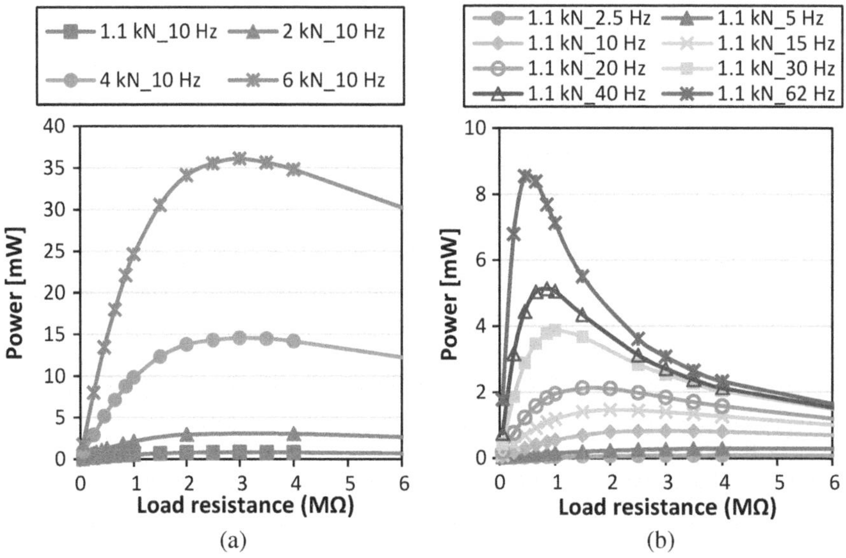

Fig. 3 RMS power output of one PZT stack **a** effect of mechanical load amplitude **b** effect of mechanical load frequency

matching impedance is not affected by the mechanical load amplitude but it is largely affected by the loading frequency as can be seen in Fig. 3b (power reaches to its maximum in lower impedance or resistance values).

The four PZT stacks in the PEH can be connected with series or in a parallel configuration. Figure 4 shows the experimental results for two stacks tested in the UTM at the same time with parallel and series connections. It can be seen that parallel configuration wins both in terms of amount of harvested power and lower matching impedance.

Fig. 4 Harvested RMS power from two PZT stacks for serial and parallel connections

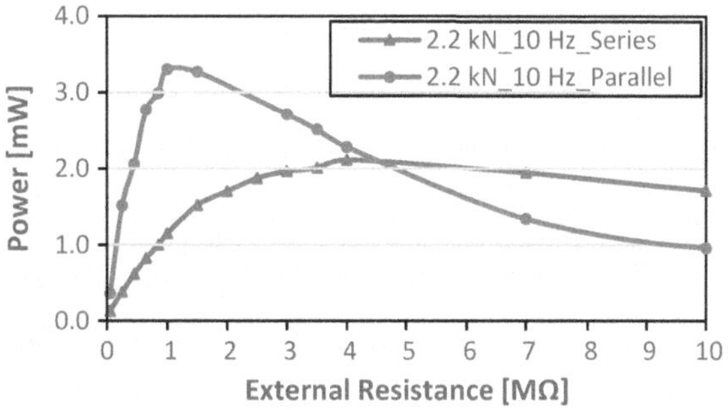

5 Electromechanical Modelling

Khalili et al. 2019 gave a through explanation of the methodology for modelling the PZT stacks used in this study.

For low frequencies, the PZT stack is modeled as a voltage source (Xu et al. 2013). The equivalent electrical circuit of the PZT is shown in Fig. 5.

As can be seen by Fig. 5, the model has five parameters: ϕ, R_{em}, C_{em}, L_{em} and C_P. These parameters were found through a calibration process in Excel solver by comparing the experimental output voltage of the PZT and the analytical output voltage of the equivalent circuit. The model was calibrated using laboratory data obtained by subjecting the PZT stacks to a load amplitude of 1.1 kN and loading frequencies ranging from 2.5 to 62 Hz. The fitted parameters are shown in Table 1.

Figure 7 suggests a very good agreement between model predictions and laboratory data. It is concluded that the electro-mechanical model fitted can reliably predict the electrical output from load input and vice versa.

As discussed earlier, the corresponding force from a car and a truck on each of the four PZT posts is 1.1 and 11 kN respectively. Using the equivalent circuit, The maximum RMS power harvested from each PZT stack is 9 and 1400 mW for 1.1

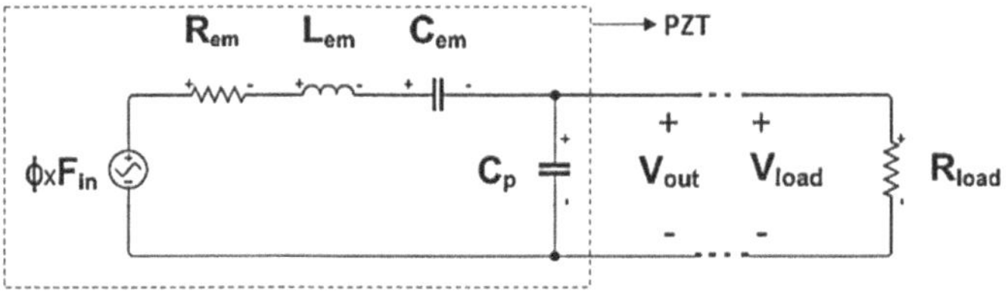

Fig. 5 The equivalent electrical circuit of the PZT. F_{in} is applied mechanical force amplitude

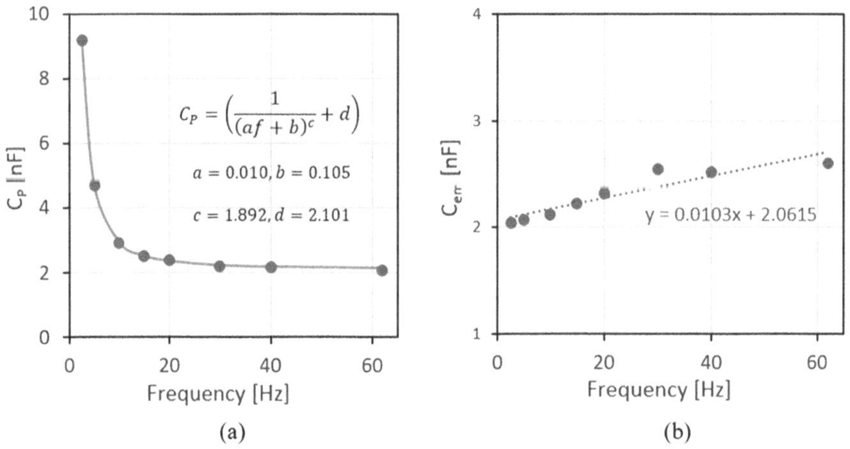

Fig. 6 Frequency dependent parameters of the model

Table 1 Fitted parameters of the model (equivalent PZT circuit)

Parameter	Values
R_{em} [Ω]	13.009
$L_{em}[H]$	0.033
C_P $[F]$	Refer to Fig. 6a
C_{em} $[F]$	Refer to Fig. 6b
$\phi[V/N]$	0.55

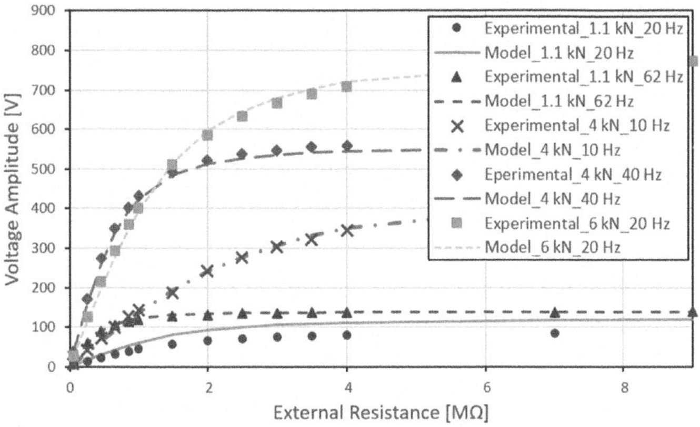

Fig. 7 Model verification using experimental data

and 11 kN, respectively for 66 Hz (which represents a vehicle speed of 96 km/h). It is noted that the power output from four of these stacks connected in parallel, as envisioned for the PEH, will be at least four times as large.

6 PEH Circuit Design and Testing

A preliminary energy harvesting circuit was designed based on computer simulations in Matlab using the electromechanical model. Figure 8 shows the lab setup for the initial power harvesting circuitry with one piezo-stack under testing connected to diode rectifier feeding a Wolfspeed/CREE CRD-15DD17P flyback dc/dc converter and a linear regulator (LT1764ET-3.3) with a resistive load.

Figure 9 depicts the experimental testing result of one piezo-stack under a continuous sinusoidal loading of 3 kN force with a frequency of 62 Hz. The output of the flyback is regulated between 5–10 V then a linear regulator is used to regulate the voltage to a constant 3.3 V.

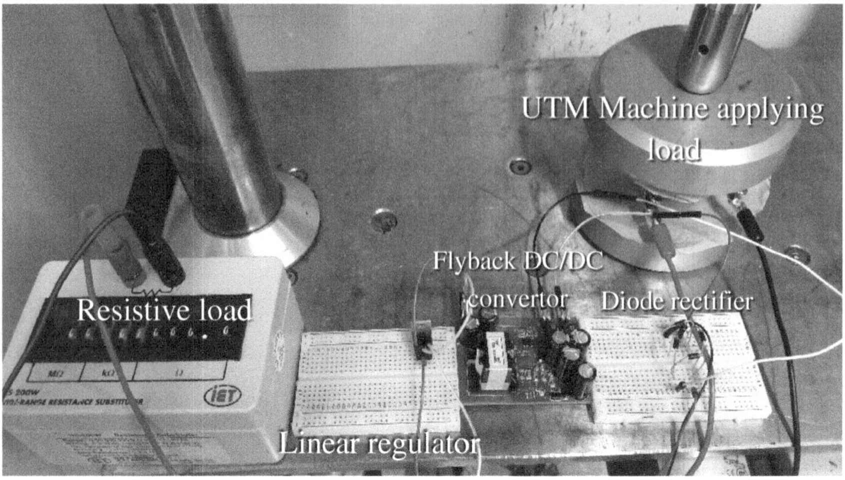

Fig. 8 Experimental Setup of one PZT stack with diode rectifier, flyback DC/DC converter and resistive load of 1.1 kΩ

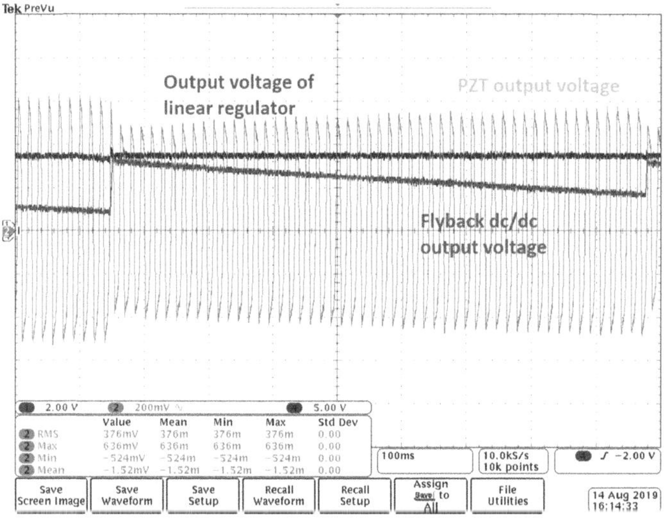

Fig. 9 Experimental results of one PZT stack with diode rectifier, flyback DC/DC converter and resistive load of 1.1 kΩ at 62 Hz and under continuous loading of 3 kN

7 Conclusions

This paper described the design and fabrication of a PEH consisting of four PZT stacks. From the extensive experimental test results on each PZT stack, the proposed electromechanical model was calibrated and then verified. Special emphasis should be given to the conditions that the PEH will experience under in-service traffic since it will determine the external impedance (i.e. resistance) of the harvesting circuit in order to maximize the harvesting power. Using the model, the PEH is estimated to produce a RMS power as high as 5.6 W.

The equivalent circuit was then used in computer simulation in Simulink in order to design a primary harvesting circuit. The circuit was tested in the lab and could provide a stable 3.3 V, DC voltage.

The output power is enough to be used for powering low consumption micro-processors which can be programmed to do the vehicle classification and counting. This study builds the foundation of a self-powered low budget WIM (Weigh-In-Motion) system alternatives. Having these cost-effective data collection systems are not only crucial to MEPDG pavement design but also provides data for a more efficient pavement management which eventually promote sustainability.

The harvested power can also be used for LED lightening promoting road safety.

Acknowledgements The authors express their sincere gratitude for the funding provided by CPS Energy of San Antonio, which made this work possible.

References

Batra AK, Alomari A (2017) Power Harvesting via Smart Materials. SPIE PRESS, Bellingham. https://doi.org/10.1117/3.2268643

Cook-Chennault KA, Thambi N, Sastry AM (2008) Powering MEMS portable devices—a review of non-regenerative and regenerative power supply systems with special emphasis on piezoelectric energy harvesting systems. Smart Mater Struct 17(4):043001. https://doi.org/10.1088/0964-1726/17/4/043001

Erturk A, Inman DJ (2011) Piezoelectric energy harvesting. Wiley, Hoboken

Feenstra J, Granstrom J, Sodano H (2008) Energy harvesting through a backpack employing a mechanically amplified piezoelectric stack. Mech Syst Signal Process 22(3):721–734. https://doi.org/10.1016/j.ymssp.2007.09.015

Gholikhani M, Nasouri R, Tahami SA, Legette S, Dessouky S, Montoya A (2019) Harvesting kinetic energy from roadway pavement through an electromagnetic speed bump. Appl Energy 250(September):503–511. https://doi.org/10.1016/j.apenergy.2019.05.060

Guilar NJ, Kleeburg TJ, Chen A, Yankelevich DR, Amirtharajah R (2009) Integrated solar energy harvesting and storage. IEEE Trans Very Large Scale Integr (VLSI) Syst 17(5):627–637. https://doi.org/10.1109/TVLSI.2008.2006792

Jiang X, Li Y, Li J, Wang J, Yao J (2014) Piezoelectric energy harvesting from traffic-induced pavement vibrations. J Renew Sustain Energy 6(4):043110. https://doi.org/10.1063/1.4891169

Jung I, Shin Y-H, Kim S, Choi J-y, Kang C-Y (2017) Flexible piezoelectric polymer-based energy harvesting system for roadway applications. Appl Energy 197(July):222–229. https://doi.org/10.1016/j.apenergy.2017.04.020

Khalili M, Biten AB, Vishwakarma G, Ahmed S, Papagiannakis AT (2019) Electro-mechanical characterization of a piezoelectric energy harvester. Appl Energy 253(November):113585. https://doi.org/10.1016/j.apenergy.2019.113585

Kour R, Charif A (2016) Piezoelectric roads: energy harvesting method using piezoelectric technology. Innov Energy Res 5(1):1–6

Platt SR, Farritor S, Garvin K, Haider H (2005a) The use of piezoelectric ceramics for electric power generation within orthopedic implants. IEEE/ASME Trans Mechatron 10(4):455–461. https://doi.org/10.1109/TMECH.2005.852482

Platt SR, Farritor S, Haider H (2005b) On low-frequency electric power generation with pzt ceramics. IEEE/ASME Trans Mechatron 10(2):240–252. https://doi.org/10.1109/TMECH.2005.844704

"Press Release: 3.2 Trillion Miles Driven On U.S. Roads In 2016, 2/21/2017| Federal Highway Administration." n.d. Accessed 23 May 2019. https://www.fhwa.dot.gov/pressroom/fhwa1704.cfm

Roshani H, Dessouky S, Montoya A, Papagiannakis AT (2016) Energy harvesting from asphalt pavement roadways vehicle-induced stresses: a feasibility study. Appl Energy 182(November):210–218. https://doi.org/10.1016/j.apenergy.2016.08.116

Sun C, Shang G, Zhu X, Tao Y, Li Z (2013) Modeling for Piezoelectric Stacks in Series and Parallel, pp. 954–957. IEEE. https://doi.org/10.1109/ISDEA.2012.228

Tahami SA, Gholikhani M, Nasouri R, Dessouky S, Papagiannakis AT (2019) Developing a new thermoelectric approach for energy harvesting from asphalt pavements. Appl Energy 238(March):786–795. https://doi.org/10.1016/j.apenergy.2019.01.152

Xiong H (2014) Piezoelectric energy harvesting for public roadways. PHD Other-Dissertation, Civil Engineering, Virginia Polytechnic Institute and State University

Xiong H, Wang L (2016) Piezoelectric energy harvester for public roadway: on-site installation and evaluation. Appl Energy 174(July):101–107. https://doi.org/10.1016/j.apenergy.2016.04.031

Xu T-B, Siochi EJ, Kang JH, Zuo L, Zhou W, Tang X, Jiang X (2013) Energy harvesting using a PZT ceramic multilayer stack. Smart Mater Struct 22(6):065015. https://doi.org/10.1088/0964-1726/22/6/065015

Yang H, Wang L, Hou Y, Guo M, Ye Z, Tong X, Wang D (2017) Development in stacked-array-type piezoelectric energy harvester in asphalt pavement. J Mater Civ Eng 29(11):04017224. https://doi.org/10.1061/(ASCE)MT.1943-5533.0002079

Performance Optimization of Warm Recycled Mixtures

F. Cardone, F. Canestrari, X. Jiang, and G. Ferrotti

Abstract Warm Mix Asphalt (WMA) technologies are becoming popular due to their ability to reduce mixing and compaction temperatures compared to the conventional hot mix asphalts (HMAs), with remarkable advantages of environment and costs. Moreover, WMA is considered as one of the most promising technology for increasing the re-use of Reclaimed Asphalt (RA) within the mixture although its effectiveness in recycling issues require more dedicated research activities. This paper describes a laboratory investigation aimed at optimizing a dense graded asphalt mixture for wearing course, produced with WMA technology and including up to 30% of RA. WMA mixtures were prepared by using two contents of a plain bitumen, two contents of RA and one chemical additive. A recycled HMA containing lower RA content, according to technical specifications currently applied in Italy, was selected as reference mixture. Strength and stiffness properties, water sensitivity, rutting and cracking resistance were investigated on shear gyratory compacted specimens. The result analysis on stiffness, rutting and fracture properties indicated the possibility to produce suitable WMA mixtures with higher RA contents without penalizing their performance compared to the reference one.

Keywords Warm Mix Asphalt · Reclaimed Asphalt · Stiffness · Rutting

1 Introduction

Compared with conventional Hot Mix Asphalt (HMA), Warm Mix Asphalt (WMA) allows the mixing and compaction temperature of mixture to be reduced by 20 to 40 °C (Rubio et al. 2012). In general, WMA are classified by the technique of production (D'Angelo et al. 2008): foaming based WMA (by adding cold water during the handling of hot binder) and additive based WMA (by adding organic additives or chemical additives to the asphalt mixture components). The application of WMA

F. Cardone (✉) · F. Canestrari · X. Jiang · G. Ferrotti
Department of Civil and Building Engineering and Architecture, Università Politecnica delle Marche, via Brecce Bianche, 60131 Ancona, Italy
e-mail: f.cardone@univpm.it

C. Raab (ed.), *Proceedings of the 9th International Conference on Maintenance and Rehabilitation of Pavements—Mairepav9*, Lecture Notes in Civil Engineering 76,
https://doi.org/10.1007/978-3-030-48679-2_22

technology could lead to both environmental and economic benefits. At lower temperature, the asphalt plant significantly reduces gas emissions including CO, CO_2, NO_x and So_2 (Capitão et al. 2012; Stimilli et al. 2017a) and compared with HMA production, a reduction in energy consumption by 20 to 75% can be gathered (Chowdhury and Button 2008). Moreover, it has been shown that, in particular conditions, WMA technology can improves workability and enhance the compaction of mixtures pavement (Mogawer et al. 2013; Stimilli et al. 2017b).

In Europe, over 80% of road pavements are constructed with asphalt mixtures (Antunes et al. 2019), therefore huge amounts of Reclaimed Asphalt (RA) material are available after the milling of old bituminous pavements. Nowadays, RA is soundly recognized as material for road construction and its use for the production of new bituminous mixtures provides remarkable economic and environmentally-friendly benefits. Because of these benefits, over recent years more and more researches focus on the performance of mixtures containing high RA contents (greater than 25%). Indeed, while the addition of RA in asphalt mixture increases its stiffness, which could increase its rutting resistance, the cracking resistance could be significantly reduced. Thus, a balanced mix design is very crucial. Actually, the simultaneous application of WMA technologies and recycling techniques constitutes a promising construction solution in terms of environment and paving cost (Putman and Xiao 2012). Moreover, plentiful investigations were carried out suggesting that mixture performance could significantly take advantage from the combination of WMA mixture and RA. In this regard, previous studies showed that high RA contents (up to 50%) could increase stiffness, improve rutting performance and moisture resistance of WMA mixture (Vargas-Nordcbeck et al. 2012; Zhao et al. 2013; Moghadas Nejad et al. 2014; Yin et al. 2014; Behbahani et al. 2017). When the WMA mixtures are produced with proper RA content and WMA additive, the findings indicate that recycled WMA mixtures have equivalent short-term and long-term performances to HMA mixtures, or slightly better at high temperature (Choubane et al. 2014; Zheng et al. 2013; Lopes et al. 2015).

Given this background, this research study aims at analyzing the mechanical properties of WMA mixtures containing different percentages of RA. Since these activities are preparatory to future works aimed at evaluating the feasibility of full-scale production, the related in situ lay-down and the compaction operations, the RA contents must be properly chosen. As the national/local specifications regarding hot recycling of mixtures for wearing courses require RA contents lower than 15%, this research study tries to extend this limit by using RA percentages up to 30%. The performance of the WMA mixtures, prepared with a chemical additive and selected RA and binder contents, were evaluated by an extensive laboratory testing aimed at determining their stiffness, cracking and rutting properties as well as moisture susceptibility.

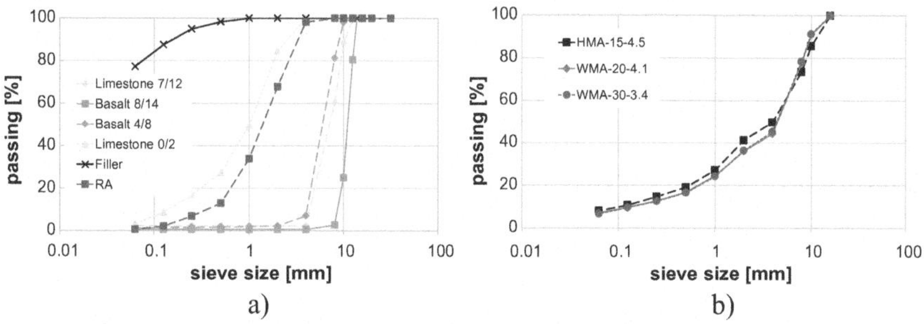

Fig. 1 **a** Virgin aggregates and RA gradations; **b** Design curves of asphalt mixtures

2 Materials

In this study, virgin (limestone, basalt, calcareous filler) and RA aggregates were used to produce the aggregate blends. The bitumen content of RA was 7.7% by aggregate weight. The recovered bitumen had a penetration of 38 dmm and a softening point of 54 °C. Figure 1.a shows the gradations of virgin and RA aggregates. A plain bitumen pen 50/70 was selected as binder. A chemical WMA additive available on the market, having an alkyl-amide polyamines-based composition and characterized by a density at 25 °C of 0.94 g/cm^3, flash point higher than 150 °C and viscosity at 25 °C of 150 cP, was chosen.

Three dense-graded mixtures for wearing courses were investigated: two WMAs and one HMA mixture as reference. The grading curves of the three asphalt mixtures are shown in Fig. 1b. All the mixes were characterized by a total bitumen content of 5.6% by the aggregate weight. The WMA mixture with 20% RA and 4.1% of added binder was coded as WMA-20-4.1, the WMA mixture with 30% RA and 3.4% of added binder was coded as WMA-30-3.4 and the reference mixture with 15% RA and 4.5% of added binder was coded as HMA-15-4.5. For the WMA mixtures, the mixing and compaction temperatures were 120 °C and 110 °C respectively, whereas for the HMA mixture they were 160 °C and 150 °C, respectively.

After mixing, each mixture was kept for one hour in an oven at a temperature equal to in situ compaction (i.e. 110 °C for WMA, 150 °C for reference mixture) in order to simulate the aging effects due to the lay-down and transportation operations. Afterwards, for each mixture a series of cylindrical specimens (D = 150 mm) were prepared by means of a gyratory compactor.

3 Testing Program

For each mixture, the compacted cylindrical specimens were analyzed to perform a mechanical characterization in terms of Indirect Tensile Strength (ITS), water sensitivity (Indirect Tensile Strength Ratio – ITSR), stiffness (Dynamic Modulus

Table 1 Testing program

Mixture	Number of replicates					
	ITS (dry)	ITS (wet)	|E*|	ITSM	ITFT	Rutting
HMA-15-4.5	3	3	2	6	6	2
WMA-20-4.1	3	3	2	6	6	2
WMA-30-3.4	3	3	2	6	6	2

|E*| and Indirect Tensile Stiffness Modulus - ITSM), fatigue (Indirect Tensile Fatigue Test – ITFT) and rutting resistance.

The summary of the testing program is reported in Table 1. The specimens used for the mechanical study were characterized by the same target air voids content (equal to 5%).

4 Results and Discussions

4.1 Indirect Tensile Strength and Water Sensitivity

ITS test was performed according to EN 12697-23 at a test temperature of 25 °C on air-conditioned specimens (i.e. dry condition) by obtaining ITS_{dry}. Moreover, the ITS test was also performed on specimens which were first submerged in water at 20 °C for 15 days to simulate wet condition and then conditioned at the test temperature of 25 °C, by obtaining ITS_{wet}. The ITSR value was calculated according to Eq. (1):

$$ITSR = \frac{ITS_{wet}}{ITS_{dry}} \times 100 \quad (1)$$

Figure 2 shows the average ITS values in both dry and wet condition along with the ITSR values for all the investigated mixtures. As expected, the comparison between WMA mixtures showed an enhanced strength for a higher RA content, even if their ITS values are lower compared to the reference HMA, in both dry and wet conditions. In terms of water sensitivity, WMA-20-4.1 showed an ITSR value comparable to the reference mixture, whereas higher RA content (WMA-30-3.4) resulted in a lower moisture resistance. However, it is worth noting that ITS and ITSR values for the WMA mixtures satisfied the local technical specification requirements reported in Fig. 2 (>0.6 MPa for ITS value and >75% for ITSR value).

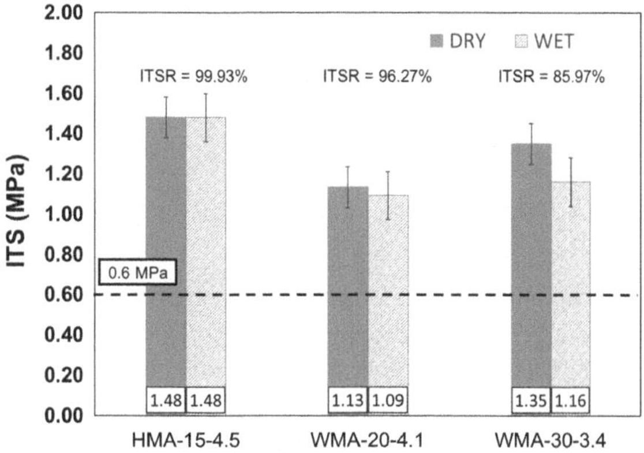

Fig. 2 ITS and ITSR results

4.2 *Stiffness*

Dynamic complex modulus was measured to evaluate the stiffness of mixtures over a wide range of temperature and loading frequency. Dynamic modulus tests were conducted at 4 temperatures (5, 20, 35 and 50 °C) and 6 frequencies (20, 10, 5, 1, 0.5 and 0.1 Hz) by means of Asphalt Mixture Performance Tester (AMPT), under haversine loading in compression mode. The test specimens have dimensions of 130 mm height and 94 mm diameter, cored from gyratory compacted 150 mm cylindrical specimens.

In this study, the Huet–Sayegh (HS) model (Sayegh 1967) was used to fit the frequency dependency of the dynamic modulus |E*| according to the equation shown in Fig. 3, where j is the imaginary unit, ω the pulsation (rad/s), t

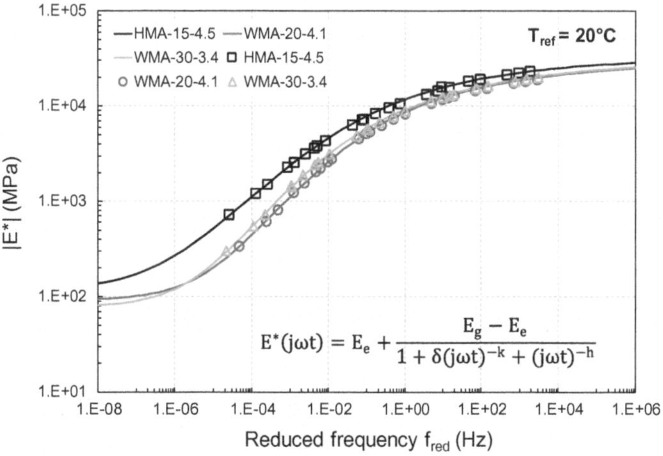

Fig. 3 Master curves of the dynamic modulus (|E*|) of mixtures at $T_{ref} = 20$ °C

Table 2 HS and WLF model parameters

	E_g (MPa)	E_e (MPa)	k (−)	h (−)	δ (−)	t (s)	C_1 (−)	C_2 (°C)
HMA-15-4.5	32056	119	0.169	0.435	1.884	1.011	53.7	412.9
WMA-20-4.1	28873	92	0.172	0.503	2.109	0.347	21.2	159.9
WMA-30-3.4	29184	78	0.182	0.518	2.510	0.793	32.1	232.7

the relaxation time (s), h and k the regression coefficients. The closed form shifting (CFS) algorithm (Gergesova et al. 2011), which is based on the minimization of the area between two successive isothermal curves, is used to translate the experimental data and to estimate the temperature shift factors. The Williams-Landel-Ferry (WLF) equation (Ferry 1961) was used to model the shift factor trend with temperature. The dynamic modulus master curves of all mixtures at the reference temperature T_{ref} = 20 °C are shown in Fig. 3 and the parameters of HS and WLF model (C1 and C2) are listed in Table 2. As can be observed, both WMA mixtures showed lower dynamic moduli with respect to the reference HMA over the entire range of loading frequency. This finding can be explained by the different short-term oxidation undergone by the mixtures. Specifically, lower production temperatures (i.e. mixing and compaction) induced less oxidation to the WMA mixtures and, consequently, a significantly reduced stiffening effect due to aging, compared to the reference mixture. Moreover, both WMA mixtures (prepared with 20 and 30% RA) showed similar trend for the master curves highlighting minor stiffening effects due to the RA content and, thus, confirming the prevalent role of aging on the development of stiffness properties.

Stiffness properties were also measured in terms of ITSM at 20 °C according to EN 12697-26. ITSM results are consistent with the dynamic modulus data obtained through uniaxial compression test (Fig. 3) and confirmed the highest stiffness of the reference mixture compared to both WMA mixtures. This similar trend can be observed in Fig. 4, which reports the comparison between ITSM values and dynamic moduli determined through the master curves at the same temperature (i.e. 20 °C) and frequency of 2 Hz (comparable with the rise time of 124 ms applied during ITSM test).

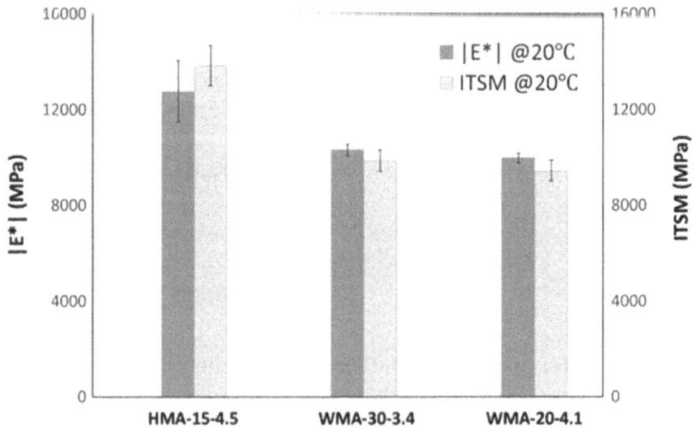

Fig. 4 ITSM vs. |E*| @ 2 Hz at 20 °C

Table 3 The rutting parameter results

Mixture	f_c (μ strain/cycle)	Δf_c (%)	ε_{1000} (%)	$\Delta\varepsilon_{1000}$ (%)
HMA-15-4.5	0.157	–	0.726	–
WMA-20-4.1	0.044	−72.1	0.453	−37.6
WMA-30-3.4	0.051	−67.4	0.323	−55.5

4.3 Rutting Performance

The cyclic compression test was performed according to EN 12697-25 to determine the resistance to permanent deformation of the investigated mixtures. The test was carried out at 60 °C by applying a cyclic axial stress with an amplitude of 0.3 MPa and a steady confining pressure of 0.2 MPa throughout 10,000 load cycles. During the test, cumulative axial strain of the specimen was recorded and the creep curve was built.

The creep rate f_c and the calculated permanent deformation after 1000 load cycles ε_{1000}, selected according to EN 12697-25, were used to characterize the resistance to permanent deformation of the mixtures under traffic loading. The parameter f_c was calculated as the slope of the least square linear fit of the (quasi) linear part of the creep curve, while ε_{1000} was calculated from the least square power fit of the (quasi) linear part at 1000th cycle. In particular, the lower the value of these parameters, the higher the rutting resistance of the mixture. The average values of the rutting parameters for all investigated mixtures are listed in Table 3.

Results clearly show that the values of both rutting parameters of WMA mixtures are remarkably lower than those of the reference mixture, denoting their improved resistance to permanent deformation. As expected, the increase in RA content within the WMA could justify their capability to be less prone to accumulate permanent deformation. The stiffness analysis previously discussed makes these results particularly interesting. Indeed, notwithstanding the mitigation stiffness due to the use of WMA technologies (i.e. reduced production temperature), the WMA mixtures would not experience any permanent deformation accumulation risk in the early period of their service life.

4.4 Fatigue Performance

Indirect Tensile Fatigue Test (ITFT) was selected to analyze cracking resistance of the mixtures. The test was performed according to the British standard BS DD ABF (1997) at a temperature of 20 °C in controlled-stress mode by applying a cyclic haversine load with a rise time of 124 ms. During the test, the number of loading cycles applied on the specimen was recorded until the complete fracture of the specimen was obtained (failure). The specimen was conditioned for at least 2 h prior the test in the climatic chamber.

Three target initial strain levels (80, 100 and 125 $\mu\varepsilon$ for the reference mixture and 100, 125 and 150 $\mu\varepsilon$ for the WMA mixtures) were selected for each mixture. The corresponding stresses to be applied were calculated by using the stiffness of the specimen in terms of ITSM (Fig. 4) measured before the fatigue tests. Thus, each specimen was firstly subjected to ITSM test at 20 °C in order to measure its stiffness modulus and then subjected to ITF test. Two specimens were tested for each strain level. Figure 5 shows the ITFT results plotting the initial horizontal strain $\varepsilon_{h,\ initial}$ as a function of the number of loading cycles to failure N_f. Test data were analyzed according to a power law regression and the fatigue curve (i.e. $\varepsilon_h = a \cdot N_f^{-b}$) was reported for each mixture. The R^2 value is 0.98 for all mixtures, which indicates good reliability to predict fatigue life by this fitting model. The average ITSM values, used to evaluate the stiffness of specimens before fatigue test, were also reported in Fig. 5.

From the comparison of fatigue curves, it is remarkable that at the same strain level WMA mixtures can withstand higher numbers of loading cycles than the reference one highlighting a better fatigue behavior for WMA mixtures. This finding is in accordance with the stiffness results that suggests a more brittle tendency of the reference mixture, due to higher stiffness. Moreover, as regards WMA mixtures higher RA content resulted in a slightly higher fatigue resistance especially at higher strain level.

Overall results highlight that lower production temperatures reduce short-term aging effects, resulting in a stiffness mitigation as well as improved fatigue performance compared to the reference mixture.

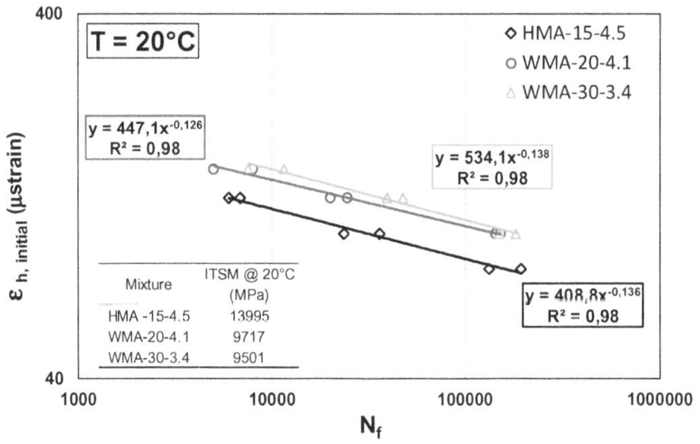

Fig. 5 Fatigue curve of mixtures at 20 °C

5 Conclusions

The present paper deals with results gathered through the performance analysis of dense graded warm recycled mixtures for wearing courses. Two WMA mixtures, including different RA amount (20 and 30%) and a chemical WMA additive, and a conventional reference mixture (HMA prepared with 15% RA) were selected for investigation. The WMA and reference mixtures were analyzed in terms of strength and stiffness properties, moisture susceptibility as well as rutting and fatigue behavior through a laboratory characterization performed on gyratory compacted cylindrical specimens.

From the test results, the following conclusions can be drawn:

1. WMA mixtures showed strength properties (i.e. ITS) lower than the reference one in both dry and wet conditions and a moisture susceptibility that tends to increase with the increase in RA content;
2. WMA mixtures showed lower stiffness properties and higher fatigue resistance compared to the reference HMA. This finding highlights that lower production temperatures (i.e. WMA technologies) reduce short-term aging effects leading to stiffness mitigation as well as improved fatigue performance. Specifically, the increase in RA content did not affect the stiffness of WMA mixtures but it allowed higher fatigue resistance to be pursued;
3. Despite the stiffness mitigation due to a reduced short-term aging effect, the WMA mixtures showed significantly higher rutting resistance than the reference one, thus excluding any permanent deformation accumulation risk in the early period of service life. No RA content effect on the rutting behavior was observed.

Overall result analysis highlighted that WMA mixtures with higher RA content with respect to a traditional recycled HMA are characterized by suitable and even improved performance. This outcome strongly encourages the use of WMA technologies in combination with higher amount of RA for the production of new bituminous mixtures denoting a promising future of these environmentally friendly materials. It is right to remark that future activities will involve the evaluation of the feasibility of full-scale production and related trial sections construction.

Acknowledgements This research was sponsored by ASFALTI s.r.l. This support is greatly acknowledged. Data analysis and opinions are those of the Authors and do not necessarily reflect those of the sponsoring agency.

References

Antunes V, Freire AC, Neves J (2019) A review on the effect of RAP recycling on bituminous mixtures properties and the viability of multi-recycling. Constr Build Mater 211:453–469

Behbahani H, Ayazi MJ, Moniri A (2017) Laboratory investigation of rutting performance of warm mix asphalt containing high content of reclaimed asphalt pavement. Pet Sci Technol 35(15):1556–1561

Capitão SD, Picado-Santos LG, Martinho F (2012) Pavement engineering materials: review on the use of warm-mix asphalt. Constr Build Mater 36:1016–1024

Choubane B, Chun S, Lee HS., Upshaw P, Greene J, Nazef A (2014) Evaluation of Warm-Mix Asphalt (WMA) performance in Florida. In: Transportation research board 93rd annual meeting, 12–16 January, Washington DC

Chowdhury A, Button JW (2008) A review of warm mix asphalt. Technical Report 473700-00080-1 Texas Transportation Institute

D'Angelo J, Harm E, Bartoszek J, Baumgardner G, Corrigan M, Cowsert J, Harman T, Jamshidi M, Jones W, Newcomb D (2008) Warm-mix asphalt: European practice. Federal Highway Administration. Office of International Programs

Ferry JD (1961) Viscoelastic Properties of Polymers. Wiley, New York

Gergesova M, Zupancic B, Saprunov I, Emri I (2011) The closed form t-T-P shifting (CFS) algorithm. J Rheol 55(1):1–16

Lopes M, Gabet T, Bernucci L, Mouillet VJM (2015) Durability of hot and warm asphalt mixtures containing high rates of reclaimed asphalt at laboratory scale. Mater Struct 48(12):3937–3948

Mogawer W, Austerman A, Mohammad L, Kutay ME (2013) Evaluation of high RAP-WMA asphalt rubber mixtures. Road Mater Pavement Des 14(2):129–147

Moghadas Nejad F, Azarhoosh A, Hamedi GH, Roshani H (2014) Rutting performance prediction of warm mix asphalt containing reclaimed asphalt pavements. Road Mater Pavement Des 15(1): 207–219

Putman BJ, Xiao F (2012) Investigation of warm mix asphalt (WMA) technologies and increased percentages of reclaimed asphalt pavement (RAP) in asphalt mixtures. FHWA-SC-12-05 Federal Highway Administration

Rubio MC, Martínez G, Baena L, Moreno F (2012) Warm mix asphalt: an overview. J Clean Prod 24:76–84

Sayegh G (1967) Viscoelastic properties of bituminous mixtures. In: 2nd international conference on structural design of asphalt pavements, 7–11 August, Washington D.C

Stimilli A, Virgili A, Canestrari F (2017a) Warm recycling of flexible pavements: effectiveness of WMA additives on SBS modified bi-tumen and mixture performance. J Clean Prod 156:911–922

Stimilli A, Frigio F, Canestrari F, Sciolette S (2017b) In-plant production of warm recycled mixtures: a case study. In: International conference on transport infrastructure and systems TIS2017, 10–12 April, Rome, Italy

Vargas-Nordcbeck A, Timm DH (2012) Rutting characterization of warm mix asphalt and high RAP mixtures. Road Mater Pavement Des 13(1):1–20

Yin F, Arambula E, Lytton R, Martin AE, Cucalon LG (2014) Novel method for moisture susceptibility and rutting evaluation using hamburg wheel tracking test. Transp Res Rec 2446(1):1–7

Zhao S, Huang B, Shu X, Woods M (2013) Comparative evaluation of warm mix asphalt containing high percentages of reclaimed asphalt pavement. Constr Build Mater 44:92–100

Zheng J, Zhao S, Huang B (2013) Rut resistance of foamed warm mix asphalt containing RAP. In: Second international conference on sustainable construction materials: design, performance, and application, 18–22 October, Wuhan, China

Development of Durable Pavement in Japan

Shigeki Takahashi, Shouichi Kanno, Yu Shirai, and Tamotsu Yoshinaka

Abstract As the existing expressways age, structural damage including bottom-up fatigue cracking occurs in the asphalt pavement of expressways in Japan. Therefore, a detailed field study was conducted to identify the damage status of the inner pavement under heavy traffic, and the weaknesses of the existing pavement were revealed. This paper proposes new ideas that can effectively contribute to the creation of exceptionally durable pavement.

Keywords Fatigue crack · Field study · Asphalt treated subbase · High modulus asphalt mixture · Long-term durability

1 Introduction

Approximately 9000 km of expressway has been constructed in Japan since 1963. The length of these expressway networks is only 1% of the nationwide road assets, but as shown in Fig. 1, they support 48% of domestic inland freight (ton-km basis), and therefore make an even bigger contribution than railway cargo.

In addition, 40% of the existing expressway is now older than 30 years of age (Fig. 2). Such heavy traffic load over many years severely damages the pavement, not only the surface but also the layers underneath. This imposes higher cost and longer closures of traffic lanes when the road operator conducts extensive repair of the damaged pavement.

Under such circumstances, an expressway operator of Japan (RI-NEXCO: Nippon Expressway Research Institute) has sought effective countermeasures against such damage in order to prolong the durability and service life of pavement. The first important step is to identify the real cause of the damage that has actually occurred in the field.

S. Takahashi (✉) · S. Kanno
Pavement Division, Nippon Expressway Research Institute Co., Ltd., 1-4-1 Tadao, Machida, Tokyo 194-8508, Japan
e-mail: s.takahashi.ai@ri-nexco.co.jp

Y. Shirai (✉) · T. Yoshinaka
Research Institute, NIPPO Corporation, 6-70, Mihashi, Nishi-ku, Saitama 331-0052, Japan

C. Raab (ed.), *Proceedings of the 9th International Conference on Maintenance and Rehabilitation of Pavements—Mairepav9*, Lecture Notes in Civil Engineering 76,
https://doi.org/10.1007/978-3-030-48679-2_23

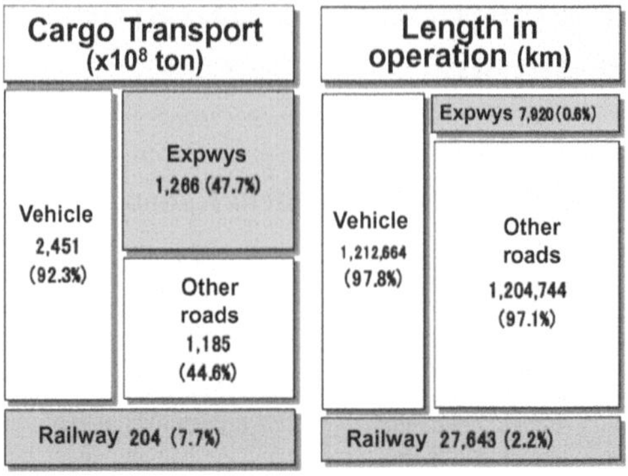

Fig. 1 Current status of inland transport in Japan

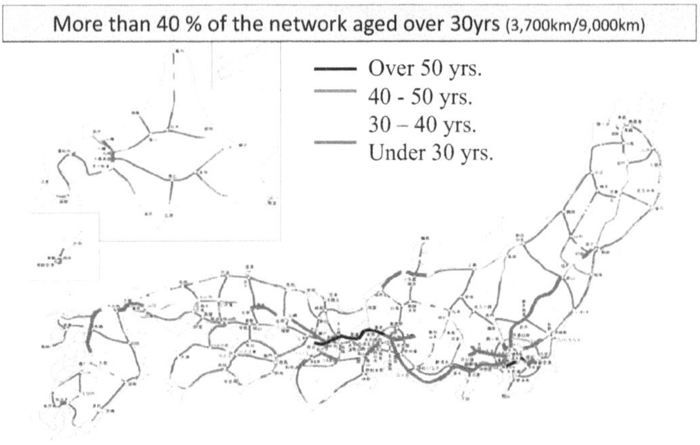

Fig. 2 Expressway network in Japan (As of 2012)

2 Field Study

The study was conducted by comparing the in situ status of damaged and non-damaged sections after almost 20 years of exposure to traffic. Specimens for lab tests were taken by block sampling of both asphalt layers and the granular subbase course. Based on the field observation of the damaged part and various testing results, several important key factors were revealed for the long-term durability of asphalt pavement. Those points, such as total thickness of the asphalt layers, durability of the asphalt treated subbase (third layer), and the bonding property of each layer are vitally important to prevent fatigue cracking. Good compaction of the granular subbase and subgrade are also key issues to support traffic over a long period without permanent deformation of the base. Preventing water penetration through the pavement into its base is another important factor to prevent severe deterioration in the field.

These findings were described in detail in a previous research paper (Takahashi 2017), but this paper highlights one of the major types of damage observed in the above field study projects.

This study focuses on the issue of fatigue cracking under wheel paths, which pass through the entire asphalt layers.

Pavement engineers in Japan believed that most of the cracks on Japanese expressways thus far were recognized as top-down cracking (Matsuno and Nishizawa 1992).

Except for the poor structure of pavements in rural roads, it was believed that bottom-up cracks rarely occur in the expressways because the pavement has sufficient thickness to counteract even the heaviest of traffic.

However, our 20-year field survey reveals that at least the structure, which has a thinner asphalt treated subbase (8–10 cm) with minimum thickness of the asphalt mix (less than 40 cm), allows bottom-up cracking in an expressway (Figs. 3 and 4, respectively). It is probably the first report of fatigue cracking observed in the field in the entire 50-year history of expressways in Japan.

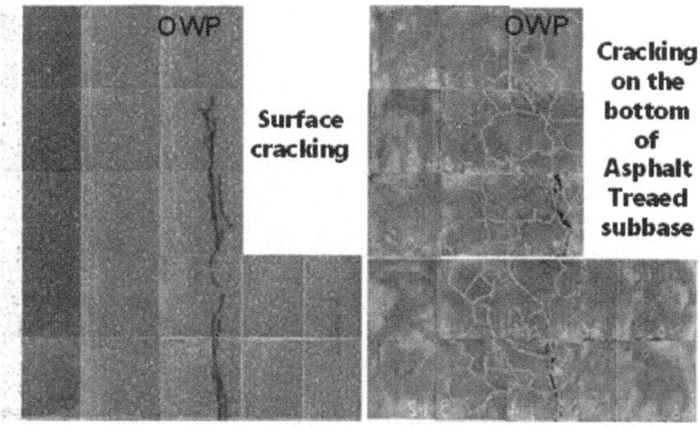

Fig. 3 Observed fatigue cracking on an expressway (Left & right are at the same location)

Fig. 4 Cut section of asphalt pavement (another expressway)

3 Countermeasures

3.1 Conventional Subbase

The first expressway of Japan was constructed in 1963 near Kyoto. At that time, the pavement had only two layers of asphalt mixture on the granular base course. A few years after its construction, severe cracks occurred, finally becoming alligator cracks as the traffic increased (Fig. 5). Based on this negative experience, the second project modified its structure by introducing an asphalt treated subbase (Fig. 6). This became the standard structure for asphalt pavement after that, and no cracks occurred.

The asphalt treated base means it is not a proper asphalt mixture, in terms of asphalt content and air voids (designed OAC is less than 4%). It is categorized as stabilized materials, such as cement or lime soil treated materials and it is cheaper than the asphalt mixture. Moreover, this asphalt treated subbase is allowed to use local materials from the cut sections of construction sites. Some of those materials

Fig. 5 Severe alligator cracks on the Meishin Expressway

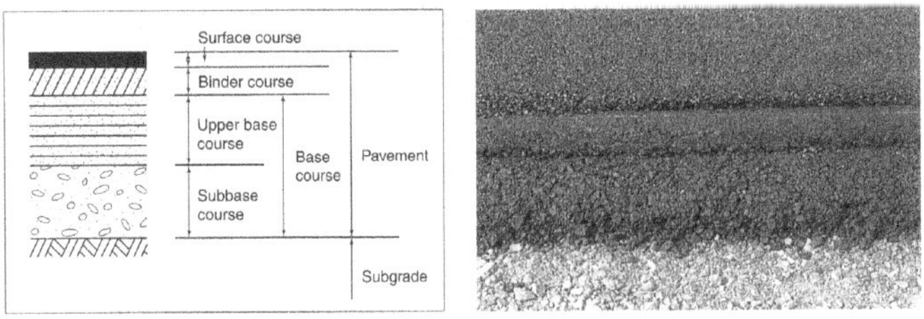

Fig. 6 Standard structure of asphalt pavement in Japan (Upper base was initially granular materials, but is now asphalt treated mixture)

have relatively poor quality, and thus it is difficult to use as it is, without any stabilizer. However, for the construction projects of expressways, it is a very beneficial way to use such local materials to minimize the construction cost.

Under the economical construction policy, cement stabilized materials became popular for the lower subbase and the asphalt treated type became widely commonplace for the upper subbase in Japan. Certain thicknesses of the pavement can support even the heaviest traffic over decades, but eventually the third layer (asphalt treated base) will succumb to the above mentioned bottom-up cracks. Currently, the structural life of the pavements is approaching its end. In fact, many of the site offices are starting to consider, or have actually started to replace this third layer on our expressways.

If time and budget permit, the existing layer can be replaced by the same material, because the initial one has performed well for more than 20 years. However, in depth repair work requires much more time to replace the layer (more than one day), compared to the process of just scarifying and overlaying the surface layer (one night work). On heavy trafficked routes, lane closure during busy hours creates severe traffic jams that unfortunately test the patience of drivers.

Therefore, a more durable mixture should be developed and introduced to prevent fatigue cracking, thus prolonging the structural life of pavements. It is thus effective to reduce the maintenance cost and improve the road user satisfaction.

3.2 High Modulus Asphalt

We studied high modulus asphalt (HiMA) as a highly durable subbase alternative mixture to that of the conventional asphalt treated subbase. HiMA was developed in France in the 1990s (Corte 2001). It has since been used globally, including in the UK, (Bankowski et al. 2009), (Nkgapel 2012) with widespread reports of use of HiMA in Australia and South Africa. There have also been cases of its use in South Africa to improve rutting resistance (Nkgapel 2012). The most significant characteristic of HiMA is that asphalt with a penetration grade as low as 10 to 30 is used as the binder.

HiMA is a mixture with (1) high stiffness, (2) high stripping resistance, (3) high water-tightness, and (4) high fatigue resistance. Compared to conventional mixtures, it has an excellent ability to distribute load to the underlying layer. It also has high fatigue resistance in the asphalt-concrete layer and high-water resistance. Compared to conventional pavement, its use is expected to prolong the pavement service life and significantly reduce the pavement's thickness. In this study, we developed a Japanese version of HiMA (HiMA (20)) based on the compositions of mixtures that have been used overseas, as well as on the standard values.

3.3 *Concept of Japanese HiMA*

Based on the results of the previous survey, stiffness, resistance to cracking, and water resistance were selected as properties required for HiMA (20). Furthermore, since the air temperature drops to about 10 °C below zero in some areas of Japan during the winter, we determined the compositions of mixtures with consideration of brittleness caused by low-temperature. In this study, we decided to use asphalt with a mixture of straight asphalt 60/80 (STAS 60/80) and a hard additive as the binder for HiMA (20). Here, a petroleum resin having a penetration of 1– 2 1/10 mm and a softening point of 90–150 °C was used as the hard additive. The binder of HiMA was adjusted to have a permeability of about 25–40 1/10 mm and a softening point of about 45–60 °C.

The size of the sieve was determined up to 20 mm, referring to information from other countries (Nkgapel 2012). The asphalt content was set to give a void of 2–3% to ensure water tightness.

4 Mix Design in Lab

4.1 *Method*

The properties of mixtures for HiMA (20) were studied in the lab. In this study, stiffness, fatigue resistance, separation resistance, and water resistance properties, including water-tightness, of a dense graded asphalt mixture with a maximum sieve size of 13 mm (DGM (13)), a coarse graded asphalt mixture with a maximum sieve size of 20 mm (CGM (20)), an asphalt stabilized subbase with a maximum sieve size of 30 mm (ASB (30)), and a large sieve-size asphalt mixture with a maximum sieve size of 30 mm (LPS (30)), which are all commonly used in Japan, were compared. Figure 7 shows the sieve size of HiMA (20) and the sieve sizes of mixtures commonly used in Japan.

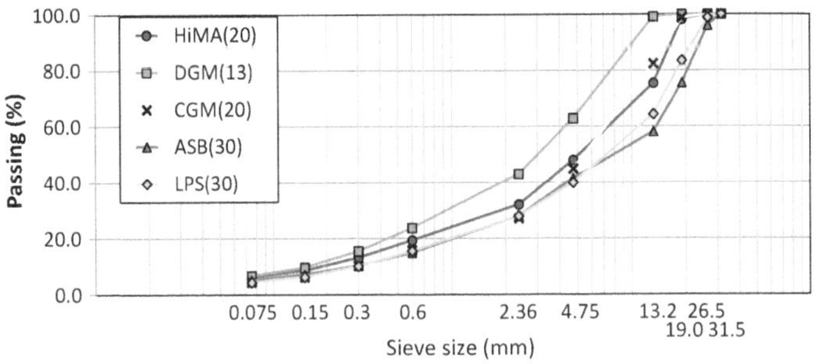

Fig. 7 Sieve sizes of HiMA (20)

Table 1 shows the properties of the mixture and Table 2 shows a list of the evaluation requirements used in this study. As the evaluation method of stiffness and fatigue resistance is a test method that can be used in Japan, the method used in the United Kingdom (McHale et al. 2012) and Poland (Petho 2014) was followed. Water resistance was evaluated using the peel rate and permeability provided in the NEXCO design procedure. In Poland, the loading frequency in fatigue testing was set to 10 Hz. However, in this study, it was set to 5 Hz due to testing equipment restrictions.

4.2 *Stiffness*

The Indirect Tensile Stiffness Modulus (ITSM) test method, using the Nottingham Asphalt Tester (NAT), was used to measure stiffness. This test method is compliant with BS EN12697. In this method, stiffness is measured by repeating an indirect tensile test of a cylindrical specimen. The test conditions were as follows: temperature: 20 °C; loading waveform: haversine; horizontal displacement: 5×10^{-6} m; Poisson's ratio: 0.35; specimen: Marshall specimen diameter 100 mm. Figure 8 shows the results of the measurement. The stiffness of HiMA (20) is 8,200 MPa, which is 1.9 times that of ASB (30), and meets the requirements shown in Table 2.

4.3 *Fatigue*

The number of cycles to fatigue failure for the mixtures was measured by the four-point bending test. The test conditions were as follows: temperature: 10 °C; loading waveform: sine; frequency: 5 Hz. The test results are shown in Fig. 9. The fatigue resistance of HiMA (20) is higher than that of other mixtures. On average, the number of cycles to fatigue failure for HiMA (20), at a strain of 200 μ, is about 1.4 to 3 times that of the comparative mixture. Figure 9 also shows that the number of cycles to fatigue failure for HiMA (20), at a strain of 130 μ, is over 1,000,000 cycles, which exceeds the requirement specified in Table 2.

4.4 *Water Resistance*

An immersed wheel tracking test was performed in accordance with the NEXCO Test Method, and a high-pressure permeability test was performed in accordance with the Pavement Survey and Test Method Handbook in Japan. Table 3a shows the test conditions, and Table 3b shows the test results. It can be seen that the stripping coefficient of HiMA (20) is less than the requirements in Table 2 and that other mixtures exceed the requirements of 13.7–38.9%. The permeability coefficient is

Table 1 The properties of the mixture

Mixture	Binder	Passing (%)							Binder content (mass%)	Air content (vol.%)	Marshal stability (kN)
		31.5 mm	26.5 mm	19.0 mm	13.2 mm	4.75 mm	2.36 mm	0.075 mm			
HiMA (20)	Binder of HiMA	100.0	100.0	98.2	75.4	48.0	32.1	6.3	5.3	2.7	13.1
DGM (13)	Straight asphalt 60/80	100.0	100.0	100.0	99.2	62.9	43.1	6.7	5.7	3.4	11.8
CGM (20)		100.0	100.0	98.7	82.4	44.9	27.2	4.5	5.3	4.1	8.5
ASB (30)		100.0	96.0	75.6	58.2	41.6	28.1	5.2	3.8	5.2	10.2
LPS (30)		100.0	98.7	83.5	64.4	40.0	28.0	4.4	4.5	4.1	9.5

Table 2 Evaluation requirements for HiMA (20)

Item	Method	Requirement	Reference source
Stiffness	Indirect Tensile test (20 °C, 124 ms)	≧5,500 MPa	UK
Fatigue	4 point bending (10 °C, 10 Hz, strain 130 μ)	≧1×10^6	Poland
Stripping coefficient	Immersion wheel tracking test (60 °C, 6 h)	≦5.0%	Nippon Expressway Co., Ltd.
Permeability coefficient	Pressurization permeability test (150 kPa, 10 min)	≦1×10^{-7} cm/s	Nippon Expressway Co., Ltd.

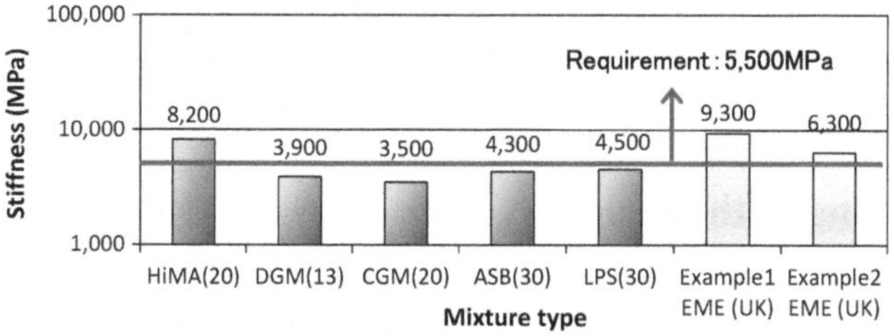

※Example 1, 2 EME (M. McHale et al.

Fig. 8 Stiffness test result

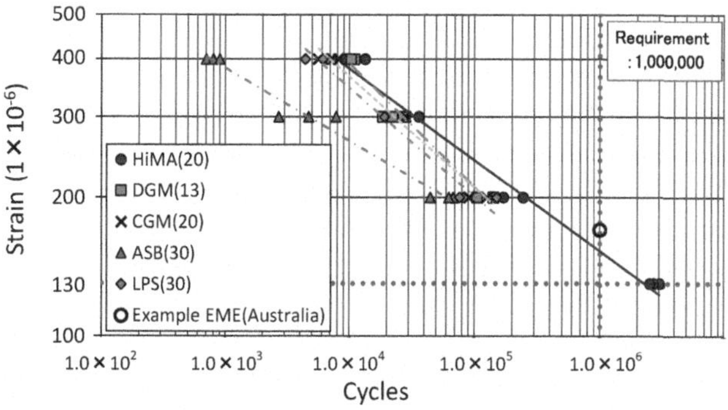

※ Example EME (Petho et al, 2014)

Fig. 9 Fatigue test result

Table 3 Water-resistance test condition (a) and results (b)

(a) Water-resistance test conditions

Item		Condition
Stripping coefficient (%)	Temperature	60℃
	Test time	6 hrs
	Curing	60℃ ,12hrs to dry , 1hrs to wet
	Water level	Under surface +5mm
Permeability coefficient (cm/s)	Temperature	Ordinary temperature
	Pressure time	24hrs
	Water pressure	150KPa
	Test time	10mins

(b) Water-resistance test results

Miture type	HiMA(20)	DGM(13)	CGM(20)	ASB(30)	LPS(30)
Stripping coefficient (%)	4.8	13.7	26.0	38.9	30.9
Requirement	≦5.0				
Permeability coefficient (cm/s)	0	0	Unmeasurable		
Requirement	$\leqq 1\times10^{-7}$				

below the requirement in Table 2 for HiMA (20) and for the DGM (13). The other three mixtures exhibited poor water tightness. Based on the above results, it was confirmed that HiMA (20) has a higher water resistance than that of other mixtures.

5 Trial Paving in the Field

In October 2018, trial paving was conducted on a highway to study the applicability of HiMA (20) to roads in service. Table 4 shows the properties of the mixture used in the trial paving. Recycled aggregate was not used in the trial paving. Figure 10a shows the sectional divisions and the cross section of the pavement in the trial paving site. An ordinary ASB was used for the upper subbase course in the comparison section. Figure 10b shows the calculation results of the estimated stiffness of each section, which was calculated by performing a dynamic back calculation of FWD measurement data. The FWD measurements were taken at six points on the surface of the upper subbase in each section to identify any difference in the performance of the upper subbase course. Wave-BALM was used for the back calculation. The analysis model consisted of three layers: the upper subbase course, the lower subbase course, and the subgrade. Figure 10b shows that the estimated stiffness averaged over the measurement points in the HiMA section is nearly twice that of the ASB section.

Table 4 Properties of the mixture

Item	Value	Requirement
Optimum binder content (mass%)	5.3	4.5– 6.5
Additive amount (mass% of binder)	10.0	–
Density (g/mm^3)	2.405	–
Air content (vol.%)	2.3	2–3
Stiffness (MPa)	10,020	≧5,500
Stripping coefficient (%)	4.2	≦5.0
Permeability coefficient (cm/s)	0	$\leqq 1 \times 10^{-7}$

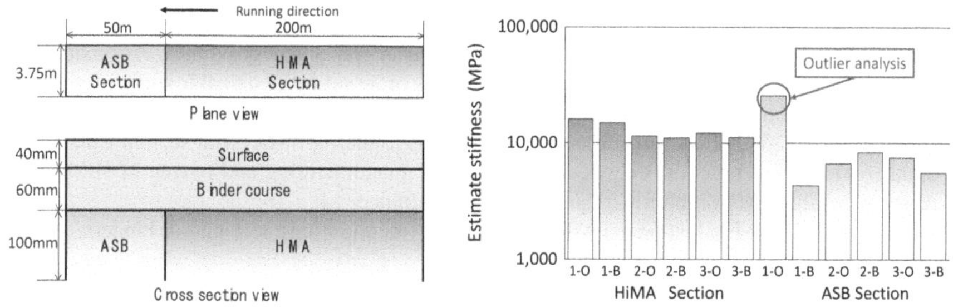

(a) Paving site plane view and cross-section

(b) Stiffness estimated from FWD

Fig. 10 The trial paving cross-section **a** and stiffness estimated from FWD **b**

It can therefore be concluded that the estimated stiffness immediately after paving is sufficiently high, and that HiMA exhibits the required performance. We plan to evaluate the long-term durability of HiMA in the future.

6 Conclusion

The results of this study are summarized below:

(1) Field inspections have identified damage to the pavement, particularly bottom-up fatigue cracks in the asphalt base course.
(2) We have developed a Japanese version of HiMA to act as a more durable asphalt base, and proposed its use.
(3) In trial paving, we evaluated the ease of paving and its performance immediately after paving. In the future, a follow-up survey will be conducted at the trial paving site to verify durability.

In the process of working on the development of HiMA from this survey, we studied paving materials to further extend the pavement service life in Japan. We hope that the knowledge and new technologies obtained through this study will help improve paving technology in the future.

References

Takahashi S (2017) Field survey project to identify the important key factors for long term durability of asphalt pavement on the expressways in Japan. In: 7th European asphalt technology association (EATA) conference, Switzerland

Matsuno S, Nishizawa T (1992) Mechanism of longitudinal surface cracking in asphalt pavement. In: Proceedings of 7th international conference on asphalt pavement, Nottingham

Corte JF (2001) Development and uses of hard - grade asphalt and of high modulus asphalt mixes in France. Transportation Research Circular No. 503

Bankowski W, Tusar M, Wiman LG (2009) Laboratory and field implementation of high modulus asphalt concrete. Requirements for HMAC mix design and pavement design. In: Proceedings of the European Commission DG research, 6th framework programme, sustainable surface transport, Sustainable Pavements for European New Members States, SPENS
Nkgapel M et al (2012) Construction of a high modulus asphalt (HiMA) trial section Ethekwini: South Africa's first practical experience with design, manufacturing and paving of HiMA. In: Abstracts of the 31st Southern African transport conference
McHale M et al (2012) Evaluation of EME2 type mixtures incorporating softer grade binders – Phase I progress report. Transport Research Laboratory
Petho L et al (2014) High modulus high fatigue resistance asphalt (EME2) technology transfer. Austroads Ltd.

Laboratory Evaluation of Recycled Asphalt Pavement Material in Warm-Mix Asphalt

Haritha Malladi, Abhilash Kusam, and Akhtarhusein A. Tayebali

Abstract This research study focuses on results from laboratory tests on Warm-Mix Asphalt (WMA) mixtures prepared using two methods—Evotherm® 3G additive and "The Foamer" device by Pavement Technology Inc.—with varying percentages of Reclaimed Asphalt Pavement (RAP). The resulting mixture combinations were evaluated in the laboratory for their performance characteristics and compared with a control Hot Mix Asphalt (HMA) with the same amount of RAP. Tensile Strength Ratio (TSR) was used to evaluate the moisture susceptibility. Dynamic modulus (|E*|) tests were conducted to obtain master curves for all mixtures. They were also used to compute the |E*| Stiffness Ratio, i.e. the ratio of dynamic modulus values of moisture-conditioned specimens to that of unconditioned specimens, analogous to the Tensile Strength Ratio. Only the Foamer mixture with the highest amount of RAP did not pass the TSR test. No correlation was found between TSR and |E*| Stiffness Ratio. Despite lower production temperatures, WMA-RAP mixtures show similar |E*| behavior to HMA-RAP. Thus, WMA technologies can be used to help incorporate higher amounts of RAP rather than using a softer virgin binder grade in high-RAP HMA.

Keywords RAP · Warm-mix · Moisture-susceptibility · Workability · Asphalt

H. Malladi (✉)
Civil and Environmental Engineering, University of Delaware, 301 DuPont Hall, Newark, DE 19716, USA
e-mail: malladi@udel.edu

A. Kusam
Trimat Materials Testing, Inc., 1 Triangle Drive, Durham, NC 27709, USA
e-mail: akusam@trimattesting.com

A. A. Tayebali
Civil, Construction, and Environmental Engineering, NC State University, 2501 Stinson Drive, Raleigh, NC 27695-7908, USA
e-mail: tayebali@ncsu.edu

C. Raab (ed.), *Proceedings of the 9th International Conference on Maintenance and Rehabilitation of Pavements—Mairepav9*, Lecture Notes in Civil Engineering 76,
https://doi.org/10.1007/978-3-030-48679-2_24

1 Introduction

Warm-mix asphalt (WMA) and Reclaimed Asphalt Pavement (RAP) material are popular green technologies used in the pavement industry. Recycling asphalt provides considerable economic benefits, which is a great incentive for its continued use at even higher percentages. Cost savings and environmental benefits from the use of WMA depend on the type of technology that is used to achieve lower production temperatures.

Material performance issues related to fatigue, thermal cracking and moisture-susceptibility arise from the presence of stiff aged binder in RAP materials. The economic benefits of using RAP can be offset by reduced workability and performance issues if appropriate measures such as the use of softer binders or rejuvenating agents are not taken (West et al. 2009; Kaseer et al. 2019). Similarly, in WMA mixtures, there are concerns with permanent deformation, fatigue, and moisture-susceptibility, especially in mixtures produced with water-based technologies such as foaming devices and additives (Chowdhury and Button 2008; Bower et al. 2012). While the use of a softer binder grade is normally considered for high RAP mixtures to overcome workability issues (McDaniel and Anderson 2001; Willis et al. 2012), the use of WMA technologies may also help compensate for the deleterious effects of stiff binder in RAP on workability (Copeland et al. 2010; Mallick et al. 2008).

Various studies by state highway agencies have observed similar workability, stiffness and in-place densities for WMA-RAP as compared to HMA mixtures (Copeland et al. 2010; Hurley and Prowell 2005). In order to promote the use of RAP, it is necessary to study WMA as a viable tool that can help incorporate higher amounts of RAP. It is also important to ensure that WMA-RAP mixtures are resistant to moisture susceptibility.

2 Research Objectives

The scope of this study was to evaluate the material characteristics of WMA mixtures produced using two different methods of WMA production—Evotherm 3G additive and "The Foamer" device by Pavement Technology Inc. with varying percentages of RAP. Mixture performance was compared with the corresponding HMA RAP mixture in the laboratory. The specific research goals were to: (1) evaluate moisture susceptibility using AASHTO T283 Tensile Strength Ratio (TSR) test; (2) compare the stiffness of the mixtures using dynamic modulus master curves; and (3) assess whether dynamic moduli of the mixtures can be used to evaluate moisture damage.

3 Materials and Methods

All mixtures used in this study were designed with 9.5 mm (12.5 mm nominal maximum aggregate size) surface course using granite aggregate (commonly used in the state of North Carolina in USA) with the same aggregate gradation. In addition to the aggregates, pond fines passing 75 μm sieve size at 1.5% by weight of aggregates were incorporated into all the test mixtures. Anti-strip additive, AD-here® LOF 6500, was added at 0.75% by weight of binder for all mixtures.

RAP material was obtained from a single local source and stockpile to maintain uniformity. RAP was incorporated at 0, 20 and 40% by weight of the mixture. To control variability in RAP aggregate gradation in the test specimens, RAP material used in the study was first separated into two size fractions using a 4.75 mm (US No. 4) sieve. Binder content and aggregate gradation were separately determined for each fraction using ignition oven tests. A blend of one-third coarse fraction (retained on 4.75 mm sieve) and two-thirds fine fraction (passing 4.75 mm sieve) was found to be ideal for mixing with the virgin aggregates to meet the design aggregate gradation. The default virgin binder grade for all mixtures was PG 64-22. However, the test samples for HMA mixtures containing 40% RAP were prepared using a softer PG 58-28 binder grade to account for workability concerns. The tests for workability of the mixtures were elaborated in a previously published paper (Kusam et al. 2017).

Two WMA technologies were used in this study—MeadWestvaco's Evotherm 3G (chemical additive) and The PTI Foamer device (water-based laboratory foaming device). Evotherm was added to the asphalt binder at a rate of 0.5% by weight of total binder in the mixture. PTI Foamer device was set to use 2% water by weight of binder for the foaming process. Mixing and compaction temperatures for all HMA mixtures were fixed at 163 °C (325 °F) and 149 °C (300 °F), respectively. All WMA mixtures were mixed and compacted at 135 °C (275 °F) and 120 °C (248 °F), respectively. The production temperatures were not changed with varying RAP contents. The mixing and compaction temperatures, and the additive dosages were chosen to reflect plant production practices at the time of study.

A two-step RAP heating procedure as recommended by the Texas Transportation Institute was followed (Zhou et al. 2011). Sample RAP fractions for each test specimen were weighed and heated at 60 °C for 12 h in individual trays. Then, they were heated to the target mixing temperature for two hours. After the two hours of heating at mixing temperature, the two fractions were mixed with the virgin aggregate and binder to prepare the test specimens.

All mixtures were designed using 65 gyrations of the Superpave gyratory compactor. Optimum asphalt content was determined for the HMA mixture at each of the three RAP contents. The same asphalt content was used in the corresponding WMA mixtures. For 0, 20 and 40% RAP mixtures, the design binder content was determined to be 6, 5.9 and 5.8%, respectively; for each 20% increase in RAP content, the optimum binder content reduced by 0.1% (Kusam et al. 2017). The amount of virgin binder was adjusted accordingly while preparing RAP mixtures.

4 Tensile Strength Ratio

Modified AASHTO T 283 (2014), "Standard Method of Test for Resistance of Compacted Hot Mix Asphalt (HMA) to Moisture-Induced Damage" as followed by North Carolina Department of Transportation (NCDOT) was used to determine the TSR. As per standard specifications, specimens were compacted to a height of 95 ± 5 mm and a target air void content of 7 ± 0.5%. For each mixture, unconditioned and moisture-conditioned sets with four replicates were prepared as per the standard specifications. NCDOT does not specify freeze-thaw cycles for TSR tests. The tensile strength ratio was determined for all mixtures as a percentage of the conditioned to unconditioned Indirect Tensile Strength (IDT).

The required minimum passing ratio specified by NCDOT is 85%. The IDT and TSR values for all mixtures are presented in Fig. 1. Figure 2 shows an interaction plot between the effects of mixture type (HMA/Evotherm/Foamer), the amount of RAP, and moisture conditioning on IDT values. Except for Foamer mixtures with 40% RAP, all mixtures pass the NCDOT acceptance criterion of 85% TSR. It should be noted that all mixtures contain an LOF 6500 anti-strip additive.

The results indicate that the addition of RAP material generally increases the IDT values in both conditioned and unconditioned states (Figs. 1 and 2b). However, the IDT values of 40% RAP HMA mixtures were slightly lower than the IDT values of HMA mixtures with 20% RAP (Fig. 2f). This may be due to the use of a softer virgin binder grade of PG 58-28 in 40% RAP HMA. TSR values for each mixture type decreased with increasing RAP content. In Fig. 2e, the slopes of the plot lines increase as the amount of RAP increases, showing a larger difference between IDT

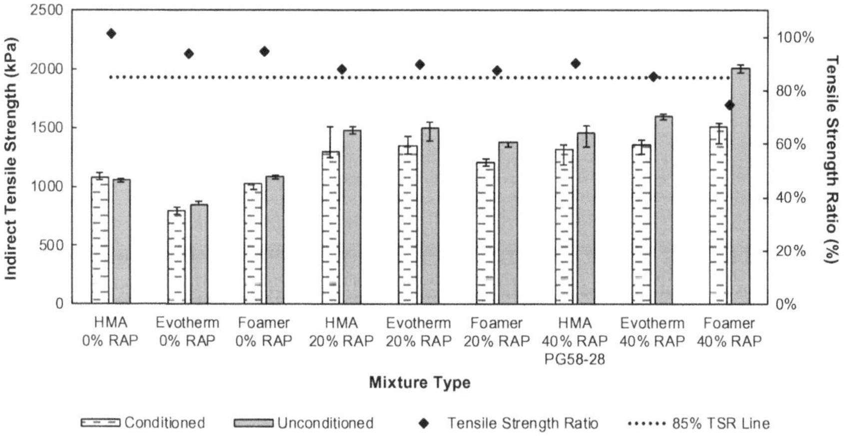

Fig. 1 Indirect tensile strength values and tensile strength ratios along with acceptance criterion. The bar graphs indicate the median IDT values. Error bars show the range of values. For HMA with 40% RAP, a softer binder grade PG 58-28 was used. All other mixtures were prepared with PG 64-22 binder

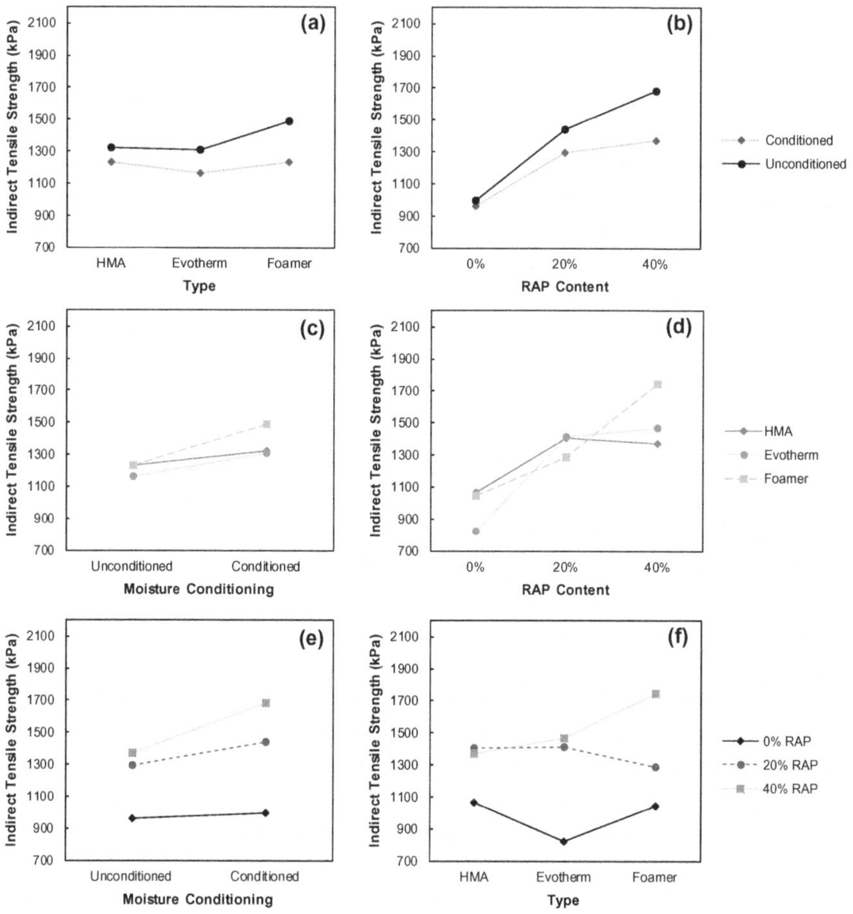

Fig. 2 Interaction plots showing relationships between mixture type, RAP content, and moisture conditioning and their effect on average indirect tensile strength values

values in the unconditioned and moisture-conditioned states with the addition of RAP.

Evotherm mixtures with no RAP had the lowest IDT values of all mixtures—even in the unconditioned state, its IDT values were lower than all measured IDT values for other mixtures (Figs. 1 and 2d, f). With 20% RAP, there was no significant difference between the IDT values of HMA and WMA mixtures (Fig. 2d). However, 40% RAP mixtures exhibited significant differences between all mixture types (Fig. 2e). Foamer mixture with 40% RAP had the highest unconditioned IDT value of all mixtures. Evotherm and Foamer mixtures with 40% RAP and PG 64-22 binder exhibited higher conditioned and unconditioned IDT values than the HMA mixture with 40% RAP and softer binder PG 58-28.

5 Dynamic Modulus

For all mixtures, dynamic modulus (|E*|) was measured at three temperatures: 4, 20 and 40 °C, and at six frequencies: 25, 10, 5, 1, 0.5 and 0.1 Hz, at each temperature, respectively. Specimens test were prepared according to the procedure described in AASHTO TP 79 (2015), "Standard Method of Test for Determining the Dynamic Modulus and Flow Number for Hot Mix Asphalt (HMA) Using the Asphalt Mixture Performance Tester (AMPT)". The specimens were initially compacted to a height of 178 mm with a diameter of 150 mm and were then cut and cored to dimensions of 150 ± 2.5 mm height and 100 ± 1 mm diameter for testing. Three replicates were tested for each mixture combination.

5.1 *|E*| Stiffness Ratio*

Since |E*| is a fundamental material property, some studies have focused on using |E*| stiffness ratio (ESR) to evaluate moisture susceptibility (Hajj et al. 2013; Nadkarni et al. 2013). To obtain ESR, samples were prepared with target air void content of 7 ± 0.5%. The ESR value is obtained from conditioned and unconditioned subsets of dynamic modulus specimens, which are subjected to a conditioning procedure similar to the TSR test. ESR is defined as the ratio of average dynamic modulus of conditioned specimens to the average dynamic modulus of unconditioned specimens. Since dynamic modulus is measured at different test temperatures and loading frequencies, ESR values can be obtained as averages for each test temperature. However, for the purposes of comparing ESR with TSR results, the |E*| values at 20 °C and 1 Hz loading frequency were used.

Since the dynamic modulus test is a non-destructive test unlike the AASHTO T 283 test for TSR, the same specimens were used for testing in both unconditioned and conditioned states. First, the dynamic moduli of unconditioned specimens were measured. Tests on conditioned specimens were conducted exactly one week later to allow recovery of residual viscoelastic strains in specimens from the unconditioned testing. Specimens were then vacuum-saturated to fill 35–45% of the air-voids with water. The saturated specimens were placed in a water bath at 60 °C for 24 h. After removal, the specimens were left to air-dry at room temperature for a period of 24 h to facilitate proper mounting of LVDTs on a dried specimen surface.

Figure 3 shows the average dynamic moduli values of unconditioned and conditioned specimens at a test temperature of 20 °C and 1 Hz loading frequency. The error bars indicate the range of values. As in the case of IDT, 0% RAP mixtures had the lowest |E*| values; and dynamic moduli values increased with the addition of RAP.

The 40% RAP HMA mixture had the highest unconditioned dynamic modulus value (as well as a high variability in the values), resulting in a low ESR. It can therefore be inferred that even with a softer binder grade, the HMA mixtures with higher

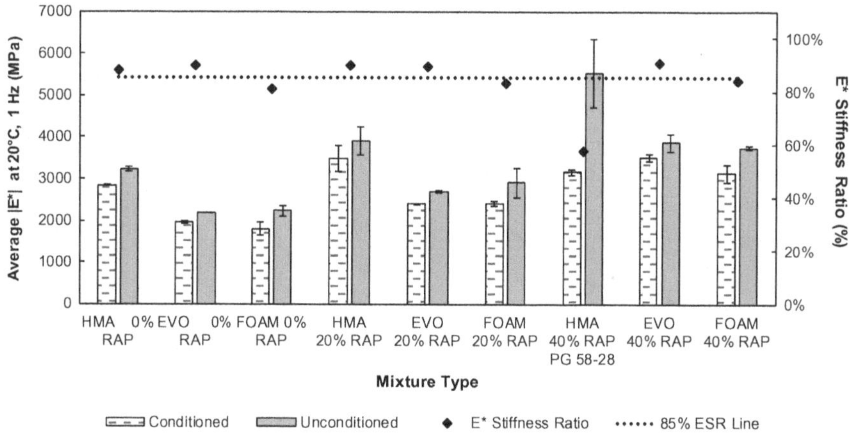

Fig. 3 Average dynamic modulus at 20 °C and 1 Hz as tested on specimens with 7% air voids along with |E*| Stiffness Ratio. Error bars show the range of values. For HMA with 40% RAP, a softer binder grade PG 58-28 was used. All other mixtures were prepared with PG 64-22 binder

amounts of RAP may have an undesirably high dynamic modulus value that could exacerbate fatigue damage. Upon moisture conditioning, these mixtures experience a large loss in dynamic modulus. WMA mixtures work well in this scenario where the need for binder grade change is eliminated while simultaneously preventing loss of stiffness in the presence of moisture.

5.2 Linear Viscoelastic Master Curves

To obtain the asphalt mixture master curves, specimens were prepared and tested at a target air void content of $4 \pm 0.5\%$. The dynamic modulus master curves for the HMA and WMA mixtures containing 0%, 20%, and 40% RAP are shown in Fig. 4. Without the presence of RAP, HMA mixtures exhibited higher dynamic moduli values than those produced with Evotherm and Foamer (Fig. 4a). With the addition of RAP, the dynamic modulus master curves generally show the same trends for all three mixture types. From the high degree of overlap between the |E*| master curves of all 40% RAP mixtures, it can be inferred that the material performance of WMA-RAP mixtures is similar to HMA-RAP even though they were produced at lower temperature without the use of a softer binder grade.

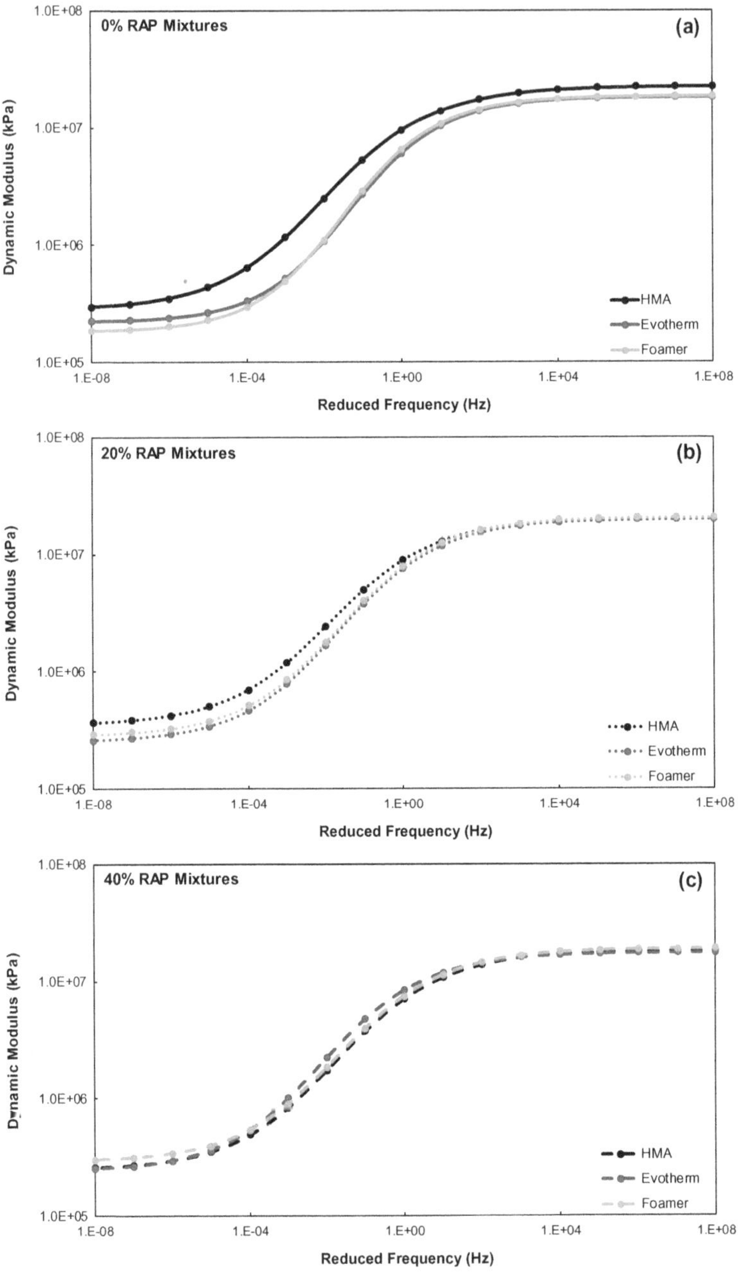

Fig. 4 Dynamic modulus master curves for HMA and WMA mixtures with **a** 0% RAP, **b** 20% RAP, and **c** 40% RAP

6 Comparison of TSR and ESR

While TSR values decreased with the addition of RAP, ESR values did not exhibit such a trend. The results indicate that there is no correlation between ESR and TSR values. TSR test measures properties in indirect tension, whereas the |E*| value is measured in compression mode of loading in AMPT. The primary cause of moisture damage is adhesive failure between aggregate and binder—subjecting their interface to tensile stress is the appropriate test for adhesion (Lee et al. 2013). Therefore, the |E*| ratio (ESR) may not be appropriate in evaluating the moisture sensitivity of mixtures, because compression testing measures more of aggregate structure properties as opposed to the binder properties in tensile or flexural mode of loading. Further investigation is needed with |E*| values obtained from tensile or flexural tests.

7 Summary and Conclusions

In this study, two WMA technologies, Evotherm 3G and PTI Foamer were used to produce mixtures with varying amounts of RAP. The mixtures' moisture susceptibility was compared with that of corresponding HMA RAP mixtures using Tensile Strength Ratio and |E*| Stiffness Ratio. The dynamic modulus master curves for all mixtures were compared as an indicator of mixture performance. Based on the results, the following conclusions can be made:

1. Non water-based WMA (additives like Evotherm) are more resistant to moisture susceptibility, and thus, preferable over water-based foaming WMA technologies.
2. The dynamic modulus master curves for HMA and WMA RAP mixtures show the same trend. Thus, WMA technologies can be used to incorporate higher amounts of RAP at lower production temperatures in lieu of using a softer binder grade in HMA RAP mixtures.
3. E* Stiffness Ratio (ESR) and Tensile Strength Ratio (TSR) do not show any correlation as they test different aspects of material behavior vis-à-vis tensile versus compressive properties.

Acknowledgements The authors would like to thank the North Carolina Department of Transportation for funding this project and their generous help and insights.

References

AASHTO (2014) Standard method of test for resistance of compacted asphalt mixtures to moisture-induced damage, Washington, D.C.

AASHTO (2015) Standard method of test for determining the dynamic modulus and flow number for hot mix asphalt (HMA) using the asphalt mixture performance tester (AMPT), Washington, D.C.

Bower N, Wen H, Willoughby K, Weston J, DeVol J (2012) Evaluation of the performance of warm mix asphalt in Washington state. Report No. WA-RD 789.1. Washington Department of Transportation, Olympia, Washington

Chowdhury A, Button JW (2008) A review of warm mix asphalt. Report No. SWUTC/08/473700-00080-1. Texas Transportation Institute, College Station, Texas

Copeland A, D'Angelo J, Dongre R, Belagutti S, Sholar G (2010) Field evaluation of high reclaimed asphalt pavement-warm-mix asphalt project in Florida. Transportation research record: journal of the transportation research board, No. 2179. Transportation Research Board of the National Academies, Washington, D.C., pp 93–101

Hajj E, Souliman M, Alavi M, Salazar LL (2013) Influence of hydrogreen bio-asphalt on viscoelastic properties of reclaimed asphalt mixtures. Transportation research record: journal of the transportation research board, No. 2371. Transportation Research Board of the National Academies, Washington, D.C.

Hurley GC, Prowell BD (2005) NCAT Report 05-04: evaluation of Aspha-Min® zeolite for use in warm mix asphalt. National Center for Asphalt Technology. National Center for Asphalt Technology, Auburn, Alabama

Kaseer F, Bajaj A, Epps Martin A, Arambula-Mercado E, Hajj E (2019) Strategies for producing asphalt mixtures with high RAP content. J Mater Civ Eng 31(11):05019002

Kusam A, Malladi H, Tayebali AA, Khosla NP (2017) Laboratory evaluation of workability and moisture susceptibility of WMA mixtures containing RAP. J Mater Civ Eng 29(5):04016276

Lee J, Lee J, Kwon S, Kim Y (2013) Use of cyclic direct tension tests and digital imaging analysis to evaluate moisture susceptibility of warm-mix asphalt concrete. Transportation research record: journal of the transportation research board, 2372. National Research Council, Washington, D.C.

Mallick RB, Kandhal PS, Bradbury RL (2008) Using warm-mix asphalt technology to incorporate high percentage of reclaimed asphalt pavement material in asphalt mixtures. Transportation research record: journal of the transportation research board, No. 2051. Transportation Research Board of the National Academies, Washington, D.C., pp 71–79

McDaniel R, Anderson MR (2001) NCHRP Report 452: recommended use of reclaimed asphalt pavement in the superpave mix design method: technician's manual. Transportation Research Board of the National Academies, Washington D.C.

Nadkarni A, Kaloush K, Zeiada WA, Biligiri KP (2013) Using dynamic modulus test to evaluate moisture susceptibility of asphalt mixtures. Transportation research record: journal of the transportation research board, No. 2127. Transportation Research Board of the National Academies, Washington, D.C., pp 29–35

West R, Kvasnak A, Tran N, Powell B, Turner P (2009) Testing of moderate and high reclaimed asphalt pavement content mixes. Transportation research record: journal of the transportation research board, No. 2126. Transportation Research Board of the National Academies, Washington, D.C., 2009, pp 100–108

Willis R, Turner P, Padula FG, Tran NH, Julian G (2012) NCAT Report 12-03: effects of changing virgin binder grade and content on RAP mixture properties. National Center for Asphalt Technology, Auburn, Alabama

Zhou F, Hu S, Das G, Scullion T (2011) High RAP mixes design methodology with balanced performance. Report No. FHWA/TX-11/0-6092-2. Texas Transportation Institute, College Station, Texas

Assessing Self-healing Asphalt by the Heating of Asphalt Mixtures

Caio Santos, Marina Cabette, Jorge Pais, Vitor Carvalho, and Paulo Pereira

Abstract Roads are the essential transport modal worldwide since it facilitates the movement of people and goods. Therefore, it is a concern to reduce costs and be sustainable for maintenance and rehabilitation. So self-healing has been studied. The self-healing properties of the asphalt mixtures had already been studied. It is previously known that asphalt mixtures can recover its strength entirely autonomously, or just with a small external stimulus such as heat. The main objective of this paper is to study the recovery of asphalt mixtures only by heating them. After heating, the complex modulus and fatigue resistance were assessed, and the capacity of recovering each mixture was evaluated. AC14 asphalt mixtures (14-mm maximum aggregate size) were produced with conventional bitumen, varying the asphalt content, namely 3.7, 4.0, 4.3, and 4.6%. Fatigued beams were heated from 90–150 °C after fatigue failure. Fatigue resistance tests were conducted at 20 °C temperature and 10 Hz frequency. It was verified that certain conditions of the heating process allow the mixtures to recover some of the initial resistance. Also, it was noted that the higher the temperature and the percentage of asphalt content, it was higher the values of complex modulus and fatigue resistance as well.

Keywords Asphalt mixtures · Self-healing · Complex modulus · Fatigue resistance · Recovery

C. Santos (✉) · M. Cabette · J. Pais · V. Carvalho · P. Pereira
Universidade do Minho, Guimarães, Portugal
e-mail: caiorubens@maua.br

M. Cabette
e-mail: marinacabette@gmail.com

J. Pais
e-mail: jpais@civil.uminho.pt

P. Pereira
e-mail: ppereira@civil.uminho.pt

C. Santos
Mauá Institute of Technology, São Caetano do Sul, Brazil

C. Raab (ed.), *Proceedings of the 9th International Conference on Maintenance and Rehabilitation of Pavements—Mairepav9*, Lecture Notes in Civil Engineering 76,
https://doi.org/10.1007/978-3-030-48679-2_25

1 Introduction

Roads are the essential transport modal worldwide since it facilitates the movement of people and goods. Therefore, there is a lot of effort and concern in the maintenance and rehabilitation of pavements. These processes generally require significant investments and demand considerable energy expenditure, even more for asphalt pavements. Due to this fact, it is common to observe a large part of the road network requiring maintenance interventions. Therefore, there is a concern to find cheaper and more sustainable solutions than the conventional technique. The primary degradation that affects the quality of the pavements is cracking, so, researchers have study ways to eliminate the cracking of pavements in their initial phase.

It was noticed that asphalt mixtures have an intrinsic characteristic after deformation. They can selfheal micro-cracks when subjected to rest periods (Xu et al. 2018a). Recovery does not occur only due to material characteristics, but it is also influenced by temperature, curing time, degree of damage, load condition, among others (Xu et al. 2018b). This ability was shown in laboratory tests and field since the 60s (Bazin and Saunier 1967; Van Dijk 1972). Although healing can restore material properties, the recovery becomes quite limited if an additional process is not applied, such as heating induction or capsules (Al-Mansoori et al. 2018).

In recent years, much research proved that when the temperature is increased, the asphalt mixtures and mastics could have their strength recovered. Carvalho (2016) investigated the self-healing only by heating fatigued asphaltic beams and concluded that asphaltic mixtures did not fully recover their strength, however as very reasonable strength recovery was verified. Thus, the self-healing by heating can be a cheaper and sustainable alternative to maintain and extend the life period of asphalt pavements. Additionally, to make a conductive asphalt mixtures, materials such as graphite or fibres (Brovelli et al. 2014, 2013) can be used to facilitate heating of the bituminous mixture through two methods of heating, namely, electromagnetic induction heating (Dai et al. 2013; García et al. 2013; Liu et al. 2012) and microwave induction heating (Qiu 2012). However, merely heating the upper layers of the asphalt pavement as a self-healing technique could be potentially used in damaged pavements (Minhoto et al. 2005; Sousa et al. 2005).

Many laboratory and in situ tests have been used to characterize the healing of bitumen, mortar, mastic, asphaltic mixtures and full road pavements. Overall, the methods to assess the self-healing potential of bituminous materials differ on the way the rest periods are introduced, and different properties and parameters of the mixture can be taken to establish recovered performance. Accordingly, there have been used different methods and properties to analyze the data of the healing tests, such as stiffness recovery, dissipated energy approach, number of loads to failure on fatigue tests, and others. (Carpenter and Shen 2006; Wistuba et al. 2017).

The main objective of this paper is to study the recovery of asphalt mixtures after heating them. Stiffness and fatigue life of asphalt concrete beams were used to evaluate the performance of the mixtures using the four-point bending test. Additionally, the effect of long resting periods was studied by performing the analysis after the production of the beam and after one year. The influence of heating temperature was also considered. To evaluate the recovery, a potential healing index calculated by measuring the stiffness and fatigue strength of the tested asphalt concrete beams.

2 Materials and Methods

The study carried out is based on four different asphalt mixtures. A dense-graded asphalt mixture, AC14 (14-mm maximum aggregate size) and a 35/50 penetration grade asphalt binder were used in all asphalt mixtures, but varying the binder content in 3.7, 4.0, 4.3, and 4.6%. Table 1 presents the properties of the asphalt binder.

In order to verify the healing potential, the effects of heating in the complex modulus and fatigue life varying the temperature of heat were investigated. A rest period of one year was assured between the production of the beams and heating procedure. Overall, 8 beams of each mixture of the four binder contents (3.7, 4.0, 4.3, and 4.6%) were produced, a total of 32 beams. The tests were conducts also on the production year (Y0) and one year later (Y1).

At year Y0, the beams were separated into two groups of 16 beams each (Groups 1 and 2). The complex modulus at temperatures of 0, 10, 20 and 30 °C was determined for group 1, and the beams into group 2 conducted the complex modulus teste and the fatigue test at 20 °C.

At year Y1, to determine the effects of the rest period of one year, the complex modulus and fatigue life tests were carried out for all the 32 beams. Afterwards, the heating procedure in the oven for 4 h was conducted in order to recovery the stiffness of the fatigued beams. For a day the beams were rest after the heating and then they were tested. The heating temperatures varied from 90–15 °C and were carried in up 3 phases, as presented in Table 2. Both complex modulus and fatigue tests at 20 °C were conducted after each heating procedure.

The complex modulus tests were conducted according to the EN 12697-26 standard, testing prismatic specimens in four-point bending at 20 °C. However, for the beams of group 1, the temperatures used in the tests were the following: 0, 10, 20, and 30 to produce master curves. The influence of the loading condition of the asphalt mixtures was investigated. Thus, the tests were carried out using the following frequencies: 0.1, 0.2, 0.5, 1, 2, 5, 8 and 10 Hz. The phase angle was also determined in the same range of frequency and temperature conditions.

The fatigue tests were carried out according to the EN 12697-24, at a temperature of 20 °C with a frequency of 10 Hz in a controlled strain mode of loading. Since

Table 1 Properties of 35/50 penetration grade asphalt (EN 1426; EN 1427; EN 13302)

Properties	Units	Limit	Test results
Penetration at 25 °C	0.1 mm	35–50	41.03
Softening point	°C	50–58	50.10
Complex viscosity at 135 °C	Pa s	–	0.49

Table 2 Heating temperatures for the beams

Beams	Tests conducted at Y0	Heating temperatures at Y1 in phase		
		1st	2nd	3rd
Group 1a (8 beams)	Complex modulus	90 °C[a]	110 °C[a]	110 °C[a]
Group 1b (8 beams)		130 °C[a]	150 °C[a]	–
Group 2a (8 beams)	Complex modulus and fatigue resistance	90 °C[a]	110 °C[a]	110 °C[a]
Group 2b (8 beams)		130 °C[a]	150 °C[a]	–

[a]Complex modulus and fatigue (strain levels for the fatigue tests were 200E-6 to 800E-6)

the breaking of the specimens is not visible, a 50% reduction of its initial complex modulus was considered to determine fatigue life.

3 Results Analysis

3.1 Complex Modulus

The complex modulus of the beams of group 1 was measured before (Y0) and after (Y1) the rest period of one year, which they were stored at room temperature. The master curves for the asphalt mixtures are presented in Fig. 1 (37 means 3.7% of asphalt content, and so on for 4.0, 4.3 and 4.6%). It was verified that, after the rest period of one year, the complex modulus was increased for all asphaltic mixtures tested.

After that, the complex modulus tests at 20 °C and 10 Hz were conducted for all specimens at year 0 and year 1 before and after the heating procedures. Figure 2 on the left presents the average of the complex modulus obtained for the specimens into

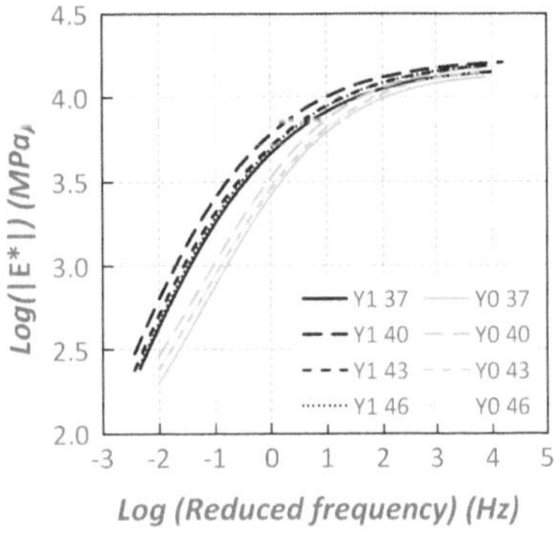

Fig. 1 Complex modulus at year 0 (Y0) and year 1(Y1) after the rest period for beams of group 1

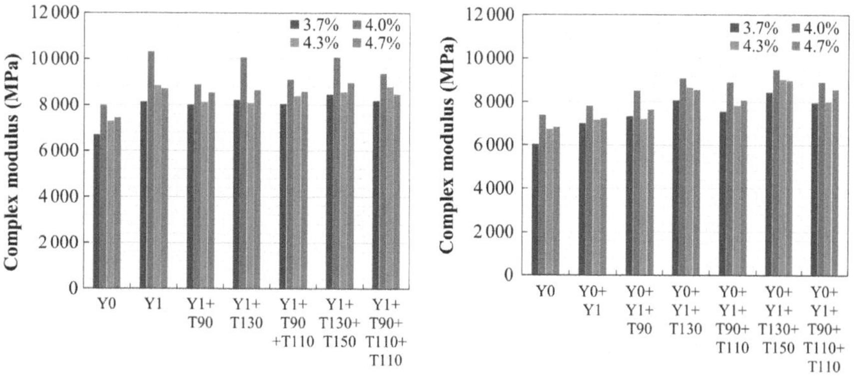

Fig. 2 Complex modulus at 20 °C and 10 Hz of Group 1 (left) and Group 2 (right), |E*| in MPa

group 1 for each asphalt content. Fatigue tests were conducted only after complex modulus testing in year 1.

Therefore, comparing the values obtained at year 0 and year 1, an increase in complex modulus of approximately 20% was verified due to the rest period.

After that, the fatigue tests were conducted, and after each heating procedure of the fatigued specimens, the complex modulus was carried out. It was verified an average recovery of about 92% on the complex modulus values when the beams were heated at 90 °C and about 98% when the heating procedure was conducted at 130 °C. Two additional proceedings of heating and fatigue tests were conducted. Finally, the total recovery from 92 to 106% was observed in the complex modulus of group 1 specimens.

Figure 2 on the right presents the results of complex modulus for beams in group 2. The group 2 beam fatigue tests were conducted in all stages of the experiment, including before the rest period of one year. Therefore, the observed increase was lower than the observed in beams of group 1. The verified increase was about 6% for 4.0, 4.3 and 4.6% asphalt content beams. For 3.7% asphalt content beams an increase yup to 20% was observed.

It was conducted the heating procedures as same as conducted to group 1 specimens. At the first heating procedure, the complex modulus of all specimens was recovered, and the specimens become stiffer as the following heating procedures were conducted. Finally, a stiffening of approximately 15% in relation to the complex modules of year Y0 was observed for group 2 beams.

3.2 *Fatigue Life*

The fatigue tests of the beams of group 2 were carried out at year 0 (Y0) and year 1 (Y1), after the rest period of one year. The four points bending test was used in a control strain mode of loading at 20 °C. Figure 3 presents the results for the fatigue

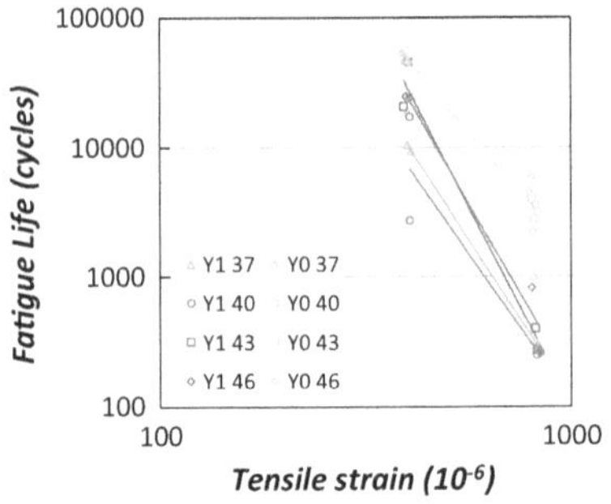

Fig. 3 Fatigue life year 0 (Y0) and year 1 (Y1) after the rest period for beams of group 2

life for beams at year 0 and year 1. It was verified that after the rest period of one year, the fatigue life of the beams for all mixtures decreased.

In order to assess the healing potential of the asphalt mixtures by heating the specimens, the fatigue tests were carried out varying the heating temperature. Figure 4

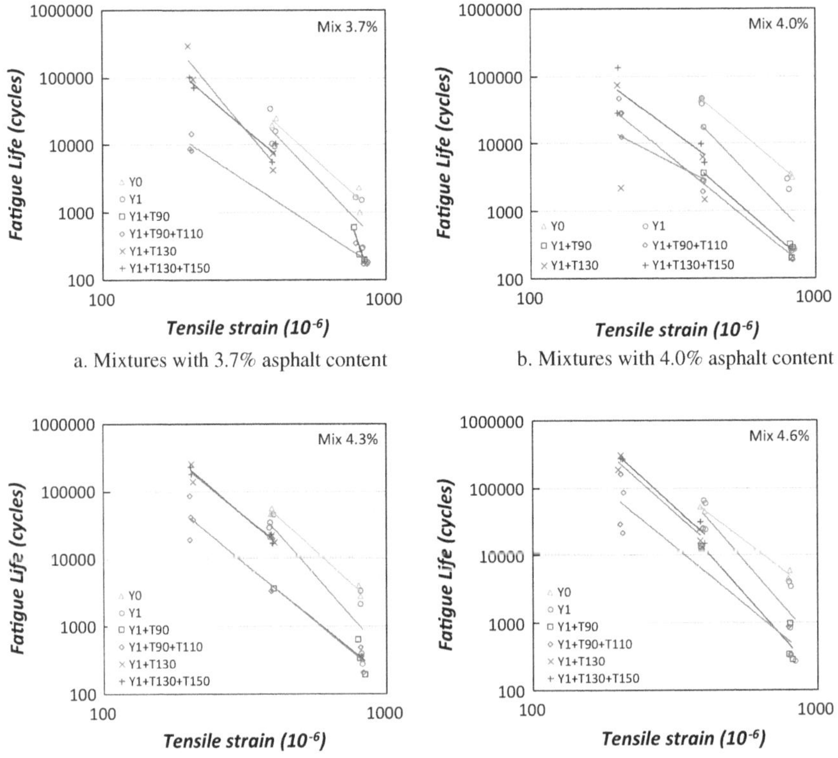

a. Mixtures with 3.7% asphalt content

b. Mixtures with 4.0% asphalt content

c. Mixtures with 4.3% asphalt content

d. Mixtures with 4.6% asphalt content

Fig. 4 Fatigue life for all mixtures asphalt content

presents the fatigue test results. Two heating procedures, considering different temperatures, were conducted for both groups of beams. One heating procedure considered the first heating of 90 °C followed by 110 °C and the other used the first heating of 130 °C and then 150 °C.

A decrease in fatigue life is observed in the mixtures. Since the mixtures became stiffer and the fatigue life became lower, the recovery is not verified. In the fatigue tests, the dissipated energy was recorded, making it possible to analyze the problem under the dissipated energy approach, as presented in the next section.

3.3 *Healing Potential Index (HPI)*

It was verified that the rest period and the heating procedures affect the performance of the mixtures, turning them stiffer and reducing fatigue life. Therefore, to better understand the recovery capacity of mixtures, it is proposed to use an index, the HPI (Healing Potential Index). This index is defined by Eq. 1.

$$HPI = \frac{DE}{a.N^b} \tag{1}$$

where DE represents the dissipated energy at the failure of the tested sample, *a* (intercept) and *b* (slope) are the coefficients of the dissipated energy law of the bituminous mixture of each beam and *N* is the fatigue life of the tested sample. Figure 5 show the average of the HPI for each mixture (3.7, 4.0, 4.3, and 4.6% asphalt content). On the left are represented the results of the first heating procedure (90 and 110 °C) and on the right are the results for the other heating procedure (130 and 150°C).

The beams of group 1 were submitted to fatigue tests only after the one-year rest period, while group 2 beams were tested to fatigue at year 0 (right after production) and after the one-year rest period. Despite this, the average initial HPI resulted in about 1.0 to all mixtures.

Regarding group 1 tests, it was observed that the lower the heating temperatures are, the lower resulted in the HPI. After heating the beams at 90 °C, a decrease of the HPI was observed, and partial recovery was verified after the last heating procedure of 110 °C. The recovery was more pronounced as the asphalt content in the mixture is increased. In the specimens of group 1, where the heating temperatures were 130 and 150 °C, the highest values of HPI were observed.

The beams of group 2 were subjected to fatigue tests before and after the one-year rest period. Therefore, the HPI calculated at Y0 + Y1 is only due to the rest period. After the last heating procedure of 110 °C, a partial recovery was observed. For the specimens subjected to the higher temperatures, a higher recovery was observed.

Regardless of the heating procedures and the two groups of tests, it was observed that the higher the asphalt content in the mixture is the higher is the HPI. Thus, mixtures with 4.3 and 4.6% of asphalt content present a better response after heating.

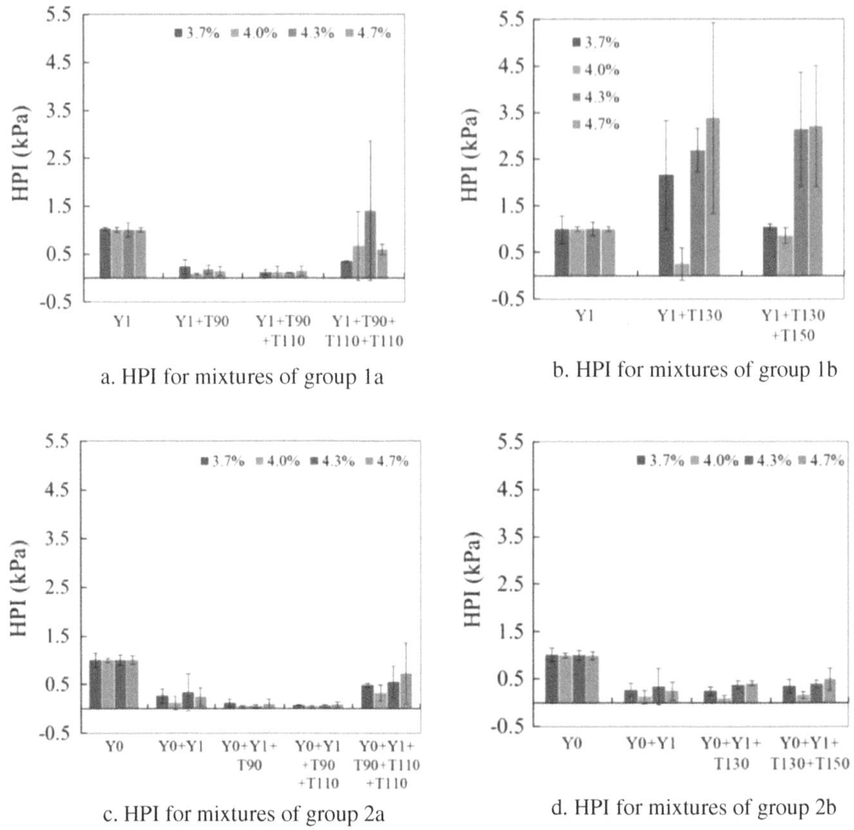

Fig. 5 Healing Potential Index (HPI) for mixtures of group 1 and 2 for both heating procedures

4 Conclusions

In the present study, laboratory tests were carried out to evaluate the healing potential of four asphaltic mixtures. The results obtained showed that the tested beams did not fully recover their load capacity, but achieved a very reasonable load capacity recovery, as the only treatment used in the fatigued beams was to heat them.

The stiffness and fatigue life of the specimens were evaluated after the one-year rest period. It was verified that the mixtures became stiffer and fatigue life decreased after the rest period even heating the beams. However, it was noted that the higher the temperature and the percentage of asphalt content, it was higher the values of complex modulus and fatigue resistance as well.

It should also be noted that there were beams tested to fatigue five times, and even in the fifth test, the results showed that the beams recovered some load capacity. It should also be noted that the temperature at which the beams were heated increased according to the tests, and the higher the temperature was, the higher the HPI.

Overall, the tests carried out in this study have presented entirely satisfactory results, showing that it was possible to recover load capacity in fatigued beams until the end of life, by only heating them, which in theory demonstrates that it is possible to recover end-of-life pavements by merely heating the surface layer.

References

Al-Mansoori T, Norambuena-Contreras J, Garcia A (2018) Effect of capsule addition and healing temperature on the self-healing potential of asphalt mixtures. Mater Struct/Materiaux et Constructions 51(2):1–12. https://doi.org/10.1617/s11527-018-1172-5

Bazin P, Saunier J (1967) Deformability, fatigue, and healing properties of asphalt mixes. In: Second international conference on the structural design of asphalt pavements, pp 2–3

Brovelli C, Crispino M, Pais JC, Pereira PAA (2014) Assessment of fatigue resistance of additivated asphalt concrete incorporating fibers and polymers. J Mater Civ Eng 26(3):554–558. https://doi.org/10.1061/(ASCE)MT.1943-5533.0000837

Brovelli C, Hilliou L, Hemar Y, Pais J, Pereira P, Crispino M (2013) Rheological characteristics of EVA modified bitumen and their correlations with bitumen concrete properties. Constr Build Mater 48:1202–1208. https://doi.org/10.1016/j.conbuildmat.2013.07.032

Carpenter SH, Shen S (2006) Dissipated energy approach to study hot-mix asphalt healing in fatigue, no. 1970, pp 178–185

Carvalho VHG (2016) Verificação Do 'Self-Healing Asphalt' Através Do Aquecimento Das Misturas Betuminosas. Guimarães: Universidade do Minho

Dai Q, Wang Z, Hasan MRM (2013) Investigation of induction healing effects on electrically conductive asphalt mastic and asphalt concrete beams through fracture-healing tests. Constr Build Mater 49:729–737. https://doi.org/10.1016/j.conbuildmat.2013.08.089

Van Dijk W et al (1972) Fatigue of bitumen and bituminous mixes. In: 3rd international conference on the structure design of pavement, pp 354–366

García A, Bueno M, Norambuena-Contreras J, Partl MN (2013) Induction healing of dense asphalt concrete. Constr Build Mater 49:1–7. https://doi.org/10.1016/j.conbuildmat.2013.07.105

Liu Q, Schlangen E, Van De Ven M, Van Bochove G, Van Montfort J (2012) Evaluation of the induction healing effect of porous asphalt concrete through four point bending fatigue test. Constr Build Mater 29:403–409. https://doi.org/10.1016/j.conbuildmat.2011.10.058

Minhoto MJC, Pais JC, Pereira PAA, Picado-Santos LG (2005) The influence of temperature variation in the prediction of the pavement overlay life. Road Mater Pavement Des 6(3):365–384. https://doi.org/10.1080/14680629.2005.9690012

Qiu J (2012) Self healing of asphalt mixtures. Delft University of Technology

Sousa JB, Pais JC, Way GB (2005) A mechanistic-empirical based overlay design method for reflective cracking. Road Mater Pavement Des 6(3):339–363. https://doi.org/10.1080/14680629.2005.9690011

Wistuba MP, Falchetto AC, Isailovic I (2017) Influence of rest period on asphalt recovery considering nonlinearity and self-heating. Constr Build Mater 140:321–327. https://doi.org/10.1016/j.conbuildmat.2017.02.122

Xu S, Tabaković A, Liu X, Schlangen E (2018a) Calcium alginate capsules encapsulating rejuvenator as healing system for asphalt mastic. Constr Build Mater 169:379–387. https://doi.org/10.1016/j.conbuildmat.2018.01.046

Xu S, García A, Junfeng S, Liu Q, Tabaković A, Schlangen E (2018b) Self-healing asphalt review: from idea to practice. Adv Mater Interfaces 5(17):1–21. https://doi.org/10.1002/admi.201800536

Ultrasound Monitoring and Microwave Self-healing of Top-Down Cracks in Asphalt Pavements

Miguel A. Franesqui, Jorge Yepes, and Juan Gallego

Abstract Surface-initiated cracking with top-down propagation (TDC) is one of the most frequent and important failure modes of asphalt pavements. In order to achieve long-lasting pavements, it is necessary to control the evolution of these cracks and so repair them before they become deeper and deteriorate the lower layers. Self-healing of asphalt mixtures is possible if the temperature is raised near the softening point of the binder, thus allowing the fusion of the cracks. For this purpose, conductive additions can be used to promote induction heating when applying electromagnetic fields. This laboratory work shows the self-healing results of TDC on bituminous mixtures after microwaves exposure. Different mixtures (semi-dense asphalt concrete AC-S, gap-graded asphalt concrete for very thin layers AC-VTL and porous asphalt PA) with diverse types, sizes and proportions of metallic additions from industrial waste were tested. Three aspects were studied: (a) analysis of the type, particle size and content of each addition on the heating speed; (b) temperature increase with the specific energy; (c) monitoring of the healing process by using ultrasounds. Microwave exposure allowed the total closure of cracks using an industrial waste, with reduced exposure times and applied energies. The results validate the microwave healing capacity, as well as the use of ultrasounds for tracking the crack depth.

Keywords Self-healing · Microwave · Metallic addition · Industrial waste · Crack depth · Ultrasound

M. A. Franesqui (✉) · J. Yepes
Research Group of Integral and Advanced Manufacturing, Department of Civil Engineering, School of Engineering (EIIC), University of Las Palmas de Gran Canaria (ULPGC), 35017 Las Palmas de Gran Canaria, Canary Islands, Spain
e-mail: miguel.franesqui@ulpgc.es

J. Gallego
Research Group of Highway Engineering, Department of Civil Engineering: Transport and Territory, School of Civil Engineering (ETSICCP), Technical University of Madrid (UPM), 28040 Madrid, Spain

C. Raab (ed.), *Proceedings of the 9th International Conference on Maintenance and Rehabilitation of Pavements—Mairepav9*, Lecture Notes in Civil Engineering 76,
https://doi.org/10.1007/978-3-030-48679-2_26

1 Introduction

Adequate monitoring and timely treatment of partial-depth top-down cracks (TDC) are essential to prolonging the life cycle of long-lasting pavement structures. This is one of the most frequent failure modes and cause of deterioration in asphalt pavements. Therefore, in order to achieve perpetual pavements, it is necessary to track the evolution of these cracks and so repair them before they become deeper and deteriorate the lower layers (Franesqui et al. 2017).

Self-healing of bituminous mixtures is possible if the temperature is raised high enough to reduce the binder viscosity, allowing the fusion of the crack faces. A possible technique is the electromagnetic induction heating of mixtures with inductive additions that raise electric conductivity. Decisive factors to ensure the efficiency of this method are the type, size and proportion of the additions. Studying the electrical conductivity of PA mixtures with steel wool, some researchers concluded that short length fibres provide optimal performance in comparison to longer fibres (Liu et al. 2010, 2011).

Some laboratory studies determined that it is also possible to raise the temperature of AC mixtures with steel wool by using microwaves with brief exposure times (Gallego et al. 2013). Consequently, microwaves seem to be promising for the self-healing of cracks in asphalt pavements (Norambuena-Contreras and García 2016).

However, the main limitations of the previous studies have been:

(a) Self-healing of dense asphalt concrete (AC) and porous asphalt (PA) have been studied, but up until now other common types such as asphalt concrete for very thin layers (AC-VTL)—also known in European Standards as BBTM "Béton bitumineux très mince"—have not been analysed. This is usually employed in wearing courses of only 2–3 cm thick.
(b) Due to the formation of clusters during mixing, the steel wool fibres are difficult to homogenize in order to reach a uniform heating (Gallego et al. 2013) and thus, increasing the air void content (Yang et al. 2016).
(c) The induction devices are difficult to use for field applications and require certain safety measures. Furthermore, the time required in order to heat the asphalt mixes by induction still remains excessive (García et al. 2015). Hence, microwave devices could be more manageable and risk free for this application.
(d) The evolution of the crack depth after microwave radiation and how the macro-cracks heal has yet to be experimentally examined.

Consequently, this research focused on the evolution of the crack depth with the specific energy applied by microwave equipment. At the same time, this laboratory study sought after an optimal addition in reduced proportions from industrial waste (Norambuena-Contreras et al. 2018) that would allow significant energy saving and achieve a complete self-healing of TDC.

Table 1 Percentages of steel addition (by total weight of mixture)

Type of mixture (HMA)	Type of metallic addition						
	Ref	SW10	SW5	SF1-2	SF0.5-1	SPC0.063-0.5	SPC<0.063
AC 16 surf 50/70 S	0.0	0.6	0.4	1.0	2.0	–	5.37
BBTM 11B PMB 25/55-65	0.0	0.6	0.4	1.0	2.0	15.51	5.5
PA 11 PMB 25/55-65	0.0	0.6	0.4	1.0	2.0	9.63	4.5

(Ref) Reference mixture without additions; (SW10) Steel wool of length 10 mm; (SW5) Steel wool of 5 mm; (SF1-2) Steel filing of size 1–2 mm; (SF0.5-1) Steel filing 0.5–1 mm; (SPC0.063-0.5) Steel filing with corundum powder of size 0.063–0.5 mm; (SPC<0.063) Steel and corundum powder less than 0.063 mm [metallic filler]

2 Materials

Cylindrical and slab specimens of the three different types of hot mix asphalt (HMA) were compacted in the laboratory: AC 16 surf 50/70 S (semi-dense) with 4.5% (by wt. of mixture) of conventional penetration bitumen (50/70 indicates the penetration grade in 10^{-1} mm); BBTM 11B PMB 25/55-65 with a bitumen content of 5%; and PA 11 PMB 25/55-65 with 4.5% of the same type of polymer-modified bitumen (25/55 indicates the penetration grade in 10^{-1} mm, and 65 is the softening point in °C). All the aggregate fractions came from massive phonolite of high density (a type of volcanic rock) with a bulk density of 2.62 g/cm^3.

Six types of steel additions were used to speed up the microwave heating. These varied in size, composition and proportion (Table 1). The additions were prepared from low-carbon steel profiles and sheets, all cut manually by the same operator: (a) steel wool fibre (5 and 10 mm long, 0.3–0.4 mm diameter approximately); (b) steel filing (1–2 mm and 0.5–1 mm) obtained from metal profiles cut with a metal lathe machine; (c) steel filing (90%) with corundum powder (10%, approximately) obtained from radial saw grindings (0.063–0.5 mm and #<0.063 mm). The steel filing with corundum powder was used to substitute either the finest aggregate or the mineral filler (#<0.063 mm), in this last case acting as a metallic filler. Thus, the mix design depended on the aggregate gradation of each type of mixture (Fig. 1) and the corresponding fractions. The mixtures were produced following the Spanish road specifications [PG-3] (Spanish Ministry of Infrastructures 2014).

3 Methodology

The cylindrical specimens (D = 101.6 mm; h = 63.5 mm) were compacted using a Marshall hammer according to EN 12697-30 with 50 blows/side. The prismatic specimens were obtained from slab specimens (300 × 300 × 60 mm), compacted by

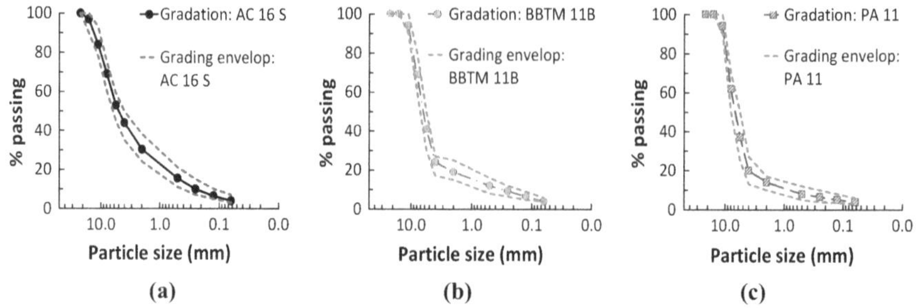

Fig. 1 Aggregate gradation and specified grading envelope for the three types of mixtures: **a** AC 16 surf 50/70 S; **b** BBTM 11B PMB 25/55-65; **c** PA 11 PMB 25/55-65

rolling according to EN 12697-33. The cylindrical specimens underwent basic characterization tests: maximum specific gravity (EN 12697-5; volumetric procedure), bulk specific gravity using the hydrostatic method (EN 12697-6; SSD procedure), air voids (EN 12697-8), moisture sensitivity (EN 12697-12) and particle loss (EN 12697-17).

Once characterized, each cylindrical specimen was cut into two halves in order to measure the temperatures inside of each compacted specimen after microwave exposure. The slab specimens were divided into three prismatic samples. A total of 110 test samples were obtained: 92 from cylindrical specimens and 18 from slab specimens.

The different test samples were initially conditioned in a heater-refrigerator at 15 °C (±0.1 °C) during 4 h, thus making this the starting temperature (T_0) from which the microwave exposure began. Exposure in the microwave oven (at 800 W and 2.45 GHz) lasted long enough to surmount the softening point of each binder (52 °C for 50/70 pen bitumen) and 67 °C for PMB 25/55-65). Using an infrared thermometer (resolution ±0.1 °C; precision ±1.5 °C) several measurements were carried out at 10 or 20 s intervals; these measurements were made at three points on the cut surface of each halved cylindrical specimen.

By cutting the slab specimens (300 × 300 × 60 mm), different prismatic beam samples were obtained to measure the evolution of the crack depth after microwave exposure (300 × 110 × 60 mm and 300 × 80 × 60 mm). With this partition of the slab specimens, a height safeguard of 50% at least was achieved for the deeper cracks studied in the laboratory in order to ensure that the notches and cracks will not fracture the samples.

The prismatic beams underwent cracking by means of three-point bending test at a low temperature (−20 °C) and deformation rate (0.5 mm/min). A notch was previously made in the centre of each beam using a radial saw in order to ensure the initiation of cracking at this point. The minimum notch depth was 20 ± 1 mm for the AC-S and 10 ± 1 mm for the BBTM and PA mixtures (according to the maximum size of each aggregate), and with a 4–5 mm slot between notch faces (Fig. 3b). The net initial crack depths (subtracting the notch depth) ranged between 11 ± 1 mm and 40 ± 5 mm.

Measurements of the initial crack depth as well as measurements of the same crack at different intervals of microwave exposure were carried out using the non-destructive method postulated by Franesqui et al. (2017), where the analytical models are founded upon a self-calibration technique based on ultrasound measurements on a single surface. This method allows the immediate determination of the depth of surface-initiated cracks in asphalt mixtures, is economically feasible and provides errors below 13% (at 95% statistical confidence level), even with micro-cracks unobservable to the naked eye. The ultrasound device was utilized with cylindrical CPC (couplant plate contact) piezoelectric transducers (26 mm diameter, 54 kHz).

In order to use the proposed models of this methodology, the mathematical functions should be previously calibrated on each specific material and with the ultrasonic equipment to be employed (see Franesqui et al. 2017). For this calibration, ultrasound propagation time measurements were carried out on the cracked surface of the beam samples at 20 °C. Figure 2 shows the functions of the calibrated models for the three types of HMA, at T = 20 °C with measurement baseline (B = linear distance between transducers) 70, 120 and 150 mm.

Before exposure, all the prismatic samples were placed in an oven for acclimatization (4 h at 20 °C). From this temperature the samples were radiated until the crack healed (total maximum exposure time of 110–210 s). The process was carried out during several cycles of microwave exposure between 20–40 s each. The ultrasound measurement of the crack depth after each gradual microwave exposure interval has allowed the assessment of the depth evolution with the exposure time and therefore, the effectiveness of the self-healing technique (Fig. 3).

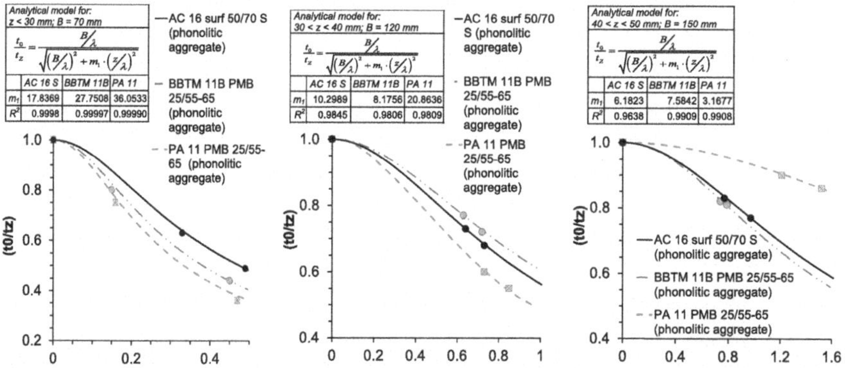

Fig. 2 Calibrated functions used to predict crack depth using ultrasounds at 20 °C: **a** For cracks up to 30 mm depth (B = 70 mm); **b** For cracks from 30 to 40 mm (B = 120 mm); **c** For cracks from 40 to 50 mm (B = 150 mm). [(t0/tz) transmission time ratio on the non-cracked HMA surface (z = 0) and on the same specimen with crack depth (z); (λ) Ultrasound wavelength; (B) distance between transducers]

Fig. 3 Prismatic sample of BBTM 11B PMB 25/55-65 with metallic filler (SPC<0.063 mm) and height 80 mm: **a** Net initial crack depth of 19 ± 1 mm, excluding the notch; **b** Detail of the same crack, completely closed in its entire depth, after an exposure time to microwaves of 110 s (starting temperature 20 °C)

4 Results and Discussion

4.1 Engineering Properties of the Mixtures

The mixtures with steel additions complied with the specifications for roads ($4 \leq V_m \leq 6\%$, for AC-S; $V_m \geq 12\%$, for BBTM; $V_m \geq 20\%$, for PA; ITSR $\geq$ 85%, for AC-S; ITSR $\geq$ 90%, for BBTM; PL $\leq$ 20-25%, for PA). However, the AC-S mixtures with steel wool fibres caused some problems during compaction which led to an increase in air voids (see Table 2).

4.2 Effect of Different Steel Additions on the Heating Speed

In order to determine the most efficient addition for each HMA, the average temperatures were calculated for each microwave exposure time. These points were fitted by linear regression functions, which allowed the assessment of the performance differences among the different metallic additions (Fig. 4). The coefficient of determination (R^2) of the fitting varied between 0.953–0.992 for AC-S; 0.958–0.997 for BBTM; and 0.907–0.998 for PA mixtures.

After examining the results, the following observations are presented:

- The filler used to substitute the mineral powder (SPC<0.063 mm) offered good results with AC-S and BBTM mixtures: exposure time reduction by 18.6% in the AC-S (with 5.4% addition) and 7.6% in the BBTM (with 5.5% addition). Nevertheless, the performance was irregular in the case of porous mixtures (PA) [4.5% of addition], being less efficient than the control specimens. On the contrary, the PA mixtures showed a good performance with the short steel wool fibres (5 mm) [0.4% of fibres], which produced a 5.6% of time reduction (González et al. 2019).
- The addition of short steel wool fibres (5 mm) [at 0.4%] proved to be more efficient than the 10 mm fibres [at 0.6%].

Table 2 Characterization properties of the mixtures (EN 12697 Standards)

Type of mixture (HMA)	Property	Type of metallic addition						
		Ref	SW10	SW5	SF1-2	SF0.5-1	SPC 0.063-0.5	SPC<0.063
AC 16 surf 50/70 S	$D_{b,SSD}$ (g/cm^3)	2.39	2.16	2.18	2.28	2.13		2.40
	V_m (%)	4.57	13.05	10.78	5.75	6.46		5.75
	ITSR (%)							85.37
BBTM 11B PMB 25/55-65	$D_{b,SSD}$ (g/cm^3)	2.19	2.14	2.07	2.20	2.13	2.19	2.06
	V_m (%)	11.99	15.58	18.59	13.26	16.36	13.69	18.63
	ITSR (%)							92.36
PA 11 PMB 25/55-65	$D_{b,SSD}$ (g/cm^3)	1.94	1.72	1.99	1.94	1.96	1.90	1.94
	V_m (%)	24.15	32.75	22.20	24.37	23.59	25.63	24.32
	PL (%)							22.0

(Ref) Reference mixture without additions; (SW10) Steel wool of length 10 mm; (SW5) Steel wool of 5 mm; (SF1-2) Steel filing of size 1–2 mm; (SF0.5-1) Steel filing 0.5–1 mm; (SPC0.063-0.5) Steel filing with corundum powder of size 0.063–0.5 mm; (SPC<0.063) Steel and corundum powder less than 0.063 mm (metallic filler); ($D_{b,SSD}$) Bulk density [saturated surface dry]; (V_m) Air void content in the mixture; (ITSR) Indirect tensile strength ratio; (PL) Particle loss

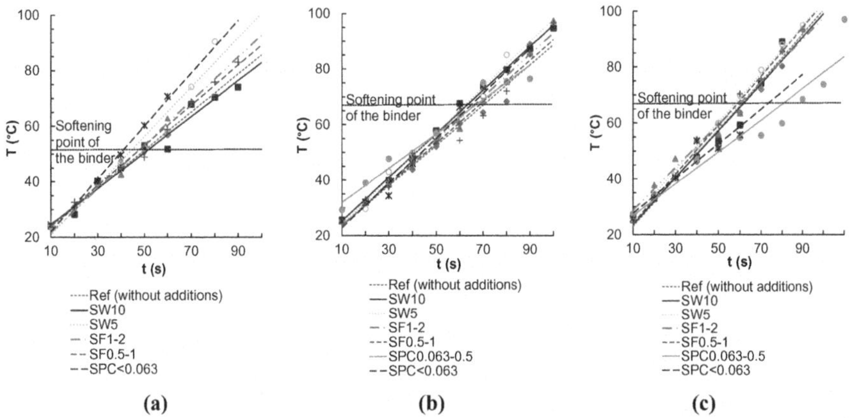

Fig. 4 Comparison of the effect of the type and size of the metallic addition on the heating speed: **a** AC 16 surf 50/70 S; **b** BBTM 11B PMB 25/55-65; **c** PA 11 PMB 25/55-65

- The intermediate size additions (steel filings 0.5<#<2 mm) offer an intermediate thermal efficiency between steel wool fibres and the finest filings, which is due to the fact that they were also used with intermediate proportions.

To summarize, the optimal addition for AC-S and BBTM, considering both practical applications (easier mix formulation and mixing) and thermal efficiency, is the metallic filler (SPC<0.063 mm) because distribution is far more homogeneous and avoids clump formation as occurs in the case of steel wool fibres, making compaction difficult. The addition of metallic filler implies significant benefits: simple production and dosage control of the mixtures; excellent homogenization and compaction with the habitual production formula used for each mixture type with similar final properties; it offers greater energy efficiency to microwaves as the smaller particles facilitate heat generated by Joule effect; furthermore, the powdery particles prevent the accumulation of charges that ionize the air, avoiding electric arcs when microwaves are applied; and finally, this filler yields environmental advantages by using up waste metal produced in the industry.

However, the best addition for PA mixtures is 5 mm steel fibres. This is because the fibres increase electric connectivity between aggregates, which counters the isolation effect produced by the high levels of porosity of PA.

4.3 Effect of Microwave Energy on Temperature

With the aim of comparing the results obtained from the different specimens (cylindrical and prismatic), the temperature was indicated as a function of the energy supplied per unit of mass (specific energy, E/m). The fitted functions were estimated by linear regression with the experimental values (Fig. 5) and can therefore be applied regardless of the mass of the mixture and the power output of the microwave device.

The analysis of these models show that the rate of temperature increase vs. the specific energy may be considered roughly constant in each HMA (Fig. 5), being 0.9 °C/(J/g) for AC-S mixtures; 1.1 (°C g)/J for BBTM; and 1.2 (°C g)/J for PA. The model obtained with BBTM offers better fit ($R^2 = 0.96$) compared to the other materials ($R^2 = 0.74$ for AC-S; $R^2 = 0.88$ for PA 11).

4.4 Monitoring of the Crack Depth

The last phase of the study made it possible to systematically study the reduction in depth of the cracks generated in the laboratory on the prismatic samples after microwave exposure. This enabled verification of the practical effectiveness of the

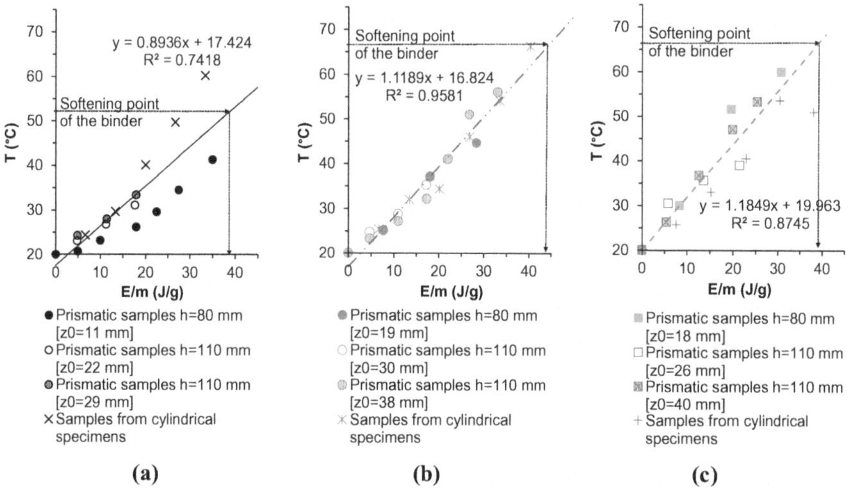

Fig. 5 Temperatures vs. specific energy supplied by microwaves on each type of HMA with steel filler (SPC<0.063 mm): **a** AC 16 surf 50/70 S; **b** BBTM 11B PMB 25/55-65; **c** PA 11 PMB 25/55-65

method and the chosen addition. In order to monitor the cracks, the previously mentioned ultrasound technique was employed. The crack depth (z) following each interval of microwave treatment was expressed according to the specific energy (E/m) supplied.

The ultrasound results show a reduction of the crack depth with the applied specific energy. This proves that the healing begins at the crack tip, where the opening is smaller, and spreads towards the surface until self-healing completion. As the experimental points demonstrate in Fig. 6, effective and complete closure was verified throughout the macro-cracks previously produced in the laboratory, including

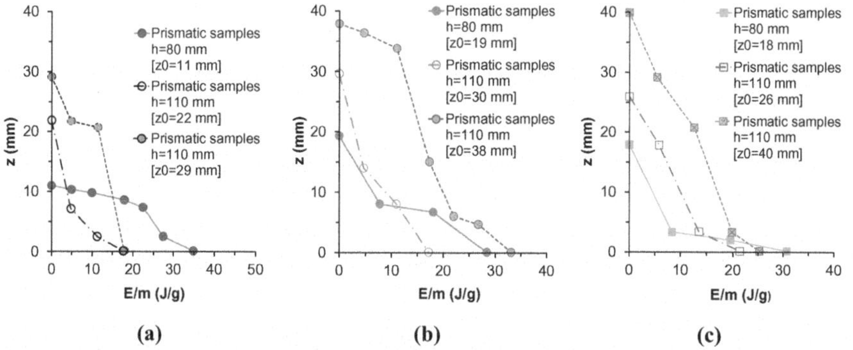

Fig. 6 Evolution of the crack depth with the microwaves specific energy on prismatic beams with steel filler (SPC<0.063 mm) from an initial temperature of 20 °C and different initial crack depths (z_0): **a** AC 16 surf 50/70 S; **b** BBTM 11B PMB 25/55-65; **c** PA 11 PMB 25/55-65

the deepest cracks (40 mm). In no event was the initial notch closed, confirming that this methodology is ineffective with wide cracks (>4–5 mm), with crack faces excessively polished and with severed aggregates. This suggests that the pavement maintenance must be periodical in order to avoid excessive deterioration.

In the laboratory, cracks up to 40 mm deep were completely self-healed after a brief microwave exposure (between 110 and 210 s), starting off at a room temperature of 20 °C. These times are equivalent to a specific energy between 17.2 and 35.0 J/g (depending on the type of HMA and crack depth). This fact demonstrates the effectiveness of microwaves for self-healing of surface-initiated macro-cracks with both types of mixtures with metallic filler. Furthermore, the crack depth measurement method by ultrasounds proved to be efficient, reasonably precise, cost effective and manageable.

5 Conclusions

Based on the results of this experimental research, the following conclusions can be drawn:

- Microwaves have proved to be an efficient way of controlled, quick heating for the three types of asphalt mixtures with metallic additions and requiring simple, safe, compact, affordable and low power equipment.
- The smaller particles (steel filings #<0.063 mm) proved to be the optimal addition for AC-S and BBTM, taking into account practical use and thermal efficiency (heating speed with lower energy consumption). Furthermore, mixing and proportioning control proves easier and the mixture is more homogeneous and compactable. Standard formulations may be used, final characteristics of the mixtures are similar, heating energy efficiency is improved and electric arcs are avoided.
- On the contrary, the PA mixtures have shown better performance with short steel fibres (5 mm) because the fibres increase electric connectivity between aggregates in these high-porosity mixtures.
- The temperature increase rate with regard to the specific energy is approximately linear and proved to be approximately 1.0 (°C g)/J.
- In the laboratory complete self-healing of surface-initiated cracks of up to 40 mm in prismatic specimens with metallic filler was achieved. The energy per unit of mass required is low (between 17 and 35 J/g, depending on the HMA type and starting at room temperature of 20 °C), and consequently a short exposure time is necessary (between 110 and 210 s with the laboratory prismatic specimens).
- The experimental results proved that healing starts at the crack tip and spreads towards the pavement surface till healing completion, providing the cracks are not excessively wide (<4–5 mm), and that the sides are not too polished nor the aggregates severed. This suggests that pavement maintenance must be periodical in order to avoid excessive deterioration.

- The use of ultrasounds for depth measurement of top-down cracks has proved to be efficient, relatively precise, cost effective and manageable.
- The proposed self-sealing technique allows not only longer service life of the pavement but also reduces rehabilitation costs. Furthermore, there are numerous environmental advantages: recovery and reuse of metal waste from industry; and prevention of new waste from milling of cracked surface layers. The time saving aspect of these procedures is noteworthy when compared to the standard pavement rehabilitation methods, allowing quick re-opening of traffic and reduced energy consumption and emissions.

Acknowledgements The activity presented in the paper is part of the research project "MW-VolcAsphalt" (Ref. BIA2017-86253-C2-2-R, "Sustainable self-healable perpetual asphalt pavements with volcanic aggregates using microwaves and additions of metallic wastes and nanoparticles") supported by the Ministry of Economy and Competitiveness (MINECO) from the Government of Spain.

References

Franesqui MA, Yepes J, García-González C (2017) Ultrasound data for laboratory calibration of an analytical model to calculate crack depth on asphalt pavements. Data Brief 13:723–730

Gallego J, Del Val MA, Contreras V, Páez A (2013) Heating asphalt mixtures with microwaves to promote self-healing. Constr Build Mater 42:1–4

García Á, Bueno M, Norambuena-Contreras J, Partl MN (2015) Single and multiple healing of porous and dense asphalt concrete. J Intell Mater Syst Struct 26(4):425–433

González A, Valderrama J, Norambuena-Contreras J (2019) Microwave crack healing on conventional and modified asphalt mixtures with different additives: an experimental approach. Road Mater Pavement Des 20(Sup1):S149–S162

Liu Q, Schlangen E, García Á, Van de Ven M (2010) Induction heating of electrically conductive porous asphalt concrete. Constr Build Mater 24:1207–1213

Liu Q, García Á, Schlangen E, Van de Ven M (2011) Induction healing of asphalt mastic and porous asphalt concrete. Constr Build Mater 25:3746–3752

Norambuena-Contreras J, Garcia Á (2016) Self-healing of asphalt mixture by microwave and induction heating. Mater Des 106:404–414

Norambuena-Contreras J, Gonzalez A, Concha JL, Gonzalez-Torre I, Schlangen E (2018) Effect of metallic waste addition on the electrical, thermophysical and microwave crack-healing properties of asphalt mixtures. Constr Build Mater 187:1039–1050

Spanish Ministry of Infrastructures (2014) Spanish specifications for roads and bridges (PG-3). Art. 542 & 543. Orden FOM/2523/2014, Madrid, Spain

Yang JM, Kim JK, Yoo DY (2016) Effects of amorphous metallic fibers on the properties of asphalt concrete. Constr Build Mater 128:176–184

Effects of Moisture and Aging on Asphalt Binder Adhesion Failure Using Pull-Off Tension Test

Muhammad Rafiq Kakar, Meor Othman Hamzah, and Christiane Raab

Abstract Moisture damage is mainly characterized by the loss of adhesion between asphalt binder and aggregate and the loss of cohesion within the asphalt binder. The binder holds the aggregates firmly together and acts as a sealant against moisture ingress. Additives are added to enhance the performance-related properties of asphalt mixtures under different climatic and environmental conditions. In this study, warm mix additive was used to modify the asphalt binder PG-64. The asphalt binder modified with chemical surfactant based additive was tested against pull-off tension force using limestone aggregate substrates. The specimens were conditioned using accelerated laboratory vacuum saturator (ALVS). The results showed that the percent adhesion failure increased when specimens were subjected to moisture conditioning. Furthermore, adhesion failure also increased with binder aging and enhanced further when the binders were long term aged.

Keywords Moisture damage · Warm mix asphalt · Adhesion failure · Bitumen bond strength

1 Introduction

Moisture damage has major concern among asphalt pavement technologists for many years. It is one of the most common causes of pavement distress that results in loss of strength, stripping, raveling, fatigue damage and permanent deformation (Hamzah et al. 2015). Moisture damage results in asphalt pavement loss of strength or durability due to the effects of moisture and can be evaluated based on the loss of mechanical properties (Little and Jones 2003). Since the 1920s, researchers have been looking for a test that can differentiate between good and poor performing stripping potential of asphalt mixtures (Kakar et al. 2015). The identification of test procedures that can

M. R. Kakar (✉) · C. Raab
EMPA, Swiss Federal Laboratories for Material Science and Technology, Ueberlandstr. 129, 8600 Dübendorf, Switzerland
e-mail: muhammad.kakar@empa.ch

M. O. Hamzah
School of Civil Engineering, Universiti Sains Malaysia, 14300 Nibong Tebal, Penang, Malaysia

C. Raab (ed.), *Proceedings of the 9th International Conference on Maintenance and Rehabilitation of Pavements—Mairepav9*, Lecture Notes in Civil Engineering 76,
https://doi.org/10.1007/978-3-030-48679-2_27

well predict mixture moisture susceptibility is still required (Solaimanian et al. 2003). Moisture damage is mainly characterized by the loss of adhesion between asphalt and aggregate and the loss of cohesion within the asphalt binder (Kakar et al. 2019; Fromm 1974). The early form of removing binder coatings from aggregate surfaces is referred to as stripping. The moisture infiltrates into the asphalt-aggregate interface and diffuses into the binder, resulting in asphalt and aggregate bond failure. Stripping is a complex phenomenon that involves the physical and chemical properties of asphalt mixtures such as chemical composition of binder/mastic and aggregate, aggregate mineralogy and surface characteristics (Kakar et al. 2017).

The bond strength of asphalt binder and aggregate is a critical parameter in evaluating a binder's ability to resist moisture damage. Bhasin et al. (2006) evaluated the adhesion between asphalt binder and aggregate using the adhesive bond energy. A test method that can effectively evaluate the influence of water in both cohesive and adhesive failure types in an asphalt–aggregate system can lead to a better understanding of the moisture sensitivity of asphalt mixtures (Kanitpong and Bahia 2003). The aim of this study is assess the adhesion failure at the asphalt-aggregate interface using pull-off tension test (setup shown in Fig. 1) under different aging and moisture conditions. The mode of failure is assessed based on image classification. A new laboratory based accelerated moisture conditioning protocol was developed to condition the asphalt-aggregate substrate specimens.

Fig. 1 Pneumatic adhesion tensile testing instrument

2 Materials and Methods

In this study, limestone aggregate substrates were prepared by cutting cores from aggregate boulders. The cores were then cut into slices using a diamond cutter. PG-64 binders was used as control adhesive materials and Table 1 summarizes the rheological properties of the binders used. Only one set of specimen was used for the test. However, all specimens were prepared using a new set of aggregate substrates. The chemical based surfactant as warm mix additive is consumed in a very limited amount of 0.2–0.5% by weight of the binder (Oliveira et al. 2013). The physical and chemical properties are given in Table 2. To assess the effects of warm mix additive, binders were blended with 0.3% additive (as per the suppliers recommendations) by mass of the binder (Hamzah et al. 2014). This additive, which is liquid at room temperature, can be easily mixed with the hot asphalt binder before the asphalt mixture production (Kakar et al. 2016). Initially, the binders were heated to the blending temperature. The mechanical overhead blender was used to blend the additive with binder. Table 4 presents the blending parameters used to blend the additive with PG-64 binder. The test sample designations are given in Table 3.

2.1 *Asphalt Binder Aging*

The modified and unmodified asphalt binders were artificially short-term and long-term aged in the laboratory. The rolling thin-film oven (RTFO) and pressure aging vessel (PAV) were used for this purpose. The short-term binder aging was performed at 163 °C for 85 min, while for the long-term aging, the binders were kept under

Table 1 Properties of PG-64 binder

Property	Penetration at 25 °C, 100 g, 5 s, (0.1 mm)	Softening point (°C)	Ductility at 25 °C (cm)	Flash and fire point (°C)	Solubility (%)	Specific gravity
Test method	ASTM D5	ASTM D36	ASTM D113	ASTM D92	ASTM D2042	ASTM D70
PG-64	86	45	<160	331 & 340	99.52	1.03

Table 2 Physical and chemical properties of WMA additive (Xiao et al. 2012)

Properties	Ingredients	Physical state	Color	Specific gravity	Flash point	Solubility in water
Cecabase RT® 975	Polymer: >45% Fatty acid amine: <50% 1,2-Ethanediamine: >1%	Liquid	Light yellow	1	>100 °C	Insoluble

Table 3 Sample designation

Binder type	Aging condition	Additive content (%)	Designation
PG-64	Unaged	0.0	AC0.0
		0.3	AC0.3
PG-64	RTFO	0.0	RAC0.0
		0.3	RAC0.3
PG-64	PAV	0.0	PAC0.0
		0.3	PAC0.3

Table 4 Dosage, blending time and temperature of additive

Additive	Dosage by weight of binder (%)	Blending time (min)	Blending temperature (°C)
Cecabase RT® 975	0.3	15	130

2.1 MPa at 100 °C for 20 h. The tests were conducted in accordance with ASTM D2872 (ASTM 2006a) and ASTM D6521 (ASTM 2006b) procedures for both RTFO and PAV, respectively.

2.2 Preparation of Aggregate Substrates

Aggregate substrates were prepared in 100 mm diameter and 15 mm thickness. The aggregate was first cored from boulder supplied by Kuad Quarry Penang using a 100 mm core cutter. Later, the cores were cut into 15 mm thicknesses slices using a diamond cutter. To minimize the discrepancy and inaccuracy of the test results, especially in terms of the uniformity, and also to avoid the loose particles on the surface of aggregate substrates these aggregate substrates were polished using a stone polishing machine. After polishing, the aggregate substrates were cleaned with distilled water to further ensure that no foreign materials were left on the surface.

2.3 Accelerated Laboratory Moisture Damage Conditioning Process

The laboratory moisture conditioning of asphalt mixtures has become more important over the past several years. The development of a quick and responsive technique that can predict the field moisture damage of an asphalt mixture is utmost important. A newly developed conditioning method for the binder aggregate substrates bond strength was used (Kakar, 2015). The device was designed to control the level of vacuum, temperature, vacuum holding time and the number of conditioning cycles. The accelerated laboratory vacuum saturator (ALVS) was used to condition the test samples. The samples were placed inside an ALVS chamber in the water, where the interface of the aggregate binder was placed at least 25 mm below surface water level. As suggested by Sulamanian (2003), the conditioning water incorporated sodium carbonate Na_2CO_3 at a concentration of 6.62 gm per liter to enhance the stripping rate. The conditioning time was set for 30 min at 35 °C.

2.4 *Pull-Off Tension Test*

An improved method for moisture damage performance based on the fundamental evaluation of moisture damage mechanisms remains a challenge, especially in terms of the adhesion of the asphalt mixtures. Copeland et al. (2007) and Kanitpong and Bahia (2003, 2004, and 2005) developed test techniques and procedures to measure directly the adhesive bond strength between bitumen and aggregates. The adhesion between asphalt binder or mastic and aggregates is considered as one of the key fundamental properties that influence asphalt mixtures performance. However, there are no established test procedures that can be used to evaluate the adhesive bond strength of the asphalt-aggregate interface. Currently, the pull-off tension test failure modes are evaluated based on visual observations, which is not precise. To overcome this, the current study applies an image analysis technique on the failed surfaces of pull-off tension test specimens to precisely quantify the percent adhesion failure. The test setup is shown in Fig. 1.

Modified pull-off tension stubs disc shapes 314.15 mm^2 surface contacts were fabricated for the test, as shown in Fig. 2. The pull-off tension test stub diameter was 20 mm. The asphalt binder was poured directly onto the substrate to develop the bonding between the stub and aggregate substrates. The thickness of asphalt film for each sample was approximately 8 mm. The excess binder from the gap in the stub after pressing the stub over the binder poured on the aggregate substrate, was trimmed. Then, one set of the specimen was exposed to moisture conditioning using ALVS. The pull-off tension test was carried out at 15 °C.

3 Results and Discussion

3.1 *Image Analysis on Binder-Aggregate Substrates Using Direct Tensile Test*

Figure 3 show the original images of PG-64 binder over limestone aggregate captured after testing the specimens using pull-off tension tests in unaged and long term

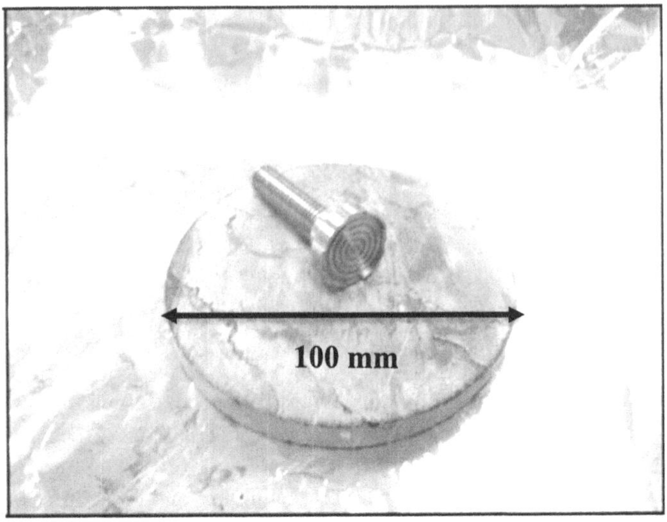

Fig. 2 Limestone aggregate substrates prepared for pull-off tension test

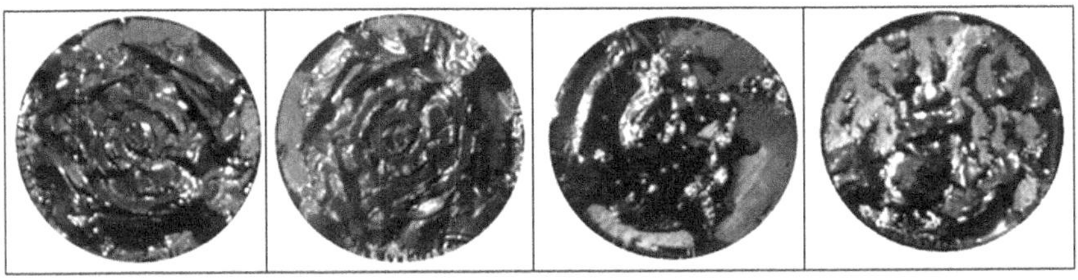

(a) AC0.0 Unconditioned (b) AC0.3 Unconditioned (c) AC0.0 Conditioned (d) AC0.3 Conditioned

Fig. 3 Original images of PG-64 binders on limestone aggregate substrate

aged conditions. The images obtained after the pull-off tension test were then processed by means of Matlab Image Processing Toolbox™ (Matlab documentation) and Environment for Visualizing Images (ENVI) software (Exelis Visual Information Solutions). As described by Hamzah et al. (2014), image processing functionality can create the color transformation structure based on Color Look Up Table (CLUT) (Matlab documentation). Figure 4 present the transformed images that clearly show the adhesion failure over the aggregate substrate.

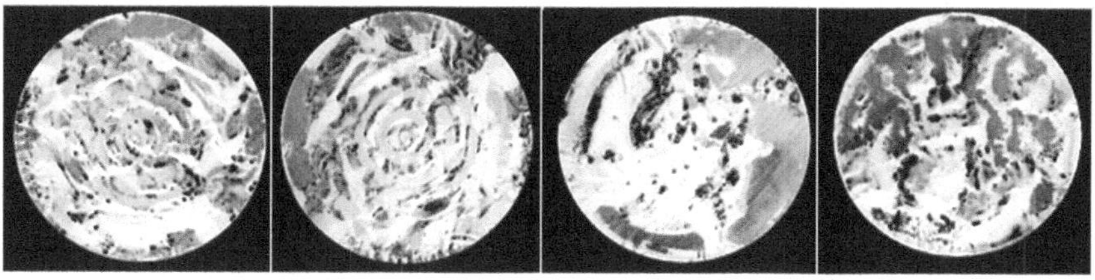

(a) AC0.0 Unconditioned (b) AC0.3 Unconditioned (c) AC0.0 Conditioned (d) AC0.3 Conditioned

Fig. 4 Transformed images of PG-64 binders on limestone aggregate substrate

3.2 *Effects of Moisture Conditioning and Binder Aging on Adhesion Failure Using Image Analysis Technique*

Figure 5 shows the percent adhesion failure of asphalt binder sandwiched between limestone substrates subjected to different aging conditions using the pull-off tension tests. The percent adhesion failure on the aggregate substrates was quantified using image analysis method as described by Hamzah et al. (2014). The results in Fig. 5 show that both aging and conditioning increases adhesion failure. The addition of chemical surfactant additive reduces the percent adhesion failure compared to the base binder irrespective of aging type and moisture conditioning. Specimen PAC0.0 subjected to conditioning exhibits (89.8%) highest percent adhesion failure. The lowest adhesion failure (7.3%) was observed for unconditioned specimen AC0.3. Hence, the results of percent adhesion failure obtained from image analysis illustrates

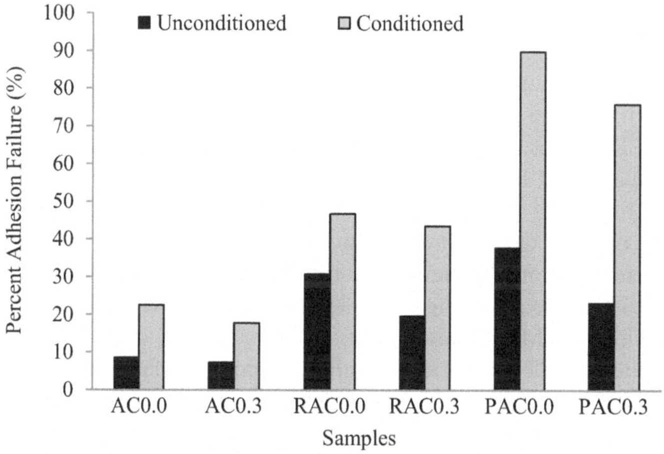

Fig. 5 Percent adhesion failure of PG-64 modified and unmodified binders on limestone substrate

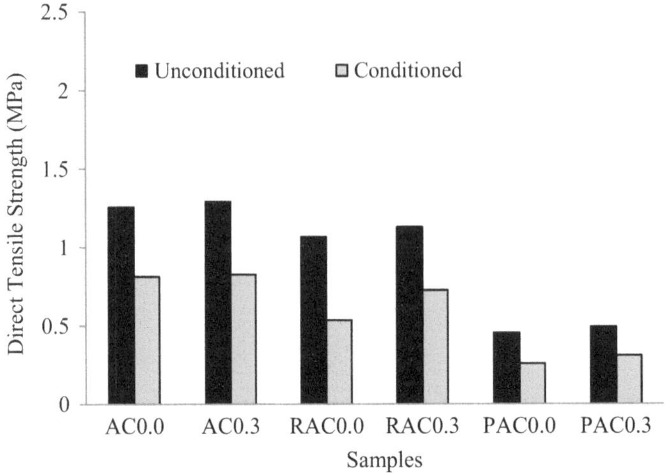

Fig. 6 Bitumen bond strength of PG-64 modified and unmodified binders on limestone substrate

that the chemical surfactant warm mix additive improves the adhesion property of PG-64 binder under moisture damage at difference aging conditions.

3.3 *Effects of Moisture Conditioning and Binder Aging on Bitumen Bond Strength*

Figure 6 shows the bitumen bond strength of PG-64 binder sandwiched between limestone aggregate substrates. The results indicate that the bond strength decreases when the specimens are exposed to moisture conditioning. The effects of aging also reduce the bitumen bond strength. The bond strength of 0.26 MPa and 1.3 MPa are obtained for conditioned specimen PAC0.0 and unconditioned specimen AC0.3, respectively. This indicates that the specimens PAC0.0 when exposed to conditioning exhibit the lowest bitumen bond strength. However, the unconditioned specimens AC0.3 registered the highest bitumen bond strength. This shows that specimen AC0.3 when exposed to moisture will have more resistance compared to specimen PAC0.0. Therefore, from the viewpoint of resistance and susceptibility to moisture damage, specimens incorporating WMA additive comparatively performs better than speicmens without WMA additives.

Overall, the pull-off tension tests results showed that adhesion failure increased with binder aging and enhanced further when the binders were long term aged. Moisture conditioning also increased the percent adhesion failure irrespective of the aging method used. The results further showed that compared to modified binder the unmodified PG-64 binder is more susceptible to moisture damage in terms of the percent adhesion failure.

4 Conclusions

The moisture damage in asphalt aggregate constituents was evaluated in terms of adhesion failure. The adhesion failure was quantified using image analysis process. Based on the results of this study the following conclusions are drawn:

1. Binder-aggregate substrate tests showed that the short term aging and long term aging of asphalt binder increased the percent adhesion failure.
2. The percent adhesion failure increased when specimens were subjected to moisture conditioning (ALVS) and binder aging.
3. Specimens PAC0.0 subjected to conditioning and the unconditioned specimens AC0.3 exhibited the maximum and minimum percent adhesion failures, respectively.
4. From the test results, it is appropriate to analyze the adhesion failure at the asphalt binder and aggregate interface. However, there are some factors that should be considered while selecting pull-off tension test method. The aggregate source, core cutting machine, diamond cutter, polishing machine and testing facility (Pull-Off tension machine) must be available.
5. Pull-off tension test is more quick and easy to test more number of specimens. However, the duration for specimen conditioning depends on the type of conditioning process. The pull-off tension test specimens are easy to prepare and test, as the modified pull stubs are used to control the film thickness of the asphalt binder.

Acknowledgements This work was supported by the Malaysian Ministry of Higher Education through the Fundamental Research Grant Scheme (grant number 203/PAWAM/6071277).

References

ASTM (2006a) ASTM D2872: standard test methods for effect of heat and air on a moving film of asphalt (rolling thin-film oven test). Annual book of ASTM standard, vol 4.03. West Conshohocken, USA

ASTM (2006b) ASTM D6521: standard practice for accelerated aging of asphalt binder using a pressurized aging vessel (PAV). Annual book of ASTM standard, vol 4.03. West Conshohocken, USA

Bhasin A (2006) Development of methods to quantify bitumen-aggregate adhesion and loss of adhesion due to water (Doctoral dissertation, Texas A&M University)

Copeland A, Youtcheff J, Shenoy A (2007) Moisture sensitivity of modified asphalt binders: Factors influencing bond strength. Transp Res Rec J Transp Res Board 1998(1):18–28

Fromm HJ (1974) Proceedings of the association of asphalt paving technologists, vol 43

Hamzah MO, Kakar MR, Hainin MR (2015) An overview of moisture damage in asphalt mixtures. Jurnal Teknologi (Sci Eng) 73(4):125–131

Hamzah MO, Kakar MR, Quadri SA, Valentin J (2014) Quantification of moisture sensitivity of warm mix asphalt using image analysis technique. J Clean Prod 68:200–208

Kakar MR, Hamzah MO, Valentin J (2015) A review on moisture damages of hot and warm mix asphalt and related investigations. J Clean Prod 99:39–58

Kakar MR, Hamzah MO, Valentin J (2017) Analyzing the stripping potential of warm mix asphalt using imaging technique. In: IOP conference series: materials science and engineering, September, vol 236, no 1. IOP Publishing, p 012013

Kakar MR, Hamzah MO, Akhtar MN, Saleh JM (2019) Evaluating the surface free energy and moisture sensitivity of warm mix asphalt binders using dynamic contact angle. Adv Civ Eng, Article ID 9153603

Kakar MR, Hamzah MO, Akhtar MN, Woodward D (2016) Surface free energy and moisture susceptibility evaluation of asphalt binders modified with surfactant-based chemical additive. J Clean Prod 112:2342–2353

Kakar MR (2015) Effects of adhesion failure on moisture damage of warm mix asphalt containing Cecabase additive, Ph.D. thesis, Universiti Sains Malaysia, Chap 3, pp 94–95

Kanitpong K, Bahia HU (2003) Role of adhesion and thin film tackiness of asphalt binders in moisture damage of HMA. In: Association of asphalt paving technologists technical sessions, 2003, Lexington, Kentucky, USA, vol 72

Kanitpong K, Bahia H (2004) Test method to determine aggregate/asphalt adhesion properties as an indication of potential susceptibility to moisture damage. The Wisconsin Highway Research program, The Wisconsin Department of Transportation

Kanitpong K, Bahia H (2005) Relating adhesion and cohesion of asphalts to the effect of moisture on laboratory performance of asphalt mixtures. Transp Res Rec: J Transp Res Board 1901(1):33–43

Little, DN, Jones, DR (2003) Chemical and mechanical processes of moisture damage in hot-mix asphalt pavements. Transportation research board national seminar, San Diego, CA, USA, pp 37–70

Oliveira JR, Silva HM, Abreu LP, Fernandes SR (2013) Use of a warm mix asphalt additive to reduce the production temperatures and to improve the performance of asphalt rubber mixtures. J Clean Prod 41:15–22

Solaimanian M, Harvey J, Tahmoressi M, Tandon V (2003) Test methods to predict moisture sensitivity of hot-mix asphalt pavements. Moisture sensitivity of asphalt pavements, a national seminar

Xiao F, Punith VS, Amirkhanian SN (2012) Effects of non-foaming WMA additives on asphalt binders at high performance temperatures. Fuel 94:144–155

Effect of Addition of Plastic Fibres on Strength Characteristics of Subgrade Soil

Ashutosh Kaushal, Rajesh Pathak, and Tanuj Chopra

Abstract In today's time, plastic materials (which otherwise pose great threat to environment), can be alternatively and smartly utilized in civil engineering as a soil stabilizing material (apart from the more traditional cement and lime) to achieve economy and reduce waste impact on environment. In present study, effect of addition of plastic fibres (shredded wrappers) on the strength characteristics of flexible pavements have been studied. Available literature after been reviewed, and then experiments were performed to compute the liquid limit, plastic limit and plasticity index of soil to categorize type of soil. OMC (Optimum moisture content) and MDD (Maximum dry density) were computed at various plastic contents (by % of dry weight of soil). The CBR (California bearing ratio) test has been performed at different percentage plastic contents. The maximum value of CBR was obtained as 4.01% at 1.5% plastic content. The design of pavement section have been carried out for different traffic volumes to find the most efficient and economical traffic condition for which this method could be most advantageous. Thickness of pavement for each layer (as per CBR corresponding to 0 and 1.5% plastic content) was determined using IRC-37:2018 design plates. The theoretical values of vertical compressive and horizontal tensile strains at critical locations have been determined. The theoretical value of modulus of elasticity (with 0 and 1.5% optimum plastic addition) were calculated using IRC 37:2018. Trial sections filled with virgin soil and reinforced soil at various percentages of waste plastic content were compacted manually to 250 mm thickness and the subgrade modulus were determined experimentally by using a Light Weight Deflectometer. The theoretical and experimental modulus values were compared, and the most optimized pavement thickness has been designed with IIT PAVE.

Keywords Plastic fibres · Flexible pavements · CBR · LWD · Subgrade modulus

A. Kaushal · R. Pathak (✉) · T. Chopra
Department of Civil Engineering, Thapar Institute of Engineering and Technology, Patiala, India
e-mail: pathak69.2000@gmail.com

A. Kaushal
e-mail: ashutoshkaushal25@gmail.com

T. Chopra
e-mail: tanuj33@gmail.com

C. Raab (ed.), *Proceedings of the 9th International Conference on Maintenance and Rehabilitation of Pavements—Mairepav9*, Lecture Notes in Civil Engineering 76,
https://doi.org/10.1007/978-3-030-48679-2_28

1 Introduction

In the past decade, plastic waste management has developed into a serious issue all across the world. With the ever rise in population there has been a growing demand of commodities manufactured using various types of plastics. Many plastics are non-biodegradable. There have been reported cases in India of animals unknowingly consuming them and dying as plastic bags cause blockage in their food pipes. In rural areas, waste plastics are still burnt today that releases harmful smoky gases that in turn cause hazardous environmental effects like smog, cancer etc. An estimated 100,000 marine mammals and up to 1 million sea birds die every year after ingesting or being tangled in plastic marine litter Soil stabilizers such as geosynthetics, fly ash, cement kiln dust, jute, blast furnace slag, high density polyethylene, copper slag, GGSB (Ground Granulated Blast-furnace Slag) etc. can be used to engineer the soil properties. From a geotechnical point of view, it was realized that waste plastic could prove beneficial if it was kept away from direct environmental contact. Especially in civil engineering, such types of plastics can be utilized as engineering materials instead of the more conventional ones. Plastics alone do not possess good engineering properties, therefore even partial replacement of standard construction materials with waste plastic will not only have really beneficial and positive impact on the environment, but also will go way forward in reducing construction costs by significant margins. This addition of waste plastic to enhance the engineering soil properties comes under a broader category, called soil stabilization. It could be defined as any process which helps improving the engineering properties of the soil, such as its shear strength, bearing capacity etc. Since plastic as a waste is available in abundance, hence it is economical to use plastic as a soil stabilizer. Also being non-biodegradable in nature it has serious impact on environment. Therefore, it can work as an ideal admixture for the subgrade.

Sobhan and Mashnad (2003) experimented with the use of recycled aggregate and waste HDPE strips for the pavement foundations using fatigue behaviour analysis. The main objective being the evaluation of a material that containing 90% recycled materials for foundation layer below the flexible or rigid pavement. Following materials were used—around 90% of recycled aggregate and cement and fly ash in very low quantity of about 4% each, and about 2% of 50 mm waste HDPE (High Density Polyethylene) strips. Different trial specimens were casted by varying the proportion of different materials. Experiments that were performed are—UCS, split tension, and static flexural tests. S–N curve is used to express the connection among stress ratio & number of cycles of failure. They reported fatigue and damage behaviour of an alternate pavement foundation material which incorporates recycled aggregate, fly ash, and waste HDPE strips.

Kolias et al. (2005) studied usefulness of high calcium fly ash and cement in the stabilisation of fine-grained clayey soil (CL, CH). While varying the percentages of fly ash (FA) and cement, various strength tests were carried out in uniaxial compression. Three type of clayey soils were used for the analysis. Cylindrical specimens were used for indirect tensile (splitting) and unconfines compressive strength tests.

Compaction for different percentages of FA shows that with increase in FA content from 5 to 20% OMC is increased while the MDD decreases. The results show that there is a advantage of stabilising clayey soil with calcium FA but it is subjected to type of soil and stabilising agent.

Choudhary et al. (2010) experimented with HDPE to be used as reinforcement in soil by improving its properties. HDPE strips were obtained from plastic waste, then mixed thoroughly with soil along with varying percentages of HDPE proportions. Lengths of strips were also varied. Large quantity of CBR tests were performed on reinforced soil ultimately concluding the beneficial effect of using HDPE strips as a soil reinforcement agent.

Hejazi et al. (2012) did a comprehensive review of the type of soil reinforcements that are being used to improve the soil characteristics by using both types of natural and synthetic fibres through referencing already published data and found out that fibers, typically at a dosage rate of 0.2–4% by weight were found to be optimum to be mixed into silt, clay, sand, or lime and cement stabilized soil. Based on the review they concluded the effect of fibre reinforcement by studying parameters like compressive and tensile strength, UCS (Unconfined Compressive Strength), shear strength, flexural strength, CBR etc. on various types of soils.

Manso et al. (2013) investigated on the use of ladle furnace slag in soil stabilization. Clayey soils have often been stabilized by addition of materials like cement, lime etc. So as to gain desirable properties for clay to be used in civil engineering projects. By-products of various industries can also be used for this objective. The properties of Ladle Furnace Slag (LFS) and the properties of different clayey soils with the addition of this slag to improve its properties have been studied. The behavior of the different soil and slag mixes were almost like that of the mixture of soil and lime. The results of the tests performed for the improvements of soil with respect to the various soil properties, like the plasticity index, expansiveness, bearing capacity and durability have been studied.

Dhatrak and Konmare (2016) studied the effect of plastic waste when mixed with soil. It was observed that inclusion of plastic bottles chips is effective in constructing improved strength subgrade of flexible pavement. In his paper author made use of various experiments predominantly CBR on soil hand blended with various proportions of plastic content. Final outcome being that usage of plastic waste strips will enhance the soil compressive strength parameter and is recommended to be used as subgrade.

Table 1 Engineering properties of soil

S. No.	Property	Result	IS code
1	% finer than 75 microns	73.22	IS: 2720 (Part 4)—1985
2	Liquid limit (%)	34.00	IS: 2720 (Part 5)—1985
3	Plastic limit (%)	16.67	IS: 2720 (Part 5)—1985
4	Plasticity index (%)	17.33	IS: 2720 (Part 5)—1985
5	Classification	CL A-6	IS: 1498—2007 AASHTO
6	Unconfined compressive strength, q_u (kN/m^2)	83.26	IS: 2720 (Part 10)—1991

2 Experimental Programme

2.1 *Materials*

2.1.1 Soil

The soil chosen for studying the effect of plastic on the strength characteristics of soil used for pavement subgrade falls in the category of fines grained soil having a 73% passing from a 75-micron sieve. Soil has been procured from Noorkhrian village located in the outskirts of the Patiala city, Punjab (India). The soil investigation was done by Indian Standard Code. Engineering properties are shown in the Table 1.

2.1.2 Waste Plastic

Plastic selected for the study was collected as form of waste shredded wrappers of fast food eatables such as chocolates, potato chips, candies etc. as shown in Fig. 1.

These plastics are made from aluminium laminated with polypropylene or low-density polyethylene film which forms an aluminium and hot plastic mass forming a multi layered type of plastic. Keeping the complex recyclable nature of the plastic in mind it was selected for the study to simultaneously reduce the waste plastic heap as well as making its sustainable use. The general specifications are provided in Table 2.

2.2 *Compaction Test*

The reduction of air voids existing in soil mass by means of applying dynamic forces (by using weights) is known as compaction test. In present study compaction test is done to compute the OMC (Optimum Moisture Content) and MDD (Maximum Dry Density) in conformity with the codal provisions of IS: 2720 (Part 7)—1980.

Fig. 1 Shredded plastic wrappers

Table 2 Waste plastic properties

Type of plastic	Shredded plastic wrappers (lays, chocolate etc.)
Length	5–10 mm
Thickness	Up to 60 μm

Table 3 Standard proctor test results

Plastic content (%)	OMC (%)	MDD (kN/m^3)
0	17.8	17.16
0.25	19	16.87
0.50	17.8	16.97
1.00	17.9	16.97
1.50	17.8	16.8

Compaction test was performed for plastic content of 0, 0.25, 0.5, 1 and 1.5%. There was a very slight change in the OMC while increasing the plastic content percentage. Due to larger size of plastic its addition in soil does not cause any reduction in voids present in between soil particles. Hence, not affecting interactions between water and soil. Therefore, there is no significant change in OMC and MDD as listed in Table 3.

2.3 California Bearing Ratio Test

California Bearing Ratio (CBR) is an empirical test method. CBR is known to be a measure of resistance offered to soil mass by the penetrating load, in this test a force

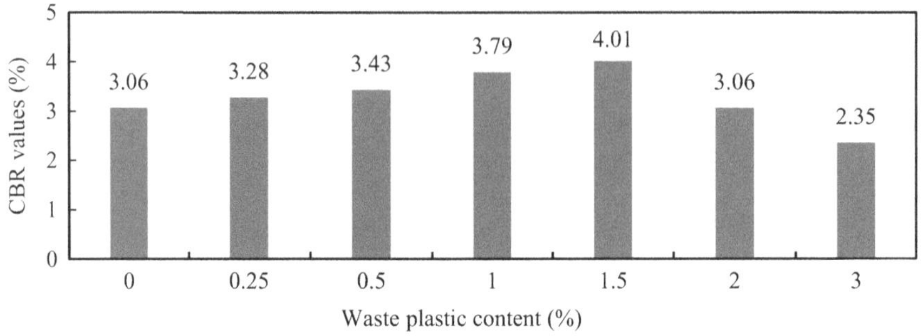

Fig. 2 CBR values with different waste plastic content

per unit area is applied to the soil sample under controlled test conditions. From Fig. 2 it is evident that with increased waste plastic content there is an increase in the CBR of soil. It reaches its maximum value at waste plastic content of 1.5% and after that decreases with further increase in plastic content. The maximum CBR obtained was 4.01%. A maximum increase of 31% in CBR was observed for reinforced soil.

2.4 Light Weight Deflectometer

Light Weight Deflectometer (LWD) is a handheld miniature version of Falling Weight Deflectometer (FWD) which is also known as the light FWD which was developed for the measurement of dynamic modulus of in situ soil. LWD can be used on directly unbound sub base and subgrade surfaces. LWD test was performed on constructed trial sections (60*45*25 cm dimensions) to determine the practical field values of the subgrade modulus of soil, it was done to simulate actual field conditions on a small scale. Figure 3 shows that the subgrade modulus determined from the empirical relations provided in the IRC:37-2018 which incorporates CBR values overstates

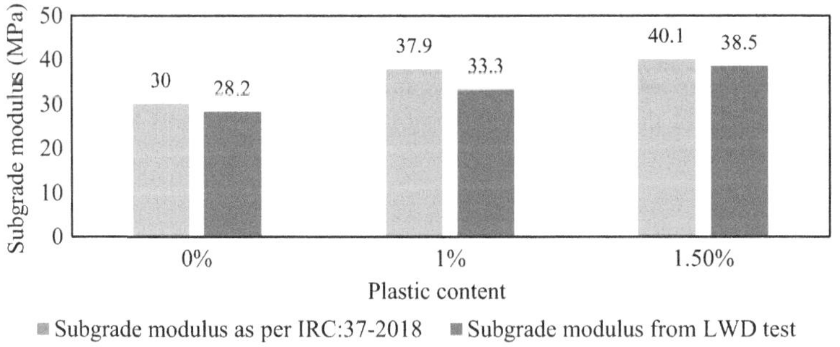

Fig. 3 Comparison of subgrade modulus

the modulus when compared to the modulus attained by LWD though the difference being very small. There is a difference of 6% for 0% plastic content and 4% for 1.5% (optimum) plastic content.

3 Analysis and Design of Flexible Pavement

The pavement is designed and compared for three cases with the traffic of 30, 75 and 125 MSA (Million Standard Axle) to study its effect on different volume roads. The maximum permissible strains are calculated for 90% reliability equations from IRC:37-2018 and actual strains are calculated from IITPAVE. Subgrade resilient modulus was calculated using equation from the IRC:37-2018 and by performing LWD test on a constructed trial section in the field. For preliminary estimate of

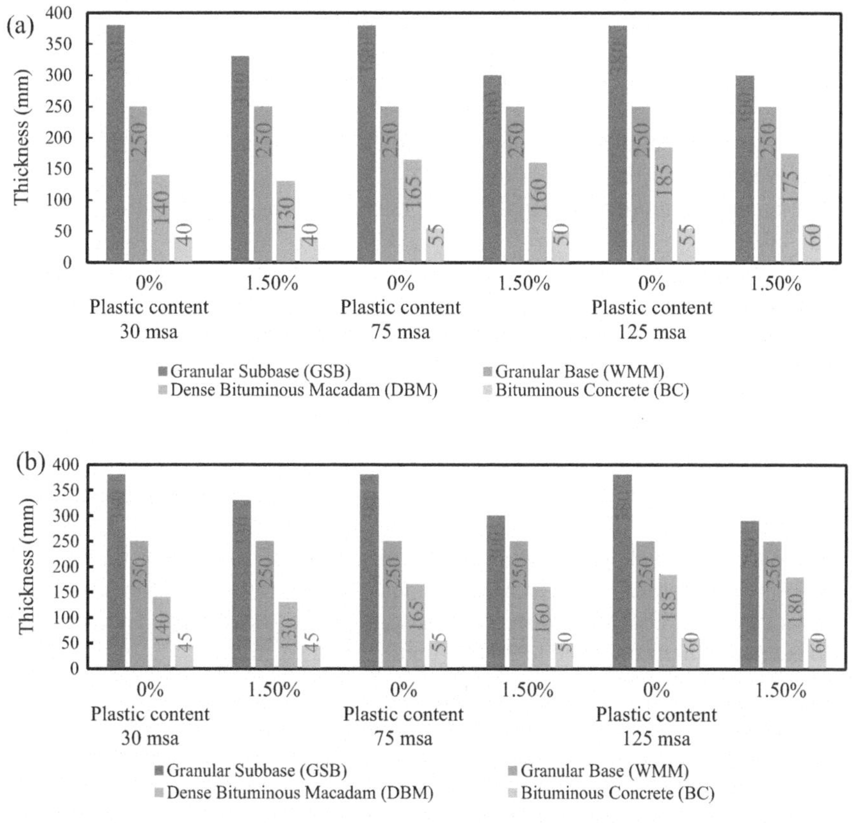

Fig. 4 Comparative thickness of pavement for reinforced and unreinforced section for different traffic (MSA) **a** while using IRC:37-2018 & **b** while using subgrade modulus determined from LWD

thickness of the pavement Design Plate 1 and 2 of IRC:37-2012 were used because of the low CBR of the soil samples. The pavement thickness was designed and compared using both subgrade moduli (from CBR and LWD). The results of design as shown in Fig. 4a and b validates that pavement thickness was reduced to a considerable extent when subgrade is reinforced with waste plastic fibres.

There is a reduction of total pavement thickness of 7.4% for 30 MSA traffic, reduction of 10.58% for 75 MSA traffic and 9.77% for 125 MSA traffic. The maximum reduction was observed for 75 MSA traffic. For the pavement designed by using experimentally determined moduli using LWD the pavement thickness yield similar results but on a slightly higher side. It can be attributed to lower subgrade modulus value derived from LWD when compared to IRC:37-2018.

4 Conclusions

In the present study the objective was to determine the influence of plastic addition on strength characteristics of fine-grained subgrade soil. From the conducted experiments following conclusions can be made:

- With the increase in plastic content (up to 1.5%) there is no significant change observed in either the OMC or the MDD.
- The maximum improvement in CBR is obtained while using 1.5% plastics content i.e. 4.01%.
- The CBR value at 2 and 3% plastic content were found to be even less than the CBR of base soil.
- The maximum CBR value of modified soil is approximately 1.3 times that of base soil.
- The difference between theoretical value of subgrade resilient modulus and LWD modulus derived from the trial sections in the field is 4% for base virgin soil and 6% for reinforced soil.
- Subgrade modulus from LWD is slightly less when compared to subgrade modulus derived from codal provisions.
- The pavement thickness is optimized by maximum of 10.58% for traffic of 75 MSA, 9.77% for traffic of 125 MSA and least for the traffic of 30 MSA by 7.4%.
- Total cost of 7.89% was reduced for the traffic of 75 MSA which was observed to be more when compared to 30 MSA and 125 MSA traffic.

References

Choudhary AK, Jha JN, Gill KS (2010) Laboratory investigation of bearing capacity behaviour of strip footing on reinforced flyash slope. Geotext Geomembr 28(4):393–402

Dhatrak AI, Konmare S (2016) Laboratory performance of randomly oriented plastic waste in subgrade of flexible pavement. Int J Innov Res Sci Eng Technol 3(1):3969–3976

Hejazi SM, Sheikhzadeh M, Abtahi SM, Zadhoush A (2012) A simple review of soil reinforcement by using natural and synthetic fibres. Constr Build Mater 30:100–116

IS 2720 (Part 10) (1991) Determination of unconfined compressive strength. BIS, New Delhi

IS 2720 (Part 16) (1987) Laboratory determination of CBR. BIS, New Delhi

IS 2720 (Part 4) (1985) Grain size analysis. BIS, New Delhi

IS 2720 (Part 5) (1985) Determination of liquid limit and plastic limit of soil. BIS, New Delhi

IS 2720 (Part 7) (1980) Determination of water content, dry density relation using light compaction. BIS, New Delhi

IRC:37-2018. Guidelines for the design of flexible pavements, fourth revision

Kolias S, Kasselouri-Rigopoulou V, Karahalios A (2005) Stabilisation of clayey soils with high calcium fly ash and cement. Cement Concr Compos 27(2):301–313

Manso JM, Ortega-López V, Polanco JA, Setién J (2013) The use of ladle furnace slag in soil stabilization. Constr Build Mater 40:126–134

Sobhan K, Mashnad M (2003) Mechanical stabilization of cemented soil–fly ash mixtures with recycled plastic strips. J Environ Eng 129(10):943–947

Modeling Rutting Behavior of Crumb Rubber Modified Binders Using Design of Experiments

Reza Azadedel, Nader Solatifar, and Maghsoud Rahbarnia

Abstract In recent decades, modifying the binder with various additives for improving its behavior has been very noticeable by researchers. Crumb rubber is one of the additives which increases the rutting resistance of binder. The main objective of this study is to identify the factors affecting the resistance characteristics of binder regarding rutting and modeling its behavior at 64 °C using crumb rubber. For this purpose, one of the Design of Experiments (DoE) methods, i.e. Response Surface Methodology (RSM) was utilized taking into account four factors, including mixing temperature, mixing time, mixing speed and content percentage (by weight) of crumb rubber, as well as one response, rutting parameter. As a result, 30 modified binder samples were produced and Dynamic Shear Rheometer (DSR) test was conducted on the samples. Modeling was done and performance of the model was evaluated. Results show that developed model has a good prediction accuracy and the content of crumb rubber is more effective factor rather than other factors and the interaction of them. In addition, in a certain content percentage of crumb rubber, considering the mixing temperature and mixing speed at a low level and mixing time at a high level, the maximum rutting parameter could be achieved.

Keywords Crumb Rubber Modified Binder (CRMB) · Binder rheology · Rutting · Design of Experiments (DoE) · Response Surface Methodology (RSM)

1 Introduction

In the recent years, due to increased traffic loads, modifying binders has drawn researchers' attention towards enhancing the performance of them. The main objective to use additives in modifying binders has been also increasing resistance of asphalt pavements against permanent deformation (rutting), fatigue cracking, as well as low-temperature cracking (Subhy 2017).

R. Azadedel · N. Solatifar (✉)
Department of Civil Engineering, Urmia University, Urmia, Iran
e-mail: n.solatifar@urmia.ac.ir

M. Rahbarnia
RK Northwest Refining and Energy Company, Tabriz, Iran

C. Raab (ed.), *Proceedings of the 9th International Conference on Maintenance and Rehabilitation of Pavements—Mairepav9*, Lecture Notes in Civil Engineering 76,
https://doi.org/10.1007/978-3-030-48679-2_29

To improve the mechanical and rheological behavior of binders, researchers have used various additives such as polymers, nano-materials, etc. Crumb Rubber, known as one of the polymer additives, is produced from car tire wastes and can improve the properties of the binder and consequently help in protecting the environment. Tire waste is formed up of three major components including rubber, steel, and fibers. So, the rubber component forming about 60% of weight of the tire can be used to modify the binders (Thodesen et al. 2009).

The results of different performance tests on the Crumb Rubber Modified Binder (CRMB) have similarly shown its improved properties; for example, the findings of Dynamic Shear Rheometer (DSR) test for adding crumb rubber to binders showed an increase in complex modulus (G^*) and a reduction in phase angle (δ). Moreover, the complex modulus increased and the phase angle diminished via adding to the crumb rubber modifier content. Increasing the parameter of $G^*/\sin\delta$ also resulted in more CRMB resistance against rutting. In addition, this parameter increased via adding to the crumb rubber modifier content (Aflaki and Tabatabaee 2009; Al-Khateeb et al. 2014; Behnood and Olek 2017; Fini et al. 2017; Venudharan and Biligiri 2017; Venudharan et al. 2018; Liu et al. 2018; Kim et al. 2010; Chen et al. 2019; Wang et al. 2018). On the other hand, decreasing the parameter of $G^*.\sin\delta$ at moderate temperatures could lead to a rising trend in modified binder resistance against fatigue cracking (Behnood and Olek 2017; Aflaki and Tabatabaee 2009; Fini et al. 2017; Liu et al. 2018).

The results of the Bending Beam Rheometer (BBR) test also showed that the stiffness of the CRMB was less than the neat binder, which could be concluded that the improved modified binder resistance to low-temperature cracking was higher than the neat binder (Aflaki and Tabatabaee 2009; Hajikarimi et al. 2013; Behnood and Olek 2017; Fini et al. 2017; Wang et al. 2012). In general, it could be observed that the addition of crumb rubber could increase the viscosity, improve the rutting resistance, enhance the fatigue resistance, and reduce the stiffness of the binder.

Rutting in asphalt pavement is the result of the combination of irreversible deformations of the asphalt layers under repeated loading at high temperatures. In order to account for the binder response to these repetitive loads, the ratio $G^*/\sin\delta$, the rutting parameter is used as a measure of the hardness of binder at high temperatures or binder resistance to permanent deformation. Increasing the rutting parameter of modified binder compared to the neat binder, indicates increased resistance of modified binder against permanent deformation.

In order to investigate the behavior of CRMB against performance parameters and failures due to loading and environmental conditions, it is necessary to conduct various experiments on samples containing different amounts of additives as well as in different testing conditions. In this study, the effect of each of the quantitative factors including mixing temperature, mixing speed, mixing time, and crumb rubber modifier content on rutting parameter was investigated. The well-known Design of Experiments (DoE) technique, i.e. Response Surface Methodology (RSM) utilized in this regard using Design-Expert software (Stat-Ease 2018); and at the end, the effect of each of the mentioned factors on the response was analyzed and the optimum rutting parameter was achieved.

2 Design of Experiments (DoE)

According to Montgomery (2017), a DoE includes a series of experiments consciously made some changes in process input variables as well as it can identify and observe the variations to be caused in the process output responses. In this method, the experiments are designed in such a way that the factors can be tested together in order to achieve the responses. The DoE model in this study was developed based on the Central Composite Design (CCD) method, which is one of the most widely used RSMs. RSM is also used to optimize the experiments. The main purpose of RSM optimization is to find levels of effective factors that cause the response of the experiment to be maximum or minimum (depending on the purpose of the experiment). The RSM can also determine factors in terms of their interactions using two methods including CCD and Box-Behnken Design (BBD) (Bezerra et al. 2008; Myers et al. 2016). In this respect; the number of required experiments could be determined according to Eq. (1).

$$N = 2^f + (2f + 1) + r \quad (1)$$

where N is the number of required experiments; f is the number of factors; and, r shows the replicate number of the central point. All the factors can be studied in five levels ($-\alpha$, -1, 0, $+1$, $+\alpha$). The α value depends on the number of variables and it can be calculated as $\alpha = \pm 2^{f/4}$. For two, three, and four variables; these values are ± 1.41, ± 1.68, and ± 2.00; respectively (Bezerra et al. 2008).

To evaluate the behavior of CRMB in terms of the four factors including mixing temperature, mixing time, mixing speed and crumb rubber modifier content, five replicates at upper and lower levels were used as illustrated in Table 1. Hence, according to Eq. (1) the number of required experiments is equal to 30.

Table 1 The factors of design of experiments

Factor	Unit	Level $+\alpha$	High-level (+1)	Center level	Low-level (-1)	Level $-\alpha$
A: Mixing temperature	°C	190	180	170	160	150
B: Mixing time	min	40	35	30	25	20
C: Modifier content	%	10	8	6	4	2
D: Mixing speed	Hz	100	86	71	57	42

3 Materials and Methods

In this study, a total number of 30 CRMB samples of 500 gr were prepared in accordance with the design specifications of the test using a CCD method. The neat binder used in this research was 85/100 pen bitumen. Classical tests including penetration point, flash point, tensile strength, softening point, and solubility in tetrachloroethylene were conducted on this binder and the results are presented in Table 2.

Crumb rubber as a polymer additive was used in this study. The characteristics of this additive including polymer and thermal analysis, ash content, as well as volatiles content are reported in Table 3. It should be noted that the crumb rubber particle size of 30 mesh (0.595 mm) was used in this study.

For mixing the binder with crumb rubber modifier, a high shear mixer was used. For this purpose, the binder in the oven was first heated up to the desired temperature, and then the modifier was added. After the required temperature of the resulting sample was obtained, the mixing was done at the desired speed and time. All the samples were produced in a temperature tolerance of ±3 °C. Some of CRMB samples produced in this study are shown in Fig. 1. For determination of rutting parameters

Table 2 Conventional characteristics of the neat binder

Parameter	Test standard	Result
Penetration point (0.1 mm)	ASTM D5 (ASTM 2013)	87
Flash point (°C)	ASTM D92 (ASTM 2018)	294
Tensile strength (cm)	ASTM D113 (ASTM 2017)	>100
Softening point (°C)	ASTM D36 (ASTM 2014a)	47.5
Solubility in tetrachloroethylene (%)	ASTM D2042 (ASTM 2015a)	99.92

Table 3 Analysis results of crumb rubber used as the binder modifier

Test	Standard	Result
Polymer analysis	ASTM D3677 (ASTM 2015b)	Natural rubber
Thermal analysis (%)	ASTM E1131 (ASTM 2014b)	Natural rubber: 52.16 ± 2 Soot: 45.36 ± 2 Ash: 2.30 ± 2
Ash (%)	–	4 ± 2
Volatiles @105 °C (%)	–	0.3

Fig. 1 Some of CRMB samples produced in this research

including G^* and δ, DSR test was conducted on unaged CRMB at the temperatures of 52, 58, 64 and 70 °C according to AASHTO T315 (AASHTO 2012) standard.

4 Results and Discussion

4.1 Developing the Rutting Model

In order to predict and select the effective factors on the rutting parameter (mixing temperature, mixing time, mixing speed, and crumb rubber modifier content); Linear, 2 Function Interaction (2FI), and Quadratic regression models were developed for responses (rutting at the temperature of 64 °C) obtained from 30 tests using Design-Expert software version 11. In Table 4, the accuracy of the developed models is presented.

Among the developed models, both linear and quadratic regression models with the p-values less than 0.05 are considered significant. The quadratic model with higher coefficient of determination (R^2) is also reported to be more accurate rather than the linear model. The general form of quadratic regression model is expressed as Eq. (2).

Table 4 Accuracy evaluation of the developed models

Model	p-value	R^2	Adjusted R^2	Predicted R^2
Linear	<0.0001	0.8699	0.8491	0.8053
2FI	0.1050	0.9214	0.88	0.7475
Quadratic	0.0077	0.9672	0.9365	0.8109

$$y = \beta_0 + \sum_{i=1}^{k} \beta_i x_i + \sum_{i=1}^{k} \beta_{ii} x_i^2 + \sum_{i=1 \le i \le j}^{k} \beta_{ij} x_i x_j + \varepsilon \tag{2}$$

where y is the response; x_i is the factor i; β_i, β_{ii} and β_{ij} are the coefficients; k is the number of the factors; and, ε is the model error (Bezerra et al. 2008).

Stepwise regression method has been used for final modeling. In this study, the results of the 64 °C testing have been discussed and analyzed. Equation (3) reflects the final stepwise quadratic regression model that was developed for modeling the rutting behavior of CRMB.

$$\mathbf{G^*/\sin\delta = (0.006208 - 0.000046 \times \mathit{A} - 0.000156 \times \mathit{B} - 0.000160 \times \mathit{C} + 0.00003 \times \mathit{D} + 5.55878\mathit{E} - 07 \times \mathit{A} \times \mathit{B} - 2.12043\mathit{E} - 07 \times \mathit{A} \times \mathit{D} + 2.50122\mathit{E} - 07 \times \mathit{B} \times \mathit{D} + 1.31909\mathit{E} - 07 \times \mathit{A}^2 + 7.25063\mathit{E} - 07 \times \mathit{B}^2 + 6.93734\mathit{E} - 06 \times \mathit{C}^2)^{-1}} \tag{3}$$

where $G^*/\sin\delta$ is the rutting parameter of CRMB, kPa; A is the mixing temperature, °C; B is the mixing time, min; C is the crumb rubber modifier content, wt.%; and, D is the mixing speed (frequency), Hz. It should be noted that increasing the rutting parameter ($G^*/\sin\delta$), indicates increasing in binder resistance against permanent deformation.

4.2 Performance Evaluation of the Model

To evaluate the performance of the developed model, results of the analysis of variance (ANOVA) of the rutting parameter, the adequacy assessment of the model and also its coefficient of determination (R^2) were utilized. The results of the ANOVA statistical analysis for model parameters are shown in Table 5, in which the p-value in the model is at 95% confidence level.

The normal plot of the residuals is shown in Fig. 2(a). As it can be seen in this figure, the straight line drawn from these points covers almost all the data, and the approximate normalization of the data is verified. The residuals versus predicted

Table 5 ANOVA results of the rutting parameter for the final developed model

Source	Sum of squares	df	Mean square	F-value	p-value	
Model	6.301E−07	10	6.301E−08	52.99	<0.0001	Significant
Residual	2.259E−08	19	1.189E−09			
Lack of fit	2.259E−08	14	1.614E−09			
Pure error	0.0000	5	0.0000			
Cor total	6.527E−07	29				

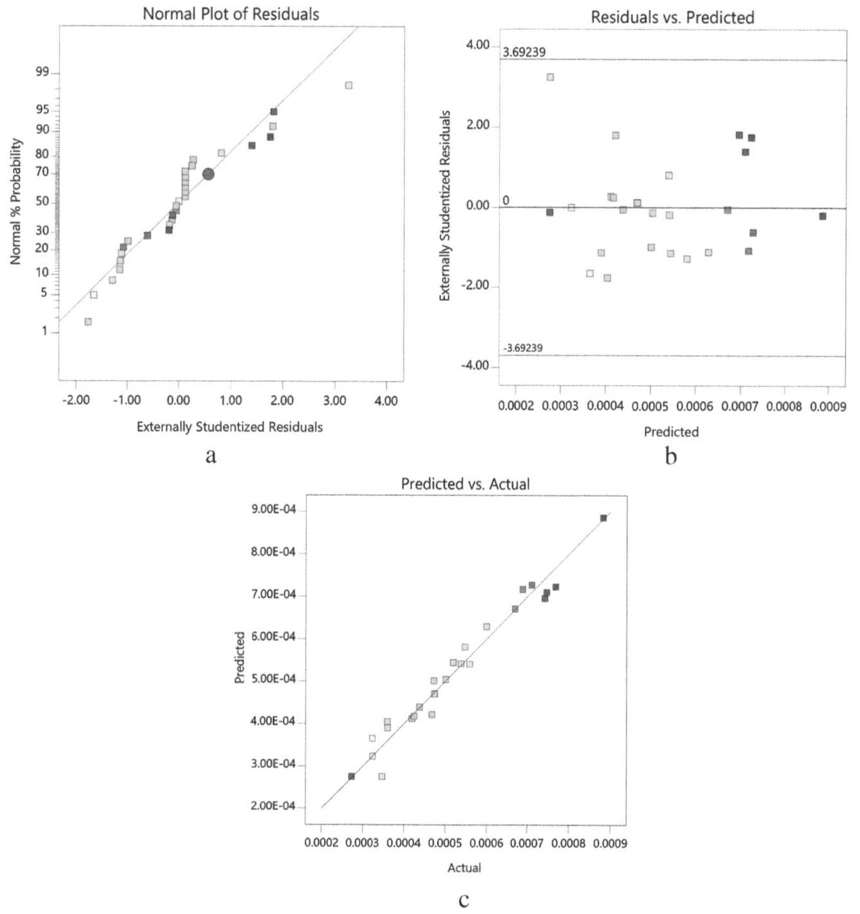

Fig. 2 Evaluation of the developed model: **a** normal plot of residuals, **b** residuals vs. predicted values, and **c** predicted vs. actual values

rutting values were also used to check the stability of the variance of the data as presented in Fig. 2(b). This figure shows that the data have not followed a particular trend and the distribution is irregular. In addition, Fig. 2(c) displays the predicted rutting versus the actual values. The data is predominantly on a straight line. This figure also indicates that the model has good accuracy despite some errors; therefore, it could be used for qualitative analysis as well as prediction of rutting parameter.

According to the adjusted R^2 value for the response variable (the rutting parameter), the whole input variables in the developed regression model could validate 94.72% of the variation of the response variable which is reported in Table 6.

Table 6 Coefficient of determination values for the final model

R^2	Adjusted R^2	Predicted R^2
0.9654	0.9472	0.9029

4.3 *Sensitivity Analysis of the Effective Factors*

Figure 3 shows the effect of variation of the four factors on the rutting parameter of the CRMB. As it can be seen in this figure, the most effective factor on the model is the crumb rubber modifier content. On the other hand, the rutting parameter is directly related to the modifier content and the interaction of the mixing temperature and the mixing speed. In addition, the relationship between the rutting parameter with the other factors, their interactions, and their quadratic order is inverse.

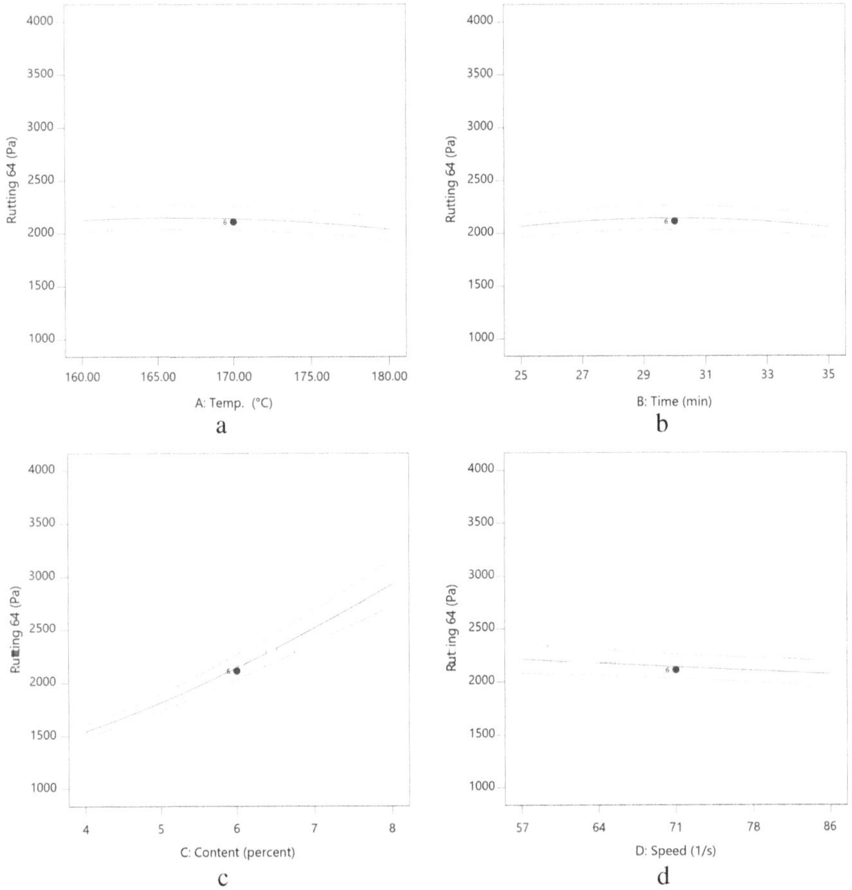

Fig. 3 Effect of variations of factors on the rutting parameter: **a** mixing temperature, **b** mixing time, **c** modifier content, and **d** mixing speed

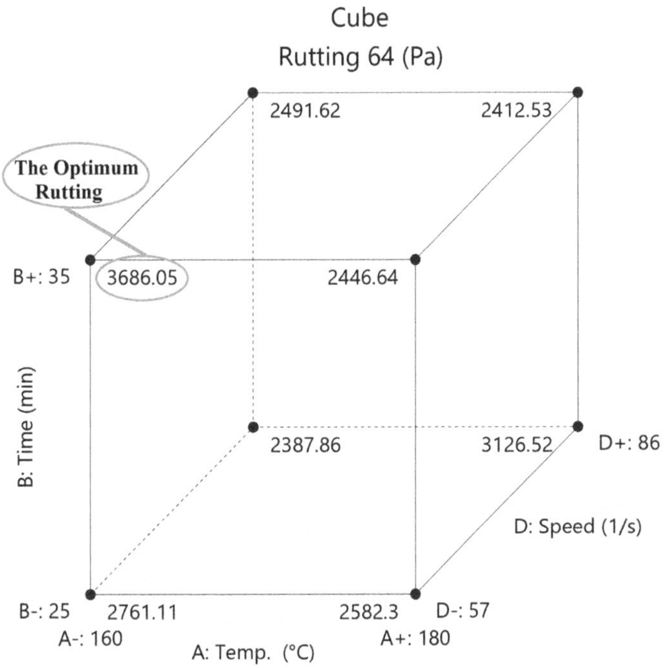

Fig. 4 Cubic diagram for analyzing effective factors on the optimum rutting parameter

4.4 Model Validation

To validate the model, the comparison between the value of the rutting parameter obtained from the test and the predicted value based on the developed model were used. For this purpose, an additional CRMB sample was produced in different conditions from that of proposed DoE. The production conditions of new sample together with the test results (responses) are shown in Table 7. Accordingly, it is found that the developed model had been able to predict the rutting parameter value for validation sample with an error of 0.43%.

Table 7 Comparison of predicted and measured rutting parameters for model validation

Sample mixing method				Response value	Rutting parameter (Pa)	Error (%)
Temperature (°C)	Time (min)	Modifier content (%)	Speed (Hz)			
170	25	8	86	Measured	2771	0.43
				Predicted	2782.86	

4.5 Optimization of Rutting Parameter

The cubic diagram of the effect of three factors including mixing temperature, mixing time and mixing speed at 8 wt.% of crumb rubber on the response, i.e. the rutting parameter is shown in Fig. 4. As it can be seen, the optimum rutting is considered as the highest response wherein in the mixing temperature is low, the mixing time is high, and the mixing speed is low.

5 Conclusions

In this study, the design of experiments was utilized to analyze the rutting behavior of the crumb rubber modified binder. Effective factors including mixing temperature, mixing time, mixing speed and crumb rubber modifier content were considered and the following results were derived:

- Stepwise regression model was found more accurate than the other models and it was selected as the final model. The developed model (Eq. (3)) can be used to predict and determine the optimum rutting parameter for the binder grade and crumb rubber type used in this study.
- Crumb rubber modifier content was the most important factor affecting the value of rutting parameter. This factor had a direct relationship with the rutting. The other factors and their interaction as well as quadratic factors could also contribute somewhat to the rutting parameter.
- The results of the sensitivity analysis showed that the crumb rubber modifier content was directly correlated with the rutting parameter. Mixing speed and mixing temperature were inversely correlated with this parameter.
- The developed model was validated to predict the rutting parameter with an error of 0.43%.
- And, considering the mixing speed and mixing temperature at a low level and the mixing time at a high level, the response value, i.e. the optimum rutting parameter, was considered as the highest level in a certain modifier content.

References

AASHTO (2012) Standard method of test for determining the rheological properties of asphalt binder using a dynamic shear rheometer (DSR). American Association of State Highway and Transportation Officials

Aflaki S, Tabatabaee N (2009) Proposals for modification of Iranian bitumen to meet the climatic requirements of Iran. Constr Build Mater 23(6):2141–2150. https://doi.org/10.1016/j.conbuildmat.2008.12.014

Al-Khateeb GG, Ramadan ZK (2014) Investigation of the effect of rubber on rheological properties of asphalt binders using superpave DSR. KSCE J Civil Eng 19(1):127–135. https://doi.org/10.1007/s12205-012-0629-2

ASTM (2013) Standard test method for penetration of bituminous materials. ASTM International, West Conshohocken. http://doi.org/10.1520/D0005_D0005M-13

ASTM (2014a) Standard test method for softening point of bitumen (ring-and-ball apparatus). ASTM International, West Conshohocken. http://doi.org/10.1520/D0036_D0036M-14E01

ASTM (2014b) Standard test method for compositional analysis by thermogravimetry, ASTM International, ASTM E1131-08, West Conshohocken. https://doi.org/10.1520/E1131-08R14

ASTM (2015a) Standard test method for solubility of asphalt materials in trichloroethylene. ASTM International, West Conshohocken. http://doi.org/10.1520/D2042-15

ASTM (2015b) Standard test methods for rubber—identification by infrared spectrophotometry ASTM International, ASTM D3677-10, West Conshohocken. https://doi.org/10.1520/D3677-10R15

ASTM (2017) Standard test method for ductility of asphalt materials. ASTM International, West Conshohocken. http://doi.org/10.1520/D0113-17

ASTM (2018) Standard test method for flash and fire points by cleveland open cup tester. ASTM International, West Conshohocken. http://doi.org/10.1520/D0092-18

Behnood A, Olek J (2017) Rheological properties of asphalt binders modified with styrene-butadiene-styrene (SBS), ground tire rubber (GTR), or polyphosphoric acid (PPA). Constr Build Mater 151:464–478. https://doi.org/10.1016/j.conbuildmat.2017.06.115

Bezerra MA, Santelli RE, Oliveira EP, Villar LS, Escaleira LA (2008) Response surface methodology (RSM) as a tool for optimization in analytical chemistry. Talanta 76(5):965–977. https://doi.org/10.1016/j.talanta.2008.05.019

Chen Z, Pei J, Wang T, Amirkhanian S (2019) High temperature rheological characteristics of activated crumb rubber modified asphalts. Constr Build Mater 194:122–131. https://doi.org/10.1016/j.conbuildmat.2018.10.223

Fini EH, Hosseinnezhad S, Oldham D, Mclaughlin Z, Alavi Z, Harvey J (2019) Bio-modification of rubberised asphalt binder to enhance its performance. Int J Pavement Eng 20(10):1216–1225. https://doi.org/10.1080/10298436.2017.1398548

Hajikarimi P, Aflaki S, Hoseini AS (2013) Implementing fractional viscoelastic model to evaluate low temperature characteristics of crumb rubber and gilsonite modified asphalt binders. Constr Build Mater 49(Supplement C):682–687. https://doi.org/10.1016/j.conbuildmat.2013.09.001

Kim H-S, Lee S-J, Amirkhanian S (2010) Rheology investigation of crumb rubber modified asphalt binders. KSCE J Civil Eng 14(6):839–843. https://doi.org/10.1007/s12205-010-1020-9

Liu W, Yan K, Ge D, Chen M (2018) Effect of APAO on the aging properties of waste tire rubber modified asphalt binder. Constr Build Mater 175:333–341. https://doi.org/10.1016/j.conbuildmat.2018.04.098

Montgomery DC (2017) Design and analysis of experiments. Wiley, Hoboken

Myers RH, Montgomery DC, Anderson-Cook CM (2016) Response surface methodology: process and product optimization using designed experiments. Wiley, Hoboken

Subhy A (2017) Advanced analytical techniques in fatigue and rutting related characterisations of modified bitumen: literature review. Constr Build Mater 156:28–45. https://doi.org/10.1016/j.conbuildmat.2017.08.147

Stat-Ease I (2018) Design-Expert (Version 11.1.0.1), Minneapolis. https://www.statease.com/dx11.html

Thodesen C, Shatanawi K, Amirkhanian S (2009) Effect of crumb rubber characteristics on crumb rubber modified (CRM) binder viscosity. Constr Build Mater 23(1):295–303. https://doi.org/10.1016/j.conbuildmat.2007.12.007

Venudharan V, Biligiri KP (2017) Effect of crumb rubber gradation on asphalt binder modification: rheological evaluation, optimization and selection. Mater Struct 50(2):129. https://doi.org/10.1617/s11527-017-0994-x

Venudharan V, Biligiri KP, Das NC (2018) Investigations on behavioral characteristics of asphalt binder with crumb rubber modification: Rheological and thermo-chemical approach. Constr Build Mater 181:455–464. https://doi.org/10.1016/j.conbuildmat.2018.06.087

Wang D, Li D, Yan J, Leng Z, Wu Y, Yu J, Yu H (2018) Rheological and chemical characteristic of warm asphalt rubber binders and their liquid phases. Constr Build Mater 193:547–556. https://doi.org/10.1016/j.conbuildmat.2018.10.199

Wang H, You Z, Mills-Beale J, Hao P (2012) Laboratory evaluation on high temperature viscosity and low temperature stiffness of asphalt binder with high percent scrap tire rubber. Constr Build Mater 26(1):583–590. https://doi.org/10.1016/j.conbuildmat.2011.06.061

Self-healing Asphalt for Road Pavements

A. Tabaković and E. Schlangen

Abstract This paper presents a unique self-healing system for asphalt pavement which employs compartmented calcium-alginate fibres encapsulating an asphalt binder healing agent (rejuvenator). This system presents a novel method of incorporating rejuvenators into asphalt pavement mixtures. The compartmented fibres are used to distribute the rejuvenator throughout the pavement mixture, thereby overcoming some of the problems associated with alternate asphalt pavement healing methods, i.e., spherical capsules and hollow fibres. The healing system performance, when embedded in Porous Asphalt (PA) mix was tested by employing: (i) Indirect Tensile Stiffness and Strength test (ii) 4 Point Bending Fatigue test. The Semi Circular Bend (SCB) test was adopted to study crack propagation and its closure (healing) in an asphalt mix. The findings demonstrate that compartmented alginate fibres have capacity to survive asphalt mixing and compaction process. The fibres can efficiently repair damage (close the cracks), increase asphalt mix stiffness and strength. However, when the asphalt mix is subjected to fatigue loading the system does not significantly improve healing properties of the asphalt mix. Nevertheless, the findings indicate that, with further enhancement, compartmented calcium alginate fibres may present a promising new approach for the development of self-healing asphalt pavement systems.

Keywords Self-healing · Alginate fibres · Asphalt pavements · Asphalt rejuvenation

---

A. Tabaković (✉)
Centre for Research in Engineering Surface Technology (CREST), FOCAS Research Institute, Technological University Dublin, 13 Camden Row, Dublin D08 CKP1, Ireland
e-mail: amir.tabakovic@tudublin.ie

School of Civil Engineering, University College Dublin, Newstead, Belfield, Dublin D04 V1W8, Ireland

A. Tabaković · E. Schlangen
Department Materials, Mechanics Management and Design (3MD), Faculty CiTG, Delft University of Technology, Stevinweg 1, 2628 CN Delft, The Netherlands

C. Raab (ed.), *Proceedings of the 9th International Conference on Maintenance and Rehabilitation of Pavements—Mairepav9*, Lecture Notes in Civil Engineering 76,
https://doi.org/10.1007/978-3-030-48679-2_30

1 Introduction

For past decade self-healing technology has been advancing asphalt pavement design (Tabaković and Schlangen 2016, Xu et al. 2018). Self-healing technology offers an alternative method for road maintenance, where the damage is repaired by an internal (implanted) healing system. The objective of self-healing technology is to enable/assist material systems to heal after damage on a local or global scale. It aims to reduce the local or global level of damage and to extend or to renew the functionality and life-time of the damaged part, system or device. To date, researchers have tested three self-healing methods for asphalt pavements as follows (Xu et al. 2018), they are: (i) induction heating, (ii) microwave heating and (iii) rejuvenator encapsulation. The rejuvenator encapsulation approach represents a more favourable method of self-healing as it allows for the rejuvenation of aged binder, i.e., enables it to return it to its original chemical, physical and mechanical properties. Researchers have demonstrated that various types of capsules containing rejuvenator can be produced and that these capsules are sufficiently thermally and mechanically stable to survive the asphalt production process (Xu et al. 2018). However, a difficulty with this approach is that large amounts of microcapsules are needed to make the process effective. The addition of large quantities of microcapsules into the asphalt mix can reduce the quality of the pavement which itself may cause premature pavement failure. Garcia et al. (2016) and Sun et al. (2015) reported that asphalt stiffness was reduced when microcapsules were added. They explained that softening of asphalt binder (viscosity reduction) was caused by the rejuvenator release. However, it is well documented (Gibney 2004) that deformation in the asphalt mix is caused by sand granulates. It is possible that the inclusion of microcapsules, sand like particles, has also contributed to increased asphalt mix deformation, i.e., rutting. Furthermore, the chemical compounds used in the production of microcapsules, such as melamine–formaldehyde (Anderson 1995), in large quantities could pose an environmental threat via leaching.

The encapsulation of rejuvenator in alginate-based compartmented fibres is explored here as a solution to these problems in asphalt mixtures. The study showed that alginate fibres have great potential as self-healing technique for asphalt pavements, i.e. they can be inserted into the asphalt mastic mix (fibres can survive asphalt mixing and compaction process) and can increase asphalt mastic mix strength by 36%. The results demonstrate that optimum rejuvenator content in the alginate fibre is of 70:30 rejuvenator/alginate ratio. The results also show that Porous Asphalt (PA) mix containing 5% of 70:30 rejuvenator/alginate ratio compartmented alginate fibres has higher strength, stiffness and better healing properties in comparison to the control asphalt mix, i.e., mix without fibres. However, 4PB fatigue test showed that self healing system does not significantly improve the healing properties of the asphalt pavement. Nevertheless, the findings indicate that, with further enhancement, compartmented calcium alginate fibres may present a promising new approach for the development of self-healing asphalt pavement systems.

2 Materials and Methods

2.1 Compartmented Alginate Fibres Production

The compartmented fibres were spun from an emulsion of rejuvenator suspended in a water solution of sodium alginate. A 6 wt% solution of sodium alginate in de-ionized water was prepared for this purpose. At the same time, a 2.5 wt% poly (ethylene-alt-maleic-anhydride) (PEMA) polymeric surfactant solution was prepared by dissolving the copolymer in water at 70 °C and mixing it for 60 min. After the PEMA has been dissolved in the water, it was allowed to cool to room temperature (20 ± 2 °C) and was combined with the rejuvenator, forming a healing agent solution, in PEMA/rejuvenator 1/1.5 proportion. Sodium alginate and PEMA/rejuvenator solutions were then combined in accordance with Tabaković et al. (2017a) with a 70/30 rejuvenator/alginate proportion, found to be optimal for the compartmented alginate fibres. Figure 1 illustrates compartmented alginate fibre encapsulating bitumen rejuvenator. All of the solutions were mixed at 200 rpm for 60 s. It is important to note that the stirring rate and stirring time can be used to control the size of the rejuvenator droplets in the solution and thus the size of the rejuvenator compartments (Prajer et al. 2015; Tabaković et al. 2016). The emulsions were spun with a plunger-based lab scale wet spinning line in a conventional wet spinning process (Mookhoek et al. 2012) to form the rejuvenator-filled compartmented fibres. More details on the fibre preparation and spinning process can be found elsewhere (Mookhoek et al. 2012). All chemicals used in the process were purchased from Sigma Aldrich, The Netherlands, except for the rejuvenator, Modesel R20, which was provided by Latexfalt B.V., Hoogewaard 183, 2396 AP Koudekerk aan den Rijn, The Netherlands.

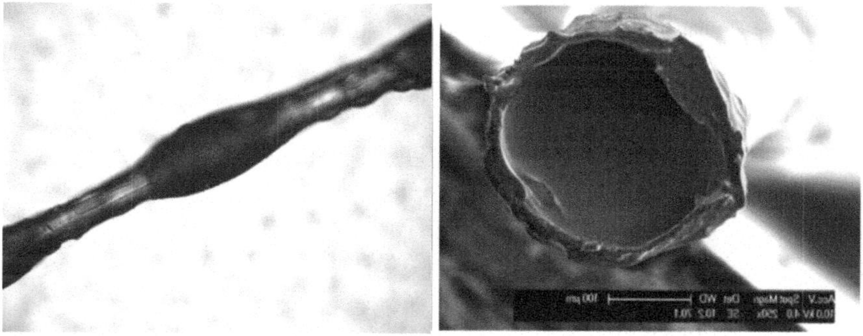

a) b)

Fig. 1 Compartmented Alginate fibre, (a) compartmented alginate fibre encapsulating rejuvenator (Tabaković et al. 2016), (b) ESEM image of fibre cross section (Tabaković et al. 2017a)

2.2 Porous Asphalt Mastic Mix Design and Mixing Procedure

In an effort to evaluate the efficiency of the rejuvenator encapsulated in compartmented calcium alginate fibres, a Porous Asphalt (PA) mix was designed. The grading envelope Rationalisatie en Automatisering Grond-, Water-en Wegenbouw (RAW) 2005 was used to produce a PA asphalt mix typical of those used for road design (Liu 2012). The limestone used originated from a quarry in Norway, the filler material was hydrated lime (Wigro 60 K), and pen 70/100 of bitumen was used. Figure 2 illustrates the mix grading curve and illustrates how the mix compares well with the grading envelope. Table 1 summarises mix constituents and shows their proportions in the mix with and without fibres. The fibres are added in amount of 5% and 10%

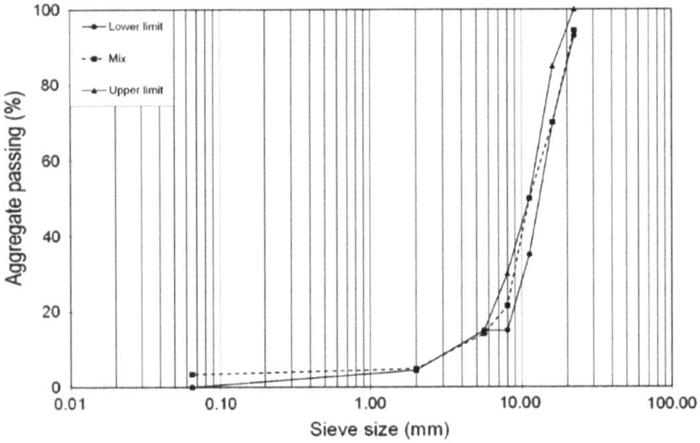

Fig. 2 Porous asphalt mix grading, grading envelope RAW 2005 (Liu 2012)

Table 1 Porous Asphalt mix design (percentage of constituent content is given by weight)

Mix constituent	Constituent size (mm)	Content in mix (%)		
		Without fibre	5% fibres	10% fibres
Gravel	22.4	20.1	20.1	20.1
	16.0	25.6	25.6	25.6
	11.2	34.8	34.8	34.8
	8.0	7.4	7.4	7.4
	6.0	6.7	6.7	6.7
Sand	2.0	0.5	0.5	0.5
Filler	<0.063	0.5	0.5	0.5
Bitumen	–	4.5	4.3	4.1
Fibre	–	0	0.23	0.45

of total bitumen content in the mix, which in total mix volume it represents 0.23% and 0.45% respectively. The aggregate mix constituent content was not changed with insertion of the fibres in the mix.

The porous asphalt mixing is described elsewhere (Tabaković et al. 2017a). In order to account for the asphalt ageing, an aging programme was developed. The ageing protocol was adopted from Kliewer et al. (1995), as follows: Long term; 15 years field ageing 4 h at 135 °C followed by 4 days at 85 °C in a forced air draft oven. Following the asphalt aging procedure, the mix was preheated to the standard asphalt mixing/compaction temperature of 160 °C and the fibres were gradually added to the mix to avoid conglomeration of the fibres within the mix. After the fibres were included in the mix, the test specimens were prepared. The air void content in the mix design was 20%, with a target density of 2.05,124 kg/cm^3 for the control mix and 2.05088 kg/cm^3 for the mix containing the fibres.

The cylindrical test specimens were compacted in accordance with IS EN 12697-31:2007 using a SERVOPAC gyratory compactor and beam test specimens were machined form an asphalt slab which was produced by using a shear box compactor in dimensions: 450 × 50 × 50 mm.

2.3 *Porous Asphalt (PA) Mix Optimisation*

The non-destructive Indirect Tensile Stiffness Modulus (ITSM) test is conducted, in accordance with EN 12697-26: 2012 and the Indirect Tensile Strength (ITS) test, in accordance with EN 12697-23: 2003 were employed in order to investigate effect of the fibres on mechanical properties of the asphalt mix and evaluate the healing efficiency of the compartmented fibres encapsulating the rejuvenator.

Healing programme is as follows:

(1) Three mixtures: control mix (0% fibre) and two mixtures containing 5% and 10% fibres.
(2) Tests: ITSM and ITS.
(3) Test temperature: 20 °C.
(4) Healing temperature: 20 °C.
(5) Healing time: 20 h and 40 h after initial test.

Testing protocol was as follows:

(1) Test samples pre-conditioning at testing temperature (20 °C).
(2) ITSM test diameter I followed by diameter II.
(3) Test specimen relaxation 2 h, followed by 1st ITS test.
(4) Positioning test specimen into healing ring and healing for 2 h.
(5) ITSM repeat–post ITS test.
(6) Positioning test specimen into healing ring and healing for additional 18 h.
(7) ITSM test-pre 2nd ITS test.
(8) 2nd ITS test.

(9) Positioning test specimen into healing ring and healing for additional 20 h.
(10) ITSM test-pre 3rd ITS test.
(11) 3rd ITS test.

2.4 SCB Test—Crack Propagation and Healing

For the SCB test, a modified NCHRP09-46 procedure was adopted. The load was applied at the centre line, above the 'V' notch, at a loading rate of 0.1 mm/s and at a temperature of 20 ± 2 °C. The support span 'S' = 80 mm, included 80% of the test specimen diameter, leaving 10 mm on each side. Figure 3 shows a schematic diagram and an actual view of the SCB test-setup. Several modifications to the test were made in the specimen geometry. The diameter of the specimen was reduced from 150 mm to 100 mm and the thickness from 50 mm to 20 mm, whereas the notch shape was changed from a straight line to the V-notch shape. These changes were employed in order to achieve full depth crack propagation throughout the depth of the test specimen, with the onset at the tip of the notch. Test samples were loaded until the crack has propagated to 25 mm or when the samples started to deform considerably. The crack propagation was measured in order to calculate the crack propagation speed. The crack propagation velocity was measured manually, by observing the crack propagation during the test, from the tip of the crack (0 mm) to the max (25 mm). For this procedure, a measuring scale was painted along the centerline of the specimen. The progression of the crack propagation was measured against the load and crosshead displacement. A special testing programme was designed, including:

- two mixtures: control mix and mix containing fibres (5%),
- test temperature of 20 ± 3 °C,
- healing temperature of 20 ± 3 °C,
- healing time of 20 and 40 h after the initial test.

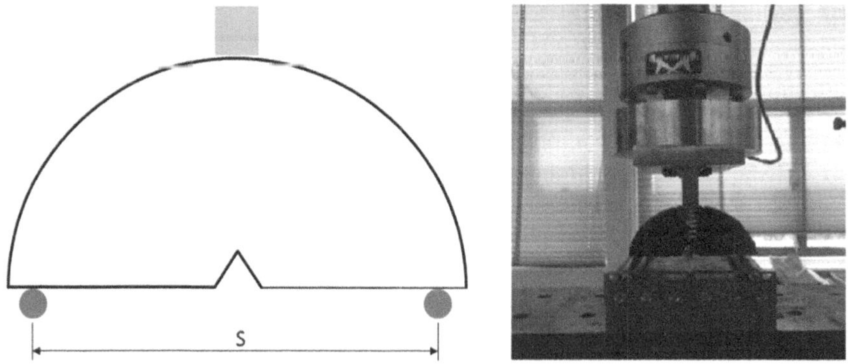

Fig. 3 Left: schematic of a SCB test specimen; right: the SCB test system set up

2.5 Four Point Bend (4PB) Test—PA Mix Fatigue Loading Performance

The Four Point Bend (4PB) Fatigue test was employed as a third test in order to evaluate of asphalt mix healing and asphalt healing system efficiency. The 4PB Fatigue test system is considered the most reliable means of evaluating asphalt pavement performance (Hartman and Gilchrist 2004). The authors believe that the healing system for an asphalt pavement is only effective for micro cracks, because large cracks, even if closed, will still deform the test specimen, thus rendering it useless and challenging the results. Hartman and Gilchrist (2004) found that microcracks result in the creation of large cracks, leading to full pavement failure. Thus, a focus of the self healing system should be the prevention of large cracks and a test system able to insert fatigue loading, simulating repetitive wheel loading, with a deflection strain control allowing small deformations and small crack formation in the test specimen is most suitable for studying the asphalt mix healing and asphalt healing system efficiency. The 4PB Fatigue test system set up is described elsewhere (Taboković et al. 2017b). Same testing programme as for SCB test was used, except that only one healing stage was performed, 20 h after the initial test.

3 Results

3.1 Porous Asphalt Mix Fibre Content Optimisation

Figure 4a, b show the effect of fibres on the stiffness, strength and their healing (strength and stiffness recovery) abilities. The results from the test support authors statement in their previous publication (Tabaković et al. 2016) "fibres increase the asphalt mix strength and stiffness". However, the results from this study show that higher fibre content does not necessarily improve asphalt mix healing properties. From the test results is clear that mixtures with lower amount of fibres (5%) and higher rejuvenator/alginate ratio (70:30) have best ability to recover its original stiffness and strength. The asphalt mixtures with higher fibre content (10%) had lower stiffness and strength recovery, between 10% and 20%. This could be simply due to the fact that fibres once broken cannot be repaired and thus the strength is not recovered. Therefore, samples with higher fibre content experience higher stiffness and strength loss. However, the test sample (asphalt mix) recovery efficiency might have depended on the test sample damage. During the test it was observed that some samples were damaged more than others perhaps this played a role in asphalt mix strength recovery. Therefore, further studies are needed in order to optimise asphalt mix design containing the compartmented alginate fibres encapsulating the bitumen

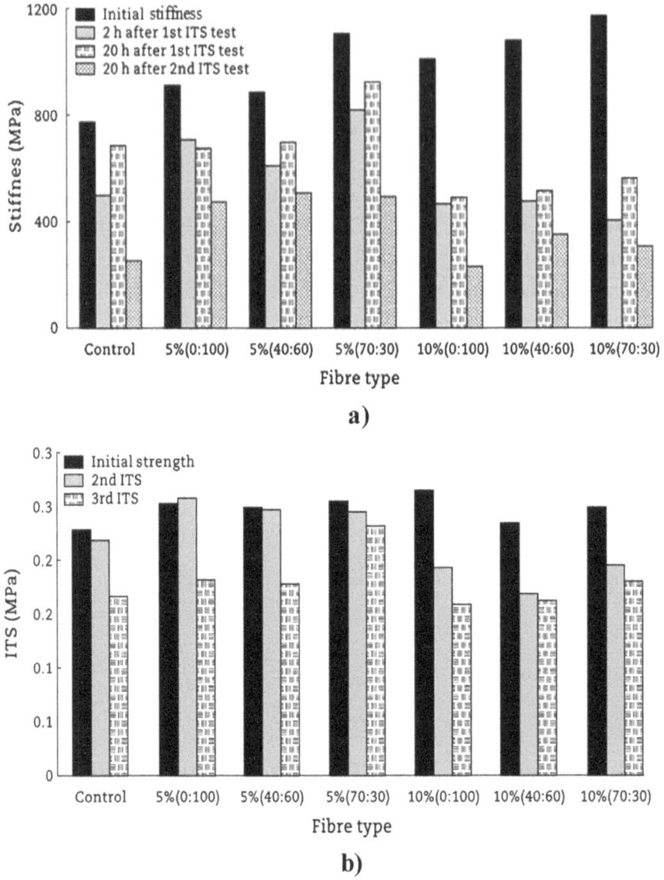

Fig. 4 Effect of fibres on: **a** asphalt mix stiffness and **b** on asphalt mix strength

rejuvenator. Nevertheless, these results confirm that compartmented fibres encapsulating the bitumen rejuvenator is viable self-healing technology for asphalt mix crack/damage repair.

3.2 SCB Test—Crack Speed

The SCB test was specifically employed in order to study crack propagation and its closure (healing) in an asphalt mix. Figure 5 illustrates successful crack closure (healing) in an asphalt test specimen containing 5% compartmented alginate fibres encapsulating the rejuvenator after 20 h of healing.

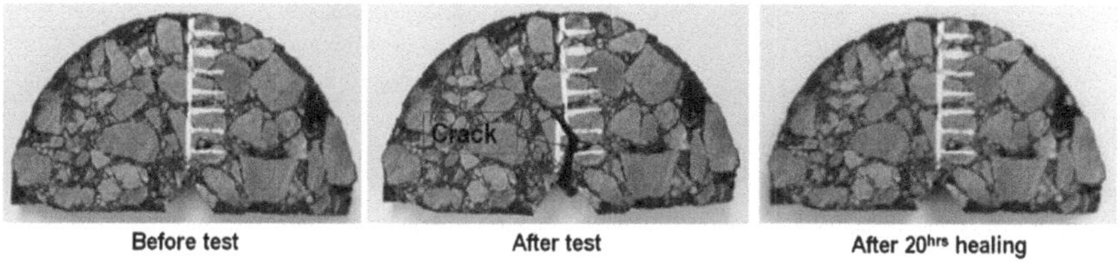

Fig. 5 SCB test specimen crack closure/healing

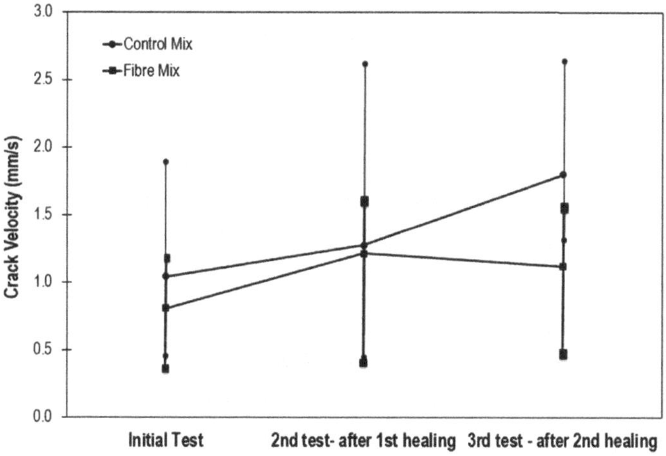

Fig. 6 Crack speed

Figure 6 shows the crack propagation speed within the test sample at the initial test and after two healing stages. It is clear from the graph that the fibre mix experiences slower crack propagation. This could be due to the softened binder and reduced fracture energy available for crack propagation. This result illustrates the benefits of the compartmented fibre encapsulating a rejuvenator mix in terms of crack propagation, where the rejuvenation (softening) of aged binder can reduce the brittleness of the aged binder, thereby reducing the energy available for crack propagation.

3.3 4PB Test-Healing Efficiency

Figure 7 shows the healing efficiency of the PA control mix and fibre mix flexural stiffness (S_{mix}). The results show a very close initial test performance (Fig. 7a). However, Fig. 7b shows a higher stiffness recovery for the PA fibre mix after 20 h healing at 20 °C. These results demonstrate the potential benefits of the PA mix containing compartmented alginate fibres encapsulating the rejuvenator.

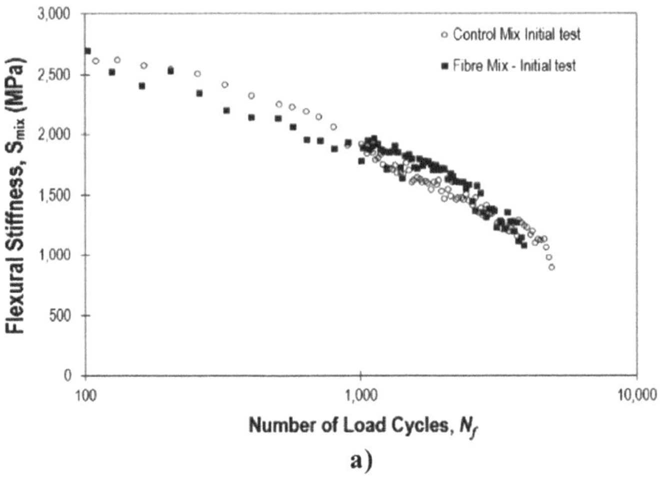

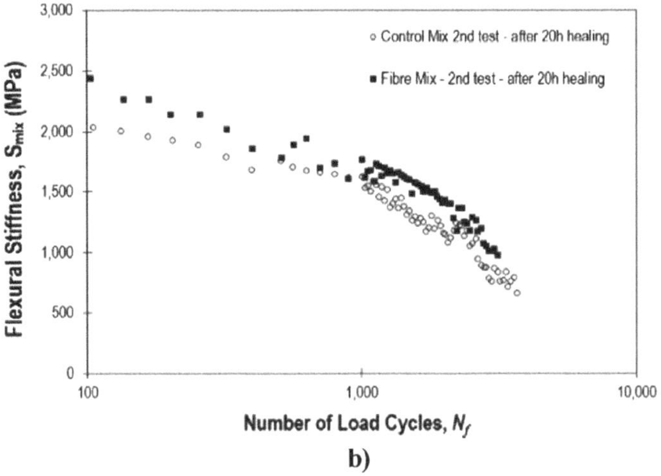

Fig. 7 4PBT results—healing efficiency of the fibre asphalt mix vs. control asphalt mix, **a** initial test results; **b** test results after 20 h healing at 20 °C

4 Conclusion

This study presents a unique concept self-healing system for asphalt pavement, where compartmented alginate fibres encapsulating rejuvenator (asphalt binder healing material) are employed to locally distribute the rejuvenator and to overcome the problems associated with spherical capsules and hollow fibres. The work presents proof of concept of the encapsulation process which involved embedding the fibres into the asphalt PA mixture and the survival rate of fibres in the asphalt mixture. The test results demonstrated that fibres have suitable thermal and mechanical strength to survive the asphalt mixing and compaction process. Furthermore, fibre optimisation process showed that PA mix containing 5% of 70:30 rejuvenator/alginate ratio compartmented alginate fibres has higher strength, stiffness and better healing properties in comparison to the control asphalt mix, i.e. mix without fibres. SCB test demonstrated that system has capacity to close (heal) crack and also to reduce crack propagation speed after the healing period. However, 4PB fatigue test showed that self-healing system does not significantly improve the healing properties of the asphalt pavement. Nevertheless, the findings indicate that, with further enhancement, compartmented calcium alginate fibres may present a promising new approach for the development of self-healing asphalt pavement systems.

Acknowledgements This research was conducted under the Marie Curie IEF research funding, research project Self-healing Asphalt for Road Pavements (SHARP), project number 622863.

References

Anderson FA (1995) Final report on the safety assessment of melamine/formaldehyde resin. J Am Coll Toxicol 14:373–385

Garcia A, Austin CJ, Jelfs J (2016) Mechanical properties of asphalt mixture containing sunflower oil capsules. J Clean Prod 118:9

Gibney A (2004) Prediction of rutting resistance of hot rolled asphalt. In: 3rd Eurasphalt & Eurobitume Congress, 12–14 May 2004, Vienna, Austria. Foundation Eurasphalt, pp 1645–1651

Hartman AM, Gilchrist MD (2004) Evaluation four-point bending fatigue of asphalt mix using image analysis. J Mater Civ Eng 16:9

Kliewer JE, Bell CA, Sasnovske DA (1995) Investigation of the relationship between field performance and laboratory ageing properties of asphalt mixtures. In: Huber GA, Decker DS (eds) Engineering properties of asphalt mixtures and the relationship to their performance

Liu Q (2012) Induction healing of porous asphalt concrete. PhD thesis, TU Delft, The Netherlands

Mookhoek SD, Fischer HR, Van Der Zwaag S (2012) Alginate fibres containing discrete liquid filled vacuoles for controlled delivery of healing agents in fibre reinforced composites. Compos A Appl Sci Manuf 43:7

Prajer M, Wu X, Garcia SJ, Van Der Zwaag S (2015) Direct and indirect observation of multiple local healing events in successively loaded fibre reinforced polymer model composites using healing agent-filled compartmented fibres. Compos Sci Technol 106:7

Sun D, Hu J, Zhu X (2015) Size optimization and self-healing evaluation of microcapsules in asphalt binder. Colloid Polym Sci 293:12

Tabaković A, Braak D, van Gerwen M, Copuroglu O, Post W, Garcia SJ, Schlangen E (2017a) The compartmented alginate fibres optimisation for bitumen rejuvenator encapsulation. J Traffic Transp Eng (Engl Ed) 4(4):347–359

Tabaković A, Post W, Cantero D, Copuroglu O, Garcia SJ, Schlangen E (2016) The reinforcement and healing of asphalt mastic mixtures by rejuvenator encapsulation in alginate compartmented fibres. Smart Mater Struct 25(8):084003

Tabaković A, Schlangen E (2016) Self-healing technology for asphalt pavements. In: Hager MD, Van Der Zwaag S, Schubert US (eds) Self-healing materials. Springer, Cham

Tabaković A, Schuyffel L, Karač A, Schlangen E (2017b) An Evaluation of the Efficiency of Compartmented Alginate Fibres Encapsulating a Rejuvenator as an Asphalt Pavement Healing System. MPDI Applied Sciences 7:16

Xu S, García A, Su J, Liu Q, Tabaković A, Schlangen E (2018) Self-healing asphalt review: from idea to practice. Adv Mater Interfaces 5:1800536

Recycling and By-products

Experimental Investigation on the Effect of Rejuvenator on the Use of a High Amount of Recycled Asphalt Binder

Di Wang, Maximilian Koziel, Augusto Cannone Falchetto, Chiara Riccardi, Martin Hugener, Laurent Porot, Yun Su Kim, Goshtasp Cheraghian, and Michael P. Wistuba

Abstract Due to economic and environmental benefits, Recycled Asphalt Pavement (RAP) has been commonly used in the asphalt pavement construction. However, the capability of existed mixing plants to incorporate a high percentage of RAP restrict the use of a high RA content. In this study, the possibility of recycling a RA binder content of 80% is experimentally investigated. First, an unmodified 50/70 binder is selected as the reference material; then, an extracted RA binder was blended with the optimal rejuvenator, and the rejuvenator content determined based on the penetration test to restore the original penetration value close to the reference material. Next, both

D. Wang · M. Koziel · C. Riccardi · Y. S. Kim · G. Cheraghian · M. P. Wistuba
Department of Civil Engineering—ISBS, Technische Universität Braunschweig, 38106 Brunswick, Germany
e-mail: di.wang@tu-bs.de

M. Koziel
e-mail: m.koziel@tu-bs.de

C. Riccardi
e-mail: chiara.riccardi@tu-bs.de

Y. S. Kim
e-mail: yunsu.kim@tu-bs.de

G. Cheraghian
e-mail: g.cheraghian@tu-bs.de

M. P. Wistuba
e-mail: m.wistuba@tu-bs.de

A. Cannone Falchetto (✉)
Department of Civil and Environmental Engineering, University of Alaska Fairbanks, Fairbanks, AK 99775-5960, USA
e-mail: afalchetto@alaska.edu

M. Hugener
EMPA—Materials Science and Technology, Dübendorf, Switzerland
e-mail: Martin.Hugener@empa.ch

L. Porot
Kraton Chemical, Almere, The Netherlands
e-mail: Laurent.Porot@Kraton.com

C. Raab (ed.), *Proceedings of the 9th International Conference on Maintenance and Rehabilitation of Pavements—Mairepav9*, Lecture Notes in Civil Engineering 76,
https://doi.org/10.1007/978-3-030-48679-2_31

materials are artificially aged to short- and long-term aging conditions. Rheological tests are performed with the Dynamic Shear Rheometer (DSR) and the Bending Beam Rheometer (BBR). Finally, the experimental results are used to evaluate the performance property of virgin and rejuvenated asphalt binders under different aging levels. Results indicate that the properties of the blend of RA and virgin binder can be only partially restored by the action of the rejuvenator.

Keywords Asphalt binder · Reclaimed Asphalt (RA) · Rejuvenator · Rheological properties · Performance tests

1 Introduction

In view of both economic and environmental factors, recycled materials such as Reclaimed Asphalt pavement (RA) (EAPA 2008; Dinis-Almeida *et al.* 2016) constructions and demolition waste (CDW) (Ossa *et al.* 2016), Recycled Asphalt Shingle (RAS) (Williams *et al.* 2019), fly ash and wasted tire rubber (Farina *et al.* 2017), and industrial by-product, steel slags (Grönniger *et al.* 2017), have been more and more commonly used in the construction of asphalt pavement (Cannone Falchetto *et al.* 2018a; Wang *et al.* 2019b).

According to several reports and researches from Europe, northern America and East Asia (McDaniel and Anderson 2001; Wei *et al.* 2007; EAPA 2017; KOSIS 2018), more than 200 million tons of RA are produced annually in the aforementioned regions. Hence, the reuse of these materials is not only a problem for the pavement industry but also a social issue. In the U.S. and Europe (including Turkey), approximately 120 million tons of RA are reclaimed each year (EAPA 2017), while more than 90% are directly reused for the asphalt pavement construction. For example, in Germany, the use of RA dates back 1980s (German Asphalt Association 2014); nowadays, the application of RA is enforced by the national standard (TL Asphalt-StB 2013), with the amount varying between 25 and 50% RA for different layers (TL Asphalt-StB 2013). In the case of South Korea and China (Wei *et al.* 2007; KOSIS 2018), the use of RA has found a substantial increase only in the recent past, with a limit between 15 and 21%. And in Japan the use of rejuvenators enables successfully to achieve up to 80% RA into new asphalt mix for more than 20 years (West *et al.* 2015).

While the reuse of RA can bring economic and environmental benefits, the application of RA may lead to negative effects on the mechanical properties of asphalt mixtures due to the oxidized and aged asphalt binder with higher stiffness and brittleness (Hansen and Newcomb 2011). Therefore, even though similar performance properties were experimentally observed in comparison with conventional Hot Mix Asphalt (HMA) when incorporating 100% of RA (Dinis-Almeida *et al.* 2016), only the use of 30% RA or lower is commonly accepted (Planche 2008). The use of rejuvenators together with RA material is an alternative method to potentially increase the amount of RA content. Several studies have already successfully investigated the idea of using rejuvenators in combination with a maximum amount of 100% recycled materials, such as RA and RAS (Mogawer *et al.* 2013; Zaumanis *et al.* 2014).

In this study, the idea of using a rejuvenator to improve the rheological properties of a blend of virgin and 80% RA binder is experimentally investigated. This work is part of the interlaboratory activity proposed by the International Union of Laboratories and Experts in Construction Materials Systems and Structures (RILEM) Technical Committee 264—Asphalt Pavement Recycling (RAP) project. For this purpose, one fresh unmodified reference asphalt binder, one extracted RA binder and a rejuvenator were selected. Then, the optimal content of the rejuvenator was determined based on the conventional penetration value test so that the blend of binders presented similar grading compared to the reference binder. This approach is the commonly adopted approach in Europe to fast determine suitable dosage of rejuvenator (EAPA 2018). Next, both materials were artificially aged to short- and long-term aging conditions. Temperature-frequency-sweep, Binder-Fast-Characterization-Test (Bitumen Typisierungs Schnell Verfahren) BTSV (AL BTSV 2017; Alisov *et al.* 2018), and low temperature creep tests were performed with DSR and BBR, respectively. Finally, the experimental results were used to evaluate the performance properties of different materials at different aging levels.

2 Materials and Testing

2.1 Materials

A single fresh unmodified 50/70 pen-graded (EN 12591, 2015) asphalt binder is used as the reference material in this study. The RA binder is extracted and recovered (EN 12697-3, 2013) from the RA material by one single laboratory. The selected rejuvenator is a bio-based rejuvenating additive. It is a liquid additive, which, with its specific amphipathic chemical structure, disperses the highly polar fractions limiting the agglomeration of asphaltenes; more detailed information on the latter can be found elsewhere (Zaumanis *et al.* 2013; Koudelka *et al.* 2018; Porot *et al.* 2020).

In the present work, a mixing ratio of 20% fresh unmodified binder together to 80% recycled material, RA binder and rejuvenator, was adopted to prepare the asphalt binder blend. Based on the commonly used penetration grading system (EN 12591, 2015), an optimal rejuvenator content of 9% was determined to add to the RA binder in a preliminary study (Cannone Falchetto *et al.* 2018b). This percentage was determined as the optimum dosage to recover the penetration value at 25 °C without harming the softening point temperature. The methodology and results are described in another publication, more detailed information can be found in the authors' previous study (Cannone Falchetto *et al.* 2018b). To better evaluate the aging effect on the performance properties of fresh binder and blended binder, both materials are artificially short-term and long-term aging using the Rolling Thin Film Oven (RTFOT) (EN 12607-1, 2014) and Pressure Aging Vessel (PAV) (EN 14769, 2012)

Table 1 Basic properties of the asphalt binder

	Fresh		RTFOT		RTFOT + PAV		RA binder
	50/70	Blend	50/70	Blend	50/70	Blend	
Softening point (°C)	52.2	49.1	57.6	55.2	63.7	63.4	69.9
Penetration (0.1 mm)	56	63	41	40	28	24	16

procedure, respectively. Then, penetration value at 25 °C (EN 1426, 2015) and softening point temperature (EN 1427, 2015) tests are performed to characterize the fresh and the aged binders according to the pen grading system (EN 12591, 2015), the results are listed in Table 1.

It can be seen that comparable penetration value and softening point temperature can be found between the reference material and the binder blend under the three different aging conditions. Additional rheological tests and parameters are further used to address the characteristics of asphalt binders as described in the next section.

2.2 Dynamic Shear Rheometer (DSR) Test

Temperature-frequency-sweep (*T-f*-sweep) tests are performed on both fresh and blend asphalt binders under different aging conditions (see Table 1) with the Dynamic Shear Rheometer (DSR) (EN 14770, 2012). The testing temperature ranges between −30 and +82 °C with an incremental step over a frequency sweep from 0.1 to 10 Hz (0.1, 0.2, 0.3, 0.4, 0.5, 0.6, 0.7, 0.8, 0.9, 1, 1.59, 2, 3, 4, 5, 6, 7, 8, 9, 10 Hz). The parallel-plate configuration with three different plate geometries (25, 8, and 4 mm) is used. At high temperatures, between +28 and +82 °C, an increasing step of 6 °C and a plate diameter of 25 mm with a gap of 1 mm was adopted. At intermediate temperatures, −6, 0, 4, 10, 16, 22, 28, 34 and 40 °C, the diameter of 8 mm with a gap of 2 mm was applied; while at low temperatures, −30, −24, −18, −12, −6, 0, 4 and 10 °C, an experimental procedure proposed by the Braunschweig Pavement Centre (Wang *et al.* 2019a) with a 4 mm diameter and the gap of 3 mm was selected. The wide range of temperature window is selected to generate master curves based on the CAM model (Christensen and Anderson 1992; Marasteanu and Anderson 1999; Zeng *et al.* 2001).

Amplitude sweep tests are initially performed to identify the Linear Viscoelastic (LVE) range and the suitable strain/stress levels to be adopted and next used for the *T-f*-sweep tests. Based on the capability of the available DSR, a mixed-mode approach was used to obtain more reliable results. Only at high temperatures, the test is performed in strain-controlled mode, while the stress-controlled mode is applied to both intermediate and low temperatures. This experimental procedure has been successfully validated by the authors' previous work (Riccardi *et al.* 2017; Wang *et al.* 2019a). At least two replicates are performed for each binder type with different aging conditions and for each plate.

Besides the conventional *T-f*-sweep test, the 'Binder-Fast-Characterization-Test BTSV' (AL BTSV 2017; Alisov *et al.* 2018), was used to characterize the binder with the DSR in place of the conventional softening point. Two rheological parameters, temperature (T_{BTSV}) and the corresponding phase angle (δ_{BTSV}) at $|G^*| = 15$ kPa can be used to represent the state of the asphalt binder. In previous studies (Alisov *et al.* 2018; Büchler *et al.* 2018) the BTSV was suggesting to determine the required amount of rejuvenator or soft asphalt binder to be blended with RA binder to achieve the selected target property, showing that higher T_{BTSV} combined with a lower δ_{BTSV} indicate a more aged material for unmodified asphalt binder. This method has been used for several years to characterize the asphalt binder and it is currently suggested to replace softening point temperature measurement as part of the German national standards (AL BTSV 2017).

2.3 Bending Beam Rheometer (BBR) Test

Low temperature three-point bending tests are conducted on the long-term aged materials for both reference and blend binders listed in Table 1 using an available Bending Beam Rheometer (BBR) (AASHTO T313, 2012). The conditioning time is set to 1 h and a constant loading force of 980 ± 50 mN is used. Four different testing temperatures −6, −12, −18 and −24 °C are imposed. The creep stiffness, $S(t)$, and relaxation parameter, m-value, are calculated for further comparison. For each material and temperature condition, at least three replicates are performed.

3 Results and Analysis

3.1 Master Curve

Master curves can provide a visual understanding of the viscoelastic behavior of bituminous materials and allow an evaluation of the rheological properties over a wide range of temperatures and frequencies. In this study, the master curves of complex shear modulus, $|G^*|$, and phase angle, δ, are generated by using the Christensen-Anderson-Marasteanu (CAM) model (Marasteanu and Anderson 1999; Zeng *et al.* 2001) together with the conventional Williams–Landel–Ferry (WLF) (Williams *et al.* 1955) equation to express the temperature-shift factor, a_T. A reference temperature of $T_0 = 16$ °C is selected for this purpose. More detailed information about the fitting of master curves can be found in the authors' previous work (Riccardi *et al.* 2017; Wang *et al.* 2019a). The fitted complex shear modulus master curves for both fresh 50/70 and blend binders under different aging conditions are presented in Fig. 1.

As shown in Fig. 1, a relatively good fitting for each asphalt binder can be observed. The lowest complex modulus curve is exhibited for the virgin binders

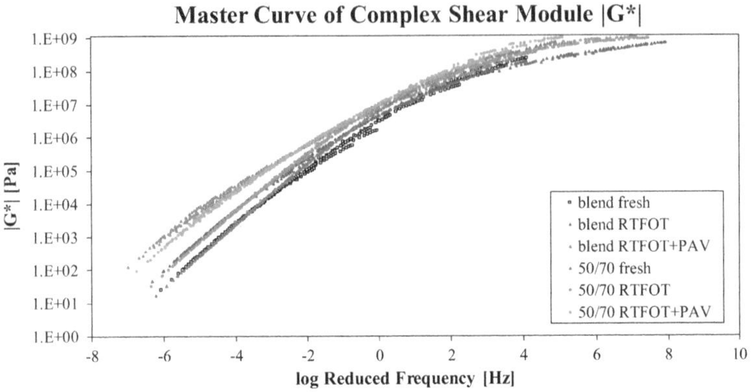

Fig. 1 Master curves at different aging conditions at a reference temperature of 16 °C

while the long-term aged binder presents the highest curves for both materials (untreated and blend). In the case of the fresh and short-term aging conditions, a very similar curve can be found between the 50/70 and blend binders. With respect to the long-term aged condition, the binder blend shows a slightly higher curve at high temperature while the differences are getting smaller when the testing temperature decreases/frequency increases. Overall, both reference and blend material have similar rheological properties over the entire spectrum of temperatures.

3.2 *Binder-Fast-Characterization-Test—BTSV Test*

The results of the BTSV tests are illustrated in Table 2 for the entire set of 50/70 and blend asphalt binders listed in Table 1. Its temperature criteria aims at correlating and with softening point temperature.

According to Table 2, the differences between the two materials are significantly different except for the BTSV temperature of the long-term aged binders. While the change in properties for the 50/70 between fresh and fully aged state is 13°, for the blend it was only 10 °C suggesting a lower aging susceptibility. This suggests a remarkable distinct behavior between the reference and rejuvenated material at high temperatures, clarifying the visual inspection of the master curves in Fig. 1. At the

Table 2 BTSV results for both reference and blend asphalt binder under different aging conditions

	Fresh		RTFOT		RTFOT+PAV	
	50/70	Blend	50/70	Blend	50/70	Blend
T_{BTSV} (°C)	50.7	54.1	55.9	58.4	63.5	64.6
δ_{BTSV} (°)	81.6	75.4	79.1	74.5	76.7	72.5

Table 3 Low temperature properties of the long-term aged asphalt binders

		−6 °C	−12 °C	−18 °C	−24 °C
50/70	S(60 s)[MPa]	61	211	499	936
	m(60 s) [–]	0.428	0.351	0.266	0.199
Blend	S(60 s) [MPa]	40	82	201	408
	m(60 s) [–]	0.437	0.367	0.313	0.258

Table 4 Low temperature properties of the long-term aged asphalt binders

	T_c (S)	T_c (m)	Continuous low PG (°C)	Low PG (°C)	ΔT_c (°C)
50/70	−15.00	−15.81	−25.00	−22	0.81
Blend	−21.51	−19.34	−29.34	−28	−2.17

same time the critical temperature is slightly higher and phase angle lower, more elastic, suggesting better performance against rutting.

3.3 Bending Beam Rheometer (BBR) Test

Creep stiffness and relaxation parameter are listed in Table 3, while the critical temperature based on creep stiffness and relaxation parameter, continuous low Performance Grade (PG) (AASHTO M320, 2017) and the actual low PG are illustrated in Table 4. The continuous low PG and the actual low PG are calculated based on the AASHTO standard M320 (2017) and T313 (2012).

It can be seen from Tables 3 and 4 that the reference asphalt binder presents a creep stiffness that is more than twice as large as the creep stiffness at temperatures lower than −12 °C, *S*(t) of the rejuvenated binder blend, while in the case of relaxation parameter, *m* parameter, closer values are found. In addition, the low PG of these two materials is also different, while the difference of the continuous low PG is more than 4 °C. In addition, more negative ΔT_c value in the blend materials indicates a more aged binder (Riccardi *et al.* 2017). Therefore, the low temperature properties of these two materials appear to be substantially different and improved for the blend.

4 Summary and Conclusions

In the present study, the possibility of using the rejuvenator to incorporate a high amount of Recycled Asphalt Pavement (RA) binder (80%) in the mix design was experimentally investigated with the Dynamic Shear Rheometer and with the Bending Beam Rheometer. Based on the experimental results and analyses conclusions can be drawn:

- Visible comparison of master curves indicate that the blend materials achieve very similar rheological properties with the 50/70 asphalt binder at different aging conditions, while BBR results show that the blend materials are more aged than the virgin unmodified 50/70 binder
- The penetration system does not provide a comprehensive vision of the performance properties of asphalt binders, but enables to determine the optimal dosage of rejuvenator at restoring flexibility without compromise of high temperature
- The rejuvenator can only partially restore the properties of a blend of RA and virgin binders for low and intermediate temperature while the high temperature is kept higher

Although the results obtained in this study are promising, additional experimental and analysis support is needed by extending the present research effort to the investigation of more types of asphalt binders and different RA binder contents.

Acknowledgements The RILEM Technical Committee on Asphalt Pavement Recycling (RAP) 264 and the contribution of Nynas AB and Kraton to the RILEM inter-laboratory activities are gratefully acknowledged.

References

AL BTSV (DSR) (2017) Arbeitsanleitung zur Bestimmung des Verformungsverhaltens von Bitumen und bitumenhaltigen Bindemitteln im Dynamischen Scherrheometer (DSR)—Durchführung des BTSV (Bitumen-Typisierungs-Schnell-Verfahren). German Technical Specification (in German)

Alisov A, Riccardi C, Schrader J, Cannone Falchetto A, Wistuba MP (2020) A novel method to characterise asphalt binder at high temperature. Road Mater Pavement Des 21(1):143–155

Büchler S, Cannone Falchetto A, Walther A, Riccardi C, Wang D, Wistuba MP (2018) Wearing course mixtures prepared with high reclaimed asphalt pavement content modified by rejuvenators. Transp Res Rec 2672(28):96–106

Cannone Falchetto A, Moon KH, Wang D, Riccardi C, You ZP, Wistuba MP (2018a) Experimental investigation on double recycling of asphalt mixture for pavement applications. In: 6th international conference on sustainable solid waste management, pp 1–10

Cannone Falchetto A, Porot L, Riccardi C, Hugener M, Tebaldi G, Dave E (2018b) Effects of rejuvenator on reclaimed asphalt binder: an exploratory study of the RILEM TC 264-RAP task group 3. In: RILEM 252-CMB-symposium on chemo mechanical characterization of bituminous materials. Springer, Cham, pp 195–200

Christensen DW, Anderson DA (1992) Interpretation of dynamic mechanical test data for paving grade asphalt cements (with discussion). J Assoc Asphalt Paving Technol 61

Deutscher Asphaltverband (DAV) e.V (2014) Wiederverwendung von Asphalt. Nachhaltigkeit auf höchstem Niveau. Bonn, Deutscher Asphaltverband (DAV) e.V., Schieffelingsweg 6, 53123 Bonn (in German)

Dinis-Almeida M, Castro-Gomes J, Sangiorgi C, Zoorob SE, Afonso ML (2016) Performance of warm mix recycled asphalt containing up to 100% RAP. Constr Build Mater 112:1–6

EAPA, European Asphalt Pavement Association (2008) Arguments to stimulate the government to promote asphalt reuse and recycling. https://eapa.org/wp-content/uploads/2018/07/arguments_stimulate_asphalt_May2008.pdf. Accessed 26 Oct 2019

EAPA, European Asphalt Pavement Association (2018) Recommendations for the use of rejuvenators in hot and warm asphalt production. https://eapa.org/wp-content/uploads/2018/08/Rejuvenators-Paper.pdf. Accessed 26 Oct 2019
EAPA, European Asphalt Pavement Association (2017) Asphalt in figures. https://eapa.org/wp-content/uploads/2018/12/AIF_2017.pdf. Accessed 11 Oct 2018
Farina A, Zanetti MC, Santagata E, Blengini GA (2017) Life cycle assessment applied to bituminous mixtures containing recycled materials: crumb rubber and reclaimed asphalt pavement. Resour Conserv Recycl 117:204–212
Grönniger J, Cannone Falchetto A, Isailović I, Wang D, Wistuba MP (2017) Experimental investigation of asphalt mixture containing Linz-Donawitz steel slag. J Traffic Transp Eng (Engl Ed) 4(4):372–379
Hansen KR, Newcomb DE (2011) Asphalt pavement mix production survey on reclaimed asphalt pavement, reclaimed asphalt shingles, and warm-mix asphalt usage: 2009–2010. Inf Ser 138:21
KOSIS, Korean Statistical Information Service (2018, 2014) http://kosis.kr/search/search.do. Accessed 21 June 2018 (in Korea)
Koudelka T, Porot L, Coufalik P, Varaus M (2018) The use of rejuvenators as an effective way to restore aged binder properties. In: Transport research arena 2018, Vienna
Marasteanu MO, Anderson DA (1999) Improved model for bitumen rheological characterization. In: Eurobitume workshop on performance related properties for bituminous binders. European Bitumen Association, Brussels
McDaniel RS, Anderson RM (2001) Recommended use of reclaimed asphalt pavement in the Superpave mix design method: technician's manual (No. Project D9-12 FY'97). National Research Council (US). Transportation Research Board
Mogawer WS, Booshehrian A, Vahidi S, Austerman AJ (2013) Evaluating the effect of rejuvenators on the degree of blending and performance of high RAP, RAS, and RAP/RAS mixtures. Road Mater Pavement Des 14(sup2):193–213
Ossa A, García JL, Botero E (2016) Use of recycled construction and demolition waste (CDW) aggregates: a sustainable alternative for the pavement construction industry. J Clean Prod 135:379–386
Planche JP (2008) European survey on the use of RAP. In: Proceeding of ISAP, asphalt and environment conference, pp 3–18
Porot L, Hugener M, Cannone Falchetto A, Wang D, Kawakami A, Tebaldi G, et al (2020) Aging of rejuvenated RAP binder—a RILEM inter-laboratory study. In: 7th EandE Congress, Eurasphalt and Eurobitume, Madrid, 12–14 May 2020
Riccardi C, Cannone Falchetto A, Wang D, Wistuba M (2017) Effect of cooling medium on low-temperature properties of asphalt binder. Road Mater Pavement Des 18(sup4):234–255. https://doi.org/10.1080/14680629.2017.1389072
TL Asphalt-StB (2013) Technische Lieferbedingung für Asphaltmischgut für den Bau von Verkehrsflächenbefestigung. FGSV Verlag (in German)
Wang D, Cannone Falchetto A, Alisov A, Schrader J, Riccardi C, Wistuba MP (2019a) An alternative experimental method for measuring the low temperature rheological properties of asphalt binder by using 4 mm parallel plates on dynamic shear rheometer. Transp Res Rec 2673(3):427–438
Wang D, Cannone Falchetto A, Moon KH, Riccardi C, Pei JZ, Wen Y (2019b) Experimental investigation on the use of re-recycled reclaimed asphalt pavement (RAP) in asphalt mixtures. Environ Sci Pollut Res, 1–9
Wei Q, Yang CH, Xiong CH, Ling TQ (2007) Recycling technology of aged asphalt pavement. J Chongqing Jianzhu Univ 29(3):128 (in Chinese)
West R, Copeland A (2015) High RAP asphalt pavements: japan practice—lessons learned. https://www.asphaltpavement.org/PDFs/EngineeringPubs/IS139_High_RAP_Asphalt_Pavements_Japan_Practice-lr.pdf. Accessed 26 Oct 2019
Williams BA, Willis JR, Hansen KR, Stanton B (2019) Guidelines for the use of reclaimed asphalt shingles in asphalt pavements (No. Information Series 136)

Williams ML, Landel RF, Ferry JD (1955) The temperature dependence of relaxation mechanisms in amorphous polymers and other glass-forming liquids. J Am Chem Soc 77(14):3701–3707

Zaumanis M, Mallick RB, Frank R (2013) Evaluation of rejuvenator's effectiveness with conventional mix testing for 100% reclaimed Asphalt pavement mixtures. Transp Res Rec 2370(1):17–25

Zaumanis M, Mallick RB, Poulikakos L, Frank R (2014) Influence of six rejuvenators on the performance properties of Reclaimed Asphalt Pavement (RAP) binder and 100% recycled asphalt mixtures. Constr Build Mater 71:538–550

Zeng M, Bahia HU, Zhai H, Anderson MR, Turner P (2001) Rheological modeling of modified asphalt binders and mixtures (with discussion). J Assoc Asphalt Paving Technol 70

Rubber-Oil Distillation Bottoms Blends as a New Recycling Solution for Bitumen Extension

G. Tarsi, C. Sangiorgi, A. Varveri, and C. Oliviero Rossi

Abstract Environmental protection is also pursued by promoting the saving of oil reserves and the proper management of waste through a circular economy. Following this objective, the present study encourages the use of recycled and by-product materials for the production of bituminous binders. The powdered rubber (R) from end-of-life tyres and the re-refined engine oil bottom (REOB), i.e. the by-product of refining mineral waste oils, were used to replace a certain amount of standard bitumen. The experimental program aims to characterize the interaction between R and REOB in order to use them as virgin bitumen extenders. The multiple variables of the production process that may affect the R-REOB blends and in turn the binders' final properties were evaluated. The R-REOB mixes were prepared by varying the type of REOB, the R content and the mixing temperature. The rheological analysis was performed by means of dynamic shear rheometer (DSR) tests with the aim to define the variables that strongly influence the response of the R-REOB compound. As a result of this preliminary study, the second phase will focus on extended bitumens that will be prepared considering some of the studied extenders.

Keywords Waste rubber · Re-refined engine oil bottom · Extended bitumen · Rheology

---

G. Tarsi (✉) · C. Sangiorgi
Department of Civil, Chemical, Environmental and Materials Engineering, University of Bologna, Viale U. Terracini 28, 40131 Bologna, BO, Italy
e-mail: giulia.tarsi2@unibo.it

A. Varveri
Department of Engineering Structures, Delft University of Technology, Stevinweg 1 – Building 23, 2628 CN Delft, The Netherlands

C. Oliviero Rossi
Department of Chemistry and Chemical Technologies, University of Calabria, Viale P. Bucci – Building 14D, 87036 Arcavacata di Rende, CS, Italy

C. Raab (ed.), *Proceedings of the 9th International Conference on Maintenance and Rehabilitation of Pavements—Mairepav9*, Lecture Notes in Civil Engineering 76,
https://doi.org/10.1007/978-3-030-48679-2_32

1 Introduction

In recent years, a growing attention to the environment and an increased awareness of the depletion of non-renewable resources led to more sustainable approaches in all construction activities included the production of materials. A sustainable approach ensures that the demands of present generations are fulfilled without compromising the needs of future generations as defined by Brundtland (1987). In this regard, the economy is moving from an open-ended system to a circular one, where a relationship between resource use and waste residuals exists (Andersen 2007).

According to the current trend and environmental strategies of European countries, the pavement engineering world has focused on eco-friendly road materials. Various studies have focused on the use of recycled and by-product materials in partial replacement of petroleum bitumen. The use of these materials turns the detriment of non-renewable resources into environmental benefit as the biologically non-degradable materials are recycled (Fernandes et al. 2018). Additives, such as polymers, resins, oils and waste materials in general can be used to decrease the proportion of bitumen in the binder mixture or can represent the alternative for bitumen themselves (Aziz et al. 2015).

The use of synthetic and natural polymers as bitumen modifiers is found as early as 1843 (Yildirim 2007). In the 20th century, bitumen modified by virgin or recycled polymers became widely used due to improvements in the service temperature range of bituminous binders, which in turn enhanced the engineering properties of asphalt mixtures (Becker et al. 2001). Moreover, virgin and recycled polymers showed similar results in improving the road performances (Kalantar et al. 2012). The possible achievement of similar performance together with the environmental benefits and, above all, the advantageous price of recycled materials in comparison to the virgin ones have promoted their use. Among all, crumb or powdered rubber obtained from end-of-life tyres (ELTs) is widely used as bitumen modifier. Various studies have demonstrated that recycled rubber represents an engineering resource and a possible solution to reuse the waste tyres in road materials contributes to mitigate the issue of their disposal. As a matter of fact, the ELTs generation represents a large and problematic source of waste; only in Europe, 3.4 million ton of ELTs are generated every year (ETRma 2019). The rubberised binder is especially used to improve the high temperature performance at binder level, which turn into improved responses at asphalt mixture level (Becker et al. 2001). Furthermore, it has been observed that the modified bitumen behaves better overall in-service temperatures with increasing rubber content. The presence of rubber does not only improve the elasticity and resilience at high temperatures, but the binder shows improved temperature susceptibility and fatigue resistance at intermediate temperatures and thermal cracking resistance at low temperatures (Becker et al. 2001). The behaviour of modified bitumen depends on size of rubber particles and base binder properties (Ziari et al. 2016). Nonetheless, the bitumen-rubber blend is also affected by processing conditions like mixing time, temperature and frequency, which influence the rheo-mechanical response of the resulting binder and asphalt concrete. Their interaction has not been fully understood yet, but can be ascribed to two simultaneous phenomena: the adsorption of the aromatic oils of bitumen that causes rubber swelling, and the digestion of the polymer into bitumen (Lo Presti 2013). However, the addition of rubber leads to an overall increase in binder viscosity that demands higher in-plant production temperature and, consequently, it produces higher greenhouse gas emissions (Amini and Imaninasab 2018). Despite of the many advantages of the use of rubberised bitumen, the high production temperature remains a drawback that should not be neglected.

Previous study has been proven that the use of additives and paraffinic waxes can be a possible solution to control the production temperature reducing the emissions (Sangiorgi et al. 2018). A more sustainable alternative can be represented by waste oils, in particular the re-refined engine oil bottom (REOB), which have been used to soften bitumen so far, reducing both the high and low continuous performance grade (Asphalt Institute 2016). The use of REOB, i.e. the residue of the refining process of waste engine oils to produce second-hand lubricants, has a twofold aim, as it increases the workability of the product and it promotes the use of recycled materials. The study of Herrington (1992) showed that the viscosity of extended bitumen by the use of REOB is slightly less than the viscosity of standard bitumen after RTFOT short-ageing. Thus, REOB addition can represent a solution to compensate the increased viscosity and the higher production temperature caused by rubber. Regardless the oil origin, the addition of REOB allows the improvement of the low temperature performance of binders and the reduction of thermal cracking (Golalipour and Bahia 2014). Conversely, a high amount of REOB seems to be detrimental for the final product, because of reduced adhesiveness to the aggregates leading to stripping and ravelling (Asphalt Institute 2016). Furthermore, the base properties of materials greatly affect the response of bitumen-REOB compounds, without any general rule (Li et al. 2017).

Recent studies have evaluated the addition of both materials as bitumen modifiers obtaining promising results. Fernandes et al. (2018) have compared the rheological and thermo-chemical results of various modified binders. The introduction of REOB reduces the temperature susceptibility of binders leading them to be more rutting resistant. Amini and Imaninasab (2018) have validated the positive effects of rubber-REOB addition on binder performance at low and high temperatures and on the Performance Grade. However, the formulation of new modified binders has to be carefully evaluated; the type and amount of additives and oils have to be balanced considering the intrinsic drawbacks of the constituents. In this regard, the present study investigates the interaction of rubber and REOB from a rheological point of view.

2 Experimental Design

2.1 Objectives

The purpose of this research is the characterisation of recycled rubber and re-refined engine oil bottom (REOB) mixes to understand their contribution as bitumen extenders. As reported in literature, the response of both rubber-bitumen and REOB-bitumen blends depends on the intrinsic properties of each constituent material, the quantity of each additive/modifier and the mixing conditions. Hence, this research considered eight binders that were made using two REOB types, at two mass proportions of the components, at two production temperatures. The binders were compared through rheological analysis in order to evaluate the mixes more suitable for neat binder extension.

2.2 Materials and Test Methods

The powdered rubber (R) was obtained by recycling ELTs of cars and trucks by double trituration processes at ambient temperature. The rubber gradation varies in the range 0–0.4 mm and its density is equal to 1.01 g/cm^3. The re-refined engine oil bottom (REOB) is the by-product from a vacuum tower in the refinery plant of exhausted motor oils to obtain recycled lubricating oils. The REOB is strongly affected by the refinery plant since it is the residue of refining processes. Two REOBs (O1 and O2) have been compared, which are produced in two distinct refinery plants of the same company. Being a petroleum-based material, the chemical composition of REOBs mainly consist of hydrocarbons; both, O1 and O2, show a predominant portion of aliphatic hydrocarbons, with some aromatic molecules. The REOBs have a similar density of about 1.00 g/cm^3, but different viscosity, i.e. O1 is less viscous than O2.

Four blends were prepared for each type of REOB, varying the R-REOB ratio and the production temperature. Two R-REOB ratios have been considered; the first adds both materials in the same quantity (R:REOB = 1:1), and the second considers a double amount of REOB with respect to R (R:REOB = 1:2). The ratios were selected based on the output of preliminary studies, which assumed that R particles can absorb the lighter fractions of REOB as they absorb the same fractions of bitumen (Tarsi et al. 2020).

All blends were prepared by means of a propeller mixer at two production temperatures, namely 130 and 160 °C. The pre-heated materials were mixed at a rate of 800 rpm for 1 h; the mixing time was set following the guidelines of asphalt rubber adopting the same process of R-bitumen interaction for R-REOB mixes (Signus Ecovalor and Ecopneus Scpa 2014). The REOBs were warmed up in oven at the established temperature for more than 1 h 30′, while R for 15′. During the mixing phase, all compounds were continuously heated at 130 or 160 °C using a heating

Table 1 Matrix of considered extenders

Temp.	O1		O2	
	R:O1 = 1:1	R:O1 = 1:2	R:O2 = 1:1	R:O2 = 1:2
130 °C	O1_1:1-130	O1_1:2-130	O2_1:1-130	O2_1:2-130
160 °C	O1_1:1-160	O1_1:2-160	O2_1:1-160	O2_1:2-160

plate; however, the temperature was not constant as it dropped of about 55–65 °C when the R particles have been incorporated to O1 or O2. The produced blends are listed in Table 1.

A 50/70 penetration grade bitumen (Pen 50/70) was chosen as the reference material. Taking into account the different nature of extenders and bitumen, the reference material was also investigated. This information may help to understand the behaviour of the R-REOB blends, which will been used to made the extended bitumens.

The rheological performance of the binders was evaluated using a dynamic shear rheometer (DSR)—Anton Paar MCR 302. The device was used to analyse the binders' response over a wide range of frequencies and/or temperatures performing the amplitude sweep test, frequency sweep test, viscosity test and multiple stress creep-recovery test. Each test has been performed twice to guarantee the reliability of obtained data.

The first analysis allows the definition of the linear visco-elastic range (LVE range) of materials, i.e. the maximum applicable strain to avoid irreversible structural changes in samples. To obtain the LVE range, the deformation changes, increasing from 0.01 up to 15% with a constant frequency of 1.59 Hz. In order to evaluate the LVE limit of materials, their rheological responses at the lowest, intermediate and the highest test temperature have been investigated. Thus, the test was performed at 10, 30 and 60 °C, which are linked to the range covered by the frequency sweep test.

The frequency sweep tests were carried out at a strain controlled mode. The samples underwent an oscillatory shear stress increasing the frequency from 0.1 up to 10 Hz with a logarithmic ramp. The tests were performed at six temperatures starting from 10 to 60 °C with an increment of 10 °C. These two tests followed the standard EN 14770 using a plate-plate configuration with 8 mm of diameter (PP08) and considering the gap equal to 2 mm.

The elastic response of R-REOB compounds were measured by performing the multiple stress creep-recovery (MSCR) test at 60 °C. The samples underwent shear creep loading and recovery at two stress levels; 10 creep and recovery cycles were run at 0.1 kPa shear creep stress followed by 10 cycles with an applied stress equal to 1.0 kPa. The test has been carried out following the standard ASTM D7405, when possible. Despite the shear stress requirements, the second stress was reduced from 3.2 to 1.0 kPa because it was not possible to perform the test on samples with R-REOB ratio equal to 1:2. The REOB that covers the R particles causes the samples to slip away. The parallel plate configuration, PP25, and the gap equal to 1 mm were adopted.

The dynamic viscosity of extenders were evaluated at three different temperatures, 60, 100 and 150 °C as suggested by the standard EN 13702. However, it was not possible to perform the test in the cone-plate configuration due to the dimension of R particles. In order to obtain representative results avoiding the friction between sample and measurement system, the tests were performed with PP08 plate and 2 mm of gap.

3 Results and Discussions

The amplitude sweep test represents the first binder characterisation that identifies the LVE range of a material. Unlike for standard bitumen, the LVE domain for the R-REOB blends decreases when the test temperature increases as can be observed in Table 2, where the average LVE limits of neat bitumen and extenders are listed. Hence, the LVE range of the R-REOB mixes that have been considered for further studies are those obtained at 60 °C as the corresponding value to the 95% of the initial storage modulus. In general, due to the higher amount of R and production temperature, both samples of O1 and O2 have wider LVE domains; it means that they can withstand larger deformations before irreversible structural changes occur. Only samples O1_1:1-160 and O1_1:2-160 do not exhibit this trend when R content increased. In addition, the specimens produced with O1 show a reduced LVE range than those made with O2, except O2_1:2-130. The high production temperature helps to enlarge the LVE limit of all extenders. The elastic behaviour dominates the viscous one as the storage modulus (G') is higher than loss modulus (G'') in all extenders as shown in Fig. 1. This response can be ascribed to the high amount of R in the R-REOB blends.

As far as the frequency sweep tests are concerned, the data did not show a thermo-rheological simple behaviour, as indicated by the Black Diagrams reported in Fig. 2 (Airey 2002). Thus, the time-temperature superposition principle (TTSP) is not valid

Table 2 Linear visco-elastic range (γ_{LVE}) of extenders and neat bitumen at 10, 30, 60 °C

Sample	γ_{LVE} @10 °C [%]	γ_{LVE} @30 °C [%]	γ_{LVE} @60 °C [%]
O1_1:1-130	0.344	0.271	0.228
O1_1:1-160	0.435	0.344	0.271
O1_1:2-130	0.271	0.271	0.203
O1_1:2-160	0.551	0.344	0.326
O2_1:1-130	1.410	0.697	0.390
O2_1:1-160	1.120	1.120	0.624
O2_1:2-130	0.697	0.308	0.129
O2_1:2-160	1.002	0.624	0.624
Pen 50/70	1.605	4.600	9.340

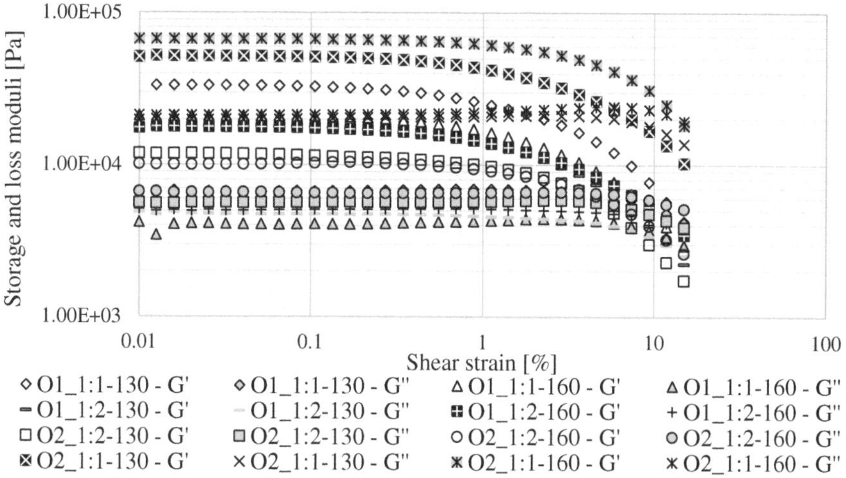

Fig. 1 Amplitude sweep test results: storage and loss moduli of all extenders at 60 °C

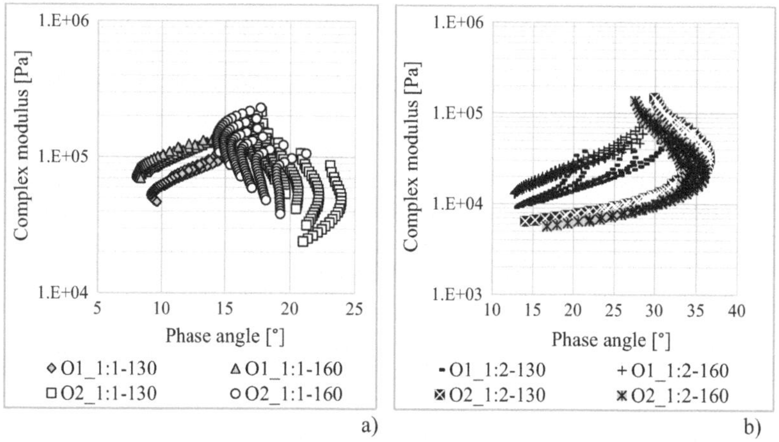

Fig. 2 Black diagrams of extenders made with: (a) R-REOB ratio 1:1; (b) R-REOB ratio 1:2

and the master curve of the rheological properties cannot be determined as done for common bituminous binders. Nevertheless, the Black Diagrams were also found to be sensitive to constituents' proportions, type of REOB and mixing temperature. In general, the six curves that made up the Black Diagram representation of each extender do not show an aligned and elongated curve. This trend may be ascribed to the less temperature-susceptibility of R and REOB. The role of R particles is crucial on the rheological response of the R-REOB compound. Increasing the amount of rubber, the extenders behave more elastically as the phase angle decreases and the G* modulus enlarges, which may reflect the polymer network of R. Moreover, the

Table 3 Percent recovery and J_{nr} parameters of extenders at two stress levels (0.1 and 1.0 kPa)

Sample	%R [%]		J_{nr} [1]	
	at 0.1 kPa	at 1.0 kPa	at 0.1 kPa	at 1.0 kPa
O1_1:1-130	89.84	84.72	0.0017	0.0061
O1_1:1-160	85.16	96.32	0.0018	0.0005
O1_1:2-130	100.00	18.28	0.0001	0.5215
O1_1:2-160	92.18	29.27	0.0045	0.2141
O2_1:1-130	90.36	95.69	0.0016	0.0008
O2_1:1-160	88.83	90.21	0.0023	0.0025
O2_1:2-130	95.19	21.61	0.0033	0.3846
O2_1:2-160	94.68	31.68	0.0042	0.2316
Pen 50/70	2.92	0.84	2.44	2.57

inverse S-curve is typical for rubberized binders. The G*-δ curves are influenced by the type of REOB, and the O2-samples show a more viscous-like behaviour generally. A higher production temperature leads to stiff the final extenders; this response may be linked to a greater oxidation process that appear during the pre-heating and mixing phases.

In addition, the multiple stress creep-recovery test was performed and the elastic response of R-REOB mixes was investigated. This test permits the evaluation of two parameters, i.e. the percent recovery (%R) and the non-recoverable creep compliance (J_{nr}). The MSCR parameters for all blends including the results of the Pen 50/70 are listed in Table 3. The extenders and bitumen behave differently as the MSCR results have a diverse order of magnitude. The presence of a polymer, the recycled R, heavily affects the elastic response of materials. The samples with R-REOB ratio equal to 1:1 show greater elastic recovery (%R) and lesser non-recoverable deformation (J_{nr}) than those specimens with 1:2 ratio; moreover, the differences are amplified when applying the higher shear creep stress (1.0 kPa). Hence, the elastic response of extenders is stress-dependent. The effect of higher production temperature on the elastic behaviour of R-REOB mixes is variable and the trend is not clear. At the highest stress level only, the extenders produced at 160 °C behave better in respect to the others. In general, the O1-samples show a greater %R and minor J_{nr} values than specimens with O2; hence, the mixes made with O1 usually better resist against rutting than extenders with O2.

The average dynamic viscosity (η) of all samples is reported in Table 4, including the value of Pen 50/70, O1 and O2. This property is fundamental for the workability of materials. The viscosity of extenders has magnitude considerable higher than that of REOBs and neat bitumen. The values seem to be influenced by both R-REOB ratio and production temperature. A larger amount of R leads to the increase of the dynamic viscosity for the most specimens over the range of test temperatures. This finding is ascribed to the higher quantity of solid particles present in the blends, which reduce the mobility of REOBs. In general, both O1 and O2 samples made at

Table 4 Average dynamic viscosity of extenders at 60, 100, 150 °C

Sample	η@60 °C [Pa · s]	η@100 °C [Pa · s]	η@150 °C [Pa · s]
O1_1:1-130	11038.15	8610.59	6034.14
O1_1:1-160	13010.97	5182.40	6817.37
O1_1:2-130	7344.05	4781.39	2945.21
O1_1:2-160	12613.58	5736.83	3706.04
O2_1:1-130	8287.13	3115.70	4225.25
O2_1:1-160	15529.46	3971.20	4834.35
O2_1:2-130	5864.40	2395.74	1480.34
O2_1:2-160	8795.13	3300.41	1975.67
O1	4.81	2.21	0.75
O2	9.39	6.38	6.89
Pen 50/70	425.31	5.47	0.40

higher production temperature show greater viscosity at the three test temperatures. The extenders with O1 exhibit a higher viscosity than those samples made with O2, except for the specimens O1_1:1-160 and O2_1:1-160. The dynamic viscosity decreases as the test temperature increases for samples with a double amount of REOB than R (1:2). Different responses can be observed in samples with 1:1 ratio. Changing the test temperature from 60 to 100 °C, the values of viscosity decrease; on the other hand a viscosity increment is observed when the test temperature was raised from 100 to 150 °C.

4 Conclusions

In the present study, a rheological analysis of R-REOB blends has been carried out to evaluate their suitability to be used as bitumen extenders considering three variables: the R-REOB ratio, the type of REOB and the temperature of production. Based on the obtained results the following conclusions can be drawn:

- The quantity of R strongly influences the rheological response of R-REOB blends. Being a polymer, the introduction of R leads to enlarge the LVE range and improve the elastic response of materials; the latter was confirmed by MSCR and FS tests. A high amount of R allows the R-REOB blends to recover more elastic deformation and this behaviour is influenced by the applied stress. Moreover, the content of R affects the dynamic viscosity of specimens reducing the mobility of REOB molecules when the amount increases.
- The rheological properties of extenders are affected by the type of REOB. The O1-mixes showed a more elastic-like behaviour as they better resist against the application of cyclic creep loads and have lower values of phase angle. On the

other hand, the specimens made with O2 exhibit a wider γ_{LVE}, thus they can withstand larger deformations before structural changes occur than O1-extenders. In addition, the dynamic viscosity of O1-bends are usually greater than those of O2.
- Among all variables, the production temperature leads to lesser changes in the rheological properties of the R-REOB blends. The influence of temperature on the MSCR results is not clear; further studies are suggested to better understand this behaviour. Nevertheless, the production at 160 °C instead of 130 °C allows the increment of LVE domains, complex modulus and dynamic viscosity values.

The R-REOB blends made with 1:1 ratio at 160 °C are the candidates to produce extended bitumens with improved elastic responses. The type of REOB leads the extenders to behave differently. It should be highlighted that the final binders will be affected by the base bitumen properties, the amount of extenders and the production conditions, which have to be carefully investigated. Nonetheless, this research foresees the assessment of emissions and fumes from extender productions to its inclusion in the bitumen.

Acknowledgements This research is funded by the Italian recycling consortium of ELTs, Ecopneus Scpa. The authors would like to thank this company, which supplies the rubber and the company Itelyum Regeneration Srl, which provided the REOBs.

References

Airey GD (2002) Use of black diagrams to identify inconsistencies in rheological data. Road Mater Pavement Des 3(4):403–424. https://doi.org/10.1080/14680629.2002.9689933

Amini A, Imaninasab R (2018) Investigating the effectiveness of vacuum tower bottoms for asphalt rubber binder based on performance properties and statistical analysis. J Clean Prod 171:1101–1110. https://doi.org/10.1016/j.jclepro.2017.10.103

Andersen MS (2007) An introductory note on the environmental economics of the circular economy. Sustain Sci 2(1):133–140. https://doi.org/10.1007/s11625-006-0013-6

Asphalt Institute (2016) State of the knowledge - the use of REOB/VTAE in asphalt - IS-235. First edit. Edited by Asphalt Institute, USA

Aziz MMA, Rahman MT, Hainin MR, Bakar WAWA (2015) Alternative binders for flexible pavement. Constr Build Mater 84:315–319. https://doi.org/10.1016/j.conbuildmat.2015.03.068

Becker Y, Méndez MP, Rodríguez Y (2001) Polymer modified asphalt. Vis Tecnol 9(1):39–50

Brundtland GH (1987) Our common future - report of the world commission on environment and development. Oxford University Press, Oxford. https://doi.org/10.9774/gleaf.978-1-907643-44-6_12

ETRma - European Tyre & Rubber manufacturers' association (2019) Europe - 92% of all End of Life Tyres collected and treated in 2017

Fernandes SRM, Silva HMRD, Oliveira JRM (2018) Developing enhanced modified bitumens with waste engine oil products combined with polymers. Constr Build Mater 160:714–724. https://doi.org/10.1016/j.conbuildmat.2017.11.112

Golalipour A, Bahia H (2014) Evaluation of oil modification effect on asphalt binder thermal cracking and aging properties. In: Proceedings of the 59th annual conference of the canadian technical asphalt association, Victoria, BC

Herrington PR (1992) Use of rerefined oil distillation bottoms as extenders for roading bitumens. J Mater Sci 27(24):6615–6626. https://doi.org/10.1007/BF01165945
Kalantar ZN, Karim MR, Mahrez A (2012) A review of using waste and virgin polymer in pavement. Constr Build Mater 33:55–62. https://doi.org/10.1016/j.conbuildmat.2012.01.009
Li X et al (2017) Performance evaluation of REOB-modified asphalt binders and mixtures. Road Mater Pavement Des 18:128–153. https://doi.org/10.1080/14680629.2016.1266754
Lo Presti D (2013) Recycled tyre rubber modified bitumens for road asphalt mixtures: a literature review. Constr Build Mater 49:863–881. https://doi.org/10.1016/j.conbuildmat.2013.09.007
Sangiorgi C et al (2018) Stone mastic asphalt (SMA) with crumb rubber according to a new dry-hybrid technology: a laboratory and trial field evaluation. Constr Build Mater 182:200–209. https://doi.org/10.1016/j.conbuildmat.2018.06.128
Signus Ecovalor; Ecopneus Scpa (2014) Guida per la produzione di bitumi con polverino di gomma da Pneumatico Fuori Uso
Tarsi G et al (2020) A study of rubber-REOB extender to produce sustainable modified bitumens. Appl Sci 10(4):1204. https://doi.org/10.3390/app10041204
Yildirim Y (2007) Polymer modified asphalt binders. Constr Build Mater 21(1):66–72. https://doi.org/10.1016/j.conbuildmat.2005.07.007
Ziari H, Goli A, Amini A (2016) Effect of crumb rubber modifier on the performance properties of rubberized binders. J Mater Civil Eng 28(12):04016156. https://doi.org/10.1016/j.conbuildmat.2007.04.010

Improving the Sustainability of Semi-Dense Asphalt Pavements by Replacement of Recycled Concrete Aggregate Fractions

Peter Mikhailenko, Muhammad Rafiq Kakar, Zhengyin Piao, Moises Bueno, and Lily Poulikakos

Abstract Waste concrete is one of the most highly produced types of waste in the urban environment and finding a means for its re-use is crucial to making infrastructure sustainable. Semi-Dense Asphalt (SDA) is a type of asphalt mixture, which is commonly used in Switzerland to reduce pavement noise. This study examined the use of various fractions of Recycled Concrete Aggregates (RCA) into SDA mixtures. The virgin aggregates were replaced by RCA in selected fractions of 2/4 and 0.125/2 at 100% and 50%, with only one fraction being replaced for any single mixture. The mixtures were evaluated by volumetrics, indirect tensile strength and water sensitivity (EN 12697-12) in order to assess the effects of each RCA fraction. The results showed that RCA coarse aggregates absorb high amounts of binder, which is not the case for the RCA sand. The ITS results showed increase peak load for the RCA replacement samples but also increased brittleness. The ITSR% was similar to the control for lower fraction of RCA replacement samples, but significantly lower with higher replacement.

Keywords Indirect tensile strength · Water sensitivity · Recycled Concrete Aggregates (RCA) · Semi-Dense Asphalt (SDA)

1 Introduction

Some of the biggest challenges to construction sustainability is the production of waste. Construction and demolition (C&D) waste accounts for 46% of total waste in the EU (Gálvez-Martos et al. 2018) and is more than 80% in Switzerland (FOEN 2018). Portland cement concrete composes 85% of the C&D waste in the EU (Gálvez-Martos et al. 2018). This concrete can be reused in the form of Recycled Concrete Aggregates (RCA), but currently, only the RCA from precast concrete, which accounts for 15 to 19% of the total concrete, can be reused (Wahlström et al. 2014).

P. Mikhailenko (✉) · M. R. Kakar · Z. Piao · M. Bueno · L. Poulikakos
Road Engineering/Sealing Components, Empa, Swiss Federal Laboratories for Material Science and Technology, Uberlandstrasse 129, 8600 Dübendorf, Switzerland
e-mail: peter.mikhailenko@empa.ch

C. Raab (ed.), *Proceedings of the 9th International Conference on Maintenance and Rehabilitation of Pavements—Mairepav9*, Lecture Notes in Civil Engineering 76,
https://doi.org/10.1007/978-3-030-48679-2_33

Therefore, finding uses for RCA is important in maintaining the construction industry sustainable.

RCA is prepared through the crushing and grading of waste concretes, which can be partially used as the raw material of virgin concrete or as back fill (McNeil and Kang 2013). Studies have also been carried out using RCA in asphalt - another widely produced material, by replacing the virgin aggregates in asphalt pavements. Usually, this has been done by partially replacing the bulk virgin aggregates with RCA aggregates of a similar gradation at a certain percentage (Pasandín and Pérez 2015).

The main benefit of using RCA is reducing bulk landfilled waste. However, the performance of RCA in asphalt mixtures has given mixed results. Some studies have found improvements in terms of asphalt mixture rutting (Ossa et al. 2016) and fatigue resistance (Pasandín and Pérez 2017) with RCA replacement, while other studies have come to the opposite conclusions (Wu et al. 2017; Albayati et al. 2018).

Semi-Dense Asphalt (SDA) is a type of asphalt mixture, which is commonly used in Switzerland to reduce pavement noise (Raab and Partl 2012). SDA is composed of relatively high quality aggregates and polymer modified binder to compensate for the strength loss from the elevated air void content. Therefore, the addition of waste materials that could potentially be of lower quality, presents challenges. However, the special gradation of the mixture means that it has to consist of selected gap graded aggregate fractions. This presents an opportunity to observe how selected fractions of RCA can perform in an SDA mixture.

The experimental program was designed to replace the virgin aggregates by RCA in selected fractions of 2/4 and 0.125/2 mm at 100% and 50% replacement, with only one fraction being replaced for any single mixture. All other parameters have been kept constant. These mixes are then evaluated for their volumetric properties and their mechanical properties/durability by indirect tensile strength and water sensitivity (EN 12697-12). The experimental program allows to observe the effect of each fraction on the volumetric and mechanical properties of the resulting mixture, potentially being able to take advantage of the benefits of single fractions economically as well as mechanically.

2 Materials and Methods

2.1 Materials

The binder used for the asphalt mixture was an SBS polymer modified binder (PmB) graded at 45/80 – 65 according to EN 1426. Polymer modified binder is part of the requirements for SDA in the Swiss standards (SN 640 436). The control aggregates were quarried limestone from the company FAMSA while the RCA was concrete plant waste from the company FBB, coming from excess of various mixes washed out of the drum. The mixture used for this testing was Semi-Dense Asphalt (SDA

Table 1 Physical properties of aggregates

Sample	Apparent density (Mg/m^3)	Bulk density (Mg/m^3)	Water abs. 24 h (%)	Sand eq. (%)	Flow coeff. (s)
	EN 1097-6			EN 933-8	EN 933-6
FAMSA Filler	2.700[a]	N/A	N/A	N/A	N/A
FAMSA 0.063/4	2.699	2.657	0.57	54	35.7
FAMSA 2/4	2.711	2.647	0.89	N/A	N/A
RCA 0.125/2	2.654	2.439	3.32	89	35.0
RCA 2/4	2.609	2.297	5.20	N/A	N/A

[a]Determine by gas pycnometer

4-16), currently a commonly used gap graded mixture in Switzerland, primarily as a noise abatement measure. The maximum aggregate size was 4 mm and air voids content was 16 ± 2% (SN 640 436).

The aggregates were divided into 3 fractions corresponding to coarse aggregates (washed), fine aggregates and filler: 2/4, 0.063/4, Filler from the control aggregate; 2/4, 0.125/2 for the RCA, respectively. The apparent and relative aggregate densities were determined by EN 1097-6 (by gas pycnometer for the filler) and are shown in Table 1, along with the sand equivalent and flow indices for the fine aggregates. The bulk densities of the RCA was significantly lower than for the control aggregates, especially with regard to the coarse aggregates. For the sand, the Sand Eq. of the RCA was higher, indicating a lower presence of clay particles, which is likely due to the fact that the RCA sand was sieved in the lab while the control sand was received directly from the asphalt plant. The flow coefficients of the sands were similar, indicating similar shape characteristics.

The aggregate replacement was conducted satisfying the gradation prescribed in SN 640 436, so the RCA aggregates were replaced by weight (Fig. 1), with the gradation of the individual fractions determined according to EN 933-1 and an asphalt binder content of 5.7%, and slightly higher for the 100 RCA 2/4 due to increased binder absorption expected. One fraction of the mixture (coarse or fine aggregate) was replaced at a time, with 50 and 100% replacement rates, for 1 control mixture and a total of 4 mixes with partial RCA replacement (Table 2).

2.2 *Methods*

The binder and aggregates were heated to 170 °C before mixing. The aggregates including the filler were added to the drum first and mixed for 5 min followed by the addition of the binder and another 5 min of mixing. The maximum relative density

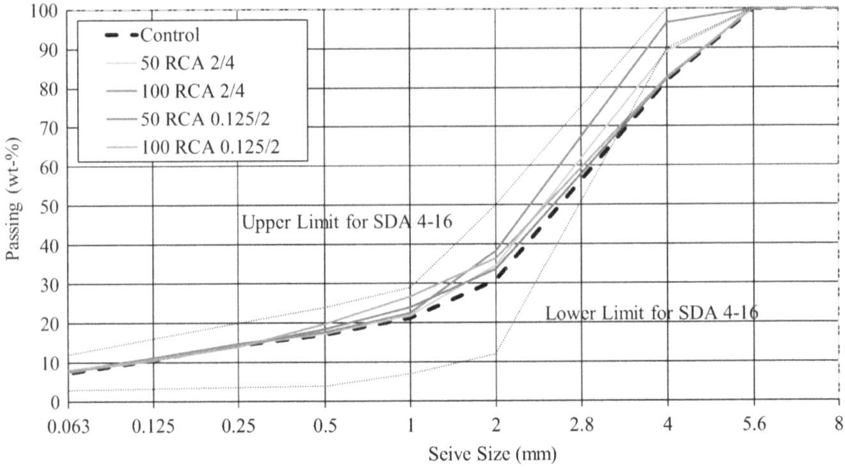

Fig. 1 Gradation curves for RCA modified mixes compared with the limits for SDA 4-16 (SN 640 436)

Table 2 SDA 4 mixture composition

Mixture type	Control sandstone %			RCA %		Binder %
	2/4	0.063/4	Filler	2/4	0.125/2	PmB 45/80 – 65
SDA 4 Control	63.1	23.9	7.3	0.0	0.0	5.7
50 RCA 2/4	31.6	23.9	7.3	31.5	0.0	5.7
100 RCA 2/4	0.0	23.9	7.2	62.9	0.0	6.0
50 RCA 0.125/2	63.1	12.0	7.3	0.0	12.0	5.7
100 RCA 0.125/2	63.1	0.0	7.3	0.0	23.9	5.7

of the loose mixture was determined by EN 12697-5 from two samples. The mixture was then compacted by gyratory compactor under a temperature of 155 °C to a sample size of 99.5 ± 0.5 mm diameter and 64 ± 1 mm height with 16 ± 2% air voids. The bulk density of the sample was determined geometrically according to EN 12697-29 as recommended for SDA. Finally, the indirect tensile strength and water sensitivity were determined according to EN 12697-12 where three samples each were used for the dry and wet conditions.

3 Results and Discussion

3.1 *Asphalt Mixture Volumetrics*

The volumetric properties of the asphalt mixtures are shown in Table 3. The mixture maximum density is reduced with RCA replacement, which is roughly proportional to the RCA replacement rate, whether for the coarse or the fines. The compactability of the asphalt mixture was much more difficult with a higher amount of the coarse 2/4 RCA replacement, and so the mixture was only able to be compacted to 17% air voids. The reason for this is likely in the higher absorbed binder content, which resulted in less lubrication between the stones from the effective binder, but also due to the higher relative volume of RCA coarse aggregates and their lower density.

The voids in mineral aggregate (VMA), which refers to the inter-granular void space of the compacted mixture, was lower for the mixtures with coarse RCA, likely due to the lower compactability of the coarse RCA and the higher amount of gyrations needed to reach the target air void content of 16%, reducing the void space. Low VMA values indicate that there may be a lower binder film thickness between the aggregates (Sengoz and Topal 2007).

The voids filled with asphalt (VFA), which is a measure of the proportion of the VMA filled with binder, was significantly lower with the coarse RCA, especially at

Table 3 Volumetric properties of asphalt mixtures

Sample	Max. density (kg/m^3)	Bulk density (kg/m^3)	Air voids %	VMA %	VFA %	Abs. binder %	Eff. binder %
	EN 12697-5	EN 12697-29	EN 12697-5				
Control	2475.2	2087.1	15.7	25.9	39.5	0.7	5.0
Std Dev	*7.0*	*2.8*	*0.1*	*0.1*	*0.2*		
50 RCA 2/4	2432.6	2041.3	16.1	23.9	32.6	1.9	3.9
Std Dev	*9.2*	*4.5*	*0.2*	*0.2*	*0.3*		
100 RCA 2/4	2399.5	1991.8	17.0	22.3	23.7	3.5	2.7
Std Dev	*0.0*	*4.1*	*0.2*	*0.2*	*0.2*		
50 RCA 0.125/2	2445.8	2053.5	16.0	26.3	39.0	0.6	5.1
Std Dev	*3.9*	*10.7*	*0.4*	*0.4*	*0.8*		
100 RCA 0.125/2	2436.8	2041.5	16.2	25.9	37.4	0.9	4.9
Std Dev	*1.2*	*3.8*	*0.2*	*0.1*	*0.3*		

maximum replacement. This was because of the lower availability of binder intergranularly as more of it was absorbed in the aggregates (Al-Bayati et al. 2018). Both the VFA results and the high absorbed binder content indicate that more binder needs to be added to mixtures with coarse RCA. The RCA sand performance was much more similar to the control sand by the VMA and VFA measures, as the binder absorption was the same as for the control.

3.2 *Indirect Tensile Strength and Water Sensitivity*

The water sensitivity results are shown in Table 4, where the maximum indirect tensile strength (ITS) loads adjusted for the sample size are shown for the dry samples and for those after wet conditioning. In terms of ITS, the dry RCA samples had a higher strength than the control. For the coarse RCA, this is likely from the higher compaction energy that went into the samples. However, the RCA sand attained these results without higher compaction energy. It should also be noted that while the maximum ITS is higher, all of the RCA samples are less ductile than the control as failure occurs at a lower strain (Fig. 2), especially for the 100% RCA sample.

In terms of the indirect tensile strength ratio (ITSR%), the RCA mixture with 50% replacement of the coarse RCA performed almost as well as the control, the 100% coarse RCA sample was much more sensitive to water damage. The RCA sand samples also showed a decrease in ITSR% with higher replacement. This is similar to the results previously found in studies where both coarse and fine RCA was replaced (Mills-Beale and You 2010; Ossa et al. 2016). The Swiss SN 640 431-1c prescribes a minimum of 70% ITSR% for asphalt mixtures. The poor performance of the RCA

Table 4 ITS results for Water Sensitivity EN 12697-12 of asphalt mixtures

Sample	Max load dry (kPa)	Max load wet (kPa)	ITSR %
SDA 4 Control	1127.3	915.0	81.2%
Std Dev	*43.9*	*15.7*	
SDA 4 50 RCA 2/4	1453.5	1157.8	79.6%
Std Dev	*66.8*	*8.7*	
SDA 4 100 RCA 2/4	1820.3	1063.0	58.4%
Std Dev	*22.6*	*25.5*	
SDA 4 50 RCA 0.125/2	1339.3	1001.3	74.8%
Std Dev	*13.3*	*10.8*	
SDA 4 100 RCA 0.125/2	1292.9	811.2	62.7%
Std Dev	*10.1*	*83.9*	

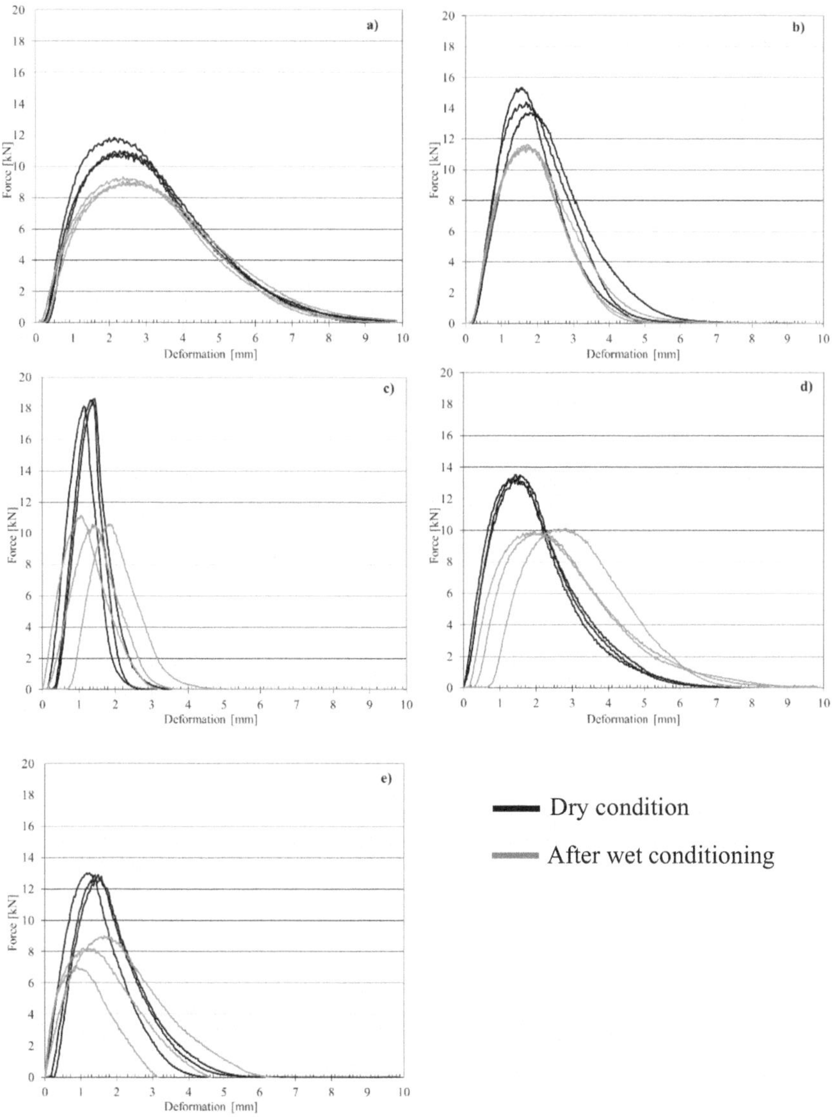

Fig. 2 Loading Curves for Water Sensitivity test: **a** Control, **b** 50 RCA 2/4, **c** 100 RCA 2/4, **d** 50 RCA 0.125/2, **e** 100 RCA 0.125/2

in ITSR% is due to the lower effective binder combined by the higher aggregate absorption (Table 1) allowing for more saturation of the mixture by the water during conditioning. Because of the brittleness in ITS and the lower ITSR%, more advanced testing is needed to show the extent of RCA asphalt mixture viability.

4 Conclusions

This paper studied the volumetric and mechanical properties of asphalt mixtures incorporation coarse and fine RCA. The conclusions are as follows:

- RCA replacement lowers the density of asphalt mixtures.
- RCA coarse aggregates absorb significant amount of binder, whereas the fine fraction do not do so significantly more relative to their virgin aggregate counterparts.
- The replacement by weight of RCA aggregates can result in a higher maximum ITS, but also result in a comparatively less ductile mixture.
- A limited amount of RCA replacement can provide acceptable performance in water sensitivity, but higher replacement with significantly reduce this property.

Acknowledgements The activity presented in this paper is part of Swiss National Science Foundation (SNF) grant No. 5211.01569 for a project titled "Urban Mining for Low Noise Urban Roads and Optimized Design of Street Canyons". Thank you to the Empa 301 technical staff for performing a part of the sample preparation and testing, notably Roland Takacs and Hans Kienast. Thank you to FAMSA (Massongex, Switzerland) for providing the asphalt mixture materials and FBB (Hinwil, Switzerland) for providing the RCA.

References

Albayati A, Wang Y, Wang Y, Haynes J (2018) A sustainable pavement concrete using warm mix asphalt and hydrated lime treated recycled concrete aggregates. Sustain Mater Technol 18:e00081

Al-Bayati HKA, Tighe SL, Achebe J (2018) Influence of recycled concrete aggregate on volumetric properties of hot mix asphalt. Resour Conserv Recycl 130:200–214

EN 933-1 (2012) Tests for geometrical properties of aggregates - Part 1: Determination of particle size distribution - Sieving method

EN 933-6 (2014) Tests for geometrical properties of aggregates - Part 6: Assessment of surface characteristics - Flow coefficient of aggregates

EN 933-8 (2012) Tests for geometrical properties of aggregates Part 8: Assessment of fines - Sand equivalent test

EN 1097-6 (2013) Tests for mechanical and physical properties of aggregates - Part 6: Determination of particle density and water absorption

EN 1426 (2015) Bitumen and bituminous binders. Determination of needle penetration

EN 12697-5 (2019) Bituminous mixtures - Test methods - Part 5: Determination of the maximum density

EN 12697-12 (2008) Bituminous mixtures - Test methods for hot mix asphalt - Part 12: Determination of the water sensitivity of bituminous specimens

EN 12697-29 (2002) Bituminous mixtures - Test method for hot mix asphalt - Part 29: Determination of the dimensions of bituminous specimen

FOEN (2018) Waste and raw materials: In brief [online]. Federal Office for the Environment. https://www.bafu.admin.ch/bafu/en/home/themen/thema-abfall/abfall–das-wichtigste-in-kuerze.html. Accessed 10 Sept 2019

Gálvez-Martos J-L, Styles D, Schoenberger H, Zeschmar-Lahl B (2018) Construction and demolition waste best management practice in Europe. Resour Conserv Recycl 136:166–178

McNeil K, Kang TH-K (2013) Recycled concrete aggregates: a review. Int J Concr Struct Mater 7(1):61–69

Mills-Beale J, You Z (2010) The mechanical properties of asphalt mixtures with Recycled Concrete Aggregates. Constr Build Mater 24(3):230–235

Ossa A, García JL, Botero E (2016) Use of recycled construction and demolition waste (CDW) aggregates: a sustainable alternative for the pavement construction industry. J Clean Prod 135:379–386

Pasandín AR, Pérez I (2015) Overview of bituminous mixtures made with recycled concrete aggregates. Constr Build Mater 74:151–161

Pasandín AR, Pérez I (2017) Fatigue performance of bituminous mixtures made with recycled concrete aggregates and waste tire rubber. Constr Build Mater 157:26–33

Raab C, Partl MN (2012) Stripping of low noise surface courses during laboratory scaled wheel tracking. In: Seventh international conference on maintenance and rehabilitation of pavements and technological control, Auckland, New Zealand

Sengoz B, Topal A (2007) Minimum voids in mineral aggregate in hot-mix asphalt based on asphalt film thickness. Build Environ 42(10):3629–3635

Swiss Standard SN 640 431-1c-NA (2014) Asphaltmischgut Mischgutanforderungen – Teil 1: Asphaltbeton

Swiss Standard SN 640 436 (2013) Semidichtes Mischgut und Deckschichten Festlegungen, Anforderungen, Konzeption und Ausführung

Wahlström M, Laine-Yliijoki J, Kaartinen T, Erlandsson M, Cousins AP, Wik O, Suer P, Oberender A, Hjelmar O, Birgisdottir H (2014) Environmentally sustainable construction products and materials–assessment of release. Nordic Innovation, Oslo

Wu S, Muhunthan B, Wen H (2017) Investigation of effectiveness of prediction of fatigue life for hot mix asphalt blended with recycled concrete aggregate using monotonic fracture testing. Constr Build Mater 131:50–56

Combined Effect of Warm Mix Processes and Multi-recycling on the Main Criteria of the French Asphalt Mix Design Method

P. Marsac, C. Petiteau, O. Burban, J. P. Terrier, G. Didelet, J. Demoncheaux, T. Lorino, and S. Pouget

Abstract The study reported here aims to precise the combined effect of warm mix processes and multi-recycling on the main criteria of the French asphalt mix design method. Asphalt concretes (AC) 0/10 mm are produced at a semi-industrial scale for a combination of 3 recycling cycles, 3 production processes (hot, warm with additive and warm with foamed bitumen) and 2 recycling rates (40 and 70%). This combination is completed by a cycle without recycling for the 3 production processes, giving a total of 21 modalities. Compactibility, water sensitivity, stiffness modulus and fatigue tests are performed on the 21 AC produced. A principal component analysis (PCA) is then performed on the set of data obtained to extract the main trends.

No drastic change is detected in the properties as all the AC produced comply with the French requirements for water sensitivity, stiffness modulus and fatigue. The main significant trends observed through PCA are a decrease of the workability and the compactibility and an increase of the water sensitivity with reclaimed asphalt (RA) addition and multi-recycling. It is also observed that high recycling rate (70% RA) tends to increase the constant of the fatigue line but also its slope, reflecting an increased sensitivity of the fatigue resistance to the strain level.

Keywords Asphalt mixes · Multi-recycling · Warm mix processes · High rate recycling · Mix design criteria · Principal component analysis

---

P. Marsac (✉) · C. Petiteau · O. Burban · J. P. Terrier · G. Didelet · J. Demoncheaux
Laboratoire Matériaux pour infrastructures de transports MAST-MIT, Univ Gustave Eiffel, IFSTTAR, 44344 Bouguenais, France
e-mail: paul.marsac@univ-gustave-eiffel.fr

T. Lorino
Laboratoire Environnement, Aménagement, Sécurité et Eco-conception AME-EASE, Univ Gustave Eiffel, IFSTTAR, 44344 Bouguenais, France

S. Pouget
EIFFAGE Infrastructures, Centre d'Etudes et de Recherches, 8 rue du Dauphiné, 69960 Corbas Cedex, France

C. Raab (ed.), *Proceedings of the 9th International Conference on Maintenance and Rehabilitation of Pavements—Mairepav9*, Lecture Notes in Civil Engineering 76,
https://doi.org/10.1007/978-3-030-48679-2_34

1 Introduction

As in many other countries, the asphalt mix design method currently used in France is based on past experience almost exclusively acquired on hot mix asphalts. In this framework, the effects of the design parameters on the AC properties were widely studied and a compilation of the knowledge accumulated is readily available for road material designer. However, in a new context of growing environmental awareness, new production processes with reduced environmental impacts, namely high rate recycling and warm mixing, tend to become standard practice. Additionally, the gradual generalization of high rate recycling tends to induce repeated recycling of the same material (multi-recycling). In continuation of other international research (Su et al. 2008, Heneash 2013, Hugener et al. 2017), the aim of this study is to gather extra knowledge about the effects of these changes of practices on the main criteria of the French asphalt mix design method in order to assist road material designers in their choices.

2 Experimental Plan

The methodology adopted is to simulate at a semi-industrial scale 3 successive recycling cycles. Each cycle includes laying, milling, ageing and screening. This multi-recycling is applied to 3 production processes (hot, warm with additive, warm with foam) with 2 recycling rates (40% and 70%). This combination is completed by a cycle without recycling for the 3 production processes, giving a total of 21 modalities. An overview of these modalities is given in Table 1 with the labeling adopted.

Table 1 Overview of the 21 modalities of the experimental plan and their labelling

Process and RA content	Without recycling	1st recycling	2nd recycling	3rd recycling
Hot 0% RA (reference)	LH0-0	–	–	–
Warm additive 0% RA	LWA-0	–	–	–
Warm foam 0% RA	LWF0-0	–	–	–
Hot 40%	–	LH40-1	LH40-2	LH40-3
Hot 70%	–	LH70-1	LH70-2	LH70-3
Warm additive 40%	–	LWA40-1	LWA40-2	LWA40-3
Warm additive 70%	–	LWA70-1	LWA70-2	LWA70-3
Warm foam 40%	–	LWF40-1	LWF40-2	LWF40-3
Warm foam 70%	–	LWF70-1	LWF70-2	LWF70-3

Table 2 Properties of the pure bitumen binders and of the binder extracted from the RA0

Test standard	Measured property	35/50	50/70	160/220	RA0 binder (extracted)
EN 1426	Penetration (0.1 mm)	40	58	170	11
EN 1427	Softening point (°C)	52.6	49.0	40.2	74.6

2.1 *Material*

The material tested is an AC10 surf/bin 35/50 according to the standard EN 13108-1 commonly used in France. For all the modalities, the proportions of the constituents are adjusted to match as closely as possible a target binder content of 5.4% mass and the same target grading curve. The virgin aggregate is a micro-granite. For the first recycling cycle, the RA is taken from a mixing plant stock selected for the purpose of the project. This initial reclaimed asphalt (RA0) is a blend of different AC materials milled on different sites and processed (crushed, homogenized and screened at 10 mm). 3 pure bitumen binders are used (35/50, 50/70 and 160/220). The properties measured on these binders and on the binder extracted from the RA0 are given in Table 2.

2.2 *Production Processes and Mixing*

The target mixing temperatures are 170 °C for the hot process and 140 °C for the 2 warm processes. The product EVOTHERM® VM30 supplied by Ingevity is used for the warm process. It is composed of fatty amines derivatives improving the wetting between aggregates and the binder. This additive is blended with the hot binder before the introduction in the mixer. The dosage is 0.4% mass of the added binder. A laboratory foaming device using the MARINI-ERMONT AQUABlack® process was especially designed for the purpose of the project to mimic real mixing-plant foaming. The dosage of the water injected in the hot binder pipe to produce the foam is 2% mass of the added binder. All the AC are produced in a semi-industrial horizontal twin-shaft asphalt mixer with a capacity of 400 kg/batch to mimic a usual batch mixing plant.

2.3 *Milling and Ageing*

The part of each production intended to be used as RA in the next recycling cycle is laid in $600\times\ 100\times\ \approx 5$ cm sections on a trial area. Then, at the end of each cycle, all these sections are milled separately with a usual milling machine (Wirtgen 100FC) before being aged to produce the different RA for the next cycle. In order to simulate the on-site ageing of the binders, the ageing protocol defined by the RILEM TC 206

ATB TG5 for the long term ageing (de la Roche et al. 2010) is applied to the milled materials: 9 days at 85 °C in a large ventilated oven with the material spread on ≈5 cm thick layers. The material is then screened through a 25 mm grid

2.4 *Properties Measured*

The properties of all the 21 AC produced are measured according to the French asphalt mix design method. The 10 properties measured, their labelling and the test standards applied are given in Table 3.

3 Implementation

In practice, the actual conditions did not match strictly the initial plan on 2 points. Firstly, the grades of the added virgin bitumen were shifted to softer grades between cycle 1 and cycle 2. The grades actually used for each recycling cycle are given in Table 4. Secondly, the grading curves of the AC with RA are slightly different from the grading curves of the AC without RA. The actual passing at 2 mm is ≈35% for AC without RA and ≈30% for AC with RA.

Table 3 Properties measured on each of the 21 AC produced and their labelling

Labelling	Property description	Standard
V_GC	Void content with the Gyratory Compactor (GC) at 60 gyrations (%)	EN 12697-31
F_GC	Shear force in the GC at 60 gyrations (daN)	–
IC	Compressive strength ratio i/C (%)	EN 12697-12
E_Mod	2 point bending (2 PB) modulus at 15 °C 10 Hz (MPa)	EN 12697-26
V_Mod	Average void content of the 2 PB modulus specimens (%)	EN 12697-8
e6	Strain for 10^6 cycles for the 2 PB 10 °C 25 Hz fatigue test (μdef)	EN 12697-24
P_Fat	Slope of the 2 PB 10 °C 25 Hz fatigue line	EN 12697-24
E_Fat	2 PB modulus at 10 °C 25 Hz (MPa)	EN 12697-24
V_Fat	Average void content of the 2 PB fatigue specimens (%)	EN 12697-8

Table 4 Grades of the added virgin bitumen actually used

RA content	1st recycling	2nd recycling	3rd recycling
40%	35/50	50/70	50/70
70%	50/70	160/220	160/220

4 Results

All the AC produced comply with the requirements of the French national foreword of the standard EN 13108-1 for water sensitivity (IC ≥ 70%), modulus at 15 °C 10 Hz (E_Mod ≥ 7,000 MPa) and fatigue resistance ($\varepsilon 6 \geq 100$ μdef). However, 18 AC on 21 do not comply with the requirement for compactibility (V_GC < 10%), but this includes the reference AC (LH0-0). Therefore, we can already assume that the different modalities of recycling rate, process and recycling cycle did not led to drastic changes of the AC properties.

5 Statistical Analysis

Rather than studying the effects on each property separately, the choice made is to extract the main trends from the global set of data using a principal component analysis (PCA) with the free software R (R Development Core Team 2005).

5.1 Data Reduction

Modulus at 15 °C 10 Hz (E_Mod) and 10 °C 25 Hz (E_Fat) are obviously correlated and the modulus and the fatigue resistance (E_Mod, e6 and P_Fat) are correlated with the void content. In order to avoid redundant information, E_Mod, e6 and P_Fat were recalculated for the same void content using regression coefficients. Thus, 3 properties of the Table 3 can be removed without losing significant information: V_Mod, V_ Fat and E_Fat. Finally, 3 illustrative variables were added for the recycling cycle, the recycling rate and the process.

5.2 Basics on PCA

The aim of PCA is to describe graphically a set of data which crosses rows containing *n* individuals (in our case AC) against columns containing *p* quantitative variables (in our case properties). The possibly correlated *p* variables are transformed in a smaller number of uncorrelated (perpendicular) variables called principal components. The scatter plot of the *n* individuals in the *p* dimensions of the variables is represented in the reduced space (dimension < *p*) of the perpendicular principal components. The projection of this scatter plot in the plan (2 dimensions) of the first 2 principal components accounting for the maximum variability of the data is then studied as it concentrates the maximum relative amount of information.

5.3 Principal Component Analysis

The first 2 principal components calculated by the PCA explain 67.5% (Dim1: 39.6%, Dim2: 27.9%) of the variability of the set of data. This value is significant. Thus, we can assume that the scatter plot of the 21 individuals in the 6 dimensions of the variables is fairly well represented in the plan of the first 2 principal components. The PCA firstly provides a projection of the variables (properties) in a correlation circle on the plan of the first 2 principal components (Fig. 1). The nearer a variable is to the circle the better it is represented in the plan. Furthermore, the nearer the variable is from a principal component axis, the more it contributes to this axis. Through an analysis of this plot, a physical signification can be associated to the first 2 principal axes.

The variables F_GC, V_GC, IC and e6 are well represented in the plan (Fig. 1). F_GC, V_GC and IC contribute strongly to the first axis (positive correlation for F_GC and V_GC, negative correlation for IC). Therefore, a positive shift along the first axis (Dim1) reflects a decrease of the workability and the compactibility and an

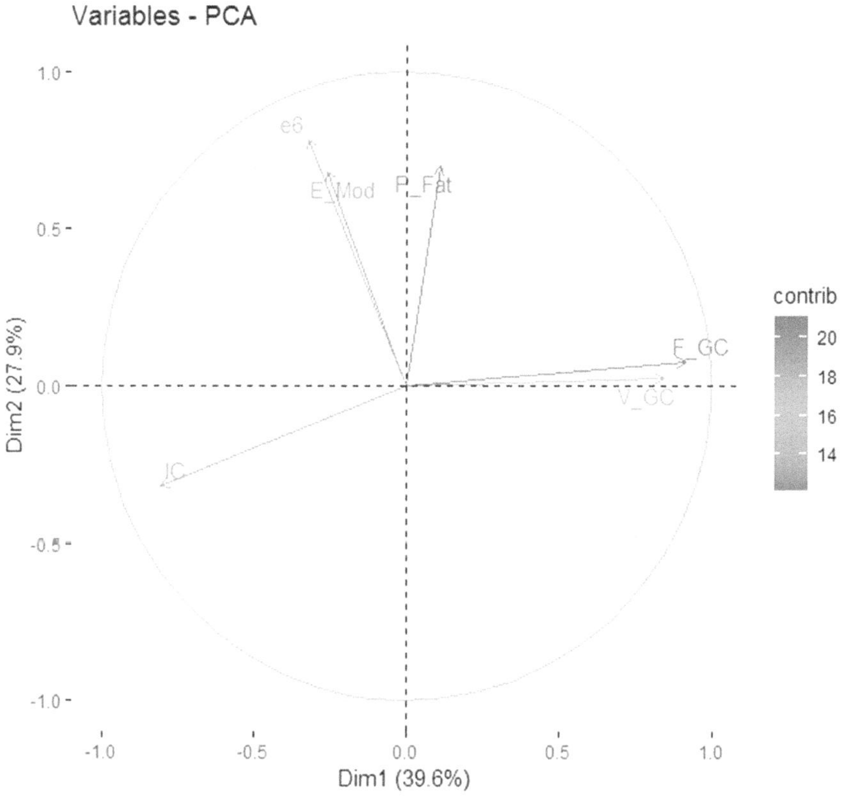

Fig. 1 Projection of the variables in the plan of the first 2 principal components

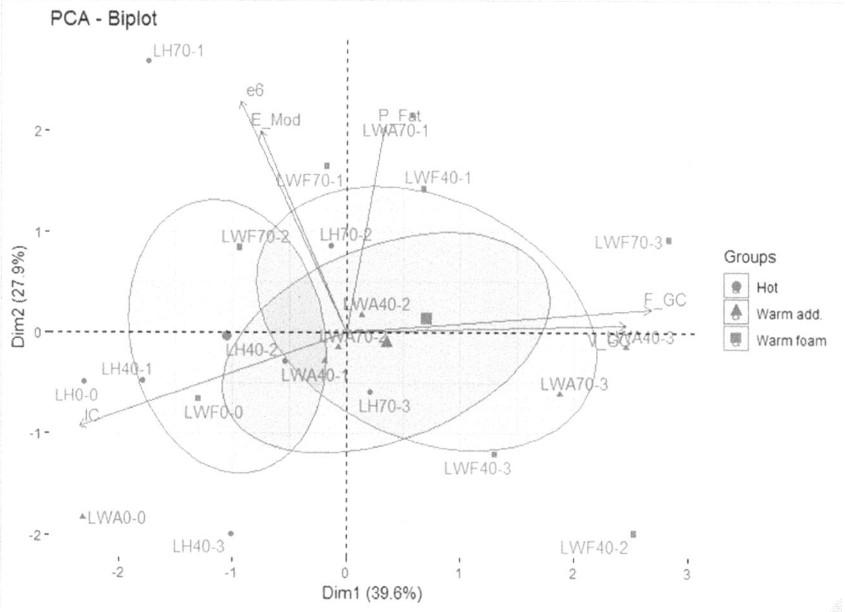

Fig. 2 Process effect in the plan of the first 2 principal components

increase of the water sensitivity. The variable e6 contributes strongly to the second axis, while the contributions of E_Mod and P_Fat are less important. Therefore, a positive shift along the second axis (Dim2) reflects an increase of fatigue resistance and a slight increase of the modulus.

The individuals can then be plotted in the plan of the first 2 principal components and clustered according to the illustrative variables (Figs. 2, 3 and 4). The small labelled points account for the individuals, the bigger points for the means of the clusters and the ellipses for the 95% confidence intervals for the means of the clusters.

The Fig. 2 illustrates the process effect. The 2 warm processes are shifted along the first axis compared to the hot process. This indicates a decrease of the workability and the compactibility and an increase of the water sensitivity for the warm processes. However, the confidence intervals of the clusters means are largely overlapping so this effect is not very significant.

The Fig. 3 illustrates the recycling rate effect. The AC with RA are shifted along the first axis compared to the AC with no RA. This indicates a decrease of the workability and the compactibility and an increase of the water sensitivity with RA addition. The AC with 70% RA are shifted along the second axis compared to the AC with 40% RA. This indicates an increase of fatigue resistance and a slight increase of the modulus with the recycling rate. These effects are significant as the overlapping of the confidence intervals is low.

The Fig. 4 illustrates the multi-recycling effect. The AC are shifted along the first axis according to the recycling cycle. This indicates a decrease of the workability

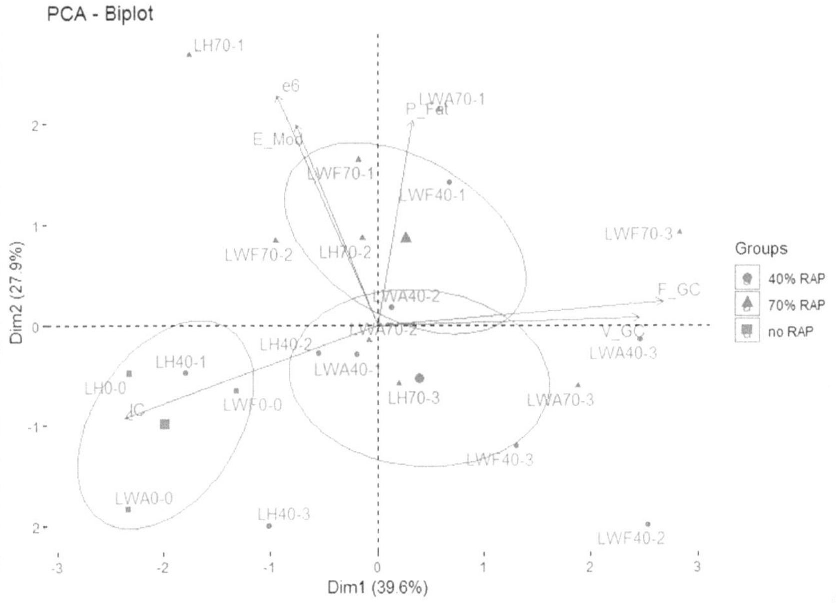

Fig. 3 Recycling rate effect in the plan of the first 2 principal components

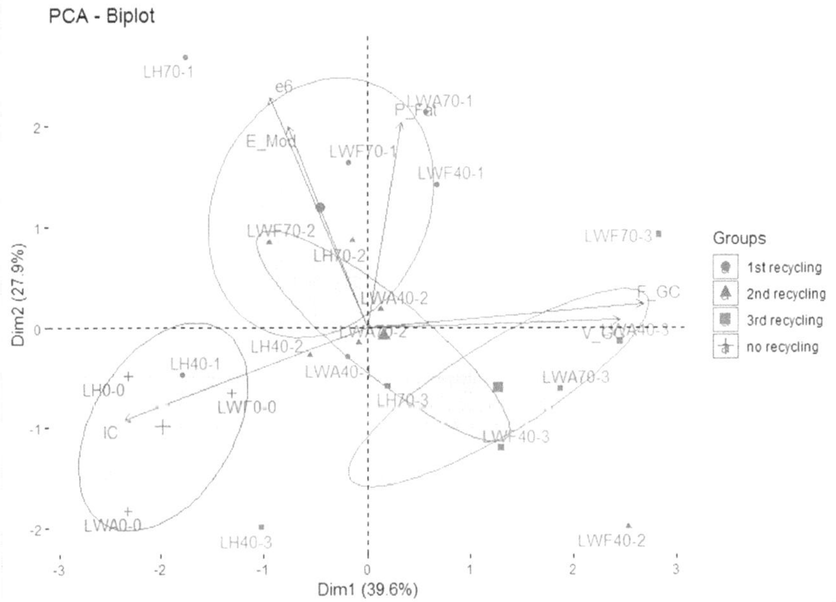

Fig. 4 Multi-recycling effect in the plan of the first 2 principal components

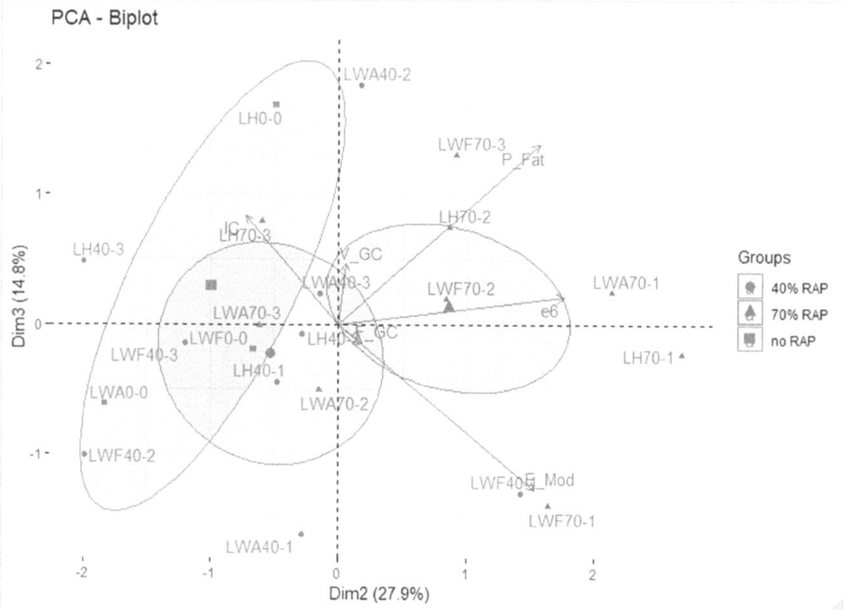

Fig. 5 Recycling rate effect in the plan of the second and the third principal components

and the compactibility and an increase of the water sensitivity with multi-recycling. The effect becomes more and more significant with the repetitions of the recycling cycles. The first recycling is also shifted along the second axis reflecting an increased fatigue resistance. This is probably linked to the harder grades of the bitumen added for the first recycling (Table 4). This shift seems similar to the shift obtained for 70% RA in Fig. 3. In order to have a deeper look on this effect, the recycling rates and the multi-recycling effects are presented in the plan of the second and the third principal components in Figs. 5 and 6.

The variable E_Mod and P_Fat are much better represented in this plan with an opposite effect on the third principal components axis (Dim 3), positive for P_Fat and negative for E_Mod. Figure 5 shows a slight positive shift for the 70% RA cluster compared to the 40% RA cluster on the third principal components axis (Dim3). This indicates that adding 70% RA tends to increase the fatigue resistance (e6 is the constant of the fatigue line) but also its sensitivity to the strain level (P_Fat is the slope of the fatigue line) with limited effect on the modulus (E_Mod). Conversely, Fig. 6 shows a negative shift for the first recycling cluster on the third principal components axis (Dim 3). Thus, the use of harder grades added binders and 70% RA addition induce similar effects on the fatigue resistance (e6), but almost opposite effects on the modulus (E_Mod) and the fatigue slope (P_Fat).

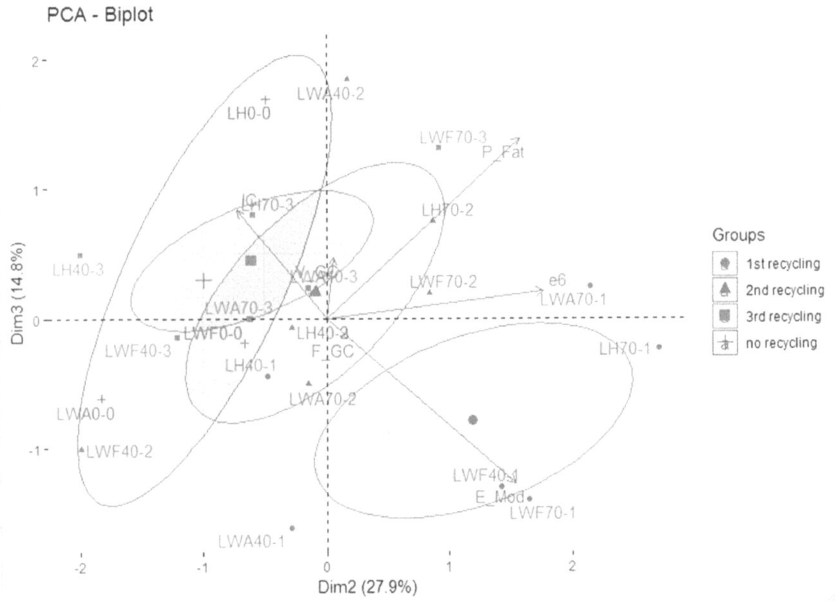

Fig. 6 Multi-recycling effect in the plan of the second and the third principal components

6 Conclusions

In order to precise the combined effect of warm mix processes and multi-recycling on the main criteria of the French asphalt mix design method, asphalt concretes (AC) are produced for a combination of 3 recycling cycles, 3 production processes (hot, warm with additive and warm with foamed bitumen) and 2 recycling rates (40% and 70%). This combination is completed by a cycle without recycling for the 3 production processes. All of the 21 AC produced comply with the requirements of the French national foreword of the standard EN 13108-1 for water sensitivity (IC $\geq$ 70%), modulus at 15 °C 10 Hz (E_Mod $\geq$ 7,000 MPa) and fatigue resistance ($\varepsilon 6 \geq$ 100 µdef). 18 AC do not comply with the requirement for compactibility (V_GC < 10%), but this includes the reference AC (hot process with no recycling). Therefore, the different modalities of recycling rate, process and recycling cycle do not led to drastic changes in the AC properties.

The main significant trends observed through principal component analysis (PCA) are a decrease of the workability and the compactibility and an increase of the water sensitivity with reclaimed asphalt (RA) addition and multi-recycling. It is also observed that high recycling rate (70% RA) tends to increase the constant of the fatigue line but also its slope, reflecting an increased sensitivity of the fatigue resistance to the strain level.

Acknowledgements Acknowledgments are dedicated to the financial support of the French National Research Agency (ANR) in the framework of the national project IMPROVMURE (Innovation for Materials and Processes for Improving the Multi-Recycling of Mixes). Project-ANR-13-RMNP-0008.

References

Heneash U (2013) Effect of the repeated recycling on Hot Mix asphalt properties faculty of engineering. University of Nottingham

Hugener M, Kawakami A (2017) Simulating repeated recycling of hot mix asphalt. Road Mater Pavement Des 18(Suppl. 2):76–90. https://doi.org/10.1080/14680629.2017.1304263

de la Roche C, Van de Ven M, Van den bergh W, Gabet T, Dubois V, Grenfell J, Porot L (2010) Development of a laboratory bituminous mixtures ageing test, ISAP, Nagoya, Japan

R Development Core Team (2005) R: A language and environment for statistical computing. R Foundation for Statistical Computing, Vienna, Austria. http://www.R-project.org. ISBN 3-900051-07-0

Su K, Hachiya Y, Maekawa R (2008) Laboratory investigation of possibility of re-recycling asphalt concretes. Proceedings of 6th ICPT conference, Sapporo, Japan

Mechanical Behaviour of Cold Recycled Asphalt Mixtures for Binder Courses Produced with Bitumen Emulsion and High Strength Cement

Chiara Mignini, Fabrizio Cardone, and Andrea Graziani

Abstract Cold recycled asphalt mixtures (CRAM) are sustainable solutions for maintenance and rehabilitation of road pavements. Notwithstanding the environmental benefits, CRAM may exhibit lower performances compared to traditional hot mix asphalt. Besides, CRAM require a curing period to reach their long-term properties, and thus, they are usually employed in base courses. This work investigates the mechanical behaviour of CRAM with bitumen emulsion and high strength cement to verify if they are suitable for binder courses. Six CRAM were produced using 80% of reclaimed asphalt, modified emulsion and two high strength cement: sulfo-aluminous and Portland-slag cement. For comparison, two additional mixtures were produced using traditional emulsion. The CRAM had an emulsion content of 4.2% or 5.0%, and residual bitumen to cement ratio of 1.0 or 1.2. After curing periods from 6 h to 1 year, the indirect tensile stiffness modulus (*ITSM*) and indirect tensile strength (*ITS*) were assessed. Their thermal sensitivity was evaluated as well, measuring *ITSM* at different temperatures. Results indicate that high strength cement leads to satisfactory mechanical behaviour in the long-term. Compared to Portland-slag cement, sulfo-aluminous cement increases the rate of development of *ITS* and *ITSM* in the early stage of about 50% and 25%, respectively.

Keywords Cold recycling · Reclaimed asphalt · High strength cement · Bitumen emulsion · Curing · Thermal sensitivity

1 Introduction

Recycling has become a major challenge in the construction field. In this context, cold recycled asphalt materials (CRAM) have been developed for roads construction and rehabilitation. CRAM are produced at ambient temperature. The lack of heating results in the reduction of emissions and energy consumption (Xiao et al. 2018) as

---

C. Mignini (✉) · F. Cardone · A. Graziani
Department of Civil and Building Engineering and Architecture, Università Politecnica delle Marche, via Brecce Bianche, 60131 Ancona, Italy
e-mail: c.mignini@pm.univpm.it

C. Raab (ed.), *Proceedings of the 9th International Conference on Maintenance and Rehabilitation of Pavements—Mairepav9*, Lecture Notes in Civil Engineering 76,
https://doi.org/10.1007/978-3-030-48679-2_35

well as material saving, allowing the re-use of high percentages of reclaimed asphalt (RA) (Lin et al. 2017; Bocci et al. 2019). Therefore, CRAM are spreading as a sustainable and cost-effective technique, alternative to hot mix asphalt (HMA).

CRAM are produced using a bituminous binder (bitumen emulsion or foamed bitumen) and a cementitious binder (e.g. Portland cement, composite cement, hydrated lime, fly ash) (Behnood et al. 2015). The two binders strongly influence the mechanical response of CRAM, lending to these materials properties deriving from both HMA and cement concrete (Du 2014; Fang et al. 2016a; Graziani et al. 2019). Despite the advantages, road agencies have still some uncertainty on CRAM usage. Indeed, CRAM are generally characterised by a higher void content and consequently lower mechanical properties compared to HMA. Besides, they require a curing period to reach long-term properties (Mignini et al. 2019; Du 2018; Cardone et al. 2015). Because of that, CRAM are mainly used for subbase and base layers (Kuchiishi et al. 2019; Raschia et al. 2019; Timm et al. 2018; Godenzoni et al. 2018).

Recent studies attempt to identify techniques to produce cold mixtures suitable for upper pavement layers. Adoption of rapid hardening cement, such as sulfo-aluminate cement (CSA) or calcium aluminate cement, has been proposed to improve the early strength of cold mixtures (Fang et al. 2016b; Saadoon et al. 2018). Supplementary blended cementitious fillers can increase the mechanical performance of cold mixtures as well (Nassar et al. 2016; Dulaimi et al. 2017). The use of blended fillers was also introduced in the UK to produce cold asphalt for surface courses (Al-Hdabi et al. 2014). However, it is important to highlight that in all these researches, natural aggregates were used. Grilli et al. (2019) used a high fineness filler and 88% RA aggregate in CRAM produced in-plant to be used in binder courses.

The present laboratory investigation aims at identifying a CRAM composition suitable for binder courses. To this goal, the use of modified bitumen emulsion and two types of high strength cement was evaluated. The evolution of the physical (water loss by evaporation, *WL*) and mechanical properties (indirect tensile stiffness modulus, *ITSM* and indirect tensile strength, *ITS*) was assessed over a year of curing. Besides, *ITSM* was measured at different temperatures to evaluate the thermal sensitivity.

2 Materials and Mixtures

The CRAM investigated were prepared using RA aggregate, natural sand, filler, bitumen emulsion, cement and water. The aggregate blend was obtained mixing 80% RA, 17% natural sand and 3% filler (Table 1). The RA aggregate had the grading distribution reported in Table 1, particle density of 2482 kg/m^3 and water absorption of 1.1%. The natural sand was a crushed limestone with an upper sieve size of 2 mm, particle density 2732 kg/m^3 and water absorption of 1.5%. The filler was a limestone powder with a particle density of 2650 kg/m^3. Two commercial cationic bitumen emulsions complying with EN 13808 were employed, both having 60% of residual bitumen. The first one (C 60 B 10) was produced using neat bitumen,

Table 1 Grading distribution of RA and mixture aggregate blend

Sieve size (mm)	31	20	16	12.5	8	4	2	1	0.5	0.25	0.125	0.063
RA0/16	100	99.7	98.4	91.3	69.2	50.4	32.9	20.6	13.2	8.5	5.3	3.5
Mixture	100	99.8	98.7	93.1	75.5	60.5	43.6	29.1	20.3	14.9	10.5	8.1

whereas the second was obtained using SBS modified bitumen (C 60 BP 10). As co-binder, two types of high strength cement were selected and provided by Italcementi S.p.A.: a CSA with compressive strength of 80 MPa after 28 days and a Portland slag cement type II/B-S (PSC), strength class 52.5N (EN 197-1). The particle density was 2900 kg/m^3 and 3090 kg/m^3, respectively. CSA is a rapid hardening cement, its compressive strength after two days is 45 MPa.

Eight CRAM were produced combining different types and dosages of the co-binders. Two reference mixtures were obtained using the traditional emulsion (coded as "B") whereas the modified emulsion ("BP") was adopted for the remaining six mixtures. The residual bitumen to cement (RB/C) ratio was equal to 1.0 and 1.2 to have a balance among the co-binders. Table 2 summarises the binder dosages (by dry aggregates mass). The total water dosage was 4.5% by dry aggregate mass, used to enhance the workability and compactability of CRAM. Total water includes water from the emulsion and added water. CRAM are coded as follows: cement code (i.e. CSA or PSC) followed by its dosage, bitumen emulsion code (i.e. B or BP) followed by its dosage. For example, CSA2.5_BP5.0 identifies the CRAM produced using 2.5% of sulfo-aluminous cement and 5.0% of modified emulsion.

Mixing was performed in the laboratory at ambient temperature using a mechanical mixer. First, the aggregate absorption water was added to the aggregates previously dried up to the constant mass. Then, the wet aggregate blend was stored in a sealed plastic bag for at least 12 h until reaching saturated and surface-dry condition.

Table 2 Summary of the testing program

Cement		Bitumen emulsion		RB/C ratio	Curing time	
Type (-)	Dosage (%)	Type (-)	Dosage (%)		*WL, ITSM, ITS* (days)	Thermal sensitivity (days)
CSA	2.5	B	5.0	1.2	6 h, 1, 3, 7, 28, 90, 365	28
CSA	2.5	BP	5.0	1.2		
CSA	2.5	BP	4.2	1.0	1, 28, 90, 365	
CSA	3.0	BP	5.0	1.0		
PSC	2.5	B	5.0	1.2	6 h, 1, 3, 7, 28, 90, 365	
PSC	2.5	BP	5.0	1.2		
PSC	2.5	BP	4.2	1.0	1, 28, 90, 365	
PSC	3.0	BP	5.0	1.0		

Then, cement, water and bitumen emulsion were gradually added and mixed to the aggregate blend until a good coating was achieved. Immediately after the mixing, the loose mix was compacted using the Superpave Gyratory Compactor (SCG) to produce cylindrical specimens with 100 mm diameter. The protocol provided constant pressure of 600 kPa, angle of inclination of 1.25° and gyration speed of 30 rpm. The fixed height compaction mode was applied to reach selected target voids of the mixture V_m (i.e. air voids and intergranular water) (Grilli et al. 2012). Two values of V_m were chosen: 8.5% for CRAM with B and 9.5% for the mixtures with BP.

Curing started right after the end of the compaction. Specimens were cured at (25 ± 2) °C and constant RH of (70 ± 5)%. Curing temperature and RH were selected to favour the development of both bituminous and cementitious bonds (Mignini et al. 2019). The specimens were cured for 6 h, 1, 3, 7, 28, 90, 365 days. To assess the thermal-sensitivity specimens were cured for 28 days (Table 2).

3 Test Methods

The evolutive behaviour of CRAM was investigated in terms of WL, $ITSM$ and ITS.

The WL was monitored by weighing the specimens immediately after the compaction (M_0) and at each curing time (M_i):

$$WL = \frac{M_0 - M_i}{M_W} \tag{1}$$

where M_W is the mass of the total water in the specimen.

The $ITSM$ was determined according to EN 12697-26, Annex C. $ITSM$ was measured applying 5 pulse loads with a rise time of (124 ± 4) ms and a horizontal target strain of $2 \cdot 10^{-6}$ mm/mm. The ITS was evaluated according to EN 12697-23, applying a constant rate of deformation equal to 50 mm/min until specimen failure. Both tests were performed at 25 °C on three replicates for each mixture and curing time. Consequently, specimens did not need further conditioning after curing.

The thermal sensitivity was assessed by measuring the $ITSM$ at 5, 15, 25, 40 °C. Two replicates for each mixture were tested with the same protocol adopted for the investigation of the evolutive behaviour. $ITSM$ was also measured by applying a horizontal target strain of $5 \cdot 10^{-6}$ mm/mm. Specimens were conditioned at each test temperature for at least 4 h before the execution of the test.

4 Results and Analysis

4.1 Evolutive Behaviour

The evolution of WL, from 6 h to 1 year of curing, of CRAM with RB/C = 1.2 is shown in Fig. 1a. The WL rate of evolution is noticeably higher in the first week, when its average value is 55 and 59% for CSA and PSC mixtures, respectively. After that, the evaporation gets slower and it is negligible after 3 months. Although the type of emulsion does not influence the WL, some differences due to the type of cement can be observed. After 6 h, CRAM with CSA lose a higher amount of water. In the long-term, CSA mixtures display a higher amount of non-evaporated water (about 25%) compared to PSC mixtures (20%). Even though a slight amount of non-evaporated water could be entrapped, such evidence points out that CSA cement bounds a higher amount of water during its hydration process.

Figure 1b depicts the average values of WL after 1, 28, 90, 365 days, for mixtures with RB/C = 1.0. Results are similar to the ones obtained when RB/C = 1.2.

Figure 2a shows the evolution of *ITSM* throughout curing when RB/C = 1.2. CRAM produced using the same cement highlight a comparable behaviour. In the early stage of curing, $ITSM$ develops rapidly using CSA. After 6 h, $ITSM$ is about 3600 MPa. Differently, $ITSM$ is about 700 MPa with PSC. After the first days, the rate of $ITSM$ evolution of CSA mixtures decreases. CRAM produced with PSC exhibit a lower $ITSM$ even in the long-term. A significative effect related to the type of bitumen is not observed. It must be underlined that CRAM with B have lower V_m.

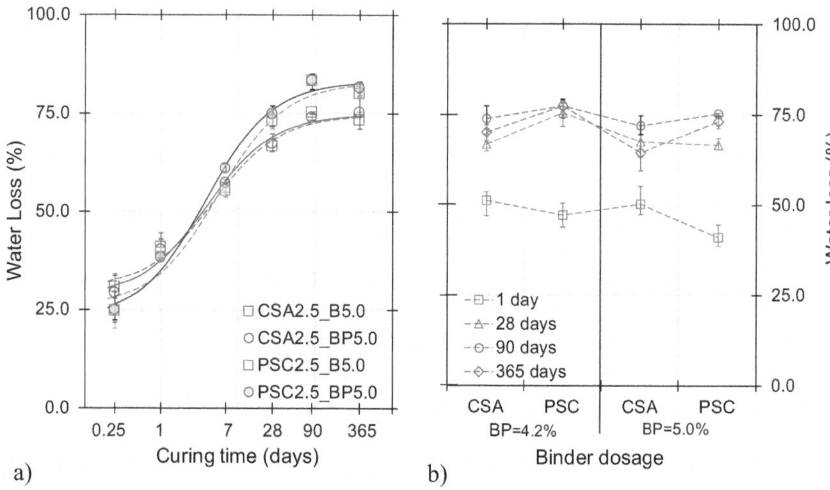

Fig. 1 Evolution of *WL*: **a** as a function of curing time of the CRAM with RB/C = 1.2, **b** of the CRAM with RB/C = 1.0 after 1, 28, 90, 365 days. Error bars show the variability of the data

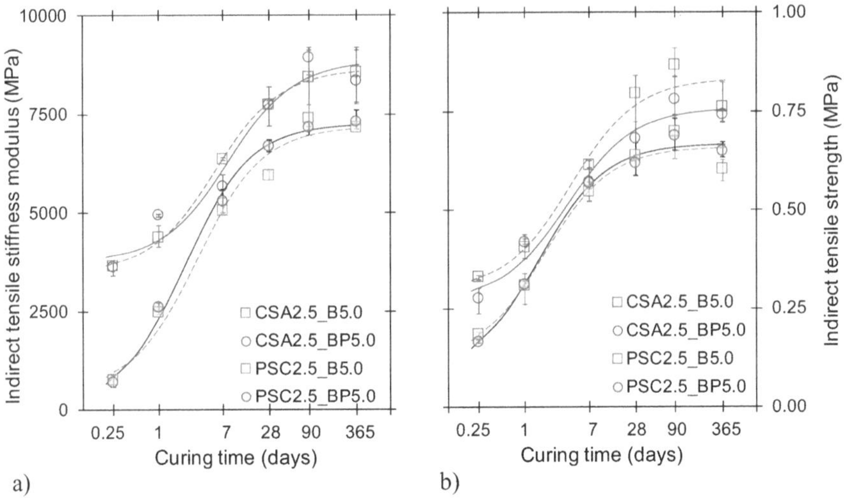

Fig. 2 Mechanical properties as a function of curing time of the CRAM with RB/C = 1.2: **a** *ITSM*, **b** *ITS*. Error bars show the variability of the data

Consequently, the effect of the modified bitumen could counterbalance the effect of the reduced V_m.

Figure 2b depicts the evolution of ITS for CRAM with RB/C = 1.2 In the early stage of curing ITS develops faster for CRAM with CSA. After 6 h, the strength is about 0.30 MPa and 0.18 MPa for mixtures with CSA and PSC, respectively. In the long-term, using CSA, ITS is 20% and 12% higher with B and BP, respectively. ITS is comparable over all the curing for PSC mixtures, whereas it is slightly higher for the mixture with B and CSA. The result can be related to the different V_m. The behaviour is comparable when the same type of cement is used.

Figure 3 depicts the average values of the mechanical properties after 1, 28, 90 and 365 days of curing, measured for CRAM with RB/C = 1.0. Results are similar to the ones measured for the higher RB/C ratio. Comparing the $ITSM$ (Figs. 3a and 2a) of CRAM produced using BP (having equal V_m) it can be observed that the different binders combinations do not have an effective influence on the $ITSM$. After the first curing day mixtures differ by up to 15% and 10% when are used CSA and PSC, respectively. After 1 year the differences are always lower than 10%. The ITS, as well as $ITSM$, does not highlight a pronounced sensitivity to the binder dosage (Figs. 3b and 2b). For CRAM produced with BP the differences due to the dosages of binders are up to 18% and 15% after 1 day and 365 days, respectively.

Figure 4 shows the relationship between $ITSM$ and ITS described by linear regression. The relationship is the same for all the CRAM, regardless of type and dosage of emulsion and cement. This outcome means that the use of a different cement influences mainly the rate of development of the mechanical properties.

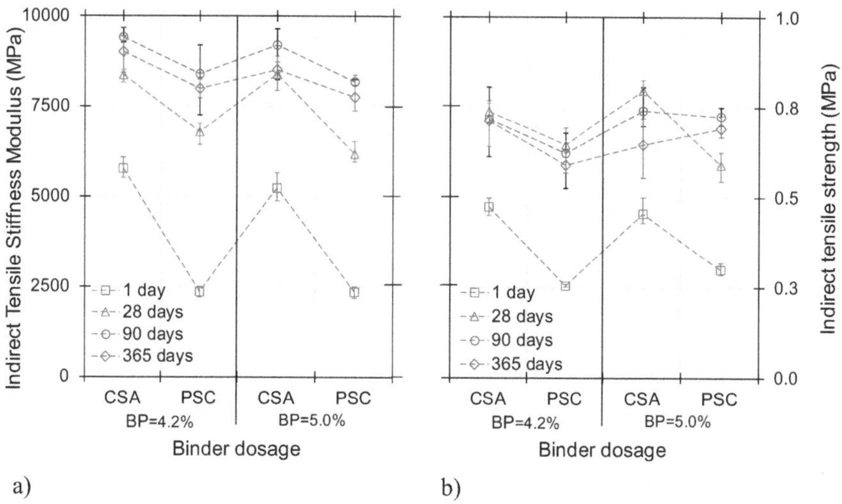

Fig. 3 Mechanical properties of CRAM with RB/C = 1.0 after 1 day, 28, 90, 365 days: **a** *ITSM*, **b** *ITS*. Error bars show the variability of the data

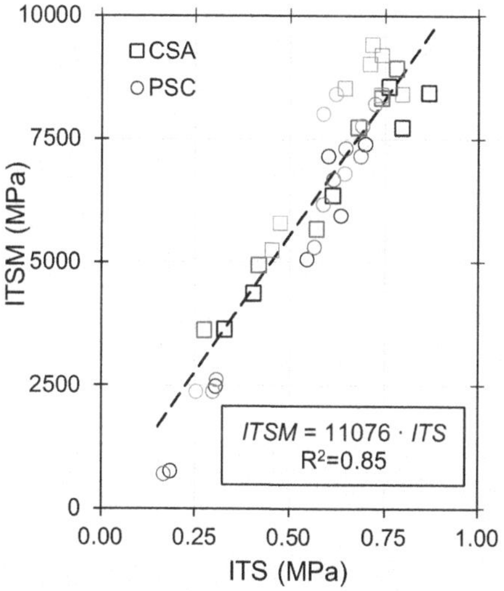

Fig. 4 Relationship between *ITSM* and *ITS*

4.2 Thermal Sensitivity

Figure 5a shows the $ITSM$ values at different temperatures obtained fixing the target horizontal deformation of $2 \cdot 10^{-6}$ mm/mm. The experimental data can be fitted using linear regression:

$$\log ITSM = -a \cdot T + b \tag{2}$$

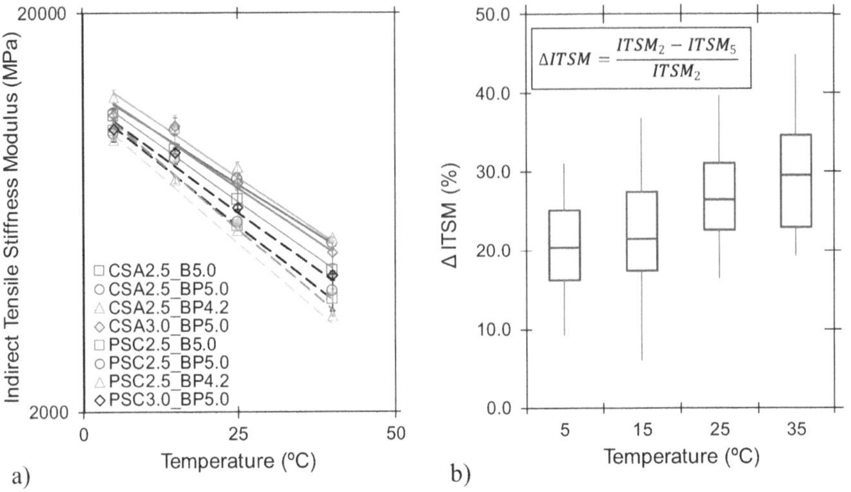

Fig. 5 CRAM thermal sensitivity: **a** *ITSM* as a function of temperature for all mixtures (horizontal deformation $2 \cdot 10^{-6}$ mm/mm), **b** variation of *ITSM* changing horizontal deformation from $2 \cdot 10^{-6}$ mm/mm ($ITSM_2$) to $5 \cdot 10^{-6}$ mm/mm ($ITSM_5$)

where T is the temperature, a and b are fitting parameters depending on CRAM characteristics (reported in Table 3). The slope of the regression line a defines the thermal sensitivity of the material.

All the investigated CRAM highlight a thermal sensitivity typical of bituminous materials: $ITSM$ decreases while temperature increases. However, thanks to the cement, the thermal sensitivity of CRAM is less noticeable compared to HMA (Cardone et al. 2015). Observing the effect of the bitumen type, a is slightly higher for mixtures with B when cement type and dosage are the same. CRAM with CSA have lower thermal sensitivity compared to CRAM with PSC and the same composition.

Table 3 Regression parameters for Eq. (2) (*ITSM*, horizontal deformation $2 \cdot 10^{-6}$ mm/mm and $5 \cdot 10^{-6}$ mm/mm)

Mixture	Hor. Def. = $2 \cdot 10^{-6}$ mm/mm			Hor. Def. = $5 \cdot 10^{-6}$ mm/mm		
	a	b	R^2	a	b	R^2
CSA2.5_B5.0	0.0113	4.112	0.98	0.0124	4.069	0.96
PSC2.5_B5.0	0.0126	4.093	0.97	0.0155	3.999	0.97
CSA2.5_BP5.0	0.0097	4.122	0.96	0.0118	4.036	0.96
PSC2.5_BP5.0	0.0118	4.077	0.96	0.0124	3.918	0.98
CSA2.5_BP4.2	0.0102	4.155	0.97	0.0115	4.074	0.96
PSC2.5_BP4.2	0.0127	4.062	0.99	0.0135	3.977	0.99
CSA3.0_BP5.0	0.0104	4.129	0.94	0.0123	4.089	0.98
PSC3.0_BP5.0	0.0109	4.087	0.98	0.0137	3.939	0.96

Figure 5b quantifies the variation of the stiffness, $\Delta\, ITSM$, when the target horizontal deformation is $5 \cdot 10^{-6}$ mm/mm instead of $2 \cdot 10^{-6}$ mm/mm. The increase of the deformation always reduces the $ITSM$. The result can be attributed to nonlinearity effects due to the horizontal deformation. Besides, the higher the temperature, the higher the $\Delta\, ITSM$.

5 Conclusions

The investigation aimed at defining CRAM composition whose mechanical properties are adequate for the application in binder courses. The influence of two types of high strength cement was evaluated, as well as the effect of traditional and modified bitumen emulsion. Outcomes showed that the use of sulfo-aluminate cement allows the rate of evolution of the mechanical properties to be considerably increased in the early stage of curing. Moreover, in the long-term stiffness and strength reach satisfactory levels. The mechanical response of CRAM produced using the Portland slag cement resulted lower all over one curing year. No particular effect related to the type of bitumen emulsion was observed, albeit the level of compaction could hide this evidence. All CRAM investigated exhibited a lower thermal sensitivity thanks to the cement. Nonlinearity effects related to the deformation were found for all the CRAM investigated.

Acknowledgements The activity presented in the paper was funded by Italcementi–HeidelbergCement Group. Furthermore, the Authors desire to thank Valli Zabban S.p.A. and Società Cooperativa Braccianti Riminese Companies for providing the bitumen emulsion and the RA aggregate.

References

Al-Hdabi A, Al Nageim H, Seton L (2014) Superior cold rolled asphalt mixtures using supplementary cementations materials. Constr Build Mater 64:95–102

Behnood A, Gharehveran MM, Asl FG, Ameri M (2015) Effects of copper slag and recycled concrete aggregate on the properties of CIR mixes with bitumen emulsion, rice husk ash, Portland cement and fly ash. Constr Build Mater 96:172–180

Bocci E, Graziani A, Bocci M (2019) Cold in-place recycling for a base layer of an Italian high-traffic highway. In: 5th international symposium on asphalt pavements and environment (APE), Padova, Italy, 11–13 September

Cardone F, Grilli A, Bocci M, Graziani A (2015) Curing and temperature sensitivity of cement–bitumen treated materials. Int J Pavement Eng 16(10):868–880

Du S (2014) Interaction mechanism of cement and asphalt emulsion in asphalt emulsion mixtures. Mater Struct 47(7):1149–1159

Du S (2018) Effect of curing conditions on properties of cement asphalt emulsion mixture. Constr Build Mater 164:84–93

Dulaimi A, Al Nageim H, Ruddock F, Seton L (2017) High performance cold asphalt concrete mixture for binder course using alkali-activated binary blended cementitious filler. Constr Build Mater 141:160–170
Fang X, Garcia A, Winnefeld F, Partl MN, Lura P (2016a) Impact of rapid-hardening cements on mechanical properties of cement bitumen emulsion asphalt. Mater Struct 49(1–2):487–498
Fang X, Garcia-Hernandez A, Lura P (2016b) Overview on cold cement bitumen emulsion asphalt. RILEM Tech Lett 1:116–121
Godenzoni C, Graziani A, Bocci E, Bocci M (2018) The evolution of the mechanical behaviour of cold recycled mixtures stabilised with cement and bitumen: field and laboratory study. Road Mater Pavement Des 19(4):856–877
Graziani A, Mignini C, Bocci E, Bocci M (2019) Complex modulus testing and rheological modeling of cold-recycled mixtures. J Test Eval 48(1)
Grilli A, Mignini C, Graziani A (2019) Field behaviour of cold-recycled asphalt mixtures for binder courses. In: International conference on sustainable materials, systems and structures, Rovinj, Croatia, 18–22 March
Grilli A, Graziani A, Bocci M (2012) Compactability and thermal sensitivity of cement–bitumen-treated materials. Road Mater Pavement Des 13(4):599–617
Kuchiishi AK, Vasconcelos K, Bariani Bernucci LL (2019) Effect of mixture composition on the mechanical behaviour of cold recycled asphalt mixtures. Int J Pavement Eng 1–11
Lin J, Hong J, Xiao Y (2017) Dynamic characteristics of 100% cold recycled asphalt mixture using asphalt emulsion and cement. J Clean Prod 156:337–344
Mignini C, Cardone F, Graziani A, Morbi A, Setti L (2019) Effect of curing on the indirect tensile failure energy of cement-bitumen treated materials. In: Bituminous mixtures and pavements VII: proceedings of the 7th international conference bituminous mixtures and pavement, Thessaloniki, Greece, 12–14 June
Nassar AI, Mohammed MK, Thom N, Parry T (2016) Mechanical, durability and microstructure properties of Cold Asphalt Emulsion Mixtures with different types of filler. Constr Build Mater 114:352–363
Raschia S, Mignini C, Graziani A, Carter A, Perraton D, Vaillancourt M (2019) Effect of gradation on volumetric and mechanical properties of cold recycled mixtures (CRM). Road Mater Pavement Des 1–15
Saadoon T, Gómez-Meijide B, Garcia A (2018) Prediction of water evaporation and stability of cold asphalt mixtures containing different types of cement. Constr Build Mater 186:751–761
Timm DH, Diefenderfer BK, Bowers BF (2018) Cold central plant recycled asphalt pavements in high traffic applications. Transp Res Rec 2672(40):291–303
Xiao F, Yao S, Wang J, Li X, Amirkhanian S (2018) A literature review on cold recycling technology of asphalt pavement. Constr Build Mater 180:579–604

Performance Evaluation of Hybrid EAF Slag and RAP in Pavement

Shih-Huang Chen, Hasnae Amal Smimine, Wei-Lun Tsai, Ching-Tsung Hung, Meng-Hsin Kuo, and Ching-Lien Zen

Abstract This study evaluated the mixture of Hot Mix Asphalt (HMA) containing Electric Arc Furnace (EAF) slag and Reclaim Asphalt Pavement (RAP). The first design contains 30% EAF slag and the second hybrid design contains mixed EAF slag and RAP, each 20%. Compared to the control group, EAF slag and EAF slag + RAP have lower stability in Marshall Test, but perform better in rutting resistance and moisture sensitivity. In the second part of the study, there are three designs mixture on the test road. The test piece was drilled for the Hamburg wheel test (HWT) after paving, showing better creep performance on EAF; hybrid EAF slag + RAP showed the best stripping results. On the Stripping Inflection Point (SIP), EAF slag has better results of the corresponding ruts and times. Besides, EAF slag + RAP does not show asphalt stripping. EAF slag and EAF slag + RAP have similar results from the performance of the rut depth of 12.5 mm. From the above study, adding one or even mixing two kinds of recycled materials can still meet the design requirements, and it is worthwhile for further study of different recycled materials using in the pavement.

Keywords EAF slag · RAP · Hybrid and pavement

1 Introduction

Most of Taiwan's road systems are made with a flexible paving structure that exhibits stability, durability, comfort and slip resistance through good design and appropriate materials. However, the exploitation of natural materials is becoming more difficult and the germination of circular economy awareness. EAF slag, for instance, is one of the currently widely used alternative materials.

EAF steel making is made by waste iron and steel. The by-product produced in this process is the EAF slag which can be divided into oxide and reduced type. In

S.-H. Chen (✉) · H. A. Smimine · W.-L. Tsai · M.-H. Kuo · C.-L. Zen
Department of Civil Engineering, National Central University, Taoyuan City, Taiwan (R.O.C.)
e-mail: shc@cc.ncu.edu.tw

C.-T. Hung
Department of Logistics and Shipping Management, Kainan University, Taoyuan City, Taiwan (R.O.C.)

C. Raab (ed.), *Proceedings of the 9th International Conference on Maintenance and Rehabilitation of Pavements—Mairepav9*, Lecture Notes in Civil Engineering 76,
https://doi.org/10.1007/978-3-030-48679-2_36

Taiwan, the output of EAF making steel was nearly 1.45 million tons in 2018, of which oxidized slag accounted for about 80% (Zhan 2019). Half of it was applied to engineering materials. For paving engineering, EAF slag can be processed into fine or coarse aggregate and applied in hot mixing, cold mixing or surface treatment of paving. Many studies compared mixtures containing EAF slag with mixtures of natural aggregates and found that the performance of EAF slag mixture is satisfactory in indirect tension, resistance to stripping, fatigue life and permanent deformation (Kim et al. 2018). Compared to virgin mixture, the testing result of elastic modulus and indirect tensile of EAF slag applied in HMA were satisfied (Asi et al. 2007). The asphalt concrete containing EAF slag also shows good rutting resistance (Hainin et al. 2012) and has better stability performance for Marshall Test.

Due to its physical properties, including rough texture, multi-angled angle and high internal friction angle, EAF slag has good results in anti-rutting, deformation, crack and water intrusion. Because of these characteristics, asphalt concrete can have better interlocking effect and improved bonding ability (Kavussi and Qazizadeh 2014). The above research results are all related to coarse EAF slag, and the characteristics of coarse EAF slag are the factors to improve the performance of asphalt concrete. Compared with coarse EAF slag, the study of fine EAF is not much. A few studies of fine EAF slag indicate that the use of fine EAF slag has lower values in Marshall Stability, indirect tension and rebound modulus. On the other hand, fine EAF slag had better resistance to water intrusion (Ameri et al. 2013). Improve the processing treatment of fine EAF slag, which amend its texture roughness and angularity in order to ameliorate the adhesion of fine EAF slag. The test results of improved fine EAF slag has better stability, rutting resistance and water intrusion resistance (Chen et al. 2015).

However, the EAF slag is prone to volume instability once it contacts with water. X-ray Diffraction studies of EAF slag show that the EAF slag has a large amount of free compounds such as MgO and CaO which lead to volume expansion when hydration resulting in volume instability (Yildirim and Prezzi 2011). Due to Volume instability problem, the EAF slag needs to be processed with weathering stabilization treatment before applied in asphalt concrete. Storing and exposing EAF slag directly to an open atmosphere for at least 3 to 18 months helps to oxidize free Ca/Mg to Ca/Mg-hydroxide (Sofilić et al. 2010).

Due to volume instability and stabilized processing time, most of the current EAF slag is ground into fine particles to reduce the stabling treatment time and volume instability in Taiwan. However, the fine EAF slag mentioned in the previous research can resist to water intrusion but cannot effectively improve Marshall Stability and indirect tension. One of the feasible solution is to mix fine EAF slag in recycled asphalt concrete, which can effectively improve the rutting resistance, moisture sensitivity and tensile strength of the mixture (Bowers et al. 2014; Huang et al. 2011; Shu et al. 2012). But the ratio of using is limited by the uncertainty of RAP. The asphalt content and viscosity of RAP are high which made influence on quality, so the upper limit of use is 40% (Chen 2012).

Table 1 Natural aggregates and oxidized slag specific gravity and water absorption

Aggregates	Stone	Sand	Fine EAF slag	RAP
Specific gravity (oven-dry)	2.575	2.609	3.292	2.574
Water absorption (%)	1.65	0.52	2.29	1.37

In the study of the combination of EAF slag and RAP in asphalt concrete, the mixture prepared by the recycled material and the slag showed better fatigue performance and low-temperature fracture performance than the virgin mixture (Wang et al. 2019). The reason for this result may be related to the material properties of EAF slag and the increments of asphalt when designing with EAF slag (Falchetto et al. 2017). In summary, the application of the combination of RAP and EAF slag to the pavement structure is feasible.

2 Materials and Methods

2.1 *Material Selection*

In this study, the nominal maximum aggregate size is designed to be 19 mm. Since the coarse EAF slag may be incomplete due to insufficient asphalt coating or insufficient asphalt film thickness, which may lead to rust spots on the EAF slag slag surface. The EAF slag used in this study is ground into fine particles (100% through No.4 sieve) replacing some natural fine aggregate. The combination of natural aggregate and EAF slag is designed with AC-20 asphalt cement. While the target viscosity of recycled asphalt concrete with EAF slag + RAP is 2000 P, the asphalt cement is AC-10 to meet design needs. The basic properties of the aggregates used in this study are shown in Table 1.

2.2 *Marshall Design*

To understand whether different recycled aggregate can be mixed or not, this study designed three mixtures. The first type is the natural aggregate control group, the second type is the design of the addition of oxidized slag, and the third type is the recycled asphalt materials mix design of adding oxidized slag. The first two use asphalt cement as AC-20, and the third uses AC-10. The asphalt concrete design is based on the Marshall Design method of the seventh edition of Asphalt Institute Manual Series No. 02. The gradation of three groups are designed according to the dense grade asphalt concrete specification table (19.0 mm). The ratio of use of each aggregate is shown in Table 2 and the gradation is shown in Fig. 1.

Table 2 Three asphalt concrete mix design aggregate ratio

Mix	Stone	Sand	Mineral filler	Fine EAF slag	RAP
Virgin	51	44	3	–	–
EAF slag	57	9.5	3.5	30	–
EAF slag + RAP	50	8.5	1.5	20	20

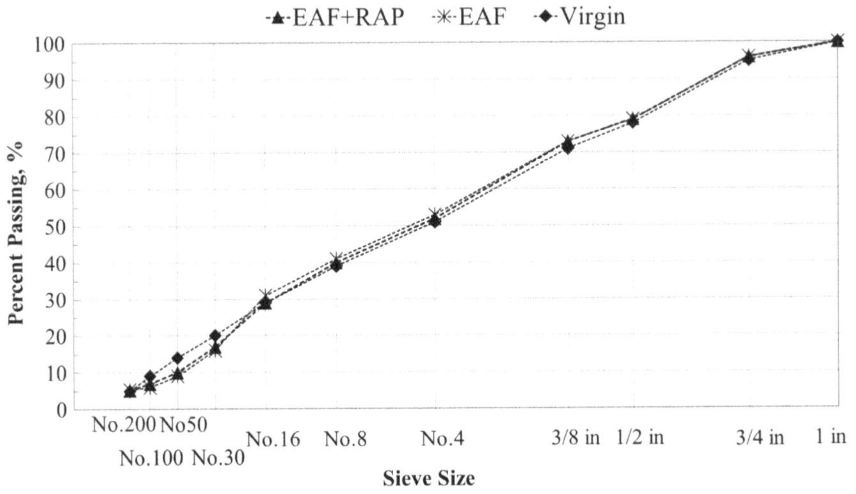

Fig. 1 Three asphalt concrete mix design gradation

A. Virgin
   Using natural aggregates and design matching the intermediate curve of the specification zone.
B. 30% EAF slag
   The recommended upper limit of EAF slag substitution is 40% as the reference, and concerning the maximum amount of natural fine aggregate is replaced by EAF slag. The gradation curve should conform to the specification zone and as close as possible to curve A.
C. 20% EAF slag + 20% RAP
   The design of recycled asphalt concrete supplemented by EAF slag and RAP is replaced with the maximum amount of natural aggregate that based on the recommended upper limit of 40%. It should meet the target viscosity of 2,000 P, so the asphalt content of RAP and the viscosity of the recycled asphalt should be concerned. The add ratio of RAP and EAF slag are 20% each and the gradation should conform to the specification zone and as close as possible to curve A.

Table 3 The test road different sections lengths

Section	Mileage	Length (m)	Material	Binder's grade
A	0K+000–0K+025	25	Virgin	AC-20
B	0K+025–0K+97	72	30% EAF slag	AC-20
C	0K+97–0K+170	73	20% EAF slag + 20% RAP	AC-10

2.3 *Method*

In order to understand whether the mixed use of different recycled materials can still meet the design requirements, this study uses Marshall design to the above three different designs to obtain the optimum asphalt content. According to the results, the indirect tensile test and the resistance of moisture-induced damage test are carried out respectively. The previous tests were carried out in order to investigate the laboratory performance for three asphaltic designs; later on, a road was paved into three different sections (each with a mixture design) for a field performance evaluation. The driveway is 170 m long and 10 m wide. According to the calculations, the traffic volume on the road for the last 20 year ESALs (Equivalent Single Axle Load) is 1.53 × 106 as a heavy traffic volume. Three sections and their lengths are in Table 3. Then, three designs were paved, and the HWT was carried out after the paving to find out whether the mixture of EAF slag and RAP was still able to meet the paving design requirements.

3 Results and Analysis

This study was designed in accordance with the heavy traffic volume in Taiwan. The specimen was made with double-sided compacting 75 blows by Marshall hammer. Finally, the most optimum asphalt content is 4% of void ratio. See Table 4 below.

Table 3 shows the three major differences in design, stability; natural aggregate has higher stability results than the other two mixtures, although they both contain fine EAF slag. From the previous research, the addition of EAF slag as coarse aggregate

Table 4 Marshall design results at 4% voids

	OBC (%)	G_{mb}	Stability (kgf)	Flow (0.25 mm)	VMA (%)	VFA (%)
Requirement	–	–	≧817	8–14	≧13	65–75
Virgin	5.0	2.334	1940	11	13.6	72
EAF slag	5.0	2.499	1504	12	14.0	72
EAF slag + RAP	5.2	2.456	1337	11	13.6	71

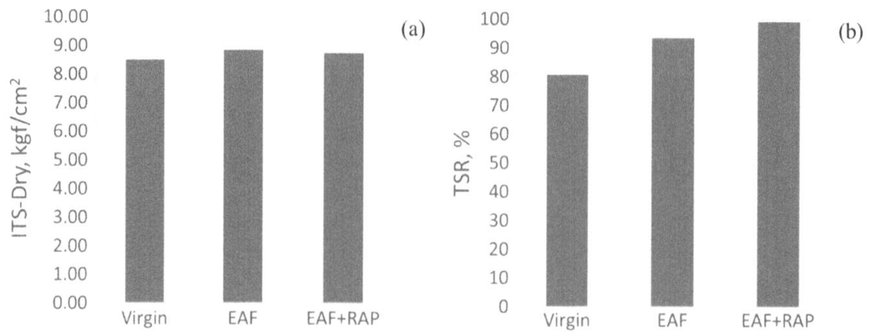

Fig. 2 Indirect tensile strength test **a** and Tensile strength ratio **b**

in hot mix asphalt concrete will increase the stability value and reduce the fluidity, and the improvement is mainly due to the physical properties of the coarse EAF slag, such as the rough surface and angularity. There is little research on the application of fine EAF slag in hot asphalt concrete, however, a few studies have shown that asphalt concrete with fine EAF slag is less stable than normal asphalt concrete, and the reason may include that fine EAF slag is not good at angularity which surface lost the interlocking effect after grinding (Chen et al. 2015).

The mixture of EAF slag and RAP has the lowest stability. Through previous studies, the aged asphalt in RAP is not easily remixed with the new asphalt, and the added ratio of new asphalt content is significantly low that only thin oil film reforming on RAP aggregate which reduced the bonding effect of the aggregate.

This study is processed with the dry curing procedure of AASHTO T283 and the average void ratio of the three mixtures was controlled between 7.1 and 7.4%. The test results are shown in Fig. 2(a), which shows no significant difference between them. Therefore, Indirect Tensile Strength (IDT), the maximum tensile stress asphalt concrete can withstand, is not affected by the addition of EAF slag.

Tensile Strength Ratio (TSR) is the result of IDT test comparing dry and wet curing after the specimen is processed with wet curing. The effects of freezing and thawing damage is not considered in wet curing due to the weather of Taiwan. Figure 2(b) shows the results of different mixtures. The virgin mixture is not ideal for moisture damage resistance. EAF slag and EAF slag + RAP show higher TSR which is 21 and 23%. The addition of EAF slag enhances the tensile strength of the mixture, indicating that EAF slag can obtain satisfactory resistance to moisture and cracking, while the mixture of EAF slag + RAP shows that there is almost no difference in tensile strength between dry curing and wet curing. EAF slag and RAP have been proved excellent in performance of moisture resistance and anti-stripping, and the results of this test indicated that mixture of EAF slag and RAP can effectively enhance the ability of resist water intrusion. The study result is worth exploring in Taiwan that is hot and rainy.

Most of the current research on water intrusion resistance is based on the strength ratio (TSR or retention strength) of the asphalt concrete specimen to evaluate its capacity. While the HWT is to immerse the specimen in 50 °C water for 45 min which is composite experiment that asphalt concrete is subjected to moisture intrusion and rutting resistance simultaneously with repeated sinusoidal motion. The HWT evaluated the sensitivity of various moisture-impacted experiments. It suggests that the creep slope and stripping slope of the HWT have a very high impact on water intrusion, so it can be used to compare the degree of moisture affected by the test (Schram and Williams 2012). This study was performed in accordance of AASHTO T324.

Standard test results include three phases: re-compacting phase, creeping phase and stripping phase. The re-compacting stage is the case where the sample void ratio is compacted. When compacted to a certain rut depth, the slope will gradually decrease to reach the latent stage. The slope at this stage is the rutting sensitivity. When the slope is large, the specimen is more likely with rutting situation. The slope of the next stage will start to increase which is the stripping stage, indicating that the asphalt starts to peel off from the surface of the aggregate and the aggregate is stripped off. This stage is greatly affected by moisture and can be used to compare the degree of moisture intrusion of the specimen. The intersection of the stripping slope and the creep slope is the stripping inflection point, which can be used to comprehensively evaluate the degree of damage caused by the rutting and asphalt stripping of the specimen under the influence of moisture.

In order to make sure the field mixture has the same performance as the laboratory, the study constructed a test road, and drilled the specimen for the HWT 3 days after the open traffic. Figure 3 compares the relationship between the rut depth (mm) and the number of passes (x1000).

From the results of the HWT in Fig. 3, we can find that many trends are similar to those in the laboratory. First, in the slope of the potential change, the natural aggregate

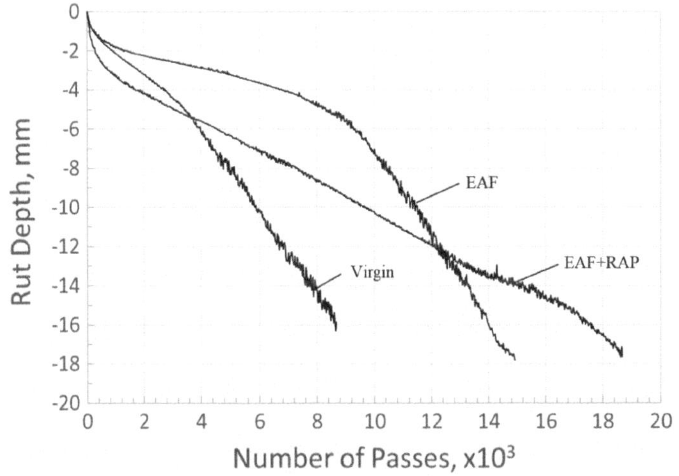

Fig. 3 Deformation (mm) versus number of passes

Table 5 HWT results

	Virgin	EAF slag	EAF slag + RAP
Creep slope	0.955	0.354	0.732
Stripping slope	2.084	2.245	N/A
SIP, (mm, passes)	(4.26, 3216)	(4.63, 9012)	N/A
Rut Depth 12.5 mm, passes	7,170	12,518	12,600

shows the maximum slope value, and EAF slag has the lowest slope value. This also has an approximate situation on the IDT but not significant. EAF slag still has a higher experimental value. In terms of exfoliation slope, EAF slag + RAP did not reach asphalt stripping from the beginning to the end of the test, while natural aggregate and EAF slag have similar slope value. Compared with the results of laboratory TSR, EAF slag + RAP is good in resisting to moisture damage.

The difference from other tests is that HWT can comprehensively evaluate the effects of anti-rutting and anti-moisture damage, so we can pay more attention to the results of SIP. Usually, if the number of passes when SIP occurs is small, it means that the mixture is more vulnerable to water damage. SIP value in Table 5 reflects that the natural aggregate does not perform well in this respect. While there is no obvious peeling point on EAF slag + RAP, indicating that it has excellent performance against moisture damage.

Finally, we compare the number of passes of the three design reaching the maximum rut depth, which is a moderate rut damage according to ASTM D6433 of 13 mm or more, so the setting rut depth is 12.5 mm. The recommended number of passes at a depth of 12.5 mm rut is not clearly defined in Taiwan; even in the United States where the numbers are not uniform. However, the recommended value for PG-64 grade asphalt cement is generally between 10,000 and 15,000. The test results show that natural aggregate has poor performance in it. In Fig. 3, the SIP of natural aggregate occurs when the number of passes is very low although the stripping slope of it is similar to EAF slag, and the corresponding rutting depth is lower than that of EAF slag. But the resistance to moisture damage of natural aggregate is unsatisfactory, which explains that EAF slag + RAP has a poor creep slope compared to EAF slag but the number of passes is almost the same when setting the rut depth of 12.5 mm.

4 Conclusions

For the test result of three mixtures, natural aggregate has higher values in stability and similar results to other mixtures in indirect tension. But when the mixture is affected by moisture, the resistance to water intrusion of natural aggregate is much lower than the other two mixtures; while EAF slag has better crack resistance and EAF slag + RAP has better results in water sensitivity.

Although this study did not discuss the effects of changes in the volume properties, the addition of fine EAF slag that the coarseness of the surface elevated and has relative small volume did increase the voids between the aggregate and the addition volume of asphalt in order to meet the design requirements of the void ratio. It is helpful for the water resistance and durability of the specimen.

The coarse EAF slag has doubts about the stability of volume, while the fine EAF slag can reduce the occurrence of this problem and has good performance not only in the laboratory but also on the field. In addition to EAF slag, the mixture of RAP and EAF slag has excellent result in resistance to moisture damage, which is a viable option for economic cost and environmental value considerations. Therefore, mixed use and the collocation ratio of different recycled materials deserve further studies.

Acknowledgements The authors acknowledge and greatly appreciate the financial support of the National Central University and Ministry of Science and Technology MOST, Taiwan.

References

AASHTO T283-14 (2018) Standard Method of Test for Resistance of Compacted Asphalt Mixtures to Moisture-Induced Damage. Part II: Tests. Washington, DC

AASHTO T324-17: Standard Method of Test for Hamburg Wheel-Track Testing of Compacted Asphalt Mixtures. Part II: Tests. Washington, DC

Asphalt Institute (2014) MS-2 Asphalt Mix Design Methods. Seventh edn. Asphalt Institute

Ameri M, Hesami S, Goli H (2013) Laboratory evaluation of warm mix asphalt mixtures containing electric arc furnace (EAF) steel slag. Constr Build Mater 49:611–617

Asi IM, Qasrawi HY, Shalabi FI (2007) Use of steel slag aggregate in asphalt concrete mixes. Can J Civ Eng 34(8):902–911

ASTM D6433-18 (2018) Standard Practice for Roads and Parking Lots Pavement Condition Index Surveys, ASTM International, West Conshohocken, PA. www.astm.org

Falchetto AC, Moon, KH, Wang D, Riccardi C, Kang MS, Wistuba MP (2017) Investigation on the combined recycling of reclaimed asphalt pavement and steel slag in asphalt mixture at low temperature. In: 5th international conference on sustainable solid waste management, At Athens, Greece

Bowers BF, Huang B, Shu X, Miller BC (2014) Investigation of reclaimed asphalt pavement blending efficiency through GPC and FTIR. Constr Build Mater 50:517–523

Chen Z, Wu S, Wen J, Zhao M, Yi M, Wan J (2015) Utilization of gneiss coarse aggregate and steel slag fine aggregate in asphalt mixture. Constr Build Mater 93:911–918

Chen CL (2012) Improvement research of reclaimed asphalt pavement in Taiwan. National Central University, Master's thesis

Wang D, Falchetto AC, Riccardi C, Wistuba MP (2019) Experimental investigation of performance properties of asphalt mixture designed with the re-recycled RAP and EAFSS. In: Proceedings of the 5th international symposium on asphalt pavements & environment (APE), pp 164–173

Hainin MR, Yusoff NIM, Sabri MFM, Aziz MAA, Hameed MAS, Reshi WF (2012) Steel slag as an aggregate replacement in Malaysian hot mix asphalt. ISRN Civil Engineering

Huang B, Shu X, Vukosavljevic D (2011) Laboratory investigation of cracking resistance of hot-mix asphalt field mixtures containing screened reclaimed asphalt pavement. J Mater Civ Eng 23(11):1535–1543
Zhan CL (2019) Laboratory performance evaluation of electric arc furnace steel slag as aggregate in dense graded asphalt concrete. National Central University, Master's thesis
Kavussi A, Qazizadeh MJ (2014) Fatigue characterization of asphalt mixes containing electric arc furnace (EAF) steel slag subjected to long term aging. Constr Build Mater 72:158–166
Kim K, Jo SH, Kim N, Kim H (2018) Characteristics of hot mix asphalt containing steel slag aggregate according to temperature and void percentage. Constr Build Mater 188:1128–1136
Schram S, Christopher Williams R (2012) Ranking of HMA moisture sensitivity tests in Iowa. In: 92nd annual meeting of the transportation research board, Washington, D.C
Shu X, Huang B, Shrum ED, Jia X (2012) Laboratory evaluation of moisture susceptibility of foamed warm mix asphalt containing high percentages of RAP. Constr Build Mater 35:125–130
Sofilić T, Merle V, Rastovčan-Mioč A, Ćosić M, Sofilić U (2010) Steel slag instead natural aggregate in asphalt mixture. Arch Metall Mater 55(3):657–668
Yildirim IZ, Prezzi M (2011) Chemical, mineralogical, and morphological properties of steel slag. Advances in Civil Engineering

Effect of Waste Fillers on the Rutting and Fatigue Behavior of Asphalt Mastic and Mixes

Jayvant Choudhary, Brind Kumar, and Ankit Gupta

Abstract This study compared the performance of rutting and fatigue resistance of asphalt mastics and mixes prepared with waste fillers (limestone sludge (LS) and glass powder (GP)) to the conventional stone dust (SD) filler counterparts. Physical and chemical properties of fillers were determined. Asphalt concrete mixes were prepared at three different filler proportions (4, 5.5 and 7%) and their optimum asphalt contents (OAC) were determined. The asphalt mastics for all mixes were prepared at their respective filler binder ratio and their rutting and fatigue analysis was conducted using multiple stress creep and recovery (MSCR) and linear amplitude sweep (LAS) analyses. At their OAC's, indirect tensile strength, Marshall quotient, resilient modulus and fatigue lives of asphalt concrete mixes were analyzed. Both LS and GP mixes displayed superior rutting resistance and fatigue resistance than conventional mixes due to their tendency to display higher stiffening and crack pinning behavior caused by the finer particle size and lower specific gravity. This trend was more or less similar in the case of mastics as well. Overall, mastics and mixes prepared with GP and LS displayed superior performance than standard SD filler and hence they could be utilized as alternative fillers.

Keywords Asphalt mixes · Waste fillers · Resilient modulus · MSCR · LAS · Sustainability

1 Introduction

Asphalt mixes are conventionally made up of non-renewable resources like aggregates and carbon-based bitumen binder. In 2017, USA and Europe together produced 640 million tonnes of hot and warm mix asphalt (EAPA 2019). Production of such large quantity of mixes has inherent distinctiveness for environmental damage due to continuous exploitation of natural resources, particularly aggregates. Exhaustive

J. Choudhary · B. Kumar · A. Gupta (✉)
Department of Civil Engineering, Indian Institute of Technology (Banaras Hindu University), Varanasi, Varanasi 221-005, India
e-mail: ankit.civ@iitbhu.ac.in

C. Raab (ed.), *Proceedings of the 9th International Conference on Maintenance and Rehabilitation of Pavements—Mairepav9*, Lecture Notes in Civil Engineering 76,
https://doi.org/10.1007/978-3-030-48679-2_37

mining of aggregates cause problems such as vegetation loss, loss of water retaining strata, lowering of groundwater table and disturbance in the existing ecosystem. However, asphalt mixes are 100% recyclable and in present scenario, researchers are emphasizing on utilizing RAP as well as various solid wastes in the construction of new pavements. Fillers are the finest part of aggregates and are defined as mineral grains most of which pass through 75 μm sieve (MORTH 2013). Inert particles of fillers fill the voids between larger aggregates while active filler articles can alter the behavior of bitumen-filler mastic, adhesion and strength of the mixture due to their physico-chemical interaction with the binder in mix. The overall performance of asphalt mixes is influenced by physical and chemical properties of fillers and due to their relative proportion in asphalt mixes (Choudhary et al. 2018; Huang et al. 2007). Stone dust, cement, and hydrated lime are being conventionally utilized in asphalt mix composition as fillers since they deliver satisfactory performances in the mix. However, several studies have observed that the solid wastes like bauxite residue (Choudhary et al. 2018); carbide lime (Choudhary et al. 2018); copper tailing (Modarres and Bengar 2017); fly ash (Chandra and Choudhary 2013) could be beneficially utilized as fillers.

Glass and stone industry are two of the largest producers of non-biodegradable solid wastes. According to the USGS 2015, the dimension limestone constitutes 45% of the USA's market (Dolly 2015). In 2007, it was estimated that the global production of glass was more than 130 million tonnes and it is estimated to increase further due to exponential growth in its demand (IEA 2007). Both limestone and glass in their respective industries undergoes finishing operations like cutting and polishing, which produces a significant amount of non biodegradable solid wastes in the form of limestone sludge (LS) and glass powder sludge (GP). This sludge largely consists of suspended dust particles and is disposed of in the landfill nearby the industrial area. Recent studies have observed that incorporation of glass powder as filler can form mixes with satisfactory stability, rutting and cracking resistance (Arabani et al. 2017; Choudhary et al. 2018). However there is a need for a much detailed study to analyze the suitability of GP and LS which not only can save significant amount of conventional fillers but also can ensure their safe disposal.

2 Materials and Experimental Investigations

2.1 Materials

Dolomite aggregates were used in the study and gradation used to prepare the asphalt concrete mix stated in Fig. 1 (MORTH 2013). VG 30 asphalt (similar to the 60/70 penetration grade binder) was utilized in this study. Dolomite SD was utilized as the conventional filler and was collected locally (24.46° N, 82.99° E). Dried LS was collected from the dump yard of dimension stone industry located in Kota city (26.91° N, 75.78°) whereas, GP was collected from the dump yard of a glass factory

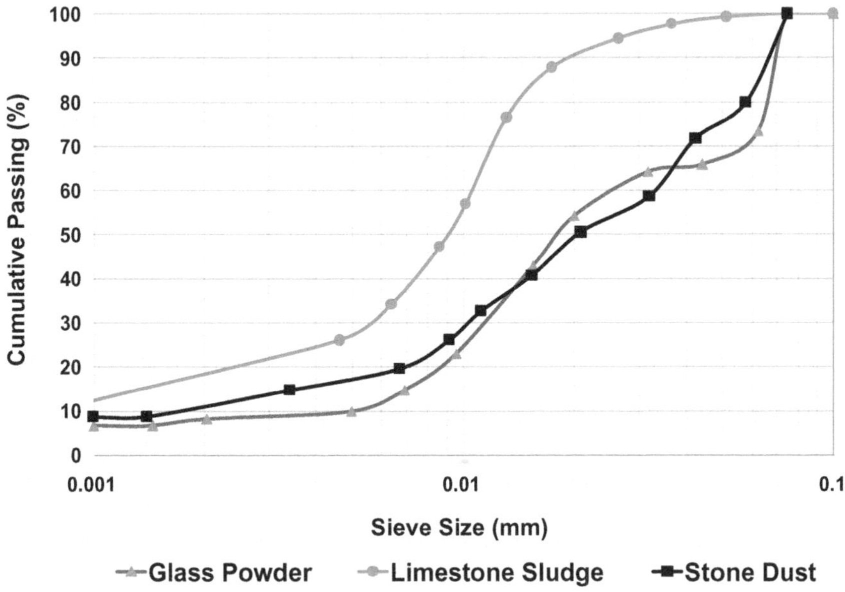

Fig. 1 Particle size distribution of fillers

Table 1 Chosen gradation of asphalt concrete mixes

Sieve sizes (mm)	19	13.2	9.5	4.75	2.36	1.18	0.6	0.3	0.15	0.075
Lower-upper limits (%)	100	90–100	70–88	53–71	42–58	34–48	26–38	19–28	12–20	4–10
Adopted gradation (%)	100	91	74	62	50	43	35	25	14	4.0, 5.5,7.0

located in Bhopal (23.26° N, 77.41° E). All wastes are obtained free of cost from their respective dump yards. Oven dried filler which passes through 0.075 mm sieve was utilized in this study (Table 1).

2.2 *Characterization Tests on Fillers*

Specific gravities of all fillers were determined as per ASTM D854 guideline. Particle size distribution curves were plotted and were characterized using fineness modulus (FM) and mean particle size (D50). Particle shape and surface texture was analyzed using scanning electron microscopy (SEM) analysis. Porosity of fillers was determined as per the German filler test (NAPA 1999). Prevalent minerals in the filler composition were evaluated using X-Ray Diffraction (XRD) investigation which was conducted using Rigaku benchtop XRD device operated with Cu Kα radiation at 1.5406 Å wavelengths. The analysis of Methylene blue values (MBV) of fillers was done as per the EN 933-9 specification to enumerate the harmful clay content and organic matters in fillers. Various results are stated in Table 2, and in Figs. 1 and 2.

Table 2 Properties of various fillers

Property	GP	LS	SD	Inferences
Specific gravity	2.370	2.650	2.698	GP and LS has lower specific gravity and thus occupies larger volume in asphalt mix
Methylene blue value	1.25	3.75	3.25	All fillers have lower MBV an harmful clay content
German filer value (g)	75	97	85	GP and LS had lowest and highest fractional voids per unit weight respectively
Fineness modulus	4.66	3.03	5.38	SD and LS found to be coarsest and finest filler respectively
Particle shape and texture (SEM)	Angular particles with smooth texture	Small size, granulous and somewhat spherical particles with rough texture	Angular particles with smooth to rough texture	Particles with rough texture may negatively affect workability and can also cause higher asphalt absorption
Primary mineralogical composition (XRD)	Quartz (SiO_2)	Calcite ($CaCO_3$), Quartz (SiO_2), Enstatite ($Mg_2Si_2O_6$)	Dolomite ($CaMg(CO_3)_2$), Quartz (SiO_2), Ertixite ($Na_2Si_4O_9$)	SD and LS has dolomite and calcite respectively which improve asphalt adhesion. GP had quartz which is associated with poor moisture sensitivity

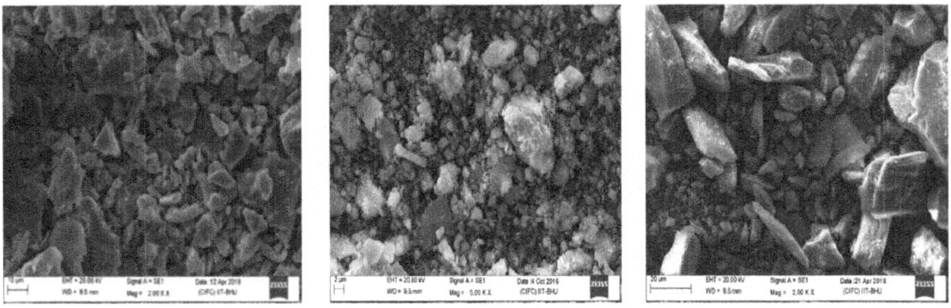

Fig. 2 SEM images of GP (left), LS (middle), and SD (right)

2.3 Design of Asphalt Concrete Mixes

The Indian mix design specification, MORTH 2013 advices to utilize Marshall mix design procedure as per MS-2 (Asphalt Institute 1997) to determine OAC of asphalt mixes. At OAC, Marshall stability (MS), flow and volumetric properties (voids in mineral aggregates (VMA), voids filled with asphalt (VFA) and air voids (VA)) were also calculated. The OAC is considered as asphalt content corresponds to 4% air voids (VA) of the mixes. The increment of filler proportion in the mix was done by reducing the proportion of fine aggregates accordingly in order to satisfy the chosen gradation. The effective filler binder ratio was then calculated and asphalt mastics were designed according to it.

2.4 Preparation and Testing of Asphalt Mastics

A total of nine types mastics were designed by mixing GP, LS, and SD fillers with bitumen in three different filler binder ratio, which was decided based on OAC of the mixes and their filler contents (4, 5.5, and 7%). The mixing was done at 163 °C using mechanical mixer operating at 2000 rpm for 30 min.

The MSCR test was conducted on short-term aged mastics at 64 °C, using a dynamic shear rheometer (DSR) with 25 mm parallel plate in diameter (1 mm gap), in accordance with the standard AASHTO TP70. During the tests, each sample was subjected to ten consecutive cycles at two stress levels (0.1 and 3.2 kPa) and every cycle underwent one second creep loading followed by a nine-second recovery without loading. The non-recoverable compliance (J_{nr}) and the percent recovery (%R) after 10 cycles at 0.1 and 3.2 kPa were studied. The J_{nr} value was calculated as the ratio between the average non-recoverable strain for 10 creep and recovery cycles, and the applied stress for those cycles.

The characterization of fatigue failure was carried out by performing Linear Amplitude Sweep (LAS) test as per AASHTO TP 101. This test was conducted using a DSR with a standard geometry of 8 mm parallel plates and a 2 mm thickness gap. It measures accelerated damage of mastic using cyclic loading by linearly increasing load amplitudes (1–30%). The relationship between the number of loading cycles to failure (N_f) and the applied initial strain amplitude (γ) can be expressed by the following equation.

$$N_f = A^{(\gamma)^{-B}} \quad (1)$$

Where A and B are the fitting coefficients

2.5 *Testing of Asphalt Mixes*

Indirect Tensile Strength

The average indirect tensile strength (ITS) of compacted asphalt concrete mix samples which was determined according to ASTM D6931 guidelines. This evaluation was performed at 25 °C in which, compressive force was imposed on compacted Marshall sample diametrically with the help of steel strips at a constant rate of 50.8 mm/min.

Rutting Resistance

Marshall Quotient (MQ) is the ratio of Marshall stability (kN) to flow (mm) of asphalt mix at OAC. Several studies have observed that asphalt mixes with higher MQ have higher stiffness as well as higher ability to spread the imposed load, which lead to their superior rutting resistance.

Resilient Modulus

Resilient modulus of all mixes was determined at 35 °C as per ASTM D4123 guideline. The UTM 14 was equipped with a temperature control chamber which maintains a constant temperature (35 °C) during specimen conditioning and testing. A haversine load pulse was applied at a frequency of 1 Hz (0.1 s load and 0.9 s rest period) vertically along the vertical diameter of specimens using curved loading strips. The stress level in this test should lie in between 10–50% of indirect tensile strength.

Fatigue Resistance

Indirect tensile fatigue test (ITFT) was performed as per EN 12697-24 specification to predict the fatigue life of asphalt mixes. The ITFT was conducted under controlled stress conditions at 25 °C with a stress level equal to 40% of the indirect tensile strength of the compacted Marshall specimen (with 4% air voids). The stress level of 40% is chosen to reduce the time of testing. The haversine load pulse with loading and rest periods of 0.1 s and 0.4 s respectively were as taken for the analysis. The specimen was continuously loaded until its complete failure.

3 Results and Discussion

3.1 *Marshall and Volumetric Properties*

In general, it is observed that almost all mixes fulfilled the requirements of strength and volumetrics as per Indian specifications (Table 3), and hence these materials can be utilized as fillers. In general Marshall stability were found to be increased with filler content. Improvement in stability might be due to toughening of asphalt mastic due to increase in filler content and a decrease in OAC. All mixes had flow values within the prescribed limit, which limit their possibility to be excessively plastic or brittle. In all cases, OAC decreases with the increase in filler content. It might be due to the asphalt "extender" behavior of fillers which suggest that less amount of binder is needed with higher filler content to make the same amount of mastic for lubrication of aggregates in the mix (Huang et al. 2007). LS displayed a greater tendency to exhibit asphalt binder extender function than GP and SD. GP also have lower specific gravity than SD and thus occupied larger volume in mix at same weight proportion which leaves lower volume for asphalt accumulation in the mixes and thus reduce the OAC of the mixes (Korayem et al. 2018).

Table 3 Marshall and volumetric properties of mixes at OAC

Type of filler	Filler content (% of the weight of aggregates)	OAC (% of the total weight of mix)	Bulk specific gravity	VMA (%)	VFA (%)	Marshall stability (kN)	Flow (mm)
Stone dust	4.0	6.20	2.430	17.02	74.22	12.22	3.43
	5.5	5.95	2.444	16.21	74.43	13.99	3.62
	7.0	5.38	2.453	15.31	74.79	15.96	3.50
Glass powder	4.0	6.03	2.427	16.51	74.85	12.98	3.38
	5.5	5.81	2.431	15.96	73.92	13.46	3.18
	7.0	5.48	2.441	14.85	72.97	14.93	3.37
Limestone sludge	4.0	5.96	2.427	16.83	74.79	12.65	3.37
	5.5	5.53	2.456	15.33	72.54	14.42	3.15
	7.0	4.98	2.469	14.27	73.84	15.60	2.95
Requirements	4–10	–	–	14.00 (min)	65–75	9.00 (min)	2–4

Table 4 Table displaying various parameters obtained from MSCR test of mastics

Filler type	Filler content in mix %	Filler binder ratio for the mastic	Abbreviation of the mastic	Jnr at 0.1 kPa (kPa^{-1})	Jnr at 3.2 kPa (kPa^{-1})	% R at 0.1 kPa (%)	% R 3.2 kPa (%)
SD	4	0.66	SD 4	1.0839	1.2343	5.15	0.78
	5.5	0.98	SD 5.5	0.7132	0.8042	5.62	0.94
	7	1.38	SD 7	0.5013	0.5733	6.55	1.70
GP	4	0.71	GP 4	0.4819	0.5491	7.86	1.00
	5.5	1.01	GP 5.5	0.2056	0.2142	13.42	1.50
	7	1.46	GP 7	0.0753	0.0597	28.66	6.10
LS	4	0.70	LS 4	0.8668	1.0184	7.09	1.10
	5.5	1.06	LS 5.5	0.5557	0.6414	5.96	1.07
	7	1.45	LS 7	0.2976	0.3282	10.93	2.99

3.2 Performance of Asphalt Mastics

The rutting resistance of mastics was found to increase with the filler binder ratio as determined from their decrease in J_{nr} values and the increase in percentage recovery values at both stress levels (Table 4). The GP mastics displayed highest rutting resistance followed by LS and SD mixes. The higher stiffening of GP mastic might be attributed to the higher volume of glass powder at each filler level. This might be due to higher stiffening of mastic caused by higher volume of glass powder per unit weight due to its lowest specific gravity as well as due to angular nature of glass particles. For all the mastics the fatigue lives were found to reduce with the increase in filler binder ratio and the increase in strain magnitude (Table 5). The decrease in fatigue lives might be due to the increase in the stiffness of the mastics. The SD mastics have the highest fatigue lives followed by LS and GP mastics. The GP mastics were found to be most strain susceptible.

3.3 Performance of Asphalt Mixes

Indirect Tensile Strength

Indirect tensile strength of all mixes increased with filler content (Table 6). It is clearly understood that filler has higher strength than the binder. So, at higher filler content, increase in the portion of filler and a simultaneous decrease in binder content in mastic will inevitably increase its strength and as well as ITS of asphalt mix (Huang et al. 2007). LS mixes has highest ITS values followed by GP and SD mixes. Recent studies have observed that finer fillers have a great potential for uniform distribution

Table 5 Table displaying various parameters obtained from LAS analysis of mastics

Type of mastic	A	B	N_f and γ relationship	N_f at different strains			
				0.1%	1%	2.5%	5%
SD 4	23903	2.50	$N_f = 23903(\gamma)^{-2.50}$	7733532	23697	2368	415
SD 5.5	20671	2.45	$N_f = 20671(\gamma)^{-2.45}$	5960708	20578	2156	391
SD 7	7719	2.73	$N_f = 7719(\gamma)^{-2.73}$	4298452	7374	585	86
GP 4	13959	2.59	$N_f = 13959(\gamma)^{-2.59}$	5554930	13904	1282	211
GP 5.5	7012	2.70	$N_f = 7012(\gamma)^{-2.70}$	3589537	6754	556	84
GP 7	2408	2.73	$N_f = 2408(\gamma)^{-2.73}$	1268580	2560	217	33
LS 4	27839	2.52	$N_f = 27839(\gamma)^{-2.52}$	9398720	27993	2766	480
LS 5.5	16440	2.51	$N_f = 16440(\gamma)^{-2.51}$	5371132	16499	1650	289
LS 7	5758	2.64	$N_f = 5758(\gamma)^{-2.64}$	2585007	5511	477	75

Table 6 Various properties of asphalt concrete mixes prepared at OAC

Filler type	Filler content in mix (%)	OAC (%)	Marshall Quotient (MQ)	Indirect tensile strength	Resilient modulus (MPa)	Fatigue life
SD	4	6.20	3.57	652	1360	2491
	5.5	5.95	3.96	718	1991	4201
	7	5.38	4.57	851	2630	6136
GP	4	6.03	3.85	727	1610	4324
	5.5	5.81	4.24	754	2134	5812
	7	5.48	4.66	892	2834	6371
LS	4	5.96	3.77	826	1429	3551
	5.5	5.53	4.62	902	2284	4985
	7	4.98	5.29	984	3037	6762

and formation of an integrated structure in the asphalt mix, which ultimately improve the ITS of the mixes (Modarres and Bengar 2017).

Marshall Quotient
MQ values were found to be increased with the filler content (Table 6). The rutting resistance of asphalt mix was found inversely proportional to their VMA (Christensen and Bonaquist 2006). They followed the trend similar to that of mastics;

however it was interesting to notice that in some cases the KS mastics (KS 5.5, and KS 7) displayed relatively higher J_{nr} values, but still their mixes displayed higher rutting resistance. It suggested that volumetric properties like VMA also dominate the stiffness of the mixes other than the mastic. Both LS and GP displayed higher MQ values than SD mixes. All mixes fulfilled the satisfactory requirements of MQ (2–5 kN/mm) specified in Indian specification.

Resilient Modulus
Resilient modulus also followed the trend similar to MQ and increased with filler content (Table 6). It was expected since OAC in both mixes decreased with the increase in the filler content and Mr usually increase with decrease in binder content (Akbulut et al. 2012). They also followed almost similar trend with the J_{nr} values of their mastics (Table 4). Mixes containing LS and GP had higher Mr values than SD mixes for each filler contents.

Fatigue Life
Fatigue life of asphalt mixes were found to increase with the filler content (Table 6). It was quite interesting since it followed almost inverse trend than that was observed by their mastics. This reason behind it would be the mode of the loading. LAS test performed on mastics was based on strain controlled, while the ITFT test performed on the mixes was stress controlled. Several studies have observed the significantly different fatigue life values when testing was done in different mode of loading. In case of stress controlled mode of testing, the fatigue life usually increases with the stiffness of mixes which is also observed here. Higher fatigue lives of LS and GP mixes might also be due to their higher tendency to show crack pinning behavior due to their finer size and higher filler volume in the mix. It was also interesting to see that trend of fatigue lives of the mixes and the percentage recovery (%R) displayed by mastics in MSCR tests are similar (Tables 4 and 6). Hence future study can focus on obtaining correlation between two test methods since they are based on similar mode of loading.

4 Conclusions

This study investigated the suitability of two industrial wastes, LS and GP as fillers on the performance of asphalt concrete mastic and mixes at three filler contents. Both waste displayed physical and chemical properties synonymous of a good filler. Both LS and GP mixes formed the asphalt mixes that delivered satisfactory strength and volumetric properties at relatively lower OAC. Lower OAC of GP and LS mixes was due to their lower specific gravity, finer size, and tendency to act as asphalt extender in the mixes. Mixes prepared with both fillers were also displayed higher indirect tensile strength, resilient modulus, rutting and fatigue resistance than conventional mixes at all filler contents. The rutting resistance of the mixes as determined from MQ test followed similar trend to that of J_{nr} values of their asphalt mastics as determined from MSCR test. However, fatigue life of mixes as obtained from ITFT test found to have inverse trend of that of their mastics (LAS test), which was attributed to the change of mode of loading in test methods. Mixes made with GP and LS also fulfilled the requirements of Marshall, volumetric properties and MQ as mentioned in Indian specifications. Hence, based on this limited study, GP and LS can be proven as viable fillers for asphalt concrete mixes. Both wastes has been obtained free of cost from their respective source and they doesn't require additional modification other than sieving. Hence their mixes can also be proven to much more economical than conventional mixes. However, this aspect needed further exploration in the future research. In addition to this performance evaluation of mixes against moisture sensitivity, raveling, low temperature cracking and long term ageing is also needed to be explored in future.

References

Akbulut H, Gürer C, Çetin S, Elmaci A (2012) Investigation of using granite sludge as filler in bituminous hot mixtures. Constr Build Mater 36:430–436

Arabani M, Seyed AT, Mohammad T (2017) Laboratory investigation of hot mix asphalt containing waste materials. Road Mater Pavement Des 18(3):713–729

Asphalt Institute (1997) Mix design methods for asphalt concrete and other hot-mix types: Manual Series No. 2 (MS-2) 6th Ed

Chandra S, Choudhary R (2013) Performance characteristics of bituminous concrete with industrial wastes as filler. J Mater Civ Eng 25(11):1666–1673

Choudhary J, Kumar B, Gupta A (2018) Application of waste materials as fillers in bituminous mixes. Waste Manag 78:417–425

Christensen DW, Bonaquist RF (2006) Volumetric requirements for superpave mix design, vol 567. Transportation Research Board, National Research Council

Dolly TP (2015) Stone (dimension), minerals commodity summaries. United States Geological Survey, Reston

Huang B, Shu X, Chen X (2007) Effects of mineral fillers on hot-mix asphalt laboratory-measured properties. Int J Pavement Eng 8(1):1–9

EAPA (European Asphalt Pavement Association) (2019). https://eapa.org/wp-content/uploads/2019/08/Asphalt-in-figures_2017.pdf. Accessed 12 Feb 2020

IEA (International Energy Agency) (2007). https://www.iea.org/publications/freepublications/publication/tracking_emissions.pdf. Accessed 12 Nov 2018

Korayem AH, Ziari H, Hajiloo M, Moniri A (2018) Rutting and fatigue performance of asphalt mixtures containing amorphous carbon as filler and binder modifier. Constr Build Mater 188:905–914

MORTH (Ministry of Road Transport and Highways) (2013) Specifications for road and bridge works (Fifth Revision). Indian Road Congress

Modarres A, Bengar PA (2017) Investigating the indirect tensile stiffness, toughness and fatigue life of hot mix asphalt containing copper slag powder. Int J Pavement Eng 20:977–985. https://doi.org/10.1080/10298436.2017.1373390

National Asphalt Pavement Association (NAPA) (1999) Evaluation of baghouse fines for hot mix asphalt. Information series, Lanham, MD, vol 127

Evaluating the Properties of Bioasphalt Produced with Bio-oil Derived from Biodiesel Production

Caio Rubens Santos, Jorge C. Pais, Jorge Ribeiro, and Paulo Pereira

Abstract Environmental concerns are conditioning all processes that use hydrocarbons products, building awareness for a more sustainable environment through products called bio. In this scenario, it is also the production of bitumen, where the substitution of part of the bitumen by bioproducts is a contribution to the sustainability of the environment. Bio-oil, which is currently used for burning with the purpose of producing energy, can be used to modify the bitumen, producing a bio-asphalt. Trying to contribute to sustainable development, this paper presents an initial analysis of the feasibility of using bio-oil in the production of pavement bitumen. One base bitumen was used, namely, residue of asphalt. The tests were carried out for physical characterization of the bio-bitumen, namely the penetration, the softening point, and the Brookfield viscosity, as well as its rheological characterization in which the dynamic shear modulus and the phase angle were evaluated. The bio-oil content, by weight, varied from 2 to 10% for the residue of asphalt. It was verified that the base bitumen, namely the residue of asphalt, can be softened significantly due to the bio-oil. The bio-oil content of 6% added to the residue of asphalt produces a bio-binder with physical behaviour similar to a conventional 35/50 pen asphalt.

Keywords Bio-oil · Bio-asphalt · Residue of asphalt · Asphalt rheological properties

C. R. Santos (✉) · J. C. Pais · P. Pereira
University of Minho, Guimarães, Portugal
e-mail: caiorubens@maua.br

J. C. Pais
e-mail: jpais@civil.uminho.pt

P. Pereira
e-mail: ppereira@civil.uminho.pt

C. R. Santos
Mauá Institute of Technology, São Caetano do Sul, Brazil

J. Ribeiro
Petrogal, Matosinhos, Portugal
e-mail: jorge.ribeiro@galp.com

C. Raab (ed.), *Proceedings of the 9th International Conference on Maintenance and Rehabilitation of Pavements—Mairepav9*, Lecture Notes in Civil Engineering 76,
https://doi.org/10.1007/978-3-030-48679-2_38

1 Introduction

Crude oil is the main source of fuel, supplying almost a quarter of the world's energy demands for the needs of the industry, transportation and housing (Hu et al. 2018). However, the depletion of crude oil reserves generates several efforts to use less petroleum asphalt binder for road pavements (Colbert et al. 2016). In the current paradigm, regarding oil-related energy policies, there is an increase to develop environmentally friendly methods and products that minimize the use of crude oil, without compromising current and even future needs of petroleum products, such as the asphalt for road paving. Nowadays, the bitumen used in road pavements is obtained by processing crude oil in a refinery. Thus, there is an awareness for a more sustainable environmental development that calls for bio-products, which are derived from different sources and which show a benefit for the environment. In this scenario, there are several initiatives to include bio-materials in asphalt production (Pérez et al. 2019). Regarding asphaltic materials, one way to contribute to environmental sustainability is the use of bio-asphalts, which are binders with a partial replacement of bitumen components by biological products such as bio-oils.

Bio-oil is a biological material generally obtained through pyrolysis from crops, cotton, straw, waste wood, animal fats, waste cooking oil, and others (Su et al. 2018). Among the bio-oils is found the Bio Heating Oil (BHO), which is a residue of the production of biodiesel from waste cooking oil and animal fats. The BHO is currently burned for energy production with the consequent greenhouse gas emissions. It also has physical and chemical properties that allow its addition to the paving bitumen with the potential to improve aggregate adhesiveness. Its use in the asphalt can reduce part of bitumen extracted from crude oil. Besides, the use of BHO in bitumen makes them more environmentally sustainable, reducing greenhouse and polluting emissions derived from the oil burning.

Currently, state of the art for the utilization of bio-oils is concentrated on its uses as renewable fuels to replace fossil fuels. However, there has been an amount of research conducted to investigate the applicability of bio-oils to produce bio-asphalt. Metwally and Raouf (2010) state that bio-asphalt can be obtained by mixing biomass heavy oil and petroleum asphalt, and it can be classified into three categories according to the content of bio-oil in blended binders. The bio-oil can be classified as a modifier of the bitumen to content less than 10%, an extender to content from 25 to 75% and an alternative binder when the bio-oil replaces the petroleum asphalt (Metwally and Raouf 2010).

A large amount of research has been conducted with the aim to evaluate the effects of the bio-oil as a rejuvenator of aged asphalts. Zargar et al. (2012) and Sun et al. (2016) investigate the use of waste cooking oil as an aged bitumen rejuvenator. Zhang et al. (2018) conduct research with wasted wood bio-oil. Zeng et al. (2018) carried out studies using residue in castor oil, and Zhang et al. (2019) used sawdust bio-oil as a bio-based rejuvenator. The ability to soften aged binders was observed in the studies. Most of the bio-asphalts can improve low-temperature performance, but high-temperature performance is reduced. With the aim to improve the bio-asphalt high-temperature performance, Zhang et al. (2017) carried out tests with SBS modified bio-asphalt, successfully improving its high-temperature performance. On the other hand, Yang and You (2015) found that with the increase of bio-oil generated from wasted woods content, asphalt high-temperature performance was improved. Additionally, it is verified that the rheological parameters have been successfully used to assess both aged bitumen and modified bitumen (Peralta et al. 2010; Brovelli et al. 2013, 2014). Overall, it was found that bio-oil derived from several biomass sources can successfully be used as an asphalt modifier, and it is expected the bio-oil to soften the asphalt binder. Therefore, it can be concluded that bio-oil can be used not only as a rejuvenator for aged binders but also to modify very hard asphalts and even residues of asphalt in order to obtain bio-asphalts with physical and chemical characteristics consistent with the European standards of paving asphalts. The use of bio-renewable materials partially substituting asphalt binder creates a more sustainable solution for pavement materials.

2 Objective

This work studied bio-asphalt binders produced from the modification of asphalt binders by the BHO, which is a bio-oil that results, as a residue, from the production of biodiesel derived from waste cooking oil. The bio-asphalts were composed with one base asphalt binder (residue of asphalt) mixed with BHO in content from 2 to 10% by weight for the residue of asphalt. The physical characteristics namely penetration, softening point, and viscosity and the rheological behaviour of the bio-asphalts were carried out to verify the effects of the modification of the base binders by adding the BHO.

3 Materials and Methods

3.1 Bitumen

The base asphalt binder used in this study was a residue of asphalt, which is a hardened material derived, as a residue, from the distillation of the crude oil. Table 1 presents some detailed technical properties of the base binder. The residue of asphalt was obtained from a Portuguese petroleum refinery; however, is not a commercial product.

Table 1 Properties of the residue of asphalt

Binder	Properties	Units	Test results
Residue of asphalt	Penetration @ 25 °C	0.1 mm	8.73
	Softening point	°C	65.65
	Dynamic Viscosity @ 135 °C	Pa s	1.96

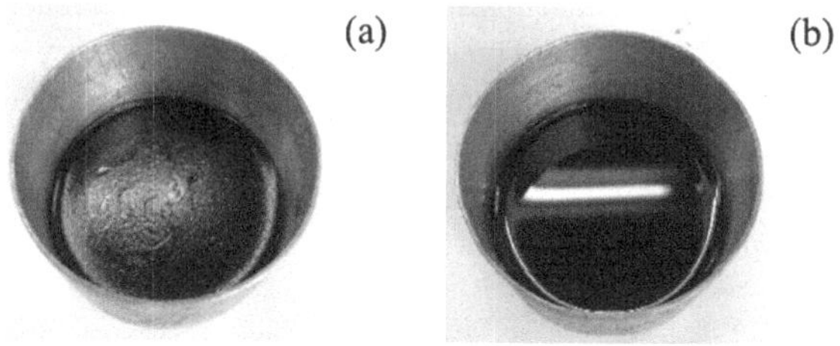

Fig. 1 Bio Heating Oil appearance at ambient temperature (**a**) and heated to 40 °C (**b**)

3.2 *Bio-oil*

The bio-oil used in this study was obtained in a plant for biofuel production. The biofuel is produced using waste cooking oil and animal fats as the main raw materials. The process involves a first step of esterifying the free fatty acids in acid medium with methanol, followed by the transesterification step using basic methanol catalysis (the catalyst is added as CH3OK potassium methoxide) and the distillation of biodiesel. BHO is the residue obtained from the distillation of the biodiesel, and it is composed of the less volatile organics, but also contains some residual Fatty Acid Methyl Ester (FAME). The calorific power of this product makes it interesting as a low-cost fuel, however generating the consequent greenhouse gas emissions. The BHO used in this study presents a density at 15 °C equal to 0.9386 g/cm^3 and a kinematic viscosity at 100 °C equal to 8,135 mm^2/s. Additionally, the chemical characteristics of BHO suggest compatibility with asphalt, with the potential to improve the interactions between bitumen and aggregate, namely the adhesion. Figure 1 shows the Bio Heating Oil appearance at ambient temperature and heated to 40 °C.

3.3 *Bio-asphalt Preparation*

The bio-asphalt was produced simply by blending the BHO into the base binder. Then the blending of the bio-oil and base asphalt binder was implemented by a mixer for 5 min at 135 °C. The rate of the mixing was 200 rotations per minute. The temperature of 135 °C and the mixing time of 5 min were chosen according to previous test experience. Five mixtures of the residue of asphalt and BHO were used. The following BHO contents were used: 2, 4, 6, 8, and 10% by weight. The base binders (residue of asphalt) was also tested as the control asphalt binder.

3.4 *Test Methods*

The physical characterization of bio-asphalt was performed in the laboratory using the following tests: dynamic viscosity at temperatures from 90 to 200 °C (EN 13302), penetration at 25 °C (EN 1426) and softening point (EN 1427). The rheological characterization was performed using the Dynamic Shear Rheometer—DSR according to the European standard EN 14770.

4 Results

4.1 *Dynamic Viscosity*

Dynamic viscosity tests were conducted using the Brookfield Viscometer for temperatures between 90 and 100 °C and for temperatures under 90 °C the Dynamic Shear Rheometer was used, so a viscosity curve (Fig. 2 left) for temperatures from 20 to 200 °C was obtained. As expected for asphalts, reduction in viscosity as a function of temperature is observed for all bio-asphalts tested in this study The addition of BHO to the asphalts promoted a reduction of the dynamic viscosity proportional to the BHO content.

Figure 2 also presents an analysis of the normalized viscosity for the bio-asphalts for 19, 46, 70, 100 and 130 °C. It is observed in this figure the effect of the BHO in

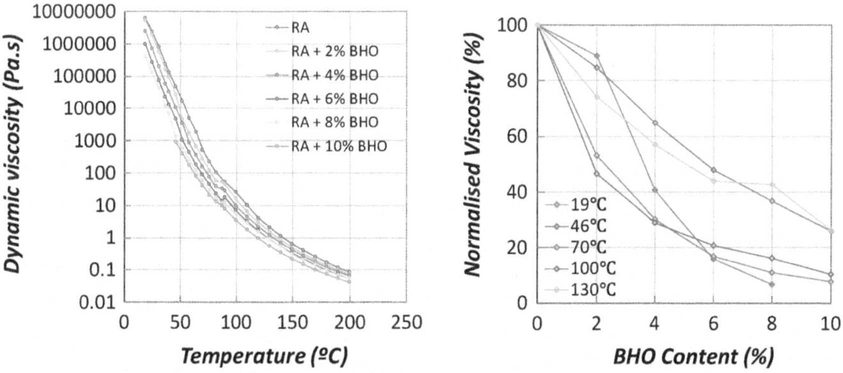

Fig. 2 Dynamic viscosity (left) and normalised viscosity (right) of bio-asphalt produced with the residue of asphalt (RA) base binder

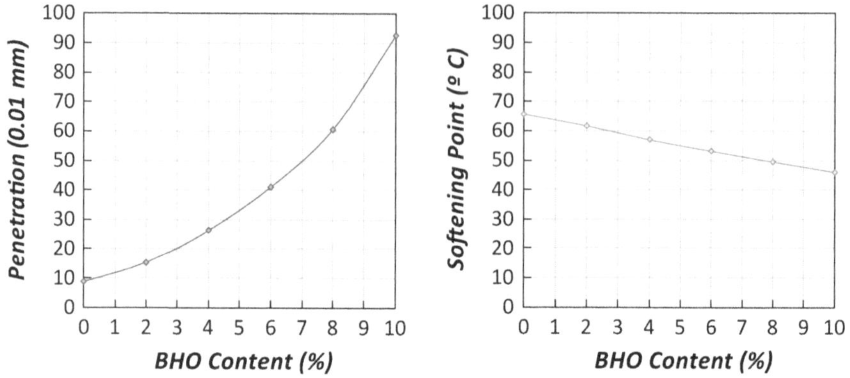

Fig. 3 Penetration at 25 °C (left) and softening point (right) for the bio-asphalt blends

the viscosity of the bio-asphalts. The greater the BHO content is greater, the decrease in the viscosity. The reduction of the viscosity due to the addition of the BHO is more significant as the temperature decreases, as can be observed to the curve for 19 °C.

4.2 Penetration Test and Softening Point

The addition of bio-oil to the base asphalt promoted a reduction in bitumen stiffness, observed with increased penetration for the bio-asphalts, as shown in Fig. 3. Regarding the softening point, the addition of the BHO to the bitumen promoted a bitumen softening observed with the decrease of the temperature determined by the ring and ball method, as shown in Fig. 3. This decrease is linear and proportional to the BHO content added to the base asphalts.

4.3 Rheological Characterization

The rheological characterization was carried out using a Dynamic Shear Rheometer to achieve the dynamic shear modulus (G*) and the phase angle of the bio-asphalt binders. In this work, the tests were conducted using a 25 mm plate and a gap of 1 mm. The temperatures of 46, 52, 58, 64, 70, 76, 82 and 88 °C and a frequency of 1.6 Hz were used. The dynamic shear modulus is used to evaluate the deformation resistance of asphalt binder under repeated shear, and the phase angle is used to evaluate the viscoelastic properties of asphalt binders. Figure 4 presents the results of the dynamic shear modulus and the phase angle of the bio-asphalts produced with the residue of asphalt.

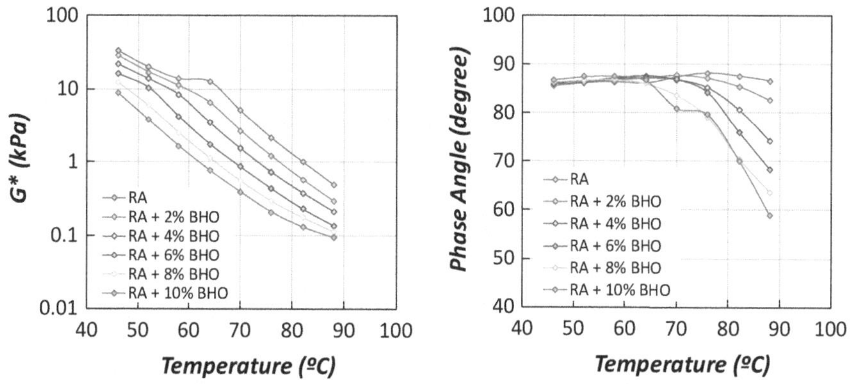

Fig. 4 Dynamic shear modulus, G* (left) and phase angle (right) for bio-asphalts produced with the residue asphalt base binder

The greater is the dynamic shear modulus better its deformation resistance to shear. From Fig. 4, it can be verified that the dynamic shear modulus (G*) of bio-asphalts had presented a decrease proportionally to the BHO content added to the binder. Therefore, the bio-oil contributes to a softening of asphalt binders. The variation is almost logarithmic for low temperatures, reducing as the temperature increases. The viscoelastic properties of the asphalt binder can be evaluated with the phase angle. The smaller the phase angle, the more elastic is the behaviour of the asphalt binder. The higher the phase angle is, the more viscous components have the asphalt binder. Figure 4 also presents the phase angle for the bio-asphalts. It is found that the BHO contents for low temperatures do not influence the phase angle. However, as temperatures and BHO contents increase, the phase angle values decrease.

The DSR test results can also be used to characterize the rutting resistance of the binder. The parameter G*/sin δ (where δ is the phase angle) is obtained based on the dynamic shear modulus and phase angle at 10 rad/s and is usually known as a rutting index of asphalt binder, which is applied to characterize the resistance to rutting at high temperature. The greater the rutting index (G*/sin δ), the stiffer is the asphalt binder, and higher is its ability to resist deformation. Figure 5 presents the effect of the BHO content on the rutting index of the binders. It is verified that the rutting index of BHO modified asphalts is much lower than the neat asphalt, indicating that the BHO can soften the base binder.

Additionally, the high critical temperature that corresponds to the temperature at which the value of the rutting index (G*/sin δ) is 1.0 kPa for both the neat residue of asphalt and the bio-asphalt binders is presented in Fig. 6.

As the residue of asphalt has a greater stiffness than the conventional asphalt, its critical temperature (rutting index equal to 1 kPa) is higher, being 82 °C for the residue of asphalt while for the conventional bitumens (i.e. 35/50 pen asphalt) it is about 70 °C. The softening effect of the bio-oil on the binders is verified by the reduction of the high critical temperatures proportionally to the BHO content.

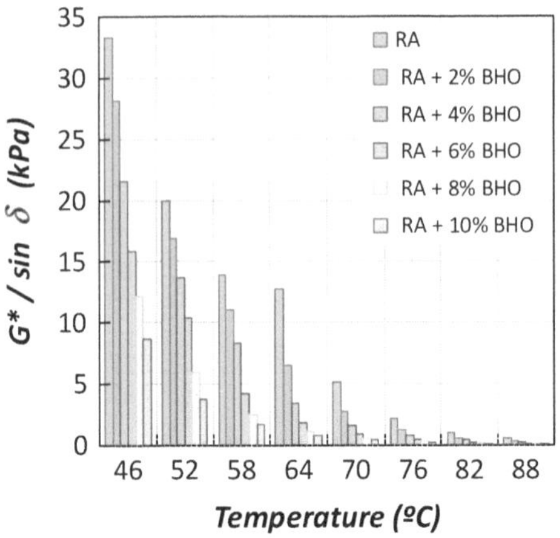

Fig. 5 Rutting index (G*/sinδ) for bio-asphalts produced with the residue asphalt base

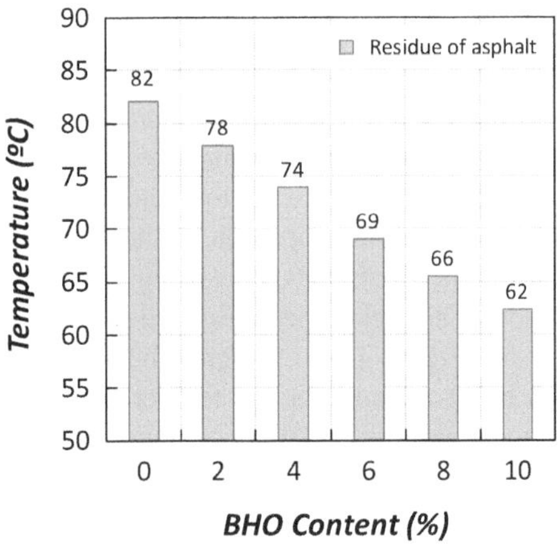

Fig. 6 High critical temperature for bio-asphalts produced with the residue asphalt base

Once the softening effects of BHO on the mechanical and physical characterization of the asphalt binders were verified, it was identified the BHO content added to the residue of asphalt (stiffer binder) to produce a bio-asphalt with same physical and mechanical characteristics of a conventional binder, namely the 35/50 pen asphalt.

It was found that 6% of BHO added to the residue of asphalt can soften the binder by making its characteristics close to the 35/50 pen asphalt.

Table 2 presents the results of the test for the neat 35/50 pen asphalt and the bio-asphalt produced with a residue of asphalt and 6% of BHO content. And in Fig. 7 it is plotted the rutting index vs temperature, highlighting the 35/50 pen asphalt and bio-asphalt (RA + 6%BHO) curves. Thus, regarding the physical and mechanical characteristics of the binders alone, it could be possible to produce bio-asphalts that attend to 35/50 pen asphalt specifications by adding BHO into the residue of asphalt.

Table 2 Characterisation of a 35/50 pen asphalt and bio-asphalt (RA + 6%BHO)

Properties	Units	35/50 pen asphalt	Bio-asphalt (RA + 6%BHO)
Penetration @ 25 °C	0.1 mm	44	41.13
Softening point	°C	48	53.05
Dynamic viscosity @ 135 °C	Pa s	0.6	0.76
High critical temperature	°C	70	69

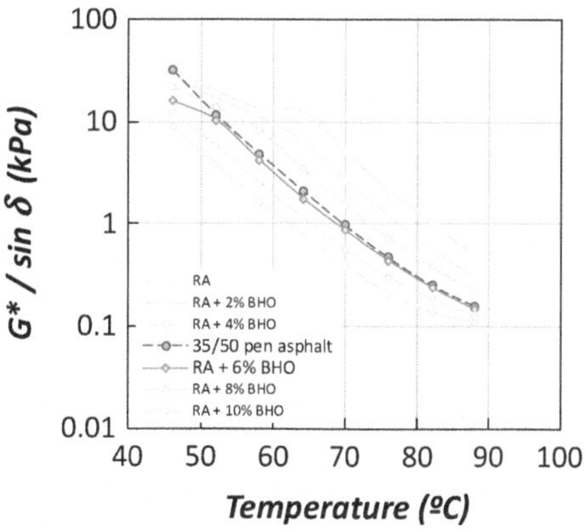

Fig. 7 Rutting index for 35/50 pen asphalt and bio-asphalt (RA + 6%BHO)

5 Conclusions

This study investigated the utilization of bio-oil to produce bio-asphalts. The residue of asphalt was used as the base asphalt binder. The bio-oil used was the BHO, which is a residue from the production of biofuel using waste cooking oil and animal fats as the main raw materials. The physical and mechanical properties of bio-asphalt were evaluated by the penetration, softening point, Brookfield viscosity and rheology tests. The BHO contents studied varied from 2 to 10% of the total binder by weight. According to the results of the tests carried out, the conclusions can be obtained:

- The addition of the BHO softens the asphalt binder. This fact was observed in all tests that were carried out.
- The penetration values increase and the softening point decrease as the BHO content increases.
- The rheology tests show that the dynamic shear moduli and the phase angle decrease as the BHO content is increased.

- The BHO content of 6% appears to be sufficient to soften the residue of asphalt on the physical and mechanical levels of a conventional asphalt binder, namely the 35/50 pen asphalt.

Overall, regarding the physical and mechanical properties, the Bio Heating Oil can be used to soften asphalt binders. However, it is still necessary to verify the effects of the bio-oil in the chemical properties of the bio-asphalts and its storage stability.

References

Brovelli C et al (2013) Rheological characteristics of EVA modified bitumen and their correlations with bitumen concrete properties. Constr Build Mater 48:1202–1208. https://doi.org/10.1016/j.conbuildmat.2013.07.032

Brovelli C et al (2014) Assessment of fatigue resistance of additivated asphalt concrete incorporating fibers and polymers. J Mater Civ Eng 26(3):554–558. https://doi.org/10.1061/(ASCE)MT.1943-5533.0000837

Colbert B, Mohd Hasan MR, You Z (2016) A hybrid strategy in selecting diverse combinations of innovative sustainable materials for asphalt pavements. J Traffic Transp Eng (English Edition) 3(2):89–103. https://doi.org/10.1016/j.jtte.2016.02.001

European Standard EN 13302. Bitumen and bituminous binders - Determination of dynamic viscosity of bituminous binder using a rotating spindle apparatus

European Standard EN 1426. Bitumen and bituminous binders - Determination of needle penetration

European Standard EN 1427. Bitumen and bituminous binders - Determination of the softening point; Ring and Ball method

European Standard EN 14770. Bitumen and bituminous binders. Determination of complex shear modulus and phase angle. Dynamic Shear Rheometer (DSR)

Hu C et al (2018) Waste packaging tape as a novel bitumen modifier for hot-mix asphalt. Constr Build Mater 193:23–31. https://doi.org/10.1016/j.conbuildmat.2018.10.170

Metwally M, Raouf MA (2010) Development of non-petroleum binders derived from fast pyrolysis bio-oils for use in flexible pavement. Iowa State University

Peralta J et al (2010) Rheological quantification of bitumen aging: definition of a new sensitive parameter. Appl Rheol 20(6):63293. https://doi.org/10.3933/ApplRheol-20-63293

Pérez IP et al (2019) Use of lignin biopolymer from industrial waste as bitumen extender for asphalt mixtures. J Clean Prod 220:87–98. https://doi.org/10.1016/j.jclepro.2019.02.082

Su N et al (2018) Productions and applications of bio-asphalts – a review. Constr Build Mater 183:578–591. https://doi.org/10.1016/j.conbuildmat.2018.06.118

Sun Z et al (2016) Properties of asphalt binder modified by bio-oil derived from waste cooking oil. Constr Build Mater 102:496 504. https://doi.org/10.1016/j.conbuildmat.2015.10.173

Yang X, You Z (2015) High temperature performance evaluation of bio-oil modified asphalt binders using the DSR and MSCR tests. Constr Build Mater 76:380–387. https://doi.org/10.1016/j.conbuildmat.2014.11.063

Zargar M et al (2012) Investigation of the possibility of using waste cooking oil as a rejuvenating agent for aged bitumen. J Hazard Mater 233-234:254–258. https://doi.org/10.1016/j.jhazmat.2012.06.021

Zeng M et al (2018) Laboratory evaluation on residue in castor oil production as rejuvenator for aged paving asphalt binder. Constr Build Mater 193:276–285. https://doi.org/10.1016/j.conbuildmat.2018.10.204

Zhang R et al (2017) High temperature performance of SBS modified bio-asphalt. Constr Build Mater 144:99–105. https://doi.org/10.1016/j.conbuildmat.2017.03.103

Zhang R et al (2018) Using bio-based rejuvenator derived from waste wood to recycle old asphalt. Constr Build Mater 189:568–575. https://doi.org/10.1016/j.conbuildmat.2018.08.201
Zhang R et al (2019) The impact of bio-oil as rejuvenator for aged asphalt binder. Constr Build Mater 196:134–143. https://doi.org/10.1016/j.conbuildmat.2018.10.168

Evaluating the Characteristics of Crumb Rubber Modified Asphalt Binders Produced with Neat Bitumen—Case of Kuwait

Taha Ahmed, Dawoud Bahzad, Abdullah Al-Marshed, Zein-Eddine Merouani, and Mohamed Omar

Abstract This research study evaluated the characteristics of crumb rubber modified (CRM) asphalt binders produced with neat bitumen. Two different percentages of crumb rubber (CR); 10 and 15% and one bitumen type with penetration (Pen) grade 60/70 were evaluated. The evaluated characteristics included viscosity, ductility, Penetration, and softening point of the produced asphalt binders. Additionally, images obtained by Scanning Electron Microscopy (SEM) were used to investigate the surface topographies and compositions of the neat and produced CRM binders. SEM images were used to check for separation between the crumb rubber and the bitumen. Furthermore, the Multiple Stress Creep Recovery (MSCR) test was conducted in order to evaluate the rheological properties of the produced CRM asphalt binders. MSCR results showed that the addition of CR to the neat bitumen significantly improved the performance grade (PG) of the produced binder as per AASHTO M332 standards. The traffic loading level for both 10 and 15% CRM asphalt binder increased from "Heavy" for neat to "Extreme" level. Both 10 and 15% CRM binders passed local Kuwaiti requirement for modified binders. Based on AASHTO M332 standards, the 15% CRM binder is said to be not modified with an elastomeric polymer, unlike the 10% CRM binder, which was found to be modified with an acceptable elastomeric polymer. This can be attributed to the lack of full blending between the CR particles and the binder's bulk due to the increase in CR percentage as shown by the SEM images

Keywords Crumb rubber · Modified asphalt · Rheological characteristics · Multiple Stress Creep Recovery · Scanning Electron Microscopy

---

T. Ahmed (✉) · Z.-E. Merouani · M. Omar
Australian College of Kuwait, P.O. Box 1411, Safat, 13015 Mishrif, Kuwait
e-mail: t.ahmed@ack.edu.kw

D. Bahzad · A. Al-Marshed
Petroleum Research Center, Kuwait Institute for Scientific Research, P.O. Box 24885, Safat, 13109 Ahmadi, Kuwait

C. Raab (ed.), *Proceedings of the 9th International Conference on Maintenance and Rehabilitation of Pavements—Mairepav9*, Lecture Notes in Civil Engineering 76,
https://doi.org/10.1007/978-3-030-48679-2_39

1 Introduction

The increasing number of vehicles on the road resulted in a huge number of scrap tires. The disposal of these waste tires have been one of the big environmental problems in the modern society. Currently, there are several industrial applications that incorporate the recycling of waste tires in their processes. Some of these applications include new tires manufacturing, tire-derived fuel production, civil engineering applications and products, molded rubber products, agricultural uses, recreational and sports applications, and crumb rubber modified asphalt applications (Presti 2013; Fontes et al. 2006).

One of main applications of waste tires is the production of crumb rubber modified (CRM) asphalt binder. According to the American Society for Testing and Materials (ASTM) D8-18 standards, crumb rubber modified asphalt, also known as asphalt-rubber binder, can be defined as "a blend of asphalt cement, reclaimed tire rubber, and certain additives in which the rubber component is at least 15% by weight of the total blend and has reacted in the hot asphalt cement sufficiently to cause swelling of the rubber particles" (ASTM D 8-18; ASTM D 6114-97).

Improving the long-term performance of hot mix asphalt (HMA) mixtures means reducing the main leading distress modes to pavement failure. These modes include permanent deformation related distresses such as rutting or shoving, fatigue and thermal cracking, and disintegration related problems such as raveling and stripping. Several studies showed that CRM asphalt binder can improve HMA performance by reducing/delaying the propagation of those main distress modes (Presti 2013; Fontes et al. 2006; Buncher 1995). The added crumb rubber can increase the elasticity of the asphalt binder and reduce the mix sensitivity to temperature changes. Increasing asphalt binder elasticity improves its fatigue characteristics while reducing binder sensitivity to temperature changes reduces its rutting potential by increasing its stiffness. Also, several studies showed that the added crumb rubber can improve the adhesion between the binder and aggregate which reduces the raveling and stripping problems (Buncher 1995; Fontes et al. 2010). Several US states are using CRM asphalt binder in their mix designs specifications such as, California, Arizona, Florida, and Washington.

In order to produce crumb rubber, scrap tires can be processed using two main technologies; ambient grinding, and cryogenic grinding. Figure 1 (left) shows a typical ambient grinding process diagram where scrap tires are processed at room temperature or higher. During the ambient grinding process, the rubber size is being reduced using a granulator by means of cutting and shearing action. The produced CR particles by this method normally have a cut surface shape and are rough in texture with similar dimensions on the cut edges. The ambient grinding method can be considered economic if the bulk of the rubber output needs to be relatively coarse materials, i.e. down to 1.2 mm CR size.

In the case of cryogenic grinding, liquid nitrogen is used to bring tires temperature down to approximately −120 °C before grinding. Below this glass transition temperature, rubber becomes nearly as brittle as glass and size reduction can be

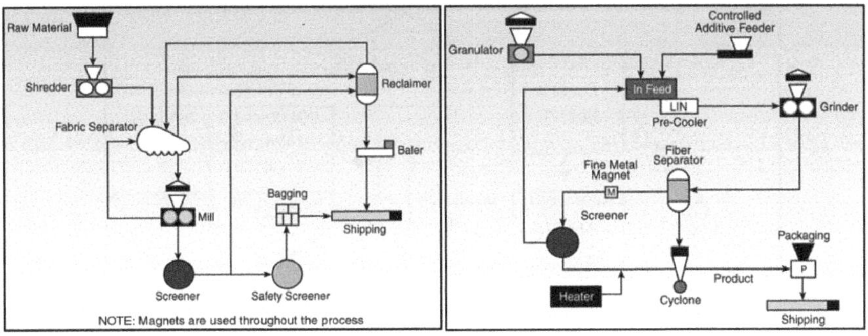

Fig. 1 Typical crumb rubber ambient grinding process (left) and cryogenic grinding process (right) diagrams (Scrap Tire News)

done by crushing and breaking. The resulting CR particles in this process are clean, shiny with fractured surfaces and low steel and fiber content due to the clean breaks between fiber, steel and rubber. Also, the CR particles can be produced with sizes less than 0.85 mm. another advantage of the cryogenic grinding process is that the steel and fiber liberation is much easier, which leads to cleaner product (Fontes et al. 2010; Baker et al. 2003; Scrap Tire News; Way et al. 2011). Figure 1 (right) shows a typical cryogenic grinding process diagram (Scrap Tire News).

Adding crumb rubber to the asphalt binder can then be done through two main methods; wet and dry processes.

1.1 Wet Process

Wet process covers all methods of blending the specified percentage of crumb rubber with the asphalt binder before mixing it with aggregate. This method requires mixing the crumb rubber in hot asphalt binder and holding the blend at a temperature of 190–230 °C (375–450 °F) for a sufficient period of time, typically 45 min, to allow a chemical interaction between the crumb rubber and the asphalt binder (ASTM D 6114-97). This method is known as McDonald's wet process (see Fig. 2) (Presti 2013).

1.2 Dry Process

In the dry process, crumb rubber is mixed with the aggregate prior adding the asphalt binder. The added crumb rubber can substitute 1.0–3.0% of aggregate weight. The crumb rubber acts as a rubber aggregate in the mixture and is added to the plant before the hot asphalt binder is introduced. This method only applies to hot mix

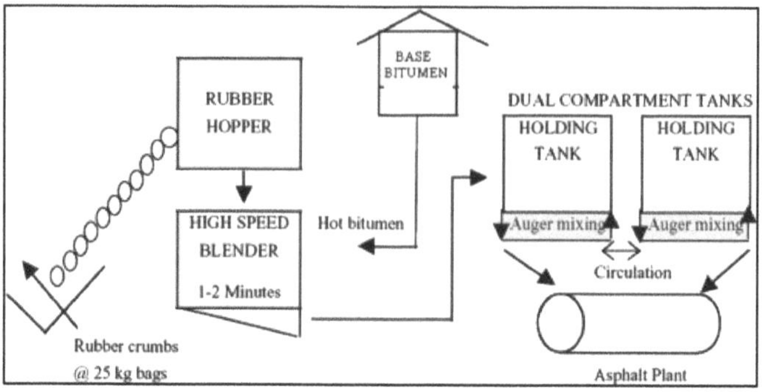

Fig. 2 Schematic diagram of McDonald's wet process (Presti 2013)

asphalt production. The asphalt binder is not considered to be modified in this process because of the limited interaction between the crumb rubber and the asphalt binder.

1.3 Crumb Rubber Modified Asphalt Binder Design

According to the rubber pavement association (RPA), mass production of CRM asphalt binder requires a laboratory approved asphalt binder design profile before the production starts (Way et al. 2011). Initially, the selected asphalt binder performance grade (PG) must be appropriate for the climatic region. Similar to the selection of asphalt binder grade, an appropriate crumb rubber gradation must be chosen to be used with the mix design (Way et al. 2011). The development of asphalt binder profile is a typical process in asphalt mix deign in order evaluate the compatibility and interaction between the used materials. Asphalt binder tests are used to evaluate the rheological characteristics of the CRM asphalt binder. These tests measures the CRM asphalt binder susceptibility to the main failure modes in HMA mixtures, which include permanent deformation related distresses such as rutting or shoving, fatigue and thermal cracking, and disintegration related problems such as raveling and stripping.

2 Research Objectives

Safe disposal of scrap tires is one of the main goals of the Environment Public Authority (EPA) in Kuwait. Therefore, most of the other public authorities in Kuwait guided by EPA are encouraged to use scrap tires in every possible safe application in Kuwait. One of the main applications is incorporating CR material in the production of Hot Mix Asphalt (HMA). This can help in reducing the negative impacts of scrap tires in addition to providing a safer and cleaner alternative to scrap tires disposal

(Green Rubber Tire Recycling Plant Kuwait). CR material can be added directly to the asphalt binder according to the wet or dry mixing process. In order to have a successful HMA with CRM asphalt binder, the properties of CRM asphalt binder should be evaluated and compared to the conventional asphalt binders before using it in the local HMA mixes in Kuwait.

Thus, the main objective of this research study is to evaluate the characteristics of the CRM asphalt binders comparing to the original asphalt binders used in the local HMA mixes in Kuwait. The project started with two percentages of CR material, which were added to the neat bitumen and investigated in order to evaluate their characteristics for potential usage to improve the performance of locally produced bitumen. The work presented in this paper is based on preliminary results derived from an ongoing research project run by the authors of this paper.

3 Materials and Experimental Work

The State of Kuwait uses one bitumen type with 60/70 Pen grade for all HMA mixtures. All HMA mixtures are designed according to Marshall mix design method for heavy traffic level. The used bitumen has an equivalent Superpave performance grade of PG 64-16 (Ministry of Public Works, Roads Administration 2012). Locally produced bitumen Pen 60/70 was mixed with 10 and 15% CR by weight. The crumb rubber materials used in this study were obtained from "Green Rubber Tire Recycling Plant" in Kuwait. The plant uses Ambient Grinding Process in order to produce the crumb rubber. The gradation of the produced CR is shown in Table 1. In order to eliminate the influence of CR gradation on the testing results, CR particles passing sieve number 30 (0.595 mm) and retained on sieve number 50 (0.297 mm) were used in this research study. Challenges were faced during the preparation of CRM asphalt binders. The original effort consisted of heating the asphalt binder to 190 °C before blending it with CR material using a regular mixer at 2,000 rpm for 45 min. However this process didn't result in a homogenous mix. Few more trial was completed and finally, CR was mixed with the bitumen at an interaction temperature of 190 °C for 90 min of reaction (Way et al. 2011). The produced CRM asphalt characteristics were evaluated according to the experimental testing plan in Table 2.

Table 1 Crumb rubber gradation produced by green rubber tire recycling plant

Sieve size (mm)	Percent passing [%]
2.36	99
2.00	91.8
1.18	31.9
0.595	0.6
0.297	0.2
0.075	0.1

Table 2 Experimental testing plan

Test method and parameters	Standards
Penetration of bituminous materials (ASTM D5/DM5 - 19 2019)	ASTM D5/D5M
Ductility of asphalt materials (ASTM D113 - 17 2019)	ASTM D113
Softening point of bitumen (ring-and-ball apparatus) (ASTM D36/D36M 2019)	ASTM D36/D36M
Asphalt binder viscosity (ASTM D4402/D4402M - 15 2019)	ASTM D4402/D4402M
Multiple Stress Creep Recovery (MSCR) of asphalt binder (ASTM D7405-15 2019)	ASTM D7405
Performance-graded asphalt binder using Multiple Stress Creep Recovery (MSCR) test (AASHTO 2017)	AASHTO M332

Additionally, images obtained by Scanning Electron Microscopy (SEM) were used to investigate the surface topography and composition of the produced CRM asphalt. SEM images were used to check for separation between the crumb rubber and the bitumen.

4 Analysis of the Results

Table 3 shows a summary of the test results for the neat and CRM binders. Apparently, the addition of CR material can change the properties of the neat binder, which can lead to not meeting some of the state requirements. Thus, the state of Kuwait requires that all asphalt cement not meeting the requirement of 60/70 Pen grade may be accepted if it complies with viscosity grade AC-40 in accordance with AASHTO M226. Therefore, AC-40 specifications were considered for the CRM binder Pen, viscosity and ductility testing results. Also, the state of Kuwait requires that the modified asphalt binders must meet the MSCR and dynamic shear rheometer (DSR) requirements at 76 °C (Ministry of Public Works, Roads Administration 2012). Therefore, MSCR test was done at 64 °C for the neat binder and 76 °C for the CRM binder. The testing results showed that, adding CR to the neat binder significantly improved the rheological characteristics of the asphalt binder.

The CRM binders met the AC-40 specifications for penetration, ductility and viscosity requirements. It should be noted that there is no requirements for ductility in the AC-40 specifications in Kuwait (Ministry of Public Works, Roads Administration 2012).

For the softening point test, the Asphalt-Rubber Standard Practice Guide by Rubber Pavements Association was used (Way et al. 2011). The softening point test results of the CRM binders were checked against the hot climatic zone requirements, which is similar to Kuwait's climate. It was found that both CRM binders meet the softening point requirements.

Table 3 Summary of the test results

Parameters	Neat	10% CR	15% CR	Specifications
Penetration, mm	70	52	40	60–70 mm for Neat (Ministry of Public Works, Roads Administration 2012) Min. 20 mm for CRM (Ministry of Public Works, Roads Administration 2012)
Ductility @ 25 °C, cm	144	12	7	Min. 100 cm for Neat (Ministry of Public Works, Roads Administration 2012) NA for CRM (Ministry of Public Works, Roads Administration 2012)
Softening point of bitumen, °C	52	58	70	Min. 57 for CRM Hot climatic zone (Way et al. 2011)
Viscosity (Brookfield) @ 135 °C, Cs	832	2,000	3,500	Min. 300 for CRM (Ministry of Public Works, Roads Administration 2012)
Mixing temperatures, °C	167–174 (330–345 °F)	191–198 (375–388 °F)	209–217 (408–422 °F)	Viscosity range (AASHTO 2017) 0.17 ± 0.02 Pa s (170 ± 20 Cs)
Compaction temperatures, °C	155–161 (311–321 °F)	177–183 (350–361 °F)	193–199 (379–390 °F)	Viscosity range (AASHTO 2017) 0.28 ± 0.03 Pa s (280 ± 30 Cs)
Multiple Stress Creep Recovery (MSCR) of asphalt binder				
Non-recoverable creep compliance, $J_{nr3.2}$, kpa^{-1}	1.22 @ 64 °C	0.23 @ 76 °C	0.08 @ 76 °C	Max. 4.0
Percent recovery, %R	19.3	74.6	35.6	Max. 75%
Traffic level	Heavy "H"	Extreme "E"	Extreme "E"	AASHTO M332 (AASHTO 2017)

Production temperatures were determined from the developed viscosity-temperature chart for each binder. Production temperatures for CRM binders were found to be approximately 30 °C higher than the neat binder which originally has high production temperatures.

The non-recoverable creep compliance, $J_{nr3.2}$, and the percentage recovery, %R of each binder were obtained from the MSCR test results and then used to determine the new asphalt binders PGs and traffic levels according to AASHTO M332 standards. As mentioned earlier, the Ministry of Public Works (MPW) in Kuwait states that all modified binders must meet the asphalt binder rheological properties requirements at 76 °C temperature (Ministry of Public Works, Roads Administration 2012), therefore the MSCR was done at 76 °C for CRM asphalt binders and at 64 °C for neat bitumen. The asphalt binder traffic level was found to be increased from “Heavy” for neat binder to “Extreme” for both 10 and 15% CRM binders with an increase to the PG high temperature from 64 to 76 °C.

Furthermore, $J_{nr3.2}$ and %R values can be used together as an indication of elastic response of the modified binder. This can be done by plotting $J_{nr3.2}$ and %R values in the chart provided by AASHTO M332 standards. If the plotted results on the chart falls above the solid line, the binder is said to be “modified with an acceptable elastomeric polymer”; and “if the point falls below the line on the graph, the indication is that the asphalt binder is not modified with an elastomeric polymer” (AASHTO 2017). Figure 3 shows the $J_{nr3.2}$ and %R values of each binder plotted on the chart provided by AASHTO M332 standards. It can be seen that only 10% CRM binder is said to be modified with an acceptable elastomeric polymer. This means that increasing the CR amount to 15% didn’t improve the elastic response of the modified binder. This can be attributed to the improper mixing between the 15% CR and the neat bitumen. Therefore, surface topographies and compositions of the neat and CRM binders were investigated using images obtained by Scanning Electron Microscopy (SEM). Figure 4 shows SEM images of the neat (a), 10% CRM (b) and 15% CRM (c) binders. The SEM images didn’t show significant differences in the surface topography between the neat and the 10% CRM binders, while the 15% CRM binder image showed several CR particles (circled in red), which are not fully blended with the binder’s bulk. This might be the reason for the reduction in elastic response when adding more CR.

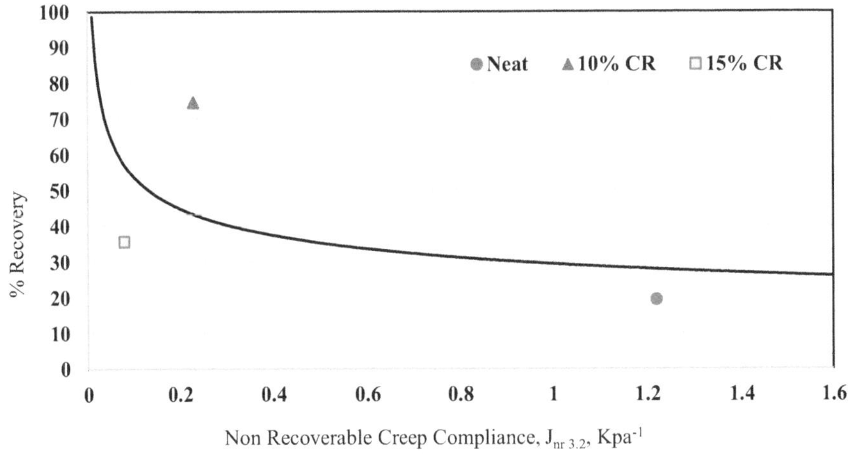

Fig. 3 Indication of elastomeric modification, from AASHTO M332 (AASHTO 2017)

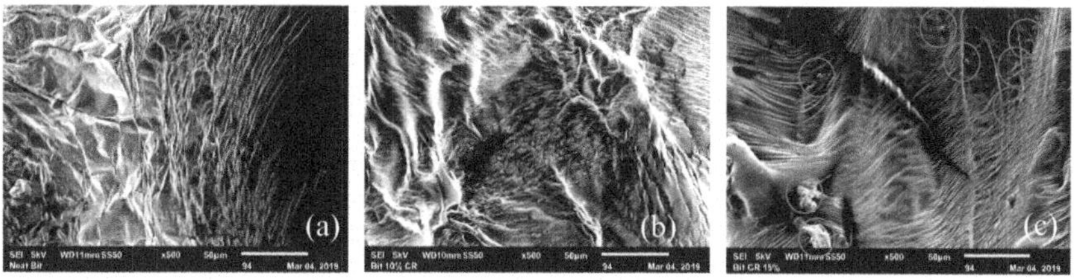

Fig. 4 SEM images of neat (**a**), 10% CRM (**b**) and 15% CRM (**c**) binders

5 Conclusions and Future Work Recommendations

The study aimed to evaluate the characteristics of the CRM asphalt binders produced with local 60/70 Pen grade bitumen. 10 and 15% CR percentages were investigated and their characteristics were checked against the local specifications. Both CRM binders met the local requirements for modified binders. MSCR test results showed a significant improvement to the binders' rheological properties and traffic level, which was increased from "Heavy" for original binder to "Extreme" level for both CRM binders. Based on AASHTO M332 standards, the 15% CRM binder is said to be not modified with an elastomeric polymer, unlike the 10% CRM binder, which was found to be modified with an acceptable elastomeric polymer. This can be attributed to the lake of full blending between the CR particles and the binder's bulk due to the increase in CR percentage as shown by the SEM images. Generally, it can be concluded that 10% CR or less should be used for this specific neat bitumen in order to obtain a CRM binder with acceptable characteristics.

For future work, problems related to 15% of crumb rubber blending process should be further investigated and worked out. A proposed approach, is to consider adding warm mix asphalt (WMA) additives to 15% and higher CR percentages in order to achieve appropriate blending and to reduce the production energy, emissions and carbon footprints. Less CR dosages (4, 6 and 8%) can be also investigated to see their performance comparing to neat bitumen.

Acknowledgements The activity presented in the paper is part of the research grant provided by Kuwait Foundation for the Advancement of Science (KFAS).

References

AASHTO (2017) Standard specifications of transportation materials and methods of sampling and testing. American Association of State Highway and Transportation Officials

ASTM D113 - 17 (2019) Standard test method for ductility of asphalt materials. American Society for Testing and Materials

ASTM D36/D36M (2019) Standard test method for softening point of bitumen (ring-and-ball apparatus). American Society for Testing and Materials

ASTM D4402/D4402M - 15 (2019) Standard test method for viscosity determination of asphalt at elevated temperatures using a rotational viscometer. American Society for Testing and Materials
ASTM D5/DM5 - 19 (2019) Standard test method for pen of bituminous materials. American Society for Testing and Materials
ASTM D 6114-97. Standard specification for asphalt-rubber binder. American Society for Testing and Materials
ASTM D7405-15 (2019) Standard test method for multiple stress creep and recovery (MSCR) of asphalt binder using a dynamic shear rheometer. American Society for Testing and Materials
ASTM D 8-18. Standard terminology relating to materials for roads and pavements. American Society for Testing and Materials
Baker TE, Allen TM, Jenkins DV, Mooney T, Pierce LM, Christie RA, Weston JT (2003) Evaluation of the use of scrap tires in transportation related applications in the State of Washington. Washington State Department of Transportation, Washington, D.C.
Buncher MS (1995) Evaluating the effects of the wet and dry processes for including crumb rubber modifier in hot mix asphalt. PhD thesis, Auburn University, Alabama, August 1995
Fontes LPTL, Pereira PAA, Pais JC, Trichês G (2006) Performance of wet process method alternatives: terminal or continuous blend. In: Proceedings of the asphalt rubber, 2006 conference, Palm Springs, CA, October 2006
Fontes LPTL, Trichês G, Pais JC, Pereira PAA (2010) Evaluating permanent deformation in asphalt rubber mixtures. Constr Build Mater 24(7):1193–1200
Green Rubber Tire Recycling Plant Kuwait. http://www.grrcq8.com/about_us.html. Accessed 25 June 2018
Ministry of Public Works, Roads Administration (2012) General Specifications for Kuwait Roads and Highways. State of Kuwait, January 2012
Presti DL (2013) Recycled tyre rubber modified bitumens for road asphalt mixtures: a literature review. Constr Build Mater 49:863–881
Scrap Tire News. Crumb rubber manufacturing technologies. http://www.scraptirenews.com/crumb.php#overview. Accessed 17 Apr 2018
Way GB, Kaloush KE, Biligiri KP (2011) Asphalt-rubber standard practice guide, 1st edn. Rubber Pavements Association, Final report, October 2011

Properties of Hot Mix Asphalt Containing Treated Recycled Concrete Aggregates Using SCB and ITS Tests

A. Kavussi, F. Kazemian, and M. Bayzidi

Abstract Recycling of Construction and Demolition Waste (CDW) with the purpose of re-using these in construction projects has substantial benefits for the environment and for conservation of natural aggregates resources. Recycled Concrete Aggregates (RCA) that are derived from buildings, concrete structures or distressed PCC pavements can be used in various applications. This paper reports results of the experimental study on the application of demolished concrete aggregates in preparing Hot Mix Asphalt (HMA). RCA materials were used both in their original form; and, treated before that these were added to HMA mix. Two treatment methods were applied to coarse RCA materials. Treatments were consisted of; one, a two-stage treatment of RCAs, using acid and a pozzolan material (calcium metasilicate, named Wollastonite); two, applying hydrated lime solution to RCAs. Physical and mechanical characteristics of the treated/untreated RCAs were determined. Various asphalt mixtures with different amounts of RCAs were prepared. Moisture susceptibility and fracture properties of mixes were determined, using Indirect Tensile and Semi-Circular Bending (SCB) testing methods. Results indicated that although treating RCAs might require more effort in production processing, significant benefits will result in reducing moisture susceptibility and increasing fracture resistance of HMA mixes containing treated RCA materials. It was also found that replacing natural aggregates with RCA, improved fracture properties of asphalt mixes appreciably.

Keywords Construction Demolition Waste (CDW) · Moisture susceptibility · SCB fracture energy

1 Background

Road construction is one of the industries that consumes huge amounts of natural resources around the world. In recent decades, with remarkable rise in rehabilitation and reconstruction projects, the amounts of Construction and Demolition Waste (CDW) materials were increased. The high cost of transportation of natural mineral

A. Kavussi (✉) · F. Kazemian · M. Bayzidi
Faculty of Civil and Environmental Engineering, Tarbiat Modares University, Tehran, Iran
e-mail: kavussia@modares.ac.ir

C. Raab (ed.), *Proceedings of the 9th International Conference on Maintenance and Rehabilitation of Pavements—Mairepav9*, Lecture Notes in Civil Engineering 76,
https://doi.org/10.1007/978-3-030-48679-2_40

aggregates, and the environmental problems of deposition of CDWs, lead to the idea of recycling these waste materials and re-using them in concrete mixes (Kazemian et al. 2019).

Aggregates are extensively used in various civil engineering projects. Over-utilization of aggregates can cause detrimental impacts on the environment. Hence, several strategies were taken by researchers to use CDWs in different types of asphalt mixtures, including hot mix asphalt (Farina et al. 2017; Kavussi et al. 2019).

CDWs are generally composed of different amounts of concrete, masonry, wood, metal, plastics and other materials (Pasandín et al. 2015). For instance, in Tehran, approximately one-fifth of all municipal CDWs are composed of demolished concretes (Asgari et al. 2017). These are produced by processing crushed demolished concrete elements. Based on various benefits from the practical and eco-friendly standpoint application of RCAs, these have attracted interests among many researchers (Hou et al. 2018; Kareem et al. 2018; Ma et al. 2019). Studies indicated that the use of simple RCAs can also cause problems, such as increased pH in surrounding water resources (Gul and Guler 2014). Therefore, development in application and technology of re-using RCAs cannot only provide a solution for re-using waste materials, but it can reduce the application of natural aggregates (NA).

Recycled concrete aggregates, compared with natural mineral aggregates, are characterized by their low apparent density, high porosity, rough surface, high water absorption, high crushing value, and many micro-cracks in their particles. This is due to existence of adhered old mortars on surfaces of RCA particles (Brasileiro et al. 2017; Fatemi and Imaninasab 2016; Ismail and Ramli 2013). Adhered cement mortar and old Interfacial Transition Zone between virgin aggregates and the mortar affect quality of RCA materials (Kazemian et al. 2019). The adhered mortar has typically greater porosity and lower density than NA. This is believed to be the most important reason behind unsatisfactory quality of RCA materials (Al-Bayati et al. 2016). Researchers have conducted several research works on the application of RCA in asphalt mixes, claiming that this incorporation might result in weakness either in terms of mechanical properties (Fatemi and Imaninasab 2016) or durability of mixes (Motter et al. 2015; Wu et al. 2017). These disadvantages have imposed some limitations to the application of recycled concrete in projects. Researchers found that with increased amounts of ordinary RCA materials in asphalt mixtures, properties of these will be dropped appreciably (Al-Bayati et al. 2018). Therefore, various approaches have been employed to enhance performance of recycled aggregates and recycled asphalt mixtures (Al-Bayati et al. 2016; Katz 2004; Spaeth and Tegguer 2013).

Many researchers have considered removing adhered mortar of RCAs by washing these with tap water, in order to remove impurities (Katz 2004). Others have proposed the approach of pre-soaking RCAs in acid (Tam et al. 2007). In a research, strengthening of adhered mortars was achieved, using nano-pozzolanic solutions. In this method, the nano-pozzolanic solution coated RCAs by impregnating these in the slurry material. These effectively enhanced quality of RCA materials (Babu et al. 2015). Some researchers, carried out a mechanical treatment of RCA materials and then applied chemicals to these. With this approach, Kazemian et al. (2019) utilized a removal treatment and then strengthened the adhered mortar, with applying some additives. Other researchers found that treatment materials can form a water-repellent coating on the surface of RCA materials, reducing the absorption of bitumen by RCA particles (Kou and Poon 2010). Other researchers impregnated RCA materials in a lime and silica fume solution (Shayan and Xu 2003). They found that silica fume improved properties of RCA materials greatly. The efficiencies of silica fume is attributed to its high surface area and the activities of its particles (Singh et al. 2013).

Pretreating RCAs with a liquid silicone resin resulted in a treatment method that improved moisture resistance, fatigue performance at low temperatures and rutting resistance of HMA mixes at high service temperatures (Zhu et al. 2012). Pasandín and Pérez indicated that RCA materials perform differently than the conventional mineral (natural aggregates), when hydrated lime is used as an anti-stripping agent. This is because mixtures containing Portland Cement display better RCA-asphalt bond than mixtures that contain hydrated lime (Pasandín et al. 2015). In another work of these researchers, it was reported that coating RCAs with 5% asphalt emulsion provide satisfactory results in terms of stripping resistance of mixes (Pasandín and Pérez 2014).

2 Materials

2.1 *Binder*

Asphalt binder was a 60/70 penetration grade bitumen from Pasargad Oil Company in Iran. Standard laboratory tests were carried out to determine its physical properties. The results of the conventional tests on the bitumen are reported in Table 1.

2.2 *Aggregates*

In this research a conventional mineral aggregate, consisting of 100% crushed particles and a recycled concrete aggregate (RCA) were used to prepare HMA mixes.

Table 1 Standard testing results of the 60/70 pen bitumen used in asphalt mixes

Test	Standard	Result
Flash point (°C)	ASTM D92	316
Softening point (°C)	ASTM D36	53
Loss on heating (%)	ASTM D1754	0.19
Specific gravity (g/cm^3)	ASTM D70	1.02
Ductility at 25 °C; (Cm)	ASTM D113	164
Penetration at 25 °C; (0.1 mm)	ASTM D5	65
Viscosity at 60 °C; (Poise)	ASTM D 2171	1900

The RCA materials were sampled from a Construction and Demolition Waste (CDW) materials site in Tehran Province. From previous research works, it was known that, coarse recycled aggregates, compared with fine ones, have a lower portion of adhered mortar materials (Pasandín and Pérez 2015). Accordingly, it is expected that the coarse RCA particles possess better qualities, compared with the fine RCA materials. In addition, their usage is likely to be more successful than the fine recycled aggregates. This investigation was primarily focused on substituting natural mineral aggregates (NA) with coarse RCA materials. In this context, RCA was defined as the fraction of materials retaining between sieves corresponding with 4.75 and 12.5 mm sizes. Three levels of RCAs, namely 0, 25, and 50% were used to prepare asphalt mixtures.

2.3 Additives

A hydrated lime filler and a calcium metasilicate material (named, Wollastonite) were used to treat the RCAs. Both of these were used in solution condition when were subjected to treat RCA materials. Performing XRF and XRD tests on the additives, their major components were determined, as shown in Table 2.

3 Treatment Methods

RCA materials were subjected to three types of treatment, as are described in the following sections.

Table 2 Chemical composition of the treatment additives

Additive	$Ca(OH)_2$ (%)	MgO (%)	SiO_2 (%)	CaO (%)	Al_2O_3/Fe_2O_3 (%)	Passing sieve no. 200 (%)	Density (g/cm^3)
Hydrated lime	91.14	3	–	72^{Active}	<0.5	99	2.34
Calcium metasilicate (Wollastonite)	–	1.4	50	45.3	1.5	99	2.86

3.1 Pre-treatment

The coarse RCA particles were washed thoroughly so that all noticeable impurities, including wood chips and other similar materials, were removed. Then, these were dried at ambient temperature for 24 h, before being subjected to stages 2 and 3 treatment methods.

3.2 Two-Stage Treatment (Acidic and Pozzolanic Treatment)

The procedure began with soaking RCA for 24 h at ambient temperature in an acidic environment (HCL 0.1 M). Next, the RCA particles were washed and were soaked in 5% calcium metasilicate solution before that these were used it the asphalt mix.

3.3 Hydrated Lime Solution Treatment

In order to reduce stripping susceptibility of the recycled asphalt mixes, hydrated lime was added as the treatment additive. The benefits of hydrated lime is its anti-stripping properties, abundance and ease of application in mixes. RCA particles were impregnated in 6% solution of hydrated lime for 24 h at ambient temperature. Then these were dried before that were added to asphalt mixes.

4 Testing

Asphalt mixtures were prepared and were compacted, applying 75 blows of Marshall Hammer on each side of the cylindrical specimens, following Standard Marshall mix design procedure (ASTM D1559). After mixing the materials and before compacting the samples, these were kept in an oven for two hours at the assigned compaction temperature. This was in order to simulate field conditions, allowing the aggregate particles to absorb asphalt binder. The mixing temperature was set at 163 °C and that of compaction was 150 °C. For RCA mixtures, three levels of 0, 25 and 50% RCA materials were used as partial substitution of coarse NA. The optimum bitumen content was 4.7% which was kept constant in all mixes. The different mix compositions and the assigned nomination of mixes are shown in Table 3.

Table 3 Various mix combinations and the assigned abbreviations

Abbreviation	Description
Control	Natural aggregates
25R	25% RCA
25MTR	25% RCA treated with calcium metasilicate
25LTR	25% RCA treated with hydrated lime
50R	50% RCA
50MTR	50% RCA treated with calcium metasilicate
25LTR	50% RCA treated with hydrated lime

4.1 Tensile Strength

In accordance with AASHTO T-283 Standard procedure, specimens with air voids of 6.5 to 7.5%, were prepared and were conditioned before determination of indirect tensile strength (ITS) and indirect tensile ratio (TSR) parameters. First, vacuum was applied to partially saturated specimens to a level between 55 and 80%. The vacuum-saturated samples were kept under freeze-thaw cycles consisting of −18 °C for 16 h and in water bath for 24 h at 60 °C. After this period, specimens were considered conditioned. The other three samples remained unconditioned. The applied loading rate was 2 in/min (50.8 mm/min). The failure load of each specimen was determined in controlled stress condition. ITS results of the specimens were determined from Eq. (1) below.

$$ITS = \frac{2F}{t\pi d} \tag{1}$$

where, ITS is tensile strength (kPa), F is the failure load (kN), t is the sample thickness (mm), and d is the sample diameter (mm). Indirect Tensile Strength Ratio (TSR) was determined from the following Equation:

$$TSR = 100\left(\frac{F_{Cond}}{F_{Uncond}}\right) \tag{2}$$

where F_{cond} and F_{uncond} are the indirect tensile strength of the conditioned and unconditioned samples, respectively.

4.2 Semi-Circular Bending Strength

Semi Circular Bending (SCB) test was performed according to ASTM D 8044 Standard procedure. This test with three different notches of 25, 32, and 38 mm on samples was conducted using UTM-25 testing machine. The loading was applied in monotonic compression condition at the speed of 0.5 mm/min. Figure 1 shows

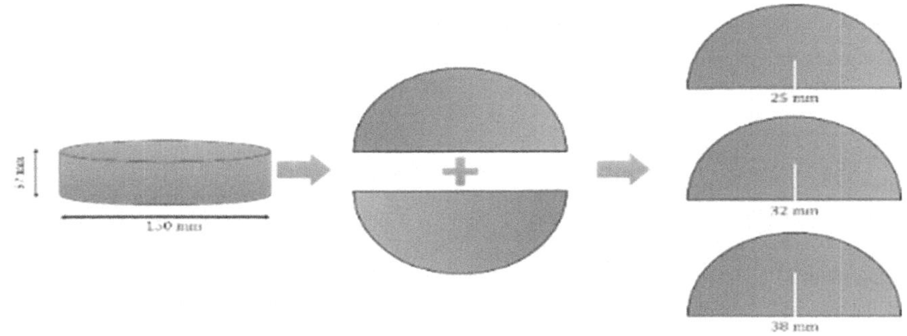

Fig. 1 Preparation of samples in Semi-Circular Bending (SCB) test

both the sample conditions with the assigned notches and the testing unit. From this testing method, J-integral parameter of the samples were determined. This parameter which is a criterion for resistance of materials to cracking and describes fracture behavior of asphalt materials in a nonlinear mode, is determined calculating area under displacement-force curve in SCB testing.

In addition, from the above test, fracture energy parameter of samples, which is defined as their "critical strain energy release rate" were determined using Eq. 3 below.

$$J_C = \frac{-1}{b}\left(\frac{dU}{da}\right) \tag{3}$$

where:

J_c = critical strain energy release rate (kJ/m^2),
b = sample thickness (m),
a = notch depth (m),
U = strain energy to failure (kJ), and
dU/da = change of strain energy with notch depth (kJ/m).

5 Results and Discussion

5.1 Characteristics of Treated RCA Materials

Physical and mechanical characteristics of both selected natural aggregates and the differently treated RCA materials were determined. The results are reported in Table 4. Based on these results, it could be observed that quality of variously treated RCA materials were enhanced appreciably. Physical characteristics of the aggregates, such as their water absorption, is related to durability of RCA materials.

Table 4 Physical properties of the natural and RCA aggregates

Test method	Standard method	Natural aggregate	Untreated RCA	Two-stage treatments	Hydrated lime treatment
Specific gravity (g/cm^3)	AASHTO T96	2.61	2.39	2.52	2.54
Water absorption (%)	ASTM C127	1.05	5.32	4.36	4.5
Abrasion loss (%)	ASTM C131	27	45	37	38
Flatness particles (%)	ASTM D4791	12	15	13	13
Elongation particles (%)	ASTM D4791	13	16	15	16
Aggregate Impact Value (AIV)	BS812	14	23	19	19

While acid solution can remove loose cemented particles from the surface of the aggregates, calcium metasilicate solution decreases water absorption properties of the RCA particles. This was described in detail in a previous research work of the authors (Kazemian et al. 2019). On the other hand, hydrated lime powder will partly fill the pores of RCA materials, resulting again in lower water absorption of the particles.

5.2 *Indirect Tensile Strength (ITS)*

Results of the indirect tensile testing of both treated and untreated RCAs are shown in Fig. 2. These are in dry, one cycle, and three cycle conditions. As it can be seen in this figure, increased level of substitution of natural aggregates with RCAs (i.e. from 25 to 50%) resulted in reduced tensile strength values. As it was mentioned before, water and bitumen absorption of RCAs are greater than those of the natural aggregates. On the other hand, the bitumen contents were constant in all specimens. Therefore, with increasing the amount of RCAs in asphalt mixtures, their effective bitumen content will be decreased. This can cause reduced adhesion between particles and a drop in tensile strength properties of mixes. The magnitude of the indirect tensile strength of asphalt mixes containing treated RCAs were greater than those untreated. This indicates that the treatment can successfully enhance tensile strength properties of the recycled asphalt mixtures. Because of the greater angularity of RCA particles, compared with the natural aggregates, indirect tensile strength of RCA25 and TRCA25 were greater than the control specimen. In contrast, in RCA50

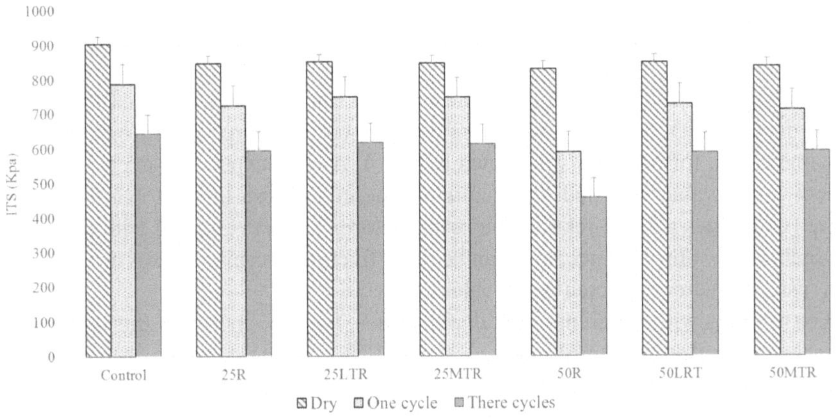

Fig. 2 Indirect tensile strength testing results of HMA mixes containing treated RCAs

and TRCA50 mixes, a significant decrease occurred in effective bitumen content of mixes. This might have resulted in the drop in indirect tensile strength of the specimens. Lime solution causes a reduction in RCAs porosities and increase in adhesion properties between bitumen and aggregate particles. This can successfully enhance tensile strength properties of mixes.

The ratio of indirect tensile strength of samples in wet and dry conditions represents moisture susceptibility of asphalt mixes (Fig. 3). For specimens that experienced one cycle treatment it resulted that all the specimens, except RCA50, met the minimum requirement of moisture resistance (based on Iranian Asphalt Road Pavement Specification). The results indicated that although specimens containing 50% coarse RCAs are prone to moisture damage, specimens containing 50% TRCAs successfully met the moisture resistance requirements.

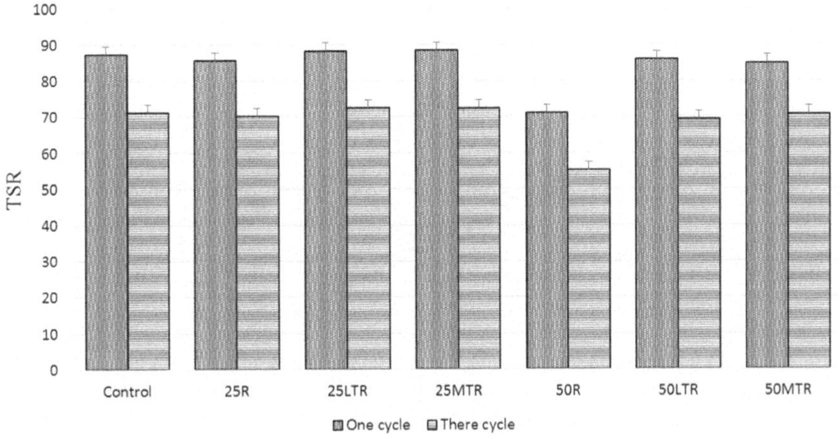

Fig. 3 TSR results of HMA mixes containing various treated RCAs

5.3 Semi-Circular Bending (SCB) Test

SCB testing was performed at 25 °C and Fracture Energy of samples that contained notch depths of 25, 32 and 38 mm were determined. The force-extension curves of specimen 25R, for example, are shown in Fig. 4. As it is expected, specimens with deeper notch values resisted lower forces. Therefore, area under force-extension curve (which has a direct relationship with facture energy) is related with the trend of notch depth values. Hence, as it can be seen, fracture energy of mixes are reduced as a result of increased notch depth values.

For samples with different notch depth values, fracture energy of the specimens were determined and the results are reported in Fig. 5. With reference to this figure, it can be seen that RCA substitution in asphalt mixtures resulted in increased fracture energy, compared with conventional mixes that contained natural mineral aggregates. Fracture energy in 25LTR specimen was decreased 8% only, while in other cases

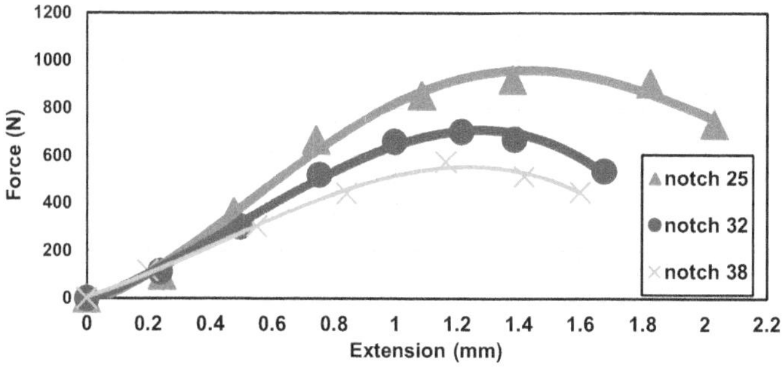

Fig. 4 Force-extension curves for Specimen 25R

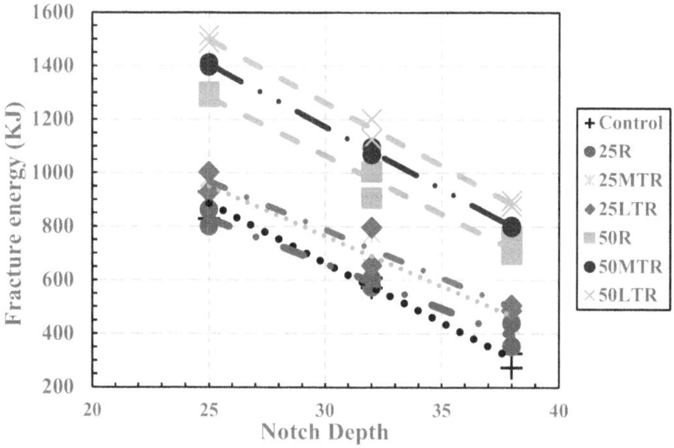

Fig. 5 Fracture energy versus notch depths of the different specimens

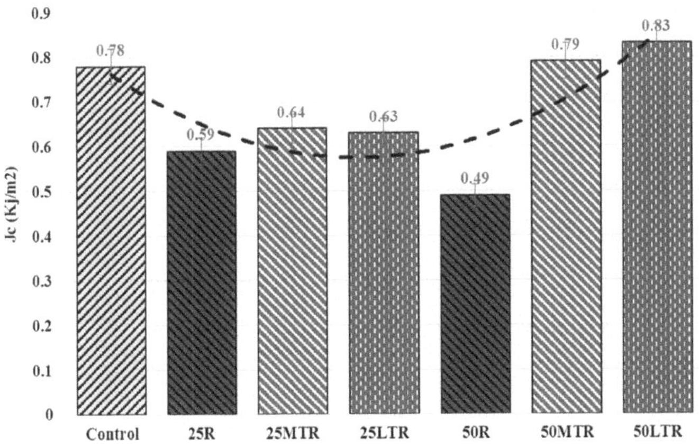

Fig. 6 Critical energy of specimens containing different treated RCAs

an increment in fracture energy was resulted as a result of utilization of RCA in asphalt mixtures. The same trend occurred on samples having different notch depth values. With the same percentages of substitution, fracture energy of asphalt mixtures containing treated RCA materials were greater than those untreated. Results reported in this figure indicates that both treatment methods successfully increased fracture energy of asphalt mixtures, with hydrated lime exhibiting more satisfactory results.

The results of critical strain energy release rate parameter (J_C) of various mixes are shown in Fig. 6. As it can be seen in this figure, J_C values confirm the results that were achieved in fracture energy analysis. This indicates that cracking resistance of asphalt mixes are improved with the addition of treated RCA materials.

6 Conclusions

Asphalt mixtures were prepared, containing 25 and 50% RCA materials of the sizes ranging from 4.75 to 12.5 mm. In order to improve quality of mixes, RCAs were treated with hydrated lime and calcium metasilicate solutions and HMA mixes were prepared. From performing various tests on mixes the following conclusions were drawn.

1. Substitution of RCA materials with conventional natural mineral aggregates in HMA mixes are beneficial both in terms of environmental issues and enhanced mechanical properties of mixes.
2. Limited percentages of RCA materials can be applied in asphalt mixtures without occurring significant changes in performance of asphalt mixes.
3. Both of the treatment methods that were imparted in this research (i.e. application of hydrated lime and calcium metasilicate solutions) resulted in increased

intrinsic properties of RCA materials. Among these reduced water absorption of the particles was quite distinct.
4. Asphalt mixes containing 50% untreated RCA materials showed some moisture susceptibility in HMA mixes. While, asphalt mixes containing 50% treated RCA showed improved moisture resistance.
5. Hot mix asphalt containing RCA exhibited greater fracture energy than conventional mixes that contained natural mineral aggregates.

References

Al-Bayati HKA et al (2016) Evaluation of various treatment methods for enhancing the physical and morphological properties of coarse recycled concrete aggregate. Constr Build Mater 112:284–298

Al-Bayati HKA, Tighe SL, Achebe J (2018) Influence of recycled concrete aggregate on volumetric properties of hot mix asphalt. Resour Conserv Recycl 130:200–214

Asgari A et al (2017) Quality and quantity of construction and demolition waste in Tehran. J Environ Health Sci Eng 15(1):14

Babu VS et al (2015) Strength and durability characteristics of high-strength concrete with recycled aggregate-influence of processing. J Sustain Cement-Based Mater 4(1):54–71

Brasileiro LL et al (2017) Concrete production of hot asphalt using recycled aggregates CDW. Mater Sci Forum 881:346–350

Farina A et al (2017) Life cycle assessment applied to bituminous mixtures containing recycled materials: crumb rubber and reclaimed asphalt pavement. Resour Conserv Recycl 117:204–212

Fatemi S, Imaninasab R (2016) Performance evaluation of recycled asphalt mixtures by construction and demolition waste materials. Constr Build Mater 120:450–456

Gul WA, Guler M (2014) Rutting susceptibility of asphalt concrete with recycled concrete aggregate using revised Marshall procedure. Constr Build Mater 55:341–349

Hou Y et al (2018) Adhesion between asphalt and recycled concrete aggregate and its impact on the properties of asphalt mixture. Materials 11(12):2528

Ismail S, Ramli M (2013) Engineering properties of treated recycled concrete aggregate (RCA) for structural applications. Constr Build Mater 44:464–476

Kareem AI, Nikraz H, Asadi H (2018) Evaluation of the double coated recycled concrete aggregates for hot mix asphalt. Constr Build Mater 172:544–552

Katz A (2004) Treatments for the improvement of recycled aggregate. J Mater Civ Eng 16(6):597–603

Kavussi A et al (2019) Laboratory evaluation of treated recycled concrete aggregate in asphalt mixtures. Int J Pavement Res Technol 12(1):26–32

Kazemian F, Rooholamini H, Hassani A (2019) Mechanical and fracture properties of concrete containing treated and untreated recycled concrete aggregates. Constr Build Mater 209:690–700

Kou S-C, Poon C-S (2010) Properties of concrete prepared with PVA-impregnated recycled concrete aggregates. Cement Concr Compos 32(8):649–654

Ma J et al (2019) Potential of recycled concrete aggregate pretreated with waste cooking oil residue for hot mix asphalt. J Clean Prod 221:469–479

Motter JS, Miranda LFR, Bernucci LLB (2015) Performance of hot mix asphalt concrete produced with coarse recycled concrete aggregate. J Mater Civ Eng 27(11):04015030

Pasandín AR, Pérez I (2014) Mechanical properties of hot-mix asphalt made with recycled concrete aggregates coated with bitumen emulsion. Constr Build Mater 55:350–358

Pasandín AR et al (2015) The effect of hydrated lime on the bond between asphalt and recycled concrete aggregates. Pet Sci Technol 33(10):1141–1148

Shayan A, Xu A (2003) Performance and properties of structural concrete made with recycled concrete aggregate. ACI Mater J Am Concrete Inst 100(5):371–380

Singh LP et al (2013) Beneficial role of nanosilica in cement based materials–A review. Constr Build Mater 47:1069–1077

Spaeth V, Tegguer AD (2013) Improvement of recycled concrete aggregate properties by polymer treatments. Int J Sustain Built Environ 2(2):143–152

Tam VWY, Tam CM, Le KN (2007) Removal of cement mortar remains from recycled aggregate using pre-soaking approaches. Resour Conserv Recycl 50(1):82–101

Wu S, Muhunthan B, Wen H (2017) Investigation of effectiveness of prediction of fatigue life for hot mix asphalt blended with recycled concrete aggregate using monotonic fracture testing. Constr Build Mater 131:50–56

Zhu J et al (2012) Investigation of asphalt mixture containing demolition waste obtained from earthquake-damaged buildings. Constr Build Mater 29:466–475

Bituminous Mixtures with High Environmental Compatibility: Laboratory Investigation on the Use of Reclaimed Asphalt and Steel Slag Aggregates

C. Nodari, M. Crispino, and E. Toraldo

Abstract Recycling waste materials is a key issue in sustainable road pavements construction. This is the reason why the efforts of the scientific and technical community are addressed in studying sustainable solutions for producing bituminous mixtures. As a result of these efforts, the use of Reclaimed Asphalt (RA) is becoming a standard practice, even if in Italy the use of RA is limited to a certain content. So the increase of RA content appears to be the new border of the research in this field. At the same time, the scientific literature demonstrates that other recycled materials, such as Steel Slags Aggregates (SSAs), can be used with satisfactory results. To go a step further, the research herein described focused on the evaluation, at the laboratory scale, of the effects of both high amounts of RA (from 15 up to 60% by weight) and RA plus SSAs (from 55 up to 71% by weight). Compaction properties, volumetric characteristics and mechanical performances were investigated. Even if the obtained results are at the laboratory scale and keeping in mind that the research needs an on-site validation, the investigated mixtures seem to be promising for a future development of bituminous mixtures with high environmental compatibility.

Keywords Reclaimed Asphalt · Steel slag aggregates · Recycled bituminous mixtures · Mechanical performance

1 Introduction and Literature Review

Recycling waste materials is a key issue in sustainable construction and rehabilitation of road pavements. This is the reason why the efforts of the scientific and technical community are addressed in studying sustainable solutions for producing bituminous mixtures (Lopes et al. 2015; Ghabchi et al. 2016; Moghaddam and Baaj 2016; Zhao et al. 2016; Ashouri Taziani et al. 2017; Fakhri and Ahmadi 2017b). As a result of these efforts, the use of Reclaimed Asphalt (RA) is becoming a standard

C. Nodari (✉) · M. Crispino · E. Toraldo
Department of Civil and Environmental Engineering, Politecnico di Milano, Piazza Leonardo da Vinci 32, 20133 Milan, Italy
e-mail: claudia.nodari@polimi.it

C. Raab (ed.), *Proceedings of the 9th International Conference on Maintenance and Rehabilitation of Pavements—Mairepav9*, Lecture Notes in Civil Engineering 76,
https://doi.org/10.1007/978-3-030-48679-2_41

practice (European standard EN 13108-8:2016), even if in several countries, including Italy, the use of RA is limited to a certain content (ANAS 2010; Moghaddam and Baaj 2016). The limitation on the use of RA into new bituminous mixtures is a consequence of the RA composition. In fact, since it derives from a process of removal, milling and crushing of old pavements (Lopes et al. 2015; Moghaddam and Baaj 2016; Fakhri and Ahmadi 2017b), both the RA selection and the rheological behavior of aged bitumen into the RA could be detrimental to the performance of the mixtures in which high contents of RA are included. As for the aged bitumen, the available literature demonstrated that mixtures in which RA is included suffer from problems of hardening, drying and brittleness (Moghaddam and Baaj 2016), this is the reason why rejuvenators and neat bitumen are used with satisfactory results (Silva et al. 2012; Izaks et al. 2015; Liphardt et al. 2015; Moghaddam and Baaj 2016; Ashouri Taziani et al. 2017; Mohammadafzali et al. 2017). Other negative factors affecting these mixtures seem to be the reduction of workability during construction (Zaumanis and Mallick 2015; Fakhri and Ahmadi 2017a), and the increase of pavements susceptibility to cracking (Barco Carrión et al. 2015; Mohammadafzali et al. 2017) resulting in a fatigue life reduction (Moghaddam and Baaj 2016; Fakhri and Ahmadi 2017b; Mohammadafzali et al. 2017). Moreover, other researchers observed an increase of both stiffness (Silva et al. 2012; Ghabchi et al. 2016; Fakhri and Ahmadi 2017a; Pasetto and Baldo 2017), tensile strength (Mohammadafzali et al. 2017) and rutting resistance (Colbert and You 2012; Visintine et al. 2013; Ghabchi et al. 2016). Considering what above reported, the increase of RA content into new bituminous mixtures with no detrimental effects appears to be the new border of the research of sustainable solution for road pavement construction. Obviously, there are other recycled materials potentially suitable for bituminous mixtures' production, such as recycled concrete, recycled asphalt shingles, crumb rubber, plastic bottles, copper slag, waste glass, steel slag, incinerator bottom ashes, etc. (Toraldo and Saponaro 2015; Zhao et al. 2016; Fakhri and Ahmadi 2017a; Lastra-Gonzàlez et al. 2017; Topini et al. 2018). Among these materials, Steel Slags Aggregates (SSAs) appear to be appropriate to replace natural aggregates into bituminous mixtures, because of their ability to increase both stability and skid resistance (Brand and Roesler 2016; Ferreira et al. 2016; Fakhri and Ahmadi 2017a); and mechanical performances (Ferreira et al. 2016), such as fatigue life and rutting resistance (Fakhri and Ahmadi 2017b). An increase of stiffness modulus has been also observed (Pasetto and Baldo 2017). However, other SSAs' properties, such as irregular shape and porous texture, appear to be detrimental in terms of mixtures' workability and binder consumption (Fakhri and Ahmadi 2017a, b). To go a step further, in the recent past some researchers addressed their efforts in studying the concurrent use of both RA and SSAs for the production of bituminous mixtures, with some satisfactory results (Pasetto and Baldo 2012; Fakhri and Ahmadi 2017b; Lastra-Gonzàlez et al. 2017; Pasetto and Baldo 2017).

Given the above mentioned scientific literature and considering that, from one side, the increase of RA content is a need in terms of sustainability, and from the other side that SSAs reveal promising results, the research described in this paper focused on the evaluation, at the laboratory scale, of the effects of both high amounts of RA (from 15 up to 60% by weight) and RA plus SSAs (from 55 up to 71% by weight) into new bituminous mixtures. The laboratory evaluation was performed by means of a set of tests selected to highlight the effects of the use of high amounts of the afore mentioned recycled materials into bituminous mixtures in terms of compaction properties, volumetric characteristics and mechanical performances.

2 Materials and Methods

2.1 Materials

The investigation included the following materials:

- natural lithic aggregates, deriving from an Italian quarry;
- Recycled Asphalt (RA) sourced by milling a highway porous wearing course, sieved in the range of 0–16 mm with an aged bitumen content of 5% by the weight of the RA (EN 12697-1:2012);
- Steel Slags Aggregates (SSAs) deriving from an Italian manufacturer of seamless steel tubes, selected in the range of 0–4, 4–8 and 8–16 mm;
- calcareous filler;
- 50/70 neat bitumen (EN 1426:2015) obtained from a refinery located in Busalla (Genoa);
- amine-base rejuvenator available on the Italian market, specifically tetraethylene-pentamine (TEPA), amber coloured and liquid at ambient temperature.

The main characteristics of natural aggregates, neat and aged bitumen, RA and SSAs are reported in Table 1.

2.2 Experimental Plan and Methods

The experimental laboratory investigation included six bituminous mixtures. Three mixtures were prepared using the only RA: the content of which was 15, 30 and 60% by the total mass, respectively. The other three mixtures were obtained adding SSAs (40, 23 and 11% by the total mass, respectively) to the mixtures with RA, progressively diminishing the virgin aggregate contents. The six mixtures were proportioned according to Table 2, considering an intermediate (binder) layer for road pavement currently used in Italy (ANAS 2010).

The mixtures were prepared using a laboratory mixer, both mixing and compaction temperatures were at 150 ± 5 °C, as suggested by the Italian Specifications (ANAS 2010). Four nominally alike cylindrical specimens (150 mm in diameter, 180 revolutions) of each mixture were prepared by Gyratory Shear Compactor GSC

Table 1 Key material main characteristics

	Bulk density [Mg/m^3] EN 1097-6:2013	LA [%][a] EN 1097-2:2010	FI [%][a] EN 933-3:2012	SI [%][a] EN 933-4:2008	SE [-][a] EN 933-8:2015
Natural aggregates	2.690	20.3	8.11	–	93
Steel slags aggregates	3.860	22.1	0.79	–	71
RA aggregates	2.685	16.2	10.1	9.49	–
	Penetration [0.1 mm] EN 1426: 2015		Softening point [°C] EN 1427: 2015		
Neat bitumen	53		48		
RA bitumen	17		75		

[a]LA: Los Angeles Index; FI: Flakiness Index; SI: Shape Index; SE: Sand Equivalent

Table 2 Mixtures composition

	Natural aggregates [%]	RA (including aged bitumen) [%]	SSA [%]	Total recycled amount [%]	Total bitumen content [%]	Rejuvenator [% RA bitumen]
RA15	80.2	15	–	15	4.8	2.0
RA30	65.3	30	–	30	4.7	2.0
RA60	35.3	60	–	60	4.7	1.2
RA15_SSA40	40.44	15	40	55	4.6	2.0
RA30_SSA23	42.52	30	23	53	4.5	2.0
RA60_SSA11	24.36	60	11	71	4.6	1.2

(EN 12697-31:2007), measuring both self-compaction c_l and workability k during the compaction process.

A complete volumetric characterization of the specimens, including the air voids content (v), the voids in mineral aggregate (VMA) and the voids filled with bitumen (VFB), was also performed.

As first stage of the mechanical performance evaluation, stiffness tests (EN 12697-26:2012—Annex C) and indirect tensile strength tests (EN 12697-23:2018) at 25 °C were carried out. The stiffness behavior was investigated considering three different temperatures (5; 20 and 40 °C) and three load frequencies (1.25; 2.00 and 4.00 Hz). Average stiffness test results are shown in Table 3. As expected, stiffness increases according to temperature decrease and load frequency increase. Moreover, the stiffness increases according to both the RA and SSAs contents. The first is due the increasing of the aged bitumen into the mixtures (according to the RA content), the

Table 3 Average stiffness test results

	Stiffness modulus [GPa]								
Temperature [°C]	5			20			40		
Frequency [Hz]	1.25	2	4	1.25	2	4	1.25	2	4
RA15	18.2	19.3	21.4	8.4	8.9	11.1	1.7	1.9	2.5
RA30	21.9	23.4	25.4	9.7	10.6	12.0	2.1	2.4	3.1
RA60	25.5	26.9	29.2	12.9	13.9	15.5	2.5	2.9	3.7
RA15_SSA40	24.8	26.1	28.4	12.4	13.4	15.2	2.2	2.5	3.1
RA30_SSA23	25.5	26.8	29.3	12.1	13.3	15.1	2.3	2.6	3.5
RA60_SSA11	30.3	30.8	33.6	17.1	18.3	20.4	2.4	2.8	3.7

latter is probably due to the angular shape of the SSAs particles (according to the FI results reported in Table 1), that produce an increase of stone-to-stone contacts into the mixtures.

Then, the stiffness test results were used to calculate the mixtures' Master Curves at reference temperature of 20 °C, according to the time-temperature superposition principle. To do that, three sigmoidal models were compared: Pellinen-Witzack, AASHTO TP-62 and Medani-Huurman. The latter allowed the best data fitting—Eq. (1).

$$\log(S_{mix}) = \log(S_{min}) + \left[\log(S_{max}) - \log(S_{min})\right] * S$$
$$S = 1 - exp\left[-\left(\frac{10 + logfr}{\beta}\right)^{\gamma}\right]. \tag{1}$$

where: S_{mix} is the stiffness modulus obtained by the model [MPa]; S_{min} is the minimum stiffness modulus by laboratory test [MPa]; S_{max} is the maximum stiffness modulus by laboratory test [MPa]; f_r is the reduced frequency [Hz] and β and γ are the shape parameters.

As second stage of the investigation, rutting proneness and fatigue performances were also evaluated on the mixtures with the higher content of recycled materials (RA60 and RA60_SSA11).

Rutting proneness (EN 12697-22:2007—method B in air at 40 °C) of each mixture was measured on two slab samples (0.3 × 0.4 × 0.08 m) obtained by a Rolling Compactor (EN 12697-33:2007).

The fatigue life (EN 12697-24:2012—Annex E) was investigated on cylindrical specimens (three GSC nominally alike replicates for each mixture). The fatigue tests were run at 20 °C, 5 Hz in controlled horizontal stress mode (σ_H equal to 1,000 kPa). The failure criterion was 50% reduction of the initial stiffness.

3 Results and Discussion

3.1 Compaction and Volumetric Characteristics

Average values (including maximum and minimum values as error bars) of both GSC compaction parameters (self-compaction c_l and workability k) and samples' volumetric characteristics (air voids content v, voids in mineral aggregate *VMA* and voids filled with bitumen *VFB*) are shown in Figs. 1 and 2, respectively.

Focusing on the compaction characteristics of the mixtures in which only RA is used, both the investigated parameters seem not to be affected by the RA content; this is reasonably because of the rejuvenator effectiveness. This is not true for the mixtures in which SSAs are used. In fact, an increase of c_l and a decrease of k are appreciable according to the increase of SSAs content.

Moreover, comparing mixture at the same RA content, those in which SSAs are included show higher c_l and lower k. This is probably due to both the high bulk density of the slags (as reported in Table 1), that increase the proneness of the mixtures to be compacted under their own weight, and their shape and angularity, that results as detrimental to the mixtures' ability to be compacted by the GSC. However, a

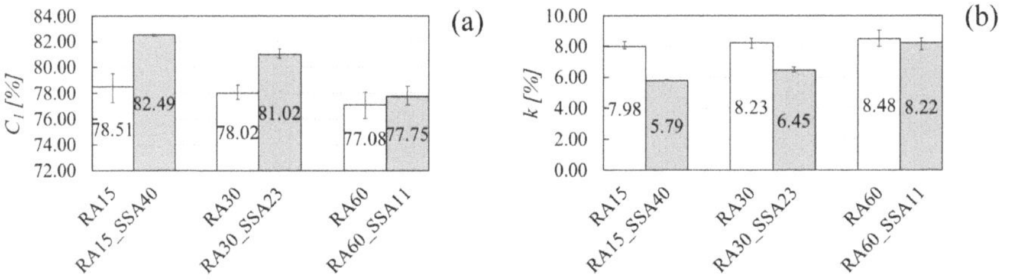

Fig. 1 Compaction test results: **a** self-compaction and **b** intrinsic workability

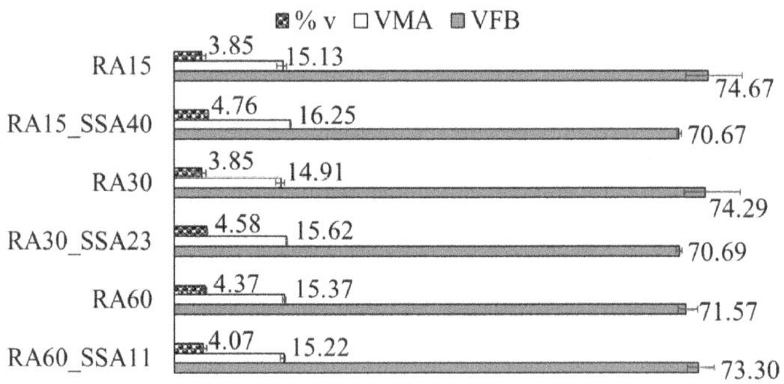

Fig. 2 Volumetric test results: voids content (%v), voids in mineral aggregates (VMA) and voids filled with bitumen (VFB)

sort of balancing effect of the compaction parameters can be expected at the end of compaction, in terms of volumetric characteristics.

The volumetric results reported in Fig. 2 confirm the expectations deriving from the compaction analysis. In fact, the investigated volumetric parameters, air voids content (v), voids in mineral aggregate (*VMA*) and voids filled with bitumen (*VFB*) are quite similar in average, regardless of RA and/or SSAs content. Also the error bars in the graph reveal an insignificant variability of the results among the samples.

3.2 Mechanical Performances

3.2.1 Stiffness Master Curves and Indirect Tensile Strength (ITS)

Mixtures' Master Curves obtained by Medani-Huurman model at reference temperature of 20 °C (including the shifted experimental values) and Indirect Tensile Strength results (average values, including error bars as for the previous compaction and volumetric results) are given in Figs. 3 and 4, respectively. The results in Fig. 3 indicate an increase of the mixtures' frequency sensitivity according to both RA and SSAs amounts, except for the RA in the range of 10–1,000 Hz, where the curves are partially overlapped. More in detail, it is possible to note a stiffer effect of the recycled materials at both high and low frequencies (that, according to the time-temperature superposition principle, means low and high in situ temperatures, respectively). If it is definitely a good result at low frequencies (or high temperatures), because it makes the mixtures able to reduce their rutting potential, it is detrimental considering in situ low temperatures (high frequencies) when the lower stiffness is desirable in order to reduce the mixtures' low temperature cracking proneness.

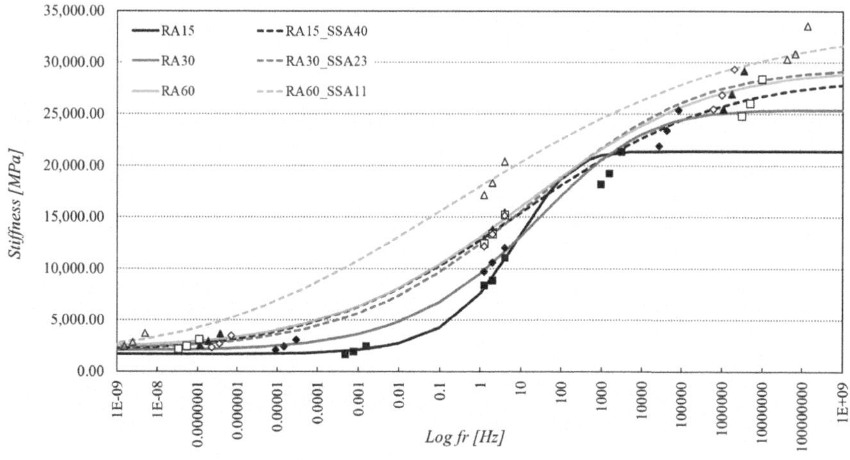

Fig. 3 Master Curves of the investigated mixtures

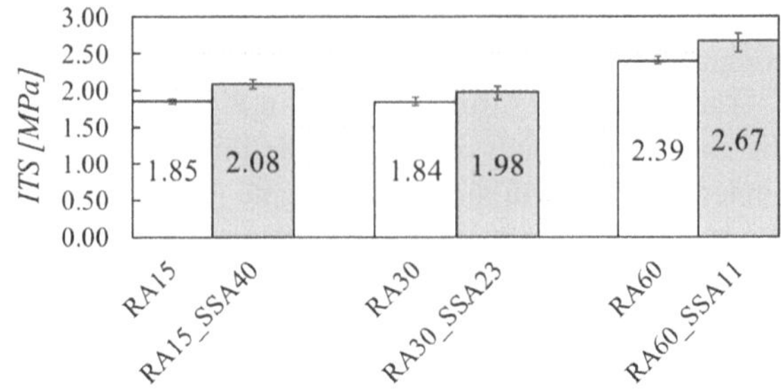

Fig. 4 Indirect tensile strength at 25 °C

Regarding the ITS results reported in Fig. 4, it is possible to note an increase of the performance according to the RA content. It is an expected result and it is due to the increase of the aged bitumen into the mixtures. A slight increase of ITS is also appreciable when the SSAs are used, at the same content of RA (e.g. comparing the mixture RA15 with RA15_SSA40). However, considering the average results and the error bars in the graph, the performance increase has the same order of magnitude as of the test variability.

3.2.2 Fatigue and Rutting Resistance

As a second stage of the investigation, fatigue and rutting tests were performed on the mixtures with the higher content of recycled materials (RA60 and RA60_SSA11). The results are given in Fig. 5. Regarding fatigue resistance (Fig. 5a), a huge variability of test results is demonstrated by the error bars amplitude, it is probably due to the intrinsic variability of the recycled materials into the mixtures. However, considering the average results, the number of cycles at which failure is reached (failure criterion: 50% reduction of the initial stiffness) is lower for the mixture containing steel slags. As regards wheel tracking test results (Fig. 5b), both mixtures present the same *WTS*, in other words the same rutting proneness, but the final rut is quite different. In fact, the mixture containing steel slags (RA60_SSA11) experienced a final rut deformation that is 50% of the one obtained on the RA60 mixture, indicating that the use of steel slags helps to prevent rutting damages in the field.

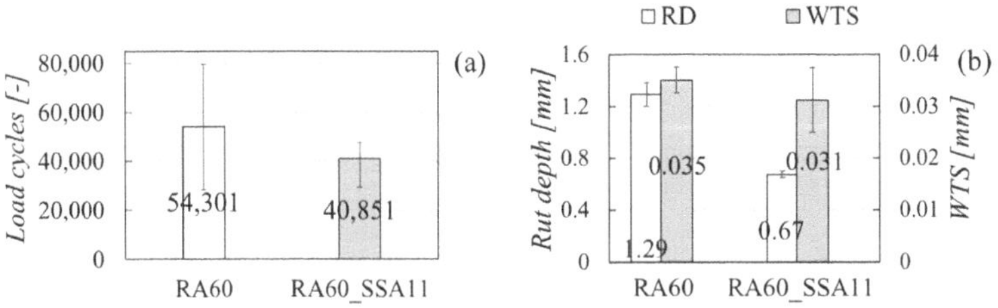

Fig. 5 **a** Fatigue results expressed in terms of number of pulse application to failure and **b** wheel tracking test results

4 Conclusions

The aim of this study was to evaluate, at the laboratory scale, the compaction properties, the volumetric characteristics and the mechanical performances of bituminous mixtures with high recycled materials content, ranging between 15 and 71% by mass, in which Reclaimed Asphalt and Steel Slags Aggregates are included.

As a summary of the obtained results, the following conclusions can be drawn:

- both compaction and volumetric characteristics seem not to be influenced by the increasing of the recycled materials; in this way a crucial role is played by the rejuvenator in reducing the viscosity of the RA aged bitumen; moreover, the compaction characteristics of the mixtures including SSAs are balanced by the bulk density of these slags (that increases the self-compaction proneness) and their shape and angularity (that reduces the workability);
- the stiffness of the mixtures is highly influenced by the recycled materials; in general terms it increases according to the recycled materials content: if it is definitely a good result at low frequencies (or high temperatures), because it could reduce the effects of rutting, it is detrimental considering on site low temperatures (high frequencies) when the lower stiffness is desirable in order to reduce the mixtures' low temperature cracking proneness;
- the ITS results demonstrated that RA increase the mixtures' performance, no significant effects deriving from the use of SSAs are appreciable;
- fatigue and rutting results on the mixtures' having the higher content of recycled materials revealed an unclear trend in terms of fatigue life and a positive role of SSAs in reducing rutting potential.

To conclude, even if the obtained results were at the laboratory scale and a small number of recycled mixtures were investigated (and keeping in mind that the research needs an on-site validation), the investigated mixtures seem to be promising for a future development of bituminous mixtures with high environmental compatibility.

Acknowledgements The Authors gratefully acknowledge Impresa Bacchi s.r.l. for their technical support. Specifically, they express their appreciation to Mr M. Bacchi, Eng. C. Panceri and Eng. L. Trussardi. The support of the Experimental Laboratory of Transportation Infrastructures of Politecnico di Milano is also gratefully acknowledged.

References

Ashouri Taziani E, Toraldo E, Crispino M, Giustozzi F (2017) Application of rejuvenators and virgin bitumen to restore physical and rheological properties of RAP binder. Aust J Civ Eng 15:73–79

Barco Carrión AJD, Lo Presti D, Airey GD (2015) Binder design of high RAP content hot and warm asphalt mixture wearing courses. Road Mater Pavement Des 16(S1):460–474

Brand AS, Roesler JR (2016) Expansive and Concrete Properties of SFS–FRAP aggregates. J Mater Civ Eng 28(2):04015126

Capitolato speciale d'appalto, Parte 2 Norme tecniche, Pavimentazioni stradali/autostradali ed (2010) ANAS S.p.A. (Italian Road Specifications)
Colbert B, You Z (2012) The determination of mechanical performance of laboratory produced hot mix asphalt mixtures using controlled RAP and virgin aggregate size fractions. Constr Build Mater 26:655–662
Fakhri M, Ahmadi A (2017a) Recycling of RAP and steel slag aggregates into the warm mix asphalt: a performance evaluation. Constr Build Mater 147:630–638
Fakhri M, Ahmadi A (2017b) Evaluation of fracture resistance of asphalt mixes involving steel slag and RAP: susceptibility to aging level and freeze and thaw cycles. Constr Build Mater 157:748–756
Ferreira VJ, Saez-De-Guinoa Vilaplana A, García-Armingol T, Aranda-Uson A, Lausín-Gonzalez C, Lopez-Sabiron AM, Ferreira G (2016) Evaluation of the steel slag incorporation as coarse aggregate for road construction: technical requirements and environmental impact assessment. J Clean Prod 130:175–186
Ghabchi R, Singh D, Zaman M, Hossain Z (2016) Laboratory characterisation of asphalt mixes containing RAP and RAS. Int J Pavement Eng 17(9):829–846
Izaks R, Haritonovs V, Klasa I, Zaumanis M (2015) Hot mix asphalt with high RAP content. Procedia Eng 114:676–684
Lastra-González P, Calzada-Pérez MA, Castro-Fresno D, Indacoechea-Vega I (2017) Asphalt mixtures with high rates of recycled aggregates and modified bitumen with rubber at reduced temperature. Road Mater Pavement Des
Liphardt A, Radziszewski P, Król J (2015) Binder blending estimation method in hot mix asphalt with reclaimed asphalt. Procedia Eng 111:502–509
Lopes M, Gabet T, Bernucci L, Mouillet V (2015) Durability of hot and warm asphalt mixtures containing high rates of reclaimed asphalt at laboratory scale. Mater Struct 48:3937–3948
Moghaddam TB, Baaj H (2016) The use of rejuvenating agents in production of recycled hot mix asphalt: a systematic review. Constr Build Mater 114:805–816
Mohammadafzali M, Ali H, Musselman JA, Sholar GA, Massahi A (2017) The effect of aging on the cracking resistance of recycled asphalt. Adv Civ Eng
Pasetto M, Baldo N (2012) Fatigue performance of asphalt concretes with rap aggregates and steel slags. In: RILEM bookseries, vol 4, pp 719–727
Pasetto M, Baldo N (2017) Dissipated energy analysis of four-point bending test on asphalt concretes made with steel slag and RAP. Int J Pavement Res Technol 10(5):446–453
Silva HMRD, Oliveira JRM, Jesus CMG (2012) Are totally recycled hot mix asphalts a sustainable alternative for road paving? Resour Conserv Recycl 60:38–48
Topini D, Toraldo E, Andena L, Mariani E (2018) Use of recycled filler in bituminous mixtures for road pavements. Constr Build MaterConstr Build Mater 159:189–197
Toraldo E, Saponaro S (2015) A road pavement full-scale test track containing stabilized bottom ashes. Environ Technol 36(9):1114–1122
Visintine B, Khosla NP, Tayebali A (2013) Effects of higher percentage of recycled asphalt pavement on pavement performance. Road Mater Pavement Des 14(2):432–437
Zhao S, Huang B, Shu X, Woods ME (2016) Quantitative evaluation of blending and diffusion in high RAP and RAS mixtures. Mater Des 89:1161–1170
Zaumanis M, Mallick RB (2015) Review of very high-content reclaimed asphalt use in plant-produced pavements: state of the art. Int J Pavement Eng 16(1):39–55

Investigation of Selected Properties of Crumb Rubber Modified Bitumens with Different Rubber Contents

E. Manthos, J. Valentin, L. Benešová, D. Giannaka, P. Gravalas, and Ch. Tsakalidis

Abstract The current paper investigates the characteristic properties of three crumb rubber modified bitumens (CRMBs) of Greek origin with different crumb rubber (CR) contents, i.e. 5, 10 and 15%, before and after aging. The virgin bitumen used for the production of the three modified bituminous binders, was a 50/70 paving grade bitumen. The characteristic properties examined were penetration, softening point, dynamic viscosity, storage stability and force ductility. Temperature sweep test was also performed, and certain rheological properties were determined. Additionally, results of characteristic and rheological properties of CRMB which was designed in the Czech Republic, by using activated rubber powder (10 and 15%), were presented. Results showed that although CR addition (5, 10 and 15%) affects the characteristic properties, only 10 and 15% of CR affects the rheological properties before and after aging. The overall effect is mainly hardening of the bitumen, rather than improvement of elastic behavior before aging. Only after aging and with respect to phase angle measurements CRMBs seem to present a more elastic behavior than the virgin bitumen. Activated rubber addition seems to lower the viscosity of the modified bitumen and at some cases (with respect to activation agent content and CR content) improves significantly the storage stability of the CR modified bitumen.

Keywords Rubber · Crumb rubber · Crumb rubber modified bitumen · Temperature sweep test · Activated crumb rubber

1 Introduction

Nowadays crumb rubber (CR) coming from waste tires has been recognized as a waste with multiple uses. CR is commonly used in the following sectors: (i) sport surfaces, (ii) automotive industry, (iii) buildings' construction, (iv) geotechnical/asphalt applications, (v) adhesives and sealants, (vi) shock absorption and safety products

E. Manthos (✉) · D. Giannaka · P. Gravalas · Ch. Tsakalidis
Civil Engineering Department, Aristotle University of Thessaloniki, 54124 Thessaloniki, Greece
e-mail: emanthos@civil.auth.gr

J. Valentin · L. Benešová
Technical University of Prague, Prague, Czech Republic

C. Raab (ed.), *Proceedings of the 9th International Conference on Maintenance and Rehabilitation of Pavements—Mairepav9*, Lecture Notes in Civil Engineering 76,
https://doi.org/10.1007/978-3-030-48679-2_42

and (vii) rubber and plastic products. Asphalt application of CR refers to its use as a constituent of rubberized asphalts. Along the last 60 years that CR is used in the pavement industry, a number of procedures with several variations have been developed by researchers worldwide for CR addition in asphalts. However, all these procedures have originated by the two basic ideas of CR incorporation in pavement construction; incorporation either in the bitumen or in the asphalt mix. In the first case we refer to the "wet process", while in the second case we refer to the "dry process".

In this paper, which presents results of an experimental study, the wet process was used to produce three crumb rubber modified bitumens (CRMBs) with different CR contents, i.e. 5, 10 and 15%. Their characteristic and rheological properties before and after aging were determined and comparison with the respective properties of the virgin bitumen was made. Additionally, results of CRMBs with activated rubber are presented, so to see the possible effect of rubber activation to the CRMB properties.

2 Literature Review

With regards to the fact that crumb rubber is used nearly from the 1960s there are many scientific papers and research reports describing and analyzing the performance either of the CRMBs or of the rubberized asphalts. Given the fact, that the performance of the rubberized asphalts depends mainly on the CRMBs used and also that this paper focuses on the CRMBs performance, literature review will be focused on CRMBs and not on rubberized asphalts.

Depicted overall review papers such as those by Papagiannakis and Lougheed (1995), Huang et al. (2002), Lo Presti and Airey (2013), Lo Presti (2013), Bennert (2013), Mturi et al. (2014), Pratico et al. (2016), Astolfi et al. (2019) and Bressi et al. (2019) give an overall perspective of the history of CR use, the interaction mechanisms between bitumen and CR, the procedures and technologies for CR production and of the research that has been done up until nowadays worldwide.

More detailed scientific papers have described the effect of rubber source, rubber particle and origin of CR. Willis et al. (2013) have investigated the effect of rubber particle size to selected CRMBs properties. They concluded that surface area and particle size were the most influential factors on the increase of critical high temperature grade of the CRMB and that the smaller the rubber size the better the results of the storage stability test. This was not fully confirmed by Soukupová and Valentin (2015). They showed that if very fine crumb rubber is used (CR size up to 0.5 mm), most of CR will float into the bitumen and reach the top of the tube after storage stability test, resulting in an inhomogeneous binder. Neutag and Beckedhal (2011) investigated the performance of CRMBs with 5, 10 and 15% of CR and base bitumens of different origins. Their research showed that bitumen origin has a substantial effect on the CRMBs properties.

Other researchers investigated the effect of CR content to the properties of CRMB. In detail, Mashaan et al. (2011) utilized a 80/100 bitumen in order to make CR blends with different CR contents up to 20%. CR addition resulted in decrease of the penetration and ductility and increase of the elastic recovery. Rheological properties results showed that CR addition resulted in increase of the complex, storage and loss modulus and decrease of phase angle and enhanced the rutting resistance of the bitumen. Similar results observed by Cong et al. (2013), Al-Khateeb and Ramadan (2015) and Nejad et al. (2012) for different base binders and CR contents. Liu et al. (2009) evaluated the performance of CRMBs with different CR types and concluded that among CR type, rubber particle size and CR content, the CR content is the foremost factor affecting the CRMB performance, followed by the type and particle size. Lucaz and Valant (2011) investigated the mechanical and rheological properties of CRMBs with 0.4% amine and 15 and 18% CR. The CRMBs showed increased softening point, and reduced penetration and Fraas breaking point compared to the base binder. Rheological characterization showed that CRMBs exhibited higher viscosity and elasticity.

Research has also been focused on the comparison of CR with other modifiers or on the incorporation of a second material, such as reclaimed asphalt binder or wax additives into the CRMB. Eldouma and Xiaoming (2019) conducted a study on the mechanical performance of bitumens and mixtures utilizing crumb rubber, tafpack super and polypropylene. CR modified binders showed the highest reduction in phase angle and the most significant elastic response (especially for 4% crumb rubber). Girimath et al. (2018) have investigated the possibility of adding reclaimed asphalt bitumen into the CRMB. They concluded that the addition of reclaimed asphalt bitumen has increased the viscosity and improved the thermal susceptibility of the final product. Aging index has also improved but fatigue resistance of the final blends has worsen. Lee et al. (2019) have added WMA wax-based additives to CRMB and found that the blends with the wax additives were effective on increasing rutting resistance and did not show a significant difference on Jnr, %Recovery and %Jnr values during MSCR tests.

3 Materials

The materials used for the production of the CRMBs of the current study were a 50/70 paving grade bitumen and CR coming from car tires (not truck tires). The gradation of the CR and the penetration and softening point values of the 50/70 bitumen are given in Table 1. As can be seen from Table 1, CR particle size is less than 1.18 mm and most of its particles are between 0.3 and 0.6 mm.

4 Experimental Work

Experimental work included the determination of characteristic and rheological properties of the 50/70 binder and of the CRMBs before and after aging (RTFOT) (CEN EN 12607-1, 2014). The characteristic properties determined were: penetration, softening point, viscosity, elastic recovery and force ductility. The rheological properties

Table 1 Gradation of CR and penetration and softening point of virgin bitumen

Crumb rubber particle distribution		50/70 bitumen	
Sieve size (mm)	Passing (%)	Penetration (dmm)	Softening point (°C)
1.18	100.0	66	47.8
0.6	87.7		
0.3	23.0		
0.125	4.6		
0.075	0.8		

determined were the dynamic shear modulus $|G^*|$, the phase angle δ, the dynamic viscosity n* and the rutting parameter G*/sinδ. Additionally by utilizing the G* and phase angle results, the bitumen transition temperature from elastic to viscous condition was determined in all cases.

The blending of the CR and 50/70 virgin bitumen has been made on a very high shear rate mixer of 36.500 rpm for 15 min. The small mixing time was selected due to the very high shear rate of mixing. After mixing, the resulted blends were left with no agitation for 45 min to 160 °C in order take place the chemical reaction between 50/70 bitumen and CR.

5 Characteristic and Rheological Properties of CRMB Before and After Aging

5.1 Characteristic Properties Before and After Aging

The test results of the bitumen characteristic properties are given in Tables 2 and 3. As it can be seen from Table 2, and with respect to penetration and softening point results, the addition of CR has hardened the 50/70 bitumen. The higher the CR content the more the reduction in penetration and increase in softening point.

Elastic recovery test has shown some increase in elasticity with the addition of CR, however the elastic recovery result of CRMB-15 is much lower than those achieved in other studies using 15% CR (Nejad et al. 2012) or with other modifiers such as SBS.

Force ductility results on the CRMBs has shown an increase in the deformation energy with CR content, which ascertains the modification of the virgin bitumen.

Storage stability results show that there was separation between bitumen and CR, which is consistent with the results of other studies and confirms the fact that CRMBs should be stored in constant agitation.

Table 2 Bitumen characteristic properties

Property	Standard	50/70	CRMB-5	CRMB-10	CRMB-15
Penetration at 25 °C (0.1 mm)	CEN EN 1426	66	47	46	43
Softening point (°C)	CEN EN 1427	47.8	51.0	51.6	52.6
Elastic recovery at 25 °C (%)	CEN EN 13398	15	19	27	41
Force ductility at 15 °C (J/cm^2)	CEN EN 13585	–	0.15	0.17	0.22
Dynamic viscosity (Pa s)	CEN EN 13302	See Table 2 and Fig. 2			
Storage stability Diff. in softening point(°C)	CEN EN 13399	–	6.6	7.6	7.8

Table 3 Viscosity results, at the rate of 20 rpm ($6.8\ s^{-1}$)

Bitumen	Testing temperature (°C)					
	130	140	150	160	170	180
	Viscosity (Pa s)					
50/70	0.495	0.305	0.205	0.150	0.115	0.095
CRMB-5	0.585	0.375	0.235	0.165	0.115	0.095
CRMB-10	0.675	0.350	0.225	0.165	0.120	0.090
CRMB-15	0.555	0.350	0.230	0.165	0.120	0.090

Dynamic viscosity has been determined with the use of a rotational viscometer. Testing temperatures were from 130 to 180 °C in steps of 10 °C. As can be seen by Table 3, the viscosity of the virgin bitumen is lower than that of the CRMBs up to 160 °C. Above 160 °C, all bitumens show almost the same viscosity values. There are discrepancies among the viscosity values of CRMBs up to 160 °C and no specific ranking is observed among them

Results of bitumen characteristic properties after aging are given in Tables 4 and 5.

Table 4 Bitumen characteristic properties after aging

Property	Standard	50/70	CRMB-5	CRMB-10	CRMB-15
Penetration at 25 °C (0.1 mm)	EN 1426	42	30	25	21
Softening point (°C)	EN 1427	56.0	60.4	61.2	63.0
Elastic recovery at 25 °C (%)	EN 13398	12	14	22	35
Force ductility at 15 °C (J/cm^2)	EN 13585	–	0.34	0.42	0.55
Dynamic viscosity (Pa s)	EN 13302	See Table 5 and Fig. 1			

Table 5 Viscosity results after aging, at the rate of 20 rpm (6.8 s^{-1})

Bitumen	Testing temperature (°C)					
	130	140	150	160	170	180
	Viscosity (Pa s)					
50/70	0.765	0.415	0.265	0.180	0.125	0.95
CRMB-5	0.995	0.540	0.355	0.220	0.165	0.130
CRMB-10	0.800	0.420	0.265	0.180	0.120	0.080
CRMB-15	0.860	0.505	0.315	0.205	0.140	0.105

As it can be seen from Table 4, and with respect to penetration and softening point results, all bituminous binders have hardened after aging. The aging hardening had an effect on the elastic recovery (reduced in all cases) and to the force ductility (increased in all cases). Dynamic viscosity was higher after aging in all cases (Table 5).

5.2 *Rheological Properties Before and After Aging – Temperature Sweep Test*

Rheological properties of all tested bituminous binders have been determined via the temperature sweep test, by using a dynamic shear rheometer (DSR) with a 25 mm diameter plate and 1 mm gap. Figures 1, 2, 3 and 4 give the results of dynamic shear modulus, phase angle, dynamic viscosity and G*/sinδ, before and after aging.

As it can be seen from dynamic shear modulus $|G^*|$, results (Fig. 1) 50/70 and CRMB-5 bitumens show similar behavior, while the CRMB-10 and CRMB-15 show almost identical behavior. At 33 °C, CRBM-10 and CRBM-15 show distinctively higher dynamic shear modulus values, than those of CRMB-5 and 50/70 binders. In the range of 33–55 °C the differences between the values of $|G^*|$ for the four compared binders are subsequently reduced. It is even visible that the drop in G* is

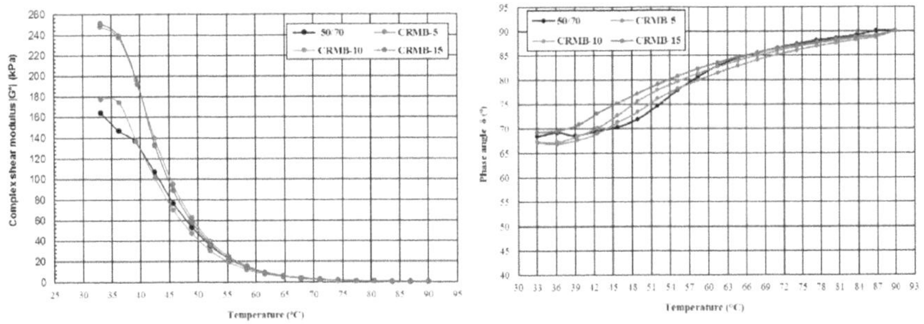

Fig. 1 Dynamic shear modulus $|G^*|$ and phase angle δ results before aging

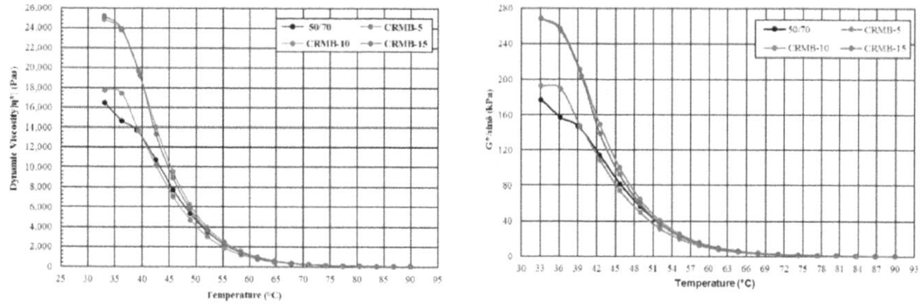

Fig. 2 Dynamic viscosity $|\eta^*|$ and $G^*/\sin\delta$ results before aging

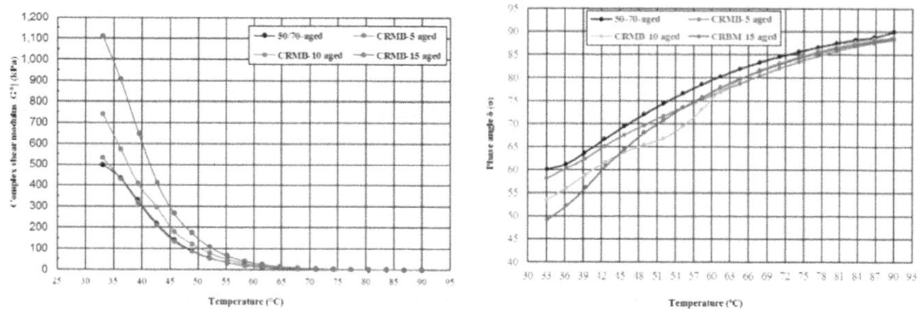

Fig. 3 Dynamic shear modulus $|G^*|$ and phase angle δ results after aging

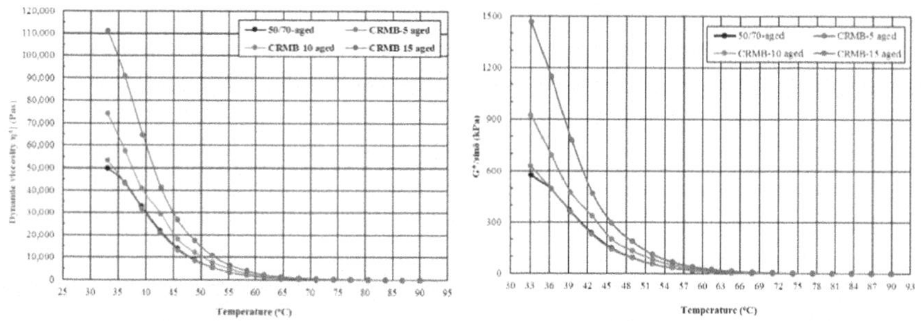

Fig. 4 Dynamic viscosity $|\eta^*|$ and $G^*/\sin\delta$ results after aging

faster for the CRMB-10 and CRMB-15. After 55 °C all tested binders showed almost the same behavior (same values of dynamic shear modulus).

As Fig. 1 shows the more the CR content the higher the phase angle. This is more distinct in temperatures up to almost 75 °C and less distinct in higher temperatures. Phase angle increase with increased CR content, implies that CR addition hardened the bitumen rather than enhanced its elastic behavior.

Phase angle of 50/70 bitumen is lower than the phase angles of CRMBs between 45 and 54 °C. At temperatures higher than 54 °C the 50/70 phase angle increases and above 63 °C reaches values higher than those of the CRMBs.

Dynamic viscosity and $G^*/\sin\delta$ curves of Fig. 2 show similar trends with those of the $|G^*|$. Hence, at 33 °C the dynamic viscosity values of CRBM-10 and CRBM-15 are distinctively higher, than those of CRBM-5 and 50/70 binders. For the temperature range between 33 and 55 °C the differences in the $|\eta^*|$ values are showing similar progress like for dynamic shear modulus.

$G^*/\sin\delta$ is the Superpave rutting parameter, which has been used to measure the rutting resistance of bitumens and to classify them with respect to their high-performance grade. Superpave specifies a minimum limit of $G^*/\sin\delta$ of 1.0 kPa for virgin bitumens and that of 2.2 kPa for aged bitumen by the RTFOT method. Considering the limit of 1.0 kPa and from the analytical DSR results (not shown in the paper), it can be noted that all bitumens (50/70 and CRMBs) satisfy the criterion for temperatures from 33 to 77 °C. At 33 °C the $G^*/\sin\delta$ values of CRMB-10 and CRBM-15 are distinctively higher, than those of CRBM-5 and 50/70 bitumens, indicating a much better rutting performance. In between 33 and 55°C the differences between the $G^*/\sin\delta$ values of the four bitumens are reduced. After 55 °C all bitumens show almost the same $G^*/\sin\delta$.

Figure 3 shows the results of $|G^*|$ and phase angle δ with increasing temperature, after aging. $|G^*|$ values are in all cases higher than those before aging. Bitumens 50/70 and CRMB-5 show similar behavior. However, the $|G^*|$ values of the CRMB-15 are distinctively higher from those of CRMB-10, up to 55 °C. The latter implies that at the certain temperature range (33–55°C) the CRMB-15 will exhibit better performance in terms of deformability.

Phase angle results after aging show that 50/70 bitumen shows the highest values throughout the testing temperature range. This implies that the CRMBs are more elastic after aging than the 50/70 bitumen. It is also noted that for all CRMBs phase angle values throughout the temperature range are lower after aging than before aging, which also shows the positive effect of CR on elasticity after aging. CRMBs show phase angle variation among them and different ranking at different temperatures.

Dynamic viscosity and $G^*/\sin\delta$ curves (Fig. 4) show the same trend with those of the $|G^*|$. Hence, at 33 °C the CRMB-10 and CRMB-15 show distinctively higher dynamic viscosity values, than those of CRMB-5 and 50/70 bitumen. Between 33 °C and 55 °C the differences between the $|\eta^*|$ values of the four bitumens are reduced. After 55 °C all bitumens show almost the same dynamic viscosity. In all tested bitumen variants the viscosity values of the aged binders are higher than those of the unaged bitumen.

The $G^*/\sin\delta$ criterion of 2.2 kPa, is satisfied for the 50/70 bitumen at temperatures lower than 71 °C, for the CRMB-5 and CRMB-10 at temperatures lower than 74.2 °C and for the CRMB-15 at temperatures lower than 77.3 °C (from the analytical results of DSR, not shown in the paper). The above shows an improvement of the resistance to rutting after aging as the CR content increases.

By utilizing the results of $|G^*|$ and phase angle, the temperature at which the bitumen transits from the elastic to the viscous region, can be determined. The transition

Table 6 Bitumen transition temperature from elastic to viscous condition

Bitumen	Transition temperature (°C)	
	Before aging	After aging
50/70	38.8	38.5
CRMB-5	38.8	37.8
CRMB-10	40.5	37.5
CRMB-15	40.0	38.3

temperature is defined as the intersection point of $|G^*|$ and phase angle curves. The results for all bituminous binders tested are given in Table 6.

As Table 6 shows before aging the transition temperature is the same for 50/70 and CRMB-5 binder. This is consistent with the $|G^*|$, $|\eta^*|$ and $G^*/\sin\delta$ curves, where the two binders showed similar behaviour. Even further, it could be said that 5% of CR has not affected the rheological behaviour of the 50/70 bitumen. CRMB-10 and CRMB-15 show higher transition temperature, almost the same for both of them.

After aging, for all bitumens tested transition temperatures are in the region of 38 °C. CRMBs transition temperatures are 1–3 °C lower compared to those before aging.

6 Use of Activated Rubber Powder

This part aims to assess the characteristics of CRMBs composed by 50/70 paving grade bitumen, mechanically activated fine-ground rubber (CR_k) with maximum particle size of 0.8 mm, as well as a new type of benzothiazol-based activating catalyst (AKH). This type of crumb rubber with activating catalyst (chemically activated crumb rubber) is developed for several years in the Czech Republic. For comparison reasons, different blends with 15% CR and one with 10% CR were produced. The resulting values of the basic empirical tests, including the dynamic viscosity, are summarized in Table 7 and can be briefly interpreted as follows.

The presence of rubber in the binder always hinders penetration but with respect to the content of activating catalyst, no clear dependence has been observed. In the case of the softening point, the presence of pulverized rubber results in a higher value; in comparison to the base bitumen (50/70), the increase may amount up to 15 °C. No clear relationship between the content of the activating catalyst and the softening point has been observed, either. Generally, when compared to the base binder, the use of activated pulverized rubber shows improved penetration index in the sense of increased binder stiffness, therefore improved resistance to deformation can be expected. In the assessment of elastic recovery of modified binders, it must primarily be noted that none of the CRMBs reached the 50% minimum value adopted by the specifications, implemented in Germany as technical requirements E GmBA (2012) or by Czech national standard ČSN 65 7222-2 (2017). Some blends exhibited storage stability results where the differences in softening points did not exceed 3 °C and

Table 7 Results of basic empirical bitumen tests and dynamic viscosity

Sample	Pen	R&B	PI	Elastic recovery	Storage stability	Dyn. Visc. @ 6,8 s^{-1}	
	[0,1 mm]	[°C]	[-]	[%]	[°C]	135 °C [Pa s]	150 °C [Pa s]
50/70	71.2	47.5	−1.0	4.6	–	0.5	0.3
15% CR_K	41.0	63.1	1.1	35.1	−8.4	4.5	3.2
15% CR_K + 10% AKH	44.8	60.2	0.8	42.8	2.7	3.9	2.9
15% CR_K + 5% AKH	46.2	60.0	0.8	41.8	1.6	4.3	3.4
15% CR_K + 2.5% AKH	42.2	62.3	1.1	46.0	−12.5	4.9	3.8
10% CR_K + 2.5% AKH	49.6	55.7	0.1	31.0	–	1.9	1.2

can therefore be considered as fully storage stable. From the dynamic viscosity test results, it is obvious that higher applied contents of the AKH reduce the viscosity value by roughly 10% in comparison to binders without such catalyst.

From measured MSCR values (Table 8), it can be noted that the bituminous binder with 15% CR_K without activating catalyst, or the bituminous binder with 15% CR_K with activating catalyst 2.5% by mass, perform the best. It is obvious that the J_{nr} value increases with the increase of catalyst quantity.

Table 8 MSCR test results for assessed CRMB binders

Sample	0.1 kPa		3.2 kPa	
	Elastic recovery [%]	J_{nr} [kPa^{-1}]	Elastic recovery [%]	J_{nr} [kPa^{-1}]
50/70	0.60	5.73	0.02	5.97
15% CR_K	34.83	0.39	9.46	0.60
15% CR_K + 10% AKH	30.35	0.61	7.90	0.89
15% CR_K + 5% AKH	33.53	0.47	9.22	0.72
15% CR_K + 2.5% AKH	38.99	0.33	11.84	0.53
10% CR_K + 2.5% AKH	28.03	0.82	3.27	1.35

Table 9 Dynamic shear modulus and phase angle values for T = 60 °C and f = 1.59 Hz

Sample	G′	G″	G*	δ	G*/sin(δ)	G*x sin(δ)
	[kPa]	[kPa]	[kPa]	[°]	[kPa]	[kPa]
50/70	0.2	3.0	3.0	86.5	3.0	3.0
15% CR_K	4.8	13.7	14.6	70.5	15.4	13.7
15% CR_K + 10% AKH	3.8	11.0	11.6	70.8	12.2	10.9
15% CR_K + 5% AKH	4.1	11.5	12.2	70.5	12.9	11.5
15% CR_K + 2.5% AKH	5.4	13.6	14.6	68.4	15.6	13.5
10% CR_K + 2.5% AKH	1.8	7.6	7.8	76.9	8.0	7.6

It is also noticeable that the content of CR_k affects the value of irreversible shear compliance and higher contents result in improved deformation characteristics By evaluating the Frequency Sweep test results, Table 9, it can be confirmed that the CR_k content affects both the dynamic shear modulus and the phase angle which indicates the degree of bitumen elasticity. Increasing content of AKH in the resulting CRMB reduces the dynamic shear modulus $|G^*|$,—this shows a slightly negative impact on deformation characteristics.

7 Conclusions

The incorporation of 5–15% CR to a 50/70 bitumen has resulted in harder CRMBs but with low elastic recovery behaviour and storage stability deficiency. Rheological properties determination also verified the bitumen hardening (as CR content increased) and the not highly improved elastic behaviour.

All CRMBs have hardened after aging. With respect to dynamic shear modulus $|G^*|$, dynamic viscosity and G*/sinδ after aging, CRMB-15 showed the highest values from 33 °C to 55 °C and thus is expected to perform better in terms of deformability than all the other bitumens tested. Phase angle determination after aging, showed increased elastic behaviour of the CRMBs and more elastic behaviour than the 50/70 bitumen, throughout the testing temperature range.

With respect to CRMBs with activated rubber, the new type of activating catalyst has, a positive effect on the resulting characteristics of the modified bitumen if appropriate quantity of such additive is used. This confirms the often-repeated finding that a higher proportion of activated rubber powder (ARP) in the bitumen improves the deformation characteristics of the material on one hand while, on the other hand, it reduces the dynamic viscosity value which, however, is an assumed dependence. It is shown that in comparison to a CRMB without the catalyst, the addition of a catalyst improves homogeneity, primarily in cases where 5–10 M% of the activating catalyst is added.

The tests carried out on DSR to predict the deformation behavior show that modified bitumens achieve very good values when using only 15% crushed rubber is used, or when the 15% rubber and 2.5% activating catalyst are mixed.

References

Al-Khateeb GG, Ramadan KZ (2015) Investigation of the effect of rubber on rheological properties of asphalt binders using superpave DSR. KSCE J Civ Eng 19(1):127–135

Astolfi A, Subhy A, Praticò F, Lo Presti D (2019) Quality-control procedure for dry-process rubberised asphalt mastics. In: Bituminous mixtures and pavements VII, proceedings of the 7th international conference 'bituminous mixtures and pavements' (7ICONFBMP), Thessaloniki, Greece, pp 560–567

Bennert T (2013) Grade determination of crumb rubber-modified performance graded asphalt binder (No. C-08-20). University Transportation Research Center

Bressi S, Fiorentini N, Huang J, Losa M (2019) Crumb rubber modifier in road asphalt pavements: state of the art and statistics. Coatings 9(6):384

Czech Agency for Standardization (2017) ČSN 65 7222-2: bitumen and bituminous binders – modified bitumens – Part 2: crumb rubber modified bitumens

Cong P, Xun P, Xing M, Chen S (2013) Investigation of asphalt binder containing various crumb rubbers and asphalts. Constr Build Mater 40:632–641

Eldouma IB, Xiaoming H (2019) Comparative study of the mechanical performance of bitumen binders and mixtures utilizing crumb rubber, tafpack super, and polypropylene. J Civil Environ Eng 9:330. https://doi.org/10.4172/2165-784X.1000330

FGSV (2012) Empfehlungen zu Gummimodifizierten Bitumen und Asphalten (E GmBA). Version 2012 and 2016. FGSV Verlag. ISBN 978-3-86446-040-1

Girimath S, Singh D, Manthos E, Mampearachchi WK (2018) Effects of reclaimed asphalt binder on rheological properties and cohesion energy of crumb rubber modified binder. Innov Infrastruct Solut 3(1):57

Huang B, Mohammad LN, Graves PS, Abadie C (2002) Louisiana experience with crumb rubber-modified hot-mix asphalt pavement. Transp Res Rec 1789(1):1–13

Lee SJ, Kwon SA, Jeon SI, Lee MS (2019) MSCR analysis of rubberized asphalt binders containing wax additives. In: Bituminous mixtures and pavements VII, proceedings of the 7th international conference 'bituminous mixtures and pavements'

Liu S, Cao W, Fang J, Shang S (2009) Variance analysis and performance evaluation of different crumb rubber modified (CRM) asphalt. Constr Build Mater 23(7):2701–2708

Lo Presti D, Airey G (2013) Tyre rubber-modified bitumens development: the effect of varying processing conditions. Road Mater Pavement Des 14(4):888–900

Lo Presti DL (2013) Recycled tyre rubber modified bitumens for road asphalt mixtures: a literature review. Constr Build Mater 49:863–881

Lukac B, Valant AZ (2011) The investigation of rubber modified bitumen as a binder for the production of asphalt mixture. In: Proceedings of the 5th international conference on bituminous mixtures and pavements. Thessaloniki, Greece

Mashaan NS, Ali AH, Karim MR, Abdelaziz M (2011) Effect of crumb rubber concentration on the physical and rheological properties of rubberised bitumen binders. Int J Phys Sci 6(4):684–690

Mturi GA, O'Connell J, Zoorob SE, De Beer M (2014) A study of crumb rubber modified bitumen used in South Africa. Road Mater Pavement Des 15(4):774–790

Nejad FM, Aghajani P, Modarres A, Firoozifar H (2012) Investigating the properties of crumb rubber modified bitumen using classic and SHRP testing methods. Constr Build Mater 26(1):481–489

Neutag L, Beckedahl HJ (2011) Performance of crumb rubber modified binders and asphalts. In: Proceedings of the 5th international conference on bituminous mixtures and pavements, Thessaloniki, Greece

Papagiannakis AT, Lougheed TJ (1995) A review of crumb-rubber modified asphalt concrete technology. Res. report for project T9902-09 "Rubber-Asphalt Study", Washington State Transp. Commission and US Department of Transp.

Pratico FG, Moro A, Noto S, Colicchio G (2016) Three-year investigation on hot and cold mixes with rubber. In: 8th International conference on maintenance and rehabilitation of pavements (MAIREPAV8), Singapore

Soukupová L, Valentin J (2015) Posouzení charakteristik CRMB pojiv s využitím aktivo-vané jemně mleté pryže a nového typu aktivačního katalyzátoru AKH. Research Report (only in Czech). Faculty of Civil Engineering CTU in Prague, 17 pages. Report No. WP1_01_2015

Willis JR, Turner P, Plemmons C, Rodezno C, Rosenmayer T, Daranga C, Carlson D (2013) Effect of rubber characteristics on asphalt binder properties. Road Mater Pavement Des 14(sup2):214–230. https://doi.org/10.1080/14680629.2013.812845

Performance Assessment of Rubberized Mixtures Containing Reclaimed Asphalt and a Viscosity Reduction Additive

Leonardo Urbano, Davide Dalmazzo, Pier Paolo Riviera, and Ezio Santagata

Abstract This paper illustrates the results obtained in a study which focused on the mix design and performance-related characterization of rubberized dense-graded mixtures containing a viscosity reduction additive and different percentages of reclaimed asphalt. Following preliminary mix design, plant-produced mixtures were employed for the construction of full-scale trial sections and were subjected to standard quality assurance tests and to advanced mechanical characterization tests. Analysis of results was performed by referring to typical acceptance thresholds and to the outcomes of simplified viscoelastic continuum damage modelling. It was concluded that reclaimed asphalt can have a non-negligible effect on performance properties and that the considered rubberized mixtures, which can be compacted at reduced temperatures, are suitable for paving applications.

Keywords Rubberized asphalt mixture · Reclaimed asphalt · Viscosity reduction additive · Dynamic modulus · Flow number · Fatigue · Viscoelastic continuum damage

1 Introduction

Recycling of waste materials and preservation of the ecosystem and of natural resources are fundamental needs which have to be taken into account in the design of today's civil infrastructures. In such a context, new asphalt pavements are designed in a framework in which sustainability is a key concept. In the specific case of bituminous mixtures employed in paving applications, the combined use of crumb rubber (CR) derived from end-of-life tires as a modifier of asphalt binders and of reclaimed asphalt (RA) as a partial substitute of virgin aggregates is a technology that can lead to significant benefits in terms of energy saving, environmental impact, human health, and preservation of ecosystems (Farina et al. 2017; Pouranian and Shishehbor 2019). Furthermore, rubberized asphalt mixtures containing RA have

L. Urbano · D. Dalmazzo · P. P. Riviera · E. Santagata (✉)
Department of Environment, Land and Infrastructure Engineering, Politecnico di Torino, 24, Corso Duca degli Abruzzi, 10129 Turin, Italy
e-mail: ezio.santagata@polito.it

C. Raab (ed.), *Proceedings of the 9th International Conference on Maintenance and Rehabilitation of Pavements—Mairepav9*, Lecture Notes in Civil Engineering 76,
https://doi.org/10.1007/978-3-030-48679-2_43

shown improved mechanical properties and performance in terms of indirect tensile strength, moisture susceptibility, resilient modulus and rutting resistance (Xiao et al. 2007; Xiao and Amirkhanian 2008). In these mixtures the viscosity of the binder is greatly increased by the presence of CR particles (Xiao and Amirkhanian 2008; Santagata et al. 2012), thus leading to a significant increase of mixing and compaction temperatures. The use of viscosity reduction additives, conventionally used for warm mix asphalt, can help to reduce these temperatures thereby leading to a reduction of emissions and to a lower energy consumption in mix production and compaction. The adoption of such a solution is encouraged by the fact that several experimental works (Behroozikhah et al. 2017; Mogawer et al. 2013; Saberi et al. 2017) have shown that rubberized mixtures containing RA and viscosity reduction additives exhibit an improved performance in terms of stiffness, rutting resistance and fatigue life.

The Authors of this paper have addressed a multitude of issues related to the use of rubberized mixtures in road pavements as part of the "TYREC4LIFE" project funded by the European Commission through its "LIFE+" program. Investigations were carried out in the laboratory and in the field, with the evaluation of several alternative solutions that entailed the production of gap-graded and dense-graded mixtures by means of dry and wet technologies (Zanetti et al. 2014, 2015; Santagata et al. 2015, 2016). Due to the encouraging results obtained during its development, the project was followed by supplementary activities, funded by the Italian Ministry of the Environment, Land and Sea, which were focused on the evaluation of rubberized mixtures of premium sustainability containing RA and viscosity reduction additives. This paper provides a synthesis of some of the activities which were undertaken in the "TYREC4LIFE" follow-up project.

2 Experimental Program

The investigation illustrated in this paper focused on dense-graded wearing course rubberized mixtures containing RA and viscosity reduction additives. The experimental program included an initial mix design phase (including characterization of component materials) and a further phase of assessment of plant-produced mixtures which were laid in full-scale field trial sections.

Mix design activities considered three different dense-graded mixtures—indicated as ARS0, ARS10 and ARS20—containing increasing quantities of RA (with dosages by weight of total aggregates equal to 0, 10 and 20%, respectively). Mixture AS0, with no RA, was included in the study for comparative purposes.

Following their mix design, the mixtures containing RA were manufactured in a batch plant (with corresponding codes ARS10-P and ARS20-P) and laid on site for the construction of a pavement wearing course of an urban road. During construction, loose mixtures were sampled with the purpose of performing standard quality assurance (QA) tests and advanced mechanical characterization tests. The QA assessment included tests for the evaluation of the volumetrics of Marshall and gyratory-compacted specimens, Marshall stability and indirect tensile strength. Advanced mechanical characterization included tests for the evaluation of the dynamic modulus master curve, flow number and fatigue life.

3 Materials and Methods

The rubberized binder considered in the study was an asphalt rubber (AR), provided by an Italian supplier, fabricated by mixing a 50/70 penetration grade bitumen with 20% CR. Particles constituting CR were entirely passing the 1.18 mm sieve in accordance to ASTM D6114.

The viscosity reduction additive employed to decrease mixing and compaction temperatures of bituminous mixtures was a synthetic microcrystalline wax produced by means of coal gasification, characterized by the presence of long chain aliphatic hydrocarbons (Jamshidi et al. 2013). Based on manufacturer's recommendations, a dosage of 3% by weight of binder was adopted during the entire investigation. In the laboratory, the additive was blended with the AR binder at 190 °C by means of a mechanical mixer operating at a constant speed of 400 rpm for 10 min.

The original AR and the AR containing the additive (ARS) were preliminarily characterized in terms of their penetration at 25 °C (EN 1426) and of their softening point (EN 1427). Furthermore, they were subjected to frequency sweep tests (EN 14770) for the construction of the master curves of the norm and phase angle of the complex modulus at a reference temperature of 34 °C. Tests were carried out with a dynamic shear rheometer equipped with parallel plates of variable diameter (25 mm or 8 mm), operated in the frequency range of 1–100 rad/s and in the temperature range of 0–82 °C. Dynamic viscosity tests (EN 13302) were carried out by making use of a rotational viscometer equipped with a cylindrical spindle rotating at constant shear rate of 6.8 s^{-1} in the temperature range comprised between 125 and 175 °C.

Bituminous mixtures were prepared by employing three locally-available aggregate fractions (10/15, 3/8 and 0/8 mm), a mineral filler and RA material obtained from the milling of wearing courses. Particle size distribution analyses of aggregates and RA, before and after binder extraction, were carried out according to EN 933-1. RA binder content was determined as per EN 12697-39. For the three mixtures considered in design (ARS0, ARS10 and ARS20), fractions were combined in order to yield the same target gradation, typical of dense-graded wearing courses.

Optimum binder content of each mixture was identified by considering different binder contents for each combination of aggregate fractions, filler and RAP. For each binder content, four Marshall specimens (EN 12697-30, 75 blows per side) and four gyratory specimens (EN 12697-31, 100 gyrations) were compacted at 155 °C. Volumetric properties (EN 12697-5, -6, -8) were assessed for both sets of specimens. Thereafter, Marshall tests (EN 12697-34) and indirect tensile strength (ITS) tests at

25 °C (EN 12697-23) were performed on Marshall specimens and gyratory specimens, respectively. Optimum binder content was identified based on acceptance criteria defined in terms of Marshall air voids content (3–6%), Marshall stability (≥9 kN) and Marshall flow (≥2 mm). ITS results were considered as supplementary information which supported the choice derived from simple Marshall design.

QA tests and advanced mechanical tests were carried out on the ARS10-P and ARS20-P mixtures manufactured in the asphalt plant and sampled during paving operations. In addition to the specimens required for the QA assessment, six specimens 180 mm thick and 150 mm in diameter were compacted at 155 °C with 40 gyrations of the gyratory compactor for each mixture. The specimens were then cored and trimmed to obtain testing samples 150 mm thick and 100 mm in diameter (AASHTO PP 60). The number of gyrations equal to 40 was defined by means of a preliminary volumetric analysis which showed that they were those necessary in order to obtain a target air voids content of 8.0% ± 0.5%.

Cyclic uniaxial compression tests (AASHTO T 378) at three temperatures (4, 20 and 40 °C) and six frequencies (25, 10, 5, 1, 0.5 and 1 Hz) were performed on all ARS10-P and ARS20-P specimens by applying an axial strain of approximately $100 \cdot 10^{-6}$ mm/mm, with the consequent evaluation of dynamic modulus and phase angle. Flow number (FN) was obtained as the average of three replicates from repeated load permanent deformation tests (AASHTO T 378) carried out at 58 °C by applying load pulses of 600 kPa with a 0.1 s duration every 1 s. Resistance to fatigue cracking was investigated by means of time sweep tests (AASHTO TP 107) performed at three strain levels, in the direct tension configuration and in the strain-controlled mode, at 18 °C and 10 Hz. The first peak-to-peak strain was set equal to $300 \cdot 10^{-6}$ mm/mm, while the second and the third strain levels were selected as function of the failure cycles obtained during the first test.

4 Results and Discussion

4.1 *Characterization of Component Materials and Mix Design*

The original AR was characterized by a penetration of 29 dmm and a softening point of 73 °C. Addition to the AR of the viscosity reduction additive led to important changes in consistency, with the consequent reduction of penetration (25 dmm) and increase of softening point (101 °C).

These results were consistent with those obtained from frequency sweep tests, which are displayed in Fig. 1a in the form of master curves of the norm of the complex modulus and of the phase angle (at 34 °C). It can be observed that binder ARS exhibited complex modulus values which were higher than those of binder AR at low and intermediate frequencies, while similar values were recorded at high frequencies. On the contrary, the phase angle was slightly reduced by the presence

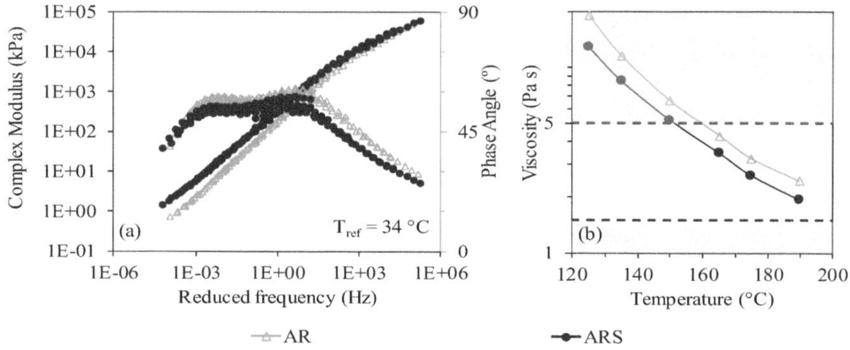

Fig. 1 Complex modulus and phase angle master curves **a** and dynamic viscosity **b** of binders AR and ARS

of the viscosity reduction additive mainly at intermediate frequencies. These effects can be explained by the presence of a crystalline structure in the additive which may provide an enhanced stiffness and elasticity to the ARS binder.

As proven by the results of viscosity tests (Fig. 1b), the abovementioned stiffening effects were lost at temperatures above the melting point of the employed wax (i.e. above 115 °C), where binder ARS displayed a viscosity which was significantly lower than that of the original AR. Based on these results, compaction temperature of the rubberized mixtures was set at 155 °C, at which the viscosity of the ARS binder was in the 1.5–5 Pa s range, often referred to in specifications for acceptance purposes.

Based on the results obtained from particle size distribution analyses (Table 1), the RA material was classified as 10RA0/8, while its binder content coming from ignition tests was found to be equal to 6.8% by the total weight of RA.

Table 1 RA gradation before and after binder extraction, gradation limits and job mix formula for design mixtures ARS0, ARS10 and ARS20

Sieve size (mm)	RA black curve (%)	RA white curve (%)	Gradation limits (%)	Job mix formula (%)
12.5	100	100	100–100	100
10	100	100	87–96	92
8	95.7	98.0	70–90	83
6.3	80.9	87.0	60–80	67
4	55.2	64.4	40–60	51
2	32.8	41.9	25–38	33
1	17.9	29.1	19–30	24
0.5	8.7	21.3	13–22	17
0.063	0.4	8.2	6–9	7.5

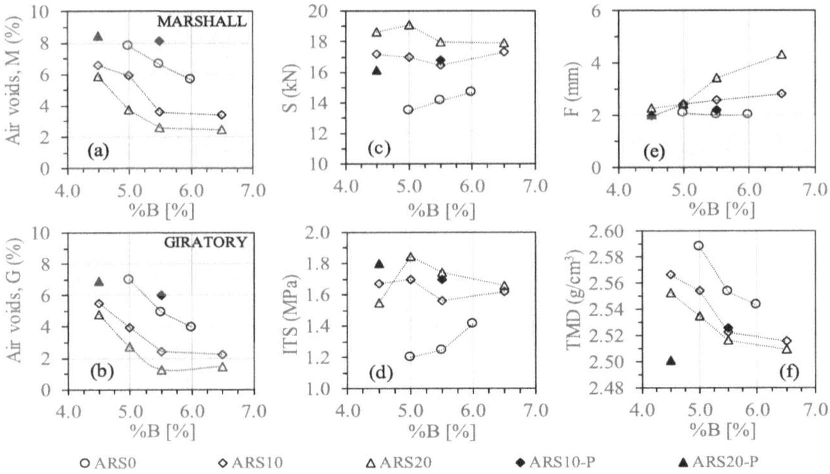

Fig. 2 Mix design and quality assurance results: Marshall air voids content **a**, gyratory air voids content **b**, Marshall stability, S **c**, ITS **d**, Marshall flow, F **e**, and theoretical maximum density, TMD **f**

Design of the three mixtures included in the study (ARS0, ARS10 and ARS20) was performed by referring to typical specification limits adopted for dense-graded wearing courses and by considering a single target gradation (Table 1).

Results collected in the mix design phase are shown in Fig. 2, where the average values of various volumetric and mechanical parameters are plotted, for each mixture, as a function of binder content (%B), expressed as the percentage of added ARS by the total weight of aggregates (virgin and RA). Additional data points which are shown in Fig. 2 (with indicators filled in black) refer to the results obtained in the QA of the plant-produced mixtures (discussed in Sect. 4.2).

It can be observed that the introduction of RA into the mixtures led to a reduction of air void contents (Fig. 2a, b), probably as a result of the presence of the additional binder covering the RA particles which may allow these to more easily adapt to the packed configuration reached during compaction. The resulting higher density and the presence of the aged RA binder also led to an overall increase of stiffness and strength, as shown by the higher Marshall stability (Fig. 2c) and ITS (Fig. 2d). The presence of the additional binder coming from the RA, which reflected in a TMD reduction (Fig. 2f), also provided the mixtures with a more ductile behaviour, demonstrated by the higher values of Marshall flow (Fig. 2e).

Based on the results discussed above, optimum binder contents selected for the mixtures containing RA were equal to 5.5 and 4.5% (by total weight of aggregates) for mixtures ARS10 and ARS20, respectively.

4.2 Quality Assurance and Performance-Related Evaluation

The two rubberized mixtures containing RA were manufactured at 175 °C in a batch asphalt plant located 15 km from the construction site. Average compaction temperatures recorded during paving, slightly lower than target, were equal to 145 and 153 °C for mixtures ARS10-P and ARS2-P, respectively.

Total binder content (by weight of aggregates) derived from ignition tests was equal to 5.9 and 5.8% for mixtures ARS10-P and ARS20-P, respectively. By assuming the presence of the design RA dosages, these values correspond to dosages of added binder equal to 5.25 and 4.5%, which are very close to optimal mix design values.

Aggregate gradation showed significant deviations from target for both mixtures, probably as a result of inconsistencies occurring in the filler dosage system at the plant. While such a problem mainly reflected in the low filler content of mixture ARS10-P (equal to 5.7%), it affected the entire fine fraction of mixture ARS20-P, with very low percentages passing the 1.0, 0.5 and 0.063 mm sieves (equal to 18.6, 12.7 and 4.5%, respectively).

As mentioned in Sect. 4.1, the black indicators displayed in Fig. 2 refer to the volumetric and mechanical properties of the two plant-produced mixtures. Thus, it can be observed that the very high values of air voids contents of both Marshall and gyratory specimens, probably due to the lack of filler, were not consistent with design values and did not satisfy specification limits. On the contrary, Marshall stability and ITS values were similar to those coming from design.

In order to derive master curves at a reference temperature of 20 °C, complex modulus data collected at different temperatures and frequencies were fitted to the sigmoidal model (Bonaquist and Christensen 2005) and to the Arrhenius equation given in Eqs. (1) and (2), respectively. The limiting maximum modulus was estimated from the composition and volumetrics of the mixtures by making use of the relationships indicated in AASHTO T 378.

$$\log\left|E^*\right| = \delta + \frac{(\mathrm{Max} - \delta)}{1 + e^{\beta + \gamma \log f_r}} \tag{1}$$

$$\log a(T) = \frac{\Delta E_a}{19.14714}\left(\frac{1}{T} - \frac{1}{T_r}\right) \tag{2}$$

where |E*| is the norm of the complex modulus (MPa), Max is the limiting maximum modulus, f_r is the reduced frequency (Hz), a(T) is the shift factor at the generic test temperature T (K), T_r is the reference temperature (K), and δ, β, γ, ΔE_a are fitting parameters.

Master curves are represented in Fig. 3, while corresponding model parameters are listed in Table 2. It can be observed that mixture ARS-20P was characterized by a higher stiffness and greater elasticity (i.e. lower phase angle) with respect to mixture ARS-10P. These outcomes are consistent with the higher RA content and lower volume of binder which differentiates mixture ARS-20P from mixture ARS-10P.

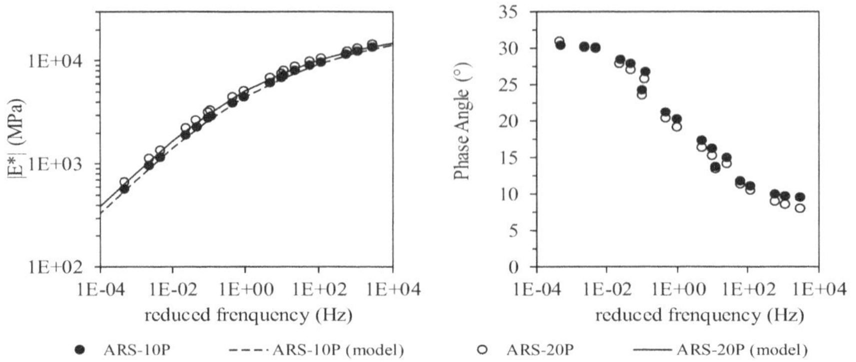

Fig. 3 Master curves of mixtures ARS-10P and ARS-20P

Table 2 Master curve parameters of mixtures ARS-10P and ARS-20P

	$\lvert E^*\rvert_{Max}$ (GPa)	δ (-)	β (-)	γ (-)	ΔE_a (kJ/mol)
ARS-10P	21.074	2.96	−1.44	−0.37	202
ARS-20P	21.145	3.03	−1.52	−0.38	211

Experimental data recorded during repeated load permanent deformation tests were analysed by making use of the model proposed by Francken, given in Eq. (3), which is capable of describing the primary, secondary and tertiary stages of the flow process (Biligiri et al. 2007).

$$\varepsilon_P(N) = AN^B + C\,(e^{DN} - 1) \tag{3}$$

where ε_P is the permanent axial strain in 10^{-6} mm/mm, N is the number of loading cycles and A, B, C and D are the fitting parameters.

Average N-ε_P curves obtained for the two mixtures are displayed in Fig. 4a in which the points corresponding to the FN values, that identify the number of load repetitions at which shear deformation under constant volume is initiated, are also highlighted. This information is supplemented by the data listed in Table 3, which contains average FN values and the corresponding model fitting parameters. As in the case of linear viscoelastic characterization (Table 2 and Fig. 3), the stiffening effect caused by increasing quantities of RA in the mixtures is clearly visible. Thus, mixture ARS20-P was found to be more rut resistant than mixture ARS10-P.

Results obtained in time sweep tests performed on specimens of the two plant-produced rubberized mixtures were modelled by making use of the so-called simplified viscoelastic continuum damage approach (S-VECD). In particular, by following the procedures suggested by Underwood et al. (2012), experimental data were fitted to Eq. (4) and to Eq. (5), thereby obtaining the damage characteristic curve and the number of cycles to failure corresponding to any given strain. Fatigue curves

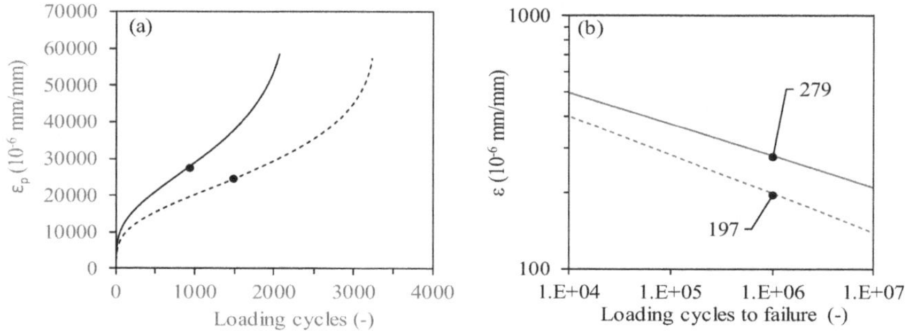

Fig. 4 Results obtained from FN tests **a** and fatigue damage modelling **b**

Table 3 FN values and model fitting parameters of mixtures ARS-10P and ARS-20P

	FN	A	B	C	D
ARS-10P	935	2,171.0	0.36269	409.62	0.001837
ARS-20P	1,498	1,726.1	0.37306	49.84	0.002020

were then derived for both mixtures as shown in Fig. 4b, where the strain level corresponding to one million loadings to failure (ε_6) is explicitly highlighted.

$$C(S) = 1 - C_{11} S^{C_{12}} \tag{4}$$

$$N_{failure} = \frac{(f_{red})(2^{3\alpha})S_f^{\alpha-\alpha C_{12}+1}}{(\alpha - \alpha C_{12} + 1)(C_{11}C_{12})^{\alpha}\left[(\beta+1)(\varepsilon_{0,pp})(|E^*|_{LVE})\right]^{2\alpha}K_1} \tag{5}$$

where S is damage, C(S) is the pseudo-secant modulus, C_{11} and C_{12} fitting parameters, f_{red} is the reduced frequency, α is the damage evolution rate, S_f is the cumulative damage at the failure point, β is the load form factor, $\varepsilon_{0,pp}$ is the peak-to-peak strain magnitude, $|E^*|_{LVE}$ is the norm of the complex modulus and k_1 is the loading shape factor. Calculated values of all the abovementioned fitting coefficients are listed in Table 4.

Table 4 Damage model fitting parameters

	C_{11} (-)	C_{12} (-)	f_{red} (Hz)	α (-)	S_f (-)	β (-)	$\|E^*\|_{LVE}$ (GPa)	K_1 (-)
ARS-10P	0.0056	0.4078	17.9	4.0	104627	0	7.1	0.27
ARS-20P	0.0034	0.4598	17.8	3.3	66961	0	7.6	0.30

It was observed that the stiffening effects related to the presence of increasing quantities of RA in the mixtures had a negative impact on their response under repeated loading in terms of their fatigue resistance. In particular, as clearly shown in Fig. 4b, mixture ARS-20P had a fatigue life which was shorter than that of mixture ARS-10P.

5 Conclusions

The experimental results obtained in the investigation described in this paper indicate that in the design of rubberized mixtures it is possible to successfully combine the use of reclaimed asphalt and of viscosity reduction additives. A proper use of the additives may allow compaction temperatures to be reduced by 15 °C, while the introduction of reclaimed asphalt in the bearing skeleton of the mixtures can simultaneously lead to an increase of their density and to an overall enhancement of stiffness, elasticity and resistance to rutting. Residual concerns may be related to the fatigue properties of these mixtures, for which further refinements in mix design may be necessary.

Although the considered rubberized mixtures seem to be suitable for paving applications, it is envisioned that more research is needed in order to fully exploit their performance potential and to develop comprehensive mix design guidelines and technical specifications. Furthermore, additional analyses are recommended for a full quantitative assessment of the sustainability benefits which they bring to pavement construction and maintenance.

References

Behroozikhah A, Morafa SH, Aflaki S (2017) Investigation of fatigue cracks on RAP mixtures containing Sasobit and crumb rubber based on fracture energy. Constr Build Mater 141:526–532

Biligiri KP, Kaloush KE, Mamlouk MS, Witczak MW (2007) Rational modeling of tertiary flow for asphalt mixtures. Transp Res Rec 2001:63–72

Bonaquist R, Christensen DW (2005) Practical procedure for developing dynamic modulus master curves for pavement structural design. Transp Res Rec 1929:208–217

Farina A, Zanetti MC, Santagata E, Blengini A (2017) Life cycle assessment applied to bituminous mixtures containing recycled materials: crumb rubber and reclaimed asphalt pavement. Resour Conserv Recycl 117:204–212

Jamshidi A, Hamzah MO, You Z (2013) Performance of warm mix asphalt containing Sasobit: state-of-the-art. Constr Build Mater 38:530–553

Mogawer W, Austerman A, Mohammad L, Kutay ME (2013) Evaluation of high RAP-WMA asphalt rubber mixtures. Road Mater Pavement 14:129–147

Pouranian MR, Shishehbor M (2019) Sustainability assessment of green asphalt mixtures: a review. Environment 6:73

Saberi KF, Fakhri M, Azami A (2017) Evaluation of warm mix asphalt mixtures containing reclaimed asphalt pavement and crumb rubber. J Clean Prod 165:1125–1132

Santagata E, Dalmazzo D, Lanotte M, Zanetti MC, Ruffino B (2012) Relationship between crumb rubber morphology and asphalt rubber viscosity. In: Proceedings, 5th asphalt rubber international conference, pp 513–532
Santagata E, Baglieri O, Alam M, Lanotte M, Riviera PP (2015) Evaluation of rutting resistance of rubberized gap-graded asphalt mixtures. In: Proceedings, 6th international conference bituminous mixtures and pavements, pp 407–412
Santagata E, Lanotte M, Baglieri O, Dalmazzo D, Zanetti MC (2016) Analysis of bitumen-crum rubber affinity for the formulation of rubberized dry mixtures. Mater Struct 49:1947–1954
Underwood BS, Baek C, Kim YR (2012) Simplified viscoelastic continuum damage model as platform for asphalt concrete fatigue analysis. Transp Res Rec 2296:36–45
Xiao F, Amirkhanian S, Juang CH (2007) Rutting resistance of rubberized asphalt concrete pavements containing reclaimed asphalt pavement mixes. J Mater Civ Eng 19:475–483
Xiao F, Amirkhanian S (2008) Resilient modulus behavior of rubberized asphalt concrete mixtures containing reclaimed asphalt pavement. Road Mater Pavement 9:633–649
Zanetti MC, Fiore S, Ruffino R, Santagata E, Lanotte M (2014) Assessment of gaseous emissions produced on site by bituminous mixtures containing crumb rubber. Constr Build Mater 67:291–296
Zanetti MC, Fiore S, Ruffino R, Santagata E, Dalmazzo D, Lanotte M (2015) Characterization of crumb rubber from end-of-life tyres for paving applications. Waste Manag 45:161–170

Evaluation of the Properties of Asphalt Concrete Modified with Crumb Rubber Using Marshall Test

Olumide M. Ogundipe, Omotola C. Aboloye, and Stephen O. Fatuase

Abstract This study examines the properties of asphalt concrete modified with crumb rubber using Marshall test. In this study, 5 different mixtures containing 3, 5, 7, 9 and 11% crumb rubber by total mix were considered. The Marshall samples were prepared, tested in the Marshall stability and flow equipment and analysed. The results from the Marshall tests indicate that the stability of the asphalt concrete increases with increasing bitumen content up to 6.2% bitumen content with the mixture containing 7% crumb rubber having the highest stability. The flow initially increased with the addition of crumb rubber up to 7%, but decreased gradually as the percentage of the crumb rubber was further increased to 9 and 11%. Generally, the study indicates that crumb rubber can be used to modify typical asphalt concrete used in Nigeria. This implies that economic gains could be derived by generating wealth from the use of discarded tyre and job creation for those who will be responsible for collecting and processing the discarded tyres and construction of roads that will stand the test of time. Also, successful utilisation will solve the problem of disposing the tyres.

Keywords Asphalt concrete · Crumb rubber · Flow · Marshall · Stability · Tyre

1 Introduction

Road transport has been identified to be the most important out of all the modes of transport, because roads are vital to economic development, as they link remote places, facilitate movement of goods and are cheaper to maintain and run than other modes. The commuters depend on it to access all other modes. An assessment of the roads in Nigeria shows they are in state of disrepair. According to TRL (1993), it is certain that road pavement will fail in one way or the other while in service, i.e., when subjected to loading over time because of the nature of the materials used for construction. However, the design life of the pavement could be elongated by modifying the mixture. Hence, the need to modify the asphalt mixture used on Nigerian roads

---

O. M. Ogundipe (✉) · O. C. Aboloye · S. O. Fatuase
Department of Civil Engineering, Ekiti State University, Ado-Ekiti, Nigeria
e-mail: momide2002@yahoo.com; olumide.ogundipe@eksu.edu.ng

C. Raab (ed.), *Proceedings of the 9th International Conference on Maintenance and Rehabilitation of Pavements—Mairepav9*, Lecture Notes in Civil Engineering 76,
https://doi.org/10.1007/978-3-030-48679-2_44

to improve the performance. As highlighted by TRL (2002), typical wearing courses of a flexible pavement should possess the characteristics highlighted as follows: (i) good deformation resistance; (ii) good fatigue resistance and the ability to withstand high strains; (iii) adequate stiffness to reduce stresses in the layers lying below the wearing course to levels, which are acceptable; (iv) sufficient durability, i.e., ability to withstand environmental degradation; (v) low permeability to prevent water from infiltrating the pavement; (vi) good workability to allow adequate compaction during construction; (vii) good surface texture that can ensure good skid resistance in wet condition and; (viii) predictable performance. All these characteristics cannot be achieved without optimising the mixture, by modifying it.

Many additives like ethylene vinyl acetate (EVA), styrene butadiene styrene (SBS), styrene-butadiene rubber (SBR), lime, sulphur, crumb rubber, etc. had been used to modify asphalt for the purpose of improving the properties. This study focuses on the use of crumb rubber, a sustainable way of improving the road pavement. Santos et al. (2018) observed that one of the main indicators for assessing the sustainability of a project is the amount of secondary material consumed, i.e., the amount of the recycled materials used in the project. Crumb rubber is derived from used tyres, which are available in large quantities in Nigeria, because of the dominance of the road transportation mode. As engineers seek to hold the two major failure modes in pavements, which are fatigue cracking and permanent deformation to acceptable limits within a pavement design life, it becomes necessary to use asphalt concrete with improved performance. It has been documented that crumb rubber had been used to great effect to improve the resistance of asphalt to fatigue and permanent deformation. Marais et al. (2017) pointed out that the major driver for the use of crumb rubber is the increased fatigue resistance. Wong and Wong (2017) reported that crumb rubber resulted in higher performance of both binders and mixtures at high temperatures. Also, Al-maamori and Hussen (2014) mentioned that desired stiffness and viscosity with high resistance to, permanent deformation, fatigue cracking and low temperature cracking of asphalt cement could be achieved by the modification of binder with a suitable quantity and proper particle size of crumb rubber. Crumb rubber could be introduced in two different forms called the wet and dry processes (Heitzman 1992). In the wet method, the CRM particles are mixed with the bitumen to produce a blend called asphalt rubber (AR); which in turn is used to produce hot-mix asphalt concrete, while in the dry process CRM particles are mixed with the aggregate (dried) before the bitumen is added (Bahia and Davies 1994). This study adopts the wet process method.

Colucci et al (1994) worked on the possibility of adding crumb rubber modifier (CRM) into hot-mix asphalt (HMA) to be used for pavements in Puerto Rico. They found that in Puerto Rico, as it was in many parts of the mainland United States of America, used motor vehicle tyres was a great problem and that using CRM in asphalt concrete would give a partial solution to the problem of heap of used worn out tyres facing Island. They concluded that the engineering benefits that could be derived from the use of asphalt rubber for road surfacing were not certain; the high cost associated with its use could not be justified by better performance and durability. Mashaan et al (2014) observed that the use of crumb rubber produced from used/worn out automobile tyres was not only advantageous in the form of cost reduction contrary to the observation of Colucci et al. (1994), but also had reduced impact on the ecology by ensuring the environment is clean and providing better balance for the natural resources. They stated further that the use of crumb rubber as reinforcement for asphalt was a smart solution for sustainable development as waste material was being reused. CDOT (2003) stated that asphalt rubber binders (ARBs) could be made to give satisfactory performance under any climatic condition. They mentioned that at intermediate and high temperatures, the physical properties of asphalt rubber differed significantly from those of neat paving grade asphalts, because, the rubber stiffened the binder and improved the elasticity (proportion of deformation that is recoverable) over operating temperatures of the pavement; which in turn reduced pavement's susceptibility to temperature and improved the permanent deformation (rutting) and fatigue resistance with small impact on the properties at cold temperature (CDOT 2003).

Bahia and Davies (1994) studied the impact of crumb rubber modifiers on performance-related properties of asphalt binders. In their study, they obtained the rubbers from whole passenger tyres. They employed three different methods to get the rubber. The first was obtained by ambient shredding (AS), the second by cryogenic grinding (CG) and the third by a special extrusion process with the use of some additives (TP). The three rubbers obtained from the tyres had similar size distribution and the maximum particle size was about 1 mm. They worked on four asphalts (two AC-10 grade asphalts, an AR-2000, and a 200/300 pen). They observed relative change in properties before and after the pressure aging vessel (PAV) aging; which showed that the use of rubber in the asphalt might result in the decreased hardening because of oxidative aging. The thin film oven test (TFOT) data showed that the use of rubber resulted in increase in the loss mass. Both the PAV and the TFOT data indicated that the aging characteristics data were not specific for the type of rubber production method employed.

Wulandari and Tjandra (2017) investigated the effect of introducing crumb rubber into asphalt mixture using wet process. They considered two rubber contents—1 and 2% according to the weight of the asphalt mixture and two different rubber sizes - #40 and #80. They observed that the using crumb rubber in asphalt mixture resulted in increased strength and quality of the asphalt mixture, but highlighted concern about the durability caused by lower bitumen content. Generally, they recommended crumb rubber as additive in asphalt mixture because all the results of the tests conducted were observed to be within the standard requirements. Mehta et al. (2005) studied the effect coarse ground tyre rubber (GTR) and crumb rubber on rutting performance of Superpave asphalt concrete. They used the locally available uniformly graded coarse rubber as replacement for aggregate particles. They examined nine different levels of rubber percentages; 0–4.0% at 0.5%, increments. They maintained the well-graded Superpave gradation. Also, in order to find the optimum percentage of coarse and fine rubber particles that could be used in the asphalt concrete mixture, they used fine crumb rubber particles (No 80 and No. 20 mesh) in proportions of 5 and 15%. They found that the coarse rubber particles improved rutting resistance when used up to 3.5%, while the rutting performance of the mixture improved significantly with addition of fine rubber up to 15% according to the total weight of binder. Also, Room et al. (2014) investigated the impact of introducing crumb rubber into asphalt concrete. The crumb of size ranged from 0.60 mm to 0.75 mm and it was heated at 150 °C at a rate of 5, 10, 15 and 20% by mass. They adopted the wet process. Their results showed that the values for resilient modulus and creep stiffness were greater than the conventional asphalt, while the accumulated strain values were less; implying implied improved performance.

Marais et al. (2017) carried out a study to compare the South African technology in crumb rubber modified bitumen (CRMB) with what was obtainable in the United States of America, where the technology originated. One of their tasks involved using the Arizona crumb size (<2 mm), which is coarser than the South African crumb size (<1 mm) to make a CRMB blend and assessing them by performing the full range of specified tests. They discovered that the viscosity of the Arizona blend with 18% crumbs (with no addition of extender oil) at 190 °C was much lower than typical South African CRMB, Again, they stated that higher blending and reaction temperature would be required to properly blend the mixture and obtain suitable reaction. Their flow values confirmed that there was poor reacted blend, as the values were extremely high for blends that were reacted at 190 °C for up to 3 h and were only within the acceptable range from 4 to 5 h reaction time. They decided it was better to use the South African crumb grading and CRMB composition specification as it has a proven track record and the technology is well understood in the local industry. Generally, the literature review indicates that results are mixed and more studies are required, especially in countries like Nigeria.

It is therefore pertinent that we employ crumb rubber to improve the performance of the asphalt mixture. This will go a long way in ensuring the use of the waste material which will positively affect the economy in terms of job creation, wealth generation and improvement in the performance of the pavement. Therefore, this study focuses on examining the effects of crumb rubber on a typical asphalt mixture used in Nigeria. A successful utilization of the crumb rubber in asphalt will not only work for the optimisation of the asphalt mixture, but will help in putting the waste tyres into good use, thereby solving the problem that may arise from the disposal.

2 Materials and Method

The materials used for the studied were sourced locally. The materials and the methodology are discussed in this section

2.1 Materials

The materials used in the preparation of the asphalt mixture in this study are:

Table 1 Mix Compositions

Materials	Specimens				
	1 (%)	2 (%)	3 (%)	4 (%)	5 (%)
12.5 mm Aggregates	45	45	45	45	45
10 mm Aggregates	25	22	20	20	20
River sand	25	25	25	23	21
Quarry Dust	2	3	3	3	3
Crumb rubber	3	5	7	9	11
60/70 Bitumen (% of total mix)	5.4	5.8	6.2	6.6	7

(i) 12.5 mm aggregates; (ii) 10 mm aggregates; (iii) Quarry dust (stone dust); (iv) River sand(fine aggregates); (v) 60/70 penetration grade bitumen; (vi) Crumb rubber in powdered form.

The 12.5 mm and 10 mm aggregates, quarry dust and the river sand were sourced from the yard of a Construction company at Ifaki—Ekiti, Ekiti State. The crumb rubber was gotten from waste tyres. The waste tyres were collected, washed and dried to remove dirt that could contaminate the crumb rubber. The steel fibres in the tyres were removed and they were cut into small pieces before being ground into powdered form using a grinding machine. The ground crumb rubber passing 0.075 mm sieve was used for the study. The particle size distributions of the aggregates were evaluated using the sieve analysis method (BS 1377 1990).

2.2 *Methods*

2.2.1 Mix Composition

The mix compositions adopted for the research are shown in Table 1. These mix compositions were selected to get the optimum mixture for the asphalt concrete modified with crumb rubber. The mixtures were prepared in conformity with the specification for roads and bridges (FGN 1997).

2.2.2 Specimen Preparation and Marshall Test

The Marshall stability equipment is shown in Fig. 1. In preparation of the specimens, 1200 g of the dry aggregates (mix composition) were measured as shown in Table 1 and mixed thoroughly. The aggregates were placed in a pan and heated to 150 °C in the oven. The crumb rubber was added to the bitumen and mixed with a stirrer. The modified bitumen (60/70 penetration grade) was placed in the oven and heated to 150 °C. It was added to the aggregates and mixed until the aggregates were fully coated. The mould, other accessories and the compaction hammer were placed in the

Fig. 1 Marshall stability and flow test equipment

oven and heated to 150 °C. The mould was set properly and the asphalt mixture was filled in the mould. The surface of the mixture was smoothened. The top and bottom of the mixture were given 75 blows of the hammer at temperature of 140 °C. The use of 75 blows was recommended by the General Specifications—Road and Bridges (FGN 1997). The sample was removed from the mould using sample extractor and allowed to cool.

Density determination: The density was determined by weighing the asphalt concrete samples both in the air and in clean water at room temperature. The difference between the measurements and the density were calculated.

Stability and flow determination: The asphalt concrete samples (cores) were placed in the water bath to bring the samples to the test temperature of 60 °C. Then the asphalt concrete cores were tested for stability and flow.

3 Results and Discussion

The particle size distribution curves for the aggregates used in the study are presented in Fig. 2. This shows that they conform with the standards specified in the General Specifications—Roads and Bridges (FGN 1997). The Marshall stability values of the asphalt concrete with and without crumb rubber are shown in Fig. 3. The results indicate that the Marshall stability of asphalt concrete increases as the bitumen content increases up to 6.2% bitumen content. The asphalt concrete modified with the crumb rubber follows the same trend as the one not modified and maximum stability

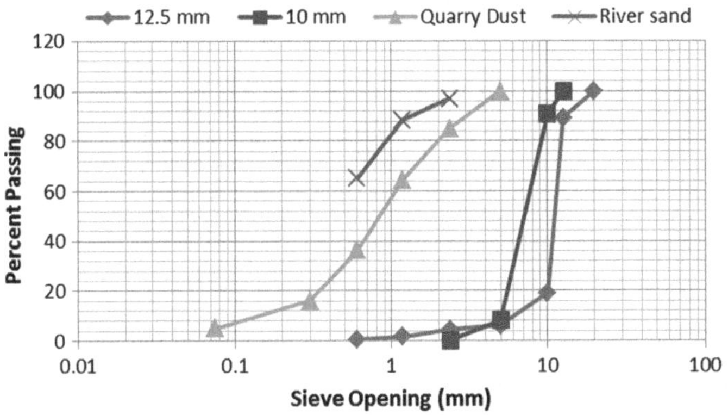

Fig. 2 Particle size distribution curve

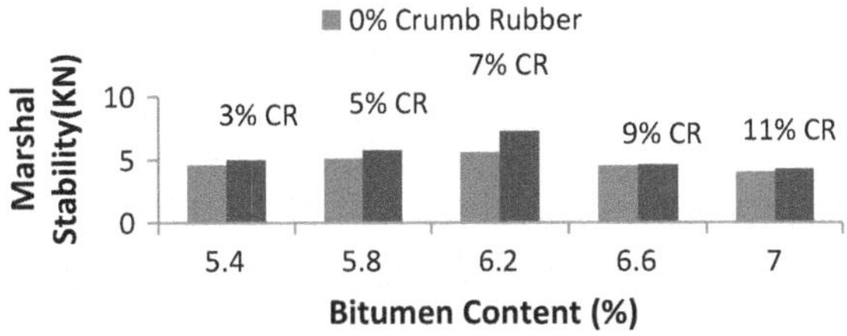

Fig. 3 Marshall stability of asphalt concrete with and without crumb rubber (CR) with varying bitumen contents

was recorded with 7% crumb rubber. Generally, the results imply that modifying the asphalt concrete with crumb rubber improves its stability. The increase in stability of the crumb rubber- modified asphalt can be linked to the improved bond (adhesion) of the asphalt concrete. However, further addition of the crumb rubber resulted in a reduction in the stability of the mix, which might be because of the reduced adhesiveness caused by inadequate binder in the mix as observed by Wulandari and Tjandra (2017).

The Marshall stability of the asphalt concrete containing crumb rubber ranges from 4.25–7.25 kN. These values are adequate and meet the specification of greater than 3.5 kN of the General specifications - Road and Bridges (FGN 1997). The optimum Marshall stability of 7.25 and 5.60 kN were obtained at 6.2% bitumen content for crumb rubber mix and conventional mix respectively.

The flow values of the asphalt concrete without crumb rubber and those with various percentages of crumb rubber are presented in Fig. 4. The flow increased initially with the increase of the crumb rubber up to 7%, but decreased gradually as the percentage of the crumb rubber was increased to 9 and 11%. The low flow values recorded for 9 and 11% crumb rubber is caused by the increase void spaces in the asphalt because of the introduction of high percentage of crumb rubber. Generally, the results show that the introduction of crumb rubber into the asphalt concrete reduces its flow. The flow value of 4% for 6.2% bitumen and 7% crumb rubber is within the range specified by the General Specifications—Roads and Bridges (FGN 1997).

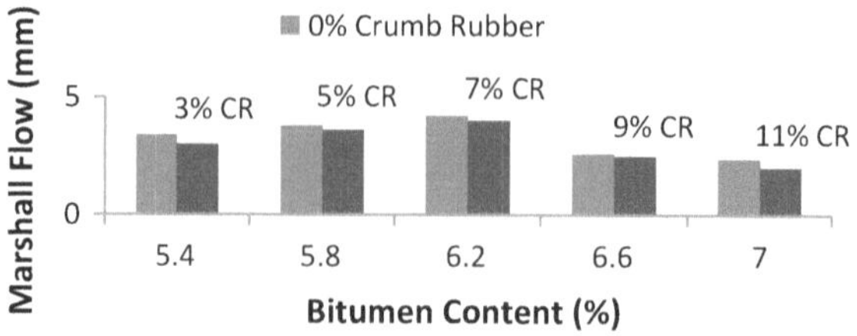

Fig. 4 Flow of the asphalt concrete with and without crumb rubber (CR) with varying bitumen contents

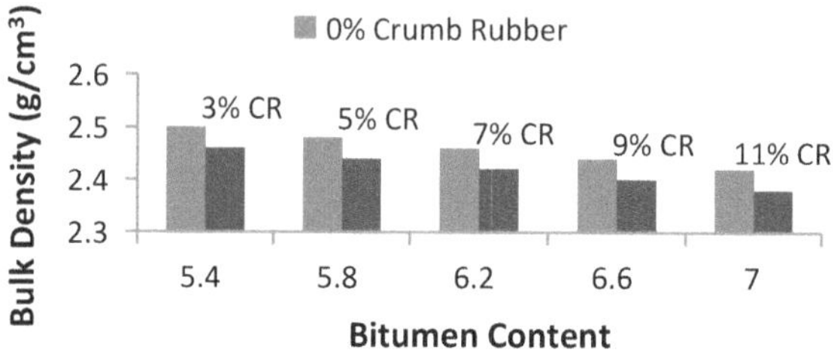

Fig. 5 Bulk density of the asphalt concrete with and without crumb rubber (CR) with varying bitumen contents

The bulk density of the asphalt concrete modified with crumb rubber was lower than that of the asphalt concrete without crumb rubber. As shown in Fig. 5, the bulk density of the asphalt reduces as the crumb rubber content increases. The resulting decrease in the bulk densities of the asphalt concrete modified with crumb rubber mixtures is because of the density of the crumb rubber which is lower than that of the mineral aggregates. However, the decrease in the density did not result in reduction of the stability of the asphalt concrete. This further confirms that the introduction of the crumb rubber improves the bond of the mixture of the aggregate, binder and the crumb rubber, which implies improved stability.

The values of the air voids for the asphalt concrete with and without crumb rubber are presented in Fig. 6. The values showed that increasing the crumb rubber content in the mixture reduces the air voids in the mixture. The decrease in air voids while the crumb rubber content is increased depends partially on the total filler content. The air voids decrease slightly due to partial replacement of mineral aggregates by crumb rubber which increases the amount of filler resulting in reduction the void spaces between the granular particles in the compacted mix.

Figure 7 shows the void filled with bitumen of the asphalt concrete with and without crumb rubber. It shows that the percentage of void filled with bitumen decreased

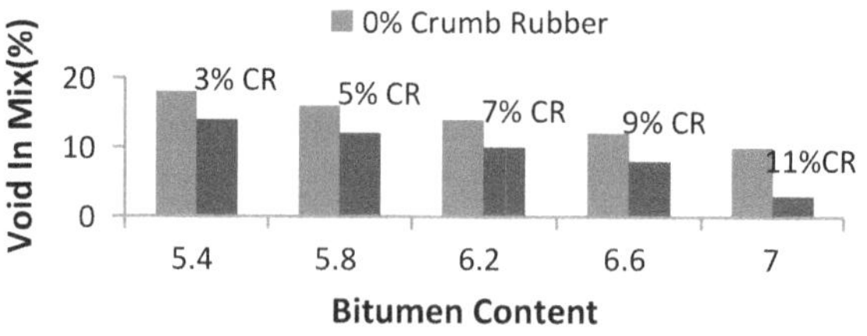

Fig. 6 Air voids of the asphalt concrete with and without crumb rubber (CR) with varying bitumen contents

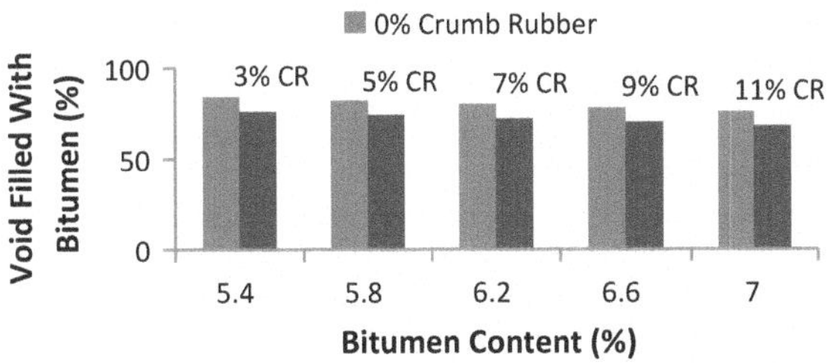

Fig. 7 Void filled with bitumen of the asphalt concrete with and without crumb rubber (CR) with varying bitumen contents

with increasing crumb rubber content. The resulting decrease in the percentage of void filled with bitumen will result in improved permanent deformation resistance of the asphalt concrete modified with crumb rubber.

4 Conclusions

This experimental study on the introduction of tyres waste (crumb rubber) into asphalt concrete indicates that:

(i) tyres waste in form of crumb rubber can be introduced into asphalt concrete to optimise its performance; (ii) introducing crumb rubber into asphalt concrete improves its resistance to deformation at high temperature (stability); (iii) the highest stability was recorded for mixture with 7% of crumb rubber; and (iv) the lowest flow was recorded for mixture with 11% crumb rubber.

These imply that economic gains could be derived by generating wealth from the use of waste tyre and job creation for those who will be responsible for collecting and processing the waste tyres and construction of roads that will stand the test of time. It is therefore recommended that further research into the performance of asphalt concrete modified with crumb rubber will be worthwhile.

Acknowledgements The authors acknowledge the financial assistance provided by Tertiary Education Trust Fund, Abuja, Nigeria (TETFUND) for the research.

References

Al-maamori MH, Hussen MM (2014) Use of crumb rubber as a way to improve performance grade for asphalt cement. Nat Appl Sci 5(1):66–77

Bahia HU, Davies R (1994) Effect of crumb rubber modifiers (crm) on performance-related properties of asphalt binders. J Assoc Asphalt Paving Technol 63:414–449

BS 1377 (1990) Method of test for soils for civil engineering purposes, British Standard Institution, London

CDOT (2003) Asphalt Rubber Usage Guide. State of California Department of Transportation

Colucci B, Colucci JA, Prieto JV, Dayton D (1994) Feasibility of using crumb rubber modifier in hot-mix asphalt pavement applications in Pueto Rico. Puerto Rico Highway and Transportation Authority

FGN (1997) Government of the Federal Republic of Nigeria, General Specification (Roads and Bridges), vol II

Heitzman MA (1992) State of the practice- design and construction of asphalt paving materials with crumb rubber modifier. U.S. Department of Transportation, Federal Highway Administration, FHWA-SA-92-022, Washington, D.C

Marais HIJ, Botha C, Hofsink W, Muller J, van Heerden J (2017) Latest developments in crumb rubber modified bitumen for use in asphalt and seals - the south african experience. In: 17th AAPA international flexible pavements conference, Melbourne, Victoria, Australia

Mashaan NS, Ali AH, Karim MR, Abdelaziz M (2014) A review using crumb rubber in reinforcement of asphalt pavement. Sci World J, 1–21. http://dx.doi.org/10.1155/2014/214612

Mehta Y, Jahan K, Laicovsky J, Miller L, Parikh D, Lozano AL (2005) Evaluate the effect of coarse and fine rubber particle on laboratory rutting performance of asphalt concrete mixtures

Room S, Aleem M, Rana MA, Awan UA, Ali S (2014) Modification of asphalt by the use of crumb rubber. Int J Eng Res Appl 4(12):44–48

Santos J, Bressi S, Cerezo V, Lo Presti (2018) D. SUP&R DST: sustainable pavement & railways decision support tool. In: Proceedings of the sixth international symposium on life-cycle civil engineering, Ghent, Belgium

TRL (1993) Overseas road note 31: a guide to structural design of bitumen-surfaced roads in tropical and sub-tropical countries. Transport Research Laboratory, Crowthorne

Wong CC, Wong WG (2017) Effect of crumb rubber modifiers on high temperature suceptibility of wearing course mixtures. Constr Build Mater 21:1741–1745

Wulandari PS, Tjandra D (2017) Use of crumb rubber as an additive in asphalt concrete. Procedia Eng 171:1384–1389

Advanced Pavement Materials and Technologies

Influence of Source and Ageing on the Rheological Properties and Fatigue and Rutting Resistance of Bitumen Using a DSR

Mrinali Rochlani, Sabine Leischner, Gustavo Canon Falla, Puneet Goudar, and Frohmut Wellner

Abstract Bitumen of penetration grade 50/70 is the most commonly used binder in asphalt pavements of Europe. This research attempts to differentiate between three bitumen of penetration grade 50/70 that have been acquired from three sources. The rheological properties, fatigue and rutting resistance were analyzed and compared using the results of Dynamic Shear Rheometer (DSR) tests. The rheological test data acquired were modelled and further approximated using the 2S2P1D model. Additionally, to understand the changes in performance due to ageing, all the DSR tests were repeated for Rolling Thin Film Oven Test (RTFOT) aged and Pressure Aging Vessel (PAV) aged bitumen. Finally, the tests results were summarized in a performance diagram to aid in overall ranking of the bitumen taking into consideration relevant parameters.

Keywords DSR tests · Rheology · Rutting · Fatigue · Ageing · 2S2P1D

1 Introduction

Owing to the large expenditure invested in construction and maintenance of road networks, it has become crucial to take into account the quality considerations for bitumen. The quality of bitumen varies not only with the refining procedure used, but also with the crude oil source (Singh and Jain 1997; Oner et al. 2017). Bitumen acquired from a particular refinery might meet the existing specifications, however, fail to provide the required performance during the in-service period under heavy traffic and changing environmental conditions, indicating the insufficiency of the specifications that exist. As of now, conventional testing methods in Germany include needle penetration, softening point ring and ball, flash point, Frass breaking point methods, etc; belonging to the empirical (catalog) approach, these fail to define the rheological properties of the bitumen (Wegmann and Vienenkötter 2017). In Germany the Dynamic Shear Rheometer (DSR) temperature sweep between 30 to

M. Rochlani (✉) · S. Leischner · G. C. Falla · P. Goudar · F. Wellner
Institute of Urban and Pavement Engineering, Technische Universität Dresden, Georg-Schumann-Straße 7, 01187, Dresden, Germany
e-mail: mrinali_rajkumar.rochlani@tu-dresden.de

C. Raab (ed.), *Proceedings of the 9th International Conference on Maintenance and Rehabilitation of Pavements—Mairepav9*, Lecture Notes in Civil Engineering 76,
https://doi.org/10.1007/978-3-030-48679-2_46

90 °C has been added recently as an additional procedure to gather information, however, it is not a requirement (Wegmann and Vienenkötter 2017).

Other than source related problems, bitumen deterioration due to ageing is of grave concern. Due to bitumen ageing, physical changes and chemical reactions occur within the structures. Physical changes include bitumen hardening and changes in rheology that could result in an increased cracking susceptibility of the pavement (Petersen 2009; Alfaqawi et al. 2018). While, chemical changes include alterations in the reactions with oxygen, evaporation of lighter molecules, polarity changes, crystallisation of paraffinic saturates to name a few affect adsorption, aggregate bonding abilities along with viscosity (Hunter 2015). Hence, it is of utmost importance to study changes in performance due to ageing; this is generally simulated in the lab using short-term (significant rate of ageing due to construction processes—mixing, transport, placement) and long-term ageing (ageing at a much lower rate during the service life of around 8–10 years) procedures to understand the changes in performance of bitumen, thereby affecting those of the final asphalt mix (Jemere 2010). Short-term and long-term ageing can be done using various methods, with Rolling Thing Film Oven (RTFOT) being a popular choice for short-term and Pressure Ageing Vessel (PAV) for long-term.

It would be therefore beneficial to have more detailed specifications that would include more than just the empirical testing procedures so as to enable in the selection of a bitumen on the basis of specific desired performance (Singh and Jain 1997; Oner et al. 2017). Accurate specifications for bitumen could include tests using just one machine - the DSR; each batch of bitumen could be tested for the rheology, ageing, rutting and fatigue performance within a reasonable time and cost frame at point of origin, i.e. as soon as it is derived from the source (Rochlani et al. 2019). This would help in differentiating between batches of superior and inferior quality and saving maintenance costs in the long run (Rochlani et al. 2019).

The desired goal of this research was to verify the performance of three bitumen of the same grade but from three different sources using the results of DSR tests. Performance includes rheology, rutting and fatigue are to be tested to be able to compare and rank. Additionally, a method to appropriately rank all the materials using different parameters from the tests would be required.

2 Materials and Methodology

2.1 Materials

A total of three materials and three ageing conditions were tested, which include three unaged bitumen, three RTFOT aged and three PAV bitumen. The table below summarizes the bitumen used along with their short descriptions. (Table 1)

Table 1 List of bitumen samples tested—each source with 3 aging conditions

Bitumen source	Sample name (unaged, RTFOT aged, PAV aged)
1	B1, B1 RTFOT, B1 PAV
2	B2, B2 RTFOT, B2 PAV
3	B3, B3 RTFOT, B3 PAV

2.2 Ageing Procedure

In order to investigate the short-term and long-term performance of the bitumen, the bitumen were artificially aged in the laboratory. For short-term ageing, the materials were put in a RTFOT at 163 °C for 75 min (EN 12607-1 2007). On the other hand, to replicate the short term + long term ageing effect, generally RTFOT procedure mentioned followed by PAV at 100 °C for 20 h and 3.2 MPa is used. However, in this research, short + long term ageing was simulated using only PAV at 100 °C for 25 h at 2.07 MPa based on the findings of by Migliori and Corte (2007), who concluded that 5 h of PAV is equivalent to the standard RTOFT aging procedure, thereby simplifying the ageing procedure.

2.3 Dynamic Shear Rheometer Tests

An Anton Paar Modular Compact Rheometer 502 was used to evaluate the rheology, fatigue and rutting performance of the bitumen. To study rheology, strain and frequency sweeps were conducted in the temperature range of −10 to 70 °C. The strain sweeps were firstly undertaken to evaluate the linear viscoelastic limit (LVE) strain, which are required to conduct the frequency sweeps. The strains were further reduced by 20% to ensure the LVE region was maintained for the frequency sweeps. A frequency range of 0.0159 to 47.7 Hz was used and the results acquired were superposed and approximated further using the time-temperature superposition principle and the 2 spring – 2 parabolic elements – 1 dashpot (2S2P1D) rheological model respectively (Benedetto et al. 2007). This model is valid for any rheologically simple bituminous material for relevant pavement conditions in terms of frequency and temperature and describes the performance of the material through seven parameters (Benedetto et al. 2007).

Secondly, for evaluating the rutting performance, single stress creep recovery (SSCR) tests were undertaken. This test follows the same procedure as Multiple Stress Creep Recovery (MSCR) test (*AL MSCR-Test (DSR) FGSV 723* 2016) where a sample is subjected to stress for one second, followed by nine seconds without stress, allowing the material to relax at a temperature of 60 °C with the 25 mm plate. This step is repeated consecutively 10 times. The only difference between MSCR and SSCR is that, in MSCR tests multiple stresses are used, while in SSCR tests only a single stress of 3.2 kPa was utilized. From each cycle two variables—non-recoverable compliance, J_{nr}, and percentage recovery, $\%Rec$ were calculated and averaged.

Lastly, fatigue performance was determined using stress-controlled tests on column cylindrical samples. The tests were carried out at a temperature of 20^0 C and frequency 10 Hz for both unaged and aged samples. For each material, a total of nine samples were tested at three different stress levels. The stress levels were determined using pre-tests so as to attempt to keep the initial strain level of each material within the range of 0.5 to 2.2% This was done to remain the LVE region.

3 Results and Discussion

3.1 Frequency Sweep Tests

The frequency sweep test data modeled using the time-temperature superposition and 2S2P1D model is presented in terms of the master curves for dynamic shear modulus and phase angle in Fig. 1a and b respectively. Black diagrams and Cole-Cole plots in Fig. 2a and b are presented for only one material (aged and unaged) to observe and appreciate the fit of the model, while 2c an d represent model data for all materials.

From Fig. 1a and b, the dynamic shear modulus and phase angle of different bitumen and the changes that occur can be visually observed. It can be seen from the

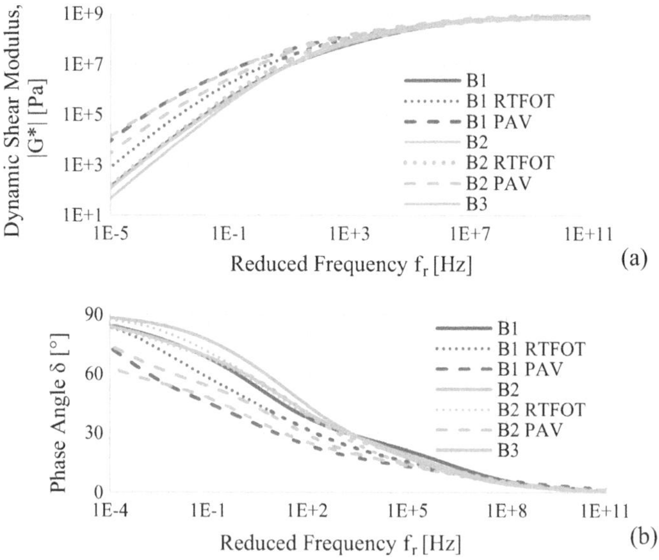

Fig. 1 Master curves **a** Dynamic Shear Modulus **b** Phase Angle using 2S2P1D model at 20 °C

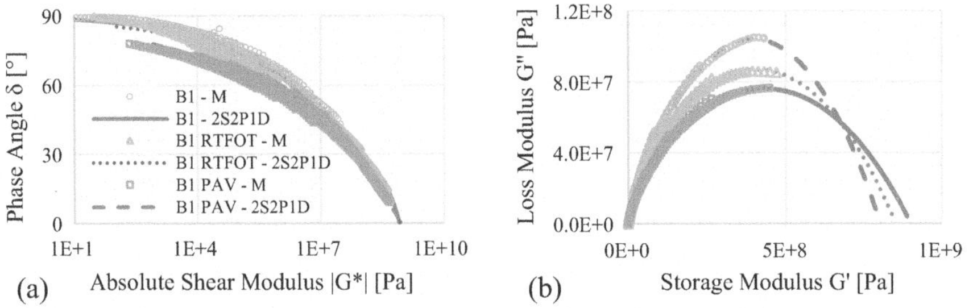

Fig. 2 **a** Black Diagram **b** Cole-Cole Diagrams for B1 with measured test data (denoted M) and 2S2P1D data for Bitumen B1

shear modulus master curve, for unaged materials, B1 and B3 tend to overlap, while B2 has slightly lower moduli values than the others at lower reduced frequencies (i.e. higher temperatures). If RTFOT aged bitumen are observed, B1 has the greatest visual shift, with an increase of approximately 600% at higher temperatures to 3% at lower temperatures (higher reduced frequencies), followed by B2 with an increase of 134% at higher temperatures and as low as 2% at lower temperatures while B3 shows almost overlapping results for unaged and RTFOT aged, with a numerical increase ranging from 20 to 6.5% as temperatures reduce. For PAV aged materials, all three bitumen, show a relatively higher increase, as expected. B1 shows a numerical increase in the range 4700 to 5% from higher to lower temperatures, while B2 showed 1940% at higher temperatures to 3% at lower temperatures and lastly the highest increase was observed with B3, with 7952% at higher temperatures and 46% at lower temperatures. On the other hand, if the master curves for phase angle are observed, with RTFOT ageing the phase angles reduce slightly, however a much higher decrease is visible for PAV aged materials. As ageing increases, the stiffness increases along with phase angle reductions, that signify a more elastic, rather than a viscous behavior. Overall, B1 and B3 with overlapping curves for PAV ageing would be expected to perform similar and better than B2 due to their increased stiffness and reduced phase angles, signifying an increased elasticity at higher temperatures. This would enable these materials to resist the stresses, and recover faster. A better numerical evaluation of the effect of ageing could be observed using the aging index. From this test, the parameters of shear modulus at 20 and 60 °C at 10 Hz in RTFOT aged conditions were chosen in order to compare and rank the materials in the later section. These conditions were chosen as these are typical conditions for permanent deformation and fatigue criterion of asphalt.

Figure 2 shows the Black diagrams and the Cole-Cole plots using the 2S2P1D model data. The Black diagrams show the relation between the phase angle and the dynamic shear modulus values. The Cole-Cole plots are used to show the model fit between the actual test data and the model data. Below in Fig. 2a and b, show the fit of the model with the test data and for clarity purposes only one material, B1, unaged and aged, is presented below. As observed, the fit is sufficiently visually accurate, therefore the model can be appropriately used for better comparison purposes to obtain smoother curves and approximate further data.

3.2 Ageing Index

The ageing of bitumen can be expressed numerically in terms of the ageing index. This index is calculated in terms of ratios corresponding to several physical grading tests and is given by the following equation:

$$A.I. = \frac{P_{aged}}{P_{unaged}} \quad (1)$$

Where, P_{unaged} is any physical parameter (e.g. viscosity, softening point, stiffness, etc.) of the unaged bituminous materials and P_{aged} is the same physical parameter as for an aged bituminous materials.

The A.I. was calculated for dynamic shear moduli (bars) and phase angles (lines) at 20 and 60 °C at a frequency of 10 Hz. The lower the ageing index value, lower is the effect of the ageing sensitivity on the material.

From Fig. 3a which represents the A.I. stiffness (with full colors bars for RTFOT and dashed for PAV). At 20 °C for RTFOT, all materials have almost the same A.I. in the range 1.35 to 1.5, while at 60 °C, B1 has the relatively higher A.I. with 2.8, than the 2.0 and 2.1 of B2 and B3. For PAV aged materials, at 20 °C, B3 shows a much higher susceptibility to ageing with the highest value of 3.9 while B1 and B2 are both around 2.5. At 60 °C however, B3 has the lowest A.I., closely followed by B1, with 60% higher than B3. From the results, it can be concluded that B3 was most affected by RTFOT ageing, with B2 being the most susceptible and B3, the least susceptible to PAV ageing at higher temperatures. For phase angles (Fig. 3b), the changes at each temperature and ageing suggest very similar values, with almost horizontal lines for

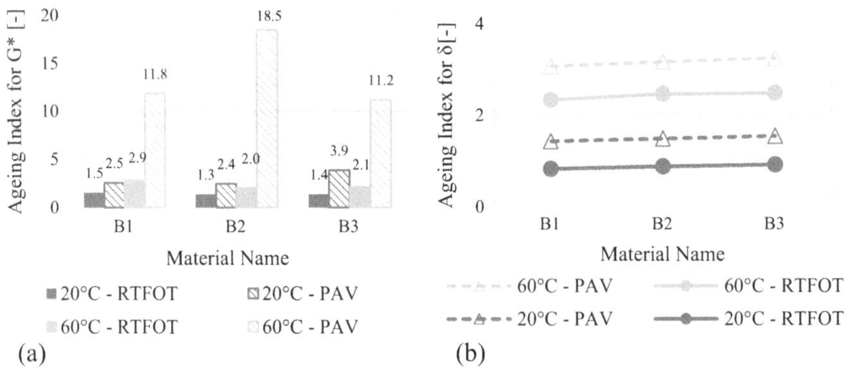

Fig. 3 Ageing index for **a** shear modulus **b** phase angle at 20 °C and 60 °C and 10 Hz

every temperature and ageing type. The phase angle A.I. for PAV are almost double of the RTFOT values for the same temperatures. The main parameters for this section include the A.I. (Stiffness) at 60 °C for RTFOT materials, as rutting occurs at earlier in the lifetime and at higher temperatures, while the A.I. (Stiffness) at 20 °C for PAV materials w.r.t fatigue. For comparison and ranking, the A.I at 60 °C for RTFOT and 20 °C for PAV were taken into consideration for purpose of rutting and fatigue evaluations.

3.3 Single-Stress Creep Recovery (SSCR) Test

SSCR tests were undertaken to evaluate the rutting performance. These tests were carried out on unaged and RTFOT aged materials. PAV aged materials were not tested, as rutting is crucial in the initial lifespan of a pavement and fatigue is more critical in the long term performance. The test results are presented in Fig. 4—showing the deformation with time for ten cycles of loading and unloading.

From Fig. 4, it can be seen that the unaged materials have much higher deformation than the RTFOT aged materials, with values for unaged being in the order of 105 to 45 to being less than 25 for aged. There are vast differences between the unaged materials, with B1 having a significantly higher cumulative deformation. After ageing however, B1 and B3 overlap, with B2 being slightly higher. In order to evaluate numerically the rutting performance, the J_{nr} and %R values were calculated and are presented in Fig. 5. Unaged B2 has the highest recovery and lowest J_{nr} value, as desired, however, for RTFOT, it has the lowest recovery and the highest J_{nr} value. With RTFOT ageing being more important, it can be concluded that bitumen B1 and B3 performs similar. It can be concluded that all three bitumen show significant different rutting performance - in particular bitumen B2 has much better rutting performance than the other two bitumen. The standard J_{nr} values were considered in RTFOT aged condition of comparison purposes presented in the later section as this is the standard value used under the German standards (*AL MSCR-Test (DSR) FGSV 723* 2016).

Fig. 4 Stress dependant fatigue functions for the bitumen tested

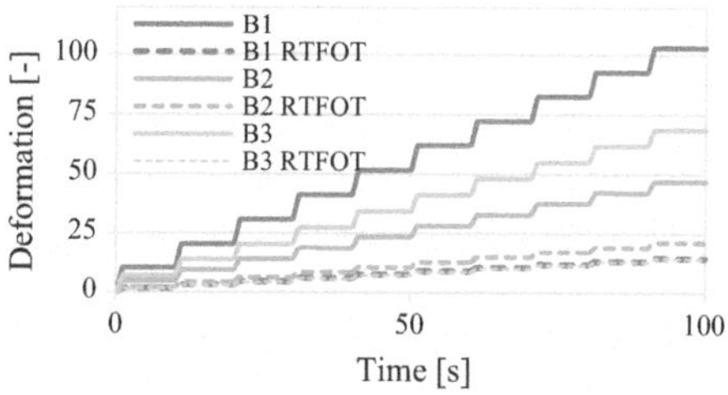

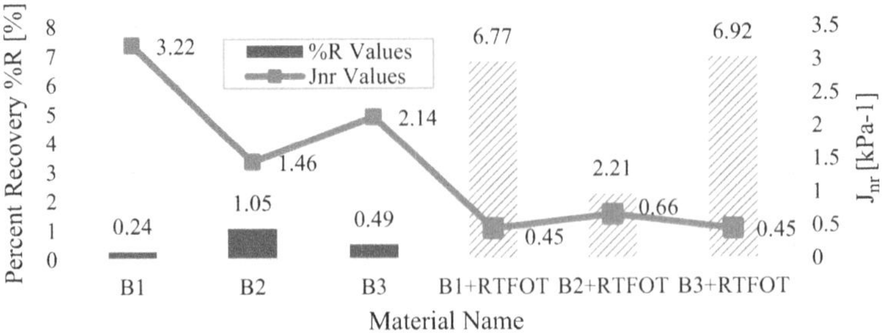

Fig. 5 %R and J_{nr} values for unaged and RTFOT aged bitumen

3.4 *Fatigue Tests*

The fatigue test results were analysed using the Dissipated Energy ratio approach. This states that the dissipated energy of bitumen at failure can be calculated as a function of the product of number of load cycles and the corresponding dynamic shear modulus. For these stress controlled tests, the stress levels were chosen in a way to maintain an initial strain level at 100 load cycles within the range of 0.5 to 2.2%. Due to this, the unaged were tested at a stress range of 100 to 200 kPa, RTFOT at 150 to 250 kPa and PAV at 250 to 400 kPa. This is expected due to an increase in stiffness with an increase of ageing, therefore, causing a lower strain for the same stress. The fatigue curves for applied stress versus the number of load cycles at failure are presented in Fig. 6.

From the figure, it can be observed that there is a clear distinction between the unaged, RTFOT and PAV aged bitumen curves, with the slope for all the materials other than B1 being relatively similar. The slope could be an indicator of the type of failure, the testing temperature and frequency. Also, all materials showed high R^2 values of at least 0.96, indicating a high accuracy of the test.

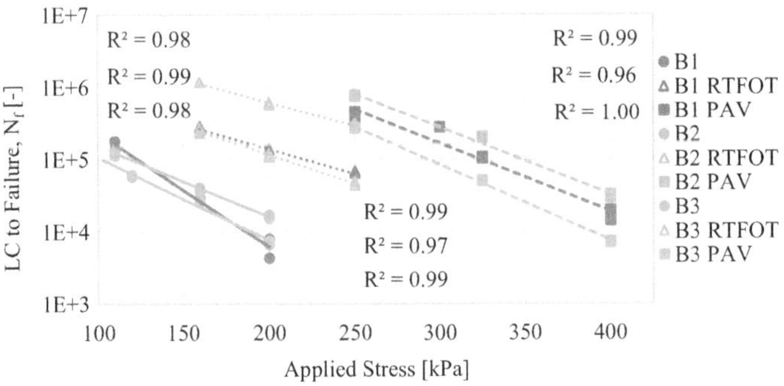

Fig. 6 Stress fatigue curves for the bitumen tested

If the ranking is considered, B2 seems to have the highest fatigue curves, with RTFOT being significantly higher than the other two materials. B3 fails the fastest for unaged and aged materials. For comparison purposes, PAV aged material can be considered the main parameter of this test for ranking purposes in terms of the stress required by the bitumen to achieve 100,000 load cycles. All the values are given in Table 2 below.

3.5 *Synthesis of the Results*

Using the characteristic parameters of the tests undertaken, the diagram presented in Fig. 7 was constructed using to pictorially rank the admixtures based on their stiffness, ageing indexes, fatigue, and rutting behavior. This diagram, which was introduced by the authors in a previous work (Rochlani et al. 2019), gives the possibility of ranking bitumen on the basis of any of the specific parameters studied in this research and obtained via the DSR instrument. From the figure, the ageing indexes and Jnr value axis is reversed, as performance relates inversely to these parameters. This gives the

Table 2 Fatigue performance parameter

Unaged	Stress [kPa] @ 1E+5 LC	RTFOT Aged	Stress [kPa] @ 1E+5 LC	PAV Aged	Stress [kPa] @ 1E+5 LC
B1	123	B1 RTFOT	222	B1 PAV	325
B2	118	B2 RTFOT	320	B2 PAV	350
B3	105	B3 RTFOT	212	B3 PAV	290

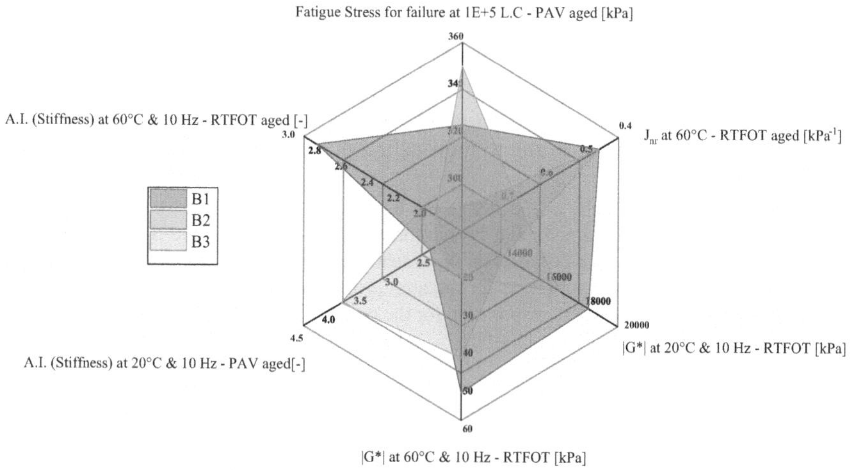

Fig. 7 Performance diagram

opportunity to rank the materials in a way that a larger enclosed area suggests the best performance overall for the parameters used. Another way to use the diagram is be able to identify the particular material based on properties required. According to the figure, B1 is the best and overall performing bitumen, while for specific properties—B1 is best where rutting is critical, while B2 where fatigue performance is of value. This material is also less sensitive to aging compared to the other two bitumen at both aging conditions. It is to be noted that these diagrams can be modified to incorporate other relevant parameters when necessary; and if properties are considered to be more relevant than others for a specific project, it could be possible to add scalar weights to each axle, which will impact the computation of the overall area.

4 Conclusion

Bitumen of grade 50/70 from three different sources were tested in terms of rheology, rutting and fatigue resistance using the DSR. Moreover, testing was also done for RTFOT and PAV aged materials. It was found that despite the similar penetration grade, there were vast differences found in terms of specific performances. While bitumen B1 and B3 provided almost similar rutting resistance, B2 had the best fatigue related performance, while B3 had the lowest performance in this criteria. While for ageing, the materials had similar phase angle ageing indexes, but their stiffness varied significantly with ageing. It can therefore be concluded that more detailed specifications are necessary for bitumen evaluation, and could be made possible using just the DSR, which provide a cost and time efficient option. Finally, to bring together all test results, important parameters were chosen from each test and synthesized in one performance diagram, to give a visual differentiation between the materials, and provide the opportunity to choose a bitumen required for specific purposes. These diagrams would be further developed at TU Dresden to provide a range of values in each specific performance for bitumen using the results of DSR tests. The test results clearly prelude that the current procedure to group materials using the results of needle penetration and ring and ball tests is not an appropriate procedure to characterize bitumen according to their actual performance.

Acknowledgements We hereby acknowledge the financial support provided by TU Dresden as part of the Scholarship Program for the Promotion of Early-Career Female Scientists for Ms. Rochlani. This paper is also partly based on the research project carried out at the request of the German Research Foundation (DFG), under research project No. WE 1642/1-2 and LE 3649/1-2 (FOR2089). The authors are solely responsible for the content.

References

AL MSCR-Test (DSR) FGSV 723 (2016): Arbeitsanleitung zur Bestimmung des Verformungsverhaltens von Bitumen und bitumenhaltigen Bindemitteln im Dynamischen Scherrheometer (DSR) (FGSV), Forschungsgesellschaft für Straßen- und Verkehrswesen

Alfaqawi R, Airey GD, Grenfell JRA (2018) Effects of mineral fillers and bitumen on ageing of asphalt mastics properties. In: bearing capacity of roads, railways and airfields - proceedings of the 10th international conference on the bearing capacity of roads, railways and airfields, BCRRA 2017

Benedetto HD, Delaporte B, Sauzéat C (2007) Three-dimensional linear behavior of bituminous materials: experiments and modeling. Int J Geomech 7(2)

EN 12607-1 (2007) EN 12607-1:2007, Bitumen and bituminous binders - Determination of the resistance to hardening under the influence of heat and air - Part 1: RTFOT method. http://www.aenor.es/

Hunter RN (2015) The shell bitumen handbook, 6th edn. ICE Publishing, London

Jemere Y (2010) Development of a laboratory ageing method for bitumen in porous Asphalt. Delft University of Technology

Migliori F, Corté J-F (2007) Comparative study of RTFOT and PAV aging simulation laboratory tests. Transp Res Rec J Transp Res Board

Oner J, Sengoz B, Maidanova N, Topal A, Malkoc G (2017) Evaluation of rheological effects of waxes on bitumen from different sources

Petersen JC (2009) A review of the fundamentals of asphalt oxidation: chemical, physicochemical, physical property, and durability relationships. Transportation Research E-Circular (E-C140), The National Academies: Washington D.C

Rochlani M, Leischner S, Falla GC, Wang D, Caro S, Wellner F (2019) Influence of filler properties on the rheological properties and performance of mastics. Submitted to Construction and Building Materials

Singh H, Jain PK (1997) Bitumen quality and manufacturing processes—past and present technological status. Indian J Chem Technol

Wegmann M, Vienenkötter L (2017) Bitumen quality recording—a German approach

Permanent Deformation Characterisation of Gap-Graded and Continuous Graded Aggregate Blends for Bituminous Mixtures

V. T. Thushara and J. Murali Krishnan

Abstract One of the major parameters which influence the performance of bituminous mixtures is a packed aggregate skeleton with minimum particle rearrangement during traffic loading. The current practice of aggregate design for bituminous mixtures is mostly based on the Fuller-Thomson model to achieve maximum density. However, this approach does not explicitly take into account the volumetric requirements of the bituminous mixture in terms of rutting and fatigue. This study investigates the mechanical behavior of bituminous mixtures prepared using aggregate gradations, designed based on binary particle packing model of Powers (1964) extended to multicomponent packing by Lees (1970). One control sample and two designed gradations are analysed, in which the first one is a continuous gradation, whereas the second one is a gap gradation. A binder content of 5% is used for the three bituminous mixtures. Samples are subjected to dry rut wheel testing and repeated creep and recovery test with a modified test protocol to quantify their rutting performance. Test results indicate superior rut resistance of the gap-graded aggregate mix. Also, the link between aggregate gradation, compactability, and rut resistance of bituminous mixtures is established for a given binder content.

Keywords Particle packing theory · Gap-graded · Continuously-graded · Aggregate gradation design · Bituminous mixtures · Rutting

1 Introduction

Bituminous concrete is a heterogeneous mixture consisting of bitumen and aggregates. Achieving the expected performance, in terms of rutting and fatigue cracking, is the major criteria in the design of bituminous mixtures. The performance of bituminous mixtures is influenced by the quality and quantity of the constituents, as well as the packing characteristics of the mixture. The aggregate skeleton forms

V. T. Thushara · J. Murali Krishnan (✉)
Department of Civil Engineering, Indian Institute of Technology Madras, Chennai 600036, India
e-mail: jmk@iitm.ac.in

V. T. Thushara
e-mail: thusharatkm@gmail.com

C. Raab (ed.), *Proceedings of the 9th International Conference on Maintenance and Rehabilitation of Pavements—Mairepav9*, Lecture Notes in Civil Engineering 76,
https://doi.org/10.1007/978-3-030-48679-2_47

the backbone of bituminous mixtures and is responsible for resisting the permanent deformation. The current gradation design methodology follows Fuller and Thomson (1907) model to achieve maximum density. Even though this model works very well for materials such as concrete, in which there is no particle reorientation while in service; it cannot explicitly take into account the requirements of Hot Mix Asphalt (HMA) in which there can be evolution of the aggregate structure during service.

Typically in bituminous mixtures, there are two stages of densification, one due to roller compaction and the other due to passage of traffic wheel loads. The existing mix design procedures for HMA, specify an initial air void requirement of 6–7% while opening for service and a final air void of 3–4% at the end of design life. The presence of high air voids in the initial stages will lead to particle rearrangement with traffic passes leading to the damage due to rutting. This densification will make the mixture stiffer, which will further aggravate damage due to fatigue. Hence, it is seen that the density restrictions are contradictory, and due to the peculiar nature of dense gradation, it is challenging to balance the performance. If the gradation is designed such that aggregates form an interlocked skeleton during the compaction phase and will not evolve substantially further with traffic loading, it is expected to result in a balanced bituminous mixture (Cooper et al. 2014) which is resistant to rutting and fatigue. In this context, the framework of particle packing approaches will facilitate gradation design in a rational manner. This knowledge, in turn, can aid designers to select the optimum combination of aggregates, which will form interlocked internal structure and will show increased resistance to accumulation of permanent deformation as well as fatigue damage.

Various particle packing approaches are reported in the literature, for discrete and continuous aggregate combination. Furnas (1928), outline the procedure to estimate the optimum proportion for maximum packing of two components based on the void values of the separate component for size ratio = 0. Here, the size ratio is defined as the ratio of the diameter of smaller particles to the diameter of larger particles in a binary mix. This method is extended by Powers (1964), in terms of specific void content graph for size ratio = 0. In real aggregates, the size ratio is never zero, and there exist interactions among aggregates. Many interactions are postulated, and these include (i) wall effect (ii) interference effect and (iii) wedge effect. If the average particle dimension of fines is small enough compared to that of coarse particles, the wall effect is linear, and the interference effect is non-linear (Olard and Perraton 2010). Olard and Perraton (2010) have applied the principles of packing based on Baron's approach for optimization of aggregate gradation for high-performance bituminous mixtures.

In this study, one control sample and two gradations designed appealing to particle packing are investigated. One of the designed gradations is a continuous gradation, whereas the other one is a gap gradation. The discrete model-based approach due to Powers (1964) is used to get the proportion of aggregate size ranges, for minimum void index. Rutting caused due to densification and shear flow, has a direct influence on aggregate gradation. The rut magnitude can be minimised if an aggregate skeleton that will not evolve much during loading could be designed. The verification/quantification of the achievement of such an aggregate skeleton is complex with the available test protocols. In this study, the formation of the designed aggregate skeleton is ascertained in a gross sense using some measures of compactability and laboratory based rut resistance measurements. The compactability of the mix with different gradations indicates the gradations which require substantial effort for particle rearrangement at compaction temperature. Such gradations are most likely to have lesser particle rearrangement due to traffic passes at lower service temperatures, leading to higher rut resistance. The rutting performance of the bituminous mixtures produced from the designed gradations in this study showed a link to compactability when quantified in terms of dry rut wheel experiments and repeated creep and recovery tests done with modified trapezoidal protocol at 60 °C.

2 Material Used

Quarry located at Madurantakam, Kanchipuram, India is the source of granite aggregates used in this investigation. The specific gravity is 2.717, 2.774, and 2.752 for coarse aggregates, fine aggregates, and filler, respectively. The selected size range of aggregates is passing 13.2 mm to 75 µm; the range for bituminous concrete grade II as per the current Indian guidelines (MoRTH 2013). The binder used is of VG 30 grade as per Indian Specifications for paving bitumen (IS 73, 2013). The binder used has a penetration of 40 dmm at 25 °C, a softening point of 51 °C and viscosity at 60 °C of 2511 Poise. Binder content of weight equal to 5% by weight of total mix is used for all the mixes.

3 Design of Gradations

Three mixtures designated as Mix A, Mix B, and Mix C with different aggregate gradations are used in this study. Mix A, the control sample, is the mid-gradation of conventional dense-graded Bituminous Concrete Grade II specification, as per the current Indian specification, MoRTH (2013). Mix B and Mix C are designed gradations based on the concepts of Powers (1964), which is an adaption of Furnas (1928) binary model. Lees (1970) notion of an extension of binary component mix to the multicomponent domain is followed for the gradation design of Mix B and Mix C. The optimum proportion of aggregate size ranges are estimated targeting a minimum void index in stages. The procedure adopted for the gradation design of Mix B and Mix C includes measurement of void ratio and stage by stage estimation of optimum proportion for minimum voids. The design steps are detailed below.

Step 1: Selection of aggregate size ranges
Mix B is designed to have continuous gradation, as a ternary blend, in which optimum proportion of coarse aggregates (CA), fine aggregates (FA), and filler is estimated. The definition of coarse aggregate and fine aggregate is taken from Bailey's method of gradation design (Vavrick et al. 2002). As per this procedure, the sieve which differentiates the coarse and fine aggregates is dependent on the Nominal Maximum Aggregate Size (NMAS) of the mix as 0.22 × NMAS. The 0.22 factor is the average condition of many different packing configurations of two dimensional and three-dimensional analysis of the packing of differently shaped particles. Also, it is stated that 0.22 is only a suggestion, and it can vary from 0.18 to 0.28 (Vavrick et al. 2002). The 0.22 factor is considered here for Mix B design. Accordingly, coarse aggregate for Mix B has NMAS 13.2 mm, fine aggregate has NMAS 1.18 mm, and filler constitutes material passing 75 μm. The ratios of mass median diameter D_{50} of consecutive components are kept below 0.2, to minimize interaction effects. The gradation details of the coarse, fine, and filler components are given in Table 1.

Mix C is designed as a gap graded mix with aggregate size ranges passing 13.2 mm. The criterion followed for gap graded design is that the size of the fine component to be mixed with the coarse component at any stage of grading is to be approximately equal to the size of voids in the coarse aggregate skeleton (Lees 1970). Therefore Mix C has only specific sizes of aggregates with the constraint of size ratio ≈0.2 between adjacent size ranges so that the fine aggregates are filled in the voids created by the coarse aggregate skeleton with minimum interference effect. In this selection process, many size ranges in between are skipped to meet the size ratio criterion,

Table 1 Gradation of coarse, fine and filler components of Mix B

Sieve size (mm)	Cumulative percentage passing		
	Coarse aggregate	Fine aggregate	Filler
19	100	100	100
13.2	75	100	100
9.5	50	100	100
4.75	25	100	100
2.36	0	100	100
1.18	0	67	100
0.6	0	33	100
0.3	0	0	100
0.15	0	0	67
0.075	0	0	33
pan	0	0	0
D_{50}	9.5	0.85	0.11
Size ratio	$\frac{D_{50}FA}{D_{50}CA} = 0.09$	$\frac{D_{50}FA}{D_{50}Filler} = 0.13$	

and hence the overall gradation is gap graded in comparison with the control mix A. Mix C gradation is quinary with aggregate sizes 9.5, 2.36, 0.6, 0.15 mm and material passing 75 μm

Step 2: Determination of void index

The void index is calculated by subjecting the known quantity of the required blend to vibratory tamping in the relative density apparatus. Vibratory tamping is continued until the height of the sample becomes constant. The final height of the compacted sample is measured, and the volume of the compacted sample is calculated. Knowing the mass and specific gravity of the components mixed, the solid volume and the corresponding void index of the blend are calculated.

Step 3: Determination of optimum proportion for a binary mixture

The void index due to 100% fine aggregate as well as 100% coarse aggregate are measured and are plotted in the void index vs. coarse aggregate volume proportion plot. The left ordinate of the graph represents 100% fines (0% coarse), and the right ordinate represents 100% coarse (100% fines). From the void index value of 100% fine aggregate, the theoretical boundary line of the fine dominant mixture is drawn to the opposite vertical ordinate value of 0. Also, the void index value of 100% coarse aggregate is connected to the value −1, in the opposite ordinate, to get the theoretical boundary line of the coarse dominant mixture. The values 0 and 1 represent the theoretical minimum void index, as discussed by Powers (1964). The proportion corresponding to the intersection point of the two theoretical boundary lines for size ratio = 0 is adopted as the optimum binary blend proportion for the minimum void index, assuming that the locus of optimum proportion for size ratio ≈0.2 will not deviate much horizontally from that for size ratio 0 (Powers 1964).

Different volume proportions of coarse and fine components are mixed, and their measured void indices are plotted in the void index graph, to compare the trend in experimental data with the theoretical estimation. The optimum proportion determination of the largest two components (Binary Stage 1) for Mix B and Mix C is given in Fig. 1a and b respectively.

Step 4: Extension to the multi-component mixture

The optimum binary blend proportion of the largest two aggregate size ranges under consideration estimated in stage 1 is combined with the next finer component, and the optimum ternary proportion is calculated in stage 2. This procedure is repeated to get an optimum multi-component mix (Lees 1970). Stage 2 details of Mix B and Mix C are given in Fig. 2a and b, respectively. Mix B blends the optimum proportion of coarse aggregates and fine aggregates obtained in stage 1 with filler passing 75 μm whereas Mix C combines the optimum blend of 9.5 mm and 2.36 mm with 0.6 mm in the second stage. Stage 3 and stage 4 of Mix C further find the optimum combination of stage 2 optimum mix with 0.15 mm and filler passing 75 μm respectively (Fig. 3a, b).

The gradation details of the three mixes chosen are given in Table 2.

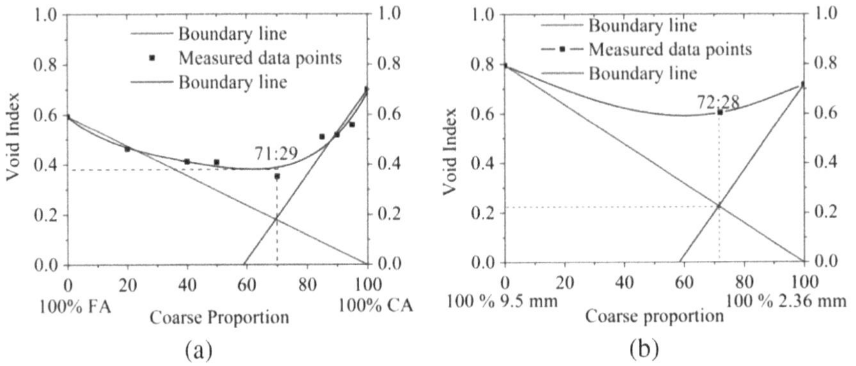

Fig. 1 Binary stage 1 (a) Mix B (b) Mix C

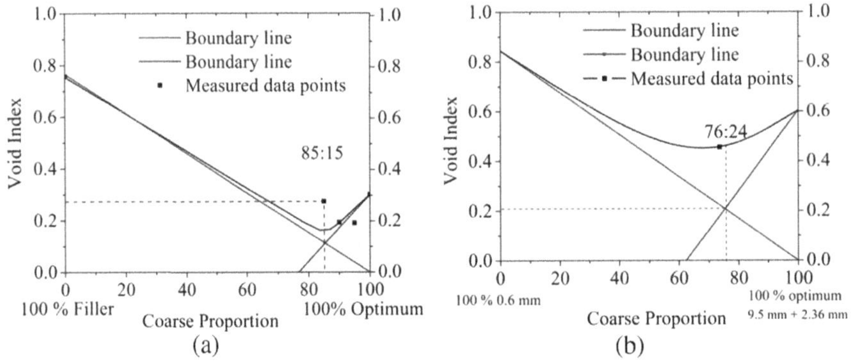

Fig. 2 Gradation design stage 2 (a) Mix B (b) Mix C

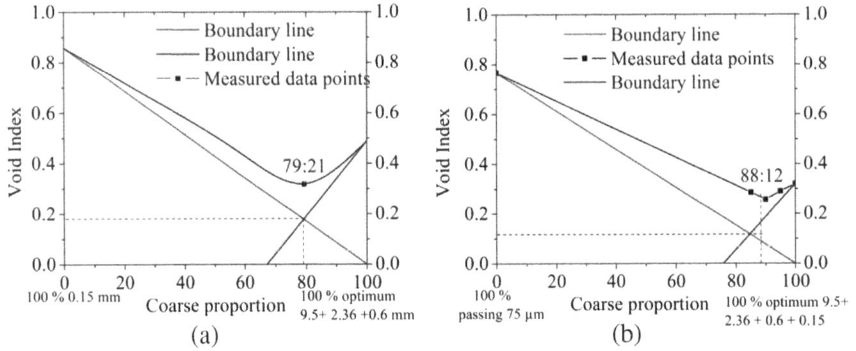

Fig. 3 Mix C gradation design stages (a) Stage 3 (b) Stage 4

Table 2 Gradation details

Sieve size (mm)	Cumulative percentage passing		
	Mix A	Mix B	Mix C
19	100	100	100
13.2	95	93.9	100
9.5	79	72.3	62.6
4.75	62	51.8	62.6
2.36	50	39.7	47.7
1.18	41	33.6	47.7
0.6	32	28.7	31
0.3	23	25.5	31
0.15	16	19.6	12.2
0.075	7	15.2	12.2
Pan	0	0	0

4 Sample Preparation

Three sets of bituminous mixtures are designed with the three gradations A, B, and C. Experiments are designed to capture the effect of designed gradations, and hence to keep the number of variables minimal the binder is kept constant as 5% by weight of the total mix. Preheated aggregates at 170 °C are mixed with binder heated at 165 °C so that a temperature of 165 °C is maintained during mixing. The prepared mixtures are kept for 4 h short-term aging at 135 °C and are compacted at 155 °C. Sample compaction is carried out in shear box compactor (ASTM: D7981-15 2015), to a beam of size 450 mm × 150 mm × 145–180 mm. All the samples are compacted in such a manner that the air voids of the cored/sliced test samples are identical in the range of 6 ± 0.5%. The shear box compactor ensures uniform compaction of the sample throughout and uniform air void distribution for the test samples (Priyadharshini and Krishnan 2014). For the rut wheel test, slab samples of 300 mm × 150 mm × 50 mm are sliced, and for flow number tests cylindrical samples of 150 mm diameter × 100 mm height are cored from the shear box compacted beam samples. The samples with an air void range of 6 ± 0.5% are used for both dry rut wheel and flow number tests. From each shear box compacted beam, three flow number samples or two rut wheel slabs can be obtained.

5 Results and Discussion

5.1 Compactability

Due to the varying nature of the gradations, different compactive effort in terms of a number of shear cycles has to be given to achieve target air voids of 6 ± 0.5%. The shear box compaction data captured during compaction of the three mixtures are analysed, and the normalized percentage air voids vs. the number of cycles of compaction are plotted in Fig. 4. This plot indicates the relative compactability of the mixtures prepared from the three gradations considered for the combination of higher load level at high temperature (during compaction) and it is assumed that as the temperature and load level reduces during service, the material which exhibited higher resistance during compaction will indeed exhibit identical levels of resistance. The number of compaction cycles required to reach the target air void, in this case, for Mix A, Mix B, and Mix C is 17, 14, and 26 cycles, respectively. From the data, it can be interpreted that the gap graded mix, Mix C is difficult to be compacted compared to the other two. On the other hand, Mix B requires a lesser number of compaction cycles to achieve the same target air voids, which is an indication of the ease of compaction. Permanent deformation measurements in this study also indicated that less compactable mixtures are more rut resistant.

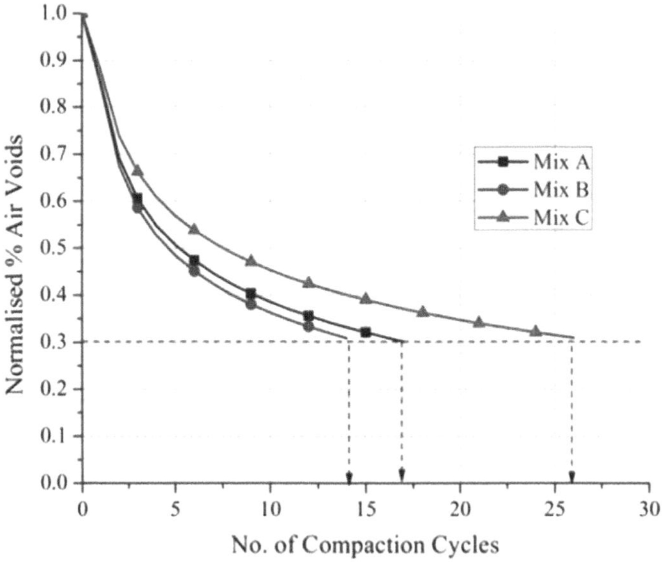

Fig. 4 Compactability of mixes

5.2 Rut Wheel Tracking Test

A dry wheel tracker is used to determine the rut depth of the bituminous mixtures. In this test, a standard loaded wheel (700 N) moves forward and backward on top of the bituminous concrete sample at a speed of 26.5 rpm at a controlled temperature. The rut depth at each cycle is calculated using the mean rut depths measured at every 4 mm interval along the wheel path. In this study, the rut wheel samples are tested at 60 °C. The termination criteria for the test are set as either 30,000 wheel passes or 10 mm rut depth, whichever occurs first.

The slab samples prepared with 6 ± 0.5% air voids are tested after 6 h of temperature conditioning, and 500 conditioning passes. The test results are reported in Fig. 5. It is observed that the gap graded Mix C is more rut resistant, followed by control sample Mix A. Mix B with continuous gradation has the least rut resistance, despite the expected coarse aggregate skeleton structure. The reason for this may be attributed to the use of the same binder content of 5% in this study, for all the mixtures, irrespective of the gradation used. Mix B is physically a rich mixture in terms of binder content. The compactability data presented in Fig. 4, indicates the same as its air voids are getting reduced rapidly in the initial cycles. In reality, the specific surface area is different for the three mixtures owing to the variation in the gradation. This, in turn, will demand varying binder content to satisfy the coating requirements as well as the packing requirements.

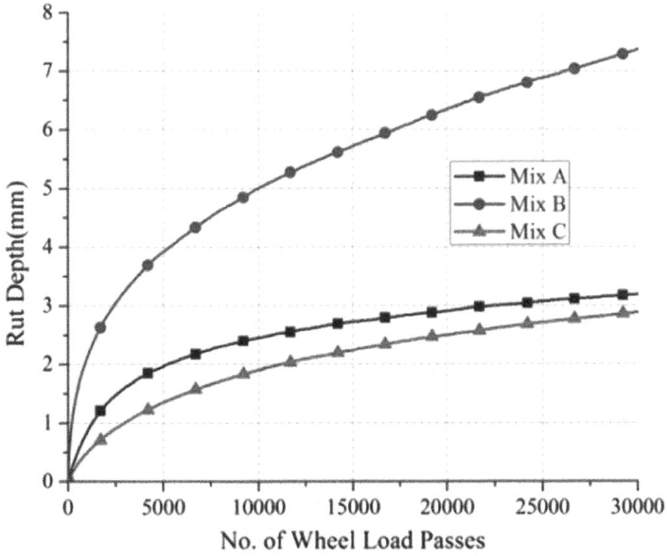

Fig. 5 Rut Depth vs. no of passes

5.3 Flow Number Test

Flow number test is a cyclic creep and recovery test in which the specimen is subjected to repeated load cycles. Different load waveforms are available in the literature (Roy et al. 2016) to simulate the field loading. The standard test protocol as per AASHTO T378-17 (2017) suggests using a haversine pulse load of duration 0.1 and 0.9 s rest under confined or unconfined conditions. However, it is reported (Gayathri et al. 2016) that the existing 0.1 s haversine loading protocol, will not lead to a steady-state response, and it captures only the transient response of the material. Also, the rest period of 0.9 s may not give enough time for the material to recover all the strains, and the residual deformation at the end of each cycle may not be a true residual strain but strain at a transient state. Hence, in order to capture the steady-state behavior in cyclic creep and recovery, a modified test protocol of trapezoidal loading with 1 s loading and 2 s rest period is given for the current work.

The loading pattern is given as a ramp load increment in 0.1 s, followed by the constant load for 0.8 s and then unloading by ramp load decrement in 0.1 s. A maximum compressive deviatoric load of 200 kPa is applied. Confinement pressure of 200 kPa is applied. After the application of 1 kPa seating load, confinement pressure is ramped up in 100 s and is held throughout the test. The loading protocol adopted is given in Fig. 6. The tests are conducted at 60 °C. Termination criteria are given as 10,000 cycles of load repetitions or a maximum actuator displacement of 25 mm. The software interface is modified to measure the applied deviatoric load and confinement pressure, as well as the corresponding sample response in terms of actuator displacement, at every 0.001 s through the data acquisition system. Therefore

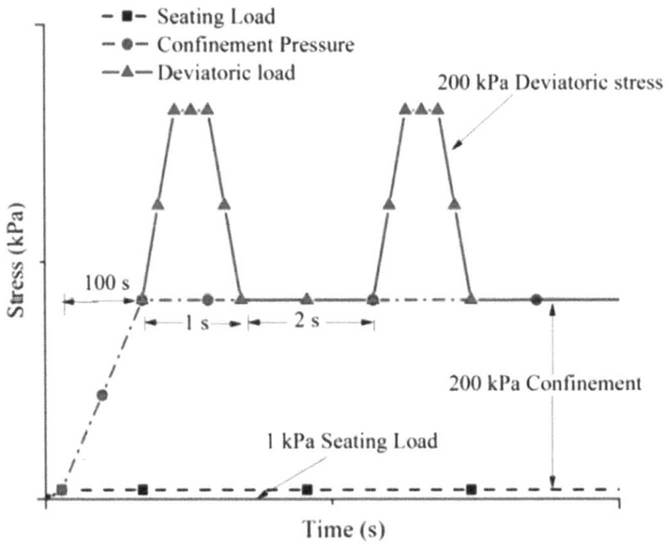

Fig. 6 Trapezoidal loading protocol (not to scale)

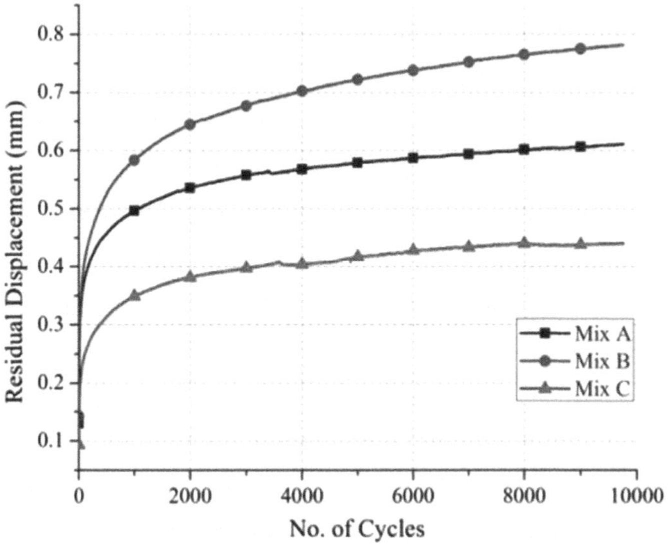

Fig. 7 Residual deformation vs. time

each cycle of loading is captured using 3000 data points, which essentially captures the complete creep and recovery response.

The creep and recovery test results can be presented in terms of cumulative irrecoverable strain value at the end of each load cycle vs. the time of loading. The test results for the three mixes are shown in Fig. 7. The results also show a similar trend as that of the rut wheel test result. The rut resistance ranking of the mixtures is in the order of Mix C, Mix A, and Mix B.

6 Conclusion

The study provides a rational gradation design approach based on particle packing method, so that well packed bituminous samples, with an interlocked coarse aggregate skeleton, is targeted with minimum void index. Three sets of gradations are considered, in which the first one is a control sample with conventional dense grading. The other two gradations are designed based on the particle packing approach given by Powers (1964) and Lees (1970), in which the first one is a continuous gradation and the second is a gap gradation. Particle sizes are selected such that a coarse aggregate skeleton structure is formed in the mix, and the finer materials will be filled in the void space in the coarse skeleton without disturbing the structure with minimum interactions. The samples are cast identically with the 5% binder content and compacted to identical air void content for all the three gradations, and are tested for capturing its rutting performance using dry rut wheel testing and repeated creep and recovery using trapezoidal loading protocol.

The mechanical test results in terms of rut depth from dry rut wheel test and residual deformation from flow number tests show the superior rut resistant behavior of gap graded mixes, followed by the control sample having conventional dense gradation. The optimal gradation design adopted may have formed an interlocked aggregate structure which in turn contributes to the higher rut resistance of Mix C. The lower rut resistance of Mix B, irrespective of the expected coarse aggregate skeleton may be attributed to two reasons. (i) gradation design rationalization is found to be more effective in gap gradation, as theoretically a telescopic fit of smaller particles into the void space created by larger aggregate skeleton is possible. (ii) the use of the same binder content of 5% in this study. The applied binder content will form different mastic volume in the mixture, due to the varying surface area of the aggregates. This will affect the coating and packing characteristics of the mixtures. The tender nature of Mix B is reflected in the compactability data, as the air void reduction rate is high in comparison to other mixtures during the initial number of cycles. This points to the need to determine optimum binder content for effective packing. Further studies are to be carried out to identify the optimum amount of binder for the different gradations to have similar binder richness, based on the surface area of aggregates.

References

AASHTO T 378 (2017) Standard method of test for determining the dynamic modulus and flow number for asphalt mixtures using the asphalt mixture performance tester (AMPT). American Association of State Highway and Transportation Officials, Washington DC, USA

ASTM D 7981-15 (2015) Standard practice for compaction of prismatic asphalt specimens by means of the shear box compactor. ASTM International, West Conshohocken, PA

Cooper SB III, Mohammad LN, Kabir S, King W Jr (2014) Balanced asphalt mixture design through specification modification: Louisiana's experience. Transp Res Rec: J Transp Res Board 2447:92–100

Furnas CC (1928) Relations between specific volume, voids and size composition in systems of broken solids of mixed sizes. Department of Commerce, U.S. Bureau of Mines. Report of Investigation, 2894

Fuller WB, Thompson S (1907) The laws of proportioning concrete. Trans Am Soc Civ Eng 59:67–172

Gayathri VG, Rajasekar YP, Lakshmi Roja K, Murali Krishnan J (2016) Influence of confinement pressure on the development of three-stage curve for bituminous mixtures. Transp Dev Econ 2(2):1–8

IS (Indian Standards). (2013). Paving bitumen: Specification. IS: 73-2013,New Delhi, India

Lees G (1970) The rational design of aggregate gradings for dense asphaltic compositions. In: The proceedings of the association of asphalt paving technologists, Kansas City, Missouri, vol 39, pp 60–90

MoRTH (2013) Specifications for roads and bridge works. Indian Roads Congress, New Delhi, India

Roy N, Veeraragavan A, Krishnan JM (2016) Influence of confinement pressure and air voids on the repeated creep and recovery of asphalt concrete mixtures. Int J Pavement Eng 17(2):133–147

Olard F, Perraton D (2010) On the Optimization of the aggregate packing characteristics for the design of high-performance asphalt concretes. Road Mater Pavement Des 11(sup1):145–169
Powers TC (1964) Geometric properties of particles and aggregates. J Portland Cement Assoc 6(1):2–15
Priyadharshini RY, Krishnan JM (2014) How workable are modified binder mixtures? A comparative study with gyratory and shear compactor. In: Colloquium on transportation systems engineering and management, CTR, CED, NIT Calicut, India
Vavrick WR, Huber G, Pine WJ, Bailey R (2002) Method summarized for gradation selection in hot-mix asphalt design. Transportation Research Circular E-C044, TRB, National Research Council, Washington, DC

Linear Viscoelastic Properties of a Half Warm Asphalt Mixture (HWMA) with Bitumen Emulsion

Silvia Angelone, Marina C. Casaux, Luis Zorzutti, and Fernando Martinez

Abstract Half Warm Asphalt Mixtures (HWAM) with bitumen emulsion are a relative new, cost effective and environmental friendly emerging technology. HWAM are produced using a combination of aggregates heated up to approximately 120 °C and a percentage of a bitumen emulsion heated to a temperature around 60 °C in order that the mixing and compaction temperature results between 80 to 90 °C. Hence, this procedure involves great savings of energy and low harmful emissions. Although there are recommendations for the formulation of this type of mixtures and references to their conventional mechanical properties, there is a lack of information regarding their rheological properties. This paper presents preliminary results concerning the design and performance of a Half Warm Asphalt Mixture using bitumen emulsion with the aim to evaluate the linear viscoelastic behaviour of this kind of materials. A HWMA was formulated using a cationic super stabilized bitumen emulsion and the Dynamic Modulus (E*) and the Phase Angle (ϕ) at different temperatures and loading frequencies were measured to produce experimental data for pavement design procedures based on mechanistic principles. It is considered that those findings could encourage greater confidence in promoting the use of these types of sustainable asphalt mixes.

Keywords Half Warm Asphalt Mixture (HWMA) · Bitumen emulsion · Linear viscoelasticity · Dynamic Modulus

1 Introduction

Sustainability and environmental care have become key concepts in our lives. In road engineering, these concepts are related to more reliable pavement design procedures based on mechanistic principles, longer pavement design lives and more efficient pavement conservation strategies. Also, the reuse or recycling of materials looking for lower consumptions of natural or non-renewable materials, lower mixing

S. Angelone · M. C. Casaux · L. Zorzutti · F. Martinez (✉)
Road Laboratory, IMAE Institute of Applied Mechanics and Structures, National University of Rosario, Riobamba and Beruti, 2000 Rosario, Argentina
e-mail: fermar@fceia.unr.edu.ar

C. Raab (ed.), *Proceedings of the 9th International Conference on Maintenance and Rehabilitation of Pavements—Mairepav9*, Lecture Notes in Civil Engineering 76,
https://doi.org/10.1007/978-3-030-48679-2_48

and compaction temperatures for asphalt materials and lower energy consumptions during the overall construction process must be considered.

Traditionally, the asphalt mixtures were classified in two groups: Hot Asphalt Mixture (HMA) and Cold Asphalt Mixture (CMA).

HMA uses bitumen heated at 150 °C or above to reduce its consistency and in this condition, is mixed with the hot aggregates and compacted in the pavement. HMA is the most common type of material used for construction and maintenance of flexible pavements with an excellent mechanical performance. However, this is a material with a high demand of energy to reach the required high temperatures with large amounts of emissions of pollutants gases during production and construction stages. These emissions pollute the air, intensify the greenhouse effect and are harmful for the health and working conditions of the workers.

CMA uses bitumen emulsions and the mixing with the aggregates and the compaction in the field are at ambient temperature. This kind of mixtures is less used and they do not require the heating of the aggregates or the binder. CMA has lower emissions of harmful gases, lower energy consumptions and an excellent environmental footprint but its mechanical performance is poorer compared to HMA mainly at early age and hence, its use is limited to low and medium traffic levels or in roads of developing countries as a first paving alternative.

To mitigate the energy and emission disadvantages of HMA but with the same mechanical performance as those, a kind of mixtures made at lower temperatures called Warm Asphalt Mixtures (WMA) has been developed where the production and placing are at temperatures between 30 to 40 °C below the normal temperatures used for HMA. Different technologies have been proposed to produce lower asphalt viscosity, better workability of the mix or lower harmful emissions (Rubio et al. 2012).

The advantages of these lower temperatures related to the environmental care are significant but still there is a necessity of achieve high temperatures to dry the aggregates and then, the energy savings are not enough important.

Looking for the optimization of this issue, a new generation of asphalt mixtures called Half Warm Asphalt Mixtures (HWMA) is in development. These mixtures are prepared and compacted at temperatures below the water evaporation temperature (100 °C) with bitumen emulsion as the binder to obtain a similar mechanical performance as for the HMA. They have gained a lot of acceptation in Spain and Portugal where road sections of low volume of traffic have been built with this technology for surface and base layers.

Lopez et al. (2017) have compared a Half Warm Asphalt Mixture with bitumen emulsion to a Hot Asphalt Mixture using samples compacted by impact (Marshall) and gyratory compaction. They conclude that it is not possible to achieve the same volumetric properties than in the HMA when the impact compaction procedure is used. When the gyratory compaction procedure is applied, it is possible to obtain the same degree of compaction but the mechanical properties of the HWMA are poorer than those of the HMA used for comparison.

Sangita et al. (2015) have carried out an experimental study comparing the mechanical properties of a HWMA and a CMA. The study concludes that HWMA have higher Marshall Stability and Indirect Tensile Strength than the CMA.

Dinis-Almeida et al. (2010) have presented a case study about the use of a Half Warm Asphalt Mixture with bitumen emulsion and Reclaimed Asphalt Pavement (RAP) for the rehabilitation of a road section in Portugal. A large number of laboratory specimens were produced for both Marshall and immersion-compression tests.

The experience obtained in this case study indicates that RAP gradation, recovered bitumen content from the RAP, mixing and compaction temperatures, as well as curing period, should be taken into account for mix design of the HWMA. Also, considering the results obtained using the Marshall mix design method compared to the immersion compression tests used for Cold Asphalt Mixtures, it can be pointed out that the Marshall procedure seems to be more adequate for mix design of HWMA.

Lizarraga et al. (2017) describe a study about the use of a Half Warm Mixture with high Reclaimed Asphalt content in wearing courses. For this purpose, HWMA with 70 and 100% Reclaimed Asphalt Pavement (RAP) and emulsion were designed in the laboratory. In a second stage, HWMA were manufactured in-plant, and laid and compacted in an Accelerated Pavement Test track. The results from the tests conducted with both laboratory specimens and cores showed that the performance of HWMRA is comparable to that of HMA. In a similar study, Garcia Santiago and Lucas Ochoa (2014) have proposed a specific methodology for the asphalt mix design of HWMA when an amount of RAP up to 100% is used.

Other study by Swaroopa et al. (2015) compared the mechanical properties of a CMA, a HWMA with bitumen emulsion and a Mild Warm Asphalt Mixture (MWMA) where the aggregates were heated up to 70 °C and the mix was compacted at 50 °C. Results from performance tests indicated that MWMA was better than CMA and comparable with HWMA.

Since there is not a regulatory framework for the HWMA (Miranda Perez et al. 2013), all of these studies were carried out using different procedures regarding the temperatures of aggregates and bitumen emulsion, the compaction methods or the curing conditions. Looking for the standardization, the Technical Association of the Bitumen Emulsion (*ATEB for Asociacion Tecnica de la Emulsion Bituminosa* in Spanish) has promoted a monographic study that could be used as a guide for the design of Half Warm Asphalt Mixtures (ATEB 2014). Also, a road authority in Spain (*Agencia de Obras Publicas de Andalucia* in Spanish 2012) has developed some recommendations about the materials, asphalt plants, placing and quality control when HWMA with bitumen emulsion are used. It can be concluded that HWMA can be used in road construction with a similar performance as for other asphalt mixtures conventionally used. Also, some of these papers have analyzed the volumetric and conventional mechanical properties (Marshall Stability and Indirect Tensile or Compressive Strength) but there is a lack of information about their rheological properties.

The objective of this paper is the evaluation of the rheological properties (dynamic modulus and phase angle) of a HWMA for different conditions of testing temperatures and loading frequencies to have experimental results that could be applied in pavement design procedures based on mechanistic principles. A description of the experimental procedures, used materials, obtained results and conclusions are presented and discussed. Other material properties like moisture sensitivity, fatigue and permanent deformation resistance will be published in future studies.

2 Viscoelastic Response of Asphalt Mixtures

In this paper, the Huet-Sayegh Model (H-S Model) was adopted for the characterization of the rheological behaviour of the asphalt materials (Sayegh 1967). This is an analogical model composed by two parabolic dampers and two springs connected in parallel and series as it is shown in Fig. 1. The analytical formulation of the H-S model is given by:

$$E^*(i\omega) = E_0 + \frac{E_\infty - E_o}{1 + \delta(i\omega\tau)^{-k} + (i\omega\tau)^{-h}} \tag{1}$$

where $E^*(i\omega)$ is the complex modulus, ω is the angular frequency, E_∞ is the glassy complex modulus for $\omega \to \infty$, E_0 is the static complex modulus for $\omega \to 0$, h and k are exponents characterizing the damper characteristics, δ is a dimensionless constant, τ is the characteristic time and i is the imaginary number ($i = \sqrt{-1}$).

The characteristic time τ varies only with temperature and its evolution is given by a shift factor aT that describes the temperature dependency of the material on the basis of the principle of frequency-temperature superposition for thermo-rheologically simple materials:

$$aT = \frac{\tau}{\tau_0} \tag{2}$$

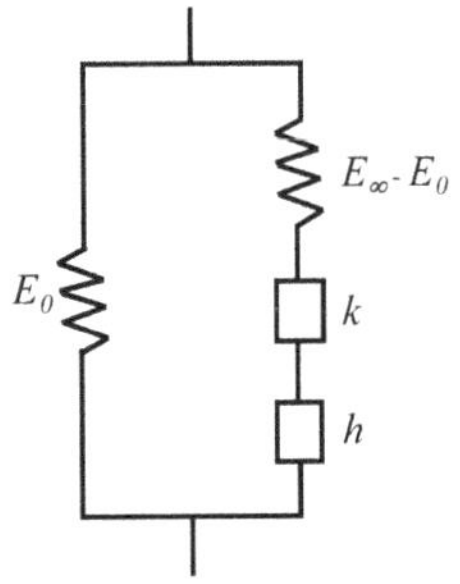

Fig. 1 The Huet-Sayegh model (H-S)

where τ is the characteristic time at the experimental temperature T_i and τ_0 is the characteristic time at the Reference Temperature T_R.

In this paper the shift factor aT was modelled according to an Arrhenius equation in the form of:

$$aT = e^{C_A\left(\frac{1}{T_i} - \frac{1}{T_R}\right)} \tag{3}$$

where C_A is the Arrhenius material constant and T_i and T_R are temperatures in °K.

Using complex algebra, it is possible to obtain the components of the complex modulus and finally, the Dynamic Modulus |E*| and the Phase Angle ϕ. Several authors have used the H-S model describing the viscoelastic behaviour of asphalt materials (Yussof et al. 2011; Olard 2003; Cauhapé et al. 2017).

3 Materials and Procedures

3.1 Half Warm Asphalt Mixture with Bitumen Emulsion (HWMA)

In this investigation a HWMA has been designed according to the recommendations developed by the Spanish road authority previously referred, for a semi-dense gradation with a maximum aggregate size equal to 19 mm. For comparison purposes, a Hot Asphalt Mixture (HMA) was also considered as is described latter. The gradation curves of both mixtures are presented in Fig. 2.

For the HWMA, a slow setting cationic emulsion with conventional bitumen that satisfied the requirements of the Argentinean standards has been used (IRAM 2019). The main characteristics of this emulsion are shown in Table 1.

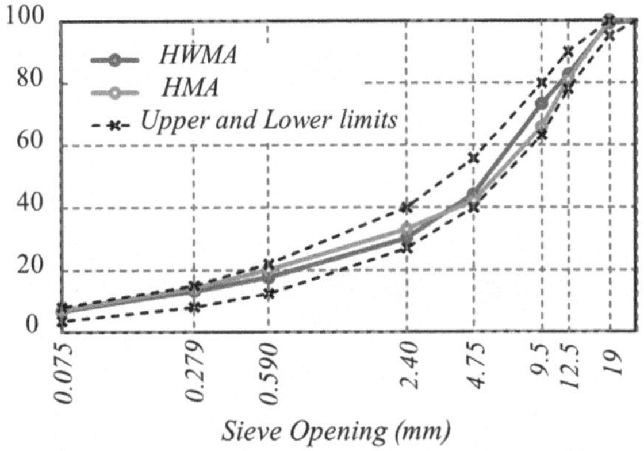

Fig. 2 Gradation curves for the considered mixtures

Table 1 Characteristics of the bitumen emulsion

Property	Test results
Polarity	Positive
Residue by evaporation (%)	61.1
Residue penetration at 25 °C (0.1 mm)	90.1
Softening Point Ring & Ball (°C)	46.3
Breaking value (g/100 g)	170

Table 2 Characteristics of the HWMA and the HMA

Mixture	Bitumen content (%)	Air voids (%)	VMA (%)	VFA (%)	S (kN)	F (mm)	S/F (kN/mm)
HWMA	5.0	4.8	16.9	71.7	3.77	3.2	1.2
HMA	5.1	3.8	15.7	72.5	9.92	3.1	3.2

S: Marshall Stability F: Marshall Flow

For the preparation of the samples, the aggregates were heated up to 120 °C and the emulsion at 60 °C to manually produce the mixture at a temperature ranging between 80 to 90 °C. Marshall samples were compacted with 75 blows per face and different contents of emulsion to determine an optimum content of residual bitumen as in the conventional procedure used for hot asphalt mixtures.

After mixing, the loose mixture was placed in an oven at 90 °C during 30 min and then, the compaction was immediately carried out at this temperature. Compacted samples were cured at room temperature during 72 h when constant weight was reached.

Finally, on the basis of these results and the recommended minimum value (ATEB 2014), the amount of emulsion that provide a residue of bitumen equal to 5% referred to the total weight of the mixture was adopted.

As it was previously stated, a HMA was also considered for comparison purposes. This mixture was formulated with a very similar aggregate gradation as the HWMA and conventional bitumen with comparable characteristics to the residue of the emulsion.

The volumetric and mechanical properties of the HWMA and the HMA finally considered in this study are listed in Table 2.

3.2 Dynamic Modulus Tests

The dynamic modulus tests were experimentally carried out with the Indirect Tension (IDT) mode with haversine loadings under controlled stress conditions using a servo-pneumatic machine. Displacements were measured along the horizontal diameter on a gauge length equal to 60 mm using LVDT's mounted on both faces of the samples. The load capable to produce a measurable displacement by the recording device was adjusted at each testing temperature. Marshall specimens were used. The test frame is enclosed into a temperature chamber where the temperature control system is able to achieve the required testing temperatures ranging from 0 to 50 °C. Two samples of each asphalt mixtures were tested and the obtained results were averaged. The |E*| and ϕ results were determined for 7 frequencies (5, 4, 2, 1, 0.5, 0.25 and 0.10 Hz) and 5 temperatures (0, 10, 20, 30 and 40 °C) to have a full rheological characterization of the asphalt mixtures.

4 Obtained Results

Based on the experimental results obtained for both samples of each mixture, the parameters that define the H-S model were adjusted by an optimization process using the Solver function of the Excel spreadsheet. A Reference Temperature $T_R = 25$ °C was adopted and the obtained values are listed in Table 3. Figure 3(a) and (b) show a comparison of measured and modelled values using the adjusted H-S models for the dynamic modulus |E*| and the Phase Angles ϕ respectively. The line of equality is also included in these figures.

A visual and qualitative evaluation of this figure shows that the points of comparison are tightly located along and on both sides of the line of equality indicating that the H-S model is able to describe the viscoelastic behaviour for the two asphalt mixtures and the conditions of frequency and temperature considered in this study.

Table 3 Parameters of the H-S model for the HWMA and the HMA

Mixture	E_∞ (MPa)	E_0 (MPa)	δ	k	h	C_A (°K)	τ_0 (s)
HWMA	17,230	22	1.01	0.48	0.48	17,644	2.043×10^{-3}
HMA	21,360	111	1.00	0.47	0.47	23,122	$6{,}710 \times 10^{-3}$

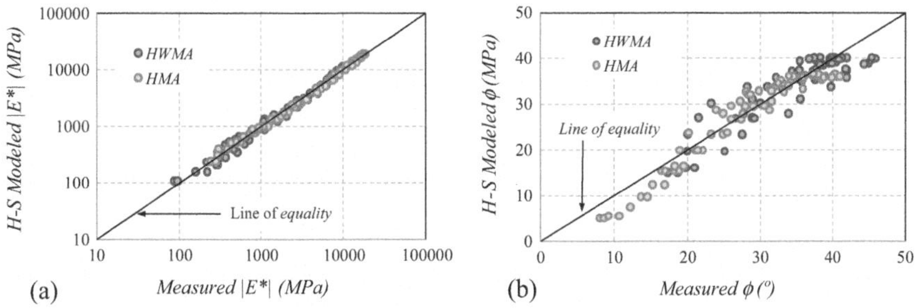

Fig. 3 Comparison of measured and modelled |E*| values (**a**) and ϕ values (**b**)

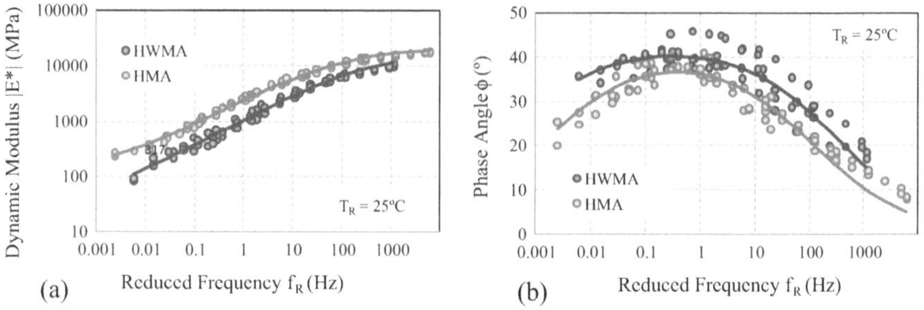

Fig. 4 Master curves of |E*| (**a**) and ϕ (**b**) for the mixtures

Figure 4(a) and (b) present the Master Curves of |E*| and ϕ at T_R = 25 °C. As expected for conventional HMA, the dynamic modulus increases with increasing frequency (and decreasing temperature). Also, the phase angle increases as the frequency vary from high to medium frequencies (from low to medium temperatures) where it reaches a maximum value and then, the phase angle decreases with decreasing frequencies from medium to low (from medium to high temperatures). For the HWMA, the |E*| values are lower and the ϕ values are higher than those of the HMA showing that the HWMA has a softer and more viscous behaviour. For the range of temperatures and frequency experimentally covered, the |E*| results for the HWMA are approximately 47% of those of the HMA and ϕ results are 11° greater compared to the HMA.

Figure 5(a) and (b) present the results for the mixtures in the Black and the Cole-Cole diagrams. Black diagram expresses the relation between the material modulus |E*| and the phase angle ϕ while the Cole-Cole diagram expresses the relation between the E_1 and E_2 components of E*.

The existence of a continuous curve in both diagrams shows that both mixtures can be considered as linear thermo-rheologically simple viscoelastic materials and validates the application of the principle of frequency-temperature superposition for them. From the Black diagram and for a given |E*| value, it can be observed that both mixtures show almost the same behaviour for the range of medium to high |E*|

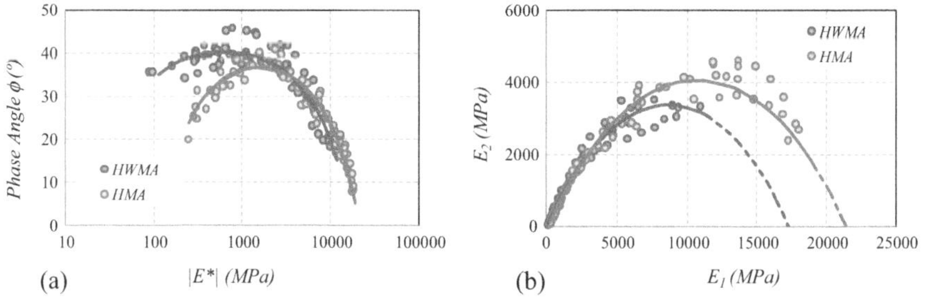

Fig. 5 Black (**a**) and Cole-Cole (**b**) diagrams for both mixtures

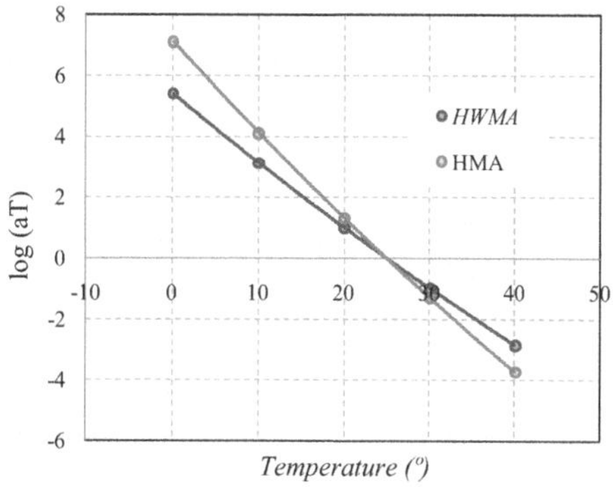

Fig. 6 Shift factors aT for the analyzed mixtures

values where both curves are superimposed while for lower |E*| values, the HWMA shows higher ϕ values than the HMA with a more viscous behaviour.

From the Cole-Cole diagram and also for a given value of the component E_1 (elastic), the same behaviour is observed because the component E_2 (viscous) of the HWMA is always greater than those of the HMA.

Figure 6 shows the shift factor aT as a function of the temperature for both mixtures. The HMA has a steeper curve than the HWMA indicating that HWMA is less sensitive to temperature changes and its stiffness has smaller variations when the mixture is subjected from low to high temperatures.

5 Conclusions

In this paper, the rheological behaviour of a Half Warm Asphalt Mixture (HWMA) with bitumen emulsion has been investigated. This mixture was formulated using a semi-dense gradation and a slow setting bitumen emulsion. For comparison purposes, a Hot Asphalt Mixture (HMA) was also formulated with the same type and a very similar aggregate gradation as the HWMA and conventional bitumen with comparable characteristics to the residue of the emulsion.

Dynamic modulus tests were carried out at different frequencies and temperatures for both asphalt mixtures. The obtained results of dynamic modulus |E*| and phase angle ϕ were modelled according to the Huet-Sayegh model. This rheological model was able to describe the viscoelastic behaviour for the two asphalt mixtures.

Master curves of |E*| and ϕ were constructed using the principle of frequency-temperature superposition. For the HWMA, |E*| values were lower and ϕ values were higher than those of the HMA showing that the HWMA has a softer and more viscous behaviour within the range of temperatures and frequency experimentally covered in this study. Experimental results were also drawn using the Black and Cole-Cole diagrams. The existence of a continuous curve in both diagrams shows that both mixtures can be considered as linear thermo-rheologically simple viscoelastic materials and validates the application of the principle of frequency-temperature superposition for them. Finally, it could be considered that those findings could encourage greater confidence in promoting the use of this kind of sustainable asphalt mixtures in mechanistic pavement design procedures.

References

Rubio MC, Martínez G, Baena L, Moreno F (2012) Warm mix asphalt: an overview. J Clean Prod 24:76–84

López C, Thenoux G, Sandoval G, Armijos V, Ramírez A, Guisado F, Moreno E (2017) Study of warm mix asphalt with super stabilized emulsion. Revista Ingeniería de Construcción RIC 32(1):57–64

Sangita D, Singh MP, Yadav A, Pandey Y, Tare V (2015) Comparative study of emulsion based half warm mix and cold mix for construction of SDBC and DBM. Int J Sci Res Dev 3(3):2933–2935

Dinis Almeida M, Castro Gomes J, Antunes M (2010) Mix design criteria for half warm asphalt recycling (HWMR)—case study. In: 16th International road federation world meeting 2010, Lisbon, Portugal

Lizárraga JM, Jiménez del Barco-Carrión A, Ramírez A, Díaz P, Moreno-Navarro F, Rubio MC (2017) Mechanical performance assessment of half warm recycled asphalt mixes containing up to 100% RAP. Materiales de Construcción 67(327):129

Garcia Santiago JL, Lucas Ochoa FJ (2014) Mezclas templadas con reutilización del RAP con tasa alta y tasa total. Aplicación, experiencias reales y resultados. Asociación Española de Fabricantes de Mezclas Asfálticas, Número 14, Volumen IV (in Spanish)

Swaroopa S, Sravani A, Jain PK (2015) Comparison of mechanistic characteristics of cold, mild warm and half warm mixes for bituminous road construction. Indian J Eng Mater Sci 22:85–92

Miranda Pérez L, García Santiago JL, Uguet N, Andaluz D, Colás Victoria, MDM, Lucas FJ, Gil S (2013) Mezclas templadas con emulsión bituminosa. Carreteras, 4ª. Época. Asociación Española de la Carretera (AEC), pp 30–36 (in Spanish)

Asociación Técnica de Emulsiones Bituminosas (ATEB) (2014) Mezclas templadas con emulsión bituminosa. Monografía (in Spanish)

Agencia de Obra Pública de la Junta de Andalucía (2012) Recomendaciones para la redacción de especificaciones técnicas para el uso de mezclas bituminosas a bajas temperaturas (in Spanish)

Sayegh G (1967) Viscoelastic properties of bituminous mixtures. In: Proceedings of the second international conferences on the structural design of asphalt pavements, Ann Arbor, Michigan, USA

Yusoff NI, Shaw MT, Airey G (2011) Modelling the linear viscoelastic rheological properties of bituminous binders. Constr Build Mater 25:2171–2189

Olard F (2003) Comportement thermomécanique des enrobes bitumineux à basses températures. Relations entre les propriétés du liant et de l'enrobé. These. Ecole Doctorale des Sciences pour l'ingenieur de Lyon, France (in French)

Cauhapé CM, Angelone S, Martínez F (2017) Evaluación del comportamiento visco elástico de mezclas asfálticas mediante modelos mecánicos. In: XIX CILA, Medellin, Colombia (in Spanish)

IRAM Instituto Argentino de Normalizacion y Certificacion (2019) IRAM 6691: Asfaltos. Emulsiones asfálticas catiónicas convencionales. Clasificación y requisitos (in Spanish)

Long Lasting Asphalt Materials with Highly Modified Asphaltic Binder

Laurent Porot, Erica Jellema, and David Bell

Abstract On mature road networks, there is a constant need for long-term pavement preservation solutions. In the 1980s, SBS polymer brought a breakthrough technology with superior performance for surface layer. In more recent years, highly modified asphaltic binder, HiMA, took a step further for structural layers with fatigue resistance. At the same time, circular economy enhances the valorisation of Reclaimed Asphalt. However, its reuse in top layers is challenging, as RA may deteriorate cracking resistance. This paper presents results of laboratory evaluation of an asphalt mix combining 25% RA and the use of highly modified bitumen with 7.5% specific SBS polymer and compared with a same mix using a normal Polymer modified Bitumen 25/55-55. The evaluation was carried out with compactability, rutting resistance, complex modulus, fatigue and low temperature cracking susceptibility. From the mixes, binders were extracted and recovered for further evaluation similarly to the asphalt mixes. This study demonstrated the benefits of both standard PmB and highly polymer modified bitumen to achieve high performance asphalt mix combined with RA. The HiMA binder was able to restore, to a greater extend, the lost properties against cracking and still maintaining the benefits in terms of rutting resistance. This emphasises the benefit for PmB usage for both structural and surface layers, even when using Reclaimed Asphalt, for sustainable modern asphalt pavements.

Keywords Asphalt mix · Polymer modified Bitumen · HiMA binder · Pavement preservation

1 Introduction

The asphalt industry is continuously improving quality and performances of the asphalt materials to answer the increasing demand of traffic and user needs. In the 1980's Polymer modified Bitumen (PmB) was successfully developed to answer

---

L. Porot (✉)
Kraton Chemical, Almere, The Netherlands
e-mail: laurent.porot@kraton.com

E. Jellema · D. Bell
Kraton Polymers, Amsterdam, The Netherlands

C. Raab (ed.), *Proceedings of the 9th International Conference on Maintenance and Rehabilitation of Pavements—Mairepav9*, Lecture Notes in Civil Engineering 76,
https://doi.org/10.1007/978-3-030-48679-2_49

to the increase of traffic in volume and loading (Lu and Isacsson 1995). It is now commonly used for binder and surface layers. It enhances the rutting resistance and ensures durable surface integrity towards ravelling and surface cracking. In the last recent years, the use of highly modified bitumen, HiMA, were further developed for structural layers and rolled out over the world as the next step in pavement engineering (Habbouche et al. 2017).

The European commission has set up a road map for the implementation of circular economy in March 2019 (European Commission 2019) which reinforces the valorisation of any material at the end of its life. This is particularly valid for asphalt materials which are 100% reusable. However the usage in surface layers is still limited due to the risk of lost characteristics especially for cracking resistance.

When PmB is used in asphalt mix containing Reclaimed Asphalt (RA), there is a need to compensate the potential deficit of polymer network from the RA. With higher polymer content, HiMA binder offer the possibility to manufacture new asphalt mix containing RA similarly to high performance mix (Porot et al. 2019).

Recent development of Styrene-Butadiene-Styrene (SBS) polymer has made possible increasing polymer content in binder and still having stable bitumen for normal use at the mix plant. When blended with bitumen, SBS polymers, swell and increase by 5 to 10 times in volume. When increasing the polymer content, phase inversion occurs between a rich-bitumen phase to a rich-polymer phase. Usually at 5% the binder displays a balanced phase between bitumen and polymer, while at 7.5%, the polymer phase is predominant and the behaviour is more rubber like (Xia et al. 2016).

When increasing the SBS content in bitumen, the binder may be more difficult to produce with storage stability issues or higher viscosity leading to excessive mixing temperatures for producing asphalt mix. Special SBS grade, with high vinyl content, can mitigate these different potential drawbacks. It has a smaller size, which keeps the binder viscosity in reasonable range.

The concept of HiMA binder has been initially investigated by Delft Technical University in 2004–2008 (Molenaar et al. 2008). The outcomes showed that its use in structural layer can lead to substantial thickness reduction under a wide range of flexible pavement structures. Later the concept was validated in full scale experiment at the National Center for Asphalt Technology, NCAT, in the US and confirmed higher service life as compared to conventional binder, using the same designed asphalt mix (West et al. 2012).

This paper presents a comparative study between asphalt mix with 25% RA made with standard PmB and HiMA. It includes asphalt mix characterisation and also properties of the extracted and recovered binder from the mixes.

2 Experimental Plan

2.1 Materials

For the purpose of the study a highly polymer modified bitumen (HiMA) was compared with a commercial PmB, graded as 25/55-55. The HiMA binder was made with a 70/100 base bitumen and modified with 7.5% of special vinyl SBS polymer, D0243. This tailored SBS polymer is especially designed to have better compatibility with a wider range of bitumen sources and thus enables high concentration in bitumen. The structure and low molecular weight enable keeping the viscosity of the PmB at a reasonable level, limiting the mixing temperature to maximum 160–170 °C. The HiMA binder was graded as PmB 45/80-85 according to EN 14023.

2.2 Binder Testing

Binder testing was conducted on the original binders and on binders recovered from the mixes themselves. They were evaluated through conventional characterisation with penetration value at 25 °C and softening point temperature. The original binders were subjected as well to laboratory aging, with Rolling Thin Film Oven Test (RTFOT) and further with Pressure Aging Vessel (PAV). Fundamental testing was done using Dynamic Shear Rheometer (DSR) and Bending Beam Rheometer, BBR.

Binder Fatigue testing was also performed using DSR in repeated shear loading stress according to the RILEM protocol (Partl et al. 2013). The test consists of applying a cyclic shear load to a bituminous binder at different strain levels and recording the shear modulus. The fatigue life is determined when the remained shear modulus is equal to 50% of the initial value. In this study, to shorten the experiment, one strain level was used. The test was performed with an 8 mm plate, 2 mm gap at 10 Hz and 10 °C on the recovered mix binders.

2.3 Asphalt Mix Testing

The mixes used were an optimised AC 16 asphalt concrete mixture with a maximal aggregate size of 16 mm according to German specifications for binder layer application. They were both similar and contained 25% Reclaimed Asphalt (RA) with 4.8% final binder content. Table 1 shows the asphalt mix composition for both mixes.

The mixes were produced according to EN 12697-35 with the same mixing temperature of 170 °C and stored in an oven at 145 °C before compaction and testing.

Table 1 Asphalt mix formulation for AC16

Sieve, mm	0.063	0.25	2	5.6	8	11.2	16	22.4
Passing, %	5.4	9.3	25.1	44.8	51.0	63.6	97.8	100

Gyratory compaction was performed according to EN 12697-31. The specimen height is recorded along the number of gyrations and a curve of void content versus number of gyrations is derived. The test was carried out up to 400 gyrations to have a wide range of compactability, while usually a compaction at 60 gyrations is indicative parameter for binder layer to be laid in 6 cm thickness.

The resistance to permanent deformation (rutting) was determined using the German wheel tracking test (WTT) according to EN 12697-22 with a rubber wheel. The WTT was carried out at 60 °C for 10,000 cycles and rut depth recorded.

Stiffness modulus was measured by 4-point bending test, 4PBT, according to EN 12697-26, where a sinusoidal load is applied to beam at an initial strain level of 50 μstrain. The test, for each sample, was performed at four different temperatures ranging from −10 to +20 °C in frequency sweep from 0.1 to 30 Hz. This enables to produce master curves in a wide range of temperatures.

Fatigue performance was determined with the 4-Point-Bending Test, 4PBT, according to EN 12697-24. Initial strains were applied in a range of 125 to 250 μstrain at temperature of 20 °C and loading frequency of 10 Hz. The resulting stiffness modulus is recorded along the test duration and fatigue performance is determined when the modulus is decreased by 50% from the initial value.

Finally, the low temperature cracking susceptibility was evaluated through Thermal Stress Restrained Specimen Tests (TSRST) on prismatic samples, according to EN 12697-46. During the test, the specimen is kept at a constant length, while temperature decreases from 20 to −40 °C at a constant rate of −10 °C/h. With temperature decrease, the specimen shrinks. During the test, the force to maintain zero deformation is recorded and the induced stress is calculated, until the specimen breaks.

3 Results

3.1 Initial Binder Evaluation

A full evaluation on the original binder, as used later in the asphalt mixes, was made with empirical properties determined as per EN 14023 for PmB. Table 2 displays the main results. The HiMA binder combined very high softening point temperature and intermediate penetration value, still keeping the viscosity in a similar range, and at the same time having a higher elastic recovery. This results from the specific features of the high vinyl SBS used in the formulation.

Table 2 Empirical binder properties for both binders PmB and HiMA

Properties	Standard	Unit	PmB 25/55-55	HiMA binder
Penetration value at 25 °C	EN 1426	0.1 mm	42	56
Softening point	EN 1427	°C	58.4 °C	90.0 °C
Penetration index			0.28	5.90
Resistance to hardening Mass loss	EN 12607-1	%	0.13%	0.09%
Viscosity at 135 °C	EN 12595	mm^2/s	1510	2270
Elastic recovery at 25 °C	EN 13398	%	77%	98%

Table 3 Change of binder properties after lab aging for 20/30 and HiMA binders

	PmB 25/55-55		HiMA binder	
	Penetration value	Softening point (°C)	Penetration value	Softening point (°C)
Original binder	42 0.1 mm	58.4	56 0.1 mm	90.0
RTFOT	29 0.1 mm	63.6	41 0.1 mm	89.0
Change *	69%	+5	73%	−1
RTFOT+PAV	19 0.1 mm	71.2	26 0.1 mm	90.0
Change *	45%	+13	46%	0

*as compared to original binder

Aging was also evaluated on the binders for short-term and long-term aging conditioning with RTFOT, and PAV. Table 3 displays the changes of properties. Usually, bitumen ages by one grade harder through short term and further another one for long term aging. On the contrary, the HiMA binder did not show important changes in the properties; especially the softening point temperature hardly changed. However, it is worth noticing that RTFOT is not always best ad hoc test for PmB due to high viscosity that limits thin film formation in the bottle.

Furthermore rheological measurements were carried out. Shear moduli were measured using DSR in temperatures ranging from −30 to +90 °C at a fixed frequency.

Figure 1 displays the shear modulus |G*| vs. temperature. For a given temperature, the HiMA binder had lower values, less stiff, although for the high temperature range both binders displayed similar values.

Figure 2 displays the DSR results in Black Space with phase angle versus shear modulus.

The lower the phase angle is, the more elastic response the material has, which is a benefit at high temperature against permanent deformation. Both binders displayed a clear rubber plateau for low shear moduli, high temperature domain. It is more pronounced for the HiMA binder starting at 10^7 vs 3.10^5 Pa for the PmB, respectively corresponding to temperature of 10 and 40 °C.

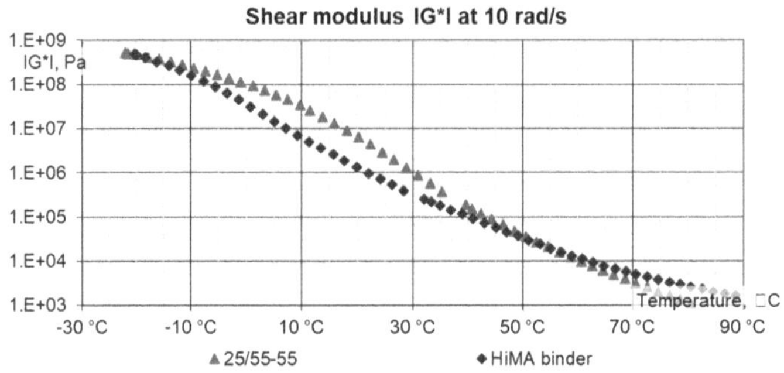

Fig. 1 DSR shear modulus results for both PmB and HiMA binders

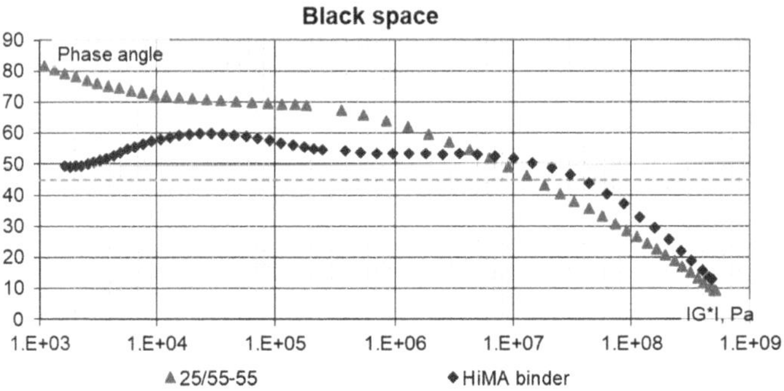

Fig. 2 DSR Black Space for both PmB and HiMA binders

Additionally, BBR was performed. This, in combination with DSR measurements, allows determining the PG grading of the binder. Table 4 displays the main criteria determining PG grading. The HiMA binder displayed a higher temperature PG, and at the same time a lower temperature PG, as compared to the 25/55-55 PmB.

Table 4 Results of PG grading for PmB and HiMA binder

	Criteria	PmB 25/55-55	HiMA binder
High temperature (original)	$\|G^*\|/\sin\delta \geq 1.00$ kPa	81.6 °C	97.3 °C
High Temperature (RTFOT)	$\|G^*\|/\sin\delta \geq 2.20$ kPa	79.5 °C	90.4 °C
Intermediate T (RTFOT+PAV)	$\|G^*\|\sin\delta \leq 5000$ kPa	25.9 °C	17.3 °C
Calculated intermediate T	(highT + lowT)/2 + 4	32 °C	34 °C
Low temperature BBR (RTFOT+PAV)	S ≤ 300 MPa m-value ≥ 0.300	−15.7 °C −13.0 °C	−19.6 °C −20.1 °C
True PG		80–23	90–30
PG grade		PG 76-22	PG 88-28

3.2 Asphalt Mix Characterisation

Asphalt mixture characterisation was carried out following the European performance approach, in place for more than 15 years. It is based on compactability using the gyratory compactor, resistance to permanent deformation through WTT, and, for structural layers, mechanical characteristics with modulus and fatigue. While it is not mandatory, low temperature cracking susceptibility was also measured.

Compactability was addressed with gyratory compaction at 145 °C shown in Fig. 3. Despite different properties especially with higher softening point temperature for the HiMA binder, both mixes did not exhibit any difference in compactability and after 60 gyrations achieved void contents between 6 and 8%.

The resistance to permanent deformation was carried out through WTT at 60 °C. Figure 4 displays the results with the rut depth in mm vs. the number of passes. Both mixes performed very well with rut depth kept below 2.5 mm.

Fig. 3 Gyratory compaction for both mixes with PmB and HiMA binders

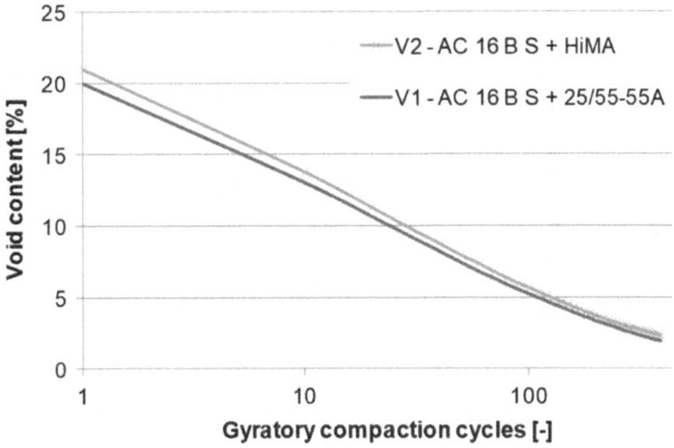

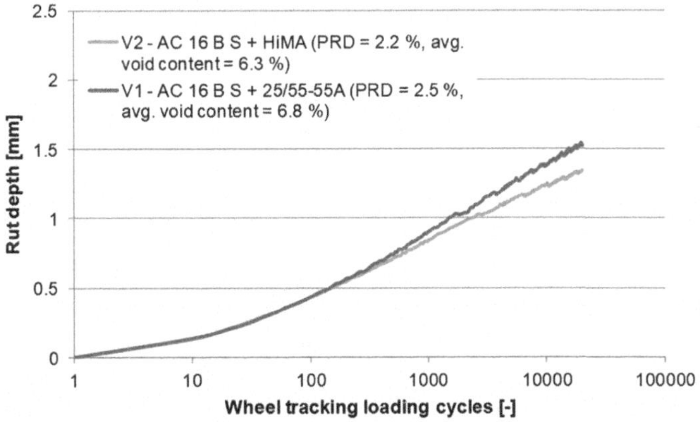

Fig. 4 Wheel Tracking Test for both mixes with PmB and HiMA binders

The mechanical characteristics were evaluated with stiffness modulus at different temperatures and frequencies and fatigue resistance at 20 °C and 10 Hz, with four point bending beam test.

For stiffness modulus, master curves were developed using sigmoid model and Arrhenius equation. Figure 5 displays these master curves at a reference frequency of 10 Hz. They both displayed high values, although slightly lower for the HiMA mix. At 20 °C in the range of 10,000 MPa, while standard asphalt mixes are more between 5000 to 8000 MPa. This may be a result of the use of the 25% RA in the mix.

Figure 6 displays the fatigue life curves for both asphalt mixes with strain level vs. number of cycles. For the same strain level, the number of repeated loading cycles was higher for the HiMA mix. With extrapolation of the curve, at one million cycles, the strain would be respectively 128 and 110 μstrain for the HiMA and PmB mix.

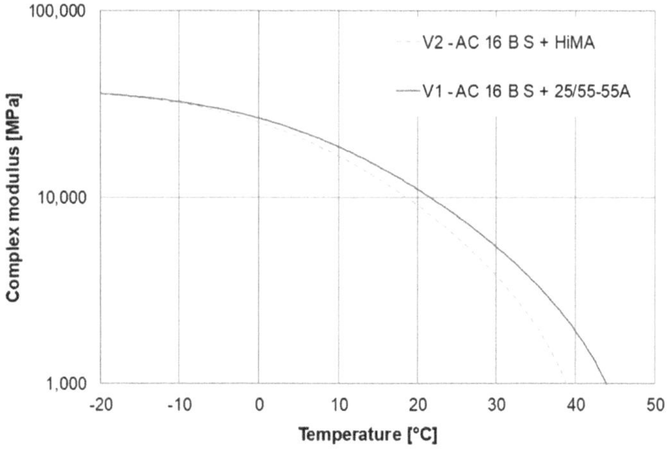

Fig. 5 Stiffness modulus for asphalt mixes with PmB and HiMA binder

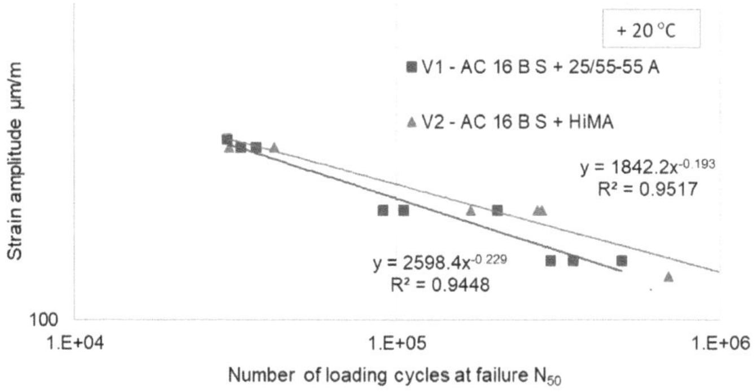

Fig. 6 Fatigue life for asphalt mixes with PmB and HiMA binder

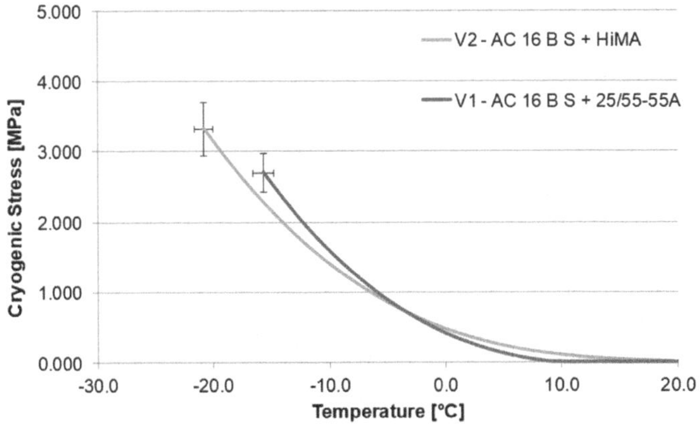

Fig. 7 TSRST results for mix with PmB and HiMA binder

Finally, the low temperature cracking susceptibility was assessed with TSRST, on three replicates for each mix. Figure 7 displays the results of the mean value of each mix in terms of stress curve vs temperature, the error bars being for min and max of failure values. The HiMA mix showed the lowest failure temperature as compared with the PmB mix. This is consistent with the binder evaluation where the HiMA binder had the best performing low temperature PG grading.

3.3 Recovered Binder Evaluation

From the asphalt mixtures, the binders were extracted and recovered according to EN 12697-3 and further tested. In addition to conventional characterisation, fundamental testing was performed and only discussed below. It is worth to notice that the mixes had 25% RA, which affected the recovered binder. RTFOT aged binders were compared with recovered binders and the original binders being for reference.

DSR measurements were investigated via the Black Space, and displayed in Fig. 8. For each binder, the curves overlapped with a contraction from left to right. The rubber plateau remained either after RTFOT or from the recovered mixes. There is a slight change in the shape, the curvature of the curve, between RTFOT and recovered binder, this is most likely due to the presence of RA binder in the latter.

In addition to DSR test, binder fatigue test was performed at 10 °C with repeated shear loading. Figure 9 displays the percentage of remained shear modulus vs. the number of load cycles at same strain level of 1.3%. For the PmB, after a slight increase of the shear modulus, it decreased after 3.10^4 cycles. With the HiMA binder, the shear modulus slowly continuously decreased but without reaching the 50% criteria.

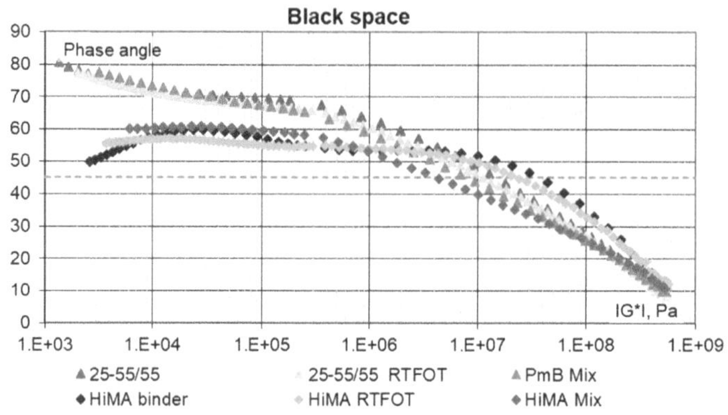

Fig. 8 Black space comparison of fresh and recovered binders, HiMA and PmB mixes

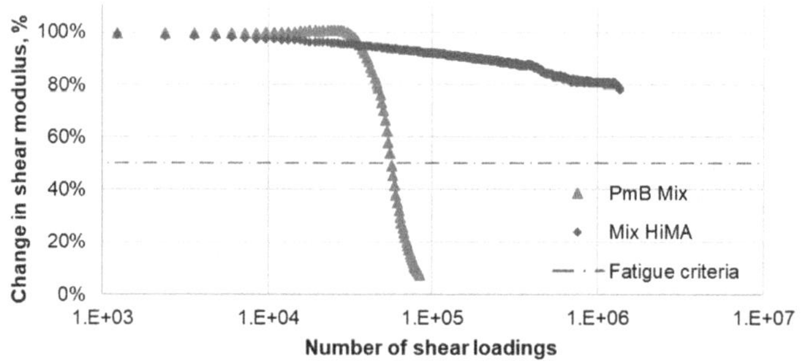

Fig. 9 Binder fatigue life for the recovered binder from PmB and HiMA, change in shear modulus

4 Conclusion

The asphalt industry is continuously developing solutions to address the current and future needs of road network. Polymer modified Bitumen, developed 40 years ago, has addressed the need for increased traffic, in volume and load. The European Commission is implementing Circular Economy across all industries. This implies for road mature network the necessity to reuse more Reclaimed Asphalt into surface layer.

The use of highly modified bitumen in combination of RA in asphalt mix can mitigate the potential risk of premature cracking failure.

This study focused on the comparison of standard PmB and HiMA binder in asphalt mixtures with 25% RA. Asphalt mix results showed that both can mitigate potential cracking risk either fatigue or thermal without compromise in rutting resistance. The HiMA binder offered even better cracking resistance than standard PmB.

The evaluation carried on the binder with fresh and recovered binder from the asphalt mixes have shown the effect of polymer modification and were consistent with mix characteristics. The HiMA binder demonstrated higher performance either on pen grading system or PG grading system with broader temperature interval. Fatigue testing on recovered binders showed that the HiMA binder displayed very long life spam.

Overall, it is possible to reuse 25% RA in asphalt mix for surface layer application with polymer modified bitumen without compromise on cracking and rutting. The use of HiMA offer an even better solution maintaining reasonable modulus and ensuring lower cracking susceptibility.

References

European Commission (2019, March) Implementation of the circular economy action plan. http://ec.europa.eu/environment/circular-economy/index_en.htm

Habbouche J, Sebaaly P, Hajj E (2017) Structural coefficient for high polymer modified asphalt mixes in Florida—literature review. University of Nevada, Reno

Lu X, Isacsson U (1995) Testing and appraisal of polymer modified road bitumens—state of the art. Mater Struct 28:139–159

Molenaar A, Van de Ven M, Liu X, Scarpas T, Medani T, Scholten E (2008) Advanced mechanical testing of polymer modified base course mixes. In: 4th Euraspahlt & Eurobitume congress, Copenhagen

Partl M, et al (2013) Advances in interlaboratory testing and evaluation of bituminous materials. State of the art report of the RILEM Technical Committee 206-ATB. Springer

Porot L, Jellema E, Bell D, Gronniger J (2019) Binder and mix evaluation of highly modified asphaltics binder. In: 18th AAPA, Sydney

West R, Timm D, Willis R, Powell B, Tran N, Watson D (2012) Phase IV NCAT pavement test track finding. NCAT, Auburn

Xia T, Huang T, He J, Zhang Y, Guo J, Li Y (2016) Viscoelastic phase behavior in SBS modified bitumen studied. Constr Build Mater 105:589–594

Mechanical Properties of Bio-Asphalt on Recycled Asphalt Pavement Binder

Atmy Verani Rouly Sihombing, Bambang Sugeng Subagio, Eri Susanto Hariyadi, and Anwar Yamin

Abstract Recycling asphalt technology is a technology that reuses of aged asphalt pavement that has been damaged and recycled as a new asphalt mixture material which in using it requires a modifier to restore the performance of aged asphalt. Bio-asphalt which is a fraction of bio-oil derived from biomass containing lignin has the potential as a modifier/rejuvenator for aged asphalt. The objective of this study is to determine the potential of the bio-asphalt as a modifier. Bio-asphalt modified aged asphalt binder is tested and evaluated based on AASHTO requirements. Best bio-asphalt content is suggested for modifying aged asphalt binder. Bio-asphalt modified aged asphalt binder were tested to know the mechanical properties by dynamic shear rheometer (DSR) temperature sweep test and frequency sweep test. In this study, the bio-asphalt used are coconut shell bio-asphalt (BioCS) and BitutechRAP, each of which was mixed with Aged asphalt binder from recycled asphalt pavement (RAP) extraction and controller asphalt binder use pen 60/70 (PG 64). The result showed that bio-asphalt can improve the mechanical performance of aged asphalt binder close to pen 60/70 as a controller.

Keywords Bio-asphalt · Aged asphalt · RAP · DSR · Mechanical properties

1 Introduction

1.1 Background

Bio-asphalt is asphalt derived from non-petroleum materials that can be renewed. The use of this bio-asphalt, in addition to being used as a modifier/rejuvenator (10%

A. V. R. Sihombing (✉) · B. S. Subagio · E. S. Hariyadi
Civil Engineering Department, ITB, Jl. Ganesha No. 10, Lb. Siliwangi, Coblong, Bandung, West Java 40132, Indonesia
e-mail: atmyvera@gmail.com

A. Yamin
Research and Development Center for Road and Bridge, Ministry of Public Works and Public Housing Bandung, Bandung, Indonesia

C. Raab (ed.), *Proceedings of the 9th International Conference on Maintenance and Rehabilitation of Pavements—Mairepav9*, Lecture Notes in Civil Engineering 76,
https://doi.org/10.1007/978-3-030-48679-2_50

usage), can also be used as an additive in binders (25–75% use) and binder use (100% use) (Peralta et al. 2012). In Indonesia, research on bio-asphalt, namely bio-asphalt from coconut shells, has begun to be carried out using for binders, while the optimal results obtained using bio-asphalt as a binder is 1% (Kusuma and Haryanto 2014; Perdana and Utomo 2013). The use of bio-asphalt as a binder, according to the authors, has not provided an optimal potential for the number of uses produced, so the authors have a hypothesis as the potential of this coconut shell bio-asphalt may be as a modifier/rejuvenator for aged asphalt. Based on previous research, from basic physical properties, chemical and morphological characteristics, the use of bio-asphalt in aged asphalt uses its potential as a rejuvenating agent. To increase confidence in this potential, further testing was carried out by looking at the mechanical properties of aged asphalt which has been equipped with bio-asphalt based on the results of DSR temperature sweep test and DSR frequency sweep test.

1.2 Objective

The purpose of this study is to discover the potential of bio-asphalt as a rejuvenating agent by looking at the mechanical characteristics of aged asphalt before and after bio-asphalt is added using the results of the DSR temperature sweep and frequency sweep test. As a comparison for the rejuvenating process in aged asphalt after the addition of bio-asphalt, virgin asphalt pen 60/70 (PG 64) is used as an asphalt controller, because this asphalt is commonly used as a binder in Indonesia.

2 Experimental Plan

The objective of this research can be achieved by developing the research plan (Fig. 1). Two types of bio-asphalt (Bio-asphalt coconut shell/BioCS and BitutechRAP) are added to the aged asphalt (extraction bitumen from RAP). Samples were prepared by mixing bio-asphalt and aged asphalt using magnetic stirrer with speed rotation 0.4 –0.6 kr/sec and mixing temperature of 120 °C for 15–20 min (Williams and McCready 2008; Williams et al. 2010). While the variation of bio-asphalt percentage to the weight of aged asphalt is 2, 4, 8, 16, 20, 25, and 30%. Penetration and softening point tests are basic tests to find out the characteristics of asphalt which have been used since the late 19th century (Read and Whiteoak 2003) thus the bio-asphalt + RAP bitumen mixture was then tested for penetration and softening point (SNI 06-2434 1991; SNI06-2456 1991) and compared to pen 60/70 to obtain the optimum content. The content of bio-asphalt are 23% for BioCS and 17% for BitutechRAP by weight of Aged Asphalt (Sihombing et al. 2019; Sugeng and Sihombing 2019).

The temperature sweep of bio-asphalt + aged asphalt was test in unaged, RTFO and PAVT conditions by using Dynamic Shear Rheometer (DSR) tools. This test was carried out based on AASHTO (AASHTO T315-10 2010) at a test frequency of

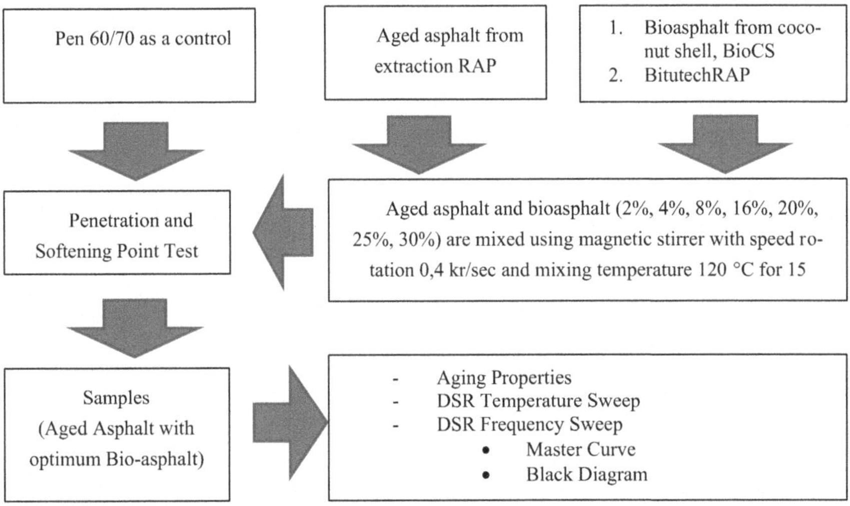

Fig. 1 Research plan

1.59 Hz, where the test on the original condition was carried out at 12% strain and a temperature range of 52 to 76 °C at 6 °C interval, test on the RTFO condition was carried out at 10% strain and temperature range of 52 to 76 at 6 °C interval, and test on the PAVT condition was carried out at 1% strain and temperature range of 22 to 28 °C at 3 °C interval. The complex shear modulus G*, phase angle δ, and the rutting factors G*/sinδ of the bio-asphalt + aged asphalt, were measured and analyzed.

DSR frequency sweep test is carried out to measure the change in rheology due to variations in loading time (frequency) at a certain temperature, testing is carried out in the frequency range of 0.2–100 Hz, at temperatures of 25, 45, and 60 °C with a 0.1% strain magnitude.

The effect of bio-asphalt addition was evaluated by looking at its effect on aged asphalt before and after bio-asphalt was added compared to the performance of virgin asphalt with DSR test results parameters.

3 Materials

The material used in this study is: PG 64-22 (pen asphalt 60/70) was chosen in this study as virgin asphalt which controls asphalt, the aged asphalt used is asphalt from RAP extraction taken from old pavement dredging in Karawang, West Java, Indonesia. Extraction is done with the help of trichloroethylene (TCE) chemical using a centrifugal extractor (Burr et al. 1990, 1991), and the bio-asphalt used is the coconut shell bio-asphalt produced by PT Nuciferra and BitutechRAP from Hydrogreen.

Table 1 Characteristics of asphalt pen 60/70, aged asphalt, bio-asphalt and aged asphalt

Test	Unit	Specification	Pen 60/70	Aged asphalt	BioCS + Aged asphalt	BitutechRAP + Aged asphalt
Penetration, 25 °C, 100 gr, 5 s	0.1 mm	60–70	65	10	65	65
Kinematic viscosity 135 °C	cSt	≥300	409.6	–	463.5	421.1
Softening point	°C	≥48	51	80	54	53.8
Ductility, 25 °C, 5 cm/minute	cm	≥100	>100	38	>100	>100
Flash point with Clevelen open cup	°C	≥232	340	240	325	335
Solubility in trichloroethylene	%	≥99	99.868	–	99.735	99.805
Specific gravity		≥1	1.024	1.061	1.015	1.022
Mass loss	% weight	≤1	0.15	0.75	0.51	0.35
Penetration RTFO	% initial	≥54	58.8	–	54	58
Softening point RTFO	°C	–	56	–	59	56
Ductility RTFO	cm	≥100	110	–	115	105

4 Test Result

4.1 Asphalt Characteristic

The optimum percentage of use of bio-asphalt on aged asphalt, then used as samples with the same mixing process as before and tested further. The rheological characteristics of aged asphalt mixed with BioCS and BitutechRAP (with optimum percentage values) are outlined in Table 1. Based on these rheological data, it can be proved that the addition of bio-asphalt on aged asphalt can meet specifications for asphalt pen 60/70.

4.2 Aging Properties

The aging properties reviewed in this study are penetration, softening point, ductility and mass loss in unaged and RTFO conditions. Figure 2a, b, and c illustrates the change in physical properties of asphalt after RTFO, where the penetration of each

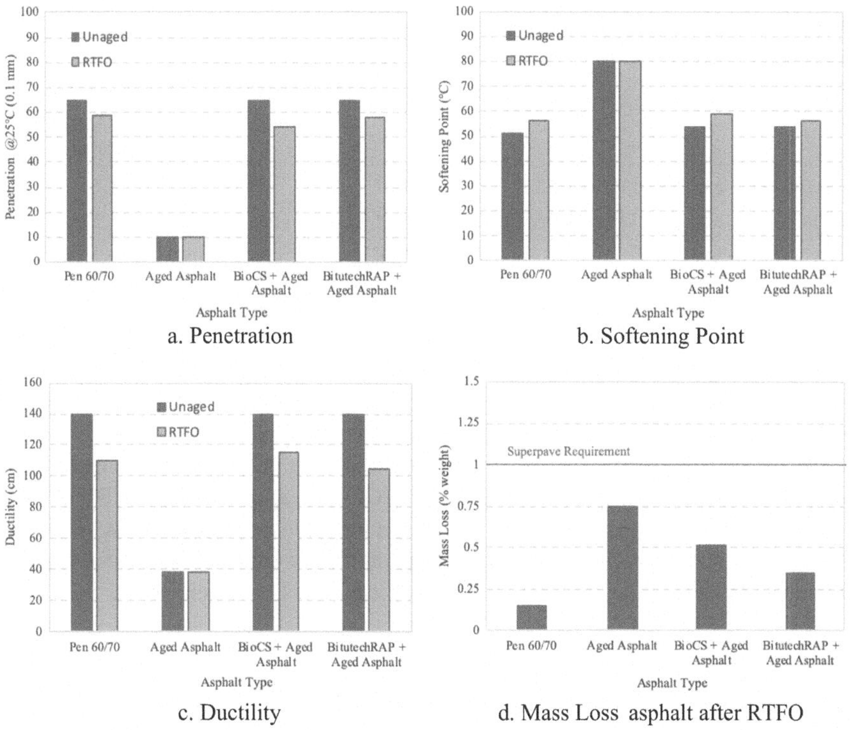

Fig. 2 Physical properties of asphalt binder before and after RTFOT and mass loss asphalt after RTFO

type of asphalt decreases, the softening point increases and the ductility decreases. The addition of bio-asphalt in RAP bitumen in RTFO condition has characteristics that are still in accordance with the specifications.

Mass losses between unaged and RTFO aged are shown in Fig. 2d. Superpave specification recommends that the mass loss of asphalt binder subjected to standard RTFO test should be lower than 1%. Pen 60/70 demonstrates only 0.15% mass loss while the extracted RAP binder exhibits significant reduction (0.75%) of volatile fractions and barely passes the requirement. During the RTFO aging any residual solvent would evaporate thus contributing to the mass loss. Both of bio-asphalt can reduce the mass loss of aged asphalt, although not the same as virgin asphalt. This shows that bio-asphalt can reduce susceptibility to aging.

4.3 Result of DSR Temperature Sweep

DSR temperature sweep test is usually done to verify the grade of asphalt in the PG system. The test was conducted at a test frequency of 10 rad/s with the test

Table 2 Critical temperature of asphalt

Conditions	Requirements	Test result (°C)			
		Virgin asphalt	Aged asphalt	BioCS + Aged asphalt	BitutechRAP + Aged asphalt
Original	Min 1,00 kPa	65.3	110.7	75.6	72.7
RTFOT	Min 2,20 kPa	63.6	120.2	111.7	85.2
PAVT	Max 5000 kPa	23.7	52.8	53.7	25.5

temperature referring to the temperature range contained in the PG AASHTO M320 (AASHTO 2013) asphalt specification table. The determination of PG was only done at the upper limit of critical temperature and was not done to the lower limit. This was because Indonesia has a tropical climate that does not have low temperatures. According to Table 2, asphalt pen 60/70 has PG of 64, aged asphalt has PG of 112, bioCs + aged asphalt has PG of 76, and bitutechRAP + aged asphalt has PG to 70. Thus, mechanically bio-asphalt + aged asphalt shows changes from aged asphalt with a smaller PG value, although it does not reach the PG of pen 60/70.

4.4 *Result of DSR Frequency Sweep*

A. Master Curve

The combined master curve in Fig. 3 shows the comparison between the aged asphalt master curve before and after mixing BioCS and BitutechRAP with the virgin asphalt master curve. The position of the aged asphalt master curve tends to be at a high modulus, indicating that the aged asphalt is elastic, this tendency is indeed found in aged asphalt. Whereas the virgin asphalt master curve is in a position with a small modulus, indicating that virgin asphalt is more viscous. The addition of bio-asphalt as a modifier/rejuvenator to the aged asphalt is expected to be close to the position of the virgin asphalt master curve. From the drawing of the merging of the master curve into the four types of asphalt, it can be seen that the addition of bio-asphalt can change the position of the aged asphalt master curve more closely to virgin asphalt, meaning that elastic aged asphalt becomes more viscous after adding bio-asphalt.

The addition of bio-asphalt to the aged asphalt changes the aged asphalt master curve sloping to become more gentle. This shows that the addition of bio-asphalt to aged asphalt can improve the performance of asphalt by increasing the temperature range and time of loading which results in the easier use of aged asphalt in the field.

The addition of BioCS to aged asphalt is more resistant to rutting due to the higher modulus in bitutechRAP at slow loading. Quantitatively, the effect of adding bio-asphalt to aged asphalt can be seen from the comparison of G* and Sbit values based on shift factors. The smaller the difference in value to the virgin asphalt, the better the performance of the asphalt produced. To get the target value of Sbit in

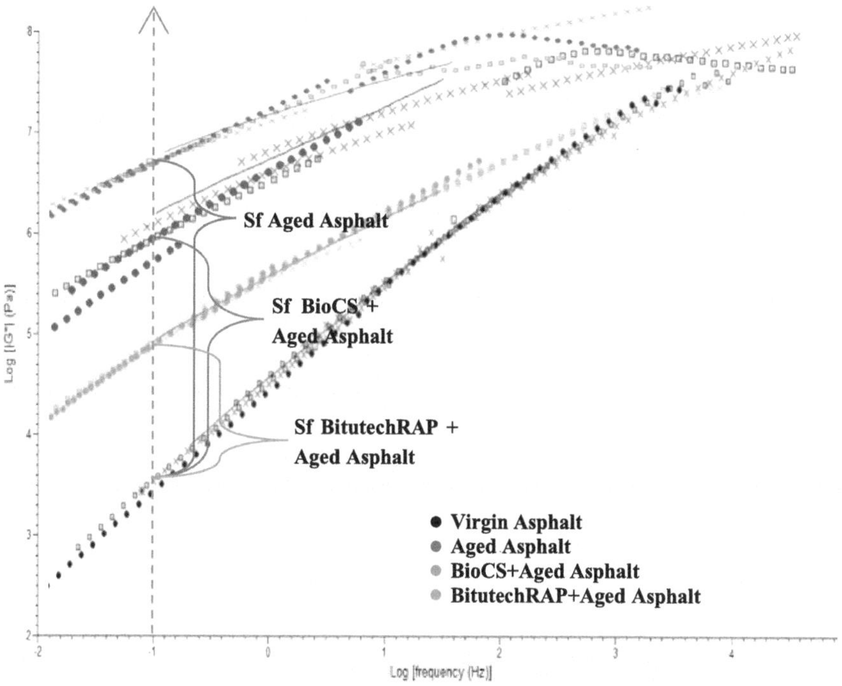

Fig. 3 Master curve by DSR frequency sweep

low frequency and high temperature conditions by using a speed value of 5.65 km/h (critical point) at a frequency of 0.10 Hz where the log value (frequency) = (−1) then a vertical line is drawn to intersect the master curve is then drawn to the left to get the value G* (complex shear modulus) which will later be converted to the target value of Sbit (stiffness modulus).

The results obtained as a minimum target (critical point) value of the stiffness modulus (Sbit) of BioCS and aged asphalt are greater than bitutechRAP and aged asphalt.

B. Black Diagram

Black Diagram illustrates the relationship of Stiffness modulus (Sbit) and Phase Angle (Francken 1998). Some things that can be conveyed, as an analysis of the Black Diagram (Fig. 4), are as follows:

1. The addition of bio-asphalt to aged asphalt affects the value of Sbit, the value of Sbit decreases after the addition of bioasphalt.
2. The value of Sbit decreases with increasing Phase Angle, showing that increasing the proportion of viscous (G″) will decrease Sbit.
3. Changes in behavior occur at the level of asphalt stiffness modulus 1000 Pa, at that level a decrease in asphalt modulus of rigidity together with an increasing in phase angle (increase in viscous portion), behavior changes at the modulus level

of 10,000,000 Pa occur with an decrease in the value of the phase angle (increase portion of elastic). This phenomenon is in accordance with the theory of black diagrams by Franken (1998).

4. The addition of bioCS to aged asphalt, aging faster than BitutechRAP to aged asphalt, this is seen in the RTFOT conditions, where asphalt modulus of stiffness at 10,000,000 Pa changes at 45 °C, and in the PAVT condition, the BioCS curve and aged asphalt are the same as the aged asphalt curve. This indicates the addition of BioCS faster stiff or faster aging.

The Sbit value of the shifting factor master curve is then plotted on the black diagram. Plotting is done to determine the value of the phase angle (δ) of each type of asphalt. The phase angle of the plotting results as described in Table 3.

According to (Francken 1998), the value of phase angle (δ) is one of the factors that influence the age of road pavement services. Based on the black diagram of each type of asphalt with the results of the plotting presented in Table 4. It can be seen that the addition of BioCS gives a smaller phase angle value than BitutechRAP, it

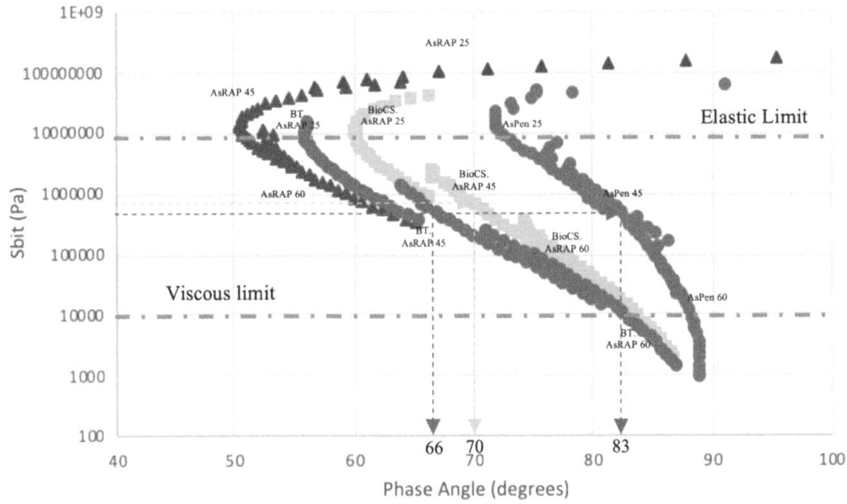

Fig. 4 Black diagram by frequency sweep

Table 3 Sbit target of bio-asphalt dan aged asphalt by *shift factor*

Asphalt type	G* target (Pa)	Sbit target (Pa)
Virgin asphalt	180300	540900
Aged asphalt	20100000	60240000
BioCS and aged asphalt	280000	838500
BitutechRAP and aged asphalt	197000	589500

Table 4 Phase angle by Sbit target master curve

Asphalt type	Phase angle (δ)
Virgin asphalt	83
Aged asphalt	(has exceeded the elastic limit)
BioCS and aged asphalt	70
BitutechRAP and aged asphalt	66

shows that the addition of BitutechRAP has more influence on the service life of road pavement.

5 Conclusions

The conclusions that can be drawn from the experiment on the effect of adding bio-asphalt to aged asphalt mechanically from the results of the DSR temperature sweep test and DSR frequency sweep are:

1. The addition of bio-asphalt can affect the PG value of aged asphalt, the high PG value of aged asphalt can change to small. The addition of BitutechRAP is more influential in reducing the value of PG compared to BioCS.
2. The addition of BioCS is more resistant to rutting as seen from Sbit which is greater than BitutechRAP at slow loading.
3. The addition of BitutechRAP is more resistant to fatigue, it can be seen from the Sbit value which is smaller than BioCS.
4. The addition of BitutechRAP has more influence on the service life of road pavement compared to BioCS.

References

AASHTO (2013) Standard specifications for transportation materials and methods of sampling and testing and AASHTO provisional standards

AASHTO T315-10 (2010) Determining the rheological properties of asphalt binder using dynamic shear rheometer (DSR)

Burr BL, Davison RR, Glover CJ, Bullin JA (1990) Solvent removal from asphalt

Burr BL, Davison RR, Jemison HB, Glover CJ, Bullin JA (1991) Asphalt hardening in extraction solvents. Transp Res Rec 1323:7

Francken L (1998) Bituminous binders and mixes. State of the art and interlaboratory tests on mechanical behaviour and mix design. New York

Kusuma RA, Haryanto I (2014) Pengujian marshall campuran laston yang mengandung bioaspal untuk lalu lintas sedang. Universitas Gajah Mada

Peralta J, Raouf MA, Tang S, Christopher Williams R (2012) Bio-renewable asphalt modifiers and asphalt substitutes. Green Energy Technol 62:89–115

Perdana S, Utomo HB (2013) Karakteristik marshall campuran aspal beton dengan menggunakan substitusi bioaspal sebagai bahan ikat untuk kategori lalu lintas berat. Universitas Gajah Mada

Read J, Whiteoak D (2003) The shell bitumen handbook

Sihombing AVR, Subagio BS, Susanto E, Yamin A (2019) The effect of bioasphalt on aged asphalt. In: IOP conference series: materials science and engineering, vol 508

SNI 06-2434 (1991) Metode Pengujian Titik Lembek Aspal Dan Ter

SNI06-2456 (1991) Metode Pengujian Penetrasi Bahan-Bahan Bitumen

Sugeng B, Sihombing AVR (2019) Utilization bioasphalt as a modifier in aged asphalt

Williams RC, McCready NS (2008) The utilization of agriculturally derived lignin as an antioxidant in asphalt binder, vol 2

Williams RC, Schaefer VR, Robert Stephenson W (2010) Asphalt modification by utilizing bio-oil ESP and tall oil additive. Major: Civil Engineering

The Effect of Fly Ash Based Geopolymer on the Strength of Problematic Subgrade Soil with High CaO Content

Nawfal Farooq Kwad, Ahmed H. Abdulkareem, and Taher M. Ahmed

Abstract The weak subgrade soil is one of the major challenges for civil engineering applications such as roads and foundations. This study aims to find out the influence of fly ash-based geopolymer on the strength of weak soil to fulfill the requirements of the subgrade layer in the pavement structure. Fly ash particles of class F was used as a raw material for geopolymer synthesis. The alkaline liquid consists of the Sodium hydroxide (NaOH) at 8 molars solution and Sodium silicate Na_2Sio_3 in liquid form and the ratio of NaOH:Na_2Sio_3 remained constant at 60:40 by weight. Low plasticity sandy silt was utilized in the study and stabilized using various proportions of fly ash (5, 10, 15, 20, 25, and 30%). Laboratory investigation involves the compaction properties of soil-fly ash mixtures in addition to the mechanical properties including the Unconfined Compressive Strength (UCS) test and the Indirect Tensile Strength (ITS) test. The UCS test results revealed that the compressive strength of the soil greatly improved after adding the fly ash-based geopolymer and 20% of fly ash content achieved the highest UCS at 28 days of curing time. The ITS test results exhibited a progressive increase in the tensile strength of the soil with fly ash geopolymer, which corresponds to a great resistance for cracking in the soil. Geopolymer gel was observed in the stabilized soil, as confirmed by the SEM analysis.

Keywords Subgrade soil · Geopolymer · Chemical stabilization · Calcium · Unconfined compressive strength

---

N. F. Kwad (✉) · A. H. Abdulkareem · T. M. Ahmed
College of Engineering, Civil Engineering Department, University of Anbar, Ramadi, Anbar, Iraq
e-mail: naw17e101@uoanbar.edu.iq; civ7977@gmail.com

A. H. Abdulkareem
e-mail: ahm1973ed@uoanbar.edu.iq; ahm1973ed@gmail.com

T. M. Ahmed
e-mail: drtaher.ahmed@uoanbar.edu.iq; alanitaher2014@gmail.com

C. Raab (ed.), *Proceedings of the 9th International Conference on Maintenance and Rehabilitation of Pavements—Mairepav9*, Lecture Notes in Civil Engineering 76,
https://doi.org/10.1007/978-3-030-48679-2_51

1 Introduction

Soil stabilization is the process of modifying the physical and mechanical properties of weak and fine-grained soils to increase the strength and long-term performance. Soil stabilization by cement or lime is the most common method to improve the strength of the soils. According to Davidovits (2015), the cement manufacturing releases 0.85–1.0 ton of Carbon dioxide CO_2 for one ton of the Ordinary Portland Cement (OPC). This harmful effect on the atmosphere has made a necessary action to find a suitable alternative material that is environmentally friendly and has similar or better properties than the OPC and to have a practical approach to utilize waste materials. Geopolymers are inorganic polymers consist of chains or 3D network of mineral molecules produced from the reaction between high alkali solution and the alumina-silicates materials at low temperatures, normally below 100 °C (Davidovits 2017). In the high calcium materials, the significant amount of calcium that exists in the raw material will produce C–S–H gel phase, which is the dominant phase, while in the low calcium mixtures, a poly-condensation process occurs by sharing the oxygen ions to form Si–O–Si and Si-O-Al bonds which define the three-dimensional structure of geopolymer. This component is very stable and durable termed as the geopolymer (Cristelo et al. 2012). Fine-grained soil with a high content of clay and silt are less proper material as a pavement subgrade.

The addition of fly ash geopolymer to the soil can develop a cementitious bond which results in strength gain, the strength of soil increases as fly ash content increases (Santos et al. 2011). The previous studies have demonstrated the effectiveness of fly ash geopolymer in soil stabilization, where the stabilized soil with fly ash and slag geopolymer passed 12 cycles of wetting and drying conditions with only 14% weight loss (Amulya et al. 2018). A study conducted by Dungca et al. (2018) concluded that the fly ash-based geopolymer is an effective technique to increase the CBR index of the embankment material by adding 30% of fly ash geopolymer to achieve a significant increase in the CBR value up to 34%. It is evident from Phani et al. (2004) that swelling pressure and swell potential of the clayey soil were decreased by approximately 50% after fly ash addition and both of compression index and secondary consolidation coefficient of the expansive and non-expansive clay were decreased by 40% after adding 20% of fly ash to the soil. Many researchers (Cristelo et al. 2011; Cristelo et al. 2013; Yaghoubi et al. 2018) have studied the effects of fly ash-based geopolymer on the soil, they concluded that the fly ash-based geopolymer is an appropriate material for soil stabilization and increasing the fly ash content to a certain limit improves the UCS and the performance of the soil. In addition to the UCS, it is important to investigate the tensile property of the stabilized soil especially after the geopolymer incorporation, as the new structure is being fully bound material within the soil. The fatigue life of the stabilized soil is an important factor to be considered in the design of pavement layers as the cracking behavior of the soil is one of the factors that affect it. Generally, the increase of the horizontal tensile strain of cement-stabilized material results in a decrease in fatigue life (González et al. 2013). Many researchers (Cristelo et al. 2012; Khater 2011; Yip et al. 2005) have studied the influence of calcium components in the fly ash on the geopolymerization products, they concluded that the high amount of calcium in the fly ash contributes in precipitation of both the calcium silicate hydrate (C-S-H) gel and the geopolymeric gel. Also, the high content of calcium in the raw material increases the short-term strength and may prevent the formation of the 3D structure of geopolymer which is important for the long-term performance of soil (Cristelo et al. 2012). C-S-H gel is likely to exist in the calcium-based materials that contain alumina and silica, alongside the geopolymer binder (Link et al. 2001).

2 Materials

2.1 *Soil Properties*

Low plasticity sandy silt was used in the study, collected from Ramadi City. The physical properties of the soil are listed in Table 1. The particle size analysis in Fig. 1 shows that the soil consists of 60% of silt, 9% of clay and 31% of fine sand. The chemical composition of the soil is listed in Table 2. It is known that the soil in the western part of Iraq was formed originally from the natural lime and silicates and consists mostly of lime rocks with marl, which is found largely in the western desert of Iraq (Muhaimeed et al. 2014). It is clear that the soil has a high content of calcium oxide (CaO), this component is known as quicklime and it is soluble in water. This study aims to predict the influence of this component on the strength of soil-fly ash geopolymer mixture.

Table 1 Geotechnical properties the soil

Property	Standard designation	Value
Liquid limit (%)	ASTM D4318-10	38
Plastic limit (%)	ASTM D4318-10	27
Plasticity index	ASTM D4318-10	11
Specific gravity	ASTM D854-14	2.66
Unified soil classification system (USCS)	ASTM D2487-11	ML-low plasticity sandy silt
Optimum moisture content (%)	ASTM D1557-12	15.5%
Maximum dry density (g/cm^3)	ASTM D1557-12	1.73 g/cm^3

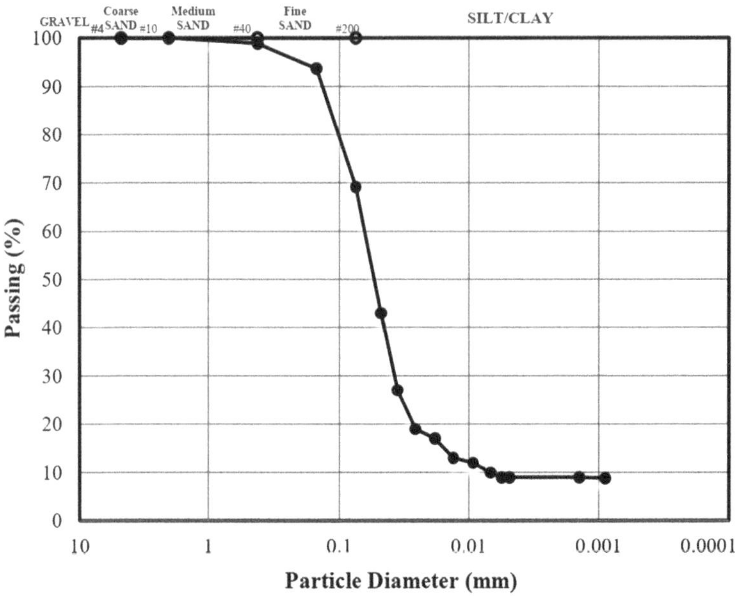

Fig. 1 Particle size analysis of the soil

Table 2 The chemical composition of the soil and fly ash

Component	SiO_2	Al_2O_3	CaO	Fe_2O_3	MgO	Na_2O	K_2O	SO_3	MnO
Soil	37.64	6.5	23.43	4.06	3.9	1.3	1.1	0.7	0.1
Fly ash	53.23	25.91	4.95	6.46	1.2	0.1	1.4	0.78	0.16

2.2 *Geopolymer Synthesis*

2.2.1 Fly Ash

According to ASTM C618-12, the fly ash is classified as Class F due to the low content of calcium. This type of fly ash is preferred in the geopolymer synthesis, where it produces higher strength and better response in Sodium-based geopolymers. The chemical composition of fly ash through X-Ray Florence is listed in Table 2. The specific gravity of the fly ash is 2.14.

2.2.2 Alkaline Solution

The alkaline solution consists of Sodium hydroxide (NaOH) pellets of 99% purity, and Sodium silicate (Na_2SiO_3). The NaOH concentration was 8 mol and the molecular weight is 40 g/mol. The Sodium silicate (Na_2SiO_3) was in liquid form and contains 32.5% of SiO_2, 13.4% of Na_2O and 54% of water. The ratio of 60:40 of $NaOH:Na_2SiO_3$ was used to activate the fly ash and soil particles. The high concentration of NaOH was used to ensure the dissolution of fly ash particles alongside the soil components that may be affected by the alkaline solution.

3 Experimental Work

3.1 Sample Preparation

The dry mixing of soil and fly ash was used to prepare the samples of the unconfined compressive strength (UCS) and the indirect tensile strength (ITS) tests. The soil was thoroughly mixed with fly ash for five minutes to provide a homogenous mixture. Various percentages of fly ash (5, 10, 15, 20, 25, and 30%) by weight of soil were used in the study. NaOH pellets were dissolved in distilled water to avoid undesirable reactions and left for 24 h prior to blending with the Na_2SiO_3 to ensure the efficiency of the dissolution. Sodium silicate is already in liquid form and can be mixed with NaOH solution to produce the alkaline solution. After getting a uniform mixture of soil and fly ash, the alkaline solution of NaOH and Na_2Sio_3 was added as a replacement of the optimum water content of soil-fly ash mixture and thoroughly mixed with the solids using a mechanical mixer for 10 min to ensure the homogeneity of the geopolymer material with the soil. The extra quantity of the alkaline solution was added to compensate NaOH solids which are dissolved in the solution based on a previous study conducted by Adhikari (2017). For the UCS and ITS tests, the stabilized samples were molded and cured in a drying oven for 48 h at 65 °C, then removed from the oven and protected in Zip-lock and kept at room temperature to be tested after 7 and 28 days. To obtain moisture density relationships, the modified proctor test was conducted according to ASTM D 1557-12. The soil was dried for 24 h at 110 °C before mixing with various proportions of fly ash. The time from mixing with water until getting the compacted specimen did not exceed 30 min to avoid the conglomerate of fly ash particles or any effectual reactions with water or soil components.

4 Results and Discussion

4.1 Compaction Test

According to the modified compaction test results, the maximum dry density (MDD) of the soil is 1.73 g/cm^3 and the optimum moisture content (OMC) is 15.5%. Figure 2 shows the relationships between the dry density and moisture content for the soil and soil-fly ash mixtures at various fly ash percentages. It can be noticed from the

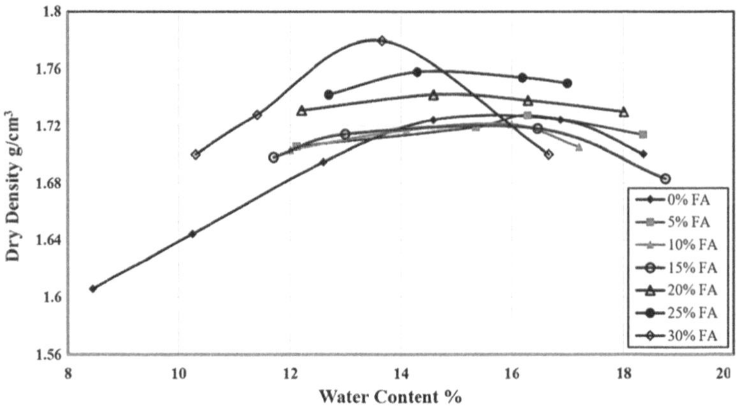

Fig. 2 Moisture-density relationship of soil-fly ash mixtures

compaction results that the maximum dry density of soil-fly ash mixtures exhibited an increasing trend from 1.73 g/cm^3 for the pure soil to the maximum dry density of 1.78 g/cm^3 at 30% of the fly ash content. While the optimum water content of soil-fly ash mixtures decreased from 15.5% for the untreated soil to the maximum dry density of 13.5% obtained at the soil with 30% fly ash mixture. However, this increase in the MDD was observed clearly beyond 15% of fly ash addition and the lower content of fly ash did not affect the MDD significantly.

The high density of soil +30% of fly ash mixture may be attributed to the interaction of fly ash particles into soil voids resulting in higher density. Furthermore, there is a noticeable fluctuation in the optimum water content where the OMC increases when fly ash increase to 15%, and decreases beyond this content. This increase in the OMC may be due to the hydration reactions of the fine fly ash particles which require more water (Santos et al. 2011). However, if the fly ash content increase to more than 15%, the OMC decrease while the MDD increases. The MDD of soil-fly mixture varying with test conditions and preconditioning period, which is the period between adding the water and compaction.

4.2 *Unconfined Compressive Strength (UCS) Test*

Unconfined compressive strength test (UCS) conducted to evaluate the effectiveness of the fly ash geopolymer as a soil stabilization agent. The subgrade layer exposes to compressive stresses imposed by the traffic loads and this test is very important to evaluate the ultimate strength of the stabilized soil under excessive loads. Soil and soil-fly ash mixtures were prepared and compacted into large size cylindrical molds, i.e. 100 mm diameter and 200 mm height according to ASTM D4219-08. The alkaline solution was used in soil-fly ash mixtures as a replacement of the optimum water content. The samples were initially cured at 65 °C for 48 h and tested after 7 and 28 days of curing at room temperature. Three samples were prepared for each mixture. The unconfined compressive test was performed by a constant loading rate of 0.1 kN/s. The results of 7 and 28 days cured specimens are shown in Fig. 3. The highest compressive strength was recorded at the optimum mixture that consists of the soil with 20% of fly ash irrespective of the curing period. The highest UCS values were 20.7 and 22.2 MPa after 7 and 28 days of curing respectively. Any further increase in the fly ash content led to strength reduction as shown in Fig. 4. The significant increase in the UCS was during the initial seven days of curing, and there is no significant variation in the results of the 7 and 28 days cured specimens. This trend may be attributed to the formation of the geopolymer binder at the early ages of curing at 65 °C and the subsequent reactions in the first 7 days which improve the strength significantly (Fig. 5).

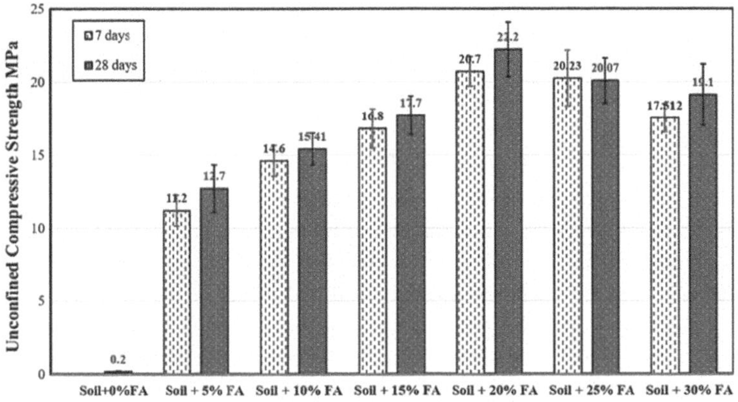

Fig. 3 Results of the UCS test for 7 and 28 days cured specimens

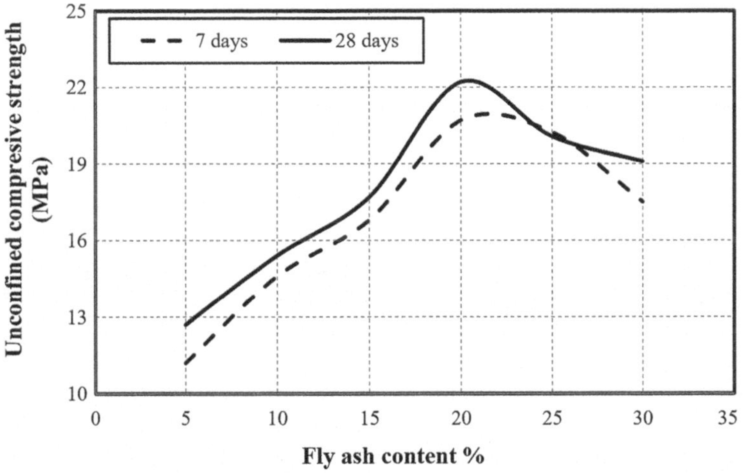

Fig. 4 Variation of UCS with fly ash increase

Fig. 5 Failure of UCS test specimen

The significant increase in the strength is related directly to the curing temperature and period, where the heat-curing has a substantial effect in accelerating the chemical reactions of geopolymer paste and longer curing period enhances the polymerization process (Rangan 2009). In addition to that, the high compressive strength of the stabilized soil may be attributed to the combined effect of the geopolymer binder and CaO in the mixture. The great increase in the compressive strength of the soil after adding the fly ash geopolymer reveals the effectiveness and feasibility of the fly ash-based geopolymer in the silty soil stabilization.

4.3 Indirect Tensile Strength (ITS) Test

The split cylinder test or known as the Brazilian test is the indirect method that determines the tensile strength of the material by applying compressive loads in the diametrical plane of a cylindrical specimen placed horizontally in the loading machine. It is a key factor in evaluating the cracking behavior of the stabilized soil when used in pavement structure or the other earth structures. There is limited information regarding the tensile strength of soil-fly ash geopolymer mixtures. Burmister (Burmister et al. 1944) concluded that the tensile stresses in a layered system will be developed at the locations of interfaces between the layers especially when the upper layers have a modulus of elasticity higher than underlying pavement layers.

The study involves preparing compacted specimens into cylindrical molds that are the same as the modified compaction molds, i.e. 101.6 mm diameter and 116.4 mm height, at alkali content corresponding to the optimum moisture content of the mixture. The influence of the cylindrical mold size was studied by Wright (1955), Thompson (1966), they concluded that the size of the cylindrical mold has a little effect on the tensile strength results and specimens with the larger size showed a smaller coefficient of variation. The indirect tensile strength at failure is given by:

$$\sigma t = \frac{2P}{\pi dt} \tag{1}$$

Where

σt = Tensile strength (MPa)
P = Maximum load at failure (N)
d = Diameter of specimen (mm)
t = thickness of specimen (mm)

The test results are shown in Fig. 8, and the variation of ITS with fly ash content is shown in Fig. 6. For the untreated soil, the indirect tensile strength was 0.04 N/mm^2 which is not effectual. When fly ash geopolymer was added to the soil and after 7 and 28 days of curing, the tensile strength increased significantly to 3.6 and 3.9 N/mm^2 respectively. These results observed at 20% of fly ash content and any further increase in the fly ash content decreased the ITS irrespective of the curing period. The reduction in the tensile strength may be attributed to the unreacted amount of fly ash which acts as a filter material at the early ages of curing. It is obvious from the test results that the stabilized soil by fly ash-based geopolymer acts as a rigid material and the failure of specimens is clear splitting in the longitudinal plane and similar to the brittle materials as shown in Fig. 7.

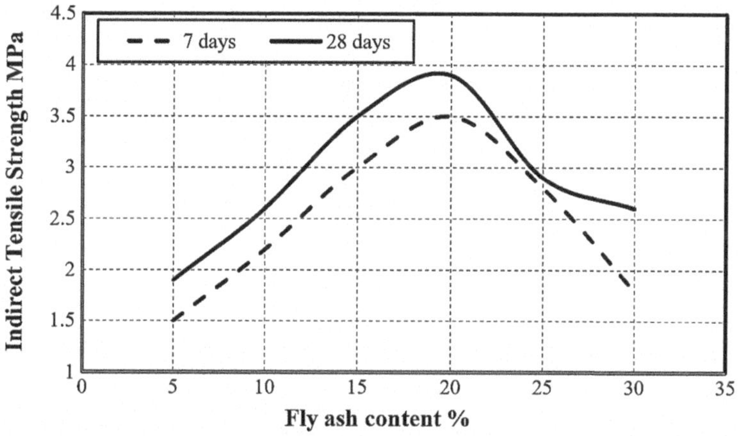

Fig. 6 Variation of ITS with fly ash increase

Fig. 7 Failure of ITS test specimen

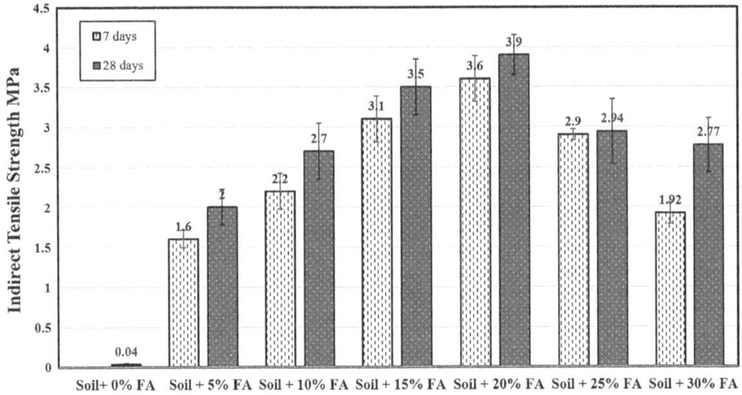

Fig. 8 Indirect tensile strength test results

4.4 *Effect of CaO Content and Curing Conditions*

The role of calcium compounds such as CaO is critical in the geopolymer synthesis. CaO effect may enhance the mechanical performance of the fly ash-based geopolymer when it is cured at low temperatures. However, the high content of calcium can affect polymerization development and change the microstructure of the product material (Temuujin et al. 2009). The previous studies showed that the alkaline activation of metakaolin and furnace slag will produce the calcium silicate hydrate (C-S-H) gel and geopolymer paste and both of them be formed simultaneously (Yip et al. 2003, 2005). Furthermore, the presence of CaO has a substantial effect on the hardening time of geopolymers. The high content of CaO in the fly ash accelerates the initial setting time of geopolymer paste (Wijaya et al. 2016).

It is proved that the hydration process of the Portland cement produces (C-S-H) gel and $Ca(OH)_2$ and the (C-S-H) gel provides the high strength to the mix. While the calcium hydroxide is unstable component and can easily react with CO_2 in the air and produce calcium carbonate which is the main reason behind concrete deterioration (Temuujin et al. 2009). According to the chemical composition, the soil contains 23% of the calcium oxide (CaO) which may participate in C-S-H gel formation alongside the geopolymer binder from the activation of fly ash (Bagheri et al. 2017). Many researchers (Temuujin et al. 2009; Cristelo et al. 2012; Canfield et al. 2014) corroborate this idea especially in the activation of the high calcium fly ash as they concluded that the C-S-H gel will be formed in addition to the geopolymer binder. Figure 9 shows the SEM analysis of the soil with 20% of fly ash geopolymer mixture. It can be seen that the geopolymer binder acts as a glue material to bond the soil particles resulting in a dense matrix in addition to the poorly crystalline C-S-H gel. Also, micro-cracks were observed in the structure and may be attributed to the reactive-silica reaction with the sodium hydroxide that produces a material absorbs the water in the mixture and increase the volume resulting in cracks formation (Slaty et al. 2015). This analysis is very comparable with strength results as the high strength of specimens indicates the precipitation of C-S-H gel and geopolymer. In addition to the CaO, heat-curing contributes positively to the chemical reactions that occur in the geopolymer paste and longer curing period enhances the polymerization development and strength gain in the fly ash geopolymer (Hardjito et al. 2005). Soil-fly ash geopolymer mixtures were cured at 65 °C for 48 h to accelerate the reaction of the alkaline activator and mixture components which is highly affected by curing temperature. Both of the UCS and ITS of soil-fly ash geopolymer increased with curing time after exposing the specimens to the heat-curing for 48 h.

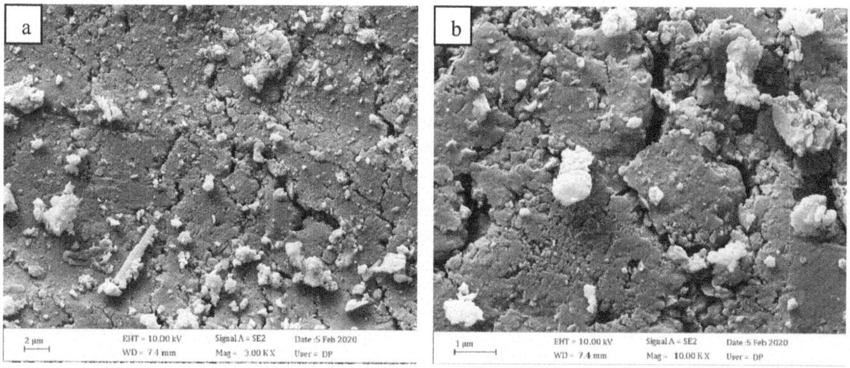

Fig. 9 SEM images of soil and fly ash geopolymer mixture with 23% of CaO content in the soil, **a** 2 μm magnification **b** 1 μm magnification

5 Conclusions

This study presents the influence of fly ash-based geopolymer on the mechanical properties of fine-grained soil and the following conclusions are estimated:

- The fly ash-based geopolymer has been demonstrated to be an efficient additive to the weak soil by satisfying the required strength and compaction properties of the subgrade layer and exhibited excellent performance in terms of the compressive and tensile strength of the soil.
- The unconfined compressive strength of the stabilized soil increased significantly after the addition of fly ash geopolymer. The stabilized soil exhibited high compressive strength at both 7 and 28 days and soil with 20% of fly ash content achieved the highest compressive strength by 22 MPa at 28 days of curing time. The indirect tensile strength increased up to 80 times and the highest ITS value was 3.9 MPa recorded also at soil +20% fly ash mixture. The UCS and ITS values are comparable to the strength of concrete, and specimen failure similar to brittle materials.
- Generally, the addition of fly ash increased the dry unit weight of the soil and reduced the optimum water content. However, the OMC increased slightly at the low content of fly ash below 15%. The maximum increase in the dry density from 1.73 to 1.78 g/cm^3 occurred after adding 30% of fly ash to the soil.
- The high content of CaO in the soil improved the short-term performance of the soil-fly ash mixtures. However, the exact role of the calcium oxide in the soil is not obvious in terms of the new products in the soil and durability.
- Durability tests are essential to predict the long-term performance and stability of the new structure in the soil which may be affected by the leaching of CaO and unstable bonds in the alkali-activated material especially after exposure to the water and severe environmental conditions.

References

Adhikari S (2017) Mechanical properties of soil-RAP-geopolymer for the stabilization of road base/subbase. University of Louisiana at Lafayette

Amulya S, Ravi Shankar AU, Praveen M (2018) Stabilisation of lithomargic clay using alkali activated fly ash and ground granulated blast furnace slag. Int J Pavement Eng 1–8

Bagheri A et al (2017) Alkali activated materials vs geopolymers: role of boron as an eco-friendly replacement. Constr Build Mater 146:297–302

Burmister DM et al (1944) The theory of stress and displacements in layered systems and applications to the design of airport runways. In: Highway research board proceedings

Canfield GM et al (2014) The role of calcium in blended fly ash geopolymers. J Mater Sci 49(17):5922–5933

Cristelo N et al (2012) Effect of calcium content on soil stabilisation with alkaline activation. Constr Build Mater 29:167–174

Cristelo N et al (2013) Effects of alkaline-activated fly ash and portland cement on soft soil stabilisation. Acta Geotech 8(4):395–405

Cristelo N, Glendinning S, Teixeira Pinto A (2011) Deep soft soil improvement by alkaline activation. Proc Inst Civil Eng-Ground Improv 164(2):73–82

Davidovits J (2015) False values on CO2 emission for geopolymer cement/concrete published in scientific papers. Technical paper, no 24

Davidovits J (2017) Geopolymers: ceramic-like inorganic polymers. J Ceram Sci Technol 8(3): 335–350

Dungca JR, II E (2018) Fly-ash-based geopolymer as stabilizer for silty sand embankment materials. Int J 14(46):143–149

González A et al (2013) Laboratory fatigue life of cemented materials in Australia. Road Mater Pavement Des 14(3):518–536. https://doi.org/10.1080/14680629.2013.779300

Hardjito D, Rangan BV (2005) Development and properties of low-calcium fly ash-based geopolymer concrete

Khater HM (2011) Effect of calcium on geopolymerization of aluminosilicate wastes. J Mater Civil Eng 24(1):92–101

Link RE et al (2001) California bearing ratio behavior of soil/fly ash mixtures. J Test Eval 29:220–226. https://doi.org/10.1520/jte12249j

Muhaimeed AS et al (2014) Classification and distribution of Iraqi soils. Int J Agric Innov Res 2(6):997–1002

Phani Kumar BR, Sharma RS (2004) Effect of fly ash on engineering properties of expansive soils. J Geotech Geoenviron Eng 130(7):764–767

Rangan B (2009) Engineering properties of geopolymer concrete. In: Geopolymers: structures, processing, properties and industrial applications, pp 211–226. https://doi.org/10.1533/9781845696382.2.211

Santos F et al (2011) Geotechnical properties of fly ash and soil mixtures for use in highway embankments. In: Proceedings of the world of coal ash (WOCA) conference, Denver, USA, pp 125–136

Slaty F et al (2015) Durability of alkali activated cement produced from kaolinitic clay. Appl Clay Sci 104:229–237

Temuujin JV, Van Riessen A, Williams R (2009) Influence of calcium compounds on the mechanical properties of fly ash geopolymer pastes. J Hazard Mater 167(1–3):82–88

Thompson MR (1966) Split-1'"ensile strength of lime-stabilized soils

Wijaya SW, Hardjito D (2016) Factors affecting the setting time of fly ash-based geopolymer. In: Materials science forum. Trans Tech Publications, pp 90–97

Wright PJF (1955) Comments on an indirect tensile test on concrete cylinders. Mag Concr Res 7(20):87–96

Yaghoubi M et al (2018) Effects of industrial by-product based geopolymers on the strength development of a soft soil. Soils Found 58(3):716–728

Yip CK, Van Deventer JSJ (2003) Microanalysis of calcium silicate hydrate gel formed within a geopolymeric binder. J Mater Sci 38(18):3851–3860

Yip CK, Lukey GC, van Deventer JSJ (2005) The coexistence of geopolymeric gel and calcium silicate hydrate at the early stage of alkaline activation. Cement Concr Res 35(9):1688–1697

ASTM C618-12 (2012) Standard specification for coal fly ash and raw or calcined natural pozzolan for use in concrete. ASTM International, West Conshohocken. www.astm.org

ASTM D4318-10 (2010) Standard test methods for liquid limit, plastic limit, and plasticity index of soils. ASTM International, West Conshohocken. www.astm.org

ASTM D854-14 (2014) Standard test methods for specific gravity of soil solids by water pycnometer. ASTM International, West Conshohocken. www.astm.org

ASTM D2487-11 (2011) Standard practice for classification of soils for engineering purposes (unified soil classification system). ASTM International, West Conshohocken. www.astm.org

ASTM D1557-12 (2012) Standard test methods for laboratory compaction characteristics of soil using modified effort (56,000 ft-lbf/ft3 (2,700 kN-m/m3)). ASTM International, West Conshohocken. www.astm.org

ASTM D4219-08 (2008) Standard test method for unconfined compressive strength index of chemical- grouted soils (withdrawn 2017). ASTM International, West Conshohocken. www.astm.org

Low Temperature Behavior of Asphaltite Modified Binders and Asphalt Concretes

Andrea Themeli, Emmanuel Chailleux, Cyrille Chazallon, and Nicolas Bueche

Abstract Natural asphalts are often used as stiffeners to obtain hard bitumens. With the decrease of the global production of hard bitumens, the use of natural asphalts as stiffeners becomes increasingly interesting. The main concern when natural asphalts are used as modifiers, is the behavior at low temperature (LT). Studies have shown a brittle behavior in the LT domain, especially when high-content-asphaltene natural asphalts, like the Gilsonite, are used. In the context of a wider study, carried out in IFSTTAR, we have tested the low-temperature behavior of asphaltite modified bitumens (AMB) and that of corresponding asphalt concretes. The Selenizza® asphaltite is used at 5, 10 and 15% to modify a pure petroleum bitumen of 50/70 penetration grade. Asphalt concretes of type EME were then produced. 3P bending tests on notched bitumen bars were performed in order to assess the cracking temperature of binders. On the asphalt concrete scale, thermal stress restrain tests and traction resistance tests at low temperatures were carried out in order to assess the low temperature behavior. Equivalent materials (pure refinery binders and EME produced with hard petroleum binder) were also tested in order to have references. Compared to these equivalent materials, the asphaltite stiffened materials (AMB and corresponding asphalt concretes) present comparable, and in some cases, better LT behavior.

Keywords Low temperature behavior of binders and asphalts · Natural bitumen · Asphaltite modified bitumens · High modulus asphalt concrete

A. Themeli (✉) · N. Bueche
Department of Architecture, Wood and Civil Engineering, Bern University of Applied Sciences, Pestalozzistr. 20, 3400 Burgdorf, Switzerland
e-mail: andrea.themeli@bfh.ch

E. Chailleux
LUNAM Univ., IFSTTAR, Route de Bouaye, BP 4129, 44341 Bouguenais, France

C. Chazallon
ICUBE (UMR 7357, CNRS, National Institute of Applied Sciences), 24, Bd. de la Victoire, 67084 Strasbourg Cedex, France

C. Raab (ed.), *Proceedings of the 9th International Conference on Maintenance and Rehabilitation of Pavements—Mairepav9*, Lecture Notes in Civil Engineering 76,
https://doi.org/10.1007/978-3-030-48679-2_52

1 Introduction

Hard bitumens are of real interest in pavement engineering nowadays. They are used in the production of high modulus asphalt concretes (HMAC). HMAC allow the reduction of thickness of structural pavement layers and/or the prolongation of the pavement lifetime (Corté 2001; Geng et al. 2013).

Hard bitumens are produced in petrol refineries by processing the residue of the vacuum distillation of petrol by means of different techniques as air blowing, oxidation, solvent deasphalting etc. (Eurobitume 2011). A shortage of hard bitumens is observed and appeals are made to a careful use (Lombardi 2012). For these reasons, several studies aim to develop alternatives for the production of hard bitumens. These alternatives very often consist in the modification of soft petroleum bitumens by various modifiers (Bardesi et al. 1999). Research is dedicated to understand the impact of various bitumen modifiers on the mechanical properties (Aflaki et al. 2014; Aflaki and Tabatabaee 2009; Ameri et al. 2011; Huang and Shu 2009; Kök et al. 2011; Yildirim 2007).

Asphaltites have a good potential for the production of hard pavement bitumens.

Asphaltites are hard natural bitumens, chemically similar to petroleum bitumens. In a wider work (Themeli 2015), the authors of this paper have shown that the asphaltite modification of pure petroleum bitumens lightly affect their chemical composition in terms of SARA fractions and molecular associations determined by GPC. But the mechanical behavior is strongly affected (Themeli et al. 2015; Themeli et al. 2017).

One of the main concerns when using asphaltites as modifiers, is the behavior at low temperatures (LT). In general, hard bitumens and the corresponding asphalt concretes are susceptible to show a brittle behavior at LT.

The natural asphaltite used in this study is mined in Albania in the region of Selenizza. The asphaltite modification of soft petroleum bitumens gives hard binders. With these binders, HMACs which satisfy the European standards can be obtained. However, we do not know how asphaltite modified bitumens (AMB) and asphaltite modified HMACs behave at LT. So, the scope of the work presented in this work is to study the LT behavior of Selenizza® modified bitumens and HMACs. The LT behavior of modified bitumens is studied by a three-point bending test performed on notched bitumen bars. The LT behavior of HMACs is studied by thermal stress restrained specimen test (TSRST) combined with uniaxial tension stress test (UTST). In addition, the results are compared with them of equivalent materials (pure petroleum bitumen and HMAC produced with hard petroleum bitumen). This allows to make relevant comparisons and to draw clear conclusions on whether asphaltite modified bitumens can substitute or not hard petroleum bitumens as binders for HMAC production.

2 Experimental

2.1 Tests on Modified Bitumens

3P bending test on notched bitumen bars are performed on aged bitumens.

2.1.1 Ageing Procedure

PAV ageing for 25 h at 100 °C and under a pressure of 2.1 MPa are applied. The additional 5 h compared to a standard PAV test (which takes 20 h) allow to skip the RTFOT tests normally performed before PAV. Indeed, it has been shown that 5 h of PAV exposition give an aged bitumen with penetrability, R&B, asphaltene percentage, creep and complex modulus equivalent to those of bitumens submitted the RTFOT ageing (Migliori and Corte 1998). Extended PAV tests are often used in the literature (Guern et al. 2010a).

2.1.2 Test for the Low Temperature Crack Behavior

The LT mechanical behavior of bitumens is studied with a 3P bending test on notched bitumen bars. The test is based on the principles of fracture mechanics adapted for bitumens (Lee and Hesp 1994; Morrison et al. 1994; Ponniah and Hesp 1996). The test geometry is optimized (Chailleux et al. 2008) (Fig. 1) and a viscoelasticity analysis is used to interpret the results (Guern et al. 2010a).

The test is carried out in a bath of water and potassium acetate (concentration 42%) at constant temperatures. A vertical displacement, of constant speed 0.01 mm/s, is applied at the mid-length of the beam and the resulting force is recorded.

Depending on the test temperature, the specimen presents a fragile crack or an excessive plastic deformation (Fig. 1b). The cracking limit temperature (CLT), which corresponds to the temperature below which the sample presents a fragile rupture,

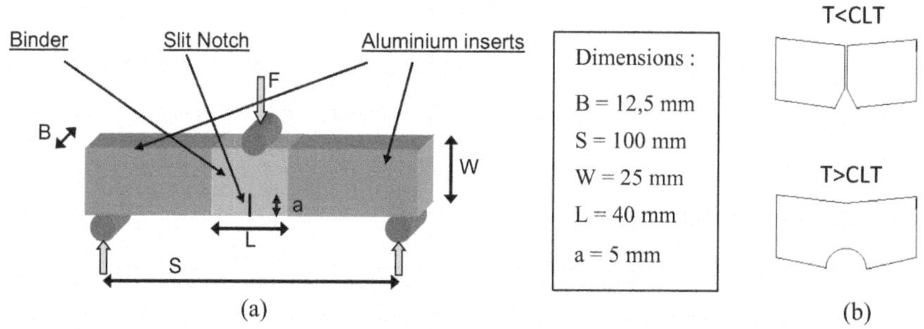

Fig. 1 3P bending test on notched bar: **a** geometry, **b** failure modes

defines the transition between the fragile and non-fragile state. The test is very sensitive to the temperature and allows to determine the CLT to the degree, i.e. if the CLT is 3 °C, the sample always cracks at 3 °C and not at 4 °C.

The test allows to calculate the critical energy recovery rate (G_C) at failure. Its value, derived from a viscoelasticity analysis (Guern et al. 2010a) is given by the expression:

$$G_C = 409.7 \cdot F_u \cdot d_u \tag{1}$$

where F_u and d_u are the force and the deformation at the failure moment.

2.2 *Tests on HMAC*

The LT behavior of HMAC is determined by thermal stress restrained specimen tests (TSRST) combined with uniaxial tension stress test (UTST) according to the European standard EN 12697-46. These tests allow the calculation of the tensile strength reserve.

2.2.1 UTST

The specimen maintained at constant temperature is subjected to a tensile stress with a constant deformation rate of 0.625 ± 0.025%/min until failure. The test results are expressed as tensile resistance $\beta_t(T)$ and deformation at rupture $\varepsilon_{failure}(T)$. Different specimens of the same material are to be tested at different temperatures. The results are reported on a temperature-resistance diagram. The test results at different temperatures are fitted by a cubic polynomial function.

2.2.2 TSRST

The sample maintained at a constant length is subjected to a constant decrease in temperature of 10 °C/min. Due to the restrained shrinkage, cryogenic stress develops in the specimen. The test results are the evolution of cryogenic stress as a function of temperature $\sigma_{cry}(T)$, the ultimate stress $\sigma_{cry,failure}$ and failure temperature, $T_{failure}$.

2.2.3 Tensile Strength Reserve

The tensile strength reserve expresses the decrease in the mechanical tensile strength due to the presence of thermal tensile stresses. At a given temperature T the strength reserve is given by the expression:

$$\Delta\beta_t(T) = \beta_t(T) - \sigma_{cry}(T) \tag{2}$$

with $\beta_t(T)$ the tensile resistance and $\sigma_{cry}(T)$ the cryogenic stress at temperature T determined respectively by means of UTST and TSRST.

3 Materials

3.1 *Bitumens*

The bitumens considered in this study are presented in the Table 1.

3.1.1 The Asphaltite

The asphaltite is mined in Albania in the region of Selenizza. In its natural state it contains 15–18% of fine mineral material. The organic phase, which is used to modify the petroleum bitumens, is isolated by dissolution in tetrachloroethylene and filter-centrifugation. The composition and some basic properties of the asphaltite are given in Table 2 in comparison with the properties of the petroleum bitumen P50/70 and the properties of the Gilsonite. The SARA fractions, the FTIR indices, the agglomerate contents and the glass transition temperatures are determined according to methods explained by Le Guern et al. (Guern et al. 2010b).

Table 1 Bitumens considered in this study

Reference	Description	Pen. (dmm)	R&B (°C)
Selenizza®	Organic phase of asphaltite extracted in deep layers of the mine	NA	120
P50/70	Petroleum bitumen of **50/70** grade to be modified	54	49.0
P35/50	Petroleum bitumen of **35/50** grade used for comparison	44	56.4
5%AS + 95%P50/70	**50/70** grade petroleum bitumen modified with **5%** of Selenizza®	38	52.6
10%AS + 90%P50/70	**50/70** grade petroleum bitumen modified with **10%** of Selenizza®	28	56.2
15%AS + 85%P50/70	**50/70** grade petroleum bitumen modified with **15%** of Selenizza®	20	61.6

Table 2 Characteristics of Selenizza® asphaltite; comparison with other materials

Test		Selenizza®	P50/70	Gilsonite (Huang et al. 2006)
c7—precipitation (NF T60-115)	Asphaltenes c7 (%)	43.8	10.2	
	Maltenes (%)	56.2	89.8	
SARA fractions	Saturates (%)	1.7 ± 0.35	6.7 ± 0.65	2–6
	Aromatics (%)	24.8 ± 2.29	50.5 ± 1.81	
	Resins (%)	35.1 ± 1.35	26.1 ± 1.64	21–37
	Asphaltenes Iatrosc. (%)	38.4 ± 1.88	16.7 ± 1.42	57–76
Oxidation (FTIR** indexes)	Sulfoxyde	6.36	–	
	Carbonyl	3.99	–	
Agglomerate content (HS-SEC*) (%)		2.4	0.92	
Glass transition temperature (°C)		−1.1	−22.9	85–107
Penetrability (0.1 mm) (EN 1427)		NA	54	0
R&B Temperature (°C) (EN 1426)		119	49	129–204
$\|E^*\|$(15 °C, 10 Hz) (Pa)		$1.23 \cdot 10^9$	$1.26 \cdot 10^8$	

*High speed size exclusion chromatography
**Fourier transform infrared spectroscopy

3.1.2 The Modified Bitumens

The modifying process consists in adding the fine grained asphaltite ($\Phi < 1$ mm) in the preheated soft petroleum bitumen P50/70. The blend is carried out by mixing both materials with a high shear mixer for 1 h at 180 °C. Modification rates of 5, 10 and 15% are chosen (Table 1). The modified binders get harder with the modification rate. They satisfy the European Norms (EN 12591, EN 13924).

3.1.3 The Petroleum Bitumen P35/50 for Comparison

The petroleum bitumen used for the comparison is found in the database of IFSTTAR. It was chosen to be compared with the 5%AS + 95%P50/70 bitumen because of the rheological similarity between the two bitumens (Fig. 2).

3.2 HMACs

The HMACs considered in this study are given in Table 3.

The comparison of EME-02 with EME-03 allows to study the impact of the mode of introduction of asphaltite on the properties of asphalt mixes. The comparison of

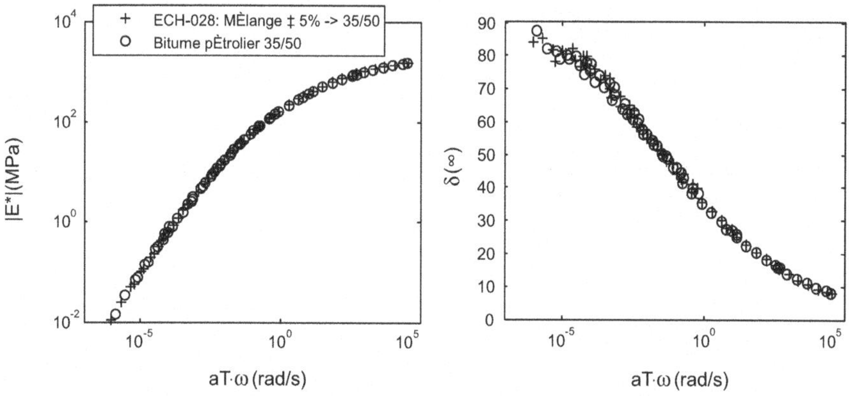

Fig. 2 Complex modulus of the 5%AS + 95%P50/70 modified bitumen compared to the petroleum bitumen P35/50 (Tref = 0 °C): **a** norm of the complex modulus, **b** phase angle

Table 3 The HMACs considered in this study

Reference	Description
EME-01	HMAC produced with hard petroleum bitumen of penetration ***grade 20/30***
EME-02	HMAC produced with ***asphaltite modified***[a] ***50/70 petroleum bitumen***
EME-03	HMAC produced with petroleum bitumen of grade **50/70**[b] **and** ***asphaltite added***[c] ***in the asphalt mixer***

[a]The quantity of asphaltite in order to obtain the $|E^*|$ at 15 °C/10 Hz of the hard petroleum bitumen of grade 20/30 of the EME-01 is determined according to a dosage calculation procedure which is developed and can be found in Themeli (2015)
[b]The 50/70 is the same petroleum bitumen which is modified and used for the production of the EME-02
[c]The quantity of the asphaltite equal to the quantity that is used to modify the bitumen 50/70 of the EME-02

these two asphalt mixes with the EME-01 allows to compare the alternative solutions produced with asphaltite, with what is produced with hard petroleum bitumens.

The EME-01 is optimized according to the French method for asphalt concrete design in order to have a HMAC of class 2. The design is based on gyratory tests, 2P complex modulus and 2P fatigue tests. The bitumen content is 5.6% of the total asphalt mix mass. The EME-02 and the EME-03 have the same constitution as the EME-01 (granulometric curve, bitumen content). The HMACs are produced at 180 °C.

The mixing time for the EME-01 and the EME-02 is as follows:

- Aggregate homogenization: 30 s; Mixing with the binder: 2 min

The mixing time for the EME-03:

- Aggregate homogenization: 30 s; Mixing with crushed asphaltite (Ø ≤ 2 mm): 30 s; Mixing time with petroleum bitumen: 2.5 min.

4 Results and Discussion

4.1 Asphaltite Modified Bitumens

The evolution of the CLT and of the G_C with the asphaltite content are given in Fig. 3. Modified bitumens become more fragile with the modification rate. The comparison of the 5% AMB with the petroleum bitumen 35/50 of Fig. 2 shows that the AMB is more resistant to cracking (Fig. 3a). It will be interesting in the future to compare the CLTs of the other modified bitumens with petroleum bitumens of equivalent viscoelastic behavior. The trend observed in Fig. 3b seems to show that 10% of asphaltite is a threshold at which the G_C begins to decrease. However, the error ranges of the measurements do not allow to validate this conclusion.

4.2 HMACs

As stated before, the LT behavior of the HMACs is studied by TSRST and UTST tests. The results of these tests are combined in order to calculate the tensile strength reserve.

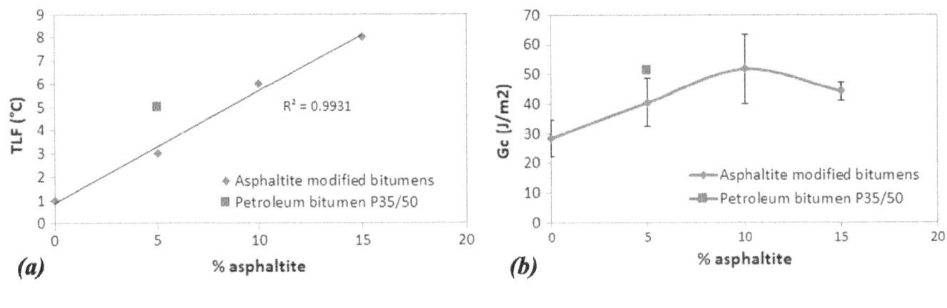

Fig. 3 Evolution of the cracking limit temperature (**a**) and of the critical energy recovery rate (**b**)

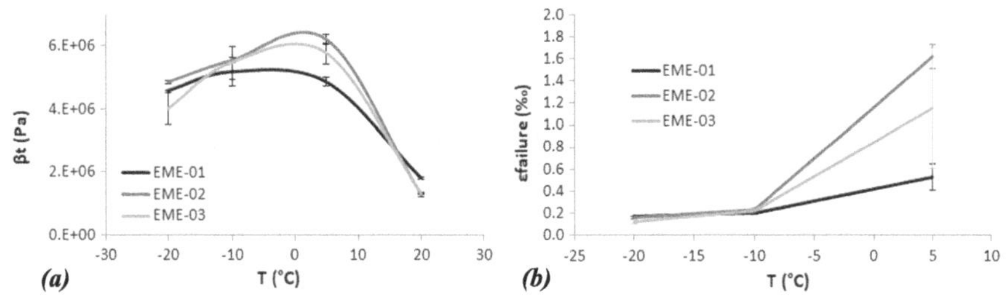

Fig. 4 Uniaxial tensile strength (**a**) and failure deformation (**b**)

4.2.1 Uniaxial Tensile Strength (UTST)

The tensile strength is determined at −20, −10, 5 and 20 °C. The results presented in Fig. 4 show that asphalt mixes with asphaltite (EME-02, EME-03), in the majority of the covered temperature range, have higher tensile strengths than the EME-01 which is produced with 20/30 petroleum bitumen. For example, at 5 °C, the tensile strength of the EME-01, is 4.87 MPa, while the tensile strength of EME-02 is 6.06 MPa. The results presented in Fig. 4b show that EMEs with asphaltite are more ductile. For example, the EME-02 has a failure deformation of 1.625‰ at 5 °C. On the other hand, the failure deformation of the EME-01 at 5 °C is 0.528‰.

4.2.2 Thermal Stress Restrained Specimen Tests (TSRST)

The results in Fig. 5 show that the EME-01 is more thermo-rigid than the EMEs with asphaltites, which perform better at LT. The failure temperature of the EMEs with asphaltite are 3.5–4.5 °C lower than the failure temperature of the EME-01 (Table 4). We observe in Table 4, that for the EME-03, in which the asphaltite is added directly to the asphalt mixer, the deviation of the results is more important. We note the same thing for the UTST results presented in Fig. 4. These observations can be explained by a non-uniform repartition of the asphaltite within the EME-03.

Fig. 5 Thermal stress during TSRST

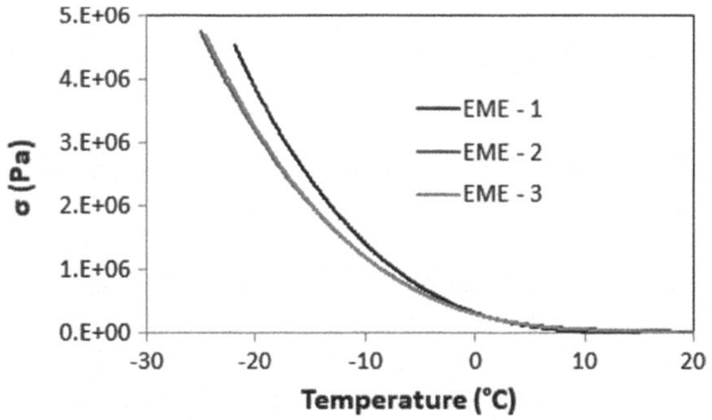

Table 4 TSRST results

Ref.	$T_{failure}$ (°C)		$\sigma_{cry,failure}$ (MPa)	
	Mean value	*Standard deviation*	*Mean value*	*Standard deviation*
EME-01	−21.4	0.27	4,523	–
EME-02	−25.1	0.48	4,752	0.13
EME-03	−24.9	1.47	4,715	0.36

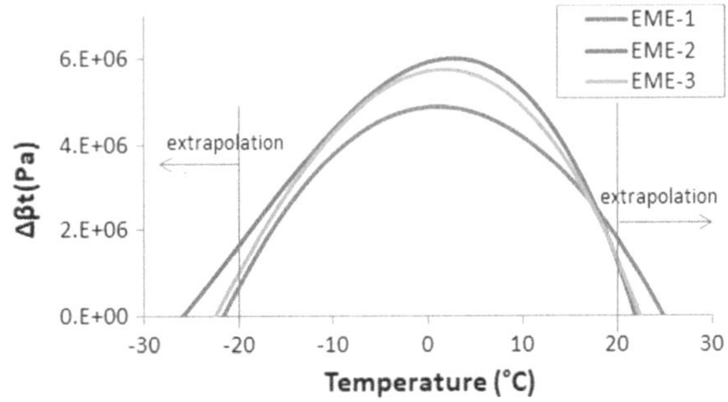

Fig. 6 Tensile strength reserve

Table 5 Tensile strength reserve characteristics

Reference	T_F (°C)	σ_F (MPa)	$T(\Delta\beta_{t,max})$	$\Delta\beta_{t,max}$
EME-01	−21.7	4,470	1.0	4,900
EME-02	−26.0	5,091	2.7	6,000
EME-03	−22.5	4,021	1.8	5,759

4.2.3 Tensile Strength Reserve

The EMEs produced with asphaltite present a higher tensile strength reserve (Fig. 6). Based on failure temperatures T_F and on the other results presented in the Table 5, the EME-02 presents the best low-temperature performance.

5 Conclusions

The scope of the present paper was to study the LT behavior of Selenizza® modified binders and of HMAC produced with the Selenizza® natural bitumen. The LT behavior of these materials was compared with the behavior of typical materials (hard petroleum bitumens and associated HMAC) actually present in the market and widely used in for road construction. In order to assure a relevant comparison, care was made to choose/produce materials for comparison which are equivalent to the materials obtained by Selenizza® modification. It resulted that:

- The Selenizza gives hard bitumens. A 50/70 bitumen gives 35/50, 20/30 and 10/20 bitumens with 5, 10 and 15% of Selenizza respectively.
- The comparison of the LT behavior of the 35/50 modified bitumen with a 35/50 petroleum bitumen with similar viscoelastic behavior, shows that the Selenizza modified 35/50 has a lower cracking limit temperature (3 vs. 5 °C).
- HMAC of type EME of class 2, can be obtained by a 50/70 bitumen modified with Selenizza or with a 50/70 bitumen and addition of Selenizza in the asphalt mix.
- TSRST, UTST results and the calculated values of the uniaxial strength reserve show that the LT performance of HMACs that contain Selenizza® is higher than that of the HMAC produced with 20/30 petroleum bitumen. The HMACs with asphaltite are more ductile and in the majority of the covered temperature range, present higher tensile strength and higher tensile strength reserve. In addition, the $T_{failure}$ determined by TSRST is 3.5–4.5 °C lower for HMACs with Selenizza.

So, these findings suggest that Selenizza can be used for the production of hard bitumens and of HMACs and can be an alternative to hard petroleum bitumens.

References

Aflaki S, Hajikarimi P, Fini EH, Zada B (2014) Comparing effects of biobinder with other asphalt modifiers on low-temperature characteristics of asphalt. J Mater Civ Eng 26(3):429–439

Aflaki S, Tabatabaee N (2009) Proposals for modification of iranian bitumen to meet the climatic requirements of iran. Constr Build Mater 23:2141–2150

Ameri M, Mansourian A, Ashani SS, Yadollahi G (2011) Technical study on the iranian gilsonite as an additive for modification of asphalt binders used in pavement construction. Constr Build Mater 25(3):1379–1387

Bardesi A, Brûlé B, François Corté J (1999) Use of modified bituminous binders, special bitumens and bitumens with additives in road pavements. Technical report, Association mondiale de la route AIPCR

Chailleux E, Mouillet V, Scholten E, Blomberg T, Dreessen S (2008) Evaluation of the three point bending test as a tool to qualify the low temperature behaviour of bituminous binder. In: 'Congrès E&E', Copenhague, pp 21–28

Corté J-F (2001) Development and uses of hard-grade asphalt and of high-modulus asphalt mixes in France. In: 'Perpetual bituminous pavements', number 503, transportation research circular

Eurobitume AI (2011) The bitumen industry - a global perspective. Production, chemistry, use, specification and occupational exposure. Asphalt Institute Inc. & European Bitumen Association-Eurobitume

Geng H, Clopotel CS, Bahia HU (2013) Effects of high modulus asphalt binders on performance of typical asphalt pavement structures. Constr Build Mater 44:207–213

Guern ML, Chailleux E, Dreessen S, Farcas F (2010a) Essai de flexion trois points sur barreau entaillé de bitume. In: XXVIIIe Rencontres Univ. de Génie Civil

Guern ML, Chailleux E, Farcas F, Dreessen S, Mabille I (2010b) Physico-chemical analysis of five hard bitumens: identification of chemical species and molecular organization before and after artificial aging. Fuel 89(11):3330–3339

Huang B, Li G, Shu X (2006) Investigation into three-layered HMA mixtures. Compos: Part B 37:679–690

Huang B, Shu X (2009) Investigation into two procedures of applying gilsonite into HMA mixtures. In: Efficient transportation and pavement systems
Kök BV, Yilmaz M, Guler M (2011) Evaluation of high temperature performance of SBS+ gilsonite modified binder. Fuel 90:3093–3099
Lee N, Hesp S (1994) Low-temperature fracture toughness of polyethylene modified asphalt binders. Transp Res Rec 1436:54–59
Lombardi B (2012) Les bitumes de grade dur et le 20/30. Bitume Info
Migliori F, Corte J-F (1998) Comparative study of RTFOT and PAV aging simulation laboratory tests. Transp Res Rec 1638:56–63
Morrison GR, Lee JK, Hesp SAM (1994) Chlorinated polyolefins for asphalt binder modification. J Appl Polym Sci 54(2):231–240
Ponniah J, Hesp S (1996) Use of fracture energy in performance-based specifications of asphalt binders. In: Proceedings of Canadian TAA, vol 41, pp 127–144
Themeli A (2015) Etude du potentiel d'emploi des bitumes naturels dans la production des liants bitumineux durs et des enrobés à module élevé. PhD thesis, Strasbourg University
Themeli A, Chailleux E, Farcas F, Chazallon C, Migault B (2015) Molecular weight distribution of asphaltic paving binders from phase-angle measurements. RMPD 16(sup1):228–244. https://doi.org/10.1080/14680629.2015.1029667
Themeli A, Chailleux E, Farcas F, Chazallon C, Migault B, Buisson N (2017) Molecular structure evolution of asphaltite-modified bitumens during ageing; comparisons with equivalent petroleum bitumens. IJPRT 10(1):75–83 http://www.sciencedirect.com/science/article/pii/S1996681416301547
Yildirim Y (2007) Polymer modified asphalt binders. Constr Build Mater 21:67–72

Repeatability Study on the Laboratory Production Process of Cement Bitumen Treated Materials with Foamed Bitumen

Simone Raschia, Amir Rahmanbeiki, Daniel Perraton, Alan Carter, Andrea Graziani, and Andrea Grilli

Abstract Cold recycled materials are mixes with high percentages of reclaimed asphalt (RA), while the binding phase is commonly composed of Portland cement and bitumen (emulsion or foam). The reproduction of such materials in the laboratory is an important issue for an accurate mix design, even more when applying a volumetric approach. Particularly, the lack of a specific control on the reliability of the foaming bitumen machine seems to hinder the potentiality of these mixtures. Hence, the goal of this study is to assess the composition variability of several batches produced in the laboratory starting from a specific cement-bitumen treated materials (CBTM) recipe and production procedure. An ignition oven was used to measure RA binder and the effective foamed bitumen amounts in produced specimens and a comparison between maximum dry density and theoretical maximum dry density was carried out. The results have shown that the volumetric approach is a suitable tool to obtain the desired volumetric properties in CBTM produced with foamed bitumen. A certain variability between the theoretical maximum dry density and the measured value can be caused by the variability of bitumen content in the mixture.

Keywords Cold recycling · Foamed bitumen · Reclaimed asphalt · Repeatability · Volumetric approach

1 Introduction

Cold recycling of bituminous pavements can be processed in-place or in-plant, and the reclaimed asphalt (RA), up to 100% of the granular mass, is usually treated with

S. Raschia (✉) · A. Rahmanbeiki · D. Perraton · A. Carter
Construction Engineering Department, École de Technologie Supérieure, Montréal, Canada
e-mail: simone.raschia.1@etsmtl.net

A. Graziani
Department ICEA of Civil and Building Engineering, and Architecture,
Polytechnic University of Marche, Ancona, Italy

A. Grilli
Department of Economics, Science and Law, University of the Republic of San Marino,
San Marino, San Marino

C. Raab (ed.), *Proceedings of the 9th International Conference on Maintenance and Rehabilitation of Pavements—Mairepav9*, Lecture Notes in Civil Engineering 76,
https://doi.org/10.1007/978-3-030-48679-2_53

a bituminous binder in form of bituminous emulsion or foamed bitumen, and cement (Bowering 1970; Jenkins 2000).

Even if many successful sections using cement-bitumen treated materials (CBTM) have been built over the past 25 years (Kowalski and Starry 2007; Bergeron 2005), the use of cold recycling seems still limited to specific project and commonly not adopted as a design tool. The gaps among scientific knowledge and relationship between the laboratory procedures and field performances and the definition of a representative curing methodology are still open challenges (Cardone et al. 2015; Gandi 2018).

2 Background and Research Objective

The intrinsic heterogeneity of the tested materials related to the sampling, preparation and testing methods cause a certain variability of the results. When using RA, the result variability is generally greater than when using virgin aggregates (FHWA 2011; Zaumamis et al. 2018). Obviously, this discrepancy can be reduced when the RA storage and treatment processes are managed properly (Nady 1997; West 2009).

For CBTM, there is limited information on the acceptable variability for the different characteristics and mix performance. For example, Carter et al. (2007) have shown a coefficient of variation (COV) of 11% on Marshall Stability measured on 106 specimens of the same field samples of cold recycled material. This is higher than the maximum 6% COV limit according to ASTM D6927-15 standard for asphalt mixtures. High variability has also been shown for laboratory measured rutting and fatigue resistance in the project "Characterization of Advanced Cold-Recycled Bitumen Stabilized Pavement Solutions" (COREPASOL) (Čížková et al. 2015) that was done specifically on cold recycled materials. According to the authors, the variability may be due to the variability of the RAP itself and also because of the different compaction method used. The variability in the maximum specific gravity was also evaluated in COREPASOL, and the highest standard variation measured was 0.288 (Batista et al. 2014) which is higher than the usual 0.016 limit used for asphalt mixtures for different laboratories according to ASTM D2041-19.

In most of the studies available in the literature, part of the variability measured can be linked to the different compaction methods, the variability of the RAP source and to the mix design methodology. Differently to hot mix asphalt (HMA), most CBTM mix design are not based on the volumetric properties. An accurate volumetric approach for CBTM in the fresh state (Grilli et al. 2012) can help in reducing the variability of the mix and results. The aim of the study is to evaluate the composition variability of CBTM with foamed bitumen and cement following a volumetric mix design procedure.

Table 1 RA aggregate properties

Property	Standard	Unit	Value
Binder content	ASTM D6307	%	4.79
Nominal maximum particle dimension	ASTM D448-03	mm	16
Maximum specific gravity	ASTM C127-128	–	2.482

Table 2 Basic properties of bitumen for foaming

Property	Standard	Unit	Value
Penetration	EN 1426	0.1 mm	64
Softening point	EN 1426	°C	52

3 Materials and Methodology

3.1 Materials

Main properties of the RA aggregate employed in the CBTM produced are listed in Table 1, whereas the characteristics of the bitumen used for foaming are collected in Table 2. Foamed bitumen was produced by a laboratory unit at a temperature of 170 °C and a foaming water content of 1.9% by bitumen weight. The cement is a general use (GU) type (CSA A3000) with compressive strength at 28 days of 43.9 MPa (ASTM C109).

3.2 Mixtures

Table 3 shows the typical mix composition for a 30 000 g batch, containing 90% of RAP, 5.7% filler (virgin aggregate), 1.4% Portland cement and 2.9% of foamed bitumen.

Six different 30 kg CBTM batches (Table 4) were prepared in this study to limit the effect of the possible variability of the foam dosage. In fact, the capacity of the pug mill is suitable for 30 kg batches and a reduced amount would probably mean a not uniform blending between the granular material and the foamed bitumen.

Each batch produced was then compacted in laboratory with different techniques (Marshall hammer, Shear Gyratory Compaction (SGC) 100 mm and 150 mm, and Proctor rammer).

After curing for 14 days at 40 ± 2 °C and RH of 55 ± 5% specimens were tested and the remaining material was crashed to conduct control tests on the mixtures, such as maximum dry density tests by pycnometer (G_{mm}) and bitumen content by ignition oven (3 repetitions of each test on each batch produced).

Table 3 CBTM mixes composition

	Components						
	RAP	Filler	Cement	Bitumen	Intergranular water	Absorbed water	Total
Mass (g)	M_{RAP}	M_{FILL}	M_{CEM}	M_{BIT}	M_{WI}	M_{WA}	
	26005.2	1659.9	415.0	830.0	830.0	260.1	30000.0
Density (g/cm^3)	2.486	2.650	3.015	1.015	1.000	1.000	
Volume (cm^3)	V_{RAP}	V_{FILL}	V_{CEM}	V_{BIT}	V_{WI}	V_{WA}	
	10486.0	626.4	137.4	817.7	830.0	260.1	482.4
Proportion (% by mass of mix)	86.7	5.5	1.4	2.8	2.8	0.9	100.0
Proportion (% by volume of mix)	79.7	4.8	1.0	6.2	6.3	2.0	100.0

Table 4 CBTM batches produced and experimental tests

Batch number	Compaction technique	Experimental tests
B1	Marshall	3 × G_{mm} (total of 18) 3 × Ignition oven (total of 18)
B2	SGC 100	
B3	SGC 150	
B4	SGC 150	
B7	Proctor 150	
B8	Proctor 150	

3.3 *Methodology*

A CBTM mixture was produced in the laboratory using fixed constituent materials and dosages. After checking the bitumen content by ignition testing, the theoretical maximum specific gravity was calculated and compared with the maximum specific gravity measured on laboratory samples.

Considering that replicate batches of a homogeneous mix have to show the same value of the maximum specific gravity, the variability of the mixes was assessed by monitoring the maximum specific gravity on the cured samples (after 14 days at 40 °C and 55% of RH) from each batch according to the ASTM D2041/D2041M-19 standard.

The total bitumen content (bitumen in the RA and added foamed bitumen) was measured with the ignition oven.

The theoretical maximum specific gravity can be calculated through Eqs. 1 and 2. With Eq. 1, the G_{mm} is calculated according to the total volume of the loose mix, with no voids, and with Eq. 2 the G_{mm} only considers the solid particles, not the bitumen.

$$G_{mm} = \frac{M_{DRY}}{V_{DRY}} = \frac{M_{RAP} + M_{FILL} + M_{CEM} + M_{BIT}}{\frac{M_{RAP}}{\rho_{RAP}} + \frac{M_{FILL}}{\rho_{FILL}} + \frac{M_{CEM}}{\rho_{CEM}} + \frac{M_{BIT}}{\rho_{BIT}}} \quad (1)$$

Theoretical $G_{mm,s}$ (with respect to the solid phase: RAP, filler and cement)

$$G_{mm,s} = \frac{M_{SOLIDS}}{V_{SOLIDS}} = \frac{M_{RAP} + M_{FILL} + M_{CEM}}{\frac{M_{RAP}}{\rho_{RAP}} + \frac{M_{FILL}}{\rho_{FILL}} + \frac{M_{CEM}}{\rho_{CEM}}} \quad (2)$$

The theoretical percentage of foamed bitumen ($\%b^T$) content can be calculated with respect to the solids volume (RAP, filler and cement) by using both G_{mm} calculated from Eqs. 1 and 2:

$$\%b^T = \frac{\rho_{BIT} \cdot \left(G_{mm,s} - G_{mm}\right)}{G_{mm} \cdot \left(G_{mm,s} - \rho_{BIT}\right)} \cdot 100 \quad (3)$$

The $\%b^T$ value obtained with Eq. 3 represents the target value for mix design. Using Eq. 3, it is possible to calculate the percentage of bitumen according to the measured G_{mm} ($\%b^{(1)}$) corresponding to each G_{mm} test, using the measured value of $G_{mm}^{(1)}$ instead of the theoretical one.

In addition to this, ignition oven test was also performed to assess the total bitumen content of the mixtures ($\%b^{(2)}$). From previous RAP characterization tests, the RAP binder is fixed at 4.79% by mass of the RAP aggregate (10 samples with standard deviation $s = 0.229$), which corresponds to 4.31% by dry mass (RAP, Filler, cement and foamed bitumen).

Using again Eq. 3, it is possible to evaluate the values for $G_{mm}^{(2)}$ by calculation knowing the foamed bitumen by ignition test.

$$G_{mm}^{(2)} = \frac{G_{mm,s} \cdot \rho_{BIT}}{\rho_{BIT} \cdot \left(1 - \%b^{(2)}\right) + \left(G_{mm,s} \cdot \%b^{(2)}\right)} \cdot 100 \quad (4)$$

4 Results and Discussion

4.1 Laboratory Tests

According to the considerations previously made, values related to Eqs. 1 and 2 for this study are $G_{mm} = 2.396$ kg/dm^3 and $G_{mm,s} = 2.496$ kg/dm^3. As a consequence, by Eq. 3, $\%b^T = 2.87\%$ by solids weight (RA, filler and cement). It is important to remark that these are theoretical (and target) values related to the mix design.

As already mentioned, G_{mm} values and the bitumen content by ignition oven were measured on eighteen samples. For both tests, average value $\bar{X}$, standard deviation s and confidence interval $\mu^{\pm}$ (Eq. 5) were calculated on a sample of 18 values, fixing a level of confidence $\alpha = 99.9\%$.

$$\mu^{\pm} = \bar{X} \pm t_{\frac{\alpha}{2};n-1} \cdot \frac{s}{\sqrt{n}} \tag{5}$$

Figures 1 and 2 show the G_{mm} and bitumen content values in relationship with their confidence intervals, respectively.

The average value of bitumen content is 8.28% with a standard deviation of 0.783 while the average value of G_{mm} is 2.420 with a standard variation of 0.013. According to these results, the G_{mm} values are 67% reproducible (12 values on 18 are in the

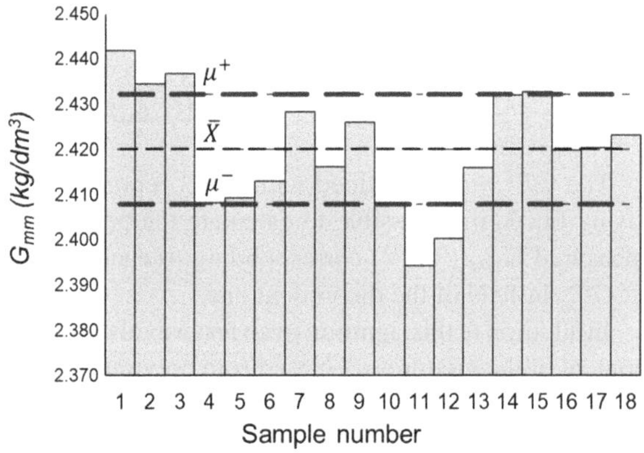

Fig. 1 Measured G_{mm} for all 18 specimens

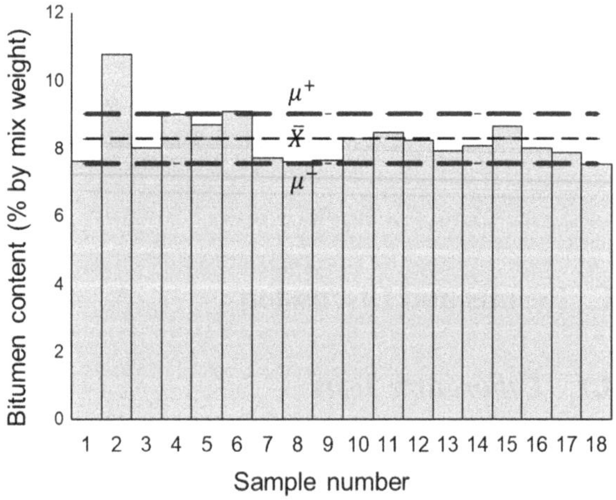

Fig. 2 Measured bitumen content in comparison with confidence interval

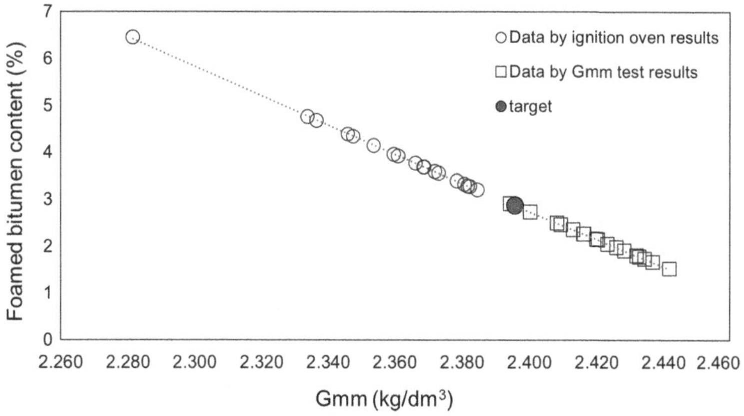

Fig. 3 Bitumen content according to G_{mm}

confidence interval). On the other hand, ignition oven tests were 83% reproducible (15 values on 18 are in the confidence interval).

Moreover, the value $\%b^{(2)}$ is linearly related to $G_{mm}^{(2)}$, whereas the value $G_{mm}^{(1)}$ is linearly related to $\%b^{(1)}$, as shown in Fig. 3. It is important to remark that such calculations are valid assuming that the solid phase composition (RAP, filler and cement), and then the $G_{mm,s}$, are consistent with the mix design.

From the results shown on Fig. 3, it can be observed that the ignition oven test overestimates the values of the added foamed bitumen in the mix, affecting in this manner the respective $G_{mm}^{(2)}$ values. This can be caused by an undesired material loss in the oven due to the high temperatures. According to Rodezno et al. (2017), using lower temperature with the ignition oven when RAP is present should reduce the variability.

At the same time, values obtained by the pycnometer test for the maximum specific gravity overestimate the $G_{mm}^{(1)}$, if compared to the theoretical value. Those higher values mean lower amount of foamed bitumen in the mix. This is, in some way realistic, because during the production and mixing of the foam mixture, it is clearly visible that not all the foamed bitumen is spread in the granular material. In fact, a significant amount tends to stick on the internal walls or the beaters of the pug mill. Moreover, a higher G_{mm} than the theoretical value (and a lower amount of foamed bitumen) can also be an indicator of bitumen absorption from the granular components. This would be unexpected in this case since the mix is made with RAP that should not absorb bitumen, and filler that also does not absorb bitumen.

The correspondence between $G_{mm}^{(1)}$ and $G_{mm}^{(2)}$, as well as between $\%b^{(1)}$ and $\%b^{(2)}$, can be observed in Fig. 4.

Having found two different ranges of values for the G_{mm}, the evaluation of the air voids in the specimens should be carried out considering separately with $G_{mm}^{(1)}$ and $G_{mm}^{(2)}$. The density of the compacted specimens has been assessed by geometrical measurements and recording the weight at the end of curing (G_{sb}). For the volume

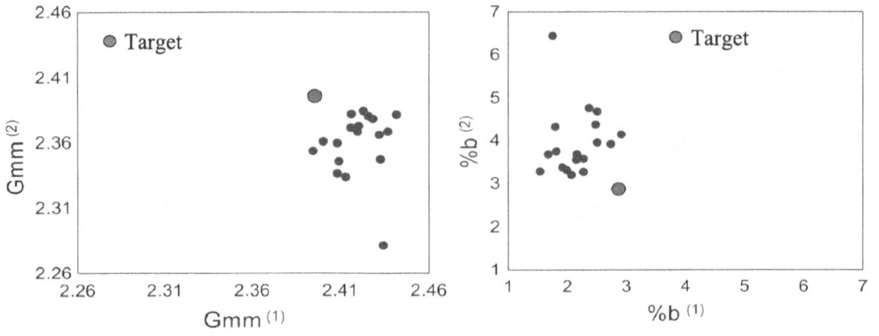

Fig. 4 Relationships between $G_{mm}^{(1)}$ and $G_{mm}^{(2)}$, and $\%b^{(1)}$ and $\%b^{(2)}$

measurements, three measures of diameter and three measures of the height of the specimens were used. The air voids are calculated using Eq. 6.

$$Air\ voids\ (\%) = \frac{G_{mm} - G_{sb}}{G_{mm}} \cdot 100 \tag{6}$$

In the analysis, three different air voids content (Fig. 5) were calculated:

- $\%V_{th}$, which refers to the theoretical value of G_{mm};
- $\%V^{(1)}$, which refers to $G_{mm}^{(1)}$ (pycnometer test);
- $\%V^{(2)}$, which refers to $G_{mm}^{(2)}$ (calculated from ignition oven results);

The air voids of the specimens made from the two batches compacted with the Proctor rammer were not measured because the height of the specimens were difficult to evaluate. The top surface of those specimens were uneven making an average height not representative. For all the compaction methods, the highest air voids is the one calculated according to the pycnometer test. It is interesting to note that even if the ranking of the different air voids remains the same for each compaction methods, the

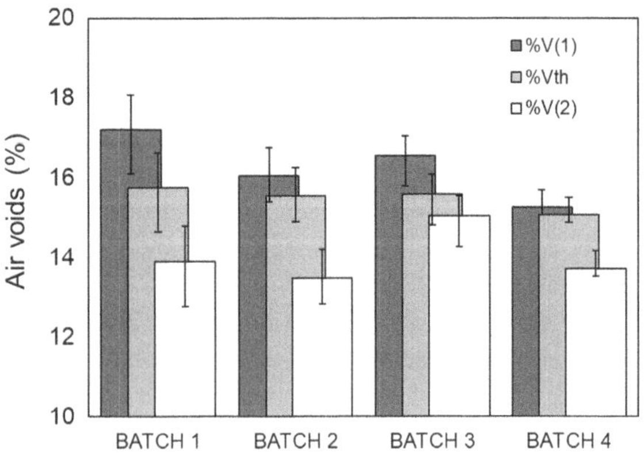

Fig. 5 The three different calculated air voids according to the different compaction methods

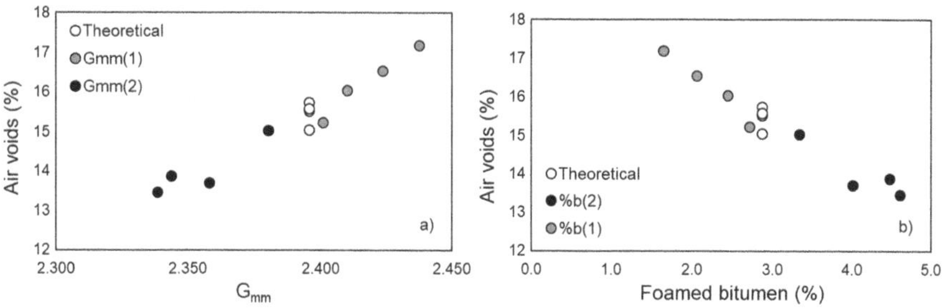

Fig. 6 Air voids vs G_{mm} and foamed bitumen content

amplitude of the differences are quite different. Higher variability can be observed for batch 1, which are the specimens compacted with the Marshall hammer.

Figure 6 shows that the air voids measured decrease with the decrease of the G_{mm} and the increase in the foamed bitumen content. The values related to the theoretical value of G_{mm} are placed in between the two calculations made using the pycnometer test and the ignition oven test.

5 Conclusion

In this study, the variability of a CBTM mix made with foamed bitumen was tested. The variability in the components was evaluated based on the maximum specific gravity, G_{mm}, and the bitumen content. The theoretical values based on volumetric calculations and the laboratory measured values were compared. The main conclusions are:

- The amount of bitumen measured with the ignition oven and the one calculated from the G_{mm} are different. The bitumen content from the ignition oven is higher than the target value. This is probably due to the loss of fine particles during the ignition test;
- The measure G_{mm} is higher than the theoretical value. This is normally explained by lower bitumen content which is not the case here according to the ignition oven results. However, as mentioned previously, we believe that the ignition results are too high.
- The volumetric approach cannot be validated by the two chosen tests, but it can be used as a tool to estimate real foam bitumen content.

The next step in this study is to test a greater number of specimens to have a better understanding of the variability. Different mix design should also be tested. The same exercise should also be carried out with mixes made with bituminous emulsion.

Acknowledgements The authors wish to thank the company Nynas for providing the bitumen used for the foamed bitumen, and Cooperativa Braccianti Riminese (CBR) for providing the reclaimed asphalt used in this study.

References

Batista F, Valentin J, Čížková Z, Valentova T, Simnofske D, Mollenhauer K, Tabakovic A, McNally C, Engels M (2014) Report on available test and mix design procedures for cold-recycled bitumen stabilised materials. CoRePaSol. Characterization of advanced cold-recycled bitumen stabilized pavement solutions. Deliverable D1.1

Bergeron G (2005) Performance des techniques de retraitement en place et de recyclage à froid au Québec. Transportation association of Canada. Congrès Annuel de l'ATC-TAC, Calgary, Canada, 18–21 September 2005

Bowering R (1970) Properties and behaviour of foamed bitumen mixtures for road building. Paper presented at the proceedings of the 5th Australian road research board conference

Cardone F, Grilli A, Bocci M, Graziani A (2015) Curing and temperature sensitivity of cement-bitumen treated materials. Int J Pavement Eng 16(10):868–880

Carter A, Fiedler J, Kominek Z, Vacin O, Barberi A, Perraton D (2007) The influence of accelerated curing on cold in-place recycling, Canadian technical asphalt association. In: CTAA annual conference, Niagara Falls, Canada, 17–21 November 2007

Čížková Z, Valentin J, Suada J, Krpalek O, Simnofske D, Mollenhauer K, Batista F (2015) Report on durability of cold-recycled mixes: test procedures for stiffness determination. Deliverable D2.1. COREPASOL. Characterization of advanced cold-recycled bitumen stabilized pavement solutions. Final report

FHWA (2011) Reclaimed asphalt pavement in asphalt mixtures: state of the practice. Publication no. FHWA-HRT-11-201. Federal Highway Administration

Gandi A (2018) Laboratory characterization of bitumen treated full depth reclamation materials. PhD thesis, École de technologie superieure, Montreal, Canada

Grilli A, Graziani A, Bocci M (2012) Compactability and thermal sensitivity of cement–bitumen-treated materials. Road Mater Pavement Des 13(4):599–617

Jenkins KJ (2000) Mix design considerations for cold and half-warm bituminous mixes with emphasis of foamed bitumen. Stellenbosch University, Stellenbosch

Kowalski TE, Starry DW (2007) Cold recycling using foam bitumen. In: Characterization and improvement of soils and materials session, transportation association of Canada, TAC annual conference, Saskatoon, Canada, 14–17 October 2007

Nady RM (1997) The quality of random RAP: separating fact from suppositions. Focus Hot Mix Asphalt Technol 2(2):14–17

Rodezno C, Brown R, Julian G, Prowell B (2017) Variability of ignition furnace correction factors. NCHRP research report 847. National Cooperative Highway Research Program

West RC (2009) Keys to managing RAP variability. Better Roads 79(10):12–15

Zaumamis M, Oga J, Haritonovs V (2018) How to reduce reclaimed asphalt variability: a full-scale study. Constr Build Mater 188(2018):546–554

Effect of Global Thresholding Algorithms on Pervious Concrete Pore Network Properties Using XRCT-Based Digital Image Processing

Ajayshankar Jagadeesh, Ghim Ping Ong, and Yu-Min Su

Abstract Digital image processing of the X-ray computed tomography images involves the crucial step of image segmentation which affects the subsequent pore structure quantitative analysis. The main objective of this study is to investigate the effect of ten different global thresholding algorithms based on the grey scale histogram, clustering, entropy and laboratory volumetric characteristics on the internal pore structure properties of the pervious concrete. The key microstructural parameters of the pervious concrete air voids such as porosity, tortuosity, throat number, pore coordination number and distributions of pore volume, throat area, pore sphericity, shape factor and throat eccentricity were analyzed for different thresholding algorithms. It was found from the analysis that the nine histogram, clustering and entropy based algorithms are found to be either under or over estimating the air void voxels compared to the volumetric segmentation method. And as the threshold value increases, effective porosity and number of throats increases and isolated porosity and tortuosity decreases due to the increase of air void voxels and pore connectivity. Overall, it is expected that the present study will help in understanding the importance of threshold segmentation in the field of pavement image processing.

Keywords Pervious concrete pores · X-ray computed tomography · Global thresholding · Volumetric segmentation

1 Introduction

Although the usage of X-ray computed tomography (XRCT) and digital image processing in pavement engineering started over the past 20 years, a brief literature review indicated that a significant importance has not been given to the usage of

A. Jagadeesh · G. P. Ong (✉)
Department of Civil and Environmental Engineering, National University of Singapore, Block E1A, #07-03, 1 Engineering Drive 2, Singapore 117576, Singapore
e-mail: ceeongr@nus.edu.sg

Y.-M. Su
Department of Civil Engineering, National Kaohsiung University of Science and Technology, 415, Chien-Kung Road, Sanmin District, Kaohsiung 80778, Taiwan

C. Raab (ed.), *Proceedings of the 9th International Conference on Maintenance and Rehabilitation of Pavements—Mairepav9*, Lecture Notes in Civil Engineering 76,
https://doi.org/10.1007/978-3-030-48679-2_54

threshold segmentation algorithms. The threshold parameter has not been actively investigated to any great extent in the field of pavement image processing; Several literatures have not included the threshold algorithm or the value used and several literatures directly provides the threshold values without mentioning the thresholding algorithm used (Iassonov et al. 2009). The current study is aimed at investigating the effect of various global thresholding algorithms on the pervious concrete pore network properties.

Pervious pavement is a special class of pavement comprising materials with sufficient continuous voids ranging from 15 to 35% to allow water to pass from the surface to the underlying layers (ACI 2010). The amount of isolated voids within the porous concrete layer is minimal because of the usage of uniform single sized coarse aggregates and the lack of fine aggregates in the mixture. The benefits of using the pervious pavements such as the increase in the water permeability, skid resistance and reduction in hydroplaning, storm water run-off, tire/road noise, splash and spray effect are derived from the size and shape characteristics of the pervious concrete air voids (ACI 2010). The pervious concrete pore network properties such as porosity, tortuosity, pore distribution, surface area and so on have been investigated in several studies using XRCT and digital image processing. The usage of XRCT enables the accurate measurement of pervious concrete pore network parameters such as volume, surface area, size and shape parameters within the voxel resolution. Manahiloh et al. (2012) studied the effect of pervious concrete clogging using the porosity versus depth profiles obtained from Otsu's global segmentation of CT images. Kuang et al. (2015) examined the pervious concrete pore parameters such as total porosity, effective porosity, pore size distribution, tortuosity and specific surface area using Bayesian segmentation of CT images. Zhang et al. (2015) studied the volumetric properties of the porous asphalt using the global minima segmentation of high resolution industrial CT images. Abera et al. (2017) studied the effect of five different global thresholding algorithms on the total porosity of pervious concrete sample. Zhang et al. (2018) studied the 2D and 3D pore size distributions of the pervious concrete samples using watershed segmentation algorithm. Chandrappa and Biligiri (2018) examined the pervious concrete pore morphology characteristics using the global minimum thresholding algorithm on the industrial CT scan images. Jagadeesh et al. (2018a, b, 2019a, b) studied the effect of global thresholding and watershed algorithms on the permeability and volumetric characteristics of the pervious concrete samples. Jagadeesh et al. (2019c) also investigated the effect of pore and solid network characteristics for different pervious concrete sample textures. This paper highlights the impact of over or under estimating the threshold value in the key microstructural parameters of the pervious concrete air voids such as effective porosity, isolated porosity, tortuosity, throat number, pore coordination number, distributions of pore volume, throat area, pore sphericity, shape factor and throat eccentricity for ten different global thresholding algorithms.

2 Global Thresholding Algorithms

In recent years, significant efforts have been made in the field of digital image processing to improve the segmentation algorithms, which is considered as the most crucial step affecting the image quantitative analysis (Iassonov et al. 2009). Image segmentation is the process of binarisation commonly used to separate images of various grey scale intensities into two discrete phases and is categorized as global and local thresholding. In global thresholding algorithm, a single grey scale value is used to divide the three dimensional voxels of varying intensities into discrete phases, whereas in the local thresholding algorithm, variable thresholding based on the local image characteristics and the histogram is carried out for each slice. The global thresholding algorithms are further classified into several categories according to the global grey scale histogram shape, entropies, higher order probability distributions, spatial correlations between image pixels and so on (Iassonov et al. 2009) and can be bi-level or multi-level. Following global threshold segmentation algorithms are used in the current study for the pervious concrete air voids: Histogram based thresholding algorithms (intermodes, miniumum), clustering based thresholding algorithms (Otsu bi-level, tri-level, isodata), entropy based thresholding algorithms (Kapur, Yen, Sahoo, Li) and volumetric based thresholding algorithm (Zelelew and Papagiannakis 2011). The detailed review of the histogram, clustering and entropy based global thresholding algorithms including the objective functions, procedures, etc. can be found in the works of Iassonov et al. (2009), Sezgin and Sankur (2004) and so on.

3 Measurement of Pore Network Properties

3.1 Sample Preparation and Obtaining XRCT Images

Pervious concrete cylinder of height 250 mm and diameter 150 mm was casted using with the siliceous aggregates (4.75–2.38 mm) from southern Taiwan and cured for 28 days in the water tank. The mixture proportions are water-cement ratio of 0.3, coarse aggregates of 1530 kg/m^3, cement of 340 kg/m^3, superplasticizer of 2 kg/m^3. The Somatom Emotion 16-channel medical XRCT scanner manufactured by Siemens Healthcare with an output voltage of 110 kV energy was used to obtain at a total of 354 section images of pervious concrete sample at the E-Da/I-Shou University Hospital with the pixel resolution of 0.326 mm × 0.326 mm and vertical resolution of 0.7 mm. The laboratory effective porosity value for the pervious concrete sample as per ASTM C1754 (2012) drying method B was found to be 26.6%. It has to be noted that in the current study, the term "volumetric" represents the segmentation algorithm and "experimental porosity" represents the ASTM C1754 (2012) drying method B.

(a) Effective pores

(b) Effective pores after watershed segmentation

(c) Isolated pores

Fig. 1 Three dimensional volumetric segmented pervious concrete pore network

3.2 Processing XRCT Images

The Simpleware ScanIP N-2018.03 software was used in the conversion of pervious concrete XRCT absorption coefficients to the 12-bit greyscale intensities ($2^{12} = 4096$) varying from 0 to 4095, followed by image pre-processing and mask segmentation using various global thresholding algorithms. The ImageJ 1.51j8 software was used to obtain the air void upper limit threshold values for the various histogram, clustering and entropy based segmentation algorithms. The global threshold segmentation of the pervious concrete voids is followed by the watershed segmentation algorithm to divide the interconnected pore structure into several pores based on the image background gradient. The lack of watershed segmentation algorithm leads to the consideration of the whole interconnected air voids as the single void structure and results in significant errors in the pore network properties (Jagadeesh et al. 2019b). Figure 1 shows the three dimensional volumetric segmented pore network structures of the pervious concrete sample such as effective pores before and after watershed segmentation, and isolated pores.

3.3 Quantification of Pore Network Properties

This step briefly discusses the various pervious concrete pore network properties examined in this research work. Throats or geometric constrictions are defined as the contacts between the connecting pores obtained from the watershed segmentation algorithm, whereas the pore coordination number is defined as the number of throats in an individual pore. The pore volume distributions for effective and isolated pores and throat area distributions are examined in this study, instead of the equivalent spherical pore radius and equivalent circular throat radius. The tortuosity of the interconnected pores is defined as the ratio of the shortest actual path of the fluid flow to the sample length in the same direction. The sphericity is the measure of 3D shape characteristics of the pore ranging from 0 (non-uniform pore dimensions along its axis) to 1 (sphere shaped pore particle) and can be divided into following ranges: low (<0.4), moderate (0.4–0.6), high (0.6–0.8) and extreme (>0.8). The oriented bounding boxes are used to quantify the pore shape factor and the throat eccentricity. Pore shape factor attempts to characterise the shape of the pores by the ratio of elongation to flatness and can be divided into elongated (0–0.99), spherical (0.99–1.01), flatted (1.01–2) and highly flatted (>2). A spherical pore will have a shape factor of 1 and elongated pores have less than 1 and flatted pores have greater than 1. The throat eccentricity is a measure of how far an ellipse differs from a circle and ranges from 0 (circular) to almost 1 (very elongated ellipse).

4 Results and Discussion

4.1 Effect of Threshold on Porosity and Air Void Distribution

Figure 2a shows the effect of threshold on total, effective and isolated porosity using ten different segmentation algorithms. The total porosity curve versus grey scale values of the pervious concrete sample was obtained from the grey scale histogram (black curve). The vertical color lines in Fig. 2a indicates the air void threshold values for the ten different segmentation algorithms. The symbols triangle and rectangle represent the effective and isolated porosities of the corresponding segmentation algorithms. It can be seen that as the threshold value increases, size of the individual pores widens resulting in the increase of total and effective porosity of the sample and the pore connectivity increases resulting in the reduction of the isolated porosity. The percentage error of the effective porosity in the commonly used segmentation algorithms such as Minimum, Otsu bi-level, Volumetric, Kapur and Yen are −38.91, −5.43, 0, 17.14 and 38.15% respectively. Figure 2b and c highlights the effect of threshold segmentation algorithms on the distributions of air void percentage and number along the depth of the sample based on the horizontal cross sectional analysis. Analyzing a single threshold graph in Fig. 2b and c, for example from 150 to 180 mm depth, the air void percent increases with reduction in the number of air voids in the slice. And also it can be seen in Fig. 2c that as the threshold value increases, number of air voids in the slice decreases due to pore connectivity as expected.

4.2 Effect of Threshold on Tortuosity, Throat Number and Pore Coordination Number

Figure 3a shows the effect of threshold on tortuosity and the total number of geometric constrictions or throats of the pervious concrete sample. The increase in the threshold

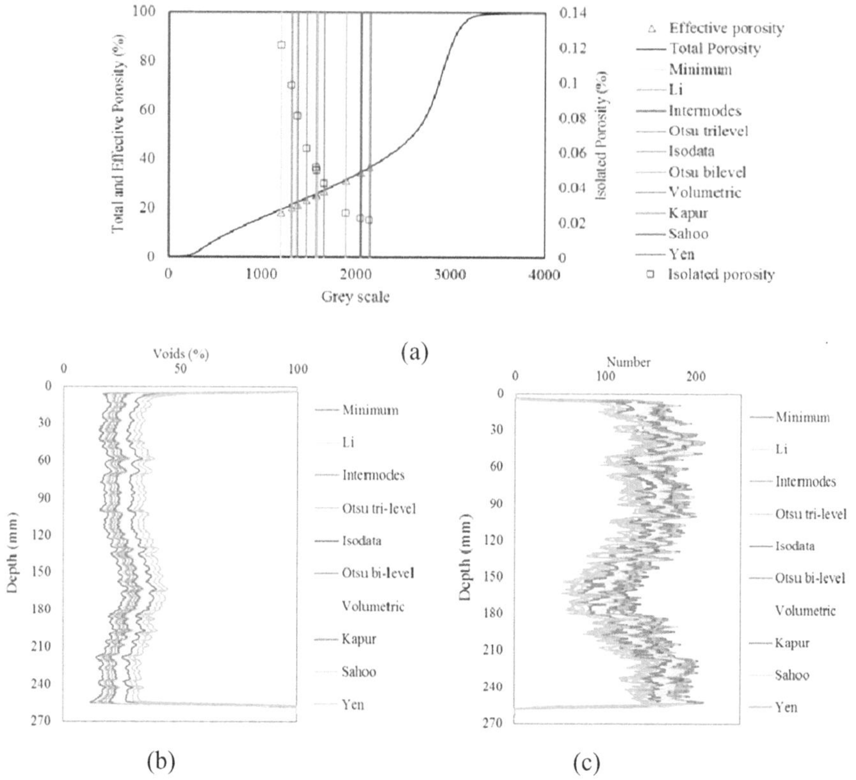

Fig. 2 Effect of threshold on porosity and air void distribution

value results in the opening of new pores and hence the tortuosity decreases and the number of throats increases. Figure 3b shows the plot of pore coordination number for the pervious concrete sample using different threshold segmentation algorithms. The pore coordination number of 0 indicates the isolated air voids and as the threshold value increases, the air voids percent in the range 0 decreases and the air voids in the higher ranges increases as expected.

4.3 Effect of Threshold on Pore Volume and Throat Area Distribution

Figure 4a and b shows the effect of threshold on effective and isolated pore volume distribution of the pervious concrete sample. It can be seen that the effective or interconnected pores lies in the range of 10 to 10000 mm^3, whereas the ineffective or isolated pores lies in the range of 0.1 to 100 mm^3. The majority of the effective and isolated pores lies in the range of 100–1000 mm^3 and less than 1 mm^3 respectively.

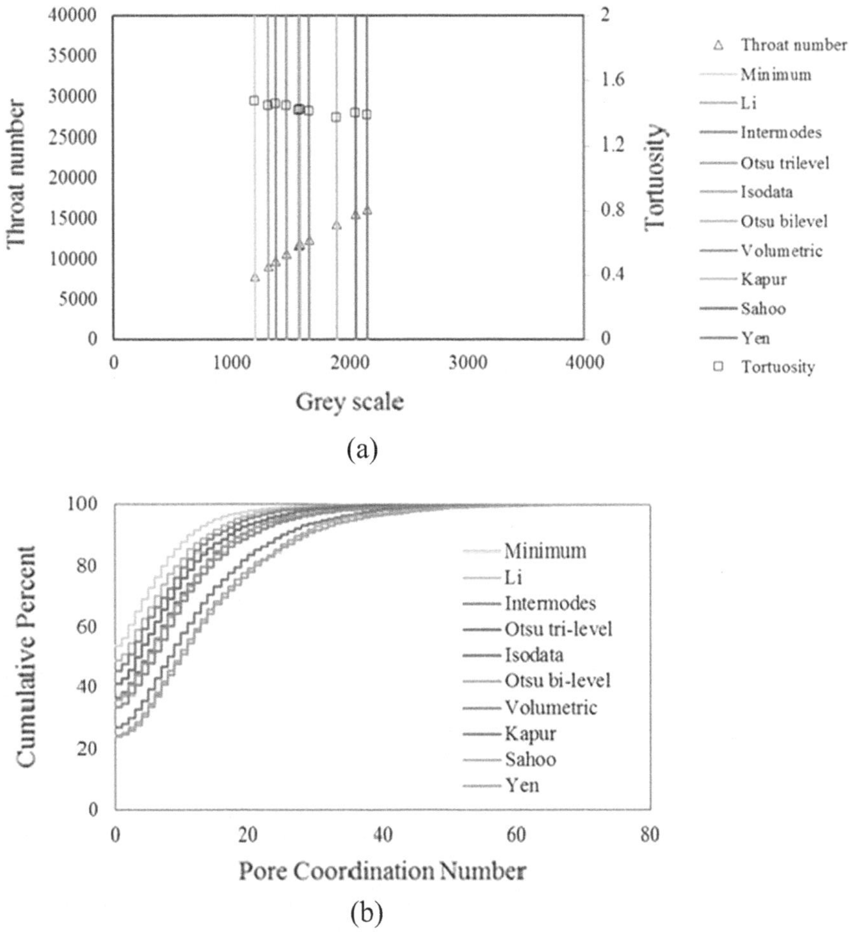

Fig. 3 Effect of threshold on tortuosity, throat number and pore coordination number

The increase in the threshold value results in the increase of effective air void percent in higher pore volume range (>1000 mm^3) and reduction in the lower pore volume range (<100 mm^3). For the isolated pores, the increase in the threshold results in the increase of isolated air voids percent in the lower range (<1 mm^3) and reduction in the higher range (>10 mm^3). Figure 4c shows the effect of threshold on throat area distribution of the pervious concrete sample. It follows the similar trend as that of the effective pore volume distribution, the increase in the threshold value results in the increase of throats in the area range greater than 10 mm^2 and decrease of throats in the lower area ranges.

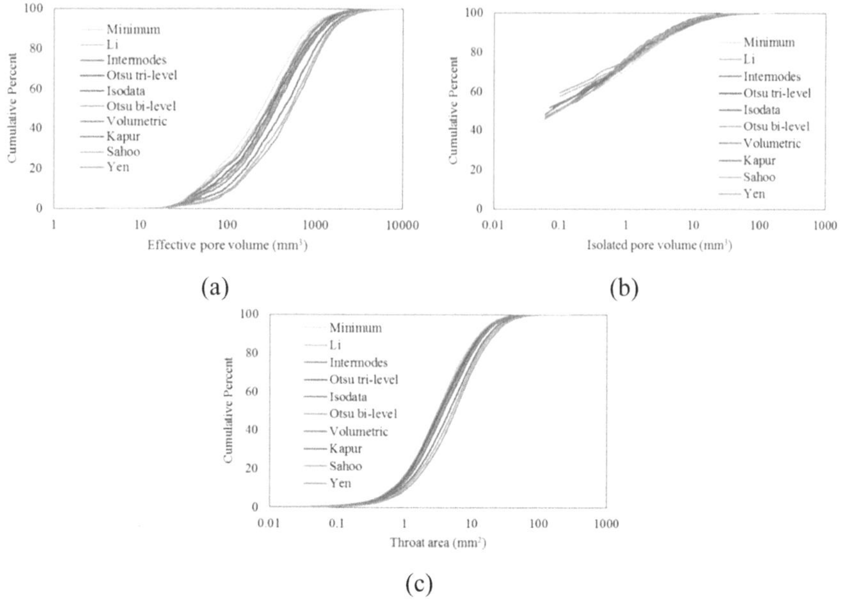

Fig. 4 Effect of threshold on pore volume and throat area distribution

4.4 Effect of Threshold on Pore Sphericity Distribution

Figure 5 shows the effect of threshold on effective and isolated pore sphericity distribution of the pervious concrete sample. The majority of the effective and isolated pore sphericity lies in the moderate and high range respectively. In Fig. 5, it can be seen that the increase in the threshold value results in the reduction of effective air voids percent in the low sphericity range and increase in the moderate sphericity range. In Fig. 5, the increase of threshold results in the increase of isolated air

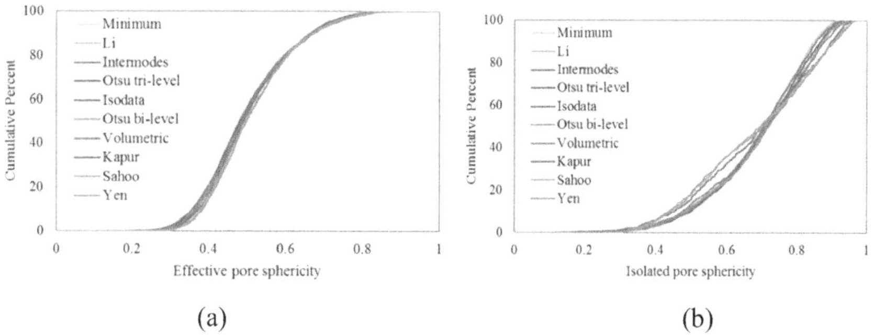

Fig. 5 Effect of threshold on pore sphericity distribution

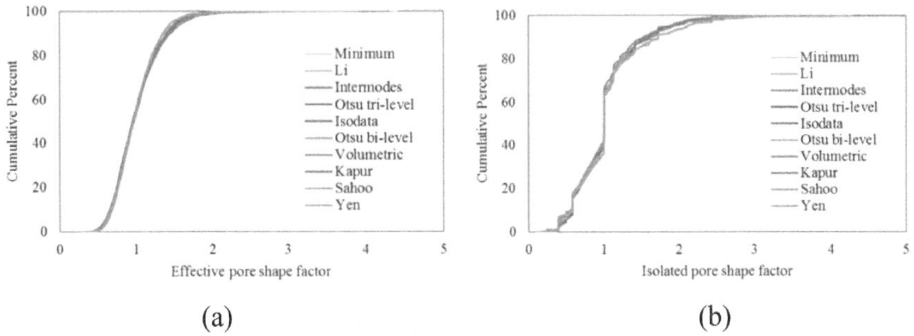

Fig. 6 Effect of threshold on pore shape factor distribution

voids percent in the low, moderate and extreme sphericity and reduction in the high sphericity range. The fluctuations in the effective and isolated air voids sphericity with increase in threshold values are attributed to the variation in the effective and isolated air voids percent as shown in Fig. 2a.

4.5 *Effect of Threshold on Pore Shape Factor Distribution*

Figure 6 shows the effect of threshold on effective and isolated pore shape factor distribution of the pervious concrete sample. The majority of the effective and isolated pore shape factor are elongated in shape and is consistent with Mahmud et al. (2017). It can be seen from Fig. 6 that the percentage of isolated pores in the spherical zone and highly flatted zone are higher compared to the effective pores and vice aversely in the elongated and flatted zone. No significant trend in the effective and isolated pores shape factor distribution with variation in threshold values can be observed due to the method of shape factor calculation.

4.6 *Effect of Threshold on Throat Eccentricity Distribution*

Figure 7 shows the effect of threshold on throat eccentricity distribution of the pervious concrete sample. No significant variation in the throat eccentricity distribution with respect to threshold values can be observed in the range of 0 to 0.5. The majority of the throats lies in the eccentricity range of 0.75 to 1. It can be seen from Fig. 7 that as the threshold value increases, the eccentricity of throats in the range 0.75 to 1 reduces and increases in the range 0.5 to 0.75.

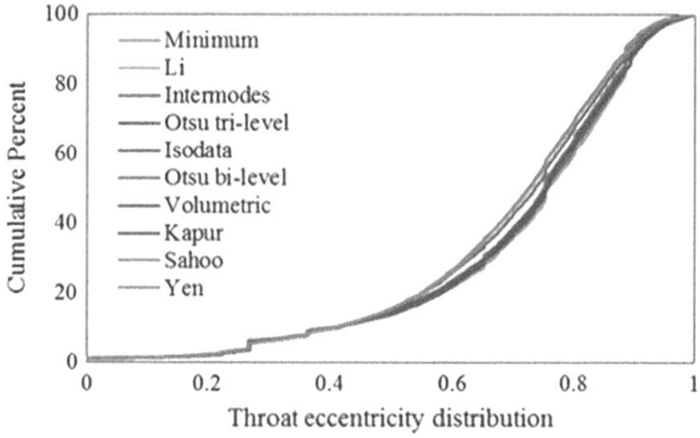

Fig. 7 Effect of threshold on throat eccentricity distribution

5 Conclusion

This study analyses the effect of over or under estimating the threshold on the volumetric and pore network characteristics of the pervious concrete mixtures using ten different global threshold segmentation algorithms. Based on the results of this study, the following conclusions may be drawn: (a) volumetric segmentation method followed by Otsu bi-level method were found to be predicting the effective porosity more closely to the experimental porosity compared to the other eight histogram and entropy based algorithms; (b) Increase in the threshold value results in the increase of air void voxels and the opening of new pore channels, resulting in the increase of effective porosity, throat number and decrease of isolated porosity and tortuosity; (c) And also the increase in threshold results in the increase of percentage of larger pores and reduction in the percentage of smaller pores; (d) The majority of the effective air voids were found to be in the moderate sphericity range and isolated air voids in the high sphericity range. Additionally, the majority of throats were found to be elliptical in shape. (e) Consistent variation of pore sphericity and throat eccentricity and the fluctuations in pore shape factor with the increase of threshold was observed. Overall, it is expected that the present research on the effect of threshold on the pervious concrete pore network characteristics will help in understanding the importance and significance of threshold segmentation in the field of pavement image processing.

References

Abera KA, Manahiloh KN, Nejad MM (2017) The effectiveness of global thresholding techniques in segmenting two-phase porous media. Constr Build Mater 142:256–267
ACI 522R (2010) Report on Pervious Concrete. American Concrete Institute
ASTM (2012) Standard test method for density and void content of hardened pervious concrete. ASTM C1754-12/C1754M. West Conshohocken. ASTM, PA
Chandrappa AK, Biligiri KP (2018) Pore structure characterization of pervious concrete using X-ray microcomputed tomography. J Mater Civ Eng 30(6):04018108

Iassonov P, Gebrenegus T, Tuller M (2009) Segmentation of X-ray computed tomography images of porous materials: a crucial step for characterization and quantitative analysis of pore structures. Water Resour Res 45(9)

Jagadeesh A, Ong GP, Su YM (2018a) Porosity and permeability evaluation of pervious concrete using three dimensional X-ray computed tomography. In: 4th international conference on transport infrastructure, pretoria, South Africa, July

Jagadeesh A, Ong GP, Su YM (2018b) Digital sieving of pervious concrete air voids using X-ray computed tomography. In: 11th Asia pacific transportation and the environment conference, Malang, Indonesia, October

Jagadeesh A, Ong GP, Su YM (2019a) Development of discharge-based thresholding algorithm for pervious concrete pavement mixtures. J Mater Civ Eng 31(9):04019179

Jagadeesh A, Ong GP, Su YM (2019b) Evaluation of pervious concrete pore network properties using watershed segmentation approach. In: Airfield and highway pavements 2019: testing and characterization of pavement materials. American Society of Civil Engineers, Reston, pp 437–446

Jagadeesh A, Ong GP, Su YM (2019c) Texture evaluation of pervious pavements using three-dimensional X-ray computed tomography. In: 11th international conference on road and airfield pavement technology, Kuala Lumpur, Malaysia, July

Kuang X, Ying G, Ranieri V, Sansalone J (2015) Examination of pervious pavement pore parameters with X-ray tomography. J Environ Eng 141(10):04015021

Mahmud MZH, Hassan NA, Hainin MR, Ismail CR (2017) Microstructural investigation on air void properties of porous asphalt using virtual cut section. Constr Build Mater 155:485–494

Manahiloh KN, Muhunthan B, Kayhanian M, Gebremariam SY (2012) X-ray computed tomography and nondestructive evaluation of clogging in porous concrete field samples. J Mater Civ Eng 24(8):1103–1109

Sezgin M, Sankur B (2004) Survey over image thresholding techniques and quantitative performance evaluation. J Electron Imaging 13(1):146–166

Zelelew HM, Papagiannakis AT (2011) A volumetrics thresholding algorithm for processing asphalt concrete X-ray CT images. Int J Pavement Eng 12(6):543–551

Zhang J, Ma G, Ming R, Cui X, Li L, Xu H (2018) Numerical study on seepage flow in pervious concrete based on 3D CT imaging. Constr Build Mater 161:468–478

Zhang Y, Verwaal W, Van de Ven MFC, Molenaar AAA, Wu SP (2015) Using high-resolution industrial CT scan to detect the distribution of rejuvenation products in porous asphalt concrete. Constr Build Mater 100:1–10

Performance Characteristics of Nano-Modified Asphalt Mixtures

Lucia Tsantilis, Giuseppe Chiappinelli, Orazio Baglieri, Pier Paolo Riviera, Fabrizio Miglietta, and Ezio Santagata

Abstract The growing need for high quality paving materials has aroused an increasing interest in innovative reinforcing agents, such as those characterized by nanometric dimensions. The experimental study presented in this paper focused on the use in asphalt mixtures of nanoclays and multiwall carbon nanotubes as bitumen modifiers. The performance characteristics of asphalt mixtures containing these nano-sized additives and those of a reference neat mixture were compared in a wide array of temperature and loading conditions. The testing program included the assessment of linear viscoelastic characteristics, anti-rutting potential and crack propagation resistance, by means of stiffness modulus, flow number and semi-circular bending tests, respectively. Results highlighted that both types of nano-additives have the potential to improve the performance properties of neat asphalt mixtures, with nanoclays yielding a superior reinforcing action.

Keywords Asphalt mixture · Nanoclays · Multiwall carbon nanotubes · Rutting · Cracking

1 Introduction

Nanotechnology has become increasingly popular in recent years in many areas of materials science and engineering, as it allows phenomena that occur at the nanometre scale to be exploited for design, characterization and production purposes (Ramsden 2011). Among the various forms and applications of nanotechnology, the fine-tuning of advanced composite materials obtained by making use of nano-additives has opened up new perspectives in the pavement research community (Gopalakrishnan et al. 2011; Li et al. 2017).

In the case of bound layers of flexible pavements, several nano-sized products have been considered in order to improve the physicochemical properties of both neat and polymer-modified bituminous matrices. In such a context, nanoclays and carbon

L. Tsantilis (✉) · G. Chiappinelli · O. Baglieri · P. P. Riviera · F. Miglietta · E. Santagata
Department of Environment, Land and Infrastructure Engineering, Politecnico di Torino, 24, Corso Duca degli Abruzzi, 10129 Turin, Italy
e-mail: lucia.tsantilis@polito.it

C. Raab (ed.), *Proceedings of the 9th International Conference on Maintenance and Rehabilitation of Pavements—Mairepav9*, Lecture Notes in Civil Engineering 76,
https://doi.org/10.1007/978-3-030-48679-2_55

nanotubes actually represent the most promising reinforcing agents against rutting and fatigue cracking distresses (Yang and Tighe 2013; Santagata et al. 2016b, c; Tsantilis et al. 2019). However, as synthesized in the following, the substantial diversity between nanoclays and carbon nanotubes in terms of origin, geometry and physicochemical nature, leads to completely different interactive mechanisms occurring at the nanoscale.

Layered silicates used as nano-additives, the so-called nanoclays, have been given widespread consideration since the 1980s, when Toyota researchers reported on the strengthening action yielded in composites by clay platelets via intercalation and/or exfoliation mechanisms. Intercalation occurs when matrix molecules penetrate within clay sheets which, nevertheless, still maintain a predefined basal spacing. On the other hand, exfoliation occurs when penetration leads to a complete breakdown of the original crystallographic structure of clay, with a consequent random distribution of platelets in the composite material. When considering organic matrices such as bitumen, detachment of platelets cannot be easily achieved as a consequence of the combined effect of layer stacking phenomena and of the hydrophilic character of clays. Therefore, specific surfactant coatings, characterized by hydrophilic heads and hydrophobic tails, are used to organically modify the pristine clay with the twofold objective of changing the overall polarity of the layered particles and of expanding clay galleries. Due to their high charge density and cation exchange capacity, montmorillonites are the most suitable minerals to be employed in the manufacture of organophilic clays. Moreover, their abundance in nature limits production costs and causes only minor environmental concerns related to the exploitation of resources (Pavlidou and Papaspyrides 2008).

Carbon nanotubes were first discovered by Iijima in 1991, as a product of carbon-arc discharge experiments in fullerene soot. They are one-dimensional carbon materials composed of rolled-up hexagonal networks of carbon atoms in sp^2 hybridization, arranged in the form of hollow cylinders of nanometric diameters and micrometric lengths. Depending on the number of coaxial tubular layers of which they are composed, that can typically vary between 1 and 50, commercial carbon nanotubes can be found in either single-wall or multi-wall configurations. These peculiar structural features lead to outstanding mechanical properties, that potentially make carbon nanotubes excellent candidates to be used as reinforcing agents in composite materials. Furthermore, sustainable technologies to produce carbon nanotubes, based on the use of waste materials or green synthesis methods, are currently being developed in addition to classical processes such as the electrical arc discharge, chemical vapour deposition, and laser ablation (Dresselhaus et al. 2001; Deng et al. 2016).

Several research studies have been carried out for the evaluation of the physicochemical properties of bituminous binders containing nanoclays and carbon nanotubes. However, scarce data have been reported on the performance characteristics of corresponding nano-reinforced asphalt mixtures.

This paper focuses on the effects of organophilic clays and carbon nanotubes used in mixtures as bitumen modifiers. The experimental investigation which is discussed in the following sections included the assessment of linear viscoelastic characteristics, anti-rutting potential and crack propagation resistance by means of stiffness modulus, flow number and semi-circular bending tests, respectively.

2 Materials and Methods

The materials used in the experimental investigation included a reference asphalt mixture (B), containing neat bitumen, and two nano-modified mixtures (NC and CNT), in which the same bitumen used for B was reinforced with two different types of nano-sized additives, nanoclays and carbon nanotubes, respectively.

The nanoclay employed in the study was a natural montmorillonite organically modified via an ion exchange reaction. The surfactant agent used to change the hydrophilic nature of the clay in its virgin state to the hydrophobic character of the employed additive, was a quaternary ammonium salt composed of two methyl groups and two alkyl chains bonded to a positively charged nitrogen atom. Carbon nanotubes were obtained by means of the catalysed chemical vapour deposition process in multiwall structures of nanometric diameter. The main characteristics of the two commercially available products are reported in Tables 1 and 2.

The neat binder was a 70/100 penetration grade bitumen belonging to Performance Grade (PG) 58-22. Nano-reinforced blends were prepared in the laboratory combining the same neat bitumen with fixed dosages of nanoclays and carbon nanotubes. Based on the results obtained from previous investigations (Santagata et al. 2016a; 2015a), percentages of nanoparticles of 3 and 0.5%, by weight of base bitumen, were chosen for nanoclays and carbon nanotubes, respectively. Blending was performed at 150 °C following a protocol which combines shear mixing and sonication. Shear mixing was carried out with a mechanical stirrer for 90 min operating at 1550 rpm, while ultrasounds were applied by means of an ultrasonic homogenizer for 60 min operating at 24 kHz with a wave amplitude of 157.5 μm (Santagata et al. 2015b).

For the preparation of asphalt mixtures, the three binders were mixed with siliceous aggregates, with the same reference gradation (Fig. 1), at a constant binder content of 4.8% by weight of dry aggregates. Cylindrical specimens of 150 mm diameter and variable height (equal to either 170 or 50 mm), with a target void content of 4% ± 0.5%, were obtained by making use of a gyratory shear compactor. Based on the different rheological properties of the employed binders, mixing and compaction temperatures were changed from one case to the other as detailed in Table 3.

Table 1 Physical characteristics of the nanoclay

Basal spacing [nm]	Cation exchange capacity (CEC) [meq/100 g]	Density [g/cm^3]	Anion
3.15	125	1.66	Chloride

Table 2 Physical characteristics of carbon nanotubes

Average diameter [nm]	Average length [μm]	Density [g/cm^3]	Carbon purity [%]
9.5	1.5	1.72	90

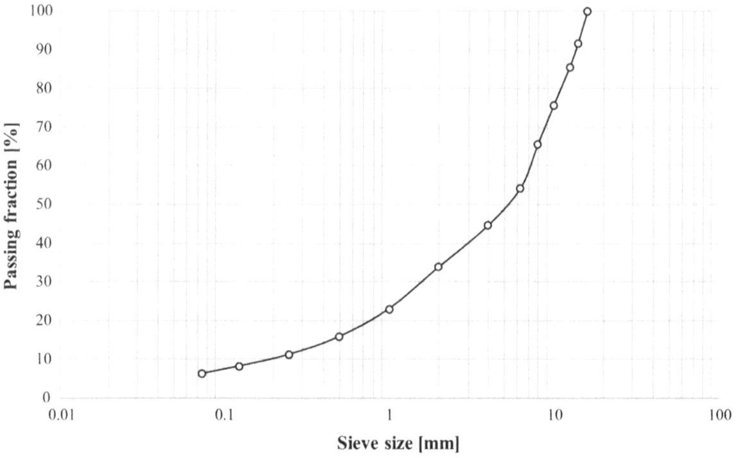

Fig. 1 Aggregate gradation of asphalt mixtures

Table 3 Mixing and compaction temperatures of asphalt mixtures

Mixture	Mixing temperature [°C]	Compaction temperature [°C]
B	150	140
NC	180	170
CNT	165	155

The testing program included the assessment of linear viscoelastic characteristics, anti-rutting potential and crack propagation resistance by means of stiffness modulus, flow number and semi-circular bending tests, respectively.

Stiffness modulus measurements (EN 12697-26) were performed at testing temperatures of 5, 20 and 30 °C on cylindrical specimens of 150 mm diameter and 50 mm height. Indirect tension was induced in the specimens by applying repeated load pulses characterized by a pulse repetition period of 3 s and by different values of rise time, selected in the range of 60–160 ms.

Flow number tests (AASHTO T378) were carried out on cylindrical specimens of 100 mm diameter and 150 mm height obtained by coring and sawing the larger gyratory specimens. Haversine axial compressive load pulses with a duration of 0.1 s were applied every 1.0 s in unconfined conditions with a fixed deviator stress equal to 600 kPa at a testing temperature set equal to 58 °C.

Semi-circular bending tests (EN 12697-44) were performed at 20 °C on half cylinder test pieces of 150 mm diameter and 50 mm height (obtained by sawing gyratory compacted specimens), with a 10 mm midspan notch. Specimens were loaded in the three-point bending configuration by imposing a constant deformation rate of 5 mm/min.

3 Results and Discussion

3.1 Stiffness Modulus Tests

The effects of nano-sized additives on the linear viscoelastic response of asphalt mixtures were assessed by means of stiffness modulus tests performed by simulating relevant in-service conditions in terms of temperature and traffic loading. The outcomes of these tests are presented in Fig. 2, which allows direct comparisons to be made among mixtures at 5, 20 and 30 °C and at several rise-time durations.

From a general overview of the results it can be seen that, regardless of the loading duration, the presence of nano-particles in the mixtures did not cause dramatic changes at 5 and 20 °C. This is proven by the values of the relative variations in stiffness, calculated with respect to the reference mixture, that never exceed 15%. On the contrary, at 30 °C such variations were found to be higher, with an increase in stiffness that reached values of 48% and 30% for NC and CNT mixtures, respectively.

When focusing on the minor changes in modulus which occurred at 5 and 20 °C, it is worth noting that, on the whole, the two additives showed a diverging effect on the response of the neat asphalt mixture. While nanoclays in most cases exhibited a stiffening action, carbon nanotubes generally induced a stiffness reduction which was greater at the lowest temperature. These findings, although unexpected, are consistent with those obtained from the linear viscoelastic characterization of carbon nanotubes-bitumen blends (Santagata et al. 2012).

When considering the effect of different loading durations, it is interesting to observe that at the highest temperature (30 °C) the viscoelastic response of the

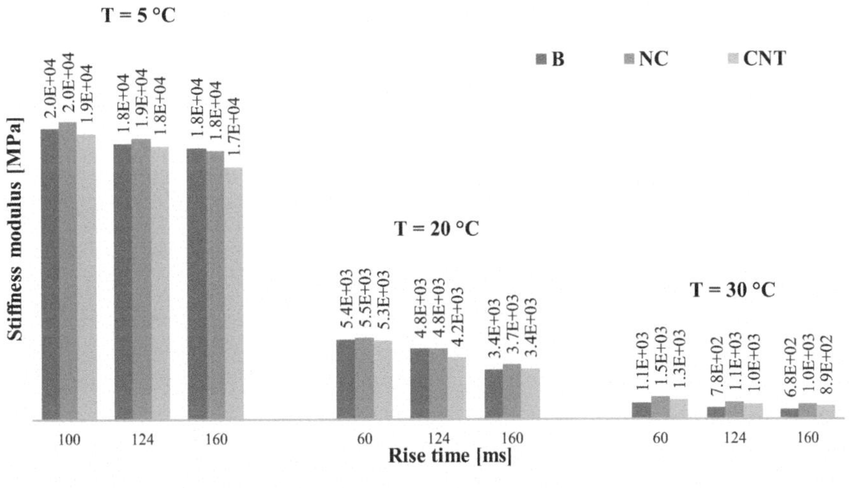

Fig. 2 Stiffness moduli of asphalt mixtures

nano-modified mixtures showed a lower time-dependency when compared with the reference mixture, thus indicating an enhancement of the degree of elasticity.

3.2 Flow Number Tests

The results of flow number tests performed at the reference temperature of 58 °C are displayed in Fig. 3, where the evolution of cumulative permanent strain is plotted versus the number of loading cycles. In line with typical outcomes reported in the literature for both neat and modified mixtures (Santagata et al. 2017), all asphalt mixtures exhibited three distinct stages of response under repeated loading: a primary stage, characterized by a progressively decreasing rate of strain, a secondary stage in which the rate of strain is almost constant, and a final stage of tertiary flow, during which the rate of strain accumulation dramatically increases with load repetitions.

The flow number is defined as the number of loading cycles which corresponds to the transition between the secondary and tertiary stage. It is also associated to the point of minimum rate of change of permanent strain. Average flow number values, calculated from experimental curves fitted to the analytical model proposed by Franken, are listed in Table 4 for the three mixtures included in the study.

In good agreement with previous studies performed at the binder scale (Santagata et al. 2013, 2015b, 2016b), also at the mixture scale nano-modified materials were

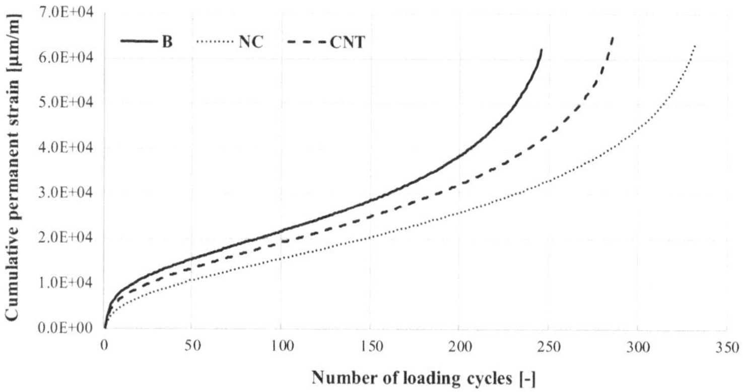

Fig. 3 Typical curves obtained in flow number tests

Table 4 Average flow number values of asphalt mixtures

Mixture	Flow number [-]
B	111
NC	146
CNT	132

found to be less susceptible to permanent deformation accumulation under repeated loadings, the best performance being exhibited by NC. These observations, which are consistent with the enhancements in stiffness and elasticity found in the linear viscoelastic response, reveal that nano-sized additives yielded a reinforcing action in the mixture that was also effective in the domain of large deformations.

3.3 Semi-circular Bending Tests

Semi-circular bending tests were carried out at 20 °C in order to evaluate the impact of nano-modification on the resistance to crack propagation. Typical outputs of tests are presented in Fig. 4 as load-versus-displacement curves. Average experimental results are summarized in Table 5 in terms of fracture toughness (K_{Ic}), strain at maximum load (ε_{max}), maximum stress at failure (σ_{max}), and strain energy to failure (U). This last parameter is calculated as the area under the load-versus-deflection curve up to peak load.

When considering the effect of nano-modification with organophilic clays, a non-negligible improvement in fracture toughness was exhibited by mixture NC with respect to reference mixture B. Moreover, changes in fracture properties were

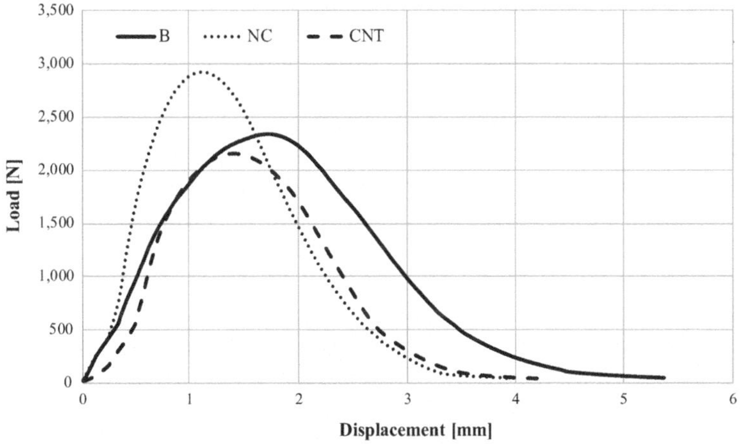

Fig. 4 Typical load-versus-displacement curves obtained in semi-circular bending tests

Table 5 Average results of semi-circular bending tests

Mixture type	K_{Ic} [N/mm$^{3/2}$]	ε_{max} [%]	σ_{max} [N/mm^2]	U [kJ]
B	7.78	2.09	1.31	2.01E+03
NC	9.56	1.73	1.61	2.05E+03
CNT	7.61	1.99	1.28	1.94E+03

revealed by the higher values of both maximum stress and strain energy to failure, and by the lower value of strain at maximum load recorded during tests. Based on these experimental outcomes and by taking into account the encouraging enhancements in fatigue resistance found in previous studies by means of indirect tension cyclic tests (Miglietta et al. 2018; Santagata et al. 2019), it can be inferred that organophilic clays can play an important role in preventing distresses related to fatigue cracking phenomena.

On the other hand, modification with carbon nanotubes was not effective in improving the fracture properties of the reference asphalt mixture. This is shown by the limited variations in fracture toughness, strain at maximum load and maximum stress at failure which were recorded during the investigation. It is interesting to note that as a result of the presence of carbon nanotubes, strain energy to failure decreased. Such a finding is in general disagreement with the results obtained by other researchers who performed semi-circular bending tests on mixtures containing the same dosage of carbon nanotubes (Ameri et al. 2016). Further work is certainly needed to clarify the cause of this apparent discrepancy in results.

4 Conclusions

Based on the results presented in this paper, it can be concluded that nano-sized additives have the potential to positively affect the performance properties of neat asphalt mixtures, with organophilic clays yielding a superior reinforcing action when compared to carbon nanotubes. However, the general effectiveness of such improvements should be carefully evaluated by means of a cost- and environmental-based analysis.

Considering the mechanical behaviour of the mixtures in their linear viscoelastic domain, no major changes were generated by both types of nano-particles. However, at the highest investigated temperature a moderate increase in stiffness and elasticity was highlighted in the case of the nano-modified mixture.

With respect to their anti-rutting performance, the two nano-modified mixtures showed a superior resistance to accumulation of permanent deformations under repeated loading when compared to the reference neat material. In particular, the best performance was exhibited by the mixture reinforced with nanoclays.

When focusing on the resistance to crack propagation, beneficial effects associated to nanomodification were found only in the case of nanoclays, thus suggesting that for the considered mixture employed carbon nanotubes may have an inadequate scale length for crack bridging contributions.

Acknowledgements The study reported in this paper is part of the FIRB research project on "Innovative nano-structured and polymer-modified bituminous materials" funded by the Italian Ministry of Education, University and Research (MIUR) (Grant RBFR10JOWO).

References

Ameri M, Nowbakht Sh, Molayem M, Aliha MRM (2016) Investigation of fatigue and fracture properties of asphalt mixtures modified with carbon nanotubes. Fatigue Fract Eng Mater Struct 39:896–906

Deng J, You Y, Sahajwalla V, Joshi RK (2016) Transforming waste into carbon-based nanomaterials. Carbon 96:105–115

Dresselhaus MS, Dresselhaus G, Avouris Ph (2001) Carbon nanotubes: synthesis, structure, properties, and applications, vol 80. Topics in applied physics. Springer, New York

Gopalakrishnan K, Birgisson B, Taylor P, Attoh-Okine N (2011) Nanotechnology in civil infrastructures: a paradigm shift. Springer, Berlin

Li R, Xiao F, Amirkhanian S, You Z, Huang J (2017) Developments of nano materials and technologies on asphalt materials – a review. Constr Build Mater 143:633–648

Miglietta F, Underwood BS, Tsantilis L, Baglieri O, Kaloush KE, Santagata E (2018) Fatigue properties of nano-reinforced bituminous mixtures: a viscoelastic continuum damage approach. Int J Pavement Res Technol 11(7):766–773

Pavlidou S, Papaspyrides CD (2008) A review on polymer – layered silicate nanocomposites. Prog Polym Sci 33:1119–1198

Ramsden JJ (2011) Nanotechnology: an introduction. Elsevier, Waltham

Santagata E, Baglieri O, Tsantilis L, Dalmazzo D (2012) Rheological characterization of bituminous binders modified with carbon nanotubes. Procedia Soc Behav Sci 53:546–555

Santagata E, Baglieri O, Tsantilis L, Chiappinelli G (2013) Effects of nano-sized additives on the high-temperature properties of bituminous binders: a comparative study. In: Kringos N, Birgisson B, Frost D, Wang L (eds) Multi-scale modeling and characterization of infrastructure materials. Springer, Dordrecht, pp 297–309

Santagata E, Baglieri O, Tsantilis L, Chiappinelli G (2015a) Fatigue properties of bituminous binders reinforced with carbon nanotubes. Int J Pavement Eng 16:80–90

Santagata E, Baglieri O, Tsantilis L, Chiappinelli G, Brignone Aimonetto I (2015b) Effect of sonication on high temperature properties of bituminous binders reinforced with nano-additives. Constr Build Mater 75:395–403

Santagata E, Baglieri O, Tsantilis L, Chiappinelli G (2016a) Storage stability of bituminous binders reinforced with nano-additives. In: 8th RILEM international symposium on testing and characterization of sustainable and innovative bituminous materials. RILEM bookseries. Springer, Dordrecht, pp 75–87

Santagata E, Baglieri O, Tsantilis L, Chiappinelli G, Dalmazzo D (2016b) Bituminous-based nanocomposites with improved high temperature properties. Compos Part B 99:9–16

Santagata E, Baglieri O, Tsantilis L, Dalmazzo D, Chiappinelli G (2016c) Fatigue and healing properties of bituminous mastics reinforced with nano-sized additives. Mech Time Depend Mater 20:367–387

Santagata E, Baglieri O, Riviera PP, Alam M (2017) Correlating creep properties of bituminous binders with anti-rutting performance of corresponding mixtures. Int J Pavement Res Technol 10:38–44

Santagata E, Baglieri O, Miglietta F, Tsantilis L, Riviera PP (2019) Impact of nano-sized additives on the fatigue damage behaviour of asphalt mixtures. Fatigue Fract Eng Mater Struct 42(11). https://doi.org/10.1111/ffe.13110

Tsantilis L, Dalmazzo D, Baglieri O, Santagata E (2019) Effect of SBS molecular structure on the rheological properties of ternary nanomodified bituminous binders. Constr Build Mater 222:183–192

Yang J, Tighe S (2013) A review of advances of nanotechnology in asphalt mixtures. Procedia Soc Behav Sci 96:1269–1276 CICTP - 13th COTA international conference of transportation professionals

Properties of Styrene-Isoprene-Styrene (SIS) Modified Asphalt Binder

Mithil Mazumder, Soon-Jae Lee, and Moon-Sup Lee

Abstract Physical and rheological properties of asphalt binder modified with Styrene-Isoprene-Styrene (SIS) were investigated through the rotational viscometer (RV), the dynamic shear rheometer (DSR), and the bending beam rheometer (BBR). In order to have a depth understanding on the SIS binder at micro level, morphological observations were conducted using optical microscopy. The result of this study showed that (1) the addition of SIS modifiers increases the viscosity and has positive effect on rutting resistance of the binder; (2) the higher the SIS content, the better the cracking resistance of the binder.

Keywords Styrene-Isoprene-Styrene · SIS · Microscopy

1 Introduction

The styrene-butadiene (SB) copolymer consists of linked blocks of polystyrene (PS) and polybutadiene (PB). According to Becker et al. (2001), it is the most appropriate and used polymer for asphalt modification, followed by reclaimed tire rubber. It is the formation of critical network between the binder and SBS that increases the complex modulus, resulting the increase in rutting resistance. In 2004, Florida Department of Transportation (FDOT) and Federal Highway Administration (FHWA) reported that SBS benefited the cracking resistance by reducing the rate of micro-damage accumulation (Roque et al. 2005). However, the addition of SBS has some drawbacks in terms of economic and technical limits. It is capable to increase the low temperature flexibility but some authors report that a decrease in strength and resistance to penetration is observed at higher temperature (Yildirim 2007; Mazumder et al. 2016; Kim et al. 2019; Ali et al. 2018).

On the other hand, SIS polymer has the potential to overcome the challenge of low temperature failure and cracking due to its high stiffness and elasticity.

M. Mazumder · S.-J. Lee (✉)
Texas State University, San Marcos, TX 78666, USA
e-mail: SL31@txstate.edu

M.-S. Lee
Korea Institute of Civil Engineering and Building Technology, Gyeonggi 10223, South Korea

C. Raab (ed.), *Proceedings of the 9th International Conference on Maintenance and Rehabilitation of Pavements—Mairepav9*, Lecture Notes in Civil Engineering 76,
https://doi.org/10.1007/978-3-030-48679-2_56

SBS (styrene-b-butadiene-b-styrene) has butadiene as rubbery mid-block whereas SIS (styrene-b-isoprene-b-styrene) has isoprene as rubbery midblock. The SIS molecule chain is composed of isoprene. It increases the complex modulus at high temperature and due to its branch methyl in the isoprene group it has better tenacity and compatibility with other materials (Zhang et al. 2018). The drawback of SIS polymer is its higher price compared to SBS polymer. However, due to its strong rheological characteristics and higher demand of durable pavement it can be used in bridges or other heavy duty structures. The study on SIS polymer mostly limited to its individual characteristics. Also, the study on modification of bitumen using SIS is limited as well in terms of rheological and microstructural properties. The study selected the different percentage of SIS (0, 5, 10, 10 and 20%) to show the change of trend of SIS modification in terms of its rheological and microstructural properties.

For this purpose, control PG 64-22 is modified with five different percentages of SIS content (0, 5, 10, 15, and 20%). Viscosity change as a function of SIS amount is evaluated through rotational viscometer (RV) test using two testing temperatures (135 and 180 °C). Dynamic shear rheometer (DSR) and bending beam rheometer (BBR) were used to investigate high temperature and low temperature rheological behavior of asphalt binder modified with SIS at original state. Microstructural properties of SIS modified binder were investigated using optical microscopy.

2 Experimental Design

2.1 *Materials*

Performance grade (PG) 64-22 asphalt binder as a base binder was used in this study. The SIS modifier is a blend of linear SIS triblock and SI diblock copolymer. It contains approximately 18% SI diblock copolymer. Figure 1 and Table 1 show the image and properties of SIS modifier.

2.2 *Superpave Asphalt Binder Tests*

The Superpave asphalt binder tests are used to measure the asphalt's performance at three stages of its life: in its original state, after mixing and construction, and after in-service aging. In this study the selected test procedures included the viscosity test (AASHTO T 316), the dynamic shear rheometer (DSR) test (AASHTO T 315), and the bending beam rheometer (BBR) test were conducted at original state. A 8.5 g sample of the control binders and a 10.5 g sample of SIS binders were tested with a number 21 spindle and with a number 27 spindle in the Brookfield rotational viscometer at 135 and at 180 °C. In the DSR test, the original binders were tested at a frequency of 10 radians per second which is equal to approximately 1.59 Hz.

Fig. 1 SIS (Styrene-isoprene-styrene)

Table 1 Properties of SIS modifier

Properties	Test method	Units	Typical value
Styrene	TSRC/DEXO method	wt%	15
Diblock content	TSRC/DEXO method	wt%	18
Melt flow rate (200 °C/5 kg)	ASTM D1238	g/10 min	11
Solution viscosity	ASTM D2196	cps	1240
Ash	ASTM D5630	wt%	0.3
Tensile strength	TSRC/DEXO method	MPa	25
300% modulus	TSRC/DEXO method	MPa	1.1
Elongation	TSRC/DEXO method	%	1250
Hardness	ASTM D2240	Shore A	33
Bulk density	ASTM D1895	g/cm^3	0.55 (4113 A)
Specific gravity	ASTM D792		0.92

Each asphalt binder at original state used to determine the G*/sin δ at 82°C. The G*sin δ at intermediate temperature was measured to evaluate the fatigue cracking property for unaged binders at 25 °C. The BBR test was conducted on original asphalt beams (125 × 6.35 × 12.7 mm) at −24 °C, and the creep stiffness (S) of the binder was measured at a loading time of 60 s. A constant load of 100 g was then applied to the beam of the binder, which was supported at both ends, and the deflection of center point was measured continuously. Testing was performed on aged samples.

2.3 Optical Microscopy

The sample morphology was observed using Hirox digital microscope (RH-2000E digital microscope). A binder sample of the mold specimen was prepared using different percentage of SIS modifier and viewed under the microscope at a magnification of 1500×.

3 Results and Discussions

3.1 Rheological Properties

3.1.1 Viscosity Property

The viscosity of asphalt binder at high temperature is considered to be an important property to decide working temperature because it reflects the binder's ability to be pumped through an asphalt plant, thoroughly coat aggregate in a HMA mixture, and be placed and compacted to form a new pavement surface (Asphalt Institute 2003). Figure 2 shows the standard RV test results for SIS binders at 135 and 180 °C. It is evident that the addition of SIS into the asphalt binder increases the binder viscosity for both testing temperatures. The viscosity of 15 and 20% SIS binder could not be measured at 135 °C. The viscosity values of PG 64-22, SIS 5% and SIS 10% at 135 °C are found to be 635, 2028, and 7008 cP, respectively. It is worth to note that the viscosity of SIS binder seems to have insignificant change after adding more than 15% of SIS content.

3.1.2 Rutting Property

The higher G*/sin δ values from the DSR test indicate that the binders are less susceptible to rutting or permanent deformation at high pavement temperature (Asphalt Institute 2003). The G*/sin δ values of unaged SIS binders at 82 °C are shown in Fig. 3. It is evident from the figure that the addition of SIS modifier significantly increases the rutting resistance of the binder. It means that the SIS has a positive effect on the rutting resistance at high temperature which causes an increase in the complex modulus of the binders. However, the percentage improvement of rutting resistance after 15% addition of SIS modifier is found to be less significant.

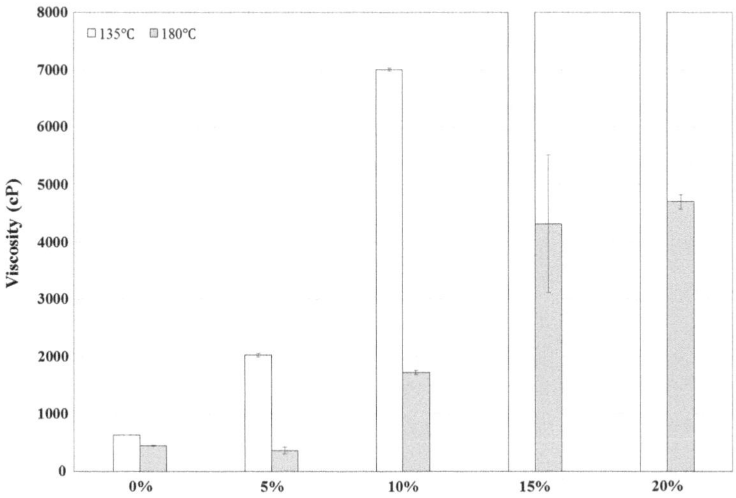

Fig. 2 Viscosity of SIS binders at 135 and 180 °C

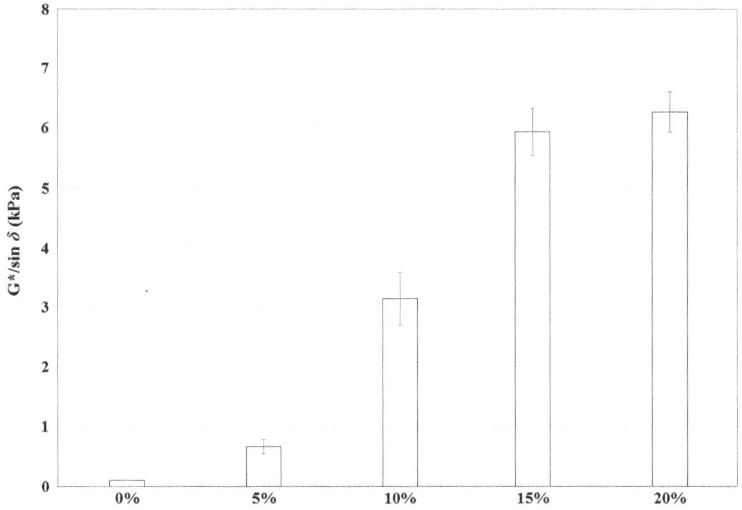

Fig. 3 G*/sin δ of SIS binders at 82 °C

3.1.3 Cracking Property

In general, the lower G*sin δ values are considered to be desirable attributes from the standpoint of fatigue cracking resistance (The Asphalt Institute 2003). The G*sin δ values of the SIS binders are determined using the DSR at 25 °C and the results are illustrated in Fig. 4. The G*sin δ values are found to be 1210, 1101, 146, 140 and

97 kPa for the binders of control (PG 64-22), SIS 5% (PG 64-22 + 5% SIS), SIS 10% (PG 64-22 + 10% SIS), SIS 15% (PG 64-22 + 15% SIS) and SIS 20% (PG 64-22 + 20% SIS), respectively. With the increase of SIS percentage the binder is found to have more cracking resistance.

From the BBR tests at −24 °C, the stiffness and m-value of SIS binders at original state are calculated and results are illustrated in Figs. 5 and 6, respectively. It is

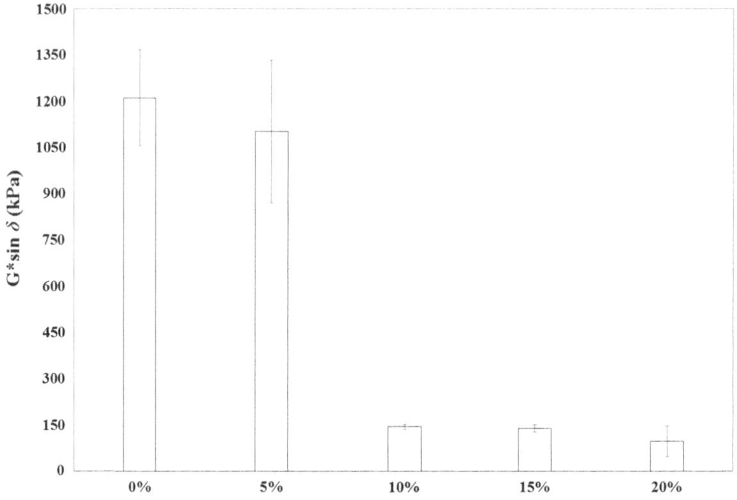

Fig. 4 G*sin δ of SIS binders at 25 °C

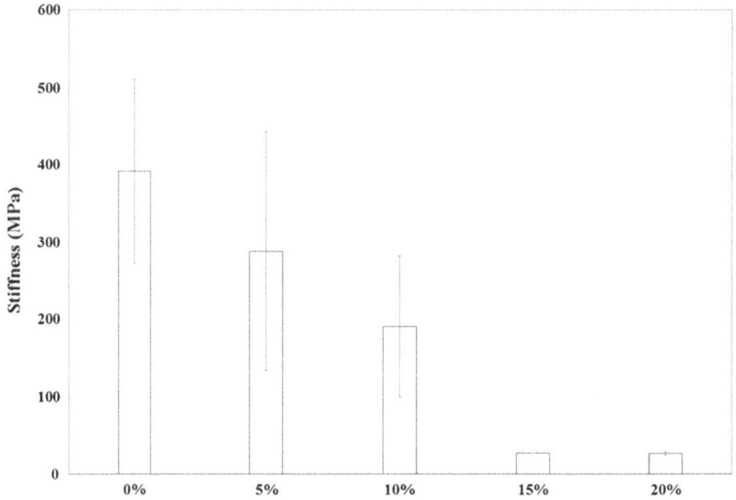

Fig. 5 Stiffness of SIS binders at −24 °C

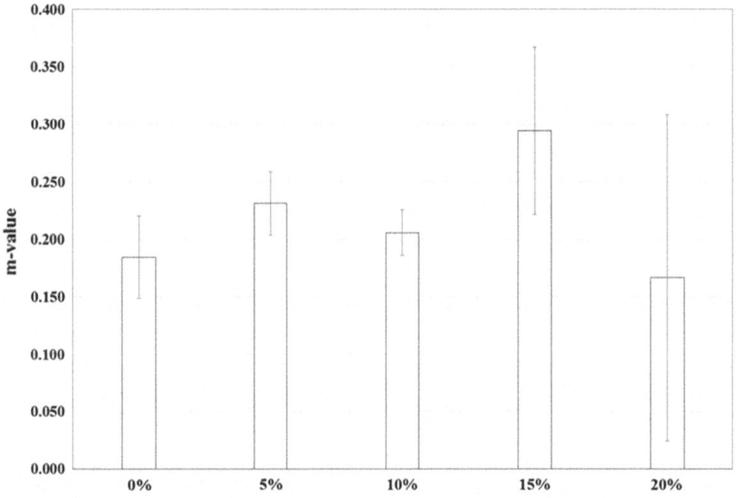

Fig. 6 m-value of SIS binders at −24 °C

found that the addition of SIS into the asphalt binder significantly decreases the low temperature stiffness. The addition of 5% SIS and 10% SIS resulted in decreasing the stiffness of control PG 64-22 binder by 26 and 51%, respectively. Also, with the percentage of SIS modifier increased, the low temperature stiffness of the binder is observed to be decreased.

3.2 Microscopy Properties

3.2.1 Optical Microscopy Analysis

Figures 7a–e illustrates the surface images of the prepared SIS binder sample at its original state. The surface image of control binder PG 64-22 or SIS 0% shows blue spots homogeneously spread all over the surface. The addition of SIS modifier exhibited a new elongated phase along with bee-structure. With further addition of SIS content of 10% the density of new phase increased and at 15% content the new oval phase evident in Fig. 7d. The addition of 20% content of SIS shows the homogeneous spread of oval phase throughout the sample surface.

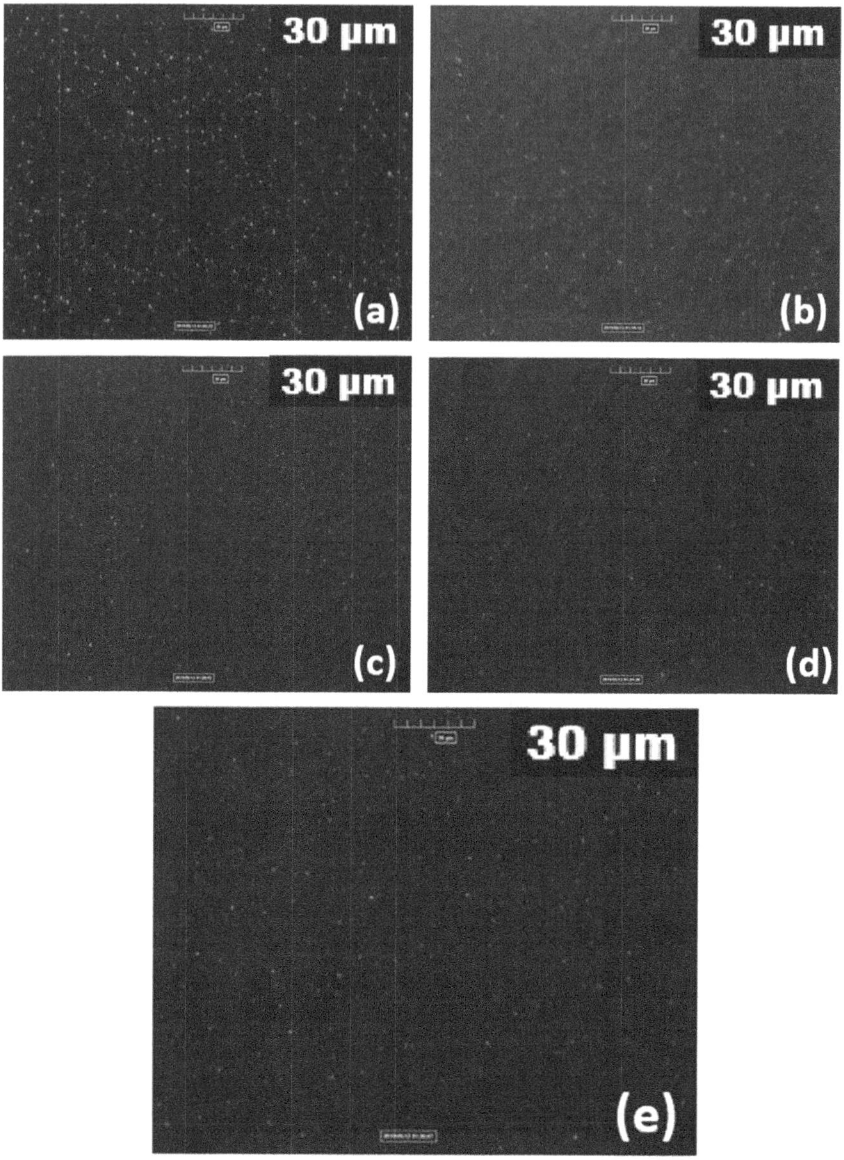

Fig. 7 Optical microscopy images at original state: **a** SIS 0%, **b** SIS 5%, **c** SIS 10%, **d** SIS 15%, and **e** SIS 20%

4 Conclusions

This study is to investigate the physical, rheological and microstructural properties of SIS binder. The binders are produced using 0, 5, 10, 15 and 20% of SIS by the weight of the control PG 64-22 binder. The viscosity property is determined by RV, rheology and stiffness properties are measured by DSR and BBR, respectively and micromorphology is investigated using optical microscopy. Based on the result of these tests, the conclusions are drawn for the materials used in this study as following,

(1) The addition of SIS into asphalt binder can significantly increase the viscosity of binders. With the SIS percentage increased, the binder viscosity increases at both testing temperatures. However, the viscosity of SIS binder seems to have insignificant change after adding more than 15% of SIS content.
(2) The amount of SIS modifier has a positive effect on the rutting resistance at high temperature.
(3) The addition of SIS into asphalt binder can significantly decrease the $G^*\sin\delta$ and creep stiffness of SIS binder at low temperature which can ensure better cracking resistance of asphalt binder.
(4) Optical microscopy images showed that the addition of SIS content introduce a new elongated phase which observed to have increased with the addition of SIS content.

Acknowledgements This study was conducted under research project "Development of High-Performance Concrete Pavement Maintenance Technology to Extend Roadway Life (Project No: 18TLRP-B146699-01)" funded by the Ministry of Land, Infrastructure and Transport (MOLIT) and the Korea Agency for Infrastructure Technology Advancement (KAIA). The authors would like to thank the members of research team, MOLIT and KAIA for their guidance and supports throughout the project.

References

Ali AW, Kim HH, Mazumder M, Lee MS, Lee SJ (2018) Multiple Stress Creep Recovery (MSCR) characterization of polymer modified asphalt binder containing wax additives. Int J Pavement Res Technol. http://www.ijprt.org.tw/reader/pdf.php?id=535
Asphalt Institute (2003) Performance graded asphalt binder specification and testing, SP-1, Lexington, KY
Becker Y, Mendez MP, Rodriguez Y (2001) Polymer modified asphalt. VisTechnol 9(1):39–50
Kim HH, Lee MS, Lee SJ (2019) Performance evaluation of polymer modified asphalt (PMA) binders containing ground tire rubber (GTR). Int J Pavement Res Technol 12(2):215–222
Mazumder M, Kim H, Lee SJ (2016) Performance properties of polymer modified asphalt binders containing wax additives. Int J Pavement Res Technol 9(2):128–139
Roque R, Birgisson B, Drakos C, Sholar G (2005) Guidelines for use of modified binders. University of Florida, Report No. 4910-45054-964-12, pp 7–13
Yildirim Y (2007) Polymer modified asphalt binders. Constr Build Mater 21(1):66–72
Zhang F, Hu C, Zhang Y (2018) The research for SIS compound modified asphalt. Mater Chem Phys 205:44–54

Effects of Binder Modification on Rutting Performance of Asphalt Binders

A. Seitllari, M. Ghazavi, and M. E. Kutay

Abstract Asphalt binder is the driving economical factor in asphalt pavement design. As a result, it is a good target to optimize the design by minimizing the overall cost while maximizing the performance. Modifying asphalt binders with polymers has been historically a successful method of optimization of pavement design. However, fluctuations in polymer prices and concerns related to sustainability lead many engineers to consider scrap tire rubber as a full or partial replacement of the polymers. Polymer coated rubber (PCR), a hybrid blend of crumb rubber and polymer, is a relatively new alternative material to improve asphalt mixture performance. The objective of the study reported in this paper is to investigate the relative performances of the neat asphalt binder, polymer modified (PM) binder and PCR modified binder using the wet process (PCR-Wet). The influence of the modifications on the viscosity, stiffness and rutting performance were explored. The experimental program included rotational viscosity measurements, linear viscoelastic characterization using frequency sweep (|G*|) test and rutting resistance using multiple stress recovery (MSCR) test on the base binder and the PCR modified binder. Findings from this study suggest that the more sustainable and rut-resistant asphalt binders can be achieved by using the PCR-Wet process.

Keywords Polymer coated rubber · Dynamic shear modulus · Rutting performance · MSCR · Viscosity · Aging effects

1 Introduction

Enhancing the performance and reducing the life cycle cost of asphalt pavements is very important for transportation agencies and asphalt paving industry (Hansen and Copeland 2015; Seitllari and Kutay 2019). Exploring possible alternatives to improve the service life of asphalt pavements without compromising the cost has always been a concern especially for agencies (Diefenderfer and Bowers 2019).

A. Seitllari · M. Ghazavi · M. E. Kutay (✉)
Department of Civil and Environmental Engineering, Michigan State University, East Lansing, MI, USA
e-mail: kutay@egr.msu.edu

C. Raab (ed.), *Proceedings of the 9th International Conference on Maintenance and Rehabilitation of Pavements—Mairepav9*, Lecture Notes in Civil Engineering 76,
https://doi.org/10.1007/978-3-030-48679-2_57

Asphalt binder is the driving economic factor in asphalt pavements, hence, it is a good target to optimize the design by minimizing the overall cost while maximizing the performance. In this context, the use of crumb rubber (CR) from the end of life (EoL) tires in paving practices have been an alternative practice to increase the overall performance of asphalt pavements without compromising the general cost (Santagata et al. 2016). Several researchers have studied the use of CR modified asphalt binders in road construction and satisfactory performance of pavements using CR modified binders has been documented (Stroup-Gardiner et al. 1996; Way 2000).

Traditionally, CR modifications are grouped into three processes: wet process, dry process and terminally blended methods (Heitzman 1992). The wet process is the most commonly used modification method. CR particles are added to the hot asphalt binder (160–220 °C) and blended (Seitllari et al. 2019a, b). Various amounts of CR are utilized in this method. For the dry process however, the CR material is directly added to the asphalt mixture. In fact, the CR material for this method is used as a replacement for fine aggregates. On the other hand, the terminally blended method is similar to the wet process except for the inclusion of polymeric additives. The performances of mixtures strongly depend on the mode of interaction between CR and the other components of the asphalt mixtures. It is also true that in some cases the interaction process results in a final product (e.g. "asphalt rubber") that is not easy to work with in the field (Kocak 2016; Seitllari et al. 2019a, b). To overcome these problems encountered in the field, many technologies and pre-treatment methods of the rubber particles have been developed. Polymer coated rubber (PCR) is a relatively new alternative to address these concerns (Kurgan and Dongre 2015). The PCR is a chemically enhanced/polymer-coated rubber particle where a polymer film partially covers the surface area of the CR particles. The PCR is made of scrap tire rubber coated with about 1.5% SBR (Styrene-Butadiene co-polymer latex liquid) emulsion. A Scanning Electron Microscope (SEM) image of the PCR particles is shown in Fig. 1, where the liquid polymer partially coats the rubber reducing the potential for absorption of light components of the asphalt binder into the rubber. The SBR

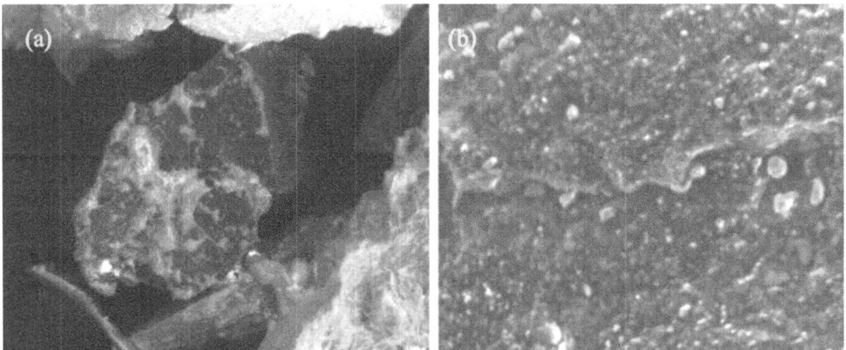

Fig. 1 Polymer coated rubber with patented coating process **a** at 180× magnification and **b** 5000× magnification

polymer is thought to promote and control the reaction with asphalt and allows the development of strong bonds between materials during the production process of the mixture (Kurgan and Dongre 2015).

The PCR particles can be added to the mix using either wet or dry technology and some of the previous field applications in the US have shown good performance in both cases (Gibson et al. 2012). However, the use of PCR in asphalt mixtures using the dry process possesses several benefits from both economic and environmental aspects. In fact, this is not a direct modification of the asphalt binder, no prior mixing between PCR and bitumen is required. Nonetheless, the presence of the PCR particles will reduce but not fully eliminate the need for polymer modification of the asphalt binder used for the production of asphalt mixes. Thus, some of the savings mentioned above will be mitigated by the fact that high performance (polymer modified) asphalt shall be produced and purchased. In response, the wet process is seen as a potential candidate to address this issue (Jamrah 2018).

As mentioned earlier, the wet process involves blending process of the PCR particles and asphalt binder at elevated temperatures. It is believed that the inclusion of PCR particles in asphalt binder contributes to softening action at lower temperatures and stiffening at higher temperatures (Jamrah and Kutay 2015).

2 Objective and Scope

The objective of this paper was to investigate the relative performance of the neat asphalt binder, PM binder and PCR modified binders using the wet process (PCR-Wet). The scope of this study included a neat base binder (PG 58-28), one type of polymer (styrene-butadiene-styrene (SBS)) and one type of PCR obtained from commercial manufacturers. PM binder was prepared at 3% SBS by the weight of the binder. The production process was performed in general accordance with the procedure developed at the Asphalt Advanced Characterization Laboratory (AACL) at Michigan State University (Kocak and Kutay 2016). PCR-Wet was prepared at two percentages, 8 and 15% by total weight of the binder. Three different blending durations were attempted to optimize the blending speed of PCR-Wet. In addition, the base binder, 3% SBS and PCR-Wet were subjected to a series of testing. The tests include viscosity measurements, linear viscoelastic characterization using frequency sweep ($|G^*|$) test and rutting resistance using multiple stress recovery (MSCR) test.

3 Sample Preparation and Testing

The PCR-Wet modification process included two basic steps. As part of the first step, the base binder was heated to 163 °C and the desired amount of PCR particles was added and mixed using a rotor-stator head in the high shear mixer at a rate of 5000 revolutions per minute (rpm) for 30 min. The heat produced during the size reduction of the PCR pills allows keeping the temperature of the system approximately constant, eliminating the need for a heating mantle at this step. As part of the second step, the PCR-binder compound was transferred into an adjustable heating mantle for the low shear mixing at a rotational speed of 1000 rpm at 160–170 °C. As previously mentioned, three different low shear mixing durations of 120, 60 and 30 min were investigated. In addition, a liquid Sulphur at an amount of 0.2% by the weight of the binder was introduced to the mixture as a cross-linking agent during the last 30 min of the low shear process. The AASHTO T 316 "*Viscosity Determination of Asphalt Binder Using Rotational Viscometer*" was followed to determine the viscosity of the binders and two replicates were tested at 135 °C at a rotational speed of 20 rpm. The Brookfield Rotational Viscometer was utilized for viscosity measurements of the base binder, 3% SBS and the PCR-Wet combinations. The AASHTO T 240 "*Effect of Heat and Air on a Moving Film of Asphalt Binder (Rolling Thin-Film Oven Test)*" and AASHTO T 350 "*Standard Method of Test for Multiple Stress Creep Recovery (MSCR) Test of Asphalt Binder Using a Dynamic Shear Rheometer (DSR)*" tests were conducted to investigate the rutting performance of the binders. Frequency sweep (|G*|) tests were also performed on unaged and short-term aged samples in general accordance with AASHTO T 315 "*Determining the Rheological Properties of Asphalt Binder Using a Dynamic Shear Rheometer (DSR)*". Three test replicates were used to generate the data at 14, 24, 34, 44, 54, 64 and 74 °C and a frequency of 0.02–16 Hz. The 25 mm plate geometry with a 1 mm gap was used for |G*| measurements at 34–74 °C. And the 8 mm plate geometry with a 2 mm gap was used for measurements at 14–34 °C. A minimum of two binder batches was prepared with two replicates per each batch that were sampled and tested.

4 Results and Discussion

Initially, the viscosity test was performed on PCR-Wet 15% and PCR-Wet 8%. According to the Superpave binder specification, PCR-Wet 15% failed the viscosity requirement. On the other hand, PCR-Wet 8% met the requirements of Superpave binder specification. Therefore, a decision was made to exclude PCR-Wet 15% from further evaluations and consider only PCR-Wet 8%. Figure 2 presents the viscosity values for PCR-Wet 8%, 3% SBS and base binder. As shown in this figure, a considerable increase in viscosity is obtained for the asphalt binder modified with PCR when compared to 3% SBS and base binder. The PCR-Wet 8% subjected to 1/2 h and 2 h low shear resulted in very similar viscosities followed by PCR-Wet 8% at 1 h low shear and base binder. The difference of PCR-Wet 8% combinations is probably due to sample variability.

Percent recovery (%R) and non-recoverable creep compliance (J_{nr}) of the selected binders were determined according to AASHTO T 350. Initially, the binders were subjected to short-term aging as recommended in AASHTO T 20. The short-term aged binders were then molded and prepared for testing. The MSCR testing temperature was selected according to the regional high surface temperature, which is

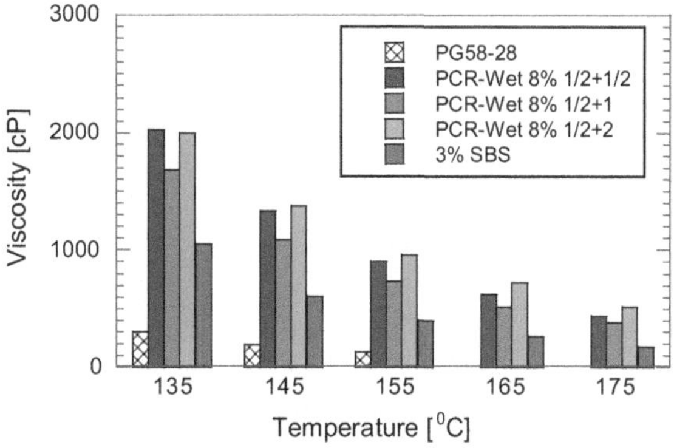

Fig. 2 Rotational Viscometer data for three different blending combinations

58 °C for the State of Michigan. The primary objective of this test is to investigate the rutting performance of asphalt binders. The MSCR test results are also used to correlate the performance of asphalt mixtures. Researchers have successfully used J_{nr} for estimating the rutting performance of asphalt binders. Studies show that the MSCR test is well correlated to pavement rutting (Bukowski et al. 2011).

Figure 3 shows the multiple stress creep and recovery (MSCR) test results of the PCR modified binders mixed at three durations, as well as the 3% SBS and base binder (PG 58-28). J_{nr} percent difference for all extracted binders was less than 75% according to AASHTO T 350. As shown, all mixing durations produced a V-Grade binder except the base binder which was an S-Grade (Fig. 3a). In general, a lower $J_{nr3.2}$ value indicates better rutting resistance. Thus, the rutting performance of the PCR-Wet 8% and 3% SBS are anticipated to outperform the base binder. All PCR-Wet 8% passed the elastic recovery requirement by demonstrating similar performance while the 3% SBS performed best. On the other hand, the base binder failed the elastic recovery requirement (Fig. 3b). In terms of low shear mixing duration, it was concluded that all mixing durations produce similar PCR-Wet 8% modified binders. Therefore, it was decided to use the first mixing procedure (1/2 h high shear mixing plus 1/2 h low shear mixing, hereafter noted PCR-Wet 8%) for binder modification.

To further examine the mechanical behavior of the modified and base binder, the dynamic shear modulus (|G*|) test was conducted at different frequency-temperature combinations. The purpose of this exercise helped to capture the fundamental behavior of the binders under different loading conditions.

The short-term aging as recommended in AASHTO T 20 was performed on selected binders except for 3% SBS, and the linear viscoelastic behavior was measured as described above.

Figure 4a shows |G*| values of the tested binders at 58 °C at multiple frequencies, where PCR-Wet 8% modified binder is stiffer than the neat binder by almost one order of magnitude at low frequencies and to a lesser extent at high frequencies at this temperature. Nonetheless, the 3% SBS outperforms the other two binders. Figure 4b shows the |G*| master curves of the PCR-Wet 8%, 3% SBS and base binder. Overall,

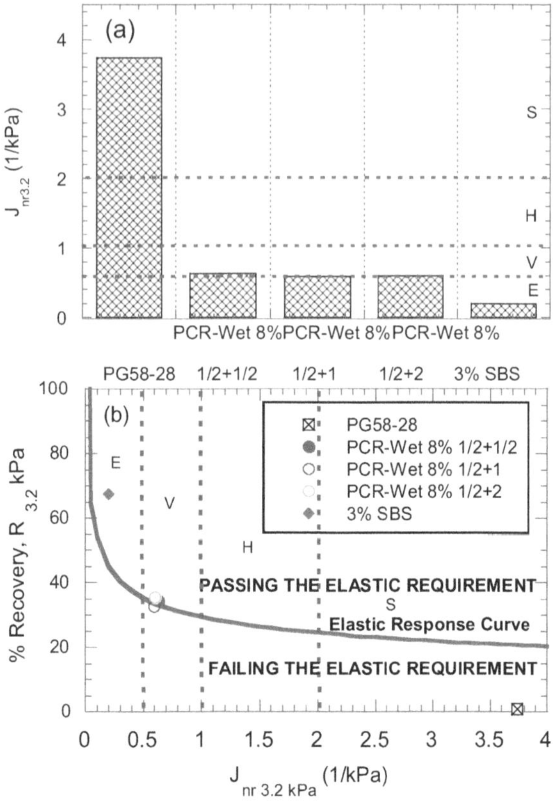

Fig. 3 MSCR test results for all the considered binders

the |G*| values of PCR-Wet 8% binder at nearly all temperature/frequency combinations are higher as compared to the neat binder but lower than 3% SBS. At high temperature (74 °C) and low frequency (0.02 Hz), the PCR-Wet 8% had higher |G*| values than a base binder and lower than 3% SBS, whereas at low temperature (14 °C) and high frequency (16 Hz), all the binders have similar stiffness. It was somewhat expected to have the PCR-Wet 8% modified binder to be softer than the base binder at low temperatures, but this was not observed. However, this distinction was more pronounced at high temperatures where the effects of PCR particles increased the overall stiffness of the modified binder relative to base binder.

In general, there is an obvious impact of aging in both PCR-Wet 8% and base binder in terms of |G*|, as expected. This is also shown in Fig. 4. Typically, the binder stiffness increases when subjected to short-term aging. This change is more pronounced at low frequencies and high temperatures. Nonetheless, the aging effects are not the same for PCR-Wet 8% and base binder. Figure 5 shows the effects of aging on the |G*| values in terms of short-term aged and unaged ratio for the tested asphalt binders. The PCR-Wet 8% experienced lower aging ration compared to the base binder. This is perhaps due to the presence of PCR particles that remain softer

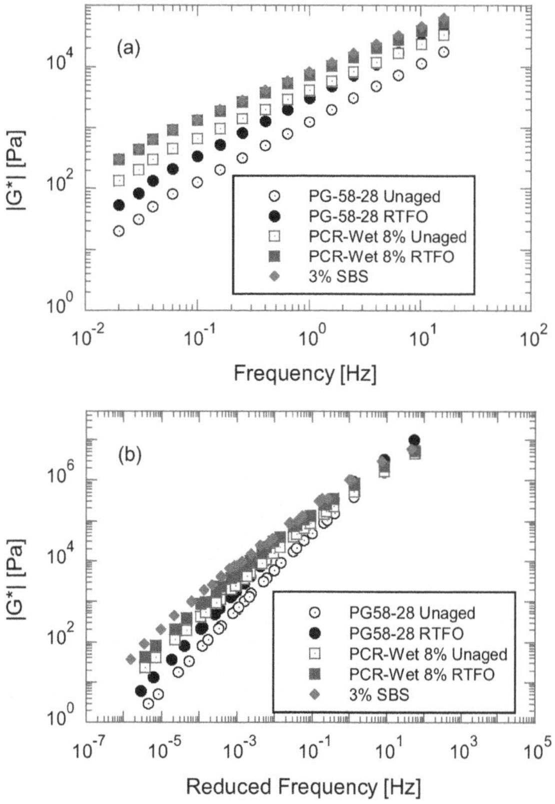

Fig. 4 **a** Dynamic shear modulus values at 58 °C and varying frequencies, **b** Dynamic shear modulus master curve for all conisdered binders

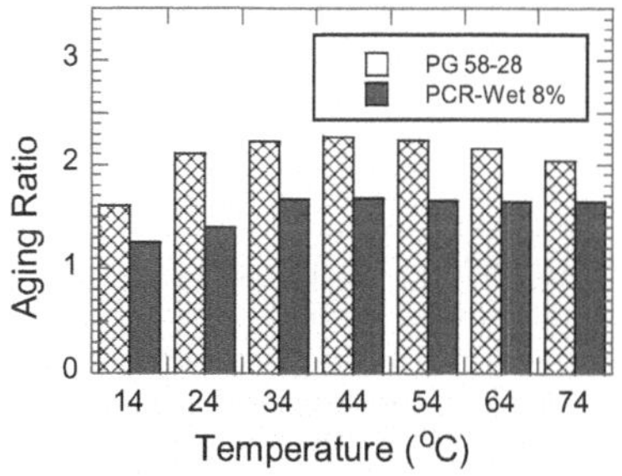

Fig. 5 The aging ratio of PCR-Wet 8% and base binder (PG 58-28)

than the asphalt binder when subjected to higher temperatures thus, mitigating the stiffening effects. This is an important fact knowing that the aging phenomenon in asphalt pavements is accepted as a key factor that can induce early deterioration on the road pavement structure.

Rheological testing of PCR modified asphalt binders provides a fundamental characterization of the material. But in order to capture the complete effect of PCR modified asphalt blends, mixture testing is needed to confirm the validity of results showed in this paper.

5 Conclusion and Recommendations

In this study, the effects of binder modification with PCR at different concentration levels and blending durations were investigated. Initially, the effects of PCR concentration on binder viscosity was explored. Then, the rutting performance of the selected concentration was studied using the MSCR procedure. It was observed that the blending duration of PCR-Wet 8% didn't affect the rutting resistance. However, in general, PCR-Wet 8% outperformed the base binder (PG 58-28) showing significantly better rutting performance while 3% SBS showed E-Grade binder. In addition, |G*| test results showed compelling differences between PCR-Wet 8% and base binder. This difference was less when compared to 3% SBS. Finally, the short-term aging ratio of the PCR modified binder was lower compared to the base binder. Further testing is needed on asphalt mixtures to confirm the validity of the results in this paper.

References

Bukowski J, Youtcheff J, Harman T (2011) The multiple stress creep recovery (MSCR) procedure

Diefenderfer SD, Bowers BF (2019) Initial approach to performance (balanced) mix design: the virginia experience. Transp Res Rec J Transp Res Board 2673(2):335–345

Gibson N, Qi X, Shenoy A, Al-Khateeb G, Kutay ME, Andriescu A, Stuart K, Youtcheff J, Harman T (2012) Performance testing for superpave and structural validation

Hansen KR, Copeland A (2015) Asphalt pavement industry survey on recycled materials and warm-mix asphalt usage: 2014

Heitzman MA (1992) State of the practice-design and construction of asphalt paving materials with crumb rubber modifier

Jamrah A (2018) Novel parametric approach to characterize crumb rubber modified asphalt binder performance based on parallel plate dynamic shear rheometry

Jamrah A, Kutay ME (2015) A method for quantifying the extent of crumb rubber pre-treatment. In: 6th conference of the european asphalt technology association (EATA), Stockholm: Sweden, 15–17 June 2015

Kocak S (2016) Interaction between recycled tire rubber, polymers and high amount of reclaimed asphalt pavements

Kocak S, Kutay ME (2016) Combined effect of SBS and devulcanized rubber (DVR) modification on performance grade and fatigue cracking resistance of asphalt binders. In: Chabot A, Buttlar W, Dave E, Petit C, Tebaldi G (eds) 8th RILEM international conference on mechanisms of cracking and debonding in pavements. Springer, Dordrecht, pp 269–274

Kurgan G, Dongre R (2015) International trends in low-carbon/low-energy pavement construction

Santagata E, Lanotte M, Baglieri O, Dalmazzo D, Zanetti MC (2016) Analysis of bitumen–crumb rubber affinity for the formulation of rubberized dry mixtures. Mater Struct/Materiaux et Constructions 49(5):1947–1954

Seitllari A, Kutay ME (2019) Development of 3-point bending beam fatigue test system and implementation of viscoelastic continuum damage (VECD) theory. J Assoc Asphalt Paving Technol 88:783–810

Seitllari A, Lanotte MA, Kutay ME (2019) Calibration of the MEPDG rutting model: issues and consequences on rutting prediction. In: Transportation research board 98th annual meeting, Washington DC: no 19–02795, 6 p.

Seitllari A, Lanotte MA, Kutay ME (2019) Comparison of uniaxial tension-compression fatigue test results with SCB test performance indicators developed for performance-based mix design procedure. In: Bituminous mixtures and pavements VII: proceedings of the 7th international conference 'bituminous mixtures and pavements' (7ICONFBMP)

Stroup-Gardiner M, Chadbourn B, Newcomb DE (1996) Babbitt, Minnesota: case study of pretreated crumb rubber modified asphalt concrete. Transp Res Rec J Transp Res Board 1530(1):34–42

Way GB (2000) Flagstaff I-40 asphalt rubber overlay project: Nine years of success. Transp Res Rec 1723:45–52

Seitllari A, Boz I, Habbouche J, Diefenderfer SD Assessment of cracking performance indices of asphalt mixtures at intermediate temperatures. Int. J. Pavement Eng:1–10

Viscoelastic Response of Bitumen Emulsion Mastic with Various Active Fillers

Ahmed Al-Mohammedawi and Konrad Mollenhauer

Abstract The performance of the Cold Bitumen Emulsion (CBE) mixture is strongly controlled by the bituminous binder (bitumen emulsion) and the hydraulic binder (active filler). CBE mixture behavior, based on the mastic (both earlier mentioned binders) properties, may vary from a stiff behavior (hydraulic binder) to an asphalt like behavior with viscoelastic deformation failure (bitumen emulsion). This research aims to assess the impact of different fillers on the viscoelastic characteristics of CBE materials and also to evaluate the suitability of the sigmoid model to describe the rheological properties of CBE mastic. To address this aim, bitumen emulsion was mixed separately with eight fillers/blended fillers. The rheological properties of the prepared mastics were evaluated by conducting oscillating temperature-frequency sweep tests, utilizing the Dynamic Shear Rheometer (DSR). The raw data were shifted to the reference temperature and modelled using the sigmoid model to get a smooth master curve. Results showed that most of the active fillers exhibited stiff behaviour. Among the eight fillers, two of them own balanced rheological characteristics while only two fillers reduced the neat CBE binder complex modules. It was found that the sigmoid model was able to satisfactorily describe the rheological properties of CBE binder and mastic over a range of temperatures and frequencies.

Keywords Cold bitumen emulsion mastic · Viscoelastic response · Active fillers · Sigmoid model · Time-temperature superposition principle

1 Introduction

Cold Bitumen Emulsion (CBE) mixture has several advantages over hot mix asphalt (HMA) in terms of conservation of materials, reducing energy consumption, preservation of the environment and reduction in cost. CBE mixture consists of an aggregate skeleton bonded by a thin-film material matrix composed of Bitumen Emulsion (BE) and mineral fillers (Turk et al. 2016). The mastic constitutes the dominant part in the

A. Al-Mohammedawi (✉) · K. Mollenhauer
Engineering and Maintenance of Road Infrastructure/Transportation Institute, University of Kassel, Mönchebergstraße 7, 34125 Kassel, Germany
e-mail: a.al-mohammedawi@uni-kassel.de

C. Raab (ed.), *Proceedings of the 9th International Conference on Maintenance and Rehabilitation of Pavements—Mairepav9*, Lecture Notes in Civil Engineering 76,
https://doi.org/10.1007/978-3-030-48679-2_58

CBE mixture; consequently, its behavior is highly governed by its mastic. The rheological response of the mastic is significantly controlled by the type of mineral fillers (Lesueur et al. 2016). In this sense, mineral fillers can be chemically categorized into active and inactive fillers based on their reactivity. Inactive fillers (i.e. basalt, lime-stone) are considered as inert fillers due to their chemical composition and normally used as a stiffness regulator by controlling the mastic stiffness throughout adding solid particles to the bitumen colloid system without or little chemical reaction, hence the viscoelastic properties are maintained (Buczyński and Iwański 2017). On the other hand, active fillers can be described as the fillers that can react with water or with bitumen itself (changing bitumen structure) due to the alkali species in its composition (Kakade et al. 2018). The most important chemical elements in the active fillers that can control the mastic properties are CaO, Al_2O_3, and SiO_2. Calcium oxide gives fillers pozzolanic properties and, generally used to accelerate the setting time by forming $Ca(OH)_2$ as a hydration product (Temuujin et al. 2009). On the other hand, the presence of Al_2O_3 in fillers can shorten the setting time of the system. However, it also has a detrimental effect on strength. Normally, the presence of SiO_2 tends to produce microstructures with low porosity and hence enhances the strength but it inhibits the setting time (Dutta et al. 2010). The presence of CaO and SiO_2 leads to the coexistence of the C-S-H phase with the geo-polymeric gel which has been shown to improve the mechanical properties of the final product (Steveson and Sagoe-Crentsil 2005). Intriguingly, the existence of Al_2O_3 with CaO and SiO_2 results in the C-A-S-H phase which provides the system with extra strength with short setting time (Criado et al. 2016). Besides, some chemical elements have been found to have a negative effect on the cementitious binder such as Fe_2O_3, as this oxide is found to hinder compressive strength (Choi and Lee 2012). Therefore, the filler should be carefully designed to achieve the desired rheological and mechanical properties, taking into account the mixture type (HMA or CBE).

In terms of strength development, the final strength of HMA mastic is achieved as the mastic cools down. For CBE mastic, the mechanism is completely different. Since the bitumen emulsion contains water (the bitumen is liquefied by dispersing in water), the strength gaining of the mastic material develops as the emulsion sets and returns to a bitumen phase and water is absorbed by the mineral, as well as evaporates from the system. In this way, the BE mastic may take several weeks to effectively to gain the required adhesion with the aggregates and thereafter to reach its full strength (Thanaya 2003). To this purpose, active fillers (normally cement) are added to CBE mixtures which in turn with water, producc a hydraulic binder that can chemically form rigid structure within the CBE mixture. The type and nature of the hydraulic binder affect the rheological properties of CBE mastic in the fully cured state and in the fresh state, achieving suitable volumetric properties (Abdo et al. 2013). Thus, the behavior and structure of BE mastic are different from that of HMA. Incorporating active fillers in the CBE mixture may result in stiffer mastic which could show poor fatigue and fracture resistance under high load frequencies. In this scenario, CBE mastic may behave like stiff materials to a flexible material (Nikolaides 2015).

This study, therefore, evaluates the feasibility of using different by-product fillers/blended fillers in improving the linear viscoelastic properties of CBE mastics, through measuring the complex modulus (G^*). The suitability of the sigmoid model to represent the rheological characteristics of the CBE binder and mastics is also assessed. Besides, statistical correlations between the measured and modelled data are evaluated.

2 Materials and Experiments

2.1 Basic Materials

An unmodified cationic slow-setting bituminous emulsion designated as C60B10 according to EN 13808 was used to produce the Bitumen Emulsion (BE) specimens and to prepare all the mastics. In order to avoid the effect of the particle to particle contact, a relatively low mass ratio of 0.21 (filler to residual bitumen ratio lower than the practical ratio) was chosen. Hereafter particle structuralizing is not expected. In this work, the mastics were designed by the mass ratio of the filler and binders in order to simulate the mix design of the whole asphalt mixtures.

The BE was mixed separately with CE (Portland cement), limestone (LS), Ladle slag (LD), silica fume (SF), fly ash (FA), ettringite (ET), geopolymer with activator (GE) and geopolymer without activator (GO). ET blended filler was prepared by using 70% LD and 30% of gypsum as recommended by Nguyen et al. (Nguyen et al. 2019), which is the optimum combination to produce ettringite binder. GO blended filler was prepared by mixing 55% LD, 35% FA, and 10% SF. GO was mixed with a 3.5% activator by its weight to prepare GE. The activator was a combination of 50% sodium hydroxide (NaOH) in the concentration of 10 M and 50% sodium silicate as an alkaline activator (da Silva Rocha et al. 2018). All used fillers are smaller than 63 μm. The properties and chemical compositions of the fillers are given in Table 1.

2.2 Mastic Preparation

In this study, a filler of 43 g was added and mixed gradually with 340 g of BE at 25 °C, using a mechanical mixer. The mixing regime was 5 min slow mixing (150 rpm) to initially agitate the filler into the BE to help disperse the particles and then 500 rpm was applied continuously for 45 min. This mixing procedure was also performed to the reference binder to avoid possible effects on results. During mixing, it was observed that there was no particle sedimentation by visual inspection during pouring the sample. However, at the end of mixing for certain fillers, breaking was noticed. In order to avoid water evaporation bubbles, the mix was poured into a shallow glass dish. After that, the specimens were cured for 3 days under 40 °C

Table 1 Properties and chemical composition (wt%) of used fillers

Oxide	SiO_2	Al_2O_3	Fe_2O_3	CaO	SO_3	Other	Density (g/cm^3)	Delta R&B	Rigden voids (%)
CE	20.86	4.97	3.86	64.74	3.3	2,26	3.143	2.5	40.5
LS	0.53	0.16	0.082	98.1	0.08	0.36	2.726	1	27.1
SF	97.57	0.06	0.06	0.71	0.12	1.46	2,271	3	70.8
FA	48.42	26.85	15.71	0.91	1.67	6.43	2.317	1.25	25.5
LD	8.33	28.91	1.39	52.41	1.83	7.13	2.577	1	39
Gypsum	1.54	0.21	0.17	41.65	55.8	0.58	2.726	–	–
GE	–	–	–	–	–	–	2.455	2	38.7
GO	–	–	–	–	–	–	2.455	2	38.7
ET	–	–	–	–	–	–	2.622	2	35.9

temperature and 65% relative humidity. These curing conditions were chosen to ensure all water evaporated and to accelerate the hydration process (Graziani et al. 2016). Small specimens (approximately 8 mm diameter) were cut for DSR testing. DSR plates were heated up to 64 °C and then the specimens were carefully placed on the bottom plate. The upper plate was lowered to achieve 2 mm thickness. Afterward, the unnecessary edges were trimmed.

2.3 *Testing Program*

Strain Sweep tests were initially performed using a dynamic shear rheometer to determine the Linear Viscoelastic (LVE) limit and define a suitable range of deformation level. Strain sweep tests were conducted at −20 °C by using an 8 mm parallel plate geometry and a 2 mm gap, conducting a constant frequency of 10 Hz. From the results, a strain of 0.05% was found appropriate as per the LVE limit. Thereafter, the rheological properties of the CBE mastics were characterized with DSR by frequency sweep tests at a constant strain amplitude of 0.05% over a range of frequencies between 0.1–10 Hz and at temperatures from −20–40 °C.

The resulted raw data were interpreted in master curves. The master curves were constructed using the sigmoidal model Eq. (1) along with shift factor by means of horizontal time-temperature superposition principle (TTSP), following Arrhenius law Eqs. (2) and (3).

$$G* = G*_0 + \frac{G*_\infty}{1 + e^{-\left(\frac{\ln fr - f0}{z}\right)}}. \quad (1)$$

$$a_T = e^{\frac{-\Delta H}{R}\left(\frac{1}{T} - \frac{1}{T_{ref}}\right)}. \quad (2)$$

$$fr = a_T * fi. \quad (3)$$

Where G^*, G^*_0, G^*_∞, fr, $f0$, z, α_T, ΔH, R, T, T_{ref} and fi are, respectively, the modelled complex modulus, the minimum complex modulus, the maximum complex modulus, the reduced frequency, the minimum frequency, the fitting parameter, the shift factor, the activation enthalpy for flow, the universal gas constant, the testing temperature, the reference temperature (20 °C) and testing frequency.

3 Results and Discussion

3.1 *Master Curves*

The dynamic master curves were constructed using the shift factor equation based on the TTSP and modelled by the sigmoidal model are illustrated in Fig. 1a and b. From Fig. 1a, LS and FA fillers reduced the stiffness of BE at the high-frequency range and slightly increased stiffness at very low frequencies compared with the neat binder BE. Besides, Fig. 1b showed that they exhibited a lower phase angle (higher elastic response) at low frequencies compared to the BE. This could be due to the flow of bitumen through the filler particles at high temperatures and fragile behaviour at low temperatures. Furthermore, it is worth to note that GE and LD fillers have a similar effect on the BE, both have higher complex modulus and lower phase angle at the whole range of frequencies compared to the BE. The reason behind that is these fillers have the main compositions that form rigid phases (C-S-H and C-A-S-H) within the mastic. However, further investigations about the coexistence of these phases are required (Češnovar et al. 2019).

Using geopolymer without activator (GO) shows higher stiffness and lower phase angle than those of LD and GE at the whole range of frequencies, thus confer higher elastic property on mastic (higher deformation resistance). As expected, CE mastic exhibited higher stiffness and lower phase angle than those of GE, LD and GO, especially at low temperatures. In addition, despite having different chemical compositions, ET and SF have a relatively similar effect on BE. They showed the highest stiffening potential and the lowest phase angle among all tested mastics over the whole frequency range. This observation, however, does not involve CE at high frequencies since it has the highest complex modulus in that zone. This inversion showed that CE mastic being slightly stiffer than ET and SF. In this context, higher complex modulus and lower phase angle are not usually preferable at low temperatures as it makes the CBE mixture very stiff (no longer viscoelastic materials at higher frequencies), thus; CBE mixture becomes very sensitive to cracks development.

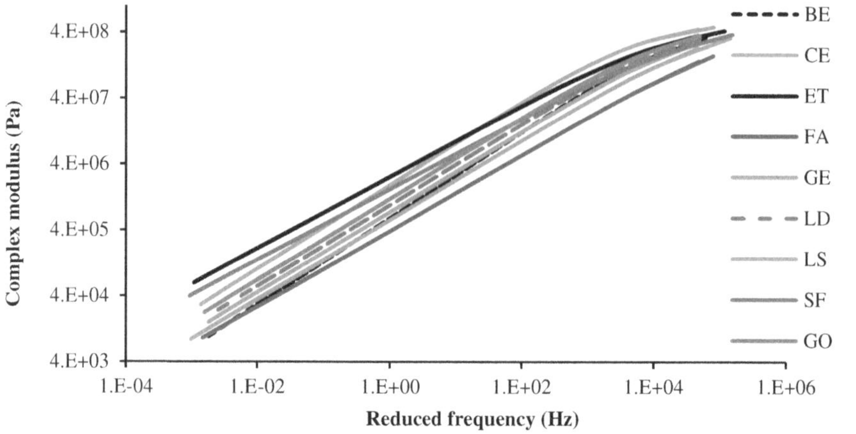

(a)

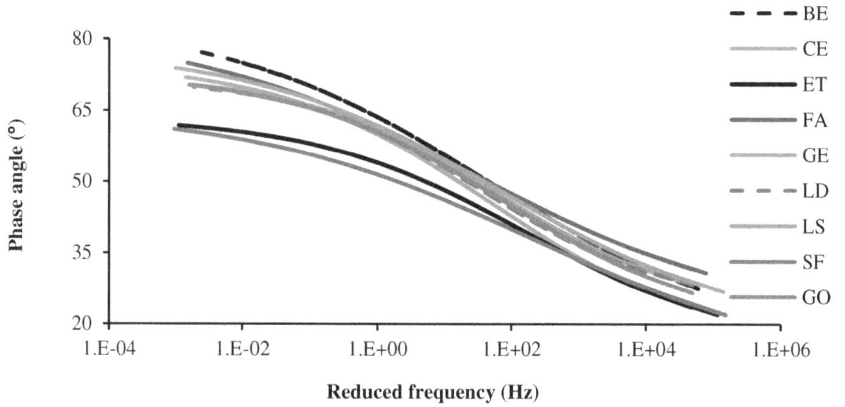

(b)

Fig. 1 Modelled master curves (a) complex modulus (b) phase angle

3.2 *Stiffing index (Si) and Difference in Phase Angle (DPA) Analysis*

Table 2 shows the effect of temperature and the volume filling on the *Si* and *DPA*. *Si* and *DPA* were computed through evaluating G^* ratios and the difference in δ to the BE respectively at a single loading frequency of 1.59 Hz and two selected temperatures. Volume filling is represented by the volume ratio of filler to residual bitumen (f/b). In general, *Si* and *DPA* values increase as temperature increases. Conspicuously, the CE, GO, ET and SF mastics have the highest values of *Si* and *DPA* at both temperatures among all mastics while FA and LS have the lowest *Si* and *DPA*. On the other hand, at low temperature, the GE and LD have *Si* values about 1 and *DPA* lower than 1°. While at high temperature, they have considerably higher values. In terms of volume

Table 2 Stiffing index (*Si*) and Difference in Phase Angle (*DPA*)

Mastic	Volume ratio (f/b)	*Si* (G^*_i/G^*_{BE})	*DPA* $(\delta_{BE}-\delta_i)$	Temperature (°C)
CE	0.068	2.36	5.01	−20
		3.40	5.21	40
ET	0.081	2.11	6.70	−20
		6.24	13.54	40
FA	0.092	0.50	−2.22	−20
		0.81	2.91	40
GE	0.087	0.88	0.10	−20
		1.49	5.24	40
LD	0.083	1.04	0.93	−20
		2.27	5.34	40
LS	0.078	0.86	1.40	−20
		1.00	2.40	40
SF	0.094	1.77	6.43	−20
		3.97	15.71	40
GO	0.087	1.45	2.19	−20
		2.32	5.81	40

filling effect, it can be observed that despite having a relatively high-volume ratio, the FA mastic has the lowest *Si* and *DPA* values. This perhaps due to the high Fe_2O_3 content that cause the poor FA particles-bitumen affinity. Alternatively, the fillers with moderate volume ratio (LD, GE and GO) increased the *Si* and *DPA*, and all of them have relatively similar volume ratios. Although SF filler has the highest volume ratio, its *Si* is not the highest. Besides, CE has the greatest *Si* value when compared with the aforementioned fillers, considering that it has the lowest volume ratio, but it has a moderate value of *DPA*. The addition of ET, which has a moderate volume ratio among all fillers, offered the highest *Si* and relatively high *DPA* value. It can be concluded here that the discrepancies in *Si* and *DPA* of different fillers not only related to the particle's density in the mastic matrix but also it is associated with the nature of the resulted hydraulic binder. Commonly, higher *Si* and *DPA* (higher elastic component) are appreciated at high temperature in road applications due to reduced susceptibility to permanent deformation on the pavement whereas lower *Si* and *DPA* are generally valued at low temperatures (less brittleness).

3.3 Statistical Analysis

Graphical Comparison

Figure 2a and b present comparisons between the measured and modelled G^* and δ data of all tested specimens. The measured and modelled values were equated and

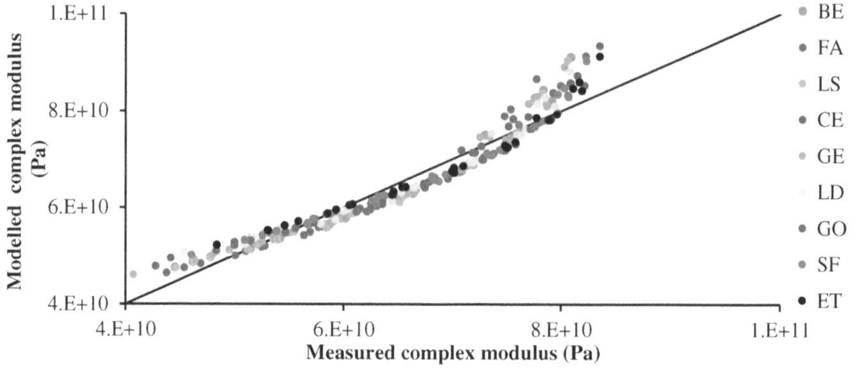

(a)

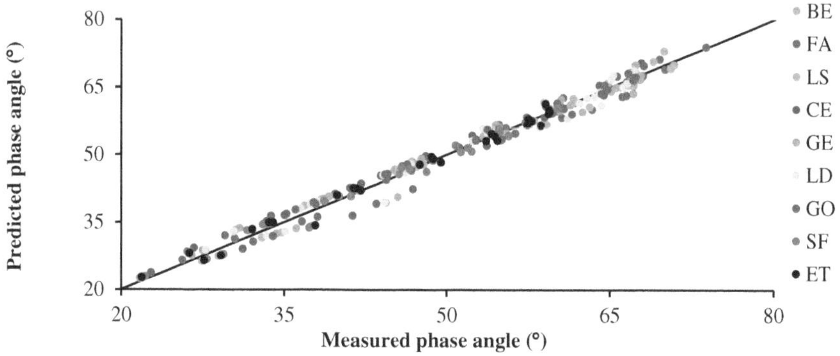

(b)

Fig. 2 Graphical comparison between measured and modelled (a) complex modulus data and (b) phase angles data

graphically represented. From Fig. 2a, it can be observed that the sigmoid model can describe G^* values for the BE and CBE mastics. However, the modelled G^* of all specimens slightly diverge from the equality line at the lowest and highest G^* values particularly at lower temperatures (higher G^* values). On the other hand, Fig. 2b shows a correlation between measured and modelled phase angle values. For all specimens, a good correlation between measured and modelled data was noticed. Nevertheless, the modelled data of the BE exhibits a slightly better correlation (less scattering) from the equality line. In addition, the R^2 between the measured and modelled G^* and δ data are illustrated in Table 3a and 3b respectively. In general, all samples have an excellent correlation between the measured and modelled data. However, δ data have higher correlations than the G^* data.

The Comparison of Sigmoid Model Parameters

The parameters of the sigmoid model of the mastics used in this study are shown in Table 3a and b. The parameters were fitted by an excel solver tool until good curves are achieved. The complex modulus and phase angle master curves were plotted

Table 3 The model parameters for (a) complex modulus and (b) phase angle

Sample	(a)					(b)				
	G^*_0	G^*_∞	f_0	z	R^2	δ_0	δ_∞	f_0	z	R^2
BE	0	5.9E+08	436.62	23.65	0.91	14.29	71.42	4.03	−5.06	0.99
CE	0	5.9E+08	487.87	26.52	0.91	15.33	59.3	4.23	−3.9	0.98
ET	0	5.9E+08	560.57	30.54	0.93	14.99	48.64	5.12	−3.65	0.99
FA	0	5.9E+08	455.88	24.65	0.92	19.45	63.76	3.48	−5.27	0.98
GE	0	5.9E+08	479.04	25.98	0.92	21.67	51.46	4.41	−3.76	0.98
LD	0	5.9E+08	544.39	29.59	0.93	24.23	46.96	3.97	−3.34	0.98
LS	0	5.9E+08	441.59	23.92	0.92	18.73	58.8	4.13	−4.1	0.98
SF	0	5.9E+08	535.55	29.14	0.93	11.24	53.51	5.2	−4.67	0.99
GO	0	5.9E+08	487.79	26.5	0.92	18.2	55.15	4.58	−3.89	0.98

using the resulted parameters. Table 3a shows that the BE has the lowest f_0 and z values. While, G^*_0 and G^*_∞ are the same for all specimens. For G^*_∞, this explains the lower effect of filler to the stiffening index *Si*, as shown in Table 2. At low temperatures, the high bitumen stiffness predominates the effect of filler particles. From Table 3b, it can be observed that all parameters are slightly different for each mastic, which explains the slightly scattered results of the modelled phase angle and that is confirmed by the observation in Fig. 2b.

4 Conclusions

In this study, the effect of different active fillers on the viscoelastic properties of CBE mastics was investigated. The rheological behavior of the tested mastics is different from that of pure BE as identified by the master curves, the *Si* and *DPA*. Each active filler offered unique rheological behavior and stiffening potential in terms of changing the LVE behavior from that of viscous behavior to that of viscoelastic behavior based on filler type. In general, CBE mastics containing ladle slag fillers exhibited higher stiffness at high temperatures. That might be due to a stable mastic matrix strengthened by the stiffness of ladle slag particles, as well as the expected chemical action between alkaline components in ladle slag fillers and water (C-S-H and C-A-S-H phases). On the contrary, LS and FA fillers decreased the stiffness of BE because of their inert nature. SF and ET corresponding mastics presented better viscoelastic properties than CE mastic. Nevertheless, at low temperatures, their higher stiffening potential (CE, ET and SF) not always preferable as they make CBE mixture very stiff (high crack susceptibility) leading to an alteration in the viscoelastic properties. From the flexibility perspective, it can be concluded that LD and GE have a balanced rheological behaviour as they have little or no effect on the BE at low temperatures, but they improve the rutting resistance at high temperatures.

In regard to data modelling, the sigmoid model and the TTSP were successfully applied and statistically verified for describing the rheological performance of CBE mastics with excellent approximation.

References

Abdo J, Serfass JP, Pellevoisin P (2013) Pavement cold in-place recycling with hydraulic binders: the state of the art in France. Road Mater Pavement Des 14:638–665

Buczyński P, Iwański M (2017) Inactive mineral filler as a stiffness modulus regulator in foamed bitumen-modified recycled base layers. In: IOP conference series: materials science and engineering

Češnovar M, Traven K, Horvat B, Ducman V (2019) The potential of ladle slag and electric arc furnace slag use in synthesizing alkali activated materials; the influence of curing on mechanical properties. Materials 12:1173

Choi SC, Lee WK (2012) Effect of Fe2O3 on the physical property of geopolymer paste. In: Advanced materials research

Criado M, Aperador W, Sobrados I (2016) Microstructural and mechanical properties of alkali activated Colombian raw materials. Materials 9:158

Dutta D, Thokchom S, Ghosh P, Ghosh S (2010) Effect of silica fume additions on porosity of fly ash geopolymers. J Eng Appl Sci 5:74–79

Graziani A, Godenzoni C, Cardone F, Bocci M (2016) Effect of curing on the physical and mechanical properties of cold-recycled bituminous mixtures. Mater Des 95:358–369

Kakade VB, Reddy MA, Reddy KS (2018) Rutting performance of hydrated lime modified bituminous mixes. Constr Build Mater 189:1–10

Lesueur D, Teixeira A, Lázaro MM, Andaluz D, Ruiz A (2016) A simple test method in order to assess the effect of mineral fillers on bitumen ageing. Constr Build Mater 117:182–189

Nguyen H, Adesanya E, Ohenoja K, Kriskova L, Pontikes Y, Kinnunen P, Illikainen M (2019) Byproduct-based ettringite binder – a synergy between ladle slag and gypsum. Constr Build Mater 197:143–151

Nikolaides AF (2015) Bituminous mixtures & pavements VI. In: Bituminous mixtures and pavements. VI Proceedings of the 6th international conference on bituminous mixtures and pavements, ICONFBMP 2015

da Silva Rocha T, Dias DP, França FCC, de Salles Guerra RR, de Oliveira LRDC (2018) Metakaolin-based geopolymer mortars with different alkaline activators (Na+ and K+). Constr Build Mater 178:453–461

Steveson M, Sagoe-Crentsil K (2005) Relationships between composition, structure and strength of inorganic polymers. J Mater Sci 40:2023–2036

Temuujin J, van Riessen A, Williams R (2009) Influence of calcium compounds on the mechanical properties of fly ash geopolymer pastes. J Hazardous Mater 167:82–88

Thanaya IN (2003) Improving the performance of cold bituminous emulsion mixtures (CBEMs): incorporating waste

Turk J, Mauko Pranjić A, Mladenovič A, Cotič Z, Jurjavčič P (2016) Environmental comparison of two alternative road pavement rehabilitation techniques: cold-in-place-recycling versus traditional reconstruction. J Cleaner Prod 121:45–55

Numerical Studies on Coir Geotextile Reinforced Flexible Pavement

V. Anusudha, V. Sunitha, Chithu Babu, Chetan R. Bhole, and Samson Mathew

Abstract Geotextiles are permeable fabrics that can separate, filter, strengthen and drain when used in association with the soil. Natural geotextiles are more advantageous than artificial geotextiles because of its cost and environmental concern. Coir geotextiles have a life of 3 to 5 years depending on the type of coir geotextile. Many roads have been built and are being built in India using coir geotextiles as reinforcement. The main aim of this study is to study the strengthening aspects of woven coir geotextiles in weak subgrades. For this study, Sobanapuram to the Uppiliyapuram road in the district of Tiruchirappalli was selected. Two different varieties of coir geotextiles H2M5 and H2M6 were used in the study. PLAXIS 3D finite element analysis software was used for analyzing the sections with and without coir geotextile with varying GSB layer thickness under static loading. It was found that the section with H2M5 has lower displacement and strain on the subgrade than the other sections.

Keywords Coir geotextile · Plate load test · Direct shear · Plaxis 3D · Finite element analysis

1 Introduction

Low volume roads play an important role in the Indian economy (Sharma et al. 2017). Traffic demand has been increasing nowadays and so as to satisfy the requirements, the road network at times has to be aligned through places where the subgrade soil is not suitable for road construction. Construction and maintenance of roads along the unsuitable soils have been problematic due to their inherent potential for volume change in the presence of water, which affects the performance of roads. The subgrade has to be properly treated at the construction stage itself. Otherwise the total transportation cost will increase substantially due to deteriorated pavement performance. One of the ways to treat subgrade is to reinforce it with Geotextiles. Geotextiles have become increasingly popular for use as reinforcement in earth structures in recent

V. Anusudha · V. Sunitha (✉) · C. Babu · C. R. Bhole · S. Mathew
Department of Civil Engineering, National Institute of Technology, Tiruchirappalli, Tamil Nadu, India
e-mail: sunitha@nitt.edu; sunitha.nitt@gmail.com

C. Raab (ed.), *Proceedings of the 9th International Conference on Maintenance and Rehabilitation of Pavements—Mairepav9*, Lecture Notes in Civil Engineering 76,
https://doi.org/10.1007/978-3-030-48679-2_59

years. They serve the functions of drainage, filtration and separation by preventing the soil from migrating while maintaining the water flow (Subaida et al. 2009). Geotextiles can be natural or synthetic. The synthetic fibres have longer life but do not generally undergo biological degradation, thus causing environmental pollution. Natural geotextile, particularly coir geotextile, is being recognized as an ideal material that is capable of offering an environment-friendly and ecologically sustainable solution to many of the environment related issues (Jayachandran et al. 2019).

Natural geotextiles can be produced from coir, jute, sisal, etc. Geotextile made of coir are ideally suited for low cost applications because coir is available in abundance in India at very low price compared to other synthetic geotexiles. Coir geotextiles are popular for erosion control, slope stabilization, reinforcement and bioengineering and has a life of 3 to 5 years depending on the type of coir geotextile. Since coir is a naturally available material, its use in pavement construction can economize the pavement construction cost and pavement life cycle cost. This study focuses on the interface behaviour of the coir geotextile between subgrade and granular subbase layer. This will be helpful in evaluating the strength properties of pavement section with and without coir geotextile and ultimately in design of an economical section for a particular geotextile. Many researches have been done till now to find out the behaviour of geotextile reinforced pavements. Faheem and Hassan (2014) has presented an axisymmetric finite element (FE) model to analyze the behavior of unreinforced and geogrid reinforced bituminous pavement under static and dynamic loads. The geogrid was kept at the base course/bitumen layer interface. For unreinforced and geogrid reinforced flexible pavement, efficient stress and vertical surface deflection have been determined. The results indicated that due to the reinforcing geogrid layer, a moderate effect on the pavement behavior was observed during static loading. The effect of the dynamic loading frequency on the pavement settlement was particularly significant for high loading amplitudes.

Finite element analysis software, ABAQUS was used effectively in analyzing stress—strain behaviour under the application of static wheel load in the modelled flexible pavement structure by Loui et al. (2014). The parametric study for unreinforced and reinforced sections showed that coir geotextiles at the interface of the asphalt layer and base course layer significantly reduced lateral strain. The inclusion of geotextile reinforcement also contributed to the reduction of the horizontal tensile load developed in the pavement. Subaida et al. (2008) carried out monotonic and repeated plate load tests on the test section consisting of subgrade and base course layers of varying thicknesses. The study showed that the plastic surface deformation under repeated loading was greatly reduced by the inclusion of coir geotextiles in the base course regardless of the thickness of the base course. A study on strength of granular sub-base (GSB) underlied by a subgrade layer in terms of CBR and bearing capacity was conducted by Naagesh et al. (2013). CBR and shear parameters were determined for different combinations of granular subbase layer and subgrade thickness. The combination of GSB and sub-grade simulated for the study shows that as GSB/total thickness ratio increases, CBR increases from 2.5 to 55% for soil-GSB specimen. The angle of internal friction increases from 12 to 40° and the corresponding shear strength from 0.91 to 1.75 kg/cm^2. Lal et al. (2017) studied the potential of geocell and planar form coir geotextiles in reinforcing shallow foundations by conducting laboratory plate load tests and reported that for a settlement of 15% of foundation width, the maximum improvement in the bearing capacity was found to be 7.92 in the case of geocell and 5.83 in the case of planar form. Most literature was based on laboratory plate load tests. It is very difficult to obtain the same field condition in a laboratory and thus results differed in different literature. Using finite element software makes the analysis precise as it involves less human error. Some literature assumes the pavement as a two-layer system, which in the actual field is not the case. The objective of this study was to carry out the total strain and total displacement analysis under static load by changing the sub base thickness.

2 Methodology

As a part of this study, Sobanapuram to Gandhipuram road (3.5 km) of Uppiliyapuram block in Tiruchirappalli district has been chosen. Sudgrade soil samples from the study area were collected to conduct experimental soil studies. Figure 1 shows the methodology adopted. Table 1 shows the properties of the soil sample. Figure 2 shows the longitudinal and plan of the section adopted in the field. Two types of coir geotextiles, i.e. H2M5 and H2M6 were used as shown in Fig. 1. Coir geotextile of minimum 6 m width is advisable to use as reinforcement for low volume roads. The total 3.5 km length is divided into three sections i.e. control section, coir geotextile Sects. 1 and 2. Control section have 175 mm thick GSB layer, coir Sect. 1 have only 125 mm thick GSB and coir Sect. 2 have no GSB layer. Coir geotextile is placed at the subbase and subgrade interface for Sects. 1 and 2. In the study, two varieties

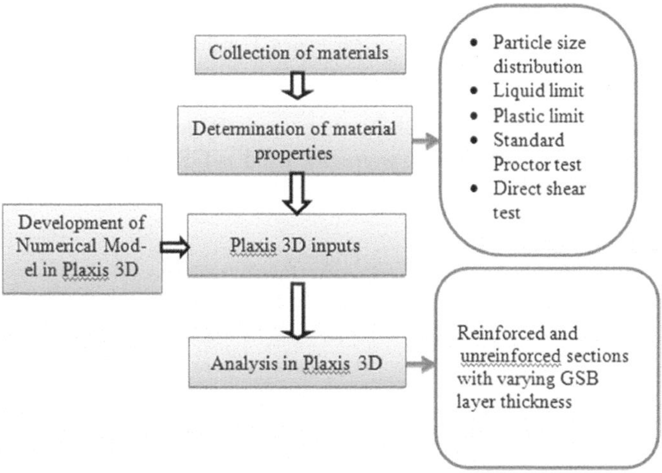

Fig. 1 Methodology

Table 1 Properties of soil sample

S. No.	Property	Value
1	Liquid limit (%)	32
2	Plasticity index (%)	11
3	Optimum moisture content (%)	12
4	Maximum dry density (g/cc)	1.47
5	Specific gravity	2.7
6	CBR (%)	3.1

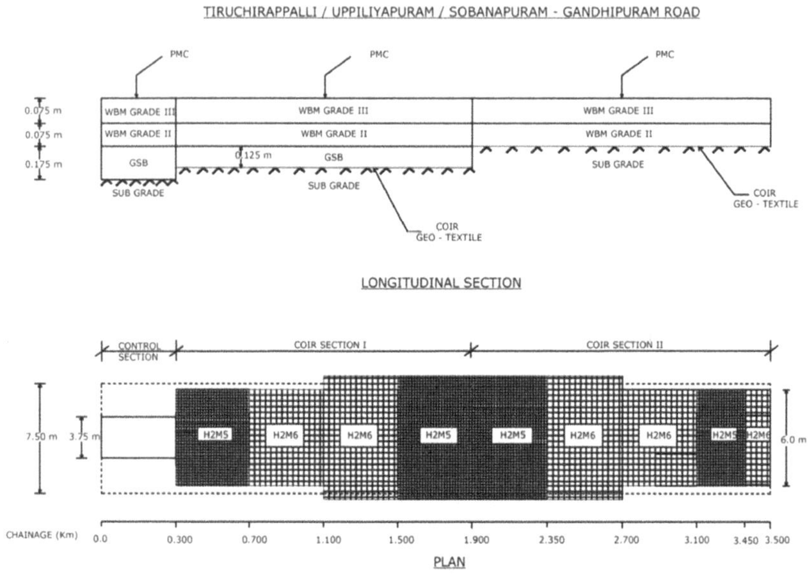

Fig. 2 Longitudinal section and plan of Sobanapuram to Gandhipuram road

of coir geotextiles, H2M5 and H2M6, were used. Table 2 shows the engineering characteristics of the crushed stone aggregate used as GSB material.

Table 2 Properties of GSB material

S. No.	Property	Value
1	Specific gravity	2.73
2	Aggregate impact value (%)	23.8
3	Optimum moisture content (%)	6
4	Maximum dry density (g/cc)	2.21
5	California Bearing Ratio (CBR) (%)	29

3 Analysis of Pavement with Varying Thickness of GSB Under Static Load

The PLAXIS 3D is a special purpose finite element computer program used to perform deformation and stability analysis for various types of geotechnical applications. A typical pavement layer consists of subgrade, sub base, base course and bituminous layer as well as reinforcement layers. The pavement is modelled as a multilayer structure subjected to static loading. The model was developed using PLAXIS 3D finite element software to analyse the unreinforced and geotextile reinforced flexible pavement structure. The load is applied as uniform pressure acting on a square area of 0.15 m and the applied pressure was 550 kPa (to represent an equivalent standard axle wheel load of 80 kN). Linear elastic model was assigned to PMC, base course and the geogrid layers whereas the base course and the sub base layers were modelled using Mohr-coulomb constitutive model. Dimensions of the control section of Sobanapuram to Gandhipuram road of Upilliyapuram block of Tiruchirappalli district were used for the model. The bottom of model was fixed in both vertical and horizontal directions and both edges of the models were restricted against horizontal movement. Table 3 shows the properties of the each layer material and Table 4 shows the mechanical properties of geotextile reinforcement. Young's modulus values from geogauge readings were used. For all layers, the Poissons ratio is 0.35. Dry density was obtained from the standard Proctor test and the cohesion and friction angle from the direct shear test were used. Figure 3 shows the model of the pavement. Analysis was done by reducing the thickness of GSB layer.

Table 3 Properties of materials in each layer

Material	Surface course	Base course WBM 3	Base course WBM 2	Sub base	Subgrade
Model	Linear elastic	Linear elastic	Linear elastic	Mohr-Coulomb	Mohr-Coulomb
Thickness (mm)	25	75	75	changes	300
Young's modulus (MPa)	3000	154	123	85	63
Poisson's ratio	0.35	0.35	0.35	0.35	0.35
Dry density (kN/m^3)	20	19	19	18.5	17.4
Cohesion (kPa)	NA	NA	NA	5	12
Friction angle (°)	NA	NA	NA	27	21

Table 4 Mechanical properties of geotextile reinforcement

Coir geotextile	Tensile strength (kN/m) warp	Tensile strength (kN/m) weft	Poisson's ratio
H2M5	20	18	0.25
H2M6	18	15	0.25

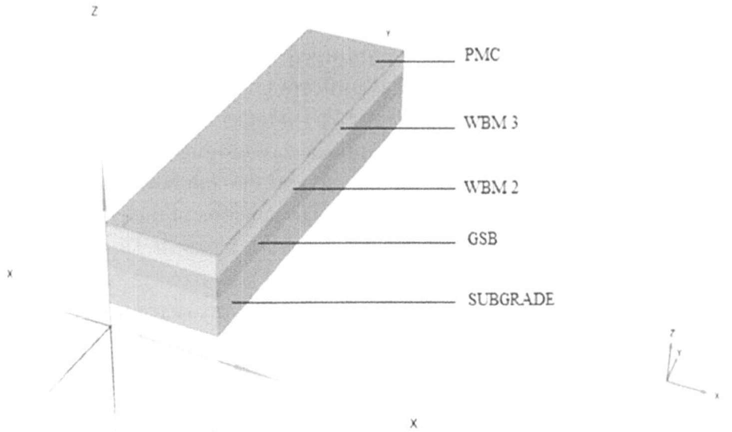

Fig. 3 Model of pavement in PLAXIS 3D

4 Results

Static loading was simulated for both unreinforced and geotextile reinforced section. Applied pressure was 550 kPa and geotextile was placed at the interface of sub base and sub grade. Critical pavement responses i.e. total displacement and strain at the top of the sub grade of unreinforced and geotextile reinforced pavements are determined under static loading. Table 5 shows the maximum total displacement values and Table 6 shows maximum strain values

Table 5 Maximum total displacement values

Thickness of GSB (mm)	Unreinforced (mm)	With H2M6 (mm)	With H2M5 (mm)
175	0.6444	0.5412	0.5319
150	0.6982	0.6103	0.5993
125	0.7490	0.6936	0.6869
100	0.7981	0.7532	0.7514
75	0.8439	0.8226	0.8189
Without GSB	1.506	1.483	1.468

Table 6 Maximum total strain values

Thickness of GSB (mm)	Unreinforced ($\times10^{-3}$)	With H2M6 ($\times10^{-3}$)	With H2M5 ($\times10^{-3}$)
175	0.7241	0.6593	0.6459
150	0.8064	0.7468	0.7316
125	0.8839	0.8317	0.8249
100	0.9643	0.9238	0.9181
75	1.037	0.9953	0.9835
Without GSB	2.236	2.098	2.088

Figure 4 illustrates the total displacement contour for applied pressure of 550 kPa for case of unreinforced pavement. Figure 5 shows the total displacement contour for H2M6 and H2M5 reinforced pavement of sections with 175 mm thickness of GSB layer. It can be observed that the total settlement for reinforced pavements has been significantly reduced. It is observed that maximum total displacement is 0.64 mm for case of unreinforced pavement, while it is 0.54 mm for pavement reinforced with H2M6, while it is 0.53 mm for pavement reinforced with H2M5 in the pavement section when the GSB layer thickness was 175 mm.

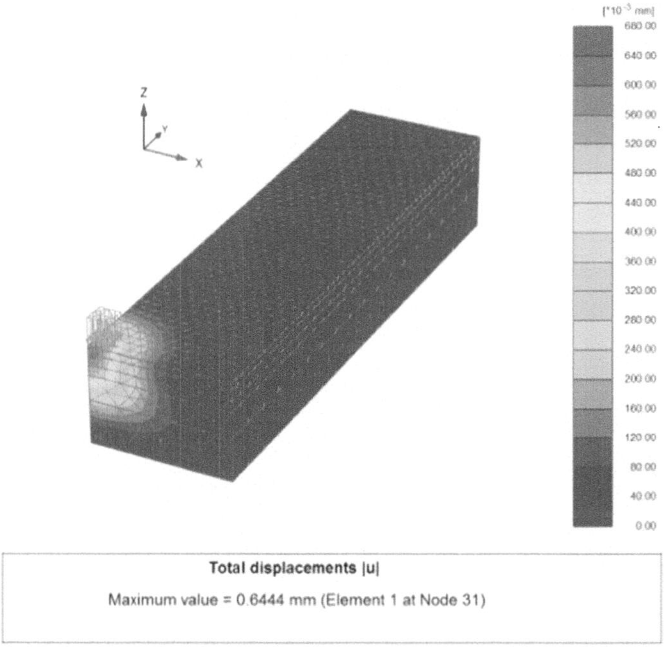

Fig. 4 Total displacement contour for unreinforced pavement with 175 mm GSB

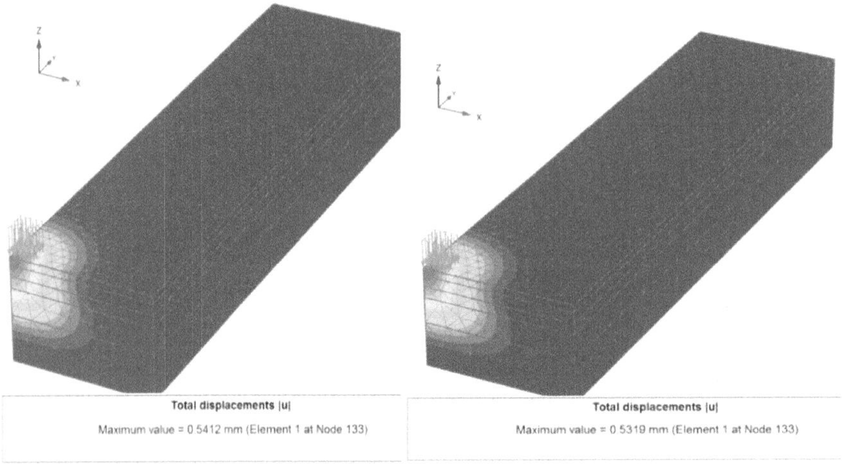

a. Total displacement contour for H2M6 b. Total displacement contour for H2M5

Fig. 5 Total displacement contour for reinforced pavement with 175 mm GSB

4.1 Outputs for 175 mm GSB

When the GSB layer thickness is 150 mm, it is observed that the maximum total displacement is 0.7 mm for unreinforced pavement, while 0.61 mm for H2M6-reinforced pavement and 0.6 mm for H2M5-reinforced pavement. Same way the displacement and strain values so obtained for sections with varying GSB thickness has been found out using the analysis. Figure 6 and 7 shows the variation in displacement and strain values for various sections.

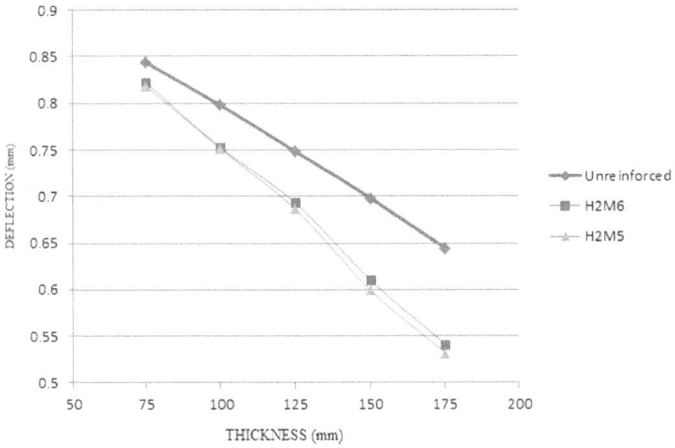

Fig. 6 Variation of deflection with varying thickness of GSB

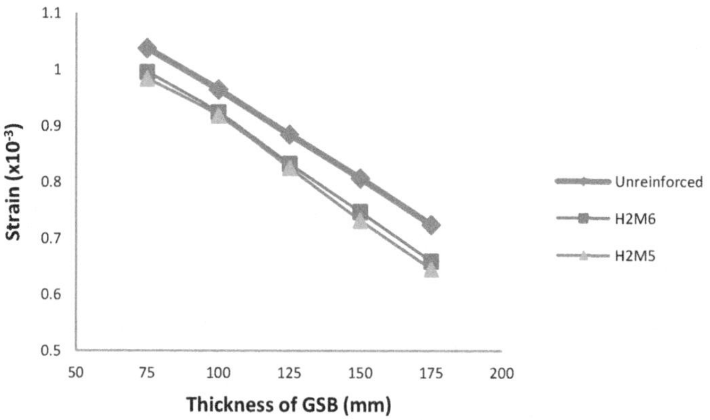

Fig. 7 Variation of strain with varying thickness of GSB

It has been observed that deflection increases as the thickness decreases. More reduction in deflection values was observed when the thickness is higher.

5 Conclusions

The following conclusions were drawn from the present study:

1. A significant improvement in the pavement behavior is achieved through inclusion of geotextile. The maximum total displacement was observed to be 0.64 mm for unreinforced pavement, while 0.54 and 0.53 mm for pavement reinforced with H2M6 and H2M5 respectively. The section with H2M5 coir geotextile has shown 18% less settlement, while it is 15% with H2M6, when compared with unreinforced section under static loading.
2. Under static loading, total strain on the pavement is reduced by 10% when reinforced by H2M5 coir geotextile, while it is 7% with H2M6 coir geotextile.
3. Unreinforced pavement with 175 mm GSB thickness has maximum total deflection value of 0.64 mm which is more than the deflections of H2M6 reinforced pavement (0.61 mm) and H2M5 reinforced pavement (0.6 mm) with 150 mm GSB thickness. Thus 25 mm thickness can be reduced by using the coir geotextile. Sub base layer can be reduced from 175 to 150 mm with H2M6 or H2M5 coir geotextile.

   The results of the FEM analysis can be validated by conducting plate load tests with instrumented strain gauges.

References

Faheem H, Hassan AM (2014) 2D PLAXIS finite element modeling of asphalt-concrete pavement reinforced with geogrid. J Eng Sci 42:122–126

Jayachandran MC, Evangeline YS, Rao GV, Sayida MK, Krishna A, Vijayan A (2019) Study on the coir geotextile reinforced flexible pavements using DCP. In: Evangeline S, Rajkumar MR, Parambath SG (eds) Recent advances in materials, mechanics and management. CRC Press, Boca Raton

Lal D, Sankar N, Chandrakaran S (2017) Effect of reinforcement form on the behaviour of coir geotextile reinforced sand beds. Soils Found 57(2):227–236

Loui TR, Satyakumar M, Padmakumar R, Aavani P (2014) Finite element analysis of coir geotextiles. J Eng Dev 5(1):112–122

Naagesh S, Sathyamurthy R, Sudhanve S (2013) Laboratory studies of strength and bearing capacity of GSB-soil subgrade composites. Int J Innov Eng Technol 2(2):245–254

Sharma M, Trivedi AS, Sahu R (2017) Pavement evaluation studies on low volume roads using plastic coated aggregate and bituminous mix. Int J Appl Environ Sci 12(5):953–966

Subaida EA, Chandrakaran S, Sankar N (2009) Laboratory performance of unpaved roads reinforced with woven coir geotextiles. Geotext Geomembr 27(3):204–210

Evaluation of Pavement Performance

Relating Asphalt Mixture Performance to Asphalt Mastic Rheology

Johannes Büchner and Michael P. Wistuba

Abstract In this study, usual asphalt mixtures are tested using known and established performance testing methods. In a second step, the asphalt mastics (mastic considered as composite of asphalt binder and aggregate filler) that correspond to the asphalt mixtures are investigated with regard to their rheological properties. All asphalt mixtures and corresponding asphalt mastics are subjected to (i) creep tests in the high temperature range, (ii) fatigue tests in the intermediate temperature range, and (iii) relaxation tests in the low temperature range. Hence, test results are obtained on both scales, the asphalt mixture scale and the corresponding asphalt mastic scale. At the mastic scale, a new set of rheological tests is composed. This set covers the full temperature range, but all results are obtained from the same Dynamic Shear Rheometer (DSR). In the last step, data from both scales are correlated, and finally, sound regressions are identified between asphalt mixture performance parameters and asphalt mastic rheological parameters. It is concluded, that the identified set of rheological tests at the asphalt mastic scale provides a rapid and significant estimate of asphalt mixture performance.

Keywords Asphalt performance testing · Asphalt mastic · Dynamic Shear Rheometer · Creep test · Fatigue test · Relaxation test

1 Background

Laboratory testing of asphalt mixture performance is generally considered appropriate and helpful in order to better understand material behaviour. In comparison to most empirical parameters derived from conventional tests (e.g. softening point ring and ball, void content, compaction degree), data obtained from performance tests are more reliable, especially when complex asphalt binders are concerned.

J. Büchner (✉) · M. P. Wistuba
Braunschweig Pavement Engineering Centre (ISBS), Technische Universität Braunschweig, Beethovenstr. 51 b, 38106 Braunschweig, Germany
e-mail: j.buechner@tu-braunschweig.de

M. P. Wistuba
e-mail: m.wistuba@tu-braunschweig.de

C. Raab (ed.), *Proceedings of the 9th International Conference on Maintenance and Rehabilitation of Pavements—Mairepav9*, Lecture Notes in Civil Engineering 76,
https://doi.org/10.1007/978-3-030-48679-2_60

Whereas most conventional tests are quick and cheap, performance tests require cyclic testing instruments and sophisticated analysis tools, and therefore they are usually considered cost and time consuming. However, if performance test methods are carried out carefully, and data are analysed correctly, they can reliably evaluate asphalt mixture performance and provide a sound basis for estimating the asphalt pavement behaviour and durability during its service life.

As stated by many authors, the performance of asphalt mixture is significantly influenced by the quality of the asphalt binder, and asphalt mastic (see, e.g. Airey et al. 2006), where mastic is considered as the composite of asphalt binder and aggregate filler. Therefore, key material properties like stiffness, strength, viscosity, creep behaviour, fatigue resistance, rutting resistance and aging behaviour, are driven by the properties of the asphalt binder, and the asphalt mastic respectively.

Various authors also tried to replace complex asphalt binder testing by simplified asphalt binder testing (e.g. Farrar et al. 2016; Johnson et al. 2009; Pérez-Jiménez et al. 2008).

2 Objective and Work Plan

The objective of this study is to help establishing laboratory testing techniques to estimate asphalt mixture performance, which are based on rheology rather than on empirical analysis. These testing techniques shall be less elaborate then conventional cyclic performance tests of asphalt mixtures, and therefore applicable for day-to-day routine testing purposes.

For this reason, a set of rheological tests based on measurements with a Dynamic Shear Rheometer (DSR) is presented in this study, which enables to identify key rheological properties of asphalt mastic in the full in-service temperature range of asphalt pavements. In this context, asphalt mastic is considered as the composite of asphalt binder and aggregate filler (<0.063 mm). Due to its variability the DSR is fully capable of testing asphalt mastic as long as the specimen thickness is higher than 10 time the particle size, which is been accounted for.

Even though this set of rheological tests covers the full temperature range, all results are obtained from the same DSR. Hence, it is intended to serve asphalt pavement engineers as a rapid and sufficiently accurate alternative to estimate asphalt mixture performance for routine purposes, such as quality control, selection of material components and mix design.

In the first step of this study, usual asphalt mixtures are tested, using known testing methods. The obtained parameters describe key performance properties of the asphalt mixtures, such as low temperature resistance, stiffness, resistance to fatigue and resistance to permanent deformation.

In the second step, the asphalt mastics that correspond to the asphalt mixtures, are investigated with regard to their rheological properties.

All asphalt mixtures and all corresponding asphalt mastics are subjected to (i) relaxation tests in the low temperature range, (ii) fatigue tests in the intermediate temperature range, and (iii) creep tests in the high temperature range.

Hence, test results are obtained on both scales, the asphalt mixture scale and the corresponding asphalt mastic scale, and they cover the full in-service temperature range of asphalt mixtures.

In the last step, data from both scales are correlated, and the relation is described between asphalt mixture performance parameters and asphalt mastic rheological parameters.

3 Materials and Testing

3.1 *Asphalt Mixture Variants*

Asphalt mixtures are produced in the laboratory, according to an asphalt concrete AC 11, with 11 mm maximum aggregate grain size and typically used for surface layers of asphalt pavements.

18 different asphalt mix compositions are distinguished through variation of the asphalt binder (two plain and two modified binders are considered, i.e. 50/70, 70/100, 25/55-55, and 45/80-65), of four asphalt binder manufacturers (A to D), and of three types of aggregate (gabbro, porphyry, silicate), as summarized in Table 1. The composition of the asphalt mixture variant is always adjusted such, that volumetric mix properties of all variants are more or less in the same range (with regard to void content, binder content, and grain size distribution).

For all the variants of asphalt mixture specified in Table 1, the individual fractions of filler are determined and used to produce asphalt mastic test samples. In this context, filler is understood as the aggregate fraction of grain size smaller than 0.063 mm. Hence, for each asphalt mixture variant, the corresponding asphalt mastic is determined, and composed in the laboratory from the relevant asphalt binder and filler (reclaimed filler, and new filler). The so-obtained asphalt mastic test samples are used for the subsequent DSR tests.

3.2 *Performance Testing of Asphalt Mixtures*

Because the material behaviour of visco-elastic complex materials like asphalt mixtures changes significantly with temperature, the assessment of asphalt mixture performance in laboratory requires differentiation of temperature ranges. Therefore, a combination of different tests is inevitable to study the individual temperature-dependent domains of material behaviour: Cracking resistance at low-temperatures (roughly ranging from −40 to 0 °C), fatigue resistance at intermediate temperatures

Table 1 Asphalt binders and aggregates considered in this study to compose 18 different variants of asphalt mixture

Asphalt mixture AC 11 variants	Asphalt binder type (plain and polymer modified)		Aggregate type (gabbro, porphyry, silicate)
1	50/70	Manufacturer A	Coarse aggregate: gabbro Added filler: limestone Provenience: Germany
2		Manufacturer B	
3		Manufacturer C	
4		Manufacturer D	
5	25/55-55	Manufacturer A	
6		Manufacturer B	
7		Manufacturer C	
8		Manufacturer D	
9	70/100	Manufacturer B	
10	45/80-65		
11	50/70		Coarse aggregate: porphyry Added filler: limestone Provenience: Austria
12	25/55-55		
13	70/100		
14	45/80-65		
15	50/70		Coarse aggregate: silicate Added filler: limestone Provenience: Switzerland
16	25/55-55		
17	70/100		
18	45/80-65		

(roughly ranging from −10 to +40 °C), and resistance to permanent deformation at high temperatures (roughly ranging from +30 to +90 °C).

In this study, all 18 variants of asphalt mixtures are subjected to known test procedures, i.e.

(i) Relaxation tests at low temperatures:
It is assumed, that low-temperature performance of asphalt materials can be characterized through relaxation behaviour at low temperature (cp. Laukkanen 2018). In this study, relaxation of prismatic test specimen (40 × 40 × 160 mm) is studied through the Thermal Stress Restrained Specimen Test (TSRST) according to European Standards (EN 12697-46 2012).

(ii) Fatigue tests at intermediate temperatures:
Prismatic test specimen (40 × 40 × 160 mm) are subjected to Uniaxial Tension Compression Tests (UTCT) according to Isailović and Wistuba (2018).

(iii) Creep tests at high temperatures:
Cylindrical test specimen (Ø 200 mm; h = 40 mm) are subjected to Cyclic Compression Tests (CCT) according to German standards (TP Asphalt-StB 2010).

3.3 *Rheological Testing of Asphalt Mastic*

Differentiation of testing techniques in function of temperature is also inevitable for asphalt binders and asphalt mastics. However, for asphalt binders and mastics most relevant tests can be run in the same testing device, i.e. the Dynamic Shear Rheometer (DSR). The DSR is generally capable of testing rheological properties of asphalt binder and mastic in the entire range of in-service temperatures.

Therefore, all relevant rheological parameters are derived in the DSR in this study. Small samples are tested in shear mode between two parallel plates. Because of the temperature-dependent change in material stiffness and the limited torque of the instrument, different plate diameters are used (according to Wistuba et al. 2019), as specified in Fig. 1.

For each of the different plate diameters (Ø 4, 8, and 25 mm) a temperature-specific testing technique is used to study mastic performance (acc. to Wistuba et al. 2019). In best accordance with asphalt mixture testing, all 18 variants of asphalt mastics corresponding to the asphalt mixtures are subjected to this set of DSR tests, i.e.

(i) Relaxation tests at low temperatures:
The cylindrical asphalt mastic sample (Ø 4 mm, $h = 3.1 \pm 0.15$ mm) is loaded in the DSR by a constant shear deformation of 0.1%, at a temperature corresponding to a complex shear modulus of the asphalt binder of $T_{G^*(binder)} = 400$ MPa, for a test duration of 60 min. After loading, the relaxation of the shear stress is monitored, and the proportional stress relaxation after 60 min is calculated.

(ii) Fatigue tests at intermediate temperatures:
The cylindrical asphalt mastic sample (Ø 8 mm, $h = 2$ mm) is loaded in the DSR by an initial shear stress of 150 kPa in oscillation mode at a frequency of 10 Hz, at a temperature of 10 °C, 20 °C, and a temperature corresponding to a complex shear modulus of the asphalt binder of $T_{G^*(binder)} = 15$ MPa. In order to reduce the testing time, the shear stress is increased every 4000 load cycles by +50 kPa, as similarly proposed by Johnson (2010).

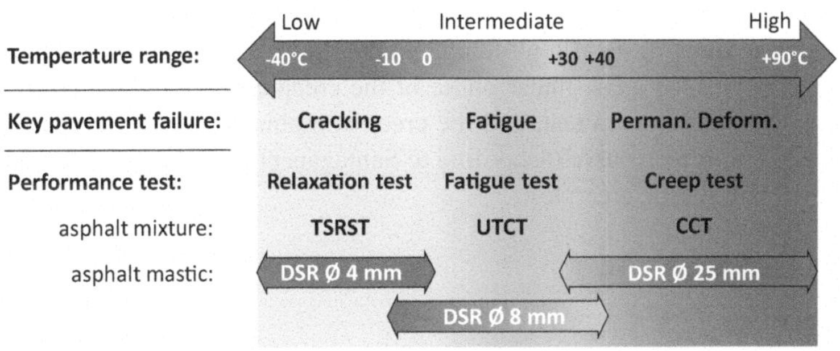

Fig. 1 Key performance properties of asphalt pavements and appropriate DSR test types in function of temperature (TSRST…Thermal Stress Restrained Specimen Test; UTCT…Uniaxial Tension Compression Test; CCT…Cyclic Compression Test)

(iii) Creep tests at high temperatures:
The cylindrical asphalt mastic sample (Ø 25 mm, h = 1) is loaded in the DSR by a shear stress of 0.1 kPa, at a temperature corresponding to a complex shear modulus of the asphalt binder of $T_{G*(binder)} = 15$ kPa, and for a test duration of 5 h in order to guarantee stationary creep behaviour. The creep compliance rate in the quasi-linear creep phase is derived according to Santagata et al. (2014).

3.4 Correlating Asphalt Mixture and Asphalt Mastic Tests

Tests on both scales, the asphalt mixture scale and the corresponding asphalt mastic scale, deliver material characteristic parameters, which are assumed to be mainly driven by the asphalt binder properties. Therefore, a good correlation between the parameters is assumed, if the individual temperature ranges are adequately distinguished. In this study, correlation is analysed separately for the relaxation tests, the fatigue tests, and the creep tests:

(i) As to the relaxation tests at low temperatures, the correlated parameters are the cryogenic stress at a temperature corresponding to a complex shear modulus of the asphalt binder of $T_{G*(binder)} = 400$ MPa as obtained from TSRST, and the proportional stress relaxation after 60 min as obtained from the asphalt mastic test.
It is assumed, that for high cryogenic stress of the asphalt mixture, which indicates unfavourable relaxation behaviour, the asphalt mastic also shows unfavourable low stress relaxation.
(ii) As to the fatigue tests at intermediate temperatures, the correlated parameters are the number of load cycles until specimen failure (according to Rowe and Bouldin 2000) as obtained from asphalt mixture fatigue testing at three different temperatures, and the number of load cycles until specimen failure (according to Rowe and Bouldin 2000) as obtained from the asphalt mastic test at three different temperatures.
(iii) As to the creep tests at high temperatures, the correlated parameters are the creep rate in the quasi-linear phase of the creep curve, as obtained from the asphalt mixture creep test, and the creep compliance rate in the quasi-linear phase of the creep curve (according to Santagata et al. 2014), as obtained from the asphalt mastic creep test.

4 Results

In the following, the correlations obtained from all tests on asphalt mixtures and mastics are presented. In the Figures, the results from tests at the asphalt mixture scale are always plotted over the related results at the asphalt mastic scale. Individual test results are always displayed as small transparent circles, while the corresponding mean values for each parameter are displayed in colour and with individual symbols. In addition, the total number of underlying results for asphalt mixture and mastic (n [-]) is indicated, as well as the test temperature (T [°C]).

In Fig. 2, the results are shown from relaxation testing of 18 asphalt material variants for a total number of n = 77 test results for asphalt mixture and mastic. A reasonable linear correlation is obtained with a coefficient of determination of 0.87.

Even though the results are quite different for the 18 asphalt material variants investigated, the influence of aggregate type is of minor importance, as for identical asphalt binders the results are very similar (cp. 3 red rectangles and 3 red triangles in Fig. 2). It appears that the low temperature relaxation behaviour of asphalt mixture is mainly driven by the type of the asphalt binder and there is no difference between plain and polymer modified asphalt binders.

In Fig. 3, the results are shown from fatigue testing of 18 asphalt material variants for a total number of n = 337 test results for asphalt mixture and mastic. Again, a reasonable linear correlation is obtained with a coefficient of determination of 0.74.

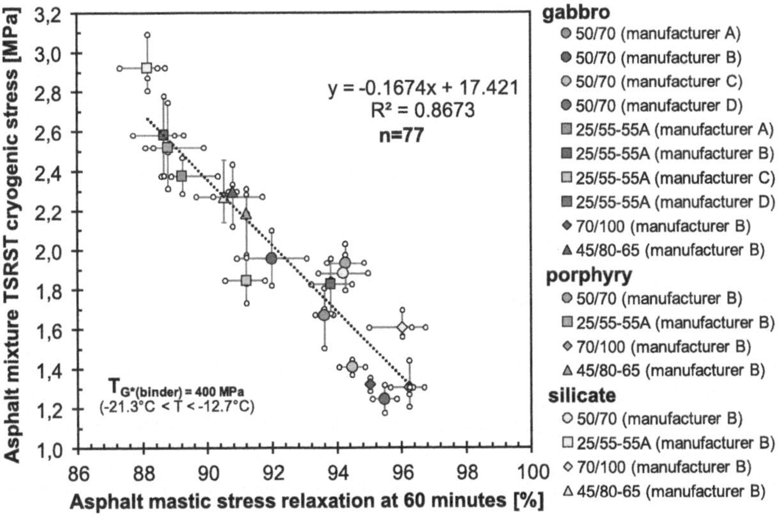

Fig. 2 Correlating results of 18 asphalt material variants, as obtained from relaxation testing of asphalt mixture based on Thermal Stress Restrained Specimen Tests (TSRST) and of asphalt mastic using DSR tests (Ø 4 mm) (Color figure online)

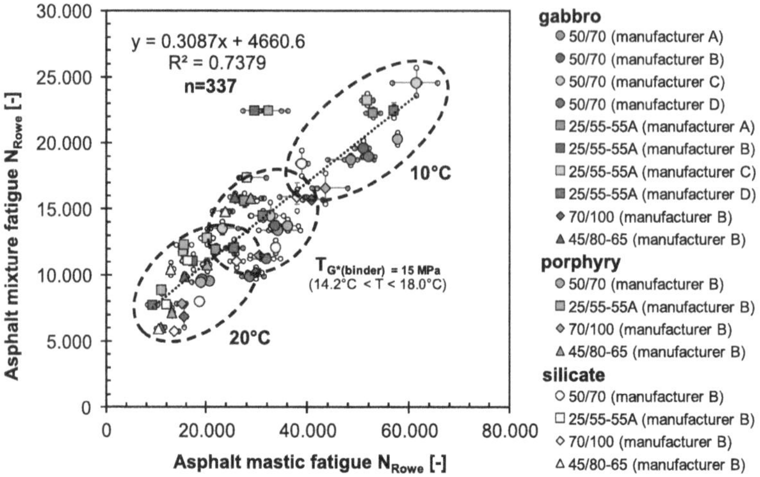

Fig. 3 Correlating results of 18 asphalt material variants, as obtained from fatigue testing of asphalt mixture based on Uniaxial Tension Compression Tests (UTCT) and of asphalt mastic using DSR tests (Ø 8 mm) (Color figure online)

(The red rectangles outside the dashed, oval shaped lines in Fig. 3 represent two outliers, which are further analysed in an ongoing study and are therefore not discussed in this paper).

As can be seen from the data obtained at three different test temperatures, the fatigue life of asphalt materials is significantly depending on temperature. Surprisingly, polymer modified binders and plain binders cannot be clearly differentiated.

Even though the results are quite different for the 18 asphalt material variants investigated, the influence of aggregate type is of minor importance. It appears that also the intermediate temperature fatigue behaviour of asphalt mixture is mainly driven by the type of the asphalt binder.

In Fig. 4, the results are shown from creep testing of 18 asphalt material variants for a total number of $n = 63$ test results for asphalt mixture and mastic. A reasonable exponential correlation is obtained with a coefficient of determination of 0.95.

Even though the results are quite different for the 18 asphalt material variants investigated, the influence of aggregate type is of minor importance. It appears that also the high temperature creep behaviour of asphalt mixture is mainly driven by the type of the asphalt binder. For both asphalt mixture and asphalt mastic scale, creep rates are smaller for polymer modified asphalt binders than for plain asphalt binders, most probably due to the elastic influence of the polymers (cp. rectangles and triangles in Fig. 4).

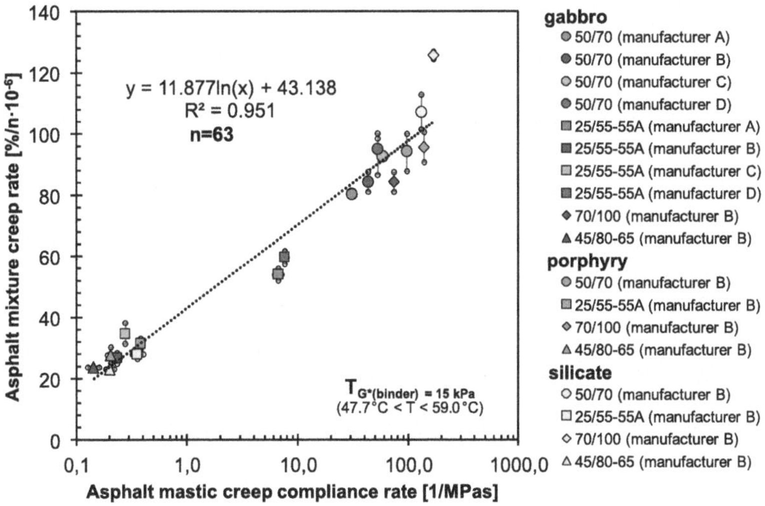

Fig. 4 Correlating results of 18 asphalt material variants, as obtained from creep testing of asphalt mixture based on Cyclic Compression Tests (CCT) and of asphalt mastic using DSR tests (Ø 25 mm) (Color figure online)

5 Summary and Conclusions

In this study, a number of 18 different asphalt mixtures and the corresponding asphalt mastics were subjected to (i) creep tests in the high temperature range, (ii) fatigue tests in the intermediate temperature range, and (iii) relaxation tests in the low temperature range. Hence, test results were obtained on both scales, the asphalt mixture scale and the corresponding asphalt mastic scale. While known test procedures were used at the asphalt mixture scale, a new set of rheological tests at the mastic scale was composed, that covers the full temperature range, but is run on one and the same Dynamic Shear Rheometer.

Finally, data from both scales were correlated, and sound regressions were identified between asphalt mixture performance parameters and asphalt mastic rheological parameters. Based on the results for 18 different variants of asphalt mixture, the following conclusions can be drawn:

- Asphalt mixture performance and asphalt mastic performance are linked, due to the viscoelastic properties of the asphalt binder. This is exemplarily shown in this study, considering the full in-service temperature range of asphalt pavements from −40 to +90 °C, and presenting correlations between the asphalt mixtures and the corresponding asphalt mastics for relaxation, fatigue and creep behaviour, with a coefficient of determination of at least 0.74.
- The identified set of rheological DSR tests at the asphalt mastic scale, as presented in this study, is a suitable tool to assess performance parameters of asphalt mastics, and it provides a rapid and significant estimate of asphalt mixture performance.

These findings may help to establish appropriate tools for day-to-day routine testing purposes of asphalt materials, such as for deriving limiting values to ensure sufficient asphalt mixture performance.
- The influence of aggregate type on performance of investigated AC 11 asphalt mixture and of asphalt mastic is less dominant than the influence of asphalt binder type.
- In the high temperature range, the influence of plain asphalt binder and of polymer modified binders can be differentiated clearly by means of creep tests. For intermediate and low-temperatures such differentiation was not distinct.

Acknowledgements This study was partly financed through the D-A-CH infrastructure research program of Germany, Austria and Switzerland. The partners of FFG-project Nr. 863063 (VEGAS) are kindly acknowledged for accompanying support. The authors further thank Malvern Panalytical for providing the DSR used to determine the rheological material properties.

References

Airey GD, Liao M-C, Thom NH (2006) Fatigue behaviour of bitumen-filler mastics. In: Proceedings of the 10th international conference on asphalt pavements, Quebec, 12–17 August 2006. International Society for Asphalt Pavements (ISAP)

EN 12697-46 (2012) Bituminous mixtures - test methods for hot mix asphalt - part 46: low temperature cracking and properties by uniaxial tension tests. European Committee for Standardization (CEN), Brussels

Farrar MJ, Kim S-S, Pauli T, Planche JP (2016) An advanced low temperature rheological and fracture test method for bitumen purchase specifications and pavement performance prediction: 4-mm DSR/ABCD. In: Canestrari F, Partl MN (eds) 8th RILEM international symposium on testing and characterization of sustainable and innovative bituminous materials, vol 11. RILEM bookseries. Springer, pp 25–36. https://doi.org/10.1007/978-94-017-7342-3_3

Isailović I, Wistuba MP (2018) Sweep test protocol for fatigue evaluation of asphalt mixtures. Road Mater Pavement Des 20(5):1131–1144. https://doi.org/10.1080/14680629.2018.1438305

Johnson CM (2010) Estimating asphalt binder fatigue resistance using an accelerated test method. Dissertation, University of Visconsin at Madison, Madison, Wisconsin

Johnson CM, Bahia HU, Coenen A (2009) Comparison of bitumen fatigue testing procedures measured in shear and correlations with four-point bending mixture fatigue. In: Proceedings of the 2nd workshop in four point bending, Guimarães, Portugal, 24–25 September. EAPA & Eurobitume, Guimarães

Laukkanen O-V (2018) Rheology of complex glass-forming liquids, vol 198. Aalto University publication series, Doctoral dissertations, Aalto University, Helsinki

Pérez-Jiménez FE, Reyes-Ortiz OJ, Miró R, Hernández-Noguera A (2008) Relation between the fatigue behaviour of asphalt binders and bituminous mixtures. In: Proceedings of the 4th eurasphalt & eurobitume congress, Kopenhagen, 21–23 May 2008. EAPA & Eurobitume

Rowe GM, Bouldin MG (2000) Improved techniques to evaluate the fatigue resistance of asphalt mixtures. In: Proceedings of the 2nd eurasphalt & eurobitume congress, Barcelona, 20–22 September 2000. EAPA & Eurobitume

Santagata E, Baglieri O, Alam M, Dalmazzo D (2014) A novel procedure for the evaluation of anti-rutting potential of asphalt binders. Int J Pavement Eng 16(4):287–296. https://doi.org/10.1080/10298436.2014.942859

TP Asphalt-StB, Teil 25 A 2 (2010) Technische Prüfvorschriften für Asphalt, Teil 25 A 2: Dynamischer Stempeleindringversuch an Walzwasphalt. Forschungsgesellschaft für Straßen- und Verkehrswesen (FGSV). FGSV Verlag, Köln

Wistuba MP, Büchner J, Hilmer T, Steineder M, Eberhardsteiner L, Donev V, Hofko B, Arraigada M, Raab C (2019) Simplification of asphalt performance testing. Final Report Projekt VEGAS. D-A-CH (Germany-Austria-Switzerland) Infrastructure Research, FFG-Project Nr. 863063. Austrian Research Promotion Agency (FFG)

Shear Bonding Performance of Reinforced Asphalt Pavements by Using Polyester Grids

Fabiana Leite-Gembus and Andreas Elsing

Abstract The conventional method for rehabilitation of cracked asphalt pavements is the installation of new asphalt layers. This resurfacing is often not an effective solution, as the existing cracks in the old asphalt layers can propagate rapidly to the top of the new overlay. In order to delay the development of reflective cracking, asphalt reinforcement grids have been used all over the world for many years, showing outstanding results. The grid changes the response of the pavement to loading, increasing the resistance of the asphalt overlay to high tensile stresses and distributing them over a larger area. Thereby the peak shear stresses are reduced at the edges of the existing cracks and thus the crack propagation can be retarded. Nevertheless, the performance of reinforced pavements depends strongly on the adhesion between the interlayer and the asphalt courses. The purpose of this paper is to report shear bonding performance verified in laboratory and in the practice by using a high modulus polyester grid as asphalt reinforcement. Shear bonding tests according to Leutner method have been carried out for different core samples, produced in laboratory as well extracted from test sections or from pavements in operation. The results show that a high bond strength between the asphalt layers can be reached when using an asphalt reinforcement grid, inclusive considering the long-term performance. In this way, the condition of repaired roads can be maintained at high levels over extended periods of time.

Keywords Reinforced asphalt pavement · Polyester grid · Shear bond strength · Pavement rehabilitation

1 Introduction

Construction of asphalt concrete overlays above existing cracked bound pavement surfaces, leads to problems of further migration of cracks to the surface of the new

F. Leite-Gembus (✉) · A. Elsing
Huesker Synthetic GmbH, Fabrikstraße 13-15, 48712 Gescher, Germany
e-mail: leite-gembus@huesker.de

A. Elsing
e-mail: elsing@huesker.de

C. Raab (ed.), *Proceedings of the 9th International Conference on Maintenance and Rehabilitation of Pavements—Mairepav9*, Lecture Notes in Civil Engineering 76,
https://doi.org/10.1007/978-3-030-48679-2_61

pavement surface. Traditional approaches to extending the lifetime of an overlay are to either provide a more flexible mix or to just increase the thickness of the resurfacing. Problems with increasing the overlay thickness related to cost and also potential tolerances on kerb height, for example by inner city roads. The phenomenon of reflective cracking can arguably lead to premature pavement distress and increased maintenance costs.

To delay or even prevent the development of reflective cracks in asphalt layers, asphalt interlayers have been largely used worldwide. Their application can clearly extend the pavement service life and therefore increase the maintenance intervals of rehabilitated asphalt pavements (Montestruque et al. 2004; Monser et al. 2010). Therefore the use of appropriate interlayer products, as asphalt reinforcement grids for example, should ideally be considered for future rehabilitation surfacing contracts on bound pavements, which have a history of reflective cracking at surface (Hessing and Thesseling 2013).

The lifetime of an asphalt concrete pavement is directly influenced by the adhesion between its different courses. As found by De Bondt (1999), if the reinforcement is not able to sufficiently adopt the high strains from the peak of a crack, the reinforcement cannot be effective. In his research, an equivalent "bond stiffness" in reinforcement pull-out tests was determined on asphalt cores taken from a trial road section. The equivalent bond stiffness of a bituminous coated polyester grid was found to be the best of all the commercial products investigated.

This paper focuses on the shear bonding performance of reinforced asphalt pavements. The adhesion between asphalt layers and asphalt reinforcement grid was determined according to Leutner for different cores samples, from laboratory produced slabs and from trafficked roads.

2 Reflective Cracking

Reflective cracking consists on the propagation of cracks from a deteriorated layer to the surface of a new overlay and is the major mode of failure in rehabilitated pavements (Elseifi 2015). It is well known that cracks appear in asphalt pavements due to external forces, such as traffic loads and temperature variations. The temperature influence leads to the effect that the binder content in the asphalt becomes brittle and cracking commences at the top of a pavement and propagates down (top-down cracking). On the other hand, high stresses at the bottom of a pavement, from external dynamic loads such as traffic, lead to cracks which propagate from the bottom to the top of a pavement (bottom-up cracking). The problem of continued reflective cracking can result in shortened maintenance periods which have an economic impact on maintenance budgets.

A conventional rehabilitation of a cracked flexible pavement involves milling off the existing top layer and installing a new asphalt course, but cracks are still present in the existing (old) asphalt layers. As a result of the horizontal and vertical movements at the crack tip, the cracks will propagate rapidly to the top of the rehabilitated pavement.

In order to delay the propagation of cracks is the use of interlayer systems between the old pavement and the new overlay, such as synthetic reinforcements, one of the most popular method among new techniques recommended (Nejad et al. 2016).

3 Asphalt Reinforcement Grids

The main function of geosynthetic products used in the construction and rehabilitation of roads and pavements is to reduce the amount of cracking in a new pavement or asphalt overlay. The idea of a reinforcing fabric for asphalt road construction first emerged in Germany as far back as the 1970s. Follow initial problems, manufactures have continually further developed their products and specialist contractors have continually refined their installation techniques.

Basically, there are three types of geosynthetics designed for pavement rehabilitation: geotextiles (nonwovens), geogrids (grids) and geocomposites. To further improve confidence in the utilization of these materials, there has been mandatory CE marking for the products since 2008. The functions of the products are outlined in the BS EN 15381 "Geotextiles and geotextile-related products—Characteristics required for use in pavements and asphalt overlays". According to this European Standard, an asphalt interlayer can fulfill three tasks: reinforcement, stress relief and/or interlayer barrier.

While the stress relief function concerns to soft products (as nonwovens) to dissipate strain energy by deforming itself, the reinforcement function regards stiff products (as grids) to compensate the lack of asphalt concrete tensile strength (Elseifi 2015). In providing reinforcement, the grid structurally strengthens the pavement section by changing the response of the pavement to loading (Koerner 2012). The reinforcement increases the resistance of the overlay to high tensile stresses and distributes them over a larger area, thereby reducing the peak shear stresses at the edges of the cracks in the existing old pavement.

Many products have been promoted as a reinforcement when in fact these products serve only as moisture barrier. Designers should have a clear understanding of the limitations all the different asphalt interlayer products offer in terms of position and stress-strain characteristics within the pavement structure (Asphalt Academy 2008).

For the maintenance of existing cracked roads there are currently a number of different asphalt reinforcement products made of different raw materials available in the market (e.g. polyester, recycled polyester, glass fiber, polyvinyl alcohol, carbon fiber, polypropylene), presenting different behavior and effectiveness. It is not disputed that each of these products has a positive effect in the battle against reflective cracking (Norambuena-Contreras and Gonzalez-Torre 2015); however, it is indispensable to choose the suitable product and assure a proper surface preparation, installation and asphalt construction.

Additionally detailed information about asphalt reinforcement grids can be found in the official document FGSV 770 "Working paper for the utilization of nonwovens, grids and compound materials in asphalt road construction" from the German Road and Transportation Research Association (FGSV 2013). In detail, this official document outlines important definitions of terms, application fundamentals, operating principles, types of asphalt interlayers as well as their properties, testing procedures and reference values. This official document may be used by designers and other interested parties for definition and specification of relevant functions, installation aspects and performance.

4 Interlayer Bonding

In order an asphalt reinforcement can improve the asphalt construction tensile strength behavior, it is particularly important to have a proper, expert installation of the product. Only in this case can an effective bond between asphalt courses and interlayer be achieved and, consequently, the tensile stress be transferred from one material to the next.

Interlayer bonding in asphalt surfaces is a result of combined effects from friction, interlocking and adhesion. High-quality asphalt reinforcements are specifically developed to not disturb the bond and contribute to these effects by means of appropriate mesh sizes and bituminous coating.

The draft European Standard on interlayer bonding prEN 12697-48 "Bituminous mixtures—test methods for hot mix asphalt—part 48: interlayer bonding" purpose three principal test methods to measure the adhesion of asphalt concrete courses: torque bond test, tensile adhesion test and shear bond test. This article describes the conditions of application, measurements made and results of this last one, the shear bond test according to Leutner.

4.1 Leutner Shear Test

The Leutner shear test is a static testing method developed in Germany in the late 1970s to assess the asphalt layer bond in road construction. The bonding effect of the layer surface is determined on core samples, either taken from a pavement or produced in the laboratory.

The process core sampling of compacted bituminous material from constructed or rehabilitated roads for the laboratory testing can be carried analogous to described in the European standard EN 12697-27 "Bituminous mixtures—Test methods—Part 27: Sampling". For the Leutner test the cylindrical asphalt specimens have to present a diameter of 150 ± 2 mm and the direction of traffic has to be marked. The core samples must be at least double-layered and the minimum thickness of the asphalt overlying layer has to be 20 mm and the subjacent asphalt layer 70 mm. The cores for shear bond strength investigation cannot be used for bulk density or other analysis. Before testing, the specimens must be conditioned at 20 °C for 12 h.

The equipment consists in a device able to apply a shear displacement rate (50 mm/min) across the interface between two layers of the cylindrical core, monitoring the resulting shear force (kN). Recording of displacements allows a Shear load (kN)—Displacement (mm) graph to be plotted.

Based on the German specification ZTV Asphalt-StB 07/13 the shear force within the testing procedure according to Leutner should not be lower than 15.0 kN between the binder course and the surface layer. For an interface between any other asphalt layers the shear force should be at least 12.0 kN.

For the installation of asphalt interlayers, higher spray rates of tack coat are required, when comparing to conventional pavement overlays. The use of precise amount of bonding agents, as bitumen emulsions for example, enables a proper adhesion between the reinforcement and the asphalt layers. Conversely, they have a tension-reducing effect, enabling viscose deformation. This system is known as "flexible bond" and is not equivalent to a layer bond in the usual meaning. For this reason, the German Working FGSV 770 (FGSV 2013) stipulates a shear bond force of 10 kN between the asphalt layers, when using an asphalt reinforcement.

5 Laboratory Test

In order to analyze and quantify the shear bond strength of reinforced asphalt compared to unreinforced one, different studies have been developed during the last years. The research results presented in this paper correspond to a polyester grid attached to an ultralight polypropylene nonwoven geotextile, incorporating a polymer modified bituminous coating with a bitumen content of 65%. Table 1 presents more detailed information about this product and following we would like to present some test results.

5.1 Different Asphalt Layers

In 2012, a series of tests were carried out at the EMPA Institute in Switzerland to verify the shear bond performance of a polyester asphalt reinforcement grid between different asphalt layers. Three concrete asphalt slabs were produced in laboratory and interlayer bond testing according to Leutner have been conducted. The asphalt

Table 1 Characteristics of the asphalt reinforcement grid used in all tests presented

Raw material	Type of coating	Nominal tensile strength [kN/m]	Aperture size [mm]	Mass [g/m^2]
Polyester (PET)	Bituminous (>65% content)	$\geq$50/50	40 × 40	270

Table 2 Test results of interlayer bond testing

Core number	Between base and binder course		Between binder and surface course	
	Reference (unreinforced)	Reinforced (PET grid)	Reference (unreinforced)	Reinforced (PET grid)
	Shear force [kN]	Shear force [kN]	Shear force [kN]	Shear force [kN]
1	36.7	30.7	34.4	29.9
2	35.5	26.5	43.4	30.7
3	42.7	33.6	44.1	40.6
4	32.7	35.6	34.4	29.5
5	32.8	36.5	37.9	31.1
6	39.5	36.0	39.3	36.8
Mean value	36.6	33.2	38.9	33.1

slabs presented three layers: base layer (AC T 22, 60 mm thick), binder layer (AC B 11, 50 mm thick) and surface layer (AC 8, 40 mm thick). Two slabs were reinforced, one having the reinforcement installed between base and binder layer and the other with reinforcement installed between binder and surface layer. For the installation of the polyester reinforcement grid a polymer modified bitumen emulsion with 60% bitumen content was used in a spray rate of 700 g/m^2. The reference slab was let unreinforced to enable an analysis of the reinforcement influence in the bonding performance. For the unreinforced slabs it was used a bitumen emulsion with approximately 50% bitumen content in a spray rate of 200 g/m^2. From each slab 6 cores with 150 mm diameter were taken. Table 2 shows a summary of the test results.

The results indicate a good interlayer bond in the reinforced specimens between binder and surface course, as well as between base and binder course. The measure shear forces by the reinforced specimens have shown slightly lower values, when comparing to the reference specimen. However, the values obtained comply by far with the requirements of the German specifications.

5.2 *Test Section in Aachen*

In 2018 was built at the Institute of Highway Engineering of RWTH Aachen University a real scale road segment. A new binder course (AC 16 BS) with a totally thickness of approximately 120 mm was installed. Then a polyester asphalt reinforcement grid with bituminous coating was installed on half of the binder surface. For the installation a bitumen emulsion was sprayed in a rate of 600 g/m^2, according to the manufacturer guidelines. Following this, an asphalt wearing course (AC 11 DS) with a thickness of 40 mm was placed, completing the asphalt superstructure. As reference, the other half of the binder surface was let unreinforced and the same bituminous emulsion was used and a spray rate of 200 g/m^2 was applied.

Six cores were drilled out from the real scale test road segment in order to be tested in the laboratory according to the Leutner shear method. The results are presented in Table 3.

According to the obtained results, the interlayer bonding observed for the reinforced pavement by using a PET grid can be similar to a conventional one. If tack coat spray rate, installation and asphalt works are conducted properly, it is possible to reach an adequate bond performance by using a bituminous coated polyester grid.

In comparison to an unreinforced section, the application of an asphalt reinforcement grid slightly reduces the interface strength. This is probably caused by the higher amount of tack coat applied for the grid installation and the area occupied by the fiber mesh in the specimen. The obtained values comply with the German specification ZTV Asphalt-StB 07/13, though.

As can be seen on Fig. 1, the polyester reinforcement grid was after testing still strong bonded to the asphalt layer. The large aperture size of the PET grid (40 × 40 mm) enables a good contact area between the beneath and upper asphalt layer, improving interlocking and friction.

Comparing the results obtained by EMPA with Aachen it is possible to observe a difference of approximately 25%. This fact is probably related to the specimens preparation. By the EMPA the tested cores were extracted from asphalt concrete slabs

Table 3 Interlayer bond test results from the test section in Aachen

Test temperature	Reference (unreinforced)		Reinforced (PET grid)	
	Shear force [kN]	Shear displ. [mm]	Shear force [kN]	Shear displ. [mm]
20 °C	27.6	3.9	25.3	3.5
	27.7	4.1	27.8	3.3
	23.0	3.6	21.7	4.5
Mean value	26.1	3.9	24.9	3.8

Fig. 1 Drill core after testing with reinforcement grid

produced in laboratory. Conversely, in Aachen the cores were extracted from a real scale trial section, submitted to typical site condition influences, as large equipment, asphalt mix provided from a plant, variable air and pavement surface temperature.

6 Practical Experiences: Investigation by Trafficked Roads

6.1 Schleswig Air Base, Jagel

The Schleswig Air Base in Jagel, in the northern part of Germany, has been built in 1916 and is in military use since then. Different sections of the asphalt concrete pavement have been rehabilitated in 1998, 2001 and 2005 by using a polyester asphalt reinforcement with a bituminous coating. The reinforced area corresponds to approximately 85,000 m^2 and the reinforcement solution was chosen at that time to prevent existing cracks from penetrating from the binder course into the new wearing course. The reinforcement was installed directly on the milled surface of the old existing asphalt layer and covered with approximately 50 mm of concrete asphalt.

In the year of 1998, after the first segment installation of the polyester reinforcement, cores were drilled in order to verify the interlayer bonding. For a direct performance comparison, sampling of reinforced and unreinforced pavement sections were carried out. Table 4 shows the obtained results according to Leutner method.

Comparing the results, no significantly difference in the shear bond performance could be verified for the reinforced and unreinforced areas. In 2008 a visual inspection has indicated that at this time no reflective cracks could be identified and the reinforcement solution proved to be an effective solution (Fig. 2).

6.2 Long Term Bonding Performance

In order to provide an example of long term bonding strength using polyester asphalt reinforcement grid, the project Rosenstrasse located in the Northwest German town of Ochtrup is presented. The Rosentrasse is a highly trafficked road, being one of the main connections to the nearby border of the Netherlands. Before its rehabilitation in 1996 the road revealed severe alligator cracking, longitudinal and transverse cracking

Table 4 Results of Leutner test for the Schleswig Air Base

Sample number	Reinforcement	Shear force [kN]
I	Polyester grid	36.4
II	No reinforcement	30.2
III	Polyester grid	37.5
IV	Polyester grid	36.7

Fig. 2 Schleswig Air Base pavement overview in 2008

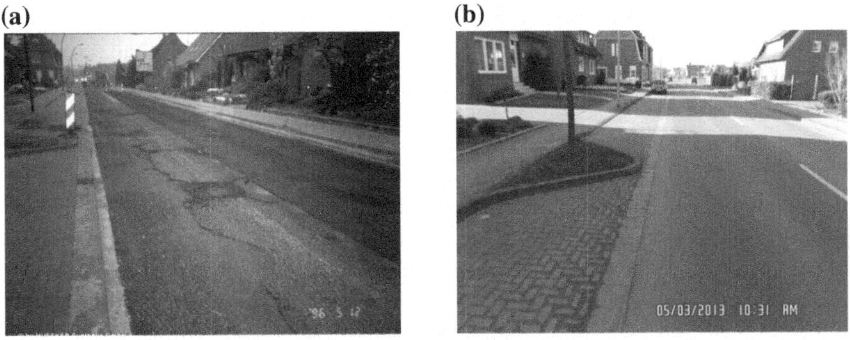

Fig. 3 **a** Condition after milling in May 1996, **b** Condition in March 2013

in large scale. For the rehabilitation the old wearing course was milled off, a PET grid as asphalt reinforcement was installed over the cracked binder course and covered again with a 50 mm thick asphalt layer. The whole project was finished in summer 1996. Directly from the beginning on this construction project has been investigated intensively (Elsing and Schröer 2005).

In context of a masters-thesis in 2013 the Rosenstrasse have again been evaluated (Quiel 2013). Drill cores have been taken and the interlayer bond according to Leutner had been checked. Between the asphalt bearing course—Polyester reinforcement grid—and upper asphalt layer a maximum shear force of 24 kN was measured. After evaluating the whole data record of the Rosenstraße it was found, that the condition, after a lifetime of 17 years, was still very good (Fig. 3).

7 Conclusions

Reflective cracking can occur in cracked pavements rehabilitated with a simple asphalt overlay. To delay the development of reflective cracks, an asphalt reinforcement grid can be placed before the new asphalt wearing course. In order to choose the proper product for a road rehabilitation, construction conditions and material characteristics must be chosen taking into account.

To mobilize tensile forces in the asphalt reinforcement grid a good bonding between the asphalt layers and the interlayer is essential. Shear bond tests according to Leutner were carried out for different core samples, produced in laboratory as well extracted from pavements in operation. The presented results have shown that it is possible to reach a good bonding performance when using a bituminous coated polyester asphalt reinforcement grid.

In comparison to an unreinforced section, the application of an asphalt reinforcement grid slightly reduces the interface strength. This is probably caused by the higher amount of tack coat applied for the grid installation and the area occupied by the fiber mesh in the specimen. However, it is possible to comply with the requirements of the German specification. Quality of all asphalt works must be carefully controlled as per relevant local guidelines and the reinforcement installation have to follow the detailed manufacturer instructions.

Practice example have shown that a high bond strength between the asphalt layers can be reached when using an asphalt reinforcement grid, inclusive considering the long-term performance. The combination of high reinforcement stiffness (polyester) and high bond stiffness (bitumen impregnation) can create such an improvement for the overlay life of an asphalt pavement, extending the service life of a rehabilitated pavement.

References

Asphalt Academy (2008) Technical guideline: asphalt reinforcement for road construction. TG 3, First Edition, South Africa

De Bondt AH (1999) Anti-reflective cracking design of (reinforced) asphaltic overlays. PhD-thesis, Delft, Netherlands

Elseifi M (2015) Presentation 2: mitigation strategies for reflective cracking in pavements. TRB Webinar – Mechanisms and Mitigation Strategies for Reflective Cracking in Rehabilitated Pavements, 24 August 2015

Elsing A, Schröer S (2005) Experience from more than 30 years of asphalt reinforcement with polyester grids. In: 15th international road federation world meeting, Bangkok, Thailand

FGSV 770 (2013) Arbeitspapier für die Anwendung von Vliesstoffen, Gittern und Verbundstoffen im Asphaltstrassenbau (in german). Forschungsgesellschaft für Straßen- und Verkehrswesen, Köln, Germany

Hessing C, Thesseling B (2013) Polyester asphalt reinforcement grids – the answer to reflective cracking and the basis for sustainable road maintenance. In: Proceedings of the XXVIII international baltic road conference, Vilnius, Lithuania

Koerner RM (2012) Designing with geosynthetics. Xlibris Corporation

Monser CA, Montestruque GE, Silva AEF (2010) Evaluation of an airport pavement after almost 8 years of overlay rehabilitation with a polyester geogrid asphalt reinforcement. In: Proceedings of the 9th international conference on geosynthetics, Brazil

Montestruque GE, Rodrigues RM, Nods M, Elsing A (2004) Stop of reflective crack propagation with the use of PET geogrid as asphalt overlay reinforcement. In: Fifth international RILEM conference on cracking in pavements: mitigation, risk assessment and prevention, Limoges, France

Nejad FM, Asadi S, Fallah S, Vadood M (2016) Statistical-experimental study of geosynthetics performance on reflection cracking phenomenon. Geotext Geomembr 44:178–187

Norambuena-Contreras J, Gonzalez-Torre I (2015) Influence of geosynthetic type on retarding cracking in asphalt pavements. Constr Build Mater 78:421–429

Quiel M (2013) Einsatz von asphalteinlagen im asphaltstrassenbau. Master thesis, Münster University of Applied Sciences, Münster, Germany

Discrete Element Simulations of 4-Point Bending Fatigue Tests of Asphalt Concrete Samples Reinforced by Fiberglass Grids

G. Liu, G. Koval, and C. Chazallon

Abstract This paper studies the fatigue damage of asphalt concrete (AC) under strain controlled load. The 3-phase fatigue life is reproduced by the proposed simulations combining a 2-phase fatigue damage law and the discrete element method (DEM) characterized by discrete force and displacement on each contact. The simulation results present good agreements with experiments after the parameter calibrations. The reinforcement effect of fiberglass grids on the fatigue life of asphalt concrete is studied with 4-point bending (4 PB) fatigue test. These preliminary results comparing simulations and experiments show the consistency of the proposed 2D model on describing part of the interactions between grids and asphalt concrete.

Keywords Fatigue · Damage · Discrete element · Four-point bending · Asphalt concrete

1 Introduction

Repeated traffic loads induce fatigue solicitation of the pavement structure. Accurate predictions of the effect of a very large number of loading cycles ($10^5 - 10^6$) over the components of the pavements is a major requirement for the optimization and design of transport infrastructures.

At laboratory scale, alternate cyclic bending tests intend to quantify deflection effects. The results of stiffness degradation (E_n/E_0, E_n is the present sample modulus and E_0 is the original modulus) with respect to the cycle numbers obtained in a typical cyclic loading test can be divided into three phases (Zheng et al. 2019; Arsenie et al. 2017). In the first phase, the defects exist and micro cracks are created in the matrix between the aggregates. In the second phase, the cracks grow and connect. In the third phase, the cracks interconnect. The sample loses its homogeneity and large cracks appear and propagate unstably with an increasing rate.

G. Liu · G. Koval (✉) · C. Chazallon
ICUBE Laboratory, CNRS, National Institute of Applied Sciences of Strasbourg, 24 Boulevard de la Victoire, 67084 Strasbourg, France
e-mail: georg.koval@insa-strasbourg.fr

C. Raab (ed.), *Proceedings of the 9th International Conference on Maintenance and Rehabilitation of Pavements—Mairepav9*, Lecture Notes in Civil Engineering 76,
https://doi.org/10.1007/978-3-030-48679-2_62

The fatigue damage causes large amount of maintenance cost. Thus, the reinforcement of the asphalt concrete (AC) is widely studied by researchers (Guo et al. 2015; Ge et al. 2015; Nguyen et al. 2013). In the last decade, the reinforcement by fiberglass grids has been applied to improve the mechanical response of pavement structure materials. Hu and Walubita (2009) indicate that fiberglass grid helps to distribute the stress and prevent crack propagation from top to bottom and the opposite direction. They may dissipate stress concentrations due to reflection and fatigue cracks, and work as moisture barriers (Nguyen et al. 2013), which significantly reduces the damage and improves the service life. However, there is still a lack of research describing how the grid works in AC under fatigue load.

The aim of this research is to analyze laboratory experimental results of Arsenie (2013, 2017), which deals with 4-point bending (4 PB) tests of reinforced and non-reinforced AC samples by means of computer simulations. The originality of the study is to implement a simple law of 2 regimes proposed by Bodin (2002, 2004) for AC as a contact law in discrete element method (DEM). This modelling approach takes advantage of DEM to reproduce the heterogeneous microstructure and crack development in cemented materials (Nguyen et al. 2019).

Section 2 presents the experimental setup. The third section covers the fatigue model description and its implementation in DEM. Section 4 presents the material calibration obtained by the comparison with experimental results. In Sect. 5, the reinforcements are introduced on the model and the numerical results are discussed. Finally, in Sect. 6, the conclusions of the paper are summarized.

2 Experimental Setup

During her study, Arsenie (2013, 2017) performed 4-point bending (4 PB) fatigue tests intending to quantify the contribution of the fiberglass grids on the fatigue life of asphalt concrete samples.

The geometry of the specimen is presented in Fig. 1. The standard beam dimensions have been adapted in order to have three warp yarns in the width of the beam. Therefore, the standard beam dimensions have been increased in width, length and

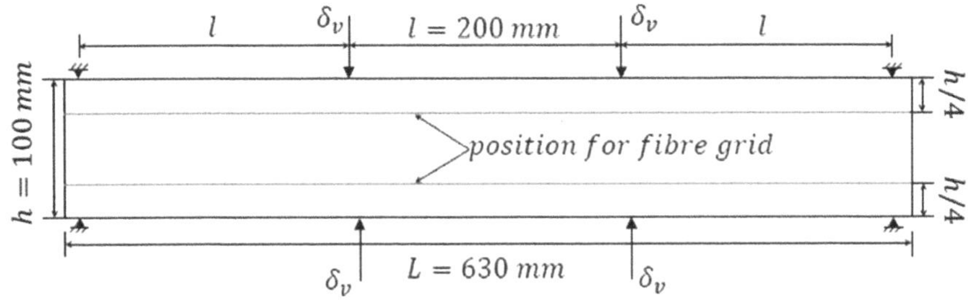

Fig. 1 Geometry of specimen for 4 PB test performed by Arsenie (2013, 2017)

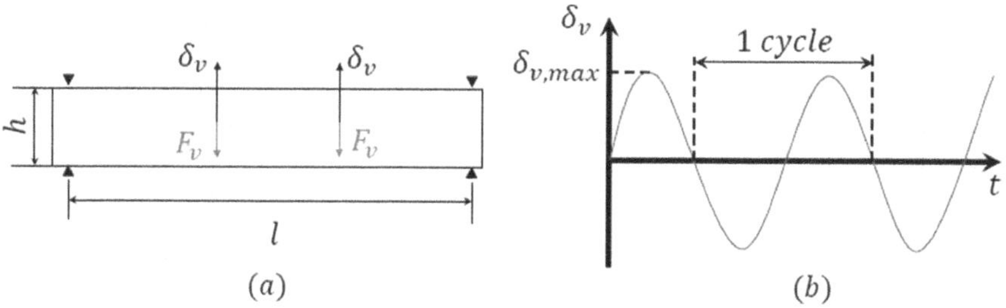

Fig. 2 Loading setup of the 4-point bending (4 PB) tests performed by Arsenie (2013, 2017)

thickness. Each beam has a length $L = 630\,\text{mm}$, and a square cross section with height $h = 100\,\text{mm}$ and width $w = 100\,\text{mm}$. The layers are bonded with a bitumen emulsion. The specimens of experiments are separated into two categories: reinforced asphalt concrete and non-reinforced asphalt concrete. The fiber grids lay between the AC layers in the reinforced beams.

A controlled strain condition is applied by the sinusoidal motion of the central supports as described in Fig. 2, which is the typical loading setup in fatigue of asphalt concrete. The reaction force on the supports F_v also presents a sinusoidal response.

3 Fatigue Model and Discrete Element Implementation

3.1 *Local Fatigue Model*

The material fatigue damage D is a state variable which characterizes the material mechanical condition. Its value ranges from 0 to 1 ($D = 0$ indicates an intact material, whilst $D = 1$ defines its complete failure).

3.1.1 Rate of Damage Growth $\dot{D}$

The rate of damage growth $\dot{D}$ is defined as

$$\dot{D} = f(D)\widetilde{\varepsilon}^{\beta}\left(\dot{\widetilde{\varepsilon}}\right)_{+}, \tag{1}$$

where $\left(\dot{\widetilde{\varepsilon}}\right)_{+}$ is the positive value of the rate of increment of the local strain $\widetilde{\varepsilon}$ (more details refer to Bodin 2002) and β is a variable related to the fatigue slope $-1-\beta$ in log-log scale. $f(D)$ is a function of the damage factor, which was proposed by Paas (1990) as

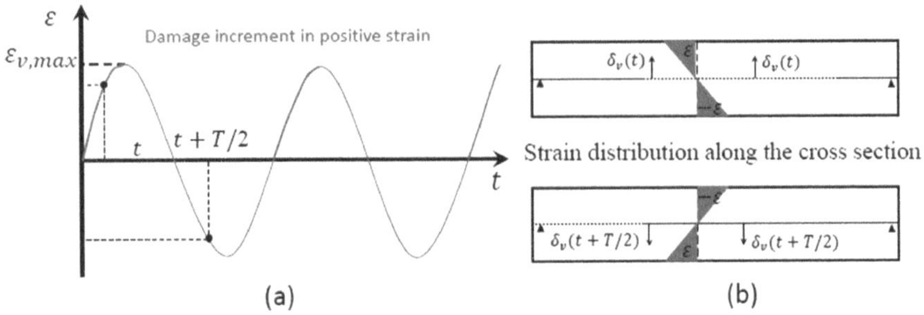

Fig. 3 Loading cycle contribution on damage increment

$$f(D) = CD^{\alpha}, \tag{2}$$

where C and α is a scalar parameter. According to Eq. (1), only a positive increment of strain may induce damage.

3.1.2 Increment of Damage δD Per Cycle

According to Eqs. (1) and (2), the damage is considered to be incremented only in tension and for positive strain rates. For a cyclic loading centered at zero, it represents the segments in Fig. 3 for $N_C \times T \leq t \leq N_C \times T + T/4$, where N_C is the number of cycles and T is the period of the loading cycles.

Equation (1) associated to Eq. (2) can be rewritten as

$$\dot{D}/D^{\alpha} = C\widetilde{\varepsilon}^{\beta}\left(\dot{\widetilde{\varepsilon}}\right)_{+}. \tag{3}$$

The local strain $\widetilde{\varepsilon}$ induces damage increment only on the first quarter of the cycle, which means $\left(\dot{\widetilde{\varepsilon}}\right)_{+} = \dot{\widetilde{\varepsilon}}$ for $0 \leq t \leq T/4$. As shown more in detail in Liu (2019), for constant strain cycles, the evolution of the damage as a function of the number of cycles N_C can be obtained by integration of Eq. (3) over the range of time $0 \leq t \leq N_C \times T$, which leads to

$$D(N_C)^{1-\alpha}/(1-\alpha) = CN_C\varepsilon_a^{1+\beta}/(1+\beta), \tag{4}$$

where $D(N_C)$ is the damage at $t = N_C \times T$ and ε_a corresponds to the amplitude of the local strain for a sinusoidal cycle. An incremental expression for damage can be obtained by derivation of Eq. (4) with respect to the number of cycles N_C

$$\delta D/\delta N_C = CD^{\alpha}\varepsilon_a^{1+\beta}/(1+\beta). \tag{5}$$

The value of damage D must be initialized ($D = 0$). The value of the damage after the first cycle $D(1)$ can be obtained from Eq. (4) for $N_C = 1$:

$$D(1) = \left[C(1-\alpha)\varepsilon_a^{1+\beta}/(1+\beta)\right]^{1/(1-\alpha)}. \quad (6)$$

3.2 DEM Implementation of the Fatigue Model—Damage and Force Calculation

In this study, the equivalent strain $\tilde{\varepsilon}$ of the simulation is obtained from the combination of the normal and tangential relative displacements (δ_n and δ_s) at the contact (for more details, see Liu 2019). The amplitude of the local strain is consequently the maximum value of the local strain under cyclic loading $\varepsilon_a = \max(\tilde{\varepsilon})$.

In the first cycle, the damage factor D is initialized, being calculated for each contact by Eq. (6). At each new cycle, D is modified incrementally, based on a first order solution of Eq. (5) as

$$D(N_C + \Delta N_C) = D(N_C) + \Delta N_C\left[C D^{\alpha}\varepsilon_a^{1+\beta}/(1+\beta)\right], \quad (7)$$

where ΔN_C is the cycle increment parameter. Any value of $\Delta N_C > 1$ allows a gain of calculation time, but must be carefully chosen not to induce cumulative errors. After the damage identification, normal and tangential forces can be updated as following

$$\begin{aligned} f_n &= k_n(1-D)\delta_n, \\ f_s &= k_s(1-D)\delta_s, \end{aligned} \quad (8)$$

where k_n and k_s are the normal and tangential stiffness of the intact contacts, respectively. The fatigue model was programmed and compiled in C++ language as part of the contact constitutive model implemented on the software PFC. For more information see reference Itasca (2014).

3.3 Geometry and Boundary Conditions of 4-Point Bending (4 PB) Tests

4 PB tests are simulated by discrete element method (DEM). The geometry follows the dimensions of the experiments performed by Arsenie (2013, 2017) as indicated in Fig. 4a with length $L = 630$ mm, height $h = 100$ mm and an equal spacing between supports $A = 200$ mm. Experimentally (and numerically), supports 1 and 2 do not move, whilst supports 3 and 4 are driven by a cyclic centered displacement with an imposed amplitude $\delta_{v,max}$. The maximum normal strain level ε_{max} is observed in the

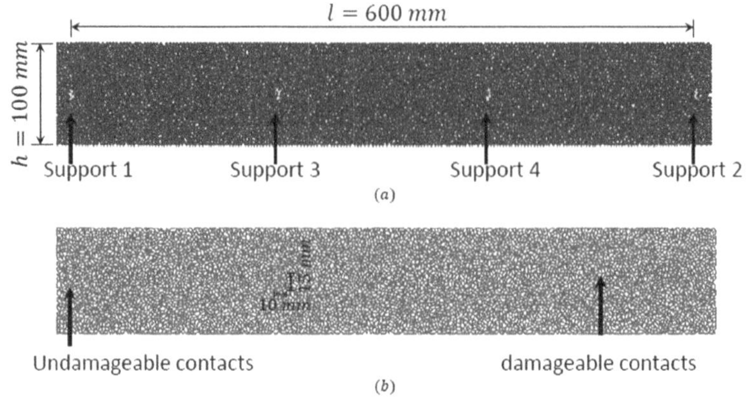

Fig. 4 Geometry of the 4 PB samples in DEM. **a** Indication of the dimensions over the particle packing and **b** contact properties by zones

middle section of the beam in upper and bottom positions and can be calculated by the expression (Arsenie 2013)

$$\delta_{v,max} = (5/3)(A^2/h)\varepsilon_{max}. \tag{9}$$

The reduced size of the supports, associated to free rotation prevent any undesired bending moment. The elastic distribution of efforts was verified comparing the results of the simulations to bending theory. However, the unrealistic dimensions of the support may induce a concentration of efforts which may lead to an inappropriate acceleration of the damage of the structure. This effect is avoided by the deactivation of the damage calculation ($D = 0$) inside a rectangular zone (10 mm × 15 mm) around the supports (see Fig. 4b). This assumption does not cause any nonphysical behavior, such as a discontinuity on the damage field, because the supports are located in an area with very low damage during the simulations, as discussed in Sect. 4.

The vertical force F_v, which correspond to the supports reactions, is measured during the fatigue tests.

4 Calibration of the Material Parameters with Experiments

In the 4 PB tests of Arsenie (2013, 2017), the strain level is driven by the value of maximum amplitude of the normal strain ε_{max} (obtained on the top and the bottom of the middle section). ε_{max} is theoretically related to the vertical displacement of the central supports δ_v by Eq. (9). Respectively, on the simulations the values of δ_v of 76.7 and 90 μm are adopted. For the discrete element material description, the radius of particles follows a uniform distribution. The average value is $\bar{R} = 2.9$ mm and the size ratio $R_{max}/R_{min} = 1.6$. The contact stiffness is $k_n = 2.0 \times 10^{10}$ N/m, $k_r = 4.5$, corresponding to a Young's modulus $E = 11$ GPa and a Poisson's ratio $\nu = 0.34$. The numerical results are obtained with an increment of cycles $\Delta N_C = 100$, which leads to accurate and efficient calculations as presented in Liu (2019).

In Fig. 5, the experimental results of the stiffness fraction F_v/F_{v0} as a function of the number of cycles N_C for $\varepsilon_{max} = 135$ and $150\,\mu$m/m are presented. Here F_{v0} is the initial value of the reaction force F_v. The envelope curves with minimum and maximum values were obtained considering 6 different samples and show the relatively large variation of the experimental results. The average curves are the references for the calibration of the parameters of the fatigue model, as follows: $C = 8.7 \times 10^8$, $\alpha = -2.0$ and $\beta = 3.0$. The average results of the simulations of the 3 different samples are in very good agreement with the experimental results with

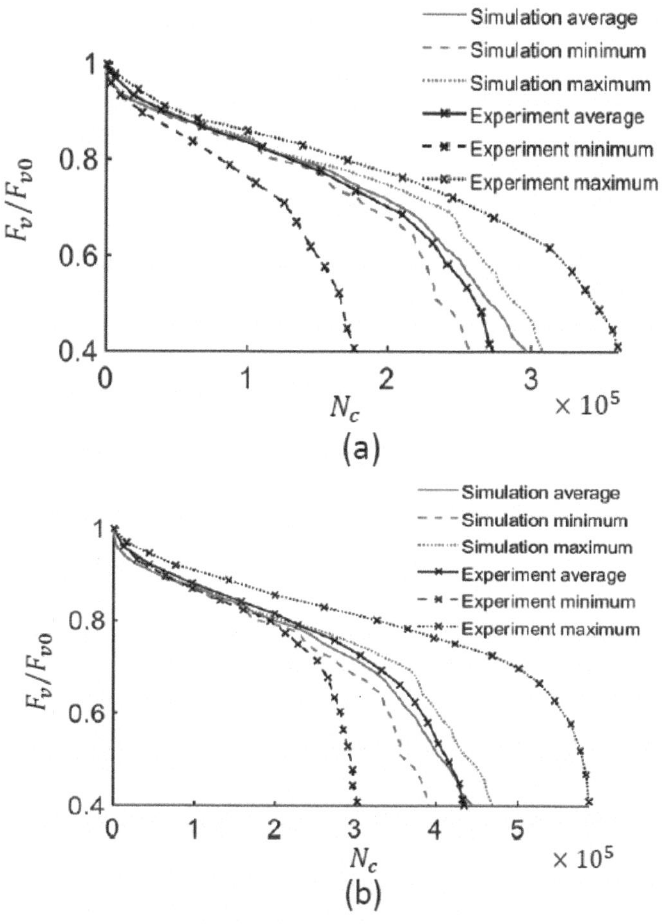

Fig. 5 Stiffness fraction F_v/F_{v0} as a function of the number of cycles N_C for **a** $\varepsilon_{max} = 135\,\mu$m/m and **b** $\varepsilon_{max} = 150\,\mu$m/m. Comparison between DEM simulations of 4 PB tests and experiments of Arsenie (2013, 2017)

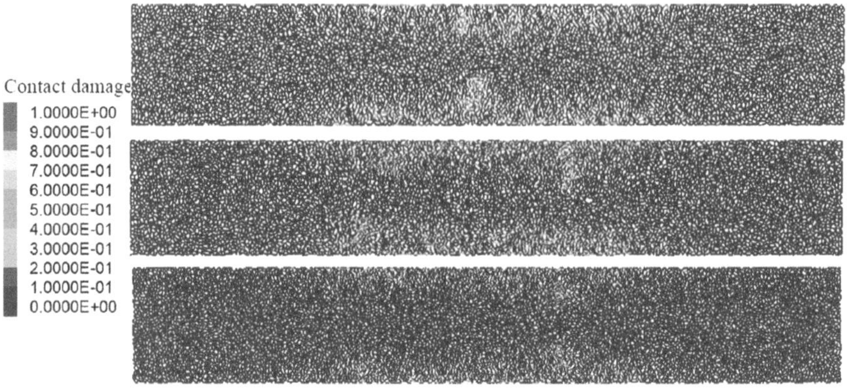

Fig. 6 Damage map for all the three samples for a stiffness fraction $F_v/F_{v0} = 0.4$

above parameters. The envelopes of results of the DEM simulations depends on the microstructure of the ensembles (granulometry, void ratio, etc.), which is an important feature of discrete approaches. However, the optimization of these microstructural parameters were not object of the present study.

Each of the numerical samples presents different damage evolutions and strain localizations due to different particle distribution. The damage maps of each analyzed sample (Fig. 6) bring up some physical elements related to the variation of results. After a roughly homogeneous damage, mostly localized at the central span, the irregularity of the inner structure of the material induces different weak points. These points give rise to concentrated damage zones, physically behaving as cracks. The localization of these damage zones affects the sample response which explains most of the visible differences between the curves F_v/F_{v0} as a function of N_C.

All damage maps seem continuous near the supports which may indicate a neglectful effect of the deactivation of the damage around them as suggested in Sect. 3.3.

5 Effect of the Glass Fiber Grids in 4 PB Tests

5.1 Model of the Reinforcements

The cross section of the bending samples is presented in Fig. 7a. Considering the lack of precise information about the fatigue behavior of the contact between asphalt concrete and fiber glass, a perfect adhesion hypothesis is adopted. Consequently, the contribution of the yarns in 2D is taken as elastic axial reinforcements working under tension and compression (bars). The elastic stiffness k_f of one segment i of a fiber glass reinforcement is determined by the expression:

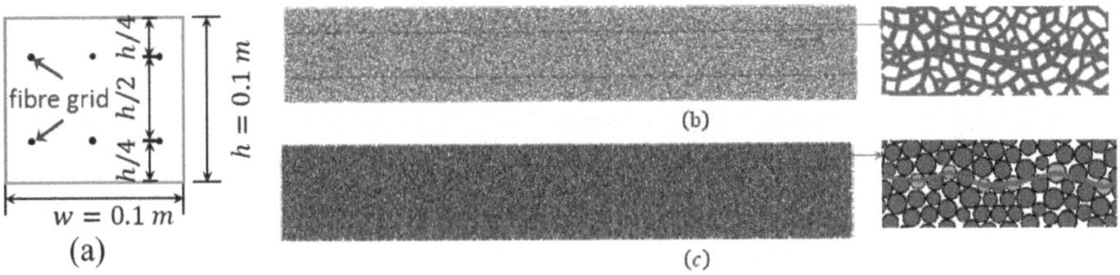

Fig. 7 **a** Cross section of the 4 PB samples. Representation of the reinforcement bars **b** as additional contacts **c** connecting particles in the same layer

$$k_{fi} = E_f A_f / l_i, \tag{10}$$

where E_f is the elastic modulus of the fiber grid and A_f is the total cross section of fiber grid per layer and l_i is the length of the segment. The effect of the bars are then taken into account on the discrete element model as additional contacts (Fig. 7b) connecting particles located at each layer, as shown in Fig. 7c. The contact properties are simply E_f and A_f, whilst the length l_i are automatically calculated based on the distance between the connected particles.

5.2 *Effect of the Fiber Reinforcements on the Fatigue Behavior Tests*

The average evolution of the stiffness factor $F_\upsilon/F_{\upsilon 0}$ as a function of the number of cycles N_C is shown in Fig. 8 for a strain amplitude of $\varepsilon_{max} = 150\,\mu\text{m/m}$. The results of 3 samples with fibers compared to the previous results without reinforcements indicates an improvement of the fatigue life induced by the reinforcements. However, the global effect of the grids shown by the experiences of Arsenie (2013, 2017) is underestimated by the simulations. The contribution of the reinforcements seems to be more effective for $F_\upsilon/F_{\upsilon 0} < 0.7$, when damage localized zones cross the fiber layer as observed by Liu (2019). On the experiments, the fiber grid improves

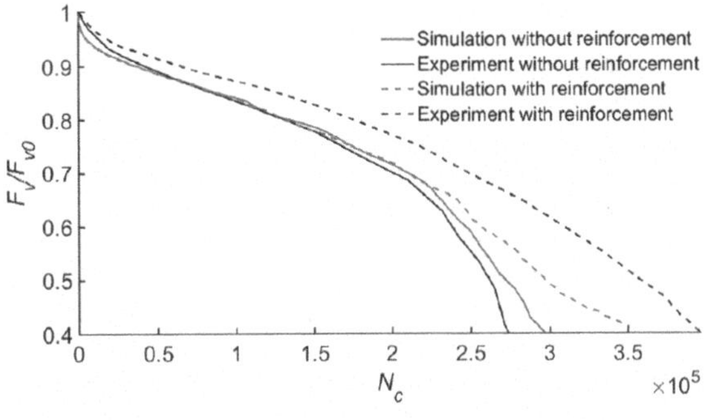

Fig. 8 Stiffness fraction $F_\upsilon/F_{\upsilon 0}$ as a function of the number of cycle N_C in 4 PB tests ($\varepsilon_{max} = 150\,\mu\text{m/m}$). Comparison between DEM simulations and the experiments of Arsenie (2013, 2017)

the fatigue response since the beginning of the test, independently to any eventual damage level.

This preliminary results suggests that a different mechanism may be associated to the fiber grid behavior, other than a simple axial reinforcement.

6 Conclusions

A fatigue damage model for asphalt concrete is adapted and implemented in a discrete element environment. A simple model describing material bulk behavior (phases I and II) associated to the natural disordered microstructure of the discrete model is shown to be able to describe the entire fatigue behavior (phases I, II and III) in a sample scale. Phase III appears as a structure effect, which is characterized by a loss of homogeneity induced by the concentration of defects and a localization of the strain.

The contribution of fiber grids as axial reinforcements on 4 PB fatigue behavior is investigated. Experimental results points to a general improvement on the fatigue curves, whilst simulations seems to underestimate this effect of the fiber grids. The numerical results indicate that reinforcements are mostly activated when crossed by localized damaged zones.

The differences on experimental and numerical results suggest that grids do not act simply as axial reinforcements in asphalt concrete. Further analysis of the interface between asphalt concrete and fiber grids may provide more information and improve the modeling as shown in Liu (2019).

Acknowledgements The study is funded by China Scholarship Council.

References

Arsenie IM (2013) Etude et modélisation des renforcements de chaussées à l'aide de grilles en fibre de verre sous sollicitations de fatigue. PhD thesis, Université de Strasbourg

Arsenie IM, Chazallon C, Duchez JL (2017) Laboratory characterisation of the fatigue behaviour of a glass fibre grid-reinforced asphalt concrete using 4 PB tests. Road Mater Pavement Des 18(1):168–180

Bodin D, Pijaudier-Cabot G, De La Roche C, Piau JM, Chabot A (2004) Continuum damage approach to asphalt concrete fatigue modeling. J Eng Mech 130(6):700–708

Bodin D (2002) Modele d'endommagement cyclique: application à la fatigue des enrobés bitumineux. PhD, University of Nantes (in French)

Ge Z, Wang H, Zhang Q, Xiong C (2015) Glass fiber reinforced asphalt membrane for interlayer bonding between asphalt overlay and concrete pavement. Constr Build Mater 101:918–925

Guo Q, Li L, Cheng Y, Jiao Y, Xu C (2015) Laboratory evaluation on performance of diatomite and glass fiber compound modified asphalt mixture. Mater Des 1980–2015(66):51–59

Hu X, Walubita LF (2009) Modelling tensile strain response in asphalt pavements: bottom-up and/or top-down fatigue crack initiation. Road Mater Pavement Des 10(1):125–154

Liu G (2019) Discrete element modelling of asphalt concrete reinforced with fiber glass grids. PhD thesis, University of Strasbourg, France

Nguyen ML, Blanc J, Kerzreho JP, Hornych P (2013) Review of glass fibre grid use for pavement reinforcement and apt experiments at ifsttar. Road Mater Pavement Des 14(sup1):287–308

Nguyen NHT, Bui HH, Kodikara J, Arooran S, Darve F (2019) A discrete element modelling approach for fatigue damage growth in cemented materials. Int J Plast 112:68–88

Paas MHJW (1990) Continuum damage mechanics with an application to fatigue. PhD thesis, Eindhoven University of Technology, The Netherlands

PFC 5.0 Itasca (2014) User manual. In Itasca Consulting group, Inc. Washington DC

Zheng M, Li P, Yang J (2019) Fatigue character comparison between high modulus asphalt concrete and matrix asphalt concrete. Constr Build Mater 206:655–664

Non-destructive Pavement Testing for Sustainable Road Management

G. Kneib

Abstract Sustainable road management asks for reliable and economic investigation methods to assess the surface and the structural substance of roads. A connection exists between properties of mixture, binder, carrots, surface texture and acoustic parameters. Noise emission and surface texture are particularly sensitive to minor variations in road engineering properties. Among the later grain size distribution is the most important quantity. Time series of drill cores, in situ surface and acoustical measurements draw a consistent picture of the process of top asphalt layer aging, including grain movements and void clogging. Georadar allows for fast microwave acquisition and mapping of electromagnetic properties as function of wave travel time. Ultrasonic measurements take more time but provide elastic parameters of road material which relate to structural substance. Measurements of mechanical road properties are sensitive to the temperature distribution inside the road body yielding temperature-dependent layer models. Depth profiles should be corrected to standard temperatures. Linear velocity-temperature relationships for ultrasonic waves yield gradients around 11 m/s/°C in a four-year old semi-dense asphalt top layer and for a temperature range between −17 and 65 °C. Heated asphalt displays strong wave absorption particularly at high frequencies which limits penetration depths and resolution.

Keywords Non-destructive pavement testing · Surface texture · Noise · Ultrasound · Georadar · Temperature

1 Introduction

Road infrastructure is under stress by increases in the amount of traffic and traffic loads. Sustainable road management aims at producing a road with the desired properties and keeping the road in proper condition for as long as possible. A prerequisite is periodical monitoring of mechanical road condition.

G. Kneib (✉)
Müller-BBM Schweiz AG, Bahnhofstrasse 48, 4132 Muttenz, Switzerland
e-mail: guido.kneib@mbbm.com

C. Raab (ed.), *Proceedings of the 9th International Conference on Maintenance and Rehabilitation of Pavements—Mairepav9*, Lecture Notes in Civil Engineering 76,
https://doi.org/10.1007/978-3-030-48679-2_63

Different countries employ different schemes and methods to quantify and to evaluate the state of the road surface and the road body. The procedure in Germany is summarized by Jansen (2017). Among other parameters the properties of drill cores representing homogeneous sections are analyzed in the laboratory to derive layer thicknesses, stiffnesses and fatigue functions. Substance values based on layer thicknesses and elastic moduli are computed via deterministic or probabilistic modelling. In Switzerland the conditions of roads are classified and indexed periodically according to the criteria: (1) surface damage, (2) surface slopes along longitudinal and cross directions, (3) grip, (4) bearing capacity (Swiss standard SN 640925b). In general, for each road section a rating for different criteria and sub-criteria is given which sum up to an overall rating. The condition classification of a road section serves as input to prediction models that are key factors for maintenance management together with catalogues of maintenance measures and cost models.

Non-destructive testing is playing an increasingly important role in road condition monitoring. Therefore, an overview is given here of existing evaluation methods and the benefit of combining different measurement methods is illustrated and the temperature-dependence of mechanical road properties is studied via ultrasound.

2 Investigation Methods for Roads

Testing of asphalt roads can be destructive or non-destructive. Destructive methods take away sections of the road for analysis in a laboratory typically in form of drill cores. Investigations based on carrots are well standardized and provide reliable information. But destructive methods require to destroy the road locally, are comparably expensive to apply and provide information only where carrots where taken. Per definition destructive road investigation is not really sustainable.

Non-destructive testing methods for roads can be subdivided into methods that investigate the surface and others that aim at the road body, i.e. mechanical and structural properties of the layering (Fig. 1). Slopes of road surfaces are measured in driving direction and perpendicular and may indicate the development of track troughs or dipping of the road due to voids in the underground. Visual inspection

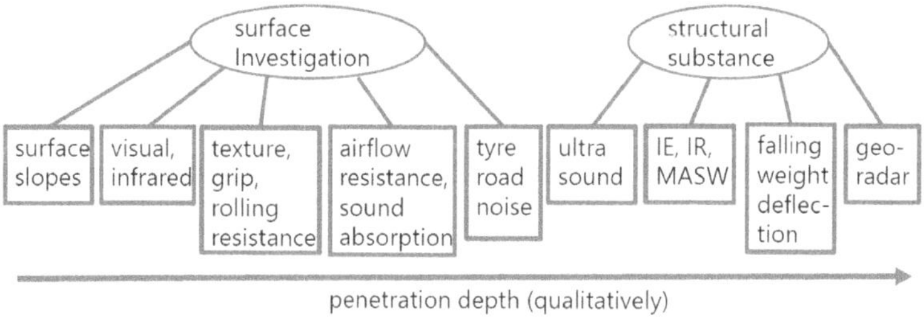

Fig. 1 Non-destructive road investigation with increasing penetration depth from left to right

is traditionally the most important assessment method for roads. Nowadays it is sometimes supported by video scanning and automated image processing to locate fissures. Grip is important for road safety and the coefficient of friction is measured by the side force method or the skid resistance tester pendulum or in Switzerland mostly via Skiddometer (SN 640925b 2003). Rolling resistance is a relevant parameter to reduce fuel consumption and traffic emissions. Surface texture is related to grip and mostly scanned by laser systems deployed on the road surface or on driving vehicles. Surface roughness spectra and other texture parameters also characterize the excitation of tyre-road noise. The later also depends on airflow resistance and sound absorption, two properties linked to geometries and sizes of connected pore spaces in the top layer. Tyre-road noise can be measured by several methods, most prominent are the statistical pass-by method and near-field close proximity (CPX) measurements. The later employ a trailer with mounted microphones that is towed along the road (Fig. 2) and provides continuous noise level profiles (ISO 11819-2 2017).

If the structural substance of the road is of interest exploration depths of several dm to m are required. The falling weight deflection (FWD) method puts a load on the road surface and measures the deflection curve (bending) due to this load (Horak et al. 2015) to indicate bearing capacity. Advanced systems make use of the load of a fast driving truck and map the resulting surface deflection by a laser system mounted on the truck. Back-calculation algorithms allow to estimate the elastic moduli of the road layers as function of known load and measured bending curve. FWD has low spatial resolution and back-calculation (as any inversion algorithm) requires certain assumptions and/or a priori knowledge to obtain (numerically) stable results.

A more direct approach to explore the road body employs body and surface seismic or ultrasonic waves that propagate in the road body and interact with the material and its internal inhomogeneities. The impact echo (IE) method employs an impulsive source at the road surface and a nearby receiver that records resonances at frequencies that are related to layer thicknesses (Gucunski and Maher 2002). A similar acquisition geometry is used by the impulse response (IR) method which identifies major voids or cracks by the presence of high-amplitude reverberations (Gucunski

Fig. 2 CPX system measuring in flowing traffic (**a**) and view on microphone installation (**b**)

and Maher 2002). Spectral analysis of surface waves (SASW) or multichannel analysis of surface waves (MASW) employ impulsive sources, record Rayleigh surface waves and evaluate frequency-dependent propagation speed (Park et al. 1999). The dispersion curves are inverted into depth-dependent shear wave velocity profiles. Spatial resolution of such investigation methods should be of the order of few cm because this is the minimum thickness of road layers. Spatial resolution is a fraction of the wavelength applied and the wavelengths equal the ratio of propagation speed to signal frequencies. Mechanical waves in asphalt display propagation velocities from about 1000–3500 m/s depending on wave type, material and its condition (like temperature). Therefore, for given target wavelengths in the cm range one has to employ frequencies between about 20 and 150 kHz, i.e. low-frequency ultrasound. Elastic waves have the advantage that they probe directly the mechanical material properties. Acquisition speed is fundamentally limited by the speed of the ultrasonic waves (Kneib et al. 2018). Electromagnetic waves in asphalt propagate with about 40–45% of the speed of light in vacuum such that frequencies in the 1 GHz range (microwaves) are needed to obtain the required spatial resolution. Application of microwaves to investigate the upper layers of the earth or civil engineering objects is called georadar or ground penetrating radar (GPR) probing. Since georadar wave propagation is very fast, acquisition is also very fast such that GPR-antennas can be mounted on a vehicle and measurements can be taken in flowing traffic delivering reflecting layer structures as function of surface position and wave travel time (Hugenschmidt et al. 1998). However, since electromagnetic waves can only probe electromagnetic material properties they cannot map mechanical road properties directly (Kneib et al. 2018).

3 Integrating Different Non-destructive Investigation Methods

3.1 Sensitivity of Road Noise Emission Due to Variability in Construction

When calls for tenders are published for road construction they specify expected road properties by demanding that the construction should follow certain specifications (usually national standards). Unfortunately, it is known from experience that quite often road construction does not exactly deliver what was wanted. The main reasons for this are related to the complexity of the building process and the many factors and uncertainties that influence the final construction. In a research project in Switzerland (Kneib et al. 2017) a dozen intra-city road surface constructions of semi-dense asphalt with 4 and 8 mm maximum grain size were monitored closely. More than 50 variables were observed at each of twelve construction sites and during an extensive measurement campaign in the weeks following the construction. Among the parameters investigated were the delivery of the asphalt mix (works, distance and mean of transport, etc.), the road construction work (company, machinery, rolling scheme, temperatures, etc.), binder (needle penetration, etc.) and mixing properties (Marshall test, binder content, grading curve, etc.), carrots (layer thickness, void content, etc.), and acoustics-related properties (surface roughness, sound absorption, airflow resistance and near-field tyre-road rolling noise). Tyre-road rolling noise was a key parameter for the project. It was measured by the close-proximity method (CPX method) which employs a purpose-built trailer with standardized tyres and closely mounted microphones (Fig. 2) and provides continuous noise emission profiles along roads.

In a first step the parameter variations were checked across construction sites with same specified grain sizes. The asphalt road construction with a maximum grain size of 4 mm (SDA 4) displayed larger parameter spans than the one with 8 mm (SDA 8). Among the road engineering properties, the largest variances were found among the grading curves although all were consistent with the standards that were to be applied (SNR 640436). Minor variances in road engineering properties often yielded major variances in properties related to acoustics. In a second step of the investigation the significance of all observation variables with respect to low acoustic noise emission was checked. Here near-field rolling noise served as a measure of noise emission. Neither the delivery of the mixtures nor the construction process or the binder properties proved to be significant. The grading curve parameter "accumulated percentage at half maximum grain size" and the void content of the carrots were the most significant road engineering properties. The former determines the proportion of large grains at the surface, the later may stand for the percentage of connected pore space. Root mean square roughness and texture shape factor (geometrical measure relating the portions of concave (peak-like) to convex (trough-like) features along the surface) also proved highly relevant as well as air flow resistivity and the degree of sound absorption, all measured in situ. Taking these six significant quantities ranges were defined within which the quantities had to be in order to obtain a silent SDA 4 or SDA 8 road. Note that the mentioned six parameters are related to each other. For example, higher roughness implies lower air-flow resistivity which implies larger sound absorption which goes along with higher void content in the drilling cores. Another example illustrates the systematic relation between properties of mixture, built road and rolling noise emission: A low accumulated portion of grains at half the maximum grain size implies a high proportion of large grains in the mixture which implies larger road surface roughness (Fig. 3a) and results in stronger excitation of tyre vibratons, i.e. higher CPX noise levels (Fig. 3b). These relations hold for both mixture types. Note that surface roughnesses and noise levels fall into two separate categories with little overlap for SDA 4 (red symbols) and SDA 8 (blue symbols). The less rough SDA 4 yields lower noise levels.

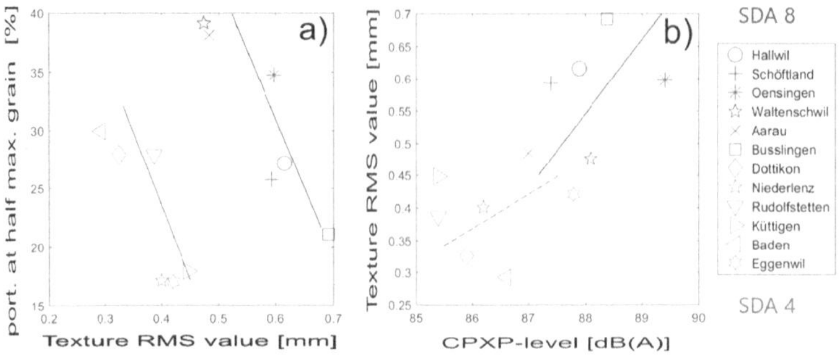

Fig. 3 Accumulated portion of grains at half of maximum grain size vs. root mean square roughness (**a**) and the later as function of CPXP rolling noise level (**b**) for asphalt surfaces SDA 4 in red and SDA 8 in blue. Different symbols denote different construction sites (Color figure online)

3.2 Time Functions of Road Engineering, Surface and Acoustical Properties

On a highway in Switzerland two pavement types were built one after another such that the same traffic passed over them. They were investigated over a time period of several years by an extensive monitoring program. The two pavements were semi-dense asphalt mixtures in accordance with Swiss regulation SNR 640436 of type SDA 8-16 (maximum grain sizes of 8 mm and typical void contents of 16%) and a mixture called SDA 8-12 (typical void contents of 12%). The higher void content of the SDA 8-16 resulted from doubling the percentage of grains larger than 4 mm compared to the SDA 8-12. Apart from this the receipts of the mixtures were identical.

A large-scale investigation program (Kneib et al. 2016) comprised permanent noise level recordings, CPX-measurements, statistical pass-by acoustic measurements and in situ measurements of surface texture, sound absorption and air flow resistance. Carrots were extracted repeatedly over time and material testing was performed on the carrots. A summary of the most significant findings shows Fig. 4. Since the same binder had been used needle penetration started at the same level but decreased faster during the following five years for the SDA 8-16 because the later offers a larger surface for oxidation processes which led to increased stiffening compared to the SDA 8-12 (Fig. 4a). The higher mass percentage of large grains in the SDA 8-16 yielded the anticipated higher void content compared to the SDA 8-12 but the void content difference between both pavement types reduced from 3 to 1.5% over the years (Fig. 4b). The shape factors of the surface roughness started also at same levels but the decrease over time was stronger for the SDA 8-16 which indicates that the grain matrix was less stable near the surface allowing for grain movements and possibly a larger loss of binder (Fig. 4c). The changes in binder, void content and surface texture led to a degradation of the acoustic emission properties (Fig. 4d). Acoustic emission is expressed here as CPX-noise levels converted into deviations from a standard covering which is indicated by a value of zero (Swiss StL86 + model,

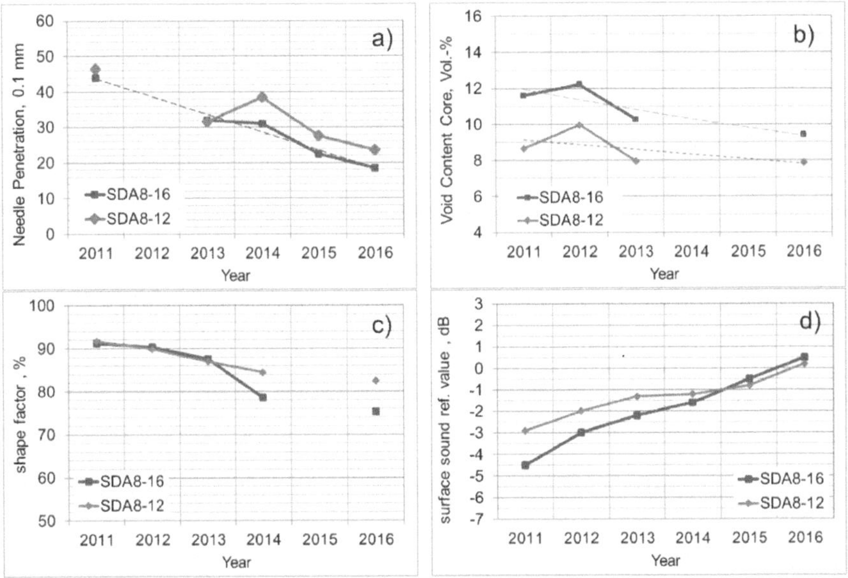

Fig. 4 Needle penetration for the binder (**a**), void content of drill cores (**b**), shape factor of texture profile (**c**) and sound emission based on CPX-measurements expressed by difference to the STL86 + reference pavement (**d**) as function of time for asphalt surfaces SDA 8-16 in red and SDA 8-12 in blue (Color figure online)

see Federal Guideline Road Noise, Switzerland). Large negative values indicate a silent road. Note that the acoustic advantage of the SDA 8-16 shortly after construction was at 1.5 dB and was gradually lost such that four years later both pavements were on same level (Fig. 4).

4 Ultrasonic Waves in Asphalt at Different Temperatures

Elastic parameters of the road body determined via mechanical waves have to be comparable to earlier and future measurements to provide meaningful time functions for maintenance management. Comparability poses a challenge for asphalt roads because (1) their mechanical properties strongly depend on material temperature and (2) large temperature gradients may occur in real roads depending on climatic conditions during the measurement and at least one day before. Fortunately, simple practical models exist to predict temperature-depth functions based only on surface measurements (Lukanen et al. 2000). Therefore, in principle the following procedure may be implemented: (1) determine depth functions of elastic parameters (for example via ultrasonic testing) in a given temperature-depth field, (2) estimate the temperature-depth field with a minimum of time-dependent surface temperature measurements, (3) convert the measured depth profiles of mechanical properties to standard temperatures (for example 20 °C). In order to perform step (3) it is necessary to establish temperature dependencies of mechanical properties, for example for ultrasonic wave speed.

In a series of experiments ultrasonic wavefields were investigated at various temperatures of asphalt probes. The probes were carrots with diameters of 100 mm and lengths of about 700 mm and consisted of semi-dense asphalt of type SDA 8 taken out from the top of a highway after four years of traffic (same project as in Sect. 3.2). The ultrasonic probes excited a pulse in longitudinal direction with frequencies between 20 and 150 kHz and were pressed by hand at the shell surface of the cylinder with sources and receivers just opposite to each other 2 cm below the top surface (Fig. 5a). The transmitted wavefield was registered as the positions of the probes were rotating along the cylinder surface in steps of 15° and the experiment was repeated. As a result, a gather of 24 ultrasonic time series was recorded for each experimental setup. At least 24 h prior to each measurement series the carrots were stored at a constant temperature to make sure the whole core was at temperature equilibrium. The cores were taken out of the temperature reservoir and the ultrasonic measurements were taken immediately within a couple of minutes at room temperatures. Infrared surface temperatures indicated only a moderate change during the measurements. The investigation temperatures of the cores were: −17, −1, +7, +18, +30, +45, +65 °C (Fig. 5).

The ultrasonic raw trace gathers measured at identical carrots show a different appearance depending on material temperature (Fig. 6). In these plots the relative amplitude scaling was the same for all traces in a gather but has been determined separately for each constant temperature gather. Absolute amplitudes decrease with temperature. Note that a 180° periodicity can be seen in each gather due to source-receiver reciprocity. Amplitude differences in the gathers are related to probe-surface coupling and variations of scattering along the wave paths, i.e. due to the random grain distribution between source and receiver. As temperature increases the binder

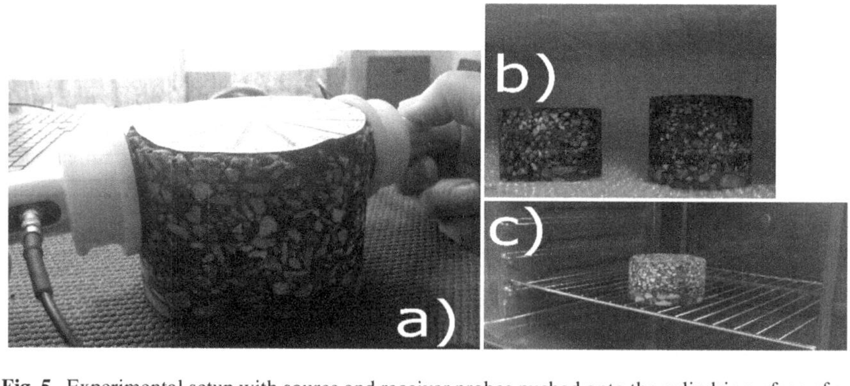

Fig. 5 Experimental setup with source and receiver probes pushed onto the cylindric surface of an asphalt carrot (**a**), probes in a refrigerator (**b**) and in an oven (**c**)

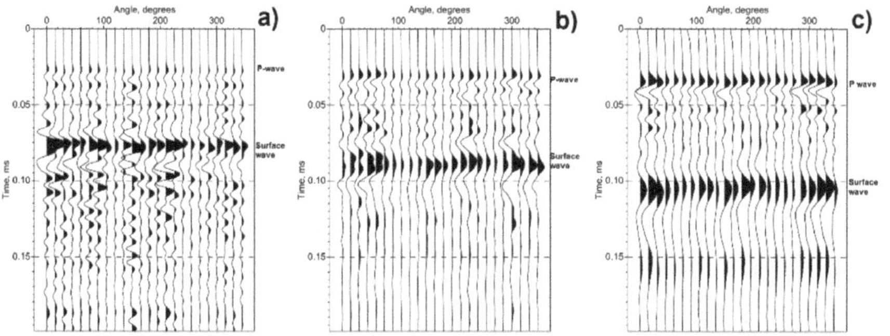

Fig. 6 Ultrasonic wavefields as function of recording time transmitted through a SDA8-16 carrot at different radial angles and material temperatures of −17 °C (**a**), 18 °C (**b**) and 45 °C (**c**)

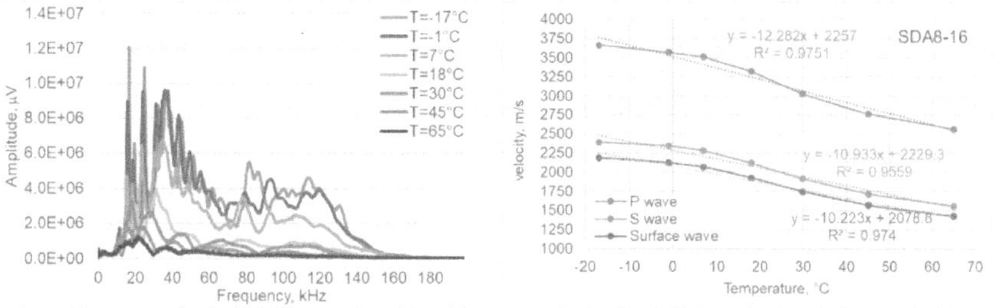

Fig. 7 Average amplitude spectra of the ultrasonic recordings at different material temperatures (**a**) and corresponding averaged propagation speeds for compressional and shear waves (**b**)

viscosity decreases which leads to stronger wave absorption and to time delays, i.e. lower propagation velocities.

Overall energy loss with temperature and frequency-dependent absorption is illustrated by the average amplitude spectra taken over the gathers and displayed in Fig. 7a. The pronounced undulations in the spectra are due to wave resonances in the core. The speed of compressional waves (P), Rayleigh surface waves and shear waves (S) can be estimated for each source-receiver pair and be averaged over each constant temperature gather. This leads to the temperature dependence of propagation speed for each wave type with velocity-temperature gradients around 11 m/s/°C (Fig. 7b). The shear propagation velocity displays a larger relative decrease than the compressional wave velocity, Poisson ratio increases by about 50% and bulk modulus nearly halves if the cores are heated from −17 to 65 °C.

5 Discussion and Conclusions

Non-destructive investigation methods have great potential to assist in road condition classification. On the one hand a connection exists between properties of mixture, binder, drill cores, surface texture and acoustic parameters and on the other hand acoustic noise emission measured via CPX levels and certain surface texture factors are particularly sensitive to changes in road engineering parameters. As a result, the former may be suitable to control the construction process. Time series of CPX levels and road surface texture may serve as early indicators of the road surface decay.

The mapping of road layering via georadar reflection time sections is an asset but it would be much better to map elastic properties as function of depth. This may be achieved in an economic way by combining georadar and ultrasonic acquisition (Kneib et al. 2018). In situ derived seismic and ultrasonic velocity-depth profiles must be converted to a standard temperature to represent true depths and comparable elastic moduli for meaningful road condition monitoring. The presented drill core transmission measurements may lead a path towards obtaining a catalogue of required correction formulas for different mixtures, frequency bands and wave types. Note that as binder viscosity quickly decreases with temperature the propagation speeds become strongly dispersive as function of temperature which poses a particular challenge for methods that use dispersion curves to invert for layer depths, like surface wave methods (Park et al. 1999).

References

Federal Office of the Environment and Federal Road Administration (2006) Guideline Road Noise – Execution support for rehabilitation, Bern, Switzerland (French and German)

Gucunski N, Maher A (2002) Evaluation of seismic pavement analyzer for pavement condition monitoring. Federal Highway Administration, Report FHWA-NJ-2002-012

Horak E, Emery S, Maina J (2015) Review of falling weight deflectometer deflection benchmark analysis on roads and airfields. In: Conference on asphalt pavements southern African (CAPSA), Suncity

Hugenschmidt J, Partl MN, de Witte H (1998) GPR inspection of a mountain motorway in Switzerland. J Appl Geophys 40:95–104

ISO EN 11819-2 (2017) Acoustics – measurement of the influence of road surfaces on traffic noise – part 2: the close-proximity method

Jansen D (2017) Methods of substance evaluation. Strasse und Autobahn, pp 958–967. (in German)

Kneib G, Belcher D, Beckenbauer T, Radojkovic N (2017) Research package: low-noise road surfaces in urban situations/EP 10: sensitivity of acoustic properties of low-noise pavements related to production variability. Federal Road Administration of Switzerland, Publication 1616 (in German). www.mobility-platform.ch

Kneib G, Belcher D, Beckenbauer T, Beyeler H-P (2016) Continuous road traffic noise monitoring and aging of asphalt surfaces. In: Proceedings of the inter-noise 2016, paper 001021, Hamburg, Germany, 21–24 August 2016

Kneib G, Jansen D, Wacker B (2018) Investigation of the structural substance of roads via georadar and ultrasound. In: Proceedings of 17th international conference on ground penetrating radar GPR 2018, Rapperswil, Switzerland, 18–21 June 2018, pp 215–220

Lukanen EO, Stubstad R, Briggs R (2000) Temperature predictions and adjustment factors for asphalt pavement. Federal Highway Administration Report FHWA-RD-98-085

Park CB, Miller RD, Xia J (1999) Multichannel analysis of surface waves (MASW). Geophysics 64:800–808

SN 640 925b (2003) Maintenance management of roadways – condition ascertaining and index evaluation. Swiss association of road and traffic experts (VSS), Zurich, Switzerland (French and German)

SNR 640 436 (2015) Semi-dense mixings and top layers SDA – rules, requirements, concepts and construction. Swiss association of road and traffic experts (VSS), Zurich, Switzerland (French and German)

Precision Assessment of the Modified Wheel Tracking Device Based on Small-Scale Testing of New Zealand Hot Mix Asphalt

Abhirup Basu Roy-Chowdhury, Mofreh Saleh, and Miguel Moyers-Gonzalez

Abstract The Wheel Tracking Device (WTD) has been widely used for laboratory characterisation of permanent deformation or rutting in Hot Mix Asphalt (HMA), however, the conventional setup of the device has limitations in terms of capturing the tertiary zone of the permanent deformation curve, and subsequently, the Flow Number (FN). This makes the Hot Mix Asphalt (HMA) characteristics related to the permanent deformation resistance difficult to analyse. Hence, a newly modified setup of WTD has been proposed for better characterisation of HMA mixes. The study focuses on the investigation of the repeatability of the new test-setup to eliminate any arbitrary calculations. The study utilised three different kinds of mixes with Nominal Maximum Aggregate Size (NMAS), and Void in Total Mix (VTM) as the factors, each with three replicates. For the experimental part and analysis, both confined and unconfined setup of the device will be used for each kind of mix. For the unconfined setup, the Flow Number (FN) was used as the parameter for the analysis. For the confined setup, the rut depths at 12000, 25000, and 50000 wheel passes or the final rut depth were utilised. This study is expected to give an insight to the modified Wheel Tracking Device and its repeatability for the purpose of robust characterisation of asphalt concrete mixes, which will in turn be helpful in developing precision estimates required for making it a standard practice in the near future.

Keywords Permanent deformation · Wheel Tracking Device (WTD) · Precision · Hot Mix Asphalt (HMA)

1 Introduction

Permanent deformation in the form of rutting is a noteworthy failure in flexible pavements. Saleh (2018), and Ebrahimi (2015) stressed on the fact that the degree of

A. B. Roy-Chowdhury (✉) · M. Saleh
Department of Civil and Natural Resources Engineering, University of Canterbury, Christchurch, New Zealand
e-mail: abhirup.basuroychowdhury@pg.canterbury.ac.nz

M. Moyers-Gonzalez
Department of Mathematics and Statistics, University of Canterbury, Christchurch, New Zealand

C. Raab (ed.), *Proceedings of the 9th International Conference on Maintenance and Rehabilitation of Pavements—Mairepav9*, Lecture Notes in Civil Engineering 76,
https://doi.org/10.1007/978-3-030-48679-2_64

compaction, composition of the asphalt mixture, and rate of loading, and temperature affect the permanent deformation resistance of asphalt mixtures, and also concluded that the shear deformation and densification cause the build-up of permanent deformation in asphalt concrete pavements. Therefore, it is of utmost importance that the susceptibility of the mixtures to shear deformation be evaluated. It has been found that undertaking a one-dimensional analysis involving only vertical deformation or rut depth is not accurate enough, and the results may be misleading regarding the permanent deformation resistance of mixtures as it does not consider the lateral deformation. As a result, for evaluating the mixtures, a two-dimensional analysis which includes both lateral strain, obtained from horizontal deformation and the vertical deformation or rut depth, are proposed for use.

Walubita et al. (2012) reported that Wheel Tracking Test is the simplest practical test available for the purpose of rutting characterisation of asphalt mixes in laboratory, which correlates well with the field performance. The conventional setup of the wheel tracking device utilises a steel mould, confined at all four sides, which, Saleh (2018) showed that no significant difference in the total permanent deformation could be observed for different mixes tested at different temperatures, which indicates the inability of the test to differentiate and rank asphalt mixes. The research used several samples of asphalt concrete slabs and were tested in the conventional setup (fully confined) setup to investigate the effect of the lateral confining stresses on the Wheel Tracking Test results. Different binder types were used along with different air voids in total mix (%VTM) and different aggregate gradations. The fully confined test was run at higher temperatures but no significant difference in the total permanent deformation was recorded, despite of different kinds of mixes. Saleh (2018) also gathered that there was no sign of inflection point for any of the test samples, therefore, it was concluded that in full-confinement, it is unlikely that the tertiary stage of permanent deformation could be reached. It is due to this reason and the absence of FN that the results from the fully-confined setup of the Wheel Tracker Test are difficult to analyse, making it not useful for ranking the mixes in terms of rutting resistance. Finally it was concluded that it is primarily because of the unrealistic lateral confining stresses in the confined setup, which causes much smaller final permanent deformation as opposed to what a realistic confinement would result in. Walubita et al. (2012) too report that high sample confinement during testing is an issue and can potentially affect the rutting performance of asphalt mixes.

The steel mould at the four sides of the asphalt sample in the confined setup causes a build-up of large reactive pressure as the sample attempts to spread laterally upon loading. This fact provides enough evidence that there exists a variable state of stress at the edges of the slab in confined wheel tracking test, which renders the test fundamentally inaccurate. This lateral confining stress unrealistically boosts the permanent deformation resistance of the mixes and thereby misguides research largely. Moreover, no valuable information about the effect of mixture parameters on the permanent deformation can be obtained. Therefore, measuring the deformation in the confined setup would merely mean measuring deformation caused by densification only, without considering the shear-related permanent deformation.

The modification of the device could simply be done by removing the side steel barriers of the mould along the lateral sides, and thereby allowing the slab to move laterally under the action of the vertical load of the tracking wheel. This simple process eases the confining pressure around the slab, which accounts for the shear deformations. This could potentially have a significant impact as to the quality assurance purposes in pavement engineering industry, as addressed by Saleh (2018), supported by the results achieved using unconfined setup. To limit the friction, paper is used to envelope the underside and the sides of the asphalt specimen. Hence, a very minimal confinement is induced in the form of friction.

The immediate advantage of using the modified setup of the wheel tracker is that, the permanent deformation curve clearly shows three distinct zones, which can in turn be utilised in quantifying the creep slope, tertiary slope, and flow number. This makes the modified setup capture the fundamental aspect of permanent deformation, i.e., the exact point when the shear deformation takes place. Although the Hamburg Wheel Tracker Test (HWTT) is conducted in submerged condition, the modified wheel tracker utilised in this study, developed in the Transportation Laboratory of the University of Canterbury, utilises the dry test condition and the tertiary zone resulting from the testing of the asphalt specimens is deemed equivalent to the "stripping zone" resulting from the Hamburg Wheel Tracker Test (HWTT).

Francken model, shown in Eq. 1, was used to determine the flow numbers of the modified wheel tracker test data, as it is reportedly the best mathematical model that addresses all three zones of the permanent deformation curve (Biligiri et al. 2007).

$$\varepsilon_p(N) = AN^B + C\left(e^{DN} - 1\right) \tag{1}$$

Where,

$\varepsilon_p(N) =$ Permanent strain or permanent deformation,
$N =$ Cycle number,
$A, B, C \ and \ D =$ Regression constants.

The first derivative of Eq. 1 defines the rate of change in the permanent strain, as shown in Eq. 2:

$$\varepsilon'_p(N) = \left(A * B * N^{(B-1)}\right) + \left(C * D * e^{(D*N)}\right) \tag{2}$$

The second derivative is determined to find the inflection point, which is the initiation of the tertiary stage or flow number (FN), as shown in Eq. 3. The flow number can be determined as the cycle number at which, the second derivative function switches its sign from negative to positive.

$$\varepsilon''_p(N) = A * B * (B-1) * N^{(B-2)} + \left(C * D^2 * e^{(D*N)}\right) \tag{3}$$

Another major advantage of the modified wheel tracker is the ability to measure both vertical and horizontal permanent deformation with the loading cycles.

Fig. 1 Modified setup of the wheel tracker

Figure 1 shows the modified wheel tracker setup in which the horizontal deformations are recorded by the Digital Transducer Interface (DTI). Therefore, in this research, the permanent deformation curves for both vertical and horizontal permanent deformations will be determined.

As the ability and applicability of the modified wheel tracker setup in characterising the rutting susceptibility of asphalt concrete specimens has been established from the research by Ebrahimi (2015), and Saleh (2018), the need for the investigation of repeatability and precision statement was identified by the authors of this paper as the next step towards validating the method. The test is currently based on AG: PT/T231 standard, however, no statement of precision has been mentioned for that. Azari (2014) pointed out that the correct design of asphalt mixtures is attributed to the precise and accurate measurement of the asphalt mixture properties, which led to his investigation of the precision statement of the Hamburg Wheel Tracking Test (HWTT). Likewise, the current study aims to investigate the repeatability of the modified wheel tracker through small-scale testing of the New Zealand Hot Mix Asphalt (HMA), which, not only will help in developing the precision statement and bias, but also prove and evaluate the performance of the conventional heavy-duty asphalt mixes in New Zealand sampled with Performance Graded (PG) binders.

2 Materials and Experimental Programme

For the purpose of the current study, two dense graded asphalt concrete mixtures AC 14 and AC 20 were used. The AC 20 is a heavy-duty dense graded asphalt mix with 20-mm nominal maximum aggregate size (NMAS), while the AC 14 has an aggregate size of 14-mm. A mixing and compaction temperature of 172°C was chosen according to AS/NZS 2891.2.1:2014 and AS/NZS 2891.2.2:2014. The loose HMA was then short-term aged for one hour at 142 °C before the compaction, according to Austroads standard. A roller compactor was used for compacting the asphalt concrete mix slab specimens. The procedure for compaction was chosen according to the New Zealand and Australia standards, and the asphalt concrete mix specimens were compacted at air voids content of 7 ± 1% as shown in Table 1. The

Table 1 Test matrix for the current study

Mix	Replicates	Binder	Mix NMAS	VTM (%)	Temperature °C	Slab thickness (mm)
1	1	PG 76-16	AC 14	7 ± 1	60	50
	2	PG 76-16	AC 14	7 ± 1	60	50
	3	PG 76-16	AC 14	7 ± 1	60	50
2	1	PG 70-16	AC 14	7 ± 1	60	75
	2	PG 70-16	AC 14	7 ± 1	60	75
	3	PG 70-16	AC 14	7 ± 1	60	75
3	1	PG 76-16	AC 20	7 ± 1	60	75
	2	PG 76-16	AC 20	7 ± 1	60	75
	3	PG 76-16	AC 20	7 ± 1	60	75

wheel tracker test was conducted at 60 °C. A total of 3 combinations of mix types, binders, and slab thickness with three replicates were considered for each kind of mix.

The wheel tracker test, based on the standard AG: PT/T231, utilises a 305-mm * 305-mm * 75-mm and 305-mm * 305-mm * 50-mm compacted HMA-mixture slabs. A vertical load of 0.7 kN is applied on the samples with a wheel tracking rate of approximately 26.3 passes/min. The test can be conducted at high temperatures such as at 50 and 60 °C, in dry condition. The test is terminated at 50,000 wheel passes or 15-mm rut depth, whichever occurs first. For the unconfined test, Francken model is fitted to the permanent deformation data to obtain the Flow Number (FN) as the data show the tertiary stage, i.e., inflection. For the confined test, the rut depth at 50,000 wheel passes or the number of wheel passes for a 15-mm rut depth is recorded.

3 Results and Analysis

For the confined condition, the rut depths at 12000, 25000, and 50000 wheel passes (or the final rut depth, in mm) were recorded and utilised for the precision estimates of the confined setup. The results are presented in Table 2. This was done to ensure the identification of more variables for the purpose of precision estimates, and to primarily acknowledge the fact that the permanent deformation in a confined mode is principally due to the mix densification, and hence, the rate of deformation and the rut depth value at increasing wheel passes with the progress of the test need to be evaluated.

The results in Table 2 indicate that the confined setup of the wheel tracker has a promising repeatability in terms of rut depth at 12000, 25000, and the final rut depth at 50000 wheel passes, as observed from the coefficient of variation (%CV) values. The results also show that the repeatability is comparable to the HWTT

Table 2 Modified wheel tracker results for confined samples

Mix	Replicates	RD at 12000 wheel passes (in mm)	Avg.	SD	CV (%)	RD at 25000 wheel passes (in mm)	Avg.	SD	CV (%)	RD at 50000 wheel passes (final rut depth. in mm)	Avg.	SD	CV (%)
1	1	2.84	2.69	0.13	**4.92**	3.34	3.13	0.19	**6.21**	3.82	3.55	0.26	**7.36**
	2	2.64				3.08				3.52			
	3	2.59				2.96				3.3			
2	1	5.08	5.45	0.48	**8.81**	5.68	6.28	0.68	**10.86**	6.37	7.14	0.82	**11.5**
	2	5.27				6.13				7.04			
	3	5.99				7.02				8.00			
3	1	4.03	4.35	0.33	**7.47**	4.54	4.85	0.33	**6.84**	5.01	5.51	0.64	**11.6**
	2	4.68				5.2				6.23			
	3	4.35				4.81				5.29			

Note: RD = Rut depth, Avg. = Average, SD = Standard Deviation, CV = Coefficient of Variation repeatability as reported by Azari (2014), and Cox et al. (2013) in their research. For the purpose of analysing the results of unconfined condition, the flow numbers (FN) were determined for the samples that were tested, and were used to determine the precision of the unconfined setup of the modified wheel tracker test. The results are presented in Table 3.

As can be observed from Fig. 2, the Flow Numbers based on vertical and horizontal deformation of the mixes and their replicates studied under the unconfined wheel tracker correlated significantly well with each other with a considerably high R^2 value. Research by Saleh (2018) suggests a higher correlation of the aforementioned parameters, which is likely due to the fact that the research used a wider range of mixes and incorporated a wider range of factors. Nevertheless, the correlation shown in Fig. 2 indicates the validation of the provision of recording the horizontal deformation and subsequently the calculation of the Flow Numbers based on that.

Figure 3a, b and c depict the representative permanent deformation curve, the vertical deformation and the horizontal deformation of three mixes utilised in the current study, studied under confined and unconfined setup of the wheel tracker respectively.

4 Discussion

Figure 3b and c clearly show that both vertical and horizontal permanent deformation curve reveal three distinct zones—the primary, secondary and the tertiary zones, as can observed from other fundamental permanent deformation tests, such as the Flow Tests (Flow Time and Flow Number). On the other hand, none of the mixes revealed such a curve (or did not fail) when tested in the confined setup, observed from Fig. 3a.

In terms of the permanent deformation resistance of the mixes studied in this paper, Mix 3 proved to be the best performing mix observed from the average Flow Numbers based on vertical and horizontal deformations. This is expected and convincing, as the mix utilised a bigger NMAS of 20-mm and stiffer binder of PG 76-16, when compared to Mix 2, which utilised 14-mm NMAS and a softer binder of PG 70-16. It was also noted that the performance of Mix 1 could not be compared with the other two as the thickness of Mix 1 slabs was 50-mm, however, the rut that accumulated has been proportional to the slab thickness. This could be observed from Table 2, where, in confined mode, the mix has an average final rut depth of 3.55-mm, which is smaller than the average final rut depths of the other two mixes which had the slab thickness of 75-mm.

The repeatability of the modified wheel tracker was investigated in this paper and it can be said that the test method showed promising repeatability in both confined and unconfined setup, observed from the coefficient of variation (%CV) values for each mix (Tables 2 and 3). The confined setup showed better repeatability compared to the unconfined setup possibly because none of the mixes and their replicates failed (i.e., did not exhibit tertiary flow or inflection point) in the confined setup. Thus, the mixes tended to behave more uniformly as the side barriers of the mould restrained

Table 3 Modified wheel tracker results for unconfined samples

Mix	Replicate	FN_V	Avg.	SD	CV (%)	FN_H	Avg.	SD	CV (%)
1	1	781	888	129	**15**	529	645	142	**22**
	2	851				803			
	3	1031				604			
2	1	791	834	218	**26**	533	644	116	**18**
	2	1071				764			
	3	641				636			
3	1	1321	1364	646	**47**	775	884	326	**37**
	2	2031				1251			
	3	741				627			

Note: FN_V = Vertical Flow Number, FN_H = Horizontal Flow Number

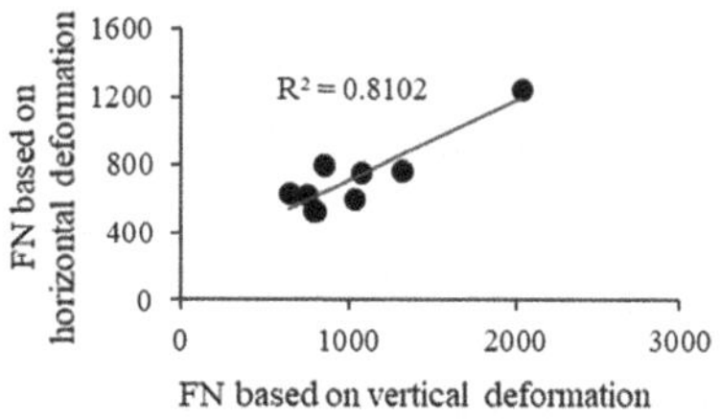

Fig. 2 Relationship of flow numbers based on vertical and horizontal deformations

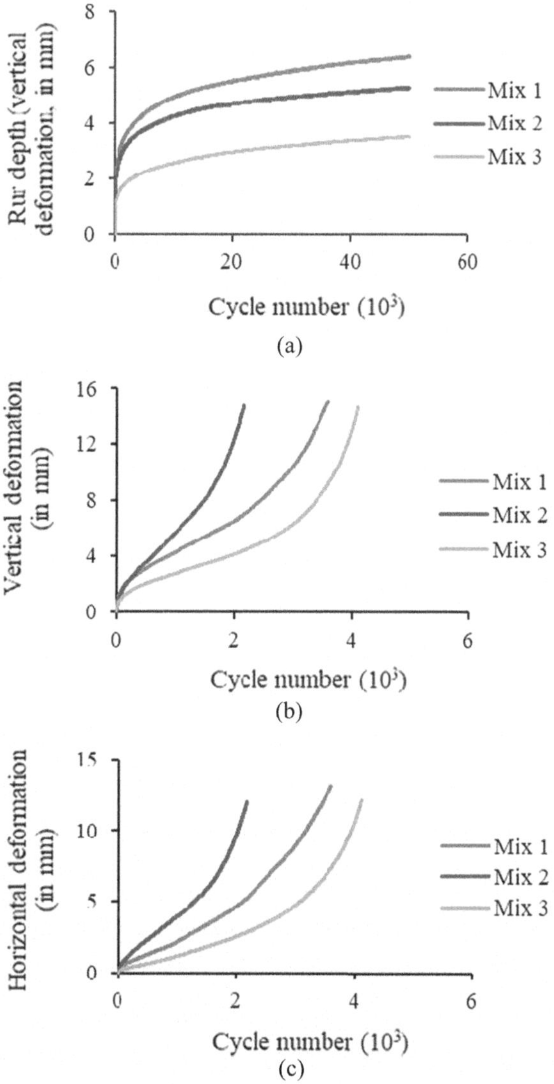

Fig. 3 Representative permanent deformation curves of the three mixes studied in **a** confined setup of the modified wheel tracker, **b** vertical deformation, and **c** horizontal deformation of three mixes studied in unconfined setup of the modified wheel tracker

the horizontal movement of the slabs, thereby leaving minimal room for variation. On the other hand, the unconfined setup is likely to experience more variation in Flow Numbers based on vertical and horizontal deformations, because the slab is free to move laterally in this setup, and this open-ended condition does leave room for variability in mix performance. However, the first two of the three mixes studied in this paper showed promising repeatability that can be compared with the CV (%) value of 23.9% for the number of passes to inflection point, as summarised by Beercroft and Petho (2015) while referring to the findings of the research by Azari (2014) on AASHTO T34 precision estimates.

Mix 3 showed higher variability of the Flow Numbers based on vertical and horizontal deformation, and this can be attributed to the added consequence of using the aggregates from two different quarries, which may have contributed to the change of specific gravity values. Moreover, the testing of the third mix was done over a longer period of time compared to the continuous testing of Mix 1 and 2. It was also noted that the wheel tracker was dismantled and re-assembled a number of times in between the testing of Mix 3 replicates, which induced handling error (such as tightening of bolts of the mould, and transducer placement), that in turn quite apparently contributed to the higher variability in Mix 3.

The variability of results across the replicates of the mixes can be attributed to several reasons, such as, the lack of homogeneity of the material used in each slab, the variability due to compaction and air voids, in addition to the variability due to the test itself. A full factorial design of experiments for the ruggedness testing of the factors associated with the precision is recommended for further study.

5 Conclusions

Based on the current study, the following conclusions can be drawn:

1. The Flow Numbers based on the corresponding vertical and horizontal deformations correlated exceptionally well each other. This proves that the horizontal deformation of the asphalt mixes is not an isolated phenomenon, but a highly relevant one, defined as the lateral deformation relative to the vertical deformation when the slabs studied under the wheel tracker experiences a lateral movement, thereby getting rid of the unrealistic full-confinement.
2. The repeatability of the modified wheel tracker was investigated in this paper, and it can be seen that it has a promising repeatability when compared to the precision estimates of the AASHTO 324 HWTT, evident from existing literature. However, an Inter-laboratory Study (ILS) is recommended for further study.
3. The unconfined wheel tracker proved much more useful over the confined one, as not only the rut depth, but a number of parameters, namely the Flow Number based on the vertical deformation, and Flow Number based on the horizontal deformation could be obtained and used for the precision estimates and also for distinguishing the mixes in terms of permanent deformation resistance. This

makes the unconfined wheel tracker much more reliable, as a wider range of parameters can be obtained and utilised for analysis.

4. From the unconfined setup results, both vertical and horizontal permanent deformation curve reveal three distinct zones—the primary, secondary and the tertiary zones, as can observed from other fundamental permanent deformation tests, such as the Flow Tests (Flow Time and Flow Number). On the other hand, none of the mixes revealed such a curve when tested in the confined setup. It can therefore be concluded that the unconfined setup is much more realistic and scientifically convincing in terms of assessing the permanent deformation performance of asphalt concrete mixes. It can also be noted from Fig. 3b and c that it takes significantly lesser time (indicated by lesser wheel-passes) for the test to end in unconfined setup when compared to the confined setup (Fig. 3a), which takes much longer time to end (indicated by longer wheel-passes). Thus, the modified setup of the wheel tracker is time-saving in the sense that the characterisation of the permanent deformation performance of HMA can be done in much shorter time.
5. For future research, a deeper statistical analysis, such as a full (or at least half) factorial ruggedness testing is recommended for the investigation of the effect of the parameters that effect the variability of test results, as well as the mix performance.

Acknowledgements The authors would like to acknowledge the staff from Fulton Hogan in Christchurch for supplying the materials utilised in this study. The authors also owe considerable thanks to the colleagues at the University of Canterbury, Christchurch, New Zealand, especially the ones in Pavement Research wing. Mr. John Kooloos, the technical officer of the Pavement Research Laboratory is particularly acknowledged for his constant help in the production of the samples, and also for useful advices.

References

AS/NZS 2891.2.1:2014 (2014) Methods of sampling and testing asphalt. Part 2.1: Sample preparation – Mixing, quartering and conditioning of asphalt in the laboratory

AS/NZS 2891.2.2:2014 (2014) Methods of sampling and testing asphalt. Part 2.2: Sample preparation-Compaction of asphalt test specimens using a gyratory compactor

Australian Standard (2005) Commentary to AG: PT/T220 - Sample Preparation - Compaction of Asphalt Slabs Suitable for Characterisation. Standards Australia

Azari H (2014) Precision estimates for AASHTO T 324, 'Hamburg wheel-track testing of compacted hot mix asphalt (HMA)', National Cooperative Highway Research Program, Research Results Digest 390

Beercroft A, Petho L (2015) Commissioning of hamburg wheel tracking device (HWDT). Final report. National Asset Centre of Excellence (NACOE)

Biligiri KP, Kaloush KE, Mamlouk MS, Witczack MW (2007) Rational modeling of tertiary flow for asphalt mixtures. Transp Res Rec: J Transp Res Board 2001(1):63–72

Cox JA, Van Frank KM, Romero P (2013) On the variability of results from the Hamburg wheel tracker device. In: Sulbaran T (ed) Proceedings of the 49th ASC annual international conference, San Luis Obispo, California, USA, 11 p
Ebrahimi MG (2015) Investigation of viscoelastic behaviour and permanent deformation modelling for New Zealand hot mix asphalts. Doctoral thesis, University of Canterbury, Christchurch, New Zealand
Saleh M (2018) Modified wheel tracker as a potential replacement for the current conventional wheel trackers. Int J Pavement Eng 21:1–9
Walubita LF, Zhang F, Das G, Scullion T (2012) Hot-mix asphalt permanent deformation evaluated by hamburg wheel tracking, dynamic modulus, and repeated load tests. Transp Res Rec: J Transp Res Board 2012:46–56

Simulation of Heavy Weight Deflectometer Test: Spectral Element Method vs Finite Element Method

Jean-Marie Roussel, Hervé Di Benedetto, Cédric Sauzéat, and Michaël Broutin

Abstract The Heavy Weight Deflectometer (HWD) is a nondestructive testing device widely used for airfield pavement assessment. The HWD test consists in applying a wheel-representative load, generated by a falling mass, at the pavement surface. Deflections time histories are recorded by geophones located at several distances from the applied load. Inverse analysis procedures are needed to estimate layers mechanical properties, such as Young's moduli for Linear Elastic (LE) materials or complex moduli for Linear Viscoelastic (LVE) materials. Most of inverse analysis processes consider (i) LE behaviour for pavement layers and (ii) static HWD loading. However, it has been shown that these hypotheses are leading to misestimate bituminous layers modulus and, in most cases, to poorly optimized maintenance. In order to better take into account bituminous layers behaviour, it is required to use LVE constitutive model and adapted numerical methods. This study presents a comparison between measured deflections and deflections computed using the Spectral Element Method (SEM) which was implemented during this work. The tested pavement was built within the test facility of the French Civil Aviation Technical Center (STAC). It is equipped with temperature probes at several depths within bituminous layers. The Spectral Element Method is a semi-analytical method in frequency domain. The pavement response is computed for unit load and for each frequency (from 1–200 Hz, every 1 Hz) at each geophone location, which leads to the Frequency Response Function (FRF). Then, frequency-domain deflections are obtained by weighting the FRF when calculating the Fast Fourier Transform (FFT) of the measured HWD load. This method is numerically efficient and shows good potential to solve axisymmetric multilayered problems. In particular, it allows to take into account frequency-dependent complex modulus analytical expressions. Spectral Element Method results have been validated with time-domain Finite Element

J.-M. Roussel (✉) · H. Di Benedetto · C. Sauzéat
LTDS (UMR CNRS 5513), University of Lyon/ENTPE, Rue M. Audin, 69518 Vaulx-en-Velin, France
e-mail: jean-marie.roussel@entpe.fr

J.-M. Roussel · M. Broutin
STAC, French Civil Aviation Technical Centre, 31 Avenue du Maréchal Leclerc, 94380 Bonneuil-sur-Marne, France

C. Raab (ed.), *Proceedings of the 9th International Conference on Maintenance and Rehabilitation of Pavements—Mairepav9*, Lecture Notes in Civil Engineering 76,
https://doi.org/10.1007/978-3-030-48679-2_65

Method data computed on a common pavement structure. SEM which is much faster than FEM shows great potential for further backcalculations application.

Keywords Heavy Weight Deflectometer · Inertia effects · Spectral Element Method · Finite Element Method · Linear viscoelasticity

1 Introduction

Pavement assessment is a major decision-making tool for road network and airport managers It provides structural information about bearing capacity of the pavement and allows for optimized maintenance and rehabilitation operations. A good structural monitoring is therefore environmentally and economically profitable.

The Heavy Weight Deflectometer (HWD) is the international reference tool for airfield pavement assessment. This device generates a load by the mean of a falling mass. This load is applied to the pavement surface through a load plate (see Fig. 1). HWD load is assumed to be representative of a high speed moving wheel load (Bretonnière 1963). A load cell is placed above the load plate and deflections (vertical surface displacements) are recorded with geophones. The studied HWD is equipped with a falling 720 kg mass which can generate a 300 kN dynamic load on a 45 cm diameter load plate. Nine geophones are positioned at several distances from the load center, until 210 cm. A typical set of data is presented on Fig. 2. The pseudo-static deflection basin is obtained by retrieving the maximum deflection from each geophone and plotting it as a function of radial distance (STAC 2014).

The most classical way to analyse HWD data is the backcalculation. It consists in a mechanical model for which parameters are calculated at each iteration of a fitting algorithm. The final set of parameters is obtained when the difference between numerical and experimental data is lower than a specified value. The most critical step of this backcalculation is the mechanical model. The quality of backcalculated parameter largely relies on the chosen hypothesis. The most common mechanical models consider a static loading and purely elastic materials. However, it has been shown that, in the case of HWD test, (i) dynamic effects are non-negligible and (ii) bituminous materials have a linear viscoelastic behavior. Hence, most of currently used backcalculation software lead to significant approximations (Broutin 2010).

This article presents and evaluates an advanced method for HWD forward simulation. This includes a dynamic and Linear ViscoElastic (LVE) mechanical model solved in the frequency-domain by the Spectral Element Method (SEM).

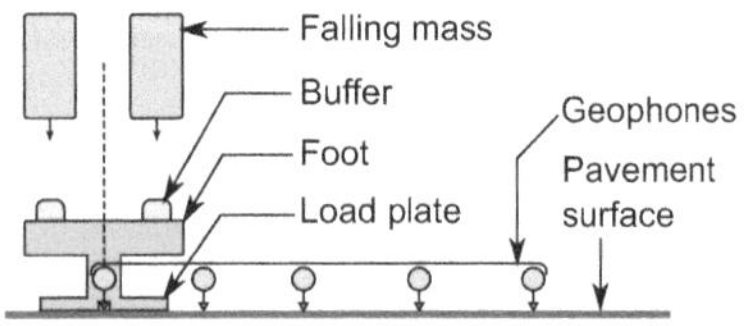

Fig. 1 Simplified scheme of a HWD with the loading system and geophones

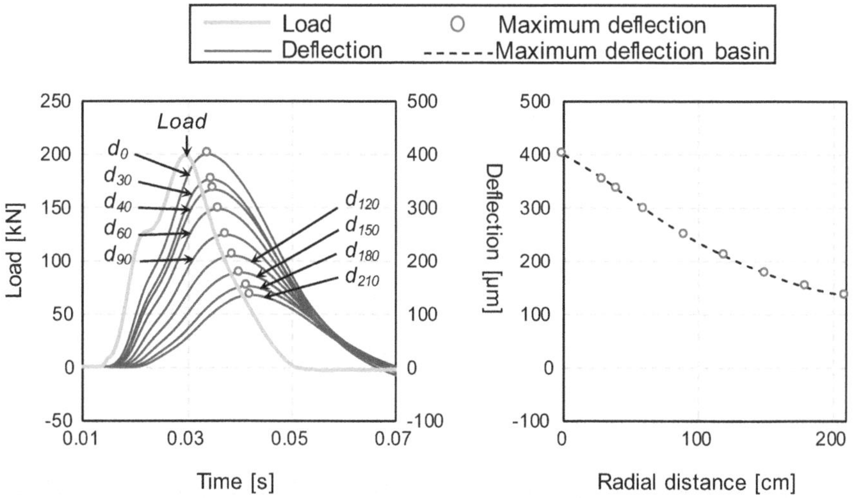

Fig. 2 Typical HWD data. Load and deflections time histories (left) and pseudo-static deflection basin (right). d_r is the deflection at a radial distance r from the load center (in cm)

2 General Framework

The mechanical model is based on an equilibrium equation describing physic phenomena. Materials behavior is taken into account with constitutive equations derived from rheological models, which give the stress-strain relationship. Finally, boundary conditions specify layers interactions and infinite conditions in-plane and in-depth.

2.1 General Equations with Inertia Effects

HWD test geometry consists in a circular load applied at the surface of a pavement made of horizontal layers. This geometry is therefore axisymmetric. Moreover, there is no torsion, meaning that shear stress components $\sigma_{r\theta}$, $\sigma_{\theta r}$ and $\sigma_{\theta z}$, $\sigma_{z\theta}$ are equal to zero. As the HWD test has to be considered as dynamic, inertia effects are taken into account by the mean of the acceleration vector $\boldsymbol{\gamma} = \left(\gamma_r\ \gamma_\theta\ \gamma_z\right)^t$ and the bulk density ρ, leading to Eq. (1).

$$\begin{cases} \frac{\partial \sigma_{rr}}{\partial r} + \frac{\partial \sigma_{rz}}{\partial z} + \frac{\sigma_{rr} - \sigma_{\theta\theta}}{r} = \rho \gamma_r \\ \frac{\partial \sigma_{zz}}{\partial z} + \frac{\partial \sigma_{rz}}{\partial r} + \frac{\sigma_{rz}}{r} = \rho \gamma_z \end{cases} \tag{1}$$

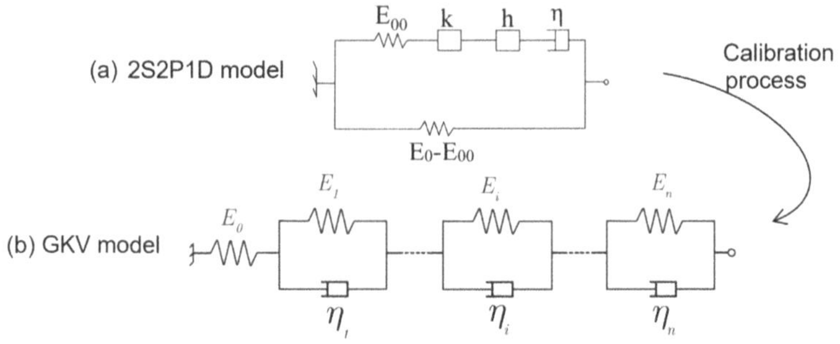

Fig. 3 Analogical representations of the 2S2P1D **a** and Generalized Kelvin-Voigt **b** models

2.2 Rheological Models

It has been shown that, under small strain amplitude, bituminous materials have a linear viscoelastic behavior (Olard 2003). This means that their complex modulus, that can be separated in terms of norm and phase angle, is a function of loading frequency. Numerous rheological models have been developed for linear viscoelastic materials. In particular, the 2S2P1D, a continuous spectrum model, has been shown to simulate very accurately bituminous materials behavior all over the frequency range. The Generalized Kelvin-Voigt (GKV) model, with n Kelvin-Voigt bodies is a discrete spectrum model which also well simulates bituminous materials when n becomes large. A calibration procedure developed by Tiouajni (2011) allows to calibrate E_i and η_i from 2S2P1D constants. Analogical representations of these two models are presented in Fig. 3.

$$E^*_{2S2P1D}(\omega) = E_{00} + \frac{E_0 - E_{00}}{1 + \delta(i\omega\tau)^{-k} + (i\omega\tau)^{-h} + (i\omega\beta\tau)^{-1}} \tag{2}$$

$$E^*_{GKV}(\omega) = \left(\frac{1}{E_0} + \sum_{i=1}^{n} \frac{1}{E_i + i\omega\eta_i}\right)^{-1} \tag{3}$$

In order to better take into account bituminous behavior, linear viscoelasticity has been used with the presented Spectral Element Method which has the advantage to directly consider the complex modulus $E^*(\omega)$ of rheological models (Eqs. (2) and (3)).

2.3 Boundary Conditions

For the HWD test, the load Q is applied on the load plate area on the pavement surface as a uniform stress. Pavement layers are assumed to be perfectly bonded by ensuring the continuity of vertical and horizontal displacements and normal and shear stresses.

Infinite conditions are analytically applied in-depth and in-place far from the load center.

3 Numerical Methods

3.1 Finite Element Method

A Finite Element Method (FEM) simulation has been set up for the HWD test on the software Comsol®. It consists in a 2D axisymmetric geometry meshed with 197 quadrangular elements. Time-domain linear viscoelasticity calculation has been implemented in the Finite Element Method (Roussel 2019). Only the discrete GKV model has been used due to the fact that continuous models are not suitable for those simulations (Pouget 2010, 2012). The final simulation has 15 000 degrees of freedom. This method will be used further to evaluate and validate an advanced Spectral Element Method.

3.2 Spectral Element Method

The Spectral Element Method is a semi-analytical method in frequency domain. It has first been introduced by Rizzi and Doyle (1992) and then applied to HWD test by Al-Khoury et al. (2001), Grenier et al. (2009), El Ayadi et al. (2012) and Zhao et al. (2015). In this method, each layer is represented by a single element in which displacements, strains and stresses are described using equilibrium and constitutive equations. Boundary conditions are then applied. The pavement response is computed for each frequency (every 1 Hz, from 1 to 200 Hz) at each geophone location for unit load, which gives the Frequency Response Function (FRF). Then, frequency-domain deflections are obtained by weighting the FRF by Fast Fourier Transform (FFT) of the measured HWD load. This method is numerically efficient and shows good potential to solve axisymmetric multilayered problems. In particular, it allows to take into account frequency-dependent complex modulus analytical expressions. In the frame of this work, the SEM has been coded in the commercial software Matlab®.

4 Validation and Application

4.1 Studied Pavement Structure

The studied pavement structure is from the STAC test facility, located at Bonneuil-sur-Marne (France). It is made of four layers detailed in Table 1. The two granular layers, GNT and SOL are assumed to be purely linear elastic. The two bituminous layers, BBA and GB have been laboratory tested with the complex modulus tension-compression test on cylindrical specimens (Gudmarsson et al. 2015). Result examples are given in Fig. 4 in terms of norm and phase angle of the complex modulus for the material BBA. 2S2P1D model has been fitted on experimental data and the GKV model with 14 kV bodies has been calibrated on the 2S2P1D model.

Table 1 Pavement structure of the STAC test facility

Layer	Behavior	Thickness [cm]	Bulk density [kg/m^3]	Modulus [MPa]	Poisson's ratio [-]
Airfield type bituminous mixture (BBA)	LVE	14.7	2540	E^*_{BBA}	0.30
Base type bituminous mixture (GB)	LVE	15.6	2406	E^*_{GB}	0.30
Untreated graded aggregate (GNT)	LE	51.5	2100	275	0.30
Subgrade soil (SOL)	LE	□inf.	1800	165	0.30

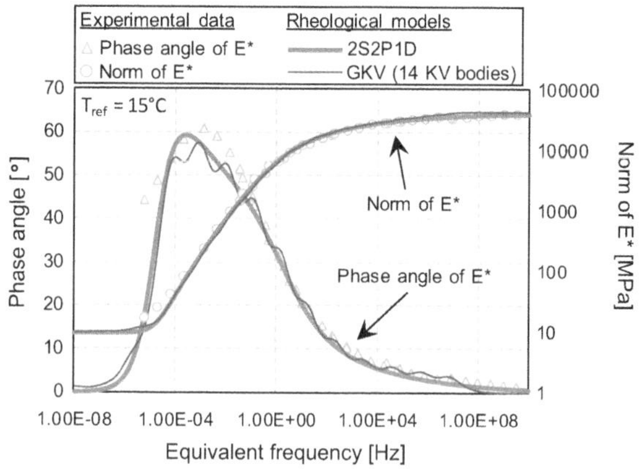

Fig. 4 Master curves of the norm and phase angle of E* used for materials BBA

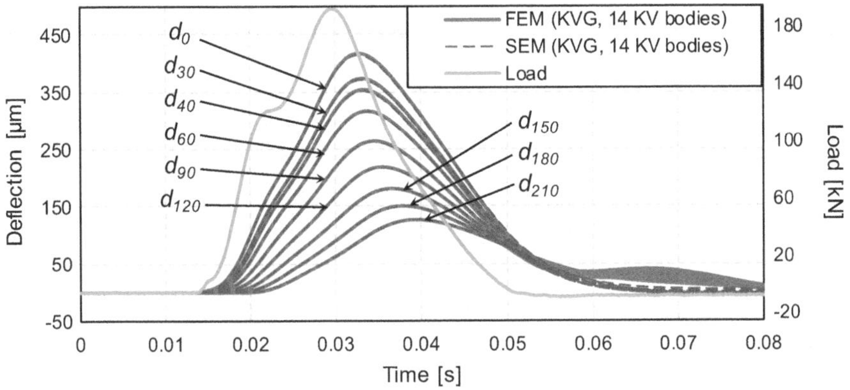

Fig. 5 Comparison between FEM and SEM calculated time histories

4.2 Comparison with Finite Element Method (FEM)

In order to validate the SEM code, a straight comparison has been done with the Finite Element Method. It has been done considering the previously defined pavement structure. FEM and SEM computations have been performed with the same mechanical properties and the same load signal. For LVE materials, the GKV model has been used since it is the only model implemented both in SEM and FEM.

The Fig. 5 shows the considered load and the computed deflections with the two methods. There is a very close correlation between deflections calculated from SEM and from FEM. The Table 2 contains the pseudo-static deflection basins and it can be observed that there is less than 1% of difference on the peak values. The main difference is seen during the free vibration phase ($t > 0.06$ s) where the deflections from FEM are still positive while deflections from SEM have returned to zero. This can be assigned to the reflection phenomenon. Indeed, although low reflective boundaries have been used in the FEM model, reflected waves may not be totally attenuated. This phenomenon seems to decrease when the mesh size increases.

The comparison with FEM validates the SEM code in terms of deflections. Moreover, it shows that the SEM better simulates infinite conditions. The computation times are 40″ for SEM and 2′30″ for FEM on a laptop computer.

4.3 Comparison with HWD Test Measurements

The SEM data have then been compared to HWD test measurements. This data set has been logged on the STAC test facility while pavement mean temperature was 13.5 °C with a weak thermal gradient. This temperature has been taken into account in bituminous mixtures behaviour. Pseudo-static deflection basins are presented in Fig. 6. Maximum deflections are of the same order of magnitude. A better match between

Table 2 Comparison between maximal deflections basins computed with FEM and SEM

Radial distance [m]	0	0.30	0.40	0.60	0.90	1.20	1.50	1.80	2.10
FEM max. deflection [μm]	417.2	373.6	354.3	316.9	289.8	219.1	181.5	151.1	127.0
SEM max. deflection [μm]	414.5	371.4	351.7	315.4	288.3	217.7	180.4	150.6	127.5
Relative difference	0.64%	0.57%	0.73%	0.48%	0.51%	0.64%	0.61%	0.29%	0.36%

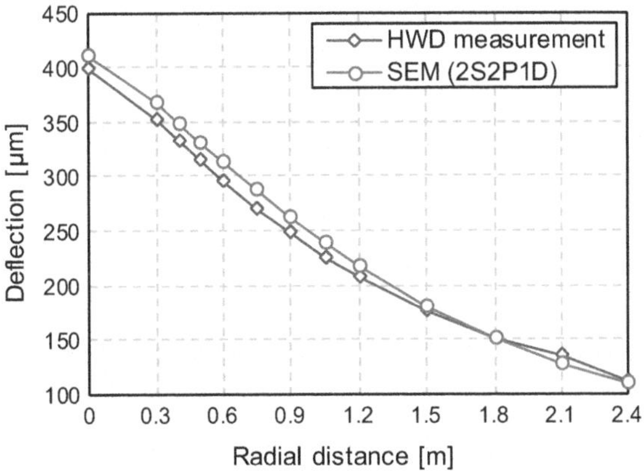

Fig. 6 Comparison between calculated and measured deflection basins

numerical and experimental data could be obtained by backcalculating materials mechanical properties.

5 Conclusion and Outlooks

A the Spectral Element Method, which integrates dynamic effects and linear elastic as well as linear viscoelastic constitutive equations has been presented in this paper. Results have been validated with time-domain Finite Element Method data computed on a common pavement structure. SEM method shows good potential to simulate HWD test and for being used in backcalculation. Comparison with experimental results from a test section is also proposed.

SEM will be further applied to analyse HWD tests and to determinate layers mechanical properties, including viscoelastic parameters. Further work on this direct simulation model will also consist in improving interfaces conditions between layers. Indeed, it has been shown that interfaces are actually not perfectly bonded and that their behavior can have significant effect on computed deflections (Chupin et al. 2010; Grellet et al. 2018; Sadoun et al. 2016).

References

Al-Khoury R, Kasbergen C, Scarpas A, Blaauwendraad J (2001) Spectral element technique for efficient parameter identification of layered media Part I: forward calculation. Int J Solids Struct 38(48–49):8753–8772

Bretonnière S (1963) Les déflectomètres à boulet pour l'étude des déflexions des chaussées sous charges dynamiques. Bulletin Des Laboratoires Des Ponts et Chaussées

Broutin M (2010) Assessment of flexible airfield pavements using Heavy Weight Deflectometers, development of a FEM dynamical time-domain analysis for the backcalculation of structural properties. Ecole des Ponts ParisTech
Chupin O, Chabot A, Piau JM, Duhamel D (2010) Influence of sliding interfaces on the response of a layered viscoelastic medium under a moving load. Int J Solids Struct 47(25–26):3435–3446
El Ayadi A, Picoux B, Lefeuve-Mesgouez G, Mesgouez A, Petit C (2012) An improved dynamic model for the study of a flexible pavement. In: Advances in engineering software, vol 44. Elsevier, pp 44–53
Grellet D, Doré G, Chupin O, Piau JM (2018) Highlighting of the viscoelastic behaviour of interfaces in asphalt pavements–a possible origin to top-down cracking. Road Mater Pavement Des 19(3):581–590
Grenier S, Konrad J-M, LeBœuf D (2009) Dynamic simulation of falling weight deflectometer tests on flexible pavements using the spectral element method: forward calculations. Can J Civ Eng 36(6):944–956
Gudmarsson A, Ryden N, Di Benedetto H, Sauzéat C (2015) Complex modulus and complex Poisson's ratio from cyclic and dynamic modal testing of asphalt concrete. Constr Build Mater 88:20–31
Olard F, Di Benedetto H (2003) General "2S2P1D" model and relation between the linear viscoelastic behaviours of bituminous binders and mixes. Road Mater Pavement Des 4(2):185–224
Pouget S, Sauzéat C, Di Benedetto H, Olard F (2010) From the behavior of constituent materials to the calculation and design of orthotropic bridge structures. Road Mater Pavement Des 11(sup1):111–144
Pouget S, Sauzéat C, Di Benedetto H, Olard F (2012) Modeling of viscous bituminous wearing course materials on orthotropic steel deck. Mater Struct 45(7):1115–1125
Rizzi SA, Doyle JF (1992) A spectral element approach to wave motion in layered solids. J Vib Acoust 114:569–577
Roussel JM, Sauzéat C, Di Benedetto H, Broutin M (2019) Numerical simulation of falling/heavy weight deflectometer test considering linear viscoelastic behaviour in bituminous layers and inertia effects. Road Mater Pavement Des 20(sup1):S64–S78
Sadoun A, Broutin M, Simonin JM (2016) Assessment of HWD ability to detect debonding of pavement layer interfaces. In: 8th RILEM international conference on mechanisms of cracking and debonding in pavements, vol 13. Springer, Dordrecht, pp 763–769
STAC (2014) Auscultation des chaussées souples aéronautiques au HWD. Technical Guide
Tiouajni S, Di Benedetto H, Sauzéat C, Pouget S (2011) Approximation of linear viscoelastic model in the 3 dimensional case with mechanical analogues of finite size. Road Mater Pavement Des 12(4):897–930
Zhao Y, Cao D, Chen P (2015) Dynamic backcalculation of asphalt pavement layer properties using spectral element method. Road Mater Pavement Des 16(4):870–888

Three Dimensional Finite Element Model for Active Crack Control in Continuously Reinforced Concrete Pavement

Muhammad Kashif, Pieter De Winne, Ahsan Naseem, Nouman Iqbal, and Hans De Backer

Abstract Crack patterns in continuously reinforced concrete pavements (CRCPs) traditionally have been controlled by continuous longitudinal reinforcement steel. This passive crack control has resulted in the formation of an unfavorable crack pattern with a high probability of clusters of closely spaced cracks, which has eventually led to premature distresses such as spalling and punch-out in the later age of CRCP. In an effort to eliminate the cluster cracking and crack meandering, the standard design concept for CRCP in Belgium underwent several changes over time, mainly addressing the longitudinal reinforcement rate, depth of the reinforcement steel, and thickness of the concrete slab. In the current design concept, the active crack control method in the form of partial surface saw-cuts on side of the concrete slab perpendicular to the axis of the road within 16–24 h after the placement of concrete is being employed to facilitate the formation of a regular spaced crack pattern in CRCP. However, this area needs further investigation and validation. The present study investigates the early-age crack pattern induced by active crack control method under typical Belgian conditions. Therefore, a 3D finite element model of CRCP with induced partial surface saw-cuts is developed using FE tool Diana 10.2. Findings show that active crack control method exhibits the cracking sooner than passive crack control method. Moreover, a more controlled and regular spaced crack pattern is produced.

Keywords Active crack control · Passive crack control · CRCP · 3D FE analysis

1 Introduction

Continuously reinforced concrete pavement (CRCP) is a rigid pavement structure that consists of a concrete slab reinforced throughout its entire length by longitudinal and transverse reinforcement. The continuous steel reinforcement eliminates the need for transverse expansion or contraction joints. Transverse cracking is allowed to form

M. Kashif (✉) · P. De Winne · A. Naseem · N. Iqbal · H. De Backer
Department of Civil Engineering, Ghent University, Technologiepark 60, B9052 Zwijnaarde, Ghent, Belgium
e-mail: muhammad.kashif@ugent.be

C. Raab (ed.), *Proceedings of the 9th International Conference on Maintenance and Rehabilitation of Pavements—Mairepav9*, Lecture Notes in Civil Engineering 76,
https://doi.org/10.1007/978-3-030-48679-2_66

freely as a result of longitudinal dry shrinkage of concrete, but the cracks are held tightly together with the continuous longitudinal reinforcement steel, which ensures that the crack opening remains relatively small and thus prevents further deterioration of the pavement structure. This type of pavement structure is mostly preferred for high priority routes due to its durability, stability and low maintenance requirements (Verhoeven 1993; Hall et al. 2007; Rasmussen et al. 2009).

For more than 40 years, Belgium has been using CRCP to build the motor-way network and other high priority roads. The proponents of CRCPs cite durability, sustainability and low maintenance cost which made it as long-lasting applications especially in Belgium. However, the crack cluster formation still occurs. Field observations of recently constructed CRCPs in Belgium following the current design concept, characterize the developed transverse crack pattern as low mean crack spacing along with a high percentage of clusters of closely spaced cracks (Rens and Beeldens 2010; Ren et al. 2013, Rens et al. 2013; Ren et al. 2014).

Initially, the idea of active crack control or induced cracking was applied to jointed plain concrete pavement (JPCP) and jointed reinforced concrete pavement (JRCP). The idea of active crack control was firstly adopted in full-scale CRCP field test sections in the United States (McCullough and Dossey 1999; Kohler and Roesler 2004). Their studies revealed that the active crack control method could be an effective attempt to induce the transverse cracks sooner, straighter and at the designated locations to decrease the risk of severe distress such as punch-out in CRCP. In the United States, the active crack control technique is applied to CRCP by either automated tape insertion or transverse saw-cut through the whole width of a concrete slab.

A new modified active crack control method in the form of partial surface saw-cuts was proposed in 2012 for Belgian conditions (Rens et al. 2013; Ren et al. 2013). It was first applied in the reconstruction project of Motorway E313 near the city Herentals, Belgium. The partial surface saw-cuts of 400 mm long were made on one side of the pavement slab at spacing of 1200 mm by using a cutting disk. The present research is conducted to study the cracking behavior of CRCP under typical Belgian field conditions. For this purpose, the 3D finite element (FE) models of a CRCP segment with active and passive control methods are developed by using FE tool Diana 10.2. The transverse crack pattern obtained from FE models is compared and discussed with respect to the field observations of CRCP in Belgium.

2 Finite Element Modeling of CRCP

The FE tool DIANA 10.2 is used to develop 3D FE models of a CRCP segment in this study, following the current standard design for CRCP in Belgium. The spacing between saw-cuts is taken equal to 1200 mm (Ren 2015). The length, width and depth of a CRCP segment are assumed as 2400, 1800 and 250 mm respectively by taking an advantage of symmetry. The behavior of CRCP can be assumed to be symmetric with respect to the center of the two adjacent transverse cracks as well as with respect to the center of the lane under environmental loading. Therefore, one-half of the slab is considered on either side of saw-cuts and other half of the lane is taken into account by considering the proper boundary conditions (Al-Qadi and Elseifi 2006; Choi et al. 2011). The geometry of the CRCP segment with the position of saw-cuts and the reinforcing steel layout is illustrated in Fig. 1.

The Euro-code model is used as a reference constitutive concrete model for real mimicking of the early-age crack development in a CRCP segment under the typical Belgium conditions. The input parameters used in the FE model are summarized in Table 1. Thermal conductivity of concrete is assumed to be 3.0 W/m °C. The heat transfer mechanism between the concrete slab and its surrounding is defined through convection only. The heat transfer coefficient is taken as 8.55 W/m^2 °C. Reinforcing steel bars are modeled using the truss elements. Full-bonding contact is assumed between the reinforcing bars and concrete. Although, the calculated stress field is highly dependent on the assumed bond stress and slip, full-bonding is considered in an effort to ensure convergence of the staggered structural-flow analysis. After a mesh sensitivity analysis, the size of element is taken as 50 mm. Two aspects of boundary conditions are considered: (1) structural restraint (2) thermal boundary.

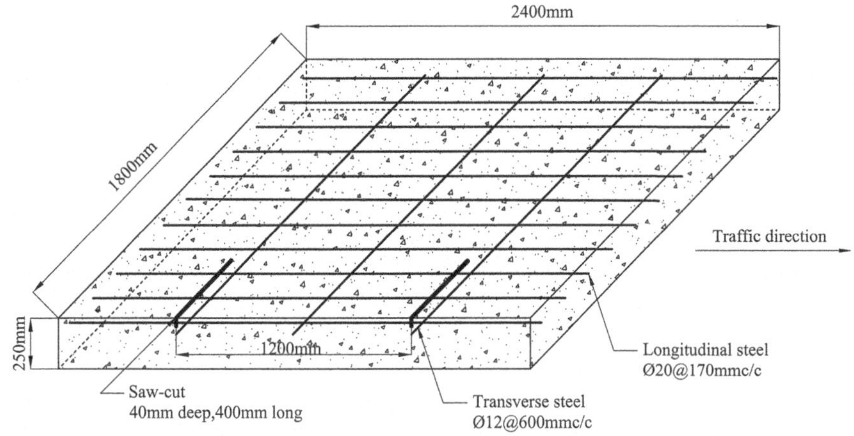

Fig. 1 Geometry of the CRCP segment

Table 1 Material input parameters

Concrete class	C40/50
Cement type	Blast furnace slag cement (CEM III/A)
Aggregate type	Sandstone
Coefficient of thermal expansion of concrete	10.0×10^{-6} (1/°C)
Relative humidity	80%
Thermal conductivity	3.0 W/m °C
Heat transfer coefficient	8.55 W/m^2 °C

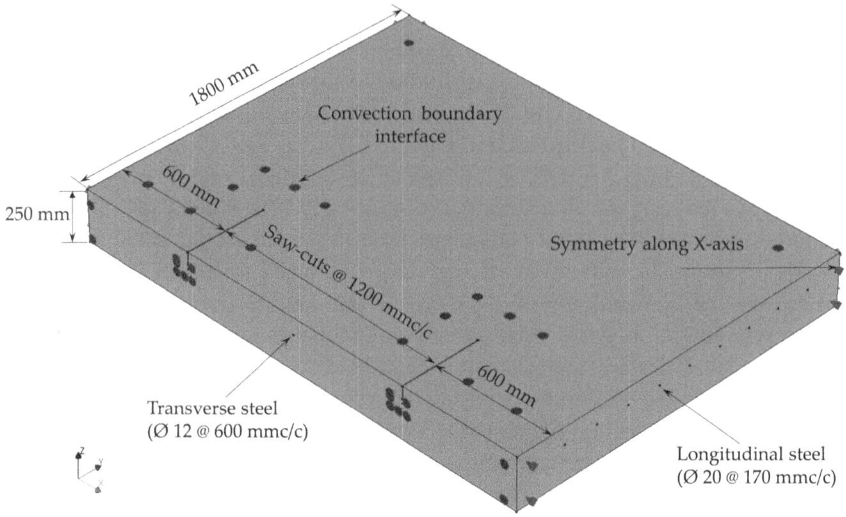

Fig. 2 Boundary conditions of 3D FE model

The structural translation restraint in an upward direction (Uz = 0) is applied at the bottom face of the concrete slab assuming that a stiffer base layer is lying below the concrete slab.

The double symmetry of the geometry is taken along X-axis and Y-axis in order to reduce the computational time and the disk space. Along the width of CRCP segment, the translation in X-direction (Ux = 0) is applied at the symmetry faces. For thermal problem, the top surface and outer edge face of the concrete slab are considered to be directly exposed to external temperature field condition. The structural and thermal boundary conditions of 3D FE model are illustrated in Fig. 2.

The 3D staggered structural-flow analysis is performed in two parts. In the first part of analysis, the structural elements are transformed into flow elements. Then in the second part, temperature output from transient heat flow analysis is used as input for structural nonlinear analysis acting as thermal strains. The external temperature filed applied to the FE model is illustrated in Fig. 3.

Fig. 3 External temperature field

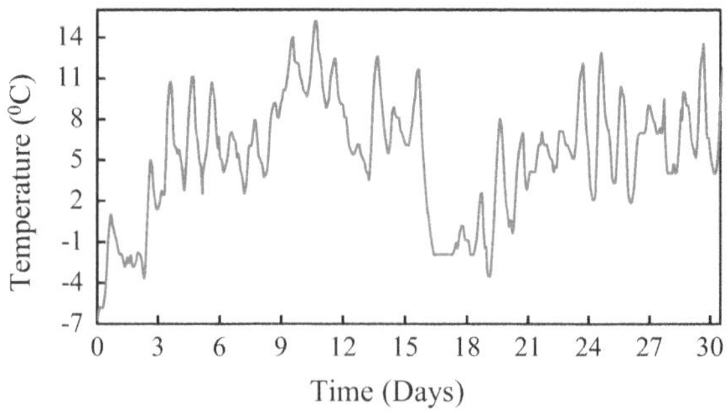

3 Results and Discussions

The development of crack pattern, steel stress in the 3rd longitudinal bar and concrete stress in the vicinity of longitudinal reinforcing steel were studied.

3.1 *Early-Age Crack Pattern*

The present study is mainly focused on the development of early-age crack pattern in a CRCP segment under external temperature field. Cracking occurs when the stress exceeds the tensile strength of concrete. In order to identify the location in the model where the first cracking is expected, the development of tensile strength is plotted against the development of tensile stress along the width of the concrete slab following the location of saw-cuts. The crack initiation in active and passive crack control FE models is demonstrated in Fig. 4. It can clearly be observed that cracking occurs all over the outer edge of the concrete slab in passive nature.

While in case of active crack control, cracking occurs from the saw-cuts. The partial surface saw-cuts in active crack control method act as the most weakest area with respect to the rest of the concrete slab from which cracking occurs. In passive crack control method, cracking occurs naturally in accordance with the boundary conditions applied in FE model. The stress vs strength plot illustrated in Fig. 5 indicates the location of cracking in the concrete slab. The transverse crack pattern produced at the age of 30 days is demonstrated in Fig. 6. Active crack control produces a more regular spaced crack pattern. The cracks initiate from the saw-cut tips and propagate along the width of the concrete slab. No cracking occurs in between the saw-cuts which indicates that the mean crack spacing will be equivalent to the design spacing of partial surface notches in CRCP. On the other hand, cracking takes place along the three transverse steel bars in CRCP segment with passive crack control method as shown in Fig. 6b.

3.2 *Stress Distribution*

In order to see the stress distribution in the longitudinal steel bars and concrete elements near to the steel, the 3rd longitudinal steel bar with respect to the entire model of CRCP is selected as a reference. For passive crack control method, three peaks of steel stress in the reference bar are observed exactly above the location of the transverse steel bars (500, 1200, 1900 mm) where the transverse cracking occurred as shown in Fig. 7. While in case of active crack control method, two stress peaks occurs exactly over the location of induced saw-cuts in CRCP.

The stress distribution in concrete elements near longitudinal steel (3rd longitudinal steel bar) is illustrated in Fig. 8. It can clearly be observed that the maximum

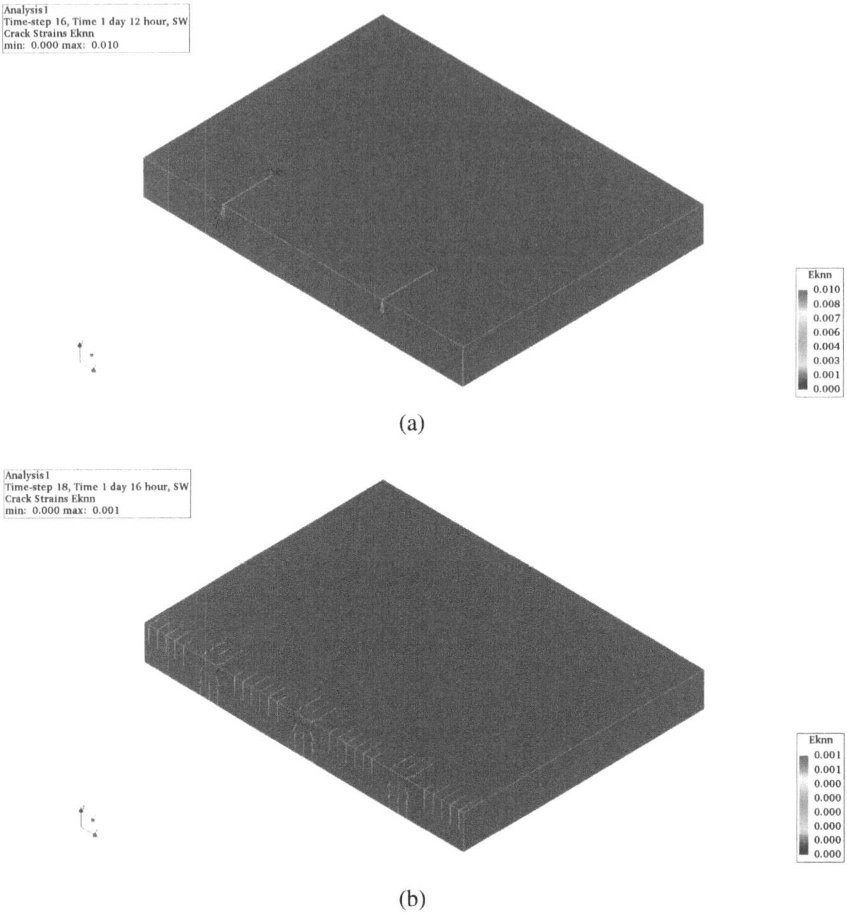

Fig. 4 Crack initiation in CRCP **a** Active crack control **b** Passive crack control

stress occurs over the saw-cuts with respect to the rest of length of the concrete slab. This could be the possible reason for initiating the cracks effectively from the saw-cuts in active crack control method. Therefore, the cluster of cracks initiates from all over the saw-cuts and propagate along the width of the pavement slab as shown in Fig. 6a. Passive crack control method exhibits the smaller stress distribution in comparison to active crack control method as there is no cluster of cracks as shown in Fig. 6b. In center of the concrete slab, the concrete stress becomes constant.

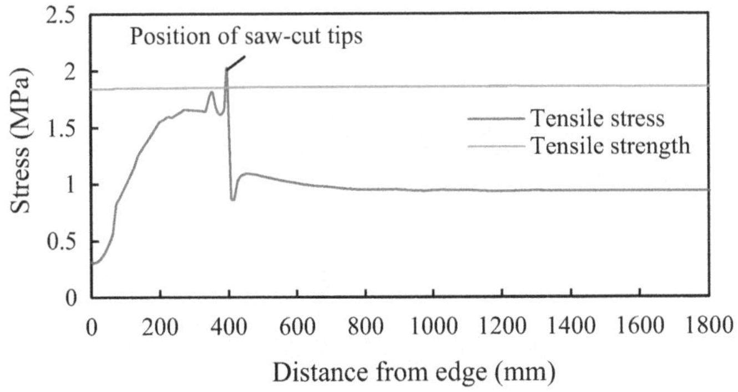

(a)

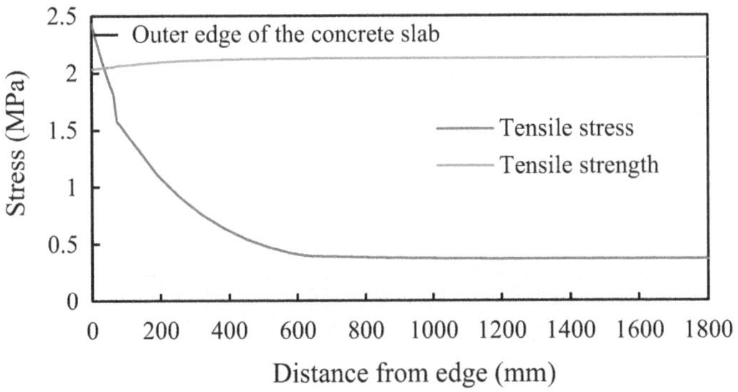

(b)

Fig. 5 Stress vs strength plot along the width of the concrete slab **a** Active crack control **b** passive crack control

4 Findings

1. Cracks imitate from the saw-cut tips and propagate along the width of the concrete slab.
2. Passive crack control method exhibits the crack initiation from the outer edge of the concrete slab and crack propagation all along the transverse steel.
3. Active crack control method exhibits the cracking sooner than passive crack control method.
4. Maximum average stress produced in 3rd longitudinal reinforcing bar by active crack control method is 15% greater than that produced by passive crack control method.
5. Active crack control produces a significant amount of stress in concrete elements in the vicinity of longitudinal steel at the crack interface in comparison to passive crack control.

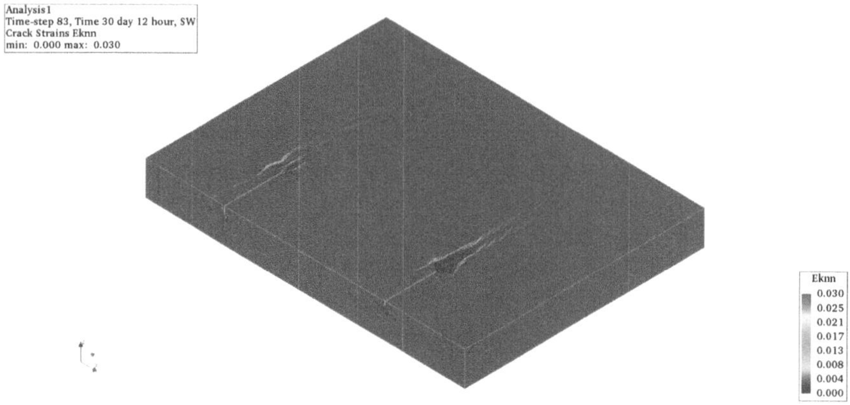

(a)

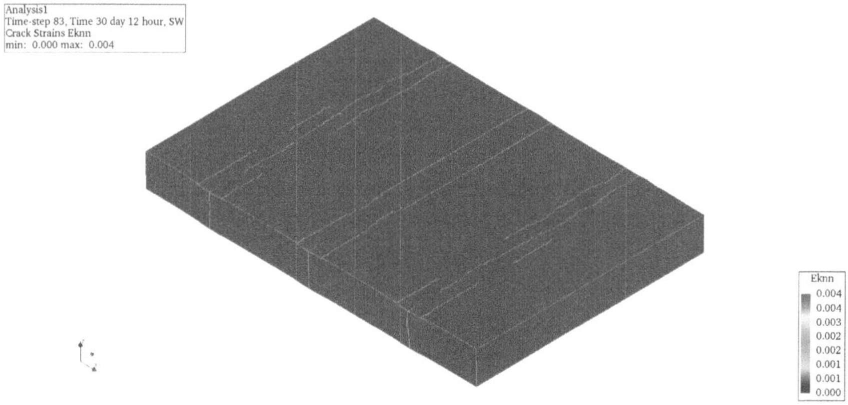

(b)

Fig. 6 Transverse crack pattern in CRCP **a** Active crack control **b** Passive crack control

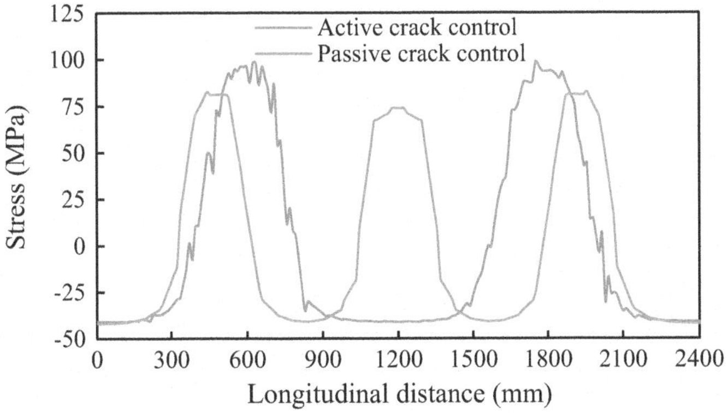

Fig. 7 Stress distribution in 3rd longitudinal steel bar

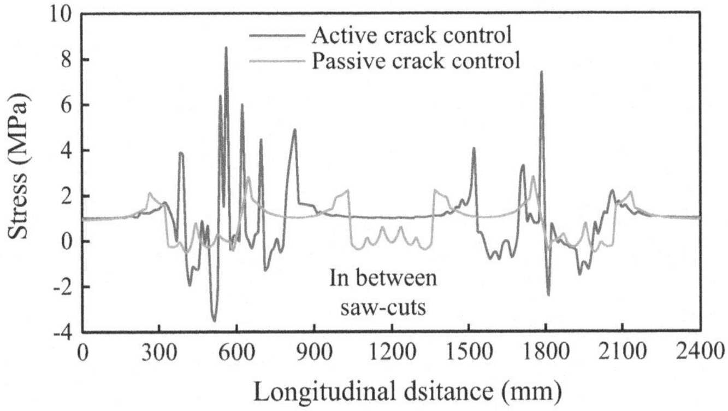

Fig. 8 Stress distribution in concrete elements near 3rd longitudinal steel bar

References

Al-Qadi IL, Elseifi M (2006) Mechanism and modeling of transverse cracking development in continuously reinforced concrete pavement. Int J Pavement Eng 7(4):341–349

Choi S, Ha S, Won MC (2011) Horizontal cracking of continuously reinforced concrete pavement under environmental loadings. Constr Build Mater 25(11):4250–4262

Hall K, Dawood D, Vanikar S, Tally Jr, R, Cackler T, Correa A, Deem P, Duit J, Geary G, Gisi A (2007) Long-life concrete pavements in Europe and Canada

Kohler E, Roesler J (2004) Active crack control for continuously reinforced concrete pavements. Transp Res Rec 1900(1):19–29

McCullough BF, Dossey T (1999) Considerations for high-performance concrete paving: recommendations from 20 years of field experience in Texas. Transp Res Rec 1684(1):17–24

Rasmussen RO, Rogers R, Ferragut TR (2009) Continuously reinforced concrete pavement design & construction guidelines. Technical Manual, Federal_Highway Administration and Concrete Reinforcing Steel Institute, Washington, DC and Schaumburg, IL. http://www.crsi.org/transportation/fhwa_crcp/pdf/CRCP_Design-Const_Manual[draft].pdf

Ren D (2015) Optimisation of the crack pattern in continuously reinforced concrete pavements
Ren D, Houben L, Rens L (2013) Cracking behavior of continuously reinforced concrete pavements in Belgium: characterization of current design concept. Transp Res Rec 2367(1):97–106
Ren D, Houben L, Rens L, Beeldens A (2014) Active crack control for continuously reinforced concrete pavements in Belgium through partial surface notches. Transp Res Rec 2456(1):33–41
Rens L, Beeldens A (2010) The behaviour of CRCP in Belgium: observation and measurement of crack pattern, bond and thermal movement. In: Proceedings of the 7th international DUT-workshop on design and performance of sustainable and durable concrete pavements
Rens L, Winne P, Beeldens A (2013) Continuously reinforced concrete pavement: new development for a sustainable concept. Belgian Road Congress, Liege, Belgium
Verhoeven K (1993) Cracking and corrosion in continuously reinforced concrete pavements. In: Fifth international conference on concrete pavement design and rehabilitation Purdue University, School of Civil Engineering; Federal Highway Administration; Portland Cement Association; Transportation Research Board; Indiana Department of Transportation; Federal Aviation Administration; and American Concrete Pavement Association

Inference of Pavement Properties with Roadside Accelerometers

Julius Nielsen, Eyal Levenberg, and Asmus Skar

Abstract An array of four synchronized single-axis accelerometers was fixed to the surface of an asphalt pavement. Vertical acceleration traces triggered by several nearby passes of a truck with known characteristics were recorded. The work focused on presenting and demonstrating an interpretation method for inferring the mechanical properties of the pavement system based on the recorded accelerations. In general terms, the method was based on careful low-pass filtering the field-measured acceleration traces, and then best-matching them with a corresponding set of calculated acceleration traces. For this purpose, the pavement system was modeled as a two-layered linear elastic half-space, and a model-guided signal filtering approach was devised to ensure that irrelevant signal content is removed prior to the matching. Based on the analysis of six separate truck passes it was noticed that the inferred upper layer modulus exhibited medium variability (coefficient of variation of 45%) while the lower (subgrade) modulus showed little variability (coefficient of variation of 8%). The moduli values displayed fair agreement with those independently estimated from non-destructive and semi-destructive tests. By analyzing many more passes inferred moduli are expected to become more representative. Overall, the method seems workable and scalable, with capacity to handle any number of acceleration sensors as well as other sensor types.

Keywords Pavement condition evaluation · Pavement sensing · Pavement accelerations · Mechanical pavement properties · Model-guided filtering

1 Introduction

Recent years has seen marked increase in public dependency on the service level and overall functionality provided by pavements. This increased dependency is accompanied by user intolerance to service interruptions, part of which are due to activities

J. Nielsen · E. Levenberg (✉) · A. Skar
Department of Civil Engineering, Technical University of Denmark, Nordvej, Building 119, 2800 Kgs. Lyngby, Denmark
e-mail: eylev@byg.dtu.dk

C. Raab (ed.), *Proceedings of the 9th International Conference on Maintenance and Rehabilitation of Pavements—Mairepav9*, Lecture Notes in Civil Engineering 76,
https://doi.org/10.1007/978-3-030-48679-2_67

taken for evaluating mechanical condition. Such condition evaluation data are essential for optimizing maintenance and repair decisions; they are also important for revisiting design decisions, for assessing the performance of non-standard materials or construction technologies, and for estimating loss of structural integrity after natural disasters. Thus the engineering community is challenged to develop non-disruptive technologies for mechanical pavement evaluation.

One approach for addressing the challenge is building and operating mobile sensing platforms that can carry out condition evaluation while moving alongside the users (Harr and Elton 1983; Hildebrand and Rasmussen 2002; Andersen et al. 2017). Another approach, which is the focus of this paper, calls for instrumenting pavements with sensing gear—effectively transforming these traditional systems into smart self-aware constructions (e.g., Lajnef et al. 2011; Klar et al. 2016; Skar et al. 2019). Specifically advocated herein is the use of accelerometers, which of all sensor types, are deemed most suited for installation over wide areas (Ryynäenen et al. 2014; Levenberg et al. 2014; Arraigada et al. 2009).

The aim of the current contribution is to further advance the idea and techniques for inferring pavement mechanical condition based on measured acceleration traces. The overall approach taken is similar to that in Levenberg (2012) and in Levenberg (2015) where measured signals are not double integrated, but carefully filtered and then directly matched against model-generated signals. The new advances being explored herein are: (i) utilizing measurements obtained from an array of synchronized accelerometers and not from a single sensor; (ii) investigating an improved filter concept that is based on truncating a Fourier series and not on a convolution with a smoothing kernel; and (iii) placement of the sensors roadside at the pavement surface and not implanting them below the ride surface. All aforementioned advancement points are closely linked to the goal of developing a roadside sensing platform for non-disruptive condition evaluation of pavements.

2 Experimental Setup and Field Measurements

The experimental investigation was performed in Henrik Dams Allé (Kongens Lyngby, Denmark), which is a single-carriageway straight asphalt road that is 610 m long. This road has a paved width of 8 m and includes shallow block-paved drainage ditches on both sides. The pavement structure is 650 mm thick, composed of 120 mm asphalt concrete and 530 mm of combined unbound base and subbase. The subgrade is a relatively stiff sandy clay extending to a large depth. The test section was located in the center of a nominally flat stretch; it was instrumented with four single-axis accelerometers (named Acc#1...Acc#4) that were fixed to the asphalt surface via 60 mm long screws. The sensors were spaced 5 m apart in the travel direction, with varying transverse offsets from the road centerline: 2.5 m for Acc#1 and Acc#4, 2.7 m for Acc#2, and 2.9 m for Acc#3. Figure 1a offers a plan view of the test section, displaying the positions of the sensors. Also shown are three nearby locations where asphalt concrete core samples were obtained for further laboratory investigation, and

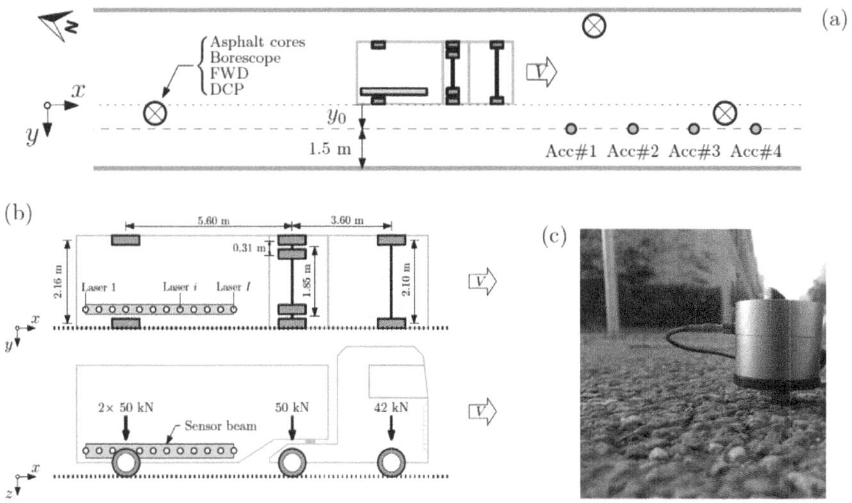

Fig. 1 Experimental setup: **a** top view (x-y plane) of the test section, **b** sketch of the Raptor truck, and **c** photograph of one accelerometer fixed to the pavement surface

where in situ tests were carried out to assess layer thicknesses, record dynamic cone penetration resistance, and collect falling weight deflections.

The accelerometers were all identical, model KB12VD manufactured by Metra Mess (Germany). A picture of one such accelerometer attached to the pavement surface is offered in Fig. 1c. The KB12VD sensors are analog, characterized by a measurement range of ±0.6 g and measurement sensitivity better than 1 μg. Two interlinked M312 units (also manufactured by Metra Mess) were employed for powering the sensors, data acquisition, and analog to digital conversion (24 bit). The M312 units were connected to the accelerometers via long cables and also to a laptop via two USB cables for data logging; they offered synchronized acquisition at a native rate of 96 kHz. Given that the KB12VD sensors are not suited for monitoring static or very slow-occurring responses, native measurements were preprocessed by high-pass filtering with a cutoff frequency of 1 Hz. Moreover, in order to produce a workable data size, and without loss of relevant information, measurements were down-converted to 1 kHz by application of a low-pass filter with 500 Hz cutoff and subsequent downsampling. Acceleration responses were generated by the Dynatest Raptor, a heavy truck that hosts a beam of lasers (as well as other sensors) designed to perform pavement evaluation while driving. Figure 1a superposes a silhouette of the Raptor over the experimental arrangement; the travel speed is assumed constant and denoted by V. Figure 1b offers side and top view sketches of the truck, showing tire configuration and indicating axle loads. A right-handed Cartesian coordinate system is introduced with the x-axis pointing in the vehicle travel direction, the z-axis pointing downward into the pavement system, and the y-axis pointing transversely to the travel direction with its origin coinciding with the edge of the rear right tire of the track. Assuming the truck is traveling in a straight line parallel to the x-axis, the

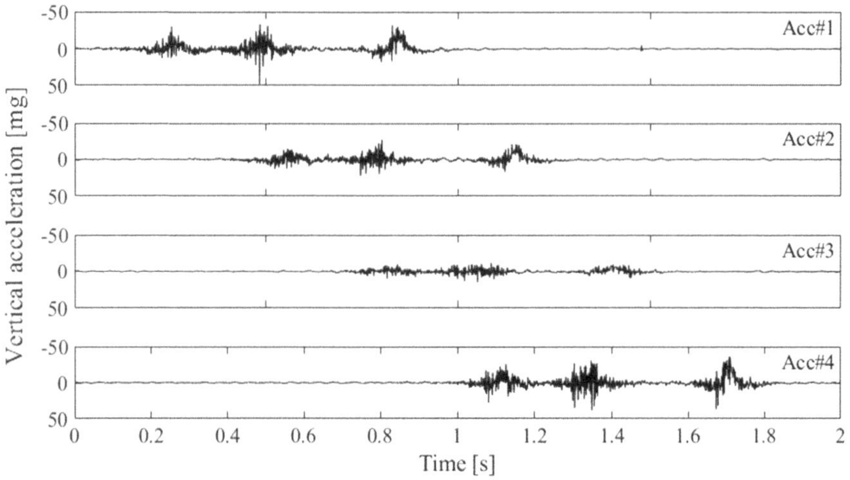

Fig. 2 Measured accelerations in the field experiment for Pass#1 ($y_0 = 250$ mm)

symbol y_0 is introduced to denote the offset distance from the rear right tire edge to Acc#1 or equivalently to Acc#4.

Testing was done while the road was partially closed to traffic. The Raptor truck was instructed to carry out several passes near the sensing array at different speeds. All passes were recorded on video, and the lateral offsets, i.e. the y_0's, were evaluated based on the video footage. To facilitate the offset determination, equidistant lines with 100 mm spacing were marked on the pavement surface prior to the truck passes. The accuracy level at which lateral offsets were determined with this procedure is estimated to be about ±50 mm. Hereafter, data from one of the passes with $y_0 = 250$ mm (Pass#1) are presented and subsequently utilized to demonstrate the interpretation method. This choice is inconsequential because the underlying procedure is essentially generic. The measured accelerations are presented in Fig. 2 with a common timeline. As can be seen, the different sensors recorded the pass with some time lag representing the actual driving speed (Levenberg 2014). Given the known vehicle dimensions, the truck passing speed was assessed to be $V = 57$ km/h. Moreover, it can also be seen that the acceleration intensities vary in accordance with the lateral distance from the truck, e.g., the signal peaks of Acc#1 and Acc#4 are relatively similar to one another, while the signal of Acc#3 seems weakest. As expected (Levenberg 2015), the lateral offset is seen to have a strong influence on the signal strength; a transverse offset of merely 200 mm considerably attenuates the acceleration signal.

3 Data Interpretation Method

The interpretation method described hereafter is focused on assessing the mechanical properties of a tested pavement based on roadside accelerations such as those shown in Fig. 2. In general terms, the assessment is sought by best fitting all measured acceleration traces with calculated acceleration traces obtained from a computational pavement model. In this connection, the classic two-layered linear elastic half-space model was chosen (Burmister 1945), with two unknowns to be inferred: E_1 Young's modulus of the top layer representing the structure, and E_2 Young's modulus of the half-space representing the subgrade. To expedite the numerical computations required for the modeling, the acceleration technique proposed in Andersen et al. (2018) was employed. Taken as known in the fitting procedure are: the thickness of the structure layer (650 mm), the Poisson's ratios of the two layers ($\nu_1 = 0.35$, and $\nu_2 = 0.40$), the travel speed of the truck (V), the truck's tire configuration and axle loads (Fig. 1b), and the lateral offset y_0 (Fig. 1a).

The fitting procedure was carried out in several steps: (i) for a trial values of E_1 and E_2 the layered pavement model was engaged to calculate vertical surface displacement traces $u_z(t)$'s at the sensor locations; (ii) a double derivative with respect to time is applied to the displacement traces (finite difference formula) to arrive at calculated accelerations corresponding to the measured signals; (iii) an individual objective function was defined to quantify the matching error between a given measured acceleration trace and it's corresponding calculated trace; (iv) a global objective function was defined, combining all individual objective functions into a single scalar entity that represents the matching error between all measured and calculated accelerations; and (v) a search was performed to find optimal moduli that minimize the global objective function. Ultimately, the optimal moduli set was deemed representative of the in situ mechanical pavement layer properties.

The above described procedure requires an additional (intermediate) step prior to defining an individual error function (Levenberg 2012, 2015). This step entails low-pass filtering the measured traces to remove high-frequency signal content that originates from vehicle dynamics and from the interaction between rolling tires and a rough textured surface. To graphically illustrate this requirement, Fig. 3a offers a

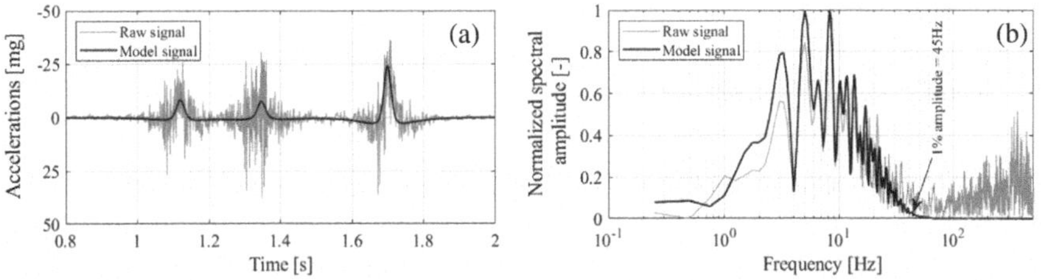

Fig. 3 Synthetic data produced with layer moduli of $E_1 = 300$ MPa and $E_2 = 100$ MPa compared to raw measurements of Acc#4: **a** time domain, and **b** frequency domain

magnified view of Acc#4 signal superposed over a corresponding calculated acceleration trace with an assumed moduli set of $E_1 = 300$ MPa and $E_2 = 100$ MPa. As can be seen, the calculated trace is smooth, clearly reflecting the effects of the different truck axles, both in terms of acceleration directions and magnitudes. On the other hand, the measured trace is very noisy, with acceleration levels abruptly changing sign producing a myriad of peaks. Because of this behavior the measured acceleration traces cannot be directly matched against calculated traces. Doing so will result in ill-defined objective functions that cannot facilitate the search for optimal E_1 and E_2.

Low-pass filtering of measured traces requires an a priori selection of a certain cutoff frequency. The chosen cutoff is critical for insuring a correct outcome. Choosing a cutoff that is too high retains irrelevant signal content that may contaminate and bias the objective function. On the other hand, choosing a cutoff frequency that is too low removes signal content that holds relevant pavement response information needed for obtaining a correct solution. To resolve this issue, a model-guided approach was taken, in which the cutoff frequency is selected based on the frequency content of the calculated acceleration trace that is being used for the matching (Levenberg 2012, 2015; Drori and Levenberg 2016). To illustrate this point, Fig. 3b presents the spectral amplitudes of both the measured and calculated acceleration traces of Fig. 3a. These spectral magnitudes are normalized by their corresponding peaks and are therefore unitless—ranging between zero and unity. As can be seen, the calculated acceleration trace is essentially band-limited with an effective high-frequency of about 45 Hz. This frequency level corresponds to a normalized spectral amplitude of approximately one percent; it essentially means that a Fourier series with frequencies up to 45 Hz can reproduce the calculated acceleration signal almost perfectly. On the other hand, and as can also be seen in Fig. 3b, the field-measured acceleration signal contains frequencies that are much higher than 45 Hz. Thus, when matching two traces for evaluating their individual objective function, the measured acceleration signal should first undergo low-pass filtering with a cutoff frequency corresponding to a normalized spectral amplitude of 1%. This cutoff is simply applied by first representing the measured signal for the entire duration of the truck passing event as a Fourier series, and then reconstructing the series while excluding all terms associated with frequencies higher than the desired cutoff. This procedure is equivalent to the application of a near-perfect filter.

4 Analysis of Properties

The interpretation method outlined in the previous section is herein applied to infer the layer properties of the tested pavement. The starting point is choosing a trial pair of layer moduli E_1 and E_2. Next, acceleration traces are calculated for each of the four sensors; these are concurrently investigated in the frequency domain to obtain a suitable set of four different cutoff frequencies based on the one percent rule—refer to Fig. 3b. Then, the cutoffs are applied to filter the measured signals—after which four individual objective functions are defined:

$$\varphi_j = \frac{1}{N}\sum_{n=1}^{N}\left|a_{j,n}^{\mathrm{mf}} - a_{j,n}^{\mathrm{c}}\right| \tag{1}$$

where $\varphi_j = \varphi_j(E_1, E_2)$ is the individual objective function for Acc#*j*, $a_{j,n}^{\mathrm{mf}}$ is the measured and filtered acceleration signal of Acc#*j* at time t_n, $a_{j,n}^{\mathrm{c}}$ is the corresponding calculated (model-simulated) signal for Acc#*j* at time t_n, and N is the total number of data points utilized for the matching (based on 1 kHz sampling rate). As a means of simultaneously considering the signals from all four sensors, the following so-called min-max multicriterion formulation is employed (Osyczka 1978; Levenberg 2013):

$$\Phi = \sum_{j=1}^{4}\left(\frac{\varphi_j}{\varphi_j^0} - 1\right) \tag{2}$$

wherein $\Phi = \Phi(E_1, E_2)$ is a global objective function, φ_j is the individual objective function defined in Eq. (1), and φ_j^0 is the smallest value the objective function φ_j can attain when minimized with respect to E_1 and E_2.

The last analysis step is minimizing the global objective function with respect to E_1 and E_2 to arrive at the sought pavement properties. This can be attempted with a nonlinear search algorithm (e.g. Lagarias et al. 1998). However, since nonlinear search algorithms cannot guarantee converges to a global minimum, especially if the search path is not smooth or contains local minima (or both), the solution space is densely gridded, and the global minimum is simply picked out.

Shown in Fig. 4a is the outcome of the above described approach, corresponding to the acceleration traces in Fig. 2 (i.e., Pass#1). It provides a contour plot of Φ for different combinations of E_1 in the range 50…500 MPa and E_2 in the range 10…200 MPa. This plot is composed of more than 2000 function evaluations with 10 MPa intervals for E_1 and 5 MPa intervals for E_2. Once a region encapsulating the global minimum was identified, with values in the range 380…400 MPa for E_1 and 80…90 MPa for E_2, the solution grid was further refined (in this region only) to intervals of 1 MPa for both E_1 and E_2. The final, globally optimal, solution is E_1 =

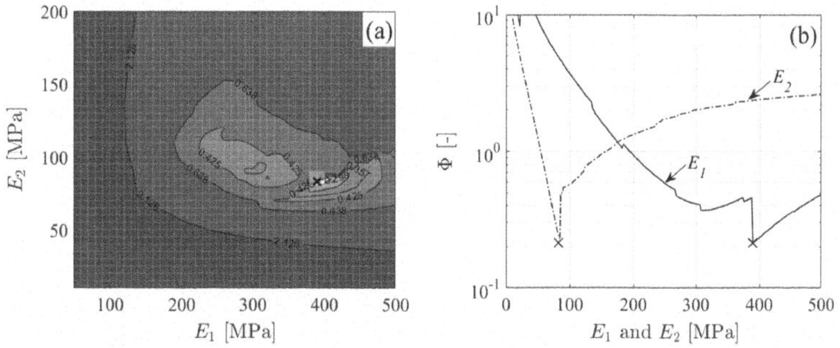

Fig. 4 Plot of the global objective function in Eq. (2) for Pass#1: **a** contour plot of Φ as a function of E_1 and E_2, and **b** sections through the minimum point (denoted with a cross marker)

390 MPa and $E_1 = 83$ MPa. Figure 4b provides two sections of the objective function surface taken through this optimal point. As can be graphically deduced, the behavior of Φ in the moduli space can be challenging for a nonlinear search algorithm, given the tortuous behavior and existence of local minima. This unfavorable behavior can be ascribed, at least in part, to the fact that the method is based on digital signals, and that the frequency cutoff changes when different trial set of moduli values are assessed.

Figure 5 shows the optimal match in the time domain for Pass#1, superposing calculated acceleration traces (with $E_1 = 390$ MPa and $E_2 = 83$ MPa) over their corresponding measured and filtered signals. As can be seen, the match is fairly good across all four sensors. Table 1 presents results from six different passes, listing the lateral offset (from video analysis), the travel speed of the truck, the inferred moduli assuming a two-layered half-space, and the minimal (optimal) value of the corresponding global objective function. To arrive at this table the entire interpretation

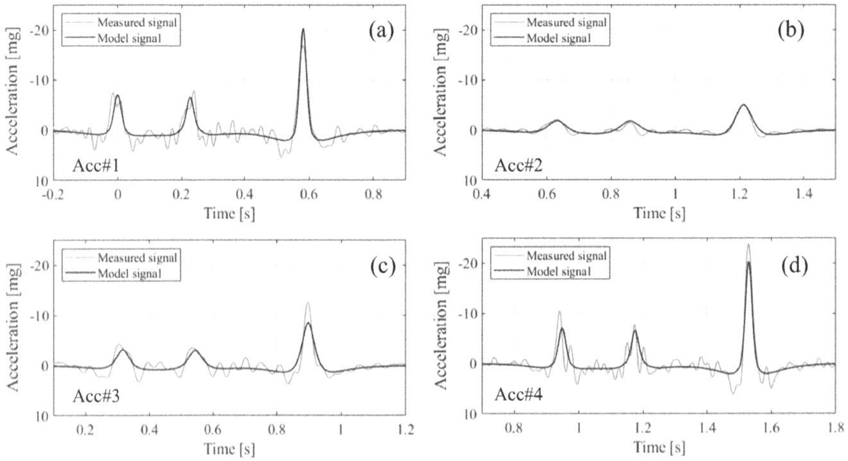

Fig. 5 Optimal match obtained for Pass#1 between measured (and filtered) accelerations and calculated accelerations in a two-layered half-space model. The time origin $t = 0$ denotes when the front axle of the truck is closest to Acc#1

Table 1 Layer moduli of the test pavement inferred from roadside accelerometer readings during different truck passes

Pass #	y_0 [mm]	V [km/h]	E_1 [MPa]	E_2 [MPa]	Φ
1	250	57	390	83	0.2126
2	300	35	189	104	0.3086
3	300	38	267	91	0.2020
4	400	58	347	92	0.2913
5	500	23	163	94	0.3698
6	400	38	548	87	0.4777

method was essentially repeated. As can be observed, the top layer modulus (E_1) varies within a wide range 163...548 MPa while the lower layer (half-space) modulus (E_2) varies within a narrow range 83...104 MPa. It is important to note that all six passes were carried out within a short time window of 20 min during which the asphalt temperature level was constant, about 7 °C.

5 Summary and Discussion

An array of four roadside accelerometers was utilized in a field experiment to record several nearby passes of a truck. Assuming a two-layered pavement model, an interpretation method was proposed and applied (separately for each pass) to infer moduli corresponding to in situ values. Upper layer moduli exhibited a coefficient of variation of 45% while the lower (subgrade) moduli exhibited a coefficient of variation of 8%. These variabilities seemed unrelated to lateral offsets, travel speeds, and optimal (minimal) value of their respective global error functions. Combining the results from the six different passes into a single moduli set, utilizing a weighted average based on Φ^{-1} (see Table 1), gives: $E_1 = 310$ MPa and $E_2 = 91$ MPa. These values do not compare well with elastostatic backcalculation of falling weight defelectometer results, which gave a pavement structure modulus of 640 MPa and a subgrade modulus of 170 MPa. However, they do compare very well with dynamic cone penetrometer results (Chen et al. 2005), which gave a pavement structure modulus of 340 MPa and a subgrade modulus of 120 MPa. Aspects that may influence the variability and inaccuracy of the interpretation results include: (i) incorrect assessment of the lateral offset y_0 (see Fig. 1); (ii) truck not driving perfectly parallel to the road centerline; (iii) truck not driving at constant speed while passing next to the sensor array; and (iii) imposition of an improper pavement model on the measurements.

Overall, the proposed method seems to yield a workable and rational solution that avoids the difficulties associated with double integration of acceleration signals. The method is also deemed scalable, with capacity to handle any number of acceleration sensors as well as other sensor types. In this connection, the envisioned application is wide-area placement of accelerometers inside, e.g., road studs.

References

Andersen S, Levenberg E, Andersen MB (2017) Inferring pavement layer properties from a moving measurement platform. In: Loizos A, Al-Qadi IL, Scarpas T (eds) Proceedings of the 10th international conference on the bearing capacity of roads, railways and air-fields. CRC Press/Taylor & Francis Group, pp 675–682

Andersen S, Levenberg E, Andersen MB (2018) Efficient reevaluation of surface displacements in a layered elastic half-space. Int J Pavement Eng. https://doi.org/10.1080/10298436.2018.1483502

Arraigada M, Partl MN, Angelone SM, Martinez F (2009) Evaluation of accelerometers to determine pavement deflections under traffic loads. Mater Struct 42(6):779–790

Burmister DM (1945) The general theory of stresses and displacements in layered systems. J Appl Phys 16(2), 89–94 (Part I), 126–127 (Part II), 296–302 (Part III)
Chen DH, Lin DF, Liau PH, Bilyeu J (2005) A correlation between dynamic cone penetrometer values and pavement layer moduli. Geotech Test J 28(1):42–49
Drori O, Levenberg E (2016) Characterization of a traveling object with an underground cluster of accelerometers. In: Transforming the future of infrastructure through smarter information - proceedings of the international conference on smart infrastructure and construction, ICSIC 2016, pp 349–356
Harr ME, Elton DJ (1983) Non-contact, non-destructive airport pavement profile, texture and deflection measurements. Report DOT/FAA/PM-83/14, U.S. Department of Transportation, Federal Aviation Administration
Hildebrand G, Rasmussen S (2002) Development of a high-speed deflectograph. Danish Road Institute, Report 117, Road Directorate, Roskilde, Denmark
Klar A, Levenberg E, Tur M, Zadok A (2016) Sensing for smart infrastructure: prospective engineering applications. In: Transforming the future of infrastructure through smarter information - proceedings of the international conference on smart infrastructure and construction, ICSIC 2016, pp 89–295
Lagarias JC, Reeds JA, Wright MH, Wright PE (1998) Convergence properties of the Nelder-Mead simplex method in low dimensions. SIAM J Optim 9(1):112–147
Lajnef N, Rhimi M, Chatti K, Mhamdi L, Faridazar F (2011) Toward an integrated smart sensing system and data interpretation techniques for pavement fatigue monitoring. Comput-Aided Civ Infrastruct Eng 26(7):513–523
Levenberg E (2012) Inferring pavement properties using an embedded accelerometer. Int J Transp Sci Technol 1(3):229–246
Levenberg E (2013) Inverse analysis of viscoelastic pavement properties using data from embedded instrumentation. Int J Numer Anal Methods Geomech 37(9):1016–1033
Levenberg E (2014) Estimating vehicle speed with embedded inertial sensors. Transp Res Part C Emerg Technol 46:300–308
Levenberg E (2015) Backcalculation with an implanted inertial sensor. Transp Res Rec 2525:3–12
Levenberg E, Shmuel I, Orbach M, Mizrachi B (2014) Wireless pavement sensors for wide-area instrumentation. In: Sustainability, eco-efficiency and conservation in transportation infrastructure asset management - Proceedings of the 3rd international conference on transportation infrastructure, ICTI 2014, pp 307–319
Osyczka A (1978) Approach to multicriterion optimization problems for engineering design. Comput Methods Appl Mech Eng 15(3):309–333
Ryynänen T, Pellinen T, Belt J (2014) The use of accelerometers in the pavement performance monitoring and analysis. IOP Conf Ser Mater Sci Eng 10(1):012110
Skar A, Klar A, Levenberg E (2019) Load-independent characterization of plate foundation support using high-resolution distributed fiber-optic sensing. Sensors 19(16):3518

Behaviour of the Interface Bonding Between Asphalt Overlays and Rigid Pavements

K. Bayraktarova, M. Dimitrov, B. Hofko, and L. Eberhardsteiner

Abstract The application of an asphalt mix overlay over old rigid pavement is a way to extend the service life of a pavement construction up to 10 years. The effectiveness of this maintenance measure strongly depends on the bonding conditions between the asphalt and the concrete layer and the effective transfer of the stresses and strains from the upper to the lower layer under traffic loads. The bond characteristics were investigated using laboratory tests and numerical simulations. Static testing equipment was used for the determination of the shear and tensile adhesion strength of cores from six sections of one highway. All six sections include three different pre-treatment methods and two types of asphalt mixtures. To evaluate resulting stresses and strains in the pavement construction under traffic load, finite elements simulations were carried out. Thereby, a simulation of the laboratory shear test setup provides information about the bond conditions as defined in the finite element model, which is used as input to simulate stresses and strains under different loading conditions. This approach is not only used to identify significant resulting stresses and strains due to traffic load but also to derive threshold values for future laboratory shear adhesion tests.

Keywords Interface · Surface · Bond · Concrete · Asphalt overlay

1 Introduction

Europe has more than 80 years history in building concrete highways. Initiated in Germany in 1929, there is currently about 25% of its freeway network made of rigid pavements. In Austria, about one third of the highway network consists of rigid pavements. The extended age of many concrete pavements requires expensive maintenance or reconstruction. A way to extend the lifespan for 5 to 10 years is to build a new asphalt surface layer over the existing concrete construction (Thouret et al. 1990; Herbst et al. 1993; Baek and Al-Qadi 2009). The interface behaviour between the

K. Bayraktarova (✉) · M. Dimitrov · B. Hofko · L. Eberhardsteiner
Institute of Transportation, Research Center of Road Engineering, Vienna University of Technology, Gußhausstraße 28, E230-3, Vienna, Austria
e-mail: kristina.bayraktarova@tuwien.ac.at

C. Raab (ed.), *Proceedings of the 9th International Conference on Maintenance and Rehabilitation of Pavements—Mairepav9*, Lecture Notes in Civil Engineering 76,
https://doi.org/10.1007/978-3-030-48679-2_68

asphalt and the concrete layer is of crucial importance for this maintenance measure (Ozer et al. 2012; Khazanovich et al. 2013). Therefore, there are different types of preparation of the concrete surface prior to the application of the asphalt mixture. Among these, high pressure water jetting, shot peening and a combination of both are commonly applied to rigid pavements before building an asphalt overlay (Kathleen Hall 2007).

There are three primary aims of this study: (i) to investigate alternative methods of surface treatment of the old jointed concrete pavement before the placement of a new asphalt overlay, (ii) to evaluate the performance of the contact interface between the concrete and the asphalt layer using static testing methods for determining the tensile and the shear resistance and (iii) to determine the occurring maximum stresses on the bonding interface caused by rolling and decelerating (braking) traffic using FE-analysis.

2 Test Sections and Sampling

To increase the knowledge for the practical feasibility of different preparation techniques, a test section was built in Austria, in April 2015 as part of a larger rehabilitation measure on a freeway. The rigid pavement on the first lane was replaced with a new asphalt binder layer. Different test sections on the second and third lane with length of 90 m were constructed.

Before the layering, the concrete surface was pre-treated in a different way (shot-peening, high pressure water jetting and combination of both) in order to ensure adequate bond between the concrete and the new asphalt layer. These pre-treatment methods were applied on the 24 years old rigid pavement before layering it with dense (type S2, mean air void content between 2 and 6%) and porous (type S3, mean void content between 6 and 12%) stone mastic asphalt SMA 11. This results in the following six test sections:

- Shot peening layered with SMA S2 (S2S)
- High pressure water jetting 2500 bar layered with SMA S2 (S2H)
- Shot-peening and high-pressure water jetting layered with SMA S2 (S2SH)
- Shot peening layered with SMA S3 (S3S)
- High pressure water jetting 2500 bar layered with SMA S3 (S3H)
- Shot-peening and high-pressure water jetting layered with SMA S3 (S3SH)

Cores were taken from the right wheel trace of the second lane before opening to traffic for each of the six test sections. One year after the rehabilitation, coring and testing was repeated to analyse the long-term evolution of the interface behaviour.

On one hand the investigation of the bond at the interface between the asphalt and concrete layer is performed using experimental tests. On the other hand, the bond conditions in the FE-analysis of the occurring maximum stresses and strains are described by the Cohesive Zone Model (CZM) (ABAQUS/Standard 2014).

3 Experimental Testing

According to ÖNORM EN 12697-48 (ÖNI 2013) there are different laboratory test methods to assess the interface behaviour. Since the shear resistance test and the pull-off resistance test are standard tests in Austria, they have been employed to examine the bond at the interface between the asphalt and the concrete layers.

3.1 *Shear Resistance Test*

The testing equipment for the shear resistance test allows direct shear testing of cylindrical samples (diameter 100 mm) without normal stress, at temperature of 20 ± 1 °C and deformation rate of 50 ± 3 mm/min. The shear strength β_S in N/mm^2 can be calculated by using the following equation:

$$\beta_s = \frac{F_{s,max}}{A} \quad (1)$$

where $F_{s,max}$ is the maximum shear force in N and A is the area of the cross section of the specimen in mm^2

3.2 *Pull-off Resistance Test*

Pull-off resistance tests are carried out at a test temperature of 0 °C and a centric tensile force applied with a constant rate of 200 N/s. The maximum reached tensile force $F_{t,max}$ in regard to the area of the specimen cross section A in mm^2 determines the pull-off resistance. The tensile adhesion strength β_{HS} in N/mm^2 is calculated by Eq. (2).

$$\beta_{HS} = \frac{F_{t,max}}{A} \quad (2)$$

4 FEM—Simulation of the Stress Conditions on the Interface Asphalt-Concrete

Finite element simulations were carried out using the software ABAQUS (ABAQUS/Standard 2014) in order to investigate the stress conditions at the interface between the existing concrete layer and the new asphalt overlay. Therefore, two finite element models were created (see Fig. 1).

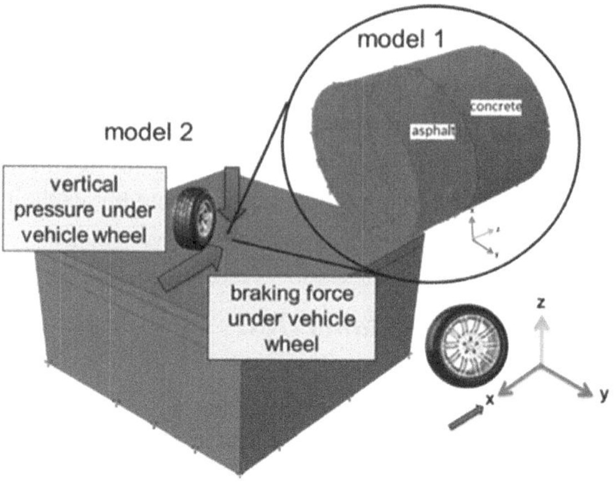

Fig. 1 Abaqus model of pavement construction and traffic loads

Model 1 simulates the static shear resistance test in order to determine the interface bond properties. It is assembled according to the test setup in ÖNORM B 3639-1 (ÖNI 1997b) and consist of a 30 mm thick SMA layer on the top of a 60 mm thick concrete layer. Both assemblies have a diameter of 100 mm. The material properties of the layers at 20 °C are given in Table 1. While the concrete layer is fixed in each direction, the asphalt layer is fixed only in y- and z-direction so that it can move in direction x. In the x-direction, which corresponds to the horizontal direction from model 2, a measured displacement from the static laboratory test is applied.

To characterise the bonding conditions on the interface between the asphalt and the concrete layer the Cohesive Zone Model (CZM) has been used as an interaction property in the FE-Model. As Fig. 2 shows, the CZM can be described by two independent parameters—the normal work of separation or the fracture energy G_{TC} and either the shear strength $t_{s=t}^{max}$ or the complete separation length $\delta_{s=t}^{f}$ (Papanastasiou and Sarris 2017). CZM is able to describe the bonding behaviour between the two different layers considering the material behaviour and the failure in the fracture process zone (traction-separation relationship across fracture surface). The bond is damaged as soon as the occurring stresses exceed the limit stress or in this case the

Table 1 Material properties of each construction layer

Material	Asphalt overlay	Concrete slabs	Asphalt layer	Unbound base	Subgrade
Layer thickness [cm]	3	25	5	45	300
E-module [N/mm^2]	4800	30000	9500	140	70
Poisson's ratio [-]	0.3	0.5	0.3	0.3	0.35

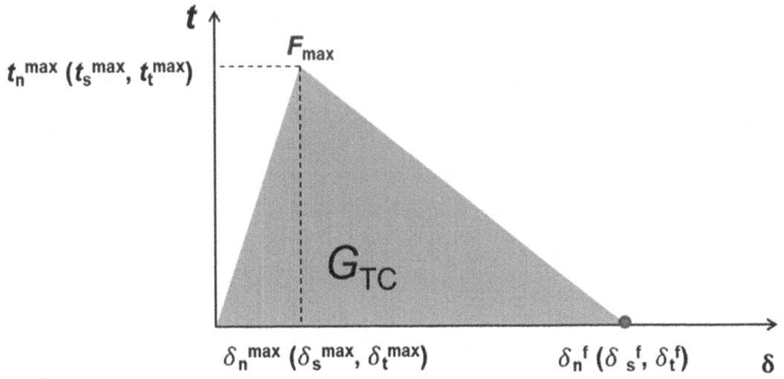

Fig. 2 Constitute law for fracture behaviour (ABAQUS/Standard 2014)

maximum shear stress $\beta_S = t_s = t_t$. The acting stresses t_n, t_s and t_t on the interface are related to the cohesive stiffness K and the displacement d. The stiffness matrix is formed from the shear stiffness components K_s and K_t and the normal stiffness component K_n.

Using the experimental data from the shear resistance test, it is also possible to estimate the shear stiffness $K_s = K_t$ with Eq. (3)

$$K_s = K_t = \frac{\beta_s}{\delta_{s=t}^{max}(F_{max})} \tag{3}$$

where $\delta_{s=t}^{max}$ is the measured displacement at the measured maximal shear force F_{max}. Thus, the experimentally obtained data serves as an input parameter to the CZM.

Figure 3 shows a comparison of the force-displacement curves from the shear resistance test and from the simulation. Both curves show a good agreement. The difference between the experimentally conducted maximal shear force and simulated maximal shear force is 2.4%. It can be concluded that the actual interface bonding behaviour of the tested specimen was reliably modelled.

A second model of an actual pavement construction was created to determine the occurring maximum stresses on the bonding interface caused by rolling and decelerating (braking) traffic (see Model 2 on Fig. 1). The CZM was used also in this model to characterise the real interface bonding conditions. The material properties of each construction layer are given in Table 1.

As the model 1 in Fig. 1 shows, the vertical and the horizontal loads induced by rolling and decelerating vehicles were simulated. Two load cases were observed; load case 1—free rolling (with standard vertical pressure 0.8 N/mm^2 for heavy vehicle trucks) and load case 2—decelerating (vertical pressure of 0.8 N/mm^2 and horizontal pressure 0.56 N/mm^2 estimated for a coefficient of friction between tyre and road of 0.7). The simulations were performed at 20 °C.

As result from the simulations with model 2 with different interface bond characteristics from the shear resistance tests can be summarized that the maximum shear stresses arise at load case deceleration.

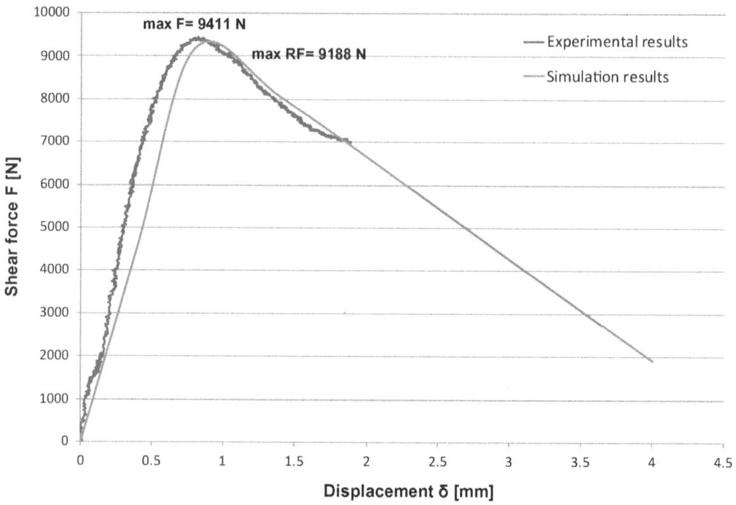

Fig. 3 Comparison of the force-displacement curves from the shear resistance test and from the simulation

Table 2 Results of the simulation in regard to the shear stresses on the interface between the asphalt overlay and the rigid pavement

Test specimen	Simulated maximum shear stress [N/mm^2]
S2KH3	0.70
S2KH4	0.70
S2KH6	0.61
S2KH6b	0.71
S2KH7	0.71
Mean value	0.69
Standard deviation	0.04
95%—quantile	0.75

With the help of this methodology, shear resistance tests on test specimens of the main test track were evaluated and the relevant shear stresses occurring at the interface between the concrete and the asphalt layer were determined. Table 2 shows the individual results of the simulation as well as the mean value, standard deviation and 95% quantile.

5 Results and Analysis

5.1 Results from the Shear Resistance Testing

The results from the test section before traffic release have been analysed using statistical evaluation (t-test). It shows that there are no significant differences between the pre-treatment methods or asphalt types (Fig. 4).

The bar chart in Fig. 5 shows the results from the shear resistance tests one year after traffic release. It becomes obvious that most of the methods lead to an increase of the shear strength. This effect is much more distinct in the sections with dense graded SMA 11 S2. However, only the specimens, pre-treated with the combination of shot peening and high pressure water jetting (S2SH), reach the required value of 1.2 N/mm^2 for the shear resistance of asphalt overlays according to national requirements (RVS 08.16.06) (FSV 2015). On the other hand, the open graded SMA 11 S3 contribution to the shear strength in the tested section is much less significant. The specimens, pre-treated with high pressure water jetting (S3H), have a shear strength of 0.65 N/mm^2 after one year, which is even a decrease of 14%.

The shear resistance results of the test section show that the overlaying with SMA 11 S2 leads to better development of the long-term strength.

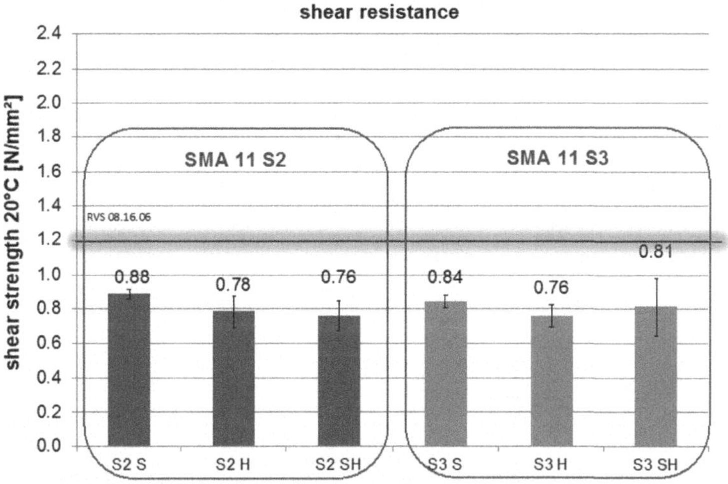

Fig. 4 Shear resistance results before traffic release

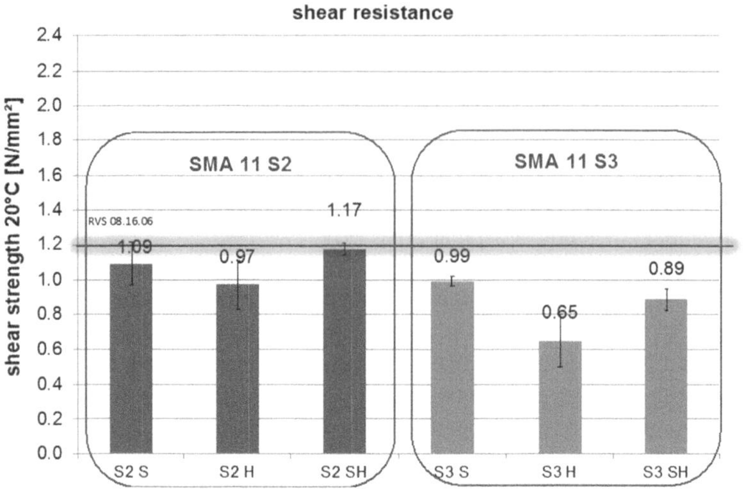

Fig. 5 Shear resistance results 1 year after traffic release

5.2 *Results from the Pull-off Resistance Testing*

The results of the pull-off resistance tests of the test section are displayed in Figs. 6 and 7. The increase of the tensile adhesion strength in the sections with SMA 11 S2 is much more obvious than the one with SMA 11 S3. The specimens from section SMA 1 S2, pre-treated with high pressure water jetting, have by far higher pull-off strengths. The average is 2.90 and 3.13 N/mm^2 respectively.

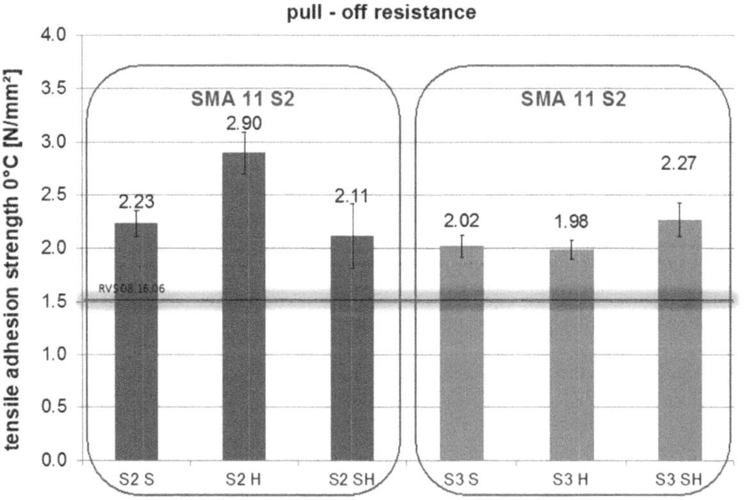

Fig. 6 Pull-off resistance results before traffic release

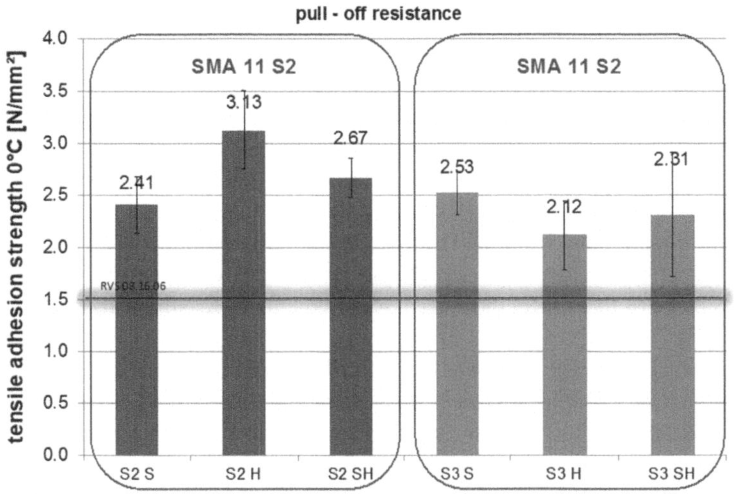

Fig. 7 Pull-off resistance results 1 year after traffic release

The sections layered with SMA 11 S3 appear to show different kind of results by comparison. The highest pull-off resistance before traffic release is determined for those samples that were pre-treated with the combination of shot-peening and high-pressure water jetting. The highest value one year after traffic release is for the specimens pre-treated with shot-peening (S3S). However, all of the tensile adhesion strengths differ slightly from each other and meet the requirement of 1.5 N/mm^2 (RVS 08.16.06) (FSV 2015).

6 Conclusion

A test section was built on a three-lane freeway. Core samples were taken before and 1 year after traffic release and afterwards tested trough shear and pull-off resistance tests. Additionally, finite elements simulations were carried out to evaluate resulting stresses and strains in the interface under traffic load.

Two finite element models were developed using the software ABAQUS (ABAQUS/Standard 2014). The first one recreates the used static shear resistance test. With the second one the relevant load and bond conditions on the interface between the asphalt layer and the rigid base were determined. The resulting maximum shear stresses from the simulations with input data from laboratory tests were statistically analysed. The mean value of the maximum shear stresses is 0.69 N/mm^2, the standard deviation 0.04 N/mm^2 and the 95% quantile 0.75 N/mm^2. Based on these results and considering a safety factor of 1.2, a threshold for the shear strength according to ÖNORM B 3639-1 (ÖNI 1997b) of 0.9 N/mm^2 can be recommended. The application of such coefficient is necessary in order to sufficiently secure the results due to the small sample size. With appropriate further investigations and simulations, this coefficient could be reduced in the future and the requirement further specified. According to this recommendation, sections with a shear strength between 0.9 N/mm^2 and 0.6 N/mm^2 can be accepted with a quality deduction. Sections with a shear strength below 0.6 N/mm^2 are not acceptable. With regard to the requirements for the adhesive bond according to ÖNORM B 3639-2 (ÖNI 1997a), no changes are recommended based on the results of this project.

Based on the results from the shear resistance tests before traffic release, it can be concluded that there are no significant differences between the pre-treatment methods. None of them meet the current requirement of 1.2 N/mm^2 in RVS 08.16.06 (FSV 2015).

The results from the pull-off tests before traffic release do not indicate a clear trend. The combination of high-pressure water jetting and the dense stone mastic asphalt SMA 11 S2 leads to the highest pull-off strength. However, the average values of the other repairing methods slightly differ from each other. Therefore, no specific statement can be made which surface treatment or layering has the most positive or negative effect on the bond between the rigid pavement and the asphalt layer. All of the specimens meet the requirement of 1.5 N/mm^2 in RVS 08.16.06 (FSV 2015).

The tested core samples, taken 1 year after traffic release, show an increase of their shear resistance. The outcome of this research indicates a clearer improvement in the sections with dense stone mastic asphalt SMA 11 S2 than with porous stone mastic asphalt SMA 11 S3. However, only the specimens pre-treated with the combination of shot peening and high-pressure water jetting reach the required value of 1.2 N/mm^2 for the shear resistance of asphalt top layers according to RVS 08.16.06 (FSV 2015). The sections with SMA 11 S2 clearly show a more significant increase in the pull-off strength in comparison to the sections with SMA 11 S3. All of the specimen meets the requirement of 1.5 N/mm^2 RVS 08.16.06 (FSV 2015).

The findings of this research are convincing that the repairing method "combination of shot-peening and high-pressure water jetting with SMA 11 S2" is recommended for future maintenance of rigid pavements.

References

ABAQUS/Standard (2014) User's manual, version 6.14. Dassault Systèmes Simulia Corp

Baek J, Al-Qadi IL (2009) Effects of bearing capacity and load transfer efficiency of jointed concrete pavements on reflective cracking in hot-mix asphalt overlays. In: Bearing capacity of roads, railways and airfields, vols 1 and 2, pp 1039–1049

FSV (2015) RVS 08.16.01: Anforderungen an Asphaltschichten. Wien, Österreich

Herbst G, Kirchknopf H, Litzka J (1993) Asphalt overlay on crack-sealed concrete pavements using stress distributing media. Reflective Crack Pavements 20:425–432

Hall K, Dawood D, Venikar S (2007) Long-life concrete pavements in Europe and Canada

Khazanovich L, Tompkins D, Wu RZ, Harvey JT (2013) Investigation and modification of available mechanistic-empirical procedures for reflective cracking in asphalt overlays of concrete pavements. Transp Res Rec 2368:126–132

ÖNI (1997a) ÖNORM B3639-2 Technische Asphalte für den Straßenbau und verwandte Gebiete - Prüfung - Teil 2: Haftverbund von Asphaltschichten. Österreichisches Normungsinstitut, Wien

ÖNI (1997b) ÖNORM B 3639-1 Technische Asphalte für den Straßenbau und verwandte Gebiete - Prüfung - Teil 1: Schubverbund von Asphaltschichten. Österreichisches Normungsinstitut, Wien

ÖNI (2013) ÖNORM EN 12697-48 Asphalt - Prüfverfahren für Heißasphalt - Teil 48: Schichtenverbund. Österreichisches Normungsinstitut, Wien

Ozer H, Al-Qadi IL, Wang H, Leng Z (2012) Characterisation of interface bonding between hot-mix asphalt overlay and concrete pavements: modelling and in-situ response to accelerated loading. Int J Pavement Eng 13(2):181–196

Papanastasiou P, Sarris E (2017) 6 - Cohesive zone models. Porous rock fracture mechanics. Woodhead Publishing, pp 119–144

Thouret D, Puig J, Martin C (1990) Repairing a cracked concrete pavement with a thin asphalt rubber concrete overlay on a geotextile. In: Geotextiles, geomembranes and related products, vol 2, p 874

Estimation of Resilient Modulus for Fine-Grained Soils Using Ground Penetrating Radar

Logan Tihey and S. Sonny Kim

Abstract As state Departments of Transportation (DOT) continue to adopt the Mechanistic-Empirical Pavement Design Guide (MEPDG), the ability to readily obtain higher quality input parameters becomes an ever-growing need. One such parameter is the Resilient Modulus (M_R), which is defined as the ratio of cyclic deviatoric stress to recoverable strain. The ensuing investigation was conducted in an attempt to replace the use of the repeated load triaxial test in determining the M_R of soils with modulus estimations obtained from Light Weight Deflectometer (LWD) and Ground Penetrating Radar (GPR) testing. The constructed test setup consisted of a test pit subdivided into six sections containing an elastic silt (MH) at different water content and density levels. LWD and GPR testing was conducted at each of the six test sections, and the results were correlated to the different soil water content and density levels. Through a regression analysis, the preliminary model was developed the E_{LWD} of fine-grained soils was estimated using the soil's water content, bulk density, and the measured dielectric constant. A strong correlation was observed (R^2 of 0.858) between the predicted and measured E_{LWD} values, indicating its potential for future use in predicting stiffness of fine-grained soils using GPR.

Keywords LWD · GPR · Resilient modulus · Dielectric constant

1 Introduction

The Resilient Modulus (M_R) is currently used as in input parameter for the Mechanistic-empirical Pavement Design Guide (MEPDG). The M_R of a soil is defined as the ratio of deviatoric cyclic stress to the recoverable strain. The repeated load triaxial test used to measure M_R is expensive, arduous, and time-consuming. This led researchers to investigate alternative methods such as nondestructive technologies for estimating the M_R. The following chapter presents a discussion of various studies conducted on the topic of estimating the M_R of a soil via stiffness parameters

L. Tihey · S. S. Kim (✉)
School of Environmental, Civil, Agricultural and Mechanical Engineering, College of Engineering, The University of Georgia, Athens, GA 30602, USA
e-mail: kims@uga.edu

C. Raab (ed.), *Proceedings of the 9th International Conference on Maintenance and Rehabilitation of Pavements—Mairepav9*, Lecture Notes in Civil Engineering 76,
https://doi.org/10.1007/978-3-030-48679-2_69

obtained from the Light Weight Deflectometer (LWD) and Ground Penetrating Radar (GPR) non-destructive technologies. The studies referenced herein concentrate on correlating the elastic modulus (E_{LWD}) obtained from the LWD with M_R test results obtained from the repeated load triaxial test. The effect of various soil parameters (i.e., water content, density, and soil type) on measured moduli is also investigated in this paper.

Mousavi et al. (2017) inform that researchers have utilized the Light Weight Deflectometer (LWD) to predict the M_R of subgrade soil at a specific confining pressure and deviatoric stress. However, these methods have failed to be sufficient in that the M_R is dependent on various stress states, and the majority of research objectives pertaining to estimating the M_R using the LWD have focused on predicting the M_R at a singular stress state.

2 Literature Review

Since the M_R was required as an input parameter for the MEPDG, researchers have sought a means to estimate the M_R of a soil. Some have attempted to utilize the elastic modulus (E_{LWD}) obtained from LWD testing as an efficient estimate for the M_R. In a study conducted by Park et al. (2018), the efficacy of estimating the M_R of untreated cohesive soils using the light weight deflectometer, falling weight deflectometers (FWD), and dynamic cone penetrometers (DCP) was investigated. It was determined that LWD, FWD, and DCP testing conducted on top of the subgrade did not show significant correlation with the laboratory determined M_R value. Relatedly, Schwartz et al. (2017) conducted a study in an attempt to standardize LWD use in compaction quality assurance (QA) of unbound geomaterials. In the study, nine resilient modulus constitutive and empirical predictive models were employed in an attempt to define a target modulus for the developed QA procedure, bypassing the impractical need for extensive M_R testing. After comparing the results from the constitutive and empirical predictive models with laboratory M_R test results, it was determined that the best of the nine models provided results at a level of accuracy considered insufficient for estimating the field LWD target modulus. Among others, both of these studies suggest that a method of accurately predicting the M_R at various stress conditions is of significant interest to both the design and constructing of pavements.

In a study conducted by Mousavi et al. (2017), a method for predicting the M_R at various stress states using the LWD is presented. The proposed method provides the ability to estimate M_R values at varying stress states, and it eliminates the need to assume a Poisson's ratio and shape factor, which can account for significant variation in measured modulus values. Mousavi et al. present a model for estimating the M_R of A-4 and A-7-5 soils from field LWD data by correlating the ratio of applied stress to measured surface deflection $\left(\frac{\sigma}{\delta}\right)$ to the MEPDG elastic modulus (k_1), stiffness hardening (k_2), and strain-softening (k_3) coefficients. It was found that the model predicted k_1, k_2, and k_3 coefficients with R^2 values of 0.71, 0.82, and 0.55 respectively. Additionally, a comparison between the laboratory measured M_R and model-predicted M_R values resulted in an R^2 value of 0.83. The model was verified by the remaining portion of the dataset and by available data from two studies with R^2 values of 0.83 and 0.96 respectively. The study showed good correlation between the model-predicted and laboratory determined M_R values for soil types A4, A4-a, and A-7-5, leaving the need for further validation of untested soil types. The proposed model eliminates significant uncertainties encountered in conventional models while displaying strong correlation with laboratory-determined M_R values, suggesting its promise for future use.

The promise displayed by the LWD in compaction quality assurance and M_R estimation has guided researchers to investigate the relationship between various soil parameters and E_{LWD}. A recent study conducted by Kavussi et al. (2019) explored the relationship and interdependency of the soil compaction level, moisture content, and stress state of unbound granular material (UGM) on the E_{LWD} using the Response Surface Method (RSM). In the study UGM was compacted in a text box (1530-mm × 1530-mm × 700-mm) in two 200-mm lifts at various moisture contents and compaction levels. The stress state of the UGM was manipulated through the changing of the LWD drop weights (10, 15, and 20 kg). Kavussi et al. discovered that, at low moisture levels, the E_{LWD} was sensitive to compaction variations. Oppositely, lower E_{LWD} sensitivities was observed at higher moisture levels with compaction variation, albeit higher moisture contents resulted in lower overall E_{LWD} values. UGM at lower compaction levels displayed lower E_{LWD} sensitivities to changes in moisture content and higher sensitivities at higher compaction levels. Kavussi et al. also verified the findings of Tatsuoka and Correia (2018) by confirming that E_{LWD} sensitivity to moisture content was dependent on compaction, and E_{LWD} sensitivity to degree of saturation was independent of compaction level. This indicates that the degree of saturation is the most significant moisture parameter as it relates to soil stiffness. The study also showed slight increases in E_{LWD} with the use of higher LWD weights. The developed model resulted in an R^2 value of 0.99. The proposed model was verified though subsequent testing of identical aggregate and gradation with new combinations of compaction, moisture, and stress levels. Some of the testing combinations of the validated data include compaction levels within and outside the range of the compaction levels used to develop the model. Upon validation, R^2 values of 0.94 and 0.985 were determined for within-range and out-of-range data respectively. Abdelmawla et al. (2019) suggested a model to predict soil physical properties (water content and soil bulk density) using Air-coupled GPR systems scan data. They also assert that the GPR technique discussed illustrates the potential for a new era in roads and geotechnical inspection methodologies for subgrade soil assessment and prediction of different properties, physical, stiffness and strength.

Non-destructive testing (NDT) methods are preferred over traditional methods–i.e., coring and destructive testing–as traditional practices are time-consuming, pose safety concerns, and are less cost-effective. Ground Penetrating Radar (GPR) is one of the most efficient NDT for subsurface monitoring. GPR can perform pavement health analysis with the advantage of scanning at traffic speed, which removes the need to disrupt traffic. The GPR's primary applications in pavement engineering are the determination of layer thickness and detection of subsurface defects.

The primary material property obtained from GPR testing is the dielectric constant. The dielectric constant, also known as the relative permittivity, ε_r, of a homogeneous media relates the relative EM velocity in a material to the speed of light in free space, (c). Leng et al. (2009):

$$\varepsilon r = (c/\upsilon)^2 \quad (1)$$

where,

ε_r dielectric constant,
c speed of light in free space of $3 * 10^8$ m/s,
ν EM velocity in the material

This dielectric constant is a material specific parameter, which will change depending on the soil condition, stiffness, physical properties, and strength. As a result, it is believed that the soil dielectric constant could be related to soil properties, with different numerical approaches.

3 Materials and Preliminary Test

In an effort to develop a relationship between the soil stiffness and the dielectric constant obtained from GPR testing, a laboratory test setup was constructed. The setup was comprised of a (6-foot by 6-foot by 1-foot) test-box subdivided into six sections with varied water content levels. Water content levels at the time of compaction were determined through the use of a microwave. Prior to placing the soil in the six sections, the distribution of grain size, Atterberg limits, and optimum moisture content were determined from a sieve analysis, Atterberg limit tests, and Standard Proctor test respectively. The results in addition to the USCS Soil Classification can be viewed in Table 1. Each of section was compacted in two 6-in. lifts. The six sections were provided sample identification numbers of D1 through D6. Their moisture content at the time of compaction can be viewed in Table 2. After compaction, sand cone test was conducted to determine the compacted density and water content at the time of testing. The results can be viewed in Table 3.

After characterizing the soil and determining its compacted properties, LWD and GPR testing was conducted. LWD tests were conducted with a 10 kg weight at drop heights of 6, 12, 18, and 24-in. Four drops were conducted at each drop height: two seating drops followed by two measurement drops. LWD testing was conducted with

Table 1 Soil classification and properties

Specific	USCS	Plastic	Liquid	Plasticity
Gravity	Classification	Limit	Limit	Index
2.76	MH	37.44	57.1	19.7

Table 2 Soil water content at compaction

Sample ID	Sample No.	Dry Plate wt. (gm)	Plate + wet sample (gm)	Plate + dry sample (gm)	Water Content	Average w%
D1	1-1	83.28	180.22	162.65	0.221368275	22.5%
	1-2	83.65	158.89	144.86	0.229210913	
	1-3	93.15	170.36	156.22	0.224195339	
D2	2-1	85.17	172.01	157.06	0.2079566	21.0%
	2-2	83.25	161.68	147.64	0.21804628	
	2-3	91.52	173.05	159.16	0.205351863	
D3	3-1	88.39	170.36	157.48	0.186423506	20.4%
	3-2	85.47	168.56	154.02	0.21210795	
	3-3	91.68	164.02	151.25	0.214369649	
D4	4-1	86.10	164.29	148.94	0.244271165	24.4%
	4-2	84.23	171.41	154.51	0.240466705	
	4-3	84.02	172.83	155.17	0.248208011	
D5	5-1	117.55	227.32	204.42	0.263612294	26.6%
	5-2	121.97	219.09	198.23	0.273537897	
	5-3	118.05	237.44	212.75	0.260718057	
D6	6-1	120.84	233.92	216.25	0.185200713	18.7%
	6-2	118.05	250.61	229.61	0.188239512	
	6-3	122.67	215.00	200.44	0.187218722	

Table 3 In situ soil compaction test results on test day

Location	Water content at testing	Density (pcf)
D1	19.0%	123.935
D2	18.0%	127.822
D3	16.0%	113.42
D4	20.0%	139.813
D5	21.0%	125.287
D6	16.0%	106.975

the 300 mm plate followed by the 150 mm plate–totaling 32 drops at each section. Variations in drop height and plate diameter were tested to investigate modulus dependency on changes in impact velocity and plate stress distribution. The LWD test setup can be viewed in Fig. 1. Air-coupled GPR testing was conducted on the six different sections following the completion of LWD testing, A picture of the GPR test setup can be viewed in Fig. 2.

Average modulus values (E_{LWD}) were developed from the two measurement drops that followed the two seating drops using a shape factor and Poisson's ratio of 1.8 and 0.35 respectively. The drops were broken down into different locations with the following nomenclature: the first part of the location refers to the respective soil section while the second part refers to the drop height and plate size. For example, D1.1 referred to the set of drops that included the 300 mm-plate at the 6-in. drop height, and D1.4 refers to the group of drops that included the 300 mm-plate at the 24-in. drop height. Thus, D1.6 would include the drops at the 12-in. drop height for the 150 mm-plate. The average E_{LWD} from the two measurement drops was determined,

Fig. 1 LWD test prep on D2

Fig. 2 Air-coupled GPR testing on D1

and the average moduli values for all of the various drop heights were averaged to determine the final E_{LWD} value for the test section with respect to its plate size as shown in Table 4.

GPR testing was conducted with a time range of 12 ns, resulting in 254 reflections per scan. Upon completing the GPR setup, the system was calibrated using a metal

Table 4 Determination of D1 E_{LWD} for 300 and 150 mm plate sizes

Location	Plate Size (mm)	Drop Height (in.)	ELWD Drop 1 (MPa)	ELWD Drop 2 (MPa)	Average Location ELWD	Average Plate ELWD	ELWD Plate Std. Dev.
D1.1	300	6	53.64392621	54.24256427	53.9	54.1	0.42544494
D1.2	300	12	53.88171663	53.95545428	53.9		
D1.3	300	18	52.73442639	54.83064705	53.8		
D1.4	300	24	54.80315719	54.63789412	54.7		
D1.5	150	6	101.5118875	100.8439522	101.2	94.0	4.818260992
D1.6	150	12	90.52735093	92.83583384	91.7		
D1.7	150	18	93.41838512	91.22147387	92.3		
D1.8	150	24	91.98404922	89.71454154	90.8		

plate, which was done to determine the amplitude of the wave reflected from the metal plate (A_m). The amplitude of the wave reflected from the metal plate was used in the analysis to determine the dielectric constant (ε_1) of each section. Considering the antenna of the GPR was stationary, each section (D1–D6) received hundreds of scans, which is more than enough to determine the dielectric constant for each section.

4 Test Results and Discussions

Test results showed no clear indication of increased modulus values as the drop height of the LWD increased. Conversely, the te results showed a clear increase in E_{LWD} collected from the 300 mm-plate as compared with the 150 mm-plate, as expected. Additionally, the 150 mm-plate displayed more variability with an average modulus standard deviation across the six sections of approximately 3.01 compared to the 300 mm-plate, which had an average standard deviation of 2.06. Figure 3a, b show a plot of E_{LWD} measured with the 300 mm-plate against the soil water content and density respectively. Upon fitting the curves, R^2 values of 0.99 and 0.80 respectively, which shows a strong relationship between the parameters. Figures 3c, d show a plot of E_{LWD} measured with the 150 mm-plate against the soil water content and density respectively. When fit, the results displayed significantly lower levels of correlation with R^2 values of 0.68 and 0.46 respectively. The lower levels of correlation and higher standard deviation observed with E_{LWD} values obtained using the 150 mm plate led to its exclusion in model development. The dielectric constant of each soil layer was determined using the following equation Maser in 1996 (Al-Qadi et al. 2003):

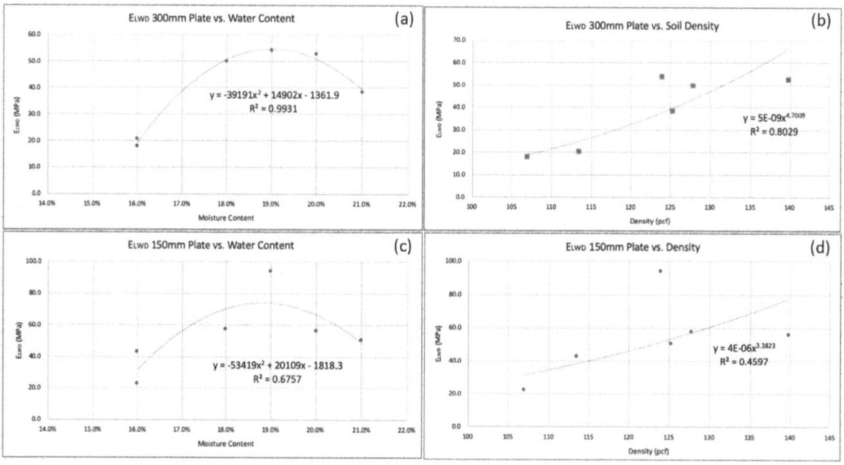

Fig. 3 E_{LWD} 300 mm as a function of Moisture Content **a** and Density **b**; E_{LWD} 150 mm as a function of Moisture Content **c** and Density **d**

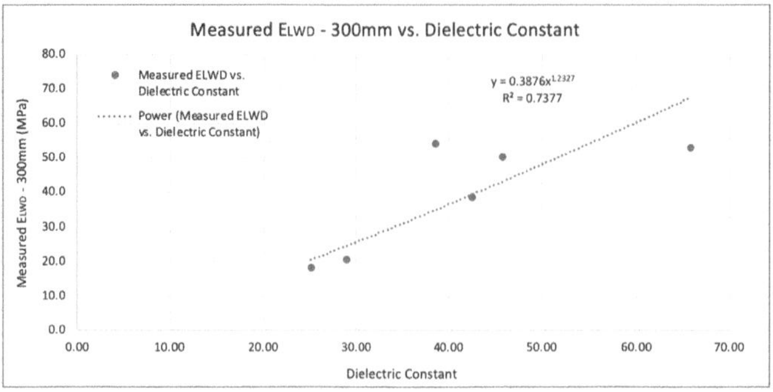

Fig. 4 Measured E_{LWD}—300 mm vs. dielectric constant

$$\varepsilon_1 = \left(\frac{1 + \frac{A_0}{A_m}}{1 - \frac{A_0}{A_m}} \right)^2 \tag{2}$$

where,

ε_1 dielectric constant of the sample,
A_0 amplitude of the soil surface reflection,
A_m amplitude of the wave reflected from the metal plate

The dielectric constant for each section was then plotted against the its respective E_{LWD} value to ascertain the relationship between the two parameters as seen in Fig. 4.

5 LWD Model Development

A preliminary model was developed through a regression analysis in an attempt to approximate the E_{LWD} of the elastic silt (MH) using the water content, dielectric constant, and bulk density. Model development included applying various trendlines to the plot of the measured E_{LWD} values versus ε_1. In doing so, a fairly significant coefficient of determination ($R^2 = 0.738$ as seen in Fig. 4) was observed when the power function trendline was applied. Thereafter, a power model was developed using the parameters listed along with three coefficients: a, b, and c. Using the solver function on excel, the error between the predicted and measured E_{LWD} values was minimized through manipulation of the a, b, and c coefficients. The proposed model can be described by the following equation:

$$E_{LWD} = \left(\frac{1.674 \cdot 10^{-5}}{W_C^{7.981}} \right) \cdot \varepsilon_1^{(0.266) \cdot \gamma_{bulk}} \tag{3}$$

where,

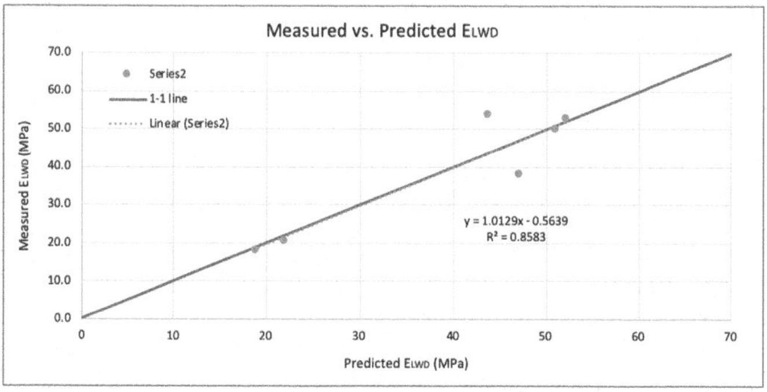

Fig. 5 Measured E_{LWD} vs. predicted E_{LWD}

E_{LWD} Predicted E_{LWD} (MPa),
W_C Water Content of Sample (decimal),
ε_1 dielectric constant of sample (predicted using Eq. (2)),
γ_{bulk} bulk density of soil sample

As previously mentioned, Fig. 5 displays a plot of the measured versus predicted E_{LWD}, which recorded an R^2 of 0.858. The strong correlation between the predicted and measured values indicates a relationship between the measured E_{LWD} values and the parameters used in the model as it relates to elastic silts. This points toward the model's promise in predicting the E_{LWD} of elastic silts and the potential of using the dielectric constant obtained from GPR testing in predicting soil stiffness.

6 Conclusions

The aforementioned investigation was conducted to ascertain, if any, the relationship between the elastic modulus obtained from LWD testing and the dielectric constant obtained from GPR. The modulus values obtained from testing showed a stronger relationship with the water content than the compaction level (R^2 values of 0.99 versus 0.80), which coincides with the findings of Kavussi et al. (2019). Results from the 150 mm-plate testing were excluded from model development due to the lower levels of correlation found between the elastic modulus and moisture content and the higher E_{LWD} variability. A preliminary model to predict E_{LWD} values based on the water content, bulk density, and dielectric constant was developed through a regression analysis. Using the model, estimated E_{LWD} values were developed and subsequently plotted against the measured E_{LWD} values, in which an R^2 value of 0.86 was obtained. This preliminary study has displayed the potential of estimating the stiffness of elastic silt using the GPR. Validation of the proposed model through additional testing is needed.

References

Abdelmawla AM, Durham SA, Kim S (2019) Estimation of subgrade soil density using ground penetrating radar. In: ICSC conference proceedings, Seoul, July 2019, Article no. A.5.3

Al-Qadi IL, Lahouar S, Loulizi A (2003) Successful application of ground-penetrating radar for quality assurance–quality control of new pavements. Transp Res Rec 1861:86–97 Paper No. 03-3512

Kavussi A, Qorbaninik M, Hassani A (2019) The influence of moisture content and compaction level on LWD modulus of unbound granular base layers. Transp Geotech 20:100252 https://doi-org.proxy-remote.galib.uga.edu/10.1016/j.trgeo.2019.100252

Leng Z, Al-Qadi IL, Baek J, Lahouar S (2009) Selection of antenna type and frequency for pavement surveys using ground penetrating radar (GPR). In: Presented at 88th Annual Meeting of the Transportation Research Board, Washington, D.C.

Mousavi SH, Gabr MA, Borden RH (2017) Subgrade resilient modulus prediction using light-weight deflectometer data. Can Geotech J 54(3):304 https://doi-org.proxy-remote.galib.uga.edu/10.1139/cgj-2016-0062

Park SS, Bobet A, Nantung T (2018) Correlation between resilient modulus (MR) of soil, light weight deflectometer (LWD), and falling weight deflectometer (FWD) (FHWA/IN/JTRP-2018/08). Purdue University, West Lafayette

Schwartz C, Afsharikia Z, Khosravifar S (2017) Standardizing lightweight deflectometer modulus measurements for compaction quality assurance

Tatsuoka F, Correia A (2018) Importance of controlling the degree of saturation in soil compaction linked to soil structure design. Transp Geotech 17 https://doi.org/10.1016/j.trgeo.2018.06.004

Impact of Construction Practices on Air Voids and Permeability of Asphalt Mixtures

Syed W. Haider, Michele Lanotte, Khurram Malik, and Aftab Quadri

Abstract Several isolated premature failures on the rehabilitated portion of the Lahore-Islamabad M-2 Motorway were observed, within a year of opening to traffic. Based on the visual evaluation of the distresses, failure causes could be attributed to poor in-pavement and surface drainage on the rehabilitated sections, segregation of HMA mixture, HMA mix-design, and the presence of heavy axle loads in the outer lanes. Cores in the distressed and normal areas were taken in different sections along the south and north bounds direction and several destructive tests were conducted on the HMA materials to determine the field compacted air voids. Also, field permeability in-situ tests were conducted on the same locations where cores were taken. Results showed that the mix segregation occurred in the fast lanes resulted in higher field air voids than the plant mixed-lab compacted samples. Thus, the permeability of mixes in these locations was significantly higher than expected, which caused water to seep through the surface mixes in all lanes. Additionally, due to the cross-fall of the HMA and aggregate base layers, the moisture entering in the fast and middle lanes accumulated under the outer lanes in both directions.

Keywords Field compaction · Permeability · Premature distresses

S. W. Haider (✉)
EMPA, Department of Civil and Environmental Engineering, Michigan State University, 428 S. Shaw Lane, Room 3546, East Lansing, MI 48824, USA
e-mail: syedwaqa@msu.edu

M. Lanotte
Civil Infrastructure and Environmental Engineering, Khalifa University, P.O. Box 127788, Abu Dhabi, UAE
e-mail: michele.lanotte@ku.ac.ae

K. Malik
Academia-Industry Linkage Centre, Military College of Engineering, National University of Sciences and Technology (NUST), Islamabad, Pakistan
e-mail: khurram.malik@mce.nust.edu.pk

A. Quadri
Zeeruk International Pvt. Ltd., 3rdfloor, Time Square Plaza, I-8 Markaz, Islamabad, Pakistan
e-mail: aftab@zeeruk.com

C. Raab (ed.), *Proceedings of the 9th International Conference on Maintenance and Rehabilitation of Pavements—Mairepav9*, Lecture Notes in Civil Engineering 76,
https://doi.org/10.1007/978-3-030-48679-2_70

1 Introduction

The permeability of a pavement surface layer is generally assumed to be proportional to its air void content. However, the lack of void interconnection and size dimensions of the individual voids may result in a watertight pavement of relatively high void content (Arambula et al. 2007; Choubane et al. 1998; Hunter and Ksaibati 2002; Masad et al. 2002). It is also well known that asphalt mixtures with different air void contents will present variable drainage capacity. The presence of water in asphalt concrete pavements may affect their mechanical properties, reducing their strength, stiffness, and durability. Water damage is generally manifested as raveling or stripping, commonly attributed to water infiltration into the asphalt mixture. Therefore, the hydraulic conductivity or coefficient of permeability of asphalt mixtures is a valuable property that should be known, in order to evaluate the water infiltration capacity of asphalt concrete pavements. Hydraulic conductivity in a saturated asphalt mixture is defined as the rate of discharge of water under laminar flow conditions through a cross-sectional unit area of a porous medium. Its value provides an indicator of the drainage capacity of asphalt concrete pavements under initial construction conditions (Norambuena-Contreras et al. 2013).

Proper compaction of hot-mix asphalt (HMA) mixtures is vital to ensure that a stable and durable pavement is built (Mohammad et al. 2016). For dense-graded mixes, numerous studies have shown that initial in-place air voids should range between 3–8%. Lower percentages of in-place air voids can result in rutting and shoving, while higher percentages allow water and air to penetrate the pavement, leading to an increased potential for water damage, oxidation, raveling, and cracking. Low in-place air voids are generally the result of a mix problem, while high in-place voids are typically caused by inadequate compaction (Brown et al. 2004). The density that can be obtained under normal rolling conditions is related to the ratio of lift thickness and nominal maximum aggregate size (t/NMAS). For improved compatibility, it is recommended that the t/NMAS be at least three for fine-graded mixes and at least four for coarse-graded mixes (Brown et al. 2004). Ratios less than these suggested numbers could be used, but more compactive effort would be required to obtain the desired density.

In most cases, a ratio (t/NMAS) of 5 does not result in the need for a more compactive effort to obtain maximum density. However, care must be exercised when the thickness gets too large to ensure that adequate density is obtained. The effect of mix temperature on the relationship between density and t/NMAS also indicates that one of the reasons for low density at thinner sections (i.e., lower t/NMAS) is the more rapid cooling of the mixture. Hence, for thinner layers, it is even more critical that rollers stay very close to the paver so that rolling can be accomplished prior to excessive cooling. For thicker sections (i.e., larger t/NMAS), the rate of cooling is typically not a problem. The in-place void content is the most significant factor impacting the permeability of HMA mixtures. This is followed by a coarse aggregate ratio and VMA. As the values of the coarse aggregate ratio increases, permeability increases. Permeability decreases as VMA increases for constant air voids. The variability of permeability between various mixtures is generally high. Some mixtures are permeable at the 8–10% void range, and others do not seem to be permeable at these higher voids. However, to ensure that permeability is not a problem, the in-place air voids should be between six and seven percent or lower. This appears to be valid for a wide range of mixtures regardless of NMAS and grading. Hydraulic conductivity of asphalt mixtures is known to be affected significantly by air void content, but may also by other factors such as lift thickness, aggregate shape, and aggregate gradation. Results of another study indicate that air void content is the predominant factor controlling hydraulic conductivity, but the aggregate shape (source) and gradation also have a statistically significant influence (Harris 2007).

2 Materials and Methods

Sixty (60) and thirty (30) HMA cores were extracted from the south and northbound direction, respectively. Coring locations were selected in both failed or normal areas of the pavement structure. The following factors were considered while identifying the locations of cores:

a. Segregation,
b. Lane type (Fast (F) innermost lane, Middle (M), and Slow (S) outer most lane),
c. Design thicknesses (3.5, 4, 5 and 10 in.),
d. Binder type (PG vs. 60–70 pen)

Cores were used to estimate the in-situ field compacted air voids. Also, plant-mixed lab compacted (PL) air voids were obtained from construction records. Statistical analyses (ANOVA and *t*-test) were performed on the data to identify significant factors impacting the field air voids. Also, field permeability tests were conducted near the core locations.

According to Darcy's Law (Eq. (1)), the coefficient of permeability (k) can be estimated by two methodologies: constant head and variable head. In the constant head method, the flow rate through a porous material is measured by sustaining a constant head of water. The coefficient of permeability can be determined by Eq. (2).

$$Q = kiA = k\frac{\Delta H}{L}A \tag{1}$$

$$k = \frac{QL}{hAt} \tag{2}$$

where: Q is the rate of flow (cm^3/sec), k is the coefficient of permeability (cm/sec), i is the hydraulic gradient, A the cross-sectional area of the specimen perpendicular to the direction of the flow (cm^2), ΔH the head loss across the specimen (cm), L is the length of the specimen (cm), t is the time of measurement (sec), and h is the pressure head (cm).

The constant head method is more appropriate for highly permeable materials with k values higher than 10^{-3} cm/s. The falling head method is preferred for materials with low permeability coefficients; k values less than 10^{-3} cm/s. Asphalt concrete generally has a k value between 10^{-3} and 10^{-5} cm/s, making the falling head method the best choice. In this study, a constant head of 850 mL was maintained to measure the time for water flow through an area of 182 cm^2.

3 Results and Discussion

Figure 1 shows the 95% confidence interval of field air voids (normal vs. segregated area) in both directions. The results show that there is a significant difference between air voids based on the area of coring. Significantly higher air voids were observed in segregated areas as compared to normal areas, especially in the southbound direction. Average air voids in south-bound were about 8%.

Table 1 shows the results of the mean air void comparison for southbound direction. The p-value less than 0.05 shows that the difference is statistically meaningful with 95% confidence. While average air voids are higher in the areas with segregation along northbound, the difference is statistically insignificant.

Table 2 shows the average air voids along different lanes. Generally, the air voids are the highest in the fast lane and lowest in the slow lanes. This observation is valid

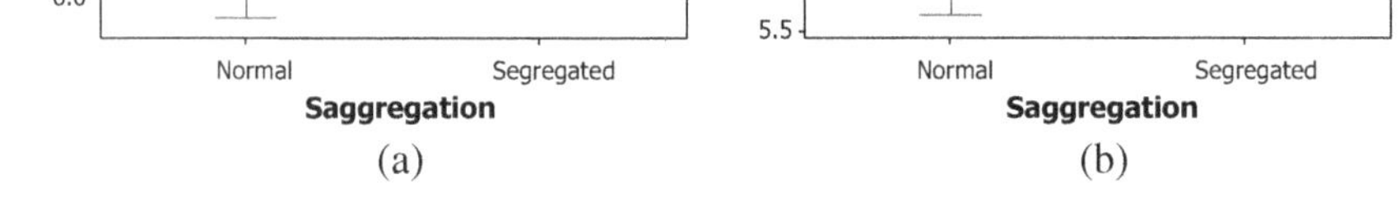

Fig. 1 Field air voids differences between normal and segregated areas **a** southbound **b** northbound

Table 1 Comparison of field air voids (normal vs. segregated areas) in southbound (SB)

Source	DF	SS	MS	F	p-value
Sag code	1	20.63	20.63	7.07	0.010
Error	58	169.18	2.92		
Total	59	189.81			

Note: DF = degrees of freedom, SS = sum of squares, MS = mean sum of squares, F = F-value

Table 2 Comparison of field air voids (normal vs. segregated areas) in different lanes (SB)

Segregation	Lane		
	F	M	S
Yes	7.42	6.33	6.08
No	8.44	7.98	6.82

Note: F = fast lane (innermost lane), M = middle lane, S = slow lane (outer most lanes)

for northbound direction as well (see Fig. 2). Table 3 shows the results of the one-way ANOVA. There is a significant difference in air voids between the inner and outer lanes.

Although no significant difference in air voids was found between different design thicknesses (see Table 4), it seems that 4 cm exhibited the highest average air voids (i.e., 8.2%) in the southbound direction.

Figure 3 shows the comparison of air voids between different binder types used in the asphalt mix along with the project. While higher air voids are observed in the asphalt mix with PG binder, the difference is not statistically significant (see Table 5). Figure 4 shows the variations of the mix and field air voids. There is a significant difference between mix and field air voids, as shown by the 95% confidence intervals in Fig. 5.

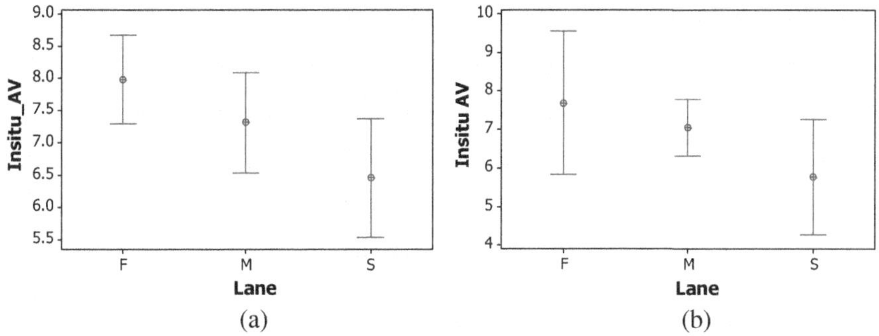

Fig. 2 Difference in field air voids between lanes **a** southbound (SB), **b** northbound (NB)

Table 3 Comparison of field air voids between different lanes (SB)

Source	DF	SS	MS	F	p-value
Lane type	2	23.54	11.77	4.04	0.023
Error	57	166.27	2.92		
Total	59	189.81			

Note: DF = degrees of freedom, SS = sum of squares, MS = mean sum of squares, F = F-value

Table 4 Comparison of field air voids between different design thicknesses

Source	DF	SS	MS	F	p-value
Design thickness	3	13.49	4.5	1.43	0.244
Error	56	176.32	3.15		
Total	59	189.81			

Note: DF = degrees of freedom, SS = sum of squares, MS = mean sum of squares, F = F-value

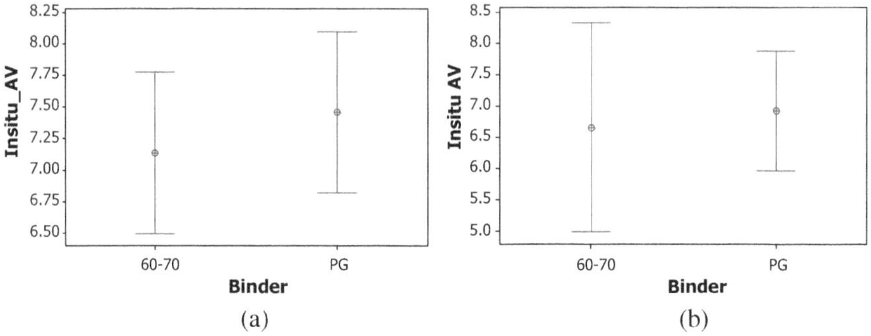

Fig. 3 The difference in field air voids between binder type, **a** SB, **b** NB

Table 5 Comparison of field air voids between binder types

Source	DF	SS	MS	F	p-value
Binder types	1	1.41	1.41	0.43	0.513
Error	58	188.4	3.25		
Total	59	189.81			

Note: DF = degrees of freedom, SS = sum of squares, MS = mean sum of squares, F = F-value

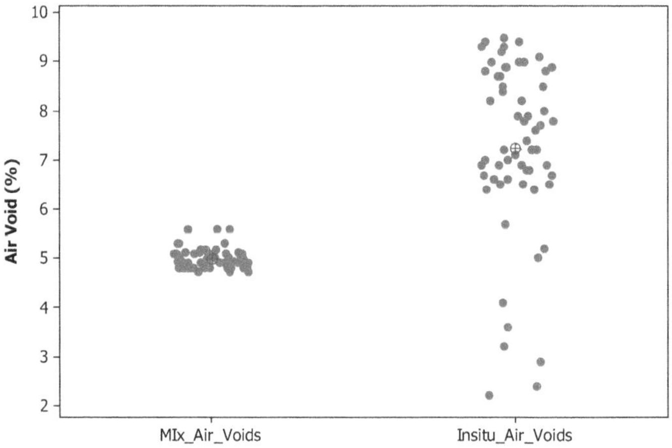

Fig. 4 Variations in air voids—SB

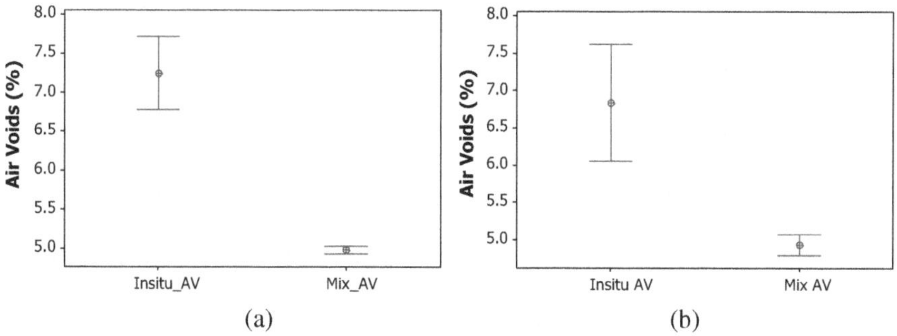

Fig. 5 Difference between as-compacted and HMA mix air voids, **a** SB, **b** NB

Figure 6 shows the relationship of air voids with asphalt mixtures permeability. The overall results of field permeability showed that asphalt mixture is very permeable (i.e., $k > 10^{-3}$ cm/s), especially in the areas of high air voids (>7%).

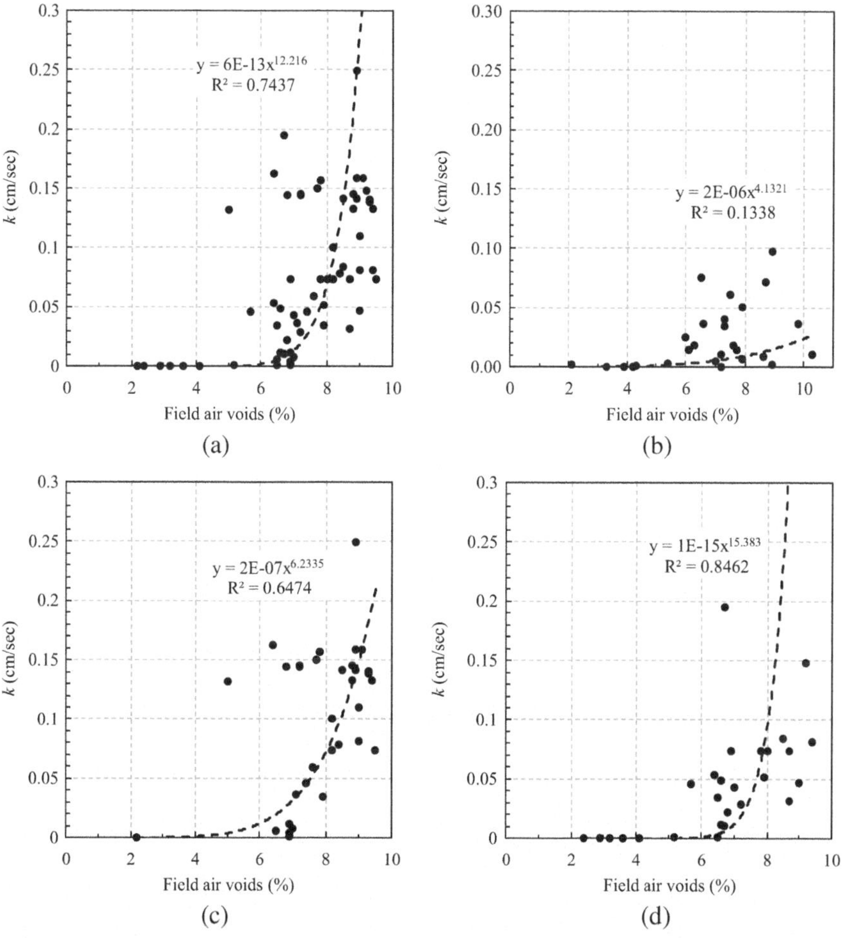

Fig. 6 Permeability of asphalt mix versus field air voids, **a** SB, **b** NB, **c** SB segregated areas, **d** SB normal areas

The permeability of asphalt mixtures is significantly lower in the northbound as compared to those in the southbound direction (see Figs. 6a, b). The permeability of mixes in segregated areas can be very high even at air voids of less than 7% (see Fig. 6c). In general, when the field air voids are more than 6%, the permeability of the mixes i.e., k > 10^{-3} cm/s. Figure 7 shows the permeability of surface mixes by different lanes, a clear trend between field air voids and permeability coefficients (k) can be observed within fast and middle lanes. However, the k-values showed high variability for the slow lane. The cause of this variability may be attributed to variations in the field air voids in the slow lanes mainly because of traffic compaction.

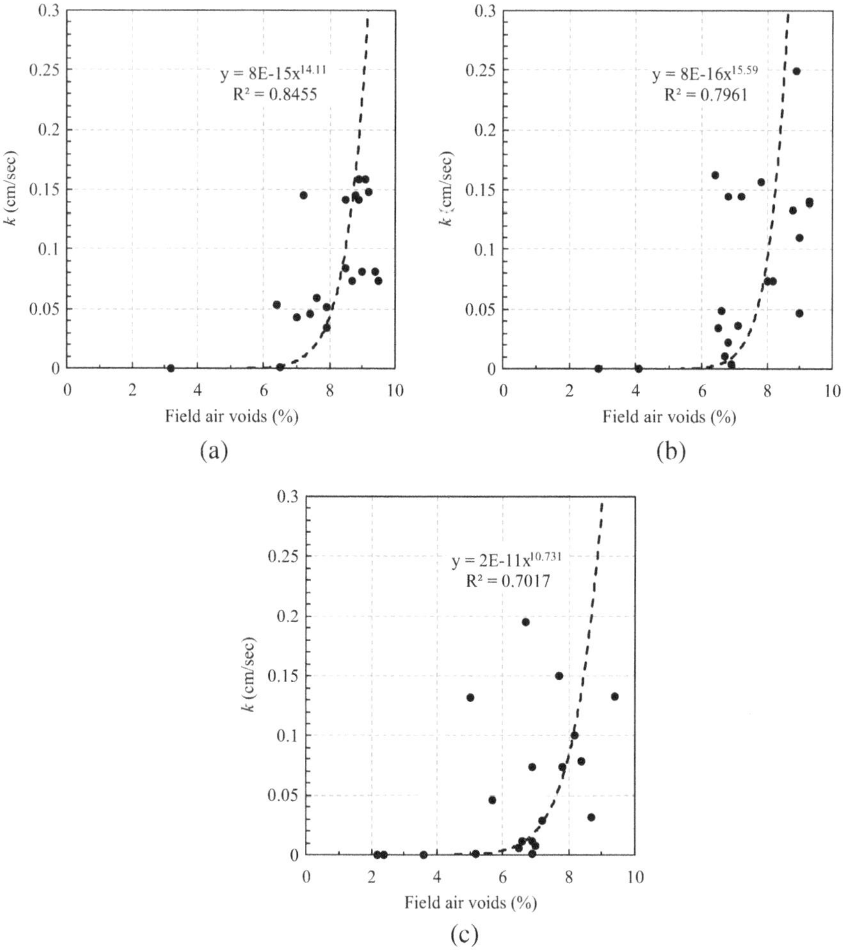

Fig. 7 Permeability of asphalt mix versus field air voids by lane type in SB direction, **a** Fast lanes(F), **b** Middle lane (M), **c** Slow lane (S)

Figure 8 presents the relationship between field air voids and permeability by binder type. More variations in air voids can be observed in the 60–70 binder (Fig. 8a), while a strong correlation can be observed for PG binder between air voids and the mix permeability (Fig. 8b). Figure 9 shows the relationship by design thickness, it seems the air voids of 6% or more can make the mix permeable irrespective of the design thickness.

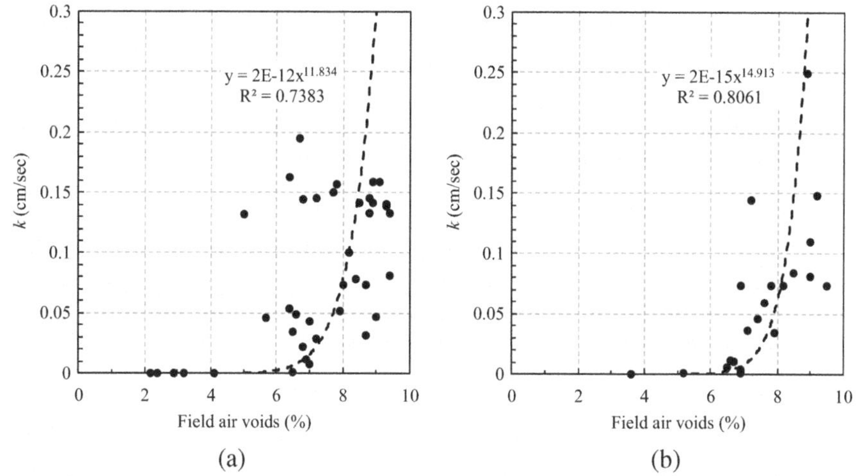

Fig. 8 Permeability of asphalt mix versus field air voids by binder type in SB direction, **a** Pen 60–70, **b** PG

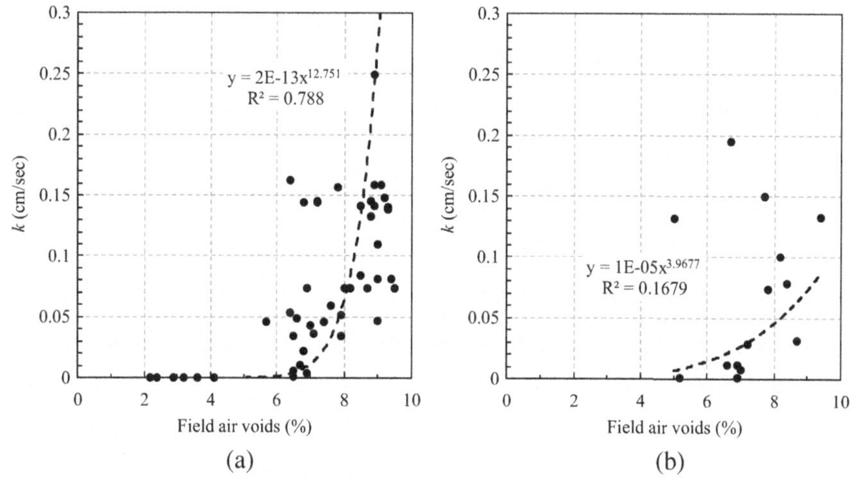

Fig. 9 Permeability of asphalt mix versus field air voids by design thickness in SB direction, **a** lift thickness $\leq$ 5 cm, **b** lift thickness = 10 cm

4 Conclusions

The following are findings from this case study based on in-situ and material testing results:

- The field air voids were found to be higher than 7%, especially in areas where the asphalt mixture segregation was observed.
- Field air voids are significantly higher in the fast lanes than slow lanes irrespective of factors considered in the material testing, especially in the southbound direction.
- Field air voids (PF) are significantly higher than the plant mixed-lab compacted (PL) air voids.
- As a consequence of above findings, the permeability of mixes is significantly higher than the standard (k value range 10^{-3} and 10^{-5} cm/s) when the field air voids are more than 6%. Because of higher permeability water can seep through the mixes in all lanes.
- Since the cross-fall of 1.8–2.2% (causing an average head of about 9 inches between fast and slow lanes) is maintained (in HMA and aggregate base layers) to cater for surface runoff, the moisture entering in the fast and middle lanes will accumulate under the outer lanes in both directions.

References

Arambula E, Masad E, Martin AE (2007) Influence of air void distribution on the moisture susceptibility of asphalt mixes. J Mater Civ Eng 19:655–664

Brown ER, Hainin MR, Cooley A, Hurley G (2004) NCHRP report 531 - relationship of air voids, lift thickness, and permeability in hot mix asphalt pavements. Washington D.C., USA

Choubane B, Page GC, Musselman JA (1998) Investigation of water permeability of coarse graded Superpave pavements. J Assoc Asph Paving Technol 67

Harris CH (2007) Hot mix asphalt permeability: tester size effects and anisotropy. Virginia Polytechnic Institute and State University

Hunter ER, Ksaibati K (2002) Evaluating moisture susceptibility of asphalt mixes. Fargo, ND

Masad E, Jandhyala V, Dasgupta N, Somadevan N, Shashidhar N (2002) Characterization of air void distribution in asphalt mixes using X-ray computed tomography. J Mater Civ Eng 14:122–129

Mohammad LN, Elseifi MA, Cooper III, SB, Hughes CS, Button JW, Dukatz Jr, EL (2016) NCHRP report 818 - comparing the volumetric and mechanical properties of laboratory and field specimens of asphalt concrete. Washington D.C., USA

Norambuena-Contreras J, Izquierdo EA, Castro-Fresno D, Partl MN, Garcia Á (2013) A new model on the hydraulic conductivity of asphalt mixtures. Int J Pavement Res Technol 6:488–495

Application of Dynamic Creep Testing to Investigate Permanent Deformation Characteristics of Asphalt Mixes

Amir Kavussi and Seyed Mohsen Motevalizadeh

Abstract Permanent deformation of asphalt mixes is caused as a result of several parameters, including; asphalt mix characteristics, climatic conditions, and traffic loading extent and configuration. For evaluating this distress mode, international institutions and standard codes of practices suggest different processes and testing methods. However, each of these investigates rutting resistance of asphalt mixes at specific conditions. The aim of this research was to investigate the effects of different loading configurations on Hot Mix Asphalt (HMA) in laboratory, applying Repeated Load Permanent Deformation (RLPD) testing method. With this purpose, various loading configuration and load durations and rest periods were applied. In order to simulate the impacts of both light and heavy weight traffic loading on permanent deformation resistance of mixes, two different stress levels of low and high extents were applied in RLPD testing. Based on the findings of this research, it resulted that asphalt mixes under low stress level (which corresponds with light weight loading traffic) in RLPD testing, do not get to "Flow Number" values. In addition, it resulted that performance characteristics of asphalt mixes cannot properly be evaluated applying one loading condition only. Among the various testing parameters, it resulted that "Flow Number" is the best indicator of resistance of asphalt mixes against permanent deformation.

Keywords Dynamic creep test · Hot Mix Asphalt · Loading combination · Rest time

1 Introduction

Permanent deformation is one of the most occurring load-associated distresses in flexible pavements (Witczak and Fonseca 1996; Qi and Witczak 1998; Ameri et al. 2014). This distress mode is associated with rutting in wheel paths, which develops

A. Kavussi (✉) · S. M. Motevalizadeh
Faculty of Civil and Environmental Engineering, Tarbiat Modares University, Tehran, Iran
e-mail: kavussia@modares.ac.ir

S. M. Motevalizadeh
e-mail: mohsen.motevalizadeh@modares.ac.ir

C. Raab (ed.), *Proceedings of the 9th International Conference on Maintenance and Rehabilitation of Pavements—Mairepav9*, Lecture Notes in Civil Engineering 76,
https://doi.org/10.1007/978-3-030-48679-2_71

gradually as the number of load repetitions accumulates (Qi and Witczak 1998). Parameters, such as traffic, environment, materials, and design considerations affect pavement performance over time (Mansourkhaki et al. 2015). Permanent deformation is revealed by longitudinal profile irregularities and particularly, by transverse profile deformations in wheel paths (Huang 2004). These distresses affect riding comfort and safety of road users to a great extent (Barksdale 1971).

Many theories and methods have been employed; including, phenomenological empirical methods and mechanistic theories, consisting of viscoelastic damage and visco-elasto-plastic damage models (BS DD 226 1996). It is well recognized that asphalt binders exhibit non-linear behavior under high strains or stresses. The evolution of micro-cracks and micro-voids and rate-dependent plastic or visco-plastic hardening are other major sources of non-linearity in thermo-mechanical response of HMA (Qi and Witczak 1998). Numerous experimental studies have shown that the response of HMA mixes is time, rate, and temperature dependent and exhibits both recoverable (visco-elastic) and non-recoverable (visco-plastic) deformations (Standard et al. 1995; Qi and Witczak 1998, Salama et al. 2006). Because of the time-dependent behavior of asphalt mixes, distresses in asphalt pavements are function of loading pulse duration and rest time (Xu et al. 2014). Several laboratory testing methods are carried out to measure rutting resistance of asphalt pavements. Most of these are based on constant load duration and resting time.

2 Permanent Deformation and Rutting

Rutting is defined as the progressive accumulation of permanent deformation of each layer of pavement structure under repetitive traffic loading (Tayfur et al. 2007; Motevalizadeh et al. 2018). It was reported that permanent deformation in asphalt mixtures includes two different modes; namely, compacted deformation that can be described by means of layer consolidation and plastic deformation that occurs as a result of asphalt shear flow (Khodaii ad Mehrara 2009).

Permanent deformation is composed of primary, secondary and tertiary zones (Zhou et al. 2004). In the primary zone, accumulated permanent deformation grows rather rapidly until it reaches to the secondary zone. In fact, in zone 1, the rate of permanent deformation per cycle is dramatically decreased until initiation of the secondary zone. In the secondary zone, rate of permanent deformation per cycle will be constant until it gets to a minimum value. At this point, the accumulated permanent deformation starts to increase rapidly. This point was introduced as the initiation of tertiary zone, named as Flow Number (FN) (Qi and Witczak 1998).

Various procedures, including "Three-Stage Method" and "Stepwise Method" have been recommended to properly calculate FN (Goh and You 2009). Moreover, another method was suggested in which each stage of rutting curve was modelled separately (Zhou and Scullion 2002). The power law, linear and exponential models, which are presented in Eqs. 1–3, respectively, were applied to describe the primary, secondary and tertiary stages.

$$\varepsilon_P = aN^b, N \leq N_{PS} \tag{1}$$

$$\varepsilon_P = \varepsilon_{ps} + c(N - N_{PS}), N_{PS} \leq N \leq N_{ST}, \textit{and}\, \varepsilon_{PS} = aN_{PS}^b \tag{2}$$

$$\varepsilon_P = \varepsilon_{ST} + d\left(e^{f(N-N_{ST})} - 1\right), N \geq N_{ST} \tag{3}$$

In these equations, ε_P is the plastic strain at each cycle; ε_{PS} and ε_{ST} are plastic strains at onset of the secondary and the tertiary stages; N is the number of load repetition; N_{PS} and N_{ST} are numbers of load cycles corresponding with the onset of the secondary and the tertiary stages. Furthermore, a, b and c are the regression constants. It is worth to consider that ε_{PS} and ε_{ST} can be obtained using $\varepsilon_{PS} = aN_{PS}^b$ and, $\varepsilon_{ST} = \varepsilon_{PS} + c(N_{ST} - N_{PS})$ equations. Stepwise Method was based on traditional methodology of the application of smoothening method to determine FN (Goh and You 2009).

3 Objective of the Research

The main objective of this research was to evaluate permanent deformation resistance of asphalt mixes using RLPD testing method. With this aim, parameters such as loading pattern and load duration, rest time and stress levels were taken into consideration and the responses of the HMA mix against permanent deformation were determined.

4 Materials and Testing

4.1 Materials

The aggregates and filler were sampled from a limestone quarry in Tehran province in Iran. The aggregate grading was continuously graded with maximum nominal size of 12.5 mm. The binder was a 60–70 penetration grade bitumen from Refinery of Tehran. Marshall Method of mix design (ASTM D-1559) was carried out and the optimum bitumen content was determined. This was determined to be %5 of the total weight of the mix.

Table 1 The selected loading patterns in RLPD testing

Variable	Corresponding traffic speed	Loading pattern					
		HT1-100	HT-20	HT-10	LT2-100	LT-20	LT-10
Loading duration (ms)	100 km/h	100	–	–	100	–	–
	20 km/h	–	500	–	–	500	–
	10 km/h	–	–	1000	–	–	1000
Loading and rest time ratio		1 (Loading)			9 (Rest time)		
Stress level (kPa)		200					
		100					
Temperature (°C)		55					
Loading wave		Haversine					
Replicate		2					

4.2 *Testing Method*

Repeated Load Permanent Deformation (RLPD) test was carried out on HMA mizes at 55 °C. Based on previous studies (Barksdale 1971) and British Standard method of practice (BS DD 226 1996), haversine loading wave was applied at different loading configurations. RLPD testing was performed at two deviator stress levels of 100 and 200 kPa. Loading durations of 100, 500 and 1000 ms were applied. With each loading duration, the rest time was selected to be 1 or 9 times of loading time. For example, at 100 ms loading, the rest times were 900 ms. It should be noted that in accordance with visco-elasto-plastic theory, resilient strain consists of two portions, namely time-dependent and time-independent parts (Witczak and Fonseca 1996). The time-dependent part that is affected by the rest time, consists of elastic and plastic deformations. Accordingly, different magnitudes of resting times result in different visco elastic and visco plastic strains. Therefore, different values of rest times would have significant impacts on permanent deformation characteristics of asphalt mixes. Table 1 reports summary of the various loading patterns that were adopted.

5 Results and Discussion

5.1 *Impact of the Low Stress Level*

Stress level of 100 kPa can be considered as low stress level, corresponding with light weight loading on road pavements. In RLPD testing of HMA mixes, this stress level was applied at different loading patterns. At this stress level, the results of the corresponding permanent deformation strains were determined, as it is shown in Fig. 1. As it can be seen from this figure, the maximum and minimum permanent deformations occurred at LT-10 and LT-100 loading patters, respectively. In addition, LT-10 and LT-20 configurations, compared with HT-10 and HT-20, resulted in greater permanent deformations. These indicate that, the increase in rest time in these conditions, does not result in lower permanent deformation values. At increased loading durations, disregarding the impact of the rest period, greater permanent deformations were resulted. The minimum permanent deformation of the HMA mix, occurred at LT-100 which is smaller than that of HT-100. On the other hand, HT-10 resulted in smaller permanent deformations, compared with LT-10. From these, it can be resulted that permanent deformation in asphalt mixes strongly depends on the applied loading and resting times.

Permanent strain results in HMA mixes under cyclic loading at stress level of 100 kPa are presented in Fig. 2. As it can be seen in this figure, combinations of loading duration and resting times affected permanent deformations appreciably. In fact, with reference to this figure, the effect of passing vehicle speed (that was simulated by loading durations) on permanent deformation of HMA mixes, can distinctively be observed. Comparing the curves for accumulated strains at loading times and those at resting times in Fig. 2a, b, shows that the differences of the effects of traffic speed are more pronounced on curves related to the accumulated strains at resting times. In other words, increased resting time resulted in greater permanent deformations, when the specimens were exposed to greater loading durations. Therefore, the above testing procedure confirms the field conditions that low speed vehicles result in more severe rutting on pavements.

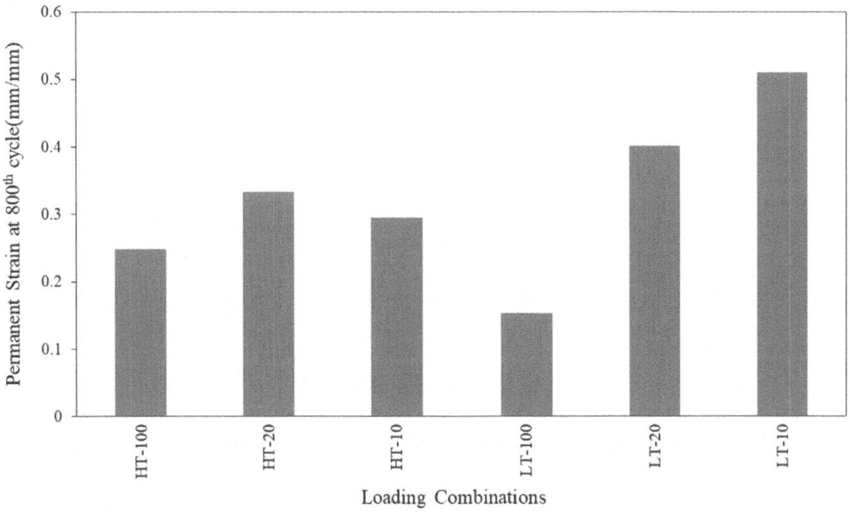

Fig. 1 Effect of loading pattern and combinations on permanent strains of HMA

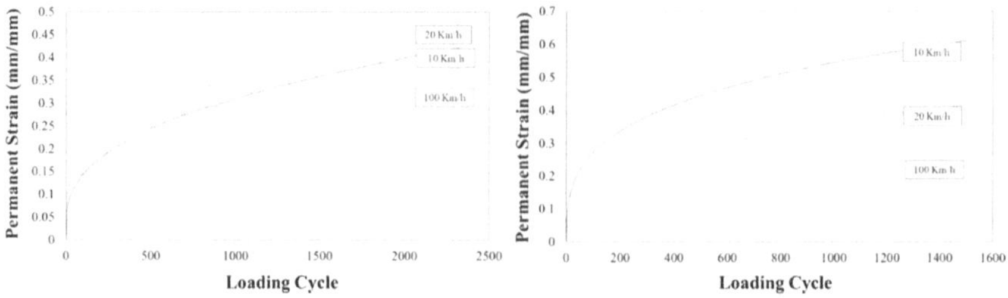

a) Rest ratio of 1/1 (loading time over rest time)

b) Rest ratio of 1/9 (loading time over resting time)

Fig. 2 Effects of vehicle speed on permanent deformation of HMA mix in RLPD test

In analyzing the testing results of RLPD test at greater loading cycles, at the initial stages of the secondary zone (Fig. 3), it can be noticed that increased loading duration resulted in early initiation of the secondary zone than the other patterns. For instance, it was observed that the mix that was subjected to HT-100 loading condition, reached to initiation point of the secondary zone much later than mixes that were subjected to HT-20 and HT-10 conditions. In contrast, initiation of the secondary zone would occur earlier as the resting time is increased. This phenomenon shows that initiation of the secondary zone strongly depends on the rest time durations.

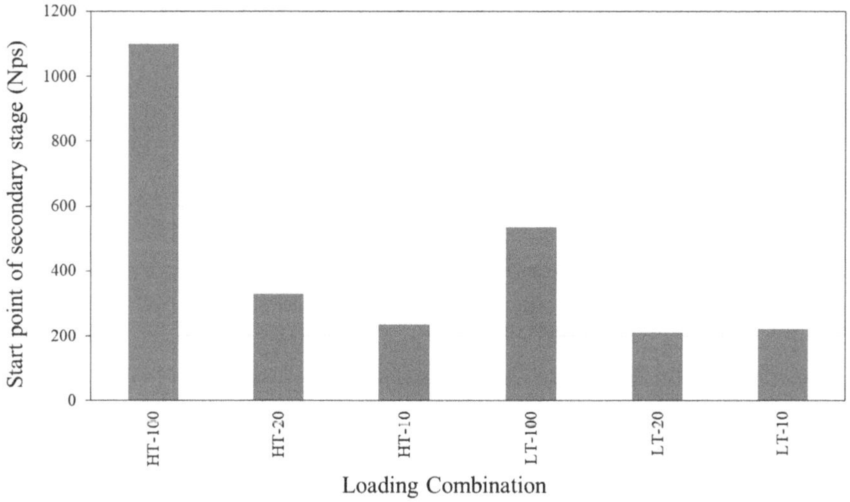

Fig. 3 Effect of loading combinations on initiation of the secondary zone in RPLD test

5.2 *Impact of High Stress Level*

The maximum permanent strains at 800th cycle was observed under the influence of LT-10 pattern and stress level of 200 kPa. The results for various loading combinations are reported in Fig. 4. The lowest strains occurred under HT-100 condition. Similar to what was mentioned in the previous section, it can be stated that increased rest periods resulted in greater permanent deformations. For instance, permanent deformation of the specimens that were subjected to HT-100 were 25% lower than those subjected to LT-100 loading. In addition, LT-10 pattern, compared with HT-10, resulted in greater permanent deformations.

Permanent deformations of asphalt mixes at stress level of 200 kPa and over a wide range of cyclic loading are presented in Fig. 5. Results reported in this figure

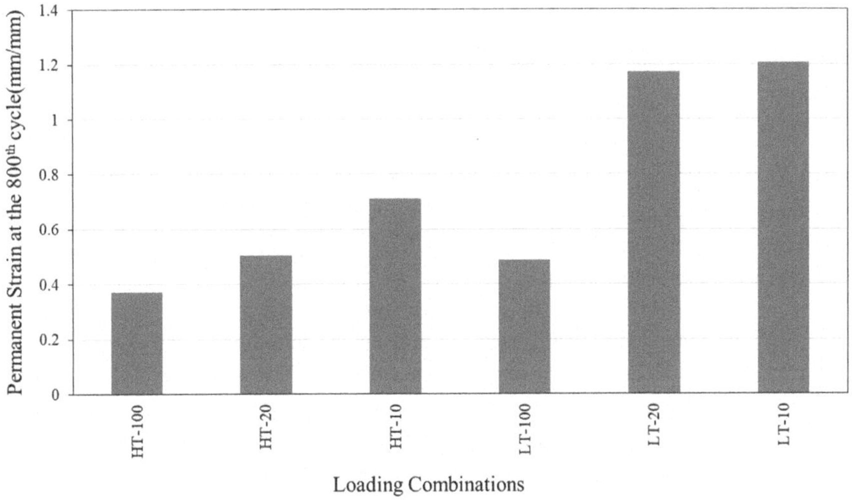

Fig. 4 Effects of loading combinations on permanent deformation of HMA mixes in RPDL test

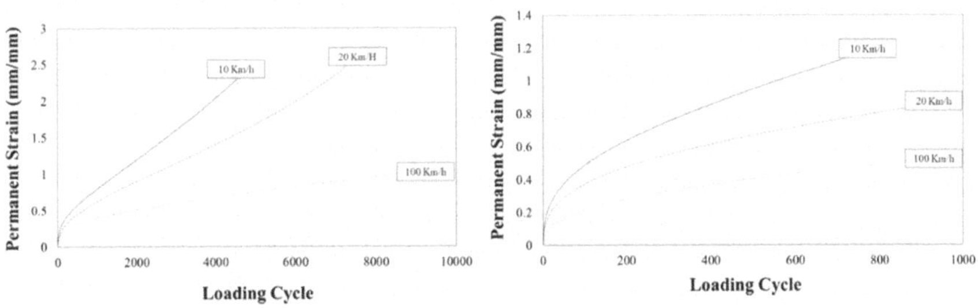

a) Rest ratio of 1/1 (loading time over resting time)

b) Rest ratio of 1/9 (loading time over resting time)

Fig. 5 Effects of vehicle speed on permanent deformation of HMA mixes

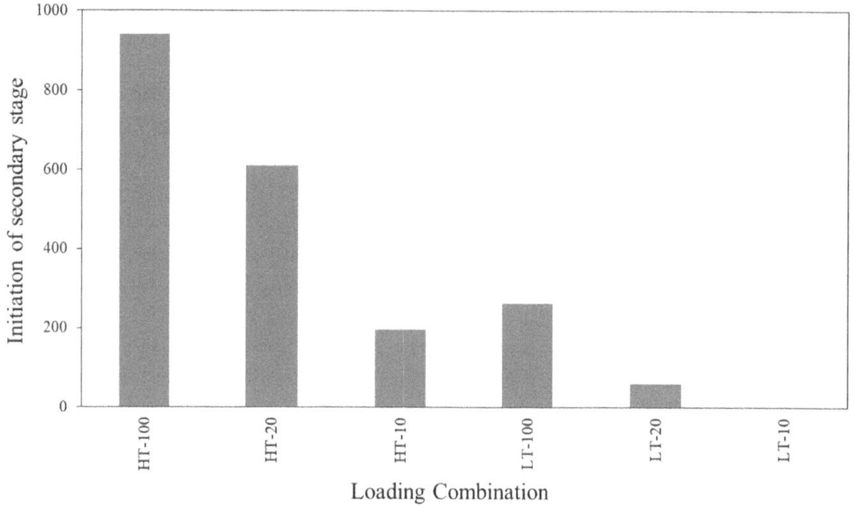

Fig. 6 Effect of loading combinations on start point of the secondary zone

and those in Fig. 2 indicate that permanent deformation resistance of HMA mixes is greatly affected by stress level. Eventually, it can be concluded that the application of various stress levels, will result in better understanding of behavior of asphalt mixes against permanent deformation.

As it is illustrated in Fig. 6, the secondary zone of HMA mixes that were exposed to HT-100 loading configuration, was initiated earlier than those at HT-20 and HT-10 conditions. The differences were approximately 50 and 35%, respectively. Moreover, number of loading cycles related to the onset of this zone were decreased with increased resting time. In addition, it can be observed that for the loading patterns in which loading duration was 1000 ms, the increased resting time resulted in greater permanent deformations (almost 95% increase). These results were in accordance with those in a previous research work (Ayar et al. 2018). As it was mentioned previously, viscous portion of bitumen's strain is a time-dependent phenomenon that would be influenced by variation of resting time. In fact, time-dependent portions of asphalt mixes would have enough time to develop if greater resting time is provided. Eventually, it can be stated that in RLDP testing, the initiation of the secondary zone will be reached earlier, if resting times are assigned on samples.

5.3 *Impact of Loading Patterns on Flow Number*

Since the adopted stress levels in this research were selected from the low and moderate levels, most of HMA specimens did not reach to Flow Number (FN) point, except at two loading patterns of HT-10 and HT-20. This indicates that HMA mixtures meet FN point only if these are subjected to heavy loading conditions; which occur in conditions such as moderate and high stress levels, heavy traffic with low resting time and slow speed vehicles (i.e. at great loading durations).

Comparison of the results of FN and permanent deformations shows an interesting contrast. In fact, although increased resting time results in greater permanent deformations, however, their related FN values would be increased. Eventually, a thorough research is needed to investigate interaction between the rest time and the loading duration at greater stress levels than those applied in this research.

6 Conclusions

From investigation of the effects of various loading parameters on permanent deformation characteristics of an HMA mix in RLPD testing, the following conclusions were achieved:

1. At stress level of 100 kPa which corresponds with light traffic loading, the testing does not get to FN point. Under the influence of this stress level, changes in loading cycles did not have significant impacts on permanent deformation characteristics of mixes.
2. The impacts of different resting times on permanent resistance of HMA mixes, strongly depends on the loading pattern. At high and moderate stress levels, greater resting times resulted in increased FN values. While, at low stress levels, where the specimens did not get to FN point, increased stress level resulted in greater permanent deformations.
3. Greater loading durations and resting times accelerated initiation of the secondary zone. In fact, greater rest periods allowed enough time for formation of viscous strains in asphalt mixes.
4. Permanent deformation in the secondary zone and initiation of the secondary zone reflect the misleading results in predicting permanent deformation characteristics of HMA mixes. Eventually, the best parameter that can be employed will be FN value.
5. Since permanent deformation of an HMA mix depends on the interaction between the loading parameters, it is not possible to introduce a certain loading pattern as a critical regime for investigating their permanent deformation resistance.

References

Ameri M, Sheikhmotevali AH, Fasihpour A (2014) Evaluation and comparison of flow number calculation methods. Road Mater Pavement Des 15(1):182–206

Ayar P, Moreno-Navarro F, Sol-Sánchez M, Rubio-Gámez MC (2018) Exploring the recovery of fatigue damage in bituminous mixtures: the role of rest periods. Mater Struct/Materiaux et Constructions 51(1)
Barksdale RD (1971) Compressive stress pulse times in flexible pavements for use in dynamic testing. Highway Res Rec 345:32–44
BS DD 226 (1996) Method for determining resistance to permanent deformation of bituminous mixtures subject to unconfined dynamic loading, p 12
Goh SW, You Z (2009) A simple stepwise method to determine and evaluate the initiation of tertiary flow for asphalt mixtures under dynamic creep test. Constr Build Mater 23(11):3398–3405
Huang YH (2004) Pavement design and analysis, p 785
Khodaii A, Mehrara A (2009) Evaluation of permanent deformation of unmodified and SBS modified asphalt mixtures using dynamic creep test. Constr Build Mater 23(7):2586–2592
Mansourkhaki A, Sarkar A, Ameri M (2015) Impact of different loading patterns with short duration on the permanent strain of asphalt mixture. J Test Eval 43(4):20130222
Motevalizadeh SM, Ayar P, Motevalizadeh SH, Yeganeh S, Ameri M, Bemana K (2018) Investigating the impact of different loading patterns on the permanent deformation behaviour in hot mix asphalt. Constr Build Mater 167:707–715
Qi X, Witczak MW (1998) Time-dependent permanent deformation models for asphaltic mixtures. Transp Res Rec 1639:83–93
Salama HK, Chatti K, Lyles RW (2006) Effect of heavy multiple axle trucks on flexible pavement damage using in-service pavement performance data. J Transp Eng 132(10):763–770
Standard ST, The RD, For D (1995) Methods of sampling and testing asphalt Method 12.1: Determination of the permanent compressive strain characteristics of asphalt — Dynamic creep test
Tayfur S, Ozen H, Aksoy A (2007) Investigation of rutting performance of asphalt mixtures containing polymer modifiers. Constr Build Mater 21(2):328–337
Witczak MW, Fonseca OA (1996) Revised predictive model for dynamic (complex) modulus of asphalt mixtures. Transp Res Rec 1540:15–23
Xu T, Wang H, Li Z, Zhao Y (2014) Evaluation of permanent deformation of asphalt mixtures using different laboratory performance tests. Constr Build Mater 53:561–567
Zhou F, Scullion T (2002) Discussion: three stages of permanent deformation curve and rutting model. Int J Pavement Eng 3(4):251–260
Zhou F, Scullion T, Sun L (2004) Verification and modeling of three-stage permanent deformation behavior of asphalt mixes. J Transp Eng 130(4):486–494

A Performance Prediction Model for Continuously Reinforced Concrete Pavement Using Artificial Neural Network

Hakan Yasarer, Mohammad Najmush Sakib Oyan, and Yacoub Najjar

Abstract The pavement performance prediction models are key components of pavement management systems. Predictions models are used for determining the future condition of pavement as well as types of maintenance and rehabilitation actions needed to keep the operation of transportation systems continuous. Accordingly, maintenance and rehabilitation actions must be prioritized due to limited budget allocations. Mississippi Department of Transportation (MDOT) utilizes Hidden Markov probability models. However, varying traffic conditions and unusual events in the operation are not reflected in these models due to the nature of probabilistic modeling. A new model using a more inclusive and powerful approach to predict the future condition of the pavement is needed. In this study, the distress data from Continuously Reinforced Concrete Pavement (CRCP) sections in Mississippi was used to develop a performance prediction model using Artificial Neural Networks (ANNs) approach. Additionally, rehabilitation actions were included as part of the model inputs to study the impact of rehabilitation actions on the model. The performance of all the developed models showed a good agreement between observed and predicted condition measures. However, only one model with the best statistical accuracy was selected to be utilized as the best performing model, which can be used for the prediction of CRCP performance within the allowable time range.

Keywords Pavement performance model · Continuous Reinforced Concrete Pavement · Artificial Neural Network

---

H. Yasarer (✉)
The Department of Civil Engineering, The University of Mississippi, 208 Carrier Hall, University, Oxford, MS 38677, USA
e-mail: hy@go.olemiss.edu

M. N. S. Oyan · Y. Najjar
The Department of Civil Engineering, The University of Mississippi, 106 Carrier Hall, University, Oxford, MS 38677, USA
e-mail: moyan@go.olemiss.edu

Y. Najjar
e-mail: ymnajjar@go.olemiss.edu

C. Raab (ed.), *Proceedings of the 9th International Conference on Maintenance and Rehabilitation of Pavements—Mairepav9*, Lecture Notes in Civil Engineering 76,
https://doi.org/10.1007/978-3-030-48679-2_72

1 Introduction

Mississippi Department of Transportation (MDOT) maintains a total of 200 miles continuously reinforced concrete pavement (CRCP) statewide. Most CRCP pavements are used in interstate or principal arterial systems (Pathway Services Inc. 2017). The pavement management system (PMS) data has been collected continuously in terms of distress type, severity, and frequency. The evaluation of these parameters is essential to pavement performance and monitoring. Pavement performance generally is an indicator of the pavement condition over its lifespan, and as a measure of ride quality. The ability to predict pavement performance over its life span is a key aspect of the pavement management system. Accordingly, prediction models are a vital tool for the network-level decision-makers to prioritize the projects based on the level of importance due to budgetary constraints. Pavement Condition Rating (PCR) and International Roughness Index (IRI) are the terms that are used to evaluate the condition of a pavement section. PCR is a mathematical expression reflecting the combined effect of varying distress type, their severity and extent[1]. IRI is a profile-based statistic, which is calculated using a "quarter car model" and is used as the index for comparing the smoothness of a pavement[2]. IRI data used in this study was collected by a data collection vehicle that is equipped with class-1 South Dakota inertial profiler class 1-ASTM E950[3]. Maintaining any pavement system is a routine and preventive process to keep roads' condition up to the standards for public use. Minor maintenance can go year-round with the routine checks for expanding the life-span of the section. Major maintenance is costly but necessary when the road condition no longer meets the quality standards. Typically, CRCP pavements have higher longevity and a higher construction cost. The maintenance and rehabilitation activities are not planned on an annual basis for every section and accordingly must be scheduled for the critical pavement sections as needed. For planning and scheduling, probabilistic models have been utilized for a long time. However, probabilistic based models are not effective tools for varying conditions. Also, the effects of rehabilitation and maintenance actions on future pavement conditions are difficult to incorporate into probabilistic models, yet they are very important in the decision-making process.

Artificial Neural Networks (ANNs) approach is a widely used tool that can learn from historical data to predict the future. ANNs approach is proven to be the best function approximation method, especially on highly nonlinear relations. Pavement performance prediction is known to be highly complex phenomena and the ANNs approach is a very suitable tool to explore the complexity of this phenomenon. In this study, 424 datasets from 53 CRCP sections in Mississippi were utilized to develop ANN prediction models. The datasets from 2010 to 2018 were included to develop

---

[1] http://www.dot.state.oh.us/Divisions/Planning/TechServ/TIM/Documents/PCRManual/2006PCRIntro.pdf.

[2] http://1204075.sites.myregisteredsite.com/downloads/RT/RT3.07.pdf.

[3] https://static1.squarespace.com/static/57978fefcd0f688f77aa40b3/t/5a092ebdec212d1131b09b38/1510551231995/pathrunnersmall.pdf.

five ANN models. Even though Equivalent Single-Axle Load (ESAL) is one of the design parameters according to the design manual by Asphalt Institute along with other parameters such as layer thickness, pavement materials, structural number, etc. (Huang 2004), none of the models included ESAL and rainfall data as inputs due to unavailability at the time of this study. Five models and their development stages are presented in the following sections.

2 Literature Review

Different optimization techniques (i.e. probabilistic and deterministic) such as ANN and regression have been utilized to predict pavement performance to support PMS. All developed models considered different aspects of the pavement system but they all had a single purpose in common: maximizing the cost-benefit actions. Probabilistic models introduce randomness in the model and the outcomes follow a probabilistic distribution. One of the popular probabilistic models in the 1990s is the Hidden Markov model which has been commonly utilized in pavement management systems. Elhadidy et al. (2015) used the Markov chain model for predicting pavement performance over the life span and developed a genetic algorithm to achieve optimum cost-benefit actions. Abaza et al. (2004) used a discrete-time Markovian model in decision-making-level to schedule maintenance and rehabilitation based on the pavement's future performance. In another study, mechanistic analysis, a type of deterministic approach, has been performed on Continuously Reinforced Concrete Pavement (CRCP) to estimate pavement life in terms of punchout frequency considering wheel load, environmental effect, and stochastic nature of material (Won et al. 1991). Another deterministic performance prediction model was developed for the rehabilitation and management of flexible pavement using AASHTO's serviceability concept (Abaza 2004). George et al. (1989) developed an empirical mechanistic model using SAS software for asphalt concrete pavement to predict Pavement Condition Rating (PCR) by gathering historical information and monitoring about 2000 miles of roads over two years. An error back-propagation neural network was constructed to predict the roughness progression in flexible pavement using road distresses, structural number, and environmental and traffic impact (Attoh-Okine 1994). Wang et al. (2003) proposed an integer linear programming model using the optimizer program, GAMS to aid network-level decision-makers for planning, maintenance, and rehabilitation activities. The pavement performance curves were generated for long sections of pavement using AASHTO's serviceability concept to establish pavement rehabilitation policies (Abaza et al. 2001). A knowledge-based computer simulation model, CA4PRS (Construction Analysis for Pavement Rehabilitation Strategies) was developed to support highway agencies for the rehabilitation and reconstruction of highways. This simulation model is inputted with schedule interfaces, pavement design and materials, resource constraints, and lane closure scheme, and contractor's logistics. The maximum distance of the highway to be rehabilitated is determined based on "What-If" scenarios (Lee and Ibbs 2005).

3 Methodology

3.1 Data Collection

The data used in this study is collected as a part of a survey in Mississippi by MDOT. Mississippi has four different pavement types: Flexible, Jointed Concrete Pavement (JCP), Continuous Reinforced Concrete Pavement (CRCP), and Composite Pavement (CP). According to the 2018 survey, a total of 69 sections were marked as CRCP in the database. The longitude and latitude and landmarks at the section beginnings and endings are used as identifiers. The section length varies from 0.028 miles to 10.055 miles and thickness varies from 152.4 to 558.8 mm. The age of the CRCP pavement has a range of 2 to 64 years until 2010. Like all other states, MDOT operates a pavement management system that includes PCR, International Roughness Index (IRI), and distress data. Usually, PCR values decrease and IRI values increase with the deterioration of pavement over time. In the CRCP database, the distresses recorded include longitudinal and transverse cracking, patching, lane-to-shoulder drop-off, durability cracking, longitudinal spalling, punchouts, and punchout repair. All of these distresses are classified into three categories based on the field condition: low, medium, and high. Besides these distresses, IRI is measured as Mean Roughness Index (MRI) and also classified as low, medium and high severity level. PCR value is calculated based on measured IRI and the deduction factors that are determined as functions of distresses and severity level. Another variable known as Distress Rating (DR) is calculated in the same manner without including the measured IRI. Finally, the ratio of PCR and DR is expressed as a percentage and designated as Roughness Rating (RR).

3.2 Data Processing

The database for ANN model development is obtained after cleansing and reorganizing the raw data files. The data collected from 2010 to 2018 for each section are included in the model development as data acquisition methods, materials, and tools have changed before 2010. Any section with missing or illogical data has been excluded as the ANN model development process needs a complete dataset for training. These exclusions reduced the number of sections from the initial 69 to the final 53. Each section is comprised of four different datasets based on PCR, IRI, and rehabilitation action, which resulted in 212 datasets. In Mississippi, the distresses data is collected every even year. To develop prediction models that are applicable for a 1-year increment, the odd year data was needed to be generated by averaging consecutive years from 2010 to 2018. This resulted in a total of 424 datasets used for model development. Concrete pavement has a rigid structure and does not deteriorate significantly over time, so only two documented major rehabilitation records were found. It is known that not all rehabilitation actions were properly recorded.

Due to the insufficient amount, another approach to assign rehabilitation actions was proposed based on the discussions with the state agency. Improvement of PCR and IRI values without any rehabilitation was found to be irrational. Some uncertainty due to the calibration of profilometer, systematic errors, and the environmental conditions on the day of the survey may have resulted in some of the irrational condition measures. To incorporate the effect of the rehabilitation on PCR and IRI, artificial rehabilitation actions based on the significant changes in PCR and IRI have been assigned categorically (i.e. Assigned "1" if the rehabilitation took place and "0" if no rehabilitation occurred). Based on the evaluation of data history, the threshold values for PCR and IRI were assumed to be 2.5 and 0.06, respectively. For example, if the PCR increased at least 2.5 points in a year compared to the previous year, the rehabilitation was assumed to take place in that year. Otherwise, it was considered as no rehabilitation. Similarly, if IRI decreased 0.06 points in a year, the rehabilitation was assumed to take place in that year. By using this methodology, two sets of assignment categories were generated: rehabilitation based on PCR and rehabilitation based on IRI. The models with two outputs (i.e. PCR and IRI) were modified to be used with the complementary PCR (i.e. 100-PCR) because of the fact that two outputs must be directly proportional in ANN modeling. It is known that PCR and IRI usually change inversely over time without any rehabilitation actions, but the complementary PCR and IRI both change proportionally over time. Thus, the utilization of complementary PCR and IRI helped the network to optimize the model with higher accuracy and to establish a better correlation between actual and predicted outputs.

3.3 ANN Structure and Modeling Process

The ANNs approach used in this study is a single layer feed-forward error back-propagation neural network. A typical ANN structure consists of an input layer, a hidden layer, an output layer, and connection weights. The input layer contains independent variables and the output layer contains dependent variable(s). The hidden layer connects the independent and dependent variables by a series of nodes and connection weights and is placed between the input and output layers. The hidden layer allows the flow of information from the input layer to the output layer through the connection links that are weighted by numerical values known as connection weights. The hidden layer is the main part of the ANN structure, as it defines the accuracy of the model. The error between the observed and predicted values is backpropagated through the connection links to update the connection weights and threshold values. ANN approach has been extensively utilized in many studies. More information can be found in the literature (i.e. Attoh-Okine (1994) and Raab et al. (2013)).

A total of five models were developed using the same CRCP database. In these models, some of the input parameters were varied to understand their effects on the outputs' prediction accuracy. All the inputs and outputs used in these models are listed in Table 1. In Table 1, age is the elapsed time between the construction year and 2010. PCR and IRI used in the inputs are the measured values in 2010. ESAL and cumulative annual rainfall were not included in these models as the data were not available at the time of the study. All ANN models were trained with 50%, tested with 25%, validated with 25% of the data and finally, trained with the full dataset. The training sets were chosen in a way that ANN models were trained with wider ranges. The first stage included training and testing with the predetermined datasets. Once the first stage was complete, the best networks were selected and compared based on minimum ASE (Averaged Squared Error), minimum MARE (Mean Absolute Relative Error) and maximum of coefficient of determination (R^2). The ASE and MARE were calculated by the Equations (1) and (2) correspondingly.

$$ASE = \frac{\sum_{1=1}^{N}\sum_{j=1}^{n}\left(Y_{ij}^{P} - Y_{ij}^{O}\right)^2}{N \cdot n} \quad (1)$$

$$MARE = \frac{\sum_{1=1}^{N}\sum_{j=1}^{n}\left|\frac{Y_{ij}^{P} - Y_{ij}^{O}}{Y_{ij}^{O}}\right|}{N \cdot n} \quad (2)$$

Here, Y_{ij}^{P} = Predicted output, Y_{ij}^{O} = Actual output, N = number of dataset, and n = number of outputs.

The best networks were validated by the validation datasets reserved solely for this step and finally, trained with the full dataset, which was called all data in Table 2. The network that had the minimum ASE and maximum R^2 in testing, validation, and in all data stages was selected for the final ANN model architecture.

The final architecture of each model is presented in the bottom row of Table 2 as "X-Y-Z", where X is the number of inputs, Y is the number of hidden nodes, and Z is the number of outputs. The correlation plots in Fig. 1 present the correlation between the actual and predicted values for all data stages for all the developed models. In

Table 1 Inputs and outputs of the developed models

Model no	Common inputs	Additional input(s)	Output(s)
Model 1	• Location • Thickness • Age • Length • Time (t) • PCR • IRI	Rehabilitation based on PCR	PCR (t)
Model 2		Rehabilitation based on IRI	IRI (t)
Model 3		Rehabilitation based on both PCR and IRI	Complementary PCR (t) and IRI (t)
Model 4		Rehabilitation based on PCR	Complementary PCR (t) and IRI (t)
Model 5		Rehabilitation based on IRI	Complementary PCR (t) and IRI (t)

Table 2 Statistical accuracy measures of all models

Model		Model 1	Model 2	Model 3	Model 4	Model 5
Training	MARE (%)	4.685	**5.865**	7.487	8.499	7.797
	R^2	0.615	**0.902**	0.825	0.721	0.798
	ASE	0.0020	**0.0009**	0.0013	0.0019	0.0014
Testing	MARE (%)	5.125	**7.719**	10.407	10.329	10.716
	R^2	0.476	**0.837**	0.681	0.639	0.654
	ASE	0.0025	**0.0016**	0.0020	0.0022	0.0022
Validation	MARE (%)	4.788	**8.119**	11.722	10.431	11.098
	R^2	0.553	**0.692**	0.589	0.623	0.599
	ASE	0.0020	**0.0026**	0.0025	0.0021	0.0022
All data	MARE (%)	4.212	**5.923**	7.365	8.734	7.969
	R^2	0.666	**0.872**	0.817	0.699	0.758
	ASE	0.0016	**0.0012**	0.0012	0.0019	0.0015
Final architecture		11-6-1	**11-8-1**	12-8-2	11-4-2	11-6-2

Fig. 1, each plot depicts only a single output (i.e. PCR or IRI). Thus, the models with the single output are shown with a single plot (i.e. model 1 in Fig. 1(a)), and the models with the two outputs are presented in two plots (i.e. model 3 in Figs. 1(c) and (d)).

4 Results

4.1 Model 1

The inputs for Model 1 are location, thickness, age, time (t), length, PCR, IRI and rehabilitation based on PCR and the output is PCR (t) as shown in Table 1. The ASE values for training, testing, validation, and all data are 0.0020, 0.0025, 0.0020 and 0.0016 respectively. R^2 values for this model are as follows: 0.615 for the training, 0.476 for the testing, 0.553 for the validation, and 0.666 for all data. The correlation of PCR in the all data stage for this model is shown in Fig. 1(a). The final architecture of this model is 11-6-1, where 11 is the number inputs, 6 is the number of hidden nodes and 1 is the number of output.

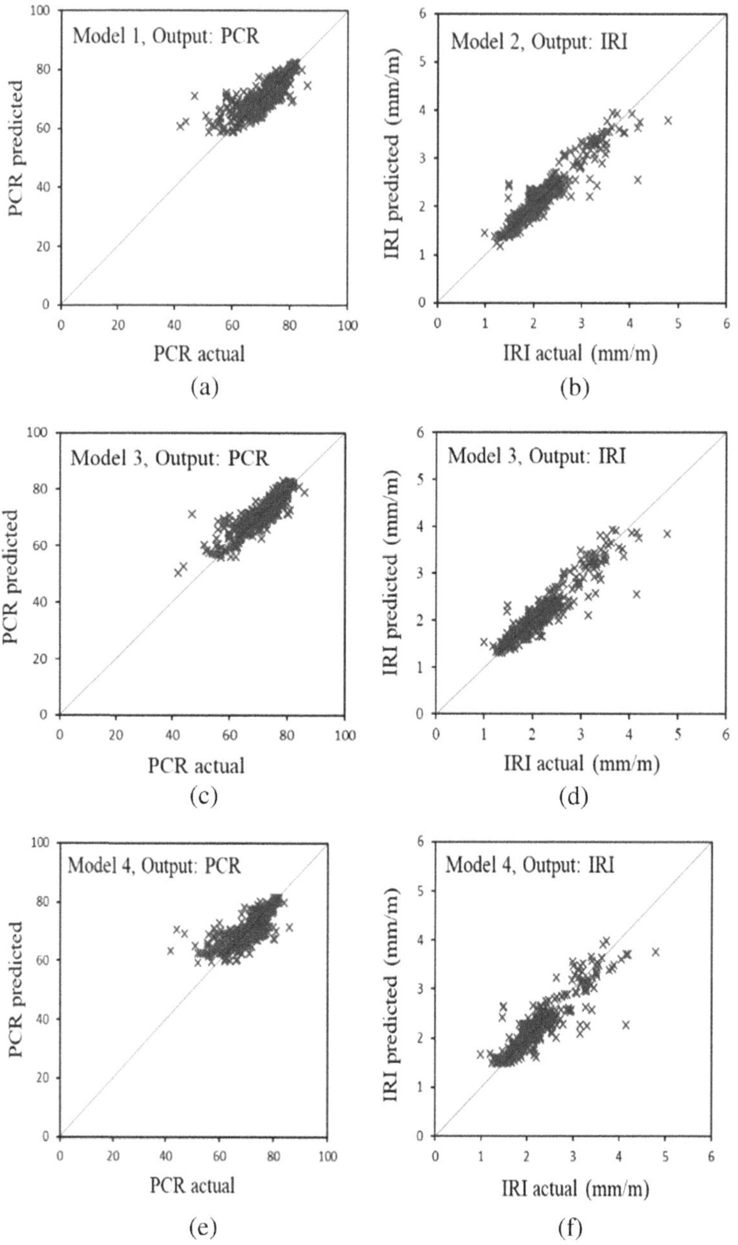

Fig. 1 Regression plots of full dataset of all models

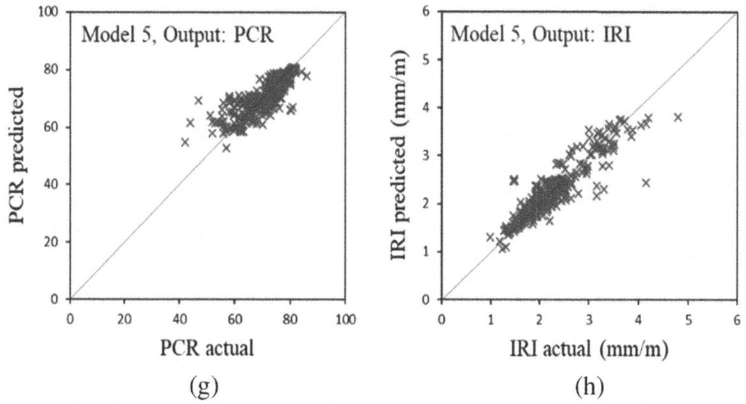

Fig. 1 (continued)

4.2 *Model 2*

Model 2 includes location, thickness, age, length, PCR, IRI and rehabilitation based on IRI as the inputs and IRI (t) as the output. The ASE values for training, testing, validation, and all data are 0.0009, 0.0016, 0.0026 and 0.0012 respectively. R^2 values are 0.902 for the training, 0.837 for the testing, 0.692 for the validation, and 0.872 for all data. The correlation between the actual and predicted IRI in all data stage for this model is shown in Fig. 1(b). This network structure is 11 inputs, 8 hidden nodes, and 1 output.

4.3 *Model 3*

The inputs and outputs of this model are very similar to model 2. In addition to model 2 inputs, rehabilitation based on PCR as an additional input and complementary PCR (t) as an additional output were considered. The ASE values for the selected network are as follows: 0.0013 for the training, 0.0020 for the testing, 0.0025 for the validation, and 0.0012 for all data. The R^2 for training, testing, validation, and all data are 0.825, 0.681, 0.589 and 0.817, respectively. Figures 1(c) and (d) show the correlation between the actual and predicted values of the PCR and IRI in all data stage, respectively. This network has 12 inputs, 8 hidden nodes, and 2 outputs.

4.4 Model 4

To develop this model, all the inputs except the rehabilitation based on IRI and all the outputs of model 3 were utilized. The ASE values are as follows: 0.0019 for the training, 0.0022 for the testing, 0.0021 for the validation and 0.0019 for all data. The R^2 values represent a moderate correlation among the datasets and the values are 0.721 for the training, 0.639 for the testing, 0.623 for the validation, and 0.699 for all data. Figures 1(e) and (f) show the correlation between the actual and predicted values of the PCR and IRI in all data stage, respectively. The final network has 11 inputs, 4 hidden nodes, and 2 outputs.

4.5 Model 5

This model utilized location, thickness, age, length, PCR, IRI and rehabilitation based on IRI as inputs and complementary PCR (t) and IRI (t) as outputs which are shown in Table 1. The ASE values for training, testing, validation, and all data were 0.0014, 0.0022, 0.0022 and 0.0015 respectively. R^2 values for this model are 0.798 for the training, 0.654 for the testing, 0.599 for the validation, and 0.758 for all data. The R^2 values indicated a moderate to a high correlation between the actual and predicted outputs. The correlations between actual and predicted values of both PCR and IRI in all data stage are shown in Figs. 1(g) and (h), respectively. This network had 11 inputs, 6 hidden nodes and 2 outputs shown in the bottom row of Table 2.

5 Sensitivity Analysis

A sensitivity analysis was carried out to evaluate the significance of the inputs on the output(s). A graphical user interface was developed by importing all the final ANN model parameters into an excel spreadsheet. To examine the models, one input was changed while keeping the other inputs constant. In this paper, the sensitivity analysis of model 2 is presented since it is the best model in terms of statistical accuracy measures. All model 2 inputs shown in Table 1 were kept constant except Time (t) that was changed from one to eight to generate predictions of IRI. Figure 2 shows two representative pavement section predictions by model 2. In each plot, the predicted IRI value follows the trend embedded within the actual IRI data. It can be inferred from Fig. 2 that IRI increased over time up to the sixth year and decreased in the latter two years without any rehabilitation action. Usually, without any rehabilitation, pavement deteriorates over time, therefore, IRI increases. Even though these two sections did not agree with the expected behaviour, the model still followed a similar trend. The CRCP has long life-span and does not deteriorate much with time; the changes in predicted and actual values of IRI are accordingly small.

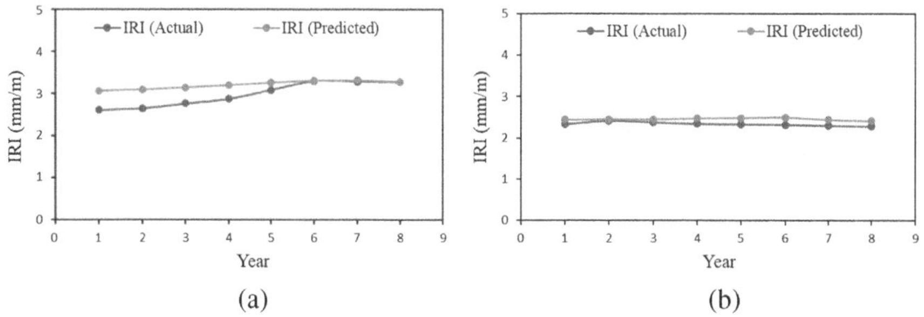

Fig. 2 Sensitivity analysis of Model 2, **a** Section ID: 1956 and **b** Section ID: 3076

6 Discussion and Conclusion

In this study, the distress data from Continuously Reinforced Concrete Pavement (CRCP) sections in Mississippi was used to develop the performance prediction models using Artificial Neural Networks (ANNs) approach with a backpropagation algorithm. IRI and PCR are the common output for all the developed models. Rehabilitation actions were categorized based on IRI and PCR. All five models with common and varying input(s) and output(s) are represented in Table 1. All the future predictions, up to 8 years, were based on the data in 2010. All the developed models have significant model statistics, which are depicted in Table 2. Model 2 was found to be the most accurate among the five models. This model had the least ASE and highest R^2 in all stages of modeling among all the models. In the sensitivity analysis shown in Fig. 2, the two randomly selected sections performed logically and followed a very similar trend as the actual measurements. At the time of conducting this study, traffic and environmental data were not available, which could potentially improve the statistical accuracy measures. It should be mentioned that the data size used in this study was small considering the life span of CRCP. To capture the ultimate response of PCR and IRI rehabilitation actions properly, more data that contain the total life span of CRCP must be included. Also, within the 8 years of data, the change in PCR and IRI was erratic which may be due to the calibration of profilometer, time and day of the survey, and weather conditions. Future studies should include cumulative ESAL, rainfall, and temperature data, which may improve the accuracy of the models and incorporate more performance logic into the existing models. The data from the upcoming years should be used for retraining the existing models for potential improvement and reliable projections. As the pavement condition in the current year is highly dependent on the previous year's condition, a new modeling approach should look into utilizing the condition of the pavement section over the years as dependent variables.

References

Abaza KA (2004) Deterministic performance prediction model for rehabilitation and management of flexible pavement. Int J Pavement Eng 5(2):111–121

Abaza KA, Ashur SA, Abu-Eisheh SA, Rabay'a A (2001) Macroscopic optimum system for management of pavement rehabilitation. J Transp Eng 127(6):493–500

Abaza KA, Ashur SA, Al-Khatib IA (2004) Integrated pavement management system with a Markovian prediction model. J Transp Eng 130(1):24–33

Attoh-Okine NO (1994) Predicting roughness progression in flexible pavements using artificial neural networks. In: Transportation research board conference proceedings (1994)

Elhadidy AA, Elbeltagi EE, Ammar MA (2015) Optimum analysis of pavement maintenance using multi-objective genetic algorithms. HBRC J 11(1):107–113

George KP, Rajagopal AS, Lim LK (1989) Models for predicting pavement deterioration. Transp Res Rec 1215:1–7

Huang YH (2004) Pavement analysis and design. Pearson/Prentice Hall, Upper Saddle River

Lee EB, Ibbs CW (2005) Computer simulation model: Construction analysis for pavement rehabilitation strategies. J Constr Eng Manag 131(4):449–458

Pathway Services Inc. (2017) Scope of the work: Survey 2017–2020. Research and Planning Division: Mississippi Department of Transportation

Raab C, Abd El Halim AO, Partl MN (2013) Utilization of artificial neural network for the analysis of interlayer shear properties. Baltic J Road Bridge Eng (BJRBE) 8(2):107–116

Wang F, Zhang Z, Machemehl RB (2003) Decision-making problem for managing pavement maintenance and rehabilitation projects. Transp Res Rec 1853(1):21–28

Won M, Hankins K, McCullough BF (1991) Mechanistic Analysis of Continuously Reinforced Concrete Pavement Considering Material Characteristics, Variability, and Fatigue (No. FHWA/TX-92+ 1169-2). Center for Transportation Research, Bureau of Engineering Research, University of Texas at Austin

Full Scale Studies Accelerated Pavement Testing

Parameter Identification of Asphalt Pavements Subjected to Moving Loads

Zhaojie Sun, Cor Kasbergen, Karel N. van Dalen, Kumar Anupam, Athanasios Skarpas, and Sandra M. J. G. Erkens

Abstract Pavements generally demand necessary maintenance and rehabilitation to maintain their service performance in the whole lifespan. The maintenance and rehabilitation strategies are usually formulated based on the results of non-destructive testing, in which the traffic speed deflectometer (TSD) test is an efficient tool for pavement structural evaluation at network level. In this paper, the TSD test on asphalt pavements is simulated by a spectral element method-based theoretical model, which is further combined with a nonlinear minimisation algorithm to achieve parameter identification. After conducting parameter sensitivity analysis, a case study is used to demonstrate the ability of the proposed parameter identification technique. The results show that this technique is able to deal with TSD measurements to effectively identify the structural parameters of asphalt pavements. The presented TSD test-based parameter identification technique is a promising tool for asphalt pavement structural evaluation at network level, which is beneficial to formulate cost-effective maintenance and rehabilitation strategies.

Keywords Parameter identification · Asphalt pavement · Moving load · Spectral element method · Traffic speed deflectometer

1 Introduction

The maintenance and rehabilitation of pavements are essential to extend their service life. In order to minimise unnecessary costs, the maintenance and rehabilitation strategies should be accurate enough, which requires reliable techniques for pavement functional and structural evaluation. An elegant method for pavement structural evaluation is the non-destructive testing, in which the falling weight deflectometer

Z. Sun (✉) · C. Kasbergen · K. N. van Dalen · K. Anupam · A. Skarpas · S. M. J. G. Erkens
Department of Engineering Structures, Faculty of Civil Engineering and Geosciences, Delft University of Technology, Stevinweg 1, 2628 CN Delft, The Netherlands
e-mail: zhaojie.sun@tudelft.nl

A. Skarpas
Department of Civil Infrastructure and Environmental Engineering, College of Engineering, Khalifa University, P. O. Box 127788, Abu Dhabi, United Arab Emirates

C. Raab (ed.), *Proceedings of the 9th International Conference on Maintenance and Rehabilitation of Pavements—Mairepav9*, Lecture Notes in Civil Engineering 76,
https://doi.org/10.1007/978-3-030-48679-2_73

(FWD) test is a widely used approach (Li et al. 2017; Elbagalati et al. 2018). In the FWD test, the pavement response caused by an impact load is measured, which can be further analysed to identify structural parameters of pavements (Al-Khoury et al. 2001). However, the FWD test is time consuming and resource intensive for network-level structural evaluation because of the stop-and-go testing process (Keenahan and OBrien 2018). The limitations of the FWD test encourage the development of another non-destructive testing method, which is called the traffic speed deflectometer (TSD) test (Levenberg et al. 2018; Nasimifar et al. 2018). This test can measure the response of pavement surface caused by a wheel loading at normal driving speeds, so it is more suitable for pavement structural evaluation at network level (Maser et al. 2017). Although the TSD vehicle has been well developed, it still needs a proper parameter identification technique to deal with corresponding measurements. Hence, this paper aims to formulate a parameter identification technique suitable for TSD measurement analysis. The presented work is beneficial to the development of engineering techniques for pavement structural evaluation at network level.

2 Model Formulation

With the assumptions that the pavement surface is smooth, the driving speed is constant, and the measurements are only caused by the right rear wheel pair of the TSD vehicle, a theoretical model for the TSD test on asphalt pavements is a layered system subjected to a uniformly moving constant surface force, which is evenly distributed over a pair of rectangular areas, as shown in Fig. 1. In order to solve the moving load problem, both a stationary Cartesian coordinate system (*OXYZ*) and a moving Cartesian coordinate system (*oxyz*) are introduced. The stationary coordinate system is stationary with respect to the layered system and its origin is located in the centre of the initial loading area, while the moving coordinate system is moving with the load and its origin is located in the centre of the moving

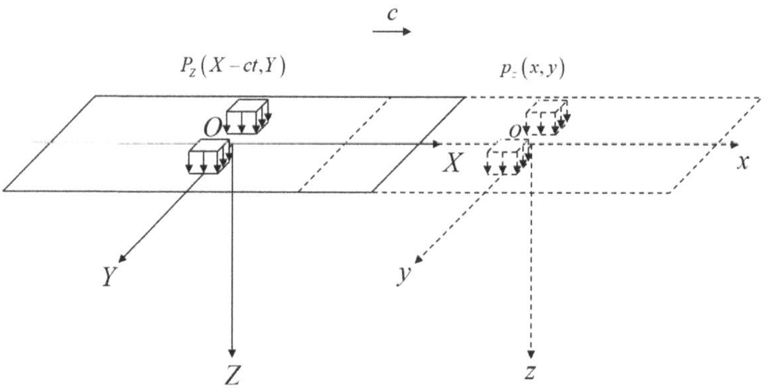

Fig. 1 A theoretical model for the TSD test

loading area. The stationary coordinate vector and the moving coordinate vector are notated as $\underline{\mathbf{X}} = \begin{bmatrix} X & Y & Z \end{bmatrix}^{\mathrm{T}}$ and $\underline{\mathbf{x}} = \begin{bmatrix} x & y & z \end{bmatrix}^{\mathrm{T}}$, respectively. If the load moves in the positive direction of the X-axis with a constant speed c, the relationships between the coordinates of the two coordinate systems are:

$$x = X - ct, \quad y = Y, \quad z = Z \tag{1}$$

where t is time, and these two coordinate systems are coincident when t is zero.

2.1 *Response of Elastic Layered Systems Under Moving Surface Loads*

As presented in Sun et al. (2019), a spectral element method-based procedure can be used to calculate the response of elastic layered systems under moving surface loads. In this procedure, a layer spectral element and a semi-infinite spectral element are developed to respectively simulate a layer and a half-space, and the combinations of these two spectral elements can model layered systems with different configurations. For the case of the TSD test, both the loading pressure and loading area are considered to be constant over time. In addition, the contact area between the right rear wheel pair of the TSD vehicle and pavement surface is assumed to be two rectangular areas with a certain distance between them, and the centre of the contact area coincides with the origin of the moving coordinate system. This configuration of the contact area can be described by a spatial distribution function $h_0(x, y)$ defined as follows:

$$h_0(x, y) = H(x_0 - |x|)\left[H\left(\frac{y_0}{2} - \left|y + \frac{y_0 + d}{2}\right|\right) + H\left(\frac{y_0}{2} - \left|y - \frac{y_0 + d}{2}\right|\right)\right] \tag{2}$$

in which $H(\cdot)$ is the Heaviside step function, $2x_0$ is the length of one rectangular area in x-direction, y_0 is corresponding length in y-direction, and d is the distance between two rectangular areas in y-direction.

2.2 *Simulation of Material Damping*

Numerically, the material damping can be simulated by replacing the Young's modulus with a complex Young's modulus derived from a certain damping model. For the moving load problem, the complex Young's modulus should be expressed in the wavenumber-frequency domain related to the moving coordinate system for further application. The asphalt layer in asphalt pavements has viscoelastic properties, which are simulated by the Zener model in this study. As shown in Fig. 2, the Zener

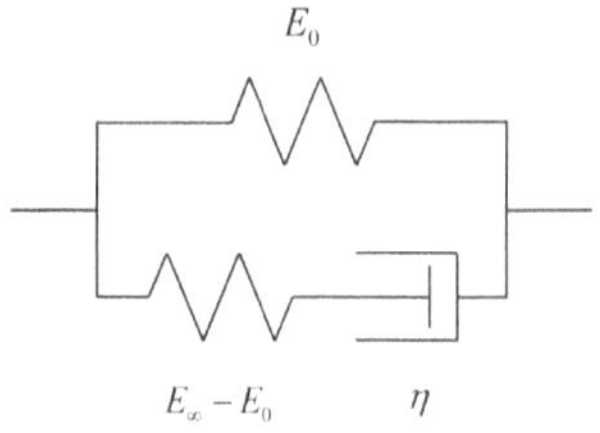

Fig. 2 Schematic representation of the Zener model

model consists of a Maxwell component and a Hookean element (with spring constant E_0) connected in parallel. The Maxwell component consists of a Hookean element (with spring constant $E_\infty - E_0$) and a Newtonian element (with viscosity constant η) connected in series. Physically, E_0 is called the static modulus because it corresponds to the complex Young's modulus when loading angular frequency equals zero, and E_∞ is called the glassy modulus because it corresponds to the complex Young's modulus when loading angular frequency equals infinity.

In the stationary coordinate system, the stress-strain relationship of the Zener model under uniaxial tension/compression can be expressed as follows:

$$[(E_\infty - E_0) + \eta\partial_t]\sigma_0(\underline{\mathbf{X}}, t) = [E_0(E_\infty - E_0) + \eta E_\infty \partial_t]\varepsilon_0(\underline{\mathbf{X}}, t) \tag{3}$$

where ∂_t means derivative with respect to t, $\sigma_0(\underline{\mathbf{X}}, t)$ and $\varepsilon_0(\underline{\mathbf{X}}, t)$ are the tensile/compressive stress and strain in the stationary coordinate system, respectively. According to Eq. (1), the stress-strain relationship of the Zener model has the following form in the moving coordinate system:

$$[(E_\infty - E_0) + \eta(\partial_t - c\partial_x)]\sigma(\underline{\mathbf{x}}, t) = [E_0(E_\infty - E_0) + \eta E_\infty(\partial_t - c\partial_x)]\varepsilon(\underline{\mathbf{x}}, t) \tag{4}$$

in which ∂_x means derivative with respect to x, $\sigma(\underline{\mathbf{x}}, t)$ and $\varepsilon(\underline{\mathbf{x}}, t)$ are the tensile/compressive stress and strain in the moving coordinate system, respectively.

By following the same Fourier transform convention shown in Sun et al. (2019), the expression of the complex Young's modulus for the Zener model in the wavenumber-frequency domain can be obtained by replacing ∂_t with $\mathrm{i}\omega$, replacing ∂_x with $-\mathrm{i}k_x$, and using the analogy relation similar to that of linear elasticity:

$$\tilde{E}(k_x, \omega) = \frac{E_0(E_\infty - E_0) + \mathrm{i}\eta E_\infty(\omega + ck_x)}{(E_\infty - E_0) + \mathrm{i}\eta(\omega + ck_x)} \tag{5}$$

where i is the imaginary unit satisfying $\mathrm{i}^2 = -1$, $\tilde{E}(k_x, \omega)$ is the complex Young's modulus in the wavenumber-frequency domain, k_x is the wavenumber in the x-direction, and ω is the angular frequency. It can be concluded that, for a certain damping model, the expression of the complex Young's modulus in the wavenumber-frequency domain related to the moving coordinate system can be obtained from its traditional expression in the frequency domain related to the stationary coordinate system by replacing ω with $\omega + ck_x$. The base layer and subgrade in asphalt pavements are considered to be elastic with hysteretic damping, which corresponds to the following expression of complex Young's modulus in the wavenumber-frequency domain related to the moving coordinate system:

$$\tilde{E}(k_x, \omega) = E\left[1 + 2\mathrm{i}\xi\,\mathrm{sgn}(\omega + ck_x)\right] \tag{6}$$

in which E is the Young's modulus, ξ is the damping ratio, and $\mathrm{sgn}(\cdot)$ is the signum function.

3 Parameter Sensitivity Analysis

The more sensitive the response is to a certain parameter, the easier it is to identify this parameter. Hence, the parameter sensitivity analysis is necessary to have an insight into the feasibility and accuracy of the parameter identification process. In practice, the TSD vehicle measures the vertical deflection slopes of surface points along the midline of right rear wheel pair. Hence, the vertical deflection slope curve along the x-axis observed on the pavement surface is the response of interest. The sensitivity of the slope curve to different structural parameters is investigated based on single factor analysis, in which the variation of a certain parameter is 50% of its reference value. According to the actual loading conditions of the TSD test, the following loading parameters are used for simulation:

- The speed of the moving load c is 13.9 m/s (50 km/h);
- The magnitude of loading pressure p_0 is 707 kPa;
- The distance between two rectangular areas d is 0.15 m;
- The parameters of the loading area are x_0 is 0.06316 m and y_0 is 0.27432 m;
- The dimensions of the space window of interest are 400 m by 400 m.

In addition, the reference structural parameters of the considered asphalt pavement are shown in Table 1. The reference parameters of the Zener model used to simu-

Table 1 Reference structural parameters of the asphalt pavement

Layers	E (MPa)	ξ	ν	ρ (kg/m^3)	h (m)
Asphalt	–	–	0.3	2300	0.1
Base	110	0.05	0.3	1900	0.3
Subgrade	60	0.05	0.3	1600	Infinite

Note: E is the Young's modulus, ξ is the damping ratio, ν is the Poisson's ratio, ρ is the density, and h is the layer thickness.

late the asphalt layer are: $E_0 = 6.4$ MPa, $E_\infty = 19500$ MPa, and $\eta = 7000$ MPa·s. The response corresponds to the reference structural parameters is shown in solid lines. In order to distinguish the sensitivity degree to different parameters, the following five levels are used for description: hardly sensitive, slightly sensitive, moderately sensitive, relatively sensitive, and highly sensitive.

3.1 Sensitivity to Parameters of the Zener Model

The sensitivity of the vertical deflection slope curve to parameters of the Zener model is shown in Fig. 3. The results indicate that the slope curve is hardly sensitive to the static modulus (E_0), highly sensitive to the glassy modulus (E_∞), and slightly sensitive to the viscosity constant (η).

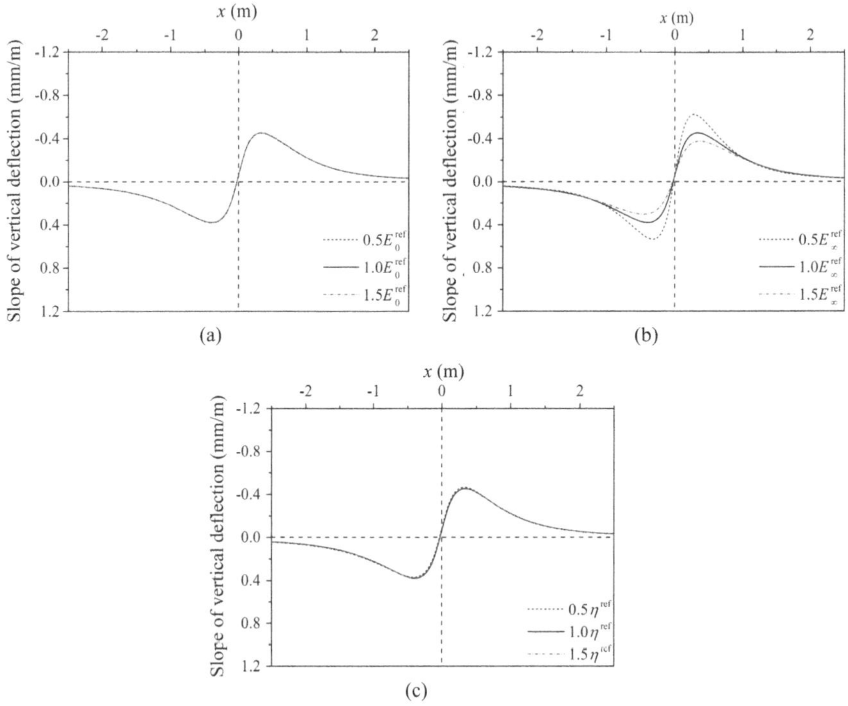

Fig. 3 Sensitivity of vertical deflection slope curve to parameters of the Zener model: (a) static modulus, (b) glassy modulus, and (c) viscosity constant

3.2 Sensitivity to Young's Modulus

The sensitivity of the vertical deflection slope curve to the Young's moduli of base layer (E_2) and subgrade (E_3) is shown in Fig. 4. The results indicate that the slope curve is relatively sensitive to the Young's modulus of base layer and highly sensitive to the Young's modulus of subgrade.

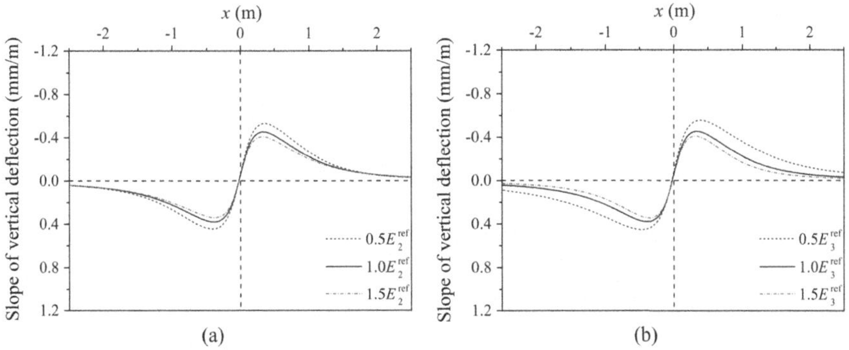

Fig. 4 Sensitivity of vertical deflection slope curve to Young's moduli of: (a) base layer and (b) subgrade

3.3 Sensitivity to Damping Ratio

The sensitivity of the vertical deflection slope curve to the damping ratios of base layer (ξ_2) and subgrade (ξ_3) is shown in Fig. 5. The results indicate that the slope curve is slightly sensitive to the damping ratios of base layer and subgrade.

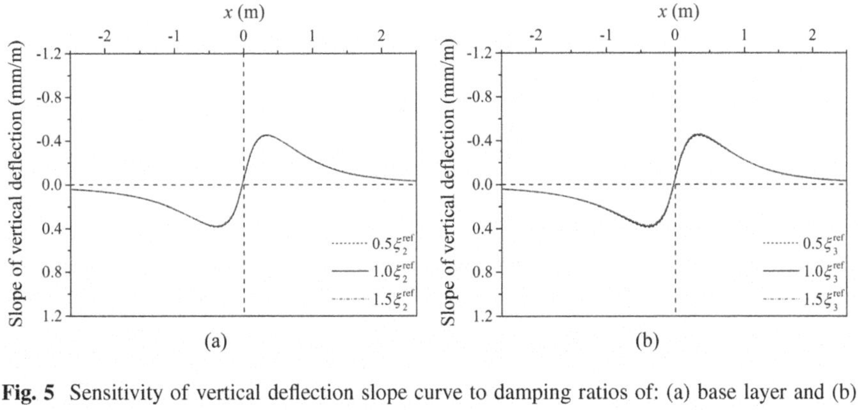

Fig. 5 Sensitivity of vertical deflection slope curve to damping ratios of: (a) base layer and (b) subgrade

3.4 *Sensitivity to Thickness*

The sensitivity of the vertical deflection slope curve to the thicknesses of asphalt layer (h_1) and base layer (h_2) is shown in Fig. 6. The results indicate that the slope curve is highly sensitive to the thickness of asphalt layer and moderately sensitive to the thickness of base layer.

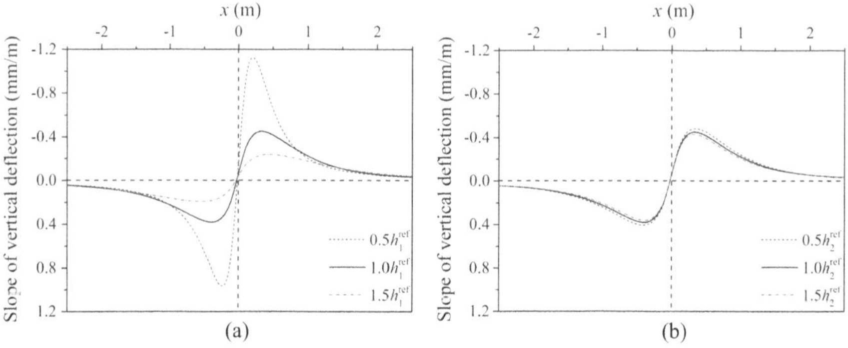

Fig. 6 Sensitivity of vertical deflection slope curve to thicknesses of: (a) asphalt layer and (b) base layer

4 Structural Parameter Identification

The response of a structure is determined by the loading conditions and structural parameters. Theoretically, if the response and loading conditions are known, it should be possible to identify the structural parameters. This parameter identification process is normally achieved by combining a forward calculation model with a proper nonlinear minimisation algorithm. A set of most likely parameters can be sought by minimising the differences between the modelled and measured response. For the case of the TSD test, the objective function $f(\underline{\mathbf{p}})$ used for the nonlinear minimisation algorithm can be defined as follows:

$$f\left(\underline{\mathbf{p}}\right) = \sum_{n=1}^{N} \left| \frac{s^{\text{modelled}}\left(x_n, y_n; \underline{\mathbf{p}}\right)}{s^{\text{measured}}(x_n, y_n)} - 1 \right| \tag{7}$$

where $\underline{\mathbf{p}}$ is a vector contains all the unknown parameters, N is the total number of measuring points, $s^{\text{modelled}}(x_n, y_n; \underline{\mathbf{p}})$ and $s^{\text{measured}}(x_n, y_n)$ are the modelled and measured vertical deflection slopes at measuring point (x_n, y_n), respectively. In general, a smaller value of the objective function corresponds to a better match between the modelled and measured response. Hence, for a set of specific measured response, the most likely parameters can be identified by minimising the objective function.

The nonlinear minimisation algorithm used in this paper is the Powell hybrid algorithm, which can solve a non-constrained system of nonlinear simultaneous equations via a finite-difference Jacobian. One feature of the Powell hybrid algorithm is that the number of unknowns equals to the number of equations. Hence, for the case of the TSD test, the number of unknown parameters should be no more than the number of measuring points. In reality, the TSD vehicle can only measure the slopes of about 9 points. Hence, some structural parameters should be fixed to make the problem solvable. With considering the results of sensitivity analysis, the following parameters are chosen for identification: E_∞, E_2, E_3, η, ξ_2, ξ_3, h_1, and h_2. For the reference asphalt pavement structure shown in the previous section, the slopes of eight points along the x-axis on the pavement surface calculated by the forward model are taken as synthetic measurements. Then, the unknown parameters are identified by using the proposed parameter identification technique. It is found that good initial guesses of the unknown parameters are important to improve the accuracy and efficiency of the parameter identification process. The initial guesses used in this case are: $E_\infty =$ 18000 MPa, $E_2 = 60$ MPa, $E_3 = 40$ MPa, $\eta = 6000$ MPa·s, $\xi_2 = 0.04$, $\xi_3 = 0.04$, $h_1 = 0.05$ m, and $h_2 = 0.2$ m. Correspondingly, the identified parameters are: E_∞ $=$ 19500.5 MPa, $E_2 = 110.0$ MPa, $E_3 = 60.0$ MPa, $\eta = 6996.3$ MPa·s, $\xi_2 = 0.050$, $\xi_3 = 0.050$, $h_1 = 0.10$ m, and $h_2 = 0.30$ m. These identified parameters have good agreement with the true values (i.e. reference parameters), which confirms the ability of the proposed technique.

5 Conclusions

This paper proposes a parameter identification technique for asphalt pavements subjected to moving loads, which can be used to deal with traffic speed deflectometer (TSD) measurements to identify the structural parameters of asphalt pavements. The proposed TSD test associated parameter identification technique may become a promising tool for pavement structural evaluation at network level, and help to formulate cost-effective maintenance and rehabilitation strategies. In the future work, more case studies will be conducted to verify the robustness of the proposed parameter identification technique.

Acknowledgements This work is financially supported by the China Scholarship Council (No. 201608230114).

References

Al-Khoury R, Kasbergen C, Scarpas A, Blaauwendraad J (2001) Spectral element technique for efficient parameter identification of layered media: part II: inverse calculation. Int J Solids Struct 38(48–49): 8753–8772

Elbagalati O, Elseifi M, Gaspard K, Zhang Z (2018) Development of the pavement structural health index based on falling weight deflectometer testing. Int J Pavement Eng 19(1): 1–8

Keenahan JC, OBrien EJ (2018) Drive-by damage detection with a TSD and time-shifted curvature. J Civil Struct Health Monit 8(3): 383–394

Levenberg E, Pettinari M, Baltzer S, Christensen BML (2018) Comparing traffic speed deflectometer and falling weight deflectometer data. Transp Res Rec 2672(40): 22–31

Li M, Wang H, Xu G, Xie P (2017) Finite element modeling and parametric analysis of viscoelastic and nonlinear pavement responses under dynamic FWD loading. Constr Build Mater 141: 23–35

Maser K, Schmalzer P, Shaw W, Carmichael A (2017) Integration of traffic speed deflectometer and ground-penetrating radar for network-level roadway structure evaluation. Transp Res Rec 2639(1): 55–63

Nasimifar M, Thyagarajan S, Sivaneswaran N (2018) Computation of pavement vertical surface deflections from traffic speed deflectometer data: evaluation of current methods. J Transp Eng Part B: Pavements 144(1): 04018001

Sun Z, Kasbergen C, Skarpas A, Anupam K, van Dalen KN, Erkens SMJG (2019) Dynamic analysis of layered systems under a moving harmonic rectangular load based on the spectral element method. Int J Solids Struct 180–181: 45–61

Effects of Field Compaction Method on Water Permeability and Performance of Asphalt Concrete Pavements

Chinecherem Agbo Igboke, Eslam Elsayed, Yasser Hassan, and Abd El Halim Omar Abd El Halim

Abstract This paper presents the results of field studies of hot mix asphalt pavements compaction using conventional compaction technologies (train of vibratory steel drum, pneumatic tired roller) and the Asphalt Multi-Integrated Roller (AMIR) to examine the effects of different compaction methods on the water permeability of asphalt concrete pavements as a surrogate measure of durability and eventual long-term performance. Nine different pavement construction projects were used in this paper where laboratory and field properties of pavements compacted using conventional and AMIR compaction were measured and evaluated. Field measurements of water permeability showed higher mean value and higher variation of the permeability coefficient for conventional compaction than the AMIR-compacted pavements, even though the air voids and relative compaction were almost the same. Compared to the conventional compaction, statistical analysis showed that AMIR compaction reduced water permeability of compacted surfaces on stiff bases (such as an overlay on top of a milled asphalt concrete pavement) and also reduced the rate of permeability increase due to increase in air voids.

Keywords Asphalt concrete pavement · Compaction · Permeability · Pavement performance

1 Introduction

The structural performance and durability of asphalt concrete pavements are controlled by a number of factors including proper mix design and adequate compaction. Hughes (1989) suggested that neither of these two factors alone can guarantee a satisfactorily performing pavement. Properties such as strength, resistance to ageing, resistance to deformation, resistance to moisture damage, permeability, and

A. Abd El Halim—Deceased

C. A. Igboke (✉) · E. Elsayed · Y. Hassan · A. E. H. O. Abd El Halim
Civil and Environmental Engineering, Carleton University, 1125 Colonel By Drive, Ottawa, ON K1S 5B6, Canada
e-mail: chinecheremigboke@cmail.carleton.ca

C. Raab (ed.), *Proceedings of the 9th International Conference on Maintenance and Rehabilitation of Pavements—Mairepav9*, Lecture Notes in Civil Engineering 76,
https://doi.org/10.1007/978-3-030-48679-2_74

skid resistance are stated to be affected by compaction (Hughes 1989). Most highway jurisdictions emphasize the quality of pavement compaction through setting a minimum compaction level or relative compaction (the percentage of compaction achieved in the field relative to maximum compaction of the mix) expected of the contractor. Such requirements are based on the premise that a higher relative compaction would ensure better performing and more durable pavements. As these requirements focus on the end result, less emphasis is placed on the method of pavement construction. However, research findings have shown that the different compaction methods impart different properties to the asphalt concrete with some more prone to distresses than others at comparable compaction levels (Tarefder and Ahmad 2016; Airey and Collop 2016). Therefore, this paper examines the effects of asphalt pavement compaction on the expected pavement's long-term performance. The paper uses the results of field trials involving two compaction methods: the conventional compaction using a train of vibratory steel, pneumatic, and static steel rollers and compaction using AMIR (Asphalt Multi-Integrated Roller) as a single roller. The pavements in the study are compared based on relative compaction, air voids, and water permeability.

2 Background of Asphalt Pavement Field Compaction

From the late 19th century when the first steam roller was invented till the late 1950s, compaction of asphalt concrete was performed by the static steel drum roller with different axle configurations (Hughes 1989; Geller 1984; Parker 1960). The 1960s saw the introduction of the vibratory steel drum roller, and the oscillatory steel drum roller was introduced between 1980s and 1990s to improve the efficiency of field compaction and provide better performing asphalt concrete pavements. Figure 1 shows the three main technologies of rotary steel drum compactors according to Kearney (2006). However, neither the shape nor the material of the drum has fundamentally changed among these three steel drum roller types (Hughes 1989; Parker 1960; Geller 1984). Hence, the current compaction technology is referred to in this paper as conventional compaction.

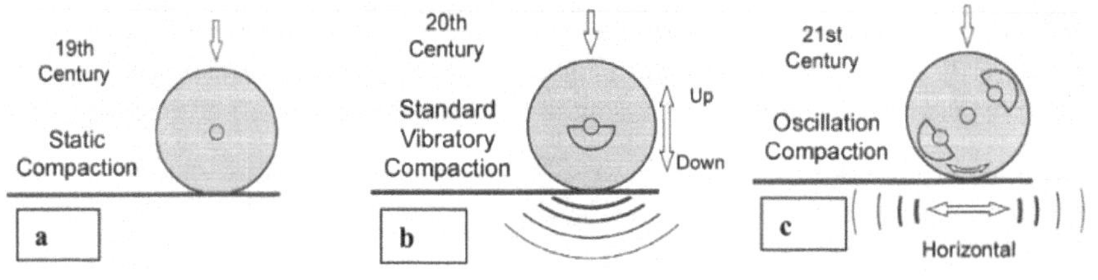

Fig. 1 Different types of rotary steel drum compaction (Kearney 2006)

For over a century now, the problems of "bow wave" phenomenon, pavement surface irregularities and transverse cracks otherwise known as roller checking, have been well documented in the literature (Hughes 1989; Parker 1960; Geller 1984; Ramezani et al. 2018). Researchers such as Geller (1984) and Parker (1960) attributed these problems to mix instability, and hence emphasized the importance of training and discipline of roller operators. Increasing the roller drum radius was also recommended to address these problems. But on the shape of the compactor equipment, Eq. (1), according to Parker (1960), shows that an infinite radius of the drum roller is required to eliminate the drawbar pull, which is a considerable factor contributing to the construction problems. Obviously, this indicates that a circular roller drum will always generate a drawbar pull and, in turn, much of the associated compaction problems.

$$P = \frac{WG}{R - H} \tag{1}$$

Where $\boldsymbol{P}$ = drawbar pull which is the engine track force that drives the roller, $\boldsymbol{W}$ = roller weight, $\boldsymbol{G}$ = the horizontal distance of the roller-pavement contact, $\boldsymbol{R}$ = drum radius, and H = depth of drum penetration into the hot asphalt during compaction.

Figure 2 shows the illustration of the roller drum dimensions, roller drum-pavement contact area, and the drawbar pull (Parker 1960).

F = resultant force

$$tan\theta = \frac{G}{R - H} = \frac{P}{W}$$

Based on tensile strength testing of field-compacted slabs, Ramezani et al. (2018) concluded that construction cracks at the top of a pavement layer would make the top portion of the layer weaker than the lower portion. Several other problems associated with the current conventional compaction of asphalt pavements have been reported in the literature including lower tensile strength in the roller direction (hence perpendicular to construction cracks) and higher susceptibility to moisture damage and other forms of asphalt pavement distress (Abd El Halim et al. 2015; Abd El Halim et al.

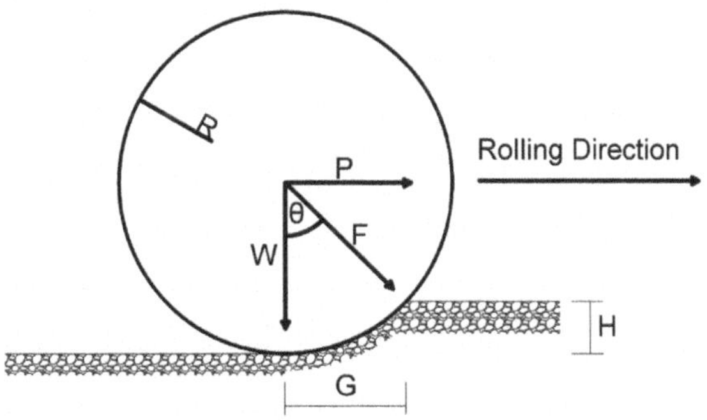

Fig. 2 Influence of roller drum radius on surface irregularities (Parker 1960)

1994; Abd El Halim and Haas 2004). A recent study concluded that the formation of bumps in asphalt concrete pavements increases with greater vibration and steeper downgrade in the direction of rolling (Shuler 2017). Unfortunately, most highway jurisdictions have not considered in their job acceptance schemes the method of compaction used in asphalt pavement construction. It is generally believed that once the required compaction level has been achieved, other properties and long-term performance of the asphalt pavement would be assured regardless of the method of compaction. However, findings by Williams (2011), Kandhal and Mallick (2007), and Fleckenstein et al. (2002) reported that the performance of asphalt pavements does not depend on the level of compaction alone but rather on other characteristics, such as permeability.

As mentioned earlier, Eq. (1) indicated that removal of the effect of the drawbar pull, which causes surface cracks, entails having an infinite radius of the steel drum roller. This condition can be achieved by replacing the drum with a flat surface such that the pavement is compacted without initiating cracks. The Asphalt Multi-Integrated Roller (AMIR) compaction technology has been developed under this premise to solve the problems of asphalt surface cracking during compaction operations. As shown in Fig. 3, AMIR changes the two parameters that have been fixed features of all steel rollers. First, the radius of the compaction equipment is changed from a finite radius in all other rollers to a flat surface or infinite radius. Second, the roller material in direct contact with the pavement is changed to rubber, which is considerably closer to the stiffness of the hot asphalt concrete than the steel used in conventional rollers. AMIR's flat rubber belt provides a greater contact area and time of compaction allowing proper densification of the mat using considerably less vertical pressure exerted on the pavement surface (Abd El Halim and Mostafa 2006; El Hussein et al. 1993). AMIR efficiently compacts asphalt concrete pavements and produces surfaces that are free of cracks and irregularities, with lower

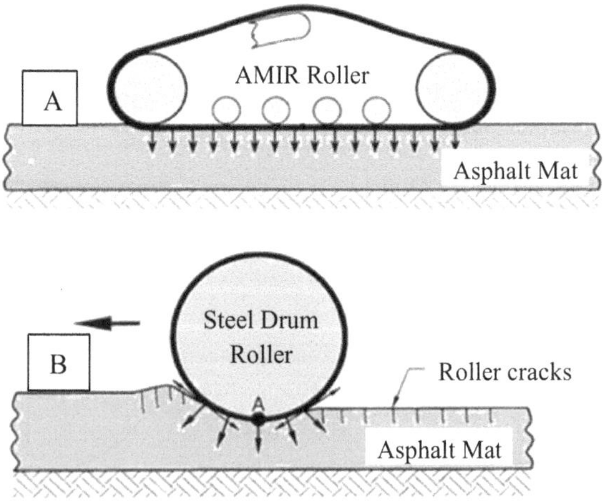

Fig. 3 The schematic diagram of **a** AMIR and **b** Steel drum rollers

water permeability, and with improved engineering properties; all of which lead to better durability and long-term performance (Abd El Halim et al. 2013; Abd El Halim et al. 1988).

3 Research Methodology

To study the comparative effects of the conventional and AMIR compaction methods on the properties of asphalt concrete, data were collected from several construction projects. The data included parameters related to the quality of the finished pavement such as density and water permeability measured in the field (referred to as field permeability) as well as parameters related to the compaction process such as compaction level, base type (compacted granular base, Portland cement concrete, or milled asphalt concrete surface), roller type, and number of roller passes. The data used in this paper comprise information from 162 cores collected from nine paving projects across Eastern Ontario, Canada. All sites have similar climatic conditions, and all fall within the same zone for performance grade requirement of asphalt cement (OHMPA 1999). The test locations were on relatively straight sections with moderate grades.

The cores were extracted at the same points where field permeability measurements were taken on the pavement surface. The cores were transported to Carleton University, where they underwent tests to determine the volumetric parameters (namely, density and air voids) and the unconditioned indirect tensile strength of each core. The volumetric properties were determined using the saturated surface dry method according to AASHTO T166 (AASHTO 2016), while the unconditioned indirect tensile strength was determined as part of AASHTO T283 method (AASHTO 2014) using an INSTRON series 5583 loading machine at 50.8 mm/minute loading rate. Field permeability tests were performed using the NCAT permeameter according to the operating manual provided by the manufacturer (Gilson 2013). Thus, a database was established for the coefficient of water permeability as measured in the field (referred to as permeability coefficient) along with other pavement characteristics for the same location of permeability measurement.

Mix types for all projects were the Ministry of Transportation Ontario's (MTO) Superpave 12.5 Nominal Maximum Aggregate Size (NMAS) with different designs for the projects. The binder content for these projects ranged from 4.7 to 5.1%. Paving operations followed the relevant MTO standards with mixing temperature and compaction temperature around 165 and 147 °C, respectively, and lift thickness of about 50 mm. Eight of the nine projects involved two side-by-side test sections at each site that corresponded to the conventional and AMIR compaction. A nuclear density gauge was used to monitor the density during compaction and ensure that enough compaction has been applied to each section to achieve the required compaction level. At the end of compaction, the two sections in each project had achieved approximately the same relative compaction. However, the conventional compaction required 22–24 passes of different rollers to achieve the required compaction, while

AMIR achieved the same compaction level in 4–8 passes by a single roller depending on the project and layer thickness.

The ninth project (Site 9) was conducted in the yard of a local contractor and involved using each of the conventional and AMIR compaction to apply varying compaction efforts on three sections for each compaction method in the yard of a local contractor. The number of roller passes for each AMIR section was one-third that of the corresponding conventional compaction section. More specifically, the three AMIR-compacted sections A1, A2, and A3 were compacted using 4, 6, and 8 passes, respectively. On the other hand, the three conventional compaction sections S1, S2, and S3 were compacted using 12, 18, and 24 total roller passes, respectively. All other parameters, including mix design and construction process, were similar to the other eight projects. A total of 30 field permeability measurements were performed, and 30 cores were recovered from the test sections of Site 9.

4 Results

First, Fig. 4 shows the compaction and permeability for Site 9 for the three AMIR-compacted sections (A1–A3) and the corresponding conventional compaction sections (C1–C3). It could be seen that the conventional compaction yielded marginally higher mean relative compaction, which was achieved with three times the number of roller passes of AMIR. However, each section of A1–A3 had a lower mean permeability coefficient and lower variance than the corresponding values of C1–C3.

To investigate the effects of the two compaction methods on the properties of asphalt pavement, regression analysis was used with the dependent variable being the coefficient of permeability (K) as the performance indicator variable, while the independent variables are those commonly used in quality control including maximum theoretical specific gravity (Gmm), relative compaction ($Comp$), bulk specific

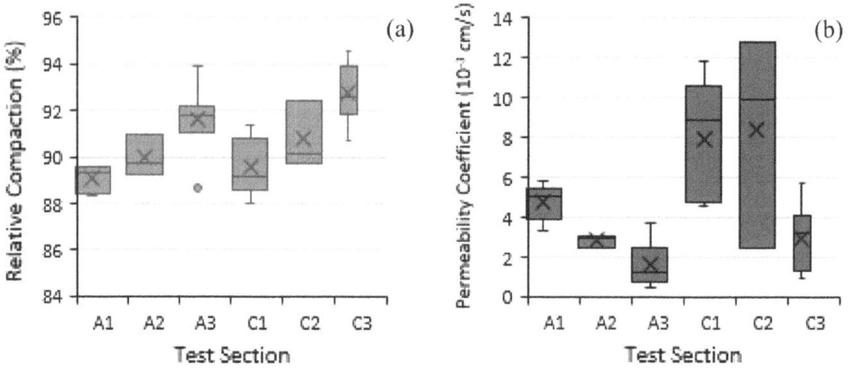

Fig. 4 Summary of relative compaction (**a**) and Permeability (**b**) Test results at site 9

gravity (*Gmb*), air voids (*Va*), indirect tensile strength (*IDS*), base type (*Pavt*: granular base = 0; milled/concrete base = 1), number of roller passes (*Pass*), lift thickness (*Thick*), roller or compaction type (*Rol*: conventional = 0; AMIR = 1), and project site. Compared to the results presented by Igboke et al. (2018), this paper adds the data from the ninth control project to provide better insight on the effect of compaction effort (*Pass*) and interaction with the other variables.

First, Table 1 shows a summary of the main characteristics of pavements compacted by the two compaction methods for all sites. As shown in the table, the relative compaction and air voids were very close for both compaction types as the construction controlled for this variable. That is, compaction continued with the conventional rollers until an acceptable compaction level was achieved. Despite the almost similar air voids and density, AMIR-compacted pavements had lower permeability with less variability (in terms of standard deviation).

Table 2 summarizes the main variables used in the regression analysis. In addition to these variables, to account for the potential of site-specific conditions that could have affected the pavement's permeability, nine dummy variables (Site 1 to Site 9) were used to represent the nine project sites. Each of these variables was set equal to 1 for the data belonging to this site, and all other site variables were set equal to 0. Finally, the interaction of the different variables was considered using independent variables that are the multiplication of two or more variables. These interaction variables account for the multiplicative effects of the independent variables on *K*.

Table 1 Permeability and compaction values of two roller types

Compaction type		*Comp* (%)	*Va* (%)	*K* (10^{-3} cm/sec)
AMIR	Mean	92.60	7.40	3.97
	Std deviation	2.89	2.89	5.47
Conventional	Mean	92.11	7.89	4.49
	Std deviation	1.92	1.92	7.90

Table 2 Summary of the main dependent and independent variables (number of samples = 162)

Variable	Mean	Std deviation	Minimum	Maximum
K (10^{-3} cm/sec)	4.28	7.00	0.005	45.3
Comp (%)	92.35	2.41	86.18	98.23
Va (%)	7.65	2.41	1.77	13.82
Gmm	2.53	0.049	2.38	2.66
Gmb	2.34	0.055	2.17	2.48
IDS (kPa)	267.23	186.37	42.25	1024.21
Pavt	0.093	0.29	0	1
Rol	0.41	0.49	0	1
Pass	15.84	7.13	4	24
Thick (mm)	52.43	7.39	42	95

Table 3 Summary of regression analysis

Variable	Regression coefficient	Standard error	*p*-value	Standard estimate
Intercept	0.0070	0.0013	<.0001	0
Comp	−0.00070	0.00022	0.0022	−0.24
Pavt	0.0063	0.0022	0.0041	0.263
Pass	−0.000202	0.000067	0.0028	−0.21
VaRol	−0.00070	0.00028	0.013	−0.23
PavtRol	−0.011	0.0040	0.0053	−0.302
sqVaRolPavt	−0.00027	0.00011	0.019	−0.21
Site 2	−0.0026	0.0010	0.013	−0.12
Site 7	0.02040	0.0028	<.0001	0.74

Table 3 summarizes the results of the significant model accounting for the independent variables and their interaction. The regression model had an *F*-statistic value of 38.29 (p-value < 0.001) indicating that the model is significant at 5% level of significance. The model's coefficient of determination (R^2) is 0.667 indicating that the model explained 66.7% of the observed variation in the permeability coefficient of the test sections.

As shown in the table, relative compaction (*Comp*), base type (*Pavt*), and number of passes (*Pass*) are significant variables at 5% level of significance. Expectedly, *K* decreases with the increase of *Comp* and *Pass*. Compared to a pavement layer over an aggregate base, *K* would increase for overlays on top of a stiff base of milled asphalt concrete or Portland cement concrete. Two sites (2 and 7) were also found to be significant with lower *K* at Site 2 and higher *K* at Site 7 compared to pavements with the same characteristics at all other sites. Three interaction terms involving *Rol*, *Pavt*, and *Va* were also significant with negative regression coefficients. First, *VaRol* (interaction term for *Va* and *Rol*) indicates that an increase in the pavement's air voids would cause a smaller increment in *K* if the pavement is compacted with AMIR than the case of conventional compaction. Similarly, *PavtRol* (interaction term for *Pavt* and *Rol*) indicates that the increase in *K* over a stiff base is lower for AMIR compaction than conventional compaction. Finally, *sqVaRolPavt* (interaction term for square of *Va*, *Rol*, and *Pavt*) indicates a further reduced rate of increase in *K* for AMIR-compacted pavements compared to conventional compaction when the percentage of air voids increases and/or overlay over a stiff base.

5 Conclusions

This paper has explored the influence of conventional vibratory steel drum compaction train and the AMIR single roller compaction technology on the asphalt pavement properties using field permeability as the property of interest to study the effect of the conventional and AMIR compaction methods on asphalt pavement properties. Based on the findings of this paper, the following conclusions are made. Field compaction equipment and its process are important factors that control the resulting properties of asphalt concrete regardless of mix design. That compaction level alone is inadequate to ensure the performance and durability of the asphalt pavement system without consideration to the process of compaction used and other related properties and benchmarked with other compaction types. Major factors such as relative compaction, stiffness of the underlain base, type of field compactor and number of rollers passes affect properties of asphalt pavement such as permeability. The major conclusions of the foregoing results are that proper compaction improves and enhances the mechanical and physical properties of asphalt pavement. Also, as-constructed asphalt concrete quality should not be based on density alone, and that the inclusion of permeability testing as part of a quality control scheme is expedient conditioned on the process of compaction used.

Acknowledgements The authors acknowledge the financial support by the Ministry of Transportation of Ontario (MTO) and the Natural Sciences and Engineering Research Council (NSERC). Worthy of mention is the support of R.W. Tomlinson Ltd and MTO for technical and logistical support.

References

AASHTO T166 (2016) Standard method of test for bulk specific gravity (Gmb) of compacted hot mix asphalt (HMA) using saturated surface-dry specimens. American Association of State Highway and Transportation Officials Provincial Standards, Washington, D.C.

AASHTO T 283 (2014) Standard method of test for resistance of compacted asphalt mixtures to moisture-induced damage. American Association of State Highway and Transportation Officials Provincial Standards, Washington, D.C.

Abd El Halim AO, Haas R (2004) Process and case illustration of construction innovation. J Constr Eng Manag 130(4):570–575

Abd El Halim AO, Mostafa A (2006) Asphalt multi-integrated rollers and steel drum compactors. Transp Res Rec 1967:173–180

Abd El Halim AO, Phang W, El Gindy M (1988) Extending the service life of asphalt pavements through the prevention of construction cracks. Transp Res Rec 1178:1–8

Abd El Halim AO, Razaqpur AG, El Kashef AH (1994) Effects of construction cracks on the design of asphalt pavements. Can J Civ Eng 21(3):410–418

Abd El Halim AO, Pinder F, Chelliah AR, Abdelalim O (2013) Reducing maintenance and rehabilitation costs through the use of AMIR compaction. Civil Eng Archit 1(3):51–60

Abd El Halim AO, Abd El Halim AO, Awadalla M, Adel Hassanin H (2015) Development of the asphalt multi-integrated roller field and experimental studies. J Constr Eng 2015:1–8

Airey GD, Collop AC (2016) Mechanical and structural assessment of laboratory and field compacted asphalt mixtures. Int J Pavement Eng 17(1):50–63

Fleckenstein LJ, Allen DL, Schultz DB (2002) Compaction at the Longitudinal Construction Joint in Asphalt Pavements. Kentucky Transportation Centre, University of Kentucky, Lexington

Geller M (1984) Compaction equipment for asphalt mixtures. In: Wagner F (ed) Placement and compaction of asphalt mixtures, ASTM STP 829, vol 54, no 6, pp 28–47

Gilson (2013) Operating manual NCAT asphalt field permeameter kit AP-1B. Gilson, Inc., Lewis Center, Ohio. www.globalgilson.com

Hughes CS (1989) Compaction of asphalt pavement. National cooperative highway research program. Synthesis of highway practice 152. Transportation Research Board, Washington DC

El Hussein HM, Abd El Halim AO, Gerhard JK (1993) Assessment of the influence of compaction method on asphalt concrete resistance to moisture damage. Constr Build Mater 7(3):149–156

Igboke CA, El Harake T, Hassan Y, Abd El Halim AO, Goubran R (2018) Field and laboratory studies for the determination of highway permeability. In: 63rd proceeding of Canadian technical asphalt association, Regina, SK, 11–14 November 2018

Kandhal PS, Mallick RB (2007) Evaluation of various longitudinal joint techniques for asphalt airfield pavements. In: FAA worldwide airport technology transfer conference, New Jersey, USA

Kearney JE (2006) Oscillatory compaction of hot-mix asphalt. In: Factors affecting compaction of asphalt pavements, circular E-C105. Transportation Research Board, Washington, DC, pp 49–53

OHMPA (1999) The ABCs of PGAC: the use of performance graded asphalt cements in Ontario. Ontario Hot Mix Producers Association (OHMPA), Toronto

Parker CF (1960) Steel-tired rollers. Highw Res Board Bull 246:1–40

Ramezani AM, Girardi G, Abd El Halim AO (2018) Effect of construction cracks on the tensile and bond strength of asphalt pavements. J Mater Civil Eng 30(7):04018121-1-7

Shuler S (2017) Preventing transverse bumps and cracks in new asphalt overlays over crack sealants. CDOT-2017-02, Colorado Department of Transportation, Arkansas

Tarefder RA, Ahmad M (2016) Effect of compaction procedure on air void structure of asphalt concrete. J Int Meas Confed 90:151–157

Williams SG (2011) Density, permeability, infiltration, and absorption used to assess quality of hot-mix asphalt longitudinal joints. Transp Res Rec 2228:120–127

Cooling Time Requirements for Asphalt Pavement Repairs

L. Chu and T. F. Fwa

Abstract In repairing asphalt pavements of busy airports or highways carrying heavy traffic, a common constraint encountered is tight time windows available for the repair work. A consideration is to provide sufficient cooling time before opening to traffic. Opening too soon to traffic with insufficient cooling time may lead to early damage to the repaired pavement. This is due to the relatively high temperatures in the asphalt layer as well as the interfacial bonding layer, resulting in insufficient strength in both layers not strong enough to resist heavy wheel loads. This paper applies a validated finite-element simulation model to analyze the time history of temperature cooling in a newly compacted asphalt pavement layer. The simulation model makes use of the thermodynamics and heat transfer theory to predict the temperature changes with time after a newly paved pavement section has been compacted. This information will enable the maintenance team to determine the appropriate time that the repaired pavement section can be opened to traffic. This paper analyzes two common thicknesses of pavement repairs, predicts time variations of temperature within the newly paved asphalt pavement materials for different weather conditions. A summary of the cooling time needed for each case of pavement repair under different weather conditions is presented. The results highlights that the commonly adopted practice of relying on the top surface temperature to open a newly repaired pavement section to traffic is inappropriate and may lead to early damages to the repaired section.

Keywords Cooling time · Pavement temperature · Finite-element simulation · Asphalt pavement repair

L. Chu · T. F. Fwa (✉)
School of Highway, Chang'an University South Erhuan Middle Section, Xi'an 710064, China
e-mail: ceefwatf@nus.edu.sg

L. Chu
e-mail: longjiachu@chd.edu.cn

T. F. Fwa
National University of Singapore, Singapore, Singapore

C. Raab (ed.), *Proceedings of the 9th International Conference on Maintenance and Rehabilitation of Pavements—Mairepav9*, Lecture Notes in Civil Engineering 76,
https://doi.org/10.1007/978-3-030-48679-2_75

1 Introduction

Temperature control is vital to all phases of asphalt mixture construction, including mix production, transportation delivery, laying and compaction, and also the earliest time for opening to traffic (Chadbourn et al. 1996). Due to high traffic volume, most busy airports in the world can only perform repairs of asphalt pavement late at night or in the early hours, and re-open the repaired pavement sections to traffic in the morning (Hampton and Hodgkinson 2017; Yeung and Fwa 2017). Opening to traffic too soon with insufficient cooling time may lead to early damages to a repaired pavement. This is due to the relatively high temperatures in the asphalt layers as well as the interfacial bonding layer, resulting in low strength in both layers not sufficiently strong to resist heavy traffic loading. Sufficient time must be allowed for the temperature to cool down and the asphalt material to gain strength to resist traffic loading. As a result, temperature monitoring and control of asphalt mixtures is important in the repair of asphalt pavements in order to determine when the newly repaired asphalt pavement section can safely open to traffic without causing any early damages.

This paper studies the temperature cooling trends of the newly-laid repaired layer on an old existing pavement using finite element simulation model. The cooling trend of asphalt mixtures is influenced by different factors, including the thermal properties of the newly laid pavement materials and the existing asphalt pavement materials, and the environmental conditions such as solar influx intensity, cloud cover, ambient air temperature and wind speed. These factors are considered in this paper. This paper also discusses the effect of different paving strategies on the temperature cooling trends of a newly laid asphalt layer. The objective for this paper is to study the temperature cooling trends under various conditions using a simulation model, and determine the cooling time period needed. This would provide the pavement maintenance team with valuable information for the determination of when the repaired pavement section can be opened to traffic. This study provides a quantitative solution for determining the time to open to traffic, instead of the traditional practice of relying on experience or the top surface temperature of the repaired layer. The improved procedure can help to reduce the occurrences of early damages of newly paved asphalt sections of busy airports or heavily trafficked highways.

2 Theoretical Basis of Heat Transfer in Action for Newly Laid Asphalt Pavement

In order to predict temperature cooling trends of newly laid asphalt layer over an old asphalt pavement, all forms of heat transfer between the newly laid layer and the surrounding must be considered. Figure 1 presents schematically different heat transfer processes involved when laying a new repaired asphalt layer on an old asphalt pavement.

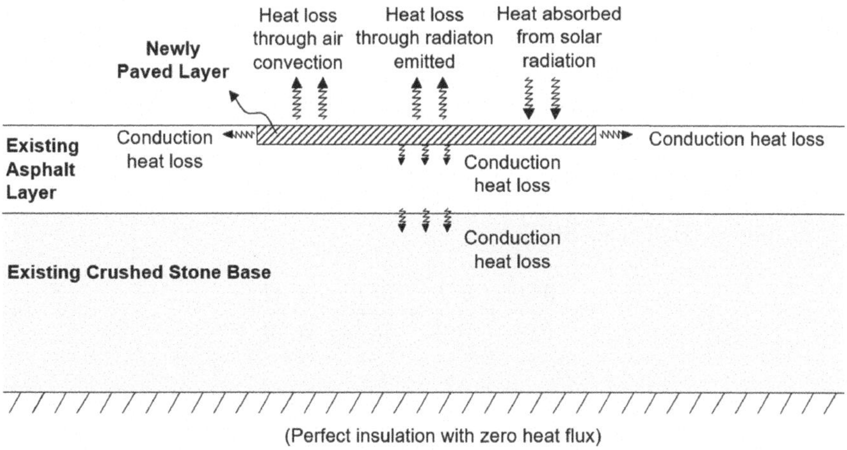

Fig. 1 Heat transfer processes considered in thermal cooling simulation model

(1) Heat transfer between the newly laid layer and the surrounding atmosphere:

- The newly laid asphalt layer will gain heat from solar influx. The intensity of solar influx varies according to the time of the day and the cloud cover. The quantity of heat absorbed can be obtained as follows (Ozisik 1979):

$$q_s = aH_s \tag{1}$$

where q_s = solar radiation energy absorbed, H_s is the net solar flux at the pavement surface, and a = solar radiation absorptivity coefficient which is a dimensionless quantity.

- The newly laid layer will also lose heat to the atmosphere through radiation. The amount of radiation energy lost can be expressed by the following equation (Hall et al. 2012):

$$qr = \varepsilon\sigma\left(T_s^4 - T_a^4\right) \tag{2}$$

where q_r = radiation heat flux emitted, ε = emissivity coefficient of asphalt layer, $\sigma = 5.669 \times 10^{-8}$ W/m^{2}K^4 is the Stefan-Boltzman constant, T_s = surface temperature of the newly laid pavement layer, and T_a = ambient air temperature.

- Heat loss to the surrounding air through convection occurred between the air and the pavement surface. This heat loss is a function of wind speed and can be expressed by the following equation (Hall et al. 2012):

$$h = 5.8 + 4.1v \tag{3}$$

where v = wind speed in m/s, and h is convective heat transfer coefficient for air flow in w/m^2.

(2) Heat transfer occurred through conduction between the newly laid layer and the old pavement structure beside and underneath:

Heat conduction occurred between the newly laid asphalt layer and the existing surrounding materials is governed by Fourier's law which can be described as follows (Hall et al. 2012):

$$\frac{\partial^2 T}{\partial x^2} + \frac{\partial^2 T}{\partial y^2} + \frac{\partial^2 T}{\partial z^2} = \frac{1}{\alpha}\frac{\partial T}{\partial t} \tag{4}$$

$$\alpha = \frac{k}{\rho c} \tag{5}$$

where T = temperature, °C, α = thermal diffusivity, m^2/s, k = thermal conductivity, W/m K, ρ = density, kg/m^3, c = specific heat, J/kg K, and x, y, z are the three coordinate directions in a three-dimensional space.

3 Finite Element Modeling

Finite element modeling is an effective technique for simulating pavement cooling (Timm et al. 2001; Kassem et al. 2015; Wang et al. 2014). The commercial software ABAQUS (2010) was used in the present study. The pavement simulated in this paper is 12 m by 21 m in plan dimensions. Preliminary analyses indicated that during cooling, the heat transfer within the central portion of surface asphalt layer was essentially 2-dimensional in nature. A simple 2-dimension finite-element model was adequate for the present study without considering the effect of vertical boundaries along the four sides of the repaired layer. Preliminary analyses were also conducted to determine the appropriate finite element mesh sizes, balancing the computation time and the required accuracy of the predicted temperatures of the repaired layer. The finite element mesh design adopted for different pavement layers are shown in Fig. 2. The newly paved repaired asphalt layer adopts the smallest mesh size of 2 mm which is the most critical part affecting the earliest time for opening to traffic. Validation of the model is found in a separate work of the authors (Chu et al. 2019).

The simulation model considers different affecting factors including wind speed, ambient temperature, solar influx intensity and radiation loss, according to the Eqs. (1) to (5). In the following sections, the finite element simulation model is applied to study the cooling patterns of two common thicknesses of asphalt pavement repairs, i.e. 100 and 150 mm thick repairs constructed in two lifts.

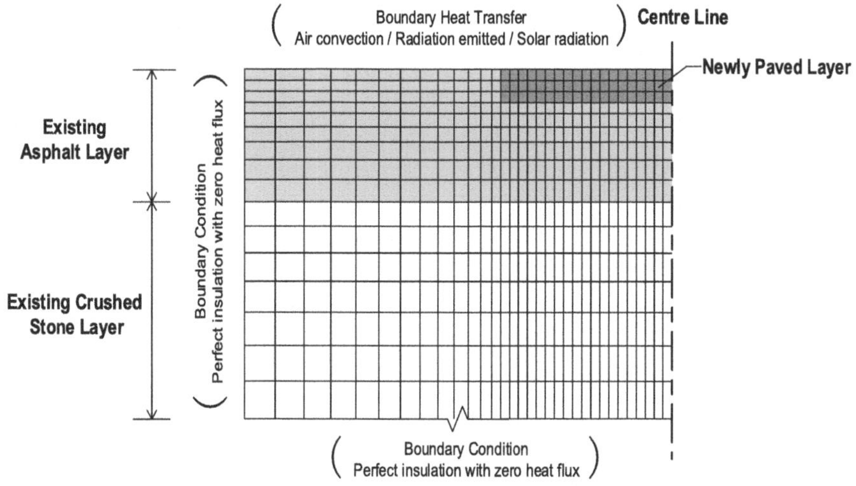

Fig. 2 Schematic representation of finite-element mesh for proposed simulation model

4 Analysis of Thermal Cooling Trends in Repaired Layer

This section presents the results of analysis considering the effect of the following three factors on the cooling time needed for the temperature in a newly paved repaired layer to reach 60 °C: (i) wind speed; (ii) ambient air temperature; and (iii) the initial compaction temperature. Traffic opening temperature of 60 °C has been a common temperature used by many airport and highway agencies. Using 60 °C as the reference temperature for opening a repaired section to traffic, the effects of these factors on the cooling trend and opening time of a newly paved layer are considered in two different ways of construction: (1) A 150 mm repair thickness constructed in two lifts of 75 mm each; (2) A 100 mm repair thickness constructed in two lifts of 50 mm each. The highest temperatures in the bottom and top lifts are identified in the analyses. The following asphalt thermal properties were applicable: thermal diffusivity = 0.98 × 10 − 6 m^2/s, absorptivity = 0.9, and emissivity = 0.8. Solar radiation flux was zero for night paving in busy airports.

4.1 Effects of Wind Speed

The effect of wind speeds, ranging from 1 to 10 m/s, on the cooling time duration needed to reach to 60 °C is presented in this section. The results of the wind speed effects are shown in Fig. 3. The plot illustrates that with the increase of wind speed, the time needed to reach 60 °C decreases for both the bottom and surface lift. Stronger wind speed helps to reduce the time needed to cool down to 60 °C. It is noted that the effect of wind tends to level off when its speed is higher than 8 m/s.

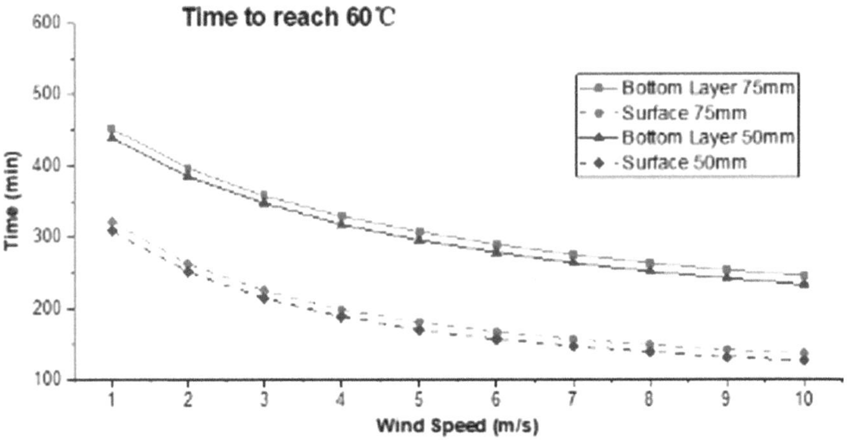

Fig. 3 Effect of wind speed on time to reach to 60 °C after compaction

Based on the results of analysis, it can be seen that the bottom lift in each case needs longer time to reach 60 °C than the surface lift for each of the paving strategies. As expected, the case with a thinner thickness takes shorter time to reach 60 °C.

4.2 *Effects of Ambient Air Temperature*

The effect of ambient air temperature ranging from 24 to 33 °C, on the cooling time duration needed to reach 60 °C is presented in this section. Figure 4 presents the

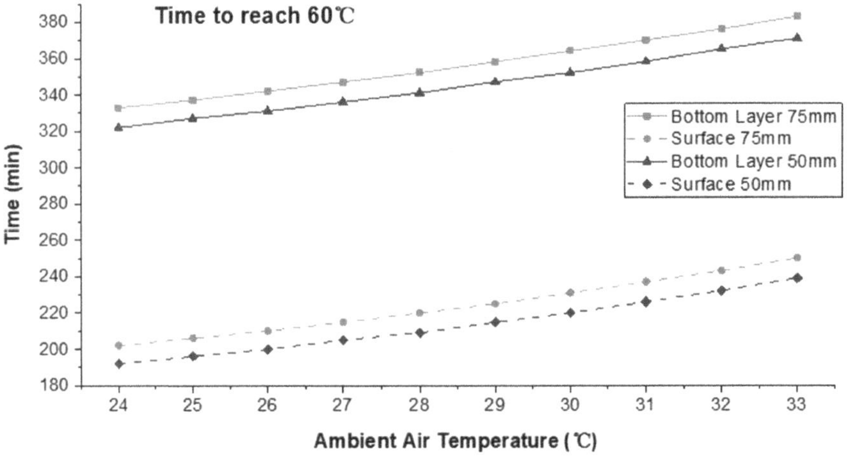

Fig. 4 Effect of air temperature on time to reach 60 °C after compaction

results of cooling trends under different ambient temperatures for the two paving strategies studied. It can be seen that as the ambient air temperature rises, longer cooling time is needed to reach to 60 °C for both the bottom and surface lift. It is noted that the bottom lift needs longer time to reach to 60 °C as compared with the surface lift. As for the effect of repair thickness, the thinner repair thickness of 100 mm takes shorter time to cool down to 60 °C. The difference roughly remains 10 min and appears to stay the same for the range of ambient air temperature studied.

4.3 *Effect of Compaction Temperature*

The initial compaction temperature can significantly affect the time duration to cool down to 60 °C. Figure 5 shows the cooling trends for the two repair thicknesses under different compaction temperatures. The cooling time needed to reach 60 °C increases with the magnitude of the initial compaction temperature. A difference of 20 °C in the initial compaction temperature, i.e. 140 and 160 °C, leads to 60 min difference in the length of cooling time to 60 °C of the bottom layers for the two repair thicknesses. When the initial compaction temperature is 140 °C, the bottom lift takes roughly 120 min longer to reach to 60 °C than that of the surface lift, while when the initial compaction temperature is 160 °C, the bottom lift requires roughly 140 min longer to cool down to the desired temperature. The case with thinner repair thickness takes about 12 min shorter time to cool down to 60 °C than the case with thicker lift thickness.

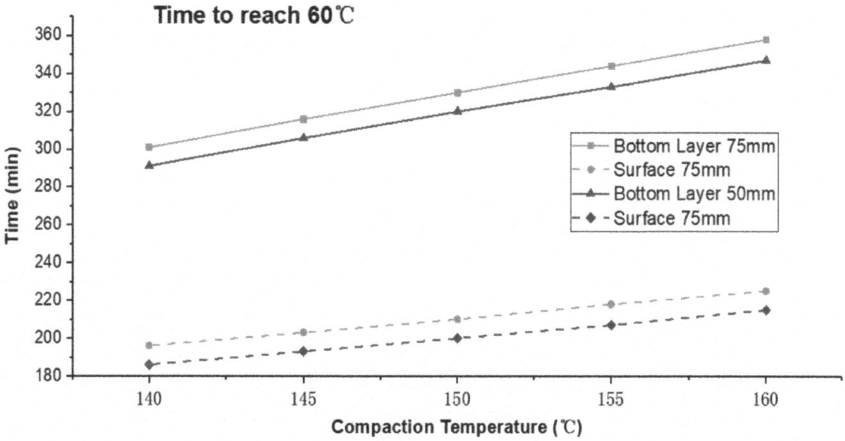

Fig. 5 Effect of compaction temperature on time to reach 60 °C after compaction

5 Conclusion

A 2-dimentional finite thermal cooling simulation model has been developed in this paper to study the temperature cooling trend of a newly laid repaired surface layer on an old asphalt pavement. Initial compaction temperature and two different environmental factors (i.e. wind speed and ambient air temperature) were the main input variables considered to study the cooling time length needed to reach to 60 °C. The three factors considered have significant effect on the length of cooling time needed to reach the target temperature of 60 °C. In the different cases studied, the temperature in the bottom lift is always higher than that in the surface lift. This is of high practical significance as it means that it is inappropriate to open a newly laid pavement section to traffic based solely on the surface temperature surface lift. The established simulation model can provide pavement engineers with information which would be useful for planning and scheduling of asphalt pavement repairs for busy airports and highways.

References

ABAQUS (2010) ABAQUS User's Manual, Pawatucket, RI

Chadbourn BA, Luoma JA, Newcomb DE, Voller VR (1996) Consideration of hot mix asphalt thermal properties during compaction. In: Decker DS (ed) ASTM special technical report STP 1299, quality management of hot-mix asphalt. American Society for Testing and Materials

Chu L, Zhu B, Fwa TF (2019) Temperature control planning tool for multi-lift resurfacing of airport pavements. Transp Res Rec 2673(7):380–389

Hall MR, Dehdezi PK, Dawson AR, Grenfell J, Isola R (2012) Influence of the thermophysical properties of pavement materials on the evolution of temperature depth profiles in different climatic regions. J Mater Civ Eng 24(1):32–47

Hampton I, Hodgkinson J (2017) Replacing concrete airport pavement slabs without loss of aircraft operations. In: Proceedings of 4th ASCP conference, Australian society for concrete pavements, Kingscliff, NSW, Australia, 17–18 July 2017

Kassem H, Chehab G, Saad G (2015) An FEM-predictive tool for simulating the cooling characteristics of freshly paved asphalt concrete layers. Int J Pavement Eng 16(2):157–167

Ozisik MN (1979) Basic heat transfer. McGraw-Hill Book Company, New York

Timm D, Voller V, Lee E, Harvey J (2001) Calcool: a multi-layer asphalt pavement cooling tool for temperature prediction during construction. Int J Pavement Eng 2(3):169–185. https://doi.org/10.1080/10298430108901725

Wang Y, Zhu S, Wong AST (2014) Cooling time estimation of newly placed hot-mix asphalt pavement in different weather conditions. J Constr Eng Manag 140(5):04014009

Yeung C, Fwa TF (2017) Airport operation-centered airfield pavement maintenance management. In: Proceedings of 10th international conference on road and airfield pavement technology, Hong Kong, 8–10 August 2017

Cold Recycling in Germany—Current Experiences and Future Projects

B. Wacker, M. Kalantari, and M. Diekmann

Abstract To guarantee the smooth transportation of goods and passengers over a highly loaded road network, more efficient and even innovative maintenance and rehabilitation measures are needed. During the last decade, the Federal Highway Research Institute (BASt) has accompanied in a rehabilitation project on a federal roadway with different cold recycling construction types and has regularly monitored it with non-destructive test methods. In recent years, different aspects of the technology have also been investigated on laboratory scale through different research programs. In view of the current positive international experience with bitumen-stabilized materials (BSM), foamed bitumen mixtures in particular have been met with great interest in Germany. In the first phase of a new research project in cooperation of BASt and Wirtgen GmbH, a 100 m long test section will be tested in the demonstration, investigation and reference areal of BASt (duraBASt) in order to extract further experience with the method and adopt it for Germany. After completion of the construction process, one cold-recycling and one reference pavement type will be analyzed with the standardized BASt Accelerated Pavement Testing Program and further laboratory tests. The collected findings of the test section in the network as well as on duraBASt will be utilized in design procedures and further development of national regulations and specifications in Germany.

Keywords Cold-recycling · Foam bitumen · duraBASt · Accelerated Pavement Testing

1 Introduction

One crucial message of the 2030 Transport Infrastructure Plan of Germany is "Preservation before new construction" (BMVI 2016). This can lead to problems because

B. Wacker (✉) · M. Kalantari
Design and Structure of Pavements, BASt, Federal Highway Research Institute, Bruederstr. 53, 51427 Bergisch Gladbach, Germany
e-mail: APT@bast.de

M. Diekmann
Wirtgen GmbH, Reinhard-Wirtgen-Str. 2, 53578 Windhagen, Germany

C. Raab (ed.), *Proceedings of the 9th International Conference on Maintenance and Rehabilitation of Pavements—Mairepav9*, Lecture Notes in Civil Engineering 76,
https://doi.org/10.1007/978-3-030-48679-2_76

there are no sufficient regulations yet to the reuse of reclaimed asphalt. To guarantee the smooth transportation of goods and passengers over a highly loaded roadway network, more efficient and even innovative maintenance and rehabilitation measures are needed. In addition, it is becoming more and more important to implement environmental-friendly and resource-saving methods as much as possible. The recycled asphalt is often used in new hot asphalt formulations in order to comply with the Closed Substance Cycle Waste Management Act. However, the question arises of how to deal with the repeated use of recycled asphalt (second-/third generation asphalt granulate) because bitumen characteristics will degrade in the long term (Gogolin 2019). This means that the reclaimed asphalt aggregates may not be suitable for hot recycling process with high rates of usage.

In this regard, cold recycling and stabilization techniques with the advantages like higher rates of recycled materials utilization, lower energy consumption and sometimes faster construction, has attracted much attention. After getting positive results from implementing the technology on low traffic volume roads, it was decided to assess that for high traffic volume roads. During the last decade, the Federal Highway Research Institute (BASt) has accompanied in a rehabilitation project on a federal highway with different cold recycling construction types and has regularly monitored it with non-destructive methods. In recent years, different aspects of the technology have also been investigated on laboratory scale through different research programs. In view of the current positive international experience with bitumen-stabilized materials (BSM) (Diefenderfer 2019; West 2018), foamed bitumen mixtures in particular have met with great interest in Germany. In Germany, only a few experiences and performance data are available. In the first phase of a new research project in cooperation with Wirtgen GmbH, a 100 m long test section will be tested in the demonstration, investigation and reference areal of BASt (duraBASt) in order to extract further experience with the method and adopt it for Germany.

2 Current Use of Cold-Recycling in Germany

There are two main national guidelines in Germany regarding to the topic of cold recycling for in-place and in-plant methods (FGSV 2005, 2007). There is a catalogue which is used to determine the thickness of the pavement layers on lower traffic classes when cold recycled layer is used (FGSV 2005). Because of the imposed limitation by the guideline and the limitation of performance data, almost all projects done in Germany are in the range of low to medium traffic classes sections.

In 2006, within the scope of a renewal project of a federal road, the Federal Highway Research Institute (BASt) was asked to supervise the planning and execution of the rehabilitation measure. The federal road was a 2.4 km long stretch with an average daily traffic of approx. 26,000 vehicles (15% heavy traffic, approx. 3,800 vehicles) based on the traffic data from 2006. Due to favorable framework conditions, it was decided to design the section with a CR layer. The section had a higher traffic level which was not included in the existing regulations therefore considered as an opportunity to earn experience and performance data about this material and construction type.

Prior to planning, a comprehensive survey was performed on the existing section by means of Ground Penetration Radar (GPR) and Falling Weight Deflectometer (FWD). The main goal was to determine the layers' thicknesses, assessing the bearing capacity of the pavement structure and to be able to define homogeneous sections (Zander et al. 2007). Based on bearing capacity measurements the section could roughly divided into two areas with different bearing capacities. It was found that the existing structure was not adequate for the future traffic and therefore a fundamental rehabilitation was recommended.

For the new construction, the road administration requested a design that would withstand for 30 million equivalent 10 t axle transitions within a service life of 20 years (Zander et al. 2007). The regulations and guidelines valid at that time did not provide a standardized construction method with CR layer for the required amount of traffic so the CR layer was arranged relatively thick under a conventional asphalt construction. However, the knowledge about the material behavior was very limited and therefore the results of the analytical design were not meaningful enough. Therefore, further experiences were required to gain more knowledge (Zander et al. 2007). Based on the existing experience on the standard pavement sections from the RStO (FGSV 2001), expected strains and stresses could be used for further analytical design for experimental purposes. Due to the higher modulus of elasticity of the CR layer (around 8,000 MPa), it was possible to calculate almost identical fatigue service lives for the asphalt package even in a design suggested by the road administration with a relatively thin CR layer (4 cm SMA, 6 cm asphalt binder, 8 cm asphalt base course and 12 cm CR). However, it was seen from the analysis results that the tensile stresses ratio (the ratio of tensile stress at the bottom of the CR layer to the tensile strength of the CR material) of the CR layer increased significantly compared to the standard pavement types (selected from RStO for the same traffic class). Due to this property, the design team expected the occurrence of early tensile fatigue cracks in the construction with thin CR thickness (Zander et al. 2007). Therefore limiting the tensile stress at the bottom of the CR layer was considered during the structural design and selection of the final thicknesses.

Three different variants were selected and implemented for the renewal measure, thus creating three sections in Fig. 1 (Golkowski 2010a). The division of the sections is mainly based on different types of CR layer construction methods (4% bitumen emulsion with 4% cement in Sects. 1 and 3 as well as 2.5% foamed bitumen with 3% cement in Sect. 2) and between Sect. 1 with a 10 cm asphalt base layer and Sects. 2 and 3 with an 8 cm asphalt base layer (Golkowski 2010a). The CR layers were laid with a thickness of 20 and 6.5 cm asphalt binder course with 3.5 cm asphalt surface course of SMA 0/8 S in all sections (Golkowski 2010a). Due to the traffic accommodation requirements, two main lanes (each direction) and one passing lane (uphill direction) were available which, resulted in a total of nine observation fields for the cold recycled sections. Because a paving problem, the cold recycled layer in Sect. 2 of the main lane was partially replaced by an asphalt base layer with the same

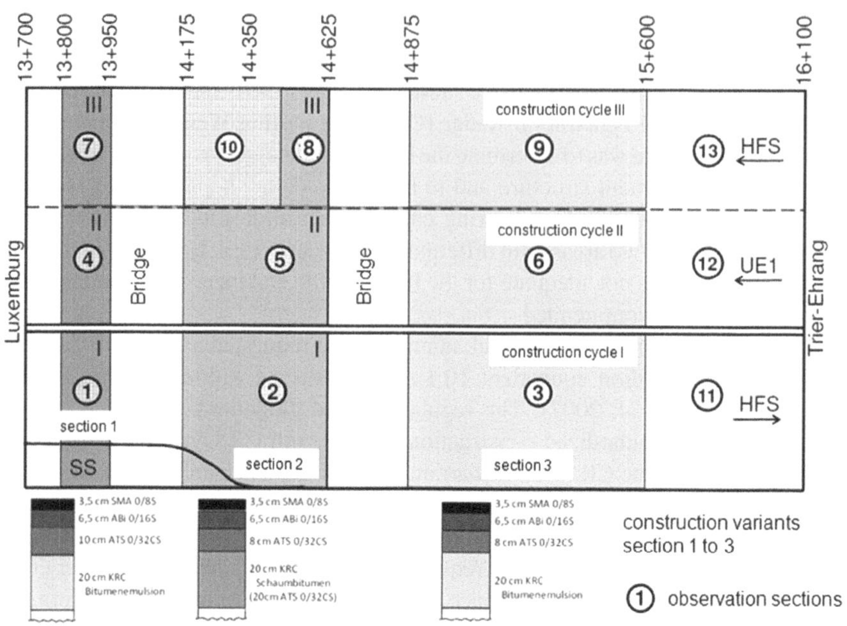

Fig. 1 Position of different construction variants (Golkowski 2010a)

thickness. The remaining observation fields (11–13) refer to a traditional renewal measure.

The investigation program provided that every year in spring and autumn the complete section of the route (2.4 km) was to be measured. In these investigations, only the main traffic lanes with two measuring lanes (right wheel path and between the wheel paths) were tested with the FWD and the GPR (Golkowski 2010a).

Two measurement campaigns were carried out within the first year after the traffic release (autumn 1st year, spring/summer 2nd year) (Golkowski 2010b). In principle, unsteady bearing capacity values could be determined in the sections with bitumen emulsion, whereas homogeneous curves could be detected in the section with foamed bitumen. In the areas with foamed bitumen, a clear difference in the level of the measured values (100–150 μm) between the two directions could be detected. It is assumed that the different stresses may lead to this difference (Golkowski 2010b). During the first measurements, no improvement or deterioration of the load-bearing behavior was found. This can be related to the curing mechanism of the bitumen recycled layers as the material gains strength and stiffness over time therefore the deflection results are expected to decrease over the time by progress of the curing. On the reference section with conventional asphalt construction (traditional renewal), lower bearing capacity parameters tended to be determined during the measurement periods. However, an assessment based on the first measurement results is presented as premature and further measurement results needs to be considered (Golkowski

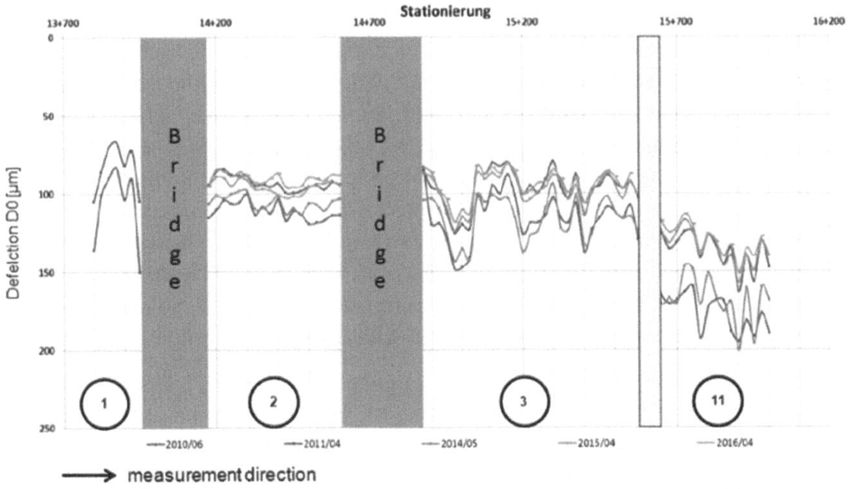

Fig. 2 Deflection data – sections 1 to 3 and reference section 11 (Tillmann 2016)

2010b). In general, difficulties in evaluating the load-bearing capacity and the load-bearing behavior with conventional methods in the field of cold recycling construction were pointed out in the evaluation report (Golkowski 2010b).

After a total of seven years of service and nine measuring actions, the pavement and the measurements were analyzed as part of a bachelor thesis. It became clear that the bearing capacity in the sections with CR layer did not deteriorate over time, but even improved in some cases (Tillmann 2016). Tillmann attributes this to post-compaction by traffic in the wheel path (Tillmann 2016). Particularly in the section of the CR layer with foamed bitumen, uniform measured values without extreme picks can be observed over the entire period. Moreover, after seven years of use, no deficiencies in the bearing capacity were found (Tillmann 2016) Fig. 2 shows the results.

3 Current State of the Research on Cold Recycling in Germany

Understanding the material's behavior is the key to get the best performance out of that and to be able to deliver optimum characteristics based on the requirements of each project. The existing guidelines in Germany on cold recycling (FGSV 2005, 2007), are not covering all aspects related to mixture production and characterization. There are some documents from different sources (contractors, consultants and state road offices) (Johannsen 2017; Weinert 2007, 2009) mainly based on the mentioned two guidelines. They aimed to improve some open sections like: methods and instructions to produce CR mixture and specimens, tests, acceptance limits and structural design. Due to the open questions from the current guidelines (FGSV 2005, 2007) and the existing experiences of the last 20 years (Mollenhauer 2017; Radenberg 2015; Hähn 2004) as well as international research, the need for further research in the field of cold recycling with foamed bitumen in Germany became necessary. This should also lead to a revision of the existing guidelines.

In this regard a fundamental research program on foamed bitumen cold mixes was planned and started from 2014 in the Institute for Road Engineering of Siegen University (ifs Siegen) and continued at BASt (Kalantari 2019). The research will try to gain a better understanding from the behavior of the foamed bitumen and cement stabilized mixes (FCSM) with the aim to integrate them by the pavement types in Germany. The main research tasks will deal with the below questions:

1. How to produce and prepare FCSM samples in laboratory? Compaction and curing are the main issues in production and preparation of the specimens.
2. How to evaluate physical properties of FCSM samples?
3. How to determine mechanical parameters of FCSM and integrate them in a pavement model?
4. How to assess material's performance in the laboratory?
5. How to perform structural design of pavements with FCSM?

The project approach is to apply as far as possible the existing and available knowledge in Germany (FGSV 2005, 2007; Wirtgen 2012) to find out up to what extent they are applicable to this material type.

Generally, during the stabilization or cold recycling process, different amounts of bitumen and cement (as hydraulic binder) can be applied as binding agents. This leads to variable material behaviors ranging from enhanced unbound granular material to bonded mixtures with bituminous or hydraulically dominant behaviors. Considering the bitumen and cement ranges proposed in the existing guideline (FGSV 2005), three different amounts of 2.5, 3.5 and 4.5% for foamed bitumen (Nyfoam 60 as the base bitumen) and three different amounts of 1, 2 and 3% for cement (type 1,42.5 N) were selected. To have a compact gradation and a good dispersion of bitumen, a material gradation with $n = 0.45$ (Fuller method) was adapted by mixing 6 different fraction sizes (0.063–2, 2–5.6, 5.6–8, 8–11.2, 11.2–16 and 16–22.4 mm) together. For filler, limestone powder was used.

Considering the compaction method in laboratory as one of the main issues in production of test specimens, the static compaction method recommended by the existing national guideline (FGSV 2005) was compared with Marshall (75 blows each side) and modified Proctor methods. Air voids content of compacted specimens showed that the static compaction with 2.8 MPa pressure is not appropriate for mixes with common ranges of foam bitumen (Fig. 3).

The results of the ITS (indirect tensile strength) tests of specimens with different bitumen contents and consistent cement content show a temperature dependency behavior (Fig. 4).

At a constant bitumen quantity of 3.5% by varying the cement content, higher ITS amounts were observed at the respective temperature levels (Fig. 5). Comparing the graphs of variable bitumen with the graphs of variable cement content, shows that

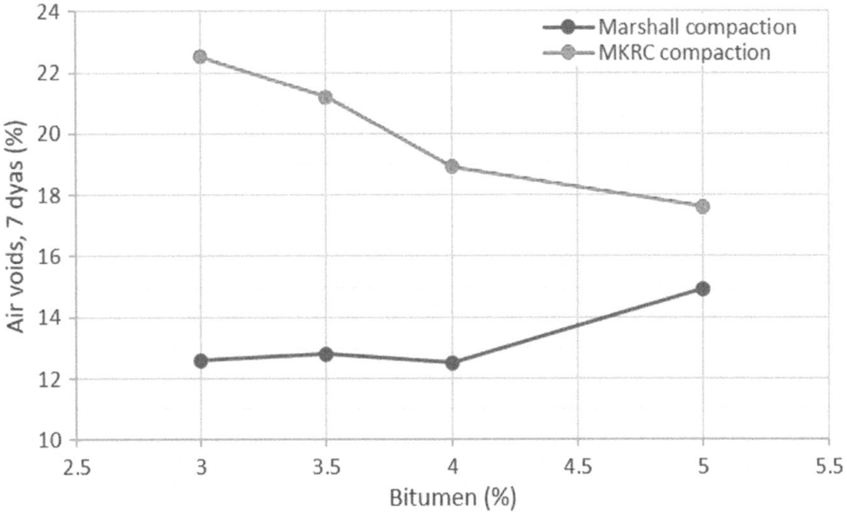

Fig. 3 The effect of bitumen content on air voids content of compacted specimens (Kalantari 2019)

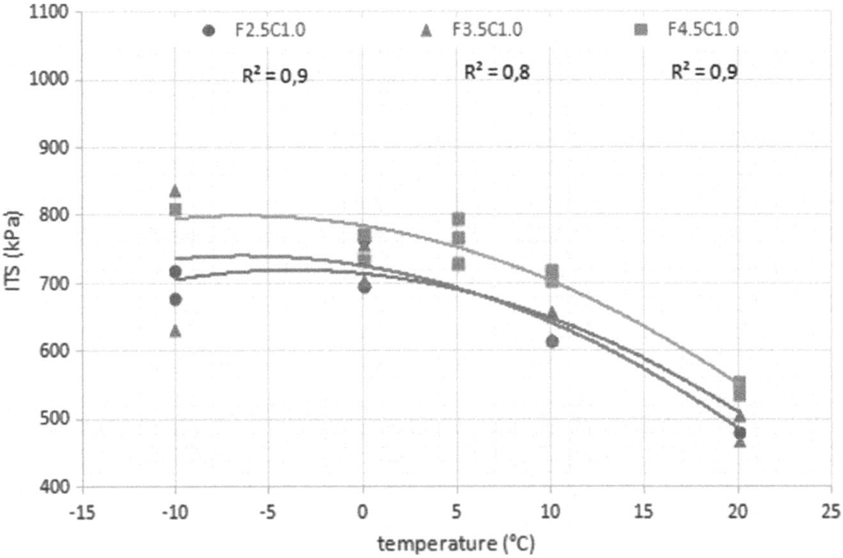

Fig. 4 The influence of temperature on ITS for different bitumen contents (abbreviations: F2.5 means 2.5% bitumen C1.0 means 1% cement), (Kalantari 2019)

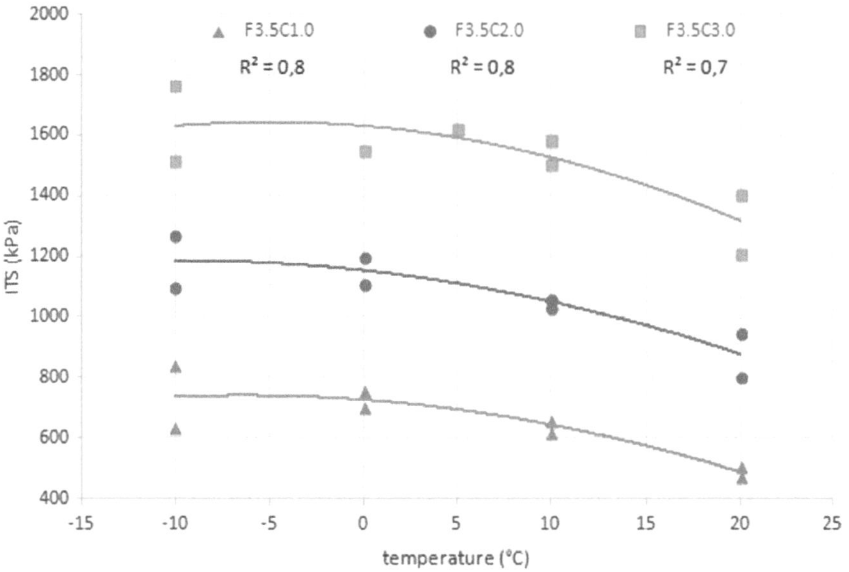

Fig. 5 The influence of temperature and cement content on ITS (Kalantari 2019)

the cement has a greater effect on the ITS results than the foamed bitumen which may easily lead to higher rigidity of the material.

Beside the ITS tests, indirect tensile cyclic tests applied on the samples to determine the stiffness at different temperatures (−10, 0, 10 and 20 °C) and different frequencies (10, 5, 1 and 0.1 Hz) based on national code AL Sp-Asphalt 09 (FGSV (2009)—similar to DIN EN12697-26:2012-06 (annex F)). The results showed less temperature dependency of stiffness in these mixes in comparison with hot mix asphalt with the same bitumen content of 3.5% (reference mix). Increasing the cement content resulted higher stiffness and lower flexibility of the material. It is important to keep a balance between the amount of bitumen and cement to profit the advantageous of both.

The results of these fundamental researches revealed the complicated behavior of this group of materials and paved the way for the next researches in this field.

4 duraBASt Investigation

Based on the experience gained in Germany, but also on the positive findings in the international research area (West 2018), BASt has initiated a project together with Wirtgen GmbH to investigate the cold recycling construction method with foamed bitumen for German conditions. At present, the potential of the construction method

Fig. 6 Mobile Load SimulatorMLS30 on duraBASt

in Germany is not yet exhausted; this could be due to the lack of methods for assessing the performance, but also to the lack of regulations for the design of the mix as well as design and quality control. For this purpose, a test track was built on the demonstration, investigation and reference area of BASt (duraBASt). Extensive measurements will carried out on this test section and realistic loads will be applied by means of Accelerated Pavement Testing (APT) tests with the Mobile Load Simulator MLS30 (Fig. 6) during the year 2020.

The main objective of the project is to determine empirical values for cold recycling construction with foamed bitumen and cement under real conditions in a large-scale APT program using the Mobile Load Simulator MLS30.

This will allow an assessment and improvement of existing material development processes using foamed bitumen and recycled material and a better understanding of the material in terms of stiffness over time. In addition, a comparison with a standard design will be made and the short and long term behavior will be investigated. As well as a relationship between laboratory tests and large-scale mixtures will be evaluated. Different mix combinations of RAP and virgin aggregates were tested during the mix design process. The final mix consisted of 75% RAP and 25% (0–2 mm sand) with 2.2% foamed bitumen and 1% cement. During the APT program in 2020, a super-single tire with a 50 kN wheel load will be used and the pavement responses will be documented with different types of installed sensors and non-destructive measuring systems.

5 Conclusions

In summary, it can be stated that within the Federal Republic of Germany various scientific empirical values with CR layers are available and are to be further analyzed. Refering to the long-term observation of CR section in a federal road under traffic as well as the extensive laboratory investigations, there are significant expectations in this technology. For a controlled and detailed evaluation of current possibilities the project on duraBASt is an essential step to generate further application possibilities. Considering the ever increasing quantities of reclaimed asphalt being produced by the maintenance measures, and the need to rethink on the use of resources, the cold recycling method is a possible and useful solution.

References

BMVI (2016) Bundesverkehrswegeplan 2030 (Federal Transport Infrastructure Plan 2030). Federal Ministry of Transport and Digital Infrastructure, Berlin

Diefenderfer BK, Timm DH, Bowers BF (2019) Structural Study of Cold Central Plant Recycling Sections at the National Center for Asphalt Technology (NCAT) Test Track: Phase II, Viginia Transportation Research Council, Charlottesville

FGSV (2001) Richtlinien für die Standardisierung des Oberbaus von Verkehrsflächen (Guidelines for the standardization of pavement structures of traffic areas). FGSV, Cologne

FGSV (2005) Merkblatt für Kaltrecycling in situ im Straßenoberbau (MKRC) – (Instruction sheet for cold recycling in situ in road pavements). FGSV, Cologne

FGSV (2007) Merkblatt für die Verwendung von pechhaltigen Straßenausbaustoffen und von Asphaltgranulat in bitumengebundenen Tragschichten durch Kaltaufbereitung in Mischanlagen (M VB-K) (Instruction sheet for the use of pitch-containing road construction materials and asphalt granulate in bitumen-bound base layers by cold processing in mixing plants). FGSV, Cologne

FGSV (2009) Arbeitsanleitung zur Bestimmung ders Steifigkeits- und ermüdungsverhaltens von Asphalten mit dem Spaltzug-Schwellversuch als Eingangsgröße in die Dimensionierung (Working instructions for determining the stiffness and fatigue behaviour of asphalts using the Indirect Tensile Strength Tests as an input variable in design). FGSV, Cologne

FGSV (2018) Technische Prüfvorschrift für Asphalt, TP Asphalt-StB, Teil 26: Spaltzug-Schwellversuch – Bestimmung der Steifigkeit (Technical test specification for asphalt, TP Asphalt-StB, Part 26: Indirect Tensile Strength Tests - Determination of rigidity). FGSV, Cologne

Gogolin D (2019) Wirksamkeit und Performance von Rejuvenatoren – Teil 1: Laboruntersuchungen (Efficacy and performance of rejuvenators - Part 1: Laboratory investigations). Asphalt 2/2019, 12–17

Golkowski G (2010a) Konzept KRC Versuchsstrecke B52 - (Concept CR test track B52). Federal Highway Research Institute (BASt), Bergisch Gladbach, unpublished

Golkowski G, Ritter J (2010b) KRC Versuchsstrecke B52 – Zwischenbericht ENTWURF - (CR test track B52 - Interim report DRAFT). Federal Highway Research Institute (BASt), Bergisch Gladbach, unpublished

Hähn G, Krass K, Drews P Kirchknopf H (2004) Prozess- und Verfahrentechnik für die umweltschonende Straßensanierung durch Kaltrecycling mit Schaumbitumen (Process and process engineering for environmentally friendly road rehabilitation by cold recycling with foamed bitumen). Federal Ministry of Education and Research, Berlin

Johannsen K, Willmeroth K (2017) Technische Spezifikation für die Dimensionierung, Produktion und Bauausführung von Fundationsschichten Typ RECYVIA CMEUROVIA. Services, Berlin

Kalantari M (2019) Assessing the effect of binding agents content on resulted mechanical and Fatigue characteristics of foamed bitumen and cement stabilized mixes, draft Manuscript for PHD, University Siegen, unpublished

Mollenhauer K (2017) Studie zum Anwendungspotenzial von werksgemischten Kaltbauweisen – Asphalt (Study on the application potential of factory-mixed cold construction methods - Asphalt). Bundesanstalt für Straßenwesen, Bergisch Gladbach

Radenberg M, Miljkovic M, Schäfer V (2015) Einfluss des Asphaltgranulates auf die bemessungs- und ermüdungsrelevanten Materialeigenschaften einer zwangsgemischten, kaltgebundenen und bitumendominanten Tragschicht (Influence of asphalt granulate on the design and fatigue-relevant material properties of a mixed, cold bound and bitumen-dominant base layer). Bundesnastalt für Straßenwesen, Bergisch Gladbach

Tillmann T (2016) Bewertung der KRC Beobachtungsfläche B52 bei Trier mittels zerstörungsfreier Prüftechnik – Bachelorarbeit (Evaluation of the CR observation plot B52 near Trier using non-destructive testing technology - Bachelor thesis). Universität Siegen, Siegen

Weinert IF, et al (2007) Sächsische Technische Richtlinie für Kaltrecycling in plan für den Straßenoberbau SN TR CR in plant (Saxon Technical Guideline for Cold Recycling in plan for the road pavement SN TR CR in plan). LIST Gesellschaft für Verkehrswesen und ingenuertechnische Dienstleistungen mbH, Rochlitz

Weinert IF (2009) Leitfaden Kaltrecycling, Bauverfahren - CR in situ und Zentralmischverfahren - CR in plant (Guide to cold recycling, construction methods - CR in situ and central mixing methods - CR in plant). Landesbetrieb Mobilität Rheinland-Pfalz, Böhlen

Wirtgen (2012) Wirtgen Cold Recycling Technology. Wirtgen GmbH, Windhagen

West R, Timm D, Powell B et al (2018) NCAT Report 16-04 Phase V (2012-2014) NCAT Test Track findings, National Center for Asphalt Technology, Auburn

Zander U, Golkowski G, Wolf A (2007) Stellungnahme zur Dimensionierung der Oberbauerneuerung auf der B52 Trier-Ehrang (Statement on the dimensioning of the renewal of the superstructure on the B52 Trier-Ehrang). Bundesanstalt für Straßenwesen, Bergisch Gladbach, unpublished

Distributed Fiber Optic Strain Measurements in an Airfield Pavement

D. Hauswirth, F. Fischli, C. Rabaiotti, and A. M. Puzrin

Abstract Distributed fiber optic strain sensors represent a powerful tool for gathering continuous strain data along an optical fiber within engineering structures under loads. On the other hand, pavements represent a harsh environment for optical fibers, in particular during construction process. Hence, optical fibers which are adequately protected by a robust cable sheath were used in this project for the purpose of measuring strains within a pavement. Strains were measured along the sensor cables with high spatial resolution while an aircraft was placed on the instrumented area, subjecting the pavement to static loads. The procedure was carried out in warm and cold conditions in order to study the effect of temperature conditions on the strains measured in the pavement. The measurements gave a detailed insight into the strain distribution in the loaded pavement section. The fiber optic sensors proved to be robust enough to be applied on a real construction project. Furthermore they may provide, in general, useful strain data e.g. for testing new pavement materials, assessing new design methods or investigating structural health of existing structures.

Keywords Full scale tests · Fiber optic strain sensors · Pavement testing · Structural health monitoring

1 Introduction

Distributed fiber optic sensor technologies provide the possibility to measure continuously external measurands as e.g. strain or temperature along an optical fiber. This makes this class of sensors in particular interesting for measuring and monitoring the performance of structures. In several studies, the application of fiber optic sensors

D. Hauswirth (✉) · A. M. Puzrin
Institute for Geotechnical Engineering, ETH Zürich, Stefano-Franscini-Platz 5, 8093 Zürich, Switzerland
e-mail: dominik.hauswirth@igt.baug.ethz.ch

F. Fischli
Marmota Engineering AG, Technoparkstrasse 1, 8005 Zürich, Switzerland

C. Rabaiotti
Institut für Bauingenieurwesen, HSR, Oberseestrasse 10, 8640 Rapperswil, Switzerland

C. Raab (ed.), *Proceedings of the 9th International Conference on Maintenance and Rehabilitation of Pavements—Mairepav9*, Lecture Notes in Civil Engineering 76,
https://doi.org/10.1007/978-3-030-48679-2_77

in pavements is reported. Often these applications adopt fiber optic point or quasi-distributed sensors as e.g. Fiber Bragg Grating sensors. Regarding the application of distributed fiber optic sensors in pavements e.g. Chapeleau et al. (2017) using swept wavelength interferometry (SWI), Bao and Chen (2016) applying Pulse Pre-Pump Brillouin Optical Time Domain Analysis (PPP-BOTDA) or Xiang and Wang (2016) adopting BOTDA showed some test application with a particular emphasis on detection of cracks. Chapeleau et al. (2017) show strain distributions using fiber optic cables embedded in an accelerated pavement testing facility. Liu et al. (2015) tested different FBG and BOTDA sensor designs with respect to their robustness and resistance against crushing during the process of road construction.

In the present study, distributed fiber optic cables were embedded in a pavement in order to measure the short term response of the pavement owing to static loading. Hot Mix Asphalt (HMA) and its construction process represent a harsh environment. Hence, optical fibers protected by a robust cable sheath were applied. The cable type versions V9 (d = 3.2 mm) and V3 (d = 7.2 mm), manufactured by Solifos AG, were used in this project. In both cable versions the optical fiber is guided, in a tight buffered design, within a thin metallic tube surrounded by a plastic cable sheath. Cable version V3 contains additionally a metal wire armoring under the outer plastic cable sheath. A further cable version (T) in loose tube design was embedded for the purpose of a potential study of temperature effects, but not used in the framework of this study. Beside of investigating the potential of distributed fiber optic sensors for quantifying strain distributions in asphalt, the trial application was foreseen to contribute also to gain further knowledge about embedment techniques and provide insights into the performance of these cables. The prior testing of the applied fiber optic cables, as well as the detailed process of cable embedment in this project, is described in Rabaiotti et al. (2017).

In order to gather strain distributions along the fiber optic cables, a distributed measurement technology (SWI), based on Rayleigh backscattering (e.g. Gifford et al. 2007), was applied in the form an OBR 4600 device from Luna Inc. This technology is characterized by high spatial resolution of around 10 mm and sensor lengths of around 70 m in a typical configuration.

2 Site Conditions and Test Procedure

In summer 2015, the construction of a temporary aircraft stand at Zurich Airport was chosen as an opportunity to instrument a pavement in a real construction project with distributed fiber optic sensors. The fiber optic cables were embedded in a loop of slightly more than 140 m length, starting and ending in an access shaft, close to the aircraft stand. Measurements were conducted by connecting the fiber optic cable ends to the measurement device within this access shaft. On the first part of the cable loop, the cables were guided in a protection tube in the subgrade (dashed line in Fig. 1), afterwards the cables were running within the pavement. Close to the planned position of the aircraft main gears, the cables were guided in a way

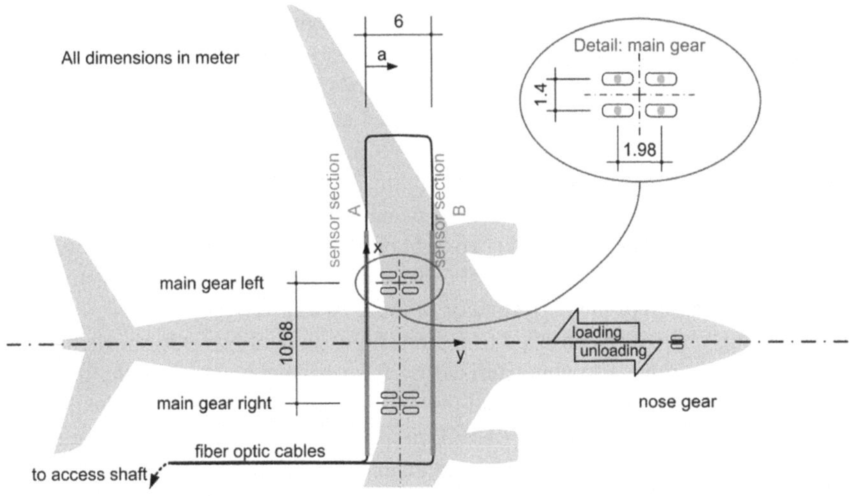

Fig. 1 Test setup

that the cables were running in two sections perpendicular to the axis of the aircraft stand. These two sections have a length of around 29 m and a distance of 6 m to each other. The measurement results discussed in this study origin from two 20 m long cable sections located centrically around the axis of the aircraft stand and are denoted with sensor section A and B (Fig. 1). An A330-300 aircraft was rolled onto the instrumented area in order to load the pavement and induce strains. The main gears of this aircraft type consist of 4 tires (main gear left and right, Fig. 1) and have a distance of 10.68 m to each other.

A cross section through the temporary aircraft stand is shown in Fig. 2a. Two layers of HMA (asphalt concrete) were placed above two layers of a hydraulically

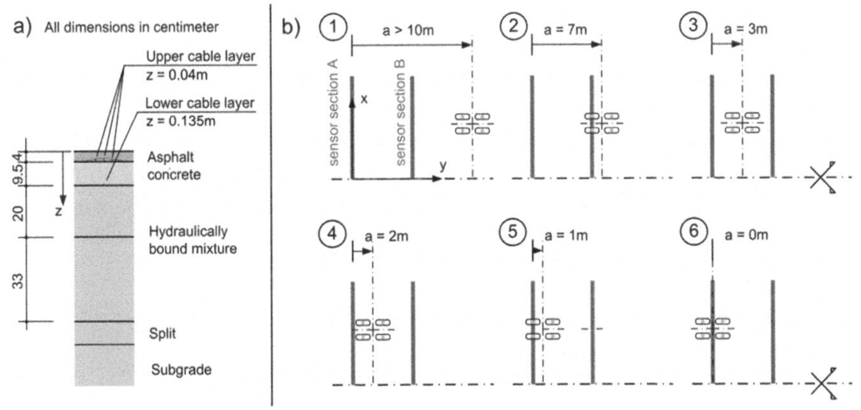

Fig. 2 Cross section through the pavement (**a**) and main gear position during test procedure (**b**)

bound mixture (cement treated material). The fiber optic cables were embedded in two depths in the pavement: an upper and a lower cable layer. The lower cable layer is located at the bottom of the two HMA layers in 13.5 cm depth and consists only of a V3 cable. The upper cable layer is located at the bottom of the upper HMA layer in 4 cm depth and contains a V3, a V9 and a T cable. These cables have a spacing of a few centimeters and are running in parallel along the instrumented area. Whereas the loose tube cable T was installed in order to have a possibility to compensate temperature effects, the two tight buffered cables V3 and V9 allowed for some redundancy in case of a potential cable rupture during pavement construction and for a comparison of the performance of these two cable types.

The testing procedure consisted of 6 aircraft positions. Figure 2b shows the left half of these positions 1–6 characterized by the variable a, which is the distance between the rear sensor section A and the axis of the main gear. When the aircraft was outside the zone of influence ($a > 10$ m, position 1), a reference measurement was carried out along each cable from both cable ends. Afterwards, the aircraft was rolled into position 2 ($a = 7$ m), where the rear tires of the main gear were resting directly above sensor section B and measurements were launched again from both cable ends. This procedure was repeated in the further aircraft positions 3 and 4, until the rear tires of the main gear were resting in position 5 ($a = 1$ m) directly above the sensor section A. Afterwards the aircraft was moved into the final position 6 ($a =$ 0 m), where the axis of the main gear is located directly above the sensor section A, but the tires are beside the cables. Whereas sensor section B was loaded immediately, sensor section A was loaded in several steps, using a decreasing distance a, in order to gather strain distributions not only directly under the tires, but also in the vicinity of the tires of the main gear. The whole test procedure lasted around 40–50 min and the aircraft rested in each position typically around 5 min while measurements were conducted. The usage of an aircraft in order to apply load to an aircraft stand in operation imposes some limits on the conduction of the tests. However, thanks to the courtesy of the airport authority, several measurement campaigns could be conducted on this test site. Owing to technical difficulties, the measurement results gathered from early measurement campaigns were not satisfying. Some first results from a technically sound measurement campaign (using a different type of aircraft and a slightly modified test procedure) are reported in Rabaiotti et al. (2017). However, an imperfect positioning of the aircraft caused presumably an asymmetric strain distribution during that test. The measurement results reported in this study originate from a warm autumn day in 2017 and a cold winter day in 2018. This allows for studying the effect of temperature in two tests, carried out by using the same test procedure, aircraft type as well as comparable aircraft weight.

3 Measurement Results

The following two sections comprise strains measured along sensor section A which were gathered in warm conditions (late autumn afternoon) and in cold conditions (early winter morning). In order to compute these strains, the frequency shift relative to position 1 (2 for section A, V3, $z = 0.04$ m), measured along the cable, was fully converted to strain. Other potential sources, which could have also contributed to the frequency shift (e.g. temperature change), were ignored. The strains were computed with a spatial resolution of 10 mm and an output was generated every 5 mm. A positive sign of the strain denotes expansion, a negative sign stands for contraction.

3.1 Measurements in Warm Conditions

Temperature readings indicated an average air temperature of around 19 °C during the test. The pavement surface was around 21 °C warm. Measurements in the protection tube in the subgrade gave temperatures of around 22 °C during the measurement period. The strains ε_x transversely to the aircraft stand axis, measured along the sensor section A, are shown in Fig. 3. Figure 3a contains the results from the V3 cable in the lower cable layer, Fig. 3b (V3) and 3c (V9) show the results gathered with the cables in the upper cable layer. The axis of the aircraft stand and the aircraft, is located at around $x = 0$ m, the position of the main gear tires are marked in grey. As Fig. 3 shows, the strains measured owing to the loading of the pavement by the aircraft are very small for position 3 ($a = 3$ m) and 4 ($a = 2$ m) compared to the distribution measured while the tires of the aircraft are resting directly above the cables (position 5, $a =$ 1 m). In position 5, the tires of the aircraft are clearly recognizable as zones of large expansive transverse strains ε_x. These zones are accompanied by a zone of contractive strains with a sharp change of the sign of the strain in between. Significant strains are occurring only in an area closely around the main gear: Already in a distance of 1 m aside the tires, the measured strains are becoming relatively small. However, closely around the tires, the strains are varying strongly within short distances. As e.g. the strain distribution around $x = 6$ m in Fig. 3c shows, spatially very high strain gradients are present in the cable below the edges of the tires (>200 $\mu\varepsilon$/cm), pushing the measurement device in this field test towards its limits, resulting locally in difficulties to compute strains in these zones. Still relatively large strains were measured while the aircraft was in position 6, where the center of the main gear but no tire was directly above the sensor section A. However, these strains are regarded to be rather the time dependent part of strains from aircraft position 5 a few minutes before, than to be the strains caused by the current aircraft position 6, as a comparison to the small strains measured during position 4 clearly underlines.

3.2 Measurements in Cold Conditions

During this test, an average air temperature of around −10 °C and pavement surface temperatures of around −9 °C were measured. Within the protection tube, temperatures of around 4 °C were registered. Figure 4 shows again the transverse strain ε_x

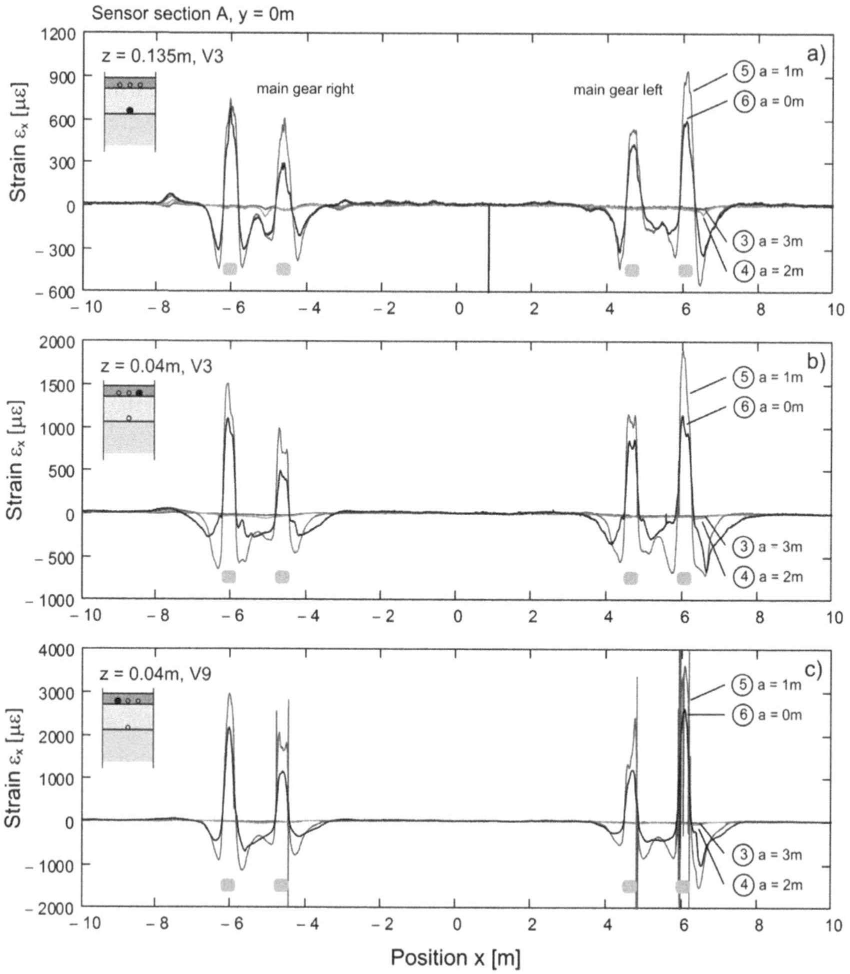

Fig. 3 Measured horizontal transverse strain ε_x in the lower cable layer using V3 (**a**), the upper cable layer using V3 (**b**) and the upper cable layer using V9 (**c**) in warm conditions

measured along sensor section A for different cables and depths. However, possibly owing to the challenging temperature and humidity conditions in the access shaft, the quality of measurements along some cables suffered and Fig. 4 contains only some measurements which could be processed successfully. The strains in the pavement are now clearly smaller compared to those observed in warm conditions. Whereas in the lower cable layer for the aircraft position 5 still some expansive strains were registered directly under the tire, in the upper cable layer the strains remained contractive. On the other hand, next to the main gear, expansive strains become visible.

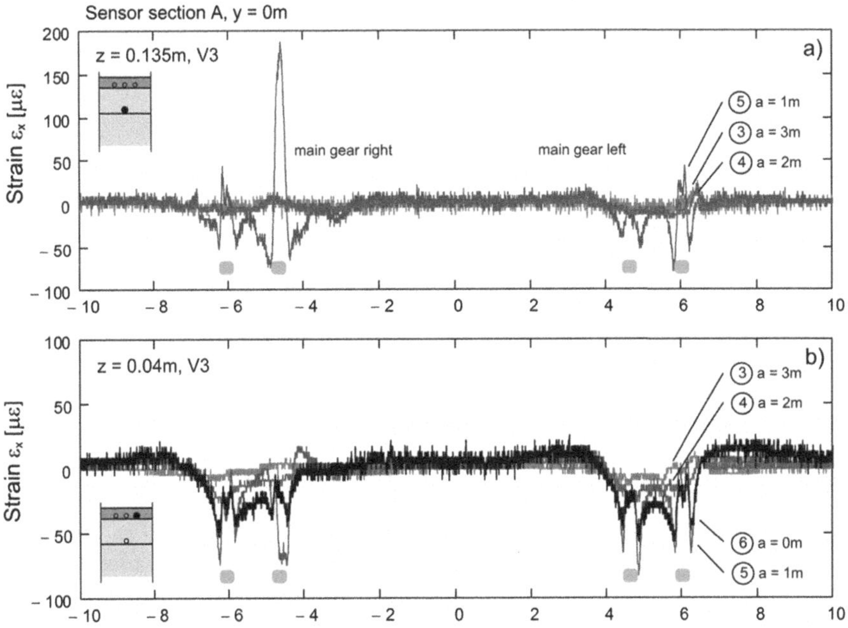

Fig. 4 Measured horizontal transverse strain ε_x in the lower cable layer using V3 (**a**) and the upper cable layer using V3 (**b**) in cold conditions

4 Discussion

The spatially high resolution of the strain data enables for a detailed insight into the strain induced around the main gear as shown in Fig. 5a and b as well as Fig. 6a. These figures are a scale up of the strain distribution around a main gear from Figs. 3 or 4 respectively. Under some tires, the strain distribution gathered with this measurement

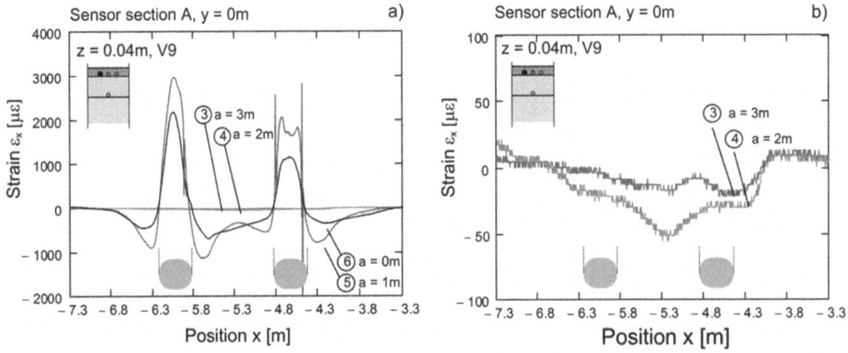

Fig. 5 Detail of strain distribution around main gear right in warm conditions (**a**) and scale up (**b**)

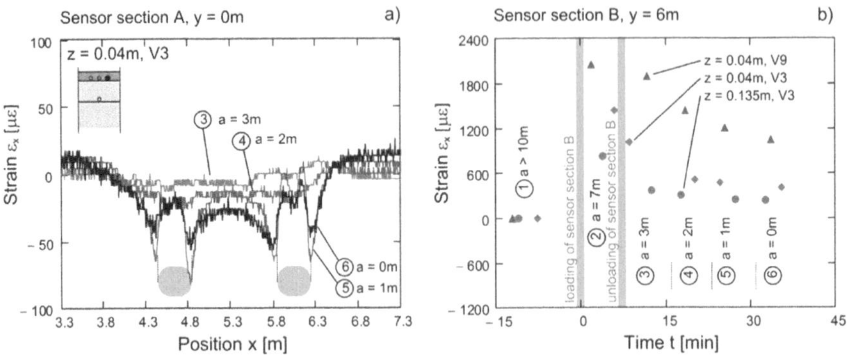

Fig. 6 Detail of strain distribution around the main gear left (**a**) in cold conditions and development of strains with time under a tire in warm conditions

technique and cables indicates two strain maxima in the upper layer instead of one as shown e.g. in Fig. 5a for warm conditions. A detailed consideration (Fig. 5b) of the strains measured around a main gear in warm conditions, while the aircraft rested in position 3 ($a = 3$ m) and 4 ($a = 2$ m), shows that the effect of the approaching aircraft can be detected. However, the strains are rather small and their magnitude differs not much from the strain magnitude observed for the same aircraft positions in cold conditions (Fig. 6a) at the same depth.

The strain distributions reported in Fig. 3 show all a similar pattern. However, regarding their magnitude, considerable differences may be observed in warm conditions, when the aircraft tires are located directly above the cables. For the upper and lower cable layer, different strain magnitudes (and also distribution) seem quite reasonable. However, the different strain magnitude within the upper cable layer, observed for the two types of cables, requires further consideration. A first reason for the observation may be the horizontal cable spacing of several centimeters between these two cables. In addition, contributions to this difference may also originate from the fact that the V3 cable has a higher longitudinal stiffness and consists of additional protective layers in the cable sheath. In case of imperfect bond, this may flatten the measured strain distribution within the core of the cable compared to the strain present in the host material. Finally, the particular test procedure applied for this study may also contribute to these differences: The fact, that the measurements could not be carried out completely simultaneous and therefore have a different time lag to the point in time, when the load was applied, also needs to be considered, owing to the time dependent strain response of the pavement under load. This circumstance is analyzed by means of Figs. 6b, 7a and 7b. Figure 7 shows the strains measured along sensor section B ($z = 0.04$ m, V9 and V3) for different aircraft positions. Since the aircraft already in position 3 only influences the cables in section B to negligible extent, the strain change is fully caused by the recovering pavement. The dependency of the strain on time in the center of the tire ($x = -6.04$ m, dot-dashed line in Fig. 7a and b) is shown in Fig. 6b. It underlines, that different time between load application and measurement conduction may indeed affect the measured strain magnitude, but

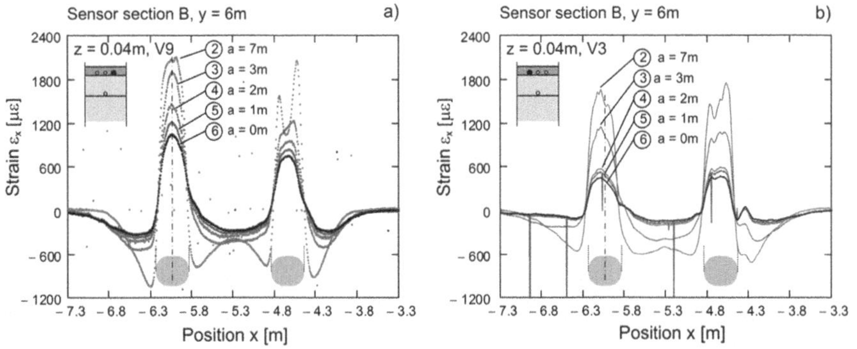

Fig. 7 Process of strain recovery in warm conditions for V9 (**a**) and V3 cable (**b**) in warm conditions (measurement data is shown as points instead of lines in Fig. 7a owing to processing difficulties)

in general the more robust V3 cable version seems not to follow the strain in the host material as closely as the V9 cable, leading to smaller strain magnitudes measured. Additionally it should be mentioned that a certain influence of the embedded cable on the strain field present in asphalt in the vicinity of the cable cannot be excluded within this study.

No attempt was made within this study to separate the potential influence of temperature on the measured frequency shift during the duration of the test. However, as the strain between the two main gears appears to be close to zero, effects of uncompensated temperature are expected to be small. In Fig. 8, strain distributions from aircraft position 5 in sensor section A were integrated to yield approximately horizontal displacement u_x. In the central part, a small average gradient of around 15 $\mu\varepsilon$ can be observed, corresponding approximately to a potentially misinterpreted temperature change of roughly 2 °C, if thermal expansion of the pavement is assumed to be completely restraint.

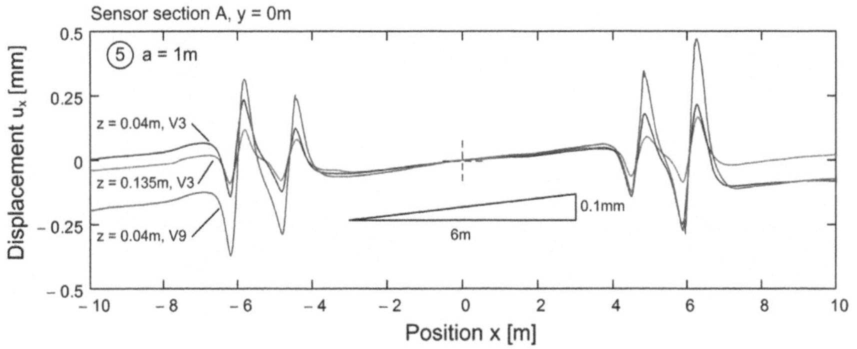

Fig. 8 Integration of strains along different cables for aircraft position 5, sensor section A in warm conditions

5 Conclusions

In the present study, robust fiber optic cables were embedded into a pavement in a real construction project for a temporary aircraft stand. This, in order to explore the potential of this sensor class to measure distributed strains within a pavement subjected to loads. The measurements under the main gears showed a spatially largely varying strain distribution with high strain gradients requiring for a high spatial resolution of the measurement device and cables with adequate strain transfer from the outer cable sheath to the core of the optical fiber. Although a comparison cannot be made straight forward, the different types of cables, applied in this study, show some differences in the measured strain magnitude for the very large strains which were observed directly under the tires of the main gear. The tests were conducted on an aircraft stand but similar measurements could also be carried e.g. in roads. The detailed insight into the behavior of the pavement provided by distributed fiber optic sensors make them in particular interesting for the testing of new pavement materials or the calibration and assessment of new design methods.

Acknowledgements The activity reported in this study is part of the research grant VSS 2014/501 of the Federal Roads Office of Switzerland. The data shown in the present study is part of a research report issued under this grant (Puzrin et al. 2019). The authors would like to thank Stefan Conrad, Hanspeter Moll, Stephan Bruderer and Thomas Hossli of Flughafen Zürich AG for their support and the possibility to carry out this measurement campaign.

References

Bao Y, Chen G (2016) Strain distribution and crack detection in thin unbonded concrete pavement overlays with fully distributed fiber optic sensors. Opt Eng 55: 011008

Chapeleau X, Blanc J, Hornych P, Gautier JL, Carroget J (2017) Assessment of cracks detection in pavement by a distributed fiber optic sensing technology. J Civ Struct Health Monit 7:459–470

Gifford DK, Kreger ST, Sang AK, Soller BJ (2007) Swept-wavelength interferometric interrogation of fiber Rayleigh scatter for distributed sensing applications. Proc SPIE 6770

Liu W, Wang H, Zhou Z, Xing X, Cao D, Jiang Z (2015) Optical fiber-based sensors with flexible encapsulation for pavement behavior monitoring. Struct Control Health Monit 22:301–313

Puzrin AM, Rabaiotti C, Hauswirth D, Fischli F, Tsirantonaki D, Iten M, Facchini M, Friedrich E (2019) Distributed fiber optic strain sensing in pavements. Research report to grant VSS 2014/501

Rabaiotti C, Hauswirth D, Fischli F, Puzrin AM (2017) Structural health monitoring of airfield pavement using distributed fiber-optic sensing. In: SMAR 2017 conference

Xiang P, Wang H (2016) Optical-fibre-based sensors for distributed strain monitoring in pavements. Int J Pavement Eng 19: 842–850

Pavement Distress from Channelized and Lateral Wandering Loads Using Accelerated Pavement Tests

Martin Arraigada and Manfred N. Partl

Abstract Stresses and strains within a road structure are affected not only by the amount and distribution of tire contact pressures as well as speed and rest periods between load passings but also by the lateral position of the wheels within the traffic lane. In this work, the effect of channelized and lateral wandering of 65 kN half axle loads on the performance of a newly constructed asphalt pavement section for low traffic volume has been studied by means of a Mobile Load Simulator MLS10 accelerated pavement loading facility. 250,000 super single wheel passings at 18 km/h were applied in each case and evaluated with visual inspections plus rutting and Falling Weight Deflectometer (FWD) measurements. Pavement response was monitored with strain gauges, accelerometers and thermocouples. The combined analysis of the results showed that channelized loading created longitudinal surface cracks in the pavement, whereas non-channelized loading did not produce any visible sign of cracking or other damage. Rutting in both loading locations showed similar depths, although the non-channelized rutting basin was wider and smoother than in the channelized case. Accelerometer and strain gauge results demonstrated the importance of the load position on the pavement's response.

Keywords Platooning · Channelized loading · Lateral wandering · Accelerated pavement testing · MLS10

1 Introduction

Autonomous driving technologies and specifically the use of platoon formations for trucks will significantly change the loading regime of the road infrastructure in near future. As a result, axle loads will be applied in a strongly channelized way with minimal lateral wandering and minimal rest periods for asphalt pavements to recover. In particular, this means that the position of the wheels relative to the pavement

M. Arraigada (✉) · M. N. Partl
Road Engineering/Sealing Components, Empa, Swiss Federal Laboratories for Material Science and Technology, Ueberlandstr. 129, 8600 Duebendorf, Switzerland
e-mail: martin.arraigada@empa.ch

C. Raab (ed.), *Proceedings of the 9th International Conference on Maintenance and Rehabilitation of Pavements—Mairepav9*, Lecture Notes in Civil Engineering 76,
https://doi.org/10.1007/978-3-030-48679-2_78

geometry will change dramatically, thus influencing stress-strain distribution and performance of the structure.

Since existing pavements were designed based on current traffic data, this kind of traffic change may require very costly strengthening or rehabilitation measures for maintaining the current level of safety and operability and for avoiding premature pavement failure. Moreover, the increasing risk of traffic jams on overloaded heavy duty roads may promote electronic alternative routing of trucks along low volume roads, thus increasing the risk of damaging the secondary road network through channelized traffic.

The influence of the wheel positions in the lane has already been studied extensively in the past, demonstrating its high impact on the durability of asphalt pavement materials. As an example, Blab and Litzka (1995) showed that fatigue failure of a pavement layer under widely spread loads will happen much later than when concentrating traffic loads in a single wheel path. Other examples of studies in that direction were reported by Villiers et al. (2005), Sirin et al. (2007), Teng et al. (2008), Bayat et al. (2010), Wu et al. (2012). Moreover, numerous design tools worldwide show that the position of the loads depends on many factors, such road geometry, vehicle speed, alignment etc.

In view of the existing research activities and the available design tools for meeting the impending future changes in the traffic loading regime, it is remarkable that direct experimental investigations on the effect of channelized and lateral wandering wheel loads on low volume roads appear comparatively scarce. In particular, the benefit of accelerated pavement test (APT) facilities for studying the different failure mechanisms by these two loading regimes in a well-defined reproducible way has not been sufficiently exploited so far (Hugo et al. 2012). This was the motivation for the Swiss research initiative sponsored by the Federal Roads Office (FEDRO). The research was performed with the Mobile Load Simulator MLS10 on a newly constructed asphalt pavement section for low traffic volume, as presented below (Arraigada et al. 2014).

2 Objectives and Experimental Approach

In order to compare the performance of the same pavement in an APT under channelized and lateral wandering loads, a new low traffic volume pavement was constructed on a special testing compound 28 × 36 m in the moderate climate region of the Swiss plateau. According to Swiss Standards, the pavement corresponded to a pavement with the designation T2-S2. This means that it was designed for a traffic load class T2 for sustaining up to 100 daily 8.6 t equivalent single axle loads during 20 years and to be built on a S2 subgrade with medium bearing capacity, defined as subgrade with a static deformation modulus E_{VI} from static plate loading at least between 11.25 and 22.5 MN/m^2.

The structure of the pavement consisted of a 60 cm thick frost-resisting unbound granular layer 0/45 mm and two asphalt concrete layers made of mixture types (N) for medium traffic. Specifically, the asphalt pavement was composed of a 30 mm thick bearing course ACT 11 N with 11 mm MAS (maximum aggregate size), topped by a 2 cm thick surface course AC 8N with 8 mm MAS. The measured binder content and maximum density of ACT 11N was 5.6% (pen 70/100) and 2.46 Mg/m^3 and for AC 8N 5.5% (pen 50/70) and 2.48 Mg/m^3. Before placing the unbound granular layer, the bearing capacity of the subgrade was verified. From light weight deflectometer LFG at 6 locations, the dynamic deformation modulus E_{vd} (in MN/m^2) was determined. Using the empirical formula proposed by Bodmer et al. (2014)

$$E_{V1} = 1.31\ E_{vd} - 15\ \mathrm{MN/m^2} \tag{1}$$

the static deformation modulus $E_{V1} = 16.42$ MN/m^2 was estimated and verified to lie within the range of an S2 medium bearing capacity.

This pavement was loaded in two positions with the MLS10 shown in Fig. 1 over a loading length of 4.2 m using 10 bar super single tires 455/40 R22.5. Each loading campaign consisted of 250,000 channelized and lateral wandering load passings at 18 km/h. Lateral wandering was performed by means of two sliding rails supporting the MLS 10 in a way that it could be slided slowly laterally with a motor in 14 steps of 21.4 mm each over a width of ±300 mm, following a normal wheel passing frequency distribution. Note that in case of lateral wandering, the central 4 cm of the wheel path were loaded 217,757 times. Both loading campaigns were performed under comparable climatic conditions in the summer period of two consecutive years, with similar temperatures during the experiments.

In addition to thermocouples, strain gauges and accelerometers were positioned as shown in Fig. 2. Based on earlier experience, longitudinal and lateral strain gauges for channelized loading were only placed on the pavement surface close to the wheel path. For non-channelized loading, additional strain gauges ware placed between the asphalt layers and in the interface between the unbound and ACT 11N layer. Longitudinal and transversal strain gauges for lateral wandering were placed at two positions within and at one position outside of the wheel path.

The location of the five accelerometer was selected such that the deformation basin in the transversal direction could be calculated and a comparison with Falling Weight Deflectometer (FWD) measurements could be made. FWD was used just

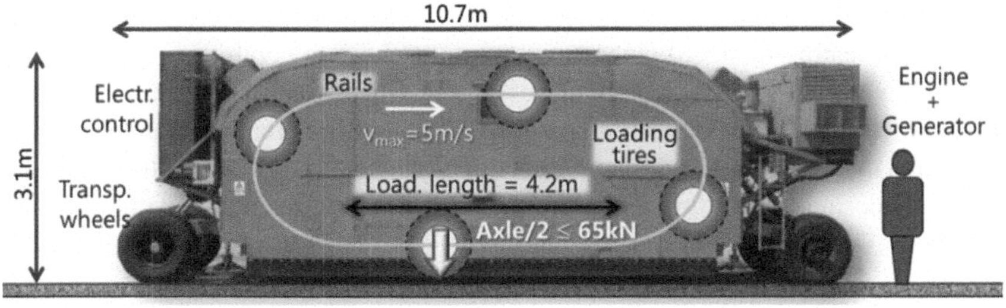

Fig. 1 View of the traffic load simulator MLS10

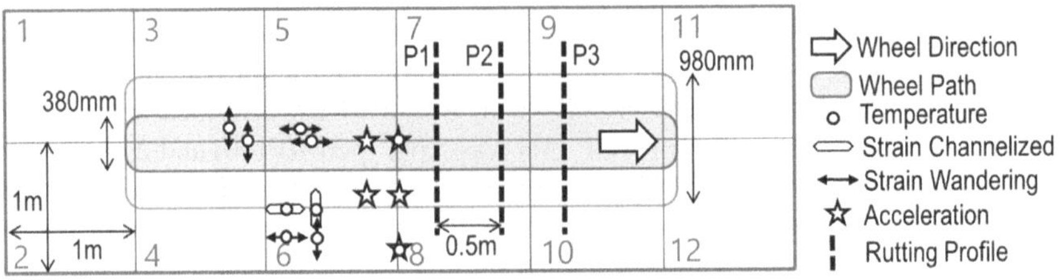

Fig. 2 View of instrumentation and measurement locations

before and after applying 250,000 wheel passings at 45 locations, following the square grid pattern in Fig. 2. In this way, five FWD measurements were taken in the middle of the wheel track and the others in 1 and 0.5 m distance in the longitudinal and transversal direction respectively. The applied FWD load was 40 kN.

Moreover, three 1.4 m long rutting profiles (P1, P2, P3) in 0.5 m distance were periodically determined with an automatic profilometer which allows registering minimal asperities of 1 mm depth. It consists of an aluminum guiding beam which serves as reference horizon for a measurement wheel that rolls slowly along the pavement surface while tracking the vertical movements.

In addition, visual inspection of crack formation was performed periodically. After testing, cores were taken and analyzed in the laboratory with respect to traffic-induced after-compaction and change in interlayer shear resistance between the asphalt layers. Interlayer shear resistance was determined at room temperature with the Linear Parallel Direct Shear (LPDS) device (Raab and Partl 2009) which is an improved version of the so-called Leutner device. Further details regarding data acquisition and performance evaluation are described in the report by Arraigada et al. (2014).

3 Results and Discussion

3.1 Temperatures

The mean temperatures in 3.5 mm depth of the pavements during testing were 23.8 °C for the channelized and 23.8 °C for the lateral wandering case. However, in case of channelized testing the local mean temperature in 3.5 mm depth directly under the wheel track was 4.7 °C higher than beside the wheel track. This was attributed to friction within the pavement, which was sufficient for creating significant heat dissipation under channelized loading but not high enough for being effective under lateral wandering. Accordingly, maximal and minimal temperatures were 37.6 and 13.5 °C for the channelized and 33.6 and 14.1 °C for the non-channelized case.

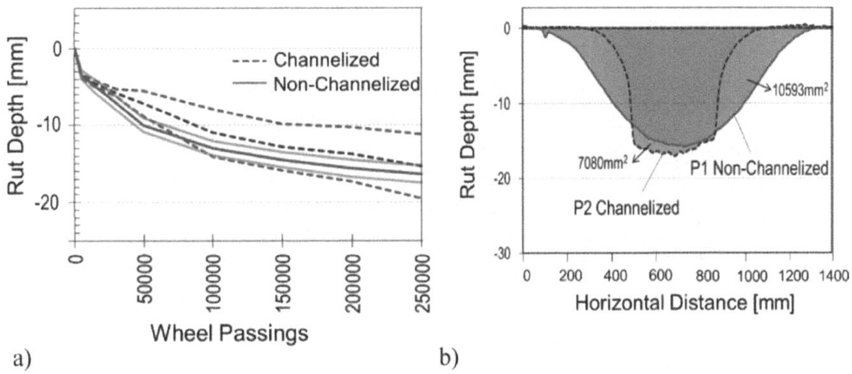

Fig. 3 **a** Evolution of rut depth in the three rutting profiles during both loading campaigns; **b** Channelized and non-channelized rut profiles P2 and P1 after 250,000 wheel passings

3.2 *Rutting*

Since the pavement was designed for low traffic, the maximum rut depth produced by the MLS10 were comparatively high as shown in Fig. 3, lying in the order of magnitude of 15 mm. Obviously, the maximum rut depths produced by both loading campaigns are comparable, with high scatter in case of channelized loading, covering a range between 17 mm in the inner profile (Profile 1) and 13 mm in the outer profile measured in sectors 9/10. The reason for this high scatter could be layer inhomogeneities below the asphalt pavement. These may have higher impact on the performance of the comparatively thin asphalt layers under concentrated channelized loading.

Figure 3b depicts the measured profiles P1 after non-channelized and P2 after channelized wheel loadings. In spite of the fact that the central part of wheel path has suffered less passings during the lateral wandering regime, the maximal rut depths after both campaigns are quite similar whereas the shape of the profiles are quite different. A reason may be the comparatively thin asphalt pavement which creates a high load transfer towards the lower unbound layer. Hence, for these low traffic volume structures, it can be concluded that, with respect to maximum rut depth requirement, channelized traffic appears to create the same endurance limit as non-channelized traffic. In view of the different profiles, the question remains, of course, if the maximum rut depths alone is a sufficiently satisfying performance criterion.

3.3 *Surface Cracks*

Periodical visual inspection revealed the formation of longitudinal surface cracks at the edge of the channelized wheel path after 150,000 wheel passings (Fig. 4). Those cracks provide openings for rain water infiltration which might be enhanced by the

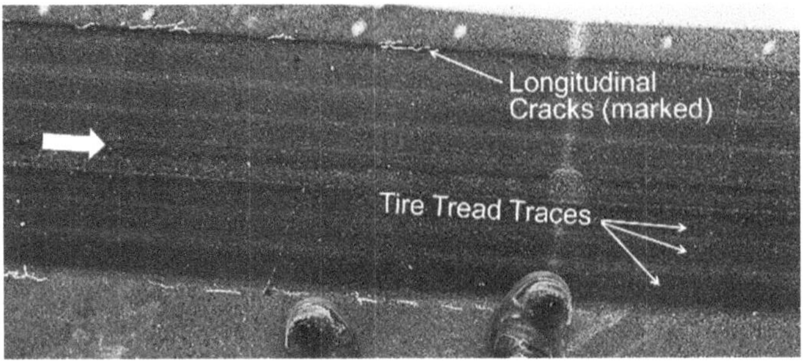

Fig. 4 Surface view of the channelized pavement showing longitudinal cracks

traffic induced pumping effects. On the other hand no visible cracks were detected during the lateral wandering regime until the end of the whole campaign. It appears that this difference in cracking behavior is a result of high shear stresses that are repeatedly imposed by the uneven tire contact stress distribution at the edge of the tire, as discussed for example by Morgan et al. (2008). In case of lateral wandering, these shear stresses induce lateral kneading of the asphalt which mitigates the risk of cracks particularly in summer time. Hence, in terms of longitudinal crack formation on low traffic roads, channelized wheel loading was found to reduce service life of low traffic roads under the investigated conditions considerably, in the order of more than 40%.

3.4 Dynamic Strains

Examples for the strain measurements in the channelized case are shown in Fig. 5. The figure shows that the strains on the pavement surface close and in the direction perpendicular to the wheel track are clearly dominated by tension during trafficking whereas compression is almost negligible, thus creating early longitudinal cracks.

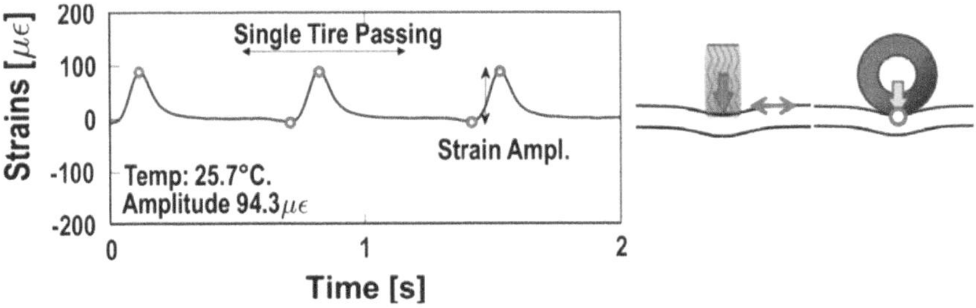

Fig. 5 Surface strains perpendicular to channelized trafficking at the beginning of the test

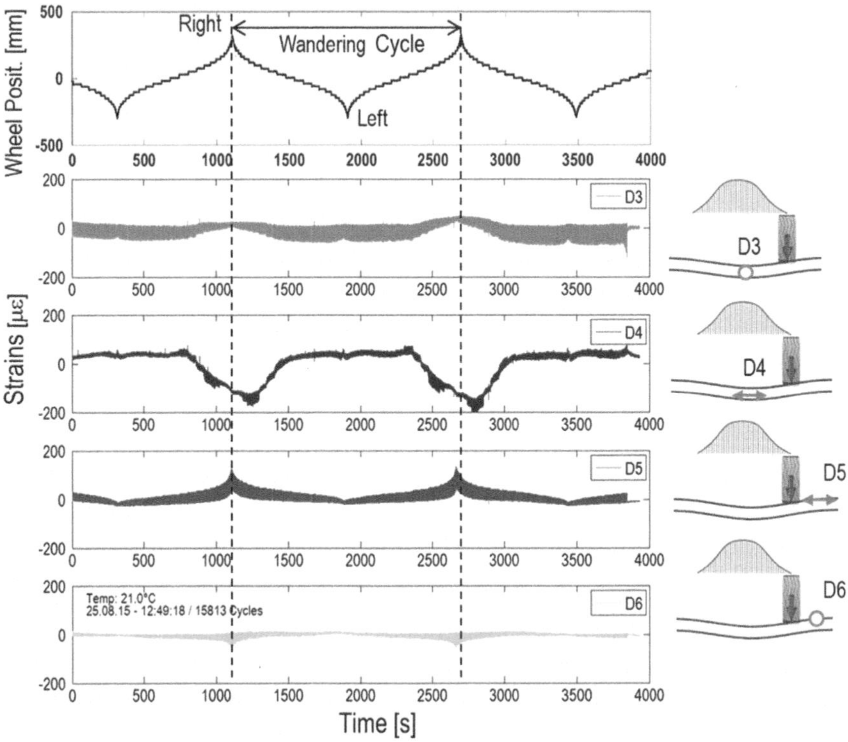

Fig. 6 Strains during non-channelized trafficking at the beginning of the test; D3, D4 strain gauges between asphalt layers, D5, D6 surface strain gauges

Stains between the asphalt layers and on the pavement surface at the beginning of non-channelized trafficking are displayed in Fig. 6 over a time period of 4000 s. Note that one complete wandering cycle takes 1560 s. Strains are much influenced by the transversal position of the wheel, in particular in case of the strains perpendicular to the wheel passings recorded by the gauge D4 between the asphalt layers. Maximum compression occurs when the wheel is on top of the sensor, which means that D4 and, hence, the interface between the asphalt layers, was located above the neutral axis of the asphalt pavement when maximum bending occurred. The signal of D4 is not ideally symmetrical, probably due to imperfections during installations. The strain sensor D3 is measuring the strains in the trafficking direction, showing highest compressive strains just when rolled over by the wheel. Note the slight tension strain maxima recorded by both sensors when the wheel is trafficking farthest away from the wheel track axis. Highest strains are recorded by D5 on the pavement surface when the wheel is closest to the location of this perpendicular sensor. In case of the longitudinal surface sensor D6, sequentially alternating tension and compression strains are observed where the compression strains appear more dominant. In case of lateral wandering compression strains certainly contribute to mitigate the effect of temporarily induced tensions strains and therefore the risk of fatigue crack failure.

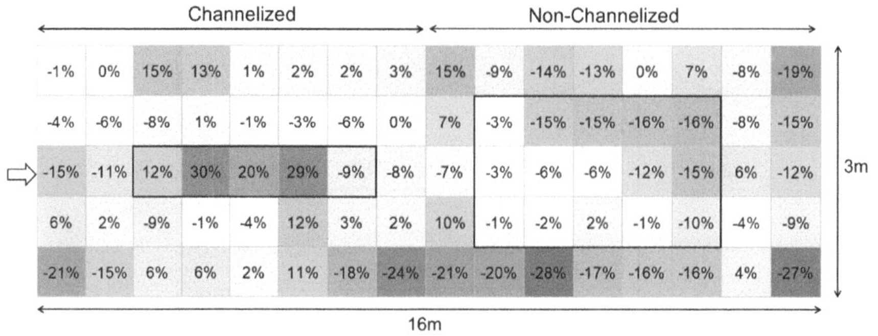

Fig. 7 Relative difference between the deflections before and after 250,000 wheel passings; positive values means deflection increase, negative values means deflection decrease (Color figure online)

3.5 Deflection Behavior

Relative change in the deflections from FWD measurements before and after applying 250,000 wheel passings are presented in Fig. 7. Red colored are zones where deflection increases were measured (positive values), whereas blue are zones with deflection decreases (negative values). Comparing the color patterns reveals that deflections clearly increased in the channelized track of the pavement whereas the deflections in the lateral wandering track changed only slightly. In fact, the deflections created by the channelized loading campaign were up to 30% higher than at the beginning of the APT campaign indicating significant damage within the asphalt layers. This difference means that the channelized loading created higher material fatigue in the pavement than the non-channelized. In the zone with lateral wandering, even locations with a reduction of the deflection could be found. However, it was not possible to attribute this reduction clearly to stiffening from to post compaction.

3.6 Properties of Pavement Cores After Trafficking

After trafficking, 100 and 150 mm cores were taken within the wheel tracks for mixture analysis of the different asphalt layers, and, in particular, for investigating after-compaction and change of interlayer shear resistance between the asphalt layers due to the different traffic regimes. From the channelized and non-channelized section respectively, 8 and 9 cores were taken. As shown in Fig. 8a, the mean bulk densities of the cores were slightly higher for the channelized than the non-channelized section due to the local after-compaction which was higher in the channelized wheel loading regime (note, that the y-axis starts at 2.2 Mg/m^3). Moreover, in this case, the scatter was higher, which corroborates the rutting tests. In addition, no significant difference between the bulk densities on the right and left side of the wheel track was visible. However, in case of lateral wandering, the bulk density in the left outer

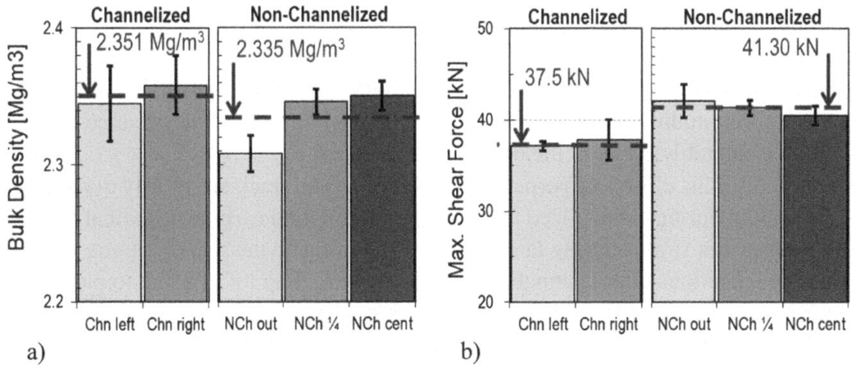

Fig. 8 **a** Bulk density of 150 mm cores taken after channelized (right and left position of wheel path) and non-channelized loading (outer left edge, ¼ wheel path towards and in center of wheel path); **b** Max. interlayer shear failure force at room temperature

zone of the wheel path was significantly lower than half way and within the center of non-channelized wheel paths (i.e. Chn out compared with Chn 1/4 and Chn center) confirming the occurrence of after-compaction effects from trafficking.

Results from quasistatic direct interlayer shear failure tests at room temperature between surface and base course are depicted in Fig. 8b (note that the y-axis starts at 20 kN). Obviously the interlayer shear failure force for the channelized loading campaign was significantly lower than for lateral wandering. This can be attributed to the higher local shear within the materials during channelized loading which also results in a weakening of the interlayer bond, confirming results reported by Partl et al. (2015). In case of lateral wandering, a slight reduction of the shear resistance between the outer part and the center of the wheel track in the order of 4% appears to occur, indicating that also with this loading regime the interlayer bond of low traffic pavements risks to be reduced.

4 Conclusions

This study addresses the effects of channelized traffic loading and lateral wandering on mechanical durability of a newly placed low traffic volume pavement. It was motivated by the expected future change in traffic regimes due to autonomous driving technologies and platooning of trucks. Tests, demonstrating the value of APT, lead to the following conclusions and findings:

- Channelized loading may lead to higher temperatures under the wheel track than lateral wandering due to higher friction within the pavement. In this study, the local mean temperature in 3.5 mm depth directly under the wheel track was 4.7 °C higher.

- Channelized loading produced high scatter in rutting, probably due to local inhomogeneities below the thin asphalt pavement. Both traffic regimes appear to create the same maximum rut depth in spite of significantly different rutting profiles.
- Due to longitudinal crack formation, channelized trafficking may reduce service life considerably; here, in the order of more than 40%.
- Surface strains close and perpendicular to the wheel track are clearly dominated by tension during channelized trafficking, thus creating early longitudinal cracks. Compression strains during lateral wandering mitigate the risk of fatigue failure.
- FWD deflections after channelized testing increased up to 30% due to pavement whereas non-channelized testing showed almost no increase.
- Cores from the wheel tracks after testing revealed high after-compaction during channelized loading, producing also high scatter in the results.
- Channelized loading resulted in clearly lower interlayer shear failure resistance between both asphalt layers.
- Finally it can be concluded that in case of low traffic volume roads channelized loading may reduce durability of pavements considerably creating special stress-strain situations relevant for future pavement design. As for planning rehabilitation measures, this means that construction related temporary traffic regimes must reduce traffic channelization to a minimum and avoid too narrow temporary traffic lanes.

References

Arraigada M, Treuholz A, Partl MN (2014) Influence of Channelized and Non-Channelized Loading with the Full-Scale Load Simulator MLS10 on the Durability of a T2 Standard Pavement. DETEC-FEDRO Federal Roads Office. Report 1615. (in German)

Bayat A, Knight M, Adedapo A (2010) Flexible pavement response under dynamic wheel loads – a CPATT full – scale instrumented test road study. In: Annual conference of the canadian society for civil engineering, vol 2, pp 1264–1273

Blab R, Litzka J (1995) Measurements of the lateral distribution of heavy vehicles and its effects on the design of road pavements. In: Proceedings of international symposium on heavy vehicle weights and dimensions, road transport technology. University of Michigan, pp 389–395

Bodmer P, Byland HH, de Witte H (2014) Use of a light weight deflectometer LFG for compaction control of unbound subbase layers. DETEC-FEDRO Federal Roads Office. Report 1459. (in German)

Hugo F, Arraigada M, Shu-Ming L, Zefeng T, Kim YR (2012) International case studies in support of successful applications of accelerated pavement testing. In: Proceedings of 4th international conference on acceleration pavement testing, Davis, pp 93–104. ISBN 978-0-415-62138-0

Morgan GCJ, Poulikakos LD, Arraigada G, Muff R, Partl MN (2008) In situ monitoring of pavement stresses on the A1 in Switzerland. J. Test. Eval. 36(4):291–299

Partl MN, Raab C, Arraigada M (2015) Innovative asphalt research using accelerated pavement testing. J Marine Sci Technol JMST **23/3**, 269–280

Raab C, Partl MN (2009) Interlayer bonding of binder, base and subbase layers of asphalt pavements: long-term performance. Constr. Build. Mat. 23(8):2926–2931

Sirin O, Tia M, Roque R, Choubane B (2007) Evaluation of performance characteristics of the heavy vehicle simulator in Florida. Build. Environ. 42:1270–1277

Teng X, Li X, Chou K (2008) Application of finite element analysis to access the rutting potential in asphalt pavements. In: Proceedings of 1st International Symposium on Transportation & Development, vol 319, pp 480–485

Villiers C, Roque R, Dietrich B (2005) Interpretation of transverse profiles to determine the source of rutting within an asphalt pavement system. Transp Res Rec 1905:73–81

Wu J, Ye F, Ling J, Qian J, Li S (2012) Rutting resistance of asphalt pavements with fine sand subgrade under full – scale trafficking at high and ambient air temperature. In: international conference on acceleration pavement testing, Davis, pp 265–276. ISBN 978-0-415-62138-0

Correlating Air Freezing Index and Frost Penetration Depth—A Case Study for Sweden

Sigurður Erlingsson and Denis Saliko

Abstract The determination of frost penetration is one of the main requirements in considering environmental effects in pavement design in cold regions. At the present time, the frost depth of pavements in Sweden is estimated computationally using computer software which approximates the heat equation by finite difference. Due to the geographical positioning of Sweden, a wide range of air freezing index and frost penetration depths were observed with lower values in the south and higher values in the north. This paper introduces a simplified design chart which is obtained by empirically correlating the air freezing index estimated from temperature measurements by 44 local meteorological stations to the maximum frost penetration depth obtained by 49 RWIS Road Weather Information Station data. The results are classified depending on their location and the climatic zones defined by the Swedish pavement design codes. Nonlinear prediction intervals are implemented to provide a range of possible frost penetration depths since local site conditions are not taken into account. Further research is required to consider local on-site effects such as frost susceptibility of pavement materials, the thermal conductivity of layers, access to water and snow covering.

Keywords Air temperature · Air freezing index · Frost penetration depth · Sweden case study

S. Erlingsson (✉) · D. Saliko (✉)
VTI the Swedish National Road and Transport Research Institute Science and Technology, Olaus Magnusväg 35, 58195 Linköping, Sweden
e-mail: sigurdur.erlingsson@vti.se

D. Saliko
e-mail: denis.saliko@vti.se

S. Erlingsson
Faculty of Civil and Environmental Engineering, University of Iceland, Hjardahagi 4-6, 107, Reykjavík, Iceland

C. Raab (ed.), *Proceedings of the 9th International Conference on Maintenance and Rehabilitation of Pavements—Mairepav9*, Lecture Notes in Civil Engineering 76,
https://doi.org/10.1007/978-3-030-48679-2_79

1 Introduction

Climate is one of the main factors influencing the performance of pavement materials with factors such as temperature, moisture, and freeze-thaw cycles known to affect the behaviour of all pavement layers (Doré and Zubeck 2009). At high temperatures, the asphalt bound surface layers become soft and viscous while with decreasing temperature, the stiffness of asphalt increases and the layers become hard and brittle (Yoder and Witczak 1975). Moisture content variations influence the behaviour of layers made of unbound granular materials. With increasing moisture content, a decrease in granular layer stiffness's is generally observed (Erlingsson et al. 2017). In the winter time frost penetrates into the pavement from the surface and penetrate downwards. Above the freezing front a zone of partially frozen material is created where both free water and ice coexist. As the portion of free water is decreasing, as larger part of the water is converted to ice, it acts like a membrane in tension and is able to cause capillary rise of underneath located water to the freezing front and therefore increase the moisture content in the material (Fredlund and Rahardjo 1993). As the material freeze it causes it's water to expand and can therefore push the aggregates away from each other. This can cause frost heave on the pavement surface and can further be observed as stiffness reduction in the unbound aggregate layers during thawing. The frost penetration depth depends mainly on the magnitude of temperatures, duration of sub-zero temperatures and the type of layers present in the pavement. The amount of heave is dependent not only on the above mentioned factors but also on the available moisture in the system and its capability to travel to the freezing front (Doré and Zubeck 2009; Hermansson and Guthrie 2005).

The freezing of pavement layers can be problematic particularly in cold regions leading to distresses and premature failure. Typical distress modes resulting from climatic factors in cold regions include transversal cracking, roughness, fatigue cracking due to frost, depressions, frost heave and thaw weakening (Doré and Zubeck 2009; Huang 2012). When the frozen mass in the cross-section is thawing, reduced stiffness of the pavement layers is observed. The reduced stiffness combined with heavy traffic loading can lead to premature failure and cracking of the asphalt bound layers (Salour and Erlingsson 2013; Simonsen and Isacsson 1999). An accurate prediction of the amount of frost penetration and the beginning of the freezing and thawing period would result in easier and more accurate decisions related to load reductions during spring thaw, a common practice in several countries.

Frost protection layers are typically added in pavements subjected to cold climate. Depending on country-specific regulations, the thickness of the frost protection layer calculations can be based on freezing index calculations, frost depth calculations or frost heave calculations (Vaitkus et al. 2016). In Sweden, the design of the frost protection layer is based on the fulfilment of frost heave requirements (Trafikverket 2011). The finite difference-based software, PMS Objekt, is used to numerically solve heat and mass transfer equations as well as predicting the frost heave and thereafter compared to the allowed values (Trafikverket 2011).

Several methods are used in various countries for the prediction of the frost penetration depth. Frost penetration depth models can be empirical, mechanistic-empirical or fully mechanistic. Mechanistic models which are typically based on computational solutions of heat and mass transfer equations possess the highest accuracy. However, due to the computational requirements they are less practical compared to other methods. Mechanistic-empirical models typically use the air freezing index and the soil properties such as thermal conductivity and latent heat of fusion (Doré and Zubeck 2009; Kersten 1959; Saarelainen 1996). Empirical models attempt to correlate the freezing index to the frost penetration depth and considering the remaining influencing factors by constants determined by statistical regressions.

The Swedish Transport Administration has operated a Road Weather Information Stations (RWIS) for more than two decades. Among other things, the system monitors the sub-surface temperatures from frost rods installed in the pavement structure down to 2 m depth. The frost rods are installed in predrilled holes and measures every half an hour the temperature with 5 cm intervals down to 2 m depth thus giving the development of frost penetration and thawing with time. The air temperature is monitored by the Swedish Metrological and Hydrological Institute (SMHI) throughout the country. This paper describes a statistical nonlinear regression model to correlate the maximum frost depth observation from the frost rods with the freezing index based on the air temperature observations.

2 Methodology and Instrumentation

The main factor which triggers temperature changes and/or freezing in pavements is the atmospheric temperature to which the structure is subjected. Air temperature is necessary to determine the surface temperature which is used as a boundary condition in any modelling. In regions subjected to cold weather, the freezing index *FI* is a commonly used method to indicate how severe a winter is. The freezing index is estimated by summing up the mean daily air temperatures (*MDAT*) for the days at which the average temperature is negative (Doré and Zubeck 2009) (Fig. 1).

$$FI = \begin{cases} \sum_{0}^{t} -MDAT & \text{if } MDAT \leq 0 \\ 0 & \text{if } MDAT > 0 \end{cases} \tag{1}$$

Additionally, the air freezing index can be estimated graphically by plotting the summed values of the mean daily temperature and calculating the difference between the upper and lower peaks of the curve (Andersland and Ladanyi 2004). It is possible to compute the freezing index mathematically as shown in Eq. 1 or graphically from the distance between the curve extrema as shown in Fig. 2.

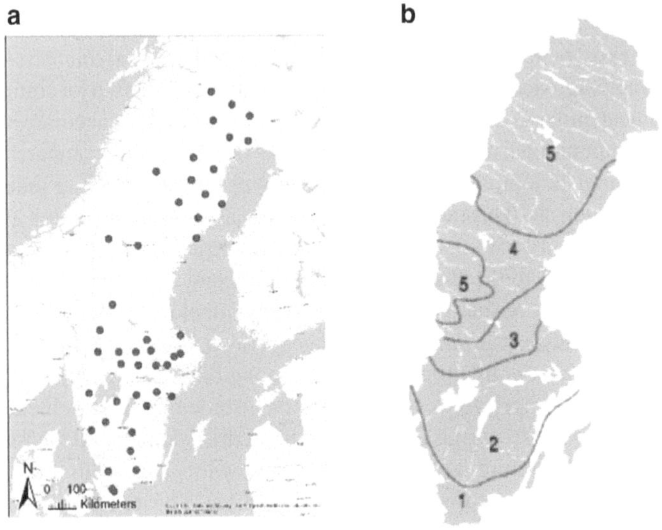

Fig. 1 Distribution of meteorological and RWIS stations (**a**) in Sweden and climatic zones (**b**)

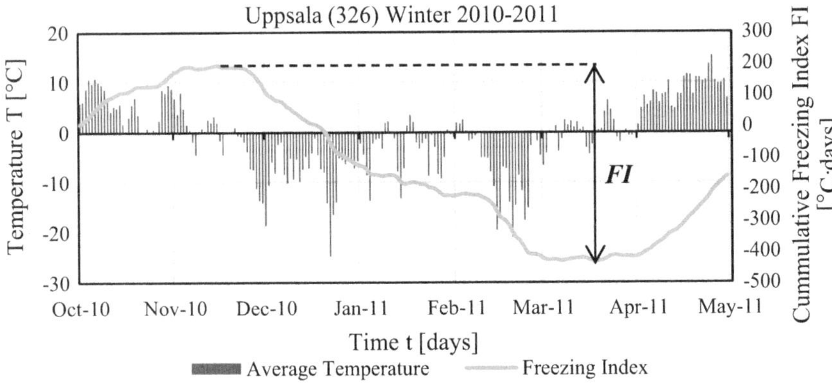

Fig. 2 Sample temperature and freezing index plot for a full winter

Air temperature registrations from 44 meteorological stations throughout Sweden were obtained from SMHI and were used to determine the air freezing index FI. The registrations covered a timespan of 10 years from 2007 to 2017 with 30 min recording intervals.

Pavement sub-surface temperature data from 49 separate frost rods from the Swedish RWIS were further provided from the Swedish Transport Administration. The average distance between the meteorological stations and RWIS was estimated to be around 6 km based on the location coordinates for each station. The positions of the stations are shown side by side in the maps below in Fig. 1.

According to the Swedish manual of technical requirements for road constructions called TVRK Väg TRV 2011:072, Sweden is divided into 5 climatic zones, with zone 1 corresponding to a milder climate and zone 5 to the regions which are more subjected to cold climate (Trafikverket 2011). The same classification system has been used in this paper to easily categorize the data and observe the differences in values between meteorological stations and RWIS located in different climatic zones.

Temperature data from RWIS were recorded using Tjälstav 2004 developed at the Swedish National Road and Transport Research Institute (VTI). It consists of 41 temperature sensors placed along a 200 cm long rod spaced with 5 cm intervals. The data is logged every 30 min and transferred remotely to a computer. Based on the sub-surface pavement temperatures from RWIS, it is possible to observe temperature changes with depth and time. The frost penetration depth also known as depth to the freezing front is obtained by interpolating for the 0 °C value between two adjacent sensors, one recording positive temperature and the other recording negative temperature.

3 Data Processing and Analysis

The recorded air temperature data from meteorological stations was plotted for complete winter periods from the beginning of October to the end of May over a 10 years long period for 44 separate meteorological stations. The period from October to May is selected to capture the winter throughout the whole country since the winter period duration varies throughout the country due to the geographical positioning with shorter winters in the southern part and longer winters in the northern part.

The air freezing index was computed and plotted over the same graph to observe its variation with changing air temperature. A typical temperature—freezing index plot for a full winter is shown on Fig. 2 for station number 326 Uppsala located in climatic zone 2. The slope of the freezing index curve changes by alternations between positive and negative air temperatures.

TVRK Väg's five climatic zones system was used to classify the pavement temperature data obtained from RWIS stations. The data was plotted into contour plots for a total of 49 stations as shown in Fig. 3 for the sample stations 1042 Putsered located in climatic zone 1, 1014 Kindjsön located in climate zone 4 and 1061 Svappavaara located in climatic zone 5 for the period 1 October 2016 to 31 May 2017.

As shown in Fig. 3, it is possible to observe the time when the freezing of the pavement starts, the rate of frost propagation, the beginning of the thawing period as well as the time when the pavement is completely unfrozen. This holds true for pavements located in climatic zones 1 to 4. Typically, lower pavement temperatures and longer thawing times have been observed for RWIS stations located to the northern zones 4 and 5.

In the case of pavements located in climatic zone 5, the frost penetration depth is larger than 200 cm. At the present time, it is not possible to measure the exact frost penetration depth for all pavements located in zone 5 due to instrumentation

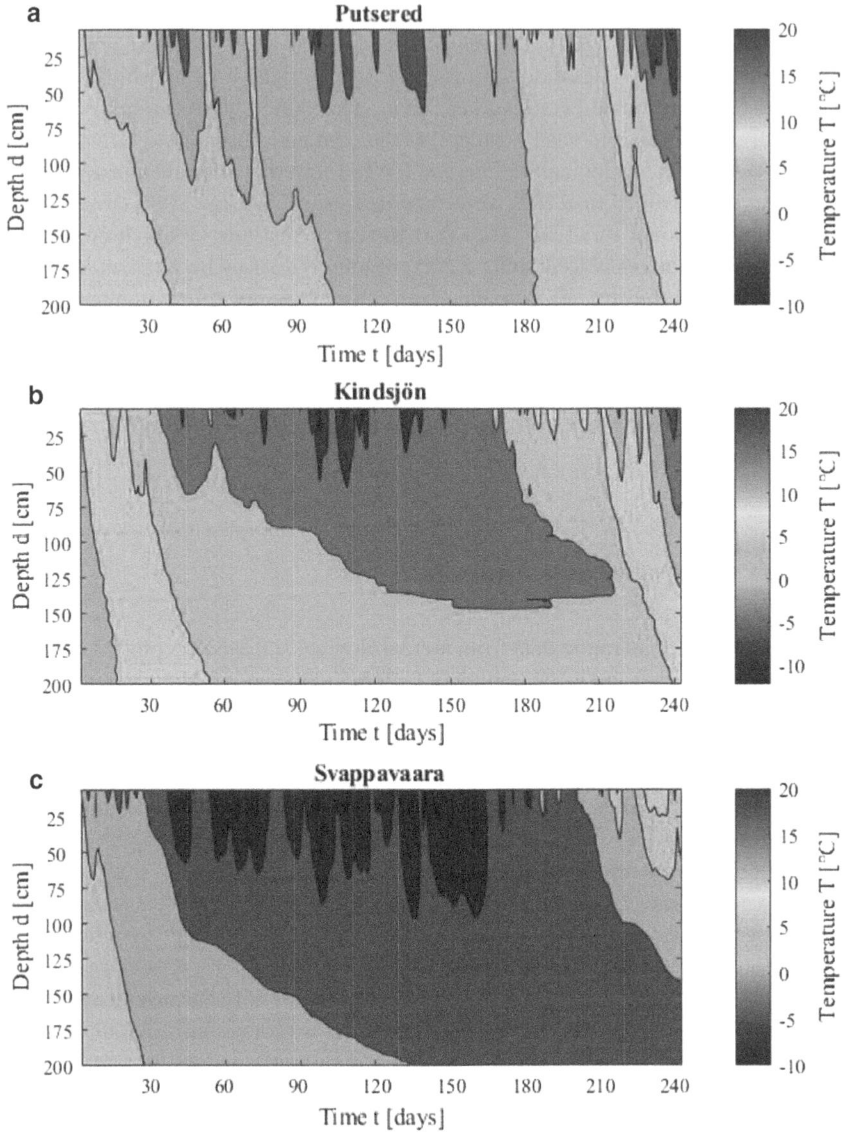

Fig. 3 Pavement temperature measurements from RWIS for **a** Putsered, **b** Kindsjön, and **c** Svappavaara

limitations. Additionally, it is possible to capture the beginning of the thawing period but not the time at which the pavement is fully thawed which occurs at a time after 31st of May.

Pavements located in climatic zones 1 are characterized by multiple intermittent freezing periods due to air temperatures alternating between positive and negative values. As seen in station 'Putsered' several small freezing fronts penetrate the pavement structure. The maximum frost penetration depth of 67.4 cm is relatively low which implies that a low thickness of frost protection layer is required and no subgrade improvement due to frost actions is needed. However multiple small freezing and thawing cycles lead to a higher moisture content and more variable stiffness of the unbound granular materials in the upper layers of the pavement. Thus, proper drainage of the upper layers should be the main focus point when designing pavements in climatic zones 1. Usually a single, full freezing front is observed in zones 2, 3, and 4 (as shown in Fig. 3b) with largest frost penetration values and latest times of thawing in zone 4. The typically expected thawing period is observed in these zones with frost penetrating the pavement continuously from top to bottom and the fully frozen zone formed at the top, partly frozen zone in the middle and unfrozen zone in the bottom of the pavement.

In order to correlate the air freezing index value to the frost penetration depth, the sub-surface temperature data from the RWIS has been interpolated to obtain the 0 °C isotherm, and the isotherm curve have been plotted for at least 10 consecutive winter periods for each station. A typical frost penetration graph for multiple winter seasons is shown in Fig. 4 for station 1030 Buskhyttan located in climatic zone 2. From Fig. 4 it is possible to observe the beginning and the end of the winter period, as well as the maximum value of the frost penetration over each winter season.

After plotting the temperature-freezing index from the meteorological data and the frost penetration depth data for each season and each station, the maximum recorded values were arranged according to the climatic zone at which each station is located. Typical graphs of the maximum freezing index and frost penetration depth of stations located in climatic zone 2 are shown in Fig. 5.

The values for the air freezing index shown in Fig. 5 were computed mathematically due to the graphical method resulting in difficulties in reading the charts in case of alternating positive and negative daily temperatures in regions of mild winter which produced flat lines of air freezing index variations.

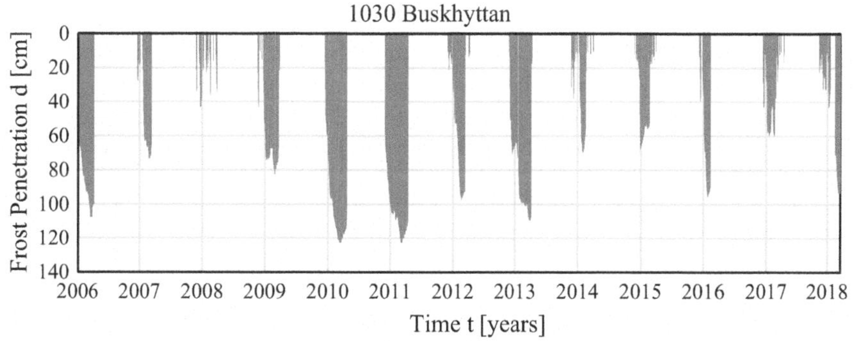

Fig. 4 Frost penetration graph for the station 1030 Buskhyttan over multiple winters

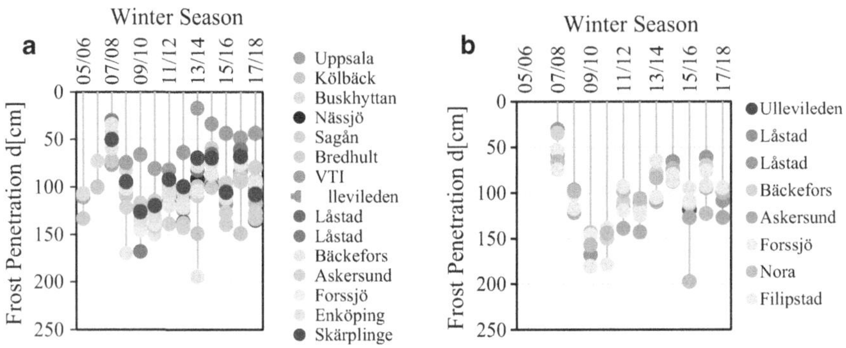

Fig. 5 Maximum values for freezing index (**a**), and frost penetration (**b**) for zone 2 pavements

Table 1 Summarized values for freezing index and frost penetration

Zone	Freezing index [°C · days]			Frost penetration [cm]		
	Avg.	Min.	Max.	Avg.	Min.	Max.
1	208.7	35.8	633.9	85.4	31.2	162.8
2	309.6	65.1	783.1	107.1	35	168.3
3	488.5	210.8	1211.8	122.8	50.7	178.3
4	732	266.5	1395.8	148.9	65.2	196.7
5	1190.7	528.7	2246.6	–	–	–

Similar trends were observed for both freezing index and frost penetration values with minimum values for climatic zone 1 and maximum values for climatic zone 5 with increasing values in the northwards of Sweden. Lower deviations between the recorded values recorded in the same season for different stations were observed in the freezing index plots compared to the frost penetration plots. This can be attributed to the plots depending only on the air temperature while the maximum frost penetration depends on multiple climatic and site-specific conditions. A summary of the recorded average, minimum, and maximum values for the air freezing index and the frost penetration values for the five climatic zones is shown in Table 1.

For the climatic zone 5, the frost rod length of 200 cm was exceeded for most of the stations on multiple seasons, thus not allowing calculation of avg., max or min values in Table 1.

4 Statistical Correlations

The frost penetration depth was plotted against the air freezing index for RWIS and meteorological stations located in the vicinity of each other.

A single point in the plot corresponds to the measured maximum frost penetration due to a computed value of air freezing index for a single station during a single winter season as shown in Fig. 6. A total of 391 points were plotted on the chart from 49 RWIS stations over a timespan of 10 years. Several points were removed from the chart due to the recorded frost penetration depth of 200 cm limited by the frost rod not being representative of the field conditions. Furthermore, for a few stations the data was not available in specific seasons, however, the number of such cases being irrelevant to the study.

Additionally, a power function given in Eq. 2 was fitted to the 391 datapoints using nonlinear least squares regression, where the freezing index FI is in [°C · days] and the frost penetration depth d_f is in [cm]

$$d_f = 12.1349 \cdot FI^{0.3861} \tag{2}$$

In order to find the limits for possible new stations, 95% prediction intervals were fitted to the nonlinear regression. This means that 95% of the new points will be located between the upper and lower limits of the 95% prediction band in Fig. 6. The upper and lower limits of the regression curve can be used to provide an estimate for the range of frost penetration based on the values of air freezing index at a new location within Sweden. This feature could be useful in the design process of new pavements or as a simple check for the predicted frost penetration by more advanced tools.

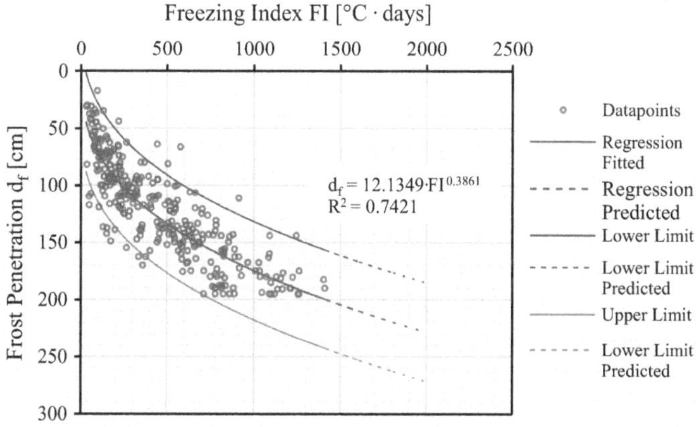

Fig. 6 Freezing index - Frost penetration depth correlation

5 Conclusions and Recommendations

In this paper, data from meteorological stations and subsurface pavement temperature data from RWIS has been used to empirically correlate the air freezing index FI to the frost penetration depth. The result of the study is a chart in which the 95% prediction intervals of a nonlinear least squares regression can be used as an estimated range to predict the frost penetration depth based on the air freezing index.

A scatter of the measurement datapoints has been observed in the freezing index—frost penetration depth plot. The scattering could be attributed to the difference in layer thermal conductivities, and in frost susceptibilities between different pavements. Possible site-specific conditions such as water ingress and snow covering could affect the frost penetration depths as well.

Several limitations apply to the design chart presented in this paper. At the present time it is not possible to measure the frost penetration in the northernmost located stations due to limited length of 200 cm of the frost rod. Due to this limitation, the data from 6 out of 49 RWIS stations were not included in the fitting of the nonlinear regression. The availability of the frost penetration depth for the northernmost stations would presumably increase the accuracy of the nonlinear regression.

Increasing the number of inputs would improve the accuracy of the model. To increase the number of inputs, more specific data for each point representing a RWIS station is needed. Such data might consist of groundwater table data, moisture distribution data through the pavement cross-section. The availability of more site-specific material properties such as unbound granular layer types, frost susceptibility of layers, subgrade types would further increase the accuracy of the model.

References

Andersland OB, Ladanyi B (2004) Frozen ground engineering, 2nd edn. Wiley, ASCE, Chichester

Doré G, Zubeck HK (2009) Cold regions pavement engineering. ASCE Press, McGraw-Hill, New York

Erlingsson S, Rahman S, Salour F (2017) Characteristic of unbound granular materials and subgrades based on multi stage RLT testing. Transp Geotech 13:28–42. https://doi.org/10.1016/j.trgeo.2017.08.009

Fredlund DG, Rahardjo H (1993) Soil mechanics for unsaturated soils. Wiley, New York

Hermansson Å, Guthrie WS (2005) Frost heave and water uptake rates in silty soil subject to variable water table height during freezing. Cold Reg Sci Technol 43(3):128–139

Huang YH (2012) Pavement analysis and design. Pearson/Prentice Hall, Upper Saddle River

Huen K, Tighe S, Mills B, Perchanok M (2006) Development of tools for improved spring load restriction policies in Ontario. In: The Annual conference of the Transportation Association of Canada (TAC), Charlottetown, Prince Edward Island

Kersten MS (1959) Frost penetration: relationship to air temperatures and other factors. Highw Res Board Bull 225:36

Saarelainen S (1996) Pavement design applying allowable frost heave. Cold Regions Engineering: The Cold Regions Infrastructure - An International Imperative for the 21st Century, pp 890–898

Salour F, Erlingsson S (2013) Investigation of a pavement structural behaviour during spring thaw using falling weight deflectometer. Road Mater Pavement Des 14(1):141–158. https://doi.org/10.1080/14680629.2012.754600

Simonsen E, Isacsson U (1999) Thaw weakening of pavement structures in cold regions. Cold Reg Sci Technol 29(2):135–151. https://doi.org/10.1016/S0165-232X(99)00020-8

Soliman H, Kass S, Fleury N (2008) A simplified model to predict frost penetration for Manitoba soils. In: 2008 Annual conference of the transportation association of Canada, Toronto, Ontario. Transportation Association of Canada

Trafikverket (2011) TRVK Väg. Trafikverkets tekniska krav vägkonstruktion (TRV 2011:072)

Vaitkus A, Gražulyte J, Skrodenis E, Kravcovas I (2016) Design of frost resistant pavement structure based on road weather stations (RWSs) data. Sustainability

Yoder EJ, Witczak MW (1975) Principles of pavement design, 2nd edn. Wiley, Hoboken

Simulating Deflection of a Jointed Rigid Pavement Under Rolling Wheel Deflectometer (RAPTOR) Loading

Pawan Deep, Mathias B. Andersen, Søren Rasmussen, Alessandro Marradi, Nick H. Thom, and Davide L. Presti

Abstract The falling weight deflectometer (FWD) has been the principal equipment used for accurate assessment of pavement structural condition for well over 30 years and back-calculation techniques are well established. In recent years however different types of rolling wheel deflectometer have been developed and they have the potential to allow the collection of structural information on a routine basis. The challenge is to interpret the data from these devices in the most meaningful way, since it is necessary that the deflections take place in different locations compared to the FWD. This paper describes an analytical approach that allows prediction of deflections in a jointed rigid pavement on a line offset from the wheel path. It outlines the assumptions that have to be made and the refinements that are potentially possible. Comparisons with both real measurements and finite element analysis are presented and it is found that the proposed approach is adequate in terms of accuracy and sufficiently rapid for practical back-calculation. The issues of vibration and noise in the deflection measurements are discussed and possible ways of handling these are put forward. Additionally, the sensitivity to model parameters of the rolling wheel deflectometer passing a joint are assessed, which can lead to strategies on back-calculation of load transfer. The conclusion reached is that the technique outlined in this paper opens the door to effective use of rolling wheel deflectometer (RWD) equipment for evaluation of load transfer efficiency in joints.

Keywords Rigid pavement · Semi-analytical model · Load transfer · FWD · RWD

1 Introduction

Structural assessment of pavement condition has been done by falling weight deflectometer (FWD) for more than 30 years and back-calculation techniques are very well established. In recent years however a different type of continuous deflection

P. Deep (✉) · M. B. Andersen · S. Rasmussen · A. Marradi
Dynatest International A/S, Tempovej 27-29, 2750 Ballerup, Denmark
e-mail: pawan.deep@nottingham.ac.uk

P. Deep · N. H. Thom · D. L. Presti
The University of Nottingham, University Park, Nottingham NG7 2RD, UK

C. Raab (ed.), *Proceedings of the 9th International Conference on Maintenance and Rehabilitation of Pavements—Mairepav9*, Lecture Notes in Civil Engineering 76,
https://doi.org/10.1007/978-3-030-48679-2_80

devices, namely the rolling wheel deflectometer (RWD) has been developed with the potential to allow collection of structural information on a routine basis. Other examples of continuous deflection devices are the traffic speed deflectometer (TSD) and rolling dynamic deflectometer (RDD). All these devices overcome the limitation of an FWD since they operate at traffic speed and require no traffic disruption. For example a continuous, contact-sensor based device such as the rolling dynamic deflectometer has been shown to be effective for jointed pavement rehabilitation (Chen 2008; Zhou et al. 2012). In a project-based study, measurement speed was limited by pavement contact nature of the sensors. Historical developments indicate devices with non-contact continuous measurement technology which has increasingly started to show its capability (Harr and Ng-A-Qui 1977; Elseifi et al. 2012). Recently, Dynatest developed a new RWD device called the Rapid Pavement Tester (Raptor) for rapid evaluation of pavements. The research presented in this study can show the potential of the Raptor for rigid pavements, since it already works for flexible pavements (Madsen and Pedersen 2019).

It is a challenge to understand and draw conclusions from these devices since the measurements are taken in different locations relative to load compared to the FWD. This happens due to the device-specific geometries which contrast with the FWD. In studies with a Rolling Dynamic deflectometer (RDD), the conclusions on joint conditions and load transfer efficiency were formed by setting threshold values for the project and thus they cannot be applied to a different maintenance project (Chen et al. 2016). RDD uses rolling geophones with limited sensor capabilities demonstrating the potential of continuous deflection measurements for jointed pavements. However, there have been no significant studies done to investigate the response of jointed rigid pavement with an array of non-contact sensors due to continuously moving load. Therefore, deflections obtained from a continuous deflection device cannot be used to draw conclusions about joints until more investigations are done.

The first step to a back-calculation is a mechanistic model which can accurately and rapidly predict the response under a moving wheel. Historically, FWD measurements use different mechanistic models to do back-calculations, but these same models cannot be applied to the RWD in the case of jointed pavements. Several FEM models of rigid pavements were able to predict deflections under all types of environmental and design loads (Davids et al. 2003). To include and characterize the load transfer mechanisms, advanced modelling strategies e.g. Enrichment of Finite elements to include discontinuities has been shown to work (Skar and Poulsen 2015; Maitra et al. 2010). Such advanced modelling strategies are challenging for a practical back-calculation method.

This paper presents a static 3D semi-analytical model that can simulate vertical deflection for jointed rigid pavement slabs due to a moving load. It can predict the deflections for a jointed rigid pavement on a line offset from the wheel path and everywhere in the vicinity of the joint. Validation is done by comparison with a FEM program. A demonstration of a potential back-calculation from an independent reference measurement setup is presented. A sensitivity study of the model reveals which quantities have the highest sensitivity to changes in the load transfer efficiency.

As such, the model could potentially be used to back-calculate and measure the load transfer efficiency.

2 Jointed Rigid Pavement Modelling

This section presents jointed pavement modelling, that can simulate a uniform pressure load with a discontinuity in the pavement structure and is able to predict vertical deflections accurately.

Rigid pavements are formed from concrete slabs. Concrete is an elastic material which can take high pressure loads. Concrete slabs are jointed together with load transfer mechanisms such as aggregate interlock, dowels and saw cuts. Their response under a load depends on structural and environmental factors. To assess the structural condition of rigid pavements, different test methods are applied depending on the type of maintenance activity under consideration. To do a structural assessment, the strength of slabs, the subgrade and joint condition are checked for their performance by instruments such FWD. Structural failures arise from the deteriorating condition of joints and weak slab joints. If slabs do not transfer loads in an optimal manner, then cracks start to appear. Deterioration in the joint's ability to transfer loads reduces the performance and life of rigid pavements. The ability to transfer load could be defined mechanically based on stresses, strains and deflections. A vertical deflection-based measure of load transfer is traditionally most used, as it is a primary kinematic unknown and easy to measure and calculate.

2.1 Mathematical Model Formulation

To predict the response of a jointed concrete pavement, a static 3D semi-analytical solution is developed. This forward model aims to be a sufficiently good approximation to real rigid pavements while being fast to calculate, e.g., in comparison with more numerically intensive approaches like finite element modelling. Figure 1 shows

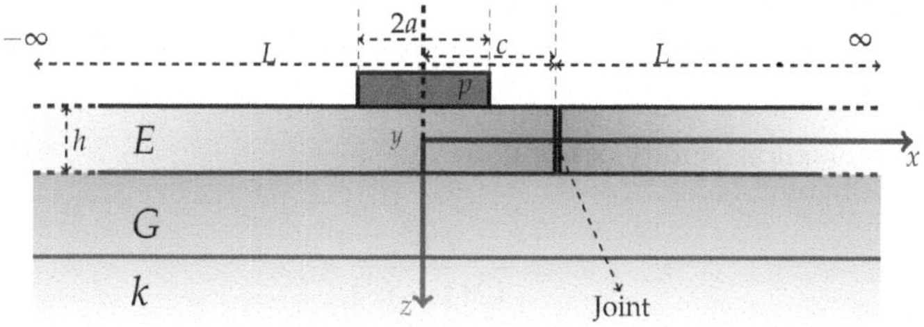

Fig. 1 Two semi-infinite jointed slabs on a Pasternak Foundation

a schematic of the model. The origin of the coordinate system is at the position of the load and x is the moving load direction, y is the transverse direction and z is the vertical direction. The origin is placed at the middle of the slab. The formulation is based on two semi-infinite jointed concrete slabs resting on a Pasternak foundation with subgrade reaction k and shear modulus G. Pasternak foundation allows for a realistic soil/subgrade behavior. The load transfer efficiency δ in Eq. (1) is the ratio of the vertical deflection on the unloaded (w_{UL}) and loaded (w_L) slab right next to the joint.

$$\delta = \frac{w_{UL}}{w_L} \tag{1}$$

This formulation has a vertical load of pressure p with a rectangular contact area $2a$ by $2b$ at a distance c from the joint. The slab is of thickness h with Young's Modulus E. A rectangular load area allows Fourier method to provide analytical solution. The model is derived from the equilibrium equation of the system. The boundary conditions imply zero vertical displacements at infinity in both x and y directions. The load pressure is assumed uniform and shear loads are not included in the model in development presented here. The solution method is presented by Van Cauwelaert (2004) but the numerically challenging implementation is done in this study.

2.1.1 Assumptions

The model follows linear elasticity and a small strain framework. A moving wheel is more static than dynamic and therefore a static loading is assumed. Thermal effects are ignored here and will make the mathematical analytical solution to not exist if included. The load transfer in the y-direction is assumed constant here.

2.1.2 Equations

The equilibrium equation in terms of the vertical deflection w is

$$\left(\frac{\partial^2}{\partial x^2} + \frac{\partial^2}{\partial y^2}\right)\left(\frac{\partial^2 w}{\partial x^2} + \frac{\partial^2 w}{\partial y^2}\right) - \frac{G}{D}\left(\frac{\partial^2 w}{\partial x^2} + \frac{\partial^2 w}{\partial y^2}\right) + \frac{kw}{D} = \frac{p}{D} \tag{2}$$

where D is the flexural rigidity of the slab

$$D = \frac{Eh^3}{12\left(1 - v^2\right)} \tag{3}$$

The relation between the radius of relative stiffness l and the flexural rigidity is

$$\frac{G}{D} = \frac{2g}{l^2} \; and \; \frac{k}{D} = \frac{1}{l^4} \tag{4}$$

To solve Eq. (2), both the load and deflection are expressed as double Fourier integrals. The ratio between the deflection on both sides of the joint, is given by the definition of the load transfer efficiency in Eq. (1). On each side of the joint, the solution is expressed as a linear combination of a particular solution w (Eq. (5)) to the inhomogeneous equation.

$$w_L = (w + A(s)w_a + B(s)w_b); \quad w_{UL} = w + C(s)w_c + D(s)w_d \tag{5}$$

$$w = \frac{p}{\pi k}\frac{1}{\sqrt{1-g^2}}\int_0^\infty \frac{\cos(sy/l)\sin(sb/l)}{s(s^4+2gs^2+1)}$$
$$\left\{e^{-(x-a)\alpha/l}\left[\sqrt{1-g^2}\cos[(x-a)\beta/l] + (s^2+g)\sin[(x-a)\beta/l]\right]\right.$$
$$\left.-e^{-(x+a)\alpha/l}\left[\sqrt{1-g^2}\cos[(x+a)\beta/l] + (s^2+g)\sin[(x+a)\beta/l]\right]\right\}ds \tag{6}$$

And the two solutions to the homogeneous equation w_a (Eq. (7)) and w_b (Eq. (8)) (w_c (Eq. (9)) and w_d (Eq. (10)) for the unloaded slab) in Eq. (5).

$$w_a = \frac{p}{\pi k}\frac{1}{\sqrt{1-g^2}}\int_0^\infty [A(s)\cos(\beta x/l)]e^{\alpha x/l}\frac{\cos(sy/l)\sin(sb/l)}{s}ds \tag{7}$$

$$w_b = \frac{p}{\pi k}\frac{1}{\sqrt{1-g^2}}\int_0^\infty [B(s)\sin(\beta x/l)]e^{\alpha x/l}\frac{\cos(sy/l)\sin(sb/l)}{s}ds \tag{8}$$

$$w_c = \frac{p}{\pi k}\frac{1}{\sqrt{1-g^2}}\int_0^\infty [C(s)\cos(\beta x/l)]e^{-\alpha x/l}\frac{\cos(sy/l)\sin(sb/l)}{s}ds \tag{9}$$

$$w_d = \frac{p}{\pi k}\frac{1}{\sqrt{1-g^2}}\int_0^\infty [D(s)\sin(\beta x/l)]e^{-\alpha x/l}\frac{\cos(sy/l)\sin(sb/l)}{s}ds \tag{10}$$

where two new auxiliary parameters have been introduced in Eqs. (11) and (12).

$$\alpha^2 = \frac{1}{2}\left[\sqrt{(s^2+g)^2+1-g^2} + (s^2+g)\right] \tag{11}$$

$$\beta^2 = \frac{1}{2}\left[\sqrt{(s^2+g)^2+1-g^2} - (s^2+g)\right] \tag{12}$$

By virtue of the fourth order partial differential equation, four conditions are required to couple the solution across the discontinuity at the joint. By relating deflections, forces and moments, Eqs. (13)–(16), can be written at the joint.

$$\delta(w + A(s)w_a + B(s)w_b) = w + C(s)w_c + D(s)w_d \tag{13}$$

Cancellation of the moment at the edge of the loaded slab

$$\left(\frac{\partial^2}{\partial x^2} + \nu\frac{\partial^2}{\partial y^2}\right)(w + A(s)w_a + B(s)w_b) = 0 \tag{14}$$

Cancellation of the moment at the edge of the unloaded slab

$$\left(\frac{\partial^2}{\partial x^2} + \nu\frac{\partial^2}{\partial y^2}\right)(w + C(s)w_c + D(s)w_d) = 0 \tag{15}$$

Equality of shear forces

$$\begin{aligned}&\left(\frac{\partial^3}{\partial x^3} + (2-\nu)\frac{\partial^3}{\partial x\partial y^2} - \frac{2g}{l^2}\right)(w + A(s)w_a + B(s)w_b)\\ &= \left(\frac{\partial^3}{\partial x^3} + (2-\nu)\frac{\partial^3}{\partial x\partial y^2} - \frac{2g}{l^2}\right)(w + C(s)w_c + D(s)w_d)\end{aligned} \tag{16}$$

Conditions in Eqs. (13)–(16) form a linear system of equations for the unknown "constants" *A(s)*, *B(s)*, *C(s)* and *D(s)* which is solved at a given coordinate by evaluating the integrals numerically for each wave number *s*.

2.2 Solution Validation

The response from the semi-analytical model is compared to the result of a FEM solution from EverFE. EverFE is a FEM tool that models the response of jointed slab systems under various load configurations (Davids et al. 1998, 2003; Davids and Mahoney 1999). Figure 2 shows a comparison of the modelled responses from the semi-analytical model and EverFE for two different slab moduli. The response from the semi-analytical model is very close to the FEM solution except under the

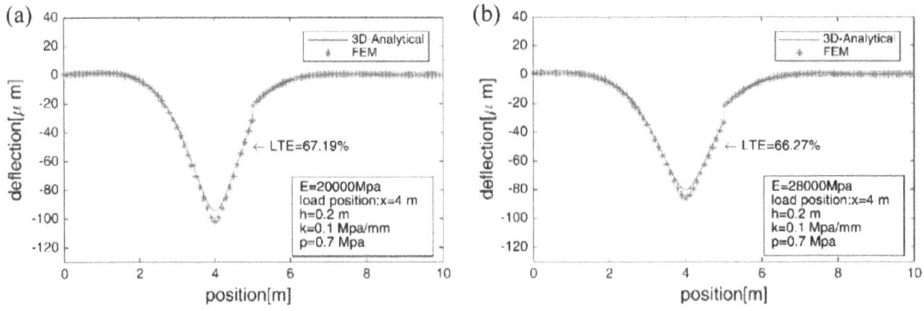

Fig. 2 Comparison of 3D semi-analytical solution and FEM solution **a** E = 20,000 Mpa **b** E = 28,000 Mpa

load (which is due to discretization error in the FEM solution). The comparison in Fig. 2 validates the 3D semi-analytical model.

With the mechanical model ready, an experiment needs to be set up to measure deflections due to a rolling wheel. From this experiment and the model presented here, information such as moduli and load transfer could be back-calculated. It is the aim that this information is sufficient to characterize the joint and predict the state of rigid pavements.

3 Field Experiment with a Moving Load

The measurement system comprises of 7 lasers, an odometer (encoder), a reference beam, an ethernet switch, a power supply, a moving load and a recording application.

To measure the edge deflection response of jointed slabs under the influence of a moving wheel load, a reference beam mounted with distance lasers is setup as shown schematically in Fig. 3. This beam is placed across joint in the center of two jointed slabs. The beam is mounted with 7 lasers such that the middle laser is placed across the joint formed by both the slabs.

The pressure load comes from a moving wheel load carrying 5 tons. The loading wheel is a part of a trailer which has an independent suspension in the rear axle. The trailer is attached to a truck and driven at a slow and controlled speed of 10 km/h for the experiment. There are two truck axles and their influence is low and ignored for this study. Initially, the trailer is moved to a far location from the joint. During the experiment, the load is moved parallel to the beam. The aim is to get as close as possible to the beam in the transverse direction to get a good signal.

The distance laser is a camera-based laser (as shown in Fig. 3a) which measures its self-distance to a ~200 mm wide line projected onto the pavement some ~300 mm

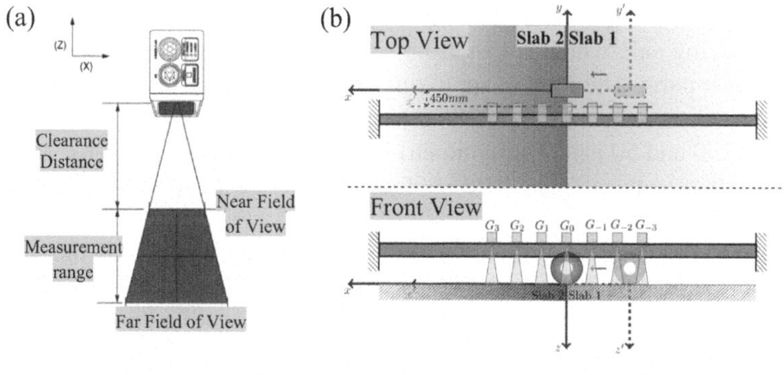

Fig. 3 **a** Coordinate system of the laser measurement system, **b** Schematic of the experiment

(a)  (b) 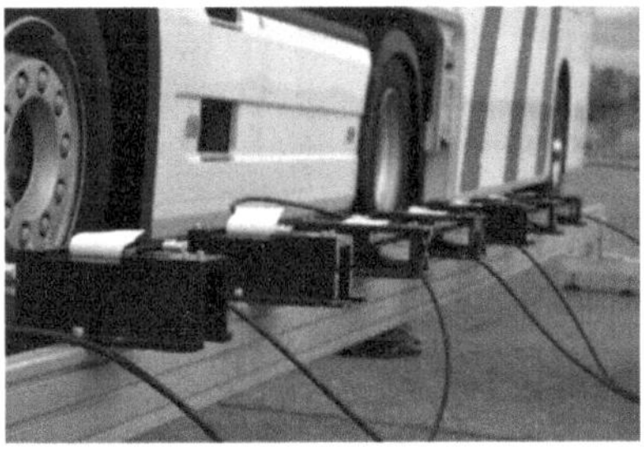

Fig. 4 a Side View of the measurement setup **b** Lasers on the beam

below it. The profile that is being scanned by the laser is parallel to the beam's length. The beam height is set up in a way so that the lasers are in their measurement range.

Experiment at the Værlose Airbase, Denmark

At the Værlose airbase, the selected site is an old apron with rigid pavement slabs. Joints to be measured are selected based on visual inspection and the setup symmetrically across the joint.

The truck is positioned 22 m (4 slabs) away from the joint with its front axle on the edge of the 4th slab. After the loaded axle of the trailer has passed the reference beam setup and the slabs on the other side, the measurement is stopped as shown in Fig. 4, and the collected data is saved. This sequence is repeated for several joints at the site. Repetitions of the experiment are done for the same joint, to check for the consistency and repeatability of the results.

Comparison of Deflection Measurements Generated from a Loaded Moving Wheel with the 3D Semi-analytical Model

Deflections measured from two different sites are presented in Fig. 5. In the plots, the x-axis represents the position of the load and the y-axis represents deflections measured by the laser over the joint. The deflection signal is marked as approaching side and leaving side which comes from separating the signal from laser over the joint into two parts. The joint is approximately positioned at $x = 0$ m in all cases. The load position ranges from -8–8 m. Measured deflections range approximately between -100 and 50 μm. The symmetric peaks in the deflection signal at $x = -4$ and 4 m is due to the moving load as it approaches the supports of the reference beam.

In order to use the model, a set of values of model parameters from the structural information at the site is chosen. The parameters are not accurate as only moduli and thickness are accurately known from FWD tests. Then the model can generate deflections in good agreement with the measured ones. A comparison of the measured signal with the modelled signal in Fig. 5 shows the match of the peaks. Modelled deflections to the left and right of the joint match the trend in the measured data.

Though the experiment was conducted in a way so as to avoid noise and with a slow speed of the moving load, the deflection values at the edges of the two slabs'

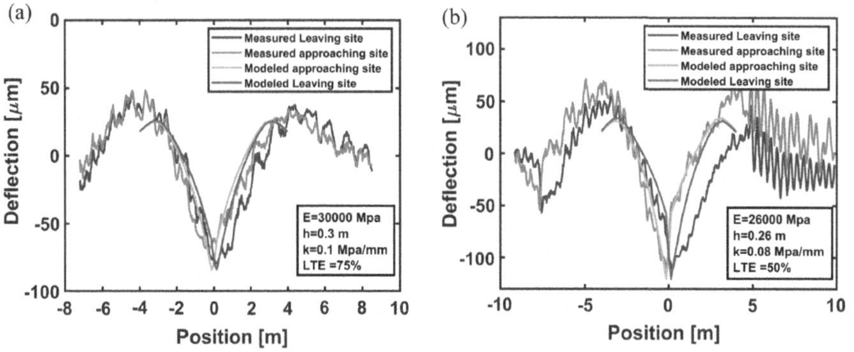

Fig. 5 Modelled vs Measured slab edge deflections under a continuously moving load from different sites

have some noise and vibration. Such behavior could be due to ground vibrations, beam vibrations and other unknown reasons. The difference between the deflections on the loaded and unloaded side of the joint are denoted as ΔP_{12}. ΔP_{12} from the experiment and the model fit was found in Fig. 6 for two different joints. Good agreement in shape and magnitude is seen between the model and the experiment. The model reveals that the load transfer efficiency is higher in case (a) than in case (b) of Fig. 6, being around 75 and 50%, respectively. Moreover, the variation in ΔP_{12} is seen to be 40 and 120 μm, respectively, which is well within the accuracy of the lasers. This confirms that the Dynatest Raptor should be able to distinguish the two joints under consideration here. This comparison demonstrates that the model can predict the response due to a moving load closer to the edge.

Sensitivity of Modelled Deflection Difference to Model Parameters

Here, a sensitivity study using the model to investigate how modelled deflection difference depends on the parameters characterizing the joint is presented. It is shown

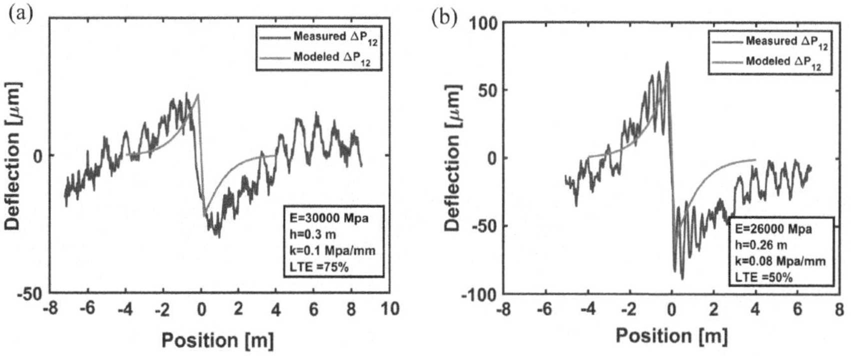

Fig. 6 Modelled vs Measured slab edge deflection difference under a continuously moving load from different sites

that modelled deflection difference has a high sensitivity to the load transfer efficiency when the unknown parameters are varied. In Fig. 7b, deflection differences across the joint increase in magnitude across the slabs as the subgrade reaction k decreases. In Fig. 7a, deflection difference values are much less sensitive to change of moduli.

In Fig. 8a Changing slab thickness affects the shape close to the joint in a pronounced way. Lastly, the load transfer parameter is varied, and it has the maximum sensitivity to predicted deflection differences which can be seen in Fig. 8b. As expected, the deflection-difference has a high and specific dependence on the load transfer and is therefore well suited as a main parameter in a backcalculation process. A higher sensitivity means that the load transfer can be calculated with a higher degree of accuracy. Furthermore, follow-up research is planned on backcalculation of Raptor data where the high sensitivity on the deflection difference is important. This study thus confirms that it is reasonable to backcalculate load transfer using Raptor data.

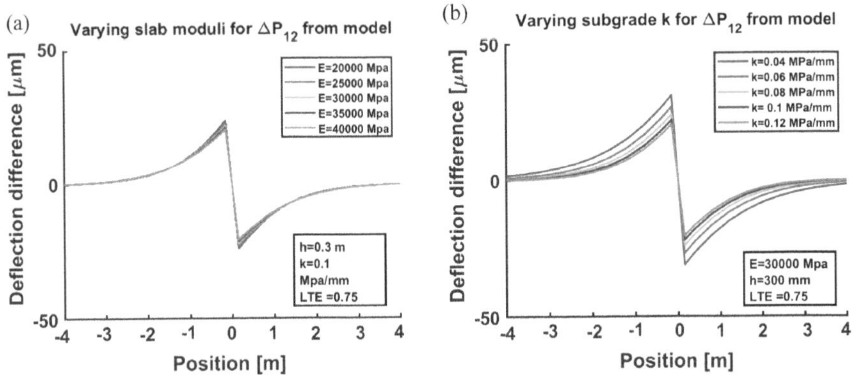

Fig. 7 Modelled deflection difference for different **a** Moduli E **b** Subgrade reaction k

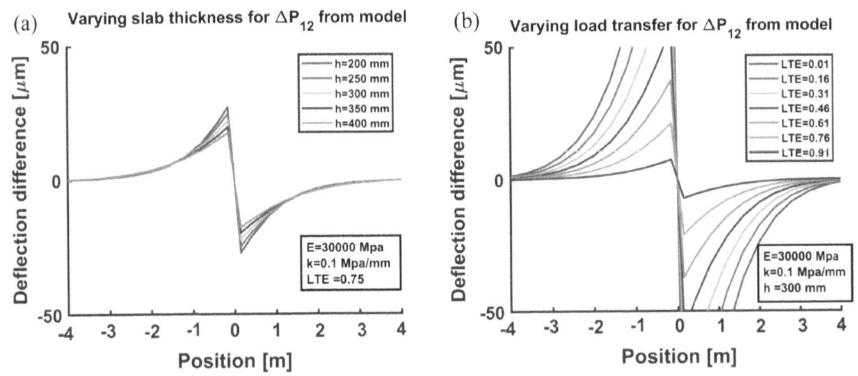

Fig. 8 Modelled Deflection difference for different **a** slab thickness h **b** load transfer δ

4 Conclusions

By using a 3D semi-analytical model and a moving wheel load to enable a continuous nondestructive measurement, it has been demonstrated that the properties of the slabs and the load transfer capability of the joints can be backcalculated using the developed model. This experiment also demonstrates the capability of the measurement system and associated sensors to capture deflections of the order of microns without embedding the sensors inside the structure. It is a simple demonstration of continuous nondestructive evaluation under controlled conditions. An advanced measurement system, which employs the same line lasers and associated setup used in this study is the Dynatest Raptor, an RWD technology platform. These results encourage the use of the developed model for continuous evaluation and investigation of jointed rigid pavement with the Dynatest Raptor in future studies.

Acknowledgements This work is carried out in collaboration with Dynatest International A/s and NTEC at the University of Nottingham. The research presented in this report/paper/deliverable was carried out as part of the H2020-MSCA-ETN-2016. This project has received funding from the European Union's H2020 Programme for research, technological development and demonstration under grant agreement number 721493.

References

Chen DH, Scullion T, Nam BH (2016) Characterization of structural conditions for pavement rehabilitations. Constr Build Mater 121:664–675. https://doi.org/10.1016/j.conbuildmat.2016.06.077

Chen D-H (2008) Field experiences with RDD and overlay tester for concrete pavement rehabilitation. J Transp Eng 134(1):24–33

Davids W, Mahoney J (1999) Experimental verification of rigid pavement joint load transfer modeling with EverFE. Transp Res Rec J Transp Res Board 1684:81–89. https://doi.org/10.3141/1684-10

Davids W, Turkiyyah G, Mahoney J (1998) EverFE: rigid pavement three-dimensional finite element analysis tool. Transp Res Rec J Transp Res Board 1629:41–49. https://doi.org/10.3141/1629-06

Davids W, Wang Z, Turkiyyah G, Mahoney J, Bush D (2003) Three-dimensional finite element analysis of jointed plain concrete pavement with EverFE2.2. Transp Res Rec J Transp Res Board 1853:92–99. https://doi.org/10.3141/1853-11

Elseifi M, Abdel-Khalek AM, Dasari K (2012) Implementation of rolling wheel deflectometer (RWD) in PMS and pavement preservation. Report FHWA/11.492, Louisiana Department of Transportation and Development

Harr ME, Ng-A-Qui N (1977) Noncontact Nondestructive Determination of Pavement Deflection under Moving Loads. Purdue Research Foundation Lafayette In

Madsen SS, Pedersen NL (2019) Backcalculation of raptor (RWD) measurements and forward prediction of FWD deflections compared with FWD measurements. In: Airfield and highway pavements 2019: design, construction, condition evaluation, and management of pavements, pp 382–391. American Society of Civil Engineers Reston, VA

Maitra SR, Reddy KS, Ramachandra LS (2010) Load transfer characteristics of aggregate interlocking in concrete pavement. J Transp Eng 136(3):190–195

Skar A, Poulsen PN (2015) 3-D cohesive finite element model for application in structural analysis of heavy duty composite pavements. Constr Build Mater 101:417–431

Van Cauwelaert F (2004) Pavement design and evaluation: the required mathematics and applications. https://trid.trb.org/view/698133

Zhou F, Hu S, Chen D-H, Scullion T (2012) RDD data interpretation and its application on evaluating concrete pavements for asphalt overlays. J Perform Constr Fac 26(5):657–667. https://doi.org/10.1061/(ASCE)CF.1943-5509.0000265

Study on the Asphalt Pavement Response in the Accelerated Pavement Testing Facility

Ruxin Jing, Aikaterini Varveri, Xueyan Liu, Athanasios Scarpas, and Sandra Erkens

Abstract Accelerated pavement testing (APT) is an effective method in evaluating pavement performance by applying wheel loading and speed under controlled conditions. This study aims to investigate the effects of wheel loading, speed and ambient temperature on the pavement responses at different directions and depths of pavement structure. A two-layer asphalt pavement structure was constructed on a base layer constructed 10 years ago. Strain gauges were installed both in the transversal and longitudinal directions of motion on the bottom of both layers. The response of the asphalt layers was monitored and the developed strains were recorded. The results show that maximum compressive strain increases with wheel load. In contrast, the maximum tensile strain decreases as load increases; this is probably due to the high confining pressure that occurs within the pavement structure when higher wheel load is applied. The maximum compressive and tensile strains decrease with wheel speed, because the asphalt mixture becomes stiffer at high wheel speed (frequency). The maximum compressive and tensile strains in the transversal direction increase with ambient temperature, because of the low stiffness of asphalt materials at high temperature, which appears to be the cause of rutting.

Keywords Asphalt · Accelerated pavement test · Cyclic Indirect Tensile Test · Mechanical response

1 Introduction

Accelerated Pavement testing (APT) was first introduced to the world in the United Kingdom in 1910s. It came to the forefront in the late 1950s with the AASHO (American Association of State Highway Officials) road test in United States and

R. Jing (✉) · A. Varveri · X. Liu · A. Scarpas · S. Erkens
Section of Pavement Engineering, Department of Engineering Structure, Delft University of Technology, 2628 CN Delft, The Netherlands
e-mail: r.jing@tudelft.nl

A. Scarpas
Civil Infrastructure and Environmental Engineering Department, Khalifa University, 127788 Abu Dhabi, United Arab Emirates

C. Raab (ed.), *Proceedings of the 9th International Conference on Maintenance and Rehabilitation of Pavements—Mairepav9*, Lecture Notes in Civil Engineering 76,
https://doi.org/10.1007/978-3-030-48679-2_81

since then has played an important role in the advancement of road construction to a largely rational process (Hugo and Ebbs 2004). APT fills the important gap between mechanistic-empirical design models using laboratory materials testing characterization and real, long-term pavement performance monitoring and data analysis. Nowadays APT is used for full-scale pavement testing, which refers to an application of the controlled wheel loading on a prototype or actual, layered, structural pavement system to determine pavement response and performance (Metcalf 1996).

To study the negative impacts of the environment and traffic on the condition and performance of pavement structures can take years under true field conditions. APT utilizes special full-scale mobile or fixed testing apparatus to simulate these effects in a shorter time period. In the early studies of APT, most of the work focused on the visible pavement performance such as fatigue cracking or permanent deformation (Xu and Meng 2004; Khan et al. 2013). Based on the test results, pavement materials and structure design was highly developed to improve the anti-fatigue and anti-rutting performance (Zhang et al. 2005; Plessis et al. 2018). New materials, epoxy asphalt, rubber asphalt, bridge deck etc. (Qian et al. 2019; Tian et al. 2017), were also evaluated and validated based on APT results.

The primary objective of this study is to investigate the effects of wheel loading, speed and ambient temperature on the mechanical responses pavement structure at different directions and depths based on APT testing. To accomplish this objective three tasks, as listed below, were performed and discussed in the following sections: 1. A series of APT tests were conducted on a porous pavement section to capture the mechanical response of the pavement structure by applying various wheel loads and speeds at two different ambient temperatures. 2. Field core samples were taken from the first and second layers of APT test sections and tested by the Cyclic Indirect Tensile Test (IT-CY) test in the laboratory. 3. The effect of wheel load, speed and ambient temperature on the transverse and longitudinal strains were discussed.

2 Test Section and Experimental Method

2.1 *Overview of the Test Section*

The construction phase of the test section started with the removal of the existing old pavement surface, which had 10 cm thickness. After the milling process, a bitumen emulsion tack coat layer was sprayed on the surface. Then the 6 cm thickness new stone asphalt concrete (STAC) layer was laid first, and on top of it a 5 cm thickness porous asphalt (PA) layer was placed. PA and STAC are commonly used as the top and base layers of pavements in the Netherlands. All layers were compacted using a roller compactor, respectively. The construction of the test sections was done in October 2014. The profile of the new pavement structures is shown in Fig. 1, in which the old STAC (17 cm thickness) and cement bound asphalt granulate base layers (AGRAC, 25 cm thickness) exist more than 10 years.

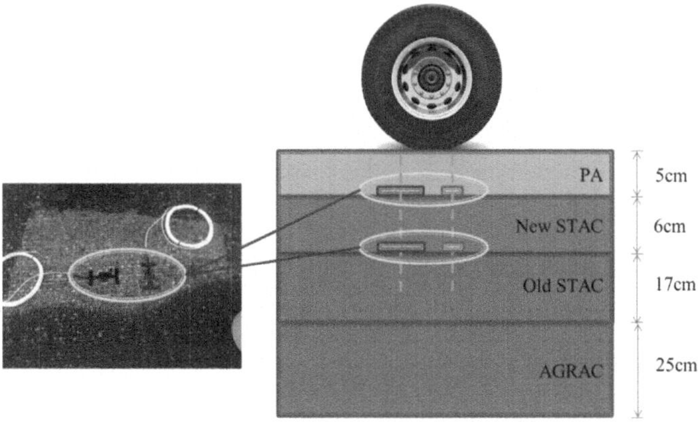

Fig. 1 Profile of test section and strain gauges instrument

In order to measure the mechanical response of pavement structure under wheel load, the test sections were instrumented with strain gauges in longitudinal and transversal directions of motion on the bottom of PA and new STAC layers, as shown in Fig. 1. The coding system in Table 1 is used for the strain at various directions and depths: X_1X_2 or $X_1X_2X_3$.

Both PA and STAC mixtures were designed using the same type of aggregate, Norwegian sandstone, with a nominal maximum size of 16 mm. Norwegian sandstone is a type of crushed stone with density 2740 kg/m^3. The aggregate gradations of both mixtures are given in Fig. 2. The target air void content was 16 and 5% for the PA and SMA mixtures, respectively. The same type of PEN 70/100 bitumen was used for both mixtures. The binder content was 5.0 and 4.6% for the PA and the STAC mixture, respectively. Table 2 shows the main physical and rheological properties of the PEN 70/100 bitumen. Moreover, a factory filler, i.e. Wigro 60K filler (with

Table 1 Coding system

Code	X_1	X_2	X_3
Indication	Capital	Number	Capital
Meaning	Direction	Depth	Maximum strain
Possibilities	T = Transversal direction L = Longitudinal direction	5 = 5 cm depth 11 = 11 cm depth	T = Tensile strain C = Compressive strain

Examples
L11 = Longitudinal direction, 11 cm depth
T5T = Transversal direction, 5 cm depth, Maximum tensile strain

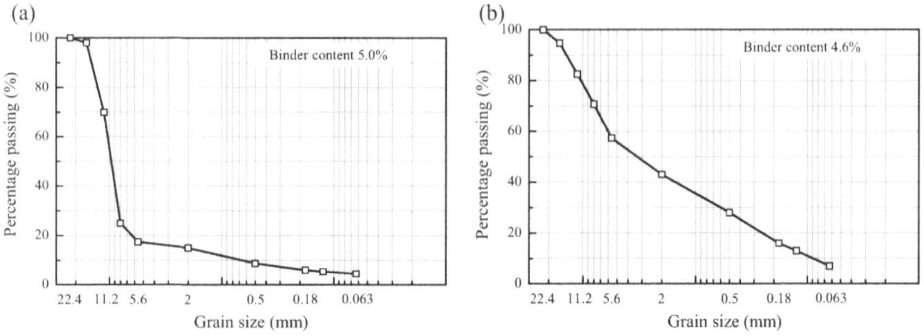

Fig. 2 Aggregate gradation of asphalt mixtures **a** PA mixture **b** STAC mixture

Table 2 Specifications of PEN 70/100

Property	Unit	PEN 70/100
Penetration at 25 °C	0.1 mm	70–100
Softening point	°C	43–51
Dynamic viscosity at 60 °C	Pa s	160
Complex shear modulus at 1.6 Hz & 60 °C	kPa	1.8
Phase angle at 1.6 Hz & 60 °C	°	88

density 2780 kg/m^3) was used for both mixtures. Wigro 60K is a limestone filler which contains 25% calcium hydroxide.

2.2 Accelerated Pavement Test

The accelerated pavement tests were performed using a linear facility, LINTRACK, at the Section of Pavement Engineering at Delft University of Technology. The facility was commissioned in 1991. In this study, the accelerated pavement tests were conducted at 5 kN single wheel load intervals in the range of 20–50 kN, at 2 km/h varying speed in the range of 5–19 km/h. Tire inflation pressure is 900 kPa. These tests were conducted in the month of February 2015 (Test-I) and August 2015 (Test-II). Ambient temperature was recorded to be 4 °C during Test-I and 22 °C during Test-II. Table 3 summarizes the testing conditions that were considered.

Table 3 Accelerated testing conditions

Test name	Wheel load (kN)	Speed (km/h)	Ambient temperature (°C)
Test-I	20, 25, 30, 35, 40, 45	5, 7, 9, 11, 13, 15, 17, 19	4
Test-II			22

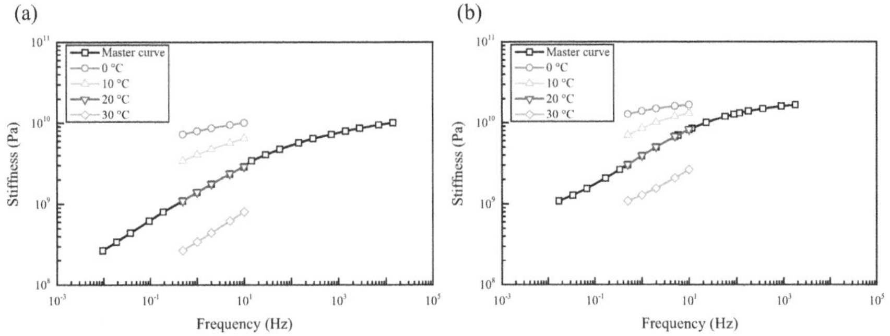

Fig. 3 Stiffness of asphalt mixtures at various temperatures **a** PA mixture **b** STAC mixture

2.3 Cyclic Indirect Tensile Test

Six samples with a diameter of 100 mm and a thickness of 50 mm were cored from the PA and the STAC layers in February 2015. The dynamic modulus of each core was determined by means of Cyclic Indirect Tensile Test (IT-CY) according to NEN-EN 12697-26. The tests were performed using the Universal Testing Machine (UTM) at five frequencies i.e. 0.5, 1, 2, 5 and 10 Hz and four testing temperature i.e. 0, 10, 20 and 30 °C. The conditioning time before testing was 4 h and three replicates were tested.

3 Results and Discussion

3.1 Cyclic Indirect Tensile Test

The stiffness of the PA and STAC mixtures at various temperatures is presented in Fig. 3. As expected, the stiffness of the mixtures significantly increases with increasing frequency and decreases as temperature rises. Moreover, the stiffness of STAC mixture is higher than that of PA mixture. To obtain the visco-elastic behavior in a wider range of frequencies, stiffness master curves were generated on the basis of the Time-Temperature Superposition (TTS) principle at a reference temperature of 20 °C and shown in Fig. 3.

3.2 Accelerated Pavement Test

The longitudinal and transversal strains at various depths were recorded during the accelerated pavement tests. Figure 4 shows the results of strain gauges at various

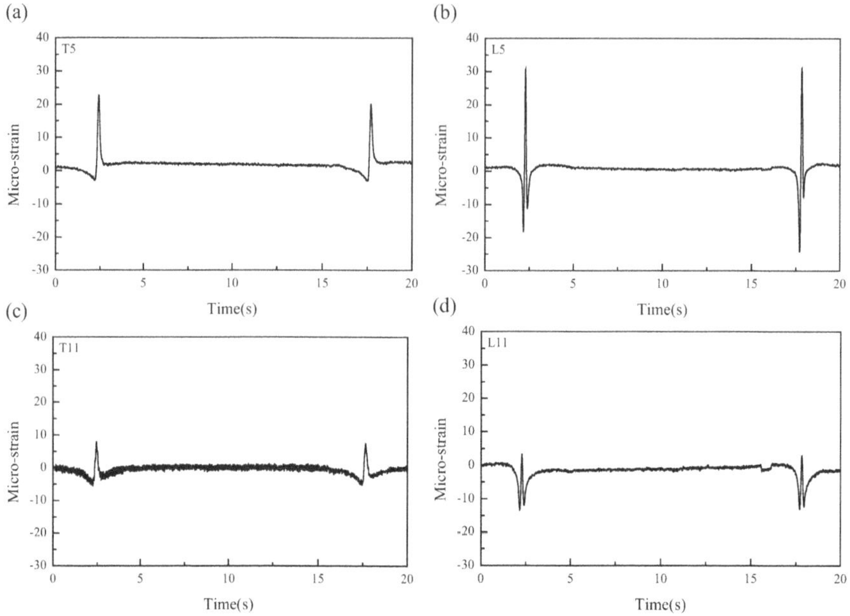

Fig. 4 Measurement of strain gauges at various directions and depths **a** T5 **b** L5 **c** T11 **d** L11 (Wheel load is 20 kN, speed is 15 km/h and ambient temperature is 4 °C)

directions and depths, when the wheel load is 20 kN, the speed is 5 km/h and the ambient temperature is 4 °C. In this figure, positive values represent tensile strains and negative values represent compressive strains.

Overall, all strain gauges experienced compression-tension-compression during the movement of the wheel load, this was true, especially for the strain gauges on the longitudinal direction, Fig. 4b, d. High tensile strains occurred when the wheel load passed, and high compressive strains occurred before and after the wheel load passed. The maximum tensile strain at the 5 cm depth (Fig. 4a, b) is higher than that at 11 cm depth (Fig. 4c, d) due to the higher stiffness of the STAC mixture. In addition, the high tensile strains exist at 5 cm are also caused by the stiffer base layer and subgrade, since they have been over compacted in the past 10 years' APT tests. A high tensile strain at the bottom of the top layer is one of the main causes of pavement cracking. From the measurement of strain data, the maximum compressive and tensile strains of each strain gauge were extracted during the tests. The effect of wheel load on strain at various both and depths is illustrated in Fig. 5.

Figure 5 shows that maximum compressive strain increases with wheel load as expected. However, the maximum tensile strain decreases with wheel load. It is probably because high confining pressure occurred in the pavement structure when heavy wheel load passed, which led to a lower measured tensile strain. In addition, it has been found that both the maximum compressive and tensile strain decrease with increasing wheel speed at the transverse and longitudinal direction for both

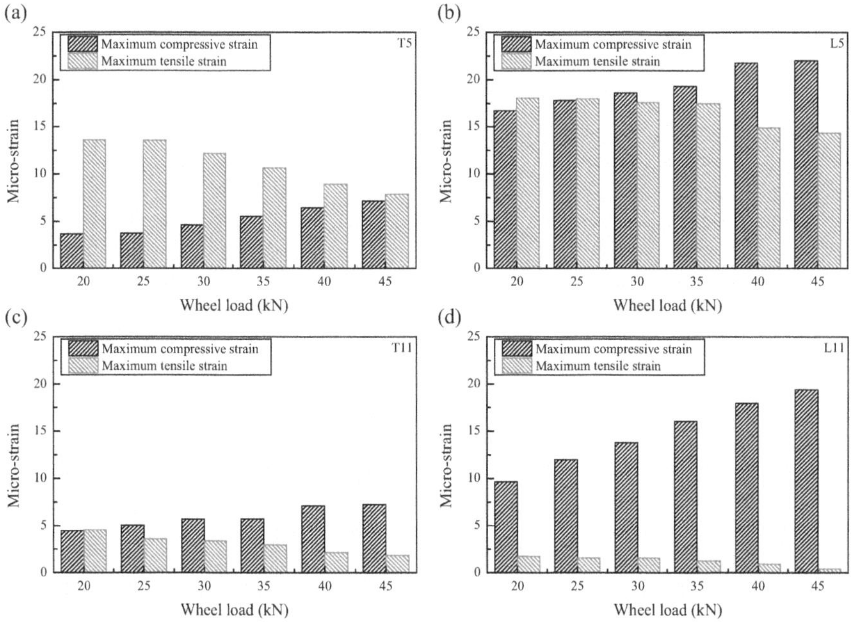

Fig. 5 Effect of wheel load on maximum compressive and tensile strains at various directions and depths **a** T5 **b** L5 **c** T11 **d** L11 (Speed is 15 km/h and ambient temperature is 4 °C)

depth locations, as shown in Fig. 6. The reason for this increases is related to the viscoelastic nature of the materials, higher speed denotes higher loading frequency and consequently higher stiffness of the mixtures, Fig. 3.

The effect of ambient temperature on the maximum compressive and tensile strains is illustrated in Fig. 7. It can be seen that the maximum compressive and tensile strain in the transversal direction increase as temperature rises, as shown in Fig. 7a, c. This is a reasonable observation considering that the material stiffness decrease with increasing temperature, Fig. 3. The occurrence of high strains in the transversal direction can lead to the accumulation of permanent deformation and significantly contribute to the rutting distress in asphalt pavements. However, the maximum compressive and tensile strain in the longitudinal direction decrease with ambient temperature. A possible reason is that the confining pressure in the longitudinal direction is higher than that in transversal direction, because the length of the pavement is much larger than its width, thus leading to the maximum compressive and tensile strains in the longitudinal direction to become smaller.

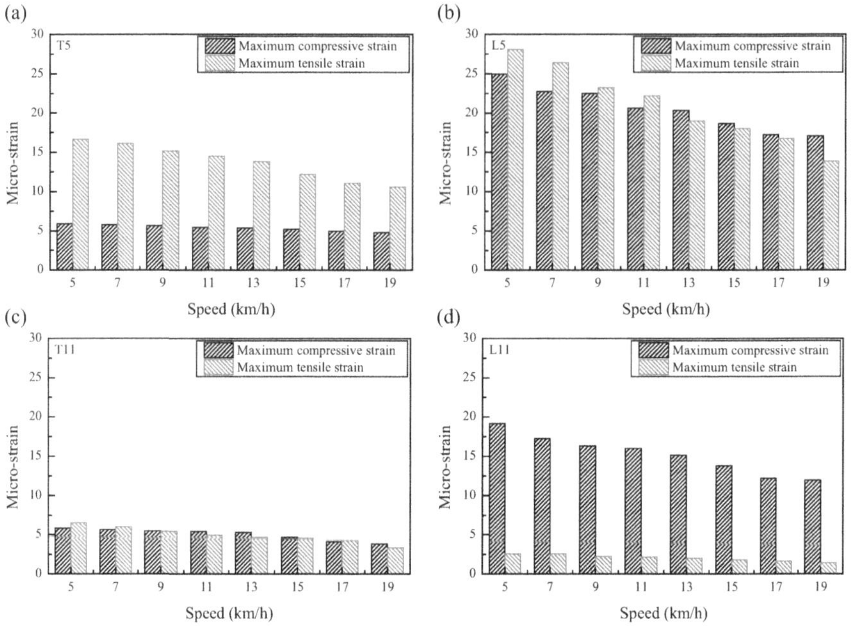

Fig. 6 Effect of speed on maximum compressive and tensile strains at various directions and depths **a** T5 **b** L5 **c** T11 **d** L11 (wheel load is 30 kN and ambient temperature is 4 °C)

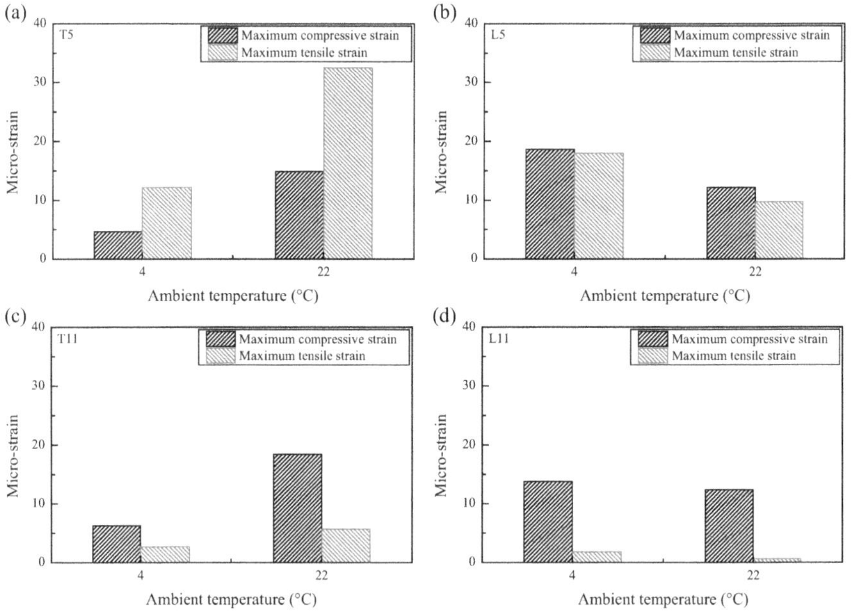

Fig. 7 Effect of temperature on maximum compressive and tensile strains at various directions and depths **a** T5 **b** L5 **c** T11 **d** L11 (wheel load is 30 kN and Speed is 15 km/h)

Table 4 between stains and loading conditions

Test		Cor_Load	Cor_Speed	Cor_Temp
APT	T5C	+	–	+
	T5T	–	–	+
	L5C	+	–	–
	L5T	–	–	–
	T11C	+	–	+
	T11T	–	–	+
	L11T	+	–	–
	L11T	–	–	–

3.3 Effect of Wheel Load, Speed and Ambient Temperature on the Transverse and Longitudinal Strains

According to the above results, the influence of wheel load, speed and ambient temperature on the transverse and longitudinal strains are concluded in Table 4. The symbols '+' and '−' denote increase and decrease in correlation. The highlight blocks denote that the mechanical behavior of the pavement structure is the same with that of laboratory evaluation results (Bayat and Knight 2012; Korkiala and Dawson 2007). To be specific, the transverse strain increases with wheel speed and decreases with wheel speed, and the longitudinal strain decreases with wheel speed as well. However, as the high confining pressure occurs within the pavement structure when higher wheel load is applied, the compressive strain increases and tensile strain decreases as wheel load increases. In addition, due to the fact that the confining pressure in the longitudinal direction is higher than that in the transversal direction, the transverse strain increases and the longitudinal strain decreases with ambient temperature. In other words, laboratory evaluation of asphalt materials cannot fully describe the field behavior of pavement structure due to the effect of confining pressure and other factors in the field pavement (Loulizi et al. 2006; Mazari et al. 2014). For example, when applying an indirect tension test or direct compression test in laboratory, the confining pressure related to the loading and the size difference at each directions normally are not considered.

4 Conclusions

The study presented in this paper shows the preliminary results of a broad study on accelerated pavement tests. To be specific, the changes in the mechanical response of pavement structure were investigated under various combinations of wheel load, speed at two different ambient temperatures. Moreover, core samples were taken from the test sections and their stiffness was determined under various combinations of frequency and temperature.

The APT results show that the maximum compressive strain increases with load amplitude, the maximum tensile strain decrease with frequency, and the maximum tensile/compressive strain in the transversal direction increase with temperature. The high compressive strain that occurs at heavy wheel loads increases the risk of pavement permanent deformation. On the other hand, the high tensile/compressive strains that occurs in the transverse direction at high temperature indicates a high risk of pavement rutting. In other words, high wheel loads, low speeds and high ambient temperatures make pavement susceptible to damage.

To fully characterize the mechanical behavior of pavement structure, a proper experimental protocol (for instance considering confining pressure) in laboratory and more verification from field measurements seem necessary.

Acknowledgements The authors gratefully acknowledge the Dutch Ministry of Transport, Public Works and Water Management for funding this project.

References

Bayat A, Knight M (2012) Field evaluation and analysis of flexible pavement structural responses under dynamic loads. Road Mater Pavement Des 13:26–37

Hugo F, Ebbs A (2004) NCHRP synthesis 325: significant findings from full-scale accelerated pavement testing. Transportation Research Board, Washington DC

Khan S, Nagabhushana MN, Tiwari D, Jain PK (2013) Rutting in flexible pavement: an approach of evaluation with accelerated pavement testing facility. Procedia Soc Behav Sci 104:149–157

Korkiala L, Dawson A (2007) Relating full-scale pavement rutting to laboratory permanent deformation testing. Int J Pavement Eng 8:19–28

Loulizi A, Alqadi I, Elseifi M (2006) Difference between in situ flexible pavement measured and calculated stresses and strains. J Transp Eng 132:574–579

Mazari M, Navarro E, Abdallah I, Nazarian S (2014) Comparison of numerical and experimental responses of pavement systems using various resilient modulus models. Soils Found 54:36–44

Metcalf JB (1996) NCHRP synthesis of highway practice 235: application of full-scale accelerated pavement testing. Transportation Research Board, Washington DC

Plessis L, Ulloa CA, Harvey JT, Coetzee NF (2018) Accelerated pavement testing efforts using the heavy vehicle simulator. Int J Pavement Res Technol 11:327–338

Qian J, Chen K, Tian Y, Zeng F, Wang L (2019) Performance evaluation of flexible pavements with a lateritic gravel base using accelerated pavement testing. Constr Build Mater 228:116790

Tian Y, Lee J, Nantung T, Haddock JE (2017) Development of a mid-depth profile monitoring system for accelerated pavement testing. Constr Build Mater 140:1–9

Xu Q, Meng S (2004) Study on accelerated pavement testing on flexible base asphalt pavement. In: Proceedings on symposium about road engineering, pp 205–210

Zhang X, Chen J, Kong F (2005) The research on heavy traffic anti- rut bituminous pavement. J Petrol Asphalt 19(6):7–11

In-Situ Measurement of Discontinuity Movements in Concrete Pavement Structures

Dongkyu Kim, Hyunsik Hwang, Christopher Jabonero, and Yoon-Ho Cho

Abstract Asphalt Concrete (AC) overlay is one of the typical pavement maintenance strategies on Portland cement concrete (PCC) pavements that enhances its service life. However, reflection cracking is one of the common distresses due to the cyclic expansion and contraction of the joints or cracks in the PCC layer that attributed to the variations in environmental conditions. The current overlay tester (OT) protocol in Texas suggested a single maximum opening displacement (MOD) in PCC pavement types. However, the displacement magnitude varies accordingly. As such, this study focuses on the field quantification of Jointed Concrete Pavement (JCP) and Continuously Reinforced Concrete Pavement's (CRCP) discontinuity movements with respect to the application of asphalt overlay and seasonal changes. Crackmeters were installed at the JCP and CRCP sections and measurements were recorded during winter and summer. Preliminary results have shown that the PCC pavement joint/crack movements significantly reduced after overlay placement. Furthermore, variations of the movement relative to PCC pavement types and seasonal variations are distinct and significant. Hence, the magnitudes of MOD according to pavement types are suggested to be modified when performing reflection cracking resistance evaluation using the asphalt overlay tester.

Keywords PCC discontinuity movement · AC overlay · Reflective cracking · Thermal insulation · Overlay tester

1 Introduction

In 1977, Darter proposed a joint opening prediction model, and Hueng assessed the prediction model as highly conservative using an FEM analysis in 1993. Armaghani (1987) measured and analyzed slab displacements corresponding to JCP temperature variations using thermocouple and linear variable differential transformer (LVDT) and suggested that weather has a strong impact on pavement temperature.

D. Kim · H. Hwang · C. Jabonero · Y.-H. Cho (✉)
Department of Civil and Environmental Engineering, Chung-Ang University, 84 Heukseok-dong, Dongjak-gu, Seoul, Republic of Korea
e-mail: yhcho@cau.ac.kr

C. Raab (ed.), *Proceedings of the 9th International Conference on Maintenance and Rehabilitation of Pavements—Mairepav9*, Lecture Notes in Civil Engineering 76,
https://doi.org/10.1007/978-3-030-48679-2_82

Later, Morian (1999) agreed and stated that the climate conditions such as air temperature have significance on joint movements as well as pavement temperatures, and Kapiri (2000) explained that pavement surface temperature plays a key role in the mechanism. Nam (2001) conducted a study on crack width of CRCP where variations in concrete volume and crack width were measured using vibrating wire strain gages (VWSG) at various locations in Texas (i.e. Austin, Cleveland, and Wichita Falls section). The study has shown that the crack width at the upper part of the slab depends on concrete temperature variations and these movements have significant effect than at the bottom part of the slab. Furthermore, crack width was observed to be inversely proportional to temperature such that it has higher opening at lower temperature. From these findings, the movements of joints at the JCP and the transverse cracks at the CRCP were predicted to exhibit varying behaviors according to seasonal changes.

Asphalt concrete (AC) overlay methods have been used as a maintenance strategy for Portland cement concrete (PCC) pavements to extend its service life. One of the most frequent distresses found in AC overlays is reflective cracking initiating from the bottom of asphalt overlay layer due to the movements at the discontinuities of PCC pavement layers. Reflective cracking reduces driving performance and the ride quality of pavements. In addition, it accelerates pavement distresses resulting to the decrease in long-term performance and durability of pavement. Reflection cracking occurs primarily due to the concrete slab contraction and expansion according to the environmental loading conditions (Chen 2007). Hence, it is imperative to define the appropriate movements at concrete discontinuities and utilize this information in laboratory simulations that evaluate the reflection cracking susceptibility of the asphalt mixture.

In Texas, thin asphalt overlays on CRCP pavement are popular, and the balanced mix design has been used to evaluate rutting and cracking resistance, where the latter is evaluated using Texas Transportation Institute Overlay Tester (TTI OT) (Zhou et al. 2003). The TTI OT subjects the specimen in a maximum opening displacement (MOD) of 0.63 mm (0.025 in.) at 25 °C to simulate movements of the discontinuities at the concrete slab due to environmental loading. This displacement magnitude was derived from the finite element analysis and theoretical calculation at a temperature gradient of 14 °C in a concrete slab having 4.5 m crack interval (Zhou et al. 2003). Later, the field measurements were also obtained from a cracked surface of AC overlays on PCC using a demountable mechanical strain (DEMEC) gauges in winter and summer. Pavement surface temperatures were measured during field measurements in Texas which ranged between 7–25 °C, and the corresponding movements were recorded to be between 0.69–1.52 mm (0.027–0.06 in.) where the variations were attributed to the intervals of the cracks (Walubita et al. 2013).

Meanwhile in Korea, a large percentage of the highway network uses PCC pavements. Through the years, there is a growing demand for the use of composite pavement systems with asphalt overlays as pavement maintenance strategy. However, current overlays have thickness between 70–100 mm which was based on prior experience and was not scientifically supported. In addition, the PCC pavements are a combination of JCP and CRCP such that the magnitude of crack width movements varies according to the pavement type and its crack spacing. Moreover, it is considered inappropriate to apply Texas experimental specifications having a 4.5 m spacing between discontinuities as well as the behavior of environmental conditions. Hence, the results of TTI OT may generate high variations and will generate conservative results; for example, cracking occurred within 10 cycles in some mixtures (Zhou et al. 2003).

Recently, a HEART Multi-modal Overlay Tester (HEART-MOT) capable of simulating a 3-axis displacement loading was developed (Jabonero et al. 2019). In order to predict a reasonable service life using cyclic horizontal displacement simulation, it is important to quantify the movement of concrete pavement discontinuities. However, due to the limited available field data to support the magnitude used in the simulation, the service life prediction model of the overlay is also compromised. As such, this study focuses on: (1) investigating the JCP and CRCP's discontinuity movements with or without overlay and (2) the effect of seasonal variations in the movements of cracks/joints. In order to achieve this, crackmeters were installed to perform field measurements on various pavement structures during summer and winter which are the period with extreme temperature fluctuations. The temperature in each pavement layer were also observed. From these results, the maximum opening displacement of 0.63 mm as suggested by TTI for the asphalt overlay testing simulations shall be compared.

2 Field Measurement of Pavement Discontinuities

To investigate movements of concrete pavement discontinuities, field measurements were conducted in three test sections as shown in Table 1. In 2005, during March and August, representative of winter and summer in Korea, surface pavement temperature variations and crack width/joint movements were measured at CRCP and JCP

Table 1 General information of measured sections

Pavement structure	JCP (Highway)	CRCP (Highway)	JCP (Airport)
Section name	Yeoju smart highway test road		Gimpo airport taxiway
Construction year	2002		1995
Steel ratio	N/A	0.7%	N/A
Slab thickness	300 mm	300 mm	500 mm
Joint/crack spacing	6 m	Variable	7.5 m
AC overlay thickness	25 mm	25 mm	100 mm
Measurement before AC overlay	Mar & Aug 2015	Mar & Aug 2015	N/A
Measurement after AC overlay	July 2016	Dec 2015 & Jul 2016	June 2016

sections in Yeoju Smart Highway Test Road. In order to maintain identical environment conditions identical, adjacent CRCP and JCP sections located around 200 m away were selected. In December 2015, dense graded AC overlay with a thickness of 25 mm was placed on the CRCP section. In July 2016, dense graded overlay with a thickness of 25 mm placed over the JCP section, and an additional Porous AC layer with a thickness of 25 mm was placed on the CRCP section which has a 25 mm dense graded AC overlay. Finally, in June 2016, the same measurements were made on AC/JCP overlay sections with a 7.5 m joint spacing at the Gimpo International Airport Taxiway. Concrete pavement thickness of the entire section was 300 mm.

Crackmeters were installed to measure crack or joint movements as well as pavement surface temperature variations. The crackmeter is made by Geokon (Model 4420). The data acquisition system consists of a datalogger (Campbell Scientific CR1000), a vibrating wire interface (AVW200), and a multiplexer (AM16/32B). According the sensor manual, the crackmeter has an accuracy of ±0.1% F.S. (Full Scale) and a temperature range between −20–80 °C.

Prior to overlay, six sensors were installed during winter and summer for CRCP section and three sensors at the JCP section. Before placing the overlay, three sensors where installed at the concrete layer protected by a steel frame and another three sensors were installed on the asphalt overlay surface and located directly above the joint/crack. Installation locations were selected based on the condition as well as the spacing of joints/cracks. The anchors of the sensors were embedded at a depth 20 mm from the surface and were fixed using epoxy. In order to protect the sensor installation from water penetration, the opening is protected with a silicon sealant. The process of sensor installation is shown in Fig. 1. Meanwhile, Fig. 2 shows the layout of: (a) Yeoju CRCP and JCP sections before overlay and (b) discontinuities in PCC/AC overlay section, while actual sensor installations.

3 Data Analysis

3.1 Movements of Discontinuities on CRCP/JCP in Extreme Seasons in Korea

Measurements were recorded for a week from 12:00 pm on the first day to 12:00 pm on the last day in winter and summer. Figure 3 illustrates pavement surface temperatures and displacements at each discontinuity location during the entire period. Some locations were excluded from data analysis due to errors such that no data were collected from the sensors.

Aligned with the previous literatures, results show that as temperature decreased, concrete slab began to contract, resulting in higher relative displacements, while as temperature increased, concrete slab expanded, leading to lower relative displacements. Specifically, the magnitude of pavement surface temperature variation was around 31 °C in a range of −6–25 °C on average in winter and around 25 °C in

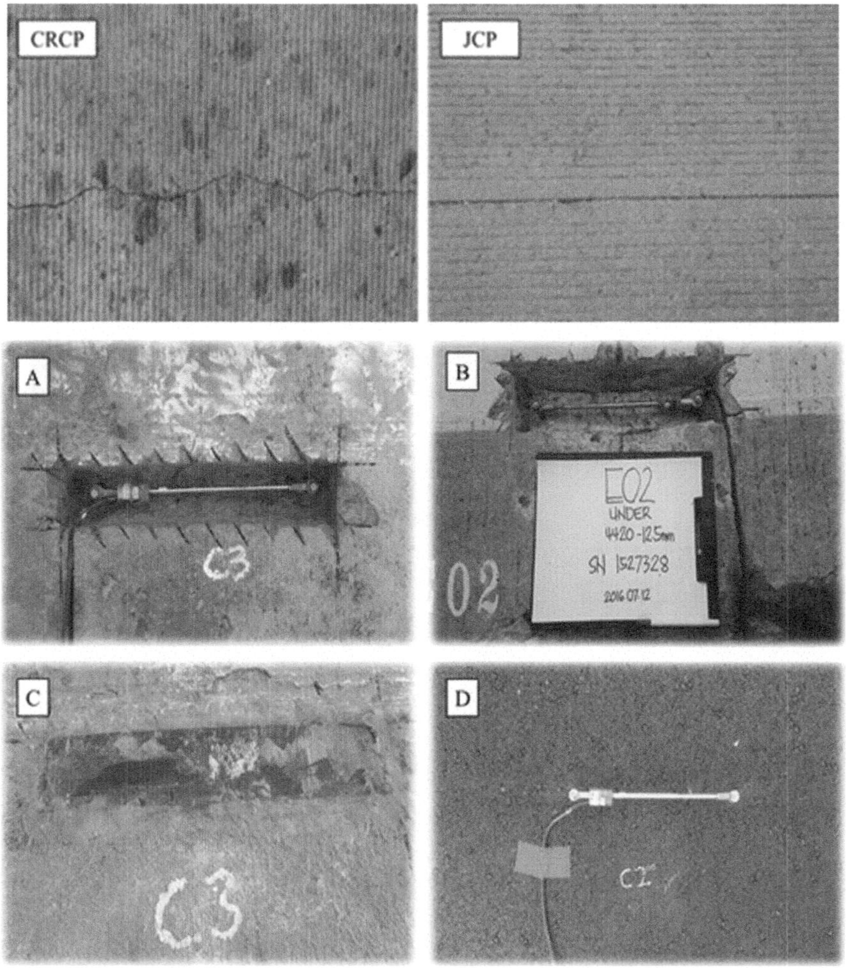

Fig. 1 Sensor installation

a range of 17–42 °C in summer. During winter, the resulting daily magnitudes of discontinuity movements were 0.54 mm on CRCP with an average crack interval of 1.5 m and 0.71 mm on JCP with a 6 m joint spacing. In summer, the daily magnitudes of discontinuity movements were found to be 0.05 mm in CRCP cracks and 0.48 mm in JCP joints. This shows that compared to summer, the discontinuity movements are significant during winter. The crack movements in CRCP during winter were around 24% smaller than joint movements in JCP, however in summer, the rate increased up to 90%. Lee (2001) defined joints in pavement that move minimally or do not move at all regardless of temperature variation as freezing joints. These are joints that are un-cracked or unmoving for another reason. Chou (2005) mentioned that slab closure occurs at higher temperatures. Kapiri (2000) and Rufino (2004) also added that joint

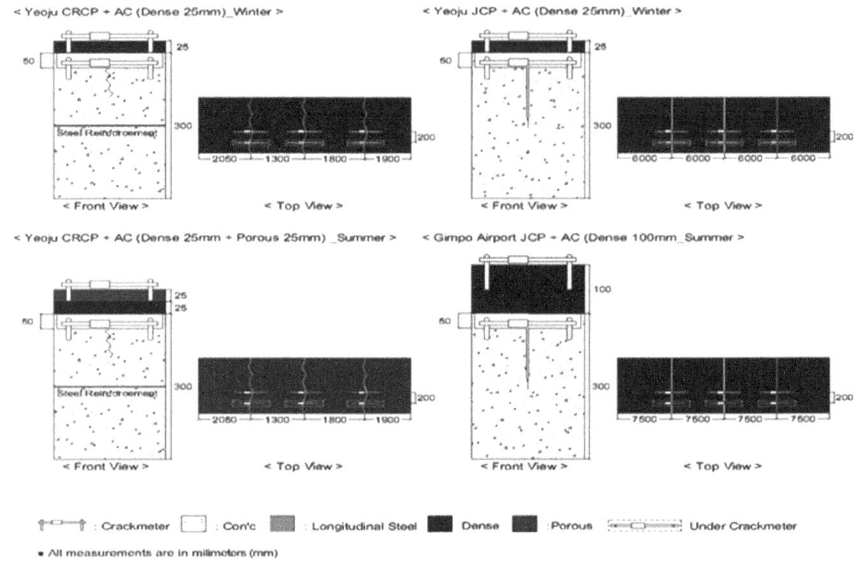

Fig. 2 Sensor installation setup

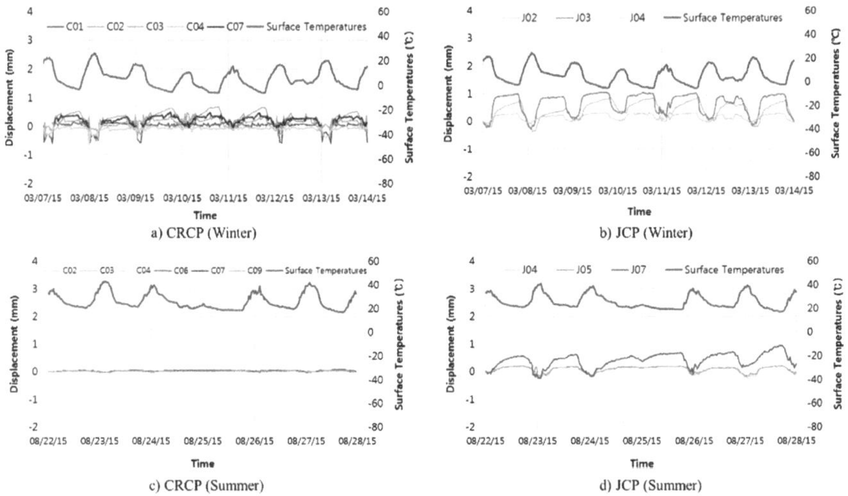

Fig. 3 Surface PCC pavement temperatures and relative displacement at discontinuities in each section: **a** CRCP/Winter, **b** JCP/Winter, **c** CRCP/Summer, **d** JCP/Summer

closing mainly occurs in summer, which is attributed to high temperatures during summer that closes the gap between discontinuities, restricting slab expansion. On the other hand, at the CRCP section, temperatures of pavement structures rose up to

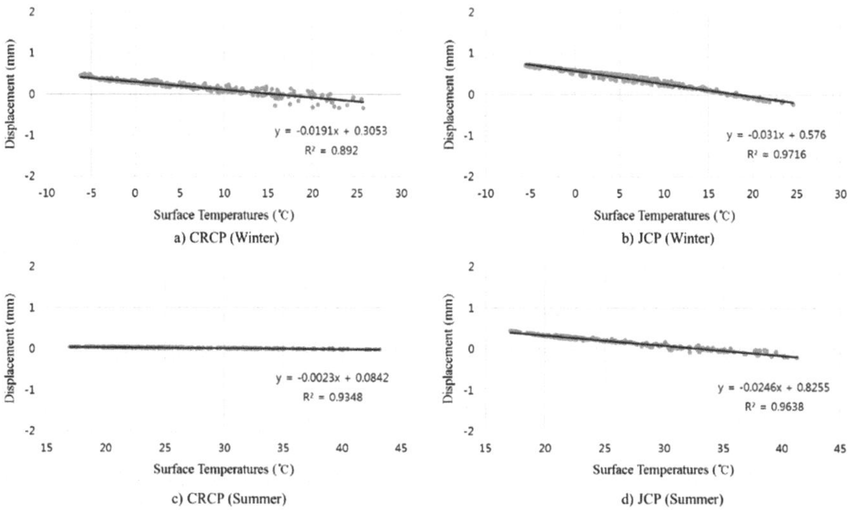

Fig. 4 Relationship between displacement and surface temperatures

about 40 °C in summer, causing joint closing of crack width, which also restrained the movements and resulted in a minimal joint movement.

In order to verify the relations between temperature variation and movement, Fig. 4 is shows the relationship between measured displacement versus pavement surface temperature. Through linear regression, a slope, that is defined in this study as DMR (Discontinuity Movement Rate) provides an information of the daily average movements at discontinuities. For CRCP, the DMR yielded 0.0023 and 0.0191 mm/°C for winter and summer, respectively. While for JCP, the DMR generated were 0.0310 and 0.0246 mm/°C for winter and summer, respectively. It can be observed that summer DMRs were smaller than those in winter. It was calculated that CRCP was around 90% lower than JCP in summer and about 38% lower during winter. As displacements in CRCP appeared somewhat sporadic at the same temperature during the measurement in winter, the coefficient of determination was relatively low. Chou (2005) measured slab movements at the Chiang Kai-Shek International Airport using optical fibers sensor and reported an average movement rate of 0.035 mm/°C for seasons of low temperatures. This result is similar to the winter results of joint movements in JCP.

3.2 *Behavior of Discontinuities on CRCP/JCP with AC Overlay*

Khazanovich (2013) discussed about the thermal insulation effects of AC overlays on concrete pavement, which means that, after AC overlay, the absolute thermal gradients of concrete slab and climate condition-dependent variability are reduced.

Furthermore, Zou (2016) found that temperatures of concrete slab under AC overlay were similar to those of asphalt pavements of similar thickness. However, the temperature at the concrete surface layer reduces significantly relative to the asphalt overlay surface when the overlay thickness increases. From this perspective, it can be assumed that the AC overlay would affect the movements of the discontinuities in the PCC layer.

Figure 5 shows the comparison of temperatures between the concrete pavement layer and the asphalt overlay surface layer. Three scenarios were investigated: (a) winter data of a 25 mm thick overlay on concrete pavement, (b) summer data of a 25 mm thick overlay, and (c) summer data of a 50 mm thick overlay on concrete pavement. For the first case, it can be observed that the temperature range of the asphalt surface temperature was about 30° (−15–15 °C) while the temperature range at concrete pavement layer was about 15° (−5–10 °C). For the second case, the temperature range of the concrete pavement layer is 20° (30–50 °C) while the temperature range at the asphalt overlay layer is 25° (22–47 °C). Gradients were computed using linear regression, and the Thermal Reduction Rate (TRR) is determined as 1 minus the slope of the regression. In the first case, around 0.31 slope was obtained implying a 69% TRR. Similarly, in the second case, the slope was about 0.54 corresponding to 46% TRR. These findings indicate that the magnitude of temperature variation in the concrete pavement layer affects in winter and summer, compared to overlay surface. In addition, the TRR was 23% smaller in summer suggesting that reduction is more significant during winter.

The second and third cases were compared to investigate the TRR relative to asphalt overlay thickness. In Fig. 5.c, it can be observed that it has about 0.32 slope which is equivalent to 68% TRR. This implies that compared to a 25 mm overlay, the 50 mm overlay increases the TRR by 22% supporting the assumption that thicker overlays generate higher thermal insulation effects. From these results, it is generalized that AC overlays will help lower concrete slab to have a thermal insulation effect which reduces joints/cracks movements. Also, given that thermal insulation effect was bigger as thickness increases, the likelihood of reflective cracking is also estimated to decrease.

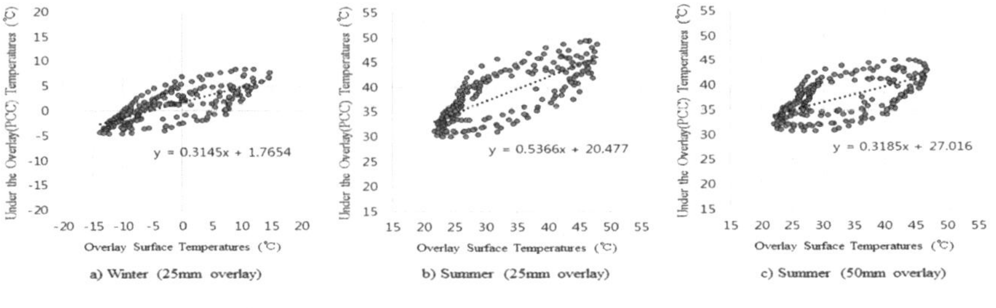

Fig. 5 Comparison of relations of overlay surface temperatures to under the overlay (PCC) temperatures

4 Conclusions

AC overlays are extensively applied to maintain PCC pavements and enhance their service life. It is important to predict and evaluate discontinuity movements of appropriate concrete pavement in order to apply a suitable composite pavement system. Focusing on the characterization of the behaviors of concrete pavement discontinuities according to concrete pavement type, weather and the application of asphalt overlays, crackmeters were installed at the transverse discontinuities in CRCP and JCP sections in order to record and analyze the pavement temperature variation and displacement which will be used determine the appropriate maximum opening displacement of the asphalt overlay tester simulations. From this study, the findings from the field measurements are as follows:

1. At the same thermal gradients, the crack width movements in CRCP during winter were around 24% smaller than JCP joints movements, but about 90% smaller in summer.
2. DMRs (Discontinuity Movement Rates) which is the slope of the displacement versus surface temperature were found to be 0.0023 in summer and 0.0191 in winter for CRCP, while, it is 0.0246 in summer and 0.031 in winter for JCP. Higher DMR values were recorded during winter regardless of pavement type.
3. Finally, the measurement data showed that compared to overlay surface, the concrete pavement layer produces lower temperature that is approximately 69% TRR (Thermal Reduction Rate) in winter and 46% in summer. This suggests that AC overlays will provide a thermal insulation effect to under the overlay concrete slab thereby reducing the magnitude of movements at the discontinuities of the concrete pavement layer.

Acknowledgements This study was conducted under research project 「Development of High-Performance Concrete Pavement Maintenance Technology to Extend Roadway Life (Project No: 18TLRP-B146707-01)」 funded by the Ministry of Land, Infrastructure and Transport (MOLIT) and the Korea Agency for Infrastructure Technology Advancement (KAIA).

References

Armaghani JM, Larsen TJ, Smith LL (1987) Temperature response of concrete pavements. Transp Res Rec TRB 1121:23–33

Chen DH, Zhou F, Lee JL, Hu S, Stokoe II KH, Yang J (2007) Threshold values for reflective cracking based on continuous deflection measurements. Can J Civ Eng 34:1257–1266

Chou C, Cheng HJ, Lin S (2005) Analysis of concrete joint movements and seasonal thermal stresses at the Chiang Kai-Shek international airport. J Eastern Asia Soc Transp Stud 6: 1217–1230

Darter MI, Barenberg EJ (1977) Design of zero maintenance plain concrete pavement, vol. II-Design Manual, FHWA-RD-77-112. Federal Highway Administration, Washington DC

Jabonero C, Park JH, Ryu SW, Lin W, Cho YH (2019) Development of a multi-modal overlay tester for the evaluation of asphalt overlay cracking behaviors. Can J Civ Eng https://doi.org/10.1139/cjce-2019-0227

Kapiri M, Tutumluer E, Barenberg EJ (2000) Analysis of temperature effects on pavement response at denver international airport. The 2020 Vision of Air Transportation. ASCE, pp 125–143
Khazanovich L, Balbo JT, Johanneck L, Lederle R, Marasteanu M, Saxena P, Tompkins D, Vancura M, Watson M, Harvey J, Santero NJ, Signore J (2013) Design and construction guidelines for thermally insulated concrete pavements, publication MN/RC 2013-02. Minnesota Department of Transportation
Lee SW, Stoffels SM (2001) Analysis of in situ horizontal joint movements in rigid pavements. Transp Res Rec TRB 1778:9–16
Nam JH, Kim DH, Choi S, Won MC (2001) Variation of crack width over time in continuously reinforced concrete pavement. Transp Res Rec TRB 2037:3–11
Morian DA, Suhahar N, Stoffels SM (1999) Evaluation of rigid pavement joint seal movement. Transp Res Rec TRB 1684:25–32
Rufino D, Roesler J, Barenberg EJ (2004) Effect of pavement temperature on concrete pavement joint responses. In: FAA worldwide airport technology transfer conference, New Jersey, USA
Walubita LF, Faruk AN, Koohi Y, Luo R, Scullion T, Lytton R (2013) The overlay tester (OT): comparison with other crack test methods and recommendations for surrogate crack tests, publication FHWA/TX-13/0-6607-2. FHWA, US Department of Transportation
Zhou F, Scullion T (2003) Upgraded overlay tester and its application to characterize reflection cracking resistance of asphalt mixtures, publication FHWA/TX-04/0-4467-1. FHWA, US Department of Transportation
Zou X (2016) Analysis of temperature gradient for cement concrete slab under the asphalt overlaying. In: BESETO international conference on concrete pavements

Surface Characteristics and Road Safety

Pavement Surface Evaluation Interacting Vibration Characteristics of an Electric Mobility Scooter

Kazuya Tomiyama and Kazushi Moriishi

Abstract Electric mobility scooters (EMSs) which are solely battery-operated personal small vehicles are becoming an increasingly common way for a walking aid. To ensure the accessibility of EMS users, pavement surface condition of sidewalks is one of the important factors in terms of traveling safety and comfort. This study examines the requirements of evaluating sidewalk pavements by focusing on the interaction between surface properties and vibration responses of an EMS. As a result, the EMS consists of two distinctive resonance frequencies of 8 and 30 Hz associated with the megatexture and microtexture in the surface properties, respectively, at a constant speed of 3 km/h. This finding proves that the EMS vibration highly correlates with the Mean Profile Depth (MPD) inducing axle excitations but not the International Roughness Index (IRI). According to the pavement-vehicle interaction analysis, this study finally suggests that measuring acceleration of an EMS is capable of estimating MPD which is a good estimator of surface friction.

Keywords Surface texture · Electric mobility scooter · IRI · MPD · Accelerometer

1 Introduction

An electric mobility scooter (EMS) which is solely battery-operated personal small vehicles has become an increasingly popular device as an alternative mobility for elder people in many countries. According to literature (Thoreau 2015), an EMS enables its users to travel distances they previously would have made by foot without any physical effort. In many countries, it is permitted to travel on sidewalks without driving license as same as a wheelchair. Secure sidewalk environment must be provided to ensure the accessibility of EMS users for their safety and comfort. In particular, pavement surface which is the interface between an EMS and sidewalk

K. Tomiyama (✉)
Division of Civil and Environmental Engineering, Kitami Institute of Technology, 165 Koen-cho, Kitami 090-8507, Japan
e-mail: tomiyama@mail.kitami-it.ac.jp

K. Moriishi
Obayashi Road Corporation, 1-2-5, Nishitenma, Kita-ku, Osaka 530-0047, Japan

C. Raab (ed.), *Proceedings of the 9th International Conference on Maintenance and Rehabilitation of Pavements—Mairepav9*, Lecture Notes in Civil Engineering 76,
https://doi.org/10.1007/978-3-030-48679-2_83

plays an important role in road management. However, very little knowledge has so far been derived for sidewalk surfaces traveled with EMSs unlike pedestrians and wheelchairs (Aghaabbasi et al. 2017 and Duvall et al. 2013).

Road surface has traditionally been characterized for the influenced performance of motor vehicles corresponding to the geometrical scale categorized by wavelengths (PIARC 1987 and Sandberg and Ejsmont 2002). As an example, the roughness which is sometimes called "unevenness" belongs to the wavelength ranges between 0.5 and 50 m. The texture is divided into three different scales such as megatexture, macrotexture, and microtexture of which wavelength ranges are 50 to 500 mm, 0.5 to 50 mm, and less than 0.5 mm, respectively. In this context, ride quality of a motor vehicle is normally associated with the roughness. Here, the term "roughness" generally has different meanings in response to the purpose of applications, which is normally recognized as a dominant source affecting motor vehicle behavior on roadways rather than sidewalks.

To better understanding how the surface characteristics are associated with an EMS traveling over sidewalks, this study examines the interaction between pavement surface properties and the vibration characteristics of an EMS. For this purpose, this study conducted a driving experiments of an EMS on ten different surfaces with various roughness levels of pavements involving dense graded, porous (open graded), and polished asphalt wearing courses as well as an interlocking block pavement. In the experiment, vertical vibration responses of the EMS were measured with accelerometers. Traveling locations of the EMS were precisely identified with a high-performance GNSS (Global Navigation Satellite System) with an accuracy of centimeter to synchronize the acceleration data with surface profiles.

2 Materials and Methods

2.1 Pavement Surfaces

The driving experiment was conducted in a test yard of Construction Machinery Center in Obayashi Road Corporation located in Saitama, Japan. The test yard consists of ten different surfaces with various roughness levels of pavements involving dense graded, porous (open graded), and polished asphalt wearing courses as well as an interlocking block pavement. as shown in Fig. 1. A series of the profile data was measured with a hand-operated low-speed profiler at an interval of 10 mm so as to acquire roughness and megatexture properties of the profile. The macrotexture was measured with the Circular Track Meter (CTM) for each pavement section shown in Fig. 1 according to the ASTM E2157 (ASTM 2015). Figure 2 shows surface measurement devices and their operations.

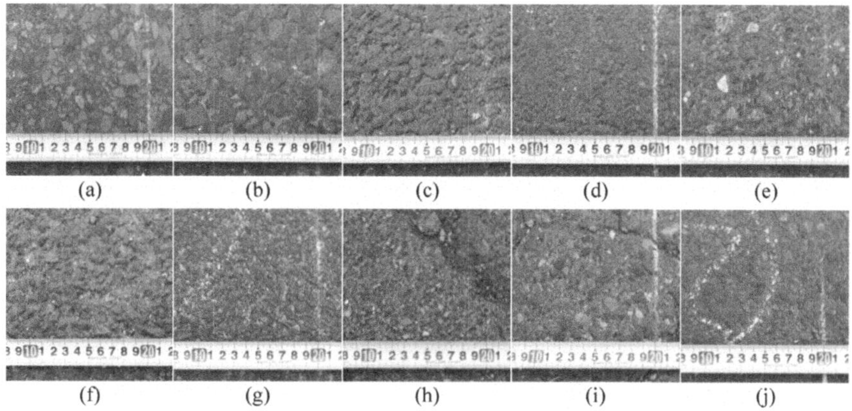

Fig. 1 Tested surfaces: **a** Polished A, **b** Polished B, **c** Porous A, **d** Dense Graded A, **e** Dense Graded B, **f** Porous B, **g** Porous C, **h** Interlocking Block (ILB), **i** Dense Graded C, and **j** Dense Graded D

Fig. 2 Surface Measurement Devices: **a** the Hand-operated Profiler and **b** the CTM

2.2 EMS Response

EMSs are solely battery-operated personal small vehicles of which dimensions have up to 1,200 mm length, 700 mm width, and 1,200 mm height according to the ministry regulation in Japan. The vibration response of the EMS can be affected by a number of factors such as the traveling speed, vehicle size, and suspension type. This study employed one of the most widely used EMSs in Japan as shown in Fig. 3. The EMS driven by an average male operator made three repeated runs at a constant speed of 3 km/h which was equivalent to the average traveling speed of wheelchair users (Duvall et al. 2013).

Figure 3 shows the EMS data collection system. The accelerometers were attached to the unsprung mass, sprung mass, and center of gravity (CG.) of the EMS as shown in the figure to record the vibration responses. The acceleration data collected at unsprung mass, sprung mass, and CG. aim to measure the direct oscillation from surface, footplate vibration, and seat vibration corresponding to the ride quality, respectively. The sampling frequency of acceleration measurement was adjusted to 100 Hz. During the trials, traveling locations were measured with a high-performance GNSS developed by i System Research, Co. Ltd. (iSRC-GNSS) and with conventional GPSs (Global Positioning Systems). The iSRC-GNSS has the capability to identify the location with the horizontal accuracy of 0.6 cm + 0.5 ppm and vertical accuracy of 1 cm + 1 ppm with an update interval of 0.05 s in real-time. This performance enables its users to synchronize accurately the acceleration data with surface profiles traveled. Figure 4 depicts an example of a traveled location identified by the positioning systems. As shown in the figure, the iSRC-GNSS accurately identified

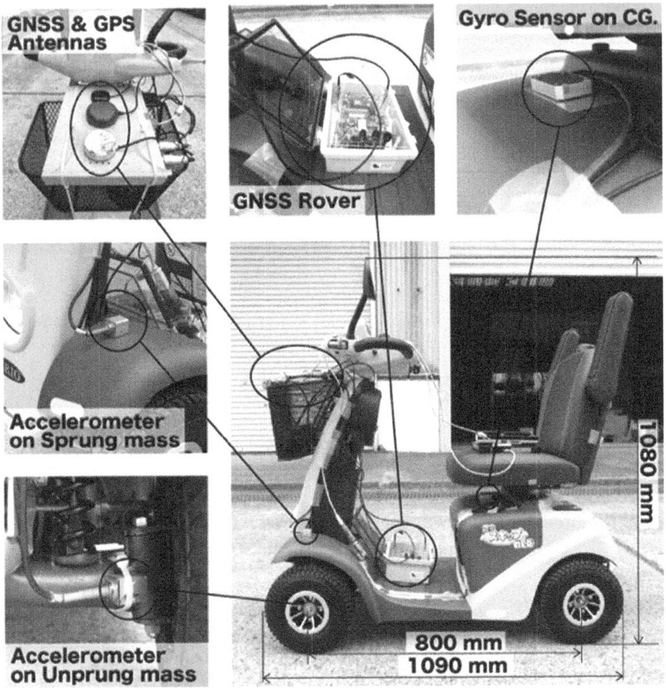

Fig. 3 EMS data collection system

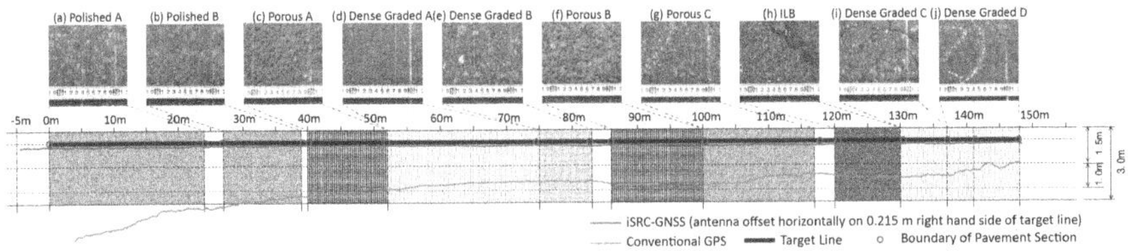

Fig. 4 Traveled location identification

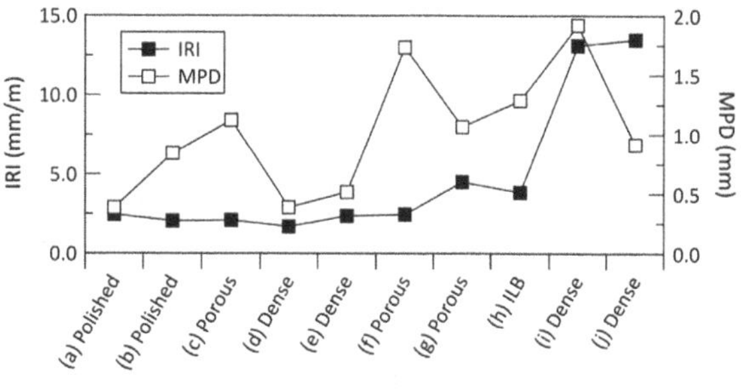

Fig. 5 Surface roughness and texture levels of the test sections

the location traveled, whereas the conventional GPSs deviated from the target path with a error of 3 m at the most. The figure also shows the location of the tested surfaces and their boundaries identified by the iSRC-GNSS corresponding to Fig. 1.

3 Analysis Result

3.1 Surface Roughness and Texture

To represent surface characteristics, the International Roughness Index (IRI) and Mean Profile Depth (MPD) are employed in this study. The IRI is the most well-known roughness index which is an accumulation of the simulated displacement between body and axle masses in the standard quarter-car model normalized by the traveled distance of the profile at a constant speed of 80 km/h (Sayers 1995). The IRIs of each tested surface were calculated form the measured profile by use of the ProVAL software (The Transtech Group 2015). The MPD is one of the best indices of macrotexture as the literature stated (Abbas et al. 2007) and is known as a good estimator of surface friction (Kouchaki et al. 2018). The MPDs of each tested surface were obtained with CTM in the present study. The surface roughness and texture levels in terms of the IRI and MPD, respectively, obtained in the test sections are shown in Fig. 5. As shown in the figure, a wide variety of roughness and texture levels were acquired for the experiment.

3.2 Vibration Response of EMS

The fundamental of vehicle ride can be recognized how vehicle components (normally a suspension system and tire damping) isolate the rider from imperfect smoothness of a road surface. It is important that the certain frequency known as resonance frequency of a mass amplifies the input motion. A frequency response plot helps to understand this phenomenon. The main application of a frequency response plot to

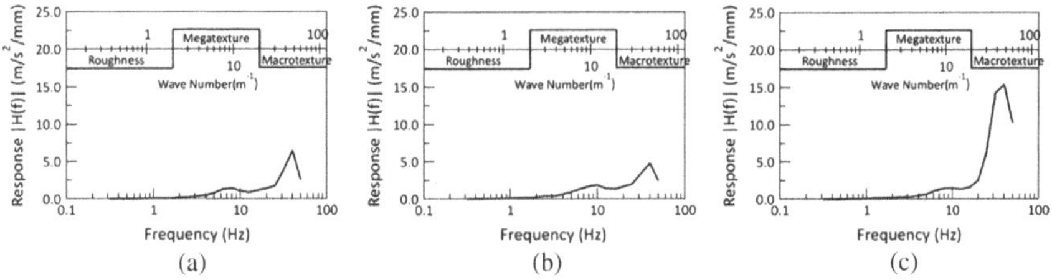

Fig. 6 Frequency Response Plots of EMS on **a** CG., **b** Sprung Mass, and **c** Unsprung Mass

vehicle ride is the way it relates the input power spectral density (PSD) of the road to the output PSD of the vehicle acceleration (Gillespie and Sayers 1981). Figure 6 shows frequency response plots of each component of which the accelerometers were attached. In the figure, the wave numbers corresponding to the traveling speed of 3 km/h are also indicated.

As shown in the Fig. 6, the EMS has two distinctive resonance frequencies of 8 and 30 Hz which can be identified as body and axle mass resonance frequencies, respectively. With respect to the surface characteristics, these frequencies are associated with the terms "megatexture" and "microtexture". Note that humans are the most sensitive to the vibration for vertical direction with a frequency range from 4 to 8 Hz according to the ISO Standard 2631-1(ISO 1997). Consequently, the megatexture plays an important role in the perception of EMS users in terms of traveling comfort. As shown in Fig. 6, the EMS resonates with the macrotexture as well. In other words, the macrotexture induces significant vibrations of the axle mass of the EMS. In contrast, the EMS does not respond to the roughness input. As already mentioned, the term "roughness" generally has different meanings in response to the purpose of applications. In case of the EMS, megatexture and microtexture become dominant sources affecting vehicle behavior instead of roughness.

3.3 Pavement-Vehicle Interaction

Summary indices of a surface profile help to monitor the performance of sidewalks. The IRI and MPD are commonly used indices to assess the roughness and texture of a surface, respectively. Good correlations between the vibration response of the EMS and profile-based indices such as the IRI and MPD contribute to the effective surface monitoring of sidewalks over time. Accordingly, the RMS values of the EMS accelerations acquired for each tested surface were calculated to investigate the correlations with the IRI and MPD. Figure 7 shows the correlation between the RMS accelerations and each profile-based index. As shown in Fig. 7 (a), poor correlations were observed between the IRI and the RMS accelerations. This result can be understandable because the IRI has been developed for a motor vehicle with the driving speed of 80 km/h. In fact, the EMS scarcely respond to the wavelengths

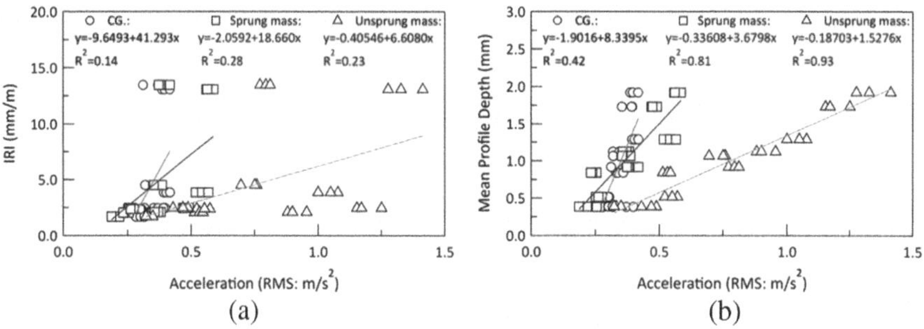

Fig. 7 Correlation of RMS Acceleration with **a** IRI and **b** MPD

of roughness which are related to the IRI as shown in Fig. 6. In contrast, the MPD has strong correlations with the RMS accelerations, in particular at the unsprung mass. This result ensures that the EMS vibration response in the unsprung mass corresponds to the macrotexture in the surface properties. It also suggests that the MPD can be estimated by the unsprung mass acceleration measured with the EMS. Since the MPD is known as a good indicator of surface friction, measuring acceleration of an EMS helps to monitor the surface friction of sidewalk pavements. However, further studies should be required in this viewpoint.

4 Discussion and Conclusions

Pavement surface condition of sidewalks plays a vital role in ensuring traveling safety and comfort of EMS users. However, very little knowledge has so far been proposed to evaluate surface characteristics interacting with EMSs traveled on sidewalks. This study examined how the surface properties interact with the vibration response of the EMS by a driving experiment on ten pavement surfaces with different roughness and texture characteristics.

As a result of the driving experiment, the EMS has two distinctive resonance frequencies for the vertical movement: one appears at 8 Hz related to the resonation of the sprung mass and CG. of the vehicle body, and the other is the excitation of 30 Hz observed at the unsprung mass. At a constant traveling speed of 3 km/h, the former corresponds to the megatexture in the surface properties, whereas the later derives from the microtexture instead.

According to the pavement-vehicle interaction analysis, the IRI has poor correlation with the EMS vibrations in terms of the RMS acceleration measured at any components. This result was easily recognized due to the different driving speed and vehicle masses. In addition, the EMS scarcely responded to the wavelengths of roughness In contrast, the MPD is strongly associated with the RMS acceleration of the unsprung mass of which resonance frequency corresponds to the macrotexture. This result accordingly suggested that measuring acceleration of an EMS helps to monitor the surface friction of sidewalk because the MPD is known as a good indicator of surface friction. The findings obtained in this study will bring the optimization of asphalt mixes including aggregate size and texture shapes for side walk pavements.

EMSs have become a popular walking aid that enables its users to travel around without any physical effort. The demand of the use of an EMS is increasing with rapid increase of the proportion of elder people. In this situation, the quality of road surfaces served for EMS users plays an important role in the traveling comfort of the sidewalk. The findings presented in this paper contribute to improving the accessibility of people with disabilities.

Acknowledgements This research was supported by the Grant-in-Aid for Scientific Research (C) Grant Number 19K04634 of the Japan Society for the Promotion of Science (JSPS). We would like to thank Dr. Keiichi Nishikawa of i System Research, Co. Ltd. for supporting the driving experiment in this study.

References

Abbas A, Kutay ME, Azari H, Rasmussen R (2007) Three-dimensional surface texture characterization of portland cement concrete pavements. J Comput-Aided Civil Infrastruct Eng 22(2007):197–209

Aghaabbasi M, Moeinaddini M, Shah MZ, Asadi-Shekarid Z (2017) A new assessment model to evaluate the microscale sidewalk design factors at the neighbourhood level. J Transp Health 5:97–112

ASTM (2015) Standard Test Method for Measuring Pavement Macrotexture Properties Using the Circular Track Meter. Publication ASTM E2157

Duvall J, Cooper R, Sinagra E, Stuckey D, Brown J, Pearlman J (2013) Development of surface roughness standards for pathways used by wheelchairs. Transp Res Rec 2387:149–156

Gillespie TD, Sayers M (1981) Role of road roughness in vehicle ride. Transp Res Rec 836:15–20

ISO (1997) Mechanical vibration and shock – Evaluation of human exposure to whole-body vibration – Part 1: General requirements. Publication ISO2631-1, International Organization for Standardization, Switzerland

Kouchaki S, Roshani H, Prozzi JA, Garcia NZ, Hernandez JB (2018) Field investigation of relationship between pavement surface texture and friction. Transp Res Rec 2672:395–407

PIARC (1987) Technical committee report on surface roughness characteristics. XIIIth World Road Congress, Brussels, pp 13–19

Sandberg U, Ejsmont JA (2002) Tyre/Road Noise Reference Book. Informex, Harg, SE-59040 Kisa, Sweden

Sayers MW (1995) On the calculation of international roughness index from longitudinal road profile. Transp Res Rec 1501:1–12

The Transtech Group (2015) ProVAL User's Guide. Manual Revisions: 6.2(3.5)

Thoreau R (2015) The impact of mobility scooters on their users. Does their usage help or hinder? A state of the art review. J Transp Health 2:269–275

Acoustic Maintenance of Pavements by Large-Scale Grinding

Françoise Beltzung and Tobias Balmer

Abstract A so-called semi-dense asphalt (SDA) is widely used in Switzerland as a low noise pavement. During 2018, our preliminary grinding tests on small SDA sections and subsequent broader tests have shown that grinding is an effective way to regain −3 to −5 dBA, measured with the close-proximity (CPX) method. The main purpose of this paper is to put forward a large-scale maintenance process which is effective in regenerating the acoustic performance of aged SDA wearing courses for entire road sections. A secondary purpose is to observe the effect of the same grinding method on dense pavements like AC, ACMR and MA. Several roads at different locations and environmental conditions have been tested in early 2019. At each location the treatment was evaluated by the measurements of grinding depth, mean texture depth and CPX rolling noise. Acoustic effectiveness of the grinding method depending on the pavement characteristics are examined. Our first results indicate that all the treated surfaces become less noisy, regardless of grain size or pore volume. The incidence of grinding on reshaping the surface texture in a way that facilitates air pumping and diminishes tyre vibration will be discussed.

Keywords Low noise asphalt · Grinding · CPX measurement

1 Introduction

Road operators worldwide have to find a way to mitigate traffic noise emission generated by the interaction between tires and the road surface. Low noise wearing courses are an effective measure to tackle the issues at its source. In Switzerland a trend of using so-called semi dense asphalt (SDA) with porosity of 12–20% and a maximum grain size 4 or 8 mm has emerged about a decade ago. While the noise reduction of −5 to −9 dBA in reference to the Swiss standard for such pavements is overall

F. Beltzung
Centre de Compétences du Domaine Routier, School of Management and Engineering Vaud, Champs-Lovats 29, 1400 Yverdon, Switzerland

T. Balmer (✉)
Weibel AG, Rehhagstrasse 3, 3018 Bern, Switzerland
e-mail: tobias.balmer@weibelag.com

C. Raab (ed.), *Proceedings of the 9th International Conference on Maintenance and Rehabilitation of Pavements—Mairepav9*, Lecture Notes in Civil Engineering 76,
https://doi.org/10.1007/978-3-030-48679-2_84

satisfying, the acoustical and mechanical lifetime is generally below expectations. As with porous asphalt (PA) the acoustic ageing is dominated by clogging of the open pore network and altering of the macrotexture due to ravelling.

Our new approach to ensure acoustic durability of SDA pavements is to lightly grind off the upper surface in a maintenance procedure at appropriate time providing two beneficial outcomes: the aged texture is restored and the clogged layer—located just under the surface (Balmer 2018)—is partially removed at the same time.

Other countries already practise the grinding of wearing courses. This is done with multiple aims such as increasing the skid resistance, or some, like us, in connection with noise maintenance (Vieira 2019; Buret 2016). Vieira et al. were principally seeking to optimise the pavement surface texture to reduce the tire/road noise. They describe their procedure as "shaving off" the peaks of the texture. It is not especially dedicated to porous low noise pavements nor to eliminate the clogged layer if any. For dense asphalt test sections, the acoustic gain is less than −1 dBA. For porous asphalt test sections, depending on the grinding depth and the state of clogging, the acoustic gain ranges between −1.6 and −3.3 dBA. Buret et al. initiated a long-term trial to investigate the acoustic performance of various asphalt pavements. After the first three years the best performing pavement was found to be a surface that was treated by horizontal diamond grinding. The "shaved" open-graded asphalt showed a consistent improvement in noise reduction of up to −3 dBA over the standard open-graded asphalt. It showed also a lesser degradation in acoustic performance over time compared to the non-treated open-graded asphalt.

Our present investigation into the relationship between surface texture and noise level includes not only SDA pavements but also dense asphalts concrete (AC), macro-rough asphalt (ACMR) and mastic asphalt (MA). The purpose behind this idea is to clarify whether the same grinding method shows also a positive effect on dense mixtures with different grains sizes. In addition this will help to distinguish the role of surface texture from the role of porosity on the overall noise gain due to grinding. Finally, it will be interesting to see how microtexture may play a relevant role as well as macrotexture in terms of noise arising from the pressure relief of trapped air.

In the long term the goal of this work is to develop a sound and economically sustainable option to rejuvenate the acoustic performance of roads at a given frequency, thus allowing road authorities to comply with noise regulations without the need for early and expensive replacements of entire wearing courses.

2 Materials and Methods

2.1 Test Sections

Table 1 provides an overview of the test sections including age at the time of maintenance, traffic load, dimension and type of tool used on the grinding machine. Among the seven locations, some have more than one type of pavement. In these cases,

Table 1 Characteristics of the test roads (all locations in canton Fribourg, Switzerland)

Location	Pavement	Age at time of maintenance [years]	Traffic load [DTV]	Maintenance length [m]	Tool type
Corserey	SDA 4	6	2400	2 × 750	Fine
Farvagny	SDA 4	2	3300	120	Fine
	AC 11	>20	3300	150	Fine
Sugiez	SDA 4	6	11600	450	Fine
La Verrerie	SDA 4	7	1500	2 × 950	Coarse
Romont	SDA 4	7	6200	2 × 500	Fine
Tafers	SDA 4	6	5200	100/100	Coarse/Fine
	MA 11	20	5200	75/75	Coarse/Fine
Villarlod	ACMR 8	8	1900	150	Fine
	ACMR 11	8	1900	150	Fine

adjacent sections to SDA 4 are AC 11 in Farvagny and MA 11 in Tafers with the numeric index being the maximum grain size in mm. Moreover, because of a specific interest, ACMR 8 and 11 pavements were selected in Villarlod to be included in the test series. The maintenance took place in two sessions with the first set of roads being treated during November 2018 and the second one during May 2019. The maintenance length is given in meters per lane. The SDA 4 in Corserey, La Verrerie and Romont was treated on both lanes of relatively long sections between 500 and 950 m in length. The other locations had smaller sections of 120 to 450 m in length. Additionally, the tool type was changed on the SDA 4 and MA 11 in Tafers on a stretch of 100 m and 75 m each. The effective treated area lies between 360 m^2 for the smallest and 7500 m^2 for the largest test section.

2.2 *Grinding Procedure*

For the grinding work, a Unimog truck with 5 HTC 950 floor grinding heads attached on a custom made rig to its front was used, having an adaptable working width of 2.4–4.2 m (Fig. 1a). We used a grinding speed of 2–3 m/min, as a compromise for grinding depth and ability to treat even the large test sections in one working day. On each grinding head four tool plates are mounted that move horizontally in a planetary motion (Fig. 1b). We used two different types of bush hammer tools with differently sized and spaced hard-metal spikes of either 30 pcs of 5 mm or 45 pcs of 8 mm spikes per single wheel. In the course of the present evaluation they are referred to as "coarse" and "fine", whereas the latter has the more densely-packed and slightly sharper spikes when in new condition. To improve skid resistance on asphalt these kind of tools would normally be used at much higher working speed than we did. To our knowledge, we are the first to use such tools on small aggregate porous asphalt

Fig. 1 Images of the large-scale grinding work. The lead vehicle **a** worked its way on the entire width of one lane. The bush hammer tools on the grinding heads **b** produced an even layer of fine-graded sand **c** that was cleaned directly by a series of street cleaners. Behind this convoy, the ground surface was left clean and ready for traffic **d**

with the purpose of accoustic maintenance. For their work on surface grinding (Veira 2019) and (Buret 2016) used diamond tools which operate differently.

Small amounts of spray water in front of the grinders is applied to bind dust. Behind the Unimog a convoy of street cleaning machines followed to sweep up the debris from the grinding process. The last vehicle did a final cleaning with high pressure water and suction, leaving a clean road ready for traffic (Fig. 1d). The overall noise and dust generation is not higher than as for other typical street construction work.

2.3 Mean Grinding Depth Measurement

The mean grinding depth MGD was determine through differential measurements from before and after grinding. A rigid metal bar was laid across the entire lane onto two supporting points that were not changed during grinding, then the distance from the bar to the road surface was measured with a ruler along the profile at spacing of 25 cm, resulting in an ensemble of 12–14 single point measures for one lane cross section. For an averaged cross-section, the accuracy of this method lies at about

±0.5 mm. The same procedure was repeated at multiple predefined locations along the worked area yielding spatially resolved grinding depth data.

2.4 Acoustic CPX Measurements

Acoustic performance of the pavement has been measured on the go with a CPX trailer (ISO 11819-2) at speed of 50 km/h. The noise reduction is calculated relative to the Swiss reference level called StL86+ which depends on the type of vehicles (car vs. truck) and the speed (BUWAL). The two values CPX_{car} and CPX_{truck} are then mathemetically mixed to reflect a standard real traffic with a certain amount of trucks called CPX_{mix}. The measurement accuracy of CPX is within ±1.0 dBA and in this work CPX_{mix} was calculated with a truck rate of 10%.

For further analysis the frequency plot is used. It allows to identify changes in the frequency curve in domains that are characteristic for certain noise sources: low frequency near 500 Hz for tire vibrations, medium frequency around 1000 Hz for sound absorption by the pavement and high frequency above 1500 Hz for air pumping from the tyre/road interaction.

2.5 Mean Texture Depth Measurements

The mean texture depth MTD or macrotexture depth is measured according to (EN 13036-1). The technique is to spread on the pavement a known volume of glass beads to form a circular patch, and then determine its surface. The MTD is calculated as the volume divided by the surface and expressed in mm.

2.6 Cross-sectional Images

To assess the state of clogging of the SDA, cores of diameter 150 mm are cut from the site. The porous layer is sawn from the rest of the core and sealed along the curved surface while the two flats remain accessible. An appropriate epoxy resin is coloured in fluorescent red or yellow. The SDA layer is fully vacuum impregnated. Finally, the specimen is sawn into two pieces to obtain a pair of sections. Each section is photographed under UV light to reveal the UV fluorescent material filling the pores. From each SDA 4 road, several cores have been impregnated including one reference from a non-grinded section.

3 Results and Discussion

3.1 MGD Results

Tables 2 and 3 provide the results of the mean grinding depth measurements. The MGD of the four dense pavements (Table 3) runs for all around 1 mm. The MGD of the six porous SDA 4 pavements fluctuates between 0.4 and 3.5 mm. The latter dispersion cannot directly be attributed to the type of grinding tool (fine or coarse), nor the age of the pavement, nor the traffic load. In Tafers where the coarse and the fine tools were applied to the same surface, it is observed that the coarse tool grinds off less material compared to the fine tool. This effect may be due to the more densely and sharper spikes of the fine tool. The main reason for the MGD dispersion can be explained by the petrography of the aggregates, namely their proportion of hard, resistant rocks. The latter observation has been previously described in (Beltzung 2019).

Table 2 Mean grinding depth and absolute acoustic improvement due to grinding; overview of the porous SDA 4 test surfaces

Location	Pavement	Tool	MGD [mm]	CPX_{car} [dBA]	CPX_{truck} [dBA]	CPX_{mix} [dBA]
Corserey	SDA 4	Fine	1–2	−2.8	−3.2	−3.0
Farvagny	SDA 4	Fine	0.3–1.3	−0.9	−1.4	−1.1
Sugiez	SDA 4	Fine	2.5–3.5	−1.3	−1.4	−1.4
La Verrerie	SDA 4	Coarse	0.4–1.7	−2.5	−2.0	−2.3
Romont	SDA 4	Fine	2–3	−3.2	−3.5	−3.3
Tafers	SDA 4	Fine	1.3–2.1	−3.0	−2.6	−2.8
	SDA 4	Coarse	1.0–1.5	−3.0	−2.5	−2.8

Table 3 Mean grinding depth MGD and absolute acoustic improvement due to grinding; overview of the dense test surfaces

Location	Pavement	Tool	MGD [mm]	CPX_{car} [dBA]	CPX_{truck} [dBA]	CPX_{mix} [dBA]
Farvagny	AC 11	Fine	0.8–1.3	−3.2	−2.3	−2.8
Tafers	MA 11	Fine	0.5–1.5	−2.7	−1.5	−2.2
		Coarse	n/a	−2.5	−1.0	−1.8
Villarlod	ACMR 8	Fine	≈ 1	−2.2	−2.0	−2.1
	ACMR 11	Fine	0.5–1.1	−3.1	−3.0	−3.1

3.2 CPX Results

The CPX values from the non-grinded sections (reference) and those measured directly after grinding are compared to determine the absolute acoustic gain due to grinding. These results are shown in Table 2 for the SDA pavements and in Table 3 for the non-porous pavements.

On the basis of the mean grinding depth there appears no direct correlation between MGD and the acoustic gain. The overall observed gain ranges between −1 and −3 dBA and is acoustically significant. We note that the observation also applies to the dense pavements which show acoustic gains of the same order of magnitude.

Figure 2 shows the frequency plots for porous SDA 4 and dense MA 11 in Tafers, including a comparison between fine and coarse grinding and a reference curve for a split mastic asphalt (SMA 11) which represents the noise standard of 0 dBA in Switzerland. The plots reveal that for an aged SDA 4 (red line) the low frequencies which are relevant for vibrational noises became louder than the SMA reference curve, while the higher frequencies (air pumping noise) still contribute to noise reduction. The curves of the aged MA 11 and the SMA 11 are very similar before grinding.

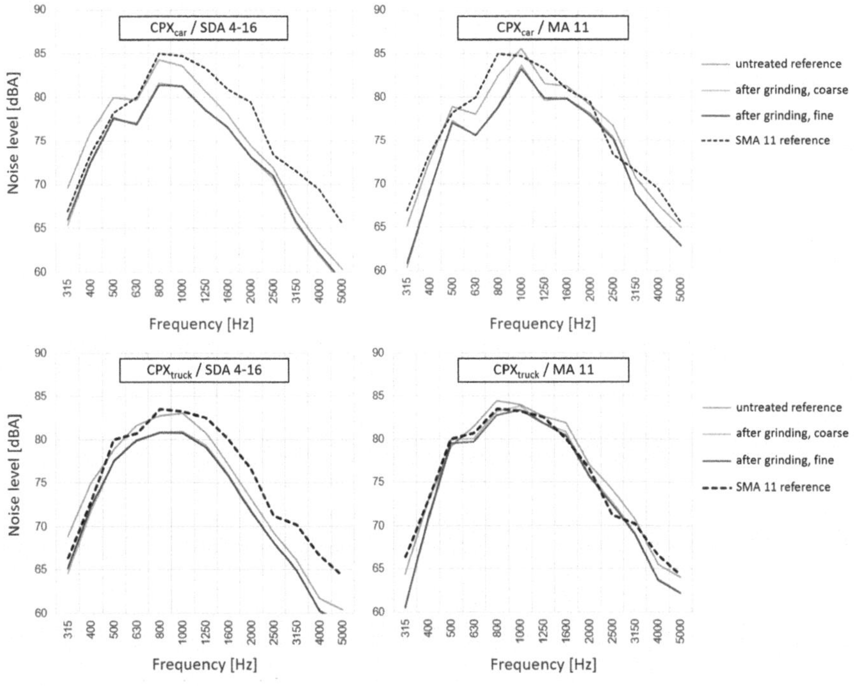

Fig. 2 Frequency plots from CPX measurements in Tafers for SDA 4 (left) and MA 11 (right). Top row shows CPX_{car} and bottom row CPX_{truck} curves

The effect of grinding on the acoustic performance is seen over the entire frequency range for both pavement types as well as for the distinctive car and truck tyres. However the acoustic gain is generally more dominant in the lower frequency domain, thus indicating that vibrational noises have been reduced by a certain improvement of the surface texture. Interestingly there is no noticeable difference between coarse and fine grinding. The only exception to this, is the truck tyre on MA 11, where the fine tool produced a slightly better effect by generating less vibrations. That can be attributed to the difference of −0.5 dBA seen in Table 3 for CPX_{truck} on MA 11 between coarse and fine grinding in Tafers.

3.3 MTD Results

Grinding has a significant effect on the roughness of the surface texture. As far as porous SDA 4 pavements are concerned, irrespective of whether they still are in a good shape or already abraded, they all tend to a unified MTD value of 0.8–0.9 mm after grinding with the fine tool (Table 4). In the specific case of La Verrerie the grinding work smoothened the surface from MTD 1.3 to 0.8 mm whereas in Farvagny and Sugiez the surface roughness increased notably from MTD 0.6 to respectively 0.8 and 0.9 mm. Note that the roughness of the SDA 4 pavement in Tafers with an initial MTD of 0.89 mm is barely affected by the grinding with only a small decrease of 0.03 mm to reach MTD 0.86 mm, both values being within the bespoken "unified" MTD range.

In all cases, whether the texture became smoother or rougher, the acoustic performance increased (Table 4). However, a closer look shows that at locations where MTD increased, the acoustic gain amounts to only 1.1 to 1.4 dBA, while locations with decreasing or stable MTD have gains as high as 2.3 to 2.8 dBA. This distinction indicates that the noise reduction due to grinding is due to a multi-fold mechanism.

At this stage of the investigation, with only four test sections, a general explanation would be premature. Instead each location is a particular case that has to be analysed separately. The most recent test surface (Farvagny) has no visible degradation and

Table 4 Comparison of non-grinded (Ref) and finely grinded SDA 4 test surfaces; roughness is expressed as a mean texture depth (MTD) and corresponding acoustic performance is measured by CPX

Location	Year of construction	Ref MTD [mm]	Post-grinding MTD [mm]	Ref CPX_{mix} [dBA]	Post-grinding CPX_{mix} [dBA]	Δ CPX_{mix} [dBA]
Farvagny	2017	0.58	0.78	−7.1	−8.2	1.1
Sugiez	2013	0.59	0.90	−3.2	−4.6	1.4
$Tafers_{fine}$	2013	0.89	0.86	−2.2	−5.0	2.8
La Verrerie	2012	1.28	0.80	−2.9	−5.2	2.3

according to the high acoustic level (−7.1 dBA) of the reference section the SDA 4 is not clogged yet (Fig. 3). The acoustic gain attributable to grinding, i.e. 1.1 dBA, may be explained by a roughening of the otherwise neat surface, i.e. which facilitates air drainage and reduces the so called air pumping. The Sugiez test surface seems to behave similarly to Farvagny, even though the pavement is older but still well preserved ("negative" texture and almost no pop-outs). Sugiez's pavement is partially clogged but from the cross-sectional image we know that removing a thin layer (3 mm) did not reopen a significant portion of new channels through the connected porosity, suggesting again that some kind of roughening of the surface must be the predominant factor to explain the acoustic improvement of 1.4 dBA. Tafers' surface texture must have changed profoundly to explain an acoustic gain of 3 dBA. Knowing that this change is not reflected by MTD, some other surface characteristics must play a role. And finally the oldest test surface (La Verrerie) was priorly in a rather bad shape, in particular numerous pop-outs are responsible for the large MTD value of 1.3 mm. Under this scenario, it is well known that to flatten and smooth the surface by grinding or by other means reduces the tire vibrations and accordingly the noise emission.

Several types of dense pavements have been analysed. They are listed in Table 1. If we now consider the family of aggregate size 11 mm, it is most interesting to notice that their acoustic performances are very sensitive to grinding (Table 5), even though they have no open-graded pores and are not intended to be used as low noise pavements.

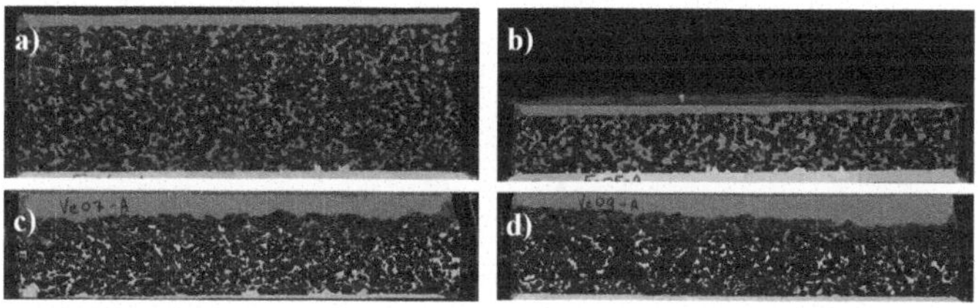

Fig. 3 Farvagny SDA 4 **a** reference Fa04 **b** Fa05 with MGD = 0.3 … 1.3 mm. The large difference in thickness between Fa04 and Fa05 is random and reflects the original layer thickness, respectively 51 and 21 mm. La Verrerie **c** reference Ve07 **d** Ve09 with MGD = 0.4 … 1.7 mm

Table 5 Comparison of non-grinded (Ref) and finely grinded test surfaces with aggregate size 11 mm; roughness is expressed as a mean texture depth (MTD) and corresponding acoustic performance is measured by CPX

Pavement	Year of construction	Ref MTD [mm]	Post-grinding MTD [mm]	Ref CPX_{mix} [dBA]	Post-grinding CPX_{mix} [dBA]	Δ CPX_{mix} [dBA]
AC 11	≈2000	0.70	0.41	−0.5	−4.0	3.5
ACMR 11	2011	1.40	1.39	0.6	−2.5	3.1
MA 11_{fine}	1999	0.37	0.31	−1.2	−3.1	1.9

Their expected acoustic performances right after laying, as a new wearing course, ranges between −1 and −2 dBA. Now that post-grinding performances of −2.5 to −4.0 dBA are observed it is permissible to infer the existence of a very special effect due to grinding. This effect improves air drainage and/or reduces tire vibrations. At this point of the investigation we assume that, besides evenness of the macrotexture, the microtexture must play a significant role too in noise reduction. Microtexture here refers to the surface abrasion of the top aggregates, a characteristic which is usually of interest for the skid resistance.

3.4 Analysis of Cross-sectional Images

The image analysis was completed to offer a visual tool which helps us to better understand the reason why a given maintenance action succeeded more ore less in terms of noise reduction. To this end, the cross-sectional images from cores drilled in Farvagny and La Verrerie are shown for comparison in Fig. 3.

In the case of Farvagny it can be seen that the porous space is entirely free from soiling. It acts as expected as an absorbent medium in a way that achieves high acoustic performances (−7.1 dBA). Further the section shows that the surface is smooth showing only "negative" cavities.

The situation in La Verrerie is quite different in many ways from that in Farvagny. Firstly, a barrier layer of soiling hinders the acoustic waves from penetrating into the porous pavement and secondly, the surface cross-sectional line is particularly rough.

4 Conclusions and Next Steps

We have demonstrated that a large-scale acoustic rejuvenation of porous SDA wearing courses is possible with a CPX_{mix} improvement of between 2 and 3 dBA after >5 years of service life. Further we observed that our method of horizontal grinding on any type of wearing course, also including dense pavements, seems to be acoustically favourable. We make the assumption that these specific qualities are to be found at the macro- *and* microtexture level. On one side, the macrotexture is transformed by grinding, especially in presence of damaged surfaces from pop-outs. Hereby the roughness is levelled and noise related to tyre vibrations diminishes. On the other hand, the acoustic performance of MA 11 and ACMR 11 pavements also increased by 2 to 3 dBA. In these cases we noticed that the noise reduction does not appear as a consequence of levelling the macrotexture. Instead we suspect a texture property that has not been taken into account yet and might be related to the microtexture. In light of these new observations we want to continue the investigations to achieve a realistic model of the low noise surface texture. These investigations may include the measurement of the skid resistance or the mean profile depth with laser-profilometry.

Acknowledgements The authors acknowledge the financial support from the Swiss innovation agency Innosuisse under agreement n° 26081.1. Further, we are grateful to the Canton of Freiburg for collaboration with test sections as well as to the people who aided in data collection, including Pascal Bangueret, Christophe Marendaz and Jonathan Brovelli from CCDR and Simon Lambert, Martin Cherey from Weibel AG.

References

Balmer T, Steiner F (2018) Akustische Alterung lärmarmer Beläge durch Verschmutzung - Grundlegende Mechanismen. Strasse und Verkehr 3:22

Beltzung F, Balmer T (2019) Rehabilitation of porous low noise pavement by large scale grinding, EEC2020 paper, under review

Buret M, Mcintosh J et al (2016) Long term asphalt trial: results of acoustic tests after three years. Acoust Aust 44:273. https://doi.org/10.1007/s40857-016-0063-6

BUWAL, Strassenlärmmodell StL-86 + , 1987, Korrekturen, 1995

EN 13036-1 Road and airfield surface characteristics - Test methods - Part 1: Measurement of pavement surface macrotexture depth using a volumetric patch technique

ISO/FDIS 11819-2:2017 Acoustics - Measurement of the influence of road surfaces on traffic noise - Part 2: The close-proximity method

Vieira T, Sandberg U et al (2019) Negative texture, positive for the environment: effect of horizontal grinding of asphalt pavement. Road Mater Pavement Des. https://doi.org/10.1080/14680629.2019.1610476

Framework for Pothole Detection, Quantification, and Maintenance System (PDQMS) for Smart Cities

Naga Siva Pavani Peraka, Krishna Prapoorna Biligiri, and Satyanarayana N. Kalidindi

Abstract Potholes in flexible asphalt pavement systems are one of the major distresses for fatal accidents. Ingress of water through the pothole disturbs the integrity of the pavement system. Delayed maintenance of potholes will adversely affect safety of road users and health of the road pavements. Therefore, detection, quantification, and maintenance of potholes are three indispensable tasks in pavement asset management. Manual collection of pothole data is time-consuming and laborious. Hence, the use of cutting-edge artificial intelligence techniques has become popular in the recent times. The major objective of this study was to develop a framework for pothole detection, quantification, and maintenance system (PDQMS) to detect and quantify potholes using pavement images collected by an automated survey vehicle; the system was also incorporated with a mechanism that calculates the amount of patching material required for maintenance. The state-of-the-art multiple-object detection algorithm, You Only Look Once version 3 (YOLOv3) was selected to detect potholes from the images. One of the salient characteristic features of the PDQMS developed in this study was to use severity-based pothole classification approach, a first-of-its-kind novel framework, which helped group the pavement sections based on severity of potholes for maintenance operations. The proposed framework is envisioned to assist the agencies in making decisions to patch potholes and reduce fatal accidents, if not maintained.

Keywords Potholes · Maintenance · Image processing · Data collection · PDQMS

1 Introduction

Pavements deteriorate over time due to the combined effect of increasing traffic, seasonal and climatic variations, and unaddressed maintenance needs. Untimely maintenance of pavements often leads to failure well before the proposed design life, which is uneconomical (Haas et al. 2015). However, due to budgetary constraints,

N. S. P. Peraka · K. P. Biligiri (✉) · S. N. Kalidindi
Department of Civil and Environmental Engineering, Indian Institute of Technology Tirupati, Tirupati, Andhra Pradesh 517 506, India
e-mail: bkp@iittp.ac.in

C. Raab (ed.), *Proceedings of the 9th International Conference on Maintenance and Rehabilitation of Pavements—Mairepav9*, Lecture Notes in Civil Engineering 76,
https://doi.org/10.1007/978-3-030-48679-2_85

most of the agencies across the globe delay the maintenance activities (Koch and Brilakis 2011). Delayed maintenance is not just a threat to the pavement, but for the road users too. A pothole, one of the distresses of flexible pavement systems frequently causes vehicle damage and fatal accidents as well. In monetary terms, the cost of accidents and compensation given to the victims was far less than the cost of maintenance activities (Eriksson et al. 2008; Koch and Brilakis 2011).

Ingress of water through cracking is the major threat for asphalt pavements, which can lead to rapid development of potholes during wet weather conditions. Furthermore, the separation of asphalt overlays due to change in permeability across the pavement system results in stripping and creates shallow potholes. A maintenance strategy must be adopted after identifying the root cause for the origin of pothole. Poor maintenance operations such as filling up the potholes with gravel in order to provide a short-term riding comfort to the road users attracts water and accelerates the deterioration rate. Similarly, poor maintenance leads to failure of pothole patches, which is consequential of pavement failure. Therefore, it is essential to address pothole maintenance as a crucial issue (Jahanshahi et al. 2013).

Automated distress detection methods are widely used to collect pavement distress data such as, cracking, potholes, patching, rutting, and roughness, etc. using cameras and sensors. Surface anomalies such as cracking, patching, and potholes are captured using cameras. The images are processed to detect and quantify distresses (Huidrom et al. 2013; Haas et al. 2015). Most of the recent studies have focused on automating the pothole detection from the images using image processing (IP) and machine learning (ML) tools in order to assist drivers to safely travel on the highways, rather than providing a potential solution. Therefore, there is a need to automate the pothole detection, and quantification that provides agencies with potential insights on treating potholes. In order to do this, as a preliminary effort, a design framework for pothole detection, quantification, and maintenance system (PDQMS) is described in this paper. The following sections detail the current state-of-the-art on pothole detection and quantification, proposed design framework, followed by conclusions and recommendations.

2 Current State-of-the-Art on Pothole Detection and Quantification

Several studies have reported pothole detection procedures based on pavement images (Koch and Brilakis 2011; Huidrom et al. 2013; Koch et al. 2013; Mathavan et al. 2015; Ryu et al. 2015; Suong et al. 2018; Ye et al. 2019). The existing methods can be broadly classified as vibration-based, three-dimensional (3D) reconstruction-based, and vision-based methods (Koch and Brilakis 2011). Accelerometers have been widely used for detecting potholes along the wheel path. The data storage required is quite low as the data is recorded in the form of signals. It has been observed that differentiating the pothole from speed breakers and manholes is a difficult task as they also cause for the change in signal recordings. In addition, removal of noise from the signals is quite a cumbersome task as finding a threshold value to segment a pothole from the whole signal is further challenging. Eriksson et al. (2008) developed a laser-based pothole detection system called Pothole Patrol (P^2), which is capable of detecting potholes on wheel paths only. Note that the P^2 system was not able to differentiate potholes from manholes.

In order to detect potholes using 3D reconstruction-based methods, highly-efficient laser scanning equipment is essential for data collection to understand both extent and severity of potholes. For instance, Chang et al. (2005) used 3D laser scanner to detect and quantify severity and extent of potholes. In another study, Yu and Salari (2011) proposed a laser-based imaging system to detect potholes. The system was designed to compare the time-of-flight of laser beam used for creating the current pavement image with the standard pavement image without potholes. The difference in time-of-flight was then used to detect potholes and other surface anomalies. The region affected due to pothole was then segmented to measure severity of the distress. However, the cost of employment of laser system is much higher than vibration-based methods and uneconomical. Stereovision-based method (Hou et al. 2007; Mathavan et al. 2015) uses simultaneous views obtained from two cameras, match the feature points to reconstruct pavement structure and detect defects such as potholes. This method requires high computational efforts, which is not practical for real-time detection. In another procedure, to obtain both depth and surface information, a low-cost Microsoft Kinect, a depth-light sensor has been used primarily equipped to a vehicle to capture pavement texture and depth information to create 3D pavement surfaces (Jahanshahi et al. 2013; Becerik-Gerber et al. 2015). Even though the cost of the sensor assembly is less, the system has two major drawbacks: i) lighting conditions greatly influence the depth readings so the system needs to be protected from the sun, and ii) the rate of data collection is quite low.

Vision-based methods use pavement images and videos collected using either auto mated distress data collection vehicles equipped with cameras, smartphones for detecting potholes, and other pavement surface anomalies using artificial intelligence and computer vision techniques such as IP, ML, and deep learning. These techniques are quite cost-effective and detect potholes from pavement images. Potholes are assumed as a near-ellipsoidal shaped object to detect it from pavement images using IP techniques. The images usually undergo four operations such as pre-processing, segmentation, feature extraction, feature selection, and detection (Zakeri et al.2016). Koch and Brilakis (2011) developed a histogram-based thresholding algorithm on MATLAB® platform. A set of 120 images were used to test the accuracy of the algorithm. Huidrom et al. (2013) devised an algorithm that is capable of detecting potholes from videoframes. A set of digital and monochrome images were used to train the algorithm for detection. In order to detect potholes in real-time, Koch et al. (2013) developed a texture-based tracking algorithm to detect and quantify potholes from video frames. The tracking-based methods were developed to alert drivers that a pothole is ahead. Hadjidemetriou et al. (2017) developed a support vector machine (SVM)-based patching detection algorithm to detect cracks from pavement images.

The aforementioned methods are localized, developed, and trained on finite datasets. Also, the computational costs of these systems are quite high, which is one of the crucial reasons for developing deep learning-based methods for detecting potholes.

Convolutional neural network (CNN) is a deep neural network designed to classify different classes such as human, animal, bag, and coin, etc. The network was trained on ImageNet, an open source dataset with millions of images of several classes. Later, researchers started adopting the concept of CNN for various domain applications such as anomaly detection, time series forecasting, video analysis, and natural language processing, etc. (LeCun and Bengio 1995). You only Look Once (YOLO) is a multi-object tracking method developed by Redmon et al. (2016) to detect multiple objects in a single image. The architecture of YOLO consists of a CNN to classify various objects (Redmon et al. 2016; Redmon and Farhadi 2017; Redmon and Farhadi 2018). Suong and Jangwoo (2018) developed a YOLO-based tracking algorithm to detect potholes from pavement images. In that study, the DarkNet architecture of YOLO version 2 (also referred to as YOLO9000) was modified to make it suitable for the proposed problem. The modified version had 27 layers and 18 million parameters to learn unlike YOLO9000, which had 31 layers and 48 million parameters. A set of 996 images were used for training the network and the precision of the algorithm was tested on 203 images. The model had achieved an average precision of 82.43%, which is better when compared with the performance of traditional YOLO9000 version. But, quantification of potholes was not performed. Ye et al. (2019) developed a CNN for pothole detection using photographs. A set of 96,000 images were used for training the network. The performance of the developed algorithm was compared with traditional IP techniques. The algorithm was capable of detecting only one pothole in an image whereas in reality, there are chances of having more than one pothole present on the pavement sections.

3 Design Framework for PDQMS Developed in This Study

The aforementioned methods are capable of detecting and quantifying potholes. An automated system to detect, quantify, and maintain potholes is essential to avoid accidents and vehicle damages. With this objective, a framework for PDQMS is necessary. In order to detect and quantify potholes from pavement images, the state-of-the-art YOLO version 3 (YOLOv3) framework was considered. YOLO is capable of detecting multiple classes in an image. Figure 1 shows the overall framework of the PDQMS developed in this study. The stages in PDQMS are given below and described in the following subsections:

- Collect and prepare image dataset
- Develop YOLO network to detect three classes
- Segment the pothole region from pavement image
- Quantify the patching area and materials

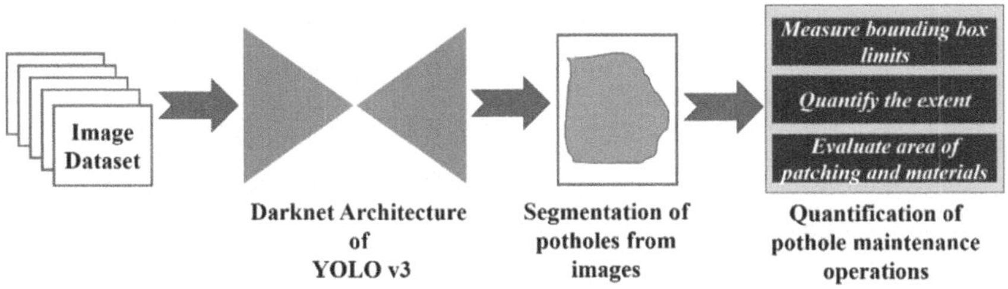

Fig. 1 Framework of PDQMS formulated in this study

3.1 Preparation of Dataset

Dataset plays a crucial role in the performance of artificial intelligence techniques. Imperfections in data would always affect the performance of the systems so it is considered as a crucial task. Differential illumination, shadows of vehicles passing nearby, oil spillage marks, and so on make it difficult for detection. Therefore, images were preprocessed to remove noise and eliminate blurring. Then, the image size that is suitable for the network was approximated to the on-site dimensions for quantification. Further, the images were classified based on the severity and extent of potholes. The American Society for Testing and Materials (ASTM) International classifies the potholes based on severity levels (ASTM D6433-18 2018), and the typical severity levels are given in Table 1. Since the images do not have the depth information, potholes were classified by observing the texture of the pothole. If the texture was dark, depth of the pothole was more, and vice versa.

Before developing the architecture, the following operations were performed on the data to make it suitable for the training process.

- Images were preprocessed using IP operations: histogram equalization (to enhance the quality of the image), filtering, Gaussian noise and removal. The aforesaid operations were performed on the image after converting it to gray scale in order to reduce blurring. This helped for better visualization of pothole features on the gray scale image when compared with the color image.
- Images were separated based on severity levels, as mentioned in Table 1.

Table 1 Severity levels of potholes as per ASTM D6433-18 (2018)

Average diameter of the pothole (mm)	Depth of pothole (mm)		
	13 to 25	25 to 50	>50
100 to 200	Low (L)	L	Medium (M)
200 to 450	L	M	M
450 to 750	M	High (H)	H

- Images with potholes were annotated by drawing a bounding box around the pothole and labeled as low severity pothole (L), medium severity pothole (M), or high severity pothole (H).

3.2 YOLO Architecture

As the name suggests, YOLO architecture reads the image only once unlike classical object detection techniques, and predicts the object class and its location. It overcomes the disadvantage of predicting background as object class in Fast Region-based Convolutional Neural Network (R-CNN) (Redmon et al. 2016; Redmon and Farhadi 2017). Therefore, YOLOv3 was chosen for detecting potholes as part of the PDQMS framework in this study. The number of classes chosen for this study were three, and the details of these classes are given in Table 2.

In order to develop the network architecture, first the dimensions of the input data were evaluated. Let the image dataset, D have N images. The mathematical representation of the dataset is given in Eq. (1).

$$D = \{(x_i, y_i) | i = [1, 2, 3, \ldots, N]\} \tag{1}$$

Where, D = image dataset; x_i = image; y_i = preprocessed tensor representation of the image; and N = number of images in the dataset

In order to label the images based on classes, a boundary box was drawn. Figure 2 represents the labeling mechanism earmarking the extent details. There are many software packages available online for labeling that can be used to mark the bounding

Table 2 Pothole classes for PDQMS developed in this study

Class description	Class label
Low severity pothole	L
Medium severity pothole	M
High severity pothole	H

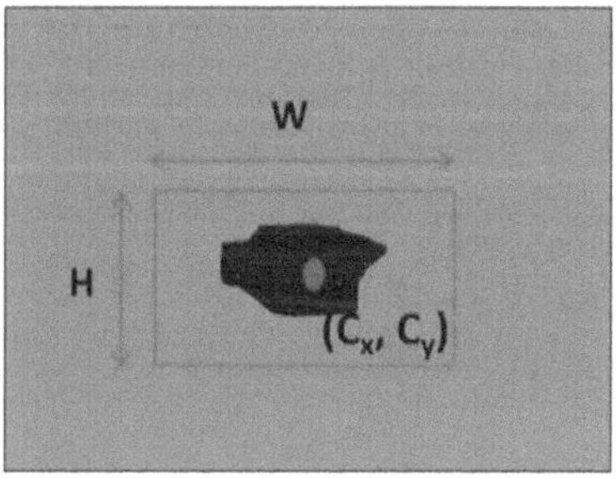

Fig. 2 A schematic representing the labeling of potholes in a typical pavement image

box and store the extent and centroid details as features to the image. Open source LabelImg tool-kit allowed the authors to label the images for object detection and classification. The annotations were saved in the.txt format to train the network.

The typical process of labelling started with identifying the severity of the pothole. Since the pavement images did not have depth information; based on the texture of pothole, the appropriate severity level was estimated. The extent of the pothole region was marked in a rectangular shape, as shown in Fig. 2. The bounding box was defined with two parameters, namely, (i) centroid (C_x, C_y), and (ii) extent of the distress class (W, H). Each of the aforementioned parameters were 2D in nature. In order to represent an image as a tensor in addition to the image resolution, dimensions per pixel were identified. For this study, the number of dimensions per pixel was the sum of dimensions of label parameters and pothole classes, as given in Eq. (2).

$$\text{pixel}_{\text{Dimensions}} = \text{centroid}_{\text{Dimensions}}(2) + \text{Extent}_{\text{Dimensions}}(2) + No.of\ classes_{Dimensions}(3) = 7 \tag{2}$$

Assume that the image resolution is A × B pixels, the dimensional tensor representation of image is resolution × dimensions per pixel, which is given in Eq. (3).

$$\text{Dimensional tensor representation of image} = \text{A} \times \text{B} \times 7 \tag{3}$$

YOLO makes use of only convolutional layers; hence it is a fully convolutional network. YOLO v3 has 75 convolutional layers, and 31 other layers such as up-sampling and route layers (Redmon and Farhadi 2018). The typical input image size used for YOLO is 416 × 416 pixels, but it can be trained on images with different aspect ratios to make it versatile for detection. Therefore, the YOLO v3 architecture was found suitable for PDQMS as well.

Training the network is a tedious task as it runs for days to initiate weights for the neurons in the entire network. In order to optimize the program duration, pretrained network on PASCALVOC dataset was used (Redmon and Farhadi 2018), which calculated initial weights to the neurons of the network. Further, the configuration files and class object files were modified to fine tune the model for predicting the three-class pothole detection problem. The dataset along with the annotations were fed into the network. The dataset was then split into train and validation sets to train the network for updating the weights of the neurons. A two-stage training process was adopted to reduce the total loss, i.e., mean squared error. In the first stage, YOLOv3 with the pothole image dataset was trained with pretrained weights of darknet53 on a batch size of 64 and subdivision of 64 for 1000 iterations. The updated weights after the training were used to retrain the model with the same configurations for 1000 iterations to reduce the average loss. The average loss after two-stage training was 0.243, which was quite low. The updated network was then used to detect and predict the class of the pothole with a bounding box.

The configurations of the network are: (i) size of the dataset: 41 images, (ii) learning rate: 0.001, (iii) momentum: 0.9, and (iv) decay: 0.0005. The performance

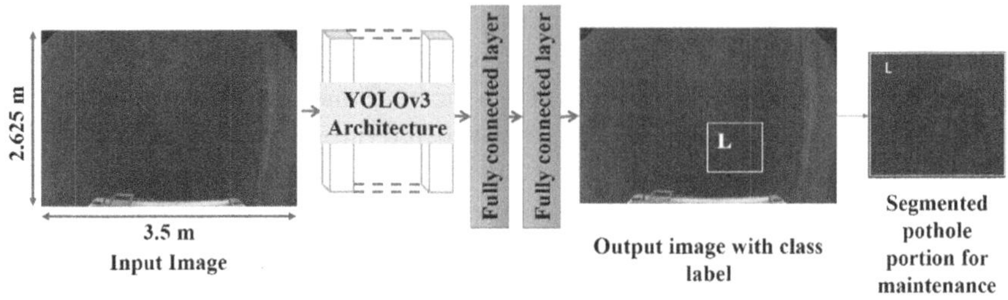

Fig. 3 Schematic representation of segmentation process of pothole from the input image for maintenance

of the model was evaluated in terms of mean average precision (mAP), a metric widely used to evaluate the performance of the object detection algorithms. In order to validate the model, mAP of the proposed network was compared with the YOLOv3 Spatial Pyramid Pooling (SPP) by Ukhwah et al. (2019) for pothole detection. The mAP reported by the authors was 88.93%, whereas the mAP of the architecture of this study was found to be 89.23% representing the performance of the proposed architecture for detecting pothole classes to be high.

3.3 Segmentation of Pothole Region from the Image

The output of the YOLO network is an image with objects marked as per their classes. In order to quantify the amount of patching material required, first the detected classes were segmented. The typical procedure of detection and segmentation accomplished in this study is shown in Fig. 3.

3.4 Quantification of Patching Area

The extent of the segmented pothole portion was correlated with the input size to calculate the patching area. As the pothole class was defined based on severity, which is dependent on the pothole depth, it helped quantify the patching material required for maintaining the pothole. The aforementioned tasks accomplished the overall framework of PDQMS, which was aimed to provide a solution to challenges in pothole maintenance. An example of the series of operations in the developed PDQMS is shown in Fig. 4. The output of the final phase of the framework is envisioned to help the agencies to calculate the cost of pothole maintenance and make decisions on maintenance activities through the future.

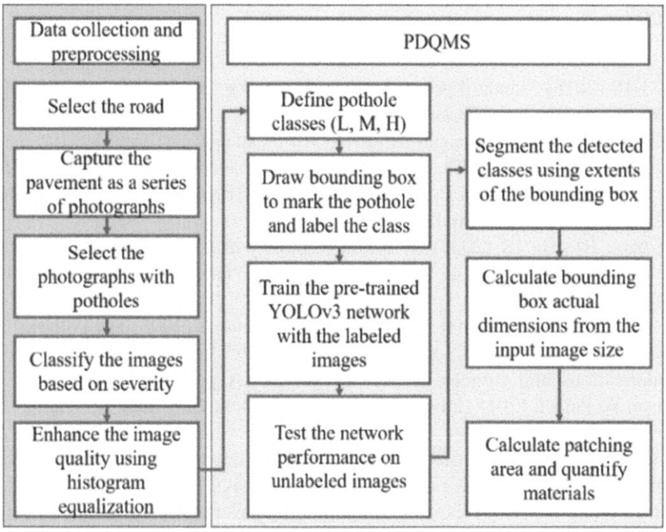

Fig. 4 An example of a typical working process of PDQMS for pothole detection and maintenance

4 Conclusions and Recommendations

In this study, a framework to manage potholes in asphalt pavements was developed and documented. The state-of-the-art object detection algorithm YOLOv3 was utilized to detect potholes based on severity class from the given image. The performance of the network measured in terms of mAP was 89.23% representing the capability of the proposed architecture to detect pothole classes much more accurate than the existing methods. The proposed framework is envisaged to help the agencies in identifying the potholes that are categorized as low, medium, and high severity levels, and make a decision on how to maintain them to avoid potential risks of occurrence of accidents and vehicle damages. However, the efficiency of the defined framework needs to be evaluated on a diverse image dataset to enhance the performance of the proposed system. Nonetheless, the proposed methodology can be comfortably adopted by agencies to provide and maintain pothole-free roadway infrastructure, which is one of the core objectives of smart cities mission.

Acknowledgements The authors would like to acknowledge Andhra Pradesh Road Development Corporation, India for sharing the image dataset for research activities. Further, special thanks to Dr. Kalidas Yeturu, Department of Computer Science and Engineering of IIT Tirupati for helping in the initial formulation of the framework.

References

ASTM D6433-18 (2018) Standard practice for roads and parking lots pavement condition index surveys. West Conshohocken, Pennsylvania, USA

Becerik-Gerber B, Masri S, Jahanshahi M (2015) An inexpensive vision-based approach for the autonomous detection, localization, and quantification of pavement defects. NCHRP Innovations Deserving Exploratory Analysis Programs Project 169, Transportation Research Board of the National Academies, Washington, DC. USA

Chang KT, Chang JR, Liu JK (2005) Detection of pavement distresses using 3D laser scanning technology. In: Proceedings of the 2005 ASCE international conference on computing in civil engineering, ASCE, Reston, VA, vol 105

Eriksson J, Girod L Hull B, Newton R, Madden S, Balakrishnan H (2008) The Pothole Patrol: using a mobile sensor network for road surface monitoring. In: 6th international conference on Mobile systems, applications, and services, ACM, New York, USA, pp 29–39

Haas R, Hudson W, Falls L (2015) Pavement asset management. Scrivener Publishing LLC, Wiley, New Jersey, USA

Hadjidemetriou GM, Vela PA, Christodoulou SE (2017) Automated pavement patch detection and quantification using support vector machines. J Comput Civil Eng ASCE 32(1):04017073

Hou Z, Wang K, Gong W (2007) Experimentation of 3D pavement imaging through stereovision. In: International conference on transportation engineering, ASCE, Chengdu, China, pp 376–381

Huidrom L, Das L, Sud S (2013) Method for automated assessment of potholes, cracks and patches from road surface video clips. Procedia - Soc Behav Sci 104:312–321

Jahanshahi MR, Jazizadeh F, Masri SF, Becerik-Gerber B (2013) Unsupervised approach for autonomous pavement-defect detection and quantification using an inexpensive depth sensor. J Comput Civil Eng ASCE 26(6):743–754

Koch C, Brilakis I (2011) Pothole detection in asphalt pavement images. Adv Eng Inform 25:507–515

Koch C, Jog G, Brilakis I (2013) Automated pothole distress assessment using asphalt pavement video data. J Comput Civil Eng ASCE 27(4):370–378

LeCun Y, Bengio Y (1995) Convolutional neural networks for images, speech, and time series. In: The handbook of brain theory and neural networks, pp 276–278

Mathavan S, Kamal K, Rahman M (2015) A review of three-dimensional imaging technologies for pavement distress detection and measurements. IEEE Trans Intell Transp Syst 16(5):2353–2362

Redmon J, Farhadi A (2017) YOLO9000: Better, Faster, Stronger. In: 2017 IEEE conference on computer vision and pattern recognition (CVPR), Honolulu, HI, pp 6517–6525

Redmon J, Farhadi A (2018) YOLOv3: An Incremental Improvement. arXiv *2018*

Redmon J, Divvala S, Girshick R, Farhadi A (2016) You only look once: unified, real-time object detection. IEEE Conf Comput Vision Pattern Recogn (CVPR) 2016:779–788

Ryu S, Kim T, Kim Y (2015) Image-based pothole detection system for its service and road management system. Math Probl Eng 1–11

Salari E, Yu X (2011) Pavement distress detection and classification using a genetic algorithm. In: Proceedings - Applied Imagery Pattern Recognition Workshop. https://doi.org/10.1109/AIPR.2011.6176378

Suong LK, Jangwoo K (2018) Detection of potholes using a deep convolutional neural network. J Univ Comput Sci 24(9):1244–1257

Ye W, Jiang W, Tong Z, Yuan D, Xiao J (2019) Convolutional neural network for pothole detection in asphalt pavement. Road Mater Pavement Des. https://doi.org/10.1080/14680629.2019.1615533

Zakeri H, Nejad F, Fahimifar A (2016) Image based techniques for crack detection, classification and quantification in asphalt pavement: a review. Archives Comput Methods Eng 24(4):935–977

Ukhwah EN, Yuniarno EM, Suprapto YK (2019) Asphalt pavement pothole detection using deep learning method based on YOLO neural network. In: 2019 international seminar on intelligent technology and its applications (ISITIA), Surabaya, Indonesia, pp 35–40

Stochastic Prediction of Short-Term Friction Loss of Asphalt Pavements: A Traffic Dependent Approach

Christina Plati, Maria Pomoni, Andreas Loizos, and George Yannis

Abstract Pavement friction significantly affects the road safety. Over years, researchers have developed multiple models in an attempt to estimate pavement friction performance. The present study aims at developing a stochastic model for the prediction of short-term friction loss based on field-friction data, which adapts the survival probability analysis in terms of a Kaplan - Meier survival curve. The traffic volume expressed through the Annual Average Daily Traffic (AADT) is the main variable of the developed model, assumed to affect more the friction loss within the short-term period of the investigation. The friction level of the preceding year is the second variable of the model that is assumed to embody other factors affecting short-term friction deterioration in the field. Following the assumptions made, the impact of other factors are deemed to be incorporated in the variable of the friction level of the preceding year. However, this assumption is discussed in the constraints of the proposed methodology. All in all, the results of the study are encouraging and can be a useful tool for timely scheduling future maintenance actions in the framework of proactive asset management.

Keywords Asphalt pavements · Stochastic model · Surface friction · Annual Average Daily Traffic (AADT)

1 Introduction

One of the most significant and determinant performance indicators for pavement serviceability condition, as well as for road safety is the surface friction. The tyre-pavement friction mainly contributes to the driver's ability to maintain vehicle control during braking. The higher the level of pavement friction is, the greater the potential

C. Plati (✉) · M. Pomoni · A. Loizos
Laboratory of Pavement Engineering, National Technical University of Athens (NTUA), 5 Iroon Polytechniou, 15773 Athens, Greece
e-mail: cplati@central.ntua.gr

G. Yannis
Laboratory of Traffic Engineering, National Technical University of Athens (NTUA), 5 Iroon Polytechniou, 15773 Athens, Greece

C. Raab (ed.), *Proceedings of the 9th International Conference on Maintenance and Rehabilitation of Pavements—Mairepav9*, Lecture Notes in Civil Engineering 76,
https://doi.org/10.1007/978-3-030-48679-2_86

vehicle control becomes Hall et al. (2009). In multiple studies, a relationship between pavement friction and accident frequency within a road network has been presented and it appears that accident rate can be reduced significantly by an improvement in pavement friction levels (Omar et al. 2017; Hall et al. 2009).

Pavement friction of asphalt surfaces is mainly affected by the surface characteristics and more specifically the texture (Plati and Pomoni 2019; Rajaei et al. 2017). In addition, friction is affected by the tyre characteristics, vehicle operation and climatic conditions (Goulias and Awoke 2020).

In general, aggregate polishing due to traffic wear affects pavement friction performance. Initially, there is an increase in pavement friction due to removal of the binder material covering surface aggregates thus exposing the aggregate rough asperities. Thereafter, pavement friction tends to decrease over the remaining pavement life as the exposed aggregates become increasingly more polished and smooth (Kane et al. 2010). Higher traffic volume and increases in heavy vehicles flow are known to have a significant effect in the expedited deterioration of the pavement surface properties and thus, the friction deterioration (Hall et al. 2009).

The modern Pavement Management Systems (PMS) comprise established procedures to collect, analyze and maintain quantitative data on road pavements. The identification of areas with friction deficiencies is of vital importance considering their impact on road decision making policies that target at optimum investments for maintaining ride quality, safety and cost-efficiency (Omar et al. 2017).

On these grounds, a multitude of different models for predicting pavement friction has been developed; these models have significant variations and depend directly on the variables examined, the approach in modelling and the initial assumptions (Omar et al. 2017). Other researchers have suggested prediction models for estimating friction loss mainly based on laboratory test data and have subsequently attempted to extrapolate their findings to the field (Hofko et al. 2019; Li et al. 2017). However, many laboratory - tailor made approaches present some limitations, as it is not possible to capture the effect of a multitude of factors in the field, including the impact of traffic, climatic conditions and road geometrics. On the other hand, friction modeling by taking into account field data appears to be a more promising and challenging analysis tool, although there are many obstacles towards achieving a univocally accepted model through this approach. As far as the modeling methods are concerned, so far, the majority of researchers have used deterministic models to predict future friction levels. Nevertheless, according to Li et al. (2017), it seems that friction modeling may be more effective for practical actions as a stochastic process rather than a deterministic one resulting in a model.

2 Objective

In light of the above, the current study is concentrated on the short-term prediction of friction deterioration, by investigating the effect of traffic volume on future friction levels. Within this scope, the ultimate goal is to propose a useful and practical approach for road agencies with the view to assessing friction condition through Kaplan—Meier survival analysis based on their pre-defined investigatory friction level and second, to predicting friction degradation for the upcoming years. Given this, a stochastic approach for the prediction of short-term friction loss is followed based on real friction data from an in-service asphalt pavement. A model is developed that takes into consideration the traffic volume and the friction level of the preceding year. The traffic volume is expressed through the Annual Average Daily Traffic (AADT) and assumed to be the main influential factor of pavement friction loss within a short-term period of investigation. Whereas, the friction level of the preceding year is expected to inherently incorporate others factors affecting friction deterioration in the field. Through the methodology presented, it is deemed that road authorities may appropriately schedule and allocate funding for potential upcoming maintenance actions, based on a condition-based preventive maintenance perspective in order to improve or even maintain road safety within a road network.

3 Field Study

3.1 Site Characteristics

The investigation was undertaken on an urban highway which has been divided into 32 sub-sections depending on their direction, location, traffic volume and structural characteristics.

The wearing course of all the investigated pavement sections was an open-graded Hot Mix Asphalt (HMA) mixture with interconnecting voids that provides increased safety in wet weather, through reduced surface water and spray during rain as well as limited noise levels. The open graded friction course mixture corresponded to Mix Designation Type II (ASTM D3515 2001) and was produced using a polymer modified bitumen 25–55/75 that ranged between 3–5%. The asphalt mix contained 4% bitumen binder by mass of the mixture and an 11.5% air void content. In addition, steel slag was incorporated into the mix design for the upper wearing course layer.

3.2 Data Acquisition

Field friction and traffic data were acquired through a database and analyzed. The traffic data was acquired in terms of AADT. The field friction data that was considered for the purpose of the present analysis were along the outer wheel path of the right lane considering that faster deterioration occurs in this lane, mainly because of heavy vehicle traffic moving. In addition, only friction data after the wet period (an extended rain period) were utilized in order to limit the effect of summer contaminants that negatively affect friction measurements (Plati et al. 2020). Importantly, based on the structural data the considered pavements were in a structurally sound condition.

The friction database consists of GripTester measurements, which is a fixed-slip device feed (Findlay Irvine 2002). The device consists of a three-wheeled system where the central-test wheel has a smooth-tread tire according to ASTM E1844 standard (2015). The axle of the test wheel is connected to a chain-system that controls the test wheel's slip speed so that a constant slip of 14% to be maintained. Also, GripTester includes a system for depositing a standard amount of water (0.5 mm water-film) in front of the test tyre so that it then passes between the tyre and the surface being measured. The GripTester system continuously measures friction reporting of Grip Numbers (GN) to 0–1 range. For the study, the GN values are the average measurements at 10 m intervals. Hence, the following data analysis was performed using a big set of data elements.

4 Data Analysis and Results

A survival analysis is applied to GN data for the description of friction deterioration. In general, survival analysis is defined as a set of methods for analyzing data, where the outcome variable is the time until the occurrence of an event of interest. It suggests a practical technique regarding hazard-based duration modeling and therefore, it has been utilized in pavement engineering analysis to model cracking (Reger et al. 2013; Loizos and Karlaftis 2005), roughness (Meegoda and Gao 2014), assess the effectiveness of pavement overlays or evaluate pavement performance (Anastasopoulos and Mannering 2015). Survival analysis method assumes an underlying failure distribution of the data (Prozzi and Madanat 2000). As survival function can be expressed explicitly in terms of a parametric distribution function, it is potentially feasible to estimate the coefficients of those parameters, or else, the influence of the affecting factors. However, the need to assume an underlying distribution introduces an obstacle. The shape of the data might not be described by a well-known specific distribution, and this constitutes a major limitation of the method.

Taking into consideration the aforementioned limitation, the authors use the Kaplan-Meier method (Donev and Hoffmann 2017) to overcome it and calculate the probability distribution of the survival random GN variable. According to Kaplan-Meier method, the survival probability at a given year can be calculated by the multiplication of each of the preceding years' conditional probability of failure. The failure criterion for the friction degradation is the level of surface friction that is below a pre-defined threshold value (Li et al. 2017). Concerning the asphalt pavement sections under investigation, this threshold value is an Investigatory Level (IL) equal to 0.41 GN. It is worth mentioning that the objective of setting an IL is to assign a level of surface friction appropriate for the risk state on the site, at or below, which further investigation is required to evaluate the site-specific risks in more detail (Design Manual for Roads and Bridges 2015).

Hence, the conditional probability of failure in a given year is defined by dividing the number of failures occurring in that year to the number of pavements sections at risk of failure (the risk to be GN ≤ IL) at the beginning of that year. The conditional probability p_t of surviving t^{th} year after having survived t − 1 years is calculated as (Balla 2010):

$$p_t = 1 - \frac{Number\ of\ pavement\ sections\ of\ GN \leq IL\ in\ interval\ t^{th}\ year\ and\ t^{th} + 1\ year}{Number\ of\ pavements\ at\ risk\ \text{of GN} \leq \text{IL}\ beginning\ of\ t^{th}\ year} \quad (1)$$

The probability of survival to time t, S(t) is calculated as:

$$S(t) = p_1 \times p_2 \times p_3 \ldots \times p_t \quad (2)$$

The graph of S(t) versus years (t) gives the Kaplan-Meier survival curve. In the current study, survival analysis is used to determine the percentage of the number of sections of which their average surface friction level is GN ≤ IL. Figure 1 depicts the survival curve for the particular highway. From years 1 to 13, the percentage of the 32 sub-sections at or above the IL is presented. The use of survival curve illustrates the evolution of the pavement survival probability in terms of surface friction deterioration (i.e. % sections ≥ IL). Noticeably, the measured GN values have been utilized for calculating the survival probability each year, up to year 13. The fourth point matches to the 4th year after opening to traffic when database started. The dashed lines after the 13th year indicate potential variability of the evolution in the survival probability.

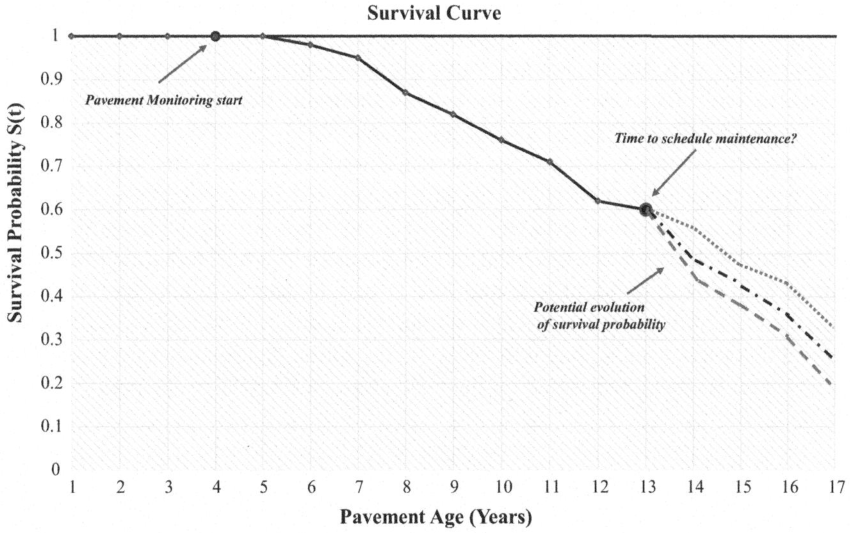

Fig. 1 Survival curve regarding friction levels in regards to IL

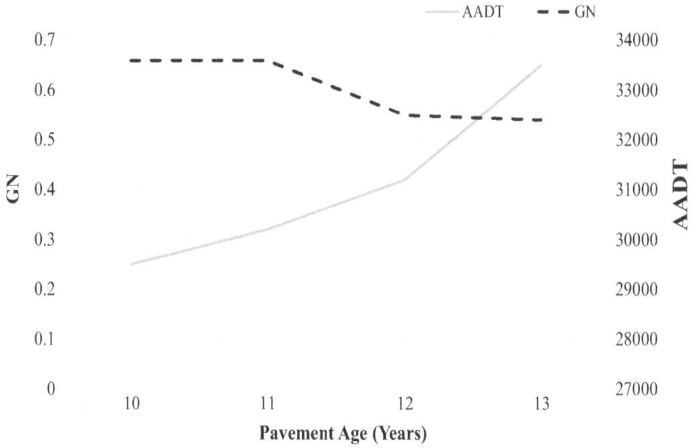

Fig. 2 GN values and AADT—Examples from the sample of the 60% test sub-sections

Thus, it seems that Kaplan-Meier survival curve can give a description on the friction levels which are below or still above IL. Thereafter, an approach is developed for providing an estimation of the surface friction level (the potential one) for the upcoming years (i.e. over year 13) in order to effectively complete the survival curve and gain further insight into the friction condition at a network-level analysis. To accomplish this, randomly 60% of the sections (i.e. 19 sub-sections) are utilized for testing, while the remaining 40% (i.e. 13 sub-sections) are utilized for validating the developed short-term degradation model, impartial of the sections selected (Shmueli et al. 2018). It is worthwhile mentioning that the traffic evolution (trend) cannot be considered stable for the entire period of the pavement monitoring period (10 years). Hence, only the last four years (from year 10 to year 13) are considered representative to describe the most recent trend of traffic evolution (that of a slight increase).

Subsequently, an investigation into friction and traffic trend between years 10-13 was conducted. Figure 2 illustrates the aforementioned trend of the two data sets.

Thereafter, a correlation between the measured friction and traffic data was performed for those four years when the aforementioned trend was observed ($R^2 = 0.85$). Hence, a linear trend between the two components was attempted which may lead to the development of a practical and simple model for the prediction of friction values. With this in mind, all GN and AADT testing data were appropriately correlated and the following empirical form described in Eq. 3 was developed for the calculation of friction coefficient in terms of GN. Through this stochastic model GN for a year (i) is calculated, based on the GN values and AADT of the previous year (i − 1).

$$GNi = GN_{i-1} - \left(AADT_{i-1} * 10^{-5}\right)/A \tag{3}$$

A = 6, (AADT > 30,000)
A = 2, (AADT ≈ 15,000–30,000, with greater heavy traffic volume)
A = 8, (AADT ≈ 5,000)

where:

GNi	= predicted GN values for a year (i)
GN_{i-1}	= GN values of the previous year (i − 1)
AADT $_{i-1}$	= AADT of the previous year (i − 1)
A	= case-adjusted factor

It is to be recalled that a range of factors affect surface friction (Rajaei et al. 2017). However, the proposed stochastic model includes only two variables assuming that other variables are somehow interrelated to GN_{i-1} for a short-term period so, there would be probably a limited benefit from considering them separately. More specifically, macrotexture level, microtexture level, asphalt mix characteristics, aging of the asphalt-mixture and the long-term effect of climatic conditions are deemed to be embodied in the variable GN_{i-1}, that describes the current friction condition. In other words, those surface characteristics may be considered roughly stable within the short-term period of the model applicability. The developed model is subsequently utilized for validation purposes on the rest sample of GN data.

4.1 Validation Procedure

For the validation process, year 11 was used as the reference year for investigating the applicability of the proposed model and predicting GN values for years 12 and 13, for which traffic data presented a trend of increase. Hence, the developed equation is initially applied on the measured GN data for year 11 (GN_{i-1}) to predict GN values for year 12 (GN_i) and on GN_{12} (predicted values for year 12) to predict GN_{13}. AADT is assumed to increase between the years 11–12 and 12–13 with a rational, stable rate. Thereafter, the measured GN values of these pavement sections for both years 12 and 13 are normally correlated with the corresponding predicted values. The goodness of fit is assessed based on the Root Mean Square Percent of Error (RMSPE) criterion (Eq. 4).

$$RMSPE = \sqrt{\frac{1}{n} * \left[\sum_{i=1}^{n} \left(\frac{GNi^{estimated} - GNi^{measured}}{GNi^{measured}} \right)^2 \right]} * 100 \quad (4)$$

Figure 3 illustrates indicatively the graphic adaptation of RMSPE criterion on GN measured and GN predictive values. All the results of the validation process are presented in Table 1, where it appears that the proposed two-variables model leads to predictions of acceptable accuracy (RMSPE < 15%).

It seems that the proposed stochastic model can accommodate the specific friction data. In addition, it is observed that the RMSPE for year 13 is mainly higher than for year 12, an issue that will be further discussed below. Overall, the proposed methodology could be considered as an efficient and practical tool for road transport agencies

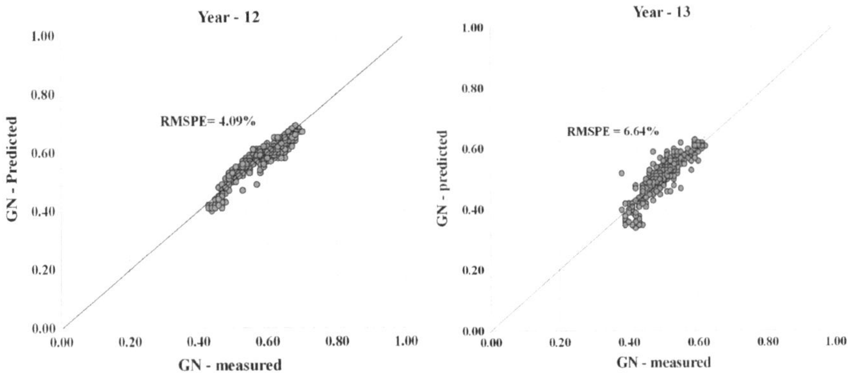

Fig. 3 RMSPE for a section where A = 8, (AADT ≈ 5,000)

Table 1 Total validation results of RMSPE criterion for the 40% of the subsections

Section	AADT	RMPSE %–year 12	RMPSE %–year 13
1	15000	8.54	9.67
2	20000	14.36	12.28
3	30000	9.02	9.04
4	20000	4.90	7.53
5	30000	4.80	6.10
6	>30000	8.33	9.80
7	>30000	6.07	12.58
8	30000	9.98	10.18
9	30000	12.97	14.74
10	15000	9.48	14.75
11	15000	13.69	14.64
12	15000	8.02	12.43
13	5000	4.09	6.64

in estimating the degradation in surface friction levels for the short—upcoming years to more efficiently plan routine maintenance actions and upgrade surface friction, if levels are expected to fall under the IL.

5 Discussion

A range of factors affect surface friction but road authorities need simple and effective tools to draw strategic planning of routine maintenance activities based on rational priorities and budgets assignment. Thus, the use of models with a multitude of variables may not be an effective approach for them. On these grounds, the presented methodology assumes that within a specific highway, yearly friction is mainly affected by traffic volume for a short-term period.

However, on a long-term basis the variables assumed to be stable in this approach will not remain constant, while traffic volumes will face changes that may affect the particular case-adjusted factor—A. Hence, a re-calibration may be needed after a period to incorporate these changes. Given this, a suggestion could be to define the calibration time based on sharp increases on RMSPE values. Also, it has to be pinpointed that the model is based on GripTester data and some discrepancies may occur in case of another measuring device. Another constraint concerning the suggested methodological approach is the inability to consider the effect of intermediate maintenance activities for surface improvement. The consideration of such an issue would raise the complexity of friction modeling. Thus, it means that the developed stochastic model can be adapted only between subsequent periods of major maintenance activities with no further intermediate maintenance in the meantime.

6 Concluding Remarks

In this study, a methodology that captures the short-term friction deterioration for an in-service pavement was presented. The developed stochastic approach adapted the survival probability analysis in terms of a Kaplan—Meier survival curve and then, a model development was proposed. The presented methodology was considered to be useful for road agencies that need to assess and predict the degradation in pavement friction and schedule routine maintenance actions. The effect of traffic volume was deemed to be critical and was embodied in the developed empirical two-variable model in terms of AADT. The basic assumption of the model was that other factors that potentially influence the yearly loss of pavement friction within a highway do not significantly change over a short-period of time. Also, it was assumed that the effect of those factors can be expressed through the level of the friction of the previous year which was the second variable of the model. The different magnitude of traffic volumes was incorporated through a case-adjusted factor—A, which is a traffic dependent parameter. Potential improvements however, may be achieved through proper adjustments to factor A in order to consider significant changes in traffic volume and the heavy vehicle traffic as well, an issue that seeks further investigation.

Acknowledgements This research is co-financed by Greece and the European Union (European Social Fund- ESF) through the Operational Programme «Human Resources Development, Education and Lifelong Learning» in the context of the project "Strengthening Human Resources Research Potential via Doctorate Research" (MIS-5000432), implemented by the State Scholarships Foundation (IKY).

References

Anastasopoulos P, Mannering F (2015) Analysis of pavement overlay and replacement performance using random parameters hazard-based duration models. J Infrastruct Syst 21(1):04014024

ASTM (2001) Standard Specification for Hot-Mixed, Hot-Laid Bituminous Paving Mixtures (Withdrawn 2009). ASTM D3515-01, West Conshohocken, PA

ASTM (2015) Standard Specification for a Size 10 × 4–5 Smooth-Tread Friction Test Tire. ASTM E1844-08, West Conshohocken, PA

Balla CK (2010) Prediction of remaining service life of pavements. Master Thesis, University of Toledo, Digital Repository

Design Manual for Roads and Bridges (2015) "Skidding Resistance". HD 28/15, 7, (3), Part 1

Donev V, Hoffmann M (2019) Condition prediction and estimation of service life in the presence of data censoring and dependent competing risks. Int J Pavement Eng 20(3):313–331

Findlay Irvine Ltd (2002) Road base grip tester survey software for roads. User manual

Goulias D, Awoke GS (2020) Novel approach to pavement friction analysis with advanced statistical methods using structural equation modelling. Int J Pavement Eng 21(2):236–245. https://doi.org/10.1080/10298436.2018.1454922

Hall JW, Smith KL, Titus-Glover L, Wambold JC, Yager TJ, Rado Z (2009) Guide for pavement friction. Contractor's Final Report for NCHRP Project 01-43, NCHRP, Transportation Research Board of the National Academies, Washington, DC, pp 19–37

Hofko B, Kugler H, Chankov G, Spielhofer R (2019) A laboratory procedure for predicting skid and polishing resistance of road surface. Int J Pavement Eng 20(4):439–447

Kane M, Do MT, Piau JM (2010) On the study of polishing of road surface under traffic load. J Transp Eng 136(1):45–51

Li L, Guler SI, Donnell ET (2017) Pavement friction degradation based on pennsylvania field test data. Transp Res Rec 2639:11–19

Loizos A, Karlaftis M (2005) Prediction of pavement crack initiation from in-service pavements: a duration model approach. Transp Res Rec 1940:38–42

Meegoda JN, Gao S (2014) Roughness progression model for asphalt pavements using long-term pavement performance data. J Transp Eng 140(8): 04014037, 7 p

Omar LG, Halim AE, Ismail K (2017) Investigating the predictability of pavement friction on rural roads in Ontario Canada. In: 96th Annual Meeting, Transportation Research Board, Washington, D.C, 13 p

Plati C, Pomoni M (2019) Impact of traffic volume on pavement macrotexture and skid resistance long-term performance. Transp Res Rec 2673:314–322

Plati C, Pomoni M, Georgouli K (2020) Quantification of skid resistance seasonal variation in asphalt pavements. J Traffic Transp Eng 7(2):237–248. https://doi.org/10.1016/j.jtte.2018.07.003

Prozzi JA, Madanat SM (2000) Using duration models to analyze experimental pavement failure data. Transp Res Rec 1699:87–94

Rajaei S, Chatti K, Dargazany R (2017) A review: pavement surface micro-texture and its contribution to Surface Friction. In: 96th annual meeting, transportation research board, Washington, D.C., 19 p

Reger D, Christofa E, Guler I, Madanat S (2013) Estimation of pavement crack initiation models by combining experimental and field data. J Infrastruct Syst 19(4):434–441

Shmueli G, Bruce PC, Yahav I, Patel NR, Lichtendahl Jr. KC (2018) Data Mining for Business Analytics Concepts, Techniques, and Applications in R. Wiley, Hoboken

Tire Contact Stress Distribution Considering the Tire Inclination in Bend

Y. Oubahdou, E. Manyo, P. Reynaud, B. Picoux, J. Dopeux, and C. Petit

Abstract Tire—road surface contact is a complex combination of stresses and deformations that depend on tire and road factors, such as vehicle speed, weight, tire material and type, tire pressure inflation, camber and texture of the surface layer. In this paper, the problem of the tire-pavement contact is studied using a realistic description in normal section and corner. Mechanical field is validated with a press system under different inflation pressures and loads. The model requires precise tire geometry and an equivalent Young's modulus. For this purpose, an optical method based on a photogrammetry method provided a 3D field of displacement. The estimation of the equivalent Young's modulus is also proposed as a function of the inflation pressure resulting from the press load tests. Finally, a comparison is made between the measured footprint of the tire, the result of the semi-analytical calculation as well as the effect of tire inclination (corner/turn) on the distribution of surface stresses that is demonstrated.

Keywords Surface layer · Tire-pavement contact · Tire inclination · Degradations

1 Introduction

The appearance of new materials associated with a decrease in the thickness of the surface layers combined with an increase in truck loads and their frequency of passage has led to new pathologies of pavements deterioration. These are directly related to the tire - pavement contact pressure. The pavement design method considers that the tire-floor contact area as a circularly shaped and the contact pressure are uniformly distributed. However, it has been shown that under no circumstances, the pressure distribution is uniform (De Beer et al. 1997).

So, traffic loads are applied to the pavements by tires that exert efforts on the contact surface. The distribution of contact pressures between the pavement and the tires depends on several factors such as:

Y. Oubahdou (✉) · E. Manyo · P. Reynaud · B. Picoux · J. Dopeux · C. Petit
Laboratoire GC2D, Université de Limoges, EA 3178, 19300 Egletons, France
e-mail: yamina.oubahdou@unilim.fr

C. Raab (ed.), *Proceedings of the 9th International Conference on Maintenance and Rehabilitation of Pavements—Mairepav9*, Lecture Notes in Civil Engineering 76,
https://doi.org/10.1007/978-3-030-48679-2_87

- Intensity of the load
- Tire inflation pressure
- Type of tire (profile, single or twin, brands, etc.)

These factors have consequences on traffic loads, in particular on the shape of the footprints, the distribution and the intensity of the contact pressures.

The distribution of the contact pressure plays a major role in the mechanical response of the pavement under load, more precisely in the singular points when the tangential force, the dissymmetry of the axle load affect the contact constrains.

In this study, we estimated the contact pressure during a turn by simulating it with the help of a Semi-Analytical calculation code. The position of the tire in inclined turns was subjected to a static vertical force. Then, the modeling was validated by an experimental method by measuring the contact pressures under the same conditions using a pressure sensor (TekScan).

2 Objectives

In the above-mentioned context, modeling and experimentation were undertaken to study tire-road contact pressures, particularly in corners where major surface layer degradations are observed due to the importance of significant aggression induced by traffic (tangential effort, axle load dissymmetry). However, it is enough to observe the footprints of existing structures to see that they are sometimes the source of significant degradations from the first months of commissioning. These singular points then present maintenance problems to the road managers and at the same time a certain inconvenience to the users. Ambassa (Ambassa et al. 2013) undertook a study to evaluate the aggressiveness of roundabout traffic. Authors quantified the effect of normal and tangential forces on the mechanical response of the surface layer at roundabouts. The distribution of the stresses depends strongly on the geometry (Radius, slope) and the speed of circulation. The structural calculations showed the vulnerability of the roundabouts regarding the effect of repeated tangential stress and the dissymmetry of the axle loads.

To perform the tribological approach modeling of the pneumatic-road contact, a semi-analytical calculation code is proposed. This method takes into account the normal and gyratory sections and allows the determination of the contact surface as well as the distribution of its pressure. The actual geometrical shape of the tire is taken into account by digitizing the tire profile using the photogrammetry method. The tire is assumed to be homogeneous, elastic linear, isotropic and moderated by an equivalent Young's modulus. In order to validate the results of the pavement tire contact obtained from the semi-analytical model, a 295/80R22.5 tire with an inflation pressure of 900 kPa was used. In normal cross-section, it was loaded with 32.5 kN and the contact points were measured by the TekScan sensor system. In gyrations the lateral load transfer tends to amplify the loading on the outer wheel. To take into account this phenomenon, we propose 3 variants of camber angle 0, 3 and 5°

with the same load of 45 kN. The obtained pressures will be validated with the same measuring system TekScan.

3 Semi-analytical Modeling of Tire-Pavement Contact

Since experimental studies are limited by the cost and complexity of the test equipment, modeling has the advantage of being able to easily study the pavement and the tire structure, the effect of the load as well as the speed.

Modeling the interaction between the tire and the pavement from contact mechanics approach have to consider the properties of non-linear materials, the rolling contact transition phenomena, the complex behavior of the contact tire and the friction o the tire—pavement interface.

Due to the complexity of the tire-road contact problem and therefore the difficulty of solving it with a purely analytical method, numerical models using the finite element method (FEM) have been developed. Several works in the tire industry have led to a very accurate simulation using 3D finite elements. Most of these calculations are extremely time-consuming to apply them in the road industry. Moreover, they are generally limited to quasi-static or stationary solicitations (Nackenhorst 2004; Zie et al. 2008; Padovan 1987).

In the pavement area, a first study of contact pressure by a tribological approach was carried out by Reynaud (Reynaud et al. 2016). In this first approach, a Semi Analytical Model (SAM) was used for a smooth tire profile. Results were compared to measurements from the literature. Manyo (Manyo et al. 2018) then performed calculations with a scanned tire profile. The authors compared the results to measurements made by the TekScan system. The maximum measured value is 1532 kPa and the calculated maximum value is 1576 kPa.

The same semi analytical method is used in this paper. This method is based on the summation of analytical solutions using digital techniques such as the Fourier Fast Transform (FFT) and the conjugate gradient method (CGM). Those techniques are proposed respectively by Liu (2000) and Polonsky (1999).

By assuming the two materials as linear, isotropic and homogeneous elastic, the resolution of the problem is then formulated according to:

Load balancing: the load applied to the upper body (the tire in our case) must be equal to the sum of the contact pressures distributed over the contact surface.
Separation of the bodies: the final separation of the bodies after the loading, is equal to the sum of the initial separation of the bodies, the rigid body displacement and the elastic displacement in surfaces of the bodies in contact.

Those conditions leads to several equations described in (Manyo 2019). Equations are solved numerically by the FFT and the CGM method. The potentials of the theory of the semi-infinite space (Love 1982) is applied for the computation of the elastic displacements on surface u.

The geometry of the bodies in contact must be known in order to have the real appearance of the distribution of the contact pressures.

Thus, in the case of tire-pavement contact, the geometry of the tire is reproduced according to a photogrammetry technique (Manyo et al. 2019). The tire inclination is achieved by rotating the tire around its center line under a commercial software (MATLAB). The mechanical properties of the tire are determined by assimilating the tire to a solid cylinder and using a force-displacement model according to a linear contact whose parameters are obtained by calibrating the force experimental deviations results (Manyo et al. 2019). The tire is supposed to be incompressible.

4 Experimental Validation

4.1 Experimental Apparatus

In order to verify the results of SAM contact pressure calculations, the contact pressures were measured by a specific sensor (TekScan). The operating principle of the TekScan is based on a resistive polymer whose electrical resistance changes according to the applied stress. This sensor is in the form of a sheet of a certain rigidity and a given resolution. This technique makes it possible to get measurement at static and dynamic states.

The system used consists of electronic sensors embedded in a 0.2 mm thin sheet and an acquisition software named iScan. The electronic sheet is incompressible and operates at temperatures between -40–$200\,^{\circ}$C with a moisture percentage of 5–10%. Two polyester sheets are needed for the construction of the sensor. On the first sheet, piezoresistive materials (sensors) are placed at equidistance. On the second one, the same sensors are placed in the same way, but perpendicular to the first (Fig. 1(a)).

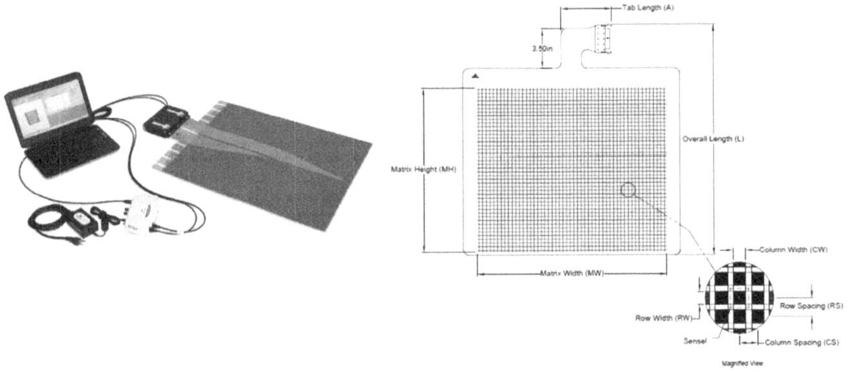

Fig. 1 (Left) TekScan piezoresistive sensors (Right) Matrix of sensitive elements formed by piezoresistive sensors (from https://www.tekscan.com/)

Sensors are semiconductors whose electrical resistance varies with the application of a force.

These rows and columns of sensors intersect forming a matrix of sensitive elements more or less tight when the two sheets are joined (Fig. 1(b)). The typical accuracy of these sensors is ±5%. In our case the area of a sensor is 25.806 mm^2 or 3 mm width and separated by 2.08 mm. Thus, in the SAM simulation, we chose a 3 mm mesh pitch in the longitudinal (x) and the lateral (y) directions.

The Tekscan pressure sensor has been calibrated with a load of 45 kN under a press system (Zwick). When calibrating a sensor, the software calculates an average applied pressure based on the area of loaded sensor and the force value appliqued.

4.2 Experimental Protocol

In the roundabouts, the truck speed is generally slow (5–30 km per hour). So the load application is longer than in a busy road section. However, because of its geometry and the effect of the centrifugal force resulting from the turn, the load distribution is unbalanced between the wheels of the same axle. It is due to the dragging of the overload of the wheels outside the turn (by 10–20% according to French measurements).

Theoretically, it is assumed that during a turn, the tires of a vehicle are inclined at a camber angle with the road surface. This angle is likely to cause a difference in shear stress and vertical distribution of stresses in the road structure compared to straight sections or when the tires must not tilt without an inclination angle. To study and verify this hypothesis, the angle of inclination of the tire was chosen to be 0, 3 and 5° as tested by Xiaodi (2017).

The tests were performed in static state. The experimental protocol used to describe cornering and normal section contact pressures is based on the measurement of the contact pressure between a heavy-duty tire (reference 295/80R22.5) with an inflation pressure of 820 and 900 kPa. For a simplicity reason, the choice of material in contact with the tire was made on a metal plate (5 mm thickness).

To simulate the position of the tire at a turn under the press, we used the principle of inclinometer (see Fig. 2).

In normal section, the tire was loaded at 32.5 kN and the contact pressures were measured by the TekScan sensor system. In the roundabout, the lateral load tends to amplify the loading on the outer wheel. To account for this phenomenon, we propose 3 variants of camber angle 0, 3 and 5° with a load of 45KN, the obtained pressures will be validated with the same measuring system TekScan.

The inclination of the tire after loading has been estimated at 1° for the different camber angles. The tire tilt angles (0, 3 and 5°), on the other hand, were chosen arbitrarily to conservatively cover the theoretical worst-case range for tilt angles of the tire when at a roundabout.

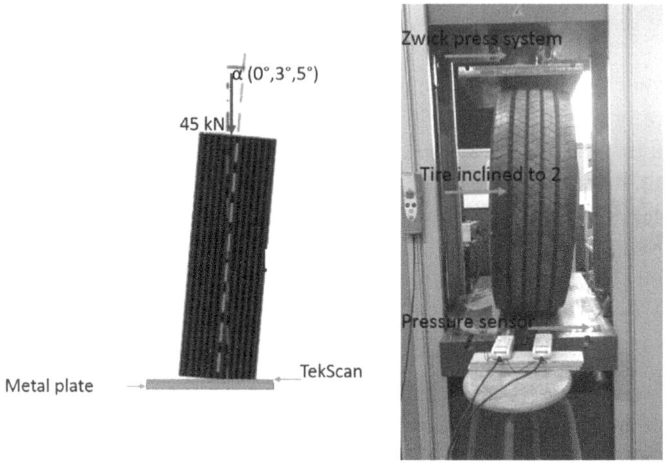

Fig. 2 Experimental protocol of inclined tire under the press

5 Analysis and Discussion

5.1 Results for 0° Inclination

Figure 3 shows the distribution of the contact pressure field for a normal load of 32.5 kN static at 0°. The size of the mesh chosen for the SAM simulation is 3×3 mm^2 in order to approach the size of the elementary sensors.

A relatively good correspondence is also observed between the numerical and experimental results, especially in terms of average maximum pressures. The average maximum measured value is 1528 kPa and the maximum value calculated is 1536 kPa. It is observed that the distributions of the contact pressures, the shape and size of the contact surface are almost similar.

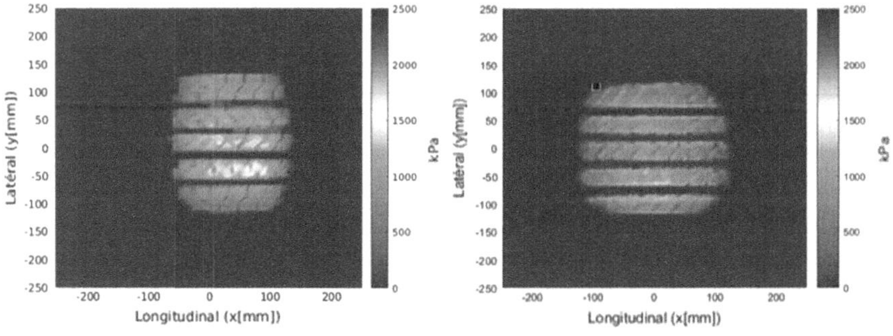

Fig. 3 Results for 0° inclinations° TekScan measurement (left) and SAM simulation (right)

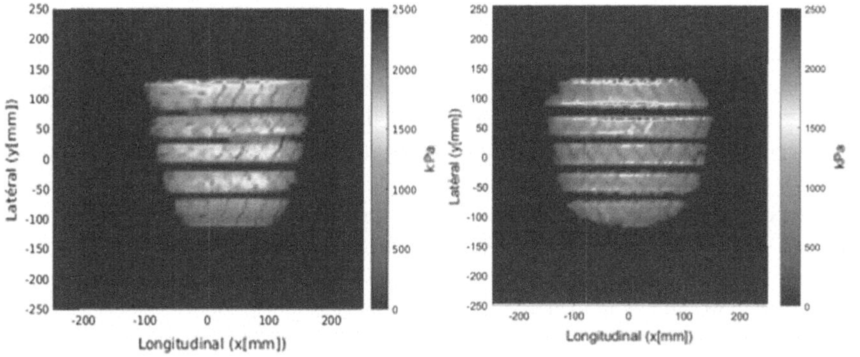

Fig. 4 Contact pressure at 3° inclination TekScan measurement (left) and SAM simulation (right)

5.2 Results for a 3° Inclination

Figure 4 shows the results for the angle of inclination of 3°.

The maximum average stress is 1788 kPa from the measurements and 1800 kPa from the calculation code. The profiles of the contact pressure for both existing inclinations seem to vary considerably.

5.3 Results for a 5° Inclination

At 5°, the maximum measured pressure is 1977 kPa against 1995 kPa calculated with SAM code (See Fig. 5). This result enhances the theoretical hypothesis that the turning point can cause greater stress on pavement. It is also observed that the distributions of the contact pressures are almost identical (calculation and measurements).

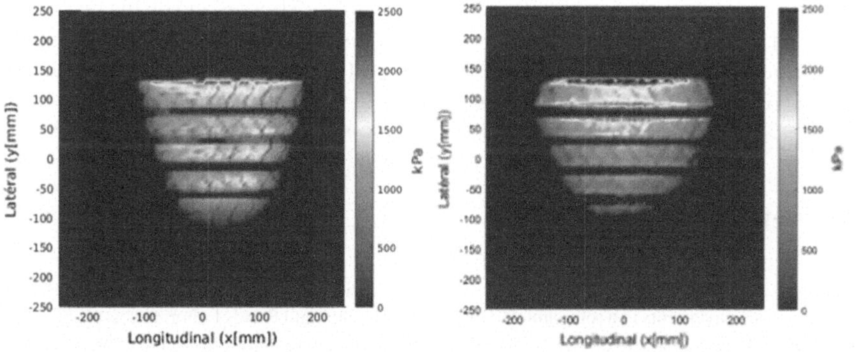

Fig. 5 Contact pressure at 5° inclination TekScan measurement (left) and SAM simulation (right)

It is clear that the contact pressure increases with an increase of the angle of inclination of the tire.

It can be concluded that the inclination of the tire in a turn causes apparent changes in tire-road contact stress. At a current section, the shape of the contact area between the pavement tire is rectangular and the maximum value of the contact stress appears more or less in the center of the contact zone. When turning (3, 5°) the contact area appears to be reduced and the location of the maximum contact stress gradually moves towards the outer edge of the contact area The reduction of the contact surface is accompanied by an increase in the contact stress and consequently the deterioration intensifies, particularly in turns.

6 Conclusions and Perspectives

This study introduced the modeling and validation of the application of a semi-analytical method for the analysis of tire-pavement rolling contact. The use of this code has the advantage of a more realistic, accurate and faster calculation than the current methods including the Finite Element Method (FEM). The developed model thus makes it possible to obtain the contact pressure while cornering when the stresses are supposed to be maximum.

Based on the simulation results and experimental validation, the key findings, conclusions, and recommendations drawn from this paper are summarized as follows:

The contact pressure increases with an increase in the angle of inclination of the tire from 0–5°, which provides analytical evidence to support theoretical hypotheses that the traffic turning on roads (for example, at intersections) can cause greater shear stress in the pavement than by traffic. This suggests that sections of a highway with a large volume of traffic in turns may be more susceptible to shearing and surfacing ruts.

Finally, our model will lead to a better understanding of the problems of the plowing and the descending cracks of the surface layer in normal section and in significant points making an analysis on the Von Mises stresses and on the maximum shear on the surface. Subsequently, we plan to integrate transversal efforts to study the phenomena of degradation in the singular road points (turns, roundabouts) where the life of the surface layers is limited.

References

Ambassa Z, Allou F, Petit C, Eko RM (2013) Evaluation of the aggressiveness of traffic on bituminous pavements in a roundabout, Bulletin des Bridges and Roads Laboratories, pp 171–188

De Beer M, Fisher C, Jooste, F (1997) Determination of pneumatic tyre/pave-ment contact stresses under moving loads and some effects on pavements with thin asphalt surfacing layers. In: 8th International Conference on Asphalt Pavements, Wash-ngton, pp. 179–227

Jacq C, Nelias D, Lormand G, Girodin D (2002) Development of a three-dimensional semi-analytical elastic-plastic contact code. J Tribol 124:653–667
Liu S, Wang Q, Liu G (2000) A versatile method of discrete convolution and FF (DC-FFT) for contact analyses. Wear 243(1–2):101–111
Love A (1892) A treatise on the mathematical theory of elasticity. The University Press, Cambridge
Manyo EY, Reynaud P, Picoux B, Tautou R, Nélias D, Allou F, Petit C (2019) Towards fast modelling of the tire-pavement contact". Eur J Environ Civil Eng. https://doi.org/10.1080/19648189.2019.1628812
Manyo EY (2019) Modélisation avancée du contact pneu chaussée pour l'étude des dégradations des chaussées en surface, Thèse de Doctorat, Université de Limoges, 14 Février 2019, 211 p
Padovan J (1987) Finite element analysis of steady and transiently moving/rolling nonlinear viscoelastic structure. Theory Comput Struct 27:249–257
Polonsky I, Keer L (1999) A numerical method for solving rough contact problemsbased on the multi-level multi-summation and conjugate gradient techniques. Wear 231(2):206–219
Reynaud P, Nasr SB, Chaise T, Nelias D, Petit C, Allou F (2016) 3D modelling of tire-pavement contact pressure. Eur J Environ Civil Eng 21(6):112–127
Xiaodi H, Abu NM, Faruk JZ, Mena IS, Lubinda FW (2017) Effects of tire inclination (turning traffic) and dynamic loading on the pavement stress–strain responses using 3-D finite element modeling. Int J Pavement Res Technol 10:304–314
Ziefle M, Nackenhorst U (2008) Numerical techniques for rolling rubber wheels: treatment of inelastic material properties and frictional contact. Comput Mech, 42(3):337–356

LCMS-2 Measurements of the Quality of Road Markings

Kars Drenth, Jun Yew Tan, Marc Drenth, and Ong Ju Kit

Abstract The assessment of the structural and functional pavement condition is nowadays a fully machine based non-destructive procedure in which equipment covers all the data input required for a modern PMS approach. A LCMS (Laser Crack Measurement System) is a high-resolution transverse profiling system based on 3D Sensors capable of real time continuous measurement of condition data in a single run. One of the images produced by a LCMS is the intensity image, which expresses the reflective properties of the pavement surface. As quality control of road marking often retro-reflectometers are used. A limitation of retro-reflectometers is that they can only measure for instance a single edge marking at any time whereas the road surface can have multiple markings. The LCMS captures the reflective properties of the surface over the total width of a lane and as such can measure the reflective properties of all road markings at any transverse location. This paper discusses the results of the correlation of the retroreflectometer measurements with the reflective properties measured by the LCMS-2. This first time study resulted in a very promising correlation showing that the LCMS-2 collected data is very useful in rating the quality of road markings and predicting the need for maintenance.

Keywords LCMS-2 · 3D-images · Markings · Intensity · Retroreflectivity

1 Introduction

The assessment of the structural and functional pavement condition is nowadays a fully machine based non-destructive procedure in which equipment covers all the data input required for a modern PMS approach. A LCMS (Laser Crack Measurement System, December 2019) is a high-resolution transverse profiling system based on 3D Sensors using custom optics and a laser line allowing the system to operate not only during daytime but at night as well. The LCMS Sensors are capable of real time

K. Drenth (✉) · M. Drenth
Samwoh Innovation Centre B.V., Kranebittenbaan 21, 3045AW Rotterdam, Netherlands
e-mail: kars.drenth@swic.international

J. Y. Tan · O. J. Kit
Samwoh Innovation Centre Pte Ltd., 51 Kranji Crescent, Singapore 728661, Singapore

C. Raab (ed.), *Proceedings of the 9th International Conference on Maintenance and Rehabilitation of Pavements—Mairepav9*, Lecture Notes in Civil Engineering 76,
https://doi.org/10.1007/978-3-030-48679-2_88

continuous measurement of condition data regarding cracking, rutting, roughness, macro-texture, ravelling, potholes, pavement marking and road geometry in a single run. The LCMS-2 is the latest upgrade of this technology having an improved vertical and horizontal resolution resulting in higher quality and reliable output.

One of the images produced by the LCMS-2 is the intensity image, which expresses the reflective properties of the pavement surface of the projected laser line (black = 0 and white = 255). The reflective quality of a road marking tells something about its visibility under different light conditions and is a safety related characteristic. As quality control of road marking often retro-reflectometers are used. A limitation of retro-reflectometers is that they can only measure for instance a single edge marking at any time whereas the road surface can have multiple markings like the centreline, arrows and other type of markings in transverse direction. The LCMS-2 captures the intensity (reflective properties) of the surface over the total width of a lane and as such can measure the intensity of all road markings at any transverse location at traffic speed. Economically there is a clear advantage as all markings can be surveyed in a single run including all other condition related surface characteristic. A retroreflectometer would require multiple runs to collect the condition of markings only as long as a traffic speed device is used. Manually it will take more time whereas less data is collected and not all roads allow a manual approach due to safety reasons.

2 Road Marking Rating and Testing Approach

There are international and local standards related to the materials to be used for road marking, its performance and the way it has to be applied. The performance-orientated standard for Singapore (Singapore Standard 2013) is rating the field measurement of the coefficient of retroreflected luminance characteristics (Q_D and R_L mcd/m^2/lx) using a portable device complying with ASTM E1710 (ASTM E1710-18). The characteristic R_L relates to nighttime visibility and the characteristic Q_D relates to daytime visibility. The minimum value of the field measurement of Q_D respectively R_L (dry condition) of thermoplastic material shall be $\geq$180/$\geq$300 mcd/m^2/lx for a white marking and $\geq$160/$\geq$160 mcd/m^2/lx for a yellow marking. The minimum value for R_L after 5 month of installation of thermoplastic material shall have a level of respectively $\geq$150/$\geq$50 mcd/m^2/lx in dry condition. The degree of wear (wear index) is based on a visual assessment and should after 5 month be $\leq$ 20 based on a specified procedure using a reference chart shown in Fig. 1 (Singapore Standard 2013).

In the Netherlands a visual quality catalogue is used for rating all elements of the public space from A+ (no damage) to D (maintenance required). Figure 2 (CROW 2018—original in Dutch) shows this rating related to road marking as well as a percentage of area to meet the rating level referenced to a NEN Standard (NEN-EN 1436:2018) based on the performance of the marking and test methods.

Condition	Wear Index	% of marking faded
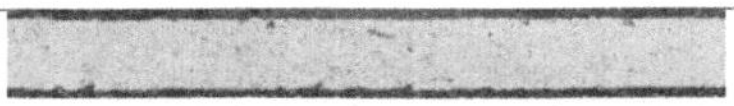	20	< 5
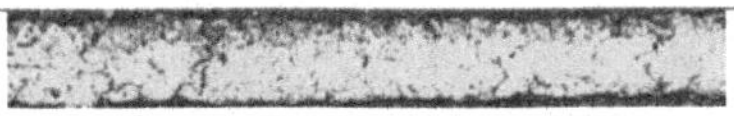	25	25
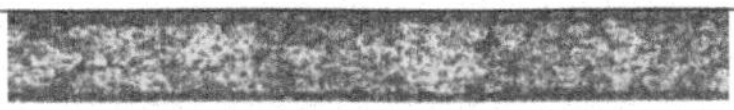	35	50
	43	60
	63	75
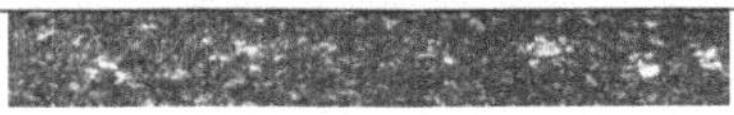	82	90

Fig. 1 Wear index reference chart

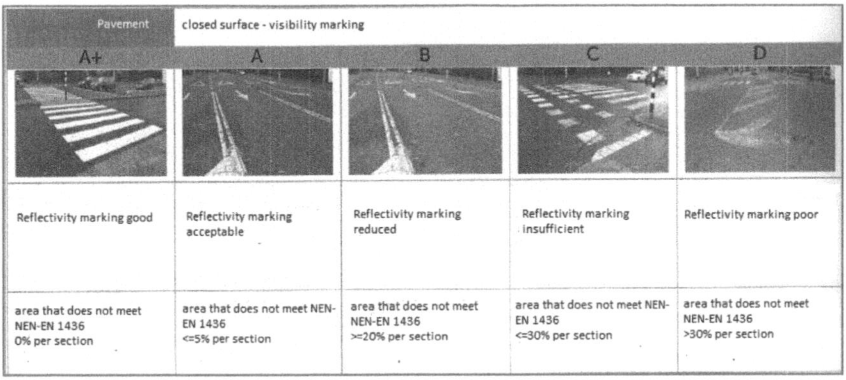

Fig. 2 Visibility rating of road markings

The listed references specify the measurement of the retroreflection of markings by means of a portable instrument as well as visual rating. The background of this research is to make the rating of markings part of the automated rating already in place for the condition of the pavement wearing course.

The reflective characteristic of a pavement surface is sampled by the LCMS, called ‘intensity’. This intensity shows a clear difference between markings and the non-marked pavement surface, for all the marking within the transverse profile of the width of a lane. Figure 3 shows the Survey Vehicle with LCMS-2 equipment.

The portable instrument used for collecting the retroreflection of different quality of markings is the Delta LTL-XL (Delta, August 2016) as shown in Fig. 4 on a relatively old surface and related aged marking.

Fig. 3 LCMS-2 Survey Vehicle with two sensors

Fig. 4 Delta LTL-XL portable retroreflectometer

3 Retroreflectometer Data

Twenty (20) retroreflectometer measurements were collected at four (4) different roads having different marking rating qualities, white as well as yellow. The retroreflectometer is factory calibrated valid till 13-09-2020 and the supplied calibration unit was used each day measurements were taken. The locations where the measurements were taken were marked for easy reference with the LCMS-2. Figure 5 shows a representative collection of the selected locations for rating the marking.

For the comparison of the reflective quality of a marking measured with the Delta LTL-XL and the LCMS-2 the daytime visibility Q_D was used as characteristic. Figure 6 shows a measurement with the Delta LTL-XL on a relative new marking and the result shown real-time.

Fig. 5 Selection of different levels of wear

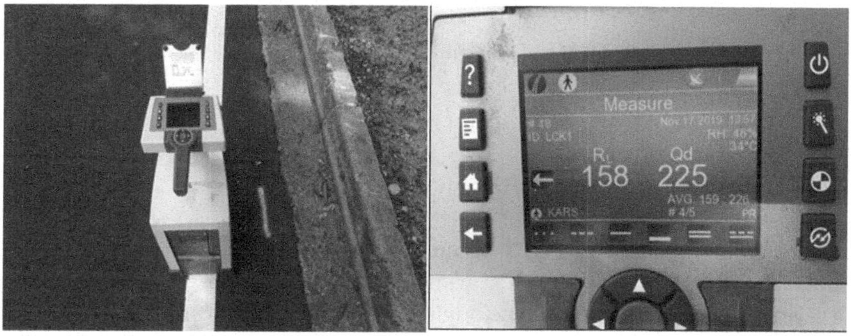

Fig. 6 Retroreflectometer result shown real-time

All the sampled retroreflectometer results for the 4 different roads and different color of the marking are listed in Table 1. The % of marking faded is based on a visual rating of the test locations in accordance with Fig. 1.

4 LCMS-2 Data

In addition to the intensity image produced by the LCMS there is as well depth map of the pavement surface (also referred to as range or XYZ image) and a 3D image. Merging the intensity and range image results in the 3D image. This continuously

Table 1 Retroreflectometer results

Road	Test	Q_D	Marking	% Marking faded
Perahu Road	1	184	White	5–25
	2	138	White	25–50
	3	87	White	60–75
	4	187	White	5–25
Lim Chu Kang Road	5	139	Yellow	25–50
	6	128	Yellow	50–60
	7	134	Yellow	50
	8	104	Yellow	75–90
Woodlands Street	9	224	White	<5
	10	189	White	<5
	11	204	White	<5
	12	177	White	<5
	13	255	White	<5
	14	226	White	<5
	15	220	Yellow	<5
	16	191	Yellow	<5
Rosewood Drive	17	189	White	5–25
	18	188	White	<5
	19	104	White	60–75
	20	112	White	60–75

sampled 3D image is showing all the condition details of a pavement surface, including the wear of all markings present within a traffic lane or the width (4 m) of the surveyed area. Figure 7 shows some brighter 3D image examples of rated locations.

The sampled LCMS-2 results for the 4 different roads and different color of the marking are listed in Table 2. The visibility is rated based on the visual condition rating of the test locations in accordance with Fig. 2.

5 Retroreflectometer Vs LCMS-2 Comparison

The LCMS automatically detects any marking independent of its transverse location or lateral wander and calculates automatically the intensity. This intensity can be calculated over a to be specified file length of 2 till 10 m, although it is foreseen to make this possible over a length of minimal 250 mm. In this comparison the LCMS-2 file length was 5 m. The retroreflectometer testing is measured over a length of 185 mm but repetitive measurements can be taken to cover a larger length by moving it forward including the calculation of the average of the number of preset tests

Fig. 7 Selection of 3D LCMS-2 images

(forward movements). The Delta LTL-XL results have been compared with those of the LCMS-2 as shown in Fig. 8. The computed R^2 of 0.8491 shows a very good polynomial correlation when taking into account the relative difference in procedure and the (slightly) larger area of the LCMS-2 results. In theory the intercept will run through zero for a black new pavement without marking and up to at least 255 for a new white marking. The highest Q_D value measured with the LTL-XL on an old pavement surface is around 110 and the lowest value measured on a relatively new black surface is around 60.

During the analysis of the LCMS-2 data the intensity results can already being visualized by moving the mouse pointer over any marking present within a lane. Figure 9 gives some examples of the intensity of typical markings like arrows and text.

Table 2 LCMS-2 results

Road	Test	Intensity	Marking	Visibility
Perahu Road	1	174	White	A
	2	157	White	B
	3	117	White	D
	4	170	White	A
Lim Chu Kang Road	5	138	Yellow	B
	6	138	Yellow	C
	7	132	Yellow	B
	8	117	Yellow	D
Woodlands Street	9	217	White	A+
	10	212	White	A+
	11	222	White	A+
	12	169	White	A+
	13	200	White	A+
	14	194	White	A+
	15	214	Yellow	A+
	16	203	Yellow	A+
Rosewood Drive	17	173	White	A
	18	174	White	A+
	19[1]	120	White	B
	20[1]	130	White	C

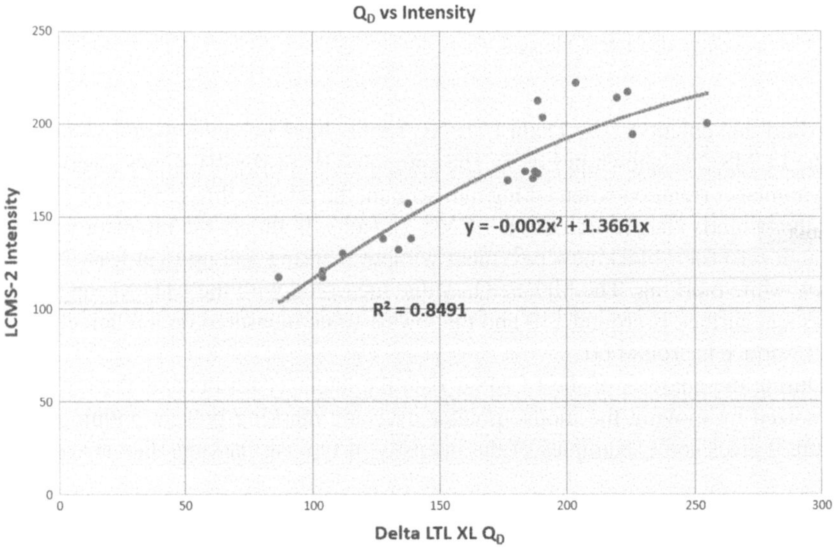

Fig. 8 Correlation of Q_D vs Intensity

Fig. 9 LCMS Intensity measurement of any marking feature within a lane

6 Conclusions

A LCMS (Laser Crack Measurement System) is a high-resolution transverse profiling system based on 3D Sensors capable of real time continuous measurement of condition data in a single run. One of the images produced by a LCMS is the intensity image, which expresses the reflective properties of the pavement surface. The LCMS captures the intensity of the surface over the total width of a lane and as such can measure the intensity of all road markings at any transverse location. The following conclusions can be drawn:

This study resulted in a very promising correlation between the Intensity of the LCMS-2 and Q_D of a manual retroreflectometer.

A LCMS survey delivers a continuous image of all markings present within a lane allowing a visual rating according to present international procedures.

The LCMS images allowing for linking a visual quality of the wear to an automated analyzed intensity in a way that the subjectivity of a human rating can be eliminated.

The availability of past LCMS data allows to qualify the performances between different marking materials.

A survey of the quality of the functional condition of the surface of a pavement can include the collection of marking data at no extra cost and is far more efficient than any other way of collecting the level of wear of markings automatically or by use of handheld equipment.

This study resulted in a very promising correlation showing that the LCMS-2 collected data is very useful in rating the quality of road markings and predicting the need for maintenance.

References

ASTM E1710-18. Standard Test Method for Measurement of Retroreflective Pavement Marking Materials with CEN-Prescribed Geometry Using a Portable Retroreflectometer

CROW Publication 380 (2018) Kwailiteitscatalogus openbare ruimte 2018—Standaardkwaliteitsniveaus voor onderhoud (in Dutch)

Delta, LTL-XL Retroreflectometer, Manual, On site quality control of road marking & road surfaces in accordance with CEN/ASTEM specifications, August 2016

NEN-EN 1436:2018. Road marking materials—road marking performance for road users and test methods

Pavemetrics, Laser Crack measuring System, LCMSAnalyser Library: User Manual (Rev 4.63.0.0), December 2019

Singapore Standard (2013) Specification for hot-applied thermoplastic road marking materials – Materials, performance and application. Singapore Standard SS, vol 589, p 2013

A Study on the Effect of Milling on Stress Distributions in Asphalt Pavements

Kaoutar Diouri, Rajae Bousselham, Anirban De, Adriana Hera, Tahar El-Korchi, and Rajib B. Mallick

Abstract Milling is an indispensable process in recycling of asphalt pavements. The process involves the fracture and removal of asphalt mix under high stress and rapid loading conditions. An understanding of milling induced stress is important for avoiding or reducing the impact on pavements below the milling line and to optimize the milling process. The objectives of this research were to develop a finite element model of the milling process and estimate stress distributions under different milling conditions. Milling of a fine and a coarse graded Hot Mix Asphalt (HMA) and a Stone Matrix Asphalt (SMA) were modeled for different milling speeds and depths. The results indicate significantly high milling induced stresses below the milling line. Gradation of mixes, interface and milling speeds have significant effect on stiffness and hence stress distributions and maximum stresses. The inference is that the depth of milling should be decided on the basis of existing conditions which include layer depths, location of interface, strength of bond between the layers, type of mix, temperature of mix and milling speed (drum rotation speed). To avoid high

K. Diouri · R. Bousselham · T. El-Korchi · R. B. Mallick (✉)
Civil and Environmental Engineering Department, Worcester Polytechnic Institute (WPI), Worcester, MA, USA
e-mail: rajib@wpi.edu

K. Diouri
e-mail: kdiouri@wpi.edu

R. Bousselham
e-mail: rbousselham@wpi.edu

T. El-Korchi
e-mail: tek@wpi.edu

A. De
Civil and Environmental Engineering Department, Manhattan College, Riverdale, NY, USA
e-mail: anirban.de@manhattan.edu

A. Hera
Mechanical Engineering Department, Worcester Polytechnic Institute (WPI), Worcester, MA, USA
e-mail: ahera@wpi.edu

C. Raab (ed.), *Proceedings of the 9th International Conference on Maintenance and Rehabilitation of Pavements—Mairepav9*, Lecture Notes in Civil Engineering 76,
https://doi.org/10.1007/978-3-030-48679-2_89

stress related damage to remaining pavement layers, milling to partial depth of a layer should be avoided. Research is needed to evaluate changes to the structural condition of remaining pavement due to milling.

Keywords Milling · Asphalt · Finite element · Von Mises stress

1 Introduction

Every year, hundreds of thousands of kilometers of pavements require maintenance and rehabilitation work. Milling of roadway layers has become almost a routine activity in asphalt pavement rehabilitation all over the world. Effective milling of the existing asphalt mix layer is the key to constructing sustainable pavements—it maintains pavement geometry, conserves aggregates, asphalt and energy, and helps us construct durable pavements at a reduced cost and environment friendly way (Brock and Richmond 2007).

In many cases, milling is not conducted at any interface and a part of the layer remains after milling (see Fig. 1). Research is needed to understand the effect of different factors on the pavement layers below the milling line. The sheer size of milling machines and the complicated machinery make it difficult to replicate milling in the laboratory, and hence, few, if any, academic research work have been conducted on milling so far. Practically in most cases, milling is conducted on the basis of empirical guidelines that have been developed by the equipment manufacturers (for example, Wirtgen GmbH, undated).

Milling involves the fracture and removal of asphalt mix under high stress and rapid loading conditions, and it is not known to what extent these forces affect layers below the milling depth. The questions that follow from this knowledge gap are as follows: How are these stresses affected by drum rotation speed, mix stiffness, depth of milling? What is the impact of the bond between different layers on the stress magnitude and distributions? If the answers to the above questions are not available, then there is a risk of damaging underlying layers during the milling

Fig. 1 Core from milled pavement showing remaining part of milled layer

process. Furthermore, this knowledge gap prevents the users from optimizing the milling process, which typically consumes a significant percentage (20%, MDOT 2019) of pavement construction budget for mill-and-fill operations.

2 Objective and Scope

The objective of this research was to develop a finite element (FE) model of the milling process, estimate the stress distributions in the pavement during milling, and infer practical significances from the model simulations. Milling of a fine and a coarse graded Superpave Hot Mix Asphalt (HMA), and a Stone Matrix Asphalt (SMA) were modeled for different types of bonding, milling speeds and depths considering elastic material behaviors.

The scope of work that is reported in this paper consists of the following steps.

1. Geometric modeling of the milling drum with teeth and pavement.
2. FE simulation of the milling process for the following different cases:

    a. A fine graded Superpave HMA mix, a coarse graded Superpave HMA mix and a Stone Matrix Asphalt (SMA);
    b. Two different types of interfaces between the HMA layers—layers that are fully bonded, and layers that are separated by a thin (3 mm) weak layer (to represent poor bonding);
    c. Three milling depths of 25, 50 and 75 mm; and
    d. Three milling speeds corresponding to drum rotation speeds of 50, 84 and 105 RPM.

Information regarding the mixes, and milling equipment were obtained from Liao (2007) and Wirtgen GmbH (2016a, b), as well as from Lyons (2019), respectively. The pavement model consisted of two 75 mm thick HMA layers (surface and binder) over a 300 mm base on top of a 600 mm subgrade.

3 Geometry Modeling

The first approach was through modeling of a three-dimensional milling drum with multiple teeth. However, this process is computationally very intensive and hence time consuming and required more than 30 h for the complete simulation, with the available hardware. Subsequently, 2D plane strain models with the drum and a single tooth were created in SolidWorks (2019) and design modeler (ANSYS 2019b). Figure 2 shows the 3D and the 2D model of the drum and milling tooth. The intent of the modeling was to evaluate the stress distribution in the pavement during the impact of one tooth, which actually results in removal of the material at the point of contact.

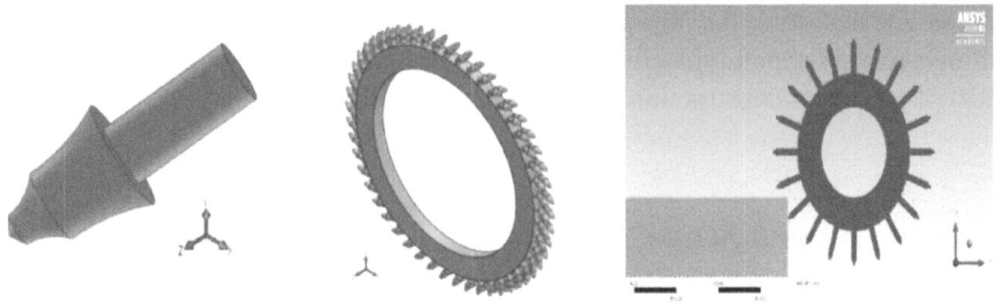

(a) 3D model of tooth, W7/20X, Wirtgen GmbH, 2016b) (b) 3D milling model (c) 2D model of milling drum and teeth

Fig. 2 3D and 2D milling models

4 Finite Element Model

The FE simulations were conducted with Autodyn (ANSYS 2019a) software, in explicit analysis mode for nonlinear dynamics, after published work on rock-cutting for tunnels (Cho et al 2013; Li and Du 2016). Lagrange-Lagrange interaction was enabled between solid parts to simulate the action of the tooth on the pavement material. The full bond layers (Fig. 3) were modeled with no-slip condition at the interface, and the weak interface was modeled with a thin layer between the two HMA layers. For meshing, the quadrilateral element sizes were chosen as 3 mm for the HMA layers (finer mesh), and as 10 mm for the milling drum, base and subgrade (coarse mesh).

Boundary conditions were defined with fixed support at the bottom of the subgrade and the left edge (opposite of the tooth contact) of the pavement structure. For the contact between the milling tooth and the pavement structure, a frictional interaction was chosen to simulate the milling process, with static and dynamic friction coefficients of 0.3 and 0.1, respectively. The drum and the milling teeth were considered to be rigid bodies, and the 2D flexible pavement structure was assumed at plane strain condition. Milling was modeled for a duration of 20 ms.

5 Material Model

Tables 1 and 2 present the material properties for the three mixes (coarse, fine Superpave and SMA) and those of the milling drum and tooth. A Drucker-Prager model (Seibi et al. 2001; Dessouky et al. 2005; Ying 2013) was used for the HMA layers. Removal of the milled materials was simulated through erosion of elements that reached a maximum tensile failure level (1 MPa) during simulation. The explanation for considering this specific tensile strength is provided in the next section.

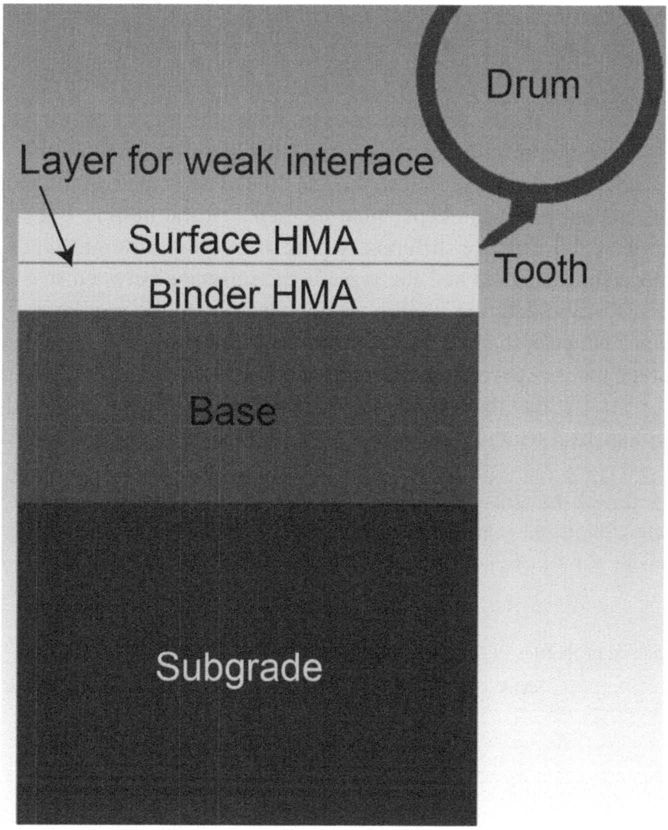

Fig. 3 Schematic of milling drum and tooth with the pavement structure

6 Results of Simulations

Von Mises stress (Eq. 1) versus depth, along the point of contact of the milling tooth with the pavement, was selected as the primary parameter of interest in the analyses. The reason is that Von Mises stress effectively combines the contributions of the different stresses that are acting concurrently, and also has been found to be effective in explaining the failure of materials under such combined action of different stresses. During the analyses, the Von Mises stress was compared against a limit of 1 MPa, because, typically, tensile strengths of Hot Mix Asphalt (HMA) mixes at temperatures of 20–25 °C have been found to be at or below 1 MPa (Arepalli and Mallick 2017; Dave et al. 2018). This means any Von Mises stress >1 MPa is indicative of a potential cracking failure at that point. It can be observed from all of the results that there is a significant depth below the milling line where this is indeed the case.

$$\sigma_{VonMises} = \left[\frac{(\sigma_1 - \sigma_2)^2 + (\sigma_2 - \sigma_3)^2 + (\sigma_3 - \sigma_1)^2}{2}\right]^{\frac{1}{2}} \quad (1)$$

Where σ_1, σ_2, and σ_3 are the principal stresses with the order of $\sigma_1 > \sigma_2 > \sigma_3$.

Figure 4 shows representative stress diagrams from the simulations. The most important observation is that the stresses at the time of the impact of the milling tooth with the pavement penetrate deeper than the depth of the milling line. This depth of penetration is different for different depths of milling, different types of mixes (dependent on the stiffness) and the type of the interface between two successive HMA layers. As expected, the stresses do not penetrate to the next layer if there is a weak interface between the two layers; however, if there is a good bond between the two layers then the stresses do penetrate into the next layer. The practical significance of this observation is that there can be a significant stress at depths below the milling line, which can lead to damage (such as the formation of micro-cracks) at those locations.

Figure 5 shows the stress versus time plots for two different mixes, at three different depths. The figure shows that the stress at the milling depth peaks within a time of 1 ms (1 * 10 – 3 s). The stress quickly dissipates as the material fails at that

Table 1 Material properties of the three mixes and the weak layer

Material	SMA	Fine graded Superpave HMA mix	Coarse graded Superpave HMA mix	Weak interlayer
Specific gravity (g/cm^3)	2.408	2.504	2.498	1.025
Reference temperature (°C)	21.1			
Poisson's ratio	0.35			0.495
Young modulus (MPa)	5,014	3,459	8,063	0.00298
Bulk modulus (MPa)	5,571	3,843	8,959	0.004966
Shear modulus (MPa)	1,857	1,281	2,986	0.001

Note Properties of the mixes were obtained from Liao (2007) Modulus values were obtained by Liao (2007) through dynamic modulus testing

Table 2 Milling drum and tooth properties

Material	Drum (steel)	W7/20X milling tooth (tungsten carbide)
Specific gravity (g/cm$^{3)}$	7.800	15.880
Reference temperature (C)	21.1	
Poisson's ratio	0.30	0.35
Young modulus (MPa)	200,000	686,000
Bulk modulus (MPa)	333,333	762,222
Shear modulus (MPa)	71,428	254,074

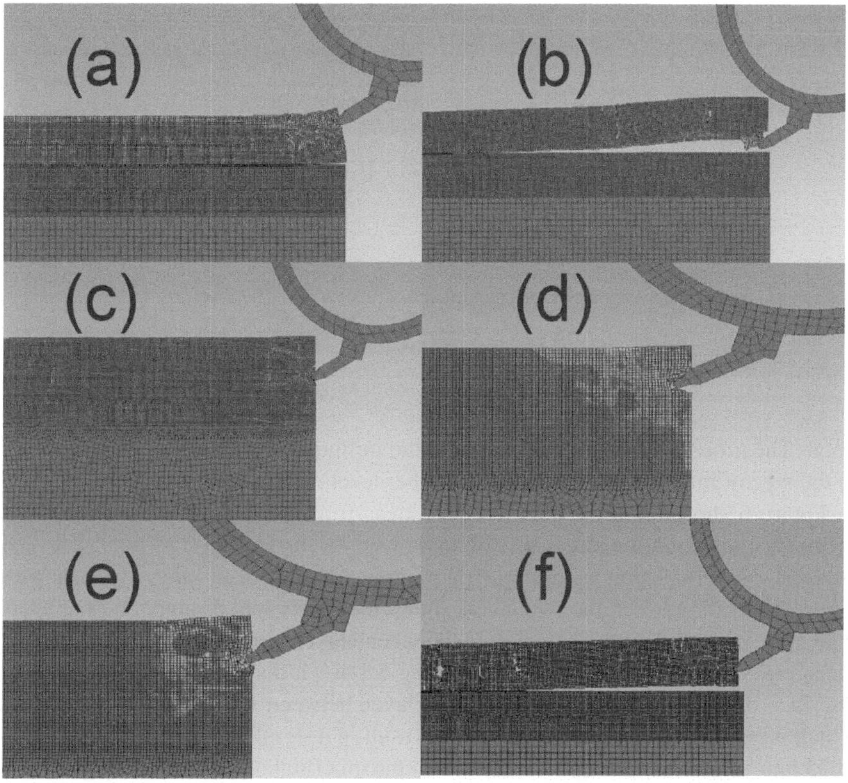

Fig. 4 Examples of stress diagrams for different milling conditions; (a) Fine mix, 25 mm milling depth, 150 RPM, weak layer at interface; (b) Fine mix, 75 mm milling depth, 150 RPM, weak layer at interface; (c) Fine mix, 75 mm milling depth, 150 RPM, fully bonded layers; (d) Coarse mix, 50 mm milling depth, 150 RPM, fully bonded layers; (e) Coarse mix, 50 mm milling depth, 50 RPM, fully bonded layers; (f) Coarse mix, 50 mm milling depth, 150 RPM, weak layer at interface. Note that RPM refers to drum rotation speed, blue represents areas of zero stress, green, yellow and red represent areas of progressively higher stresses

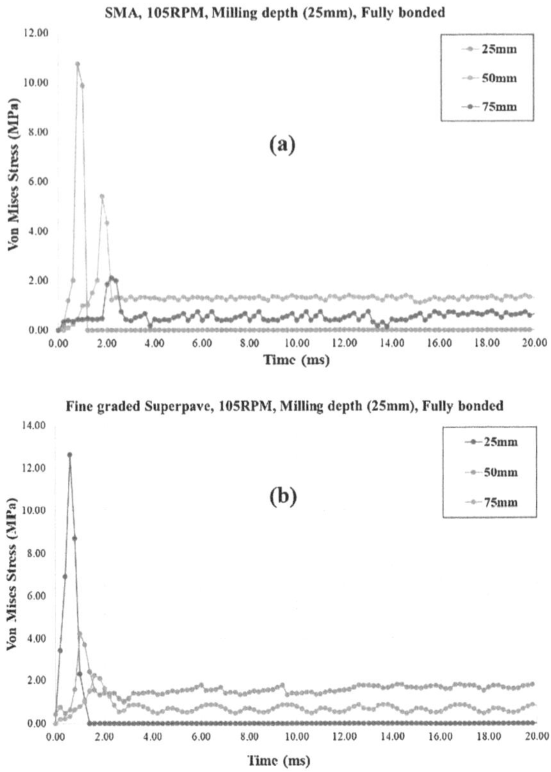

Fig. 5 Stress distribution versus time for **a** SMA and **b** Fine graded Superpave mix

point. The time to reach the peak stress at the milling depth seems to be insensitive to the type of mix and the stresses at the other levels dissipate at a slower rate.

Figure 6 shows plots of depth versus stress from the simulations. The effects of four variables on the stress distributions were evaluated: type of bond between successive layers, depth of milling, type of mix and speed of milling. The case with a weak layer in between the two layers shows the most rapid dissipation of stresses with no or very small stress penetrating the second layer. For different milling depths, the stresses penetrate well below the milling depth—it dissipates completely at the interface only in the case with the weak layer between the layers. Fine milling, which is conducted at <25 mm depth, can result in a significant stress up to a depth of 25 mm below the milling line. The stiffer the mix (higher modulus), the greater is the stress generated, and deeper is the penetration of stresses. The order in terms of stress magnitude and penetration depth is (high to low): coarse graded Superpave, SMA, and fine graded Superpave. The speed of milling also has an effect on the stresses—a higher drum rotation speed generates higher stresses and greater depth of penetration. A summary of the effect of the milling depth and the milling speed is shown in Fig. 7, for the fine graded Superpave mix.

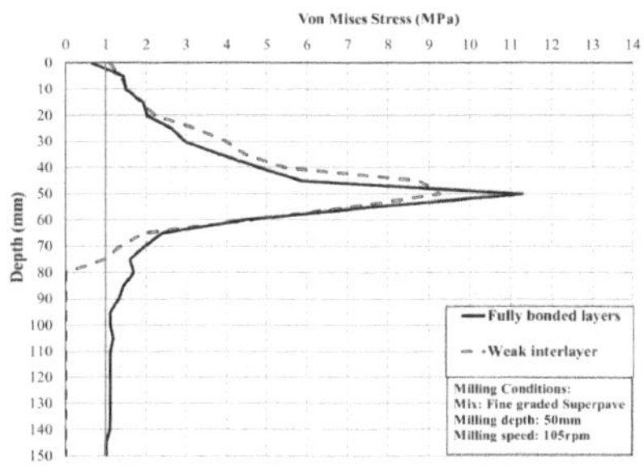

(a) Effect of type of interface

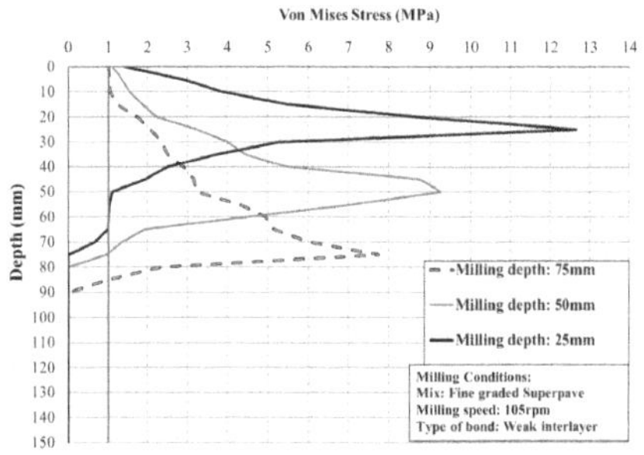

(b) Effect of milling depth

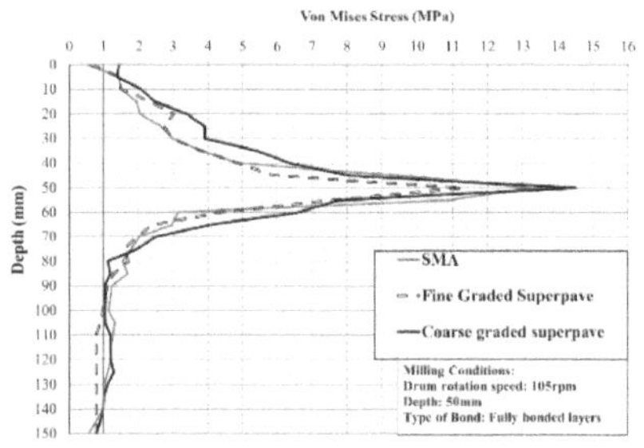

(c) Effect of mix

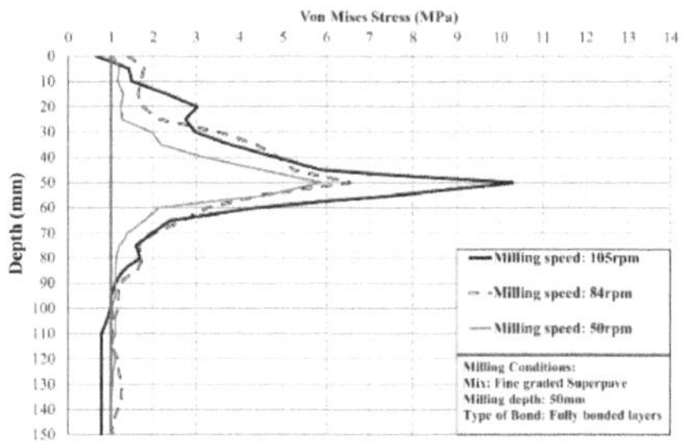

(d) Effect of milling speed

Fig. 6 Plots of stress distribution versus depths for different cases

The above observations lead to a hypothesis that damage can be expected in milled layers. Indirect evidence in support of this hypothesis is obtained from observations from cores obtained from milled pavements, as shown in Fig. 8. The figure shows different kinds of damages—cracked and crushed aggregates as well as missing aggregates and mastic from the areas that have remained from the milled layer (25 mm) on top of the next layer. These damages were very likely caused by high stresses during the milling operation. An inference is that it is not desirable to leave a part of any layer unmilled during the milling process, since cracks in this layer will trigger reflective cracking in the upper layer and lead to its premature failure. The thicker the left-over unmilled layer, the greater will be the negative impact on the overall structural strength of the rehabilitated pavement. It is interesting to note that Hossain and Wu (2002), from their studies on estimation of asphalt pavement life in Kansas, noted that "… in order to achieve a very high fatigue life, the mill-and-inlay thickness should be at least 1.25 times the thickness of the remaining AC (*asphalt concrete*) layer". They cautioned against the use of rule-of-thumb to decide on a milling thickness, and recommended the use of a rational method for the selection of a milling depth (for a mill-and-fill work).

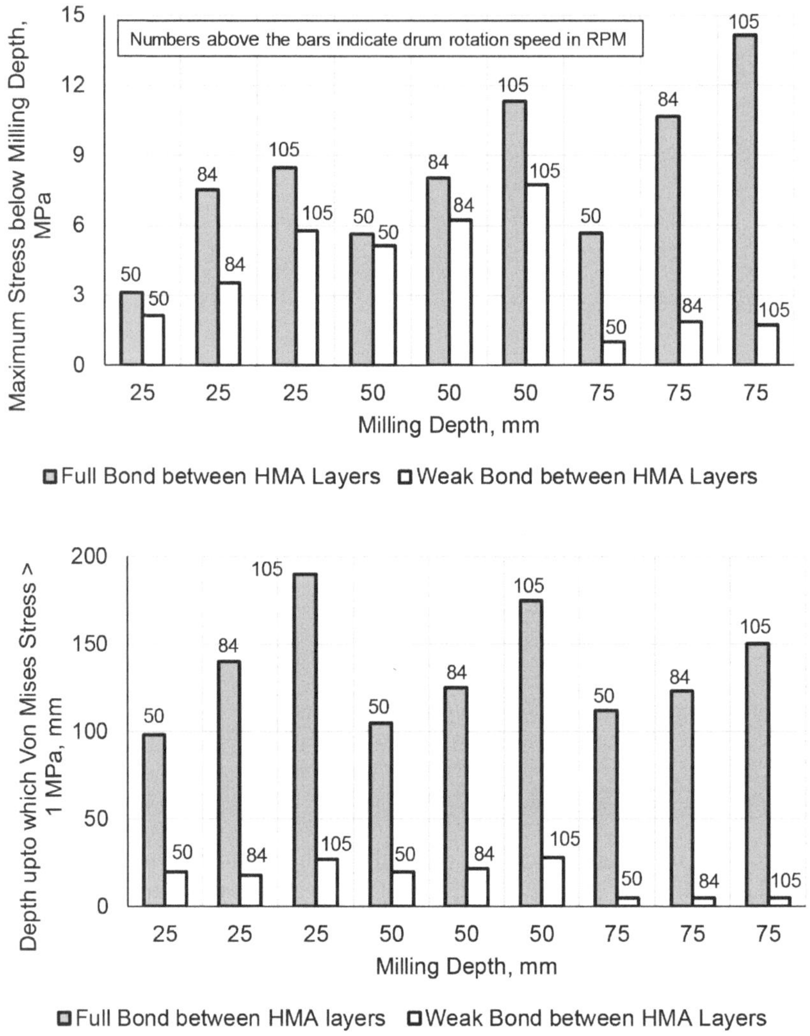

Fig. 7 Plots of maximum (max) stress below milling line and depth (below the milling line) to which stresses exceed 1 MPa versus milling depth and milling speed

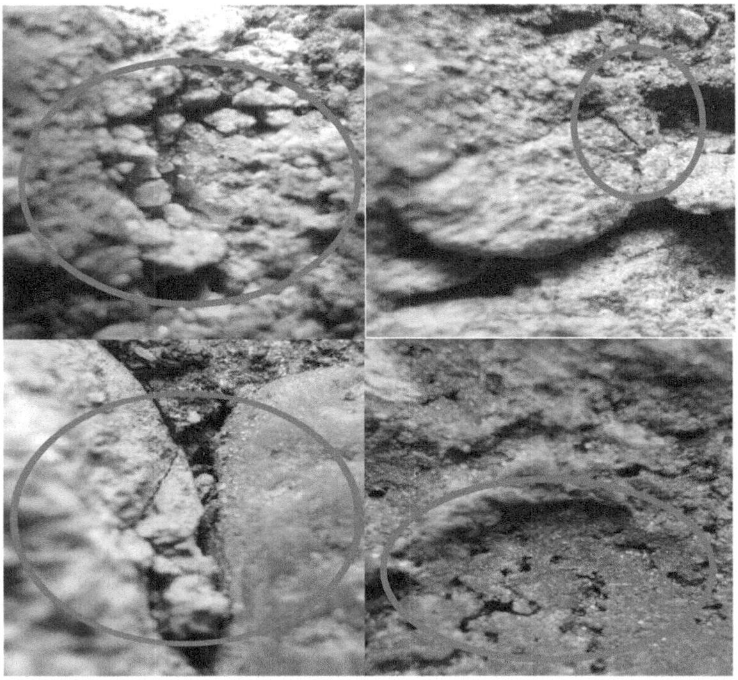

Fig. 8 Examples of crushed, cracked and missing aggregates and mastic in cores from milled pavements

7 Conclusions and Recommendations

The major conclusions from the study are as follows. In general, stresses are relatively high at the depth of milling and at the interface with the next layer. Stresses remain high up to significant depths below the milling line. Gradation of mixes have a significant effect on stiffness and hence stress distributions and maximum stresses. Stresses are also affected significantly by the milling drum speed. Cracks in aggregates and mastic are visible in cores from milled pavements.

Based on these conclusions the following recommendations are made. Depth of milling should be decided on the basis of existing conditions which include layer depths, location of interface, strength of bond between the layers, type of mix (and hence stiffness), temperature of mix (which has an effect on the material behavior) and milling speed (drum rotation speed). To avoid high stress related damage to remaining pavement layers, milling to partial depth of a layer should be avoided—the depth should match with an interface depth. Research is needed to evaluate changes to the structural condition of remaining pavement due to milling. Finally, design guidelines should be prepared to complement the empirical speed-depth-yield charts that are currently used for milling.

Acknowledgements The authors are grateful to Wirtgen GmbH and Maine Department of Transportation for their help with materials and information. This work would not have been possible without the data from Y. Liao's dissertation which is available on the Internet.

References

ANSYS (2019a) Autodyn, ANSYS, Inc., Concord, MA 01742
ANSYS (2019b) Design modeler, ANSYS, Inc., Concord, MA 01742
Arepalli UM, Mallick RB (2017) A study of moisture induced material loss of hot mix asphalt (HMA). Report submitted to Maine Department of Transportation, Augusta, ME
Brock JD, Richmond Sr JL (2007) Milling and recycling. Technical paper T-127. ASTEC INC, Chattanooga, TN, USA
Cho J, Jeon S, Jeong H, Chang S (2013) Evaluation of cutting efficiency during TBM disc cutter excavation within a Korean granitic rock using linear-cutting-machine testing and photogrammetric measurement. Tunn Under Space Technol 35(2013):37–54
Dave EV, Daniel JS, Mallick RB (2018) Moisture susceptibility testing for hot mix asphalt pavements in New England. Final Report for New England Transportation Consortium, Project 15-3
Dessouky S, Masad E, Little D (2005) Mechanistic model to predict the impact of the aggregate matrix on the permanent deformation of asphalt mixtures. Texas Transportation Institute, The Texas A&M University System, College Station, Texas
Hossain M, Wu Z (2002) Final report estimation of asphalt pavement life report, No K-Tran: KSU-97-6, Kansas State University
Li H, Du E (2016) Simulation of rock fragmentation induced by a tunnel boring machine disk cutter. Adv Mech Eng 8(6):1–11
Liao Y (2007) Viscoelastic FE modeling of asphalt pavements and its application to U.S. 30 perpetual pavement. Doctoral Dissertation, Ohio State University, Columbus, Ohio
Lyons S (2019) Personal communication. Wirtgen GmbH
Maine Department of Transportation (MDOT) (2019) Personal Communication
Seibi AC, Sharma MG, Galal AA, Kenis William J (2001) Constitutive relations for asphalt concrete under high rates of loading. Transportation Research Record 1767
SolidWorks Corporation (2019) Dassault Systems SolidWorks Corporation, Waltham, MA 02451
Wirtgen GmbH (2016a) The world of Wirtgen cold milling machines, Windhagen, Germany
Wirtgen GmbH (2016b) Parts and more compact picks, Windhagen, Germany
Ying H (2013) Finite element modeling of hot-mix asphalt performance in the laboratory. Louisiana State University Doctoral Dissertations, Baton Rouge, LA 70803

Assessment of Preformed 3D-Thermoplastic Road Markings for Long-Term Durability, Skid Resistance and Texture Functionality

Kalpesh Purohit, Mujib Rahman, Andrew Price, and Alan Woodside

Abstract Road Markings are an important safety feature for directing and guiding traffic without distracting its drivers. The erosion of the marking material as a result of prolonged exposure to rain and traffic actions, greatly reduce their performance. In this paper, results from field and laboratory investigations on innovative preformed thermoplastic road markings sheet with 3D impressions are presented. An immersion wheel tracking test was conducted in the laboratory for evaluating material effectiveness against stripping in long-term exposure to water and traffic loading. The results showed that preformed sheet installed on pre-heated and non-heated asphalt surfaces are capable of withstanding long (20,000 cycles) exposure of continuous traffic while submerged in water. Overall, the thermoplastic sheet installed in both conditions, demonstrated excellent performance against stripping, although the performance was relatively better when installed on the pre-heated surface. Furthermore, one-year field observations showed that preformed marking installed on asphalt and concrete surface retains their macrotexture, maintains excellent dry and wet frictions. The 3D impression was also found beneficial on drivers' behavior for speed reduction. Depending on the vehicle type, the reduction was 20–31% immediately after installation and 9–17% after one year. Overall, this study suggests that preformed thermoplastic road marking is not only a durable product but also 3D illusion that provides some traffic calming benefits.

Keywords 3D road marking · Preformed thermoplastic road marking · Sand patch · Skid resistance · Long term performance · Stripping resistance

1 Introduction

The underlying motivation behind road markings is to direct and manage the traffic on a highway (Highways Act 1980). They increase the function of traffic signs, fill

---

K. Purohit · M. Rahman (✉) · A. Woodside
Department of Civil and Environmental Engineering, Brunel University London, Uxbridge, UK
e-mail: mujib.rahman@brunel.ac.uk

A. Price
Preformed Markings Ltd., 6, Oyster Park, 109 Chertsey Rd., Byfleet, West Byfleet KT14 7AX, UK

C. Raab (ed.), *Proceedings of the 9th International Conference on Maintenance and Rehabilitation of Pavements—Mairepav9*, Lecture Notes in Civil Engineering 76,
https://doi.org/10.1007/978-3-030-48679-2_90

in as a mental hindrance and guide traffic to maintain lateral clearance from traffic hazards for the safe and smooth movement of vehicles. Pavement markings area unit is classified into two broad classes by service functions, particularly by removable and non-removable. The former type encompasses a comparatively short service life (less than a year) and is employed with temporary traffic patterns; whereas the later one encompasses a longer service life (more than one year) and is employed with permanent traffic patterns.

There are three key types of permanent road markings. Marking to show longitudinal lane or edge lines, stop lines or crossover markings, and other unique markings such as arrows, words, image markings, cross-hatching, speed hump markings, and automobile parking space markings. They are available in various configurations and styles, creating it visible for drivers and pedestrians to instantly acknowledge and quickly react for safe and efficient use of the road. Presently, there are a variety of road marking materials are widely used. These are solvent and water-based paints, polyester, thermoplastic, epoxy, alkyl radical methacrylate, preformed thermoplastic, and preformed tape (Traffic Signs Manual 2018; IS EN 1436 2004). The performance of road marking depends on the surface porosity, surface type and material used in the marking (Sasidharan et al. 2009; Wang 2010; Scottish Road Research Board 2015; Sheik et al. 2015; Naidoo and Steyn 2018; Brusentseva et al. 2019).

The price of road marking materials varies depending on their formulation and availability but they all must ensure two fundamental characteristics. The color of the marking must be distinguishable from the underlying surface and must be detectable even in harsh weather. Furthermore, they should apply to a range of different pavement surfaces and must be strong enough to undergo harsh traffic and climatic conditions.

The focus of this study is on permanently installed preformed 3D pavement markings. In recent years preformed thermoplastic marking is gaining popularity in an urban environment due to their flexibility in installation. The preformed marking combined with 3D that generates the illusions which appear to be more effective in controlling driving behaviors. Presently, there are some good industrial practices where 3D marking demonstrated good performances over traffic reduction and safety and better durability than conventional paint-based road marking (Sheikh et al. 2015). The future of this type of marking is also promising as road marking will be vital for the implementation of autonomous vehicles and ensuring road safety when both driver-assisted and self-driving vehicles will be sharing the same road space (Stefan et al. 2007). Moreover, if 3D marking proved effective for long-term traffic calming, this could either be used to replace and/or supplement traffic cameras as a traffic control measure.

The objective of this paper is to investigate the functional, and long-term laboratory and field performance of preformed thermoplastic material. A comprehensive laboratory and field investigation was conducted to measure dry and wet friction, macrotexture and immersion wheel tracking tests to measure stripping resistance and degradation properties over time. Furthermore, two 3D thermoplastic 15 Mph signs were installed at the campus to evaluate their on-site functional properties and effectiveness as a traffic calming measure.

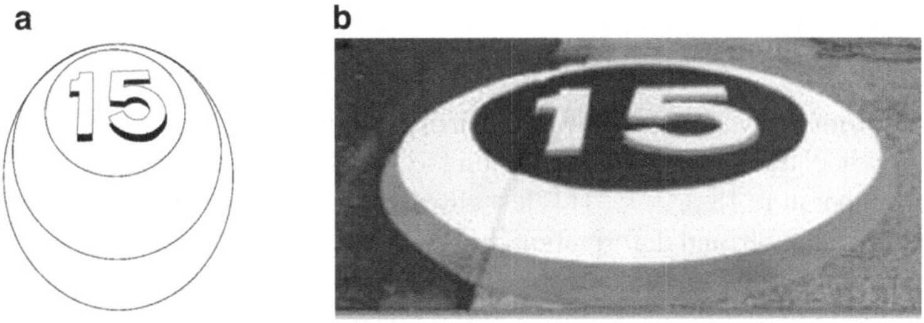

Fig. 1 **a** Schematic diagram of 3D marking **b** 3D thermoplastic marking

2 3D Preformed Thermoplastic Road Marking

3D-Preformed thermoplastic markings consist of pigments, reflective glass beads, fillers, binders and additives (PREMARK 2012). Pigments are fundamentally acquainted with imparting color to supply characteristics, like ultraviolet stability. The filler material is a combination of carbonate, sand, and different inert materials. The binder component used in the marking sheet is resin-based and does not change with chemicals on heating and application. Finally, the plasticizer is used as an additive to ensure the durability and flow of the material during the manufacturing process.

The preformed sheet is manufactured adding 30% reflective glass beads by weight mixed along with the melted thermoplastic material. Further beads are added from the top once the marking is installed. The thickness of the 3D preformed marking is between 2.29 or 5.50 mm (DMRB 2008). An image of typical 3D marking is given in Figs. 1a and b.

3 Experimental Program

3.1 Functional Properties

The testing program consisted of evaluating functional properties such as Sand patch test according to ASTM E965 for macro texture measurement, Pendulum test according to ASTM E303-93 to measure frictional properties in dry and in wet conditions (Traffic Signs Manual 2018; ASTM 2018; AASHTO 2004; Anon 2012; Praticò and Vaianab 2015). Furthermore, the marking was cut into 750 by 450 mm size and was installed on 750 mm × 450 mm × 100 mm × 20 mm asphalt slab. Two types of installation processes were followed. The first one was installed on the unheated surface and in the second one, a preheating for approximately 5 min was applied on the asphalt surface before placing the mat. Both methods then followed similar heating approximately 193 °C according to manufacturer procedure for site installation.

3.2 Stripping Resistance

Two slabs from each case were then tested in dry and in the submerged condition in the Hamburg wheel tracking machine according to AASHTO T324-04 for 20,000 passes at room temperature 18 ± 2 °C. This test allowed evaluating the stripping resistance, material degradation, and deformation due to wheel tracking (AASHTO 2004). It is important to note that, the specifications in AASHTO T324 were modified to allow large specimen (750 mm × 300 mm × 100 mm) testing instead of 150 mm diameter circular specimens. This was necessary to test the large size of the thermoplastic installation simulative to field conditions.

3.3 On-site Testing

In parallel, two 1.5 m size 3D markings were installed on campus road to study the functional properties and long-term performance in the outside environment as well as their effectiveness in traffic calming. The installation sequence is given in Fig. 2. The first marking was installed on the concrete-asphalt surface with joints in the middle and the second was on the concrete pavement. The site was chosen to evaluate the performance of the 3D marking on different surfaces including the influence of joint on long-term performance. As shown in Fig. 2, the installation procedure consisted of thoroughly cleaning the road surface and then apply preheating to dry out surface moisture (Fig. 2a–b). The discussion with the manufacturer pointed out that in a typical installation, the preheating stage is left to weather conditions and budget of the project. Following preheating, a primer was applied (Fig. 2c) before placing the marking segments on the road surface (Fig. 2d). It was ensured that all parts of the preformed markings are in contact with each other. The final stage involved applying heating up to 190–2000 °C to fuse the mat into the pavement surface (Fig. 2e). Careful attention was given in the joints between the segments and in the joints between asphalt and concrete pavement. The finished installation (Fig. 2f) was left for further 15 min to cool down to the surrounding temperature. The whole operation of installation for one marking took approximately 30 min.

4 Results and Analysis

4.1 Visual Observation of Installed 3D Marking

The visual survey was conducted at 3 and 12 months. The evidence of distress is mentioned in the table below. The road marking observed for 270 days showed no visible sign of disintegration, cracking or loss of visibility. A picture of the marking after 12 months is given in Fig. 3. The tire stain marks spotted at 3 months was found

Fig. 2 3D preformed marking installation procedure

Fig. 3 Installed 3D marking after 12 months

superficial and washed away with rain. Overall, the visual appearance did not change after 12 months. The excellent day and night visibility after 12 months indicate good retro-reflectivity of the material. Furthermore, material near the joints, as shown in Fig. 3, showed no sign of failure, indicating a good bond even at the edges.

4.2 *Macro Texture*

The macrotexture of the 3D marking before installation and periodically measured nine months after installation is given in Table 2. As expected, the texture depth in the asphalt surface was found higher than the concrete surface. On average, the change of texture from day one to after three months and 12 months is relatively constant, the reduction is less than 5–7%, indicating stable performance under traffic loading and environmental cycles (Table 1).

Table 1 Visual condition data

Observation	Condition (X = none, Y = few, Z = significant, E = excellent)		
	Day 1 (Aug 18)	3 months (Jan 19)	12 months (Sep 19)
Surfacing cracking	X	X	X
Material disintegration	X	X	X
Stain on the surface	X	Y	X
Tire mark	X	Y	X
3D impression	Z	Z	Z
Visibility at daytime	E	E	E
Visibility at night-time	E	E	E

Table 2 Macrotexture properties

Base surface	Macro texture depth (mm)				
	Statistics	Surrounding pavement	On the marking		
			Day 1	3 months	12 months
Asphalt	Ave	4.13	4.0	3.9	3.8
	Max	4.26	4.1	4.2	4.1
	Min	4.03	3.8	3.7	3.6
Concrete	Ave	3.13	2.8	2.8	2.7
	Max	3.26	3.0	3.1	3.0
	Min	3.03	2.4	2.6	2.4

Table 3 Friction in dry and wet condition

Base surface	Statistical value	Friction					
		After installation					
		Day 1		3 months		12 months	
		Dry	Wet	Dry	Wet	Dry	Wet
Asphalt	Ave	84.2	69	85	70	81.3	68
	Max	85.2	70	90	72	84.3	72
	Min	82.3	68	80	68	78.5	66
Concrete	Ave	90.4	73	90	71	82.5	69
	Max	95.1	75	95	73	84.1	72
	Min	89.0	70	85	69	79.3	64

4.3 Frictional Properties

The dry and wet pendulum test results from day one to after twelve months are given in Table 3. The installation on both asphalt-concrete and concrete surface gives average slip resistance value (SRV) as 80–90 for preformed thermoplastic marking in dry condition and 66–73 in wet condition. Which still is considerably higher than generally recommended 55 in the design standards in the UK (DMRB 2008). Looking at the results, the skid resistance value for the concrete pavement is marginally higher than that of asphalt pavement. The presence of glass bits in the thermoplastic sheet gives high frictional resistance. It is interesting to note that the frictional values change less than 10% in all surface installation even after 365 days of trafficking.

4.4 Long-Term Performance in Accelerated Laboratory Test

The test set-up for wheel tracking and tested samples after 20,000 cycles are given in Fig. 4, and in Table 4, the visual conditions of the tested samples (with and without thermoplastic sheet) are explained. The rutting overserved in the testing is related to asphalt material, therefore, it was not included to assess the performance of thermoplastic. The result showed that in both non-preheated and pre-heated cases (Table 4), there was no evidence of stripping or material disintegration in the dry test. However, there was some evidence of stripping on the non-heated case when tested submerged condition. On the other hand, no stripping or deterioration was noted in the preheated installation. This indicates that the preheating of the host surface marginally improves the adhesion.

The stripping inflexion point was calculated by subtracting the intercept of the first portion of the deformation vs load cycles plot from the 2nd portion of the same graph and then divided by the slope of the first portion minus the slope of the second

Fig. 4 Modified Hamburg wheel tracking test

Table 4 Visual observation of tested slabs after 20,000 cycles

Case	Visual condition after 20,000 cycles		Stripping inflexion point @ number of passes	
	Dry test	Submerged	Dry	Submerged
Installed on non-heated surface	No evidence of material stripping, surface crack, etc.	Evidence of some stripping	None	~18,000
Installed on pre-heated surface	No evidence of material stripping, surface crack, etc.	No evidence of stripping	None	None

portion. As reported in Table 4, no striping was evident in the dry test. However, in the submerged test, as discussed earlier, minor evidence of stripping was noted towards the end of the test in the non-heated case.

4.5 *Traffic Calming Evaluation*

The main aim of this survey was to identify the response of the driver towards the road marking installed on the university campus. A similar survey was conducted on the same location before the installation of the marking to test the average speed of the vehicles. The results showed that, despite road signs displayed at the side of the road, the vehicles run at an average speed varying from 17 to 22 mph exceeding the University premises speed limit of 15 mph.

The procedure for the survey included calculation of the speed of the vehicles going past the installed road marking. Since the average speed was in the range of 17 mph, it was assumed that the initial velocity of the vehicle is 17 mph and the final

Table 5 Average speed of the vehicle

Vehicle	Average speed at 40 m to 3D marking (mph)	Average speed near the marking (Mph)			
		Day 1	1 month	3 months	12 months
Cars (110 no)	18.6	12.90	14.3	14.2	15.3
Transit Vans (25 no)	17.5	14.07	15.3	16.7	15.8

velocity value determined the response. To calculate the final velocity, a constant distance, i.e. 40 m from start to end (standard amidst which lies the marking) was taken. After this, the time taken by the vehicle to cross that distance was observed for different types of vehicles. The measurements were done during the morning (8.30–10 am) and afternoon rush hour (between 4.30 to 5.30 pm). The average speed for cars and transit vans, the most frequent types of vehicles on the campus, are presented in Table 5.

The results showed that, depending on the vehicle type, speed reduction was 20–31% reduction after 1 day of installation and that reduced to 9–17% after 12 months. As the campus road is predominantly used by students, staff, and visitors, regular over speeding used to take place before installation of 3D marking. The 3D impression likely gives a road humps illusion, and the driver reacts even the marking location is known to them. This could be an effective and low-cost alternative to speed camera or road humps. Further large-scale trials on different roads and traffic conditions are required to verify this.

5 Conclusions

Key conclusions from this investigation are:

- 3D thermoplastic road marking does not reduce texture in asphalt and concrete surfaces. The texture depth remains relatively constant and likes the surrounding pavement even after 12 months of trafficking.
- The average wet and dry slip resistance value (SRV) of 3D marking were found between 65–90 at the start and remain relatively constant after 12 months. These values are significantly higher than conventional asphalt and concrete surfaces and indicate very good frictional properties in the installed area.
- The consistent day and night-time visibility of the marking ensure good retro-reflectivity. The vehicle speed measurement also showed speed reduction even after 12 months in service, indicating the long-term effectiveness of the marking in traffic calming operation.
- The thermoplastic material installed on non-heated and pre-heated asphalt surfaces showed a good performance against accelerated wheel tracking tests. The dry and submerged condition testing showed that the mat installed in the pre-heated surface has marginally better performance than non-heated surfaces. The preheating was also applied during the field installation.

- The overall conclusion from this investigation is that 3D preformed thermoplastic markings is a durable product and have demonstrated some benefits in traffic calming operation in a small-scale trial and can supplement traffic camera as a means for traffic control measure. Furthermore, the long-term durability of 3D marking ensures less intervention and, therefore less disruption to the traffic.

Acknowledgements The authors would like to express their appreciation to the Preformed Markings Ltd UK for providing supports during this project.

References

AASHTO T 324 (2004) Hamburg wheel-track testing of compacted hot mix asphalt (HMA)

American Society for Testing and Materials (2018) Standard designations E-303-93. Standard test method for measuring surface frictional properties using the british pendulum tester

Anon (2012) PREMARK®. https://www.preformedmarkings.co.uk/traffic-applications/premarkr/. Accessed 5 Oct 2018

Brusentseva, T, Shikalov V, Fomin V (2019) Cold spray deposition of thermoplastic powder for road marking. In: EPJ web of conferences

DMRB (2008) Standards for highways, vol 7, section 3

Highways Act (1980) Legislation.gov.uk

IS EN 1436 (2004) European standard for road marking

Naidoo S, Steyn M (2018) Performance of thermoplastic road-marking material. J South African Inst Civil Eng 60(2):P9–P22 ISSAN 1021-2019, Paper 1584

Praticò FG, Vaianab R (2015) A study on the relationship between mean texture depth and mean profile depth of asphalt pavements. Constr Build Mater 101(Part 1):P72–P79

PREMARK (2012) Specifications and MSDS sheet, preformed thermoplastic—melt down road & pavement striping. http://preformedthermoplastic.com/premark-specifications-and-msds-sheet/. Accessed 5 Oct 2018

Rehman SAU, Duggal AK (2015) Comparative analysis of pavement marking material based on retro-reflectivity. Int Res J Eng Technol (IRJET) 02(05):964–970

Sasidharan L, Karwa V, Donnell ET (2009) Use of pavement marking degradation models to develop a pavement marking management system. Public Works Manag Policy 14(2):148–173

Scottish Road Research Board (2015) Durability of white thermoplastic road markings

Traffic Signs Manual (2018) Chapter 5: Road Marking, Department of Transport, UK

Vacek S, Schimmel C, Dillmann R (2007) Road-marking analysis for autonomous vehicle guidance. In: European conference on mobile robots, 19–21 September, Freiburg, Germany

Wang S (2010) Comparative analysis of NTPEP pavement marking performance evaluation results. MSc dissertation. https://etd.ohiolink.edu/rws_etd/document/get/akron1271369205/inline

Multiple Linear Regression Models for Predicting Surface Damage Due to Repeated Dynamic Loading on Submerged Asphalt Pavement

Fauzia Saeed, Mujib Rahman, and Maher Mahmood

Abstract Asphalt surface damage due to water pumping from moving traffic is underexplored. A laboratory test has been developed to simulate the impact of moving traffic on submerged surfaces. In total 36 tests were conducted on Hot Rolled Asphalt (HRA), open-graded Stone Mastic Asphalt (SMA) and Porous Asphalt (PA). The specimens were submerged in shallow water while 5 kN repeated loading was applied at 5 and 10 Hz frequencies until failure. It was observed that irrespective of surface type, cracking, and rutting occurs simultaneously, although their magnitudes were different on different types of surfaces. The experimental data were then used to develop multi-input deterioration prediction models using regression analysis. The experimental parameters such as asphalt surface type, aggregate size, weather conditions, void contents, load magnitude and load frequencies were used as model inputs. The measured cracking and rutting were used to compare with the predicted cracking and rutting. The models yield 84 and 71.6% correlation with measured rutting and cracking respectively. Furthermore, combined distress (cracking and rutting) model for all HRA and SMA variations was developed and found 52 and 39% correlation respectively. The low correlation was believed to be due to the measurement difficulty of narrow cracks during testing. Despite this, the models showed promising results for overall distress prediction and with further development, it could be used as a screening tool to evaluate the performance of asphalt surfaces when subject to both prolong rain and traffic loading.

Keywords Deterioration prediction model · Linear regression · Cracking · Rutting · Asphalt surface damage

---

F. Saeed · M. Rahman (✉)
Department of Civil and Environmental Engineering, Brunel University, Kingston Ln, Uxbridge, Middlesex, London UB8 3PH, UK
e-mail: mujib.rahman@brunel.ac.uk

F. Saeed
Department of Civil Engineering, Engineering College, Omar Al-Mukhtar University, Al Bayda, Libya

M. Mahmood
Civil Engineering Department, University of Anbar, Ramadi, Iraq

C. Raab (ed.), *Proceedings of the 9th International Conference on Maintenance and Rehabilitation of Pavements—Mairepav9*, Lecture Notes in Civil Engineering 76,
https://doi.org/10.1007/978-3-030-48679-2_91

1 Introduction

A thick and well-constructed pavement on good foundation is considered as long-life pavement. The distress in long-life pavement is mostly limited to surface damages. If surface damage can be predicted timely and accurately, the periodic maintenance could be done more efficiently, which will ultimately lead to very long structural life of pavement. However, asphalt surfaces are complicated physical structures that react in a complicated way to environment, and traffic loading (Ozer et al. 2018; Glaoui et al. 2012; Willway et al. 2008; Liu et al. 2017). The failure of asphalt surfaces is progressive damage with time and depends on complex interaction of material, construction, traffic, environment and foundations.

Water is one of the environmental variables that can significantly reduce the performance of asphalt surfaces (Gao et al. 2015; Lei et al. 2017; Karlson 2005). It is recognised that the water on the surface or water builds up in the pavement structure due to rainfall or poor drainage allied with repetitive traffic loading accelerates surface damage, which may subsequently occur in pavement surface layer spalling or loosening, leading to localised and structural damage (Kim et al. 2008; Xiaoyong 2008). When traffic moves on the submerged pavement, the interaction of tire-water-pavement creates pore water pressure. Despite extensive studies in the last fifty years on water related material degradation, research on the impact of water pressure on pavement performance caused by the dynamic loading is still very limited.

The most common asphalt surface distresses related to mechanical properties are rutting, cracking and ravelling. There are different standard laboratory tests available to assess and quantify these distresses. However, in real pavement, in the presence of both water and traffic loading, these distresses can occur simultaneously. It is, therefore, essential to develop a test method that can simulate combined traffic-tire-water-pavement interaction, and which can be used to generate most common damages on the asphalt surfaces. The combination of load magnitude, speed, tread characteristics and depth of surface water that create maximum pore water pressure, from Saeed et al. (2018a, b), were used. By doing this, it was possible to quantify the type and amount of deterioration on different types of surfaces and compared to each other (Saeed et al. 2018a).

Once the deterioration (rutting, cracking and ravelling) has been quantified, it is useful to develop a prediction model for evaluating the relative performance of different types of surfaces, which can be utilised to optimise the selection of asphalt surface suitable for specific loading and climatic conditions.

2 Research Objectives

The primary objective of this study was to develop multi-input linear regression models to predict deterioration due to repeated traffic loading while asphalt surfaces were submerged in shallow water. Three asphalt surfaces, namely, hot rolled asphalt

(HRA), open stone mastic asphalt (SMA) and porous asphalt (PA) chosen because they are replicate the three high void content, medium void content and low void content also used as surface layers usually. Each of these mixtures was produced with different sizes of aggregates (6, 10, and 14 mm) and was tested in two environmental conditions, dry and submerged, and in two frequencies 5 and 10 Hz at 5 kN loading.

In the prediction models, both rutting and cracking were considered as the main distresses. After developing the model, model accuracy for each type of mixture was carried out.

This paper is divided into two stages (shown in Fig. 1), in stage 1, summary results from experimental study is presented. Detail explanation of the experimental set-up and analysis of results are reported elsewhere (Saeed et al. 2018a). In stage 2, a brief overview of the model generation from stage 1 experimental data, results and sensitivity analysis are reported.

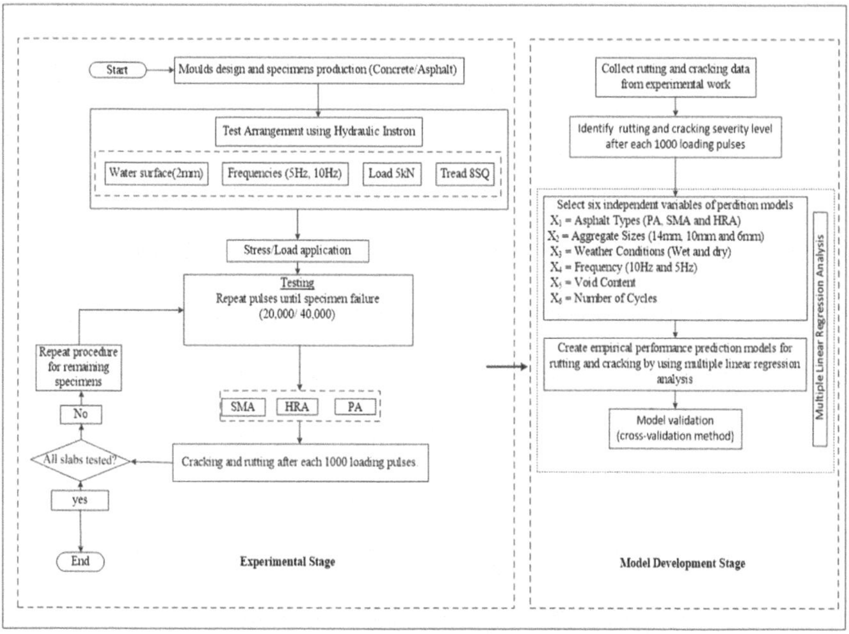

Fig. 1 Flow chart of the experimental and modeling work. *Note* Tread 8SQ refers to 8 mm deep square tread

3 Stage 1: Dynamic Compression Test for Surface Damage

3.1 Asphalt Surface Damage Testing

Stage 1 was to investigate the performance of different asphalt surfaces subjected to concurrent flooding and sinusoidal loading. For this propose, granite aggregate and a soft grade bitumen 150/100 pen were used for manufacturing test specimens. In total, 36 C40 concrete slabs (305 mm × 305 mm × 100 mm) with (200 mm × 200 mm × 50 mm) recessed area at the middle were produced. The recessed area was filled with asphalt slab. The laboratory test set-up is presented in Fig. 2. The dimensions of each asphalt specimen were measured and then tested for bulk density as detailed in BS EN 12697-6: 2003. The actual percentage of air voids of each test specimen was calculated according to BS EN 12697-8: 2003. The sample properties are given in Table 1. All mixtures showed representative VC and VMA as required by the standard.

Three asphalt mixtures, a gap graded hot rolled asphalt (HRA), and two open-graded mixtures, such as stone mastic asphalt (SMA) and porous asphalt (PA) were investigated for this research. Each of these mixtures was produced with different aggregate sizes to evaluate their impact on mixture performance.

All mixtures were tested in dry and in wet conditions by applying vertical dynamic load at 5 and 10 Hz while the asphalt surface was submerged with up to 2 mm water. All specimens went through overnight saturation before testing. The test results concerning surface cracking, rutting, and other visual distress such as ravelling, were measured using digital microscope, camera and straight plate.

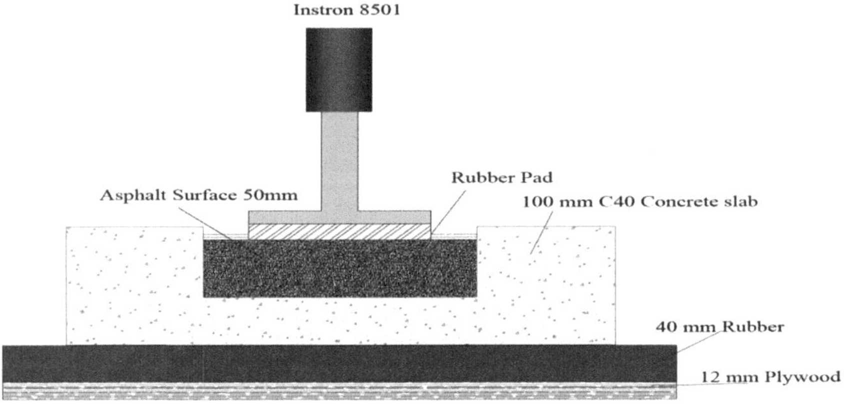

Fig. 2 The experimental setup

Table 1 Sample properties

Mix. type	Aggregate properties			No of sample	Specimen size (mm^3)	Void contents (%)				VMA %
	Nominal maximum size (mm)	Type	Filler			Max	Min	Std	AVG	
HRA	10	Granite	Granite	6	200 × 200 × 50	7.81	5.97	0.736	6.5	18.86
	14	Granite	Granite	6	200 × 200 × 50	8.00	5.20	1.017	6.31	19.28
SMA	6	Granite	Granite	6	200 × 200 × 50	12.78	8.97	1.421	10.59	15.89
	10	Granite	Granite	6	200 × 200 × 50	13.63	10.36	1.371	11.87	15.02
	14	Granite	Granite	6	200 × 200 × 50	1.50	10.33	1.024	11.13	14.13
PA	14	Granite	Granite	6	200 × 200 × 50	20.89	18.00	1.315	19.51	1.19

3.2 Test Specifications

For each surface type, three specimens were tested in dry condition and three in wet condition, i.e. specimens are submerged in 2 mm water. The samples were exposed to a total 20,000 to 40,000 load cycles, depending on the level of damage observed on the surfaces. All 6 and 10 mm HRA and all SMA mixtures were tested for 20,000 cycles, while 14 mm HRA, 14 mm SMA and 14 mm PA mixtures were tested for 40,000 cycles. The results were collected after each 1000 cycles.

After each 1000 load cycles, a picture was taken by a digital camera and a microscope with 400X magnifications to review surface condition and measure the length and width of the cracks and other distresses. A build in image processing software in the microscope was used for the measurement of cracks. The length of the crack was recorded, marked and accumulated to get a total length of the crack for subsequent 1000 cycles. Similar procedure was followed in wet condition testing; accept after lifting the load device, the asphalt surface was dried by a blotting paper prior taking any measurement. The measurement of the surface cracking was based on the distress identification manual from the LTPP study (Miller and Bellinge 2003). The rutting was also measured after each 1000 load cycles by placing the straight plate in a plane perpendicular to the load direction and the bottom surface of the plate was parallel to longitudinal slope of slab surface.

The measured cumulative cracking and rutting after each 1000 cycles are presented in Fig. 3.

Key conclusions from the experimental study were as follows:

- The presence of water accelerates surface cracking. Compared to the dry state, in a particular loading cycle, crack appearance on the wet PA was more than seven-times and was twice in the wet SMA. Least amount of cracking was observed in HRA mixtures.
- Increasing load frequency leads to accelerated damage both in terms of cracking and rutting.
- In general, presence of water accelerates rutting in all mixtures and more rapidly in 14 mm SMA. Furthermore, 14 mm HRA surface appears to perform the worst. The mechanical properties of the HRA come from the cohesion of the binder within the fine aggregates, so HRA can be less resistant to internal movement but it can be better in fatigue crack resistant.

4 Stage 2: Regression Models to Predict Combined Failures

A multi-input deterioration prediction model is generally used to find a relationship between the dependent variable, which is distress progression or a condition index, and one or more explanatory variables, such as cracking area, age and ESAL (Lytton 1987; Fwa and Sinha 1986; Kerali et al. 1996, Prozzi and Madanat 2004). To develop

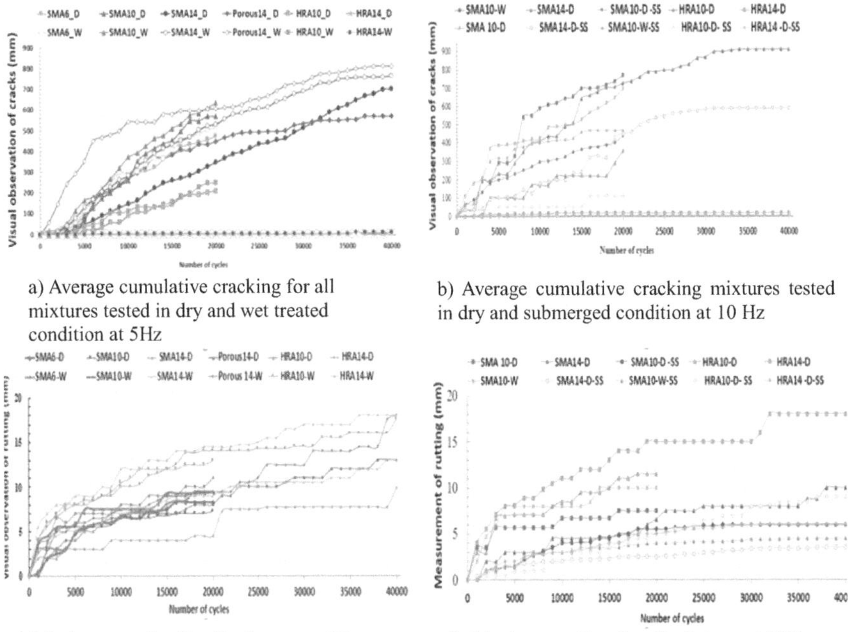

a) Average cumulative cracking for all mixtures tested in dry and wet treated condition at 5Hz

b) Average cumulative cracking mixtures tested in dry and submerged condition at 10 Hz

c) Maximum rutting for all mixtures at 5Hz

d) Maximum rutting for all mixtures at 10Hz

Fig. 3 Average cumulative cracking and rutting mixtures tested in dry and submerged condition

a multiple regression equation of deterioration prediction for cracking and rutting, the six independent variables, asphalt types, aggregate sizes, test conditions (dry and submerged), frequency, void contents and load pulses are used. The single equation of the prediction model is described consequently:

$$yy = aa_1 + aa_2x_1 + aa_3x_2 + aa_4x_3 + aa_5x_4 + aa_6x_5 + aa_7x_6$$

Where y = cracking/or rutting; x_1 = asphalt types; x_2 = aggregate sizes; x_3 = weather conditions; x_4 = Frequency; x_5 = voids content; x_6 = no of cycles; aa_1, aa_2, aa_3, aa_4, aa_5, aa_6 and aa_7 = coefficients.

Table 2 shows values used for all independent variables.

5 Results

5.1 Distress Prediction Models

In order to develop multiple regression distress model for all mixture types, the variables presented in Table 1, were used in the statistical software SPSS (Field 2009).

Table 2 X values of independent variables

Independent variables	x	Value
Asphalt types	x_1	1 = HRA, 2 = SMA, 3 = PA
Aggregate sizes (mm)	x_2	6,10,14
Weather conditions	x_3	0 = dry, 1 = submerged
Frequency	x_4	5 = 5 Hz, 10 = Hz
Void content (%)	x_5	Each mixture and aggregate size with specific void contents at the start of testing
Number of load cycle	x_6	1000, 2000, ...,40,000

Table 3 Linear equation of prediction models of cracking and rutting

Distress	Prediction model	R^2
Cracking	$y = -501.006 + 242.062x_1 - 8.0546x_2 + 147.062x_3 + 31.766x_4 - 1.827x_5 + 0.012x_6$	0.716
Rutting	$y = 7.217 - 4.277x_1 + 0.044x_2 + 1.551x_3 + 0.091x_4 + 0.268x_5 + 0.00028x_6$	0.840

The models and their corresponding coefficients of correlation (R^2) are presented in Table 3.

As shown in Table 3, 84 and 71.6% correlation were achieved for rutting and cracking respectively. It is likely that the widespread crack formation may have contributed to the model accuracy. It is interesting to note that, mixture type, voids contents, weather condition and load frequency have more impact than other variables.

In order to evaluate the accuracy of each model for cracking and rutting, the entire data set used to develop the prediction model were randomly split into training data (around 80%) and testing data (around 20%). The training data were used to generate the prediction models while the testing data were used to validate the models. The data for both models were split by using SSPS software. SPSS can automatize this selection process without requiring multiple steps. The process to split training data and testing data has been repeated to run models five times for both cracking and rutting to ensure all data were included. The observed rutting and cracking were plotted against predicted values. The results are shown in Fig. 4.

It can be seen that 84% correlation between measured and observed rutting for all mixture types. The rutting was predominantly confined in the wheel path and measurements were accurate and easier at all load cycles. On the other hand, relatively low correlation (71.6%) in cracking model was primary due to the measurement difficulty of widespread cracking in the slab. Despite great care, it was difficult to identify narrow cracks especially towards the end of the testing. Further improvement in crack measurement will enhance the model accuracy.

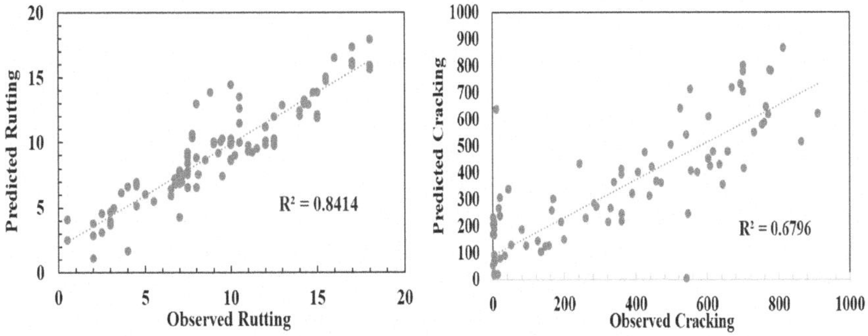

Fig. 4 Model accuracy for cracking and rutting for each type of mixture

Table 4 Combined equations

Asphalt type	Prediction model	R^2
HRA	$y = -264.9589 - 65.91269x_1 + 281.529817x_2 + 67.1378x_3 + 82.6952x_4 + 0.01789x_5$	0.52
SMA	$y = 13.5654 + 12.0760x_1 - 146.2561x_2 - 21.08433x_3 + 15.6517x_4 + 0.01252x_5$	0.39

5.2 Combined Deterioration Prediction Model

To develop a combined distress model for each mixture type, further analysis was conducted by separating each mixture type as both HRA and SMA have different mechanism of failure. The combined equations for cracking and rutting in HRA and SMA are presented in Table 4. PA was not included as only one aggregate size was tested in stage 1.

Where, x_1 = aggregate size, x_2 = weather conditions, x_3 = frequencies, x_4 = voids content and x_5 = cycle no.

R^2 for all HRA mixtures was found 52% while SMA it has reduced to 39%. This is because, as explained in the earlier section, the measurement difficulty of widespread narrow cracks during laboratory testing. Further improvement of crack measurement will enhance the model accuracy.

6 Conclusion

Key conclusions drawn from this research are:

- Stage 1: The influence of combined water and dynamic loading on surface damage was successfully simulated in the laboratory environment. The outcome of the test has demonstrated promising future for further development as a screening test for water susceptibility resistance of asphalt surfaces.

- Stage 2: Multi input linear regression models developed from stage 1 data, to predict distress on asphalt surfaces due to the combined interaction of traffic loading (frequency and load magnitude) and water. The prediction accuracy of the model was 84% for rutting, 71.6% for cracking. Combined damage prediction models yield better accuracy for HRA surface with 52% correlation and, but prediction accuracy reduced to 39% for SMA surface. The low correlation was due to the measurement difficulty of narrow cracks during testing. Despite this, the models showed promising results for overall distress prediction and with further development, it could be used as a screening tool for evaluating the performance of asphalt surfaces when subject to extreme event like prolong rain and traffic movements.

References

BS EN 12697-6:2003 (2003a) Bituminous mixtures. Test methods for hot mix asphalt. Determination of bulk density of bituminous specimens. British Standards Institute

BS EN 12697-8:2003 (2003b) Bituminous mixtures. Test methods for hot mix asphalt. Determination of void characteristics of bituminous specimens. British Standards Institute

Field A (2009) Discovering statistics using SPSS, 3rd edn. SAGE Publications Ltd, London

Fwa TF, Sinha KC (1986) Routine maintenance and pavement performance. J Transp Eng 112(4):329–344

Gao J, Guo C, Liu Y (2015) Measurement of pore water pressure in asphalt pavement and its effects on permeability. Measurement 62:81–87

Glaoui B, Merbouh M, Van De Ven M, Chailleux E, Youcefi A (2012) Thermal fatigue of polymer modified bitumen. J Thermoplast Compos Mater 25(4):469–478

Karlson TK (2005) Evaluation of cyclic pore pressure induced moisture damage in asphalt pavement. University of Florida

Kerali H, Lawrance A, Awad K (1996) Data analysis procedures for long-term pavement performance prediction. Transp Res Rec: J Transp Res Board 1524:152–159

Kim Y, Lutif JS, Bhasin A, Little DN (2008) Evaluation of moisture damage mechanisms and effects of hydrated lime in asphalt mixtures through measurements of mixture component properties and performance testing. J Mater Civ Eng 20(10):659–667

Lei Y, Hu X, Wang H, You Z, Zhou Y, Yang X (2017) Effects of vehicle speeds on the hydrodynamic pressure of pavement surface: measurement with a designed device. Measurement 98:1–9

Liu H, Hao P, Xu J (2017) Effects of nominal maximum aggregate size on the performance of stone matrix asphalt. Appl Sci 7(2):126

Lytton RL (1987) Concepts of pavement performance prediction and modeling. In: Proceedings of the 2nd north american conference on managing pavements

Miller J, Bellinger W (2003) Distress identification manual for the long-term pavement performance program. Federal Highway Administration Report: FHWA-RD-03-031

Ozer H, AL-qadi IL, Singhvi P, Bausano J, Carvalho R, Li X, Gibson N (2018) Prediction of pavement fatigue cracking at an accelerated testing section using asphalt mixture performance tests. Int J Pavement Eng 19(3):264–278

Prozzi J, Madanat S (2004) Development of pavement performance models by combining experimental and field data. J Infrastruct Syst 10(1):9–22

Saeed F, Rahman M, Chamberlain D (2018a) Impact of tire and traffic parameters on water pressure in pavement. J Transp Eng Part B: Pavements 144(4):04018042
Saeed F, Rahman M, Chamberlain D, Colins P (2018b) Asphalt surface damage due to combined action of water and dynamic loading. Constr Build Mater 196:530–538
USM. The engineering properties and performance of porous asphalt. https://www.google.co.uk/search?hl=en&q=www.civil.eng.usm.my/projekta/template/Chapter4.doc. Accessed 30 May 2018
Willway T, Baldachin L, Reeves S, Harding M (2008) The effects of climate change on highway pavements and how to minimise them: technical report, vol 1, no 1, pp 1–111
Xiaoyong LZD (2008) Axial symmetric elastic solution of pore water pressure in asphalt pavement under mobile load. J Southeast Univ (Nat Sci Ed) 5:014

The manufacturer's authorised representative in the EU is Springer Nature Customer Service Centre GmbH, Europaplatz 3, 69115 Heidelberg, Germany. If you have any concerns regarding our products, please contact ProductSafety@springernature.com

Printed and bound by CPI Group (UK) Ltd, Croydon, CR0 4YY
05/10/2026
02228149-0001